14th International Conference on Micro and Nanotechnology for Power Generation and Energy Conversion Applications

(PowerMEMS 2014)

Journal of Physics: Conference Series Volume 557

Awaji Island, Japan
18-21 November 2014

ISBN: 978-1-63439-710-0
ISSN: 1742-6588

Printed from e-media with permission by:

Curran Associates, Inc.
57 Morehouse Lane
Red Hook, NY 12571

Printed by Curran Associates, Inc. (2015)

Institute of Physics
Dirac House, Temple Back
Bristol BS1 6BE UK

Phone: 44 1 17 929 7481
Fax: 44 1 17 920 0979

techtracking@iop.org

Additional copies of this publication are available from:

Curran Associates, Inc.
57 Morehouse Lane
Red Hook, NY 12571 USA
Phone: 845-758-0400
Fax: 845-758-2634
Email: curran@proceedings.com
Web: www.proceedings.com

TABLE OF CONTENTS

POSTER SESSION W

ENGINEERED MATERIAL FOR ENERGY HARVESTING

ENERGY STORAGE

ELECTRET ENERGY HARVESTING

MICRO THERMAL DEVICES

INVITED-B

INDUSTRIAL APPLICATIONS OF ENERGY HARVESTING

MICROSCALE COMBUSTION-I

POSTER SESSION T

PIEZOELECTRIC ENERGY HARVESTING

MAGNETIC DEVICES

Journal of Physics: Conference Series **557** (2014) 011001 doi:10.1088/1742-6596/557/1/011001

WELCOME TO POWERMEMS2014

It is our great pleasure to welcome you to the 14th International Conference on Micro- and Nano-Technology for Power Generation and Energy Conversion Applications, or PowerMEMS 2014, in Awaji Island, Japan.

The aim of PowerMEM is to present the latest research results in the field of miniature, micro- and nano-scale technologies for power generation and energy conversion. The conference will also- give us the opportunity to exchange informations and new ideas in the field of Power MEMS/NEMS. The current status of the field of PowerMEMS spans the full spectrum from basic research to practical applications. We will enjoy valuable discussions not only from the viewpoint of academia but from commercial and industrial perspectives.

In the conference, three invited speakers lead the technical program. We received 172 abstracts and after a careful reviewing process by the Technical Program Committee a total of 133 papers were selected for presentation. These have been organized into 16 Oral sessions in two parallel streams and two poster sessions including some late-news papers. The oral and regular poster papers are published by the Institute of Physics (IOP). We have also organized a PowerMEMS School in Kobe-Sannomiya contiguous to the main conference. This two-day school will cover various topics of energy harvesting. World leading experts will give invited lectures on their main topics. This is a new experiment to broaden the technology remit of our conference by organizing mini symposiums that aim to gather the latest research on the following topics by the organizers: Microscale Combustion, Wideband Vibration Energy Harvesting, RF Energy Transfer and Industrial Application. We hope this, and other activities will make PowerMEMS2014 a memorable success.

One of the important programs in an international conference is the social program, and we prepare the PowerMEMS2014 banquet in the banquet room at the Westin Awaji Island Hotel. This will provide an opportunity to create strong networks between researchers. We also provide nice opportunities to experience Japanese nature and culture. The special cruise to see the magnificent whirlpool up close will definitely be one of the highlights. Additionally, we will serve Awaji's traditional performing art, Awaji Ningyo Joruri, which has a history of over 500 years and has been inherited through the generations.

There are many individuals we would like to thank for their support in organizing PowerMEMS2014. The TPC, chaired by Takayuki Fujita, have given us their valuable time and best effort in reviewing abstracts. The PowerMEMS School chair Yuji Suzuki and the expert speakers made the School possible. The local organizing committee, led by Kensuke Kanda has provided us with invaluable assistance in preparing the PowerMEMS2014 venue. The financial support from both the Tsutomu Nakauchi Foundation, the Hyogo International Association and the conference sponsors have also been gratefully appreciated. Finally, we would like to thank each of you for attending the conference and bringing your expertise. We wish you all a successful conference and an exciting and relaxing stay in Awaji Island.

Sincerely

Kazusuke Maenaka

The 14th International Conference on Micro- and Nano-Technology for Power Generation and Energy Conversion Applications

Awaji Island, Japan
18th – 21st November 2014

Conference Chair:
Kazusuke Maenaka, *University of Hyogo, Japan*

Technical Program Committee Chair:
Takayuki Fujita, *University of Hyogo, Japan*

PowerMEMS School Chair:
Yuji Suzuki, *The University of Tokyo, Japan*

CONFERENCE OFFICIALS

Conference Chair

Kazusuke Maenaka, *University of Hyogo, Japan*

Technical Program Committee Chair

Takayuki Fujita, *University of Hyogo, Japan*

PowerMEMS School Chair

Yuji Suzuki, *The University of Tokyo, Japan*

Local Organizing Committee Co-Chairs

Kazusuke Maenaka, *University of Hyogo, Japan*
Takayuki Fujita, *University of Hyogo, Japan*
Kensuke Kanda, *University of Hyogo, Japan*

International Steering Committee

Mark G. Allen, *Singh Center for Nanotechnology, University of Pennsylvania, USA*
Steve Beeby, *University of Southampton, UK*
Young-Ho Cho, *KAIST, South Korea*
Luc Fréchette, *University of Sherbrooke, Canada*
Takayuki Fujita, *University of Hyogo, Japan*
Reza Ghodssi, *University of Maryland, USA*
Hiroki Kuwano, *Tohoku University, Japan*
Jeff Lang, *Massachusetts Institute of Technology, USA*
Carol Livermore, *Northeastern University, USA*
Ryutaro Maeda, *National Institute of Advanced Industrial Science and Technology, Japan*
Kazusuke Maenaka, *University of Hyogo, Japan*
Paul Mitcheson, *Imperial College London, UK*
Albert Pisano, *University of California San Diego, USA*
Yuji Suzuki, *The University of Tokyo, Japan*
Shuji Tanaka, *Tohoku University, Japan*
Luis Velasquez-Garcia, *Massachusetts Institute of Technology, USA*
Peter Woias, *University Freiburg IMTEK, Germany*
Eric Yeatman, *Imperial College London, UK*

Mini symposium Organizer

Carlos A Fernandez-Pello, *University of California Berkeley, USA*
Shuhei Takahashi, *Gifu University, Japan*
Einar Halvorsen, *Buskerud and Vestfold University College, Norway*
Philippe Basset, *Université Paris-Est, ESYCOM, ESIEE Paris, France*
Yoshihiro Kawahara, *The University of Tokyo, Japan*
Huib Visser, *imec HOLST Centre, the Netherlands*
Shad Roundy, *University of Utah, USA*
Tomonori Seki, *OMRON Co., Japan*

International Advisory Board

Alan Epstein, *Massachusetts Institute of Technology, USA*
Masayoshi Esashi, *Tohoku University, Japan*
Susumu Sugiyama, *Ritsumeikan University, Japan*
Miwako Waga, *Susano Berkeley LLC., Japan*

Technical Program Committee

Seiji Aoyagi, *Kansai University, Japan*
David Arnold, *University of Florida, USA*
Skandar Basrour, *Tima Laboratory, France*
Philippe Basset, *Université Paris-Est, ESYCOM, ESIEE Paris, France*
Mustafa Ilker Beyaz, *Antalya International University, Turkey*
Danick Briand, *École Polytechnique Fédérale de Lausanne (EPFL), Switzerland*
Steve Burrow, *University of Bristol, UK*
Carlos Fernandez-Pello, *University of California Berkeley, USA*
Luc Fréchette, *University of Sherbrooke, Canada*
Takayuki Fujita, *University of Hyogo, Japan*
Einar Halvorsen, *Buskerud and Vestfold University College,, Norway*
Andrew Holmes, *Imperial College London, UK*
Isaku Kanno, *Kobe University, Japan*
Yoshihiro Kawahara, *The University of Tokyo, Japan*
Hiroki Kuwano, *Tohoku University, Japan*
Sejin Kwon, *KAIST, South Korea*
Jeff Lang, *Massachusetts Institute of Technology, USA*
Carol Livermore, *Northeastern University, USA*
Yiannos Manoli, *University Freiburg IMTEK, Germany*
Paul Mitcheson, *Imperial College London, UK*
Koji Miyazaki, *Kyushu Institute of Technology, Japan*
Jae Yeong Park, *Kwangwoon University, South Korea*
Michael Renaud, *imec HOLST Centre, the Netherlands*
Paul Ronney, *University of Southern California, USA*
Shad Roundy, *University of Utah, USA*
Tomonori Seki, *Omron Corp. Japan*
Junichiro Shiomi, *The University of Tokyo, Japan*
Yuji Suzuki, *The University of Tokyo, Japan*
Shuhei Takahashi, *Gifu University, Japan*
Shuji Tanaka, *Tohoku University, Japan*
Luis Velasquez-Garcia, *Massachusetts Institute of Technology, USA*
Ruud Vullers, *Eindhoven University of Technology, The Netherlands*
C. Mike Waits, *US Army Research Laboratory, USA*
Xiaohong Wang, *Tsinghua University, China*
Peter Woias, *University Freiburg IMTEK, Germany*

Peer review statement

All papers published in this volume of *Journal of Physics: Conference Series* have been peer reviewed through processes administered by the proceedings Editors. Reviews were conducted by expert referees to the professional and scientific standards expected of a proceedings journal published by IOP Publishing.

Internet of "printed" Things
low-cost fabrication of autonomous sensing nodes by inkjet printing

Yoshihiro Kawahara
The University of Tokyo / JST presto / AgIC, Inc., Japan
Georgia Institute of Technology, USA

Abstract
"What if electronics devices are printed using an inkjet printer even at home?"
"What if those devices no longer need a battery?"
I will introduce two enabling technologies for the Internet of Things concept.

1. Instant Inkjet Circuits:
A low cost, fast and accessible technology to support the rapid prototyping of electronic devices. We demonstrated that "sintering-free" silver nano particle ink with a commodity inkjet printer can be used to fabricate printed circuit board and high-frequency applications such as antennas and sensors. The technology is now commercialized by AgIC, Inc.

2. Wireless Power:
Although large amounts of data can be exchanged over a wireless communication link, mobile devices are still tethered by power cables. We are trying to solve this problem by two different approaches: energy harvesting. A simple circuitry comprised of diodes and capacitor can convert ambient radio signals into DC current. Our research revealed the signals from TV tower located 6.5km apart could be used to feed 100 microwatts to power microcontrollers.

About the Speaker
Yoshihiro Kawahara is an Associate Professor in the department of Information and Communication Engineering, The University of Tokyo.

His research interests are in the areas of Computer Networks and Ubiquitous and Mobile Computing. He is currently interested in developing energetically autonomous information communication devices. He's not only interested in academic research activities but also enjoyed designing new business and its field trial while joining IT startup companies. He is a technical advisor of AgIC, Inc.

He received his Ph.D. in Information Communication Engineering in 2005, M.E. in 2002, and B.E. in 2000. He joined the faculty in 2005. He is a member of IEICE, IPSJ, and IEEE. He's a committee member of IEEE MTT TC-24 (RFID Technologies.) He was a visiting assistant professor at Georgia Institute of Technology and MIT Media Lab.

Journal of Physics: Conference Series **557** (2014) 012002 doi:10.1088/1742-6596/557/1/012002

Development of 24GHz Rectenna for Receiving and Rectifying Modulated Waves

Naoki Shinohara[1], and Ken Hatano[1]

[1] Research Institute for Sustainable Humanosphere, Kyoto University, Japan

shino@rish.kyoto-u.ac.jp

Abstract. In this paper, we show experimental results of RF-DC conversion with modulated 24GHz waves. We have already developed class-F MMIC rectenna with resonators for higher harmonics at no modulated 24GHz microwave for RF energy transfer. Dimensions of the MMIC rectifying circuit is 1 mm × 3 mm on GaAs. Maximum RF–DC conversion efficiency is measured 47.9% for a 210 mW microwave input of 24 GHz with a 120 Ω load. The class-F rectenna is based on a single shunt full-wave rectifier. For future application of a simultaneous energy and information transfer system or an energy harvesting from broadcasting waves, input microwave will be modulated. In this paper, we show an experimental result of RF-DC conversion of the class-F rectenna with 24GHz waves modulated by 16QAM as 1st modulation and OFDM as 2nd modulation.

1. Introduction

An energy harvesting from broadcasting radio waves (RF) is considered as one of RF energy transfer. Only a difference is an RF source. The RF energy harvesting is considered as a passive RF energy transfer whose energy source is for the other purpose, broadcasting and in which RF is modulated. The development of the RF energy transfer has a long history from 1960s[1]. In the RF energy transfer, a microwave, GHz band frequency, is often used in order to focus the RF power to a receiver with high beam efficiency. It is called a microwave power transfer (MPT). The almost all systems were designed with continuous wave because we don't need any information by wave modulation for RF energy transfer. The RF itself is energy.

For the both RF energy harvesting and RF energy transfer, a receiving and rectifying antenna, rectenna, is a common key technology. The rectenna is composed of an antenna and a rectifying circuit with a diode. Conventional rectifying circuit for the RF energy harvesting is a charge pump to increase voltage. On contrary, conventional rectifying circuit for the RF energy transfer with microwave is called a single shunt full wave rectifier to increase RF-DC conversion efficiency, which composed of a shunt diode and an output filter with $\lambda/4$ distributed line and a capacitance. Theoretical RF-DC conversion efficiency is 100% caused by the output filter for processing higher harmonics[1]-[3].

2.45GHz band or 5.8GHz band, microwave band, are often used for RF energy transfer. These frequency bands are on ISM (Industrial, Science and Medical) band in radio wave regulation in the world. When the frequency increases, size of transmitting and receiving antenna can be reduced in beamed MPT system and conversion efficiency of circuits decreases on contrary. So the microwave is often used for RF energy transfer with consideration of total efficiency and system size. If the efficiency of circuits can be increased, higher frequency is suitable for RF energy transfer. Some development results of rectennas for RF energy transfer with millimeter wave have already reported[4]-[6].

2. Development of 24GHz Class-F Rectenna

In Kyoto University, we developed a rectenna for an RF energy transfer system called Fixed Wireless Access at 24GHz with NTT corporation[7]-[9]. When we use higher frequency for the RF energy transfer, we can reduce an antenna size and can realize a simultaneous energy and information transfer system in future. Our first target is 24GHz system. At first, we developed the 24GHz rectifying circuit on Teflon substrate[7] and revised it[8]. For the high frequency rectifier, we proposed a shingle shunt full-wave rectifying circuit with class-F load, which composed of a several λ/4 open stub resonators, for processing the higher harmonics instead of the capacitance (Fig.1)

Finally, we developed the 24GHz MMIC rectifying circuit as shown in Fig.2[9]. Dimensions of the MMIC rectifying circuit is 1 mm × 3 mm on GaAs. Maximum RF–DC conversion efficiency is measured 47.9% for a 210 mW microwave input of 24 GHz with a 120 Ω load[9]. Input microwave is not modulated and is continuous wave for RF energy transfer.

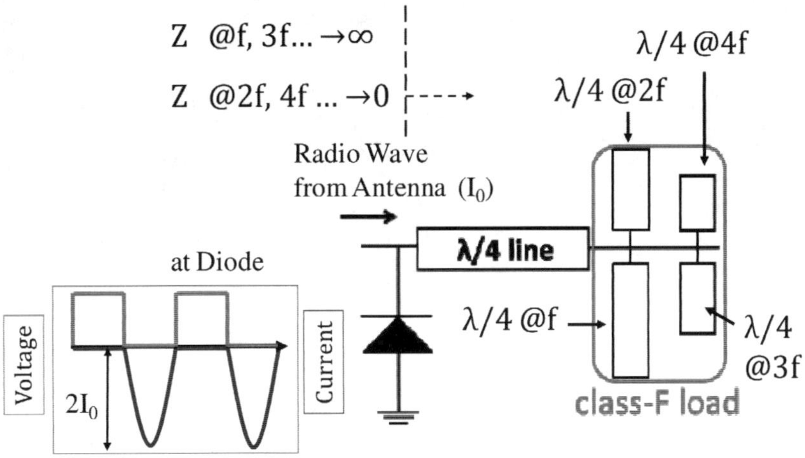

Figure 1. Circuit Design of Proposed Rectifying Circuit with Single Shunt Diode and Class-F Load as Full Wave Rectifier

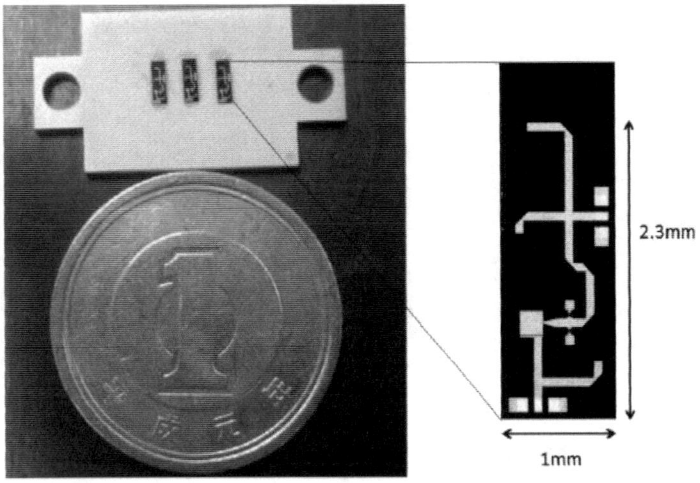

Figure 2. Developed MMIC Rectifying Circuit in 24GHz

3. Experiments of 24GHz Class-F Rectenna with Modulated Microwaves

For future application of a simultaneous energy and information transfer system or an energy harvesting from broadcasting waves, input microwave will be modulated. So we carried out an experiment of RF-DC conversion of the class-F rectenna with modulated 24GHz. Input 24GHz waves modulated by 16QAM as 1st Modulation and OFDM as 2nd Modulation are used (Fig.3). The sub carrier intervals is 312.5kHz in OFDM. It corresponds to msc3 of IEEE802.11ac. Measured results are shown in Fig.4. Connected load is 110Ω. Reflected ratio with modulated waves are higher than that with continuous wave. But RF-DC conversion efficiency with continuous wave and two modulated waves are almost same.

(a) (b)

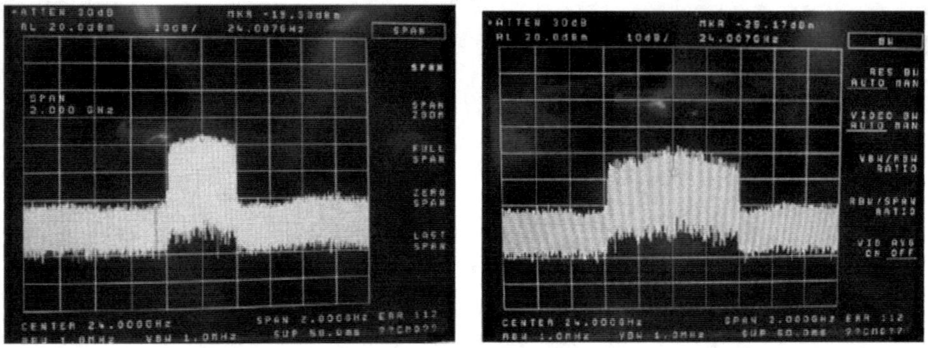

Figure 3. Wave Spectrum Modulated by 16QAM as 1st Modulation and OFDM as 2nd Modulation in 24GHz (a) Band Width 400 MHz (b) Band Width 800 MHz

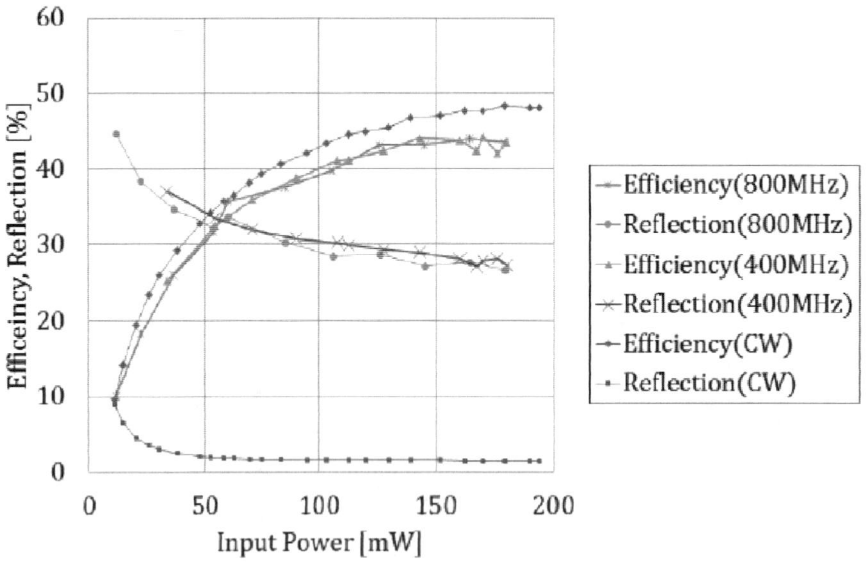

Figure 4. Experimental Results of RF-DC Conversion Efficiency and Reflection Ration of Rectifying Circuit with Modulated Waves and Continuous Wave

4. Conclusions

We show an experimental result of RF-DC conversion of the class-F rectenna with 24GHz waves modulated by 16QAM as 1st modulation and OFDM as 2nd modulation. The class-F rectenna was originally developed for RF energy transfer with no modulated wave. We conclude that our developed class-F rectenna can receive and rectify the modulated waves with high RF-DC conversion efficiency.

But the shingle shunt full wave rectifier is not theoretically suitable for rectifying the modulated waves because a λ/4 distributed line and resonators are used. So we proposed a broadband class-F rectifier with radial stab resonators and succeeded in expanding band width with high efficiency in simulations[10].

Our developed rectenna requires a several hundred mW input power to rectify it with high RF-DC conversion efficiency. For future application of a simultaneous energy and information transfer system or an energy harvesting from broadcasting waves, we should develop a revised rectenna to rectify weak microwave power with high RF-DC conversion efficiency.

References

[1] N. Shinohara, "*Wireless Power Transfer via Radiowaves*", ISTE Publishing, 2014.

[2] R. J. Gutmann, and J. M. Borrego, "Power Combining in an Array of Microwave Power Rectifier", *IEEE Trans. Microwave Theory Tech.,* Vol. 27, pp.958-968, 1979

[3] T. Ohira, "Power Efficiency and Optimum Load Formulas on RF Rectifiers Featuring Flow-Angle Equations", *IEICE Electronics Express*, Vol.10, No.11, pp.1-9, 2013

[4] T. W. Yoo and K. Chang, "Theoretical and Experimental Development of 10 and 35 GHz Rectennas", *IEEE Trans. Microwave Theory Tech.,* Vol.40, No.6, pp.1259- 1266, 1992

[5] H. Gao, M. K. Matters-Kammerer, D. Milosevic, A. van Roermund, and P. Baltus, "A 50 - 60 GHz Rectifier with -7dBm Sensitivity for 1 V DC Output Voltage and 8% Efficiency in 65-nm CMOS", *Proc. of Int. Microwave Sympo. 2014*, TH2C-4, 2014

[6] N. Weissman, S. Jameson and E. Socher, "W-Band CMOS On-Chip Energy Harvester and Rectenna", *Proc. of Int. Microwave Sympo. 2014,* TH2C-5, 2014

[7] K. Hatano, N. Shinohara, T. Mitani, K. Nishikawa, T. Seki, and K. Hiraga, "Development of Class-F Load Rectennas", *Proc. of 2011 Int. Microwave Workshop Series –Innovative Wireless Power Transmission*, pp.251-254, 2011

[8] K. Hatano, N. Shinohara, T. Mitani, T. Seki, and M. Kawashima, "Development of Improved 24GHz-Band Class-F Load Rectennas", *Proc. of 2012 Int. Microwave Workshop Series –Innovative Wireless Power Transmission,* pp.163-166, 2012

[9] K. Hatano, N. Shinohara, T. Mitani, T. Seki, and M. Kawashima, "Development of MMIC Rectenna at 24GHz", *Proc. of 2013 IEEE Radio & Wireless Symposium (RWS)*, pp.199-201, 2013

[10] K. Hatano, N. Shinohara, T. Mitani, T. Seki, and M. Kawashima, "Development of Broadband Rectenna at 24GHz", *Proc. of 6th Global Symposium on Millimeter-Waves 2013 (GSMM2013)*, CD-ROM

Experiment of Power Supply Method for WLAN Sensor Using Both Energy Harvesting and Microwave Power Transmission

K. Sakaguchi[1], S. Yamashita[1], Y. Huang[2], K. Yamamoto[1], T. Nishio[1], M. Morikura[1], and N. Shinohara[2]

[1] Graduate School of Informatics, Kyoto University, Japan
[2] Research Institute for Sustainable Humanosphere, Kyoto University, Japan

E-mail: info14@imc.cce.i.kyoto-u.ac.jp

Abstract. This paper proposes to improve effectiveness of supplying a sensor with energy using microwave power transmission (MPT) and energy harvesting (EH). The MPT duration should be as short as possible to avoid serious interference between the MPT and wireless local area network data transmission when co-channel operation of both microwave power transmission (MPT) and wireless data transmissions is performed. To shorten the MPT duration, we use multiple power sources such as an MPT source and an EH source to supply a sensor with power. Here, an overcharge or an energy shortage could occur at the sensor if the power supplied by both the MPT and EH sources is not adjusted appropriately. To solve this problem, the power supplied by multiple sources should be estimated precisely. In this paper, we propose a scheme for estimating the power supplied by multiple sources on the basis of an existing MPT scheduling system and then conducted an experiment using the scheme. From the experimental results, it is confirmed to estimate the power supplied by multiple sources successfully. In addition, the required MPT duration when the EH source is used is reduced compared to that when it is not used. Moreover, it is confirmed that the sensor station successfully estimates the power supplied by an MPT source and that by an EH source and adequately configures the MPT duration.

1. Introduction

The wireless sensor network (WSN) [1] is a technology for collecting various types of information by using many wireless sensors; it will be used in applications such as medical care or smart grids. Because the number of sensors connected to the network is expected to increase, it would become costly to replace the batteries. Thus, batteryless wirelsss sensors should be used to reduce the need for maintenance. As the technology to realize batteryless sensors, microwave power transmission (MPT) [2] and energy harvesting (EH) are cited. In this paper, we have studied how to power a sensor using both EH and MPT. We assume that a photovoltaic cell is alternately used as the EH source.

When we supply a sensor with power using MPT, the problem is serious interference between MPT and wireless local area network (WLAN) data transmission during co-channel operation of MPT and WLAN communication [3]. Therefore, a scheduling scheme for time division operation of MPT and WLAN communication has been proposed [4] to avoid the interference.

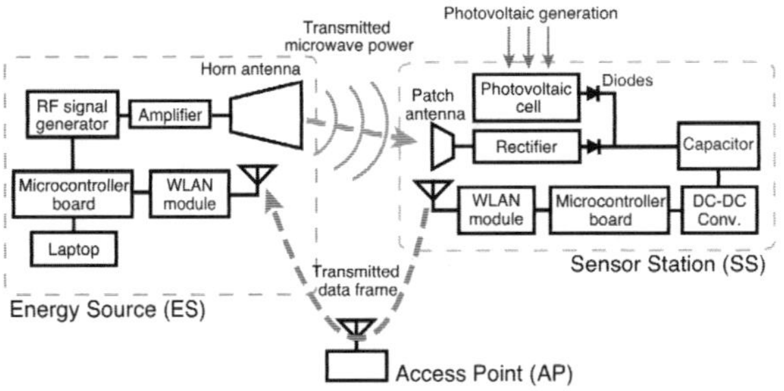

Figure 1. Experimental setup.

Figure 2. Horn antenna of ES and SS in experimental setup.

In this paper, on the basis of an existing MPT scheduling system [4], we propose a scheme for estimating the power supplied by multiple sources such as an MPT source and a photovoltaic cell and then conducted an experiment on it.

There are many studies on sensors for WSNs. In [5], a simulation of the power consumed by sensor network applications is described. In [6], the energy replenishment process and storage constraints of a rechargeable battery are discussed in terms of the design of efficient transmission strategies. In addition, a multipowered platform solution for wireless devices is presented in [7].

2. Experimental Setup

Fig. 1 illustrates the experimental setup, which consists of three devices: a sensor station (SS), an energy source (ES), and an access point (AP). In addition, Fig. 2 shows the SS and the horn antenna of the ES. We explain the detail of each device as follows.

- The SS consists of a rectenna (a patch antenna and a rectifier), a photovoltaic cell, two diodes for preventing reverse current, a capacitor, a DC–DC converter, a microcontroller board, a sensing device, and a WLAN module. The microcontroller board and the WLAN module operate on energy stored in the capacitor of the SS. The capacitor is charged with the photovoltaic cell and the rectenna.
- The AP is compatible with IEEE 802.11g WLAN. The AP periodically broadcasts beacon frames to manage the network. If the AP receives a data frame from the SS, the AP will forward the data frame to the ES.
- The ES consists of a horn antenna, an amplifier, a radio frequency signal generator, a microcontroller board, a laptop, and a WLAN module. The microcontroller board determines whether the ES transmits microwave power or not. In addition, the microcontroller board forwards data frames transmitted from the SS via the AP to the laptop. Here, the laptop is used only to display data frames transmitted from the SS via the AP, not to control the MPT.

3. Proposed Estimation Scheme for Using Multiple Sources

In this section, we explain the scheduling scheme used in this experiment and the proposed scheme of estimating the power supplied by multiple sources.

3.1. Scheduling scheme

Fig. 3 illustrates the operation of the SS, and the definitions and values of the symbols used in Fig. 3 are summarized in Table 1.

The operation of the SS is explained in detail as follows. The microcontroller board of the SS measures the capacitor voltage of the SS and then calculates the energy stored in the capacitor.

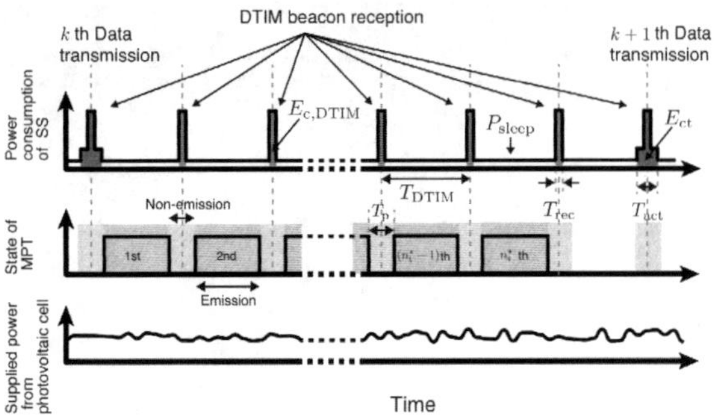

Figure 3. Operation flow.

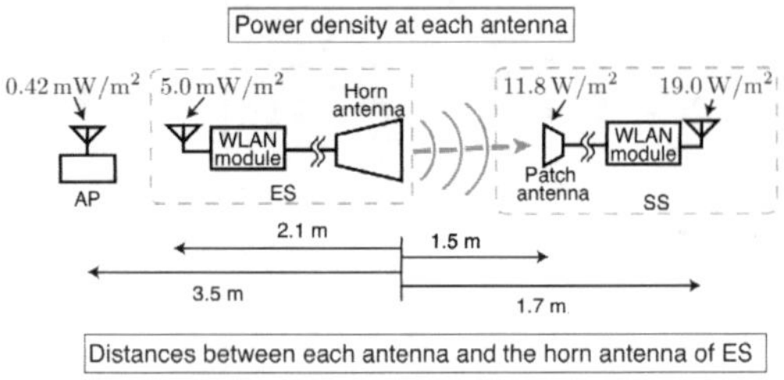

Figure 4. Position of each device and power density at each antenna in experimental setup.

Using the information on the energy, the microcontroller board determines the MPT duration using the scheme described in Sect. 3.2. The microcontroller board switches the WLAN module from sleep mode to active mode and forwards the transmitted data to the WLAN module. The data are transmitted to the ES when the WLAN module is switched to awake mode for receiving beacon frames, which contain delivery traffic indication map (DTIM) information , called DTIM beacon. The DTIM beacon is broadcasted at predetermined intervals. After transmitting the data frame and receiving the DTIM beacon, the WLAN module is switched to sleep mode. The microcontroller board is switched from sleep mode to awake mode after the MPT duration, and then the SS measures the capacitor voltage.

The operation of the ES is explained as follows. The ES receives data frames from the SS and then transmits microwave power during the MPT duration determined by the SS such that microwave for MPT does not influence on data transmission. In addition, the ES also acts as a data sink.

3.2. Estimation of power supplied by multiple sources and determination of MPT duration

In the proposed scheme, the SS operates to make the capacitor energy converge to a predetermined target value E_{target}. The reason is to prevent overcharging and energy shortage.

First, we define the symbols used in our proposed scheme. Let $V_{\text{SS},k}$ and $e_{\text{SS},k}$ denote the voltage of the capacitor and the energy stored in the capacitor at the time of the k-th data frame transmission of the SS, respectively. In addition, we let $p_{\text{micro},k}$ and $p_{\text{solar},k}$ denote the power supplied by the MPT source and EH source, respectively, estimated at the time of the k-th data transmission, and also let $n^{\star}_{\text{t},k}$ denote the coefficient for determining the MPT duration $(n^{\star}_{\text{t},k}+1)T_{\text{DTIM}}$.

Next, we explain our proposed scheme. The SS measures $V_{\text{SS},k}$ and uses it to calculate $e_{\text{SS},k}$. Here, the SS has information on $e_{\text{SS},k-1}$, $p_{\text{micro},k-1}$, $p_{\text{solar},k-1}$, and $n^{\star}_{\text{t},k-1}$. In our proposed scheme, the following equation is satisfied:

$$e_{\text{SS},k}-e_{\text{SS},k-1}=p_{\text{micro},k-1}\cdot n^{\star}_{\text{t},k-1}\cdot(T_{\text{DTIM}}-T_{\text{p}})+p_{\text{solar},k-1}\cdot(n^{\star}_{\text{t},k-1}+1)T_{\text{DTIM}}-\hat{e}_{\text{c}}(n_{\text{t},k}), \tag{1}$$

where $\hat{e}_{\text{c}}(n_{\text{t},k})$ represents the power consumption during the $(k-1)$-th and k-th data transmissions. The SS estimates the power supplied by either the MPT or photovoltaic generation assuming that the other type of supplied power has not increased or decreased since the last estimated value. In other words, we assume that if $k=0$ then $p_{\text{micro},k}=0$, if k is even and $k\neq 0$ then $p_{\text{micro},k}=p_{\text{micro},k-1}$, and if k is odd then $p_{\text{solar},k}=p_{\text{solar},k-1}$. On this assumption we calculate the value for the other type of power using (1).

Table 1. Definitions and values of parameters.

Symbols	Definitions	Values
$E_{c,DTIM}$	Energy consumption of SS for DTIM reception	17.4 mJ
E_{ct}	Energy consumption of SS for data frame transmission	21.51 mJ
T_{DTIM}	DTIM interval	10.24 s
T_{rec}	Duration of SS energy consumption to receive DTIM beacon	40 ms
T_{p}	MPT guard time for data transmission	2.0 s
T_{act}	Duration of active mode of SS	90 ms
P_{sleep}	Power consumption of SS in sleep mode	6.6 mW

Finally, we explain a method to determine MPT duration. The SS determines $n^{\star}_{t,k}$ using the following equation:

$$\Delta(n_{t,k}) = E_{target} - [e_{SS,k} + \hat{e}_t(n_{t,k}) - \hat{e}_c(n_{t,k})], \tag{2}$$

$$n^{\star}_{t,k} := \begin{cases} 0 & \text{if } k{=}0, \\ 0 & \text{if } k \neq 0 \text{ and } \Delta(n_{t,k}) \leq 0, \forall n_{t,k}, \\ \underset{n_{t,k} \in \mathcal{N}}{\arg\min}\ \Delta(n_{t,k}) & \text{otherwise}, \end{cases} \tag{3}$$

where $\mathcal{N} = \{n | \Delta(n_{t,k}) > 0, n = 1, \ldots, N\}$, $\hat{e}_t(n_{t,k})$ represents the power supplied during the $(k-1)$-th and k-th data transmissions, and N represents the predetermined maximum value of $n^{\star}_{t,k}$. The SS also informs the ES of the MPT duration $(n^{\star}_{t,k} + 1)T_{DTIM}$. Then $\hat{e}_t(n^{\star}_{t,k})$ and $\hat{e}_c(n^{\star}_{t,k})$ in the above equation are defined as

$$\hat{e}_t(n_{t,k}) = (n_{t,k} + 1) \times p_{solar,k} \times T_{DTIM} + n_{t,k} \times p_{micro,k} \times (T_{DTIM} - T_p), \tag{4}$$

$$\hat{e}_c(n_{t,k}) = E_{ct} + P_{sleep} \times (T_{DTIM} - T_{act}) + n_{t,k} \cdot [E_{c,DTIM} + P_{sleep} \times (T_{DTIM} - T_{rec})]. \tag{5}$$

4. Experiment

4.1. Experimental parameters

Fig. 4 illustrates the position of each device, the distances between each pair of antennas, and the power density at each antenna. Here, the input power of the horn antenna is 10.3 W when the ES transmits microwave power. The SS is positioned in front of the horn antenna of the ES. The WLAN modules of the ES and the AP are positioned behind the horn antenna to prevent interference with data communication.

The SS, ES, and AP transmit data using 2.446–2.468 GHz bands, while microwave for MPT uses the 2.457 GHz band. The antenna gain of the horn antenna and the patch antenna is 16.1 dBi and 7.7 dBi, respectively. The capacitance of the capacitor of the SS is 10 F, and the withstand voltage of the capacitor is 2.7 V. In addition, E_{target} is set to be 31.25 J and N is set to be 5. The DC–DC converter is used to convert the capacitor voltage to 3.3 V, which is required for operating both the microcontroller board and the WLAN module.

4.2. Experimental results

Fig. 5 shows the transition of e_{SS} when a photovoltaic cell is used, and Fig. 6 shows the results without a photovoltaic cell. The shaded areas in the figures illustrate the MPT duration. Figs. 5

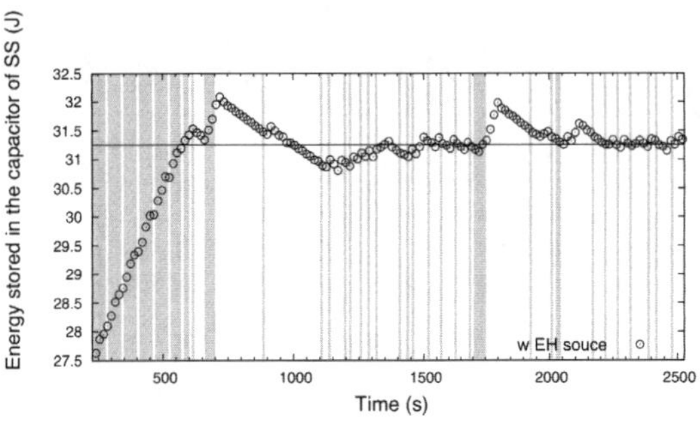

Figure 5. Transition in $e_{\rm SS}$ using both microwave power transmission and photovoltaic generation are used.

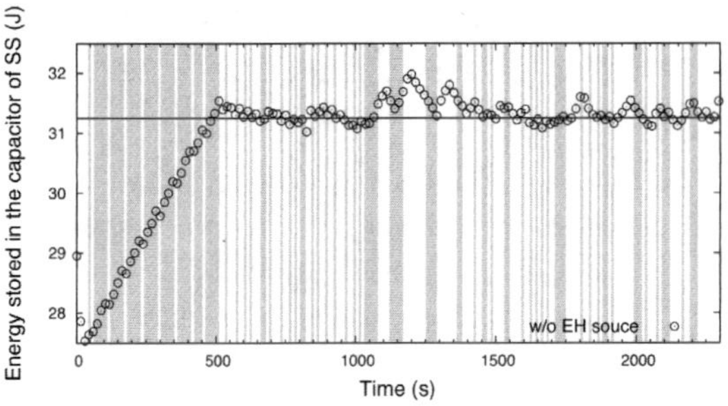

Figure 6. Transition in $e_{\rm SS}$ using only microwave power transmission is used.

and 6 show that $e_{\rm SS}$ successfully converged to $E_{\rm target}$. These results indicate that the power supplied by the MPT source and the photovoltaic cell was successfully estimated. In addition, compared with Fig. 6, the MPT duration after convergence of $e_{\rm SS}$ to $E_{\rm target}$ in Fig. 5 was reduced.

5. Conclusion

In this paper, we proposed a scheme for estimating the power supplied by an MPT source and that of an EH source on the basis of the existing MPT scheduling scheme and then conducted an experiment in which a sensor is powered with multiple sources. We confirmed the feasibility of the scheme; that is, compared with the MPT duration for a sensor without a photovoltaic cell, that for a sensor with a photovoltaic cell is reduced. Thus, by using a photovoltaic cell, serious interference between the MPT and WLAN data transmission is avoided more effectively.

Acknowledgments

This work was supported in part by JSPS KAKENHI Grant Number 24360149. This experiment was conducted using the Microwave Energy Transmission Laboratory (METLAB) system of the Research Institute for Sustainable Humanosphere at Kyoto University.

References

[1] F. L. Lewis, "Wireless sensor networks," in Smart environments: Technologies, protocols, and applications, pp. 11–46, John Wiley & Sons, 2004.
[2] N. Shinohara, "Power without wires," IEEE Microw. Mag., vol. 12, no. 7, pp. S64–S73, Dec. 2011.
[3] N. Imoto, S. Yamashita, T. Ichihara, K. Yamamoto, T. Nishio, M. Morikura, and N. Shinohara, "Experimental investigation of co-channel and adjacent channel operations of microwave power and IEEE 802.11g data transmissions," IEICE Trans. Commun., vol. E97-B, no. 9, pp. 1835–1842, Sept. 2014.
[4] S. Yamashita, N. Imoto, T. Ichihara, K. Yamamoto, T. Nishio, M. Morikura, and N. Shinohara, "Implementation and feasibility study of co-channel operation system of microwave power transmissions to IEEE 802.11-based batteryless sensor," IEICE Trans. Commun., vol. E97-B, no. 9, pp. 1843–1852, Sept. 2014.
[5] V. Shnayder, M. Hempstead, B. Chen, G. W. Allen, and M. Welsh, "Simulating the power consumption of largescale sensor network applications," Proc. SenSys'04, pp. 188–200, Nov. 2004.
[6] K. Tutuncuoglu and A. Yener, "Optimum transmission policies for battery limited energy harvesting nodes," IEEE Trans. on Wireless Commun, vol. 11, no. 3, pp. 1180–1189, Mar. 2012.
[7] R. Moraisa, S. G. Matosb, M. A. Fernandes, A. L. Valente, S. F. S. P. Soares, P. J. S. G. Ferreira, and M. J. C. S. Reis, "Sun, wind and water flow as energy supply for small stationary data acquisition platforms," Comput. Electron. Agric, vol. 64, no. 2, pp. 120–132, Dec. 2008.

Interference Cancellation for Coexisting Wireless Data and Power Transmission in the Same Frequency

Keita Yamazaki[1], Yusuke Sugiyama[1], Yoshihiro Kawahara[2], Shunsuke Saruwatari[1], Takashi Watanabe[3]

[1]Graduate School of Informatics, Shizuoka University, Japan
[2]The department of Information and Communication Engineering, The University of Tokyo, Japan
[3]Graduate School of Information Science and Technology, Osaka University, Japan

E-mail: [1]{yamazaki, sugiyama}@aurum.cs.inf.shizuoka.ac.jp, saru@inf.shizuoka.ac.jp, [2]kawahara@akg.t.u-tokyo.ac.jp, [3]watanabe@ist.osaka-u.ac.jp

Abstract. Combining wireless transmission of data and power signals enables wireless sensor networks to drive perpetually without changing batteries. To achieve the simultaneous data and power transmission, the present paper proposes power signal interference cancellation for wireless data and power transmission at the same time in the same frequency. We evaluate the performance of the proposed power signal interference cancellation using Universal Software Radio Peripheral N200 (USRP N200) software defined radio. Evaluations show that the proposed interference cancellation is feasible to receive data while transmitting power.

1. Introduction

Various devices are connected to wireless networks and exchange various information with other devices. Wireless power transmission [1] enables wireless devices to be powered permanently without the need for changing and charging batteries. The combination of wireless data and power transmission will lead to the invention of new types of wireless devices.

In this study, we use radio waves as the transmission medium as they offer two advantages: (i) long transmission distance and (ii) little calibration effort. The present paper focuses on the simultaneous data and power transmission in the same frequency to mitigate the depletion of radio resources. However, the simultaneous data and power transmission induces much interference between the data and power signals.

The present paper proposes power signal interference cancellation to achieve the simultaneous transmission even if the interference occurs. The interference cancellation subtracts a power signal from a received signal. To evaluate the performance of the proposed interference cancellation, we built an initial experiment using USRP N200 software defined radio [2] with a USRP hardware driver (UHD) [3] and C++. The evaluations show that the proposed interference cancellation reduces the power of power signal by 20 dBm at maximum.

2. Power signal interference cancellation

Figure 1 shows the final goal of our study. We assume that multiple access points are capable of transmission and reception at the same time in the same frequency. The access points cooperatively transmit data and power signals to wireless sensor nodes over a wireless network.

The wireless sensor nodes send back a data signal using the received and charged power. At each access point, the transmitted data signal from each sensor node collides with its own transmitted power signal. To extract the data signal from the collided signal, each access point needs to use interference cancellation. However, the conventional interference cancellation only assumes a collision occurred by data transmissions [4]. To handle a collision occurred by data and power transmission, this paper proposes power signal interference cancellation. The proposed interference cancellation subtracts a known power signal from the collided signal to extract a data signal.

3. Implementation of sender and receiver

3.1. Overview

Figure 2 shows the experimental setup for the proposed interference cancellation. The experimental setup consists of two senders TX_1 and TX_2, one receiver RX, three control computers, and a Fury desktop used as an external clock. The senders and the receiver were implemented by USRP N200s and connected to each control PC (Panasonic Let's note CF-B11). To reduce frequency offsets among the USRPs, a 10-MHz signal was input from the Fury desktop to all USRPs.

Figure 3 shows transmitted and received signals as a function of time. TX_1 periodically sends an Orthogonal Frequency-Division Multiplexing (OFDM) signal as a data signal. TX_2 continuously sends a sine wave as a power signal. The details of the transmission signals are described in §3.2. The OFDM signal collides with the power signal at RX. RX extracts the OFDM signal by subtracting the power signal from the collided signal and decodes the extracted OFDM signal. The details of the receiver operation are described in §3.3.

3.2. Transmission signal

TX_1 sends an OFDM signal as a data signal. The number of subcarrier is 52 and the length of Fast Fourier Transform (FFT) is 64 to consider a guard band. We used three known OFDM symbols as preambles for frame detection and channel estimation. First OFDM symbol assigns known preamble values to the odd-numbered subcarrier to detect the start position of the frame by the self-correlation function. After Inverse FFT (IFFT), the first half of the first OFDM symbol is the same as the second half of the first OFDM symbol. Second and third OFDM symbol assign known preamble values to the 52 subcarriers for channel estimation. After the preambles, we assign modulated constellations to 48 subcarriers as data symbols and Binary

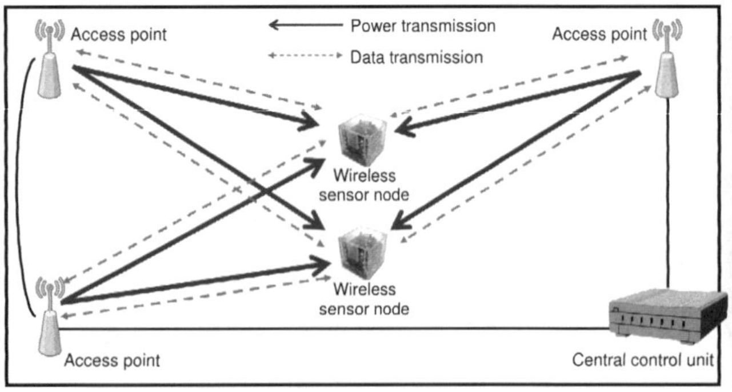

Figure 1. Final goal

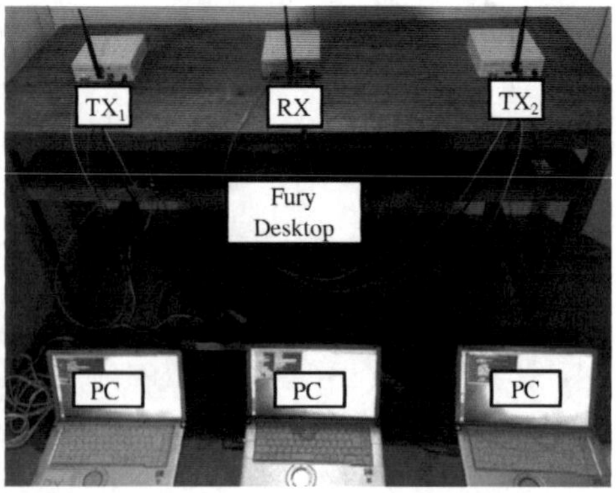

Figure 2. Experimental setup for power signal interference cancellation

Phase-Shift Keying (BPSK) signals to 4 subcarriers as pilot symbols. After IFFT, we copy 16 samples as cyclic prefix from the last to the first to mitigate Inter-Symbol Interference (ISI).

TX_2 sends a sine wave as a power signal. The sine wave $y[n]$ is described as follows:

$$y[n] = A\left(\cos\left(\frac{2\pi n}{N}\right) + i\sin\left(\frac{2\pi n}{N}\right)\right) \tag{1}$$

where n is the number of sample points, A is amplitude, and N is the number of samples in a symbol.

3.3. Receiver operation

Figure 4 shows the receiver operation to extract a data from a collided signal. The operation consists of 1) reproduction of a power signal, 2) extraction of a data signal, 3) symbol synchronization of an OFDM signal, 4) OFDM demodulation, 5) BPSK demodulation.

1) The receiver extracts samples from the received power signal and reproduces a power signal when a sender sends only a power signal. To reproduce the power signal, the receiver estimates the amplitude of the power signal by two level estimation: coarse estimation and fine estimation.

The coarse estimation extracts N samples in Eq. (1) from the received signal. The receiver regards the biggest amplitude in the samples as the estimated amplitude of the power signal. Note that we assume that N is the known value.

The fine estimation reproduces the power signal based on the estimated amplitude. The reproduced power signal approximates to the received power signal. The reproduction consists of three steps. First, the receiver adds the value from -0.01 to 0.01 in steps of 0.0001 to the estimated amplitude and lines up candidates for the amplitude of the power signal. Second, the receiver substitutes each candidate for amplitude A in Eq. (1) and generates a sine wave of one wavelength as a candidate of the power signal. Third, the receiver determines the amplitude, which achieves Minimum Mean-Squared Error (MMSE) between the generated sine wave and the received signal. To calculate MMSE, the receiver needs to detect the start position of the calculation. The receiver generates a sine wave based on Eq. (1) using the estimated amplitude in the coarse estimation. The receiver calculates MMSE between the generated sine wave and the received signal, which is shifted 0–N samples. After the calculation, the receiver detects the start position, which achieves the minimum MMSE.

2) The receiver subtracts the reproduced power signal in 1) from the received signal to obtain a subtracted signal. Figure 5 (a), (b), and (c) show the received signal, the generated power signal, and the subtracted signal, respectively.

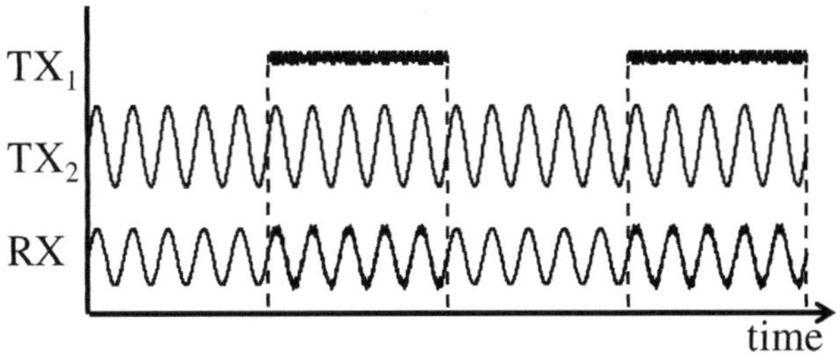

Figure 3. Transmitted and received signals as a function of time

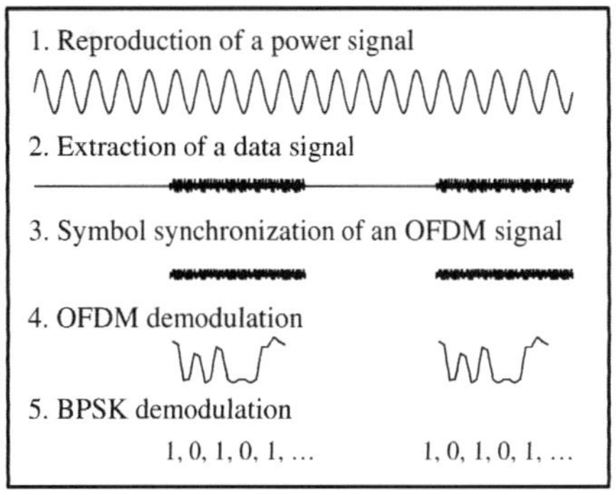

Figure 4. Receiver operation

3) After the subtraction, the receiver detects a frame. The receiver uses the first preamble of the subtracted signal to detect the frame. Since the first half of the first OFDM symbol is the same as the second half of the first OFDM symbol, 48 samples in the first OFDM signal is the same as 48 samples in the half-wave delayed OFDM signal. The receiver calculates the self-correlation of two signals for half-wave length by convolutional integral. After the calculation, the receiver detects peaks in the cyclic prefix. However, when the received power is large and the Auto-Gain Control (AGC) does not converge, the receiver detects a wrong peak. To prevent the detection of the wrong peak, the receiver normalizes the value of self-correlation by dividing by the received power. By calculating the moving average for the normalized self-correlation among 16-samples window, a peak appears in the start position of the frame. The receiver detects the peak based on a threshold obtained by experimental measurement and determines the start position of the frame.

4) The receiver removes the cyclic prefix and extracts a signal on subcarriers by FFT. The receiver removes the first 16 samples from each symbol, which are the cyclic prefix, based on the start position of the symbol synchronization of an OFDM signal.

5) The receiver obtains a bit stream from the extracted signal. When the real value of the extracted signal is greater than 0, the receiver decodes 1. On the other hand, when the real value is smaller than 0, the receiver decodes 0.

4. Evaluation

We evaluated power signal interference cancellation operating in 5.11-GHz band at the anechoic chamber in Shizuoka University, Japan. The sampling rate is 195.3125 ksps. The payload size of the OFDM signal is 1500 bytes. Power-signal samples were fixed at 500 samples.

Figure 6 shows the Bit-Error Rate (BER) as a function of Signal-to Noise Ratio (SNR). The evaluation compared theoretical values in additive white gaussian noise (AWGN) channel with experimental results. The power ratio between the OFDM signal and the power signal was -31.23–-5.57 dB. If BER is zero, we plotted 10^{-5} for convenience sake. Figure 6 shows that the digital cancellation is successful because the experimental results approach to the theoretical results.

Next, we evaluated the decoding rate as a function of the power ratio. To analyze the difference in performance between cancellation of a known power signal and an unknown data signal, the decoding rates of conventional and our implemented interference cancellation of the power signal were compared.

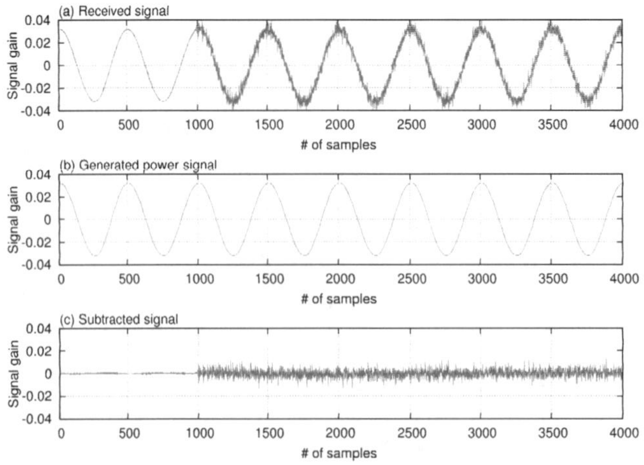

Figure 5. Signal before and after subtraction

Figure 7 shows the decoding rate as a function of the power ratio between the two signals. Data and power are proposed for power signal interference cancellation, whereas only data signals are used in conventional SIC. We calculated the power ratio with the received-signal strength indicator (RSSI) of the two signals, P_a and P_b as $10\log(P_a/P_b)$. In the data and power signal scenario, P_a and P_b are the RSSI of OFDM and power signals, respectively. In the data only scenario, P_a and P_b are the RSSI of weak and strong OFDM signals, respectively. Figure 7 shows the following:

(i) The decoding rate is small when the power ratio is small. The dynamic range of the data signal becomes smaller because AGC controls the gain based on the power signal. When the power ratio is small, the RSSI of the power signal is larger than that of the data signal. For example, when the power ratio is −30 dB, the RSSI of the power signal is 1000 times larger than that of the data signal.

(ii) The decoding rate of data and power achieved is 100% when the power ratio is higher than approximately −20 dB. This means that digital interference cancellation is successful when the RSSI of the power signal is about 100 times larger than that of the data signal. When the power ratio is smaller than −20 dB, the receiver needs to combine analog interference cancellation [5] to receive the data signal.

(iii) The decoding rate of data and power achieved is larger than that achieved with data and data. Unlike the waveform of a power signal, the waveform of a data signal is unknown. When the decoding of a strong data signal fails, the decoding of a weak signal also fails automatically.

5. Conclusions

The present paper proposed power signal interference cancellation for wireless data and power transmission at the same time in the same frequency. Currently, we study a frequency sharing protocol to improve frequency utilization by intentional collision of data transmission and electric power transmission.

References

[1] Bieler T, Perrottet M, Nguyen V and Perriard Y, *IEEE Transactions on Industry Applications* pp 1266–1272, 2002.
[2] Ettus Research, USRP N200/N210 Networked Series, 2012.
[3] Blum J, *In GNU Radio Conference (GRC'11)*, 2011.
[4] Halperin D, Anderson T and Wetherall D, *Proceedings of the 14th ACM Annual International Conference on Mobile Computing and Networking (MobiCom'08)* (San Francisco, California) pp 339–350, 2008.
[5] Jain M, Choi J I, Kim T M, Bharadia D, Seth S, Srinivasan K, Levis P, Katti S and Sinha P, *Proceedings of the 17th ACM Annual International Conference on Mobile Computing and Networking (MobiCom'11)* (Las Vegas, NV) pp 301–312, 2011.

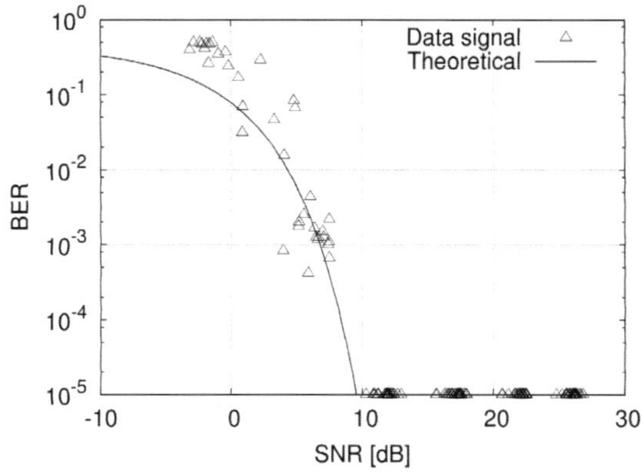

Figure 6. BER as a function of SNR

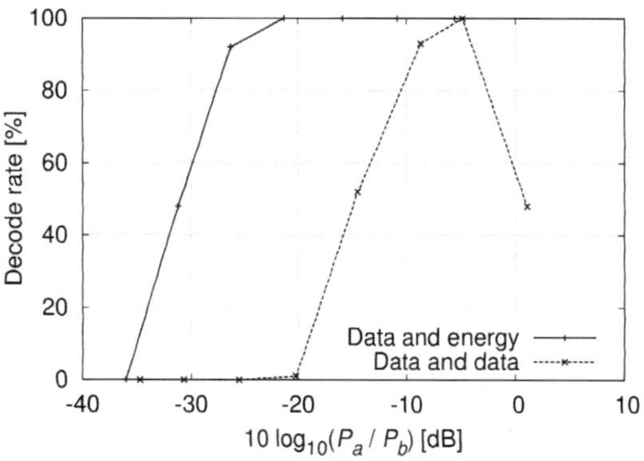

Figure 7. Decoding rate as a function of power ratio between OFDM signal and power signal

Experimental Study of RF Energy Transfer System in Indoor Environment

S.-E. Adami, P. P. Proynov, B. H. Stark, G. S. Hilton and I. J. Craddock
Faculty of Engineering, University of Bristol, Bristol BS8 1TR, UK

E-mail: salah.adami@bristol.ac.uk

Abstract: This paper presents a multi-transmitter, 2.43 GHz Radio-Frequency (RF) wireless power transfer (WPT) system for powering on-body devices. It is shown that under typical indoor conditions, the received power range spans several orders of magnitude from microwatts to milliwatts. A body-worn dual-polarised rectenna (rectifying antenna) is presented, designed for situations where the dominant polarization is unpredictable, as is the case for the on-body sensors. Power management circuitry is demonstrated that optimally loads the rectenna even under highly intermittent conditions, and boosts the voltage to charge an on-board storage capacitor.

1. Introduction

Wireless power transfer systems can use beam-forming to increase received power [1], however under standards that limit the radiated power by specifying the EIRP (Equivalent Isotropic Radiated Power) [2], focusing of the power would require the reduction of the transmitted power. This work, by contrast, demonstrates a multi-transmitter approach (Figure 1) for body-worn devices, where only the closest transmitter is powered-up at a time, thereby improving coverage without recourse to beam-forming. A rectenna (rectifying antenna) receives the electromagnetic wave and converts it to DC power, at a voltage that is low, and varies as a function of received power. A boost converter (Figure 2) provides a regulated voltage suitable for a load. An under-voltage lock-out circuit (UVLO) manages the operation of the load as function of available power [3]. The load here is a low-power RF module (74 μJ/transmission typical) which sends feedback to an access point that selects the optimal transmitter.

The paper is structured as follows: Section 2 describes the receiving patch antenna, which has two orthogonally-polarised ports, each having a separate complex-conjugate matched rectifier, in order to optimally receive the dominant transmitted polarisation. Characterisation test results are shown for the rectenna, as is the spatial power distribution for a complete system with four transmitting antennas. A high degree of power variation and intermittency is observed as the body-worn antenna location and orientation changes. Section 3 describes a self-contained and cold-starting power management circuit for highly variable power profiles. The performance of the rectenna with this power management circuit under intermittent RF power input is measured. Comparisons are drawn with commercially available power converters which are not optimised for RF power transfer.

2. RF system design and characterisation

Two rectennas have been designed and optimized for low RF power (Figure 3). The RF/DC rectifier is a passive doubler structure using HSMS2852 Schottky diode [1]. The first design uses a microstrip-based impedance matching filter which matches the complex impedance of the rectifier to 50 Ω at 2.43

GHz. The second design is based on a dual polarisation square patch antenna. Each polarisation has a voltage rectifier. The patch antenna is designed to directly match the complex impedance of each rectifier at 2.43 GHz.

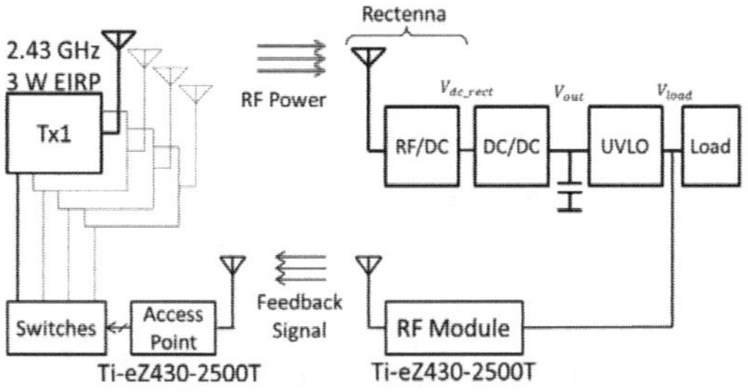

Figure 1. RF power transfer system.

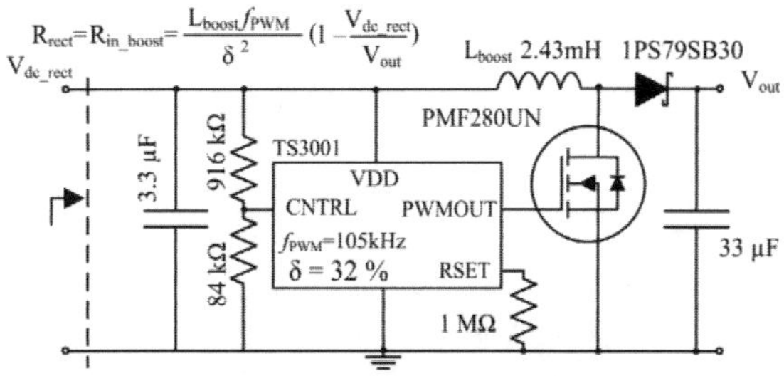

Figure 2. Boost converter.

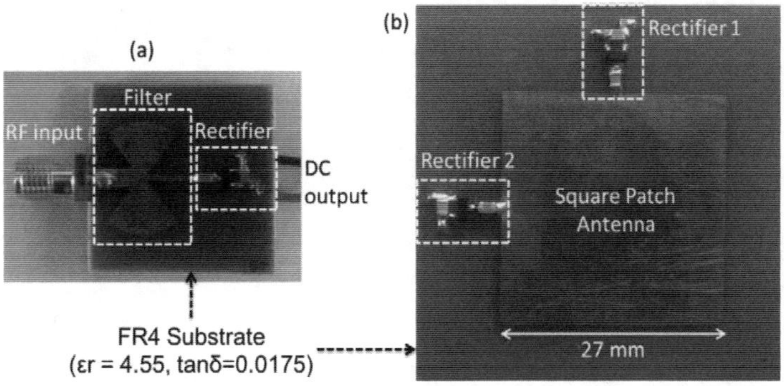

Figure 3. **(a)** Rectifier with 50 Ω matching filter, as used for Figures 4 and 5, **(b)** Dual-polarisation rectenna, as used for Figures 6, 7 and 8.

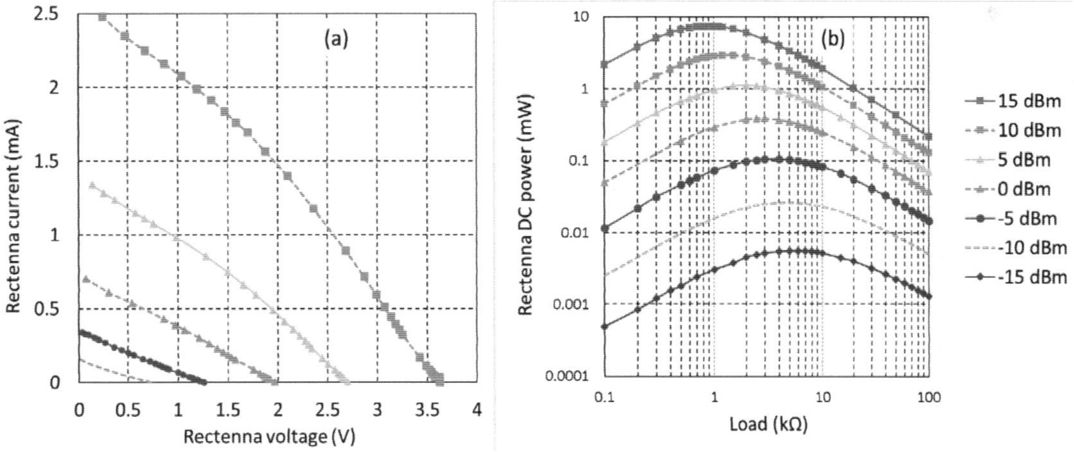

Figure 4. Output characteristics of the 50 Ω rectifier of Figure 3a: (**a**) current vs. voltage and (**b**) DC power vs. load resistance.

The 50 Ω rectifier (Figure 3a) is characterized experimentally, using an RF signal generator at the input and a variable resistive load across its DC output. Figure 4a shows rectified current against rectified voltage at specific RF power levels, and Figure 4b shows rectified output power as a function of load resistance, again as a function of RF power. A linear relation exists between output current and voltage, and the output DC power has a maximum at a certain value of the load, which is itself a function of RF power.

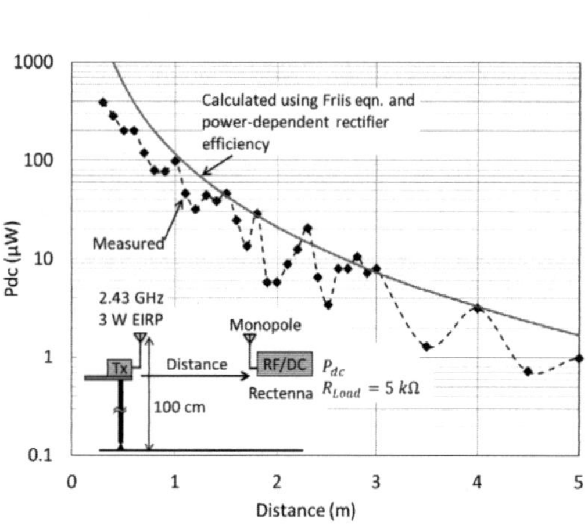

Figure 5. Measured and calculated received DC power against distance, and experimental set-up.

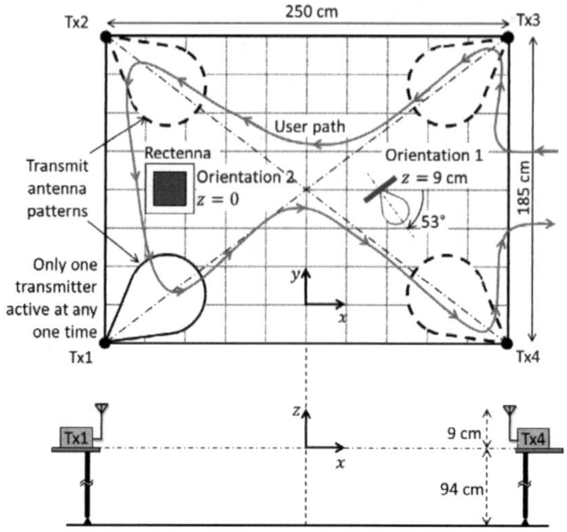

Figure 6. Plan and elevation view of experimental set-up, using four RF sources and two orientations for the dual-polarisation rectenna.

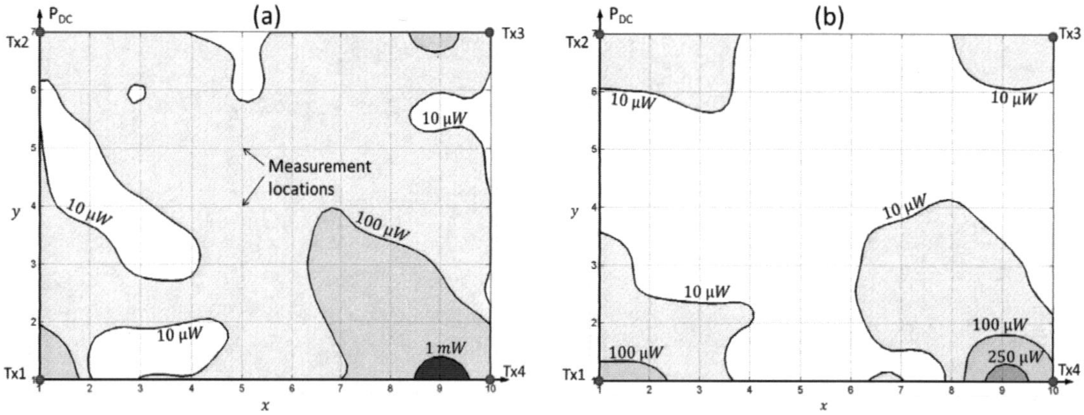

Figure 7. Measured DC power: (**a**) orientation 1 (vertical) and (**b**) orientation 2 (horizontal). (See Figure 6 for orientations). Each rectenna is connected to a 5 kΩ load.

Figure 4 indicates that the rectenna can be modelled in steady-state as a DC voltage source with an internal impedance, roughly equal to $R_{rect} = 5\ \text{k}\Omega$ at low RF power, and decreasing slightly as the RF power increases.

The 50 Ω rectifier is connected to a monopole antenna in order to evaluate the DC power obtained when the receiver is in line of sight of a transmitter. Figure 5 shows that the measured values are in a good agreement with the theoretical calculation using the Friis equation [4].

Four such RF power transmitters are placed in each corner of a rectangular laboratory indoor environment, as shown in Figure 6. The dual polarisation rectenna is used in two orientations: vertical (PCB normal to x-y plane and -53° from x-z plane) and horizontal (PCB in the x-y plane). Figure 7 shows the highest DC power obtainable from the best transmitter (only one transmitter is on at any one time). In the vertical orientation, the most of the room provided DC power levels over 10 µW, Figure

7a, however in the horizontal orientation, this is only achieved in proximity of the transmitters, Figure 7b, as the antenna's favourable radiation direction is normal to the patch (i.e. z-direction).

3. Power Management

The captured DC power for a person moving inside the test area following the path in Figure 6 has been recorded in Figure 8. The variation of the obtained DC power, ranging from a few microwatts to tens of milliwatts, sets a challenging requirement for power management circuit. The boost converter of Figure 2 is designed to operate in discontinuous conduction mode to emulate 5 kΩ at its input at large boost ratios, i.e. where $V_{out} \gg V_{dc\_rect}$. This provides impedance matching to the rectenna. The output is unregulated. The converter starts-up at 0.65 V, and has a quiescent power consumption of below 1 μW [5].

Figure 9 shows the experimental facility used to reproduce intermittent power profiles, in order to test the converter under intermittency. The RF power profile is inferred from the measured DC power of Figure 8. The converter is feeding a 2.5 V regulated rail.

In Figure 10a, the converter's useable output power is shown as P_{Boost_out}, and for reference the power in a 5 kΩ load directly connected to the rectenna. The converter is seen to follow the power profile, starting up at an RF power of around 100 μW. The converter exceeds the power of the load resistance at larger RF powers, as here the natural behaviour of the boost converter provides a better match to the rectenna than the 5 kΩ load does, despite having the quiescent power consumption. This is because, for a fixed switching frequency and duty ratio, the input impedance decreases as the ratio between output to input voltage decreases, which in turn occurs when the input voltage rises due to increasing RF power.

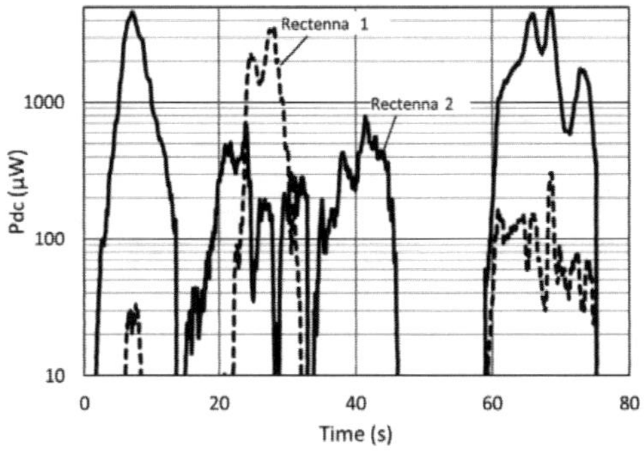

Figure 8. Measured DC power following path in Figure 6 (Blue line).

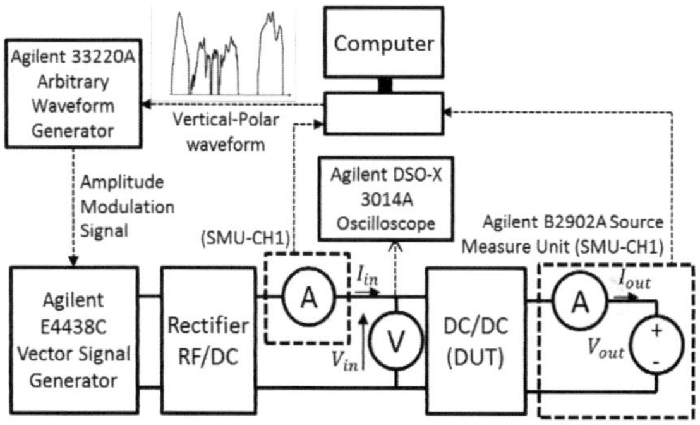

Figure 9. Experimental facility for the comparison of DC-DC converters under intermittent power.

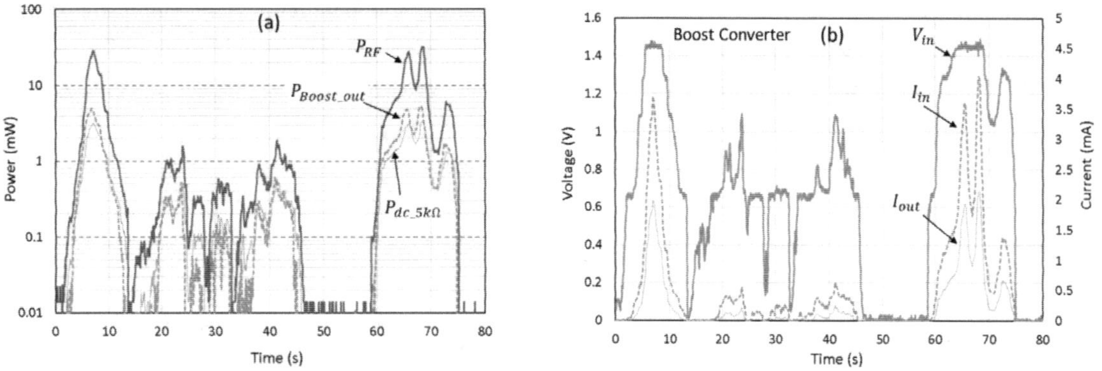

Figure 10. Performance of boost converter: **(a)** RF input power and boost converter's useful output power, power of a 5kΩ for reference. **(b)** boost converter's input current and voltage, and useful output current I_{out} into 2.5 V.

Figure 10b shows the converter input voltage following the power profile, and the converter input and output current I_{out} which charges the on-body system's energy storage. Note that the output voltage is held at 2.5 V by a source meter. The start-up voltage of 0.65 V is clearly visible.
It is therefore worth investigating the potential of commercially available converters with low start-up voltages: Touchstone TS3310 [5] and TI bq25504 [6]. These have been selected for minimum operating power, high conversion efficiency over wide range of input power levels, and low start-up voltage. The TS3310 is input-powered (as is our boost converter), whereas the bq25504 is output-powered. Estimated cold start-up input voltages are 780 mV and 330 mV, respectively.
For the experimental results of Figure 11, the two commercial converters are set to regulate the output voltage to 3 V. The bq25504, is designed for MPPT (maximum power point tracking), and regulates its input voltage to half of the source open-circuit voltage. As this is not suitable for high dynamic variations of rectenna power, this has been disabled as described in [6]. The bq25504 is seen to be drawing a high current from the rectenna which collapses the input voltage and reduces power throughput (Figure 11a). The TS3310 shows a similar input current behaviour to the bespoke converter of Figure 2, however the input voltage remains at the start-up voltage of 800 mV, reducing the effectiveness of the system (Figure 11b). It should be noted that the control systems of these converters could be adapted by the manufacturer to suit RF power transfer and provide far higher performance.

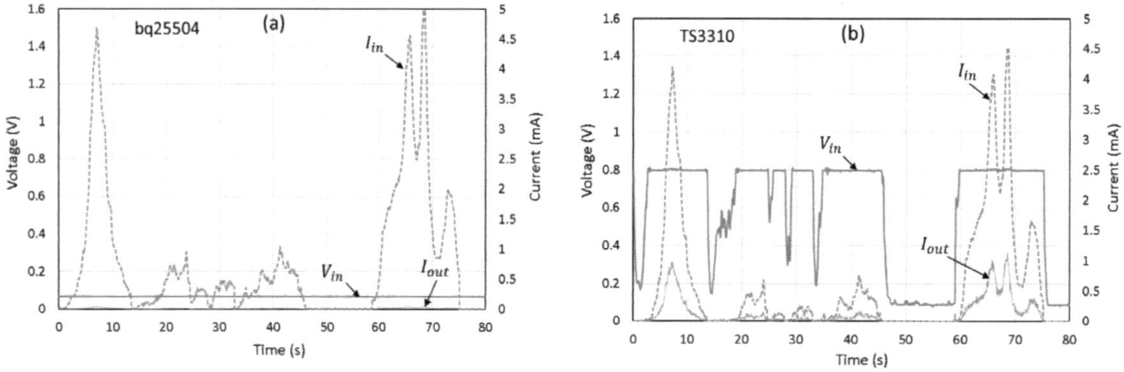

Figure 11. Performance of commercially available converters that are not optimised for RF WPT: **(a)** bq25504 and **(b)** TS3310.

4. Conclusion

Indoor RF power measurements have demonstrated the high level of intermittency that is to be expected from RF power transfer systems that involve body-worn antennas and normal user behaviour. The received power ranges from microwatts to milliwatts. A power management system which uses off-the-shelf components is demonstrated. The designed boost converter is shown to dynamically match the rectenna under these virtually unpredictable intermittent conditions. It has an efficiency of 80% at 400 μW of input DC power, and a low cold-start threshold of 0.65 V.

Acknowledgement: This work was performed under the SPHERE IRC funded by the UK Engineering and Physical Sciences Research Council (EPSRC), Grant EP/K031910/1.

5. References

[1] Visser et al. Loughborough Antennas & Propagation Conference (LAPC), 2012, pp. 1–8.
[2] CEPT, "ERC Recommendation 70-03 Relating to the Use of SRD," 2014, http://www.cept.org.
[3] Szarka et al. IEEE Trans. Power Electron., vol. 28, no. 7, pp. 3353–3362, 2013.
[4] Umeda et al. IEEE J. Solid-State Circuits, vol. 41, no. 1, pp. 35–41, 2006.
[5] TouchStone, TS3310, Datasheet, 2014, http://www.silabs.com/.
[6] Texas Instruments, bq25504, Datasheet, 2014, http://www.ti.com.

Facile 3D Metal Electrode Fabrication for Energy Applications via Inkjet Printing and Shape Memory Polymer

R C Roberts[1], J Wu[2], N Y Hau[3], Y H Chang[3], S P Feng[3] and D C Li[2]

[1] Department of Electrical and Electronic Engineering, The University of Hong Kong, Hong Kong SAR, China
[2] State Key Laboratory of Precision Measuring Technology and Instruments, Tianjin University, Tianjin, China
[3] Department of Mechanical Engineering, The University of Hong Kong, Hong Kong SAR, China

E-mail: robert.roberts@hku.hk

Abstract. This paper reports on a simple 3D metal electrode fabrication technique via inkjet printing onto a thermally contracting shape memory polymer (SMP) substrate. Inkjet printing allows for the direct patterning of structures from metal nanoparticle bearing liquid inks. After deposition, these inks require thermal curing steps to render a stable conductive film. By printing onto a SMP substrate, the metal nanoparticle ink can be cured and substrate shrunk simultaneously to create 3D metal microstructures, forming a large surface area topology well suited for energy applications. Polystyrene SMP shrinkage was characterized in a laboratory oven from 150-240°C, resulting in a size reduction of 1.97-2.58. Silver nanoparticle ink was patterned into electrodes, shrunk, and the topology characterized using scanning electron microscopy. Zinc-Silver Oxide microbatteries were fabricated to demonstrate the 3D electrodes compared to planar references. Characterization was performed using 10M potassium hydroxide electrolyte solution doped with zinc oxide (57g/L). After a 300s oxidation at 3Vdc, the 3D electrode battery demonstrated a 125% increased capacity over the reference cell. Reference cells degraded with longer oxidations, but the 3D electrodes were fully oxidized for 4 hours, and exhibited a capacity of 5.5mA-hr/cm^2 with stable metal performance.

1. Introduction

The current trend towards powerful and compact low-cost portable electronic devices has created a great need for higher performance power sources. To overcome the limits of existing technologies, development of both novel chemistries and device design are being explored. Innovations in metal electrode design have been of great interest to boost the performance, lifetime, and reliability of power generating devices including fuel cells and batteries [1]. Increasing the surface area of device electrodes to increase energy density is one popular strategy employed to boost performance including porous [2] and microfabricated structures [3]. These fabrication strategies show promising performance gains, but the techniques often require complex or time-consuming process steps, making them sub-optimal for low-cost device fabrication. To meet consumer demand for complex portable electronics, further development of high-performance low-cost power sources is paramount, necessitating the development of alternative fabrication techniques.

Inkjet printing is a promising form of additive microfabrication that allows for the direct patterning of structures from metal nanoparticle bearing liquid inks. After deposition, these inks typically require a thermal curing step to drive off the carrier solvents and render a conductive film via nanoparticle sintering [4]. As metal is selectively deposited, material waste and its associated cost can be minimized. Further, this metallization technique is compatible with large area and roll-to-roll substrate processing and can be performed at relatively low-temperatures and at atmospheric pressure, allowing the use of low-cost polymeric substrate materials. Inkjet printing has been employed to fabricate zinc-silver oxide batteries by forming arrays of silver pillars onto device electrodes to increase electrode surface area, however these pillars require time and precision to produce [5].

Shape memory polymers (SMP) are a class of material that undergoes a geometric transformation upon the application of a stimulus. Both polystyrene and polyolefin are two commercially available thermoplastics that can be manufactured to undergo a large anisotropic shape change to shrink its lateral dimensions upon the application of heat above its glass transition temperature but below its melting temperature [6]. Heat-shrink tubing to insulate electronic wiring connections is one popular application of the material. Thermally contracting SMP substrates have previously been explored for microfluidic device fabrication [7]. Upon SMP contraction, patterned areas thicken to form 3D micromold structures. Additionally, SMP substrates have previously been explored for generating wrinkle patterns in sputtered metal films [8]. Substrate contraction led to buckling of the metal thin films under the large compressive stresses to form 3D topologies. By combining inkjet printed metals with SMP contraction, a simple 3D metal electrode fabrication technique can be realized.

In this paper we combine inkjet printed silver and polystyrene shape memory polymer to realize electrode structures for microfabricated batteries. Shrinking the substrate after deposition allows one to overcome the fundamental resolution of the inkjet deposition system to yield high-resolution metal patterns in a single process. After process characterization, the resulting films are studied using scanning electron microscopy. The fabrication process is then used to realize zinc-silver oxide (Zn-AgO) microbatteries that exhibit improved performance over conventional planar thin-film devices. This work represents the first exploration of inkjet-printed metal films on SMP substrates and its application as three-dimensional Zn-AgO microbattery electrodes, demonstrating the suitability of the technique for energy applications.

2. Experimental

Low-cost consumer-grade polystyrene SMP substrate (MagicShrink) was selected for this study. As-received substrate sheets were cut into samples, rinsed in ethanol, and dried with compressed air prior to use. A commercial research-grade inkjet deposition system (Fujifilm Dimatix DMP-2831) equipped with 10pl nozzles was employed in this work using 50nm diameter silver (Ag) nanoparticle ink (ANP DGP-40LT-15C). All deposition occurred with a substrate temperature of 60°C using a nine-step interlacing technique detailed elsewhere [9]. After patterning, all nanoparticle films were placed onto a 6mm aluminium plate in a preheated laboratory oven above 150°C for 10 minutes to induce sintering and generate coherent conductive metal geometries.

Shape memory polymer behaviour was characterized by patterning nominally 3 mm by 3 mm pads onto clean substrates. The as-printed geometries were then measured using optical microscopy (Mitutoyo), before being loaded into a preheated laboratory oven. The temperature was varied from150 to 240°C for a duration of 10 minutes and then the samples removed and allowed to cool to room temperature. The resulting samples where then measured using optical microscopy and the size reduction factor computed. Finally, a scanning electron microscope (Hitachi S-3400) was employed to characterize the topology of the inkjet printed silver on the contracted SMP substrates.

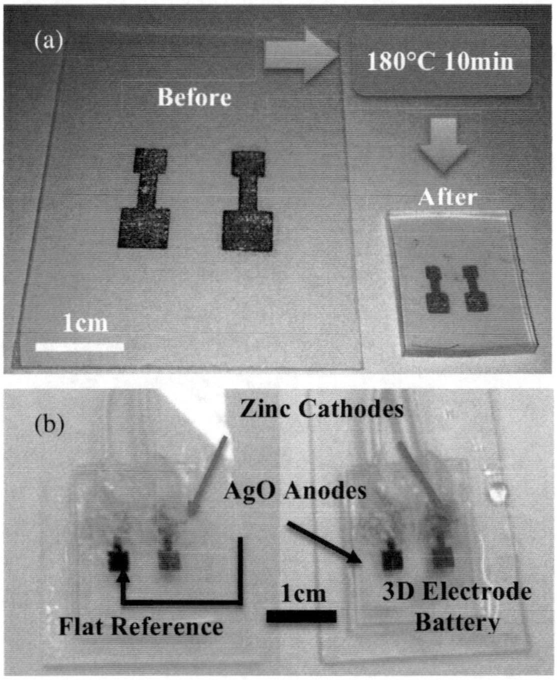

Figure 1. (a) As-printed (left) and post-heated (right) inkjet printed silver electrodes. (b) Packaged batteries after functionalization in KOH with ZnO. Black silver oxide is apparent.

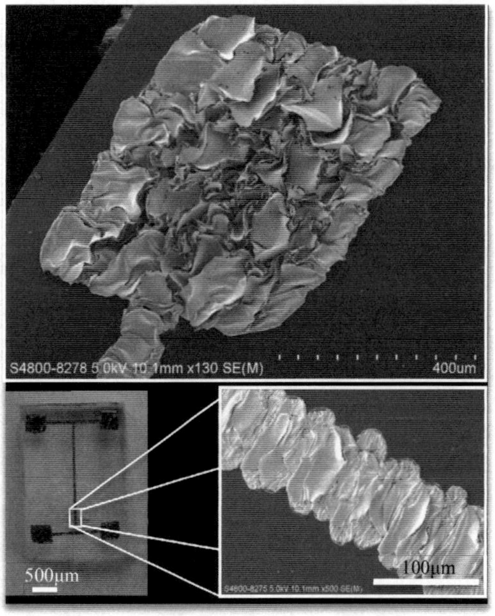

Figure 2. Scanning electron micrographs reveal the 3D topology and large surface area of the buckled printed silver after SMP substrate contraction (180°C for 10 minutes) for a pad (top) and a line (lower-right) on a bridge test structure (optical – lower-left).

Zn-AgO microbattery electrodes were fabricated by printing oversized pairs of silver electrodes onto SMP to yield an area of 3mm by 3mm after substrate contraction. 3mm by 3mm electrode pairs were also printed onto SMP substrates previously shrunk at 180°C for 10 minutes to serve as reference devices. After metal deposition, all substrates were placed in a laboratory oven at 180°C for 10 minutes to contract the substrates and sinter the Ag film. Devices were then connected to lead wires using silver epoxy (Chemtronics CW2400) and encapsulated in two-part epoxy, leaving only the square electrodes exposed. A 10M potassium hydroxide (KOH) solution doped with dissolved zinc oxide (ZnO) powder (57g/L) was selected for the battery electrolyte, and has been used previously [5]. Characterization was performed using an electrochemical workstation (CHI660E). A constant electric potential (3V) was applied across the printed Ag pads, resulting in the oxidation of the positive electrode and zinc (Zn) electrodeposition on the negative electrode to form a galvanic cell. All cells were tested in a two-electrode configuration using a 5mV/s scan rate. Measurements were performed on both 3D and planar electrode cells serving as a reference. After packaging, the resulting electrodes were characterized via optical microscopy to allow for data normalization based on true electrode area.

3. Results and Discussion

Figure 1(a) illustrates the response of shape memory polymer to the application of 180°C for 10 minutes. While the lateral dimension decrease, the substrate thickness increases proportionally. Printed silver appears dull after substrate contraction but exhibits excellent surface adhesion using a scotch-tape test. Scanning electron micrographs of samples after substrate contraction (Figure 2) reveal a highly textured 3D topology. The substrate contraction exerts large compressive stresses on the overlying patterned material, resulting in buckling of the printed Ag film. Out-of-plane buckling and ripples were observed in all features with large areas of metal exhibiting complex surface topology with crumpling and rosette-like shapes. The electrical resistance of the printed silver after SMP

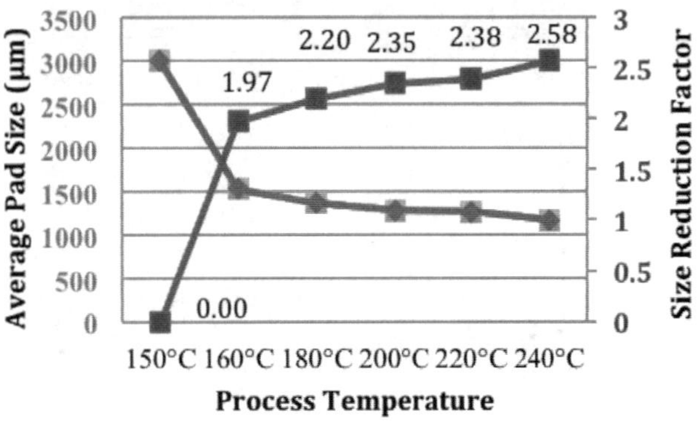

Figure 3. Average pad size (blue) and size reduction factor (red) for 10 samples at each temperature point after a 10-minute oven bake.

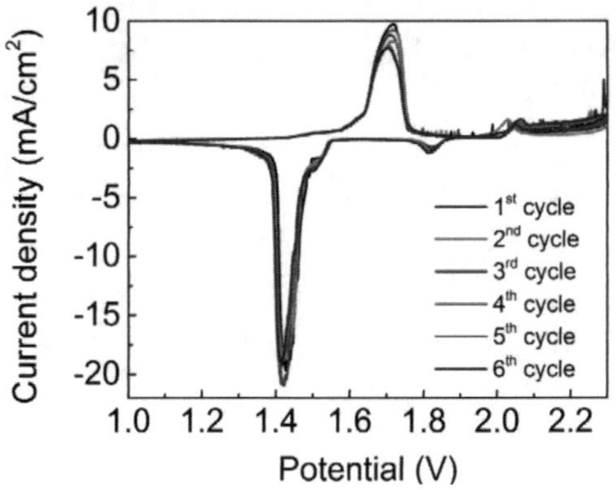

Figure 4. Cyclic voltammograms of the printed 3D silver electrodes on SMP for 6 cycles in 10M KOH electrolyte doped with ZnO (57g/L).

contraction was found to increase by a factor of 4 compared to silver films printed onto pre-shrunk SMP substrates. The average pad size of 3mm by 3mm pads and the corresponding size reduction factor is shown in Figure 3 after a 10-minute oven process for 10 samples at each temperature (error bars shown). The SMP manufacturer recommends a 160°C process temperature, and measurements were carried up to 240°C, showing an increase in SMP contraction with increasing process temperature with low variability. Above 240°C the melting point of polystyrene was reached.

Microbatteries were fabricated on both unshrunk and pre-shrunk SMP substrates, packaged, and functionalized in ZnO doped KOH as shown in Figure 1(b). Black silver oxide was apparent on functionalized microbatteries. During device functionalization, it was found that the planar reference battery electrodes delaminated from the substrate for silver oxidation/zinc electrodeposition times greater than 300s. Similar behaviour has been observed for other printed batteries and is surmised to be due to stress induced during this process [5]. However, no film delamination was observed for 3D electrodes, allowing oxidation/electrodeposition to continue for 4 hours. Enhanced metal/SMP adhesion has previously been observed for gold and should be investigated further for printed materials [8]. Cyclic voltammetry (scan rate: 5mV/s) was then used to characterize the reduction and oxidation regimes of the battery cell over six consecutive cycles of the 3D electrodes as shown in Figure 4. A strong anodic peak was observed at 1.73V with a second smaller peak at 2.05V, which correspond to the formation of monovalent and divalent silver oxide on battery anode. Upon

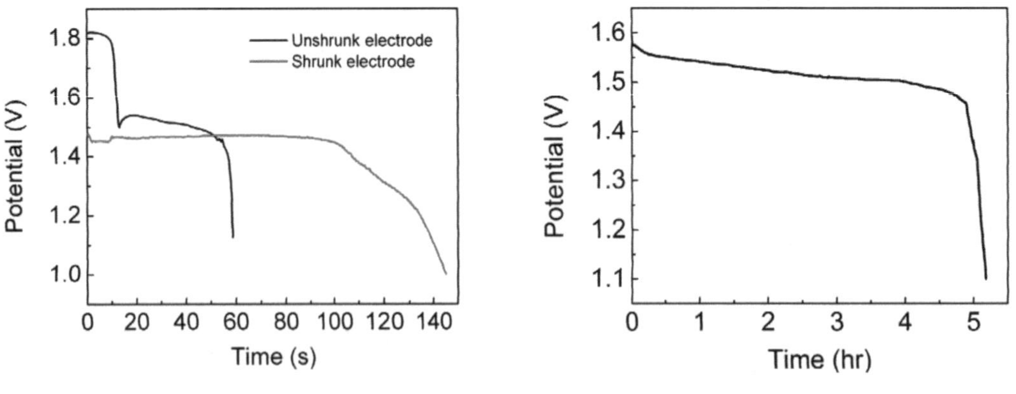

Figure 5. Discharge curves at a constant current of 1.1mA/cm^2, for (left) samples oxidized for 300s, the maximum for the reference sample, (right) and a 3D electrode battery, oxidized for 4 hours.

reversing the scan voltage direction (decreasing), a slight cathodic peak is observed at 1.81V and a strong peak at 1.42V. Over six cycles, the response of all peaks but that a 1.81V decrease in magnitude, suggesting a change in silver oxide stability over time. The microbatteries were then discharged at 1.1mA/cm^2 to test battery performance as illustrated in Figure 5. For 3D electrodes oxidized for 300 seconds to allow comparison the reference cells, the 3D cells exhibited a 125% increase in capacity. As the original electrode size was 120% larger before shrinking, the performance increase scales with electrode surface area. A 3D electrode microbattery was then discharged after a 4-hour oxidation/electrodeposition step, and exhibited an improved capacity of 5.5mA-hr/cm^2 while the underlying metal film did not show signs of delamination from the substrate.

The simple microbattery geometry used in this initial investigation has validated the concept of using inkjet printed silver on shape memory polymer as a simple method to create 3D electrodes to boost performance and stability over planar thin-film electrodes. Further study is needed to optimize the battery design using a more practical configuration such as interdigitated electrodes. The electrolyte solution ultimately degraded our packaging materials, necessitating the development of alternatives.

4. Conclusion

The use of inkjet printed silver on thermally contracting shape memory polymer to form 3D metal electrodes for energy applications has been demonstrated and validated through the realization of a Zn-AgO microbatteries. The size reduction factor of SMP films was found to be highly repeatable at multiple process temperatures, while creating 3D metal topologies. Zn-AgO microbatteries with 3D electrodes demonstrated greater capacity and better metal adhesion compared to the planar references tested. Using low-cost shape memory polymer and inkjet printed metal to more than double the electrode surface area in a simple fabrication process has the potential to improve electrode performance for energy application.

5. Acknowledgement

This work was supported by the National Natural Science Foundation of China (No. 51350110233) and the General Research Fund from Research Grants Council of Hong Kong Special Administrative Region, China (No. HKU 719512E). The authors are grateful to the staff at the Electron Microscopy Unit at The University of Hong Kong for their assistance imaging samples. We would also like to express our appreciation to Guijun Li and Yu-Ting Huang for their contributions.

6. References

[1] Baker J 2008 *Energy Policy* **36** 12 4368-4373

[2] Zhang H, Yu X and Braun P 2011 *Nature Nanotechnology* **6** 277–281

[3] Long J W, Dunn B, Rolison D R and White H S 2004 *Chem. Rev.* **104** 4463–4492

[4] Roberts R C and Tien N C 2013 Multilayer Passive RF Microfabrication using Jet-Printed Au Nanoparticle Ink and Aerosol-Deposited Dielectric *Proc. The 17th International Conf. on Solid-State Sensors, Actuators and Microsystems* (Barcelona, Spain, 16-20 June 2013) **1** 178-181

[5] Ho C C, Murata K, Steingart D A, Evans J W and Wright P K 2009 A super ink jet printed zinc–silver 3D microbattery *J. Micromech. Microeng.* **19** 094013

[6] Chen Z and Krishnaswamy S 2012 *Proc. of SPIE* **8342** 83420Z-1

[7] Grimes A, Breslauer, Long M, Pegan J, Lee L P and Khine M 2008 *Lab Chip* **8** 170-172

[8] Fu C C, Grimes A, Long M, Ferri C G L, Rich B D, Ghosh S, Ghosh S, Lee L P, Gopinathan A and Khine M 2009 *Adv. Mater.* **21** 4472-4476

[9] Li G, Roberts R C and Tien N C 2014 Interlacing Method for Micro-Patterning Silver via Inkjet Printing *Proc. The 13th International Conf. on Sensors* (Valencia, Spain, 2-5 November 2014) Preprint

Fabrication and Testing of a Bi-Conductive Polymer Membrane Fuel Cell

S Hamel[1], T Tsukamoto[2], T Tanaka[2], L G Fréchette[1]

[1]Université de Sherbrooke, Sherbrooke J1K 2R1, Québec, Canada
[2]Tohoku University, Sendai 980-8579, Japan

E-mail: Simon.Hamel@usherbrooke.ca

Abstract. This paper reports the fabrication process and testing of a bi-conductive polymer membrane (BCPM) fuel cell that integrates lateral current collectors on both sides with an ionic conductive path through the membrane. The new membrane shows major advantages over standard Nafion® membranes used in Polymer Electrolyte Fuel Cells (PEMFCs). In addition to being mechanically stable when wet, the flexible BCPM integrates efficient thin film current collectors (ICCs) on an ionic conductive membrane with a high active area ratio. Also, ICCs leave all the surface of the electrode free to eventually integrate a more efficient water and gas management system than traditional gas diffusion layers. Moreover, the fabricated membrane has shown superior volumetric power density than standard PEMFC (0.76 vs 0.47 $mW/cm^2\mu m$).

1. Introduction

Polymer electrolyte fuel cells (PEMFCs) can deliver simultaneously high energy and power densities while providing power as long as they are supplied by reactants. These characteristics give PEMFCs potential advantages compared to lithium batteries to power small portable devices. At small scales however, packaging and peripheral components tend to dominate the size of the cell while poor water management causes non optimal [1] and unstable performance of the stack [2]. These issues stem from the use of gas diffusion layers (GDLs) simultaneously as current, water and air pathway.

GDLs are an inefficient water management solution for small PEMFCs that usually work in air-breathing mode, since they can only delay flooding or drying [2] without preventing it. Moreover, current is generally drawn from the GDLs with metallic grids, called bipolar plates, which are clamped with screws, bolts or by a hot pressing process to reduce electrical contact resistance [3–6]. This approach leads to bulky packaging and leaves no room on top of the electrodes to place a more efficient water or gas management system.

A better approach is to integrate current collectors directly on the ionic conductive membrane thus eliminating bipolar plates. Photosensitive glass and polymer [7,8] have been successfully used as a substrate but the active area in these designs covered only a very small portion of the total surface. A large pattern ($1mm^2$) etched in an SOI wafer filled with ionic conductor and gold current collectors was used by Zhu et al. [9] to obtain substantial power density. In this case however, reducing the pattern size and repeating it multiple times could reduce electrical and ionic losses significantly.

To minimize low power PEMFCs and uncoupling gas, water and electrical path, a first attempt to fabricate a bi-conductive polymer membrane (BCPM) was described in previous work [10] but didn't give the expected results. That membrane used polyimide as support material and had a large ratio of active-to-total area. It was the first to integrate lateral electrical conductive layers on both sides with an ionic conductive path through the membrane. Unfortunately, the process was time consuming and lead to unrepeatable results mainly caused by poor gold adhesion and damaged current collectors during polyimide etching. Moreover, the membrane had gas hermeticity issues due to poor ionic conductor filling of the holes.

Therefore, an improved fabrication process of the BCPM is presented. The final device integrates on the same polymer sheet; mechanical support, ionic conduction and current collection while providing low ionic and electrical resistance and high active area ratio. The membrane is a 50µm thick patterned polyimide sheet (Kapton 200HN, Dupont®) between two Integrated Current Collectors (ICCs) consisting of metallic thin films (chromium, gold and titanium). The active area of the membrane is a square of 0.25cm^2 covered at 20% by through holes. Dimensions were selected to reduce ionic and electrical resistance while easing fabrication.

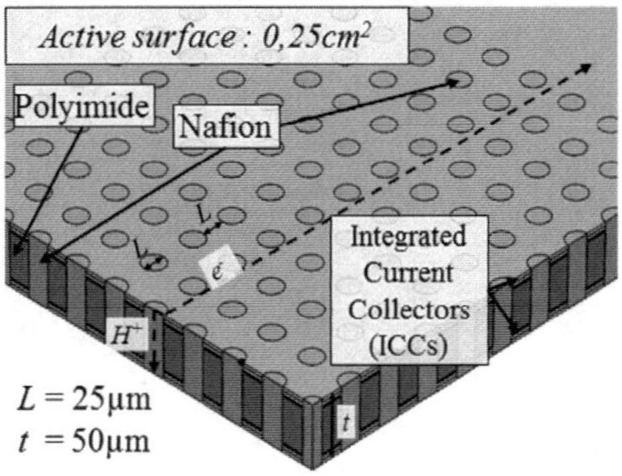

Figure 1 - Isometric cut view of the Bi-Conductive Polymer Membrane (BCPM)

In this work, we report the fabrication techniques that have been used to achieve the desired configuration and the characterization of the final device. The hermeticity, mechanical stability and efficiency of the ICCs are first tested. Then, the performances of the BCPM fuel cell are measured and compared to a standard PEMFC.

2. Fabrication process

The fabrication process of the BCPM (Figure 2) is subdivided in five major steps: ICCs deposition (Figure 2a-f), holes etching (Figure 2g), Nafion filling (Figure 2h-i), Nafion excess removal (Figure 2j-k) and electrode spray coating (Figure 2l). To solve the problems of gold adhesion, damaged ICCs and Nafion filling encountered in the first fabrication approach [10], two major changes have been made. First, metal wet etching and oxygen plasma etching of Kapton have been replaced by one quick and simple step of laser etching. This allows the use of an adhesion layer for gold in addition of being fast, low cost and repeatable. Secondly, a shear stress removal technique has been developed to remove excess Nafion and exposing the conductive layers while keeping the holes filled.

To start, a layer of αgel (Taica®) 0.5mm thick thermo-conductive rubber is laid on a silicon substrate (Figure 2a) and a Kapton 200HN sheet (50µm thick, Dupont®) is rolled on top of it (Figure 2b). Three metallic layers are then evaporated in the same step; chromium (30nm) serves as an adhesion layer for the gold conductive layer (300nm) and titanium (30nm) protects the gold in the subsequent steps. The metallized Kapton is peeled-off (Figure 2d) and cleaned with acetone and ethanol. The membrane is flipped over (Figure 2e) and the evaporation process is repeated (Figure 2f). The metallized membrane is removed from the αgel and prepared for laser etching.

Holes of 25µm in diameter separated by 25µm are etched with an excimer laser (Protolaser U3, LPKF) through the metallic and polymer layers on a 5mm x 5mm area of the metallized Kapton (Figure 2g).

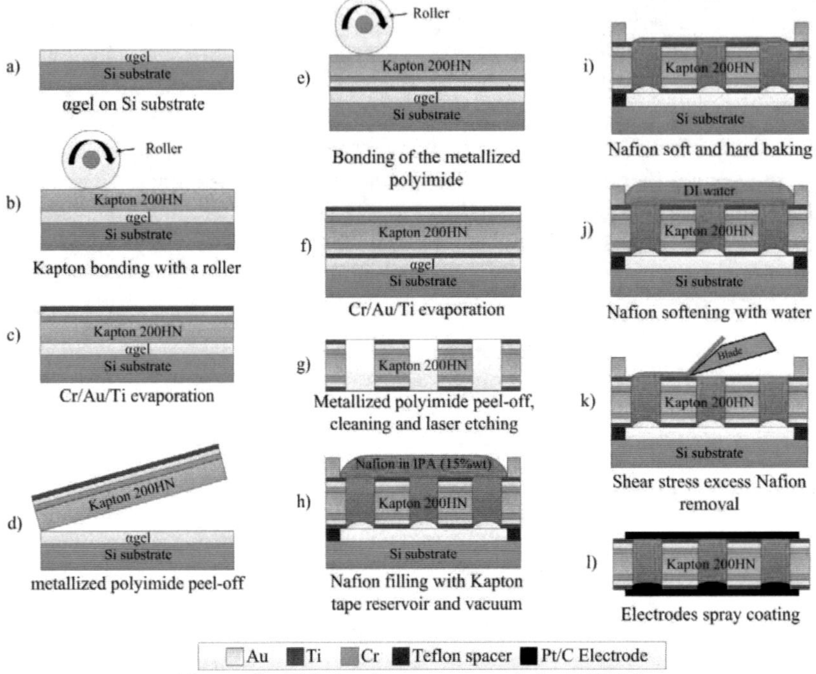

Figure 2 - Fabrication process of the BCPM

The 10 000 holes must be completely filled with Nafion to ensure ionic conductivity trough the membrane and to prevent gas leakage from the anode to the cathode. To do so, the membrane is reattached to a silicon substrate using Kapton tape and Nafion in solution (Liquion™ 15 %wt, IonPower) is deposited on the metallized Kapton surface (Figure 2h). After a 30s vacuum, the assembly is put on a hot plate at 60°C for 30min and IPA drops are periodically added to the Liquion to dilute it and prevent cracking. The temperature is then raised to 100°C for 15min to eliminate all solvent (Figure 2i).

A 10min soaking in DI water (Figure 2j) softens the Nafion and makes it easier to remove by a shear stress technique. A 23µm thick sharp blade is inserted underneath the excess Nafion layer and is slid over the etched area (Figure 2k).

To complete the fuel cell, porous electrodes are deposited both sides of the membrane. A mix of Pt/C, carbon black, Liquion and IPA is sprayed on both sides of the membrane, following standard procedures [11] (Figure 2l).

3. Experimental Characterisation

After the Nafion shear stress removal step (Figure 2k), the etched region of the membrane is observed with SEM to evaluate the efficiency of the removal technique. Figure 3a shows that the metal layer that forms the ICC is free of Nafion while the holes seem to be entirely filled. This is also confirmed with a cross-section view of a filled hole (Figure 3b). Afterwards, a differential pressure test validates that the membrane will not allow gas leakage from the anode to the cathode. A four point probing station was also used to validate the electrical conductivity of the ICCs to ensure that its surface is indeed free of Nafion.

Mechanical stability in water of the BCPM is confirmed by immersing it in DI water for 10min and visually comparing the dry and wet states. The results show (Figure 3c) that the BCPM is not mechanically affected by water which is not the case for standard PEMFC that is substantially deformed when in contact with water.

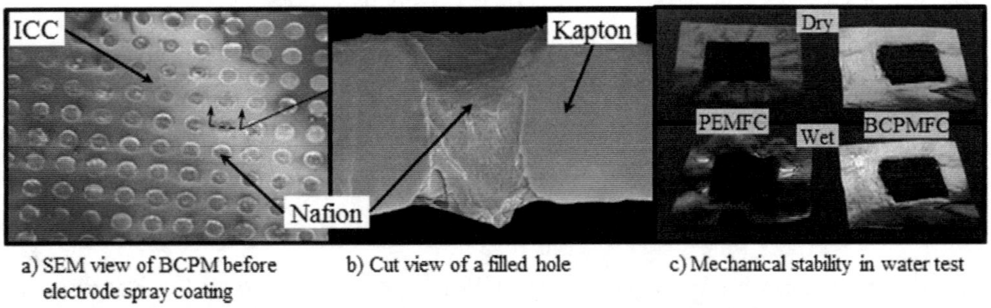

Figure 3 - Visual inspection of the BCPM and mechanical stability result.

Finally, polarization curves of the BCPM with two modes of current collection are performed to validate the efficiency of ICCs. First, traditional current collection normal to the electrodes using bipolar plates is used to draw the current from the top of the electrodes instead of using the ICCs as a reference condition (Figure 4a). Then, measurements are also taken for the same fuel cell but using the ICCs to extract the charges laterally and collect them at the periphery of the BCPM. In this case, the bipolar plates are electrically isolated from the fuel cells electrodes and are only used to ensure comparable gas flows and operating conditions (Figure 4b).

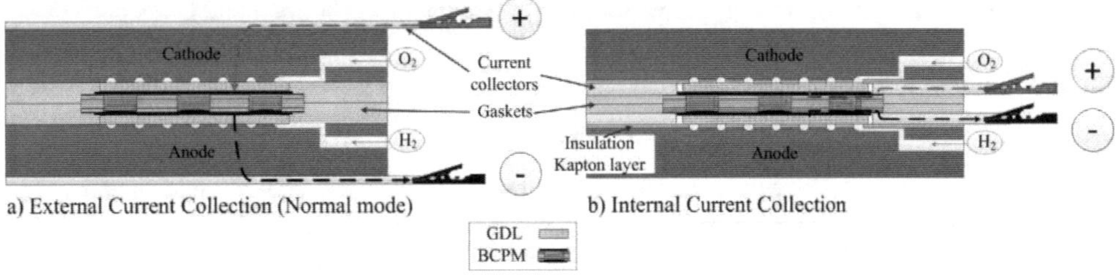

Figure 4 - Current collection modes for ICCs testing

By comparing the polarization curves of these two collection modes (Figure 5), it is possible to see that there are no major discrepancies in output voltage or power of the BCPM. Therefore, no additional ohmic losses are generated by ICCs.

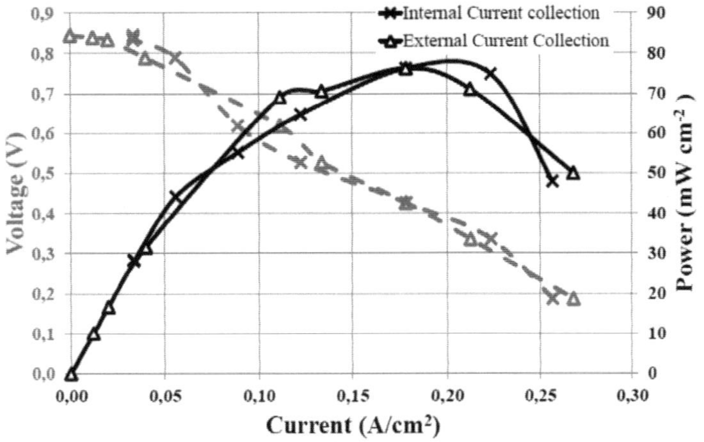

Figure 5 - Polarization curves of the BCPM with internal and external current collection; $T_{cell} = 55°C$, O_2 *flow: 10 sccm/cm*^{2}, H_2 *flow : 6 sccm/cm*^{2}

4. Comparison with standard PEMFC

The performance of the BCPM can be compared to a standard PEMFC in the same working conditions (Table 1). The PEMFC is made from a 25.4µm thick Nafion membrane (Membrane XL, Ion Power inc.) with the same electrode recipe as for the BCPM. Both membranes were tested in the test bench setup showed at figure 4a and their maximum power density was determined.

Table 1- Performance comparison between BCPM and standard PEMFC

	New BCPM	Standard PEMFC
Surface Power Density (mW/cm^2)	76	**223**
Typical Thickness (µm)	**~100**	~600
Volumetric Power Density (mW/cm^2 µm)	**0.76**	0.37

Table 1 shows that surface power density of BCPM is about 3 times lower than for PEMFC. This is mostly explained by the 10 times higher ionic resistance of the BCPM caused by the membrane thickness difference and that only 20% of BCPM cell area is covered with Nafion. Since some applications have surface limitations, surface power density of the BPCM could eventually be improved by bringing holes closer to each other thus increasing Nafion coverage. Low surface power density of BCPM is compensated by the presence of ICCs, which reduce its total thickness. For PEMFC, current collectors must be added to the thickness of the membrane to properly compare volumetric power densities. Here, current collectors were considered as two silicon wafers of 250µm bonded from each side of the membrane [5], a solution commonly used in microfabrication. In this scenario, BCPM volumetric power density is found to be twice as high as PEMFC. This validates the assumption that BCPM could help to improve the power density of a fuel cell stack.

5. Conclusion

The Bi-Conductive Polymer Membrane Fuel Cell presented in this work showed better volumetric power density than standard PEMFC (0.76 vs 0.37 mW/cm^{2}µm) while being flexible and stable in water, making it easier to integrate in a microfabrication process. Also, the fabrication process of the membrane is simple and can potentially be used in large scale production. For the first time, current collectors have been integrated directly on an ionic conductive membrane with a high active area ratio and tested successfully. Moreover, the ICCs did not add ohmic losses, leading the way to a complete re-thinking of PEMFC packaging. Since gas diffusion layers are no longer required in the current path, ICCs leave all the top surface of the electrodes available to integrate a water management system that will better suit the needs of small air-breathing fuel cells.

Acknowledgement
This work was funded by the Natural Science and Engineering Research Council (NSERC) of Canada and the Canada Research Chair program.

References

[1] S. Hamel and L. G. Fréchette, *J. Power Sources*, vol. 196, no. 15, pp. 6242–6248, Aug. 2011.
[2] M. Paquin and L. G. Fréchette, *J. Power Sources*, vol. 180, no. 1, pp. 440–451, May 2008.
[3] K.-B. Min, S. Tanaka, and M. Esashi, *J. Micromechanics Microengineering*, vol. 16, no. 3, pp. 505–511, Mar. 2006.
[4] C. Weinmueller et al., *J. Power Sources*, vol. 195, no. 12, pp. 3849–3857, Jun. 2010.
[5] D. Modroukas, Ph.D Thesis, Columbia University, 2006.
[6] J. P. Esquivel et al., *J. Power Sources*, vol. 195, no. 24, pp. 8110–8115, Dec. 2010.
[7] T. Ito and M. Kunimatsu, *Electrochem. Commun.*, vol. 8, no. 1, pp. 91–94, Jan. 2006.
[8] S. Tominaka et al., *Energy Environ. Sci.*, vol. 2, no. 10, p. 1074, 2009.
[9] L. Zhu et al., *J. Power Sources*, vol. 185, no. 2, pp. 1305–1310, Dec. 2008.
[10] S. Hamel, T. Tsukamoto, S. Tanaka, and L. G. Fréchette, *J. Phys. Conf. Ser.*, vol. 476, p. 012109, Dec. 2013.
[11] L. Sun, R. Ran, and Z. Shao, *Int. J. Hydrog. Energy*, vol. 35, no. 7, pp. 2921–2925, Apr. 2010.

Carbon Nanowalls for oxygen reduction reaction in Bio Fuel Cells

K. Lehmann[1,2], O. Yurchenko[1,2], G. Urban[2]

[1]Freiburg Materials Research Center, FMF, University of Freiburg, Germany
[2]Department of Microsystems Engineering, IMTEK, University of Freiburg, Germany

karsten.lehmann@fmf.uni-freiburg.de

Abstract. We report on the usage of Carbon Nanowalls (CNW) synthesized by a PECVD process as electrode material for oxygen reduction reaction (ORR). In order to substitute the platinum based catalysts in fuel cells, graphene is a promising candidate. Carbon Nanowalls are a graphene modification with good accessibility and a controllable morphology. By controlling height and pore size, they can be optimized for different applications. A I_D/I_G ratio around 2.5 and the SEM images indicate vertical nanocrystallin graphene sheets. Tests with ferrocene as electroactive compound verify CNW suitability as electrode material. Cyclic voltammetry measurements in oxygen saturated PBS prove the catalytic activity of CNW towards ORR. The results support the feasibility of CNW as cathode in Bio Fuel Cells.

1. Introduction

Most bio fuel cells use platinum based materials as cathode material to reduce oxygen, although depending on the field of application different problems exist. Outside of the human body a bio fuel cell is limited by its production cost which, to a significant degree, results from using platinum [1]. Replacing platinum at the cathode would therefore offer a big gain of economic competitiveness. Inside the human body bio fuel cells do not reach the lifespan necessary to be a feasible replacement for batteries or other implant energy delivering systems. One cause is the little durability due to insufficient long term stability of platinum [2].

For these reasons the search for an adequate alternative is extensive. A popular alternative are graphene based catalysts. They are catalytic active towards oxygen reduction reaction (ORR), highly selective, conductive and durable [3]. CNW are vertical grown graphene sheets with a high surface area and a good accessibility of the surface and the edges, which can be directly grown on a conductive surface [4]. Considering that the ORR is mostly limited by the diffusion of oxygen, CNW with their inherent good accessibility and ordered morphology can have an advantage over other ORR catalysts.

In this work we present the results obtained by examination of CNW for ORR at neutral conditions. The ORR kinetics are more sluggish at neutral than in alkaline conditions, but bio fuel cells only work in environments with a neutral pH. For the examination, the nanostructure, morphology and electro chemical activity of CNW were measured and discussed.

2. Methodology

The examined CNW were synthesized on top of a titanium substrate by plasma-enhanced chemical vapor deposition (PECVD) process [5]. Titanium was chosen as substrate because of its good electrochemical stability. More details about the synthesis will be presented in further publication.

Raman spectroscopy, Scanning electron microscope (SEM) and cyclic voltammetry (CV) are used to evaluate the nanostructure, morphology and electro chemical activity of the CNW.

For the Raman spectroscopy an Olympus BX40 with a laser light of 633 nm was used. The SEM Images were taken using a Quanta FEG 250.

For electro chemical measurements, a three electrode cell with Ag/AgCl as the reference electrode and platinum sheet as the counter electrode was used. In order to examine whether the CNW are conductive and therefor applicable as electrodes, the first test measurements were performed in a 5mM ferrocene acetonitrile solution with tetrabutyl ammonium hexafluorophosphate as electrolyte. Two cyclic voltammetry scans were carried out at a scan rate of 10 mV/s in the potential range of 0.1 V to 0.7 V. The second scan was used for evaluation.

The ability of CNW to enhance ORR was investigated in phosphate buffered aqueous solution (PBS) with a pH of 7.4. To confirm ORR, measurements were carried out in nitrogen and oxygen saturated electrolyte. In addition, the ORR was tested for Glassy Carbon for comparison. Two cyclic voltammetry scans were carried out at a scan rate of 5 mV/s in the potential range of -0.6 V to 0.4 V.

3. Results and discussion

Figure 1 shows a Raman spectrum of the synthesized CNW. The D peak at 1331 cm^{-1} and the D' peak at 1616 cm^{-1} are an indicator for disorder in molecular structure, while the G peak is attributed to graphitic structure with sp^2 hybridized carbon. The I_D/I_G ratio of 2.48 evidences a high amount of defects originated from the high edge density typical for CNW. The position of the G peak at 1595 cm^{-1} indicates a nanocrystalline graphite structure with little or no amorphous carbon [6][7].

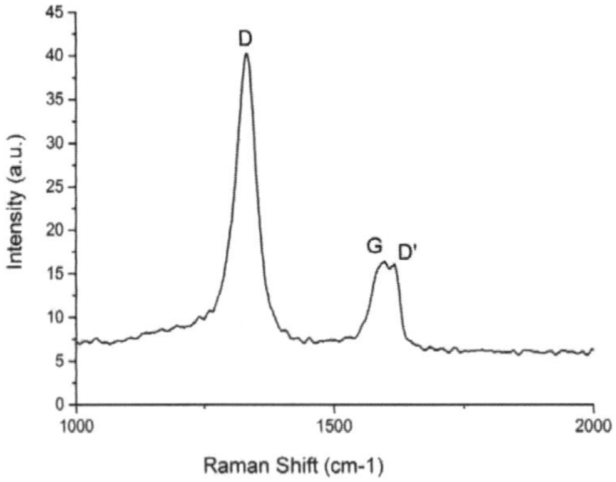

Figure 1. Raman spectrum of the CNW sample showing the peaks of the D, D' and G band.

A macroporous structure is confirmed by SEM images (Figure 2). In Figure 2a, the surface and cross section of the CNW are depicted. The CNW are homogenously distributed over the substrate and are approximately 1.2 µm in height. Figure 2b exhibits the walls with length of roughly 150 nm. The walls grew disorderly upwards creating a macroporous structure, which allow for the fast fuel transport through the CNW network.

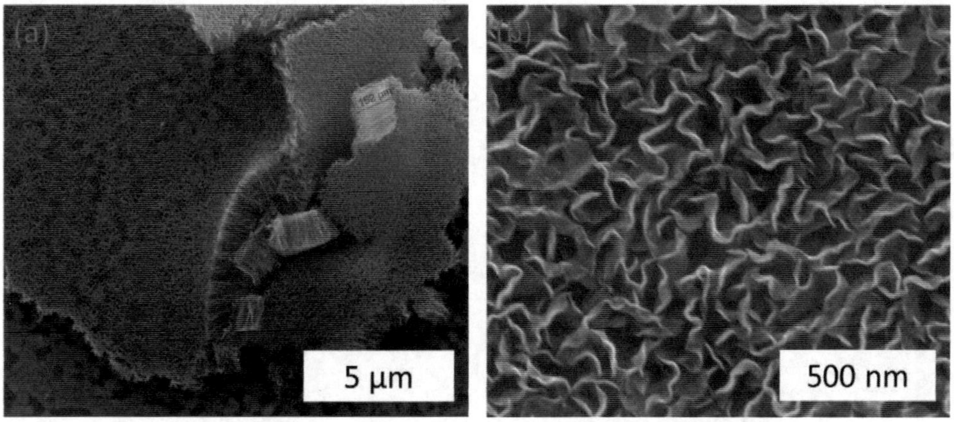

Figure 2. SEM pictures of the characterized CNW sample: (a) the surface and the cross section and (b) the surface morphology.

Figure 3 shows a cyclic voltammogram of 1 electron ferrocene oxidation performed using a CNW electrode. The high current density of the oxidation confirms the electrical conductivity and electroactivity of CNW. The potential difference ΔE_p between the anodic peak potential E_{pa} and the cathodic peak potential E_{pc} is less than 0.1 V. The theoretical value for ΔE_p for a reversible 1 electron process is 0.057 V. Large values of ΔE_p indicate slow electron transfer kinetics. Considering the large electrode area of 0.28 cm² and the high electrode surface which result in a high current density and thus high electrode resistance, the observed potential difference reveals a quite fast charge transfer on CNW electrode and proves CNW suitability for electrode application[8].

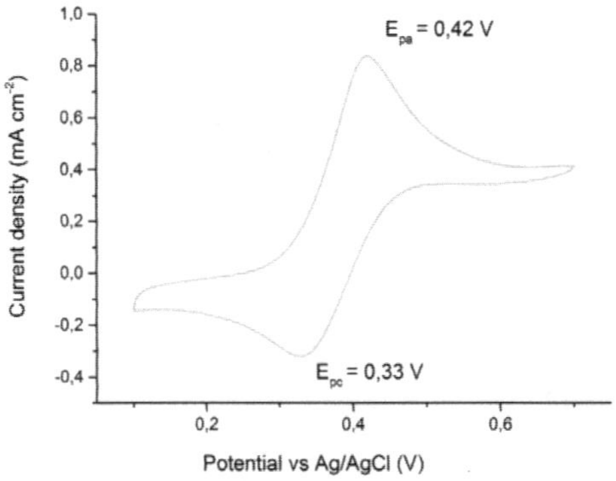

Figure 3. Cyclic voltammogram of ferrocene oxidation performed at a scan rate of 10 mV/s.

The CNW capability for oxygen reduction is shown in Figure 4a by means of CV experiments performed in nitrogen saturated (black line) and in oxygen saturated PBS solution at a pH of 7.4 (red line). At around -0.15V the oxygen reduction starts visibly increasing the cathodic current. Figure 4b compares the ORR on CNW and Glassy Carbon electrode in solution saturated only by oxygen. The onset potential for ORR is anodically shifted by 0.2 V for CNW compared to Glassy Carbon indicating catalytic activity towards oxygen reduction.

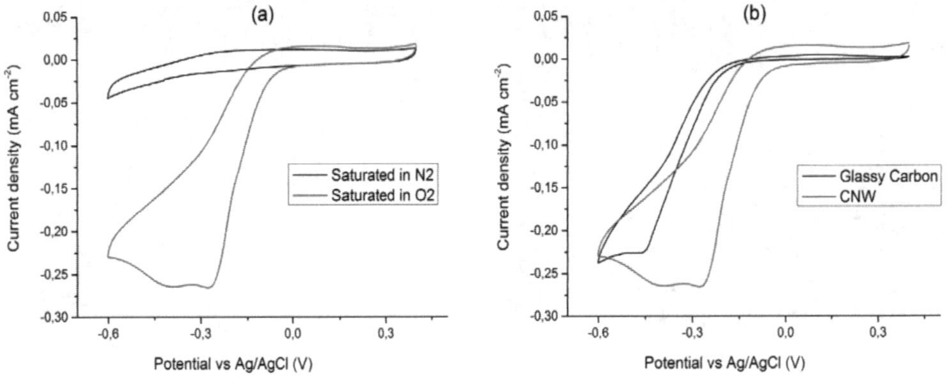

Figure 4. Cyclic voltammogram of (a) CNW electrode in a nitrogen and oxygen saturated PBS solution at a pH of 7.4 and (b) a Glassy Carbon electrode compared with a CNW electrode in an oxygen saturated PBS solution.

4. Conclusion

The results of this paper show for the first time the feasibility of CNW for ORR at neutral conditions and its potential for fuel cell application. The structure of investigated CNW was examined by RAMAN and SEM. RAMAN results indicated a nanocrystallin graphitic structure with little or no amorphous carbon. The SEM results confirm the formation of porous network from vertical aligned graphene sheets. The electro chemical applicabillity of CNW as electrode were tested by means of reversible oxidation of ferrocene. The oxidation of ferrocene was kinetically reversible confirming CNW as suitable electrode material. The catalytic activity of CNW for ORR was proved by comparison with Glassy carbon electrode. The comparison shows a shifted onset of ORR for CNW, revealing catalytic properties.

The results evidence, that CNW is a promising cathode material for bio fuel cells. In the future, the experiments for improving the electrochemical properties of CNWs by means of adjustment the material morphology and structure will be performed even further. In addition, the resulting CNW will be compared to graphene and platinum cathode.

5. Acknowledgment

We gratefully acknowledge the German Research Association DFG for the financial support within (GRK1322) and Dr. Loic Ledernez and Dr. Fethi Olcaytug for the technical support.

References

[1] Wissam A. Saidi, J. Phys. Chem. Lett. 4 (2013), pp. 4160–4165
[2] Zheng, Yao; Jiao, Yan; Jaroniec, Mietek; Jin, Yonggang; Qiao, Shi Zhang, Small 8 (2012), pp. 3550–3566
[3] Zhu, Chengzhou; Dong, Shaojun, Nanoscale 5 (2013), pp. 1753
[4] Hiramatsu, M.; Shiji, K.; Amano, H.; Hori, M., Appl. Phys. Lett. 84 (2004), pp. 4708
[5] Shiji, K.; Hiramatsu, M.; Enomoto, A.; Nakamura, M.; Amano, H.; Hori, M., Diamond and Related Materials 14 (2005), pp. 831–834
[6] Kurita, S.; Yoshimura, A.; Kawamoto, H.; Uchida, T.; Kojima, K.; Tachibana, M. et al., J. Appl. Phys. 97 (2005), pp. 104320
[7] Ferrari, A.; Robertson, J., In: Phys. Rev. B 61 (2000), pp. 14095–14107.
[8] Bond, A. M.; Henderson, T. L. E.; Mann, D. R.; Mann, T. F.; Thormann, W.; Zoski, C. G., Anal. Chem. 60 (1988), pp. 1878–1882 reference

Modified Graphene as Electrocatalyst towards Oxygen Reduction Reaction for Fuel Cells

D Qazzazie[1,2], M Beckert[3], R Mülhaupt[3], O Yurchenko[1,2], G Urban[2]

[1] Freiburg Materials Research Center, FMF, University of Freiburg, Stefan-Meier-Strasse 21, 79104 Freiburg, Germany
[2] Department of Microsystems Engineering, IMTEK, University of Freiburg, Germany
[3] Institute for Macromolecular Chemistry, University of Freibrug, Germany

E-mail: olena.yurchenko@fmf.uni-freiburg.de

Abstract. This paper reports modified graphene-based materials as metal-free electrocatalysts for oxygen reduction reaction (ORR) with outstanding electrocatalytic activity in alkaline conditions. Nitrogen-doped graphene samples are synthesized by a novel procedure. The defect density in the structure of the prepared materials is investigated by Raman spectroscopy. Further structural characterization by X-ray photoelectron spectroscopy reveals the successful nitrogen doping of graphene. The electrochemical characterization of graphene and nitrogen-doped graphene in 0.1 M KOH solution demonstrates the material's electrocatalytic activity towards ORR. For graphene an onset potential of - 0.175 V vs. Ag/AgCl reference electrode is determined, while for nitrogen-doped graphene the determined onset potential is - 0.160 V. Thus, the electrocatalytic activity of nitrogen-doped graphene towards ORR is enhanced which can be ascribed to the effect of nitrogen doping.

1. Introduction

Fuel cells generate electric power through the conversion of the chemical energy contained in a fuel into electrical energy. Within the framework of an electrochemical process the fuel such as hydrogen or methanol oxidizes at the anode and the oxidizing agent, usually oxygen is reduced to water at the cathode. In general catalysts at the respective electrode are used to increase the fuel cell's efficiency.

Due to the sluggish oxygen reduction reaction kinetics platinum-based catalyst materials are widely used as cathodes in fuel cell applications to catalyse the oxygen reduction reaction. [1] A fundamental drawback for the development of commercial fuel cells containing platinum-based electrodes is derived from high costs as well as limited reserves of platinum in nature. [2] Further problems arise from the vulnerability of platinum-based materials to time-dependent drift and poisoning effects such as CO "poisoning" leading to the catalyst deactivation. Therefore the development of novel metal-free materials has become the focus of research to substitute platinum-based catalysts. [2] [3]

Regarding to the excellent properties of carbon-based materials, different modifications of graphene are of particular interest due to easy processability, low costs, high surface and porosity, high electrical, thermal conductivity and mechanical strength. [3] The electronic and optical properties of graphene can be modified by doping with heteroatoms such as nitrogen or boron. [4] [5] In particular, nitrogen-doped carbon materials represent promising metal-free catalysts due to their enhanced electrocatalytic activity, long-term stability and environmental sustainability. [6] Compared with Pt and commercial PtC catalysts nitrogen-doped graphene is characterized by better stability. [7] [8] Besides, it is also demonstrated that the activity of nitrogen-doped graphene is not influenced by adding carbon monoxide or methanol. [4] [8]

Here, we report graphene and nitrogen-doped graphene as metal-free electrocatalysts for oxygen reduction reaction. The graphene-based materials were synthesized by a novel procedure. A detailed report about the synthesis procedure is under preparation and will be published soon. The structure of the prepared materials was characterized by Raman spectroscopy and X-ray Photoelectron Spectroscopy (XPS). Furthermore graphene and nitrogen-doped graphene were electrochemically characterized by cyclic voltammetry (CV) and linear sweep voltammetry (LSV). The electrocatalytic activity of the prepared materials towards ORR is demonstrated and compared to that of commercial 20% PtC catalyst.

2. Structural characterization

Raman spectroscopy was used to investigate the defect level in carbon lattice of the prepared materials. Figure 1 displays the Raman spectra of nitrogen-doped graphene compared with graphene and graphite. The Raman spectrum of graphite mainly shows a typical G peak at 1576 cm^{-1} corresponding to ideal single-crystal graphite. [9] The Raman spectra of graphene and nitrogen-doped graphene display G bands shifted to 1595 cm^{-1} and 1588 cm^{-1}, respectively. In addition, clearly rised D bands appear at 1344 cm^{-1} for graphene and 1365 cm^{-1} for nitrogen-doped graphene relating to the disorder-induced feature which can be attributed to a modest defective structure of the graphene-based materials. [10]

Besides the I_D/I_G ratio is taken into account to investigate the density of defects in the samples. For graphite the I_D/I_G ratio is approximately 0.15, it increases to 1 for graphene and 1.11 for nitrogen-doped graphene. The increase of I_D/I_G can be ascribed to the presence of a substantial amount of defects in the prepared materials.

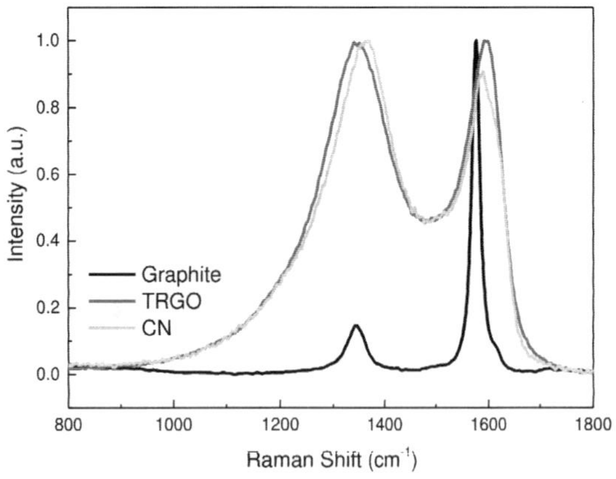

Figure 1. Raman spectra of graphite, graphene (TRGO) and nitrogen-doped graphene (CN).

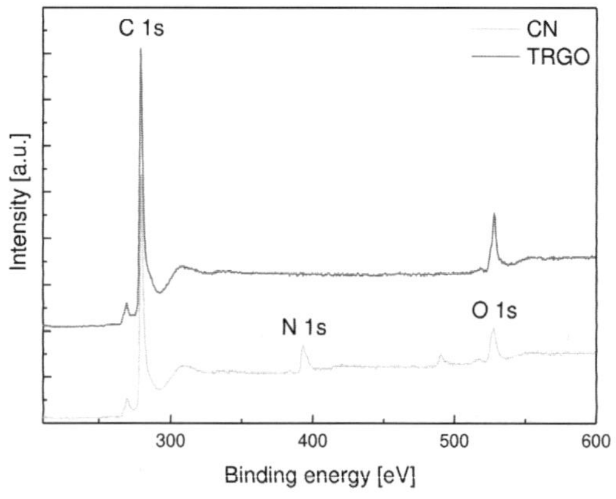

Figure 2. XPS spectra of graphene (TRGO) and nitrogen-doped graphene (CN).

X-ray Photoelectron Spectroscopy (XPS) was further used to characterize the chemical composition of the prepared samples, thus to verify the nitrogen doping. The XPS spectra of graphene and nitrogen-doped graphene are depicted in Figure 2. In the survey spectrum of nitrogen-doped graphene the peaks occurring at approximately 400 eV correspond to nitrogen and clearly confirm the successful incorporation of nitrogen atoms into carbon lattice, wherein the content of nitrogen atoms is 12 %. Furthermore oxygen functionalities are present in both graphene and nitrogen-doped graphene.

The nitrogen spectrum additionally can be deconvoluted to individual peaks corresponding to different nitrogen types in carbon lattice, usually pyridinic-, pyrrolic- and tertiary-type nitrogen functionalities. This is necessary to study the ORR mechanism in nitrogen-doped graphene and will be further investigated in future reports.

3. Electrochemical characterization

The electrochemical properties of graphene and nitrogen-doped graphene are characterized via cyclic voltammetry and linear sweep voltammetry. The electrochemical measurements are carried out at a scan rate of 5 mV/s in 0.1 M KOH solution. Each sample was dispersed in isopropanol. The working electrode, a glassy carbon (GC) electrode of 3 mm in diameter, was coated with 20 μl of the respective sample. A platinum electrode was used as counter electrode and Ag/AgCl as reference electrode.

The cyclic voltammetry was first performed in nitrogen-saturated 0.1 M KOH solution. As shown in Figure 3a, clear capacitive current background was seen within the CV. Afterwards oxygen was introduced to the solution. The measured CV curves are presented in Figure 3b featuring significant cathodic current peaks, thus showing the ORR performance of all samples in oxygen-saturated KOH solution.

The highest value of current density is obtained for nitrogen-doped graphene, followed by graphene and glassy carbon, still lower than that of PtC. Compared with glassy carbon, the onset potentials of graphene and nitrogen-doped graphene shift positively confirming better electrocatalytic activity towards ORR. The highest value of current density is obtained for nitrogen-doped graphene, followed by graphene and glassy carbon, still lower than that of PtC.

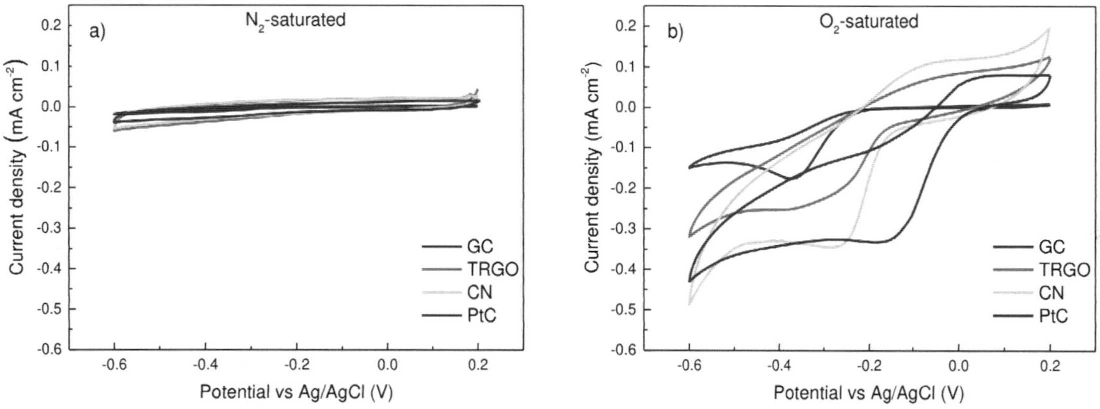

Figure 3. CV-measurement of GC, TRGO, CN, PtC in 0.1 M KOH solution in N_2-saturated ambience (a), and in O_2-saturated ambience (b) demonstrating electrocatalytic activity towards ORR.

In addition, LSV-measurements were performed on a rotating-disk electrode (RDE) in oxygen-saturated 0.1 M KOH solution at a rotation rate of 1600 rpm. The recorded LSV-curves, displayed in Figure 4, show that nitrogen-doped graphene has the most positive onset potential and highest current density among the carbon-based materials. The LSV observations accord with the CV results and demonstrate that the electrocatalytic activity of graphene towards ORR is enhanced by doping with nitrogen. Onset potentials and reduction´s current densities of glassy carbon, graphene, nitrogen-doped graphene and PtC are calculated and listed in Table 1.

Table 1. Onset potentials and reduction's current densities of glassy carbon (GC), graphene (TRGO), nitrogen-doped graphene (CN) and commercial 20% PtC derived from CV-measurements.

Material	$E_{ONSET, Red}$ vs Ag/AgCl (mV)	J_p (mA cm^{-2})
GC	- 250	- 0.175
TRGO	- 175	- 0.150
CN	- 160	- 0.256
PtC	- 20	- 0.285

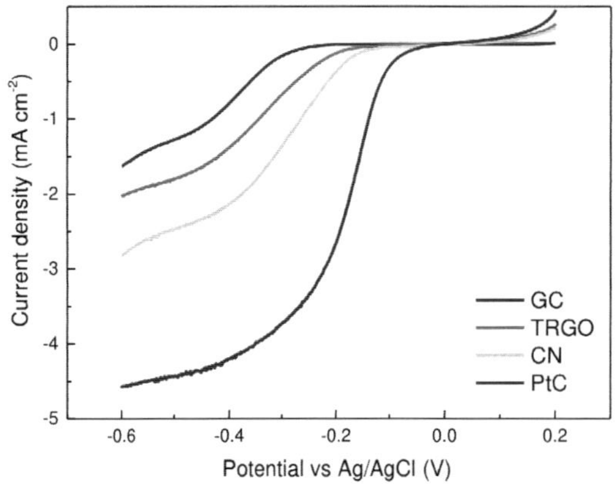

Figure 4. LSV measurements of glassy carbon (GC), graphene (TRGO), nitrogen-doped graphene (CN) and commercial 20% PtC in O_2-saturated 0.1 M KOH solution at rotation speed of 1600 rpm.

4. Conclusions

As determined by Raman spectroscopy, nitrogen-doped graphene possesses a similar defect density to that of graphene. It can therefore be concluded that the incorporation of nitrogen into graphene apparently enhances the performance towards ORR.

The successful doping process of nitrogen in carbon lattice is evidenced by X-ray Photoelectron spectroscopy. Considered together with the experimental results, we estimate a significant importance of nitrogen doping for enhancing the ORR performance. In this context, it is necessary to investigate the individual nitrogen functionalities in carbon lattice and their impact on the ORR performance which will be addressed in future reports.

In summary, prepared graphene and nitrogen-doped graphene have proven their electrocatalytic activity towards oxygen reduction reaction. The electrocatalytic activity is enhanced by nitrogen doping. First stability tests have been performed delivering good results and will be continued in more detail. With regard to further modification, nitrogen-doped graphene represent a promising metal-free electrocatalyst as cathode in fuel cell applications.

Acknowledgments
We gratefully acknowledge the German Research Association (DFG) for supporting this work within the Research Training Group (GRK 1322) "Micro Energy Harvesting".

References

[1] M. Winter, R. J. Brodd, *Chemical Reviews*, 104 (2004), pp. 4245.
[2] H. Gasteiger et al., *Applied Catalysis B*, 56 (2005), pp. 9–35.
[3] C. Zhu, S. Dong, *RSC Nanoscale*, 5 (2012), pp. 1753–1767.
[4] L. Qu et al., *ACS Nano*, 4 (2010), pp. 1321-1326.
[5] J. Liang et al., *Angew. Chem.,Int. Ed.*, 124 (2012), pp. 11664–11668.
[6] D. S. Geng et al., *Energy & Environmental Science*, 4 (2011), pp. 760–764.
[7] P. Fuping et al., *ACS Applied Materials & Interfaces*, 5 (2013), pp. 11108–11114.
[8] Y. Shao et al., *Journal of Materials Chemistry*, 20 (2010), pp. 7491.
[9] F. Tuinstra, J. L. Koenig, *The Journal of Chemical Physic,* 53 (1970), pp. 1126.
[10] M. S. Dresselhaus et al., *RSC Nano let*ters, 10 (2010), pp. 751–758.

Contacting mode operation of work function energy harvester

A Varpula, S J Laakso, T Havia, J Kyynäräinen, M Prunnila

VTT Technical Research Centre of Finland, P. O. Box 1000, 02044 VTT Espoo, Finland

Email: aapo.varpula@vtt.fi, mika.prunnila@vtt.fi

Abstract. The work function energy harvester (WFEH) is a variable capacitance vibration energy harvester where the charging of the capacitor electrodes is driven by the work function difference of the electrode materials. In this work, we investigate operation modes of the WFEH by utilizing a macroscopic parallel plate capacitor with Cu and Al electrodes and varying plate distance. We show that by charging the electrodes of the WFEH by letting the electrode plates touch during the operation a significant output power enhancement can be achieved in comparison to the case where the electrodes are charged and discharged only through a load resistor.

1. Introduction

Electrostatic energy harvesters are based on electrically charged capacitors the capacitance of which is varied by mechanical motion [1–3]. The energy of the mechanical motion is converted into electric energy of the capacitor. The electrostatic energy harvesters need either electret materials or an external power supply, such as battery, for charging of the variable capacitor. This drawback poses challenges in the fabrication and possibly lifetime of these devices. The work-function energy harvester (WFEH) is similar to the electrostatic energy harvester as it utilizes a variable capacitor. However, this energy harvester, illustrated in figure 1, is not dependent on external power supplies or electret materials [4–8]. Instead, the WFEH employs the fundamental difference in the work functions of two dissimilar materials in the charging of the variable capacitor. In this sense the WFEH is closer to a piezoelectric harvester that relies also on the fundamental properties of solid materials. Furthermore, WFEHs can generate more power than the electrostatic harvesters in many operating conditions [8].

The charging effect between two materials with different work functions was thoroughly investigated by Lord Kelvin more than 150 years ago [9, 10], but the use of work function difference in vibration energy harvesting has been, remarkably, proposed only quite recently in 2006 [4]. Since then the work has been concentrated on the modelling and design of WFEHs [5–7]. Comprehensive theory and its comparison with experiments has been performed only very recently in Ref. [8], where, for example, non-contacting operation mode of figure 1c and d was realized with a resistor load. In this work, we extend the experiments of Ref. [8] and realize an operation mode where the WFEH plates touch during the operation cycle. This type of contacting mode of operation can produce an extremely large maximum capacitance which is limited only by the native oxide layers of the capacitor plates, screening length, and parallelism of the electrode plates. As the maximum capacitance defines largely the power output of the WFEH, contacting mode is expected to lead to performance enhancements and, indeed, our experiments show that this operation mode produces significantly more power than the non-contacting mode. With our macroscopic prototype we obtain a factor of 7.4 power enhancement at 0.5 Hz.

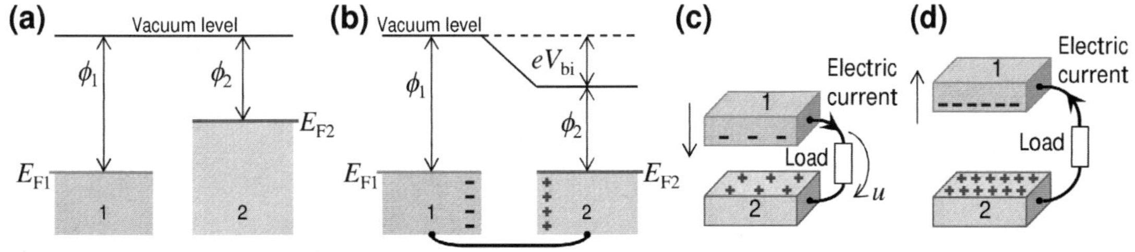

Figure 1. (a,b) Electron energy levels of the electrode materials of the work-function energy harvester (a) before and (b) in the thermodynamical equilibrium after making electrical contact. E_{F1} and E_{F2} are the Fermi levels and ϕ_1 and ϕ_2 the work-functions of the materials 1 and 2, respectively, V_{bi} is the built-in voltage, and e is the elementary charge. (c) Charging and (d) discharging phases of the operation cycle of the WFEH. u is the output voltage. Here a parallel plate WFEH with metal electrodes is illustrated. Semiconductors and other capacitor geometries can be used as well.

In addition to the high maximum capacitance, the benefit of the contacting mode is that it reproduces the key phases of the ideal operating cycle, which produces maximum output power of a WFEH. The contacting ensures that the WFEH capacitor is fully charged before the WFEH begins its work cycle. The drawback of the contacting mode is that energy cannot be harvested from the charging of the WFEH capacitor through the load as the ideal WFEH does. In most cases, however, the energy associated to the through load charging is low [8]. Figure 2 shows two circuits exploiting the contacting mode. These circuits have direct application as charge pumps for pumping electric energy into storage capacitors [8]. The circuits of figure 2 have the same principle: In figure 2a the charging of the WFEH electrodes is performed using switch S1 and the extraction of the electric energy is controlled using switch S2. In figure 2b the same switching functions are performed by the charge shuttle. The contacting mode corresponds to the case where switch S1 is closed when the electrodes 1 and 2 are in contact and open otherwise. To maximize the power output switch S2 should be closed only when the capacitance of the WFEH is at the minimum. The experiments of this work correspond to the case where switch S2 is constantly closed.

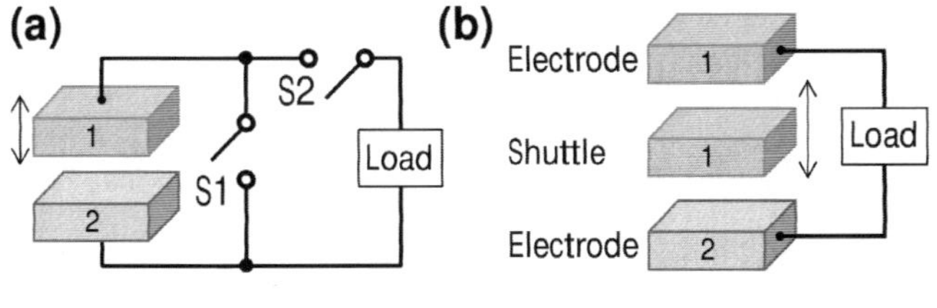

Figure 2. Schematic pictures of work-function energy harvesters based on (a) external switches S1 and S2 and (b) a shuttle. The shuttle and the upper electrode are made of the same material (1) and the lower electrode of material with dissimilar work function (2).

2. Theory

The equilibration of the Fermi levels of the electrode materials of the work-function energy harvester cause electrons to move from the electrode with higher work function to the electrode with lower work function via the circuit connecting the electrodes. In the general case the voltage across the gap between the electrodes of the work-function energy harvester is given by

$$V_{gap} = u - V_{bi}, \quad (1)$$

where u is the output voltage of the WFEH (i.e. voltage between the WFEH electrodes, measured via the connecting circuit) and the built-in voltage is given by

$$V_{\mathrm{bi}} = (\phi_1 - \phi_2)/e, \tag{2}$$

where e is the elementary charge and ϕ_1 and ϕ_2 are the work functions of the electrode materials. In the termodynamical equilibrium the voltage across the WFEH gap is V_{bi} (see figure 1b). The electric current flowing out of or into the WFEH is given by [8]

$$i = \frac{\mathrm{d}Q}{\mathrm{d}t} = \frac{\mathrm{d}}{\mathrm{d}t}(Cu) - V_{\mathrm{bi}}\frac{\mathrm{d}C}{\mathrm{d}t}, \tag{3}$$

where $C = C_{\mathrm{gap}} + C_{\mathrm{par}}$ is the total capacitance of the WFEH, C_{gap} is the capacitance between the electrodes, and C_{par} is the parasitic capacitance. The $V_{\mathrm{bi}}\cdot\mathrm{d}C/\mathrm{d}t$ term is the drive term that leads to non-zero power output of WFEH when C_{gap} is altered by an external force.

Equation (3) is the fundamental equation describing the charging phenomenon caused by the built-in voltage. The overall behaviour of the energy harvester device depends on the external circuit connected to the electrodes. It turns out that the most efficient way to operate the WFEH is to charge or discharge it only when the stored electric energy reaches a maximum or minimum, respectively. These extremes are reached when the WFEH capacitance is at maximum and minimum. This so-called charge constrained operation mode can be realized with an electric switch, for example. The maximum output power of this ideal WFEH produces the average output power of [8]

$$P_{\max} = \frac{C_{\max}V_{\mathrm{bi}}^2}{2\Delta t}\left(1-\frac{C_{\min}}{C_{\max}}\right)\left\{\frac{C_{\max}}{C_{\min}}+\frac{C_{\min}}{C_{\max}}+2-\frac{2C_{\mathrm{par}}}{C_{\min}}\left[1+\left(\frac{C_{\min}}{C_{\max}}\right)^2\right]\right\}, \tag{4}$$

where $C_{\max}$ and $C_{\min}$ are the maximum and minimum values of C during the WFEH operating cycle.

3. Experimental

In the experiments the macroscopic prototype of the work-function energy harvester based on variable-distance parallel plate capacitor was employed. The experimental setup is shown in figure 3. 63 mm × 64 mm aluminium and copper plates were used as sample electrode materials. The setup utilizes an electric motor and a gravitational counter force to move the aluminium plate sinusoidally.

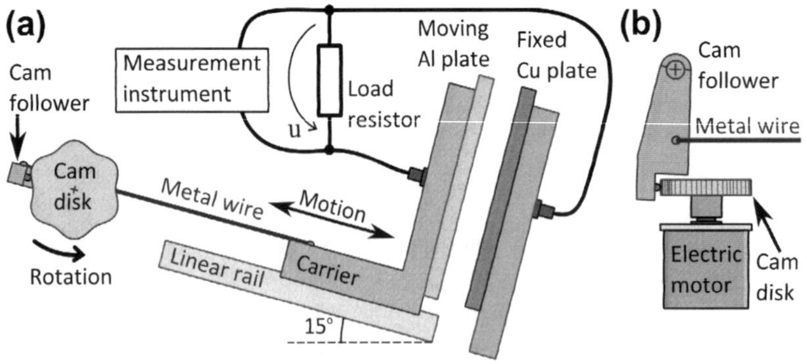

Figure 3. Schematic pictures of (a) the experimental setup (side view) and (b) the carrier drive system (top view).

In the experiments the electrode plates where connected to a load resistor and an Agilent 4156C precision semiconductor parameter analyzer, which measures the output voltage u (see figure 3a). Prior to experiments the built-in voltage of the Al/Cu system was measured with a null method to be

V_{bi} = 1.03 V. The experimental setup is the same as in Ref. [8] and further details can be found there in.

4. Results and discussion

The average output power P_{max} of two ideal WFEHs are plotted in figure 4 as functions of ratio C_{max}/C_{min}. The dependence of P_{max} on the built-in voltage V_{bi} and the operating frequency is also shown. These graphs show that high output power can be achieved with high values of C_{max}/C_{min}, V_{bi}, and the operating frequency. Achieving the high output power suggested by (4) requires that the WFEH is operated similarly as in the ideal case. As discussed above, the contacting operation mode, where the electrodes of the WFEH make periodically electric contact with each other, is close to the ideal operating cycle.

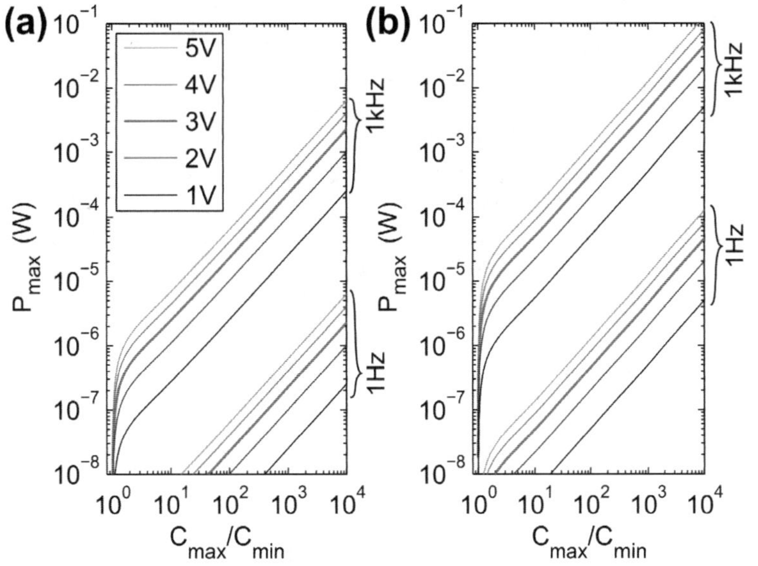

Figure 4. Calculated average ideal power P_{max} of work-function energy harvester as function of C_{max}/C_{min} with (a) C_{max} = 50 pF and (b) C_{max} = 1 nF and various values of V_{bi} and the operating frequency. Calculated using (4) with C_{par} = 0.

Figure 5a and b show the measured output voltages of a WFEH operated conventionally and a WFEH where the electrodes make electric contact with each other once during the operating cycle, respectively. The measured data shows that the peak voltage of the contacting mode WFEH is more than 3 times higher than in the non-contacting WFEH. Both the higher C_{max} and fully efficient WFEH charging contribute to the higher output voltage of the contacting mode WFEH. Since the WFEH is connected to the resistor load at all times, the WFEH capacitor begins to discharge as soon as the contact between the electrodes is lost. On the other hand, this kind of system is also able to harvest energy from the charging of the WFEH electrodes when the electrodes approach each other as can be seen in figure 5b ($t = 0.2\Delta t \ldots 0.4\Delta t$).

In the non-contacting mode, the optimal operating frequency for maximum power depends on the electric time constant of the WFEH, RC_{max}, where R is the resistance of the load [8]. Because of this the average output power of the non-contacting mode, $P_{non\text{-}cont}$, shown in figure 5c peaks around 0.2 Hz. Figure 5d confirms that the WFEH with periodic contact between plates produces much higher average output power than the non-contacting WFEH. At 0.5 Hz we obtain factor of 7.4 power enhancement.

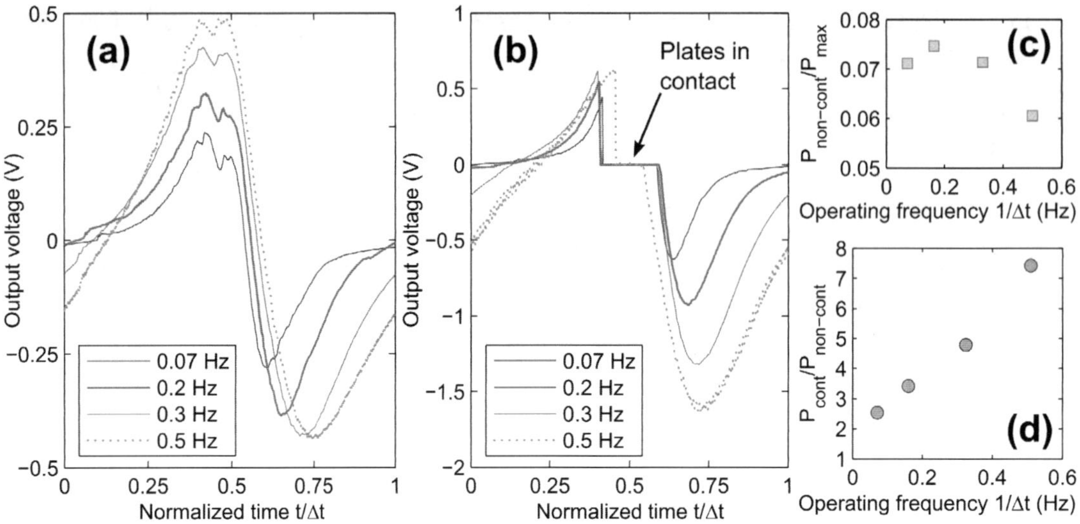

Figure 5. (a,b) Measured time-evolutions of output voltages of the Cu/Al parallel plate work-function energy harvester with a 4 GΩ load resistor at operating frequencies 0.07–0.5 Hz: (a) Operating cycle with the minimum distance between the plates of 200 μm (no contact between the plate surfaces). (b) Case where plates make contact during the operating cycle. Time is normalized with the operating period Δt. The amplitude of the sinusoidal motion in both cases is 0.75 mm. The capacitance reaches minimum at $t = 0$ and $t = \Delta t$, and maximum at $t = \Delta t/2$. (c) Ratio of measured and ideal average powers, $P_{\text{non-cont}}$ (case (a)) and P_{max}. Calculated using (4) with $C_{\text{par}} = 0.1$ pF. (d) Ratio of the measured average output powers of cases (b) P_{cont} and (a) $P_{\text{non-cont}}$ as function of the operating frequency of the WFEH.

5. Conclusions

We have demonstrated experimentally the feasibility of the contacting mode work-function energy harvester. A parallel plate WFEH with electrodes contacting each other periodically produced over 7 times higher average output power than the reference WFEH operated without contacting. Our results pave the way towards the realization of the full shuttle-based WFEH. In the microscale these devices could be realized using SiC-based MEMS switching technology [11].

References

[1] Cook-Chennault K A, Thambi N, Sastry A M 2008 *Smart Mater. Struct.* **17** 043001
[2] Beeby S P, Tudor M J, White N M 2006 *Meas. Sci. Technol.* **17** R175
[3] Roundy S 2005 *J. Int. Mat. Syst. Str.* **16** 809
[4] Kuehne I, Frey A, Eckstein G, Schmid U, Seidel H 2006 *Proc. 36th European Solid-State Device Research Conference (Montreux, 19-21 September 2006)* (IEEE) pp 138-141
[5] Kuehne I, Frey A, Marinkovic D, Eckstein G, Seidel H 2008 *Sensor. Actuat. A–Phys.* **142** 263
[6] Li L, Zhang Y, San H, Guo Y, Chen X 2008 *Proc. SPIE* **6836** 683610
[7] Dragunov V P, Lyutaeva M N 2009 *Int. School and Seminar INTERNANO 2009 (Novosibirsk, Russia, 28-31 Oct. 2009)* (IEEE) pp 93-96
[8] Varpula A, Laakso S J, Havia T, Kyynäräinen J, Prunnila M 2014 *Sci. Rep.* **4** 6799
[9] Lord Kelvin 1898 *Philos. Mag. Series 5* **46** 82
[10] Thomson W 1881 *Nature* **23** 567
[11] He T, Yang R, Ranganathan V, Rajgopal S, Tupta M A, Bhunia S, Mehregany M, Feng X-L P 2013 *IEEE Int. Electron Devices Meeting (Washington DC, 9-11 Dec. 2013)* (IEEE) 4.6.1

A Design Method of In-plane MEMS Electret Energy Harvester with Comb Drives

Qianyan Fu[1] and Yuji Suzuki[1]

[1]Dept. of Mechanical Engineering, The University of Tokyo, Tokyo, Japan

E-mail: qfu@mesl.t.u-tokyo.ac.jp

Abstract: In this report, a design method is proposed for in-plane MEMS electrostatic energy harvesters with comb drives. Dependent on the device layer thickness and the achievable aspect ratio of the DRIE process, either the overlapping-area-change or the gap-closing converter has higher output power than the other. Two prototypes of MEMS electret energy harvesters are developed to verify the present design method.

Keywords: Electrostatic generator, electret, comb drive, DRIE, aspect ratio

1. INTRODUCTION

The concept of vibration energy harvesting is to convert environmental vibration energy into electricity for powering low-power electronics such as wireless sensors. Among various energy conversion techniques, electrostatic/electret energy harvester has attracted much attention for its high output power in small volumes [1]. In electrostatic energy harvesters, the capacitance change should be maximized for enhancing output power [1]. Comb drive is the most commonly used configuration for the MEMS variable capacitor. Hoffmann et al. [2] and Nguyen et al. [3] developed in-plane overlapping-area-change (OV) electrostatic energy harvesters using deep reactive-ion etching (DRIE) process to form the device layer, and obtained output power up to 5 μW (RMS value) and 3.4 μW, respectively. Tseng et al. [4] and Guilllemet et al. [5] developed gap-closing (GC) energy harvesters and achieved up to 1.2 μW at the resonance frequency of 1870 Hz and 2.3 μW at 250 Hz, respectively. However, at the authors' best knowledge, appropriate choice of the overlapping-area-change or gap-closing converters for maximizing output power under given conditions remains unknown.

In the present study, a design method for electrostatic/electret generator with comb drives is proposed through comparison of the maximum capacitance change per unit area. In addition, we prototyped these two types of electret energy harvester and examined their performance.

2. MAXIMUM CAPACITANCE CHANGE RATIO PER UNIT AREA FOR OVERLAPPING-AREA-CHANGE AND GAP-CLOSING TYPE OF COMB DRIVES

Figure 1 shows the schematic of in-plane electrostatic converters, where $l(t)$ and $g(t)$ are the overlapping length and the gap opening between the movable and fixed fingers, respectively. The capacitance of each side of the electrodes can be expressed as:

$$C = \frac{\varepsilon_0 \cdot \varepsilon_r \cdot l(t) \cdot h}{g(t)}, \tag{1}$$

where h, ε_0 and ε_r represent the height of the comb fingers, the vacuum permittivity and the relative permittivity of the dielectric surrounding the electrodes. The total capacitance for one pair of electrodes is computed by adding the capacitance on the both sides of a finger.

For the OV type, the gap g_{ov} between the interdigitated combs is kept constant during oscillation. The maximum capacitance change ΔC_{max} per one oscillation cycle of the mass can be given by

$$\Delta C_{OV,max} = \frac{4 \cdot \varepsilon_0 \cdot h \cdot A}{g_{OV}}, \tag{2}$$

where A is the maximum travelling distance of the movable electrodes. Since g_{OV} is the minimum feature for the OV type, g_{OV} is defined by the device layer thickness h (same as the comb height) and the maximum achievable aspect ratio of the DRIE process AR ($=h/g_{OV}$).

On the other hand, for the GC type, the maximum capacitance change ΔC_{max} per one oscillation cycle can be given by

$$\Delta C_{GC,max} = \frac{4\varepsilon_0 \cdot h \cdot l_{GC} \cdot g_{GC}}{\left(2g_{GC} - g_{GC,min}\right) g_{GC,min}} - \frac{4\varepsilon_0 \cdot h \cdot l_{GC}}{g_{GC}}, \tag{3}$$

where l_{GC} and g_{GC} are the overlapping length and the initial gap opening between the movable and fixed fingers. The minimum gap $g_{GC,min}$, which is often defined by mechanical stoppers, is the key parameter for the GC type. Note that for GC type, double capacitance change occurs during one oscillation cycle.

Therefore, the capacitance change ratio per unit area, which is a measure of the power generation density, is given by

$$\Delta C_r^* = \left(\Delta C_{OV,max} / S_{OV}\right) / \left(\Delta C_{GC,max} / S_{GC}\right). \tag{4}$$

The occupied area by the electrodes S_{OV} and S_{GC} are given by

$$S_{OV} = a_{OV} \cdot b_{OV} = \left(2w + 2g_{OV}\right) \cdot \left(2A + 2d_{min} + l_{OV}\right), \tag{5}$$

$$S_{GC} = a_{GC} \cdot b_{GC} = \left(2w + 2A + 2g_{GC,min}\right) \cdot \left(2d_{min} + l_{GC}\right). \tag{6}$$

Figure 2 shows ΔC_r^* versus the maximum travelling distance A for different aspect ratios. As illustrated in Figure 2a, when AR=7, the OV type has higher capacitance change (and thus higher output power) than the GC type for $h < 22$ μm. On the other hand, the GC type is superior to the OV

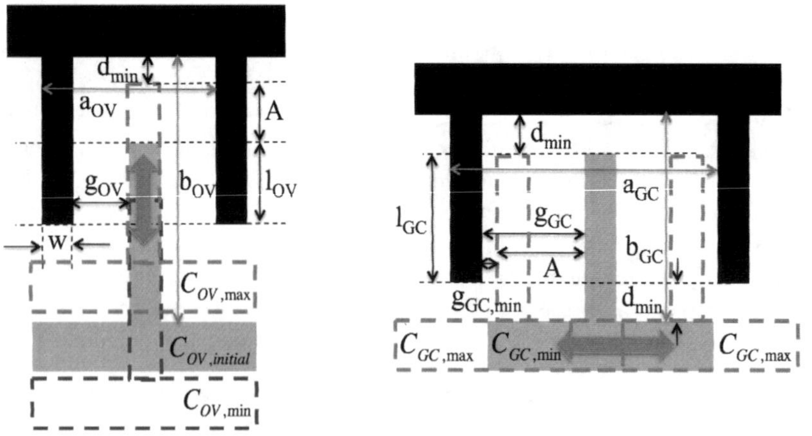

Figure 1. In-plane overlapping-area-change (OV) type (left) and gap-closing (GC) type (right). The maximum travel distance A is limited by the mechanical stoppers in the GC type.

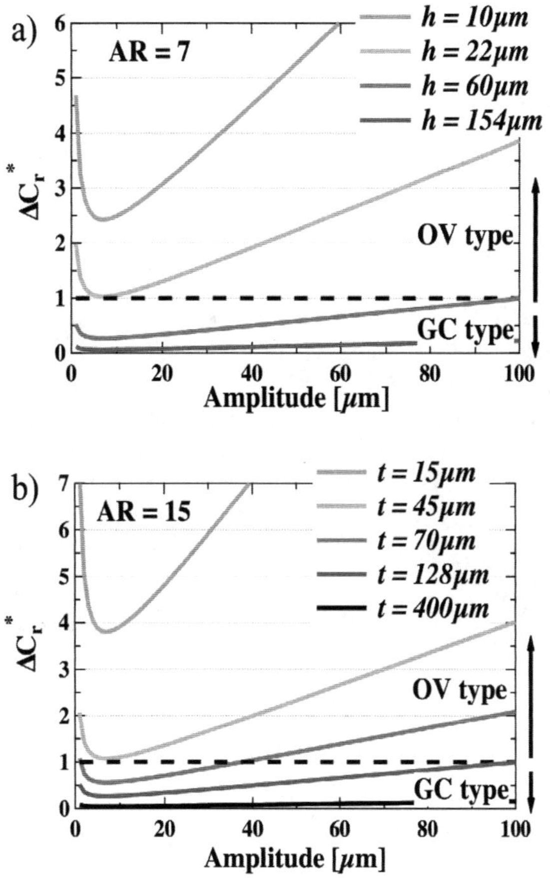

Figure 2. Maximum capacitance change ratio per unit area versus the amplitude of movable electrode for different aspect ratios. a) AR=7, b) AR=15. When $\Delta C_r^* > 1$, the overlapping-area-change (OV) type has larger capacitance change than the gap-closing (GC) type, and vice versa.

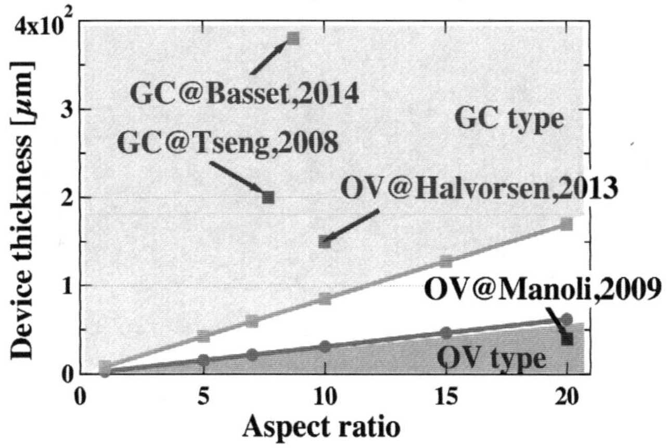

Figure 3. Thresholds of the device layer thickness versus the aspect ratio. For thinner device layer, the overlapping-area-change (OV) type has higher performance, while the threshold increases with the aspect ratio of the DRIE process.

type for $h > 60$ μm. When *AR=15*, higher output power can be achieved with the OV type for $h < 45$ μm (Fig. 2b), while the GC type is superior for $h > 128$ μm.

Figure 3 illustrates the mapping of the thresholds for different *AR*. With increasing *AR*, the region for the OV type is increased. As also shown in Fig. 3, some of the previous studies with the OV type are located in the region for the GC type. Figure 4 shows the mapping for the electret generator with the electret layer (thickness: *d*). With the electret layer, the effective aspect ratio is increased, and thus the region for the GC type is increased.

3. EXPERIMENTAL RESULTS

Figures 5 and 6 show the present prototypes of MEMS electret energy harvester [6] and their mechanical responses. The device layer thickness is 70 μm and *AR=7*. 1.5 μm-thick parylene-C layer is used as the electret layer. For the present design of $\Delta C_r^* = 0.89$ (Eq. 4), the GC type has larger capacitance change than the OV type.

Figure 7 shows the output power versus the excitation frequency for the OV and GC types. At 3 g acceleration, maximum output power of 1.19 μW and 1.71 μW are obtained, respectively; the GC type has higher output power than the OV type even with its lower resonant frequency. Although more systematic comparison is necessary, the present experimental data are in accordance with the present analysis. Note that the GC type exhibits more significant nonlinear behavior than the OV type due to its inherent nonlinear behavior of the electrostatic force.

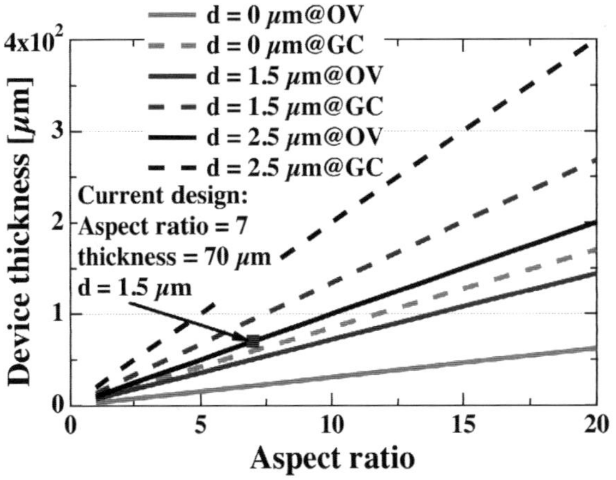

Figure 4. Thresholds versus the aspect ratio for different thicknesses of electret film on comb drives.

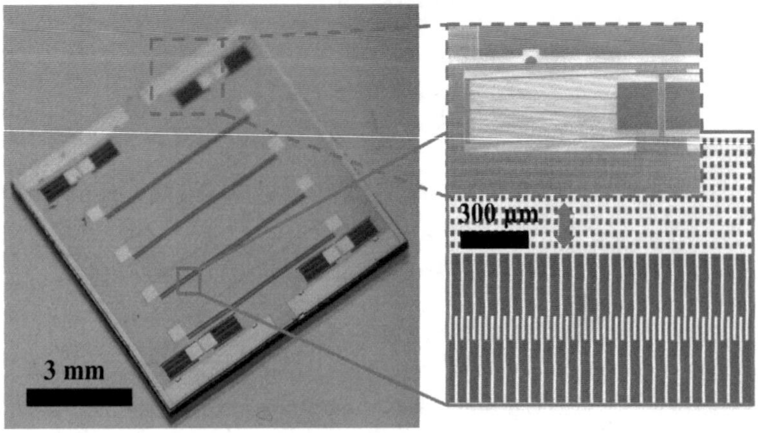

Figure 5. A prototype of MEMS electret energy harvester. Fabrication process can be found in [6].

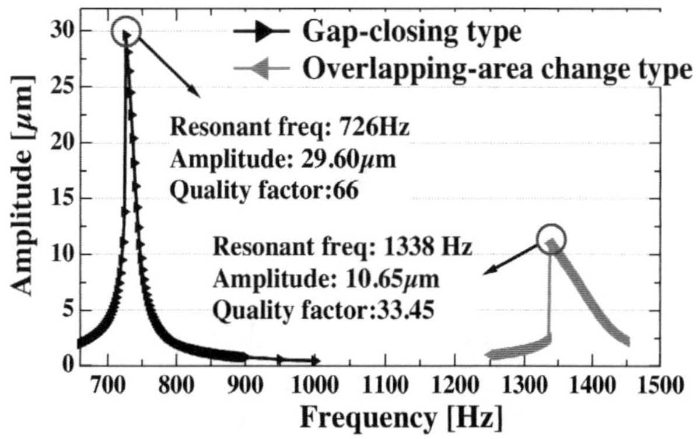

Figure 6. Frequency response for the gap-closing/overlapping-area-change type of electret energy harvesters after the parylene-C deposition.

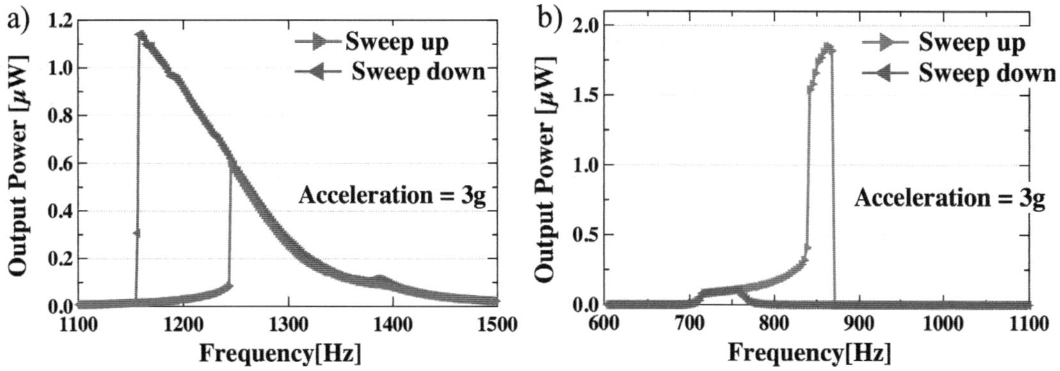

Figure 7. Output power of the present electret energy harvesters versus the excitation frequency at 3 g. a) overlapping-area-change type, b) gap-closing type. The thickness of device layer is 70 μm and AR=7. 1.5 μm-thick parylene-C layer is used as the electret layer, which corresponds to ΔC_r^*=0.89.

4. CONCLUSION

A design method for in-plane MEMS electrostatic energy harvesters is proposed. Dependent on the device layer thickness and the aspect ratio of the DRIE process, either the overlapping-area-change or the gap-closing type has higher output power than the other. Two prototypes of MEMS electret energy harvesters are developed to verify the present design method.

This work is partially supported by NEXCO Group Companies' Support Fund to "Disaster Prevention Measures on Expressways". Photo-mask is made using the University of Tokyo VLSI Design and Education Center (VDEC)'s 8-inch EB writer F5112+VD01 donated by ADVANTEST Corporation.

References

[1] Suzuki Y, 2011, *IEEJ Trans. Electr. Electr. Eng.* **6** 101-111
[2] Hoffmann D, Folkmer B, and Manoli Y 2009 *J. Micromech. Microeng.* **19**, 094001
[3] Nguyen S D, Halvorsen E, and Paprotny I 2013 *Appl. Phys. Lett.* **102** 023904
[4] Chiu Y, and Tseng V F G 2008 *J. Micromech. Microeng.* **18**, 104004
[5] Guilllemet R, Basset P, Galayko D, Cottone F, Marty F, and Bourouina T 2013 *IEEE MEMS 2013* 817-820
[6] Fu Q Y, and Suzuki Y 2014 *IEEE MEMS 2014* 409-412

Electric control of power extracting end-stop for MEMS vibration energy harvesting

Binh Duc Truong, Cuong Phu Le and Einar Halvorsen

Department of Micro- and Nanosystem Technology, Buskerud and Vestfold University College, Campus Vestfold, Raveien 215, 3184 Borre, Norway

E-mail: Einar.Halvorsen@hbv.no

Abstract. This experimental work investigates a technique to further improve performance of vibration energy harvesters under displacement-constrained operation. Previously, a device concept based on end-stops acting as additional transducers was developed so that the harvested power can be increased beyond the power obtained from a conventional harvester of the same size. However, there is a range of tested acceleration amplitudes in which the transducing end-stop device performs worse than the conventional device. In this paper, an approach using electric control is used to optimize the end-stop transducer performance and thereby further improve the system effectiveness under displacement constrained operation. For example, the maximum power increases by a factor of 2.4 compared to that of a conventional prototype under the same operating conditions and constrained displacement amplitude, while this value was about 1.3 for the previous technique.

1. Introduction

MEMS vibration energy harvesting is a technology to replace batteries in wireless sensor systems and other microscale applications. A harvester converts ambient mechanical energy into electrical energy to power the system. Typically, the energy converter is designed as a spring-mass system and is based on either of three basic mechanisms: piezoelectric, electromagnetic or electrostatic conversion [1–5]. For low loss, resonant microscale devices, the limited space available on chip can limit the proof mass displacement constraining it to a maximum amplitude X_{max}. This limitation has consequences when the vibration is strong enough. Usually, end-stops are designed to reliably define the motion range of the proof mass while preventing structural damage. To provide their function, end-stops tend to be rigid compared to the proof mass suspension. A positive effect of using the rigid end-stops is the ability to enhance system bandwidth seen in several recent works [6,7]. Nevertheless, saturated power is a major limitation observed when the proof mass hits the end-stops. In order to overcome this shortcoming, we previously introduced a device concept that uses active end-stops as additional transducers to extract extra power from internal impacts and thereby adding to the power of the main transducers. The prototypes have demonstrated improved power even when the maximum displacement amplitude is reached [8].

Despite the advantages of the end-stop transducers, there is still an undesirable gap in the tested range of acceleration amplitudes where the impact device has no benefit. A prototype in [9] with reduced mechanical stiffness for the end-stop transducers somewhat reduces this gap.

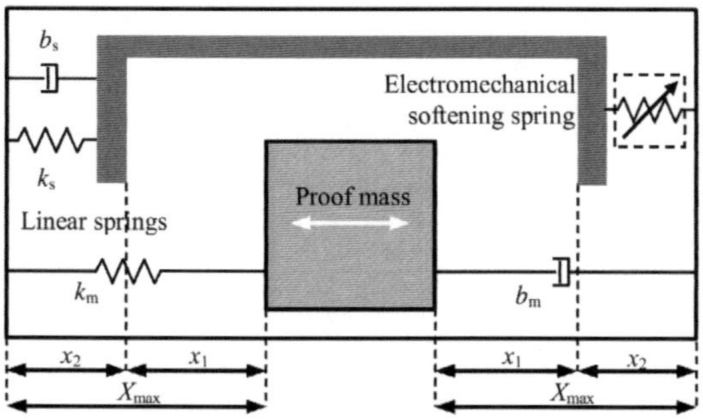

Figure 1. Device concept for utilization of movable end-stop functioning as additional transducer; the main proof mass is constrained to a maximum displacement X_{max} by rigid end-stops

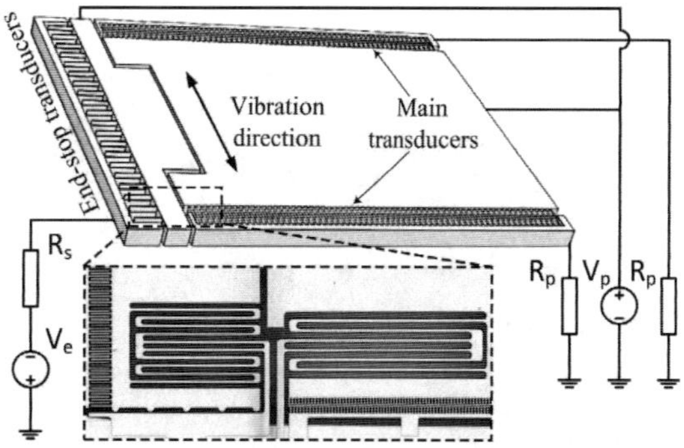

Figure 2. Schematic layout and close-up view of the fabricated impact device including main and end-stop transducers

In this work, an investigation on using electric control to reduce the effective stiffness of the end-stop transducer, and thereby further shrink the acceleration gap, is presented.

2. MEMS impact device and description

Figure 1 illustrates the device concept with the end-stop transducer that is simply represented by a spring-mass-damper system. Under sufficiently large accelerations, the main proof mass repeatedly collides with the end-stop transducer and both transducers contribute power. For small accelerations, there are no impacts and the contribution to output power from the end-stop transducer is negligible. The active area of the prototype is 4×5 mm^2 and is mostly occupied by the main proof mass and its transducer structure. The main proof mass m_m is much larger than the mass of the end-stop structure which is $m_s \sim \frac{m_m}{20}$. The maximum displacement amplitude is $X_{max} = 10$ μm. At equilibrium, the distance between the end-stop proof mass and the main proof mass is $x_1 = 6$ μm and the distance between the end-stop proof mass and the rigid end-stops is $x_2 = X_{max} - x_1 = 4$ μm.

Figure 2 shows the close-up view of the impact device, which is fabricated in the SOIMUMPS process with 25 μm device layer thickness. The main transducer is an ordinary comb-drive structure with two anti-phase overlap varying capacitances and the end-stop transducer is a gap-closing structure. All proof masses are suspended by linear folded-springs. The symmetric gap-closing structure exerts an electrostatic force corresponding to a negative stiffness, which plays a role as an electromechanical softening spring and consequently reduces the net linear stiffness of the end-stop. Further detailed parameters of the impact device can be found in [10].

A bias $V_s = V_p + V_e$ is used to control the end-stop transducer while the main-transducer bias is kept constant at V_p=9 V. When V_s increases, the net linear stiffness of the end-stop transducer decreases due to the electrostatic attraction which corresponds to a negative stiffness. Ultimately the net stiffness goes to zero at a critical value $\sim$ 22 V and a linear instability occurs. At V_s above $\sim$ 15 V, the end-stop transducer tends to dynamically pull in and stick to its own rigid end-stop, so all reported tests were conducted below this value.

In the experiments, output powers are simply measured by connecting the transducers to external resistive loads and measure the voltage across the loads. The voltages are measured through buffer amplifiers and collected by a NI-USB-6211 DAQ controlled by a computer with

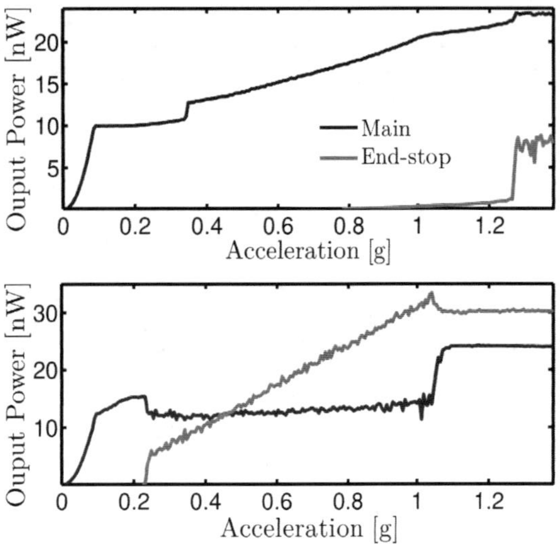

Figure 3. Measured output power of the main and end-stop transducers in sweeps of acceleration amplitude at fixed main bias voltage V_p=9 V and resonant frequency f_0=648.5 Hz for different end-stop bias voltage V_s=9 V (top) and 14.5 V (bottom)

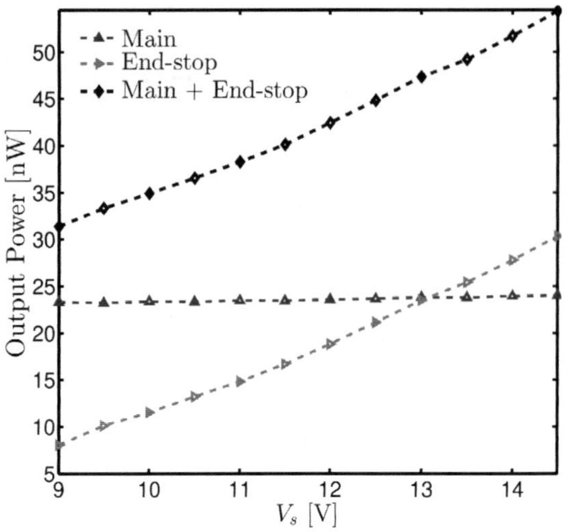

Figure 4. Measured output power of the main and end-stop transducers under increased end-stop bias voltage at fixed main bias voltage V_p=9 V, resonant frequency f_0=648.5 Hz and acceleration A=1.38 g

National Instruments LabView software. A TIRA vibration exciter is used to generate the vibrations. The total power of the impact device is the sum of powers from the main and end-stop transducers.

3. Measurements

Figure 3 shows effects of the electric control on output power of the impact device, where output powers of the main and end-stop transducers are measured in acceleration sweeps at resonant frequency f_0=648.5 Hz for different end-stop bias voltages V_s=V_p=9 V and V_s=14.5 V. The measured results indicate that the internal impact starts at an acceleration amplitude A=0.09 g. For $A > 0.09$ g, output power of the main transducers increases roughly linearly before reaching maximum value of P=23.4 nW at A=1.28 g for V_s=9 V, while for V_s=14.5 V a less acceleration amplitude of A=1.04 g is sufficient to obtain maximum power. In addition, the power extracted from the end-stop starts to increase when the acceleration amplitude exceeds A=0.24 g for V_s=14.5 V. This is a significant reduction compared to the value A=1.28 g observed when V_s=9 V. Power contribution from the transducing end-stop for V_s=14.5 V is much better than that for V_s=9 V, for example, about 3.78 times higher at their maximum levels.

Measured output powers of the main and end-stop transducers under increasing V_s are shown in Figure 4. The power extracted from the end-stop transducer is roughly linear in V_s in the impact regime while that of main transducers remains unchanged. However, in order to obtain significant benefits from the transducing end-stop compared to that of the main structure, high bias voltage for the gap-closing transducer is required while still maintain dynamic stability. The output power of end-stop transducer is equal to that of the main transducer at V_s=13 V giving P=23.8 nW and even higher for further increase provided $V_s < V_{pull-in}$.

In order to validate the technique for power improvement, the impact device performance is compared to that of a conventional reference device with rigid end-stops. The reference device is

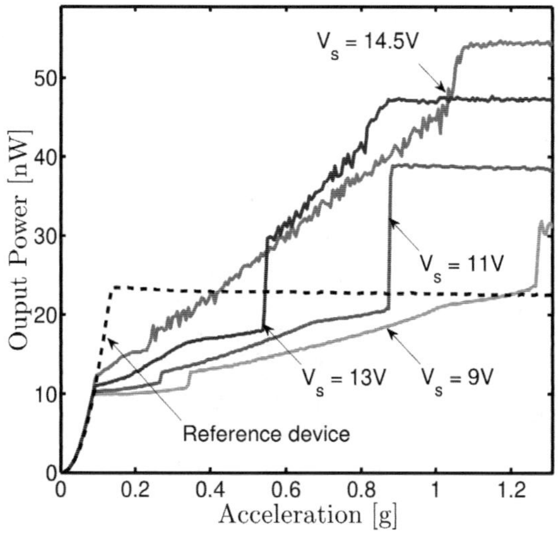

Figure 5. Measured total output power of the impact and reference devices at their resonant frequencies with different values of end-stop bias voltage. The bias is V_p=9 V for both main transducer and for reference device

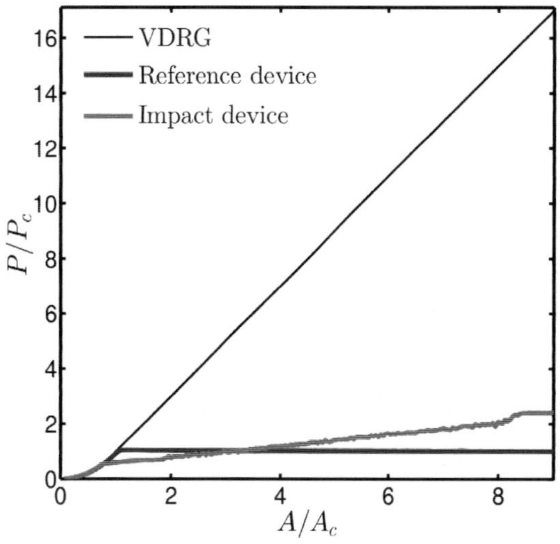

Figure 6. Measured total output power of the impact and reference devices compared to optimal velocity-damped generator VDRG

designed to have the same active area and the same maximum displacement amplitude X_{max}=10 μm as the impact device. Thus, the reference device is similar to the main transducer part of the impact device, but has space for bigger mass and larger transducers. The reference transducers are biased by the same voltage V_r=V_p=9 V. The total power of both devices at their resonant frequencies are shown in Figure 5. The conventional device displays saturated power P=23.8 nW when X=X_{max} at A=0.14 g. When increasing V_s from 9 V to 14.5 V, the output power is growing almost linearly before reaching the saturation power, instead of exhibiting jump phenomena as observed for lower bias voltages. The power of the impact device is equal to that of reference device at A =1.2 g for V_s=9 V, but this critical acceleration decreases from A =0.9 g, 0.5 g to 0.4 g when increasing from V_s=11 V, 13 V to 14.5 V respectively. Hence, the undesirable gap in the beneficial range of acceleration amplitudes for the impact device is reduced from 1.1 g to 0.3 g when V_s =14.5 V, making the impact technique useful at a significantly lower acceleration. For A > 1.05 g, the total power of impact device is 2.4 times higher than that of the reference device.

4. Discussion

The two devices are compared to the optimal velocity-damped generator (VDRG) which has nearly the maximum possible power [11] as shown in Figure 6. The output power and the input acceleration are normalized respectively by the factors P_c=$2\pi^2 f_0^2 X_{max}^2 b_r$ and A_c=$\frac{4\pi f_0 X_{max} b_r}{m_r}$ where m_r is the proof mass of the reference device and b_r is the mechanical damping of the reference device found from fit to measurement in the linear regime. The comparison demonstrates that even though the end-stop transducer is advantageous in enhancing the power under displacement-constrained operation, the obtained power is considerably lower than the maximum possible. It is interesting to consider approaches that could further boost power in the impact regime. It seems unavoidable to consider the more complicated approach of load optimization that tracks the acceleration amplitude, which is what the VDRG result corresponds to. One could also expect benefits from further developed design of the end-stop transducer, for example by combining a mechanical hardening spring with electromechanical softening while keeping the system stability in order to achieve low stiffness and high electrical damping.

5. Conclusion

A method using the electric control of the end-stop transducer has improved the performance of the impact device comparing to previous techniques in [9, 10] that caused lower power and jump phenomenon. Separate tuning the bias V_s of the end-stop transducer is able to drive its effective stiffness extremely low thanks to the attractive electrostatic force. The transducing end-stop therefore actively extracts more power from the internal impact at even lower force amplitude than the previous design. The technique displays reduction of the undesirable acceleration gap and increase of the maximum power. Nevertheless, it is clear that despite the improvements, there is still considerable room for improvement before the fundamental physical limit is reached in future works.

Acknowledgment

This work was supported by the Research Council of Norway through Grants no. 191282 and 229716/E20.

References

[1] Mitcheson P D, Yeatman E M, Rao G K, et al. (2008) *Proc. of the IEEE* **96** 1457–1486
[2] Roundy S, Wright P K and Rabaey J M (2003) *Computer Communication* **26** 1131–1144
[3] Roundy S, Wright P K and Rabaey J M (2004b), London: Kluwer Academic Publishers. ISBN: 1–4020–7663–0
[4] Moss S, Barry A, Powlesland I, Galea S and Carman G P, 2011 *Smart Mater. Struct.* **20** 045013
[5] Lei Gu and Carol Livermore, 2011 *Smart Mater. Struct.* **20** 045004
[6] Hoffmann D, Folkmer B and Manoli Y, 2009 *J. Micromech. Microeng.* **19** 094001
[7] Soliman M S M, Abdel-Rahman E M, El-Saadany E F and Mansour R R, *J. Micromech. Microeng.*, **18** (2008) 115021
[8] Le C P, Halvorsen E, Søråsen O and Yeatman E M, *J. of Int. Mat. Syst. Struc.*, **23** 1409–21
[9] Le C P, Halvorsen E, Søråsen O and Yeatman E M, *Proc. PowerMEMS* 2012: 444–7
[10] Le C P, Halvorsen E, Søråsen O and Yeatman E M, *Proc. PowerMEMS* 2011: 122–5
[11] Halvorsen E, Le C P, Mitcheson P D and Yeatman E M, 2013 *J. Phys.: Conf. Ser.* **476** 012026

Journal of Physics: Conference Series **557** (2014) 012013 doi:10.1088/1742-6596/557/1/012013

Vibrational Energy Harvester based on Electrical Double Layer of Ionic Liquid

S Yamada[1], H Mitsuya[2] and H Fujita[1]
[1]CIRMM, IIS, The University of Tokyo, 4-6-1 Komaba, Meguro-ku, Tokyo, 153-8505, Japan
[2]SAGINOMIYA SEISAKUSHO, 535 Sasai, Sayama, Saitama, 350-1395, Japan

E-mail: santa@iis.u-tokyo.ac.jp

Abstract. To boost the output of the vibration energy harvester an order of magnitude higher, we devised a high-performance energy harvester taking advantages of the two characteristics of ionic liquid, namely variable deformation of liquid and the electrical double layer between ionic liquid and metal. The electrical double layer is approximately 1nm thick and works as insulator within the voltage range of ±2.0V. Therefore, we can obtain quite high capacitance(1-10μ/cm^2). Squeezing and drawing ionic liquid between a pair of vibrating electrodes, we can obtain a μW-class energy harvester.

1. Introduction

The concept "Internet of Things[1]" is attracting much attention to realize a comfortable life with the data from wireless sensors installed in all places. Energy harvesting is a key for practical wireless sensor networks because it can eliminate periodical battery change. Energy harvesting is expected as the power source for sensor nodes because they work without maintenance and supply sensors with power until they get broken. Especially, the vibration type attracts much attention for its wide applicability.

There are 3 types of energy harvesters, a piezo type; a capacitive type; and an electromagnetic induction type. The electromagnetic induction type needs large magnets and structures and it becomes bulky. For this reason, this type is incompatible with MEMS device. To replace batteries of the wireless sensor nodes by energy harvesters, there are various difficulties such as low power generation and the narrow band width. Wireless sensor nodes consume 100μW to transmit data over 10m and the ambient vibrations associated with the bridge swing or human motion are composed of mainly low frequency under 10Hz[4][5]. Energy harvesters must generate power from such low frequency vibrations that exist abundantly in the environment. Due to its high spring constant and small mass, the resonant frequency of the piezo type tends to be high. To decrease the frequency, low spring constant and large mass are essential[2]. Thus, the structure becomes really fragile. Current devices cannot satisfy the requirement of generating large power from the low frequency source. We devised the capacitive type with ionic liquid to improve the robustness and output source. Taking the advantages of the characteristics of ionic liquid, the arbitrary deformation of liquid[6] and the high capacitance, we can fabricate a high-power energy harvester of the low resonant frequency and solve these problems. Just squeezing and drawing the ionic liquid, we can generate power. It is possible to eliminate a spring-mass structure and improves the robustness with ionic liquid. Furthermore, ionic liquid automatically forms electrical double layer of 1nm in thickness. It is unnecessary to make micro-gap like capacitive type based on the electrets[3]. The nano-gap makes the capacitance

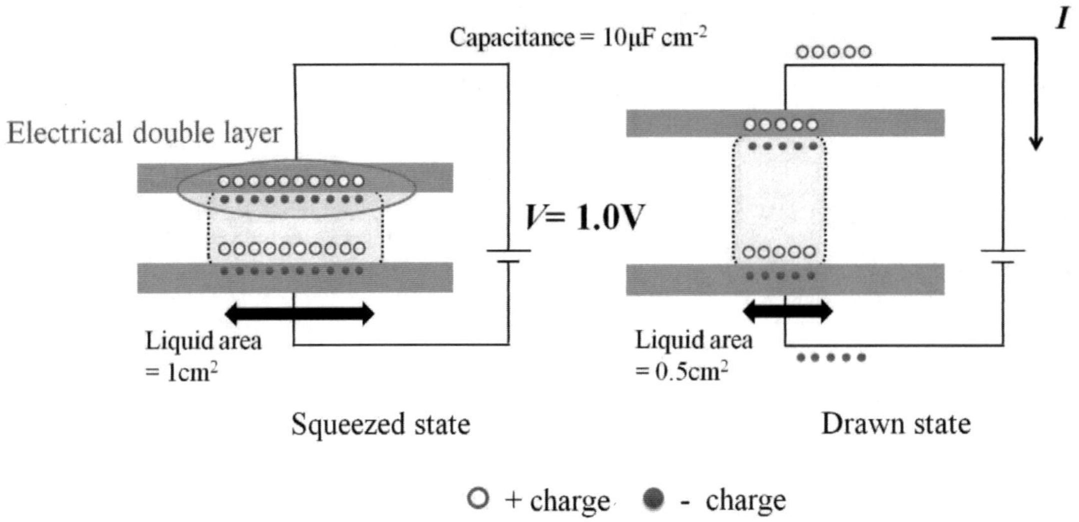

Figure 1. Schematic representation of the squeezed state and drawn state and calcuration. Narrow the gap and the contact area of the ionic liquid and electrodes changed. Thus, the charge at the electrodes changes and current occurs.

extremely high(μF) and we can obtain a large output. In this paper, we propose a new power generation principle based on ionic liquid.

2. Concept

Ionic liquid has various unique characteristics such as the formation of the electrical double layer, extremely low vapor pressure and resistance to high temperature. For this reason, remarkable attention has focused on devices using ionic liquid. The thickness of the electrical double layer is only 1nm and ionic liquid works as insulator within the voltage range called potential window. Thus, ionic liquid has quite high capacitance in the order of 10μF cm^{-2}[7]. Based on the large capacitance and its variable deformation, we devise a new energy harvester. Figure 1 shows the device structure and power

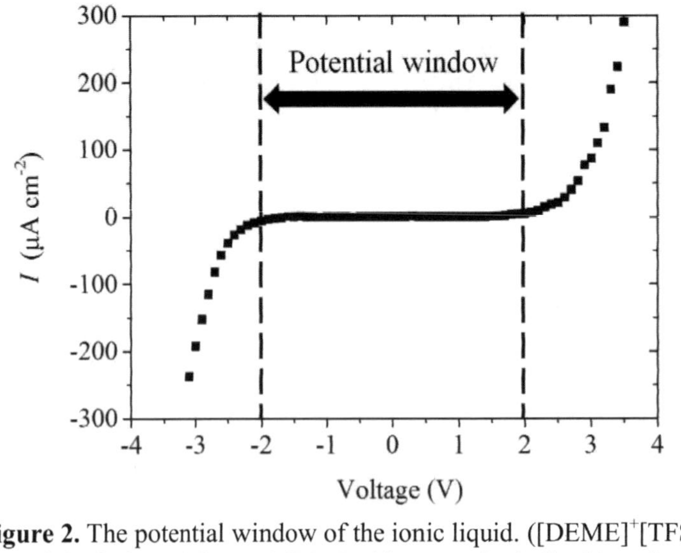

Figure 2. The potential window of the ionic liquid. ($[DEME]^+[TFSA]^-$)
potential window -1.9V to 1.9V : In this range, ionic liquid works as insulator.

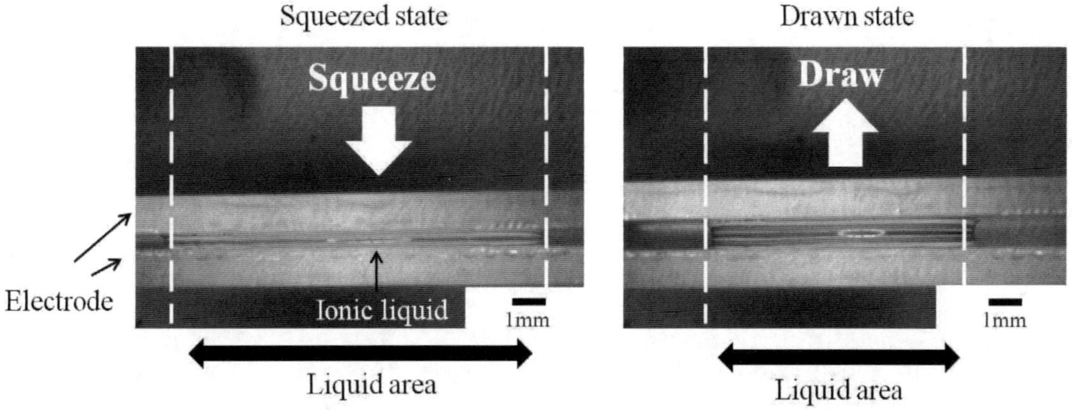

Figure 3. Photographs of the squeezed and drawn states.
Squeezed state: gap= 0.5mm, contact area= 1.6cm^2
Drawn state: gap= 1.0mm, contact area= 0.55cm^2

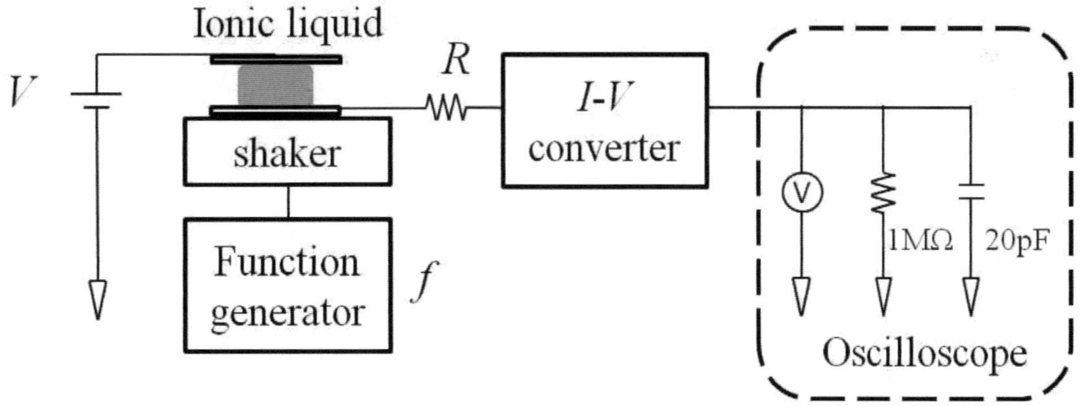

Figure 4. Schematic diagram of the experimental setup (*R* represents a loading resistance).
We applied sinusoidal movement periodically on the ionic liquid(+/- 250μm). The output current was measured by an oscilloscope and an I-V converter.

generating principle. The principle is quiet simple. Only squeezing and drawing the upper electrode, we can generate power. We put ionic liquid between electrodes and applied voltage. The voltage range is 0.5-2.0V within the potential window where the ionic liquid works as insulator as shown in Fig.2 (Please note in the real harvester the voltage source must be replaces by an electret). By drawing the upper electrode, the contact area decreases, and the capacitance between the liquid and the electrode also decreases. This change causes charge transfer and the current flows. Figure 3 shows the ionic liquid between two electrodes. To estimate the expected current ouput, we calculate the current with a simple model. In the setup in Fig. 1, let us assume the capacitance of ionic liquid in squeezed state is 10μF at V_{bias}= 1.0V. The squeezed state turns into the drawn state by the external vibration and the contact area decreases from 1.0cm^2 to 0.5cm^2. At the same time, the capacitance decreases from 10μF to 5μF. The change of the stored charge is 5μC. If this duration time is 1s, the output current will be 5μA. Based on this calculation, we prepare the setup and measure the output current. The experimental setup is shown in the Fig.4. We prepared a 90μl-droplet of ionic liquid (*[DEME]$^+$[TFSA]$^-$*) shown in Fig.3. One electrode was fixed and the other was attached to a shaker with the initial gap of 1mm. The ionic liquid was squeezed and drawn periodically by the vertical sinusoidal vibration. The gap

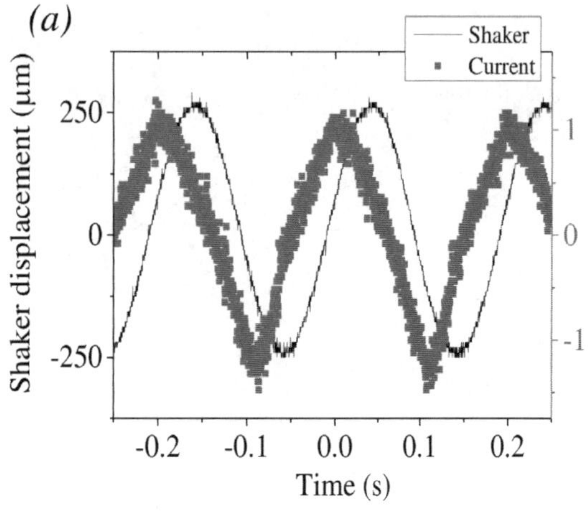

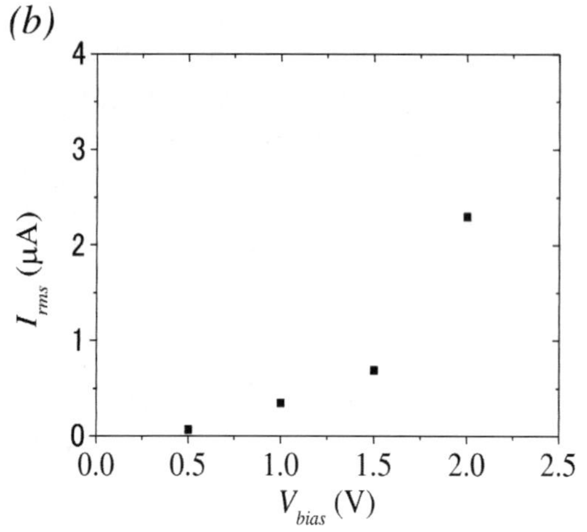

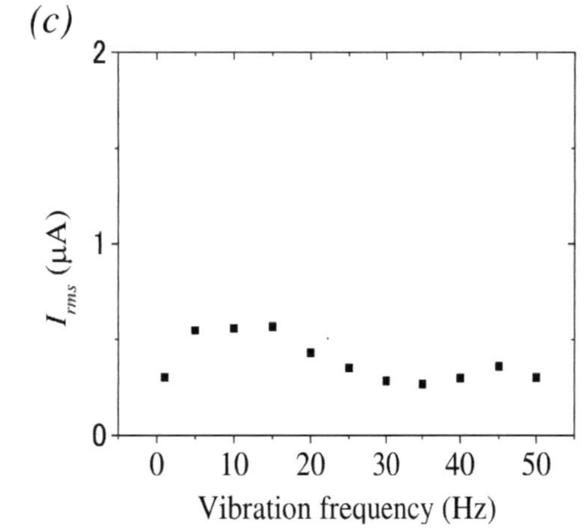

Figure 5. (a) Output current wave form at 5Hz of V_{bias} =1.5V, R=1kΩ. (b) Output at different bias voltage for sinusoidal motion at 5Hz and with R=1kΩ. The output increases as the bias Voltage increases. (c) Frequency dependence of output current at V_{bias}=1.5V, R=1kΩ, f=5Hz. The output hardly cahnged. Output must depend on the frequency, but high viscosity negatively influence on the output.

changed from 0.5mm to 1.0mm. The contact areas were 0.55cm^2 and 1.6cm^2 at squeezed and drawn states respectably. The output current was measured with an oscilloscope and an I-V converter.

3. Result & discussion

The peak value of the output current was 1μA (Fig. 5-a). We succeeded in power generation by squeezing and drawing ionic liquid and obtained sinusoidal output. The output in μA and the shaker motion in μm are expressed by 1.1sin(10πt+1.34)μA and 250sin(10πt)μm, respectively. There is a phase lag (61°) between them. Compared with the simple calculation, this output was smaller than we expected. Because of the wettablity, the spread ionic liquid could not be pulled back to the original state. This negatively influenced on the power generation.

The graphs 5-b and 5-c show output vs. bias voltage and vibration frequency, respectively. As bias voltage increased, we could obtain maximum effective current of 2.3μA at 5Hz, with a loading resistance of 1kΩ (Fig. 5-b). As the V_{bias} increased, the charge increased as well. Therefore large flow of electrons occurred in the circuit. Then, we swept the frequency from 1to 50Hz, at 1V with R of 1kΩ. We obtained output within the range (Fig. 5-c). As we swept the frequency of the shaker, the output did not increase as we expected. Because of its high viscosity, ionic liquid could not keep up with the motion of the shaker.

4. conclusion

A prototype of high-power energy harvester based on ionic liquid was reported in this paper. We succeeded in power generation with ionic liquid. The output is several nW and it was small compared with the power which wireless sensors need. However, the device generated quite high output current(μA) at this point. By improving the wettablity, the electrical circuit, and by optimizing the load resistance, it will be possible to generate a few tens of μW.

From the results, following conclusion can be drawn.

- The change of the capacitance generated power with the simple structure. Ionic liquid can eliminate complex structure from energy harvesters. It also improves the robustness of structure.
- The output depends on the bias voltage. If we utilize ionic liquid with a large potential window, we can apply high voltage on it and obtain high power generation.
- Ionic liquid driven by vibration generates power at the frequency range from 1 to 50Hz. From the results, the high viscosity negatively influenced on power generation. Ionic liquid with low viscosity must improve the frequency characteristic.

Our next steps are (1) replacement of the external power source with electrets, (2) improvement of the frequency characteristic for more efficient power generation from the environmental vibration which has wide spectrum in low frequency range, (3) optimization of the design of structures. We want to realize μW energy harvester with ionic liquid to replace the battery of wireless sensor nodes.

References

[1] Hecht J 2014 *Nat. Phys.* **10** 538
[2] Shen D, Park J H, Ajitsaria J, Choe S Y, Wikle H C and Kim D J 2008 *J. Micromech. Microeng.* **1** 055017-24
[3] Suzuki Y, Miki D, Edamoto M and Honzumi M 2010 *J. Micromech. Microeng.* **20** 104002-10
[4] Lombaert G, Conte J P and Asce M 2012 *J. Eng. Mech.* **138** 816-25
[5] Gwin J T, Gramann K, Makeig S and Ferris D P 2010 *J. Neurophysiol.* **103** 3526-34
[6] Strani M and Sabetta F 1984 *J. Fluid. Mech.* **141** 233-247
[7] Lockett V, Horne M, Sedev R, Rodopoulos T and Ralston J 2008 *J. Phys. Chem. Chem. Phys.* **12** 7486–95.

A high Reliability Module with Thermoelectric Device by Molding Technology for M2M Wireless Sensor Network

K. Nakagawa*, T. Tanaka and T. Suzuki

FUJITSU LABORATORIES LTD.
10-1 Morinosato-Wakamiya, Atsugi 243-0197, Japan

kanaka@jp.fujitsu.com

Abstract. This paper presents the fabrication of a new energy harvesting module that used the thermoelectric device (TED) by using molding technology. The output voltage per heater temperature of the TED module at 20 °C ambient temperature is 8mV/K and similar to the result with the aluminium heat sink which is almost the same fin size as the TED module. The accelerated environmental tests are performed on damp heat test that is an aging test under high temperature and high humidity, cold test and highly accelerated temperature and humidity stress test (HAST) for the purpose of evaluating the electrical reliability in harsh environments. Every result of tests indicates that the TED and circuit board can be properly protected from harsh temperature and humidity by using molding technology, because the output voltage of after tested modules is reduced by less than 5%.This study presents a novel fabrication method for a high reliability TED-installed module appropriate for Machine to Machine wireless sensor networks

1. Introduction

Energy harvesting technology allows Machine to Machine (M2M) devices to operate indefinitely without any power source [1]. M2M devices must be mechanically robust and tolerant to an external environment. However the thermocouples of thermoelectric devices (TEDs) are usually exposed and mechanically fragile[2]. There is almost no energy harvesting module with TED which can be used in external environment. In this work, we fabricated a new energy harvesting module using the thermoelectric device (TED) as an energy harvesting device.

Figure 1 is a schematic of the designed light and durable TED module with a heat transfer sheet. In this study, the TED is built into a power generation module by using molding technology [3, 4]. Through molding technology, the TED and circuit board can be properly protected and a heat-radiating fin structure can be simultaneously constructed. This is because the exterior fin structure is covered by a graphite heat transfer sheet. Good thermal contact between this sheet and the TED by thermally conductive material ensures effective heat radiation from the fin. In this paper, the fabrication of the light and durable TED modulus by using molding technology and the reliability for M2M devices is presented.

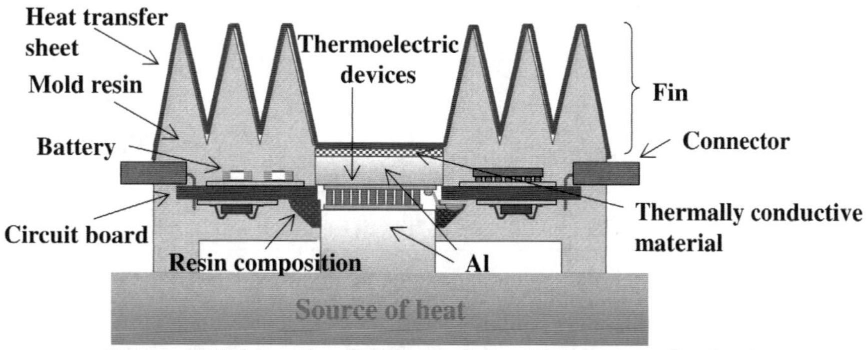

Figure 1. Schematic of TED Module with heat transfer sheet.

2. Fabrication

Figure 2 illustrates the fabrication process of the proposed TED module.

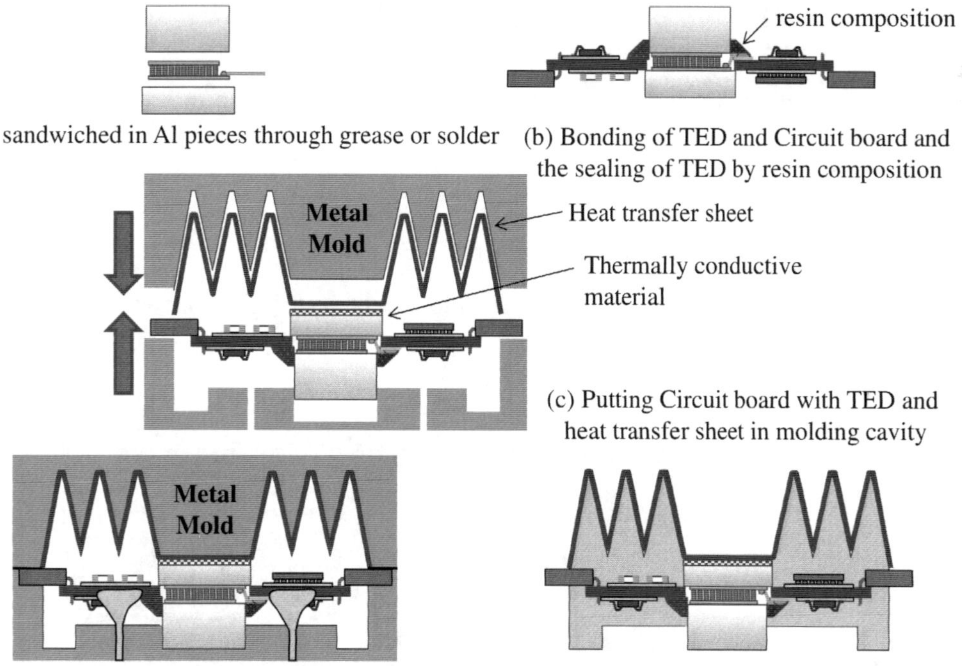

(a) TED is sandwiched in Al pieces through grease or solder (b) Bonding of TED and Circuit board and the sealing of TED by resin composition

(c) Putting Circuit board with TED and heat transfer sheet in molding cavity

(d) Injection of hot melted resin into molding cavity (e) Finished TED module after cooling

Figure 2. Fabrication process of TED module.

(a) The TED is sandwiched between Al pieces by using thermal grease or solder.(b) The TED interconnects are soldered onto the circuit board with connectors for the power supply of sensor devices and the TED is sealed by the resin composition in order to prevent of injection mold resin into the TED. (c) The circuit board installed with the TED and the heat transfer sheet are set into a metal mold. (d) The melted mold resin is injected. The injection pressure is 1.0 Pa, and the temperature of the melted resin is 230°C. After cooling, the TED module is released from a metal mold. The duration from setting parts to release of the module is a few minutes.

Molding is performed by a hotmelt process using solventless thermoplastic resin. Table 1 shows the properties of the mold resin (Henkel AG & Co. KGaA; Model OM652).

Table1 Material properties of mold resin

Operating Temperature Limits (°C)	Tg (°C)	CTE (ppm/K)	Young's Modulus (GPa) at 25°C	Thermal Conductivity (W/mK)	Flame Retardant
-40~100	-45	483	0.06	0.2	V-0

The TED is a commercial model (Model KTGS066A0: KELK Ltd.). Table2 shows the characteristics of TED.

Table 2 Characteristics of TED (KELK Ltd; Model KTGS066A00)

Demension Top side Bottom side (mm)	Thickness (mm)	Thermal Conductance (W/K) at R.T. in Dry Air	Electrical Resistamce (Ω) at 27℃	Open Circuit Voltage (V) at Th=32℃, Tc=22℃
7.16×6.57 7.57×7.16	1.32	0.035	7.11	0.26

3. Result and Discussion

Figure 3A shows the module attached to the heat transfer sheet. The module is 35 × 35 × 16.4 mm^3 in volume, and it weighs 17.5 g.

Figure 3B shows the tip point of the fin. The point angle is 3 degrees. The fin tip is filled without void with mold resin, which adheres well to the heat transfer sheet in spite of the sharp angle of the point of the fin. Through molding technology, a circuit board with the TED and a heat-radiating fin structure can be simultaneously constructed.

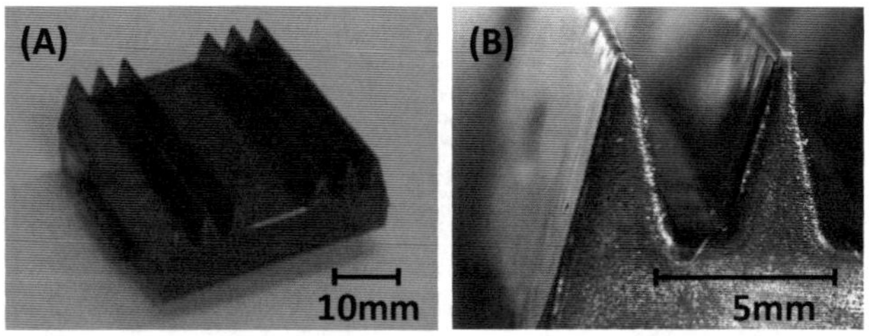

Figure 3. (A) Photograph of TED Module with heat transfer sheet. Size is 35 × 35 × 16.4 mm and weight is 17.5 g. (B) Microphotograph of fin.

Because the molding pressure of injection is more than one-hundred times smaller than conventional molding pressure, the devices on the circuit board are less damaged. In addition, because the solidification time in hotmelt molding is very short (several seconds), the turn-around time is at most only a few minutes. Therefore, the proposed fabrication process is very economical. The heat transfer sheet keeps in contact with the Al pieces with TED by molding.

Figure 4 plots the output voltage of the TED module as a function of heater temperature at 20 °C ambient temperature. For the purpose of comparison, the output voltage of the same TED with an aluminum heat sink (LPD35-7B Alpha Co. Ltd.) which is about the same fin size as the TED module is measured. The size of the heat sink is 35×35×7mm. The output voltage per heater temperature of module is 8mV/K and similar to the result with the heat sink. This result shows the fins with the graphite heat transfer sheet fabricated by molding technology have similar thermal radiation characteristics to the heat sink.

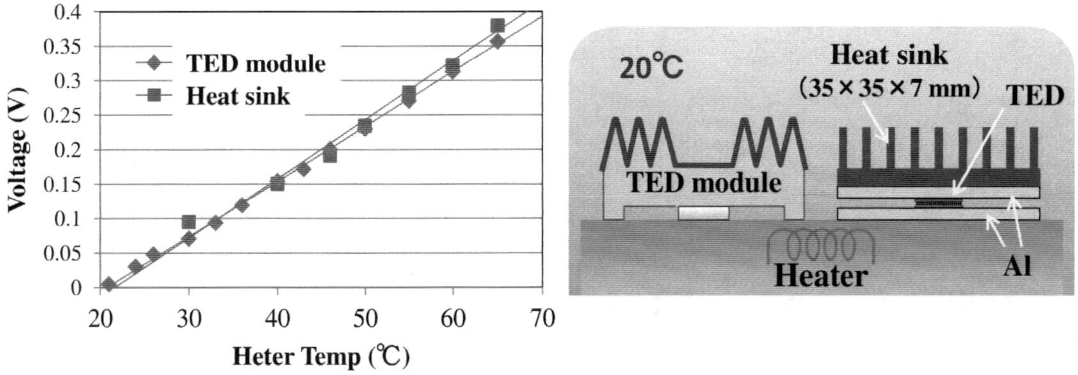

Figure 4. Output Voltage of TED Module vs. heater temperature at 20°C ambient temperature.

The accelerated environmental tests are performed on a damp heat test that is an aging test under high temperature and high humidity, a cold test and a highly accelerated temperature and humidity stress test (HAST) for the purpose of evaluating the electrical reliability in harsh environments. The damp heat test is performed in the conditions of 85 °C/85% RH/1000houers. The HAST is performed in the conditions of 110 °C/85% RH/1.2 Pa/408 h. Each result plots a ratio of the output voltage after each test V to the initial voltage V_0 (t=0), V/V_0. Figure 5 shows the damp heat test results. The damp heat test criterion is the voltage reduction within 8% (JIS C 8990). Because the output voltage of after tested modules is reduced by less than 5%, the voltage degradation is well within the pass standard. Figure 6 shows the cold test results and Figure 7 shows HAST results. Both results of the output voltage of all tested modules are reduced by less than 5%.

The thermal expansion or contraction of mold resin is a possible cause of the deterioration of power generation characteristics, because the TED is peeling off from Al pieces or the thermocouples are damaged. But the figure indicated that the power generation characteristics dose not deteriorate in hot or cold environments, because the flexibility of mold resin is considered to be retained by low Tg (glass transition temperature) -45°C.

The purpose of HAST is evaluation of the electrical reliability in humid environments. The results of HAST indicate that the cases made by hotmelt molding can protect the circuit board with TED from humid environments.

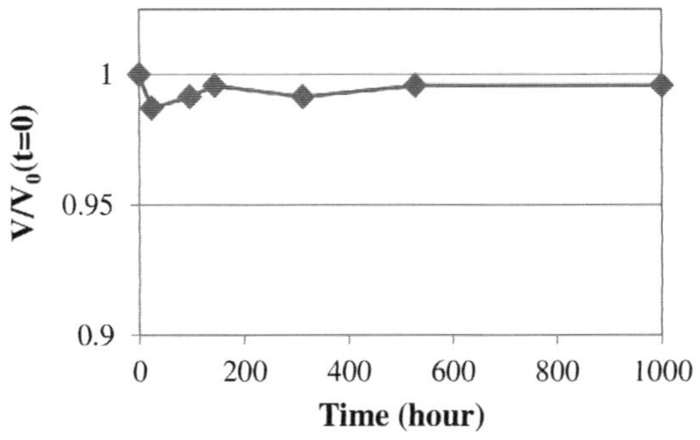

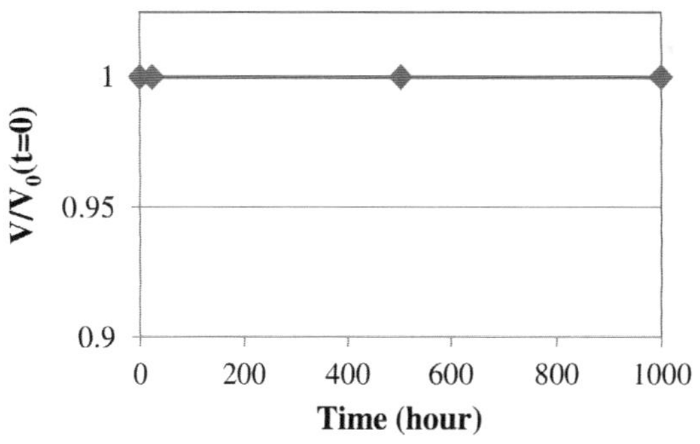

Figure 5. Result of damp heat test (85°C 85%RH1000h)**Figure 6.** Result of cold test (-45°C 1000h)

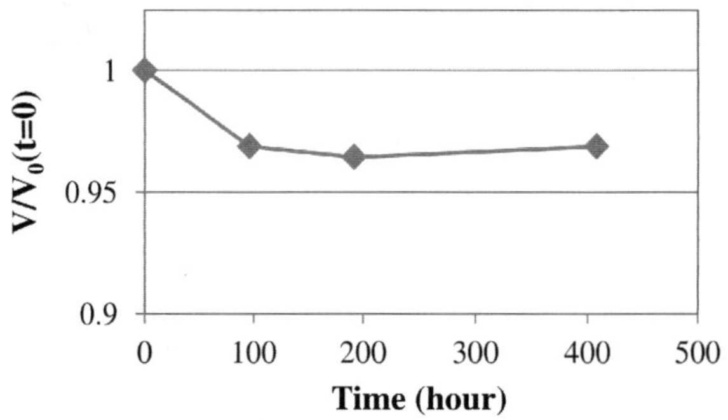

Figure 7. Result of Highly Accelerated temperature and humidity Stress test : HAST (110°C 85%RH1.2Pa408h)

Through molding technology, the TED and circuit board can be properly protected.

Considering our evaluation results described above, we conclude that the new energy harvesting module using TED as an Energy Harvesting Device by molding technology may be applicable to M2M sensor devices.

4. Conclusion

This paper presents the fabrication of a new energy harvesting module that used the thermoelectric device (TED) as an Energy Harvesting Device, for Machine to Machine (M2M) wireless sensor networks. M2M devices must be mechanically robust and tolerant to an external environment.

We propose the novel power generation module by using a molding technology enabled circuit board with the TED, the heat-radiating fin structure and heat transfer sheet to be simultaneously constructed. The light and robust power generation module with TED can be fabricated.

The output voltage per heater temperature of the TED module at 20 °C ambient temperature is 8mV/K and similar to the result with the aluminium heat sink which is about the same fin size as the TED module.

The accelerated environmental tests are performed on a damp heat test that is an aging test under high temperature and high humidity, a cold test and a highly accelerated temperature and humidity stress test (HAST) for the purpose of evaluating the electrical reliability in harsh environments. Every result of the tests indicates that the TED and circuit board can be properly protected from high and low temperature and humidity by using molding technology, because the output voltage of after tested modules is reduced by less than 5%.

We conclude that the new energy harvesting module using TED as an Energy Harvesting Device by molding technology may be applicable to M2M sensor devices.

References

[1] Tanaka T, Suzuki T and Kurihara K 2013 *FUJITSU 64,5, 557-563*
[2] Hiraishi K et al. 2009 *Jpn. J. Appl. Phys. 48 071501-1*
[3] Beaker B and Schoeneweiss A 2007 *Auto Technology 3 34-35*
[4] Yahata K 2008-09 *Materiak Stage 8(6) 14-18*

Modeling of the thermo-mechanical efficiency of the bimetal strip heat engines

A Arnaud[1,2], S Monfray[1], J. Boughaleb[1], E Trioux[1,3], F Boeuf[1], O Cugat[2], T Skotnicki[1]

1 ST Microelectronics, F38926 Crolles, France
2 G2Elab, Univ. Grenoble Alpes, F38000 Grenoble, France
3 TIMA, Univ. Grenoble Alpes, F38031 Grenoble

E-mail: arthur.arnaud@st.com

Abstract. This paper presents a theoretical demonstration of the bimetal strip heat engine working, based on the study of the thermo-mechanical instability of the pre-buckled bimetallic beams. Starting from the Euler buckling equation, this paper describes the bimetal strips like classical but non-linear thermodynamic systems, and gives the bistability criterion of such beams. Studying the thermodynamic potentials of these beams helps to evaluate the release of the kinetic energy happening during the beam snap-through, to give the Maxwell relations between each partial derivative of the thermodynamic potentials and to show that the thermal snap-through is a first-order transition according to the Ehrenfest theory. The model is then used to draw the temperature-entropy cycle of the bimetal heat engines and to evaluate the performances of these harvesters (available mechanical energy and thermodynamic cycle efficiency).

Introduction

The development of bimetal strip heat engines has been pursued as an alternative to the Seebeck thermoelectric generators based on the properties of semiconductors like bismuth tellurides, or to the conventional heat engines using fluids as a heat transfer medium. Proofs of concept have been reported in [1-6]. These devices are designed to exploit the sudden displacement of a bimetallic membrane switching from a critical equilibrium position to a stable one, to harvest a part of the thermal energy flowing through the bimetal and to convert it into kinetic energy. The existence of a hysteretic behavior, consequence of their bistability, enables to cycle beams between a hot source and a cold one (Fig.1a,b), causing then an astable behavior of the beams. The simplicity of this concept allows scaling down the devices using silicon manufacturing techniques ([4], [7]), which is expected to reduce the beam cycling duration, increasing then the heat flowing through the bimorph and finally multiplying the power generated by the harvester size scaling factor ([2]).

With this article, we aim at consolidating the theory of the bimetal strip heat engine: in a first part, we explain how the non-linear Euler buckling equation can be used to predict the bistability of bimetallic beams, whereas in a second part, the evolutions of the strain energy and entropy of the beams are studied along the equilibrium path. Finally, the model is used to evaluate the performances of the bimetallic strips as heat transfer media at the microscale.

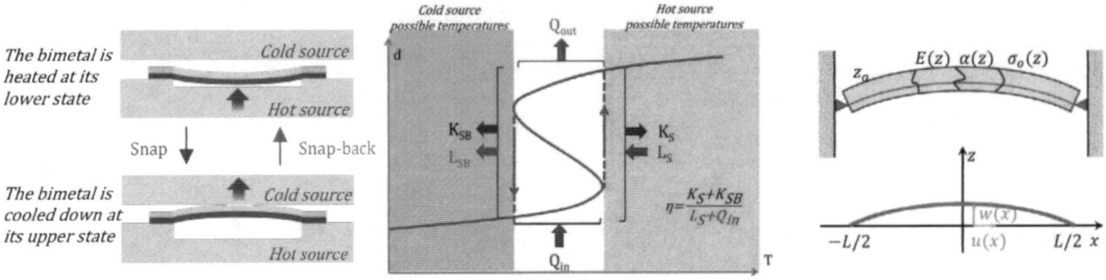

Figure 1. (a) Architecture scheme and (b) principle of the harvester. (c) Beam structure modeled

1. Snap-through modeling

The thermo-mechanical instability of the bimetallic beams is a consequence of the Euler buckling phenomenon described by equation (1), giving the dependence of the lateral displacement field on the axial thrust p_T and the beam bending stiffness i_e.

$$k^2.w_{,xx}+w_{,xxxx}=0 \tag{1a}$$

$$k^2=p_T/i_e \tag{1b}$$

Whereas the behavior of perfect and straight beam predicted by (1) looks like a pitchfork bifurcation, the behavior of an imperfect beam is sometimes a little bit more complex and can be discontinuous in some cases. This discontinuity is accompanied by a kinetic energy release and a sudden displacement of the beam called snap-through, caused by two imperfections of the beam having antagonistic effects. The bimetallic beam presented in Fig.2 is used to explain the conditions of occurrence of the thermal snap-through, by taking into account the combined effects of the residual stress and the thermal expansion. The beam's mechanical properties can be globally described through the mean of the parameters set (n_e, i_e, n_α, m_α, F_o, M_o), where n_e represents the beam stretching stiffness, n_α the mean thermal stress coefficient of the beam, and F_o the mean residual stress in the beam (measured at θ^o) which makes the beam buckle initially. The two beam's imperfections are described by m_α representing the asymmetry of the thermal expansion causing the thermal curvature of the beam in a given direction, and M_o representing the asymmetry of the residual stress in the beam's materials which orientates initially the beam buckling in the opposite direction to the thermal curvature. M_o and m_α must have the same signs to allow the thermo-mechanical instability to occur.

$$n_e=\int_{y,z} E(z).\delta z.\delta y \;;\; i_e=\int_{y,z} E(z).z^2.\delta z.\delta y \tag{2a}$$

$$n_\alpha=\int_{y,z} E(z).\alpha(z).\delta z.\delta y \;;\; m_\alpha=\int_{y,z} E(z).\alpha(z).z.\delta z.\delta y \tag{2b}$$

$$F_o=\int_{y,z} \sigma_o(z).\delta z.\delta y \;;\; M_o=\int_{y,z} \sigma_o(z).z.\delta z.\delta y \tag{2c}$$

Once these beam parameters are introduced, as we demonstrated in [8,9] it is possible to show that the thermodynamic principles, applied to the beam equilibrium, can be written as

$$dK+dF=0 \tag{3a}$$

$$(T+T^\circ).dS=\delta Q \tag{3b}$$

Where K is the kinetic energy of the beam, F its Helmholtz free energy, S its entropy and T its temperature (T° absolute temperature at which the stress is measured). Q represents the heat transferred to, or rejected from the beam during the heating and cooling periods. To understand the evolution of the internal beam efforts, one must express the constitutive equations of the beam. The Helmholtz free energy differential is given in (4a). As the volume and the temperature are the privileged variables of F in a classical thermodynamic system, we introduce here λ which is the difference between the length of the curved beam and the length of the straight beam, coming from the non-linearity of the Van Karman strain tensor (4c).

$$dF=-S.dT+(P-P_o).d\lambda \tag{4a}$$

$$dW_i=(P-P_o).d\lambda \tag{4b}$$

$$\lambda=\frac{1}{2}.\int_{-L/2}^{L/2} w_{,x}^2 .dx \tag{4c}$$

The force P_o is an invariant of the beam properties defined by (5a), which can be seen as a constant force applied by a pressure reservoir acting on the beam. The ratio of the M_o and m_α is also used to define a reference temperature T_o (5b). The force P describes the effect of the beam internal efforts (5c): the beam non-linear stretching force linking to the beam length variation λ, the mean thermal expansion, and the clamp thrust p_T which verifies the non-linear Euler equation (1).

$$P_o=n_\alpha.M_o/m_\alpha-F_o \tag{5a}$$

$$T_o=M_o/m_\alpha \tag{5b}$$

$$P=n_e.\lambda/L-n_\alpha.(T-T_o)+p_T \tag{5c}$$

Considering (4a), at a given temperature, the beam equilibrium is characterized by the equality.

$$P=P_o \tag{6}$$

As shown in [8,9], the instability appears when the external force P_o exceeds the beam Euler load P_γ (7a) which is a solution of (7b). γ, varying from 0 to ∞, models the stiffness of the clamp (0 for a simply supported beam, ∞ for a doubly clamped beam).

$$P_o>P_\gamma \Leftrightarrow n_\alpha.M_o/m_\alpha-F_o>i_e.k_\gamma^2 \tag{7a}$$

$$k_\gamma.\cot\left(k_\gamma.L/2\right)+\gamma=0 \tag{7b}$$

The evolution of the beam shape can then be plotted in function of the external force applied to the beam, showing the appearance of an unstable behavior, as shown on Fig.2a, when (7a) is verified. As seen, the bistable beams exhibit a thermal hysteresis characterized by two temperatures called snap temperature T_S and snap-back temperature T_{SB} at which the beam snaps from a critical position (or limit point) to a stable equilibrium position. Each switching is accompanied by a kinetic energy release which can be harvested if the bimetal is coupled with a piezoelectric transducer [2] for example.

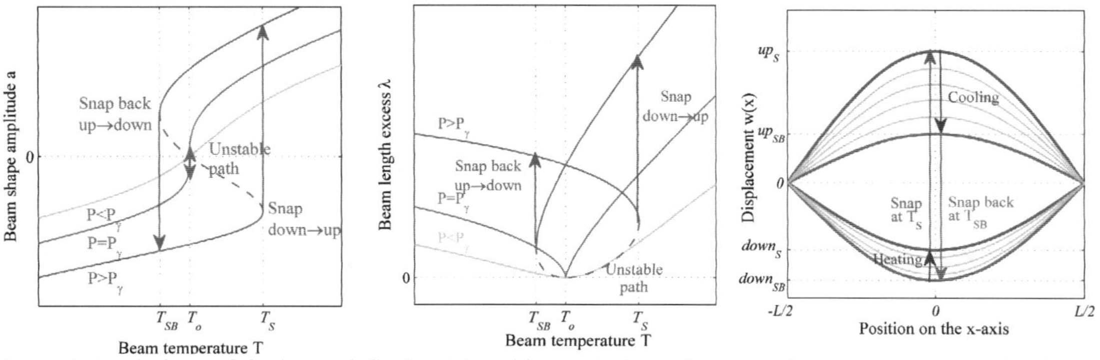

Figure 2. Evolutions of the beam deflexion (a) and length λ (b) in function of the temperature and P_o for a simply-supported beam (γ=0). (c) Evolution of the bistable beam shape in function of the temperature for a simply-supported beam (γ=0).

2. First-order transition and energy harvesting

By using (1-7), we show that the beam strain energy W_i of the harmonic functions is given by (8).

$$W_i=\frac{n_e}{2.L}.\lambda^2+i_e\left(k^2-2.\frac{k^3.\sin(k.L)}{k^2.L-k.\sin(k.L)}\right).\lambda-(n_{\alpha}.T-F_o).\lambda \tag{8}$$

Drawing the potential W_i helps understanding the energetic meaning of the thermal snap-through. As seen on Fig. 3a, for a bistable beam, W_i can exhibit two equilibrium wells separated by a potential barrier formed by an unstable equilibrium position. The beam energy evolves with the temperature and the stable equilibrium positions can become critical when reaching T_S or T_{SB}, making the beam snap to the other stable position and release its energy kinetically.

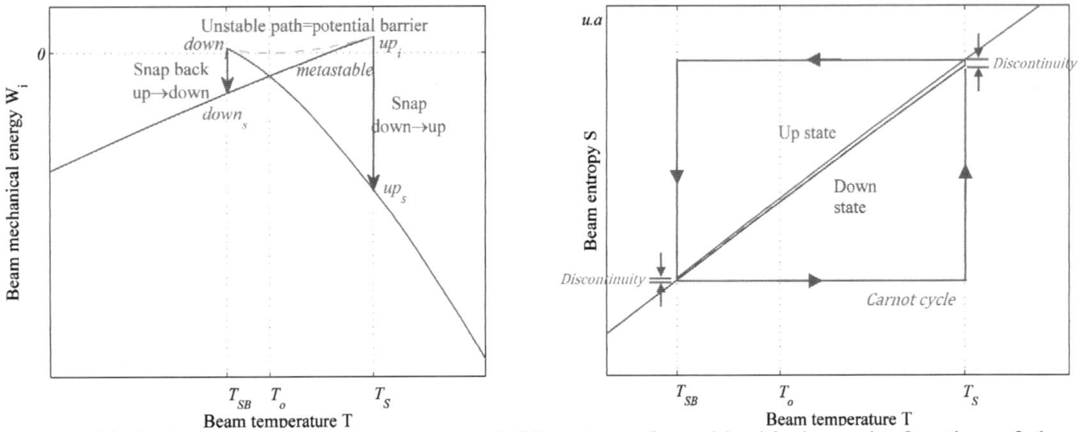

Figure 3. (a) Evolution of the strain energy and (b) entropy for a bistable beam in function of the temperature. (b) shows the thermodynamic cycle of the bimetal strip heat engine.

The Schwartz theorem is used to give the Maxwell relations between the partial derivatives of the entropy and beam force P. The bimetallic strip, as a thermodynamic system, can then be described with the matrix (9) where the partial derivative of the entropy with respect to λ expresses the thermo-mechanical coupling inside the beam materials.

$$d\begin{pmatrix}S\\P\end{pmatrix}=\begin{pmatrix}\frac{C_\lambda}{\theta} & -P^{\lambda}_{,T}\\ P^{\lambda}_{,T} & P^{T}_{,\lambda}\end{pmatrix}.d\begin{pmatrix}T\\\lambda\end{pmatrix} \tag{9}$$

The expression of the entropy (10) is found by using (9) and its evolution in function of the beam temperature is drawn on Fig.3b.

$$S\text{-}S°=\left(4.i_e.\frac{k^3.\sin(k.L)}{k^2.L-k.\sin(k.L)}+n_\alpha.(T-T_o)\right).\frac{\lambda}{T-T_o}+C_V.\ln(T+T°) \tag{10}$$

Two discontinuities of the entropy happen at T_S and T_{SB}, meaning that the thermal snap-through is a first-order transition according to the Ehrenfest and Landau theories [10]. Similarly to any other first-order phase change, a latent heat must be supplied to the beam to switch at T_S and rejected from the beam to switch back at T_{SB}. The order parameter associated to the transition is the deflection of the beam, which is either negative when the beam is at its lower state and in contact with the hot source, or positive when it is cooled at its upper state. The existence of a hysteresis on the temperature-entropy representation of the bimetals thermodynamic cycle shows the possibility to harvest energy thanks to a bimetal-based heat engines. This cycle can be compared to the classical Carnot cycle made of two isentropic and two isothermal curves and representing the maximal efficiency of a theoretically reversible heat engine.

3. Applications to micro-scale bimetals heat engine

The model can be used to evaluate the efficiency of a bimetal strip heat engine both at the macroscale and the microscale. In this paper, we study only the micro-harvesters, and take the example of a bimorph of aluminium (Al) and silicon oxide (SiO_2). This bimorph owns naturally a high difference of thermal expansion coefficients (α_{Al}=23.10^{-6}K^{-1}, α_{SiO2}=2,9.10^{-6}K^{-1}). Fig.4a gives the evolution of the beam deflexion in function of the thermal hysteresis of a 200µm-long beam (with 2µm-thick layer for each material). Fig.4b gives the energy density which can be harvested when the beam snaps and snaps back. Using (10) to evaluate the thermal exchange of the beam with its surrounding, it can be shown (Fig.4c), that a Carnot-relative efficiency of around 1.7% (up to 3.5% with a W-Al beam [9]) and an energy density of about 300µJ.cm^{-3} can be obtained with two-degree hysteresis SiO_2-Al beam.

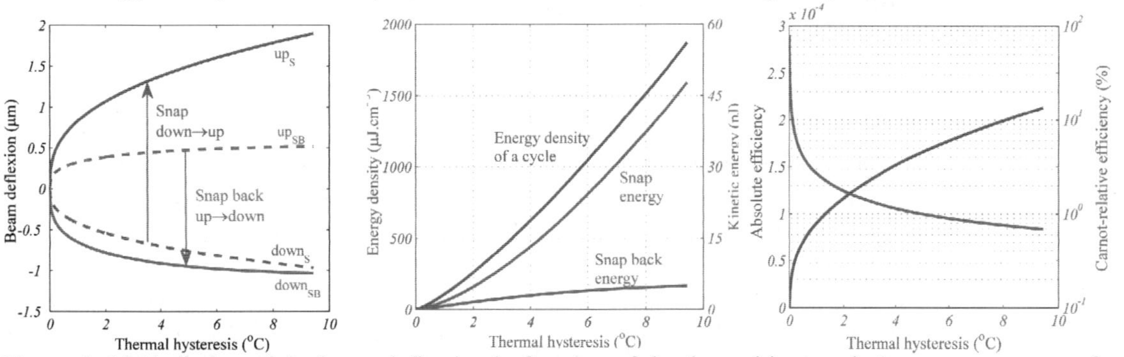

Figure 4. (a) Evolution of the beam deflection in function of the thermal hysteresis (mean temperature of the hysteresis: 75°C) (b) Evolution of the kinetic energy density of the bimetallic beam (Normalization by the volume occupied by the beam during a cycle (Fig 4.a)) (c) Absolute efficiency and Carnot-relative efficiency of the beam's thermodynamic cycle

4. Conclusion

By proposing a thermodynamic representation of the bimetallic beams, we found the necessary conditions to observe the occurrence of the thermo-mechanical bistability, and demonstrated that the thermal snap-through is a first-order transition according to the Ehrenfest theory. The hysteretic behavior of these bimorphs can then be used to harvest a part of the heat flowing through them by converting it into strain energy. Thanks to our model, we evaluated the thermo-mechanical efficiency of such heat engines both at the macro and micro-scales. A theoretical Carnot-relative efficiency of around 2~4% and a theoretical strain energy density of 300µJ/cm^3 per snap can be obtained with a two-degree-hysteresis beam, which would be sufficient to power Wireless Sensor Nodes according to the scaling laws developed in [2].

Acknowledgements
The project was partially funded by the French Inter-Ministerial Fund (FUI) through HEATec project.

5. References

[1] Ravindran S K T *et al* 2011, *Appl. Phys. Lett.* 99, 104102
[2] Puscasu O *et al* 2012, *Proc. Electron Devices Meeting (IEDM),* 12.5.1-12.5.4
[3] Boisseau S *et al* 2013, *Smart Mater. Struct.* **22**, 025021
[4] Arnaud A *et al* 2013, *J. Phys.: Conf. Ser.* **476**, 012062
[5] Ravindran S K T *et al* 2014, *Smart Mater. Struct.* **23**, 035011
[6] Boughaleb J *et al* 2014, *Sens. actuators A*, Submitted article
[7] Michael A and Kwok C Y 2006, *J. Micromech. Microeng.* **16**, 2034
[8] Arnaud A *et al* 2014, *Int. J. Therm. Sci.*, Submitted article
[9] Arnaud A *et al* 2014, *J. Appl .Phys.*, Submitted article
[10] Uzunov D I 2010, *Theory of critical phenomena second ed.*, World Scientific, Singapore

Flexible screen printed thick film thermoelectric generator with reduced material resistivity

Z. Cao[1], E. Koukharenko, R.N. Torah, J. Tudor and S.P. Beeby

[1]University of Southampton, U.K.
z.cao@soton.ac.uk

Abstract. This work presents a flexible thick-film Bismuth Tellurium/Antimony Tellurium (BiTe/SbTe) thermoelectric generator (TEG) with reduced material resistivity fabricated by screen printing technology. Cold isostatic pressing (CIP) was introduced to lower the resistivity of the printed thermoelectric materials. The Seebeck coefficient (α) and the resistivity (ρ) of printed materials were measured as a function of applied pressure. A prototype TEG with 8 thermocouples was fabricated on flexible polyimide substrate. The dimension of a single printed element was 20 mm × 2 mm × 78.4 μm. The coiled-up prototype produced a voltage of 36.4 mV and a maximum power of 40.3 nW from a temperature gradient of 20 °C.

1. Introduction

Thermoelectric generators (TEGs) can generate electrical voltage and power for autonomous microsystems from a thermal gradient. When there is heat flux along a thermoleg, a voltage will be generated from the thermocouple due to the Seebeck Effect [1]. Compared with other energy harvesting approaches, TEGs could operate in locations that lack of illumination and mechanical movement. For example, most patients or elderly people who would benefit from medical sensor systems are physically inactive in their daily life, making mechanical energy harvesting unsuitable for wearable medical sensors. Thus, flexible TEGs are an attractive approach because of their ability and suitability to power miniature electrical devices for wearable applications [2-3].

Screen printing technology has demonstrated its potential to fabricate TEGs on flexible substrate, e.g., Kapton [4]. Compared with photolithography and electrochemical deposition [5], screen printing is a low-cost process and better suited to large area fabrication, which is essential for practical applications. At room temperature, the Bismuth Tellurium ($Bi_{1.8}Te_{3.2}$)/Antimony Tellurium (Sb_2Te_3) based thermocouples have high figure of merit (ZT) values. This makes them potentially attractive for use in the screen printed TEGs [1, 4].

The efficiency of a TEG depends on the Z value of the materials ($Z=\alpha^2/(\rho\cdot\lambda)$, where α is the Seebeck coefficient, ρ is the electrical resistivity, and λ is the thermal conductivity). The printed thick film exhibits a porous structure which increases the electric resistivity and hence decreases the Z value. Cold isostatic pressing (CIP) is an approach that uniformly compresses the material over the entire surface by pressure applied through a hydraulic fluid [6]. The compressed material can be either dry particles or pastes. Figure 1 shows the basic principles of the particles densification process during pressing.

While previous work has demonstrated that the high resistivity of the BiTe and SbTe thick films is the bottle neck that limits the power output of the printed planer TEGs [4], in this paper, CIP has been introduced to densify the printed thick films in order to achieve a lower electric resistivity. The Seebeck coefficients and resistivity of thick film samples after different pressures were measured. Optimization of the printing process was also investigated. A coiled-up planar TEG on Kapton substrate was fully screen printed based on the optimized printing and pressing process condition. The voltage and power output under different temperature gradients were measured

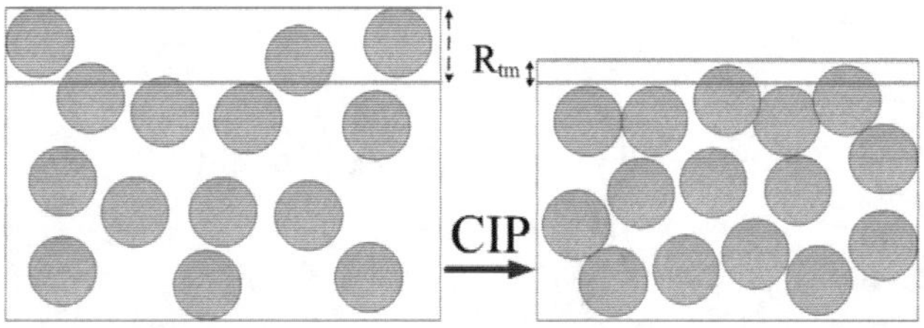

Figure 1. Illustration of particles densification in CIP process

2. Experimental section

2.1. Material processing

A screen printable thermoelectric paste basically contains 3 parts: fillers, epoxy binder system and solvent. In this experiment, the fillers were the thermoelectric material $Bi_{1.8}Te_{3.2}$ and Sb_2Te_3 particles with the size of 325 mesh and a purity of 99.99%. The particles were from Testbourne Ltd. The adjacent particles are stuck to each other and adhered onto the substrate by the epoxy binder system. The solvent was added into the paste to adjust the viscosity to a screen printable level. Finally, the paste was uniformly mixed by an EXAKT 50 triple roll mill. The viscosity of the pastes was in the range from 9000 cP to 15000 cP, measured by a Brookfield High Shear Cap-1000+ viscometer.

Initially, square thick film samples with dimensions of 1 cm × 1 cm were printed in order to measure the effect of pressure on the Seebeck coefficient and resistivity. For the CIP process, first the pastes were fully dried after printing. Then the samples were put into plastic bags, coving with a silicone film on top of the patterns. The low surface energy of silicone can prevent the uncured paste from sticking onto the plastic bag [7]. After pressing, the samples were cured under 250°C for 3 hours. The CIP machine used was CIP-20TA from MTI.

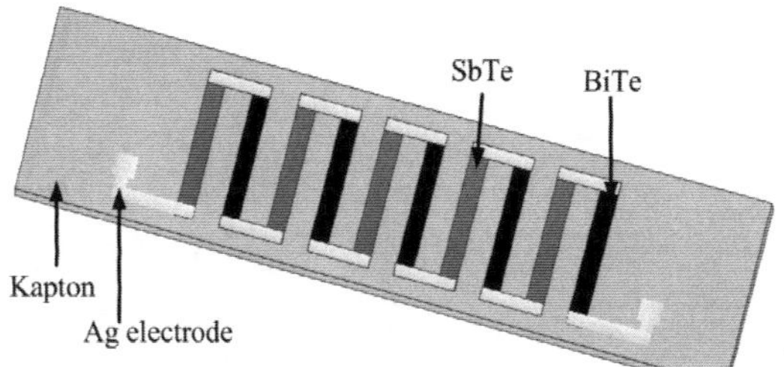

Figure 2. Illustration of screen printed planar-structured BiTe/SbTe TEGs.

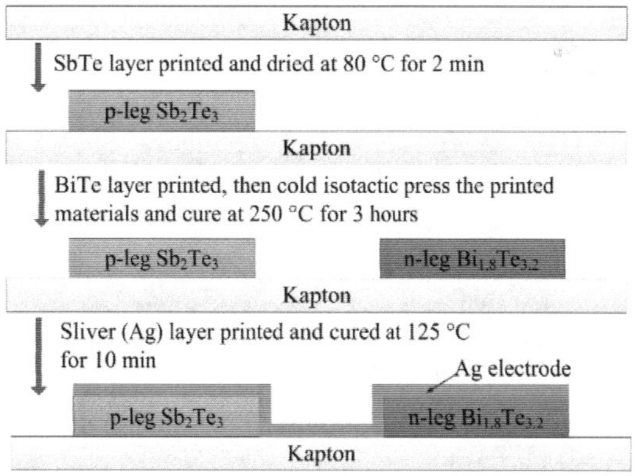

Figure 3. Fabrication process of screen printed TEGs.

2.2. Device fabrication and measurement

The schematic of the printed planar TEG was illustrated in figure 2. The thermocouples were connected electrically in series and thermally in parallel. A silver (Ag) thick film was printed on top of the thermocouples to act as electrodes. The size of an individual thermoleg was 20 mm × 2 mm × 78.4 μm (length, width and thickness) while the gap between adjacent thermolegs was 1 mm.

In order to realize this structure, a fabrication process shown in figure 3 was used. For this experiment, the volume of the CIP chamber limited the test structure to 8 thermocouples.

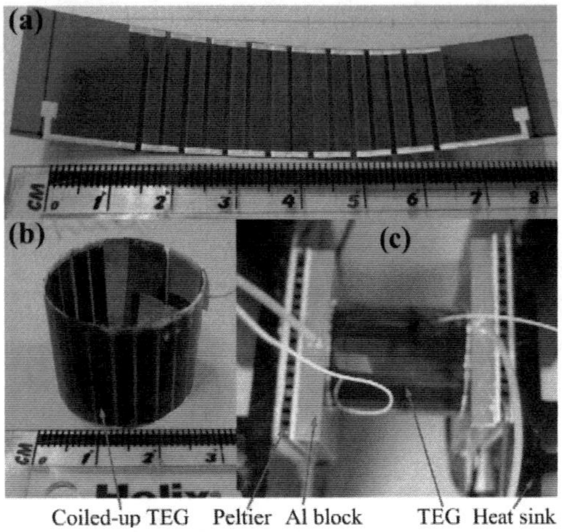

Figure 4. (a): Screen printed BiTe/SbTe TEG with Ag electrodes. (b): The coiled-up state of the TEG. (c): The setup to measure the generated thermoelectric voltage and power output when with load resistor

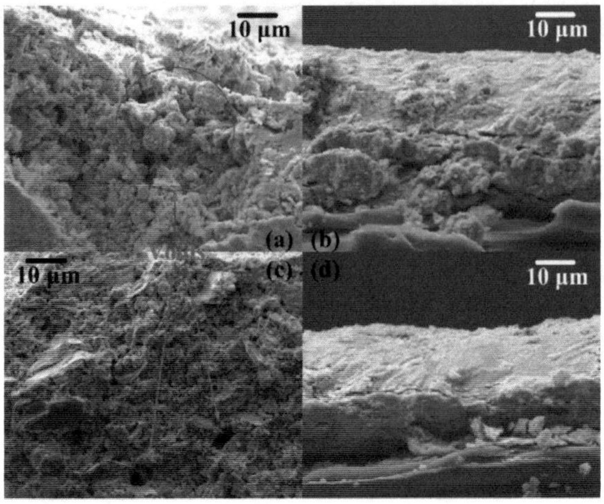

Figure 5. Cross-section SEM images of printed thermoelectric thick films. (a): BiTe thick film without pressing. (b): BiTe thick film with pressing. (c): SbTe thick film without pressing. (d): SbTe thick film with pressing.

The image of the screen printed planar TEG is shown in figure 4-(a). The strip can be rolled to a coil with a diameter of 2 cm (figure 4-(b)). In figure 4-(c), peltiers were used to supply the temperature gradient. Two square aluminum blocks were placed on the surface of each Peltier.

For the Seebeck coefficients measurement of individual materials, a bespoke test rig was used. The resistivity was measured through the Hall Effect measurement, using a commercial Hall Effect measurement system (HMS 3000 from Ecopia). The surface metrology of the samples was measured by the KLA-Tencor surface profiler.

3. Results and discussion

3.1. Material Properties

Compared with the samples without press by CIP, the pressed BiTe and SbTe thick films have a smaller surface roughness. The average distance between the highest peak and lowest valley in each sampling length is defined as a parameter (R_{tm}) and used as the measure of the surface roughness [8]. The difference of R_{tm} values for samples with and without pressing is shown in table 1. The R_{tm} values for both BiTe and SbTe thick films decreased by about 2 µm in average.

Table 1. R_{tm} values of BiTe and SbTe thick films with and without CIP.

	BiTe	SbTe
Without pressing	8000 nm	5500 nm
With pressing	6000 nm	3500 nm

Also, from the cross-section SEM images of the thick films with and without CIP in figure 5, it is observed that there were more voids in the un-pressed samples than the pressed ones. The decrease in voids created a denser film with improved electrical contact between adjacent particles. Therefore, the resistivity decreased with the densification of the thick film.

After the CIP process with an applied pressure of 250 MPa for 3 minutes, there was a reduction of 67% and 72% in resistivity compared with that of the sample without pressing for BiTe and SbTe thick films respectively. For both materials, higher CIP pressure leads to a lower resistivity. However, the variation in Seebeck coefficients for both the BiTe and SbTe under different pressure are within 5% (figure 7). The Seebeck coefficient of BiTe/SbTe thermocouple was around 248 μV/K calculated from the values after 250 MPa pressing.

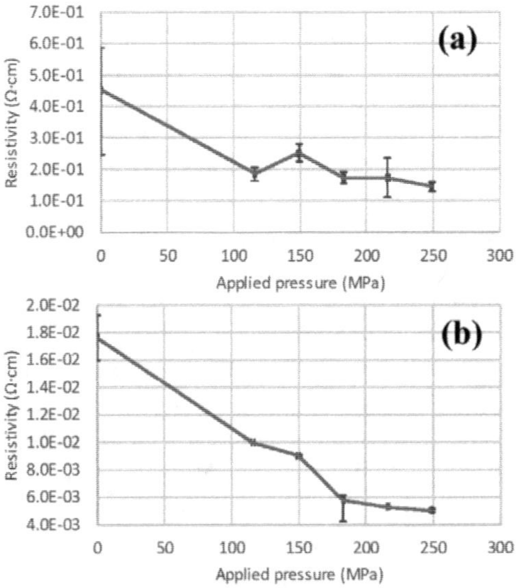

Figure 6. Electrical resistivity changing versus applied CIP pressure. (a): BiTe. (b): SbTe.

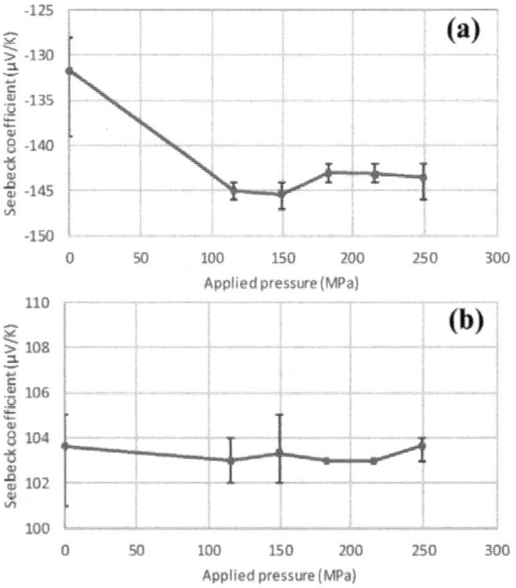

Figure 7. Seebeck coefficient changing versus applied CIP pressure. (a): BiTe. (b): SbTe.

3.2. Optimization of The Printing Process

In principle, the Ag connections can be either printed below or above the thermocouples. However, when Ag was printed as the bottom electrodes, the contact resistance between the Ag and the thermoelectric materials became unacceptably high. Table 2 lists the contact resistances between the Ag electrode and two adjacent thermolegs. The 15 junctions in figure 4-(a) had an average contact resistance of 1.3 kΩ, which is 300 times lower than the sample using Ag as bottom electrodes. Therefore, top electrodes were used in the fully screen printed Flexibly TEG.

Table 2. The resistance of single junction of BiTe/SbTe when use Ag as bottom and top electrodes

	Ag on bottom	Ag on top
Range	3.0 kΩ ~ 2.6 MΩ	0.3 kΩ ~ 2.3 kΩ
Average	437 kΩ	1.3 kΩ

3.3. Device Performance

At a temperature gradient (ΔT) of 20°C, the fully screen printed TEG generated a voltage of 36.4 mV. In figure 8, the voltage shows a linear relationship with the temperature gradient. The Seebeck coefficient of single thermocouple can be calculated out to be 227 μV/K based on the definition of the Seebeck Effect ($V=n\cdot\alpha\cdot\Delta T$, V is the open circuit voltage and n is the number of thermocouples) [1].

From figure 9, the maximum power output when ΔT=20°C was 40.3 nW. The contact resistance contributed most of the device resistance, which limited the power output. The coiled-up TEG had a resistance of 7.2 kΩ including the contact resistance. A possible reason for the high contact resistance is the formation of an insulating layer at the interface due to chemical incompatibility.

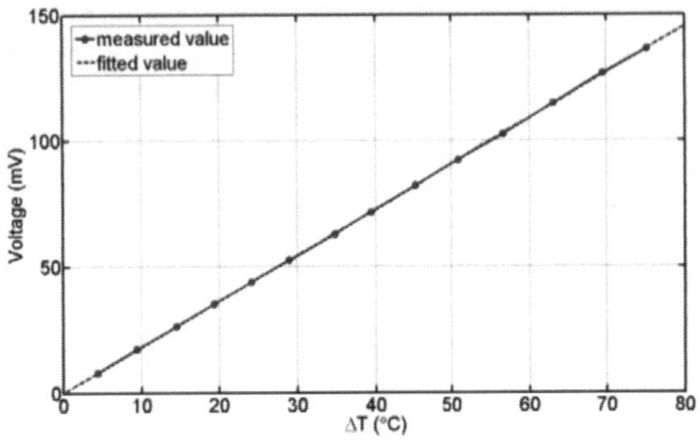

Figure 8. Generated Seebeck voltage of the coiled-up TEG at different temperature gradient

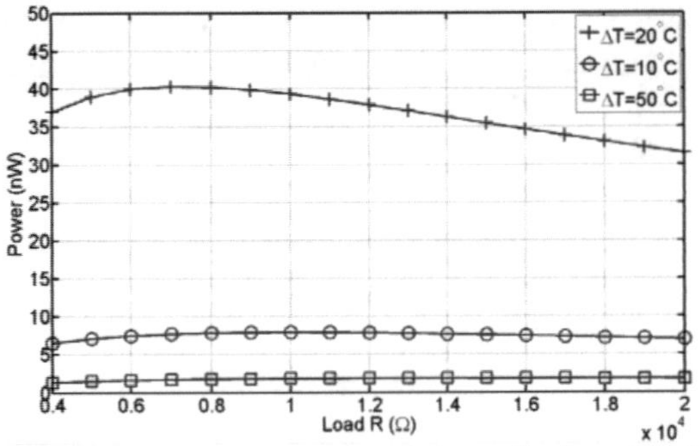

Figure 9. Power output of the coiled-up TEG varies with load resistance

4. Conclusion

The CIP process has be demonstrated to decrease the resistivity of the thick films by densifying the materials. For SbTe thick film, the lowest resistivity measured in this experiment was 5.01×10^{-3} Ω·cm, which is lower than the value of 2.0×10^{-2} Ω·cm reported in the literature [9]. However, the resistivity of BiTe thick film was still high and needs further investigation.

On the device level, a fully screen printed flexible TEG with Ag electrode connections was fabricated and tested. The Ag films was optimized to be printed as top electrodes rather than bottom electrodes. However, the contact resistance was still too high and limits the power output. In future work, approaches to reduce the contact resistance will be investigated.

5. Acknowledgements

The authors thank the EPSRC for supporting this research with grant reference EP/I005323/1. We would also like to thank Kai Yang, Yi Li and Luyi Yang for their contributions.

References

[1] S. Beeby and N. White 2010 *Energy Harvesting for Autonomous Systems*. (Norwood, MA: Artech House)
[2] S. L. Kim, K. Choi, A. Tazebay, and C. Yu 2014 *ACS Nano*, **8**, 3, 2377–86
[3] V. Leonov 2011 *ISRN Renew. Energy* **2011** 1–11
[4] Z. Cao, E. Koukharenko, M. J. Tudor, R. N. Torah, and S. P. Beeby 2013 *J. Phys. Conf. Ser.* **476** 012031
[5] W. Glatz and C. Hierold 2007 *Micro Electro Mechanical Systems, 2007. MEMS. IEEE 20th International Conference on (Hyogo, Japan, 21-25 Jan. 2007)* pp. 89-92
[6] M. Koizumi and M. Nishihara 1991 *Isostatic Pressing: Technology and Applications* (Barking, England: Springer Science & Business Media)
[7] C. Wetzel, J. Schönfelder, W. Schwarz, and R. H. W. Funk 2010 *Surf. Coatings Technol.* **205** 1618–1623
[8] A. N. S. Institute and A. S. o. M. Engineers 1996 *Surface Texture Symbols: ASME Y14.36M-1996* (New York: American Society of Mechanical Engineers)
[9] A Chen, D. Madan, P. K. Wright, and J. W. Evans 2011 *J. Micromechanics Microengineering* **21** 104006

PowerMEMS 2014 IOP Publishing
Journal of Physics: Conference Series **557** (2014) 012017 doi:10.1088/1742-6596/557/1/012017

Design of an Integrated Thermoelectric Generator Power Converter for Ultra-Low Power and Low Voltage Body Energy Harvesters aimed at EEG/ECG Active Electrodes

Milad Ataei, Christian Robert, Alexis Boegli, and Pierre-André Farine
EPFL, IMT, ESPLAB, Switzerland

milad.ataei@epfl.ch

Abstract. This paper describes a design procedure for an efficient body thermal energy harvesting integrated power converter. This procedure is based on loss examination for a self-powered medical device. All optimum system parameters are calculated respecting the transducer constraints and the application form factor. It is found that it is possible to optimize converter's working frequency with proper design of its pulse generator circuit. At selected frequency, it has been demonstrated that wide area voltage doubler can be eliminated at the expense of wider switches. With this method, more than 60% efficiency is achieved in simulation for just 20mV transducer output voltage and 30% of entire chip area is saved.

1. Introduction

In a wide range of diseases, clinicians are looking for solutions to provide an early diagnosis. Alzheimer and Epilepsy are neurological disorders that are frequently screened by physicians. Early diagnosis requires monitoring the subject's physiology and symptom development through Electroencephalography (EEG) and Electrocardiogram (ECG) signals over long periods. These examinations, done with complex devices, are currently very expensive. Furthermore they disturb normal life of patients because they have to be performed in hospital environment. Although wearable sensors can significantly benefit mankind in this long-term monitoring process, today's solutions invade the user's normal life as their platforms require removal, replacement and reconfiguration due to battery recharging. Moreover, size and weight of integrated electronic systems are usually set by batteries.

The "Zero Power" Medical devices approach can revolutionize human parameters monitoring thanks to new architectures which will lead to drastically lowering the power needs. Energy can then directly be harvested from natural sources. This makes possible to implement systems that minimize their impact on the everyday life of patients. In zero power medical applications, body energy harvesters produce unreliably small amounts of power and moreover at low voltages. On the other hand, local processing operation in body sensor nodes is power hungry. Thereby, designing an optimum architecture with ultra-low power custom integrated circuit is an inevitable choice.

Body heat can be a source for energy harvesting when autonomous medical devices have been placed directly on a human skin. In that case, miniature Thermoelectric Generators (TEG) are used as transducers. The output impedance of TEG cell is relatively constant. However, output voltage changes according to the environmental conditions and have been shown to be very small [1]. Power converter is then responsible for tapping such infinite but unreliable low potential energy from the harvesters and provide it as a high potential and reliable power source for applications. The conversion of small output voltages, in the range of a few tens of millivolts, have been achieved but with poor efficiency [2-4].

In this paper, a design procedure for an optimum integrated power converter aimed at an autonomous EEG active electrode will be presented. The procedure is based on losses examination in circuit blocks and the system. In the literature, the losses analysis of converters has been used for study of power converter characterization [2], [5]. Recently [6], this analysis has been used for optimizing specific circuit parameters such as switch sizes or frequency independently, for a fixed architecture and a transducer. However, in a body powered system different power converters are used with diverse specifications. This means that the design procedure to reach an optimum should be algorithmic, with the aim of reduce design time. Here, the design process for a simpler and more efficient converter system is discussed. Section 2 describes an autonomous active electrode as our medical platform. Section 3 introduces the design method for the converter followed by section 4 presenting the simulation results.

2. Body powered active electrode with TEG integrated

Long term EEG signal monitoring devices must be convenient while maintaining the precision compared to conventional devices. Accordingly, readout electrodes and electronic components must be tiny and lightweight and must provide high Signal to Noise Ratio (SNR) and low distortions. However, EEG signals are weak and are accompanied by strong interferences. To achieve signals with high SNR and low distortions, it is mandatory to amplify signal and reject interferences immediately after acquiring the signal. This becomes possible if active components are placed right next to the electrodes.

An autonomous active EEG electrode based on TEG harvester was introduced in [1]. It consists of a metallic plate as EEG electrode and thermal contact, TEG transducer, heat transfer structures, and electronics responsible for EEG signal conditioning and TEG power conversion. The size of entire system should be same as one electrode (i.e. 12mm×12mm in our case). This implies a hard limit form factor for the power converter. Besides area constraints, power converter must have enough efficiency to provide adequate power for the integrated EEG amplifier which consumes less than 7μW from a 1.2V supply. Table I summarizes the experimental specifications of these TEGs when transducers harvest body thermal energy [1]. This table suggested about 20% efficiency with conversion factor of 60 for low voltage TEG and 70% efficiency with a conversion factor of 15 for higher voltage transducer.

3. Power converter design

Figure 1 a) shows a modern general power converter architecture [3-4], [6] considered here. Circuit blocks including comparator are activated only when they are triggered by clock signals. Therefore, it offers the best of compromise between continuous analog control loop and power hungry precise digital one, resulting in minimizing the power consumption. In this architecture, first, the harvester charges input capacitor C_s in *tc* duration while both NMOS and PMOS switches are off. Then, NMOS transistor is turned on, PMOS is kept off, and harvested energy stored in C_s is transferred in inductor L_s, during period *t1*. After *t1*, NMOS is turned off and PMOS is switched on, so, energy will be transferred from L_s to the final *Load* during a period *t2*. This function will repeat with frequency f_s. Durations *t1* and *t2* are set by two pulse generators dedicated. To maximize energy efficiency, power transferred to the output should be maximized and converter losses be minimized. Presuming low loss on board components and big inductance L_s, the energy transferred from input to the *Load* could be written as:

$$E_T = \frac{L_s.I_P^2}{2} + \frac{V_{in}.I_P.t2}{2} \tag{1}$$

Table I: TEGs body harvester electrical characteristics at 24°C ambient temparture [1]

	Material, Size, Condition	Available Power	Output Voltage	Voltage type	Output Impedance	Impedance type
Thermal 1	Macro TEGs 15x15x4 mm	40μW	20mV	DC	2.5Ω	Constant
Thermal 2	Micro TEGs, 15x15x4 mm	10μW	80mV	DC	160Ω	Constant

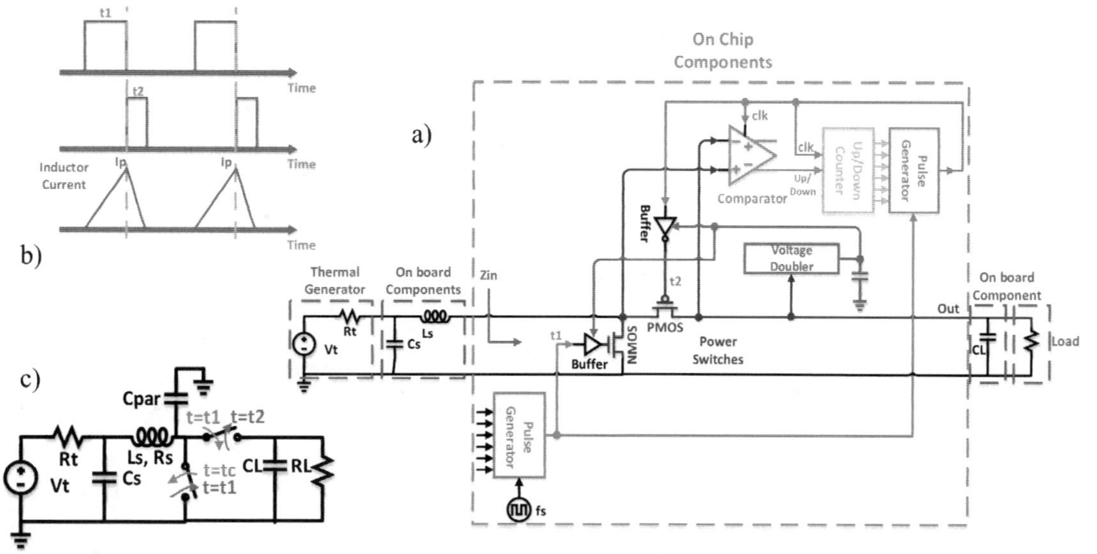

Figure 1: a) General architecture of a converter [3-4], [6], b) timing diagram, and c) its basic schematic

where the first term and second term are related to energy transmitted during *t1* and *t2*, respectively and I_p is peak current of the inductor. Simplifying equation (1), transferred power can be expressed as:

$$P_T = f_s.E_T = V_{in}^2.\left[f_s\left(\frac{t1^2}{2L_s}+\frac{t1.t2}{2L_s}\right)\right] \quad (2)$$

In (2), the statement brought in a bracket can be interpreted as the system input admittance. For maximizing power transfer, the input impedance, Z_{in}, should be matched to the harvester impedance, R_t. If it is assumed that *t1* is much longer than *t2* (it will be shown later), Z_{in} can be written as:

$$Z_{in} = \frac{2.L_s}{t1^2.f_s} \quad (3)$$

Therefore, Z_{in} is related to *t1* and it is tuned by the duty cycle of the clock frequency f_s.

Switching losses can be divided in static losses related to "On"-resistance of switches, dynamic losses including switches' gate and buffers' gate charging and discharging losses, and leakages loss. There is a trade-off in the switches sizing. Indeed, the static power losses decrease with wider switches but the dynamic power consumption increases. Based on the equations for these losses and for on board component parasitics, an optimum switches' sizes in each operating frequency can be found (Figure 2). This optimization was calculated using transducer characteristics for the worst case i.e. low voltage TEG scenario. In our micro-watts power regime, our examination showed that leakages loss does not have major effect on efficiency since it is in order of nano-watts. Inductor parasitics can also be neglected here. A simplified system loss around the optimum points presented in Figure 2 can then be stated as:

$$P_{Loss} = I_{Ls,RMS,t1}^{\ 2}.\left(R_{NMOS}+R_{PMOS}\frac{t2}{t1}\right)+f_s.V_{SW}^{\ 2}.\eta.\left(1+\frac{1}{\alpha}\right).(C_N+C_P) \quad (4)$$

where R_{NMOS}, C_N, R_{PMOS}, C_P are channel's resistances and gate's capacitance of NMOS and PMOS switches, respectively. The ratio of switch sizes to the size of the buffers which drive the switches was introduced by α. V_{SW} is switching gate source voltage and η is the efficiency of the circuit that provides this switching voltage. Root-mean-squared current through L_s during *t1*, $I_{Ls,\,RMS,\,t1}$ can be written as:

$$I_{Ls,RMS,t1} \approx \frac{V_{in}}{(L_s.f_s)^{1/4}\,Z_{in}^{\ 3/4}} \quad (5)$$

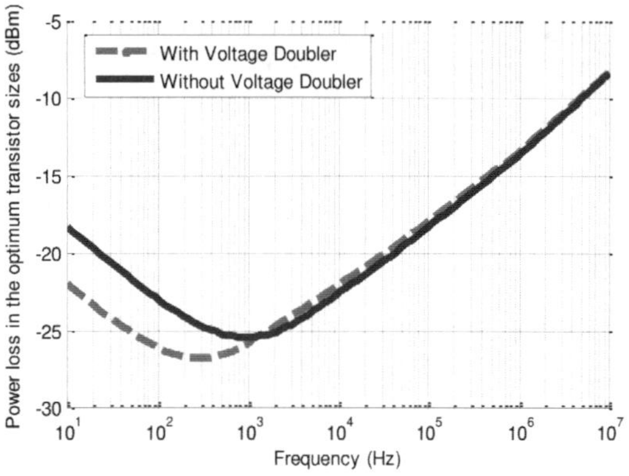

Figure 2: Optimum power loss for optimized switches' sizes versus frequency

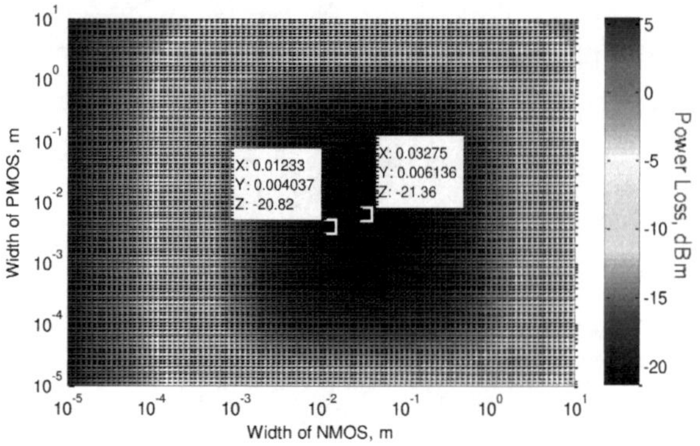

Figure 3: Power loss versus switches' sizes at f_S

From (4) and (5), it is clear that for having maximum efficiency the biggest possible inductance should be selected, however; system form factor confines inductance value. Here, the largest possible inductance with the package size of 8mm×8mm, a 47µH was selected. Simplified close form Equation (5) was written assuming that V_{in} is almost constant during inductor charging. However, V_{in} might vary during charge transfer. The exact current-voltage relations for the schematic shown in Figure 1 c) were derived and it verifies the inductance-current behavior of equation (5).

For low power applications, Discontinuous Conduction Mode (DCM) converters have a higher efficiency. In DCM, PMOS should be turned off exactly when current of the inductor reaches zero. A pulse generator and a comparator have been used to synchronize the switching off of PMOS. Comparator in Figure 1 a) detects whether PMOS is turned off early or late at the end of *t2*. Based on that pulse width of *t2* is tuned by the counter and the pulse generator dedicated to *t2*. With detail analysis of Figure 1 c) and simplifying equations during discharge of L_s in the output capacitor, *t2* can be estimated as:

$$t2 = \frac{t1.V_{in}(\mathrm{t=t1})}{V_{out} - V_{in}(\mathrm{t=t1})} \tag{6}$$

Output voltage, V_{out} must have less than 10% ripple and its fluctuations are set by C_L capacitance. $V_{in}(t=t1)$ is the input voltage at the end of charging L_s. The input voltage variation during *tc* and *t1* is inversely related to f_s, L_s, and C_s. If the system has a large V_{in} ripple, $V_{in}(t=t1)$ may reach to a small value and generating of *t2* becomes impossible. Considering that our form factor which limits the board area to the size of one EEG electrode, a C_s with also a size of 1.6 mm × 0.8mm was selected. The largest commonly available capacitance within this size is 10µF. As shown by Figure 2, maximum efficiency occurs when f_s is around 500Hz. With this frequency and the chosen C_s, our relations show that input voltage ripples become large and V_{in} reaches zero even before *t1* duration. Thus, it is not possible to create *t2* pulses. Common approach in literatures suggested increasing f_s up to have a less than 10% variation of input voltage. As Figure 2 illustrated, along the increase of frequency, losses will increase linearly. Here, frequency is increased up to the point that entire loss deviates less than 10% of input available power from the optimum point. With f_s equal to 20kHz, input voltage reaches to 5mV at end of *t1*. Therefore, by designing an apt pulse generator which provides pulses with duration of 180nsec, system can be functional with higher efficiency than a system designed with conventional approaches.

"On"-resistance and leakages loss of switches and are inversely related to switches' gate voltage. A voltage doubler is usually used to provide higher gate voltage. The output of voltage doubler is stored into on-chip capacitors. The on-chip capacitance of the voltage doubler should be large since it should supply enough energy for switching the power transistors. The voltage doubler may occupy more than 50% of area of a power converter because of its on-chip capacitors. Based on our analysis (Figure 2) it was demonstrate that if we remove the voltage doubler for f_s higher than 1kHz, power loss remains constant but the switches will be bigger. Power loss versus switches' size for the selected f_s without

Table II: Summarized this design and the prior arts

	[3][1]	[4][1]	[2][1]	This work[2]
Topology	**Inductor based**	**Inductor based**	**Inductor based**	**Inductor based**
Technology	**0.35µm**	**0.35µm**	**0.13µm**	**0.18µm**
Voltage Conversion	**26~100mV to 1.8V**	**100mV to 2V**	**20~100mV to 1V**	**20mV to 1V**
Output Power	**10µW @25mV**	**1.3mW @100mV**	**25µW @20mV**	**25µW @20mV**
Quiescent Power		**-**	**1.3µW**	**1.2µW**
Efficiency at 20mV	**<20%**	**40%**	**46%**	**60%**
Core area	***1.6mm^2**	***2.5mm^2**	****0.12mm^2**	****0.14mm^2**

[1] Measurement results [2] Layout post-simulation results * entire chip ** core

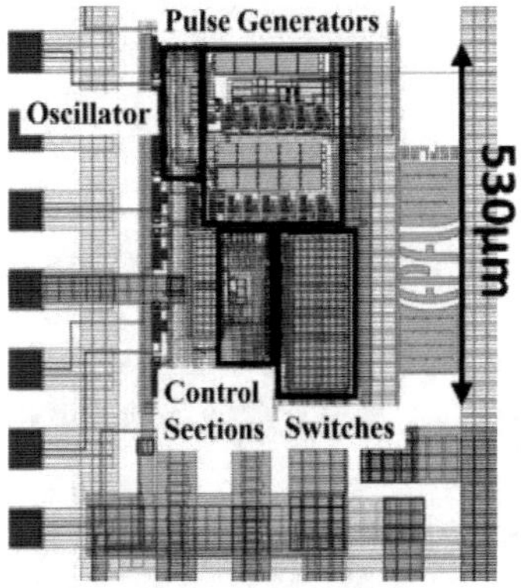

Figure 4: Layout designed in UMC 0.18µm CMOS

voltage doubler, is shown on Figure 3. Exact optimum widths supporting selected f_S for NMOS and PMOS pose large parasitics. By accepting a deviation of 10% from the optimum (Figure 3), a suitable transistor width can be achieved. With this technique, area of the chip due to switches is just 20% bigger, but overall circuit area circuit is reduced by 30%, while complexity and design procedure are drastically reduced. Also, with decreasing number of blocks the quiescent power consumption is reduced.

4. Simulation results

With the introduced algorithmic design procedure and taking into account the form factor, an ultra-low power DC-DC converter for the thermal body harvesters was designed in UMC 0.18µm CMOS technology. According to post-layout simulations, the design of 6 bit *t2* pulse generator can produce pulse durations from 44nsec to 665nsec. Such conservative circuit implementations help the system to work around its optimum point when input voltage varies depending environment or when high voltage TEG harvester characterized in Table I is used as transducer. Here, it is assumed that output energy storage has some charge during start-up that is to say that the cold start has not focused on this paper. The area of the power converter is 0.14mm^2. Its layout is shown in Figure 4. The quiescent power consumption of the system is 1.2µW and more than 60% efficiency has been achieved in post-layout simulations for 20mV TEG output voltage while the reported efficiencies in similar situation in state-of-art are below 50% [2-4]. Table II summarizes chip performances and other the state-of-the-art.

5. Conclusion

Energy transfer and loss formulas were used for a general power converter to find an optimum architecture and all the design parameters. To support calculated optimum frequency f_S, a huge input capacitance was required. With a slight deviation from the optimum f_S and design a proper pulse generator, and reasonable sizes for passive elements were achieved. Besides, the method led to a simpler design and lower power consumption since voltage doubler was eliminated without increasing loss. Presented method reduced unnecessary system complexity and area while enhancing efficiency.

Acknowledgments

This work was done under BodyPoweredSenSE project funded by NanoTera with Swiss Confederation.

References

[1] M. Thielen, M. Ataei, et al. , *International Conference on Thermoelectrics - ICT2014*, 2014
[2] E. Carlson, et al. , *Solid-State Circuits, IEEE Journal of*, **vol.45**, no.4, pp.741,750, April 2010
[3] Y. Ramadass, et al., *Solid-State Circuits, IEEE Journal* , **vol.46**, no.1, pp.333,341, Jan. 2011
[4] S. Bandyopadhyay, et al. , *Solid-State Circuits, IEEE Journal* , **vol.47**, no.9, pp.2199,2215, 2012
[5] X. Zhou, et al. , *Applied Power Electronics Conference and Exposition*, **vol.2**, pp.612, vol.2, 1997
[6] S. Bandyopadhyay, et al. , *Solid-State Circuits Conference (ISSCC), IEEE,* pp.396,397, 2014

Ignition Characteristics of Premixed Methane/air in Micro Chamber

H L Yang, J P Huo, L Q Jiang, X H Wang and D Q Zhao

Guangzhou Institute of Energy Conversion, Chinese Academy of Sciences, China

E-mail: yanghl@ms.giec.ac.cn

Abstract. The ignition process of premixed methane/air in a micro-scale chamber is simulated with the combination of surface chemistry and gas-phase reactions. The effect of different parameters on the ignition characteristics are analyzed in detail and the sensitive analysis is performed on main elementary reactions. It is found that the ignition of the mixture is inhibited by surface reaction, which mainly depends on the sticking ability of the surface on CH_3 radical. The ignition inhibition can be reduced by increasing the initial temperature and pressure. The increase of equivalence ratio has small impact on ignition delay time when it is larger than stoichiometric ratio. The ignition delay time would increase significantly with surface-area-to-volume ratio. The product of sticking coefficient and surface-area-to-volume ratio can be used to measure the intensity of surface reaction. The larger the product is, the stronger the inhibition of the surface reaction on ignition has.

1. Introduction

Due to the development of MEMS technology, the miniaturization of internal combustion engine becomes a research hotspot in recent years. The aim of maintaining stable and high efficient output is the major challenge when it is applied as a high energy power source. Micro-engines require short resident time of premixed gas in the chamber for sufficient power output, so that the short ignition delay time of fuel/air mixture is favorable [1, 2]. However, due to large surface-area-to-volume ratio of the chamber, the collision between radicals and internal wall is more frequently and the radical loss on the internal surface would not be ignored, leading to the retard of the ignition and the increase of resident time.

The study on the radical loss on the internal wall of reaction vessel was first conducted in the static-reactor experiments [3]. In this type of experiments, low temperature oxidation mechanism of hydrocarbons was studied with isothermal boundary condition in a closed quartz or silica vessel. Norrish and Frood [4] reported that the reaction rate of methane/oxygen mixture with the same operating conditions was evidently lower in the smaller vessel. The results demonstrated the existence of radical loss on the internal wall and indicated that the wall would have stronger inhibitory influence on gas phase oxidation. Bowman and Wilk [5] investigated the correlation between surface-area-to-volume ratio of the vessel and ignition delay time of methanol/air mixture. Their experimental results showed that the ignition delay time of the mixture increased with surface-area-to-volume ratio. It indicated that during the induction time, the radical loss on the wall cannot be ignored. Ignition delay time was also found sensitive to internal wall conditions in modern shock tube and jet-stirred reactor experiments [6, 7]. So it can be deducted that the homogeneous combustion assumption in micro chamber is not appropriate as the surface reaction strongly affects the radical concentration in the radical pool. Therefore, heterogeneous surface reactions should be taken into consideration for the design of micro-combustor.

In this work, the effect of heterogeneous surface reactions on ignition delay of methane/air mixture is studied using numerical simulations. The investigation on several important parameters for the combustion process including initial temperature, pressure, equivalence ratio, sticking coefficient and surface-area-to volume ratio is conducted.

2. Numerical method

2.1. Surface reaction mechanism

The surface reaction model for methane/air developed by Raimondeau *et al* [8] was used in this work, as shown in Table 1. By using this model, numerical simulation on methane/air combustion in narrow channel was conducted by Saiki *et al.* and excellent agreement to experimental result was obtained [9].

Table 1. Surface kinetic mechanism of methane/air [8]

No.	Reactions [a]	Sticking Coefficient or Pre-Exponential (s^{-1})	ΔH / (kcal/mol) [b]
R1	$CH_3 + * \rightarrow CH_3*$	0 ~ 1	0
R2	$H + * \rightarrow H*$	0 ~ 1	0
R3	$OH + * \rightarrow OH*$	0 ~ 1	0
R4	$O + * \rightarrow O*$	0 ~ 1	0
R5	$2CH_3* \rightarrow C_2H_6 + 2*$	10^{13}	-92.3
R6	$2H* \rightarrow H_2 + 2*$	10^{13}	-105.7
R7	$2O* \rightarrow O_2 + 2*$	10^{13}	-121.1
R8	$2OH* \rightarrow H_2O + O* + *$	10^{13}	-17.4
R9	$CH_3* + H* \rightarrow CH_4 + 2*$	10^{13}	-106.7
R10	$OH* + H* \rightarrow H_2O + 2*$	10^{13}	-121.1

a) * denotes reactive surface site and for example CH_3* means an adsorbed CH_3 radical.
b) ΔH means the standard molar reaction enthalpy.

2.2. Numerical setup

The CHEMKIN-Pro software [10] was used to investigate the effect of surface reactions on ignition delay time of methane/air mixture. GRI-Mech 3.0 [11] was adopted as the detail gas phase kinetic mechanism of methane/air. Adiabatic boundary condition was assumed and the surface reactions were integrated into the homogeneous combustion calculation.

The influence of five main parameters on ignition delay time, including initial temperature T_0, initial pressure P_0, equivalence ratio Φ, sticking coefficient γ and surface-area-to volume ratio *SV*, were analyzed. Series of numerical simulation were carried out and T_0 ranged from 900 K to 1500 K, P_0 from 1 atm to 30 atm, Φ from 0.2 to 1.4, γ from 0 to 1.0 and *SV* from 0.1 cm^{-1} to 100 cm^{-1}.

3. Results and discussions

3.1. Effect of initial temperature and sticking coefficient

Figure 1 shows the comparison of ignition delay time for various γ versus T_0. Larger ignition delay time is found for larger γ, demonstrating that the surface reactions have strong inhibitory effect on the induction time. However, the discrepancy of ignition delay time caused by γ becomes smaller as initial temperature increases. It indicates that the effect of surface reaction is weak at high temperature.

The ignition delay time associates with the chain branching reaction rate, which is affected by OH concentration. In figure 2, the growth rate of OH is slowest when γ=1, resulting in the longest induction time. Therefore, the wall with stronger absorption ability destroyed more radicals in the radical pool, and this is the reason why surface reaction can retard the ignition.

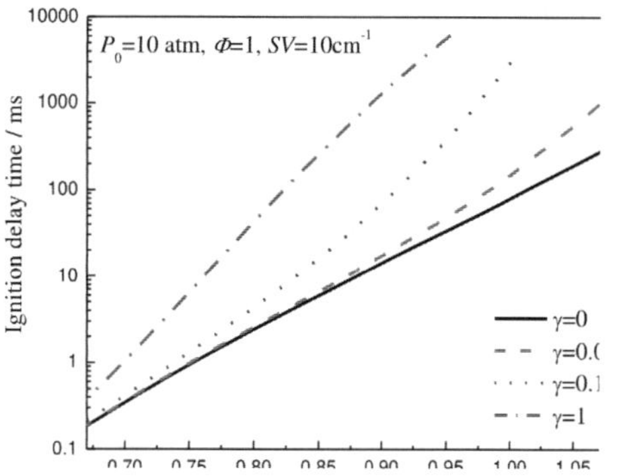

Figure 1. Comparison of ignition delay times for different γ versus T_0

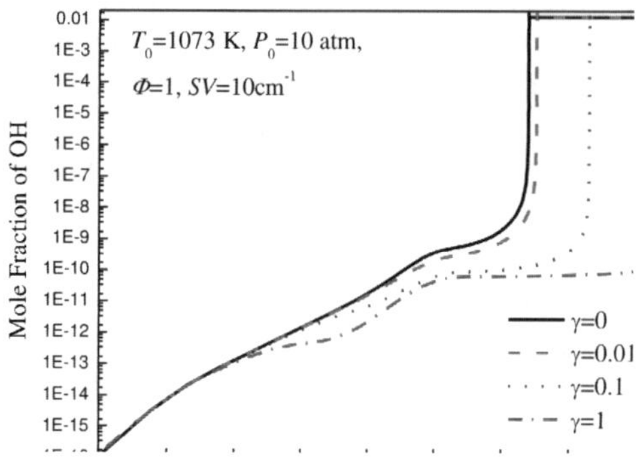

Figure 2. Comparison of OH mole fraction for different γ versus time

A sensitivity analysis was performed to study the reaction sensitivity to ignition delay time for methane oxidation as a function of sticking coefficient [12]. Figure 3 illustrates nine largest normal sensitivities among gas-phase reactions and four absorption reactions R1~R4. Both in the case of T_0=900 K and T_0=1400 K, the sensitivities for H, O and OH absorption can be neglected compared with the other reaction. Due to the rapid reaction rate of $H+O_2 \rightarrow OH+O$, $OH+H_2 \rightarrow H_2O+H$ and $O+H_2 \rightarrow OH+H$, most of the radical of H, O and OH are consumed in the gas phase and their gas-phase concentrations are relatively so low that the absorption reaction rates of R2~R4 are very small. CH_3 radical, as a relatively stable molecule, is more likely to diffuse to the wall and get absorbed. Therefore, the surface effect is mainly controlled by the absorption intensity of the wall, represented by the sticking coefficient of R1.

Noted that the sensitivity of R1 is higher than all the other reactions for T_0=900 K, shown in figure 3(a), therefore the inhibitory effect on ignition is determined by surface reaction R1. However, for T_0=1400 K, the gas-phase reaction $H+CH_4 \rightarrow CH_3+H_2$ takes the place of R1 as the largest inhibitory factor, indicating that gas-phase reactions play a dominant role during the induction period at high temperature. According to Langmuir theory [13], the absorption reaction rate can be expressed as:

$$k_{\mathrm{wall}} = \frac{\gamma}{\Gamma}\frac{P}{\sqrt{2\pi RTM}} \tag{1}$$

where Γ, R, M, P, T are the surface site density (mol/cm^2), the universal gas constant (J/(mol•K), the molecular weight, partial pressure of the radical and the gas-phase temperature. Thus, when P_0 and γ are constant, the absorption reaction rate would have slight decrease if increasing the initial temperature. However, chain branching reaction rate in the gas phase would have rapid increase even the temperature has a small increase. Therefore, a number of radical produced can compensate the radical loss on the wall, weakening the surface effect on the ignition.

3.2. Effect of initial pressure and equivalence ratio

According to the equation 1, the increase of initial pressure is able to enhance the surface reaction rate. However, the ignition delay time gets shorter when increasing P_0, but also increases with γ, shown in figure 4. In addition, the discrepancy of various γ becomes smaller as the initial pressure increase. As the density of the methane/air mixture increases with P_0, the collision frequency among the radicals increase therefore less radicals crash into the wall due to gas diffusion. It can be concluded that enhancing P_0 can effectively reduce the inhibitory effects of wall reaction.

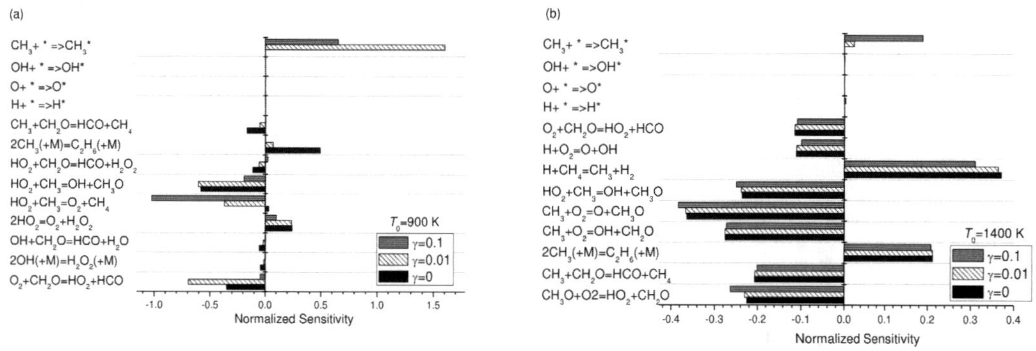

Figure 3. Normalized sensitivity of ignition delay time with respect to reaction rate (P_0=10 atm, Φ=1, SV=10 cm^{-1})

The effect of equivalence ratio Φ on ignition with various γ is illustrated in figure 5. When γ< 0.1, there is little change for the ignition delay time. When γ is up to 1.0, great increase can be seen as Φ decreases to 0.2. However, for fuel-rich condition (Φ>1), the ignition delay time changes slightly and maintains stable with a certain sticking coefficient. Therefore, high-pressure and fuel-rich condition can help weaken the surface effect.

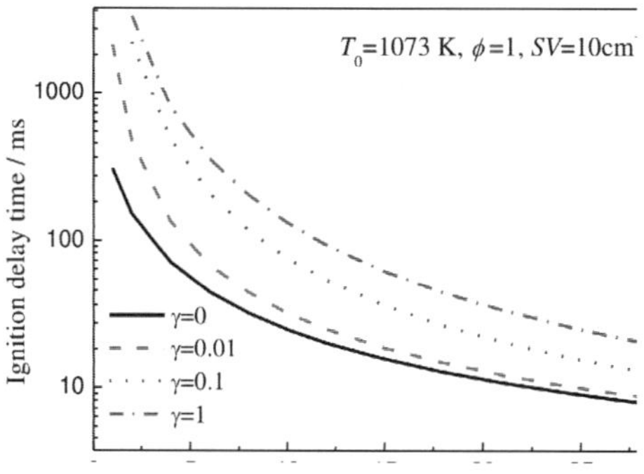

Figure 4. Comparison of ignition delay times for different γ versus P_0

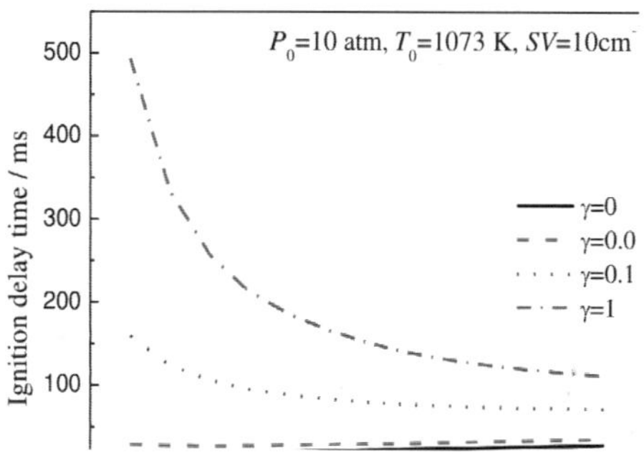

Figure 5. Comparison of ignition delay times for different γ versus Φ

3.3. *Effect of surface-area-to-volume ratio*

The investigation on SV with various γ is drawn in figure 6. The ignition delay time has a significant increase with SV for γ=1, however becomes stable for SV>10 cm^{-1}. The larger of the SV, the more possibility of radicals impinge to the surface and get absorbed, resulting in the rapid growth of ignition delay time. When SV is small, the heat released from surface reaction can be neglected. As SV increases to a certain degree, intense surface reactions release a lot of heat and enhance the gas-phase temperature due to the exothermic desorption reactions. Therefore, in an adiabatic closed system, the ignition delay time is sensitive to surface-area-to-volume ratio in a certain range for a certain sticking coefficient.

Both SV and γ affect the ignition significantly. Thus, a new parameter $\eta=\gamma\bullet SV$ is developed in this work, called sticking intensity. Figure 7 shows that the ignition delay time almost has no change for a specific η. It indicates that η can be used to measure the inhibitory effect of surface reaction. The larger of the η is, the stronger the inhibition of the surface reaction on ignition has.

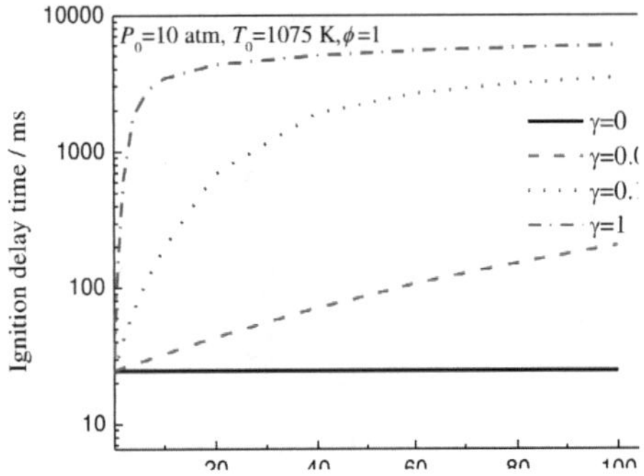

Figure 6. Comparison of ignition delay times for different γ versus *SV*

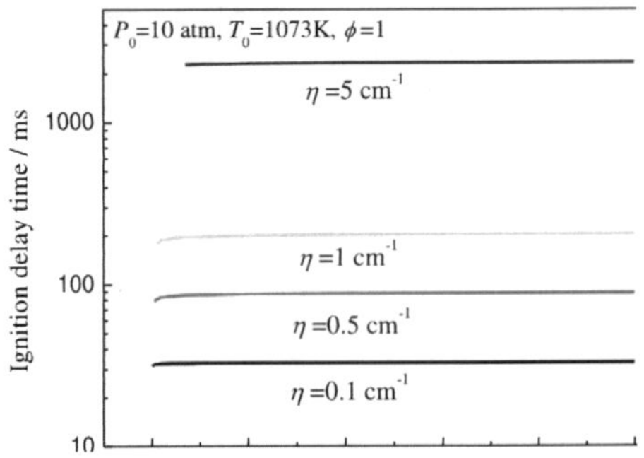

Figure 7. The ignition delay time versus *SV* with specific η

4. Conclusions

The results show the inhibitory effect caused by surface reaction mainly depends on the sticking coefficient of the absorption for CH_3 radical. Increasing the initial temperature and initial pressure can effectively weaken the surface effect. However, the increase of equivalence ratio has small impact on ignition delay time when it is larger than stoichiometric ratio. In an adiabatic system, the ignition delay time would increase significantly with surface-area-to-volume ratio in a certain range for a certain sticking coefficient. But is becomes stable when surface-area-to-volume ratio is sufficient large due to strong heat release from surface reactions. The product of sticking coefficient and surface-area-to-volume ratio can be used to measure the intensity of surface reaction. The larger the product is, the stronger the inhibition of the surface reaction on ignition has.

Acknowledgement

The current work was supported by National Basic Research Program of China (No. 2014CB239600) and National Science Foundation of China (No. 51336010).

References

[1] Sher I, Levinzon-Sher D and Sher E 2009 *Appl. Therm. Eng.* **29** 401
[2] Sher E and Sher I 2011 *Chem. Eng. Sci.* **66** 261
[3] Lewis B and Von Elbe G 1987 *Combustion, flames and explosions of gases* (Orlando: Academic Press) pp 25-76
[4] Norrish R G W and Frood S G 1936 *Proc. R. Soc. Lond.* A **157** 507
[5] Bowman M J and Wilk R D 1996 *SAE paper*
[6] Wang B L, Olivier H and Gronig H 2003 *Combust. Flame* **133** 93
[7] Dagaut P, Lecomte F, Mieritz J and Glarborg P 2003 *Int. J. Chem. Kinet.* **35** 565
[8] Raimondeau S, Norton D and Vlachos D G 2002 *Proc. Combust. Inst.* **29** 906
[9] Saiki Y and Suzuki Y 2012 *Proc. Combust. Inst.* **34** 3401
[10] *CHEMKIN-Pro* 2010 (San Diego: Reaction Design, Inc.)
[11] Smith G P,Golden D M, Frenklach M, Moriarty N W, Eiteneer B, Goldenberg M, Bowman C T, Hanson R K, Song S, Gardiner W C, Lissianski V V and Qin Z W. GRI-Mech 3.0, http://www.me.berkeley.edu/gri_mech/
[12] Zhang Y J, Huang Z H, Wei L J, Zhang J X and Law, C. K. 2012 *Combust. Flame* **159** 926
[13] Masel R I 1996 *Principles of Adsorption and Reaction on the Solid Surfaces* (New York: Wiley) pp 384-389

Numerical study of influence of surface reaction and heat-loss on flame intensity of methane–air flames

N Hayashi[1] and H Yamashita[1]
[1] Department of Mechanical Science and Engineering, Nagoya University, Furo-cho, Chikusa-ku, Nagoya, Japan

E-mail: hayashi@mech.nagoya-u.ac.jp

Abstract. This paper reports a numerical study of influence of radical quenching and heat loss on bulk flame characteristics in narrow parallel channels. Flame-wall interaction is an important phenomenon on combustors. Especially, the wall effects on the flame characteristics in a small scale combustor become larger than those on normal scale one. The wall effects are caused by heat loss and surface reaction. The surface reaction on many common non-catalytic materials may weaken or quench the flames, although those for a catalytic wall can strengthen the flames. Authors have investigated the influence of the surface reaction and the heat loss on a non-catalytic wall using numerical simulation. In this study, a two-dimensional slit burner between two parallel plates with or without surface reaction is modelled. The wall temperature is 500 and 1200 K. The flame behavior and heat release rate distributions are examined when the distance between two plates is changed.

1. Introduction

In recent years, minimization of micro electronic and mechanical devices has resulted in demands for smaller and higher power generators. One of the candidates for such generators is an ultra-micro gas turbine (UMGT), which is expected to have several times the energy density of a lithium ion battery [1]. Thus, meso- or micro-scale combustors have been investigated [2]. To further develop the UMGTs, it is important to understand the fundamental phenomena that occur in their small-scale combustors. Many researchers have studied UMGTs and related phenomena [3, 4].

One of the important features of micro-scale combustors is their large surface-to-volume ratio. This large ratio leads to enhanced heat loss and radical quenching by surface reactions on the wall. However, these phenomena are complex due to the effects of many factors, such as temperature, material, and surface texture. Heat-loss effects have been investigated by many researchers [5, 6]. Egolfopoulos et al. [6] have investigated the interaction between a chemically inert wall and a premixed flame using numerical and experimental results. Recently some researchers have made attention to chemical effects of surface [7-9]. The measurements are very difficult since reaction rates of surface reaction for radical quenching is much smaller than that for catalytic reaction and thus the S/N ratio is often small.

For numerical simulation, reaction kinetics models for the surface reaction are proposed by Aghalayam for hydrogen mixture [10] and Raimondeau for methane mixture [11]. These models are not specified for real materials, but they include important species for ignition or quenching. Thus, although results using these models may be different for real materials, it is useful to understand the radical quenching.

In the present study, a two-dimensional slit burner between two parallel plates is modeled for investigation of chemical effects. This paper reports influence of radical quenching and heat loss on bulk flame characteristics. The flame behavior and heat release rate (HRR) distributions are examined as the flame characteristics when the distance between two plates is changed.

2. Analytical model and numerical calculation method

A two-dimensional slit burner between two parallel plates was modeled using two-dimensional planer coordinates (x, y), as shown in figure 1. Stoichiometric methane-air premixed gas flows into a narrow channel from a slit nozzle of left-hand side. A symmetric condition was adopted at the center of channel of the upper boundary, and a free outflow condition was adopted at the outlet of the right boundary. The inlet velocity in the x direction, u, was assumed to be a Hagen-Poiseuille flow distribution. Here, the average velocity is 0.8 m/s and the pressure is atmospheric. The inlet temperature is 300 K. The bottom boundary (wall A) is a non-slip wall with or without surface reactions. The wall temperature, T_w, is 500 K and 1200 K, because the influence of surface reaction on flame characteristics becomes significant when wall temperature is around 900 K. The position of the wall, y_w, was varied from 3.2 mm to 1.0 mm. For each y_w condition, after ignition by high temperature region located around $x = 0$, the calculation is done until the flame becomes stable or extinct. The rim of the nozzle (wall B) is non-slip inert wall with T_w, and the inside wall of the nozzle (wall C) is non-slip adiabatic wall. The maximum size of the computational domain is 8.0 mm x 3.2 mm, and the corresponding number of grid points is 200 x 160. They are equally spaced in the y direction, and unequally spaced in the x direction. The minimum grid size is 0.02 mm.

For active wall conditions, Langmuir's adsorption model was used, and the desorption reaction rates were expressed in the Arrhenius form. The Raimondeau's surface reaction mechanism was used [11]. In this mechanism, CH_3, H, O, and OH radicals are adsorbed, and recombined to form stable species such as H_2, H_2O, O_2, CH_4 and C_2H_6. This mechanism is not considered particular material, but important radicals for flame ignition and extinction are depicted. The gas phase reaction used in this study does not contain C_2H_6, and for the C_2H_6 desorbed reaction, two CH_3 radicals are desorbed instead of C_2H_6.

The thermodynamic properties for the species were obtained from the CHEMKIN database [12, 13]. The transport properties were calculated according to Smooke's simplified transport model [14]. For the gas phase reaction, the chemical kinetics model is the Smooke's skeletal mechanism [14], consisting of 16 species and 25 elementary reactions. Details of the numerical calculation method are not shown here, but can be found in [15, 16]. Here, the gravity in the x direction was considered.

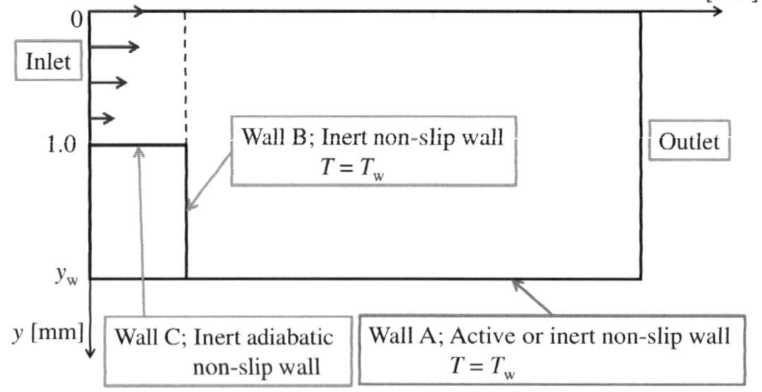

Figure 1. Analytical model and boundary conditions

3. Results and discussion

3.1. Flame Behaviors

Table 1 shows the change in flame configuration with y_w. Figure 2 shows the heat release rate distributions and velocity vectors. The attached flames (wall A) are formed in this figure. For large y_w, the flame edge forms near wall B ("Attached (Wall B)" in Table 1). When the y_w becomes smaller, the flame edge moves to near wall A (Attached (Wall A)). Unsteady flames are observed between these conditions, and they swing upstream and downstream [17]. When the y_w decreases further, blow-off occurs. This blow-off is considered as "flame quenching" in the experiments [7, 8] using the burners similar to the present study. Quenching was observed only when T_w = 500 K and y_w = 1.0 mm, which corresponds to the position of slit burner lim.

Table 1. Flame behavior

<table>
<tr><th rowspan="2">y_w [mm]</th><th colspan="2">T_w = 500 K</th><th colspan="2">T_w = 1200 K</th></tr>
<tr><th>Inert</th><th>Active</th><th>Inert</th><th>Active</th></tr>
<tr><td>3.2～2.8</td><td>Attached (Wall B)</td><td>Attached (Wall B)</td><td rowspan="4">Attached (Wall B)</td><td rowspan="4">Attached (Wall B)</td></tr>
<tr><td>2.6</td><td>Unsteady</td><td>Unsteady</td></tr>
<tr><td>2.4</td><td>Attached (Wall A)</td><td>Attached (Wall A)</td></tr>
<tr><td>2.2</td><td rowspan="6">Blow off</td><td rowspan="6">Blow off</td></tr>
<tr><td>1.8～1.6</td><td>Unsteady</td><td>Unsteady</td></tr>
<tr><td>1.4</td><td rowspan="3">Attached (Wall A)</td><td>Attached (Wall A)</td></tr>
<tr><td>1.3</td><td rowspan="4">Blow off</td></tr>
<tr><td>1.2</td></tr>
<tr><td>1.1</td><td rowspan="2">Blow off</td></tr>
<tr><td>1.0</td><td>Quenching</td><td>Quenching</td></tr>
</table>

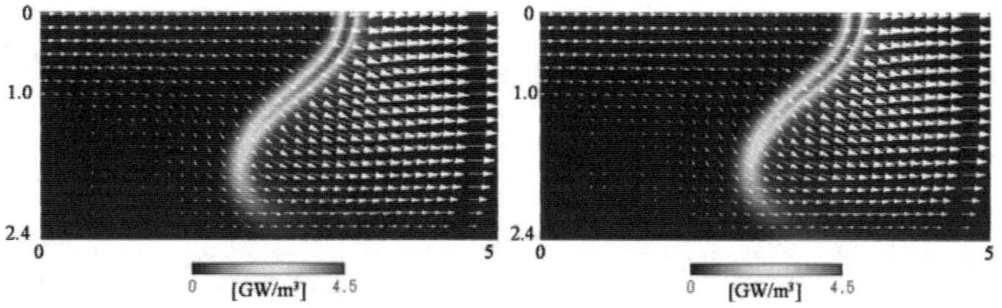

Active wall Inert wall

(a) T_w = 500 K, y_w =2.4 mm

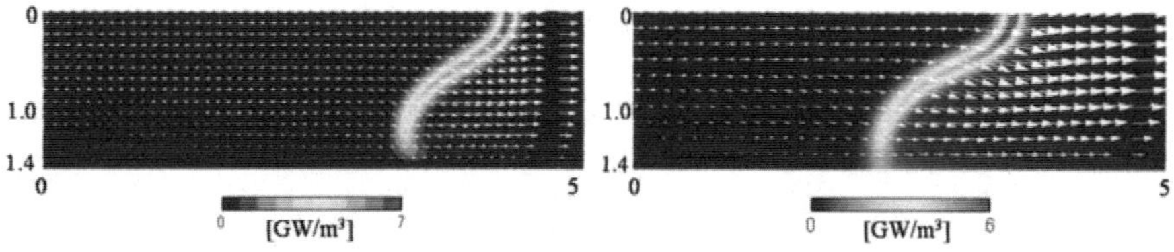

Active wall Inert wall

(b) T_w = 1200 K, y_w =1.4 mm

Figure 2. Distributions of HRR and velocity vectors

For low temperature wall, the blow-off occurs in the same y_w conditions for active and inert wall. For high temperature wall, the blow-off condition for the active wall is larger than that for the inert wall, but the difference is only 0.2 mm, and thus the influence of surface reaction on the blow-off is limited.

Figure 2 (a) shows the distribution of HRR and velocity vectors for the low temperature wall condition. The flame shapes and positions show no difference for active or inert wall conditions. This implies that the influence of heat loss is relatively large, and the influence of surface reaction is limited. The HRR distributions in figure 2 (b) show difference for active or inert wall conditions. First, the flame position for the inert wall is located in the upper stream than that for active wall. The velocity distributions in x direction at center of channel are shown in figure 3. The velocity increases with the decrease in y_w for constant inlet velocity. This tendency for the inert wall condition is more remarkable than that for the active wall condition. There is the high HRR region near the inert wall, whereas HRR is small near the active wall. In the following section, the characteristics of HRR distributions are discussed further.

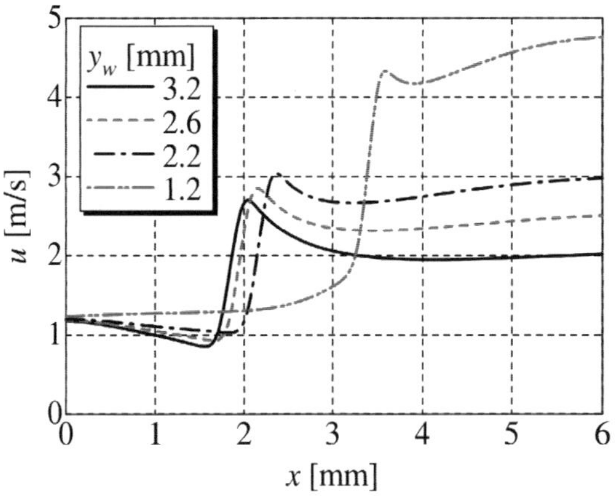

Figure 3. Distribution of velocity in x direction at center of channel;
T_w = 1200 K, Inert wall

3.2. *Influence of surface reaction on HRR*

The variation of summation of HRR, $Q_{sum,y=0}$, in the center line of channel is shown in figure 4. The $Q_{sum,y=0}$ shows no difference for active or inert wall conditions. Therefore, the influence of surface reaction on HRR is limited in center of channel, and the normal premixed flame reaction occurs around here [17].

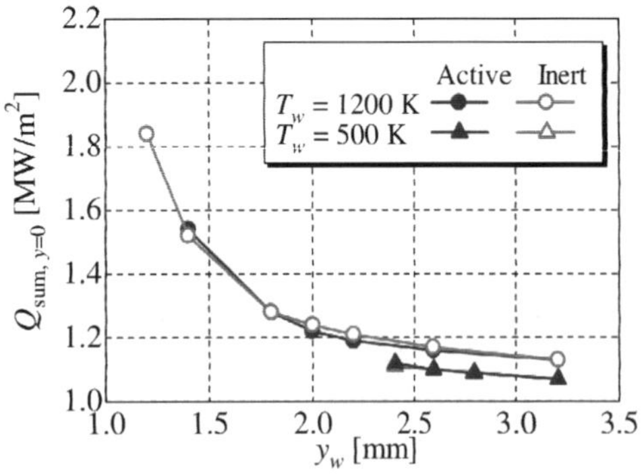

Figure 4. Relation between $Q_{sum,y=0}$ and y_w

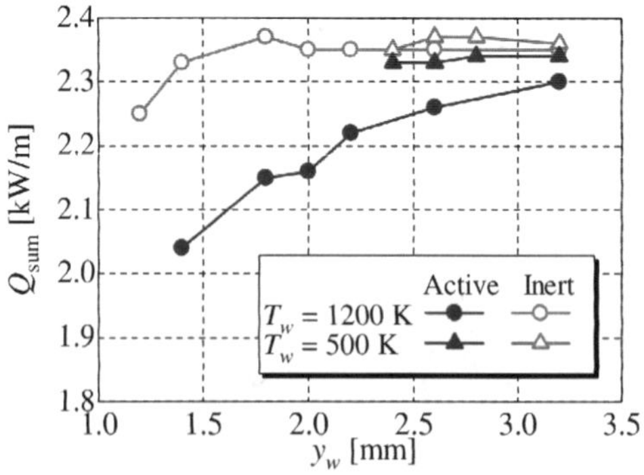

Figure 5. Relation between Q_{sum} and y_w

Figure 5 also shows the total of HRR, Q_{sum}, which is integrated over the whole calculation domain. This figure shows clear difference for active or inert wall conditions and temperature conditions. For the high temperature inert wall, Q_{sum} is almost constant except for $y_w < 1.5$ mm. The Q_{sum} for the inert wall is larger than that for the active wall in the case of high temperature wall. The difference increases with the decrease in y_w. These results clearly show the influence of surface reaction. Even when the flame forms far from wall A, the Q_{sum} is different between the active and inert wall. This is because the reaction which occurs in burned gas, such as $CO + OH = CO_2 + H$, is inhibited by the radical quenching due to the surface reaction. For both active and inert wall with low temperature, the Q_{sum} is constant for y_w and almost the same as that for the high temperature inert wall.

4. Conclusions

The numerical simulation of a two-dimensional slit burner between two parallel plates is conducted to understand the influences of surface reaction and heat-loss to the wall on quenching distance for methane-air mixture. The surface reaction causes radical quenching for active wall condition. The following conclusions are drawn.

1. With the decrease in the distance between the walls, the flame shape changes.
2. In the case of high temperature wall, the total of HRR, Q_{sum}, for the inert wall is larger than that for active wall. The difference increases with the decrease in y_w, and these results clearly show the influence of surface reactions. For low temperature wall, Q_{sum} is constant for y_w and almost the same as that for high temperature inert wall.

References

[1] Fernandez-Pello A C 2002 *Proc. Combust. Inst.* **29** 883-899
[2] Maruta K 2011 *Proc. Combust. Inst.* **33** 125-150
[3] Maruta K, Kataoka T, Kim N I, Minaev S and Fursenko R 2005 *Proc. Combust. Inst.* **30** 2429-36
[4] Yuasa S, Oshimi K, Nose H and Tennichi Y 2005 *Proc. Combust. Inst.* **30** 2455-62
[5] Norton D G and Vlachos D G 2003 *Chemical Engineering Science* **58** 4871-82
[6] Egolfopoulos F N, Zhang H and Zhang Z 1997 *Combustion and Flame* **109** 237-252
[7] Miesse C M, Masel R I, Jensen C D, Shannon M A and Short M 2004 *AIChE journal* **50** 3206-14
[8] Yang H L, Feng Y X, Wu Y Y, Wang X H, Jiang L Q, Zhao D Q and Yamashita H 2011 *Combust. Sci. Technol.* **183** 444-458
[9] Saiki Y and Suzuki Y 2013 *Proc. Combust. Inst.* **34** 3395-3402
[10] Aghalayam P, Bui P-A and Vlachos D G 1998 *Combust. Theory Modelling* **2** 515-530
[11] Raimondeau S, Norton D, Vlachos D G and Masel R I 2002 *Proc. Combust. Inst.* **29** 901-907
[12] Kee R J, Dixon-Lewis G, Warnatz J, Coltrin M E and Miller J A 1986 *Sandia Report* SAND **86**-8246
[13] Kee R J, Rupley F M and Miller J A 1989 *Sandia Report* SAND **89**-8009
[14] Smooke M D and Giovangigli V 1991 Reduced Kinetics Mechanisms and Asymptotic Approximations for Methane-Air Flames, ed Smooke M D, (Berlin: Springer-Verlag) pp 1-28
[15] Imai T, Yamashita H, Hayashi N and Zhao D Q 2011 *Journal of Combustion Society of Japan (In Japanese)* **53** 45-56
[16] Hayashi N, Miyata C, Yamashita H and Yamamoto K 2010 *Progress in Computational Fluid Dynamics* **10** 225-231
[17] Hayashi N and Yamashita H 2013 *Proceedings of 9th ASPACC* 68

Acknowledgment

This work was supported by JSPS KAKENHI Grant Number 24760163 and 26820062.

A Framework for Determining the Maximum Theoretical Power Output for a Given Vibration Energy

J Heit, S Roundy

University of Utah, Mechanical Engineering, 50 S. Central Campus Dr., Salt Lake City, UT 84121, USA

heit88@gmail.com

Abstract. This paper outlines a mathematical framework to determine the upper bound on extractable power as a function of the forcing vibrations. In addition to determining the upper bound on power output, the method described provides insight into the dynamic transducer forces required to attain the upper bound. This relationship, between input vibration parameters and transducer force gives a critical first step in determining the optimal transducer architecture for a given vibration input. The method developed is applied to two specific vibration inputs; a single sinusoid, and the sum of two sinusoids. For the single sinusoidal case, the optimal transducer force is found to be that produced by a linear spring, resonant with the input frequency, and a linear viscous damper, with matched impedance to the mechanical damper. The solution to this first case was previously known, but has been used here to validate the methodology. The resulting transducer force for the input described by a sum of two sinusoids is found to be inherently time dependent. This time dependency shows that an active system can outperform a passive system. Furthermore, the upper bound on power output is shown to be twice that obtainable from a linear harvester centred at the lower of the two frequencies.

1. Introduction

Much recent work in vibration energy harvesting has focused on structure and transducer designs to improve power output from vibration sources that are not modeled as a single sinusoidal input. Much of this work has investigated the use of nonlinearities as a way to increase energy output [1-4]. These nonlinearities are usually of the form of a nonlinear spring, such as a Duffing oscillator. These works give useful insight to the potential uses of nonlinearities for harvesting from complex vibration inputs. However these works do not give a clear relationship between the parameters that define the input vibration and the optimal transducer dynamics.

Other researchers have taken the opposite approach, starting with the vibration excitation and investigating the optimal transducer architectures to extract the maximum power. Daqaq et al. [5] showed that for Gaussian white noise the energy generation was not a function of the transducer's potential function. Halvorsen et. al. [6] proved that for an input described by a single frequency harmonic (i.e a single sinusoid) when the proof mass is subjected to viscous damping, the optimal transducer dynamics are those of a velocity damped resonant generator (VDRG) [7]. In both cases, the results are limited to very simple and specific types of forcing vibrations.

Continuing along a similar pathway, this work will present a method to find the unconstrained and globally optimal relationship between the input vibration, and force that must be produced by the transducer. This relationship will also define an upper limit for power generated for a given vibration

Journal of Physics: Conference Series **557** (2014) 012020 doi:10.1088/1742-6596/557/1/012020

input. This framework will then be applied to two case studies, a single sinusoid and the sum of two sinusoids.

2. General Model

A simple model for a kinetic harvester with a generic transducer force F_T that acts on the proof mass is shown in Figure 1. This generic transducer may contain both energy dissipative elements for power generation as well as energy conservative restoring elements. In general, the system is subject to a forcing function $F(t)$. The inherent mechanical losses that are found in any real system are approximated by a linear viscous damper described by a single coefficient b_m. This single degree of freedom system is characterized by a single displacement x. If the system is excited through base excitation, as is the normal case for an inertial generator, then $F(t)$ would be the mass (m) multiplied by the base acceleration $A(t)$. In this case the displacement x is the relative distance between the proof mass and the base. The system schematic as well as its governing equations are shown in Figure 1.

(a)

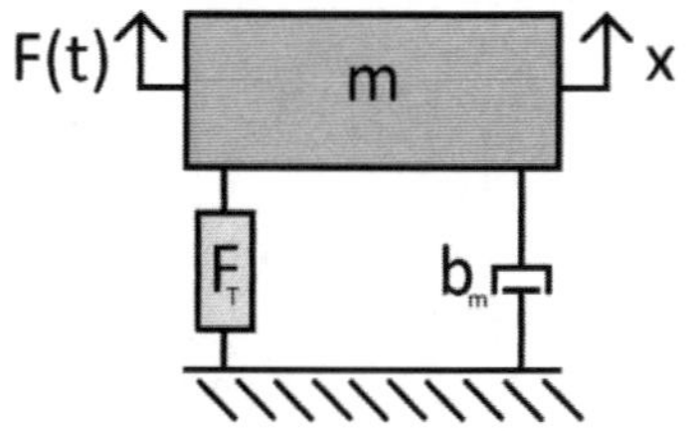

(b)

$$m\ddot{x} + b_m\dot{x} + F_T = F(t) \tag{1}$$

$$\dot{x}_1 = x_2 \tag{2}$$

$$\dot{x}_2 = \frac{1}{m}\left(-b_m x_2 - F_T + F(t)\right) \tag{3}$$

Figure 1. (a) A generic inertial generator characterized by a single displacement x. Here F_T represents the force produced by an unknown transducer architecture. b_m is the coefficient that characterizes the system's linear viscous damping due to inherent mechanical losses of the system. (b) Governing equations for the generic inertial generator.

An energy balance of the system is used in order to find an expression for the energy generated by the transducer as a functison of the input. By examining the energy balance of the system in steady state we can neglect the kinetic energy of the mass as well as the possible potential energy stored in the transducer. This is due to the fact that these energy storage elements are restorative, thus they do not represent a net energy input or output to the system while it is in steady state. The energy that can be generated as a function of system parameters and input is given in (4):

$$E_{gen} = \int [F(t)x_2 - b_m x_2^2]\, dt \tag{4}$$

For more generalized results we can look at the square of the power generated to examine a continuous positive definite functional, thereby allowing us to find the critical points in the magnitude of the energy generated.

$$J = \int [F(t)x_2 - b_m x_2^2]^2 dt \tag{5}$$

If the velocity of the proof mass x_2 is treated as the control parameter, the critical points of the functional, which represents the energy generated by the transducer, can be found through the stationary condition of the Euler-Lagrange equation [8]. Taking the derivative of I, the integrand of (5), with respect to x_2 and solving for x_2 yields three optimal velocity paths shown in Table 1. Here $\star$ denotes the critical path with respect to the energy generated. These three relationships for $x_2^\star$ represent the critical velocity paths, given a vibration input $F(t)$ to the system that will result in a minimum or maximum energy output. The second derivative determines if each solution is a maximum or minimum. By substituting the optimal velocity relationships into the governing differential equations (2) – (3), an expression for the displacement of the proof mass x_1 as well as the transducer force F_T can be expressed as a function of the system properties and the input vibration force. A summary of these solutions is given in Table 1, which shows the optimal velocity paths, position paths, and transducer forces required to produce those paths.

Table 1. Summary of critical path relationships for a general input.

Critical Velocity Path	Critical Positon Path	Critical Transducer Force	Type
$x_2^\star = \frac{F(t)}{2b_m}$	$x_1^\star = \int \frac{F(t)}{2b_m} \mathrm{d}t$	$F_T^\star = -\frac{m\dot{F}(t)}{2b_m} + \frac{F(t)}{2}$	Maximum
$x_2^\star = \frac{F(t)}{b_m}$	$x_1^\star = \int \frac{F(t)}{b_m} \mathrm{d}t$	$F_T^\star = -\frac{m\dot{F}(t)}{b_m}$	Minimum
$x_2^\star = 0$	$x_1^\star = 0$	$F_T^\star = F(t)$	Minimum

3. Model Applied to a Single Sinusoid Input

It is difficult to see the relevance of the equations in Table 1 in their general form. To help illustrate these relationships a simple example of a single frequency sinusoidal input will be examined. As the solution for this example is known, it further serves to validate the method. We will look at the relationship that maximizes the energy output of the system.

For the maximum power condition $x_2^\star = \frac{F(t)}{2b_m}$, letting $F(t) = A\, m\, sin(\omega t)$ results in the following relationships:

$$x_1^\star = -\frac{A\, m}{2b_m\omega}\cos(\omega t) \tag{6}$$

$$x_2^\star = \frac{A\, m}{2b_m}\sin(\omega t) \tag{7}$$

$$F_T{}^\star = -\frac{A\, \omega\, m^2}{2b_m}\cos(\omega t) + \frac{A\, m}{2}\sin(\omega t) \tag{8}$$

Substituting for $x_1^\star$ *and* $x_2^\star$ yields:

$$F_T^\star = \omega^2 m\, x_1 + b_m x_2 \tag{9}$$

The minimum power conditions will not be further investigated as they are not relevant to the goal of maximizing extracted power from a vibration source. Equations (6) – (9) can be interpreted as follows: the optimal transducer model is a linear spring ($k = \omega^2 m$) and a linear viscous damper which represents an electromechanical transducer. The constant of the linear spring is found to be resonant with the vibration input and the impedance of the electrical damper is found to be matched to the impedance of the mechanical damper b_m. This result is widely known and has been previously reported [6-7]. Thus, the application of this framework yielded the known optimal transducer architecture, without placing any a priori assumptions on the form of the transducer.

4. Model Applied to a Multiple Sinusoid Input

A source consisting of two sinusoids at different frequencies is a common vibration input. This type of vibration occurs in rotating machinery where two unbalanced masses rotate at different rates fixed relative to one another or in a system where multiple harmonics of a structure are well represented.

The RMS power output scales with A^2 for the standard linear system. Thus, the case where the amplitudes of the two sinusoids are equal will be examined. In the case where one sinusoid has an amplitude much greater than the other, it is reasonable to assume that the maximum power generation will be achieved by creating a linear harvester tuned to the frequency corresponding to the maximum value of A^2/ω. In the case where the two sinusoids are of similar, but different amplitudes, the following analysis is relevant. The expression for this double sinusoidal input is given as $F(t) = A\, m\, (\sin(\omega t) + \sin(n\omega t))$. Here, $n \in (0\ \infty)$ represents the multiple difference between the two frequency components.

Examining now only the vibration input which results in the maximum energy output, the optimal velocity signal for an input of two sinusoids is obtained in (10).

$$x_2^\star = \frac{A\, m}{2b_m}(\sin(\omega t) + \sin(n\omega t)) \tag{10}$$

Using the equations in Table 1, the relationships for the optimal position path and corresponding transducer force to achieve the velocity response as shown in (10) can be written as:

$$x_1^\star = -\frac{A\,m}{2\omega b_m}\left(\cos(\omega t) + \frac{1}{n}\cos(n\omega t)\right) \tag{11}$$

$$F_T^\star = \frac{A\,m}{2}\big(sin(\omega t) + sin(n\omega t)\big) - \frac{A\,\omega\,m^2}{2b_m}\big(cos(\omega t) + ncos(n\omega t)\big) \tag{12}$$

Substituting (11) and (10) for x_1 and x_2 into (12), where available, yields:

$$F_T^\star = b_m x_2 + \omega^2 m\, x_1 + TD \tag{13}$$

where *TD* is the time dependent component of the transducer force that cannot be directly substituted for by the systems states x_1 and x_2.

$$TD = \frac{A\,\omega\,m^2}{2b_m}\left(\frac{1}{n} - n\right)\cos(n\omega t) \tag{14}$$

The time dependent component of the transducer force shows that the true unconstrained optimal transducer force for an input vibration of this form cannot be realized with a passive system. Thus, in principle an active system can outperform a passive system of any type, linear or non-linear. However, this would assume that the time dependent restoring force implemented is conservative.

An energy balance is used to determine the nature of the time dependent force. It must be determined if the force does work, adding energy to the system over time, takes energy from the system, or does no net work on the system. The net energy into the proof mass from the time dependent force can be calculated by integrating the force over the displacement for a period *T* of the entire signal. Using equations (10) and (14) yields:

$$E_{TD}\left(\frac{2\pi}{\omega}\kappa\right) = \frac{A^2m^3(1-(-2+n)n-n(1+n)\mathrm{Cos}[2\kappa(-1+n)\pi]+(-1+n^2)\cos(2\kappa n\pi)^2+(-1+n)n\cos(2\kappa(1+n)\pi))}{8b^2n^2} \tag{15}$$

where κ is an integer value such that $n * \kappa \in \mathbb{Z}$. The total period for any input of this form is $T=2\pi/\omega\ \kappa$.

For the constraints of $\kappa \in \mathbb{Z}$ and $n * \kappa \in \mathbb{Z}$, (15) reduces to zero. This shows that the time dependent force acts as a conservative element, not doing any work to the system over time. The upper limit for energy output from the optimal transducer can be shown analytically. This can be accomplished in a similar manner to the derivation of the average power output for the single sinusoid case. Knowing that from the result of (13) and (15) the power output from the transducer is dissipated by the force of a linear viscous damper, the instantaneous power dissipated through this element can be written as:

$$P = F * v = b_m {x_2^\star}^2 \tag{16}$$

Here $x_2^\star$ is the optimal velocity shown in (10). Integrating the instantaneous power output over time yields the total energy generated by the transducer for all $n \in (1, \infty)$.

$$E_n = \frac{A^2\,m^2 T}{4b_m} \tag{17}$$

Note that for all $n \neq 1$, the rms value of the excitation force is Am. However, for the special case in which $n = 1$, the rms value of the excitation force is $\sqrt{2}Am$, and thus the upper bound on the energy generated would be double that shown in (17). However, if the rms value of the driving force is normalized to Am for the special case of $n = 1$, (17) will still hold. The result is that the upper bound on power output is not a function of *n*. Intuitively this means that if the transducer force given by (12) and (13) can be generated, all of the power from both sinusoids could, in theory, be captured.

In order to gain additional insight into (17) a numerical study was performed in which the energy output over a sufficiently long period was measured for various values of *n*. The output of this study is shown in Figure 2. In one case, the optimal transducer force is applied to the proof mass. In the second case, the system is characterized by a linear oscillator whose resonance is the lower of the two frequencies present in the forcing vibrations. The output is normalized to the energy generated by either system at $n = 1$. As *n* deviates from 1, the power output from the linear system quickly drops to ½. However, the power output from the optimal system remains constant at 1.

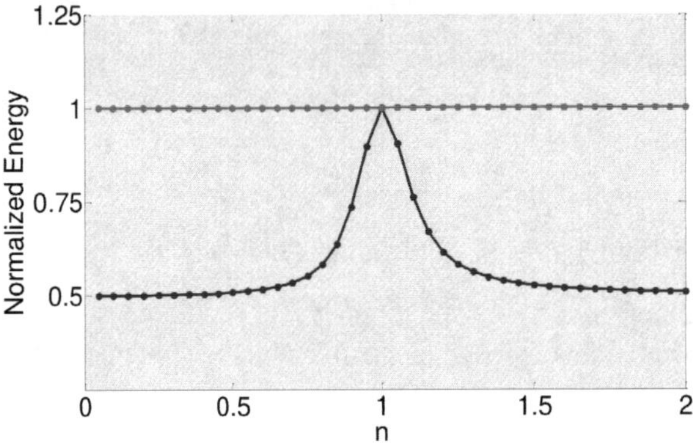

Figure 2. Numeric simulations of the energy output of the optimal transducer as compared to a linear harvester. The energy production has been normalized by the energy output of both systems at $n = 1$.

5. Conclusion

This paper has outlined a framework necessary to relate the form of an input vibration to an optimal transducer force. In creating this framework no assumptions of the transducer architecture were made. This framework was then applied to two case studies. The first was a vibration input of a single sinusoid. The optimal transducer was found to be a linear viscous damper with matched impedance, and a linear spring, resonant with the input frequency. This solution can be expressed as a function of the system's states so is considered a passive system. While the solution of this case study seems trivial, it demonstrates the method and validates it against a known case.

The second application was an input of the sum of two sinusoids at different frequencies. The optimal transducer force found was dependent on the difference between the two frequencies. In all cases the optimal transducer force consists of a linear viscous damper with matched impedance, a linear spring, and a time dependent force component. This time dependent component was found to act as a conservative force, like a time dependent spring. The framework was used to find the upper limit for power generation. This limit was found to be twice the power output of a linear system harvesting only from the lower of the two frequency components.

This basic framework could be applied to vibration inputs of various forms to determine the upper bound of power generation for that type of vibration, and the optimal transducer architecture. If a transducer architecture is assumed, a Duffing oscillator for example, this methodology can be applied to determine how close to the assumed solution is to the upper bound.

6. Acknowledgements

The authors gratefully acknowledge funding for this work from the National Science Foundation under Award Number ECCS 1342070.

7. References

[1] Daqaq MF, 2012 *Nonlinear Dynam*, vol. **69**, 1063-79.
[2] Mann BP, Barton DA and Owens BA, 2012 *J. Intell. Mater. Syst.* 1045389X12439639.
[3] Nguyen SD and Halvorsen E, 2013 *Appl. Phys. Lett.*, **102(2)**, 023904 .
[4] Hoffmann D, Folkmer B and Manoli Y, 2012 Comparative study of concepts for increasing the bandwidth of vibration based energy harvesters, *Proc. PowerMEMs, 2012.*
[5] Daqaq MF, 2014 *J. Sound. Vib.*, **333(15)**, 3479-89.
[6] Halvorsen E, Le CP, Mitcheson PD and Yeatman EM, 2013 *Journal of Physics: Conference Series*. **476(1)** 012026
[7] Mitcheson P, Green T, Yeatman E and Holmes A, 2004 *J. Microelectromech. S.*, **13(3)**, 429-40.
[8] Bryson AE and Ho Y-C, 1975 *Applied Optimal Control*, (New York: Hemisphere Publishing Corporation)

Wireless Energy Transfer Through Magnetic Reluctance Coupling

P Pillatsch

University of California – Berkeley, Advanced Manufacturing for Energy, 2111 Etcheverry Hall, Berkeley, California, 94720, USA

E-mail: ppillatsch@berkeley.edu

Abstract.
Energy harvesting from human motion for body worn or implanted devices faces the problem of the wearer being still, e.g. while asleep. Especially for medical devices this can become an issue if a patient is bed-bound for prolonged periods of time and the internal battery of a harvesting system is not recharged. This article introduces a mechanism for wireless energy transfer based on a previously presented energy harvesting device. The internal rotor of the energy harvester is made of mild steel and can be actuated through a magnetic reluctance coupling to an external motor. The internal piezoelectric transducer is consequently actuated and generates electricity. This paper successfully demonstrates energy transfer over a distance of 16 mm in air and an achieved power output of 85 μW at 25 Hz. The device functional volume is 1.85 cm^3. Furthermore, it was demonstrated that increasing the driving frequency beyond 25 Hz did not yield a further increase in power output. Future research will focus on improving the reluctance coupling, e.g. by investigating the use of multiple or stronger magnets, in order to increase transmission distance.

1. Introduction

Previously, the author presented a rotational energy harvester for human motion based on piezoelectric beam plucking [1]. Energy harvesting for body worn and implantable devices is an attractive alternative to battery powering. It can alleviate the maintenance burden and potentially eliminate the need for repeated intrusive surgery. The above-mentioned rotational energy harvester has the capability to harvest energy from low frequency motion efficiently through piezoelectric frequency up-conversion. The rotational layout of the design makes it less dependent on gravity an through the eccentric proof mass it can accept excitation in two linear and rotational directions. Figure 1 shows the different components of the harvester.

However, especially in medical applications, there is an ultimate question of what happens if a patient becomes bed-bound. This is a likely scenario and the challenge is to avoid failure of vital implants due to the energy harvester not generating any electricity at rest. Compared to thermoelectric harvesters or glucose fuel cells, this is a problem that all motion-based devices for medical applications are facing. This article demonstrates external actuation of the rotational harvester through a magnetic reluctance coupling without any changes to the initial prototype. Energy is provided by a dedicated source, in this case an electric motor, and transferred trough the magnetic coupling onto the rotor inside the harvester. This means that there is a wireless energy transfer between the source and the receiver (the harvester). The same device can thus be used for wireless energy transfer with a dedicated source and for energy harvesting and an internal battery could be charged even in the absence of an external vibration.

Typically, power transfer into the human body is performed via an inductive or ultrasonic coupling [2,3]. The novelty of this system is illustrated in figure 2. An external magnet mounted on a DC motor rotates at a distance from the micro generator. The rotor on the inside is a half disc made of magnetic mild steel. The reluctance change as the permanent magnet passes the edge of the rotor causes the rotor to latch on to the magnet and follow its rotation. The transduction mechanism is the same as for the operation as a harvester – a small permanent magnet fixed to the rotor actuates a piezoelectric beam with a second magnet on its tip each time the two pass each other.

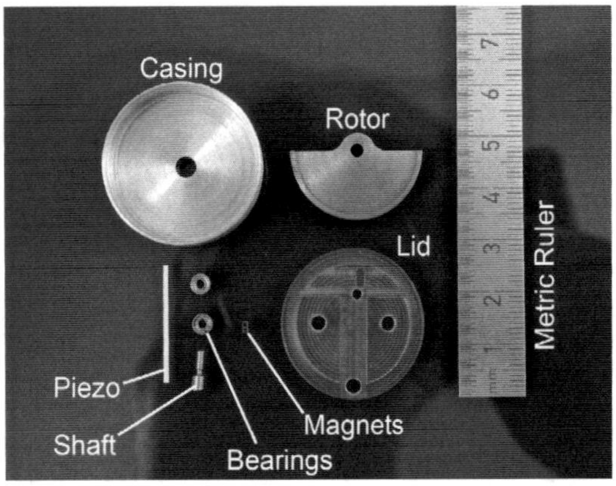

Figure 1: Energy harvester components

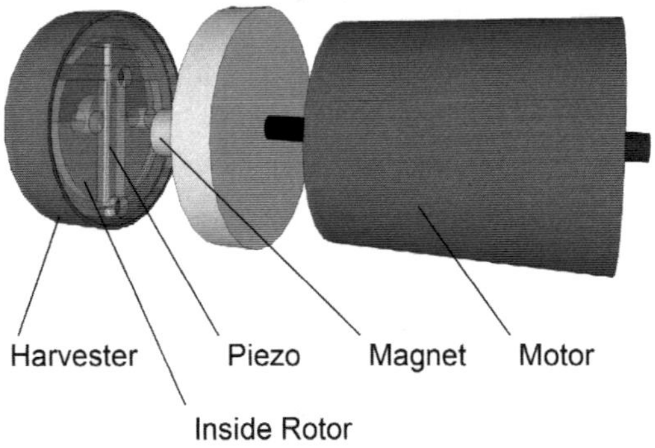

Figure 2: Reluctance coupling of the motor magnet to the discontinuity in the half disc rotor

Power transfer using permanent magnet coupling is not a novel concept as described in [4]. However, applications for the human body are little investigated. In [5] such a device is presented but the coupling magnets had a negative effect on the electromagnetic transduction mechanism. This problem does not occur in this device due to the piezoelectric transduction. In addition, the combined operation of a device as energy harvester for the general case and as a receiver for wireless energy transfer in times of need makes this a highly versatile prototype. The experimental results presented are specific to the latter case as the energy harvester operation has been shown previously.

2. Experimental Set-Up

Figure 3 depicts the experimental set-up used. The rotational energy harvester, operated as a micro-generator in this case, is held in a vertical position in a 3D-printed holder. Opposite, a DC motor is held in a similar fixture with a disc carrying permanent magnets (NdFeB disks $1/4'' \times 1/8''$). The DC motor is driven by an adjustable power supply in order to vary the rotational frequency. Both mounts are adjustable, the left one allows the gap d between the generator and the permanent magnets to be changed and the right one provides lateral displacement L to investigate misalignment between the generator and the DC motor. The piezoelectric beam inside the generator is directly connected to an impedance matched resistive load of 150 kΩ. The power output is then established by measuring the root-mean-squared voltage drop across the resistor three times in a row over a time of 10 s and taking the average.

Figure 1 already shows the components of the energy harvester. The internal series bimorph piezoelectric beam has a free length of 19.5 mm, width of 1 mm and total thickness is 0.37 mm.

The diameter of the casing is 30 mm and the functional volume without the casing is 1.85 cm^3. The total weight is 10.5 g

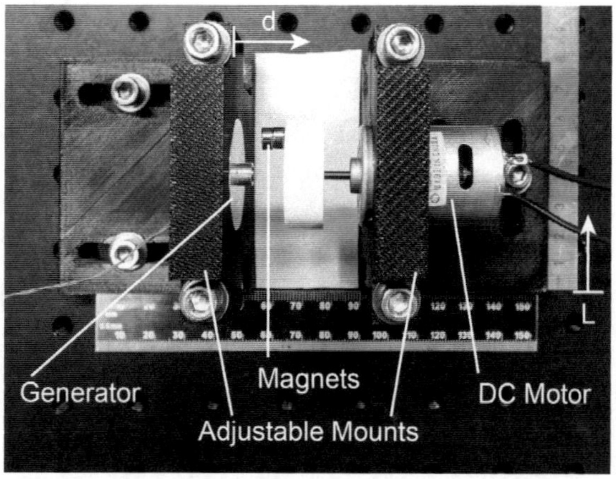

Figure 3: Experimental set-up

3. Results

Figure 4 shows the power output in relation to the distance d. The experiments are started at an initial distance of 1 mm which is progressively increased until the reluctance coupling is too weak to keep rotating the internal rotor. This happened at a maximum distance of 16 mm. The driving frequency from the DC motor was set to 10 Hz. Due to the design of the device, the power output is dependent on the driving frequency. An increased number of actuations of the internal piezoelectric beam results in an increase in power. This is why the power output stays relatively stable around 45 μW as long as the rotor keeps following the external magnet. The variations that can be seen at the smaller distances d might be caused by inconsistencies in the DC motor driving frequency. It was found that, at close proximity to the aluminium casing of the generator, Eddy currents induced by the permanent magnets can affect the driving motor. A future solution to this problem would be to use a stepper motor in order to achieve a more consistent rotation. Figure 5 shows the corresponding voltage output of the piezoelectric beam. It matches the expected behaviour from the operation as a harvester. The ring down of the beam vibration after each actuation is clearly visible.

The power output in respect to the motor rotational frequency is depicted in figure 6 at a constant distance $d = 1$ mm. As previously mentioned, higher frequencies translate to more frequent actuations of the piezoelectric beam and thus increased output. However, after about 25 Hz, the previous excitation has not fully rung down before the next actuation happens. This causes a weaker coupling between the internal magnets and explains the drop in output power beyond this point. This is further supported by figure 7, which shows the voltage output at 30 Hz being significantly lower than in figure 5. This is an interesting result as it shows that there is a limit after which it is disadvantageous to further increase the frequency. The maximum power transfer achieved was 85 μW.

Finally, the design can cope with a lateral misalignment L to a certain degree as depicted in figure 8. At a separation of $d = 1$ mm and at 10 Hz, the DC motor is displaced in 1 mm steps. Occasionally, the rotor fails to keep up with the driving magnet before latching on again. Figure 9, showing the corresponding voltage output at $L = 4$ mm, confirms an irregular actuation pattern. This explains the slightly lower power output and larger error bars with an increase in misalignment.

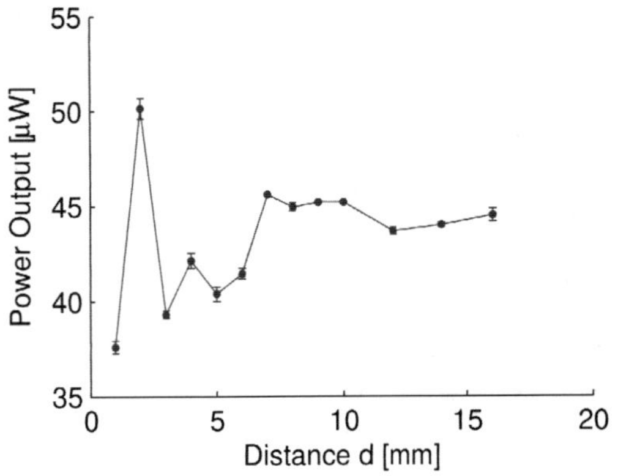

Figure 4: Power output in relation to separation distance d between motor and harvester, incl. error bars

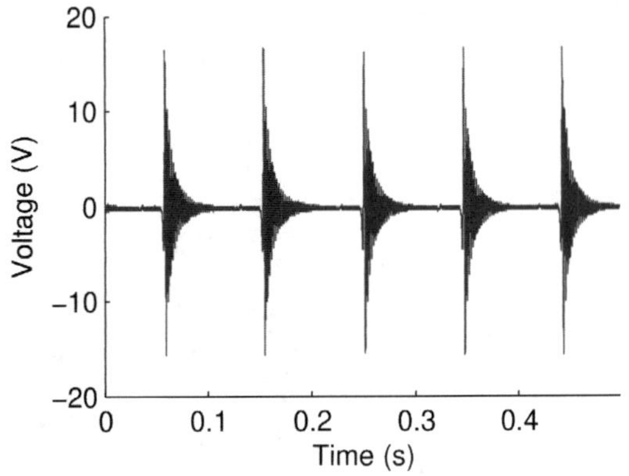

Figure 5: Voltage output at 10 Hz, $d = 1$ mm and 150 kΩ impedance matched resistive load

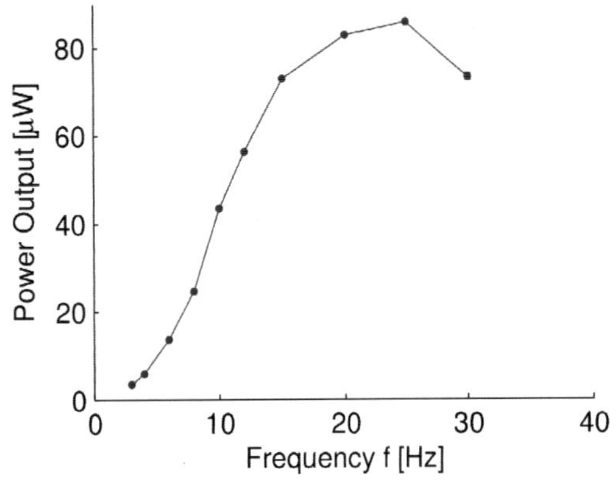

Figure 6: Power output in relation to motor driving frequency at $d = 1$ mm, error bars are insignificantly small

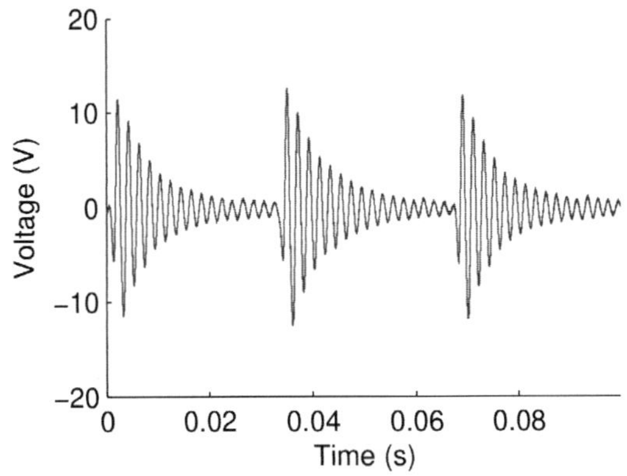

Figure 7: Output voltage waveform at $d =$ 1 mm and 30 Hz driving frequency

4. Conclusions

This paper demonstrates a mechanism for wireless energy transfer with a dedicated source to enhance the versatility of a previously demonstrated motion energy harvester. The device can be operated as a piezoelectric rotational energy harvester most of the time. However, in cases without external motion, e.g. the wearer being bed-bound for prolonged periods of time, the internal rotor of the device can be actuated through a magnetic reluctance coupling with an external permanent magnet driven by a DC or stepper motor. This makes it possible to recharge an internal battery and prevent the failure of sensors otherwise powered only when actively harvesting energy.

A novel aspect of the coupling is that it uses a permanent magnet latching onto mild steel, whereas other methods use ultrasonic or inductive coupling. A potential advantage might be reduced tissue damage and deeper penetration since a stronger coupling can be achieved by choosing a larger, stronger mechanism. This will be investigated in the future.

The energy transfer was successful over a distance of up to 16 mm and a maximum power transfer of 85 µW was achieved at a driving frequency of 25 Hz. Further increasing the driving

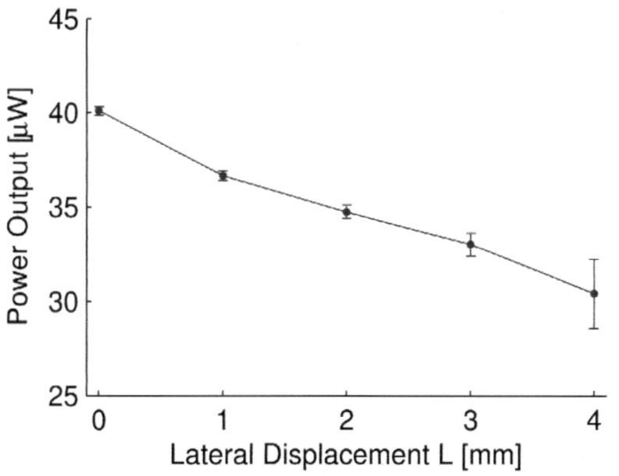

Figure 8: Power output vs. lateral misalignment L between motor and harvester at 10 Hz and $d = 1$ mm, incl. error bars

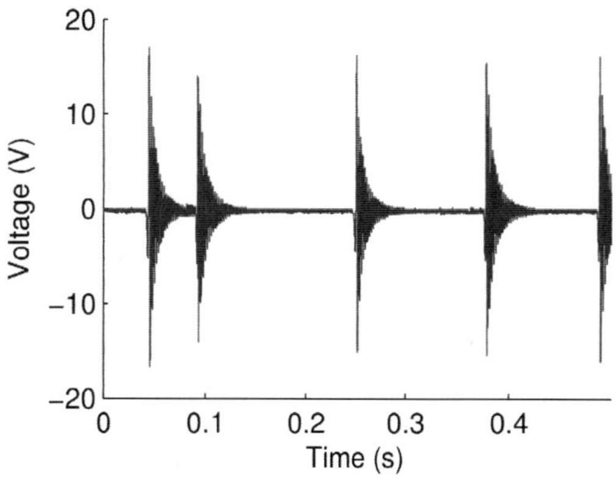

Figure 9: Output voltage waveform at $L = 4$ mm lateral displacement and 10 Hz driving frequency

frequency did not yield any more power as the actuations of the piezoelectric transducer became too closely spaced to allow a full ring down after initial deflection and release. Furthermore, it was demonstrated that the mechanism is relatively robust against lateral misalignment, showing a 25 % drop in power output for a 4 mm shift between the rotational axes of the harvester and the DC motor.

Future research will improve the consistency of the driving frequency by replacing the DC motor with a stepper motor and better controller. The reluctance coupling between the rotor and permanent driving magnet will be investigated in order to find the best combination. For instance, using multiple or stronger magnets could improve the coupling strength. Ultimately, it will be interesting to perform experiments regarding penetration through different media. This could extend the scope of the mechanism to other applications such as energy transfer in harsh environments or through water/gas pipe walls.

Acknowledgments
The author would like to thank Michael Riedel and Klaus van der Linden from Johnson Matthey Catalysts (Germany) GmbH for their help in customising the piezoelectric transducer and Dr. Richard Xu for valuable discussions. Conference attendance was made possible through a travel grant from the Institution of Engineering and Technology (IET). This research was funded in part under California Energy Commission (CEC) grant number 500-11-021.

References
[1] Pillatsch P, Yeatman E M and Holmes A S 2013 *Sensors and Actuators A: Physical* **206** 178–185 ISSN 09244247
[2] Denisov A and Yeatman E 2010 *Body Sensor Networks* pp 2–7
[3] Wei X and Liu J 2008 *Frontiers of Energy and Power Engineering in China* **2** 1–13 ISSN 1673-7393
[4] Li W 2007 *High Efficiency Wireless Power Transmission At Low Freqeuncy Using Permanent Magnet Coupling* Ph.D. thesis
[5] Suzuki S N, Ishihara M and Kobayashi Y 2011 *IEEE Transactions on Magnetics* **47** 2811–2814

Composite Piezoelectric Rubber Band for Energy Harvesting from Breathing and Limb Motion

Jhih-Jhe Wang, Huan-Jan Su, Chang-I Hsu and Yu-Chuan Su

Department of Engineering and System Science
National Tsing Hua University, Hsinchu, Taiwan

E-mail: ycsu@ess.nthu.edu.tw

Abstract. We have successfully demonstrated the design and microfabrication of piezoelectric rubber bands and their application in energy harvesting from human motions. Composite polymeric and metallic microstructures with embedded bipolar charges are employed to realize the desired stretchability and electromechanical sensitivity. In the prototype demonstration, multilayer PDMS cellular structures coated with PTFE films and stretchable gold electrodes are fabricated and implanted with bipolar charges. The composite structures show elasticity of 300~600 kPa and extreme piezoelectricity of d_{33} >2000 pC/N and d_{31} >200 pC/N. For a working volume of 2.5cm×2.5cm×0.3mm, 10% (or 2.5mm) stretch results in effective d_{31} of >17000 pC/N. It is estimated that electric charge of >0.2 μC can be collected and stored per breath (or 2.5cm deformation). As such, the composite piezoelectric rubber bands (with spring constants of ~200 N/m) can be mounted on elastic waistbands to harvest the circumferential stretch during breathing, or on pads around joints to harvest the elongation during limb motion. Furthermore, the wearable piezoelectric structures can be spread, stacked and connected to charge energy storages and power micro devices.

1. Introduction

In recent years, wearable or skin-attachable electronic sensors have been demonstrated based on a variety of detection schemes [1]. To extend the lifetime and reduce the volume of the electronics, researchers have begun investing methods of obtaining electrical energy from human body. Perhaps the most common type of human powered energy generation involves implanting piezoelectric materials into shoes [2], which harvests energy from discontinuous walking activities. For continuous activities like breathing and blood flow, very limited work has been published. The fabrication of soft piezoelectric films with high d_{33} was reported at MEMS 2013 by the authors [3]. To address the need of high coupling efficiency in the 31 mode, this paper presents novel piezoelectric composites with desired anisotropicity to achieve d_{31} >200 pC/N (or effective d_{31} >17 nC/N). As such, the soft piezo-electric composites can be worn on body to harvest the deformation during breathing and limb motion.

2. Operating principle

As illustrated in Figure 1, the structure is composed of top and bottom electrodes, and alternating two solid and one void layers in between. Upon the application of stretch, the quasi-dipoles inside the composite structures change their moments, which results in the desired piezoelectricity as illustrated. An analytical model is developed to investigate the material properties of the composite structures, with key parameters estimated by COMSOL simulation. Equation (1) and (2) show the relationships employed to optimize the design of composite structures with extreme piezoelectricity,

$$d_{33} \propto \frac{\kappa_1 t_r[\kappa_2\beta(1-s_r)+\lambda+(1-\kappa_2)](1-s_r+\alpha s_r\varepsilon_1)(1-s_r)(1-s_r+\alpha s_r)\sigma}{s_r[2(1-s_r+\alpha s_r\varepsilon_1)+(1-s_r+\alpha s_r)\varepsilon_1 t_r]^2} \tag{1}$$

$$d_{31} \propto \frac{\kappa_1 t_r^{\,2}(\beta_v-\kappa_2\beta_s)(1-s_r+\alpha s_r\varepsilon_1)(1-s_r)(1-s_r+\alpha s_r)\sigma}{x_v(2\theta_1+\theta_2 t_r s_r)[2(1-s_r+\alpha s_r\varepsilon_1)+(1-s_r+\alpha s_r)\varepsilon_1 t_r]^2} \tag{2}$$

where s_r is the ratio of A_2 divided by A_1 and t_r is the ratio of x_v divided by x_s (A_1 and A_2 are the effective areas and x_s and x_v are the thicknesses of solid and void layers, respectively). As illustrated in Figure 2(a), the piezoelectric rubber bands are designed to harvest the circumferential stretch during breathing or the elongation during limb motion. Anisotropic structures shown in Figure 2(b) and (c) are employed to further improve the piezoelectricity.

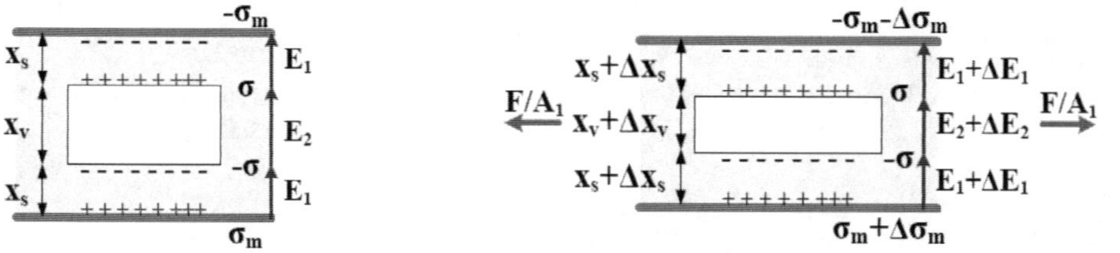

Figure 1. A model for the piezoelectricity of a charge-implanted composite microstructure.

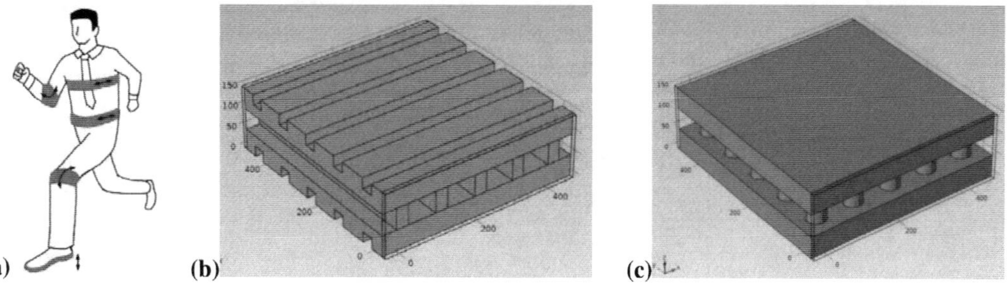

Figure 2. (a) Applications and designs for (b) anisotropic 31-mode and (c) 33-mode operations.

3. Fabrication processes

Figure 3 illustrates the fabrication processes. A layer of 50 μm thick photoresist (SU-8, MicroChem) is spin-coated and patterned on top of a blank silicon wafer to fabricate the mold used for duplicating the cellular microstructures. After the photoresist mold is fully stabilized, it is placed in a desiccator under vacuum for 3 hours with a vial containing a few drops of 1H,1H,2H,2H-perfluorooctyltrichlorosilane (Fluka) to silanize its surface for de-molding. The multilayer casting and stacking processes start with the deposition of a thin (~120 μm) PDMS-mixture layer on top of the mold, and cured for 1 hour at 85°C. Afterward, the top surface of the cured PDMS layer is treated also by 1H,1H,2H,2H-perfluoro-octyltrichlorosilane. A 2 mm thick PDMS layer, which functions as a reusable transferring carrier, is then cast on top of the surface-treated PDMS film. The thin PDMS structure is peeled off (with the carrier) from the mold, and transferred to a spin-cast, blank PDMS layer of ~70 μm in thickness. Before that, the contact surfaces of the two PDMS layers are treated with a hand-held corona treater (BD-20AC, Electro-Technic Products). The corona-treated surfaces are then pressed together and left undisturbed for at least 1 hour at 85°C for the irreversible bonding to take effect. Since the bonding strength between native PDMS surfaces is higher than that between native and silanized surfaces, the cast PDMS layer can be separated from the carrier and bonded irreversibly with another structure layer.

Once the cellular PDMS structures are fabricated, surface modification and coating of the voids are further realized. A solution of 1% (v/v) 1H,1H,2H,2H-perfluorooctyltriethoxysilane, which can promote adhesion between PDMS and PTFE surfaces, is filled into the voids and heated at 100°C for 1 hour to dry. After that, a 1% PTFE solution (Teflon AF-1600, DuPont) is filled into the voids, left at room temperature for 1 hour and then sucked out by vacuum. The PTFE-treated surfaces are then heated at 100°C for 10 minutes, 155°C for 10 minutes, and finally 175°C for 20 minutes to evaporate the solvent and anneal the PTFE films. As such, smooth PTFE films of roughly 1μm thick can be deposited inside the cellular PDMS structures. To complete the fabrication of elastic piezoelectric materials, the PDMS films are coated with stretchable gold electrodes on their top and bottom surfaces, and implanted internally with charges under strong electric fields. Before sputtering gold on the surfaces, they are silanized for 3 hours in saturated vapor of 3-mercapto-propyl-trimethoxysilane (MPTMS, Aldrich), which is utilized to promote the adhesion between sputtered gold film and PDMS substrate. Once a MPTMS layer is attached, the cellular PDMS film is heated up to 150°C and sputtered with a gold film of approximately 100 nm in thickness. The PDMS film is then cooled down slowly to ambient temperature, while compressive stress in the gold film induces spontaneous wrinkling. To achieve desired anisotropicity, a thin layer of column structures can be deposited on top of the electrodes. Figure 4 shows the SEM and photograph of fabrication results. In our trials, an alternating electric field up to 35 MV/m is employed to ionize the air within the voids. Meanwhile, the charged PDMS films are heated up to 90°C throughout the micro plasma discharge process.

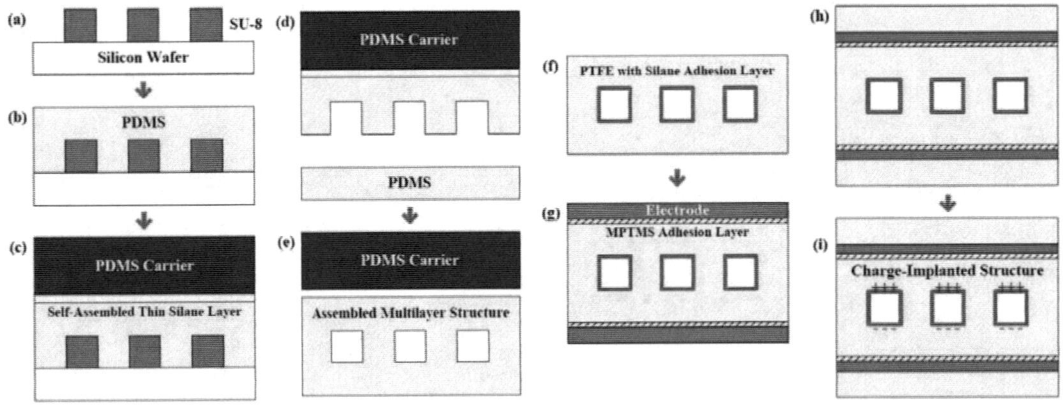

Figure 3. Schematic illustration of proposed microfabrication process for composite structures.

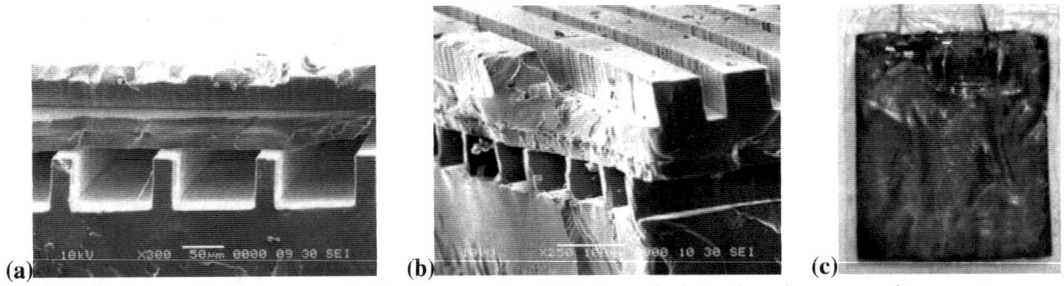

Figure 4. (a) and (b) SEM and (c) photograph of fabricated piezoelectric composite structures.

4. Results and discussion

The piezoelectricity of the fabricated and charged composite films is characterized by connecting their electrode pairs to a charge amplifier. Charge pulses (ΔQ) are converted into amplified voltage pulses (V_{out}). Figure 5(a) shows the measured d_{33} piezoelectric coefficient of a fabricated sample. The application of a 25 N force actually results in a normal strain of 10% and a charge variation of 53.6 nC. The effective d_{33} piezoelectric coefficient is estimated to be 2144 pC/N, which is >100 times higher than those of common piezoelectric polymers. When operating under 31-mode with a strain of 10%, the measured charge variation is illustrated in Figure 5(b). The working volume of the sample is 2.5cm

×2.5cm×0.3mm, and the stretch results in charge variation of 10831 pC. For the sample with a low spring constant of 220 N/m the 10% (or 2.5 mm) stretch requires only 0.63 N, so the effective d_{31} piezoelectric coefficient is estimated to be >17 nC/N. Typically, the circumferential stretch during breathing for an adult under normal conditions could be up to 2.5 cm, which is the overall elongation of 10 characterized samples connected in series. With a 0.63-N force, therefore electric charge of more than 20μC can be collected and stored per breath (or 2.5cm deformation). The experimental results are very close (with a variation less than 10%) to the values predicted by the analytical models. As such, the piezoelectric rubber bands can be mounted on human body to harvest the circumferential stretch during breathing, or the elongation during limb motion. Furthermore, the wearable piezoelectric structures can be spread, stacked and connected to charge energy storages and power micro devices.

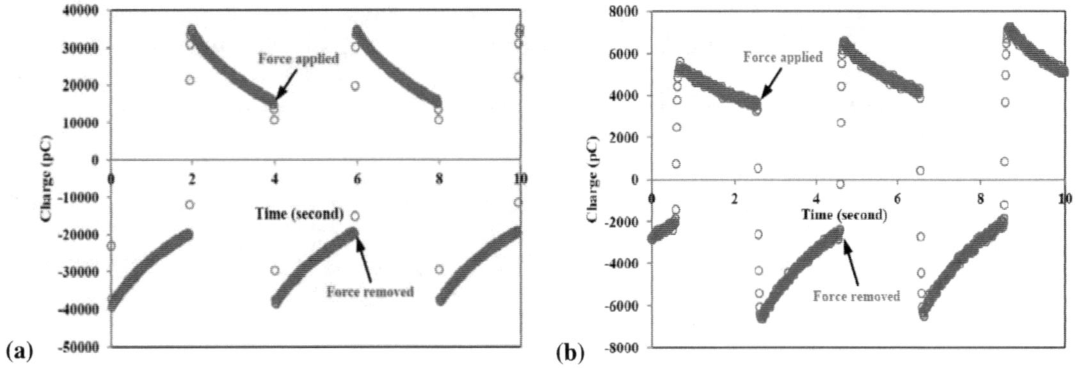

Figure 5. Measured charge variation of fabricated samples operated under (a) 33 and (b) 31 modes.

5. Conclusion

This paper presents the design optimization and microfabrication of piezoelectric rubber bands and their application in energy harvesting from human motions. Composite polymeric and metallic microstructures with embedded bipolar electric charges are employed to realize the desired stretchability and electromechanical sensitivity. In the prototype demonstration, multilayer PDMS cellular structures coated with PTFE thin films and stretchable gold electrodes are fabricated and implanted with bipolar charges internally. The composite microstructures show elasticity of 300~600 kPa and extreme piezoelectricity of d_{33} >2000 pC/N and d_{31} >200 pC/N, which are very close (with variations >10%) to the values predicted by the analytical model employed to optimize the design of composite structures. For a working volume of 2.5cm×2.5cm×0.3mm, 10% (or 2.5mm) stretch results in effective d_{31} of >17 nC/N. It is estimated that electric charge of >0.2 μC can be collected and stored per breath (or 2.5cm deformation). As such, the composite piezoelectric rubber bands (with spring constants of ~200 N/m) can be mounted on elastic waistbands to harvest the circumferential stretch during breathing, or on pads around joints to harvest the elongation during limb motion. Furthermore, the wearable structures can be spread, stacked and connected to charge energy storages and power micro devices.

6. Acknowledgements

This work was supported by the NSC of Taiwan under Contract No. 102-2221-E-007-078 and 103-3113-E-007 -006.

References

[1] D. Kim et al., "Epidermal electronics," SCIENCE, 333, pp. 838-843 (2011).
[2] N. S. Shenck and J. A. Paradiso, "Energy scavenging with shoe-mounted piezoelectrics," IEEE MICRO, 21, pp. 30-42 (2001).
[3] J. W. Tsai, J. J. Wang and Y. C. Su, "Piezoelectric rubber films for human physiological monitoring and energy harvesting," Proc. MEMS 2013, pp. 841-844.

An Innovative Controller to Increase Harvested Energy

K Makihara[1], **Y Yamamoto**[1], **C Horiguchi**[2], **H Sakaguchi**[2] **and K Fujimoto**[2]
[1]Department of Aerospace Engineering, Tohoku University, 6-6-01 Aramaki Aza Aoba, Sendai, Miyagi 980-8579, Japan
[2]Murata Manufacturing Co., Ltd., 1-10-1 Higashikoutari, Nagaokakyo, Kyoto 617-8555, Japan

E-mail: makihara@ssl.mech.tohoku.ac.jp

Abstract. This paper proposes an innovative energy-harvesting controller to increase energy harvested from vibrations. Energy harvesting is a process that removes mechanical energy from a vibrating structure, which necessarily results in damping. The damping associated with piezoelectric energy harvesting suppresses the amplitude of mechanical vibration and reduces the harvested energy. To address this critical problem, we devise an energy-harvesting controller that maintains the vibration amplitude as high as possible to increase the harvested energy. Our proposed switching controller is designed to intentionally stop the switching action intermittently. We experimentally demonstrate that the proposed control scheme successfully increases the harvested energy. The piezoelectric voltage with the proposed controller is larger than that with the original synchronized switching harvesting on inductor (SSHI) technique, which increases the harvested energy. The stored energy with our controller is up to 5.7 times greater than that with the conventional SSHI control scheme.

1. Introduction

Interest in energy-harvesting technology has been growing in recent years. Energy harvesting refers to any process that scavenges useful energy for later use from a wide variety of sources, such as wind energy, solar energy, tidal energy, piezoelectricity, thermoelectricity, and kinematic energy. Vibration-based energy harvesters convert mechanical vibration energy into electrical energy by utilizing electromagnetic materials [1], piezoelectric materials including lead zirconate titanate (PZT), macro fiber composites (MFCs), polyvinylidene difluoride (PVDF) [2, 3], and magnetostrictive materials [4]. Of these substances, here we explore the use of piezoelectric materials for energy-harvesting devices. The conventional method of energy harvesting, which uses only a diode bridge, is a simple AC-to-DC power conversion scheme [5, 6] that is commonly known as a standard rectifier. Numerous experimental and theoretical studies have been conducted with vibration-based energy harvesters that exploit piezoelectric materials. To increase the output power relative to standard energy-harvesting circuits, Guyomar *et al.* [7] proposed the technique of synchronized switch harvesting on an inductor (SSHI), in which a nonlinear electric element is appended to the conventional harvesting circuit. Specifically, their SSHI technique utilized a circuit composed of on-off switches to provide two electric states: one in which electric current flows and another in which it does not. Many proposals have been advanced and studies conducted to improve the SSHI scheme [8-10]. Badel *et al.* [8] achieved a four-fold increase in harvested energy over the standard circuit by using the SSHI circuit. Makihara *et al.* [9] used a selector switch to make the SSHI circuit more compact and more sophisticated, and enhanced harvesting performance by reducing unnecessary energy consumption. Qiu *et al.* [10] compared four different SSHI conditioning circuits and their output power equations. They proposed a simple source-less circuit to implement SSHI scheme based on the trigger circuitry design.

2. Objectives statement

Energy harvesting deprives a vibrating structure of mechanical energy, and necessarily results in damping of vibration [11]. The damping associated with piezoelectric energy harvesting suppresses the amplitude of mechanical vibration, which in turn reduces subsequently harvested energy. The reduction in harvested energy becomes clear, especially if the structural inertia or excited force is small relative to the harvesting force. From the viewpoint of vibration suppression, the reduced amplitude of the vibration is beneficial. However, in terms of energy harvesting, the reduced vibration is quite harmful because the resultant energy is also reduced. To date, energy-harvesting technologies that incorporate measures to counter the reduction of stored energy have yet to be proposed. To address this critical issue, we devise an energy-harvesting controller based on the SSHI scheme to increase the harvested energy. The SSHI scheme utilizes switching control in cases where the harvesting circuit has electric switches; the switches are controlled in synchronization with the structural vibration phase to increase the piezoelectric voltage, which eventually increases the harvested voltage. To avoid the decrease in vibration amplitude, here we manipulate the timing of switching actions for electric switches to maintain the vibration amplitude as high as possible.

3. Fabrication of energy harvester

We experimentally demonstrate that our proposed controller successfully increases the harvested energy. We fabricate a concise energy harvester comprising the following three parts: a cantilevered beam with a piezoelectric PZT patch, a switching circuit to increase PZT voltage, and a controller to govern the switching action. The cantilevered beam is made of a brass plate (38 × 8 × 0.25 mm) and a tip mass (10.3 g), as shown in figure 1. The PZT patch is attached onto the beam. The natural frequency of the beam is 18.05 Hz, with a constant-charge state.

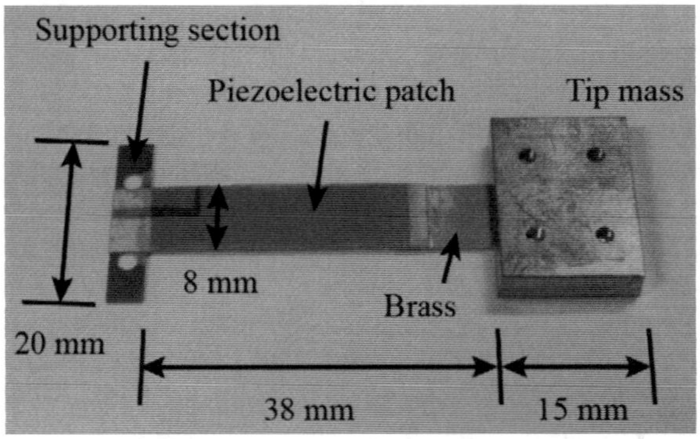

Figure 1. Energy-harvesting device.

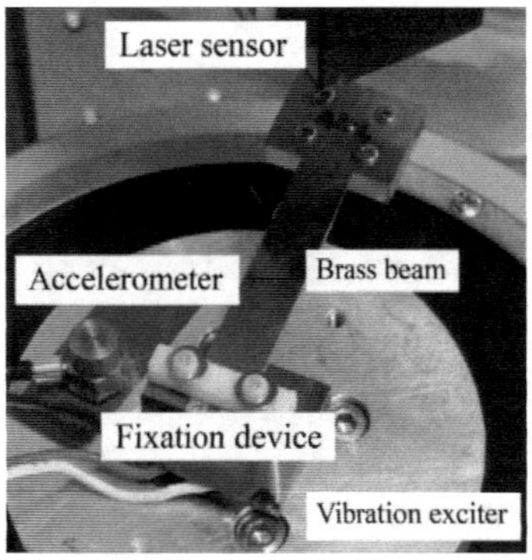

Figure 2. Experimental setup for harvesting.

Figure 2 shows the experimental setup of the concise energy harvester. The cantilevered beam is mounted on the base of a vibration exciter and is clamped with a fixation device. A function generator sends a sinusoidal wave signal for exciting the beam vibration to a vibration exciter, which moves up and down according to the signal. The motion of the exciter causes the beam to vibrate. An accelerometer is also mounted on the base to measure the vibration acceleration. It measures the base acceleration to feed it back to the function generator so that the base acceleration can be maintained at 10 mG ($0.098\,\mathrm{N/m^2}$). A laser displacement sensor measures the displacement of tip mass of the cantilevered beam. The motion of the base is negligible compared to that of the tip. The switching circuit shown in figure 3 is created to implement the SSHI scheme [7]. The circuit has a storage capacitor and two on-off switches (SW1 and SW2) to control the current direction. The circuit branch composed of the switches and the inductor boosts the value of the piezoelectric voltage with the appropriate switching action. A diode bridge comprising four diodes rectifies the boosted voltage by providing rectified current to the storage capacitor, which is connected in parallel.

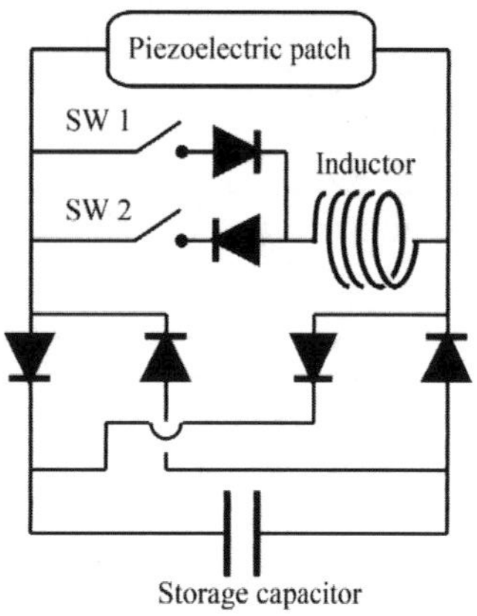

Figure 3. Switching control circuit for SSHI.

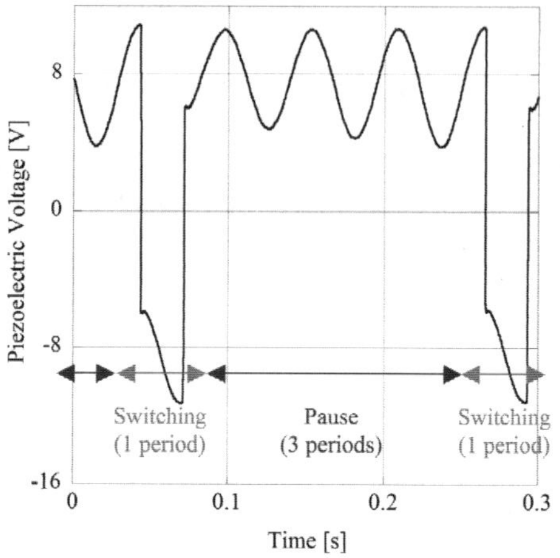

Figure 4. Concept of switching and pause durations for switching action.

4. Proposed energy-harvesting controller to increase harvested energy

According to the original SSHI scheme [7], the on-off switches are turned on every time the tip mass displacement reaches its peak value. Theoretically, the switching action at every peak of displacement results in the most efficient harvesting only when the structural inertia or excited force is large enough compared with the harvesting force. However, because the inertia of small and concise harvesters is usually small, switching at every peak suppresses the vibration and accordingly reduces the harvested energy, contrary to expectation. In other words, the switching action simultaneously imparts two effects on vibrating structures: vibration suppression and energy harvesting. Improved energy harvesting requires a new switching controller that boosts the harvesting effect and negates the suppression effect.

Our proposed switching controller is designed to intentionally stop the switching action intermittently; that is, pause durations are added to the switching durations. An example is shown in figure 4, in which the pause and switching durations are designated as three vibration periods and one vibration period, respectively. Numerous combinations of these two durations can be selected and designed. The amplitude of the vibration, which is reduced by switching, is expected to recover to its original value during the pause. This recovery of mechanical energy compensates for the vibration-suppression effect that reduces the mechanical energy. Note that our switching controller is designed so that the harvesting effect dominates the suppression effect. We have already submitted patent applications for this switching controller to contribute to the creation and development of industry [12].

5. Experimental results and discussions

We carried out experiments with the fabricated harvester to implement the proposed control scheme and investigated whether the scheme could successfully increase the piezoelectric voltage. The PZT voltage shows complex behavior caused by interference between the PZT and the storage capacitor. Therefore, the storage capacitor in the harvesting circuit was initially inactivated to focus on the behavior of the PZT voltage. Figure 5 compares the displacement of the tip mass for two switching controllers (with and without a three-period pause). The original displacement amplitude (without a pause) was $80\ \mu m$ and the maximum displacement with a three-period pause was $260\ \mu m$. The displacement with the three-period pause was larger than that without a pause. The sloped lines in figure 5 are added to easily identify the recovery of displacement during the pauses. Figure 6 shows the piezoelectric voltage history for the two switching controllers. During the switching duration, the voltage reversed from negative to positive and *vice versa*. This voltage inversion partly reflects the electrical vibration induced by the inductor and the capacitance element of the PZT. The piezoelectric voltage with the three-period pause was larger than that without the pause (the original SSHI scheme). This increase is consistent with our prediction that the new switching controller should promote recovery of the vibration amplitude. Accordingly, the voltage of the storage capacitor also increases, which increases the harvested energy.

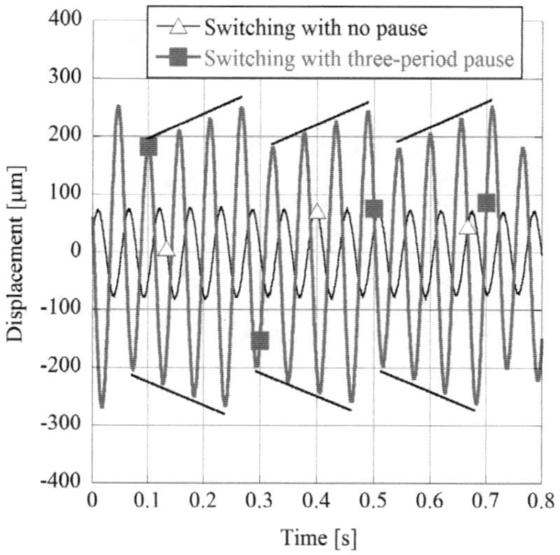

Figure 5. Tip displacement history with and without three-period pauses.

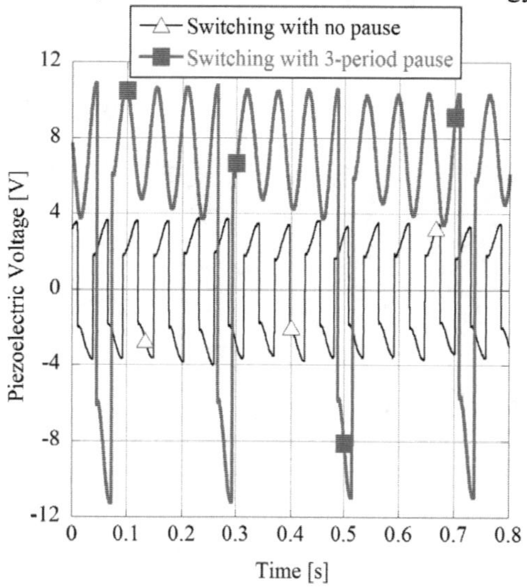

Figure 6. Piezoelectric voltage history with and without three-period pauses.

Second, by inserting a storage capacitor in the harvesting circuit as shown in figure 3, we carried out an energy-harvesting experiment to verify the proficiency of the proposed switching controller. The storage capacitance was set to $47\ \mu\text{F}$ and four types of switching controllers (no switching, no pause, one-period pause, and three-period pause) were implemented. Figure 7 compares the stored energy with the four switching controllers. The stored energy with a three-period pause was up to 2.8 times greater than that without a pause (original SSHI).

Third, we carried out extensive experiments for energy harvesting. The stored energy was normalized with respect to that of switching without a pause (original SSHI) and evaluated as a function of the number of pause periods. Figure 8 shows the normalized stored energy as a function of the number of pause periods. The controller without pause periods (number of pause periods = 0) is equivalent to the original SSHI controller proposed by Guyomar *et al.* [7], in which the switching action is activated at every peak of structural displacement. As described next, figure 8 also clearly indicates the relation between the energy-harvesting and vibration-suppression effects. Between 0 and 4 pause periods, the normalized stored energy increases as the number increases; that is, the energy-harvesting effect becomes more dominant than the suppression effect. Beyond 5 pause periods, the normalized energy decreases as the number increases; that is, the energy-harvesting mechanism of the SSHI scheme does not work well with over 5 pause periods. If the controller skips too many periods, the voltage inversion rarely occurs and the harvested energy becomes that of the standard harvester.

For our concise energy harvester, a four-period pause provides the best harvesting performance, as shown in figure 8. The stored energy with a four-period pause is up to 5.7 times larger than that without a pause (original SSHI). A factor of 5.7 is noteworthy for the future development of energy harvesters. The energy without switching (*i.e.*, standard harvesting) is indicated by the horizontal line for reference. The optimal number of pause periods for proficient harvesting may depend on the characteristics of energy harvesters, such as their mass, size, natural frequency, and PZT volume. Theoretical prediction of the optimal number will require consideration of unsteady vibrations, as seen in the sloped lines in figure 5. Investigation of this intriguing phenomenon, although promising for future work, is beyond the scope of this paper.

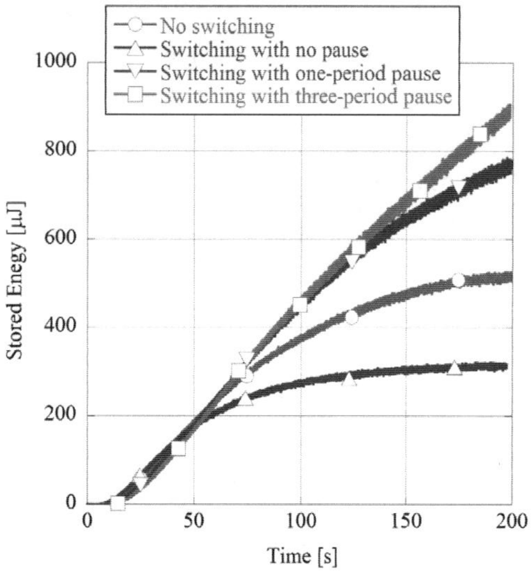

Figure 7. Stored energy history for various switching controls (Storage capacitance = 47 μF).

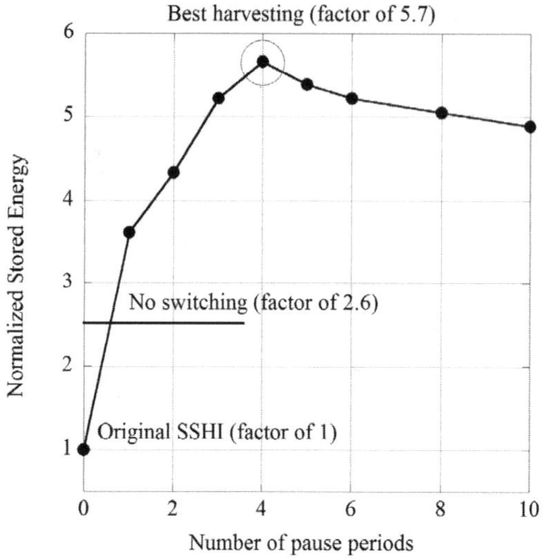

Figure 8. Normalized stored energy vs. number of pause periods.

6. Conclusions

An innovative energy-harvesting controller was proposed to increase the harvested energy from vibration. The proposed switching controller improves the original SSHI controller by intentionally stopping the switching action intermittently. Although the aim of the original SSHI scheme was to increase the harvested energy, the scheme actually suppresses the vibration amplitude and reduces the harvested output. The aim of the proposed controller was to restore the original amplitude of the vibration during the pauses. This recovery of mechanical energy was expected to boost the energy-harvesting effect and negate the vibration-suppression effect. Subsequently, we experimentally demonstrated that the proposed controller could successfully increase the harvested energy. The piezoelectric voltage with the proposed controller was greater than that with the original SSHI controller. This resulted in an increase in harvested energy. The stored energy was up to 5.7 times greater than that with the original scheme. The control principle that we proposed here can be applied to various MEMS harvesters with low voltage and power.

References

[1] Glynne-Jones P, Tudor M, J Beeby S P and White N M 2004 *Sensor Actuat. A-Phys.* **110** 344–9
[2] Stephen N G 2006 *J. Sound Vib.* **293** 409–25
[3] Shu Y C and Lien I C 2006 *Smart Mater. Struct.* **15** 1499–512
[4] Staley M E and Flatau A B 2005 *Proc. SPIE* **5764** 630–40
[5] Kazmierski T J and Beeby S 2010 *Energy Harvesting Systems: Principles Modeling and Applications* (London: Springer)
[6] Beeby S and White N 2010 *Energy Harvesting for Autonomous Systems* (London: Artech House)
[7] Guyomar D, Badel A, Lefeuvre E and Richard C 2005 *IEEE T. Ultrason. Ferr.* **52** 584–95
[8] Badel A, Guyomar D, Lefeuvre E and Richard C 2006 *J. Intel. Mat. Sys. Str.* **17** 831–9
[9] Makihara K, Onoda J and Miyakawa T 2006 *Smart Mater. Struct.* **15** 1493–8
[10] Qiu J, Jiang H, Hongli J and Kongjun Z 2009 *Front. Mech. Eng. China* **4** 153–9
[11] Lesieutre G A, Ottman G K and Hofmann H F 2003 *J. Sound Vib.* **269** 991–1001
[12] Japanese and international patent applications (No. 2013-168141, No. PCT/JP2014/063319)

Human Motion Energy Harvesting for AAL Applications

K Ylli [1,3], D Hoffmann [1], P Becker [1], A Willmann [1], B Folkmer [1] and Y Manoli [1,2]
[1] HSG-IMIT – Institute of Micromachining and Information Technology, Villingen-Schwenningen, Germany
[2] Fritz Huettinger Chair of Microelectronics, Department of Microsystems Engineering – IMTEK, University of Freiburg, Germany

E-mail: klevis.ylli@hsg-imit.de

Abstract. Research and development into the topic of ambient assisted living has led to an increasing range of devices that facilitate a person's life. The issue of the power supply of these modern mobile systems however has not been solved satisfactorily yet. In this paper a flat inductive multi-coil harvester for integration into the shoe sole is presented. The device is designed for ambient assisted living (AAL) applications and particularly to power a self-lacing shoe. The harvester exploits the horizontal swing motion of the foot to generate energy. Stacks of opposing magnets move through a number of equally spaced coils to induce a voltage. The requirement of a flat structure which can be integrated into the shoe sole is met by a reduced form factor of the magnet stack. In order to exploit the full width of the shoe sole, supporting structures are used to parallelize the harvester and therefore increase the number of active elements, i.e. magnets and coils. The development and characterization of different harvester variations is presented with the best tested design generating an average power of up to 2.14 mW at a compact device size of 75 x 41.5 x 15 mm^3 including housing.

1. Introduction

The field of electronics, including the key elements of circuits and sensors, is constantly progressing towards smaller and less power-hungry devices. This in turn has enabled the advance of mobile devices aimed at facilitating the lives of modern users on the go. Subsequently, the field of AAL applications has seen significant attention and research into body-worn systems is ever increasing. A key issue with these devices is practicality, particularly concerning their power supply which requires user intervention and maintenance when based on batteries. Advances in energy harvesting are enabling body-worn systems operating without a battery and need for maintenance as the energy to power the device is extracted directly from the human motion.

The motion of the human body consists of smooth continuous movements at a repetition frequency of approximately 1 Hz. Hence classical vibrational energy harvesting devices are not feasible for this excitation source. Instead large-amplitude devices that do not require constant vibration are being developed. The motion of the foot is amongst the most energetic of the human body as indicated in Figure 1, leading to the development of several different energy harvesting devices for the integration into the shoe sole [1–4]. As an additional advantage the shoe sole offers sufficient space to unobtrusively integrate an energy harvester.

[3] To whom any correspondence should be addressed.

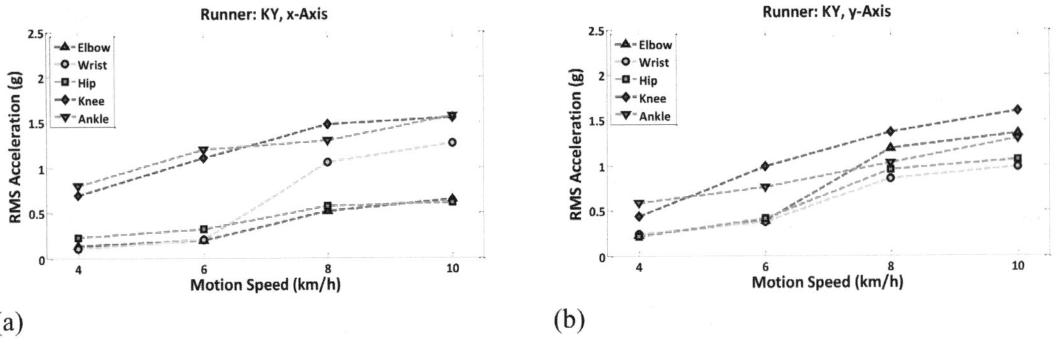

Figure 1. RMS acceleration calculated from a measurement of 60s at varying body positions. At a standstill with limbs pointing downwards, accelerations on the x-axis in **(a)** correspond to a forward (horizontal) motion and on the y-axis in **(b)** to a vertical motion.

This work is focused on the development of a linear motion harvester that exploits the swing motion of the foot during walking. Continuing the work presented in [3], the multi-coil and multi-magnet system seeks to make better use of the width available in the shoe sole. Different variations of the harvester are investigated and the key parameter to effect the generated power is shown.

2. Harvester development

In this work a multi-coil setup with a magnet stack of opposing magnets separated by steel spacers is developed. In order to meet the requirements of a flat device (height limited to 15mm), the radius of the cylindrical magnets and coils is restricted. However there clearly is more space in the width of the sole than in the height. To make use of this fact, the magnet stack is parallelised. The attractive force between magnets would cause high friction during magnet motion as parallel stacks would be pulled toward each other. To avoid this, supporting brackets are introduced which hold the stack separate at a fixed distance, thus forming a single seismic mass as shown in Figure 2.

An additional advantage of this setup is the fact that the magnet stacks can be placed in an orientation in which they simultaneously induce a signal through all coils, as they move synchronously (Figure 3). Therefore, a simple rectifying circuit is sufficient to produce a DC voltage.

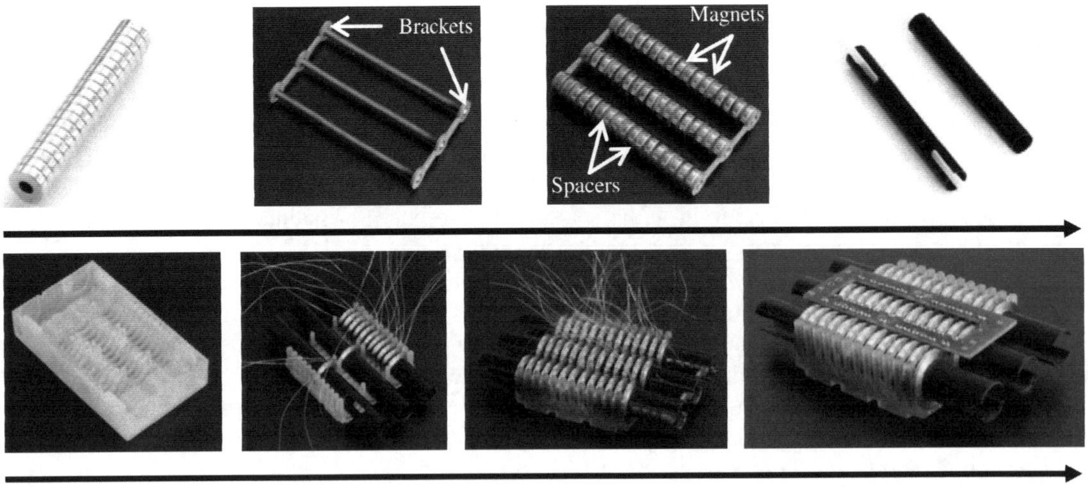

Figure 2. Depiction of the fabrication process from top left to bottom right. Placement of magnets, steel spacers, supporting steel brackets and coils in a 3D-printed housing is shown.

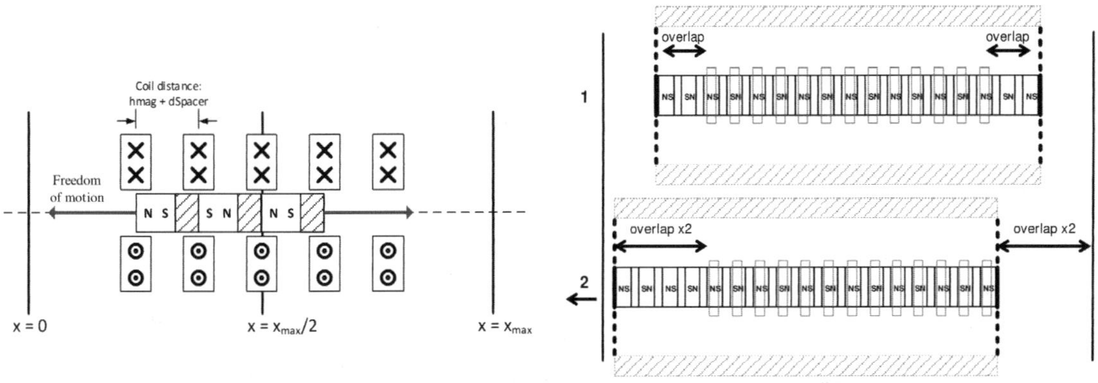

(a) (b)

Figure 3. (a) Schematic of the magnet setup. Coils are placed at intervals corresponding to the magnet height and spacer thickness to generate a synchronous signal. **(b)** Two key magnet positions. The brackets interconnecting the magnet stacks are indicated by a thick line. Coils shown in red.

The modelling of the system is done according to the work presented in [4]. The magnetic flux density around a magnet stack is extracted from an FEM simulation, providing a set of data points for positions within the considered extent of the harvester. From this set of data a coupling curve is calculated, which is used for the system model. Depending on the current position of the magnet stack, the respective coupling coefficient is extracted from a look-up table and is used to determine the voltage induced within the coils by multiplying the coupling coefficient with the velocity of the magnet stack as it moves forwards and backwards. The voltages are summed up across all coils to produce the system output.

3. Implementation and experiments

3.1. Implementation

The device was fabricated using neodymium iron boron ring magnets (NdFeB) with a height of 2 mm, an inner diameter of 2.7 mm and an outer diameter of 7 mm. The coils were fabricated with an inner diameter of 8.5 mm, an outer diameter of 12.5 mm, a thickness of 1.2 mm and 100 μm wire leading to a resistance of approximately 13 Ω per coil. The dimensions of the complete device including housing are: 75 x 41.5 x 15 mm^3 and considering only the active volume: 71 x 37.5 x 12.5 mm^3.

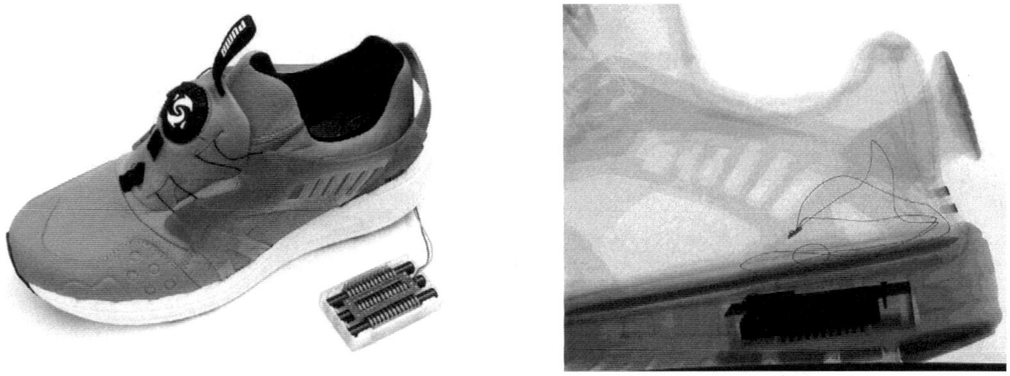

(a) (b)

Figure 4. (a) Sports shoe with integrated harvester and a second device by its side. **(b)** X-ray photography of the shoe showing the harvester system. The temperature sensor and the wireless data transmission chip from the sample application can be seen.

Table 1. Key geometrical parameters of the three investigated harvester configurations.

Configuration	Number of coils	Coil span	Number of magnets	Magnet length (incl. spacers)	Housing	Max. range of motion
V5	13	40.8 mm	17 x 2 mm	52 mm	70 x 41.5 x 15 mm^3	61.2 mm
V6	11	34.8 mm	17 x 2 mm	52 mm	70 x 41.5 x 15 mm^3	67.2 mm
V7	9	28.8 mm	17 x 2 mm	52 mm	75 x 41.5 x 15 mm^3	73.2 mm

Figure 4 (a) shows a sports shoe with a harvester system integrated into the heel part of the shoe sole, as well as a second device by its side. Apart from a simple power management circuit, a temperature sensor and a wireless transmitter are included for demonstration purposes.

In order to investigate the effect of the geometry of the magnet stack on the power output, three variations according to Table 1 were fabricated. The maximum range of motion is derived from the coil span and the overlap of a magnet stack as shown in Figure 3 (b).

3.2. Experimental data

Due to the interconnecting brackets a larger number of magnets is required for the three-stack setup than would be required otherwise. The motion of the magnet stack is stopped at the outer coils because of the brackets (see Figure 3 (b)) and therefore the number of coils has a direct influence on the power output.

Figure 5 shows the summarized results of measurements on a treadmill. Two different runners were recorded for 60 seconds at runs of 4, 6, 8 and 10 km/h. The average power for each motion speed is calculated from the measured voltage with an optimal load resistance equal to the total coil resistance.

It is instantly clear that the device with the smallest number of coils provides the best power output. In this case the motion range is the largest, which enables the magnet stacks to gain momentum and ultimately reach a higher velocity as they move through the coils.

Depending on the runner, the 9-coil harvester can generate an average of 1 mW at the slowest walking speed of 4 km/h. A maximum average power output of 2.14 mW was measured at a speed of 10 km/h. Considering the total volume (including housing) of 46.69 cm^3 and the active volume of 33.28 cm^3 this translates to a maximum power density of 45.8 $\mu W/cm^3$ and 64.3 $\mu W/cm^3$ respectively.

4. Conclusion

The integration of a harvesting device into a shoe sole is mainly limited by the height constraint. The height of the sole cannot fully be used as material is needed for cushioning. In order to maintain the level of comfort the users are familiar with, the proposed harvester was designed with a height

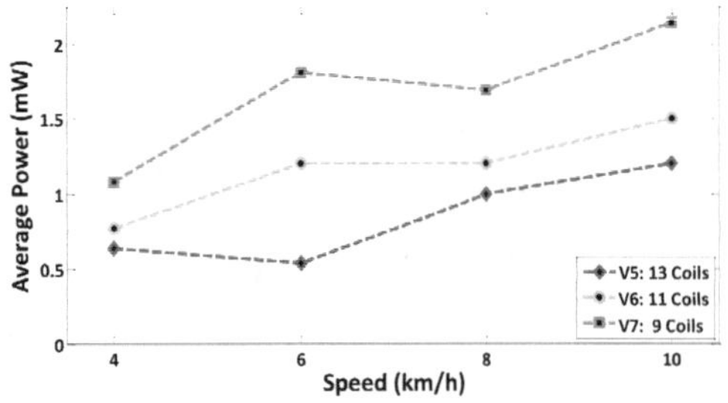

a)

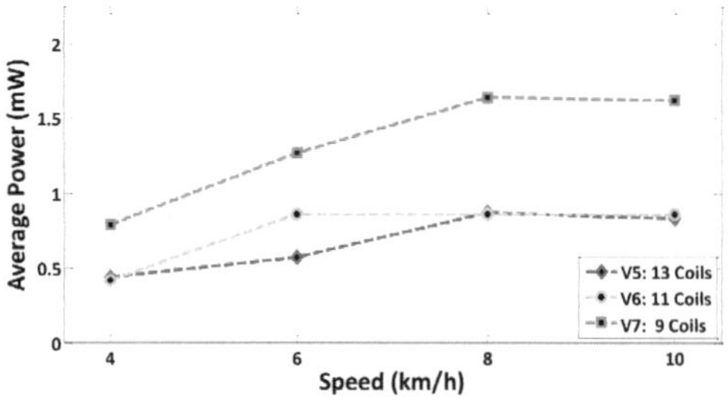

b)

Figure 5. Average power generated during treadmill runs for different configurations of the harvester. Runner KY is shown in pane **a)** and runner DH in pane **b)**.

constraint of 15 mm. The consequential reduction of the size of the magnets and coils is partially compensated by exploiting the width available in the shoe sole and parallelizing the moving magnet stacks. The best configuration out of the three fabricated ones is easily distinguished and is able to generate average powers between 1 mW and 2.14 mW depending on the walking speed.

Compared to previous work presented in [3], the triple-stack device shows a significant increase in power output. However an improvement by a factor of three was not observed. This is attributed to the limited motion range of the bracket-supported magnet stacks. In order to increase the range of motion, the length of the magnet stack must be increased, otherwise the brackets stop the stack at the coils, before the magnets reach the channel end. This in turn increases the length of the harvester housing, which should be avoided with respect to integratability.

Although the freedom of motion of the moving magnet stack was highlighted as a key factor in this work, the geometrical parameters of the setup can still be optimised further. In continuation of this work, the ongoing improvement of the simulation model is in the centre of focus. The model is being modified to be fully flexible regarding the geometrical parameters of the system and is expected to provide a new set of parameters with an improved power output.

Acknowledgments

This research (IGF-Project: 17742N) was supported by the "Programm zur Förderung der industriellen Gemeinschaftsforschung und -entwicklung (IGF, Germany)". The authors would also like to acknowledge the assistance of the "EasyFitness" gym in Villingen-Schwenningen which provided access to gym equipment.

References

[1] D. Hoffmann, B. Folkmer, and Y. Manoli, "Human Motion Energy Harvester for Biometric Data Monitoring," *J. Phys.: Conf. Ser,* vol. 476, p. 12103, 2013.

[2] D. Carroll and M. Duffy, "Modelling, design, and testing of an electromagnetic power generator optimized for integration into shoes," *Proceedings of the Institution of Mechanical Engineers, Part I: Journal of Systems and Control Engineering*, vol. 226, no. 2, pp. 256–270, 2012.

[3] K. Ylli, D. Hoffmann, P. Becker, A. Willmann, B. Folkmer, and Y. Manoli, "Energy Harvesting From Human Motion: Exploiting swing and shock excitations," *Smart Mater. Struct.* (pending).

[4] K. Ylli, D. Hoffmann, B. Folkmer, and Y. Manoli, "Design, fabrication and characterization of an inductive human motion energy harvester for application in shoes," *J. Phys.: Conf. Ser,* vol. 476, p. 12012, 2013.

Powering a Commercial Datalogger by Energy Harvesting from Generated Aeroacoustic Noise

R Monthéard[1,2], C Airiau[3], M Bafleur[1,4], V Boitier[1,5], J-M Dilhac[1,2], X Dollat[1,4], N Nolhier[1,5], E Piot[6]

[1] CNRS, LAAS, 7 avenue du colonel Roche, F-31400 Toulouse, France
[2] Univ de Toulouse, INSA, F-31400 Toulouse, France
[3] IMFT, Toulouse, France
[4] Univ de Toulouse, LAAS, F-31400 Toulouse, France
[5] Univ de Toulouse, UPS, LAAS, F-31400 Toulouse, France
[6] ONERA – The French Aerospace Lab, Toulouse, France

E-mail : dilhac@laas.fr

Abstract. This paper reports the experimental demonstration of a wireless sensor node only powered by an aeroacoustic energy harvesting device, meant to be installed on an aircraft outside skin. New results related to the physical characterization of the energy conversion process are presented. Optimized interface electronics has been designed, which allows demonstrating aeroacoustic power generation by supplying a commercial wireless datalogger in conditions representative of an actual flight.

1. Introduction

In recent years, Wireless Sensor Networks (WSN) have been considered for various aeronautical applications, including Structural Health Monitoring and flight tests [1]. Isolated from the onboard electrical network, each sensor node needs to be self-powered. Environmental concerns, together with economical and safety-related issues, often prohibit the use of electrochemical batteries. As an alternative solution, energy may be harvested in the sensor's environment, providing a reliable and sustainable electrical power source.

The energy resource in an aircraft environment is diverse yet non-uniformly distributed, which leaves room for investigating various ambient energy harvesting techniques. Motivated by the opportunity to propose an alternative and complementary solution to those already under investigation (e.g. photovoltaics, vibrations, thermal gradients, ...) [2], the goal of this work is to exploit the abundant resource associated with the relative wind.

A very well-known technique for generating electrical energy from airflow relies on the use of a wind turbine. Whereas such a device may be efficient at relatively large scale, its performance decays greatly with downscaling [3]. Alternative airflow energy harvesters have been proposed in recent years, following different approaches [4][5]. One method is based on the aeroacoustic phenomenon, which implies that a membrane is driven into vibration by acoustic waves that result from the interaction of the airflow with a particular geometry. A few contributions following this approach have been proposed recently, in which the authors employ a Helmholtz acoustic resonator, fitted with an electromagnetic [6] or a piezolectric [7] microgenerator. In [7], Matova et al. achieved 2 μW at 13 m/s airflow velocity, using a mesoscale prototype.

In this work, the aeroacoustic transduction process is investigated, with a focus on demonstrating the potential of such a technique to power a wireless sensor. Unlike earlier work, the target output power lies in the milliwatt range. Additionally, significantly higher airflow speeds are considered, as dictated by typical civil aircraft cruise speeds.

2. Aeroacoustic phenomenon

The phenomenon of cavity noise is linked to self-sustained coherent oscillations in the shear layer developing above a cavity in the presence of grazing flow. A complex feedback process can lead to important radiated noise, with both discrete and broadband components [8]. The basic geometry considered in this work is a 2-D rectangular cavity with length L and depth D, as depicted in figure 1, and figure 2 shows typical generated acoustic spectra for different airflow speeds.

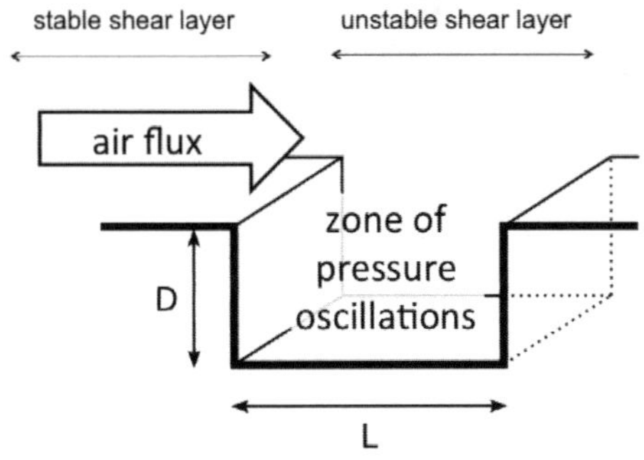

Figure 1. Schematic of a rectangular cavity of length L and depth D.

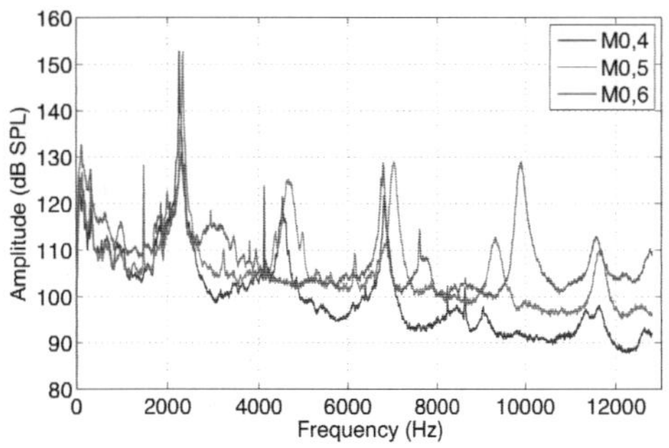

Figure 2. Generated acoustic spectra at different airflow speeds (Mach 0,4 to Mach 0,6).

Peaks in the oscillation spectrum can be empirically forecasted using Rossiter's formula [8]. Although based on a semi-empirical model, it provides a fairly accurate estimate of generated frequencies. However, amplitude and direction of the radiated noise must be determined numerically or experimentally only. Additionally, the dominant frequency can be determined according to East analysis [8], depending on the cavity's aspect ratio.

In previous work [10], experiments have been realized using multiple cavity dimensions and aspect ratios, at different airflow speeds. Experimental results confirmed Rossiter's model and East's law, and helped in the process of sizing the device described hereafter.

3. Device design and characterization

The aeroacoustic energy transducer used in this work is made of a rectangular cavity fitted with a piezoelectric membrane at its bottom (see figure 3). Table 1 gives detailed information on dimensions and references. Preliminary measurements reported in [10] show that the dominant acoustic tone produced by this device is situated at about 2300 Hz, with a sound pressure level over 152 dB.

Experiments were carried out in the aeroacoustic test bench "B2A wind tunnel" of ONERA Toulouse. This 4-meter long stainless steel tube of cross-section 50 mm x 50 mm is equipped with a quasi-anechoic outlet, to limit the amplitude of reflected acoustic waves.

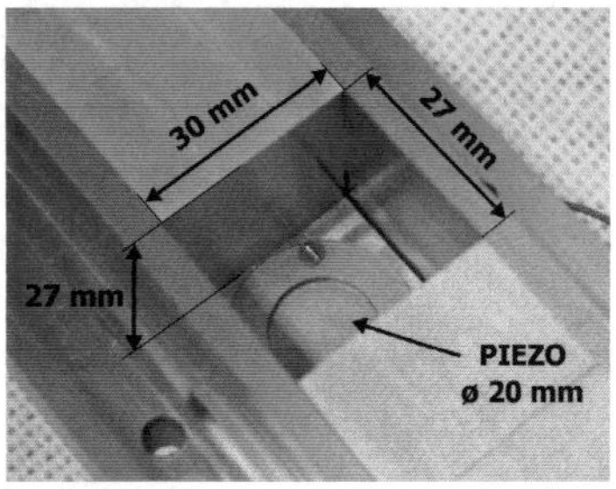

Figure 3. Fabricated aeroacoustic energy harvester.

Table 1. Dimensions of the fabricated device.

Cavity	
Length (mm)	27
Depth (mm)	27
Width (mm)	30
Piezoelectric membrane (Murata 7BB-20-6)	
Diameter (mm)	20
Thickness (μm)	420

The transducer has been flush-mounted in the wind tunnel and submitted to a grazing flow up to Mach 0.6. The AC voltage has been measured across a variable resistor directly connected to the piezoelectric membrane. Figure 4 shows the evolution of the RMS power dissipated in the load resistor with respect to the load value, for different airflow speeds. It can be seen that this energy harvester is able to deliver up to 2 mW AC power at Mach 0.5; the corresponding voltage amplitude reaches 8 Volts. Furthermore, the optimal load resistor varies very little with respect to the airflow speed. Nevertheless, as depicted in figure 5, the maximum output power significantly depends on the airflow speed.

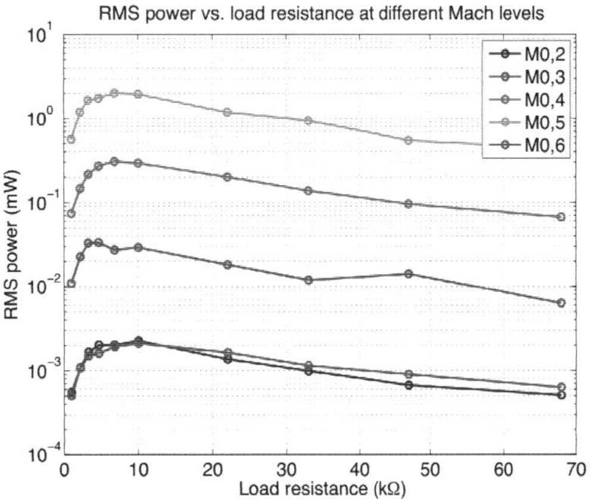

Figure 4. RMS output power vs. load resistance

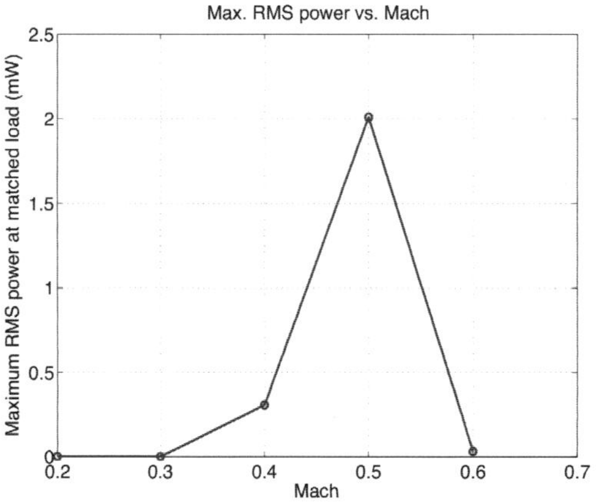

Figure 5. Max. output power vs. airflow speed

4. Power management electronics

4.1. Architecture

The power management circuit interfaces the piezoelectric membrane with the wireless sensor node. It comprises two main stages, an AC/DC converter and a DC/DC converter, followed by a storage stage based on supercapacitors as well as regulation and protection circuitry, as depicted in figure 6.

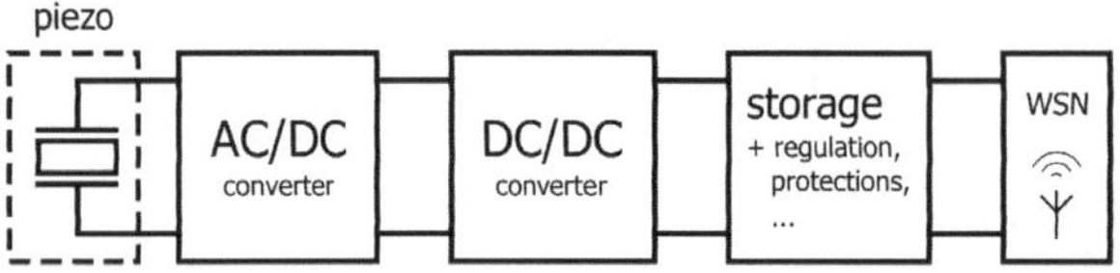

Figure 6. General architecture of the power management circuit

4.2. AC/DC converter
The first stage is used to rectify the AC voltage generated by the piezoelectric membrane. Reviews in piezoelectric power conditioning circuits highlight the performance of the SSHI (Synchronized Switch Harvesting on Inductor) interface [11][12], especially for pseudo-sinusoidal excitations. In this work, a self-powered parallel SSHI circuit has been implemented. It has been experimentally verified that the SSHI circuit allows doubling the output power compared with a standard, diode bridge rectifier and in the meantime, yields a fourfold increase of the output impedance.

4.3. DC/DC converter
The input impedance of the DC/DC converter must be controlled so as to match the output impedance of the preceding stage, in order to achieve maximum power transfer from the energy harvester. As stated previously, the optimal load varies very little with the airflow speed, which allows the use of fixed input impedance topology. Therefore, it has been chosen to implement a buck-boost converter operating in discontinuous conduction mode, as proposed in [13][14]. This converter has been designed to deal with input powers in the milliwatt range. Efficiency reaches up to 87% and is over 70% over a wide range of output currents.

4.4. Characterization
Experiments have been carried out to validate the positive impact of each stage, at an airflow speed of Mach 0.5. Figure 7 illustrates the charging of a 100 μF capacitor using different circuit configurations. It confirms the significant improvement brought by the combination of the SSHI interface and the DC/DC converter, the calculated mean output power is indeed more than six times higher over the first 20-second time interval. A picture of the realized printed circuit board is shown on figure

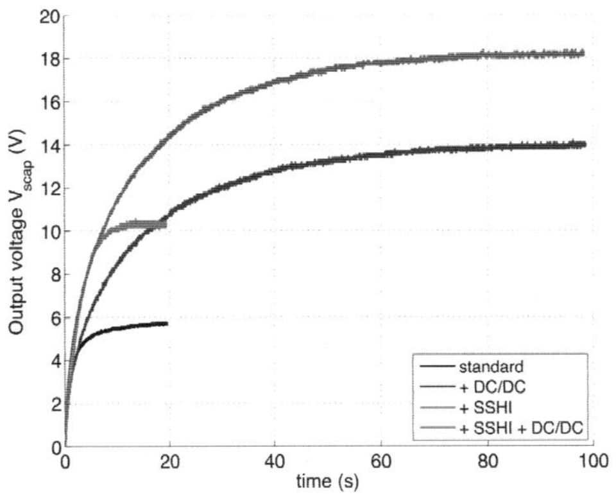

Figure 7. Charge voltage across a 100 μF capacitor, for different circuit configurations

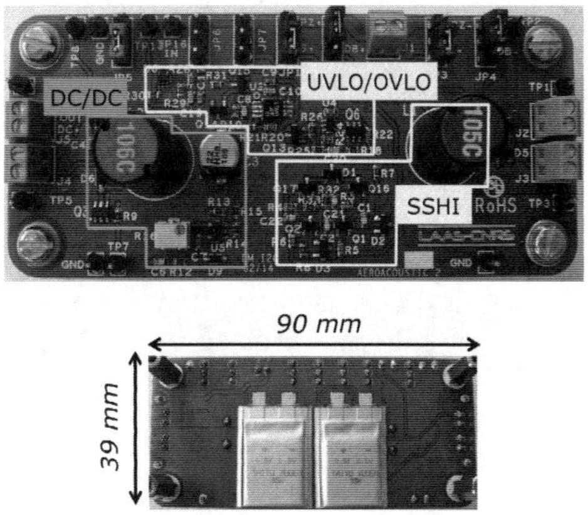

Figure 8. Top and bottom view of the printed circuit board

5. Full system test and validation
The full system includes the architecture as depicted in figure 6, that is, the DC/DC converter connected to two 0.5 F supercapacitors in series, together with an active charge balancing circuit, as well as over- and under-voltage lockout circuits. These protection stages only draw 1-3 μA so as to preserve high efficiency.

This energy harvesting system has been submitted to a Mach 0.5 airflow and used to power a wireless sensor node based on a Jennic JN5148 microcontroller[1]. Set up to measure the temperature and send the information over a 2.4 Ghz channel, once every 6 s, this module consumes a mean power

[1] Note that a video of the experiment can be iewed at the following address: http://youtu.be/1kkq-5g9D_g

of 181 μW. Figure 9 shows the evolution of the output voltage, which is applied to the datalogger. Initially fully discharged, the system initializes within 42 minutes. Because of the 0.5 J energy consumption at activation, a large capacitance and therefore a long charging time is needed. After a successful startup, the datalogger is activated whereas the storage supercapacitors keep on charging. As the output voltage reaches its maximum value, the system enters regulation. The mean charging power is about 0.8 mW.

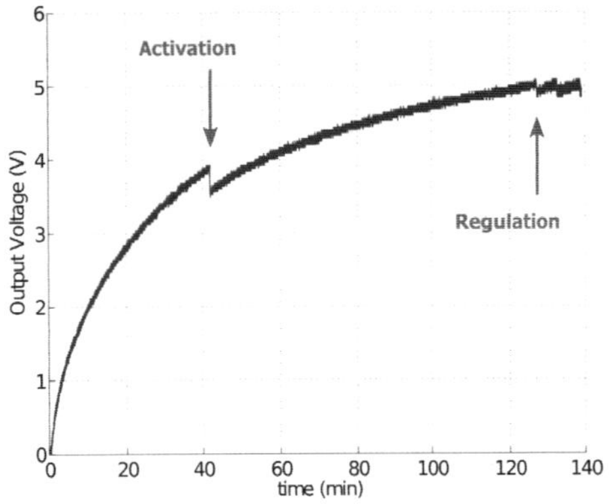

Figure 9. Full system output voltage

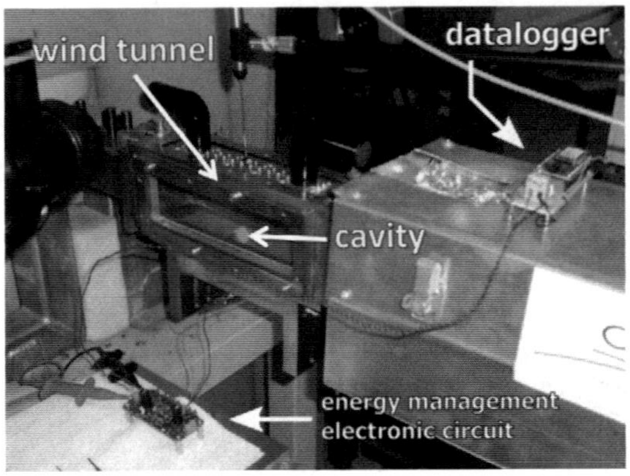

Figure 10. Picture of the experimental bench

6. Conclusion and perspectives

This work has led to the demonstration of a realistic, fully standalone system, made energy-autonomous by means of aeroacoustic energy generation. Future work will focus on refining the transducer design so as to improve its effectiveness over a wider range of airflow speeds, as well as designing a smaller scale prototype.

7. Acknowledgements

This work has been supported by the French government program *Investissements d'Avenir*, the French Defense Procurement Agency (DGA) and Intesens company.

References

[1] Dilhac JM *et al* 2009 *Proc. Int. Workshop on Structural Health Monitoring* (Stanford, USA)
[2] Dilhac JM and Bafleur M 2014 *IEEE Aerospace and Electronic Systems Magazine* **29** pp 18–22
[3] Mitcheson P *et al* 2008 *Proceedings of the IEEE* **96** pp 1457–1486
[4] Li S *et al* 2011 *Journal of Applied Physics* **109** 026104
[5] Humdinger wind energy, www.humdingerwind.com
[6] Kim SH *et al* 2009 *Journal of Micromechanics and Microengineering* **19** 094010
[7] Matova S *et al* 2010 *Proc. PowerMEMS 2010* (Leuven, Belgium)
[8] Rowley C W and Williams D R 2006 *Annual Review of Fluid Mechanics* **38** pp 251–276
[9] East L 1966 *Journal of Sound and Vibration* **3** pp 277–287
[10] Monthéard R *et al* 2012 *Proc. PowerMEMS 2012* (Atlanta, USA)
[11] Guyomar D and Lallart M 2011 *Micromachines* **2** pp 274–294
[12] Szarka G D *et al* 2012 *IEEE Transactions on Power Electronics* **27** pp 803–815
[13] Kasyap A *et al* 2002 *Proc. 9th Int. Congress on Sound and Vibration* (Orlando, USA) 271
[14] Kong N *et al* 2010 *Journal of Intelligent Materials Systems and Structures* **21** pp 1293–1302

A Skin-attachable Flexible Piezoelectric Pulse Wave Energy Harvester

Sunghyun Yoon and Young-Ho Cho

NanoSentuating Systems Laboratory, Cell Bench Research Center,
Korea Advanced Institute of Science and Technology (KAIST), Republic of Korea

E-mail: nanosys@kaist.ac.kr

Abstract. We present a flexible piezoelectric generator, capable to harvest energy from human arterial pulse wave on the human wrist. Special features and advantages of the flexible piezoelectric generator include the multi-layer device design with contact windows and the simple fabrication process for the higher flexibility with the better energy harvesting efficiency. We have demonstrated the design effectiveness and the process simplicity of our skin-attachable flexible piezoelectric pulse wave energy harvester, composed of the sensitive P(VDF-TrFE) piezoelectric layer on the flexible polyimide support layer with windows. We experimentally characterize and demonstrate the energy harvesting capability of 0.2~1.0μW in the Human heart rate range on the skin contact area of 3.71cm^2. Additional physiological and/or vital signal monitoring devices can be fabricated and integrated on the skin attachable flexible generator, covered by an insulation layer; thus demonstrating the potentials and advantages of the present device for such applications to the flexible multi-functional self-powered artificial skins, capable to detect physiological and/or vital signals on Human skin using the energy harvested from arterial pulse waves.

1. Introduction

In recent years, continuous physiological or vital signal monitoring in daily life has become conspicuously prevalent in wearable devices. Wearable devices for physiological or vital signal monitoring need power supply for the stable measurement of Human signals, such as arterial pulse wave, skin temperature, and skin conductance signals. Among the Human signals, the arterial pulsation has been considered as an energy harvesting power source for the wearable devices. The conventional energy harvesting from arterial pulsation was either invasive or non-invasive. The invasive devices [1] need surgery for device implant. Among the conventional non-invasive devices, the piezoelectric energy harvesting [2] is promising due to device simplicity. The piezoelectric energy harvesting devices, however, need rigid layers to support piezoelectric material layers; thus, reducing the device flexibility required for better energy harvesting efficiency through conformal skin contact. In this paper, we propose a non-invasive skin-attachable flexible piezoelectric generator (figure 1), where a silver electrode layer is sandwiched between a P(VDF-TrFE) piezoelectric layer and a polyimide support layer with windows for improving device flexibility, skin conformability and energy harvesting efficiency.

The originality and novelty of our device include: 1) the polyimide support layer design with the windows; 2) the simple fabrication process employing the multi-purpose silver interlayer between the piezoelectric layer and the support layer. The windows in the polyimide support layer are intended not

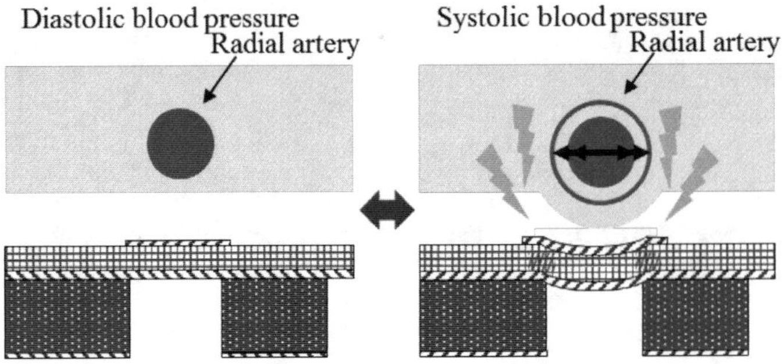

Figure 1. A skin-attachable flexible piezoelectric generator, harvesting energy from heart-beat pulse waves on Human wrist.

only to increase the sensitivity of electric generation from the arterial pulsation, but also to enhance the device flexibility and skin conformability on Human wrist. The purpose and functions of the silver electrode layer (figure 2) are three-fold: 1) an etch stop layer for the wet etching process using KOH, ethanol, and DI to pattern windows on the polyimide support layer; 2) an electrode layer for piezoelectric poling process; 3) an electrode layer for piezoelectric energy harvesting. The idea of triple-purpose silver electrode layer enables us to fabricate the flexible piezoelectric generators with the simple process of figure 3. In addition, the flexible generator devices are covered by the insulation layers, on which additional devices can be fabricated or integrated. The present skin-attachable flexible piezoelectric pulse wave energy harvester has potentials for applications to the flexible multi-functional self-powered artificial skins, capable to detect physiological and/or vital signals on Human skin using the energy harvested from arterial pulse waves.

2. Energy Harvesting Principle and Device Design

The skin-attachable flexible piezoelectric generator is comprised of a silver electrode layer sandwiched between the P(VDF-TrFE) piezoelectric layer and the polyimide support layer with the windows. Figure 1 illustrated the principle of energy harvesting from the arterial pulsation on Human wrist. Due to the systolic blood pressure, the radial artery expends and applies mechanical stress to the P(VDF-TrFE) piezoelectric layer, supported by the polyimide support layer; thus, inducing electric potential across the piezoelectric layer.

We design the skin-attachable flexible piezoelectric generator (figure 2) to have the skin contact area of 25mm x 15mm with the thickness of 70μm. The square windows in the polyimide support layer are designed in the size of 6mm x 6mm, which is large enough to cover the radial artery in the diameter of 3~5mm. We design the thickness of the piezoelectric P(VDF-TrFE) layer (figure 2) as 20μm, which is the feasible thickness [3] for poling process.

3. Fabrication

Figure 3 indicates the fabrication process of the skin-attachable flexible piezoelectric generator. A 50μm-thick polyimide film (Kapton® HN, Dupont™) is attached on the PDMS-coated silicon wafer, followed by the thermal evaporation of 0.5μm-thick silver on the polyimide film. We spin-coat the 75:25 P(VDF-TrFE) (Measurement Specialties) and Methyl Ethyl Ketone (MEK, Sigma Aldrich) solutions of 35wt% to form a 20μm-thick piezoelectric layer on the evaporated silver. We thermally evaporate 0.3μm-thick aluminum layer and patterned the aluminum layer to define the electrodes for piezoelectric energy harvesting. We detach the polyimide film from the surface of the PDMS-coated silicon wafer and attach the electrode side of the detached film to the PDMS-coated silicon wafer.

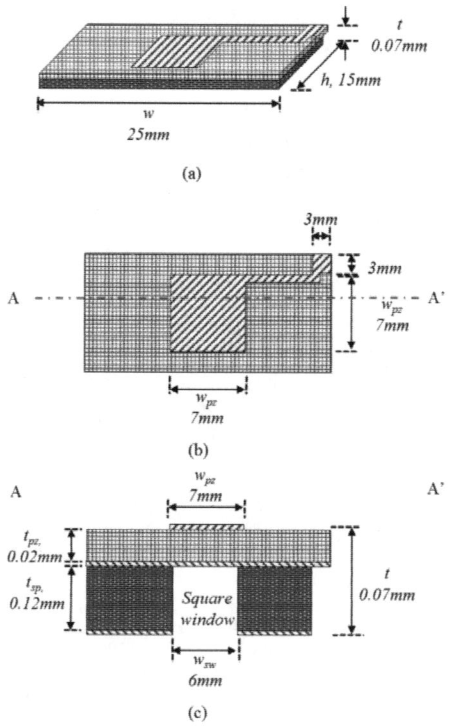

Figure 2. The skin-attachable flexible piezoelectric generator: (a) overall structure; (b) top view; (c) cross-sectional view along A-A' of figure 2b.

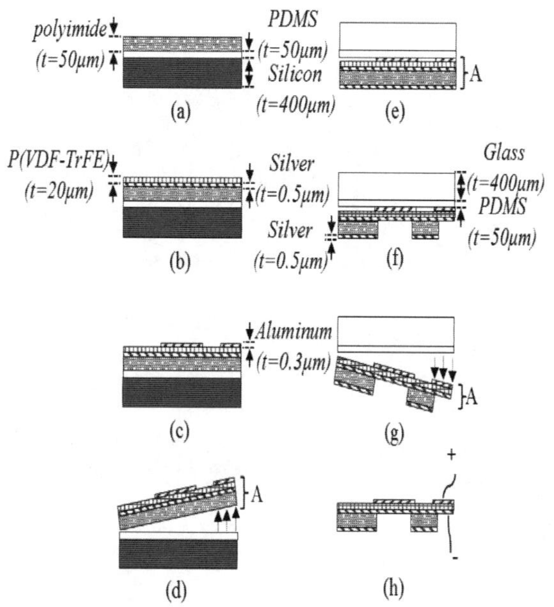

Figure 3. Fabrication process of the skin-attachable flexible piezoelectric generator.

We etch the polyimide film using the KOH, Ethanol, and DI (1:8:2) etchant [4] with the silver shadow mask for 20 minutes, followed by the 21-minute DI water dipping process to remove the high viscous residue; thus, forming the square windows through the polyimide supporting layer and the interlayer electrode on the P(VDF-TrFE) layer. Finally the fabricate device is released from the surface of the PDMS-coated silicon wafer to obtain the skin-attachable flexible piezoelectric generator.

4. Test Methods and Experimental Results

The fabricated skin-attachable flexible generator (figure 4a), having the skin contact area of 24.7mm x 15.0mm, the thickness of 70μm and the weight of 49.5mg, has been characterized by the PDMS artificial pressure actuator (figure 4b), simulating the 4mm-wide human radial artery producing Human arterial pulse waves.

Figure 5 illustrates the output response of the fabricated device for the input pressure pulse of 40mmHg (which is the human systolic and diastolic arterial pressure difference of resting period) for the Human heart rate, varying in the range of 50~220 BPM. We measure the generated energy from the flexible piezoelectric generator in terms of the RMS values of the time-dependent output voltage across 10MΩ resistor. From the 40mmHg pulse waves on the skin contact area of 3.71cm^2, the flexible piezoelectric generator is experimentally harvest the energy of 0.20±0.02μW at 50BPM, 0.70±0.05μW at 135BPM, and 1.00±0.03μW at 220BPM, respectively.

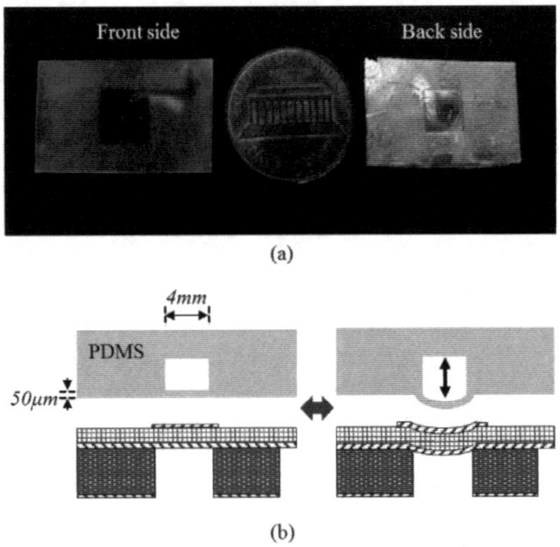

Figure 4. The fabricated flexible piezoelectric generator, (a) compared to US penny and (b) attached to an artificial pulse wave device for experimental energy harvesting performance characterization

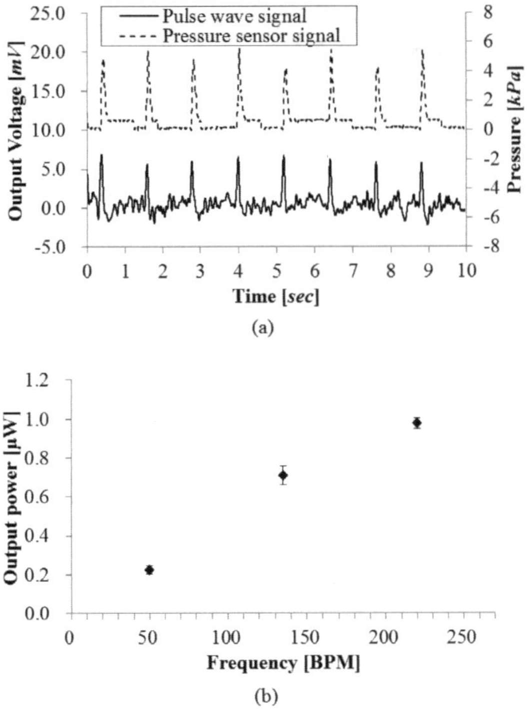

Figure 5. Experimental performance of the skin-attachable flexible piezoelectric generator for the pulse wave pressure of 40mmHg: (a) the time-dependent output voltage of the generator at the pulse wave frequency of 50BPM; (b) the frequency-dependent output power of the generator at 50, 135 and 220 BPM, respectively.

5. Conclusion

We have designed, fabricated, and characterized the skin-attachable flexible piezoelectric generator, composed of the flexible polyimide support layer with windows and the sensitive P(VDF-TrFE) piezoelectric layer, for harvesting energy from human arterial pulsation on Human wrist, We experimentally demonstrate the energy harvesting performance of the device in the range of 0.2~1.0μW from the small skin contact area of 3.71cm^2, where the pressure of 40mmHg (which is the difference of Human systolic arterial pressure and diastolic arterial pressure at Human resting period) is applied at the Human heart rate range of 50~220 BPM. The skin-attachable flexible generator is covered by the insulation layer, on which additional devices can be fabricated and integrated to measure Human physiological and/or vital signals. The flexible piezoelectric generator, capable to collect energy non-invasively from the arterial pulsation on Human wrist, has improved wearing comfort with the higher flexibility and the better conformal skin contact; thereby, demonstrating potentials for applications to the flexible multi-functional self-powered artificial skins, capable to detect physiological and/or vital signals on Human skin using the energy harvested from arterial pulse waves.

Acknowledgements
This research was supported by the Converging Research Center Program (Project No. 2014048778) funded by the Ministry of Science, ICT and Future Planning.

References

[1] Pfenniger A, Wickramarathna L, Vogel R and Koch V 2013 *Medical Engineering & physics* **35** 1256-1265
[2] H. Tseng, W. Tian and W. Wu 2013 *Sensors*, **13.5** 5478-5492
[3] Li C, Wu P, Lee S, Gorton A, Schulz M and Ahn C 2008 *Journal of micromechanical systems* **17.2** 334-341
[4] W. Choi 2010 *Transactions on Electrical and Electronic Materials* **11.3** 130-133

Comparison of the piezoelectric energy harvesters with Si-MEMS and metal-MEMS

W. J. Wu[1], C. T. Chen[1], S. C. Lin[1], C. L. Kuo[1], Y. J. Wang[1] and S.P. Yeh[1]

[1] National Taiwan University, Taipei, Taiwan

Email: wjwu@ntu.edu.tw

Abstract. This paper presents the development of piezoelectric energy harvesters based on silicon and stainless steel substrates, which have the ability to harvest mechanical energy from surrounding vibrations and transform vibration energy into useful electrical power. Our experimental results show that the silicon-based device had a maximum output power of 0.9 μW with 1.0 V_{P-P} output voltage excited at 107.9 Hz under a 0.25 g vibrating source. The metal-based device had a maximum output power of 2.7 μW with 1.5 V_{P-P} output voltage at a vibration frequency of 108.6 Hz and 0.25 g acceleration. The areal power density was 0.02 μW mm^{-2} and 0.05 μW mm-2 for the devices based on silicon and on stainless steel, respectively. The silicon-based devices broke when the device excited exceed 0.25g acceleration, while the metal-based devices can sustained for vibration level higher than 2g acceleration. The stainless steel based device is therefore proved to be much more reliable than silicon based device.

1. Introduction

In recent years, piezoelectric material has rendered a revival of intensive research activities due to their application potential and interesting physics. Once such application, energy harvesting, has become an important issue all over the world. Comparing all possible energy sources, mechanical vibration is the potential power source that is easily accessible through micro-electro-mechanical system (MEMS) technology. Most of the studies in piezoelectric MEMS are simple a cantilever beam structure based on silicon substrates. In many MEMS energy harvesting studies, to obtain high output power, it has been shown that the performance of the d_{31} mode was greater than the d_{33} mode [1]. Therefore, we used the d_{31} mode for research presented in this paper. For applications with high acceleration levels, PZT material on silicon or silicon on insulator (SOI) substrates can cause cracks owing to the brittle mechanical properties of silicon. To increase the ultimate mechanical strength of piezoelectric MEMS, finding an alternative substrate material becomes necessary.

In this paper, we show the development and construction of homemade PZT deposition equipment for the aerosol deposition method, which can deposit PZT thin films on traditional silicon substrates (silicon-based) and stainless steel substrates (metal-based) as well. We compare the two different substrate based devices by using the same laminate structure, sandwiched between top and bottom electrodes. In order to lower the resonant frequency matching with the low frequency vibrations in ambient, we designed the cantilever beam structure to operate at an appropriate resonant frequency to maximize the displacement of PZT thin film. Finally, we present a comparison of output power and lifetime between silicon-based and metal-based devices in long-term vibrating cycles.

2. Fabrication of PZT layer by aerosol deposition system

Lin et al. shows a schematic diagram of a homemade equipment to deposit PZT thin film based on aerosol deposition method [2]. The advantage of the aerosol deposition better over tradition methods is a more compact texture than that of screening printing is obtained and an easier pattern definition can be produced with simple shadow masks. However lift-off process is used in this study instead of shadow masks to define PZT patterns because of better resolution.

3. Substrate fabrication

3.1. Silicon substrate

Standard processes for silicon-based substrates have been have already been studied [3]. We therefore used a traditional process to make the devices on SOI substrates. We acquired a SOI substrate with a laminar structure of Si 60 μm/SiO_2/Si 350 μm. The SEM photograph of the cross-sectional view of the device structure is shown in Figure 3-1. Figure3-2(a)(b) shows photographs of fabricated devices.

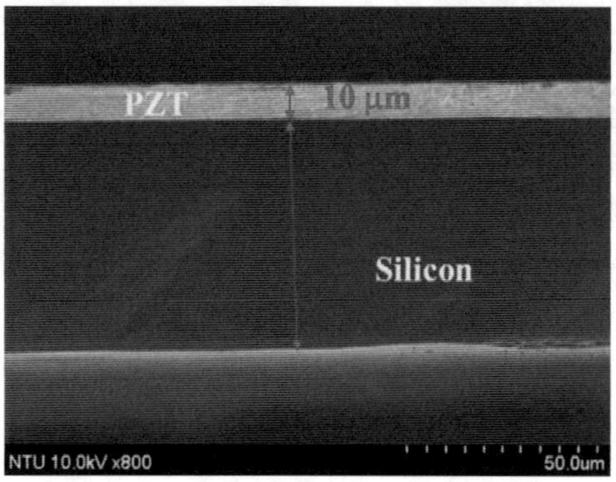

Figure 3-1 SEM photograph of a cross-sectional view of the cantilever-beam with a PZT layer of 10 μm based on a silicon substrate.

(a) (b)

Figure 3-2 Photographs of the devices: (a) after beam released structure, (b) a single device

3.2. Stainless steel substrate

In order to compare the difference between silicon-based and metal-based piezoelectric MEMS, we deposited the same PZT layer thickness of 10 μm on 60 μm stainless steel substrates. Lin et al. shows the fabrication process of metal-based devices [2]. The SEM photo of the sidewall and cross-sectional view of PZT/stainless steel samples after the lift-off process are shown in Figure3-3 (a)(b). Fabricated devices are shown in Figure3-4 (a)(b).

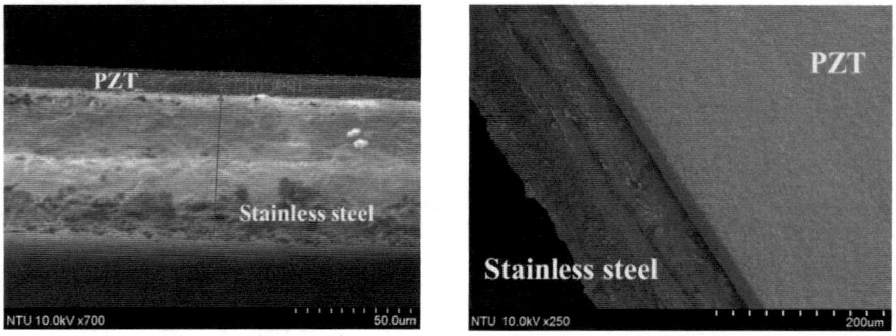

(a) (b)

Figure 3-3 SEM photograph of the device after deposition: (a) the sidewall, (b) cross-sectional view.

(a) (b)

Figure 3-4 Photographs of the devices: (a) a single device and (b) side view of the beam

4. Results and Discussion

Lin et al. shows the experimental setup [2].The output power and output voltage of the device was measured with a optimal resistive load. The output average power can be expressed by the equation [6]

$$P = \left(\frac{V_{p-p}}{2}/\sqrt{2}\right)^2/R\,,$$

where V_{p-p} is the load output voltage of peak-peak values and R the load resistance.

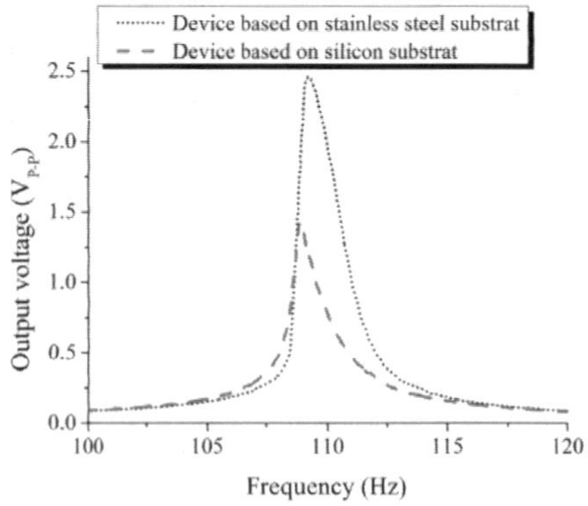

Figure 4-1 Experimental results of output voltage vs. frequency at 0.2 g acceleration

In order to compare the difference between the silicon-based and the metal-based devices, both devices were poled under 5 Vμm^{-1} for 1 hour after then annealing processes. The results of output voltage dependent frequency are shown in Fig 4-1. The resonant frequency and output voltage was observed within the temperature range investigated.

Table 4-1 Output performance of these two device at 0.25 g acceleration.

d_{31} mode with 10 μm PZT	Stainless steel base	Silicon base
Excitation Frequency	108.6 Hz	107.9 Hz
Capacitance	9.8 nF	8.6 nF
Load impedance	150 kΩ	150 kΩ
Power Output	2.7 μW	0.9 μW
Voltage Output (with load)	1.5 V_{P-P}	1.0 V_{P-P}
Quality factor Q	72.8	136

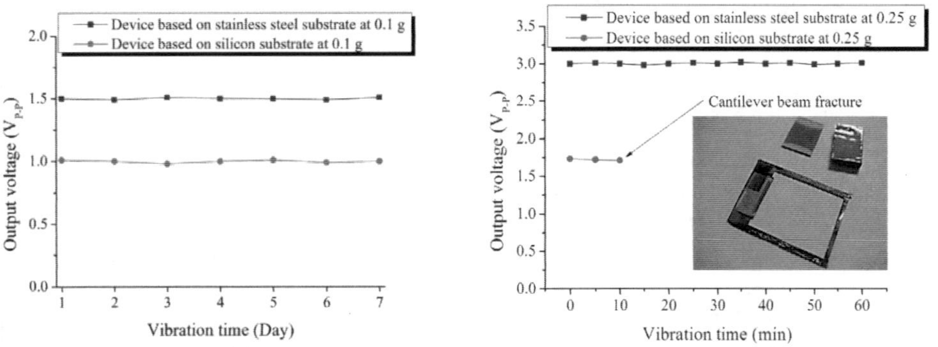

(a) (b)

Figure 4-2 Lifetime of the two devices: (a) excited at 0.1 g and (b) excited at 0.25 g.

The excitation of the silicon-based and metal-based devices was similar because the Young's modulus of silicon and stainless steel is almost the same (about 200 Gpa). Table 4-1 also shows that the capacitance of each device. Although the same laminated structure was used for each device, the total area of the laminated structure on the metal-based devices was larger than silicon-based devices. Therefore the capacitance of the metal-based devices was larger than that of the silicon-based devices. The results show the output performance of the device on stainless steel was better than the silicon-based device.

The result of the lifetime test at 0.1 g and 0.25 g acceleration level is plotted in Figure 4-2. At 0.1g low acceleration vibration excitations, both devices works continuous for more than 7 days. For slightly higher vibration level of 0.25g acceleration, the silicon-based device was observed to crack after 10 minutes.

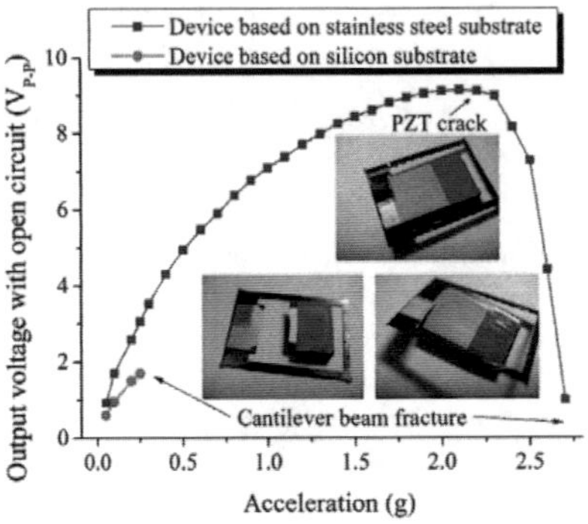

Figure 4-3 The ultimate exciting acceleration of the two devices

However, the metal-based device was still working at the 0.25 g acceleration level for hours. The cracked silicon-based device is shown in the photos of Figure 4-3. The real power density was 0.02 μWmm^{-2} and 0.05 μWmm^{-2} for the silicon-based and the metal-based device, respectively. Finally, we measured an ultimate exciting acceleration for both devices. The result was plotted in Figure4-3 and the crack behaviour was observed. The results show that the metal-based device can sustained vibration level up to 2g acceleration thanks to the flexibility of stainless steel substrates while the silicon based device all broken around 0.25g. This results shows the metal based devices have much better mechanical strength and reliability compared with silicon-based devices.

5. Conclusions

In summary, the PZT thin film was successfully deposited on silicon and stainless steel substrates by a homemade aerosol deposition machine. The devices based on traditional silicon substrates had a maximum output power of 0.9 μW with 1.0 V_{P-P} output voltage excited at 107.9 Hz under a 0.25 g vibrating source. The device based on a stainless steel substrate had a maximum output power of 2.7 μW with 1.5 V_{P-P} output voltage at a vibration frequency of 108.6 Hz at 0.25 g acceleration. Our results also shows that the silicon-based device had a longer lifetime than silicon-based device. The metal based devices can sustained vibration levels up to 2g acceleration while the silicon based devices all broken around 0.25g acceleration. All the results demonstrate that metal-based devices have better performance and also much better reliability compare with silicon-based devices.

References

[1] B. S. Lee, S. C. Lin, W. J. Wu, X. Y. Wang, P. Z. Chang, and C. K. Lee, "Piezoelectric MEMS generators fabricated with an aerosol deposition PZT thin film," *Journal of Micromechanics and Microengineering,* vol. 19, p. 065014, 2009.

[2] S. C. Lin and W. J. Wu, "Piezoelectric micro energy harvesters based on stainless-steel substrates," *Smart Materials and Structures*, vol. 22, p. 045016, 2013.

[3] S. C. Lin and W. J. Wu, " Fabrication of PZT MEMS energy harvester based on silicon and stainless-steel substrates utilizing an aerosol deposition method," *Journal of Micromechanics and Microengineering*, vol. 22, p. 045016, 2013.

[4] J. Akedo, "Microstructure and electrical properties of lead zirconate titanate (Pb (Zr_{52}/Ti_{48}) O_3) thick films deposited by aerosol deposition method," *Japanese Journal of Applied Physics,* vol. 38, p. 5397, 1999.

[5] J. Akedo, "Aerosol deposition method for fabrication of nano crystal ceramic layer - Novel ceramics coating with collision of fine powder at room temperature," *Designing, Processing and Properties of Advanced Engineering Materials, Pts 1 and 2,* vol. 449-4, pp. 43-48, 2004.

[6] Y. C. Shu and I. C. Lien, "Analysis of power output for piezoelectric energy harvesting systems," *Smart Materials & Structures,* vol. 15, p. 1499-1512, 2006.

Optimization of Passive Voltage Multipliers for Fast Start-up and Multi-voltage Power Supplies in Electromagnetic Energy Harvesting Systems

G Yang, B H Stark, S G Burrow and S J Hollis

University of Bristol, Bristol BS8 1TR, UK

E-mail: guang.yang@bristol.ac.uk

Abstract. This paper demonstrates the use of passive voltage multipliers for rapid start-up of sub-milliwatt electromagnetic energy harvesting systems. The work describes circuit optimization to make as short as possible the transition from completely depleted energy storage to the first powering-up of an actively controlled switched-mode converter. The dependency of the start-up time on component parameters and topologies is derived by simulation and experimentation. The resulting optimized multiplier design reduces the start-up time from several minutes to 1 second. An additional improvement uses the inherent cascade structure of the voltage multiplier to power sub-systems at different voltages. This multi-rail start-up is shown to reduce the circuit losses of the active converter by 72% with respect to the optimized single-rail system. The experimental results provide insight into the multiplier's transient behaviour, including circuit interactions, in a complete harvesting system, and offer important information to optimize voltage multipliers for rapid start-up.

1. Introduction

A significant challenge for an energy harvesting (EH) system with an active switched-mode converter between the harvester and main storage element is cold start [1], i.e., start-up with no previously stored energy. The most common cold-start solution is to use a parallel passive converter, e.g., a voltage multiplier, to charge the storage capacitor to the point where the converter's control circuits can be activated and supplied from this capacitor [2], as illustrated by the dashed arrows in Figure 1. This period of sub-optimum operation can take several minutes or even longer for sub-milliwatt harvesting, especially if the harvester is excited intermittently as is the case in many real environments [3].

This work presents a considerably faster start-up scheme: rather than charging the milifarad main storage capacitor to the required V_{store}, the start-up voltage multiplier output V_{cap} powers up the control of the active converter directly, as indicated by the bold arrow in Figure 1. The multiplier is optimized for rapid start-up, only having capacitances of microfarads as required to power the microwatt control circuits. Once the main converter has charged up the main energy storage, this becomes the supply for the converter's control circuits, thereby improving robustness.

For electromagnetic energy harvesting, voltage multipliers have been described in the literature for the main power extraction circuits [1], and optimized design described under specific load conditions. Typically the optimization includes minimizing the multiplier's output impedance at the harvester excitation frequency [4] and holding the output voltage constant at a value where the harvester is operating close to its maximum power point. By contrast, this paper investigates optimum multiplier design with rising output voltages, and the principal purpose is to make the charging time to reach a pre-defined supply voltage as short as possible. The multiplier is thus open-circuited at the output during this period. We demonstrate further improvements by using the inherent cascade architecture (Figure 2a) of a voltage multiplier to supply several rails at different voltages.

The electromagnetic harvester used in this work produces a nominal 110 μW into a matched (360 Ω) load under a sinusoidal excitation of 1.5 m·s^{-2}, at its resonant frequency of 44 Hz. Its full details have previously been published and can be found in [5].

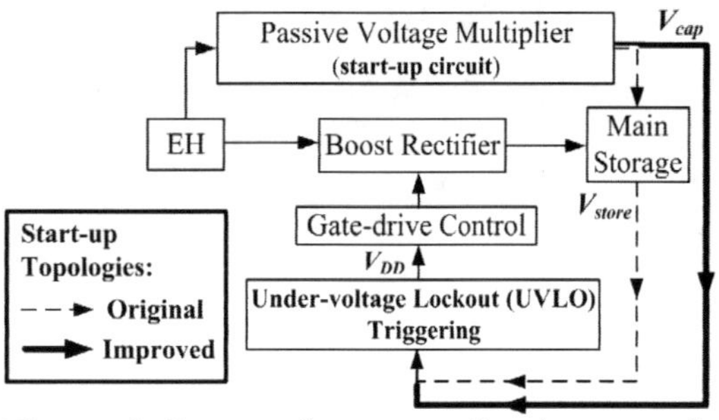

Figure 1. Improved start-up of active control of power conditioning subsystems.

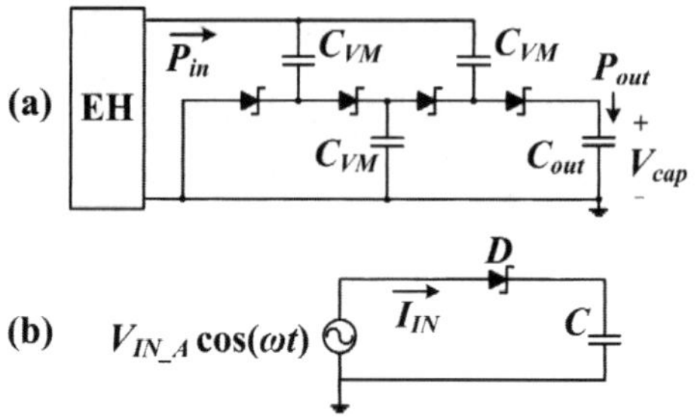

Figure 2. (a) Typical voltage multiplier, and (b) a single stage rectifier as its basic building block.

2. Analysis and optimization of voltage multipliers for fast start-up

2.1. Analysis of start-up time

For a given stored energy E_{store}, the start-up time t_{start} of a voltage multiplier (Figure 2a) is

$$E_{store} = \frac{1}{2} C_{out} V_{DD}^{\;2} = \int_0^{t_{start}} [P_{in}(t) - P_{out}(t)] \mathrm{d}t \quad , \tag{1}$$

where C_{out} is the multiplier output capacitance and V_{DD} is the required start-up voltage; the difference between the input power P_{in} and loss P_{loss} represents the charging power (i.e., P_{out} in Figure 2a). While P_{loss} is associated with the multiplier's constituent devices, e.g., the diode voltage drop and number of diodes, P_{in} is determined by the harvester loading condition, which varies continuously during start-up.

Here we present a simplified analysis of the dynamic loading by considering a single-stage rectifier as the basic building block, Figure 2b. The rectifier draws current $i_{IN}(t)$ from a sinusoidal input voltage source $v_{IN}(t) = V_{IN\_A} \cdot \cos \omega t$. Thus, the diode current (exponential) is $i_D = I_S \cdot \exp(v_D/n\phi_t - 1)$, where I_S is the diode saturation current, ϕ_t is the thermal voltage, and n is the slope factor. The diode's voltage drop is $v_D(t) = V_{IN\_A} \cdot \cos\omega t - V_{cap}(t)$, and the resistance emulated by the rectifier is governed by

$$r_e(t) = \frac{v_{IN}(t)}{i_{IN}(t)} = \frac{V_{IN\_A} \cos \omega t}{I_S \cdot [\exp{(V_{IN\_A} \cos \omega t - V_{cap}(t))}/n\phi_t - 1]} \quad . \tag{2}$$

It follows that the resistance $r_e(t)$ varies not only within every input cycle but also as the capacitor is charged during start-up and $V_{cap}(t)$ increases. This relationship becomes more complicated when the input source in Figure 2b is replaced by the actual harvester-generated voltage, since this voltage is now dependent on the loading represented by $r_e(t)$ and with time-varying amplitude. The slower the rate of change of the dynamic loading condition, the closer this condition approximates the optimum steady-state loading condition of the harvester, as demonstrated in [5]. This suggests the use of large multiplier capacitances (C_{VM} in Figure 2a) to maximise the input power, because the larger the capacitance, the more slowly $V_{cap}(t)$ increases during start-up, and thus the more slowly $r_e(t)$ increases in equation (2). Larger capacitances, however, result in longer changing time for each multiplier-stage. A good trade-off is thus required here to minimize start-up time.

2.2. Component selection to minimize start-up time

Start-up of the gate-drive control (Figure 1) occurs at 0.9 V corresponding to the minimum operating voltage of a TS3001 low-power pulse-width-modulation (PWM) generator used here. As suggested by equation (1), the lower the required rail, the shorter the start-up time. A multiplier-output capacitance of C_{out} = 47 μF is used through the optimization, as this storage is suitable for microwatt loads. The default magnitude of the harvester's excitation is 1.5 m·s^{-2} corresponding to a maximum harvestable power P_{max} = 110 μW. During low-voltage start-up, the diode's forward-voltage drop is the primary source of power loss. This points towards the use of low-threshold diodes, but typically the lower the threshold is, the higher the leakage is. Four different Schottky diodes are thus evaluated through PSpice simulation: RB411D, BAT760, BAT754, and 1PS79SB30. Despite the RB411D diode having the highest leakage (2.5 μA @ 1 V), the benefit of its low voltage drop of ~90 mV for 100-μA forward current outweighs the drawback of high leakage and results in the fastest start-up. It is thus selected.

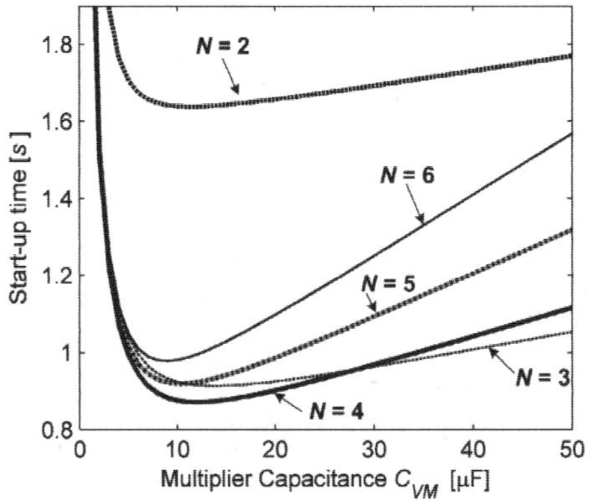

Figure 3. Simulated start-up times for V_{cap} = 0.9 V versus multiplier capacitance C_{VM}, for different stage numbers *N*.

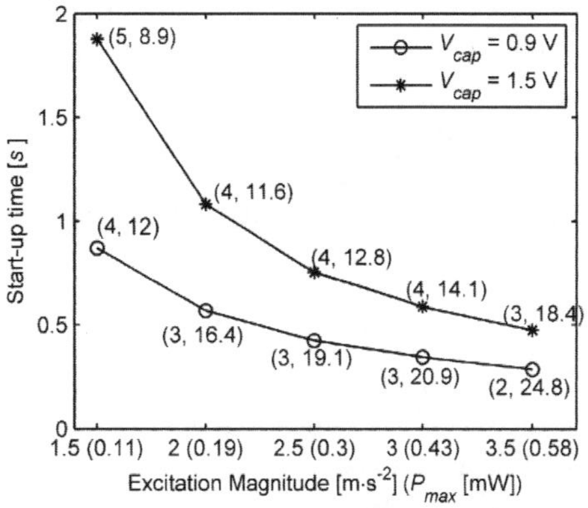

Figure 4. Simulated start-up times of optimized combinations of *N* and C_{VM} for V_{cap} = 0.9 V and 1.5 V, against harvester excitation.

Figure 3 shows the start-up time as the multiplier capacitance C_{VM} and the number of multiplier stages *N* are varied. The shortest time is found using $N = 4$ and C_{VM} = 12 μF. Larger capacitances slow down the start-up, but too small ones, which cause undesired harvester-loading, also increase the time sharply. A sufficient stage number is required for voltage boosting and offers faster start-up, whilst too many stages increase the total power losses, resulting in inefficient start-up. As the total stage number increases, varying each stage capacitance has increased effects on the total power losses and harvester loading, which in turn results in higher dependence of the start-up time on the capacitance, as seen by the increasing gradients in Figure 3. Figure 4 shows the implication of requiring a higher start-up voltage of 1.5 V. The start-up time increases, especially at low excitation magnitudes and hence low maximum harvestable power P_{max}. As the excitation increases, the start-up time shortens, with larger optimal C_{VM} and smaller optimal *N*. This is due to their impact on the harvester loading and utilisation.

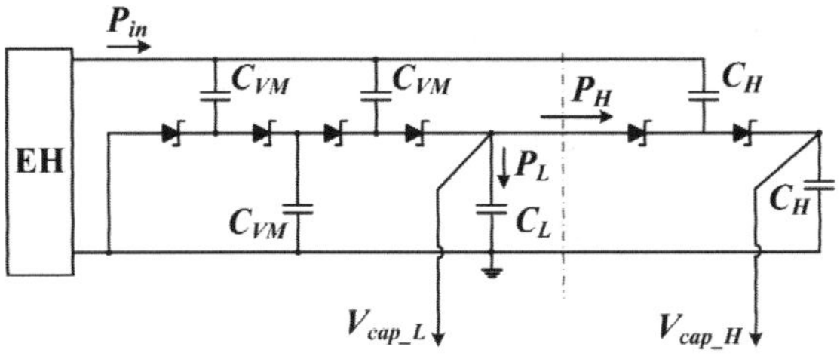

Figure 5. Dual-rail ($V_{cap\_L,\,H}$) start-up topology of a voltage multiplier.

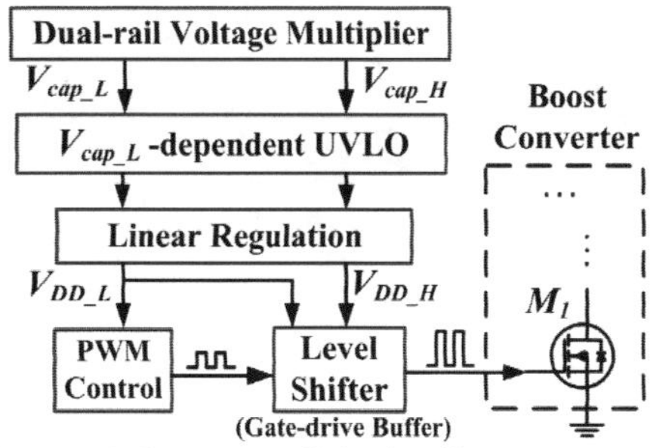

Figure 6. Prototyped dual-rail power supply for PWM control of a boost converter.

2.3. Multi-rail start-up

The single-rail start-up system suffers from a trade-off, where choosing a lower rail voltage shortens the start-up time but also increases the circuit loss of the main converter due to insufficient headroom above the gate threshold voltage. This trade-off can be improved through the powering up of different subsystems with different voltage rails. Fortunately, the voltage multiplier readily provides different internal voltages, see Figure 5, where, during start-up, the rails V_{cap_L} and V_{cap_H} rise simultaneously with $V_{cap\_H} > V_{cap\_L}$. The resulting system is illustrated in Figure 6. The majority of the control circuits are powered by the lower rail V_{DD_L}, which reduces the total quiescent power consumption, and permits these circuits to be triggered sooner via an under-voltage-lockout (UVLO). The higher rail V_{DD_H} powers only the gate-drive buffer to provide sufficient headroom above the gate threshold voltage of the main switching device M_1, thereby achieving low on-state resistance and conduction loss. Improved trade-offs between start-up speed, quiescent power overheads of active control, and power-conversion efficiency can thus be expected. In the circuit of Figure 5, the choice of C_H affects the start-up speed of V_{cap_L}, because the larger capacitance C_H is, the more power P_H is drawn from stage $N = 4$, and therefore less power P_L is available to charge C_L. Using a small C_H improves the start-up speed while the actual required C_H is small as it is only used to power the gate-drive buffer.

3. Experimental results

3.1. Measurements of transient power during start-up of single-rail system

As suggested in equation (1), the start-up time is determined by the varying input power and power loss. The input and output transient power (averaged over every excitation cycle) of the voltage multiplier of Figure 2a is measured during a start-up event and plotted in Figures 7, 8 and 9, as a function of C_{VM}, showing that these powers vary dynamically. Figure 7 shows that larger capacitances C_{VM} improve the harvester loading, as seen by the higher input powers, but also increase the power loss, which results in an optimal C_{VM} at 12 μF yielding the highest output power.

Figure 8 shows that the output powers peak at different output voltages for different stage numbers N. A higher N shifts the power peak to the higher voltage, as it provides higher voltage-boosting. This also improves the harvester loading, as each multiplier-stage draws power from its preceding stage so that charging the first stage takes more time. The resulting loading changes slowly and approaches the optimum steady-state loading condition, as noted in Section 2.1. Higher N with more diodes, however, results in higher loss. Thus, the optimal N is 4, whose corresponding output power peaks at the rail 0.9 V. With C_{VM} =12 μF, it offers the shortest start-up time for P_{max} = 110 μW as shown in Figure 3.

Figure 9 shows that different excitation magnitudes (whilst $N = 4$, C_{VM} =12 μF) not only change the power peaks but also their voltage locations. Since higher excitations and hence input power levels P_{max} result in higher voltages at which the powers peak, lower N is required to shift the peak from a low voltage level to the targeted rail, in order to achieve the fastest start-up, as shown in Figure 4.

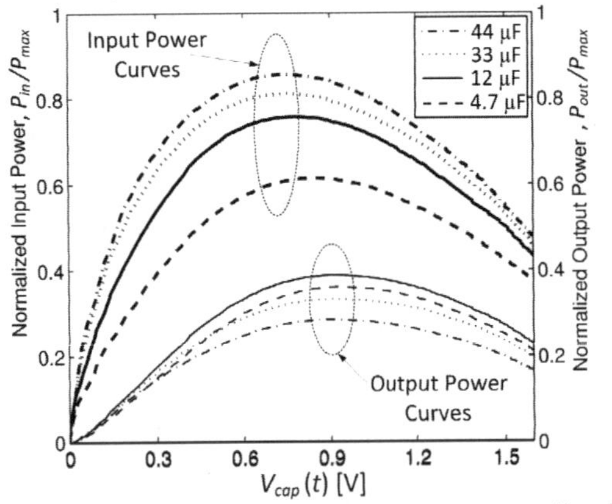

Figure 7. Input and output powers normalized to the maximum harvestable power P_{max}, and measured as the output voltage V_{cap} increases during start-up, for different C_{VM} values.

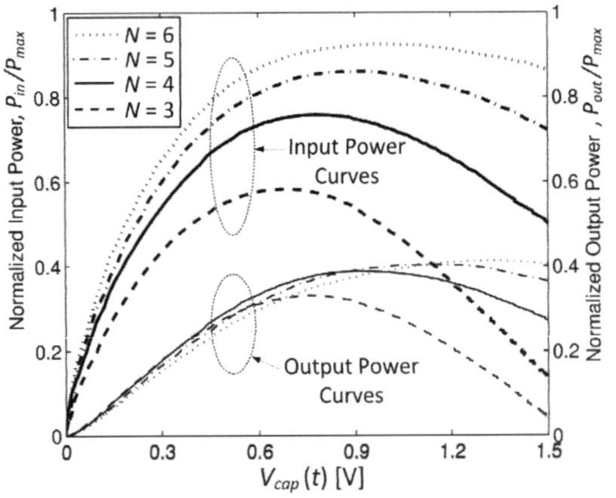

Figure 8. Normalized input and output powers, measured as the output voltage increases during start-up, for different stage numbers N.

3.2. Measurements of the dual-rail system

The dual-rail system of Figure 6 is prototyped using discrete devices including the TS12001 UVLO detector, the TS3001 PWM generator, and the NLSV1T34 level shifter (circuit details can be found in

[2] and [3]). The system is compared with its single-rail counterpart using the circuit shown in Figure 2a with $V_{DD} = V_{cap}$. Figure 10 shows the steady-state measurement results. The higher the single-rail V_{DD} is, the higher the quiescent power consumption P_q is, and the lower the converter's conduction loss P_{cond} is, since the rail regulates the gate-drive signal swing. This results in an optimal V_{DD} at 1.2 V corresponding to the lowest total power loss P_{loss}. By contrast, even lower P_{loss} is achieved by the dual-rail system with V_{DD_L} at 0.9 V for low P_q and V_{DD_H} at 1.5 V for low P_{cond}. These results, together with the measured start-up times, are summarized in Table 1. The single-rail 0.9 V start-up offers the shortest time but results in much higher P_{loss}. An improved trade-off is achieved here by the dual-rail system using small capacitances C_H in Figure 5; the total power loss is reduced by about 72%, whilst the optimized start-up time is ~1 s, as opposed to the 148 s presented in [2], under similar conditions.

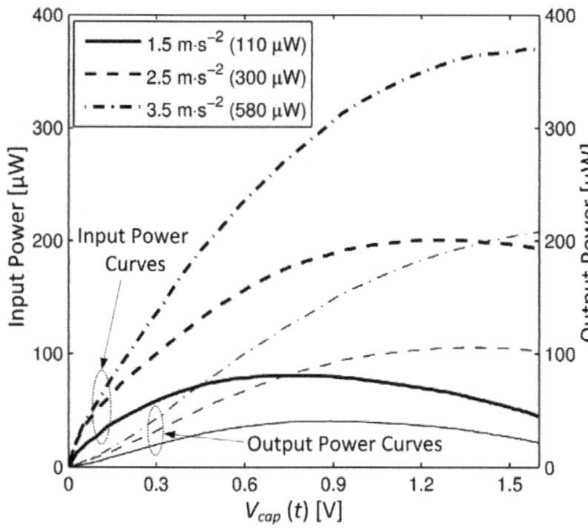

Figure 9. Input and output powers, measured as the output voltage increases during start-up, for different harvester-excitation magnitudes.

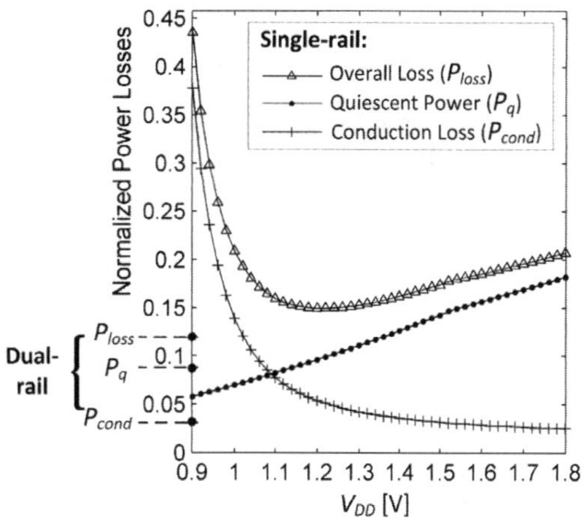

Figure 10. Steady-state measured power losses (normalized to P_{max}) of the single-rail system against the start-up rail V_{DD}, and of the dual-rail system with V_{DD_L} = 0.9 V and V_{DD_H} = 1.5 V.

Table 1. Comparative summary of the single-rail and dual-rail power-supply systems.

Power Supplies	Power Losses (relative to max. harvestable power): Quiescent, P_q	Conduction, P_{cond}	Overall, P_{loss}	Start-up Time ($C_{out} = C_L$ = 47 µF) [s]
Single-rail of V_{DD} = 0.9 V	5.8%	37.8%	43.6%	0.94 @C_{VM} = 12 µF (optimum)
Single-rail of V_{DD} = 1.2 V	9.6%	5.4%	15%	1.39 @C_{VM} = 10 µF (optimum)
Dual-rail of V_{DD_L} = 0.9 V V_{DD_H} = 1.5 V	8.8%	3.2%	12%	1.03 @C_H = 2.2 µF, C_{VM} = 12 µF 1.13 @C_H = 4.7 µF, C_{VM} = 12 µF 1.2 @C_H = 6.8 µF, C_{VM} = 12 µF

4. Conclusion

For microwatt electromagnetic harvesting systems, start-up using passive voltage multipliers can be shortened by selecting stage numbers and capacitances to achieve peak output power at the targeted rail voltage for given input power levels. Furthermore, by using the proposed multi-rail multiplier topology which distributes supply rails to various subsystems, an improved trade-off between start-up speed and active power conversion efficiency is found, resulting in a ~1 second start-up and a ~72% reduction in power loss with respect to the single-rail system.

References

[1] Torah R N, Tudor M J, Patel K, Garcia I N, and Beeby S P 2007 *Proc. IEEE Sensors (Atlanta)* p 264–7
[2] Szarka G D, Burrow S G, and Stark B H 2013 *IEEE Trans. Power Electron.* **28** 3353-62
[3] Yang G, Stark B H, Hollis S J, and Burrow S G 2014 *IEEE J. Emerg. Sel. Topics Circuits Syst.* **4** 364–74
[4] Lin P M 1977 *IEEE Trans. Circ. Syst.* **CAS-24** 517–30
[5] Szarka G D, Proynov P P, Stark B H, and Burrow S G 2014 *IEEE Trans. Power Electron.* **29** 201–12

Airflow energy harvesters of metal-based PZT thin films by self-excited vibration

E Suwa, Y Tsujiura, F Kurokawa, H Hida and I Kanno

Department of Mechanical Engineering, Kobe University, 1-1 Rokkodai-cho, Nada-ku, Kobe, Japan

E-mail: kanno@mech. kobe-u.ac.jp

Abstract. We developed self-excited vibration energy harvesters of Pb(Zr,Ti)O_3 (PZT) thin films using airflow. To enhance the self-excited vibration, we used 30-μm-thick stainless steel (SS304) foils as base cantilevers on which PZT thin films were deposited by rf-magnetron sputtering. To compensate for the initial bending of PZT/SS304 unimorph cantilever due to the thermal stress, we deposited counter PZT thin films on the back of the SS304 cantilever. We evaluated power-generation performance and vibration mode of the energy harvester in the airflow. When the angle of attack (AOA) was 20° to 30°, large vibration was generated at wind speeds over 8 m/s. By FFT analysis, we confirmed that stable self-excited vibration was generated. At the AOA of 30°, the output power reached 19 μW at wind speeds of 12 m/s.

1. Introduction

Piezoelectric vibration energy harvesters (PVEHs) have been widely studied as promising power sources for battery-free wireless sensor nodes. Conventional PVEHs are usually composed of simple cantilever structure with piezoelectric materials (bulk ceramic and thin film form), and the electric power is generated from the external mechanical vibration of cantilevers. Therefore, PVEHs have to be used in environments where relatively-large vibrations exist.

In recent years, airflow energy harvesters (AFEHs) have attracted attention as a novel type of PVEH [1-5]. The AFEHs generate electric power from the vibration induced by continuous airflow, whereas conventional PVEHs generate electric power from mechanical vibration or kinetic motion. Therefore, the AFEHs are expected to extend the range of application of PVEHs. It is known that a self-excited vibration is induced by airflow when the wind speed is above a critical value (cut-in wind speed) [6]. In the practical use of AFEHs, they are placed in airflows with higher wind speed then cut-in wind speed. However, there are no detailed studies about the mechanism responsible for self-excited vibration for MEMS-scale AFEHs, and clear design theory for the MEMS-AFEHs has not been established yet.

In this study, we fabricated PVEHs of Pb(Zr,Ti)O_3 (PZT) thin films as AFEHs driven by self-excited vibration in an airflow. We used thin stainless steel foils as base cantilevers in order to obtain a large displacement of self-excited vibration without breakdown of the AFEHs. We measured the vibration characteristics of the AFEHs and examined the effect of the various parameters (wind speed and angle of attack, etc.) to optimize the design of the MEMS-AFEHs.

2. Experiments

2.1. Device fabrication

Figure 1 shows a schematic illustration of our AFEH. We used 30-μm-thick stainless steel (SS304) foils (20 × 20 mm^2) as base cantilevers. After sputtering deposition of Pt/Ti bottom electrodes, the PZT thin films were then directly deposited onto the SS304 cantilevers by radio-frequency (rf) magnetron sputtering. Since PZT-sputtering on SS304 foils inevitably caused large initial bending due to the thermal stress, we deposited counter PZT layers on the opposite side of PZT/SS304 foils to make flat PZT/SS304 cantilevers. Subsequently, the Pt top electrodes were prepared through a shadow mask. One end of the PZT/SS304/PZT foils was fixed by a clamping jig that then looked something like a flag.

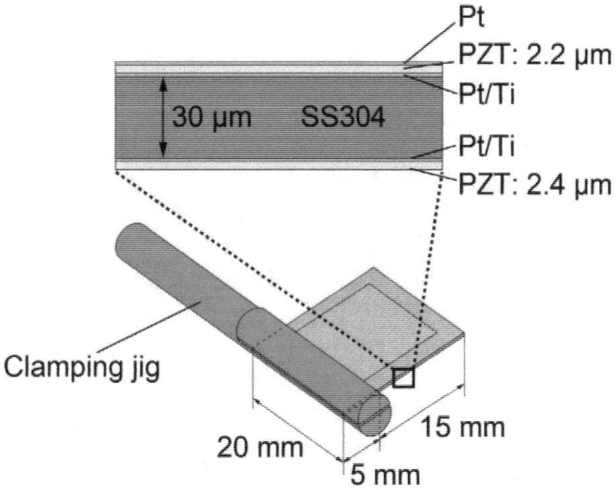

Figure 1. Schematic illustration of the AFEH.

2.2. Frequency response and optimal resistance

Prior to the evaluation of the self-excited vibration in airflow, we measured the frequency response of the output voltage and the displacement to determine the resonance frequency of the AFEHs. The AFEHs were mounted on a vibration exciter and swept the excitation frequency at an acceleration of 1.0 m/s^2. We also measured the load resistance dependence of the output power in an external load resistance at the resonance frequency of 1.0 m/s^2 acceleration. The effective electric power P was calculated using

$$P = {V_{\text{p-p}}^2}/{8R} \tag{1}$$

where, $V_{\text{p-p}}$ and R are the peak-to-peak output voltage and the load resistance, respectively.

2.3. Power-generation performance of AFEHs

Figure 2 shows the measurement setup for the power generation characteristics of AFEHs in airflow. The airflow was generated in an air channel by a fan, and the wind speed was tuned by voltage from a DC power supply. The AFEH was set at the outlet of the air channel and the angle of attack (AOA) was defined by the angle between the AFEH and the airflow direction as shown Figure 2 (c). We measured the output voltage by oscilloscope, while the cantilever displacement of the AFEHs was observed using a high speed camera. The measurements were conducted in various airflow conditions with wind speeds (0 ~ 12 m/s) and the AOA (0 ~ 180°). In order to examine the vibration mode of the AFEHs in detail, we analyzed the frequency spectrum of output voltage by fast Fourier transformation (FFT). Subsequently, we measured the effective electric power generated at the optimal load resistance and the optimal AOA.

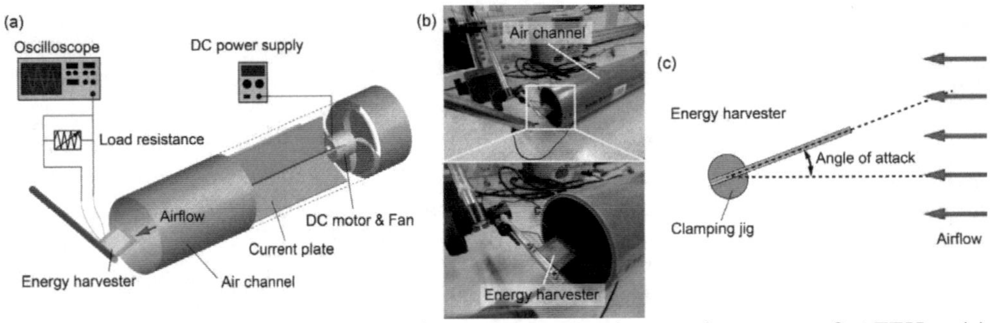

Figure 2. Measurement setup for the power-generation performance of AFEHs: (a) schematic illustration, (b) photographs, and (c) side view of the AFEH.

3. Results and discussion

Figure 3 shows the frequency response of the output voltage and the cantilever displacement at an acceleration of 1.0 m/s^2. The resonance frequency of the AFEH was measured to be 124 Hz. Figure 4 shows the output power and voltage as functions of load resistance. The maximum output power was obtained at a load resistance of 7.5 kΩ. This value is in good agreement with the theoretical value of optimal load resistance of 8.1 kΩ.

Figure 5 shows the output voltage in the open-circuit state versus wind speed at an AOA of 0° to 180°. When the free end of the AFEH faced windward, we observed large output voltage at AOA of 10° to 40° (figure 5(a)). When the AOA was 20° and 30°, the output voltage rapidly increased at the wind speed above 8 m/s, which is the cut-in wind speed of the AFEH. The output voltage reached 1.4 V_{p-p} (30°) at a wind speed of 12 m/s. One the other hand, when the free end of the AFEH faced leeward (AOA was over 90°), the output voltage was less than 50 mV_{p-p} (figure 5(b)). From these results, we considered that the AFEHs whose free end faced windward induced large self-excited vibration due to the unstable condition of the cantilever in the airflow.

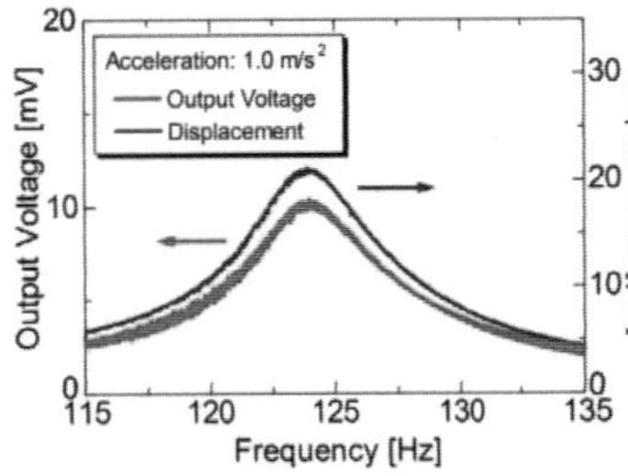

Figure 3. Output voltage in open-circuit state and cantilever displacement as functions of frequency.

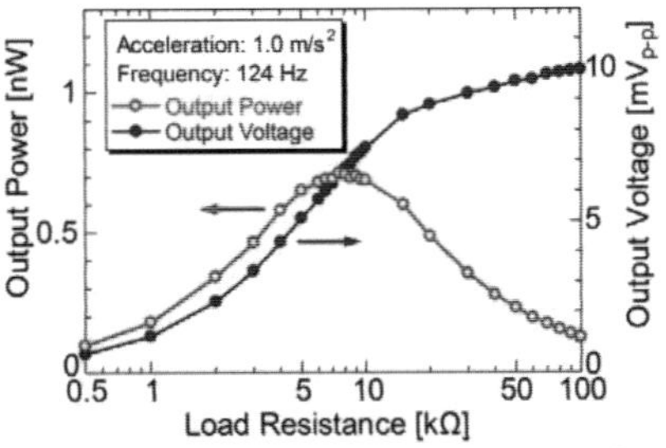

Figure 4. Output power and output voltage as functions of load resistance.

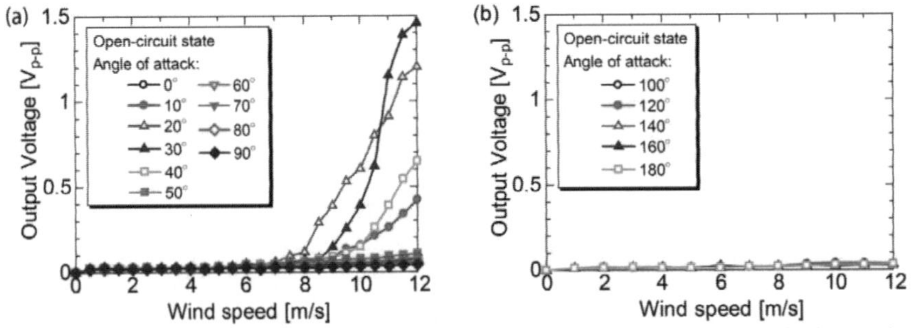

Figure 5. Output voltage in the open-circuit state as a function of wind speed at the AOA of: (a) 0° to 90°, (b) 100° to 180°.

Figure 6 shows the tip displacement (peak-to-peak amplitude) at AOA of 20° and 30° as a function of wind speed. The displacement curve is in good agreement with output voltage curve in figure 5. The tip displacement reached to 14 mm at the AOA of 30° in the wind speed of 12 m/s, however we could find no damage on the AFEHs under such large vibration. Figure 7 shows the optical images of vibrational motion of the AFEHs at the AOA of 20°. Clear vibration was not observed at wind speeds up to 8 m/s, however, large self-excited vibration was generated at the wind speeds of more than 8 m/s.

Figure 8 shows FFT spectrum of the output voltage in the open-circuit state at the AOA of 20°. With increasing the wind speed from 6 m/s to 12 m/s, the vibration frequencies of the AFEH were reduced from 119.1 Hz to 99.1 Hz. Because the vibration frequencies of the AFEH were almost the same as the resonance frequency of the AFEH measured by vibration exciter, the stable self-excited vibration was generated by the airflow. The decrease of vibration frequency with the wind speed was attributed to the softening effect of the large vibration of the cantilevers.

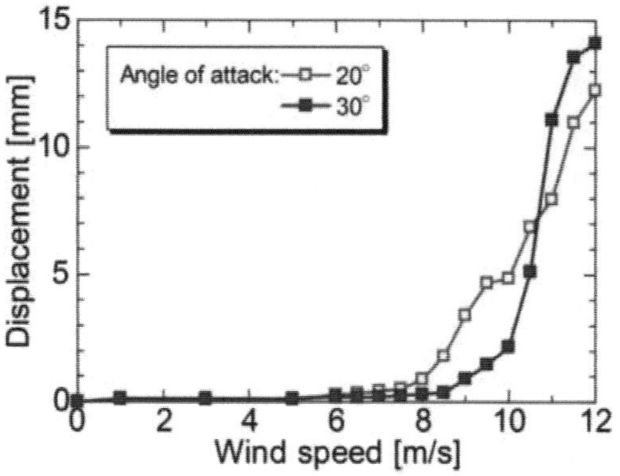

Figure 6. Tip displacement (peak-to-peak amplitude) as a function of wind speed. The AOA was 20° and 30°.

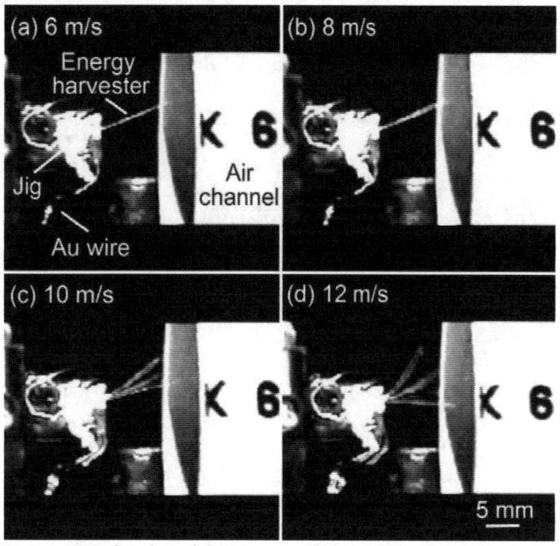

Figure 7. Optimal images of the vibrational motion at the AOA of 20° and at wind speeds of : (a) 6 m/s, (b) 8 m/s, (c) 10 m/s, and (d) 12 m/s.

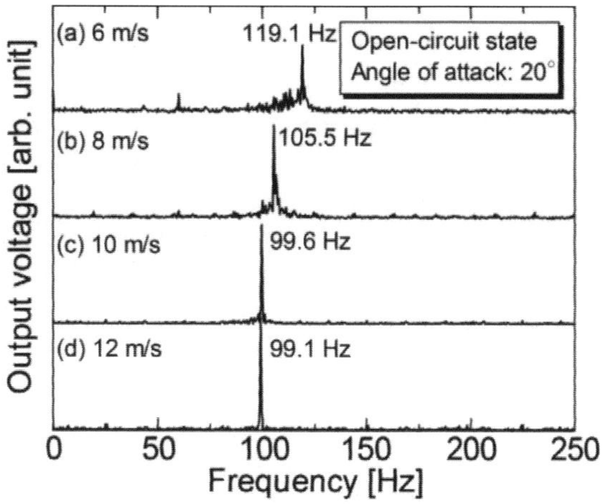

Figure 8. FFT spectrum of output voltage at the AOA of 20° at the wind speed of: (a) 6 m/s, (b) 8 m/s, (d) 10 m/s, and (d) 12 m/s.

Figure 9 shows the output power versus wind speed. When the AOA was 20° or 30° under wind speeds of 12 m/s, the maximum output power reached 12 μW (20°) and 19 μW (30°), respectively. These output powers are sufficiently large to drive MEMS devices such as wireless sensor nodes [7]. Table 1 compares the performance of our AFEH with other studies. Note that the volumes in Table 1 were calculated from the data of the references. In terms of the power density, our AFEH is comparable to that of other larger scale AFEHs. Those results indicate that MEMS-AFEHs composed of PZT thin films on metal cantilevers are an effective way to obtain large displacement from a continuous airflow.

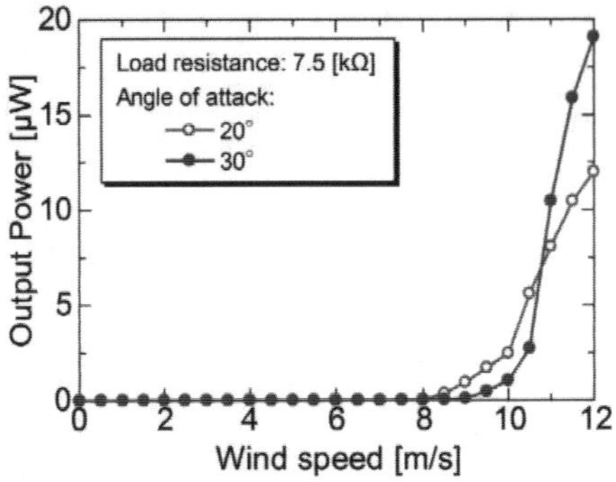

Figure 9. Output power as a function of wind speed. The load resistance was 7.5 kΩ.

Table 1. Comparison of the AFEHs performance.

Study	Piezoelectric material	Effective volume [mm^3]	Wind speed [m/s]	Output Power [μW]	Power density [μW/mm^3]
D. St. Clair (2010) [2]	PZT [3]	266 [3]	12.9	880	3.3
S. Li (2011) [4]	PVDF	472	8	615	1.3
H. D. Akaydin (2012) [5]	PZT	255747	1.192	100	4×10^{-4}
This study	PZT	10	12	19	1.9

4. Conclusion

In this study, we developed AFEHs of metal-based PZT thin films, and evaluated their power-generation characteristics. At AOA of 20° and 30°, we observed large vibration at wind speeds above 8 m/s, while we could not find any damage or breakdown of the AFEH. At the wind speed of 12 m/s, the maximum displacement reached 14 mm at an AOA of 30°. FFT analysis of output voltage revealed that a stable self-excited vibration was generated. The maximum output power at an AOA of 30° reached 19 μW at wind speeds of 12 m/s. The AFEH in this study showed that it could generate power of comparable density to conventional large-scale AFEHs.

References

[1] Truitt A and Mahmoodi S N 2013 *Int. J. Precis. Eng. Manuf.* **14** 1667-75
[2] Clair D S, Sennakesavababu V R, Daqaq M F and Li G 2010 *Appl. Phys. Lett.* **96** 144103
[3] Bibo A, Clair D S, Sennakesavababu V R, Li G and Daqaq M F 2010 *Proc. of 2010 ASME IDETC/CIE Conf.* 691-701
[4] Li S, Yuan J and Lipson H 2011 *J. Appl. Phys.* **109** 026104
[5] Akaydin H D, Elvin N and Andreopoulos Y 2012 *Smart Mater. Struct.* **21** 025007
[6] Zhao L, Tang L and Yang Y 2014 *Jpn. J. Appl. Phys.* **53** 060302
[7] Dewan A, Ay S U, Karim M N and Beyenal H 2014 *J. Power Sensor* **245** 129-43

Basic Research of Vibration Energy Harvesting Micro Device using Vinylidene Fluoride / Trifluoroethylene Copolymer Thin Film

T. Takiguchi[1], T. Sasaki[2], T. Nakajima[2], S. Yamaura[2], T. Sekiguchi[1] and S. Shoji[1]

[1]Waseda University, Japan
[2]Institute for Materials Research, Tohoku University, Japan

E-mail: takiguchi@shoji.comm.waseda.ac.jp

Abstract. Basic research of MEMS based micro devices for vibration energy harvesting using vinylidene fluoride / trifluoroethylene (VDF/TrFE) copolymer thin film was investigated. The VDF/TrFE copolymer thin film was formed by spin coating. Thickness of VDF/TrFE copolymer thin film was ranged from 375 nm to 2793 nm. Impedance of VDF/TrFE copolymer thin film was measured by LCR meter. Thin film in each thickness was fully poled by voltage based on *C-V* characteristics result. Generated power of the devices under applied vibration was observed by an oscilloscope. When the film thickness is 2793 nm, the generated power was about 0.815 μJ.

1. Introduction

Energy harvesting is innovative technology. It cropped minute energy from the environment of the circumference, and convert to electricity. Many companies and researchers focused on the field of energy harvesting. For example, Z. L. Wang at the Georgia Institute of Technology developed Radial-arrayed rotary electrification for high-performance triboelectric generator [1]. Piezoelectric polymers are one of the key technological materials for energy harvesting [2]. Especially, VDF/TrFE copolymer is one of the notable materials, because of their easy installation into mechanical sources and flexibility [3]. However, the performance and potentiality of the thin film for the energy harvesting device are still not clear. Thus, in this study, we investigated the performance of the power generation and demonstrated the practical usage of this piezoelectric polymer for energy harvesting.

2. Concept

2.1. Fabrication

Schematic image of the device fabrication is shown in figure 1. First, lower-electrode (Al) was fabricated on the polyimide substrate as shown in figure 1(a). Second, VDF/TrFE copolymer was fabricated by spin coating as shown in figure 1(b). VDF/TrFE copolymer film thickness was in a range of approximately 375 nm to 2793 nm. VDF/TrFE copolymer was coated several times to avoid the pinholes of the film. Third, upper-electrode (Al) was fabricated as shown in figure 1(c). Both electrodes were fabricated by dual ion beam sputtering method to decrease VDF/TrFE copolymer from melting by high temperature and damage of the surface of the polymer compared with RF sputtering method. The Al electrodes thickness was 230 nm.

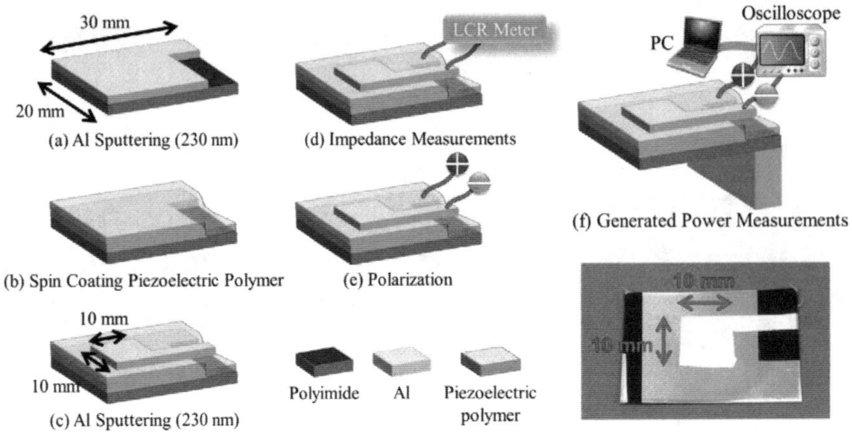

Figure 1. Schematic images of device fabrication and photographs of the fabricated device

2.2. Electric Measurement Results

Impedance of the VDF/TrFE copolymer film was measured by LCR meter as shown in figure 1(d). Frequency characteristics and *C-V* characteristics of the VDF/TrFE copolymer thin film were shown in figure 2 and figure 3. Impedance measurements results were shown in figure 4. The impedance of the film increased in proportion to film thickness. Peak electric field showing the occurrence of polarization reversal was evaluated to be 40 MV/m as shown in figure 4.

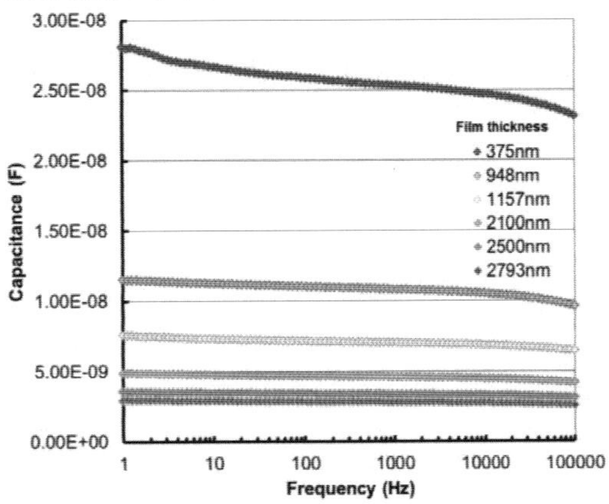

Figure 2. Frequency characteristics of the VDF/TrFE copolymer thin film

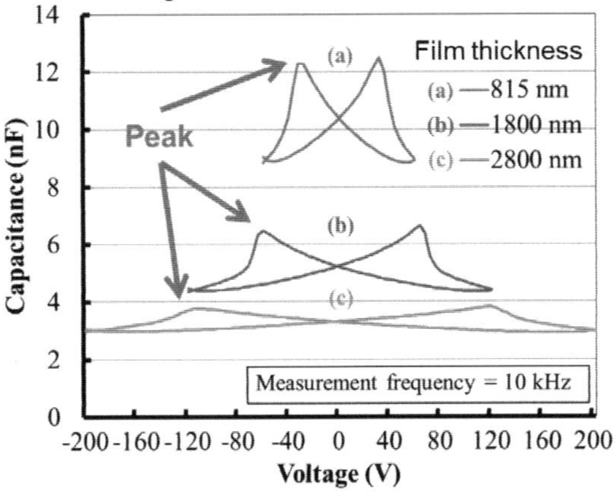

Figure 3. *C-V* characteristics of the VDF/TrFE copolymer thin film

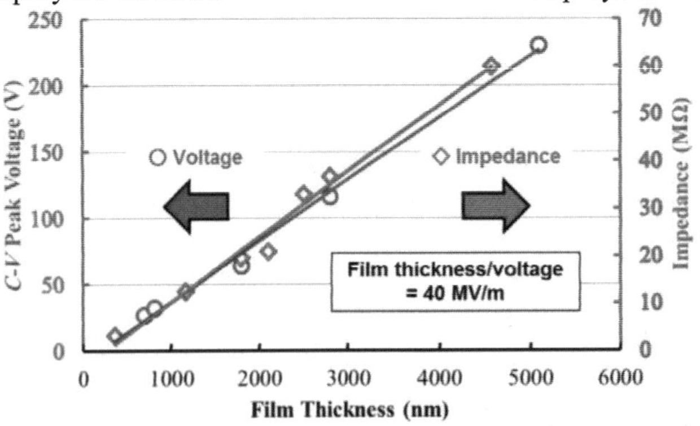

Figure 4. Film thickness vs *C-V* peak voltage and impedance

3. Results and Discussion

The device in each thickness was fully poled by applying 60 MV/m based on the result of *C-V* characteristics. Then generated power of the device by vibration was observed by an oscilloscope (probe impedance : 10MΩ+15pF). Vibration was caused by bending the device as shown in figure 5. Figure 6 shows generated power vs bending degree when the film thickness was 2793 nm. Increment of generated power depends on the bending degree. Figure 7 shows generated voltage vs vibration time when vibration was caused by bending the device to 90 degrees, and figure 8 shows relationship between film thickness and generated voltage (peak to peak: $V_{p\text{-}p}$). When film thickness was increased, $V_{p\text{-}p}$ was reduced and vibration time was extended. For example, generated power was calculated by integration of time from generated voltage. Generated power was 0.815 μJ when the film thickness is 2793 nm at 10 MΩ load resistance.

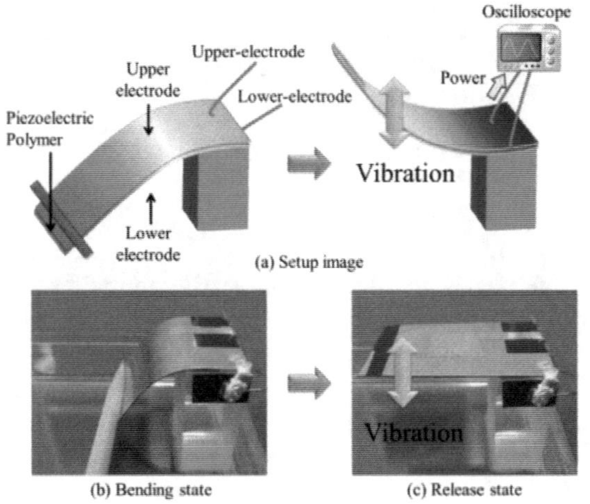

Figure 5. Experimental setup of device vibration

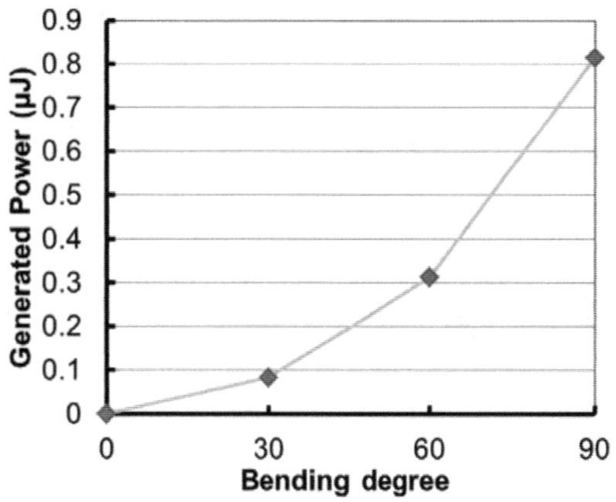

Figure 6. Generated power vs bending degree

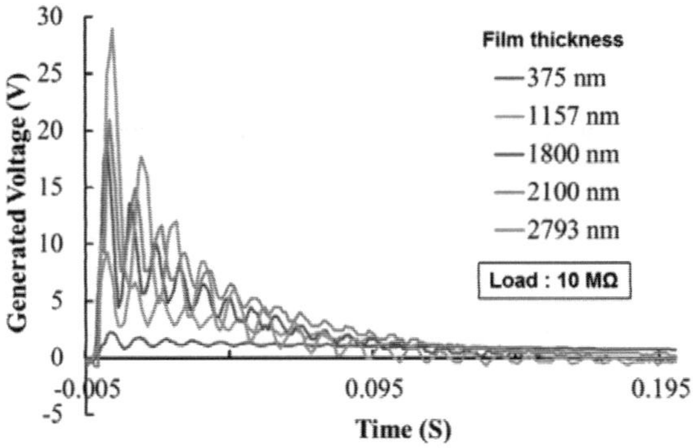

Figure 7. Vibration time vs generated voltage

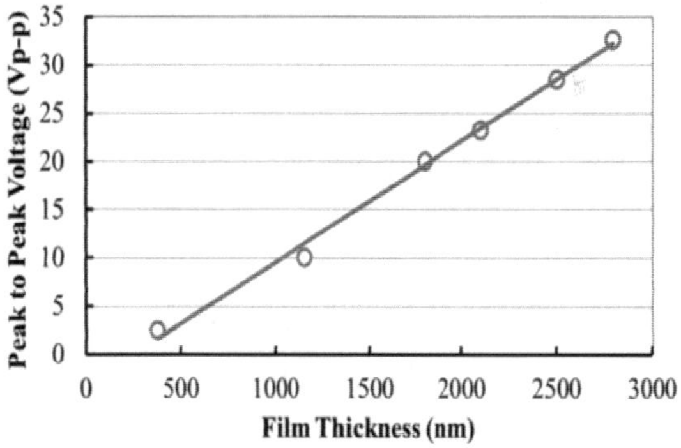

Figure 8. Film thickness vs generated voltage (Vp-p)

Generated power of the device by the vibration was observed by adding load resistance. The load resistance was in a range of approximately 1 kΩ to 1 MΩ. Setup image of the device adding load resistance shows figure 9, and the generated power results were shown in figure 10. The result investigates the generated power increased in proportion to the load resistance. When film thickness is thin, the generated power decreased with a small load resistance. In order to increase generated power, lower impedance of VDF/TrFE copolymer film is required. However, impedance of the film was limited about 3.2 MΩ when the film thickness is 375 nm as shown in figure 4. Because of electrodes were shorted through pinholes, when film thickness was thinner than 375 nm.

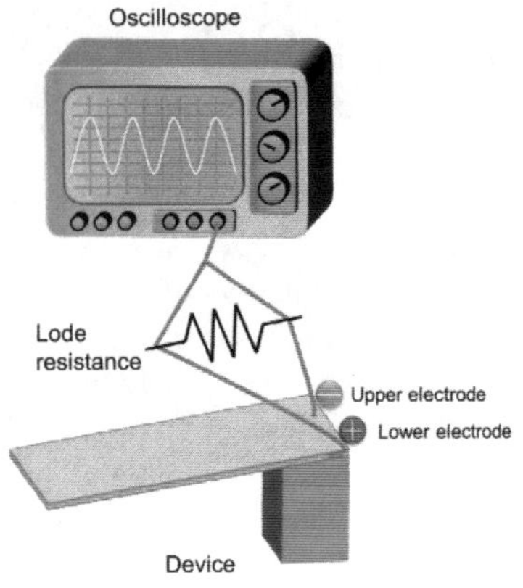

Figure 9. Setup image of device adding load resistance

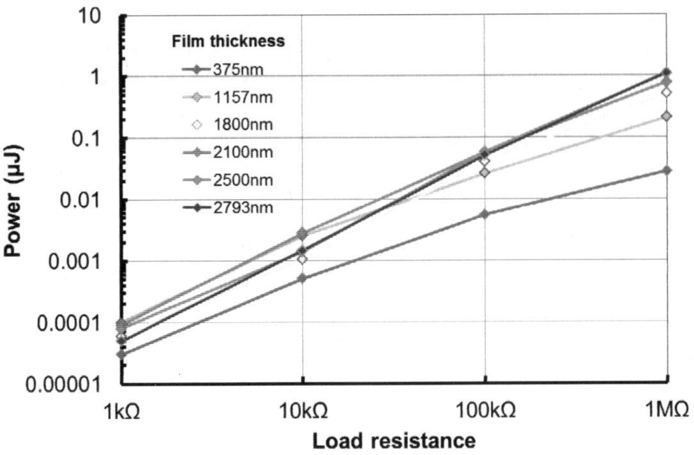

Figure 10. Generated power by adding load resistance

4. Conclusions

This paper presents the micro device for vibration energy harvesting using VDF/TrFE copolymer thin film. The VDF/TrFE copolymer thin film was fabricated by spin coating. Thickness of VDF/TrFE copolymer thin film was ranged from 375 nm to 2793 nm. The impedance of the film increased in proportion to film thickness. Peak electric field showing the occurrence of polarization reversal was evaluated to be 40 MV/m. The device in each thickness was fully poled by applying 60 MV/m based on the result of *C-V* characteristics. When the film thickness is 2793 nm, generated power was 0.815 μJ. In this research, we didn't measure or calculate d_{31} or d_{33}. It should be measured near the future. When film thickness is thin, the generated power decreased with a small load resistance. In order to increase generated power stacked, we are going to fabricate the multi-layer structures of VDF/TrFE copolymer thin films.

5. Acknowledgment

This work is partly supported by Japan Ministry of Education, Culture, Sports Science & Technology (MEXT) Grant-in-Aid for Scientific Basic Research (S) No. 23226010. The authors thank for MEXT Nanotechnology Platform Support Project of Waseda University.

References

[1] G. Zhu, J. Chen, T. Zhang, Q. Jing, and Z. L. Wang 2014 *Nat. Commun* **5** 3426
[2] S. P. Beeby *et al* 2006 *Meas. Sci. Technol.* **17** R175
[3] T. Nakajima *et al* 2011 *Jpn. J. Appl. Phys.* **50** 09ND14

A 2 DOF vibration harvester for broadband and multi-frequency harvesting using a single electro-magnetic transducer

S G Burrow and L Penrose

University of Bristol, Bristol, UK

E-mail: Stephen.burrow@bristol.ac.uk

Abstract. The narrow bandwidth of resonant vibration energy harvesters has long been seen as a drawback to exploitation. The narrow bandwidth is necessitated by the requirement to sufficiently amplify small source vibrations, but results in devices vulnerable to changes in excitation frequency, de-tuning due to ageing of components, and also makes efficient harvesting from sources with multiple frequency components difficult. In this paper a harvester based on a 2 degree of freedom oscillator is presented that not only enjoys the wider bandwidth of the higher order system, but configures the electromagnetic transducer in such a way that it requires no more components than the transducer of a typical single degree of freedom harvester. Theoretical models of the harvester system predict a range of possible frequency response functions dependent on easily-adjusted electrical parameters. These predictions are validated with experimental results.

1. Introduction

In attempts to broaden the frequency response of vibration energy harvesters, authors have proposed various ingenious modifications to the standard mass-spring-damper resonant mechanical amplifier, including devices with non-linear characteristics, electrical tuning and devices featuring multiple degrees of freedom (DOF). Non-linear devices have been well documented in the literature (e.g. [1]) over recent years and undoubtedly have some positive characteristics; however, since fundamentally superposition does not apply they are unable to display broad-band response in respect to simultaneous excitation at multiple frequencies. Electrically tuned harvesters can in theory replicate any frequency response, but in practice have limits imposed by non-ideal electrical transduction [2].

Multiple DOF (linear) devices do not have the limitations of single frequency excitation, nor the complexity of electrically tuned systems. A growing number have been reported in the literature, from the simple combinations of many 1 DOF systems combined in one package, e.g. the multiple beams described by Ferrari et al. [3], to coupled devices with higher-order behavior, such as the device described by Tang and Zuo [4], or the 3 DOF device of Wong et al. [5]. Many of the devices based on higher order responses can be derived from the classic vibration absorber described by Den Hartog [5].

In this paper a novel 2 DOF vibration harvester is described which configures a typical voice coil transducer in a non-typical way such that both coil and magnet assembly are individually sprung with respect to the base and energy is extracted by damping their relative motion. Arranging the voice coil in this manner results in the ability to harvest power over an extended frequency range, typical of higher order systems, but without the penalty of additional electromechanical transducers.

2. Theoretical 2 DOF harvester

The schematic for the 2 DOF harvester can be seen in figure 1.

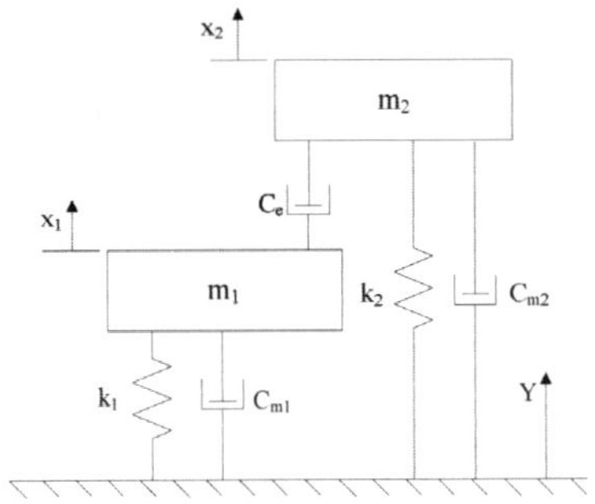

Figure 1. Schematic of the 2 DOF system, where m_1 is the coil mass with k_1 and C_{m1} as the associated compliance and parasitic loss, and where m_2 is the magnet assembly with k_2 and C_{m2} as its associated compliance and parasitic loss. C_e is the electrical damping.

The motion of the 2 DOF harvester depicted in figure 1 is described by equations 1 and 2.

$$\ddot{x}_1 = \frac{k_1}{m_1}(y - x_1) - \frac{C_{m1}}{m_1}(\dot{x}_1 - \dot{y}) - \frac{C_e}{m_1}(\dot{x}_1 - \dot{x}_2) + \ddot{y} \quad (1)$$

$$\ddot{x}_2 = \frac{k_2}{m_2}(y - x_2) - \frac{C_{m2}}{m_2}(\dot{x}_2 - \dot{y}) - \frac{C_e}{m_2}(\dot{x}_2 - \dot{x}_1) + \ddot{y} \quad (2)$$

Figure 2 shows the power frequency response of the device, from numerical simulation of equations 1 and 2, when the harvester is subject to excitation with constant displacement and output power is that developed in the electrical damping C_e. Three differing values of C_e are simulated and are classified as lightly, moderately and heavily damped conditions due to their effect on the frequency response of the system.

The device has the characteristic 'dual peak' frequency response of a third order system when the relative damping applied by the transducer is light, however when the damping is greater, the system behavior tends to that of a second order single DOF system, illustrated in figure 2. Because the system is linear, it is capable of simultaneously responding usefully to excitation from multiple frequencies (unlike non-linear systems).

For comparison, the combined maximum power frequency response of two independent 1 DOF harvesters with the same input conditions and a combined proof mass the same as the total mass of the 2 DOF system is shown in figure 3. The 'maximum power' frequency response refers to the fact that the resistive load at each particular frequency is the optimum for power generation. The wider power bandwidth of the 2 DOF system can be seen, although it will be noted that the improved response for frequencies between the two outer peaks is at the expense of frequencies outside of this range, in comparison with the dual 1 DOF devices.

The values used in these simulations were taken from the experimental device described in the next section of this paper.

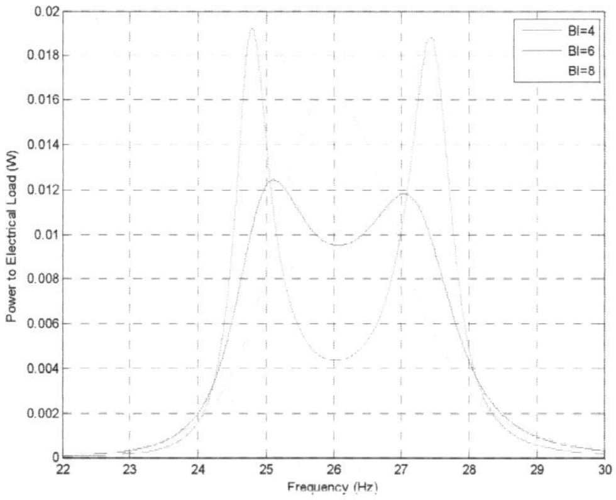

Figure 2. Power/frequency response of the harvester for various levels of damping: Red = lightly load damping, Blue = moderate load damping, Green = high load damping

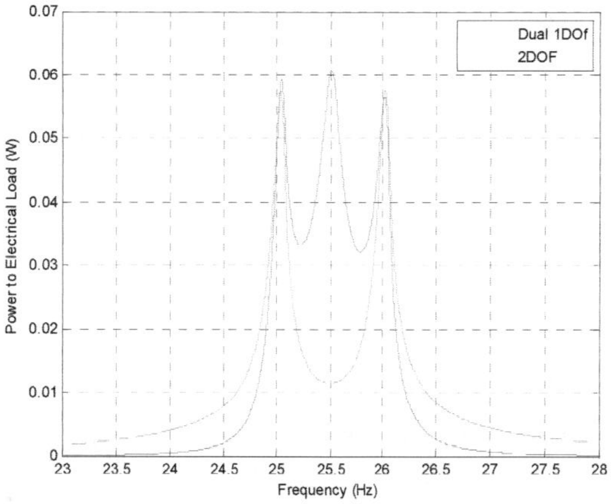

Figure 3. Comparison of maximum power/frequency response (resistive load optimised for particular frequency) of the 2 DOF device with 2, 1 DOF devices of same combined proof mass.

3. Prototype device

The device has been prototyped at a meso scale and is shown in figure 4. The assembled device is shown in figure 4a, and partially dismantled to reveal the separate springing of the magnet and coil in figure 4b. All parts were custom manufactured for the prototype including planar springs water-jet cut from beryllium copper. The voice coil and magnet assembly were numerically optimized for maximum harvested power for a fixed volume.

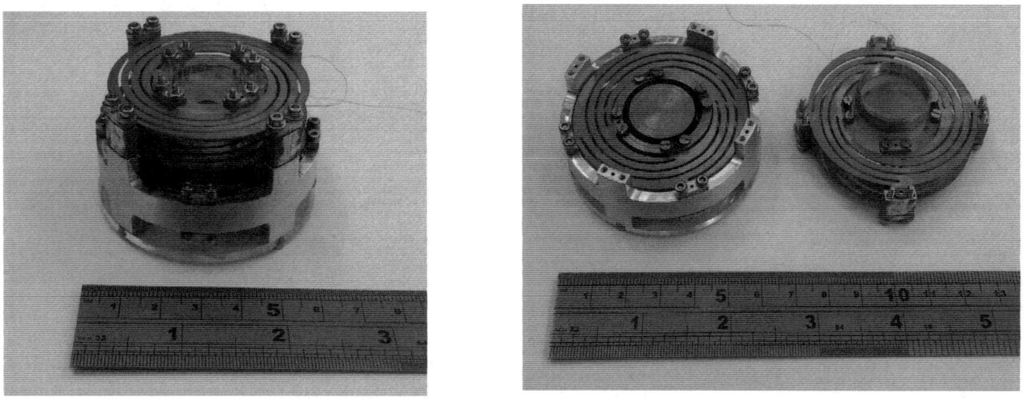

a) Device assembled b) device partially dismantled

Figure 4. Prototype 2 DOF harvesting device

The measured parameters of the device are given in table 1. The mass of each of the seismic elements was altered during testing to produce the desired resonant frequency for each individual mass/spring pair. To conditions were tested: one with resonances set at 24.7 / 27.5 Hz to reproduce a closely spaced resonant pair, and 24.7 / 40.8 Hz producing two distinct resonances.

Table 1. Properties of prototype device

Property	Value
Mechanical Damping Coefficient C_{m1} C_{m2} (Ns/m)	0.03
Length Coil (m)	14.8
Internal Resistance (Ω)	14.3
BL (Tm)	6.70
(BL)^2/Ri	3.14
Spring Stiffness K1=K2 (N/m)	1715

4. Experimental Results

The experimental harvester was tested on a rig consisting of a feedback controlled shaker driven by a Data Physics Quattro dynamic signal processor which also recorded the voltage output across the load resistance. Figure 5 gives experimental frequency response derived from swept-sine measurements. The two separate graphs correspond to the case of light and high damping which produce the differing responses in the 2 DOF system. Each graph contains 2 traces which show the frequency response of the 2 DOF system with closely spaced resonances (green, 24.7/27.5 Hz) and wide spaced resonances (red, 24.7/40.8 Hz). A third trace show the response of the system with the voice coil clamped such that the harvester behaves as a 1 DOF system with a single resonance at the lower frequency (black 24.7 Hz). An interesting phenomenon can be seen in the heavily damped case where the low value of load resistance tends to force the oscillators to behave as one. In this situation the overall damping of the system is low and the response of the coupled oscillators can be very large.

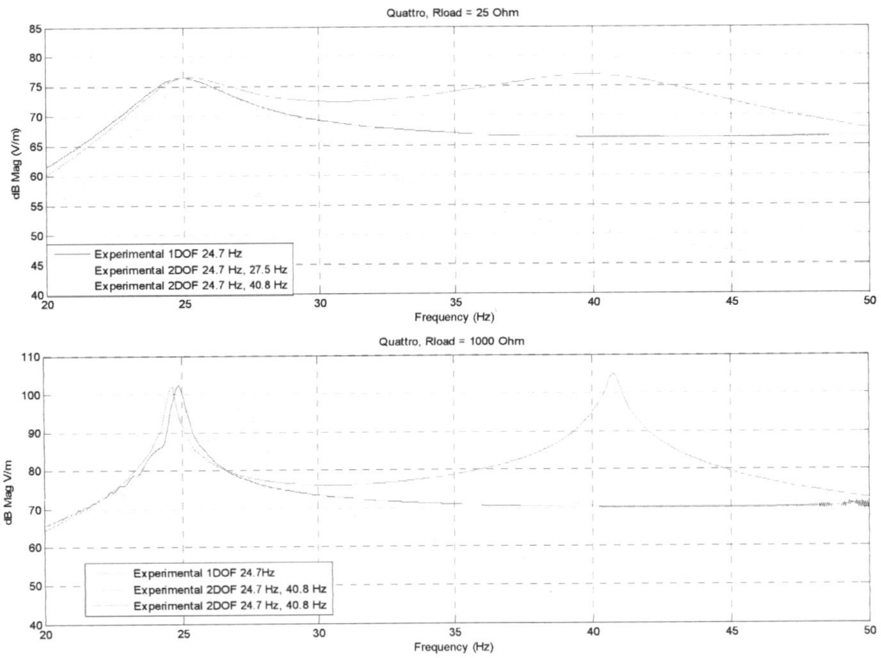

Figure 5. Experimental frequency response curves for a 1 DOF device (tuned to 24.7Hz, black) and two configurations of the 2 DOF device (tuned to 24.7 & 27.5, green, and tuned to 24.7Hz & 40.8 Hz, red). Three different electrical loads are applied corresponding to damping which illustrates the three principle behaviours of the 2 DOF system.

Figure 6 reveals the locus of the power maxima for the 24.7 / 27.5 Hz tuned condition as electrical damping is changed. This plot illustrates how the electrical damping modifies the system response at low load resistances.

In several tests the harvester was excited by two distinct frequencies and the total harvested power was found to be close to the theoretical combination of two individual oscillators. A limitation on the system used in this manner was found to be the differential displacement since the transducer must accommodate the maximum excursion of both mass/spring systems.

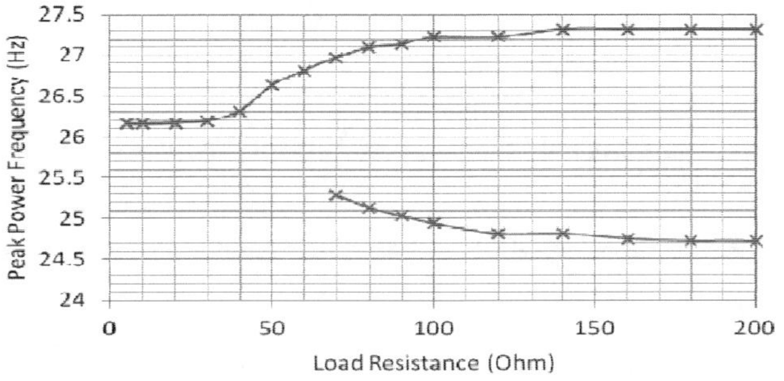

Figure 6. Locus of the peak power frequency as electrical damping is changed.

5. Conclusion

The prototype device displayed the predicted extended frequency response of a 2 DOF system and the realised design was able to efficiently couple the additional degree of freedom (over a conventional resonant harvester) by making both parts of the transducer into individual mass/spring systems.

Depending on the spacing of the resonant frequencies of the each mass-spring system, differing characteristics could be achieved, making a single broader peak or two distinct peaks. The relative motion of the two mass/spring systems is damped by the electrical load, hence the overall damping seen by the system is not a monotonic function of applied electrical load. The clearest manifestation of this is in the case of zero load resistance which results in the system approximating a lightly damped oscillator with compliance and mass equivalent to the combination of that of the individual oscillators.

In the extent of testing to-date the 2 DOF system has proved capable of harvesting from two frequencies of excitation simultaneously, demonstrating the significant advantage of higher order devices.

References

[1] Energy harvesting from vibrations with a nonlinear oscillator D.A.W. Barton, S.G. Burrow, L.R. Clare J. Vib. Acoust. 132, 021009, 2010, DOI:10.1115/1.4000809

[2] Switched-Mode Load Impedance Synthesis to Parametrically Tune Electromagnetic Vibration Energy Harvesters. James Bowden, Stephen G. Burrow, Andrea Cammarano, Paul D. Mitcheson, Lindsay R Clare. IEEE Mechatronics. Accepted publication pending

[3] Piezoelectric multi-frequency energy converter for power harvesting in autonomous microsystems. Marco Ferrari , Vittorio Ferrari, Michele Guizzetti, Daniele Marioli, Andrea Taroni,. Sensors and Actuators A: Physical. Volume 142, Issue 1, 10 March 2008

[4] Enhanced vibration energy harvesting using dual-mass systems. Xiudong Tang, Lei Zuo, Journal of Sound and Vibration Volume 330, Issue 21, 10 October 2011, Pages 5199–5209

[5] A multi-degree of freedom electrostatic MEMS power harvester. Zi Jing Wong, Jize Yan, Kenichi Soga, and Ashwin A. Seshia Proceedings of PowerMEMS 2009, Washington DC, US. December 2009

[5] Mechanical Vibrations. J.P. Den Hartog. Dover, 1985, ISBN-10:0486647854. pp93-106.

Dye sensitized photovoltaic miniaturized solar cells, used as optical sensors for line of sight detection

Cortes Torres Carlos Cesar[1,2], Kota Sampei[2], Ogawa Miho[2], Ozawa Masataka[2], Miki Norihisa[2,3]

Keio University 3-14-1 Hiyoshi, Kohoku-ku, Yokohama, Kanagawa 223-8522, Japan

cortestc@a6.keio.jp

Abstract. Dye sensitized photovoltaic devices have been studied as transparent and low-cost solar cells. Our group have miniaturized the cells and used them as transparent optical sensors. This paper reports the design and fabrication of the cells and avoids the cross talk among cells, which was found recently and such effect provokes hardware instability.

We use these optical sensors as an eye tracking device. The sensor array detects the difference in the intensity of light reflected from the pupil and the sclera and then determines the pupil position. Each sensor consists of two electrodes and electrolyte; hence our device conformed by only four semi-circular shaped sensors on eyeglasses can detect the view angle in both horizontal and vertical directions. Manufacturing process gives us freedom to easily re-arrange, add or remove sensors.

In our prior work we had good performance in stand-alone configuration. We used specialized equipment from National Instruments for our measurements. However we found that:

A cell is not 100% independent from the others, is affected by the absence or presence of light at the neighbour cells.
When our device is connected to other electronic devices (for data processing), all cells have the same voltage among them; therefore, all cells behave the same way when any of them is affected by light.

The root cause is, due to all sensors were interconnected via a micro channel and filled with electrolyte, due to its conductive properties, electrolyte does neither need electrodes nor physical paths to conduct electricity, so it creates a liquid wire between sensors, hence the gap between them become inexistent, consequently when our device is connected to other electronic devices, due to this unique channel and by sharing a common electronic ground, this connection provokes the voltage to be the same among all sensors in the array. Our device becomes four separate voltage lines in a parallel circuit.
The device was also in short circuit provoked by some overlapping paths, despite that such paths were in different layers and separated by an adhesive film of 100μm thickness, such thickness was not large enough to creates a successful dielectric to isolate the paths.

[1] To whom any correspondence should be addressed
[2] Keio University 3-14-1 Hiyoshi, Kohoku-ku, Yokohama, Kanagawa 223-8522, Japan
[3] JST PRESTO 7, Gobancho, Chiyodaku, Tokyo, 102-0076, Japan

Our proposal is to encapsulate electrodes individually and apply printed circuit board routing rules.
This improves each sensor independency without compromising the performance of the other sensors and a more reliable circuitry, stable voltage and higher accuracy can be achieved.
Results:
Electrical tests revealed improvements in voltage response. The best case for voltage dropped in prior work was a 30.5%, while in the proposed design is 11.15%.

1. Introduction

Our device is compound by 4 sensors; each sensor is compound by 2 electrodes and is interconnected by a channel filled with electrolyte which will transport the charge, the sensors react to the light and convert it into DC voltage. Our line of sight device works in the principle that based on the difference in the intensity of light reflected from the pupil and the sclera to detect pupil position.

We use dye sensitized photovoltaic technology and miniaturizing solar cells by applying pattering process of photoresist and etching process, we use them as optical sensors. Figure 1 illustrates our device.

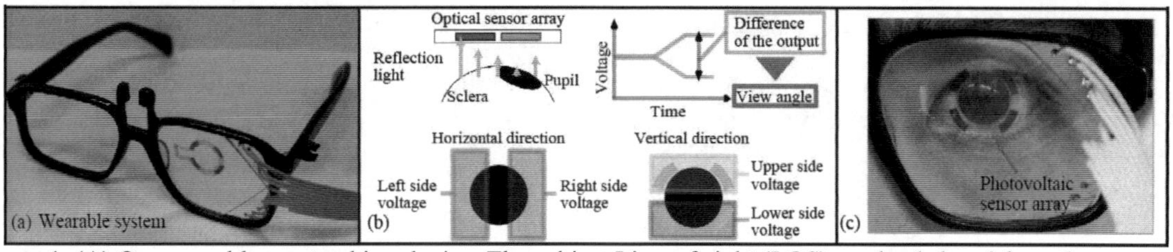

Figure 1. (A) Our wearable eye tracking device. The subject Line-of-sight (LOS) can be deduced from the sensor outputs. (B) The output of the sensor. Reflected light from the pupil is smaller than the sclera (white of the pupil.) The difference of the output is calculated to the view angle.

Because of this channel that interconnects all the 4 pairs of electrodes, thanks to the electrolyte, the charge is transported through the redox couple and this converts the device from 4 sensor array (intended design) to a single sensor (real design). Figure 2 describes the fabrication processes.

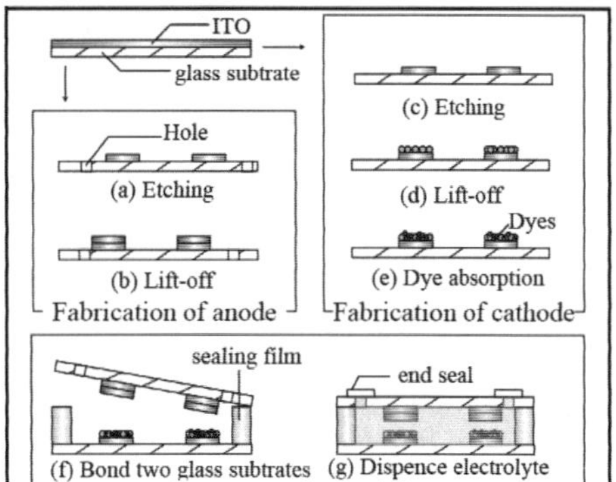

Figure 2. Fabrication and assembling process

When design electronic devices there are few things that should be consider, like: precision, fiability, robustness, etc. Also, an important point to highlight is that as the device size decreases (miniaturization) the risk of issues increases, in the case of our research we had the issue that sensors and transmission lines interference with each other (cross talk), so the a cell is not 100% independent from the others, is affected by the performance of the neighbor cells.

2. Hardware design

Several experiments were conducted, like: usage model experiments (gazed a point by some amount of time and then gazed another point, etc. trying different sequences for each experiment). However when testing the behavior of sensors individually, an experiment showed that an X sensor alters its normal performance even if the light conditions remains the same in that sensor, but it does change respect with Y sensor. The hardware set up for these experiments is: lab view software, national instrument acquisition card NI9215 and control module NI cDAQ-9178 and multimeter.

A second hardware set up was used to verify the device performance. The hardware set up for this experiment is: line of sight device coupled to an android phone through IOIO-OTG board, external circuitry and multimeter. This showed that when our device is connected to other electronic devices (for data processing), all cells have the same voltage among them; therefore, all cells behave the same way when any of them is affected by light.

To narrow down the issue we disassembled the circuit into 3 cross talk current scenarios. In the first stage of this research, we proposed a "prove of concept" design, in which we tested 3 pairs of sensors; each pair modeled a cross talk scenario.

1.-The first scenario tests the interaction of the tracks overlapping with the electrolyte channel between sensors.

2.-The second scenario tests the interaction of the tracks overlapping each other in parallel way, but separated from the electrolyte channel.

3.-The third scenario tests the interaction of the tracks by crossing between them.

These scenarios are explained in figure3.

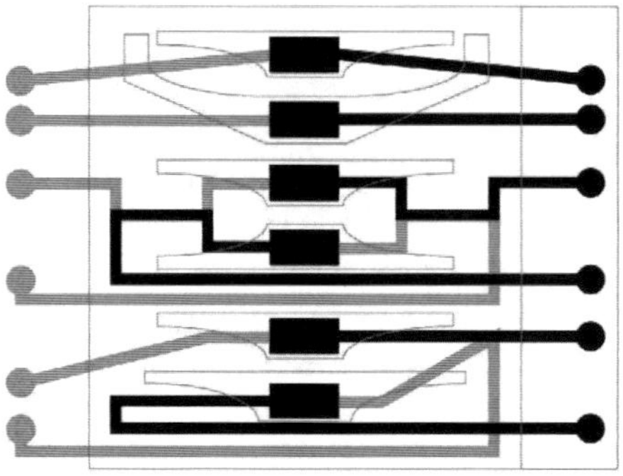

Figure 3. Cross talk basic scenarios. The color of the tracks (blue and black) only indicates that are placed at different layers in the device

3. Testing

The two experiments consisted in measure the voltage in each pair of sensors one at a time, and then covering one of them and measure the opposite cell, this shows a variation of the output voltage. This experiment was executed in each cross talk scenario described above and for each hardware set up.

In the second stage of this research, a new line of sight design was proposed, we redefine the layout of the circuit diagram and sensors, such sensors are encapsulated individually to provide more performance independency and any crossing tracks or overlapping is avoided. The experiment conducted for this second stage, was additionally to the basic voltage measurement was performed the usage model experiment, in which a user gazed marks in a whiteboard each 5 seconds and we measure the response.

4. Results

The new design showed that the percentage of dropping voltage is minor than the previous design, the

Journal of Physics: Conference Series **557** (2014) 012032 doi:10.1088/1742-6596/557/1/012032

voltage behavior is more steady, Figure 4 shows the voltage response in both cases. In our prior work, the amount of voltage dropped is larger than in our current work. i.e. In prior devices a cell will reduce its voltage by 0.1V, in our current device is only 0.02V (best case), this represents a reduction from 30.5% VS 11.15%.

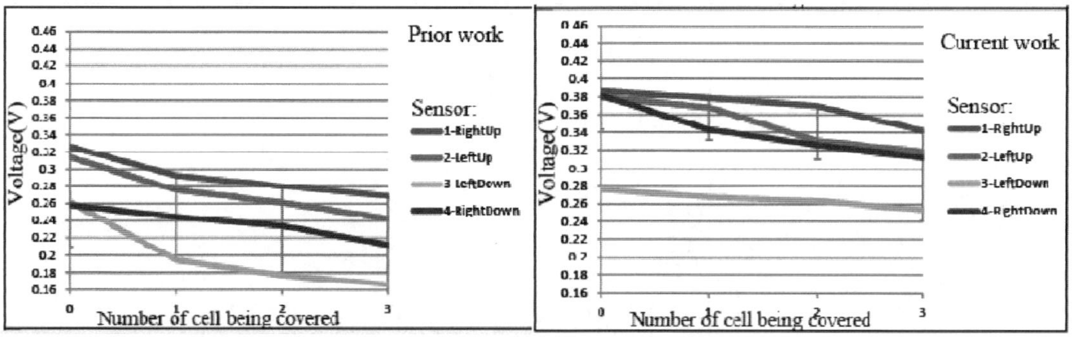

Figure 4. Voltage response. A test was conducted in which we cover all cells but one and one at a time, measuring the voltage of the uncovered cell at each step. Vertical lines, red and green respectively (standard deviation), shows the range of voltage that could be dropped in that particular point, down at vertical direction

5. Conclusions

This research shows that due to the electrolyte conductive properties creates a channel that transports the charge from point A to point B and does not need electrodes or physical paths to transport the charge, so it creates a liquid wire between any points, hence sensors should be encapsulated individually to assure its independency. Also if the electrolyte combined with the physical tracks, empower them, so any cross or overlapping provoke another electric interference despite that such paths are in different layers and separated by an adhesive film of 100μm thickness. Figure 3 illustrates this cross talk issue and illustrates our current proposal..

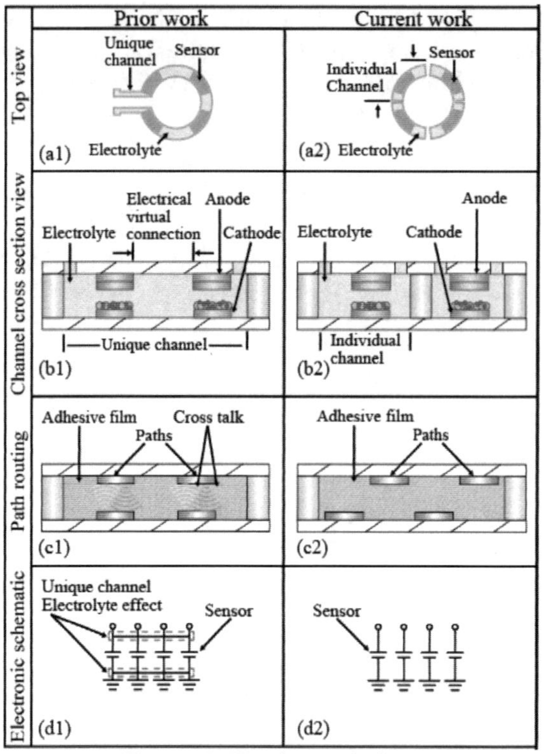

Figure 3. Hardware design.

The other findings was the difference in the results using different hardware set ups, this is due to, by using national instrument equipment, the acquisition cards have more specialized filters to isolate noise and make more precise measurements, the mobile set up was done with multipurpose board and circuits, this set up emulate the output as parallel connection, between all 4 sensors, instead of 4 individual lanes, that's why all of the sensors in the mobile experiments have the same voltage , and have the same response to light conditions, even if a sensor keep the same light conditions.

Thus by removing the cross talk, and let the sensors operates independently, a higher accuracy is achieve.

References

[1] Masataka Ozawa, Kota Sampei, Carlos Cortes, Miho Ogawa, Akira Oikawa, Norihisa Miki Wearable line-of-sight detection system using micro-fabricated transparent optical sensors on eyeglasses Sensors and Actuators A 205 (2014) 208– 214

MEMS Batch Fabrication of the Bipolar Micro Magnet Array for Electromagnetic Vibration Harvester

K Yamaguchi, T Fujita, Y Tanaka, N Takehira, K Sonoda
K Kanda and K Maenaka
Department of EECS, Graduate School of Engineering, University of Hyogo
2167 Shosha, Himeji, Hyogo 671-2280, Japan

E-mail: er14a049@steng.u-hyogo.ac.jp

Abstract. This article introduces a MEMS batch fabrication process of micro magnet array with bipolar magnetic pole for an electromagnetic vibration energy harvester. In order to obtain the large electromotive force from large magnetic flux density change, we established the fine patterned alternating magnetized bipolar magnetic structure. The batch fabrication process of bipolar magnet array is composed of two wafers processing with S-pole and N-pole magnetization and bonding process. By the prototype fabrication of bipolar magnet with the 200 μm SN-interval, we showed the usability of the batch fabrication process of the bipolar magnet array. In addition, we estimated the generated power of energy harvester with a bipolar magnet array. Compared to a harvester with monopolar magnet array, we showed the good result for bipolar one.

1. Introduction

In recent years, vibration electromagnetic energy harvesters are expected for a candidate of next generation power source for micro sensors. We have studied electromagnetic harvesters and trying to improve generated power by miniaturizing a magnet structure. Generally, implementing a minimized magnetic element into MEMS (Micro ElectroMechanical Systems) fabrication is difficult, however we have established that technique in our previous works. The miniaturized magnets on harvesting devices are fabricated by using buried and polished NdFeB magnetic material on the Si trench [1] and sputtered magnet on corrugated Si surface [2]. We have improved a change rate of a magnetic flux density by miniaturizing a magnet array, however there is a limitation to improve the power by using single direction, monopolar magnetic structure. Too fine monopolar magnet array showed decreasing of a flux density change because of the smoothing effect from the adjoining magnets.

In this paper, we propose a MEMS batch fabrication technique of a bipolar micro-magnet that is alternatively magnetized. The basic batch fabrication process and prototyping of the MEMS electromagnetic harvester is demonstrated. Furthermore, we estimate the generated power of the harvester with bipolar magnet array, and verify the usability of the bipolar magnet.

2. Bipolar magnet array

2.1. Construction of interdigitated comb structure

In our previous work, a magnetic material, NdFeB was deposited on the Si MEMS device then magnetized by pulse magnetizer with single direction strong magnetic field. Since the fine patterned magnetization with strong magnetic field is difficult, we choose the combination of two different magnetic pole fine patterned structures. Figure 1 shows a structure of prototype of a magnetic part. The N-pole and S-pole magnet comb fingers are combined after the each magnetization [3].

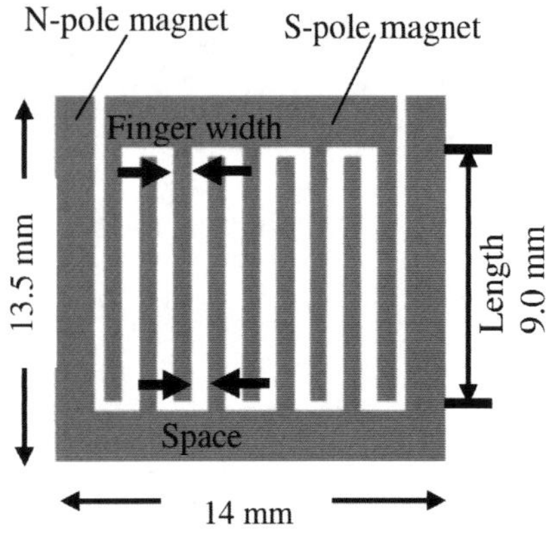

Figure 1. Combination of N- and S-pole NdFeB magnetic film on MEMS structure. The chip size is 13.5 $\times$ 14 mm^2, the finger length of the each comb is 9.0 mm, the thickness is 0.5 mm, the finger width is 100 μm. The space between fingers is equal to the finger width.

2.2. Fabrication process

In order to fabricate the micro bipolar magnet array, we have two challenges to meet. The first one is to establish high precision bonding technique of two wafers. Since each wafer has fine patterned Si comb finger structure with NdFeB magnetic film, the bonding equipment with precision alignment is essential to prevent a break of the fingers during an engaging process. The second one is use of a low temperature bonding method that can prevent the demagnetization [1] of NdFeB magnet during the fabrication step.

For the first issue, we use a wafer bonding system (Ayumi Industry Co., Ltd., Japan). By using infrared microscope, two wafers with comb finger structure can align with $\pm$ 5 μm accuracy. And for the second issue, we proposed using of the positive type photosensitive resin (OFPR-800; Tokyo Ohka Kogyo Co, Japan) for low temperature bonding. The resist can be indurated for bonding not to use a thermal treatment but to use a vacuum treatment.

Figure 2 shows the batch fabrication process flow of our bipolar magnet array. The processes are consisted of (a) S-pole magnet wafer process, (b) N-pole magnet wafer on the substrate wafer, and (c) final alignment and bonding process. As shown in Figure 2a, comb finger shape for S-pole magnet is patterned and etched by Deep-RIE. Twelve comb structures are batch fabricated on a single four-inch Si wafer simultaneously. Then the multilayer NdFeB/Ta (40 layers)[4] film was sputtered on the surface of the Si comb fingers. A pulse magnetization with magnetic field of 3.0 T are performed from bottom to top direction to make a S-pole magnetize wafer.

Figure 2b shows the process for N-pole wafer. It is similar to S-pole one, however there are additional steps for bonding with the substrate wafer. Due to the combination of the S-pole and N-pole comb fingers, an isolation structure for N-pole finger is required. At first the N-pole comb finger is patterned and etched with frame structure. Then the etched wafer is bonded on another substrate wafer by an adhesive photosensitive epoxy resin (KMPR1035; Nippon Kayaku Co, Japan). By patterning of KMPR1035, the comb finger part can be selectively bonded on to the substrate wafer. After the dicing process with appropriate cutting depth, except the comb finger are removed from the substrate. After that, the pulse magnetized with same magnetic field is performed from top to bottom direction to make a N-pole magnetized wafer.

Finally, two oppositely magnetized wafers are bonded by using wafer-bonding system in vacuum pressure (Figure 2c). The OFPR-800 photoresist is spin coated on the N-pole finger on the substrate wafer. The wafer is accurately and quickly aligned with the S-pole wafer and engaged each other. By keeping the wafer in a 1.0 mPa high vacuum chamber, two wafers are rigidly bonded. Figure 3 shows photographs on fabrication steps of S-, N-pole magnet and bipolar magnet wafer.

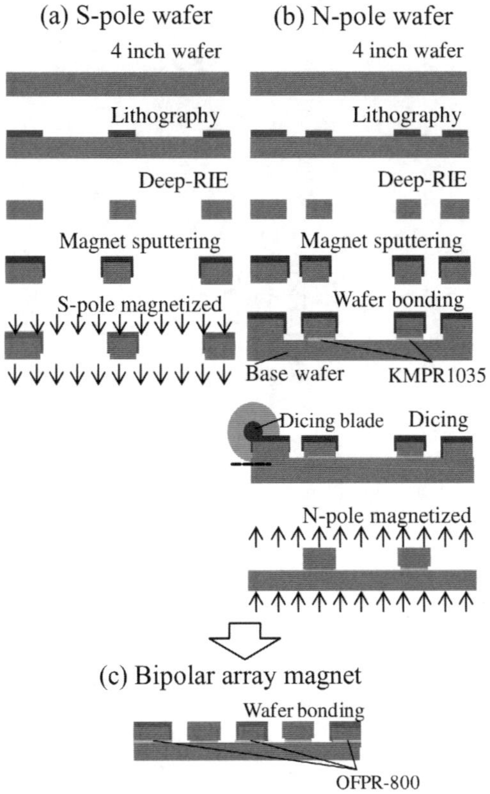

Figure 2. This is the Process flow of bipolar magnet array which are composed (a) S-pole magnet, (b) N-pole magnet and (c) bonding process.

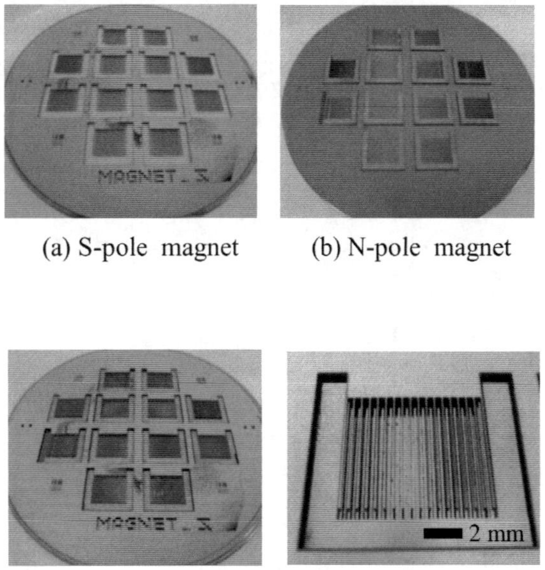

(a) S-pole magnet (b) N-pole magnet

(c) Combined S- and N-pole wafer and close-up photo of comb fingers

Figure 3. Prototype of bipolar magnet array

2.3. *Evaluation*

Figure 4 shows the magnetic flux density profile of the prototype that has 100 μm finger width and 100 μm finger space. We used small sized Hall IC with die size of 50 × 50 μm^2 (HG-0711; Asahi Kasei Co, Japan) for high-resolution measurement. From the magnetic flux density profile, the peak flux density interval is around 400 μm and is obviously similar as comb-finger size. The measurement result also shows good agreement with the FEM analysis result (ANSYS 14.5). The figure shows two-simulation result for electromagnetic flux density profiles 250 μm and 300 μm above the magnet surface. Considering the package thickness of Hall IC, the Hall IC shows reasonable value for flux density measurement.

Journal of Physics: Conference Series **557** (2014) 012033 doi:10.1088/1742-6596/557/1/012033

It is difficult to measure magnetic flux density profile of 30 μm gap that is gap between the bipolar magnet and the harvesting coil in actual device, because of the spatial resolution limitation of Hall IC. Thus, we simulated the magnetic flux density profiles on 30 μm gap for monopolar and bipolar magnetic structure as shown in Figure 5. The bipolar magnet array shows as high as 0.3 T magnetic flux density change, which is almost double of a monopolar one.

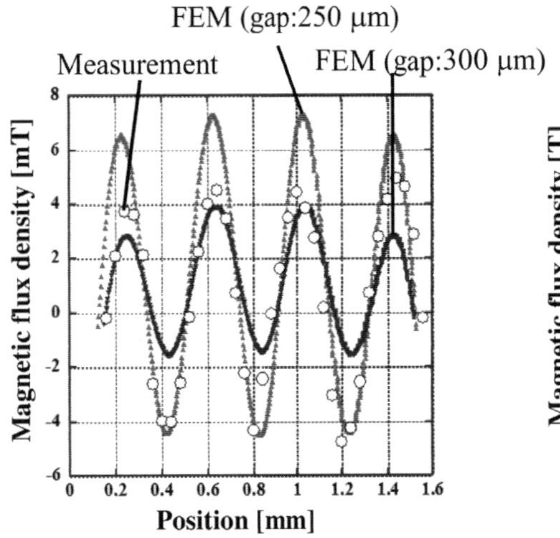

Figure 4. This shows the magnetic flux density profile of bipolar array magnet with 100 μm finger width. In addition, the FEM analysis results at 250 μm gap and 300 μm gap from the magnet surface is presented for comparison of the measurement.

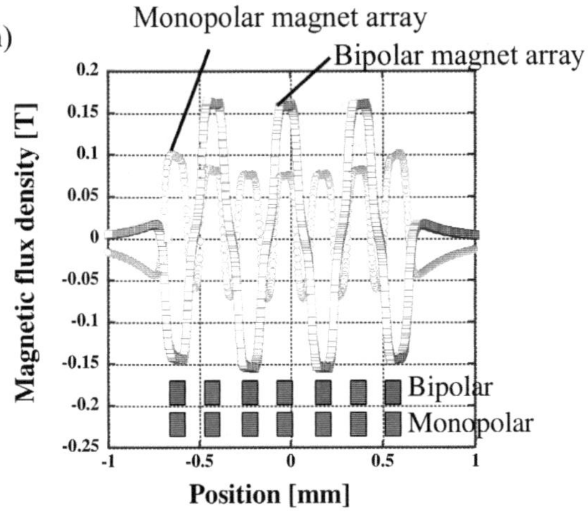

Figure 5. Analysis results of Magnetic flux density at 30 μm gap for monopolar and bipolar magnet array with 100 μm finger width.

3. Energy estimation from electromagnetic harvester

3.1. Structure of harvester

Figure 6*a* shows the cross sectional view of our electromagnetic energy harvester using the bipolar magnet array. The harvester is composed by the moving coil layer and the fixed bipolar magnet structure. The coil structure is on the spring-mass mechanical resonator. The coil with serpentine shape is made of 8 μm thickness electroplating Au [5] as shown in Figure 6b. Compare to the spiral coils, this serpentine coil is easily miniaturized according to miniaturization of magnet array.

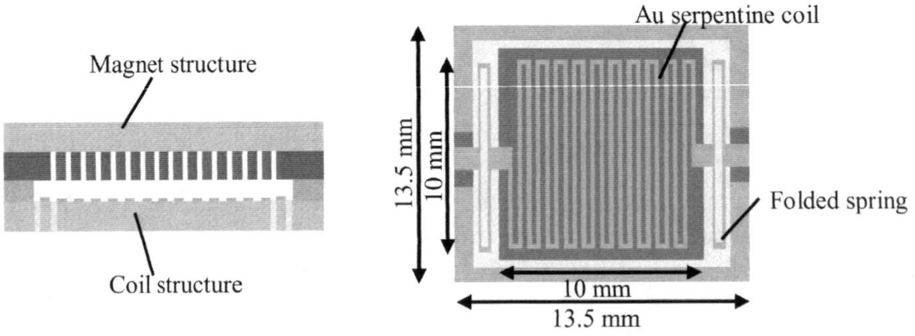

(a) Sectional view of harvester (b) Top view of coil structure

Figure 6. Electromagnetic harvester with bipolar array magnet. The device size is $13.5\times14\times1.5$ mm^3, the both of magnet area and coil area are 10×10 mm^2.

3.2. Power generation estimation

By using the structure and dimension of our electromagnetic vibration energy harvester described above, we try to estimate the harvested energy from the approximated magnetic flux density profile and numerical calculation [2]. Combining of the approximation equation for magnetic flux density from the FEM analysis and electromotive force from the serpentine coil, we can calculate a harvested energy from the various sizes of magnetic comb-fingers.

The gap between magnets and coils is set to 30 μm, the thickness of magnet film on the Si structure is measured value of 27 μm and the magnetic coercive force is 0.9 MA/m. The sectional area of Au coil line is 20×8 μm^2 and its resistivity is 2.21×10^{-8} Ωm. The Q-value of the spring-mass structure is set to 100 from the previous work. Figure 7 shows the generated power of the harvester and coil impedance with bipolar magnet array and monopolar magnet array. The bending number of the serpentine coil is depended on the rise and fall of the magnetic flux density profile. Since the bipolar magnet array makes half number of undulation with same width and space of the comb-finger for monopolar magnet array. Thus the coil impedance of bipolar magnet array shows half of the monopolar one. From the estimation result, the device using bipolar magnet array generates 3.12 μW at the optimum comb-finger width of 40 μm when applying a mass vibration with resonant frequency of 400 Hz and amplitude of 4 $\mu m_{p\text{-}p}$.

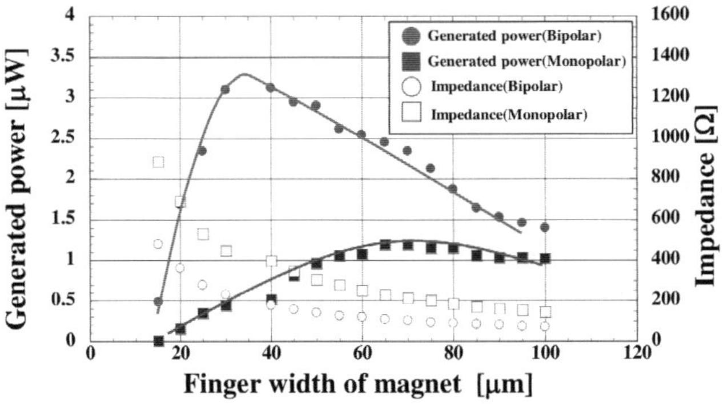

Figure 7. Generated power and impedance of the harvester versus interval of magnet comb-finger width and space. This graph shows both harvester with bipolar magnet and harvester with monopolar magnet at sin vibration(400 Hz, 4 $\mu m_{p\text{-}p}$). The size of the serpentine coil is designed by the magnetic flux density profile.

4. Conclusion

We demonstrated the MEMS batch fabrication process of bipolar magnet array. The comparison of the measurement results and simulation results, it is investigated that the simulation can be useful for magnetic flux density profile approximation. By using approximation, we found the optimum width of magnet comb finger of bipolar array. It showed 3.12 μW maximum power from the 400 Hz resonant mode vibration with 4 $\mu m_{p\text{-}p}$ amplitude.

References

[1] Miki S et al 2012 *Tech. Dig. Int. Conf. on MEMS2012 (Paris)* pp 1221-24
[2] Tanaka Y et al 2012 *Tech. Dig. Int. Conf. on PowerMEMS2013(London)*
[3] Peterson B A et al 2012 *Tech. Dig. Int. Conf. on PowerMEMS2012(London)* pp 319-322
[4] Jiang Y G et al 2011 *Fabrication of a Vibration-driven Electromagnetic Energy Harvester With Integrated NdFeB/Ta Multilayered Micro-magnets* Journal of Micromechanics and Microengineering vol. 21 095014
[5] Tanaka Y, et al 2013 *Proc. The 2nd Int. Workshop on e-Wise (Beijing)* pp 1656-58

Inductive Energy Harvesting for Rotating Sensor Platforms

T T Toh, S W Wright, M E Kiziroglou, P D Mitcheson and E M Yeatman
Department of Electrical and Electronic Engineering, Imperial College London, SW7 2AZ, U.K.

E-mail: tzern.toh02@imperial.ac.uk

Abstract. An inductive energy harvesting concept for structures rotating in proximity to a stationary body is proposed. The performance of such devices is studied analytically and numerically, and an experimental proof of concept is presented, demonstrating energy output density of 1 mW/cm^3 from a typical geometry and rotation scale. The proposed approach may be suitable for powering retrofitted wireless sensors on engine bodies and also on rotating parts where complex stator-rotor wiring solutions would otherwise be required.

1. Introduction

A range of applications for energy autonomous sensors involves rotating structures. Such applications include airborne and terrestrial vehicles as well as industrial facilities. Harvesting energy from rotary motion has been limited to locations in the vicinity of the rotation axis, mainly because the high centrifugal force that appears off-axis locks a suspended proof mass into a fixed position [1]. For the same reason, other approaches such as the combination of gravity and the Coriolis acceleration require complicated mechanical structures that are impractical to implement within the typical sizes of sensor packages. Here, an inductive device concept is introduced, for applications where relative motion between proximal structures is available. This is usually the case in rotating engines such as motors, generators, turbines etc. A schematic description of the operation principle is presented in Figure 1. The device consists of one or more permanent magnets attached to one of the structures, while one or more coils are attached to the other. During motion, as a coil passes by magnet, a voltage is induced to it. This voltage can be exploited by appropriate electronics to harvest energy from the relative motion.

2. Analysis

In order to investigate the performance of particular implementations for inductive harvesting, a formulation of the field of a permanent magnet (PM) is introduced, including the geometry of a travelling coil. A PM block of rectangular, y_m - z_m cross section, with length x_m and magnetization M along the x – axis is assumed. A rectangular, y_c - z_c cross section coil of N turns is aligned in the z-direction with the PM and can travel along the y direction at a distance x_g from the closest PM face. A schematic of this setup is given in Figure 1 (right top).
An analytical calculation of the magnetic flux density x-axis component, B_x along the central x-axis of the PM is possible by calculation of the magnetic scalar potential V_m that is imposed at a position x by each discrete PM volume element dV [2]. A vector algebraic analysis, with reference to Figure 1 (top right), gives the result:

$$V_m = \int_V \frac{1}{4\pi} \cdot \frac{M \cdot r}{|r|^3} dV = \frac{M}{4\pi} \int_{x_g}^{x_g+x_m} \int_{-y_m/2}^{y_m/2} \int_{-z_m/2}^{z_m/2} \frac{x}{\sqrt{x^2+y^2+z^2}} dxdydz \quad (1)$$

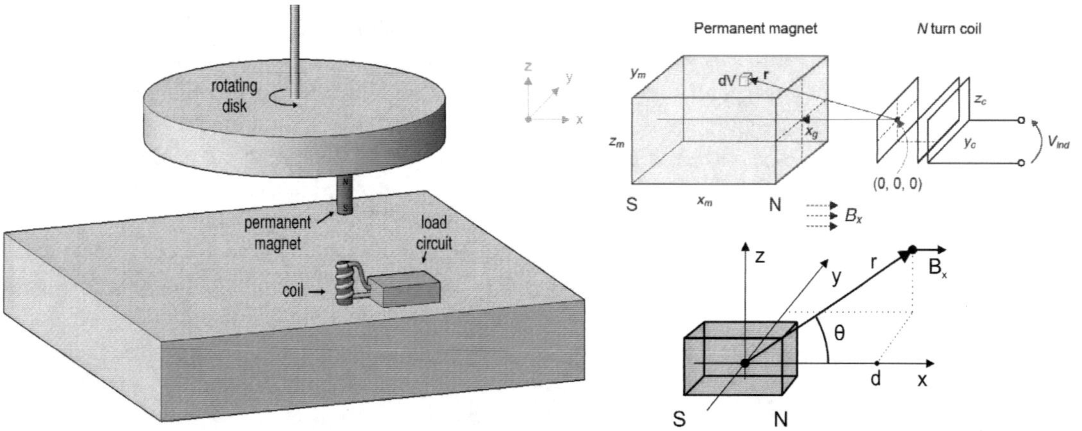

Figure 1. Left: Schematic of the device concept. Right: Vector diagrams for magnetic field calculation along the PM axis (top) and by treating the PM as a single magnetic dipole (bottom).

Then B_x can be found by calculating the integral for y and z, and differentiating V_x with respect to x:

$$B_x = -\mu \nabla V_m \cdot \hat{x} = \frac{\mu M}{\pi}\left(atan\frac{y_m \cdot z_m}{2x_g\sqrt{4x_g^2+y_m{}^2+z_m{}^2}} - atan\frac{y_m \cdot z_m}{2(x_g+x_m)\sqrt{4(x_g+x_m)^2+y_m{}^2+z_m{}^2}}\right) \tag{2}$$

where μ is the magnetic permeability of air. A first estimate of induction could be obtained by assuming a uniform B_x field abruptly appearing in front of the PM. This approach provides a reliable prediction of the induction dependence on PM – coil distance x_g and an overestimation of the induction magnitude. However, it doesn't provide information about the actual profile of the induced voltage. Another approach would be to treat the PM as a single magnetic dipole. Such an approach has low quantitative accuracy at the vicinity of a PM but provides a more detailed description of the fringing flux. In this case, the x-axis magnetic flux density at a distance r, and angle θ from the centre (Figure 1 right bottom) can be written as [3]:

$$B = \frac{\mu m_r}{4\pi r^3}(2cos\theta \hat{r}_0 + sin\theta \hat{\theta}_0) \tag{3}$$

where m_r=M·V is the total magnetic moment of the PM, V being its volume. If d is the distance between the coil and the centre of the PM, then its x-axis component in Cartesian coordinates and the corresponding induced voltage will be:

$$B_x = \frac{\mu m_r}{4\pi r^3}(2cos^2\theta - sin^2\theta)\hat{x}_0 = \frac{\mu m_r}{4\pi} \cdot \frac{2d^2-z^2-y^2}{(d^2+z^2+y^2)^{\frac{5}{2}}} \tag{4}$$

$$V = -N \cdot \frac{\partial \Phi}{\partial t} = -N \cdot U \cdot \frac{\partial \Phi}{\partial y} = -N \cdot U \cdot 2\left(F\left(\frac{z_c}{2}, y + y_c\right) - F\left(\frac{z_c}{2}, y\right)\right) \tag{5}$$

Here, U=y/t is the speed of the coil with coil-PM alignment occurring at t = 0. The last term of (5), including the factor of 2, represents the flux gradient along y. The function F is the general integral of B_x from equation (4) with respect to z:

$$\boldsymbol{F(z,y)} = \frac{\boldsymbol{\mu m_r}}{\boldsymbol{4\pi}} \cdot \frac{(d^2-y^2)z^3+(2d^2-y^2)(d^2+y^2)z}{(d^2+y^2)^2(d^2+z^2+y^2)^{\frac{3}{2}}} \tag{6}$$

From equation (5) one can calculate the induced voltage and power output for different setups. Results for a particular harvester with a 10 ×10 cm, 100 turns coil passing by a 10 × 10 × 40 mm neodymium PM at 10 mm distance with speed corresponding to 1600 rpm at 0.5 m from the rotation axis are shown in Figure 2. For the calculation of power, a resistive load matching a coil resistance of 8.56 Ω was considered, which corresponds to a 0.1 mm diameter Cu wire. As expected, a smooth alternating voltage profile is observed with amplitude exceeding 1 V for the particular device and conditions. The total energy from a single coil-magnet interaction is over 12 μJ. If one such interaction per full rotation is taken into account, an average harvesting power of 0.32 mW is obtained. Although the total PM magnetic moment was taken into account, the field is underestimated by this method, because the assumption of a single magnetic dipole leads to an effective increase of the PM-coil distance.

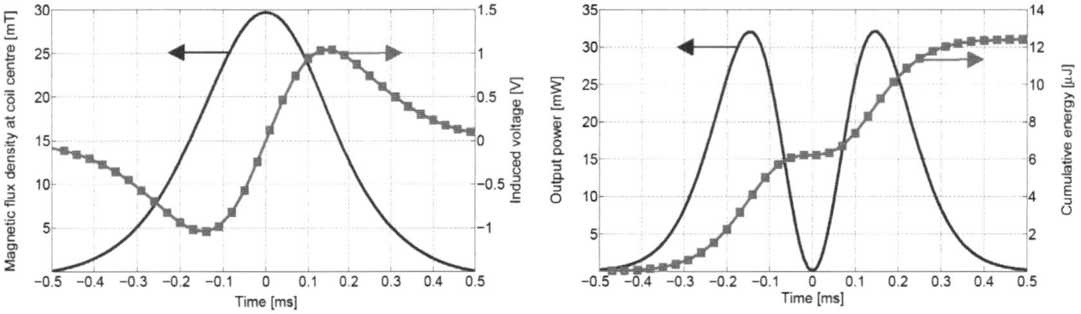

Figure 2. Analytical induction calculations for a 1×1 cm coil with 100 turns at a gap distance of 1 cm from a 1 × 1 × 4 cm neodymium PM, 0.5 m off- axis, at 1600 rpm.

3. Simulation

As a complementary method to the above analysis, the magnetic field of PMs in geometries relevant to the device design was calculated using finite element analysis in 3D. The simulation software used was COMSOL. A 10 × 10 × 40 mm neodymium PM with M = 1 MA/m and $\mu_r = 1$ was assumed. The flux through a y-z frame moving along the y-axis at a distance of 10 mm from the magnet's edge was studied, with and without a soft magnetic core. The frame size was 10 × 10 mm with negligible turn dimensions. The core size was 20 × 10 × 10 mm with its closer-to-magnet small face coinciding to the frame. A constant relative permeability of 4000 was assumed. This geometry is illustrated in Figure 3 along with x-axis flux density distribution results for two different core positions.

For a given y-axis frame velocity, the total flux through the frame was calculated as a function of time. Results are plotted in Figure 4 for the case of a 1600 rpm, 1 m diameter rotation (velocity 83.8 m/s). An over two-fold increase of the peak flux is observed with the introduction of the magnetic core. From these simulations, the open-circuit induced voltage can be calculated as the negative derivative of total flux. The corresponding results are also presented in Figure 4. In the same figure, the corresponding analytical calculation, as obtained by equation (5) is plotted for comparison. These results provide a quantitative demonstration of the performance enhancement that is achieved by employment of a soft magnetic core. In addition, they demonstrate the correct prediction of voltage profile by the proposed analytical model.

The underestimation of the magnetic field density in equation (4) is studied further by comparing the simulated B_x profile on the central PM x-axis with the analytical ones. The results are presented in Figure 5. Excellent matching between the precise expression (2) and the simulated values is observed, supporting the validity of the numerical model. As expected, the single magnetic dipole assumption of equation (4) leads to significant underestimation of B_x in the PM vicinity. This is because all the magnetic moment is assumed to be concentrated to the PM centre, effectively increasing the distance between field-source and coil. For distances larger than the PM length, all three methods converge. In the same diagram, the case of using a core at a distance of 10 mm from the PM edge is also plotted, illustrating the field concentration that is accomplished. At the core edge, the B_x maximum is increased to 135 mT, from 77 mT without core, a 75% improvement. The corresponding total flux improvement, as illustrated in Figure 4, is from 6.3 μWb to 15.2 μWb, or 141%.

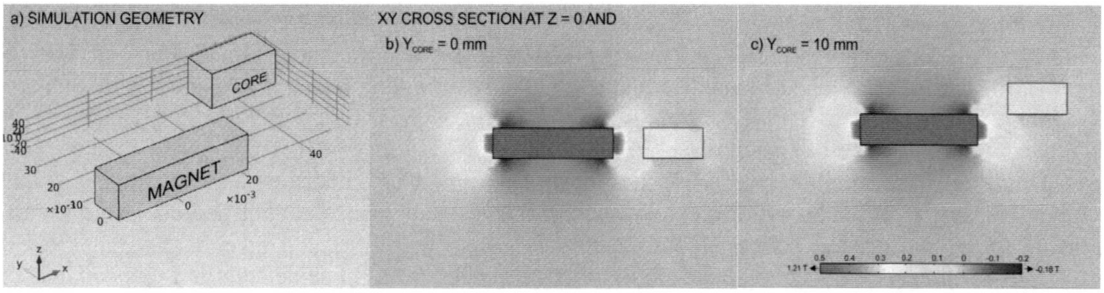

Figure 3. Geometry and x-axis flux density distribution of a magnet with M = 1 MA/m along the x-axis and a core with $\mu_r = 4000$, for 2 different core positions. Y_{CORE} is the magnet-core distance (centre-to-centre) along y.

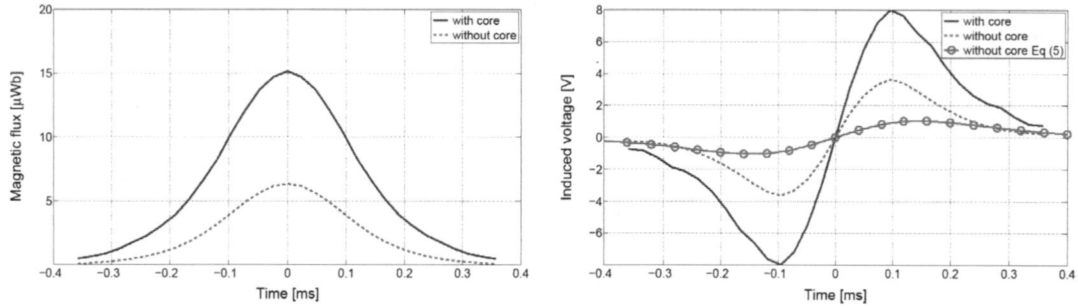

Figure 4: Left: Simulated flux through a 10 × 10 mm frame travelling at 83.8 m/s (1600 rpm, diameter: 1 m) with respect to a PM at a distance of 10 mm (Figure 3). Right: Corresponding induction on a 100 turn coil.

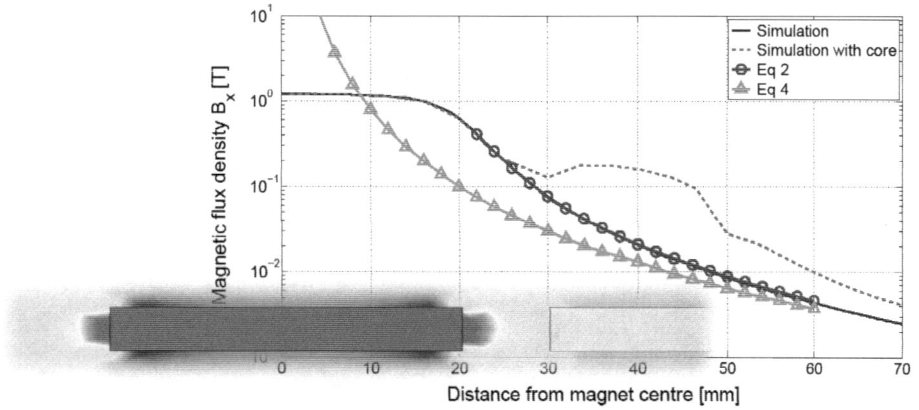

Figure 5: Simulated flux density as a function of distance from magnet along the x-axis at y = z = 0, with and without a soft magnetic core in comparison with analytical calculations. All parameters are nominal (no fitting).

4. Experimental Results

A prototype implementation was built and tested as illustrated in Figure 6. The coil – magnet gap is adjustable from 0.2 to 10 cm. and radial distance can be up to 15 cm. Two magnetic cored coils wound with 1000 turns of 0.1 mm diameter Cu wire and 79 Ω resistance are attached to the baseplate. Two 1 cm diameter, 1 cm length N42 Nd magnets are mounted on the rotor, which is spun by a motor controllable up to 1200 rpm. The voltage induced in the coil is observed on an oscilloscope and the available power into a matched resistive load calculated. The magnets are rotating and the coils are fixed in this implementation, but the converse arrangement is also possible and would allow energy harvesting on a rotating structure.

Figure 7 shows (left) a typical voltage trace observed from a 200 turn coil during rotation as a magnet passes by the coil assembly. A voltage profile shape matching that in Figure 2 is seen. On the right, measured induced open circuit voltage and the corresponding power as a function of rotation speed are shown, from a 2 PM, single coil arrangement. The linear fit seen for the voltage is in agreement with the theory of equation (5). The power levels of 4 mW at 600 rpm and 16 mW at 1200 rpm demonstrate the possibilities of harvesting significant power on a rotating system at speeds typical of much industrial and transport machinery. Such power levels are sufficient for operation of wireless sensors on moving structures where the provision of power by cabling would not be possible. Power capability can be raised by adding more magnets to the system, with the same coils and electronics.

5. Conclusion

An inductive concept for energy harvesting from proximal relative motion was introduced. The operation of such devices was studied analytically and simulated numerically, revealing an expected output exceeding the requirements of typical low-power sensor node implementations. Experimental results demonstrated several mW of continuous power supply, from a device implementation of around 2 cm^3 of magnet volume and 1000 coil turns at a 150 mm distance from the rotation axis. Further experimental results and a more detailed analysis will be presented in a following paper.

Figure 6: Inductive harvesting prototype device.

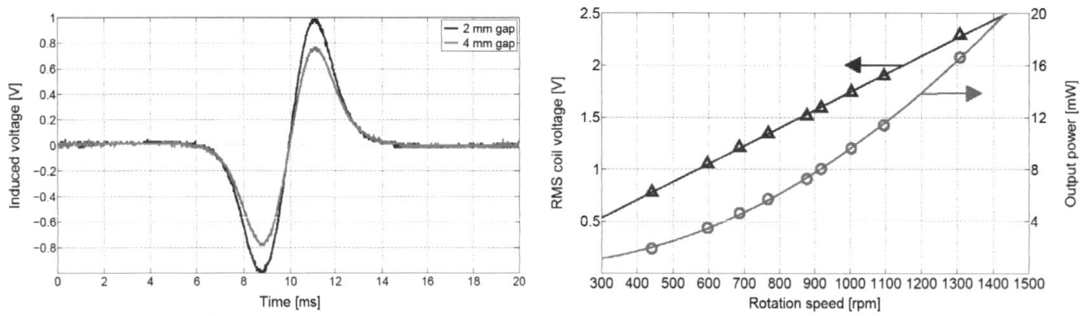

Figure 7: Left: Open-circuit induced voltage from a magnet – coil interaction at 750 rpm rotation speed, 200 turns and a rotation radius of 50 mm. Right: Open-circuit voltage amplitude as a function of rotation speed.

6. References

[1] T. T. Toh et al, *J. Micromech. Microeng.* 18 (2008) 104008
[2] J. Vanderlinde, *Classical Electromagnetic Theory* vol. Ch. 7: Wiley, 1993.
[3] M. Zahn, *Electromagnetic Field Theory* vol. 5.5.1: MIT OpenCourseWare, 2014.

Characterization of a variable reluctance harvester

M. Kroener[1], **N. Moll, S. K. T. Ravindran, P. Mehne, and P. Woias**

Group Manager, *Micro Energy Harvesting Technologies*
University of Freiburg, Department of Microsystems Engineering – IMTEK, Laboratory for Design of Microsystems, Freiburg, Germany

E-mail: michael.kroener@imtek.uni-freiburg.de

Abstract. In our last year's PowerMEMS contribution we presented a proof-of-concept of a variable reluctance harvester for the application in a railroad surveillance system. It was shown that intermittently closing a magnetic circuit could supply power output in the range of mW's. The test setup used showed unwanted energy pickup from the electro motor used. In this paper we present thorough measurements of the reluctance circuit with a compressed air motor to exclude the effects of the above mentioned magnetic stray fields. The effects of eddy currents and moment of inertia on the output power, the optimal coil position on the stators, and effects of different magnetic field strengths are studied. The gap width is set to a fixed value of 14 mm, representing a realistic scenario.

1. Introduction

In previous publications we have shown that a wireless sensor node mounted to a railroad track can be powered by vibrations of a passing train [1]. Applications for this systems lie in the field of structural health monitoring in tunnels, but also other applications are feasible and useful, such as train position or passage detection, or wheel health monitoring of the passing train [1]. In continuation of this research we have presented a contactless energy harvester tailored for the energy autonomous train passage detection [2].

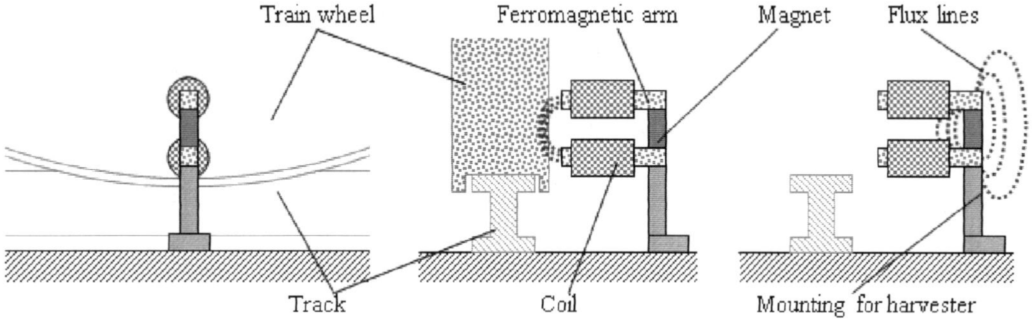

Figure 1. Schematic of the harvester mounted on a railway track. (a) shows the front view whereas (b) and (c) show the scenarios with and without the wheel, respectively [1].

[1] To whom any correspondence should be addressed.

It is based on a variable reluctance in a magnetic circuit: The train's soft magnetic wheel intermittently closes and opens the U-shaped circuit when passing by. This induces a voltage in the coil wound around one of the soft magnetic arms of the harvester (Figure 1). In literature variable reluctance circuits are used for resonance-frequency tuning of vibrational energy harvesters [3], as well as in energy autonomous anti-lock braking systems in automotive applications [4], which has a similar design as our system. The system presented in [3] is similar to our design but has a clearance between the harvester and the moving parts in the order of 100 µm. In this study the gap between stator and rotor has been increased from 10 mm to 14 mm to further increase the safety margin for the intended use in a railway monitoring application, where the position of the passing wheel is not constant (Figure 1b).

2. Design concept and experimental setup

The realized magnetic circuit is shown in Figure 2, with the concept shown on the left, and the realized test setup on the right. Two iron cores are mounted onto a rotating wheel, which is mounted to the compressed air motor. With these iron cores, the reluctance circuit is periodically closed inducing a voltage in the coil located on the static part of the magnetic circuit according to equation 1:

$$U_{ind} = -N \cdot A \cdot \frac{dB}{dt} \quad (1)$$

With N the number of turns of the coil (N = 700) and the cross-sectional area A = 7.29 cm^2. The coil is wound with a 0.22 mm enamelled copper wire. The length of the coil equals l = 10 mm, the coil resistance is measured with an *Agilent 4263B* LCR meter to be R_{coil} = 23.4 Ω. The stationary U-shaped part of the reluctance harvester is composed of a low-grade steel reluctance core as a flux guide and NdFeB (*supplier: Supermagnete.com*) magnets as the magnetic source, both with a cross-sectional area of 10×10 mm^2.

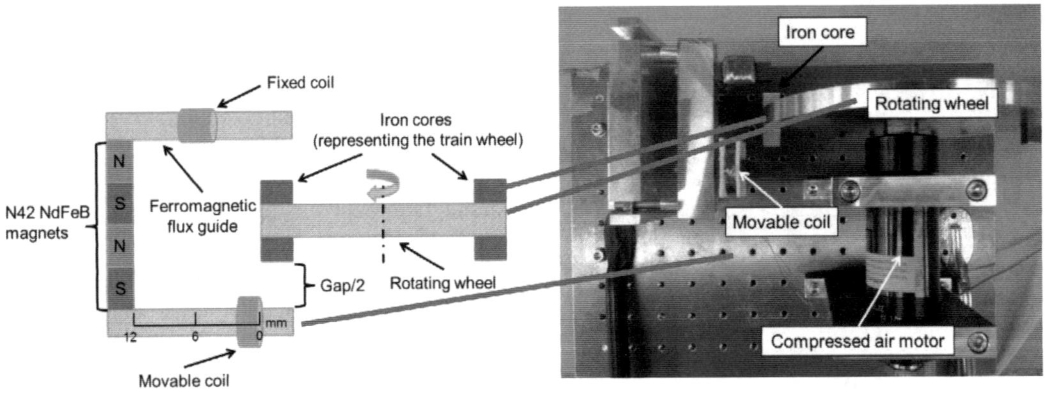

Figure 2. Concept of the variable reluctance circuit (left). The movable coil is used for all of the experiments to determine the optimal coil position and dimension. The rotating wheel carries two soft magnetic iron cores, which periodically close the magnetic circuit. In the experiment (right) the rotating wheel is driven by a compressed air motor to prevent unintended magnetic stray fields from the formerly used electromotor.

The gap between the stationary arms and the rotating iron cores is defined by magnets of different height and is set to a fixed value of 14 mm for all experiments. The relative permittivity of the low-grade steel was determined to be μ_r=666, measured with a self-built test setup according to the *DIN Norm 50 460* for soft magnetic materials. The voltage waveform is measured and recorded with a

Tektronix TDS 2002 oscilloscope, which is later used for calculation of the harvested energy per passing event.
The rotating disc has a diameter of 160 mm (Figure 2 left and right) and is fabricated from aluminium and PVC to determine the effect of eddy currents slowing down the disc. Additionally different brass weights have been used to cross correlate the effect of different moments of inertia. The electro motor used in the former study [1] was replaced by a compressed air motor *MRD 38-1460* from *Mannesmann-Demag* (Figure 2 right). This motor was used to circumvent the pickup of magnetic stray fields formerly observed from the experiments with the electro motor. The air motor could not be operated in the same rpm range (0-100 Hz) due to a limited available air pressure. By that, experiments were conducted only in the range of 5 – 25 Hz (corresponding to 10-50 Hz due to the two iron cores mounted). For rates lower than 5 Hz, the motor stalled because of internal frictional effects. Magnets with different flux density were used (N42, N45, N50) to evaluate if saturation effects in the soft magnetic core of the reluctance circuit play a role in the energy output.

3. Experimental results and discussion

3.1. Results

Figure 3 shows the measured energy per passage of the soft iron core with respect to different test setup configurations to compare the results of the electro motor and the compressed air motor. The movable coil is additionally varied in distance from the moving iron core, see Figure 2 left. For these experiments a PVC disc is used without an additional weight to exclude the effect of eddy currents in the else used Aluminium disc. The energy per passage was calculated numerically from the recorded voltage waveforms according to eq. (3), including the optimum load-matched conditions ($R_{load} = R_{coil}$).

$$E_{pulse} = \sum U^2 /(4R_{coil}) \cdot \Delta t \qquad (2)$$

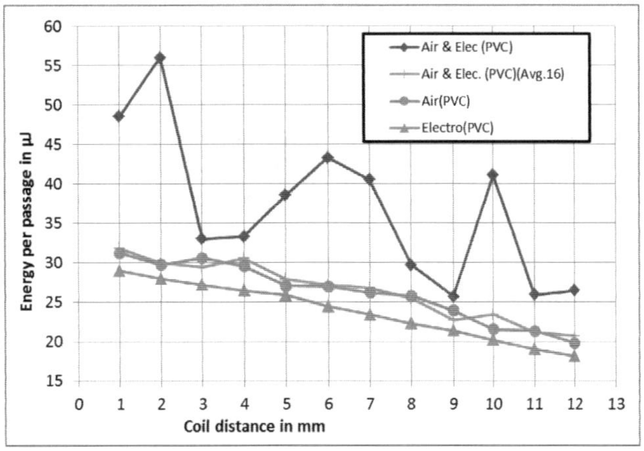

Figure 3. Measured energy per passage of the soft iron core for different test setup configurations. The two blue lines represent the setup with compressed air motor with the electromotor running close to experiment to show the effect of its magnetic stray fields. The top line is measured without averaging. It can clearly be seen that a large amount of energy is induced in the coil, which is not related to the reluctance circuit.

Four different test setups are compared in this experiment (lines from top to bottom):

1. The compressed air motor drives the disc & the electromotor is place next to the experiment, to show the effects of the magnetic stray field (top blue line). The voltage waveform is not averaged in the oscilloscope.

2. Same conditions as in one, with averaging over 16 measurements.
3. The compressed air motor drives the disc, with the electro motor being absent. No magnetic stray fields are picked up in the coil.
4. The electromotor drives the disc with averaging over 16 measurements.

These experiments clearly prove that the electro motor induces large amounts of energy into the reluctance harvester from magnetic stray fields. The experiments also shows that averaging over 16 measurements, as we did in the previous study [2], effectively eliminates these side effects. The small deviation in the harvested energy between the electro motor and the compressed air motor is related to small geometrical differences, which occur if changing the test setup.

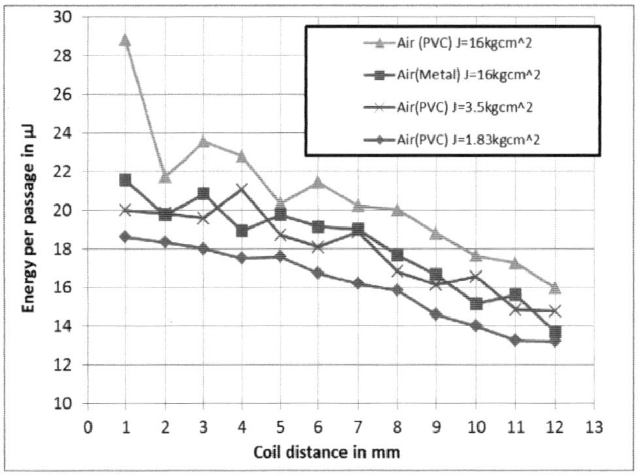

Figure 4. Measured energy harvested per passage of the moving soft iron core, and energy generated per pulse (right), with respect to the train velocity for three different clearance widths between the moving and stationary ferromagnetic parts of the reluctance circuit.

Figure 4 shows measurements of the energy harvested for different moments of inertia *J* conducted with the air motor for a fixed rotational frequency of 30 Hz. The results of the PVC disc are shown along with an experiment with an aluminium disc. It can be seen, that the higher the moment of inertia, the higher the harvested energy. It is also shown that the aluminium discs produces less energy for the same value of $J = 16\ kg{*}cm^2$. This proves the effect of eddy currents occurring in the aluminium disc, slowing down the system especially when the reluctance circuit is closed by the moving soft iron core.

Figure 5 shows measurements with magnets of different magnetization strengths (N42, N45, N50) with respect to the rotational frequency of the disc (PVC). It can be seen, that the strongest N50 magnet does not provide more energy than the weaker N45 magnet. Saturation effects in the soft magnetic material limit the maximum achievable energy. With the N42 magnets this maximum is not yet reached so a lower energy output is observed.

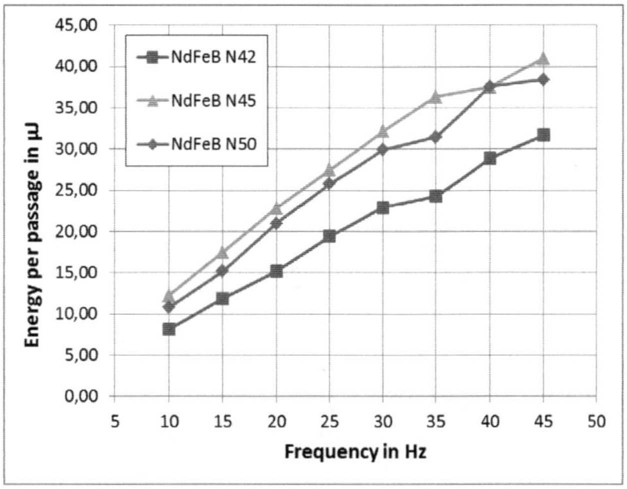

Figure 5. Measured energy per passage with respect to magnets of different strength and in relation to the rotational frequency.

3.2. Discussion

The focus in this study was set to the thorough characterization of the reluctance harvester alongside with the test setup used. Previous research revealed some magnetic stray effects supposedly coming from the electromotor which was used [2]. This side effect was proven by experiments with a compressed air motor and the electromotor four different test setup shown. While the pickup of additional energy from surrounding magnetic fields, which will occur in the application of train passage detection, is rather an opportunity than a problem, one has to consider the protection of the electric circuits behind the harvester. We could observe voltages easily going up to 10 V, which were out of the rated range of the capacitor used in the system from [5].

While the effect of eddy currents slowing down the wheel in the test setup, thereby generating less energy, certainly is not a challenge in the real application, where a very heavy train (~100's of tons) passes by. Still in other technical applications, e.g. health monitoring of tools in rotational machinery, might affect the functionality. In general, the measured energy from the test setup shown is more like a lower limit of what actually can be expected from the application.

4. Conclusions

While the applicability to use a variable reluctance was already demonstrated in the previous study, we now can derive an optimum design for the reluctance harvester and determine the limits and the boundary conditions for the electronics used to manage the harvested power. Now the harvesting principle needs to be tested in a real application to see if the technology is feasible to be used in an unclean environment, where small magnetic particles may disturb the long term use of such a harvester.

[1] Wischke M, Masur M, Kroener M, Woias P, 2011, Smart Mater. Struct. 20, 085014.

[2] M Kroener et al, 2013, J. Phys.: Conf. Ser. 476 012091, doi:10.1088/1742-6596/476/1/012091.

[3] Ayala-Gacia Ivo, Mitcheson Paul, Yeatman Eric, Zhu Dibin, Tudor M.J. and Beeby S.P., Magnetic tuning of a kinetic energy harvester using variable reluctance. *Sensors and Actuators: A Physical* **189** (2013), 266-275.

[4] Parthasarathy D., Energy harvesting wheel speed sensor, *Master thesis*, Chalmers university of technology, Goeteborg, Sweden, 2012.

[5] Woias P, Wischke M, Eichhorn C, Fuchs B, An energy-autonomous wireless temperature monitoring system powered by piezoelectric energy harvesting, *Proc. PowerMEMS* (2009), 209-212.

Micro-fabricated Liquid Encapsulated Energy Harvester with Polymer Barrier Layer as Liquid Electret Interface

L Bu[1], H Y Xu[1], B J Xu[1], L Song[2]

[1] School of Information Engineering, China University of Geosciences, Beijing, China
[2] State Grid Jibei Electric Power Co., Ltd., Beijing, China

Email: lingbu@cugb.edu.cn

Abstract. This paper addresses the electret discharge issue for liquid based electret energy harvesters. An interface structure of PDMS/PTFE polymer barrier layer between liquid and electrets is introduced, achieving 75% charge retain rate over 100h, compared with 0% without the proposed layer over 100h. Further, the PDMS/PTFE layer is introduced into liquid encapsulated energy harvester (LEEH) and is compatible with micro-fabrication process. The retain rate of device voltage is about 47%~65% over 100h. At 100h after corona charging, the device generates maximally 3.7V, 0.55μW @1Hz rotation.

1. Introduction

Liquid based electrostatic energy harvesters are recently brought forward in [1-2] with working frequency typically on the order of 0.1~10Hz, which is suitable for harvesting energy from low frequency vibrations. Such low working frequency can hardly be achieved by resonance based harvester structures such as cantilevers [3] or spring mass damper systems [4-5].

However, among the reported electrostatic energy harvesters using liquid, most only measure the capacitance change, while power output is estimated using external voltage source rather than using electrets [1,6]. One key reason is that the electrets discharge significantly when in contact with liquid, as is shown in Fig.1(a). Possibly, charges recombine with positive groups in the liquid, or transfer to sidewalls or electrode from the electrets. In either circumstance, surface potential of electrets decreases significantly, which implies very short lifespan of the harvester.

It is essential to solve the electret discharge issue for liquid based energy harvesters. In this paper, a method of polymer barrier layer is presented, which isolate electret surface from direct contact with liquid and from the infiltrative liquid vapor. The proposed polymer barrier layer is compatible with the micro-fabrication process of liquid encapsulated energy harvester (LEEH), and the power generation capability and improved reliability of LEEH are experimentally verified.

2. Concept

As is shown in Fig.1(b), polymer barrier layer is introduced at the liquid electret interface. The polydimethylsiloxane (PDMS) layer protects charges in electrets from direct contact with liquid and guarantees large contact angle for liquid motion. Further, in terms of fabrication it is feasible to bond PDMS liquid cavity with the barrier PDMS layer. Since PDMS layer is of high moisture permeation rate, it is important to introduce beneath the PDMS layer an additional polytetrafluoroethylene (PTFE) layer, which is of low moisture permeation rate and thus protect trapped charges in electrets from moisture. The combined effects of PDMS/PTFE polymer barrier layer can further improve charge storage stability of electrets in liquid environment.

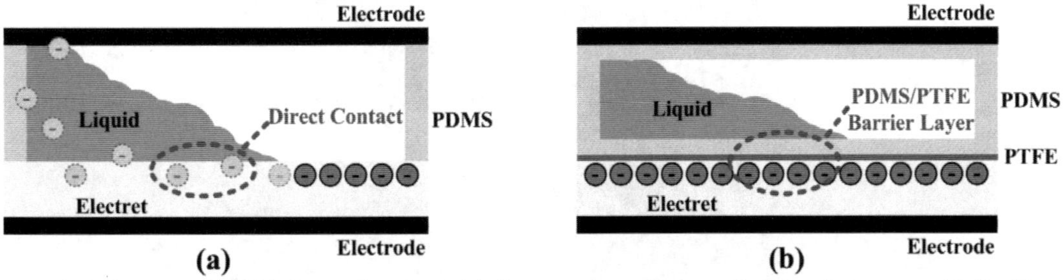

Fig.1 Concept of liquid electret interface: (a) direct contact, electrets discharge in contact with liquid; (b) this design, with PDMS/PTFE polymer barrier layer as liquid electret interface.

3. Experiment

3.1. Fabrication

The proposed PDMS/PTFE polymer barrier layer is utilized in micro-fabrication of LEEH, and the process flow is shown in Fig.2. PDMS is poured into SU-8 mold (600μm thickness) and cut into cavity units after curing and peeling-off. The PDMS cavity units are then bonded with a flat PDMS mold layer with controllable thickness (50μm-1mm). PTFE aqueous solution is curtain coated on the flat PDMS layer.

Assembly process includes three steps. First, di-methylformamide (DMF) is injected into PDMS cavity using micro-syringe. DMF is chosen due to its large dielectric constant and low viscosity [2]. The hole for injection is covered with a small amount of liquid PDMS and cured to keep watertight. Second, thin film PTFE electrets are attached on the electrode and corona charged. Surface potential of thin film PTFE electrets is recorded using electrometer (TREK 344), which is approximately -700V. Finally, the above two parts are combined by pasting the edge together, with curtain coated PTFE layer facing the electrets, so as to reduce the possible parasitic capacitance.

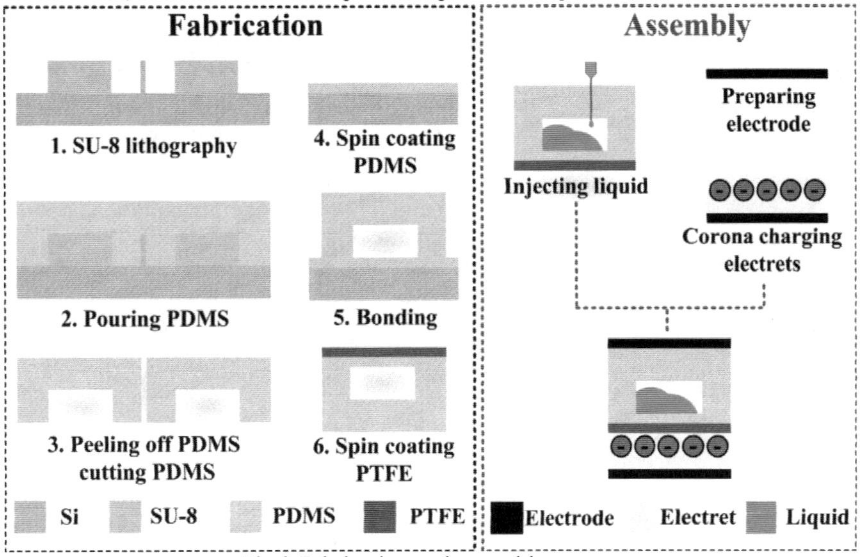

Fig.2 Fabrication and assembly process

Fig.3 shows the fabrication results. In Fig.3(a), the thickness of curtain coated PTFE layer is not uniform, achieving average thickness of 80μm. This is because the non-wetting feature of both PDMS and PTFE renders it difficult to fabricate a flat and smooth interface layer. Fig.3(b) is the photo of fabricated PDMS cavity arrays after demolding. Cavity dimension is 4.4mmx3.4mmx0.6mm. Fig.3(c) shows the assembled LEEH. The size of the entire device is 6.5mmx5.4mmx4.3mm.

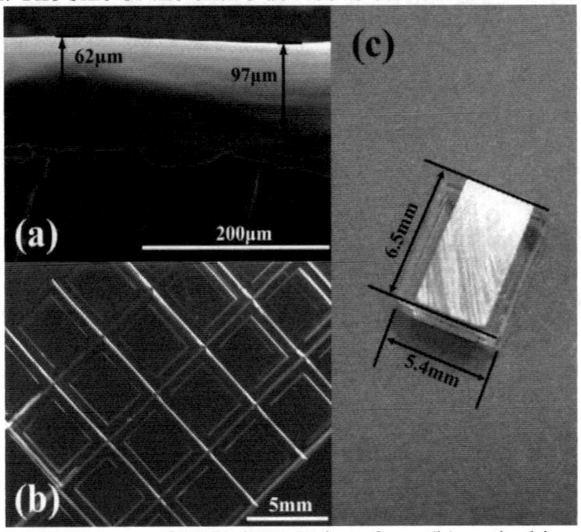

Fig.3 Fabrication result: (a) SEM image of PTFE/PDMS interface; (b) optical image of PDMS cavity arrays after demolding; (c) photo of micro-fabricated and assembled LEEH.

3.2. Testing

Our previous work shows that LEEH generates more power in rotation than in in-plane vibration [2]. Therefore, only rotary test is adopted here. LEEH is positioned on one side of the stepping motor shaft, and is driven by out of plane rotation (0~360° continuously). The rotation frequency is controlled at 1Hz. During test, output voltage is recorded using Agilent 34410A digital multimeter with 10M input resistance probe.

4. Results and discussion

Fig.4 shows that thicker PDMS/PTFE layer is conducive to more stable charge storage in liquid environment, at the cost of lower surface potential due to the dielectric loss within the thick layer. Thus thickness of PDMS/PTFE layer should be traded off between high surface potential and high charge retain rate in liquid environment. In this work, PDMS/PTFE layer with 200μm/80μm thickness is adopted, corresponding to -636V surface potential and 75% retain rate over 100h.

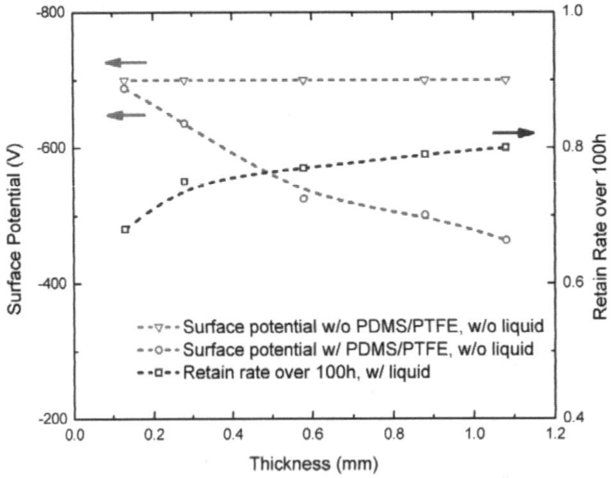

Fig.4 Surface potential and retain rate (over 100h) for different thicknesses of PDMS/PTFE layer.

Fig.5 shows the effect of the 200μm/80μm PDMS/PTFE layer in maintaining surface potential in liquid environment. Within 0.5h, the surface potential retain rate is 93% for both PDMS/PTFE layer and PDMS layer (200μm), while that of direct contact is 12%. From 0.5h to 100h, the retain rate decreases to 75% for PDMS/PTFE layer and 44% for PDMS layer, indicating the necessity of PTFE layer for shielding moisture over the long term. The retain rate of direct contact is 0% over 100h.

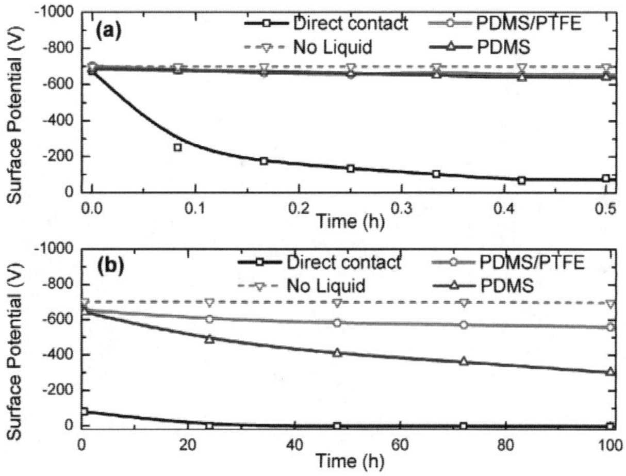

Fig.5 Surface potential decay curves for (a) short term (0~0.5h); (b) long term (0.5~100h).

Fig.6 shows the voltage-time trace @1Hz rotation on 5MΩ resistance at 0.5h and 100h after corona charging. The frequency of voltage-time traces are typically 2Hz, because during one rotation (1s) the capacitance change is symmetrical. In Fig.6(a), maximal and minimal voltage are 1.20V and -1.99V, while that in Fig.6(b) are 0.79V and -0.95V. Thus device voltage retains by 47%~65% over 100h.This lower retain rate is possibly related to degraded paste for package in liquid environment.

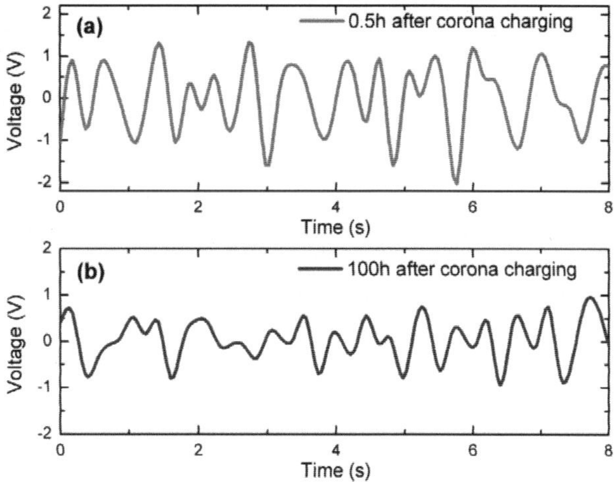

Fig.6 Voltage-time trace of LEEH with PDMS/PTFE layer: (a) 0.5h after corona charging; (b) 100h after corona charging.

Fig.7 shows power-resistance trace of micro-fabricated LEEH at 100h after corona charging. Load resistance changes from 1MΩ to 60MΩ. Maximally, 3.7V voltage and 0.55μW power are achieved on

25MΩ resistance @1Hz rotation. These results prove that the PDMS/PTFE polymer barrier layer is effective in extending the lifespan of liquid based electret energy harvester. It is believed that a thinner and more uniform polymer barrier layer will further enhance the long-term reliability and improve device output power.

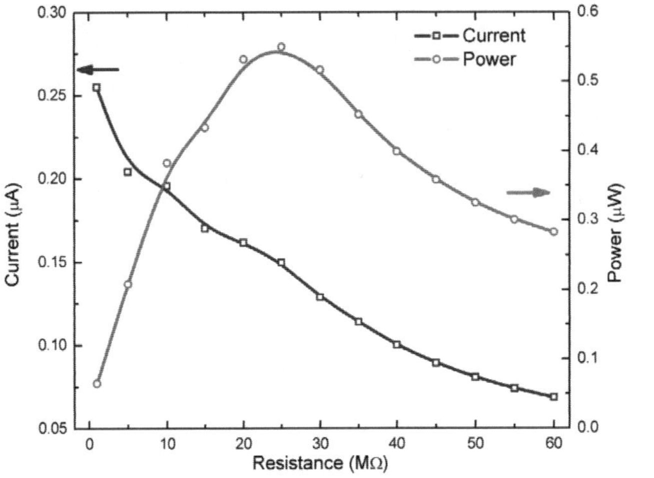

Fig.7 Power-resistance trace of LEEH with PDMS/PTFE polymer barrier layer.

Conclusion

This paper presents PDMS/PTFE polymer barrier layer to address the electret discharge issue for liquid based electret energy harvesters. the PDMS/PTFE layer is compatible with micro-fabrication process of liquid encapsulated energy harvester (LEEH). The PDMS/PTFE polymer barrier layer achieves 75% charge retain rate over 100h, compared with 0% without the proposed layer over 100h. Device voltage retain rate is about 47%~65% over 100h. At 100h after corona charging, the device generates maximally 3.7V, 0.55μW @1Hz rotation.

Acknowledgements

This work is supported by Fundamental Research Funds for Central Universities (Grant No. 2-9-2014-010) of China.

References

[1] D. Choi, et al. Liquid-based electrostatic energy harvester with high sensitivity to human physical motion. *Smart Mater. Struct.*, 2011, 20, 125012 (8pp).

[2] L. Bu, et al. Liquid encapsulated electrostatic energy harvester for low-frequency vibrations. *J. Intel. Mat. Syst. Str.*, 2013, 24(1), pp.61-69.

[3] E. Aktakka, et al. A CMOS-compatible piezoelectric vibration energy scavenger based on the integration of bulk PZT films on silicon. *Proc. IEDM*, 2010, pp.31.5.1-31.5.4.

[4] S. Beeby, et al. Energy harvesting vibration sources for microsystems applications. *Meas. Sci. Technol.*, 2006, 17(12), pp.R175-R195.

[5] Q. Zhang, et al. Microelectromagnetic energy harvester with integrated magnets. *Proc. Transducers*, 2011, pp.1657-1660.

[6] T. Galchev, et al. A gap-varying electrostatic transducer utilizing ferrofluid-based actuation for motion harvesting. *Proc. IEEE MEMS*, 2014, pp.350-353.

White Noise Responsiveness of an AlN Piezoelectric MEMS Cantilever Vibration Energy Harvester

Y Jia and A A Seshia
Nanoscience Centre, University Cambridge, Cambridge, CB3 0FF, UK
E-mail: yj252@cam.ac.uk; aas41@cam.ac.uk

Abstract. This paper reports the design, analysis and experimental characterisation of a piezoelectric MEMS cantilever vibration energy harvester, the enhancement of its power output by adding various values of end mass, as well as assessing the responsiveness towards white noise. Devices are fabricated using a 0.5 μm AlN on 10 μm doped Si process. Cantilevers with 5 mm length and 2 mm width were tested at either unloaded condition (MC0: f_n 577 Hz) or subjected to estimated end masses of 2 mg (MC2: f_n 129 Hz) and 5 mg (MC5: f_n 80 Hz). While MC0 was able to tolerate a higher drive acceleration prior to saturation (7 g with 0.7 μW), MC5 exhibited higher peak power attainable at a lower input vibration (2.56 μW at 3 ms^{-2}). MC5 was also subjected to band-limited (10 Hz to 2 kHz) white noise vibration, where the power response was only a fraction of its resonant counterpart for the same input: peak instantaneous power >1 μW was only attainable beyond 2 g of white noise, whereas single frequency resonant response only required 2.5 ms^{-2}. Both the first resonant response and the band-limited white noise response were also compared to a numerical model, showing close agreements.

1. Introduction

Amongst the various designs for piezoelectric vibration energy harvesting, cantilever-based topologies have been the most popular [1] due to simplicity of design and its ability to accumulate strain-induced electrical charges near its clamped end and room to accommodate a proof mass near its less strained free end. MEMS iterations of these piezoelectric cantilever harvesters have also been demonstrated in the literature for at least over half a decade [2].

The inclusion of an end mass has been shown, both through simulation and experiment, to drastically improve the power response of the harvester [3] compared to a plain cantilever beam. However, most of these previous studies omit an analysis on the response of these devices for broadband excitation and focuses purely on the resonant response.

This paper experimentally analysis the power responsiveness of a plain micro-cantilever (MC), a MC with 2 mg added end mass and a MC with 5 mg added end mass. Both the single frequency resonant response as well as the white noise response were investigated. A numerical model based on beam theory and piezoelectric strain-to-charge transduction model was constructed and used to verify the experimental observation.

2. Apparatus

The devices were fabricated by a 0.5 μm aluminium nitride (AlN) on 10 μm doped silicon (Si) process using the material stack shown in figure 1. A cantilever beam of 5 mm length and 2 mm width, as shown in figure 2 was employed.

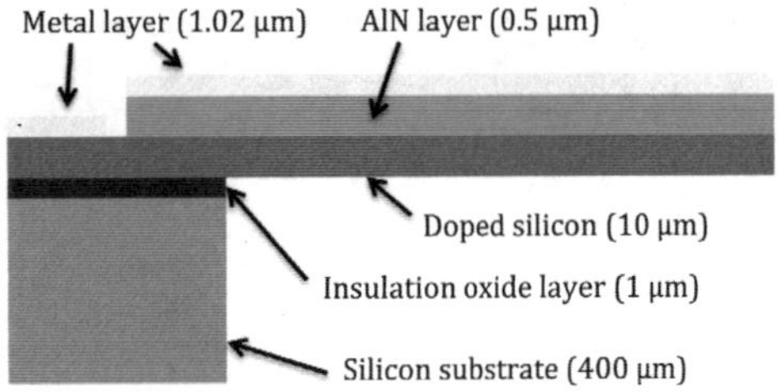

Figure 1. Stack of material used to fabricate the MEMS AlN-on-Si cantilever harvester prototypes.

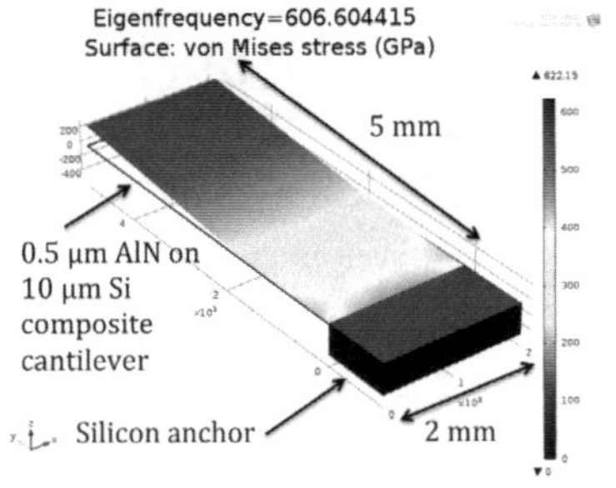

Figure 2. Plain cantilever, fn ~600 Hz.

Additional mass was added (post fabrication) to the free end of the cantilever using a non-conductive adhesive as shown in figure 3. The mass blocks were made from lead-based solder wire for its high mass density. Using shifts in the natural frequency, beam theory equations and COMSOL simulation fits, the approximate value of the added mass were estimated.

The experimental setup mounted on top of a mechanical shaker is shown in figure 4. Experimentally matched optimal resistive loads were connected in parallel for each variation of the micro-cantilevers: MC0, MC2 and MC5. The central base of the chip carrier was machined to accommodate large travel (±5 mm) in order to prevent premature saturation as a result of physical limits imposed by the package.

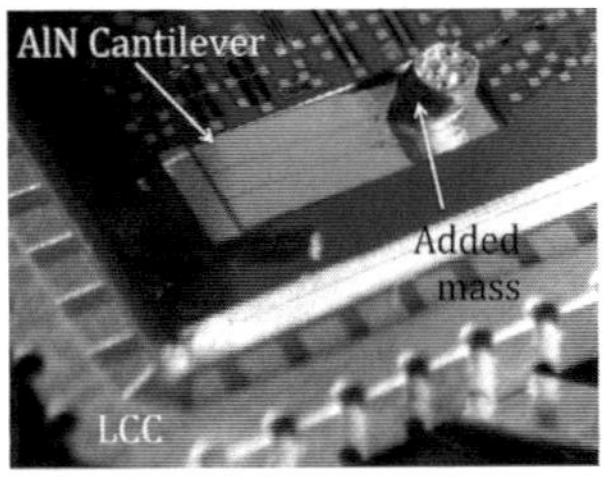

Figure 3. Photograph of micro-cantilever with added solder mass near the free end.

Figure 4. Photograph of the experimental setup on a mechanical shaker.

3. Model

To verify the experimental data measured, a MATLAB numerical model (figure 5) was constructed using the various established equations associated with a classical second order mass-spring-damper resonator (equation 1) when subjected to a periodic mechanical forcing, beam theory for extracting the dynamic bending strain value for a given deflection (equation 2) and the piezoelectric transducer model based on the mechanical strain-induced-charge generation (equation 3). The generated charge can then be assumed as a power source across an impedance equivalent to the capacitance of the piezoelectric layer (equation 4).

$$\ddot{x} + 2\zeta\omega\dot{x} + \omega^2 x + \mu x^3 = A\omega^2 \cos(\omega_0 t) \quad (1)$$

where, x is the displacement or deflection at the free end of the cantilever resonator, ζ is the damping ratio, μ is the Duffing coefficient, ω is the excitation frequency, ω_0 is the natural frequency, A is the excitation displacement, and t is the time domain.

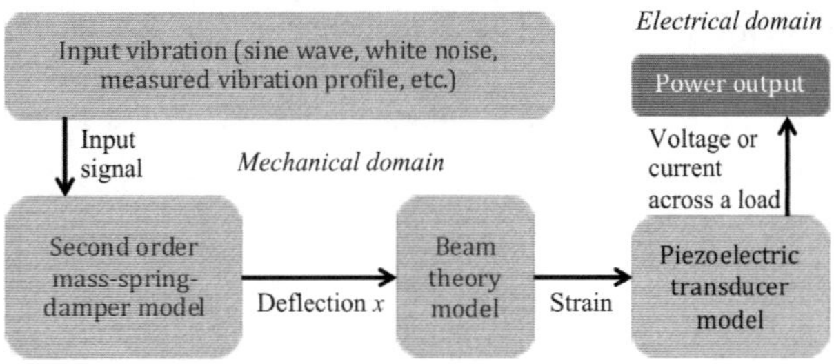

Figure 5. Overview of the MATLAB numerical model used to predict the power output for piezoelectric-on-substrate cantilever designs when subjected to a given vibration profile.

$$\epsilon_y = \frac{3x_{max}h(l-y)}{2l^3} \tag{2}$$

where, ϵ_y is the strain at any arbituary point y along the effective beam length l, x_{max} is the displacement amplitude or the maximum deflection and h is the effective thickness of the beam.

$$q = d_{31}\epsilon_y E w_p l_p \tag{3}$$

where, σ_y is the mechanical stress induced, E and is elastic modulus, q is the total charge generated, and w_p and l_p are the width and length of the active piezoelectric region.

$$P = \frac{\omega h_p q^2}{\varepsilon_0 \varepsilon_r w_p l_p} \tag{4}$$

where, P is the power, ω is the frequency, ε_0 is the permittivity of free space and ε_r is the dielectric constant of the piezo material and h_p, w_p and l_p are the thickness, width and length of the piezo-layer.

4. Result

The characteristics and performance of the three prototype iterations are summarised in table 1. The peak attainable power from the device increased with larger values of added mass, while the acceleration levels required to attain this power level decreased as expected. Driving the cantilevers beyond these acceleration levels yielded either little observable power increase or resulted in failure.

Table 1. The peak attainable power at the required acceleration for the micro-cantilevers.

Device	Added mass (mg)	Frequency (Hz)	Attainable power (μW)	Saturation acceleration (ms^{-2})
MC0	0	577	0.70	70
MC2	2	129	1.90	7
MC5	5	80	2.56	3

Figure 6 illustrates the power output per drive acceleration for the three devices. While a relatively linear relationship can be seen for the plain cantilever MC0, both MC2 and MC5 experienced diminishing returns at higher acceleration levels. The power level at which this deviation from linearity onset are all approximately in the order of few hundreds of nano-watts to a micro-watt. Additionally, MC2 and MC5 demonstrated appreciable Duffing nonlinearities in the form of spring softening (figure 7), even at relatively low acceleration levels. The unloaded MC0 on the other hand, exhibited a predominantly linear resonant peak within the scanned acceleration range.

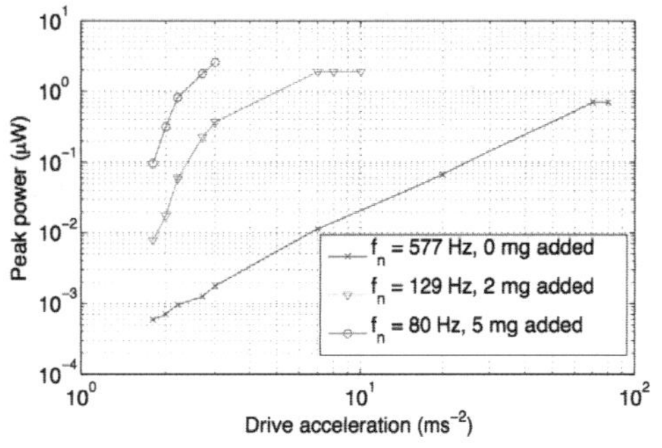

Figure 6. Power response per acceleration for 3 devices with varying added mass.

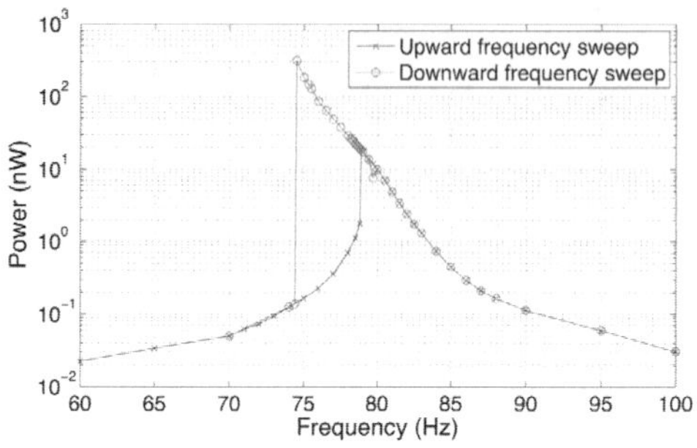

Figure 7. Frequency domain power response for MC5 when driven at 2 ms^{-2}.

The experimental measurements are compared against the numerically simulated results in table 2. Damping ratios for the numerical model of each device configuration was estimated from the measured half power bandwidth and the natural frequency of each prototype. Overall, the model deviation, for the first resonant mode, from the experimental measurement was within $\pm10\%$ in terms of the power output across a matched impedance.

Table 2. Numerical model deviation from the experimental results for the first resonant mode.

Device	Acceleration (ms^{-2})	Recorded power (μW)	Simulated power (μW)	Deviation (%)
MC0	70	0.70	0.77	+10
MC2	7	1.90	1.86	−2.1
MC5	3	2.56	2.62	+2.3

The cantilever with 5 mg added mass was also subjected to varying levels of band-limited white noise vibration with bandwidths ranging from 10 Hz to 2 kHz. Typical time domain output voltage characteristics of the harvester from amplitudes of 1 g and 4 g are shown in figure 8, which yielded average power outputs of 30.83 nW and 123.32 nW respectively. Figure 9 further details the experimentally measured and simulated peak and average power response of this device from the band-limited white noise. Experimentally, peak instantaneous power in excess of 1 μW was only attainable beyond 2 g of white noise, whereas single frequency resonant response just required $\sim$2.5 ms^{-2}. Detailed results are presented in table 3.

While the model developed here does not consider higher resonant modes, the first resonant mode dominates the power response. This is primarily due to the partial strain-charge cancellation experienced by the piezoelectric layer at these higher resonant modes when opposing strain gradients co-exist from twists or multiple curvatures. For instance, COMSOL simulation reveals that MC5 has a shear mode at $\sim$400 Hz and the second transverse mode at $\sim$1 kHz, where both possess opposing strain gradients.

Therefore, the higher average response amplitude from the experimental prototype towards wide-band excitation, as illustrated by figure 9, is expected; since higher vibrational modes are

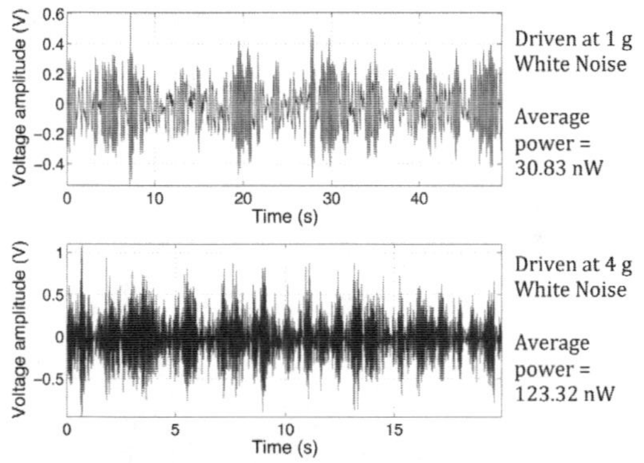

Figure 8. Voltage response from white noise (10 Hz to 2 kHz) of 1 g and 4 g for MC5.

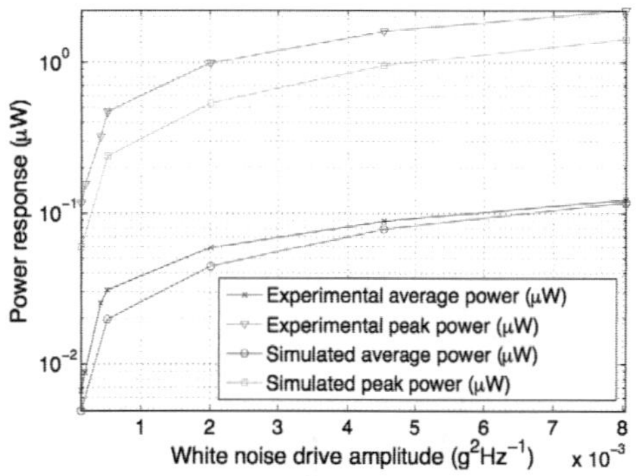

Figure 9. Power response from white noise (10 Hz to 2 kHz) for MC5.

Table 3. Numerical model deviation from the experimental result of MC5 for a band-limited white noise excitation of bandwidth 10 Hz to 2 kHz.

Acc.	Average power (μW)		Deviation	Peak power (μW)		Deviation
(g)	Experimental	Numerical	(%)	Experimental	Numerical	(%)
0.5	0.007	0.005	-28.6	0.116	0.06	-48.3
1	0.031	0.02	-35.5	0.469	0.238	-49.3
2	0.059	0.044	-25.4	0.988	0.536	-45.7
3	0.089	0.079	-11.2	1.61	0.953	-40.8
4	0.123	0.118	-4.1	2.22	1.43	-35.6

reduced to off-resonant response in the numerical simulation. However, the difference is still well within an order of magnitude, with model deviation varying from −4.1% to −35.5% for the scanned amplitude range. The model deviation from the instantaneous power peaks is slightly more significant, but remains relatively consistent, ranging from −35.7% to −49.2%.

Conclusion

Numerical models and experimental prototypes for AlN-on-Si cantilevers with varying added end mass were developed and investigated. While the first resonant mode response yielded a peak attainable 2.56 μW at 3 ms^{-2} of acceleration for an iteration with 5 mg of added end mass, the same prototype only yielded 123 nW of average power and 2.22 μW of peak instantaneous power when subjected to band-limited white noise (10 Hz to 2 kHz) of 4 g. The model deviations from the measurements were either small or relatively consistent. Future work involves further model validation and improvement towards white noise and broadband excitations by including the response of higher orders of resonance. The model can in turn be used as an optimisation tool for designing for a given vibration profile representative of a specific application.

References

[1] A. Erturk and D. Inman, 2011, *Piezoelectric energy harvesting*, New Delhi, India: Wiley.
[2] M. Renaud, K. Karakaya, T. Sterken, P. Fiorini, C.V. Hoof and R. Puers, 2008, *Sens. Actuators A*, **145-146**, pp. 380-386.
[3] R. Andosca, T.G. McDonald, V. Genova, S. Rosenberg, J. Keating, C. Benedixen and J. Wu, 2012, *Sens. Actuators A*, **178**, pp. 76-87.

A Hip Implant Energy Harvester

K Pancharoen, D Zhu, and S P Beeby
Electronics and Computer Science, University of Southampton, Southampton, UK

E-mail: kp6e12@ecs.soton.ac.uk

Abstract. This paper presents a kinetic energy harvester designed to be embedded in a hip implant which aims to operate at a low frequency associated with body motion of patients. The prototype is designed based on the constrained volume available in a hip prosthesis and the challenge is to harvest energy from low frequency movements (< 1 Hz) which is an average frequency during free walking of a patient. The concept of magnetic-force-driven energy harvesting is applied to this prototype considering the hip movements during routine activities of patients. The magnetic field within the harvester was simulated using COMSOL. The simulated resonant frequency was around 30 Hz and the voltage induced in a coil was predicted to be 47.8 mV. A prototype of the energy harvester was fabricated and tested. A maximum open circuit voltage of 39.43 mV was obtained and the resonant frequency of 28 Hz was observed. Moreover, the power output of 0.96 μW was achieved with an optimum resistive load of 250Ω.

1. Introduction

Failure of joint replacements damaging the surrounding bone results in difficult surgery and rising costs of health care [1]. In-vivo monitoring of joint replacements has been proposed to address this problem and to predict faults in advance. However, effectively electrical supplying energy to power instrumented hip implants remain a challenge.

The inductive power link has previously been used in-vivo to power the instrumented hip joint [2], [3]. The main components of this power system are a primary coil wrapped around the patient's thigh and a secondary coil embedded inside the hip prosthesis. Such components are uncomfortable for patients, limited activities of patients and place energy constraints on the operation of the instrumented implants. In addition, three kinetic energy harvesting mechanisms have been presented by Silva et al [4]: translation-based electromagnetic, rotation-based electromagnetic, and piezoelectric harvester. The amount of energy that can be harvested depends on the amplitude of hip movements, relative movement velocity between magnets and coils, and the axial force on the prosthesis during the human gait cycle. According to [4], a regular human gait frequency of 3.94 Hz is required to generate useful power for enabling operation of a Bluetooth protocol (BLE112 from Bluegiga) up to 50 seconds. This may not be possible due to poor mobility of patients who have a hip replacement. To address this difficulty, we aim at designing an energy harvester operating at low frequencies around 1 Hz which is associated with body motion during routine activities of patients.

A magnetic spring has been used to achieve a compact energy harvester and this enables the spring constant to be adjusted by varying magnet strength. In this paper, we present design of the first prototype of the energy harvester in hip implant including a comprehensive simulation and experimental results. Such results will be a benchmark to develop a practical device.

2. Harvester design

A prototype of energy harvester with magnetic spring is shown in figure 1. Two design challenges of the generator are the limited space available (volume of femoral head and femur is about 8 cm^3 and 20 cm^3 respectively) and the low frequency associated with human movements. Due to the constrained volume of the hip prosthesis, the size of the prototype was designed to be 2 cm in length with a diameter of 4 mm and the harvester is to be embedded at the top of femur as shown in figure 2.

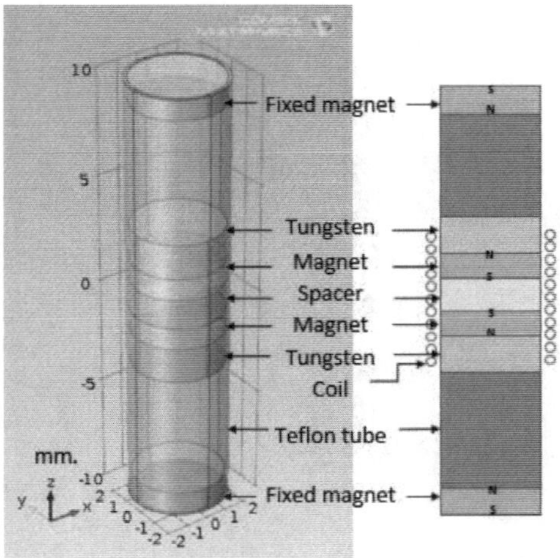

Figure 1. Schematic of magnetic spring harvester.

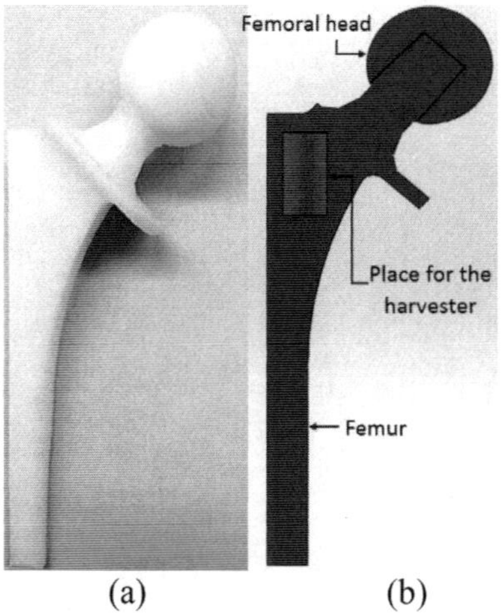

Figure 2. Schematic of hip prosthesis: (a) 3D-printing model (b) cross-section diagram.

The design consists of two fixed magnets at top and bottom of the tube and a moving mass inside the tube. A Teflon tube was used in order to reduce the friction between moving mass and inner surface of the tube. The moving mass consists of two magnets with a ferrite spacer placed between them in order to join such magnets and concentrate the flux density [5]. The thickness of the spacer was optimized for the purpose of avoiding saturation in it. As shown in figure 3, the thickness of 1.4 mm is the point that balances magnetic flux density between spacer and magnets. Moreover, two tungsten pieces were attached on moving magnets to increase inertial mass thus the output power as well as to reduce the resonant frequency of the energy harvester [5].

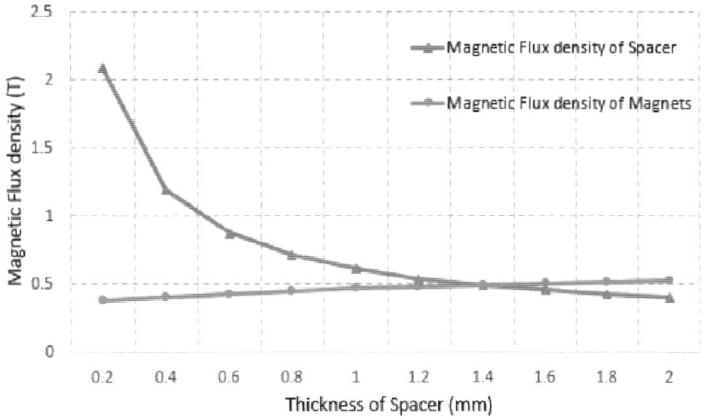

Figure 3. The optimal thickness of spacer.

3. Fabrication and Experimental arrangement

Figure 4 shows the assembled harvester. Two 4mm-diameter SmCo cylindrical magnets were attached at top and bottom of the Teflon tube which has a coil wound from 50μm-diameter Copper wire with 1100 coil turns. Two cylindrical Tungsten pieces, two SmCo magnets and a ferrite spacer have been glued together to be the moving mass and placed into the tube. After that the harvester was placed in the holder as shown in figure 5(a) and tested on the shaker which was shaken in the vertical direction.

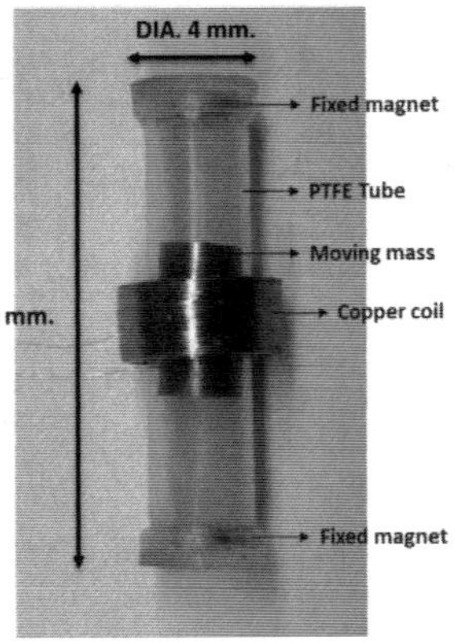

Figure 4. Fabricated energy harvester.

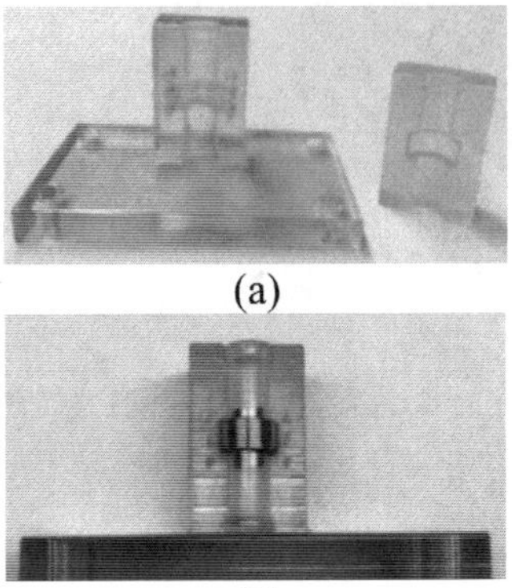

(a)

(b)

Figure 5. Experiment settings: (a) holder (b) holder with generator.

4. Results and discussion

4.1. Simulation results

The proposed model was simulated in COMSOL to study magnetic flux variation, the resonant frequency of the harvester and voltage induced in a coil. The magnetic flux density along the tube is shown in figure 6. According to COMSOL simulation, the resonant frequency of the device was found to be 30.17 Hz and a peak voltage of 47.8 mV was induced in a coil when excited at 800mg (figure 7).

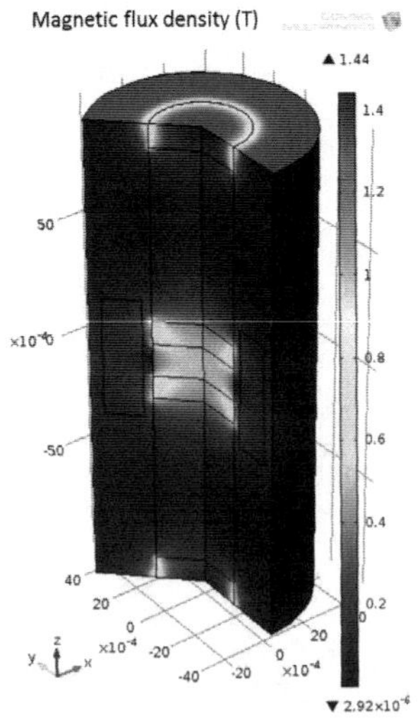

Figure 6. The simulation results of magnetic flux density along the tube

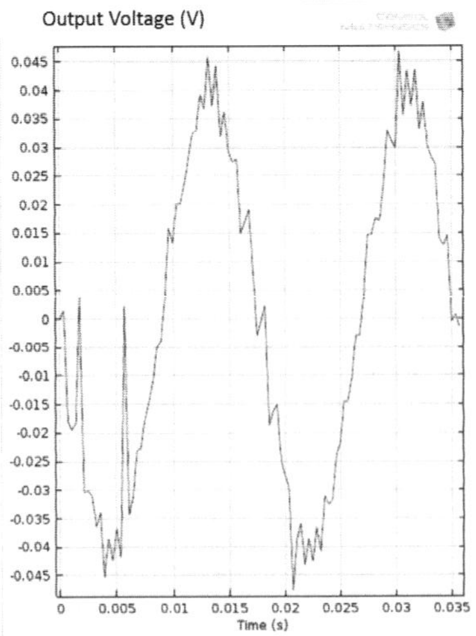

Figure 7. The simulation results of voltage induced in a coil

4.2. Experimental results

The energy harvester was tested by exciting it at 800mg on an electromagnetic shaker. According to the experiment results, the resonant frequency of 28 Hz was reached and the maximum open circuit voltage of 39.43 mV was obtained (figure 8). Moreover, the maximum output power of 0.96 μW was achieved at the optimal load of 250Ω (figure 9).

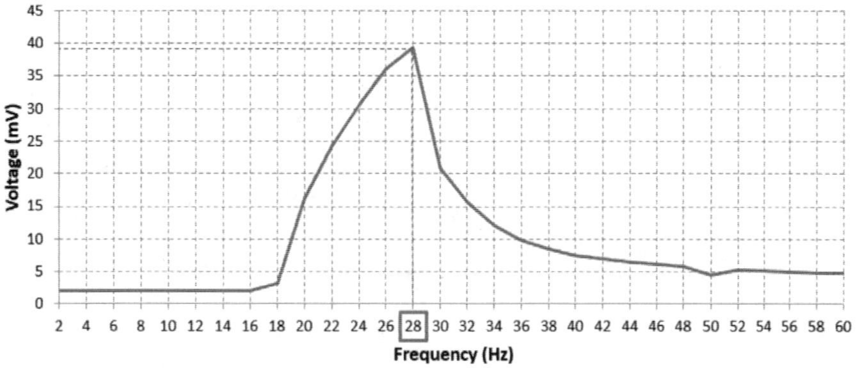

Figure 8. Experiment results: open circuit voltage vs frequency.

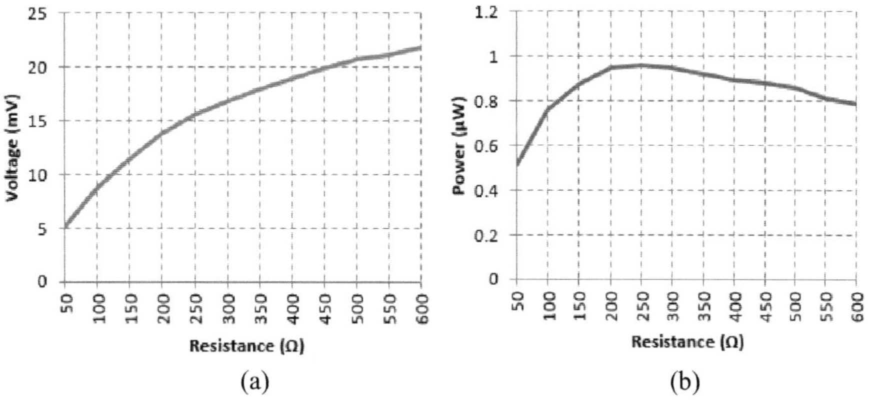

Figure 9. Experiment results: (a) voltage output and (b) power output vs. load resistance of generator at 28 Hz.

5. Conclusion and future work

There are slight differences between simulation and experimental results, i.e. 2 Hz (6.7%) difference for the resonant frequency and about 20% different for the output voltage. At the optimal load of 250Ω, the highest power output of 0.96 μW and output voltage of 15.51 mV can be achieved. According to aforementioned results, the resonant frequency of the presented energy harvester is higher than expected. In order to harvest energy from patients who have had a hip replacement, the operating frequency of the device should be decreased to working frequency which is around 1-2 Hz associated with frequency of patients during free walking [6]. Therefore, overall structure of moving mass will be redesigned, i.e. increase mass. Moreover, coil parameters such as number of coil turns or diameter of coils have to be considered as well in order to increase power output and improve efficiency of the device. Finally, the magnitude of damping effects e.g. tube friction, will be established and its effect on the power output quantified.

6. Acknowledgement
This work was performed under the SPHERE IRC funded by the UK Engineering and Physical Sciences Research Council (EPSRC), Grant EP/K031910/1.

References
[1] J. N. Katz, J. Wright, E. A. Wright, and E. Losina, "Failures of Total Hip Replacement: A Population-Based Perspective," *Orthop. J. Harvard Med. Sch.*, vol. 9, pp. 101–106.
[2] F. Graichen, G. Bergmann, and a Rohlmann, "Hip endoprosthesis for in vivo measurement of joint force and temperature.," *J. Biomech.*, vol. 32, no. 10, pp. 1113–1117, 1999.
[3] P. Damm, F. Graichen, A. Rohlmann, A. Bender, and G. Bergmann, "Total hip joint prosthesis for in vivo measurement of forces and moments.," *Med. Eng. Phys.*, vol. 32, no. 1, pp. 95–100, Jan. 2010.
[4] N. M. Silva, P. M. Santos, J. a. F. Ferreira, M. P. Soares dos Santos, a. Ramos, J. a. O. Simões, M. J. C. S. Reis, and R. Morais, "Power management architecture for smart hip prostheses comprising multiple energy harvesting systems," *Sensors Actuators A Phys.*, vol. 202, pp. 183–192, Nov. 2013.
[5] S. Beeby and N. White, *Energy Harvesting for Autonomous Systems*. Norwood: Artech House, 2010.
[6] OrthoLoad Team, "Orthoload: Loaing of Orthopaedic Implants." [Online]. Available: http://www.orthoload.com/.

Bandwidth widening in nonlinear electromagnetic vibrational generator by combined effect of bistability and stretching

P Podder, D Mallick and S Roy*

Microsystems Centre, Tyndall National Institute, Lee Maltings, Dyke Parade, Cork, Ireland

E-mail: saibal.roy@tyndall.ie

Abstract. This work reports the novel concept of frequency response widening of vibrational energy harvesters exploiting the combined effects of bistable and monostable nonlinearities in a single device. The bistability is introduced into the system by repulsive arrangement of magnets, while monostable nonlinear force through the stretching is incorporated by large deformation of two-end-fixed cantilevers. The simulation results show wider bandwidth in the bistability and stretching combined configuration compared to the bistable only or the monostable stretching only configuration.

1. Introduction

The continual reduction in power requirement of wireless sensor nodes (WSNs) and the possibility of generating the required power from ambient energy sources have made the widespread deployment of self-powered autonomous WSNs feasible in recent times [1]. In this context, the conversion of ambient vibrations into usable electrical power has become a thriving research topic over the last few years due to the abundance of vibration sources in the modern urban environment [2]. However, most of the reported vibrational energy harvesters are based on linear resonant oscillators capable of producing usable energy only within a very narrow range of vibration frequency near the resonance [3-5]. The real vibration sources, on the other hand, often produce vibration that are either random in nature, or the frequency of vibration vary considerably over time [6]. An attractive solution to this problem is the intentional incorporation of nonlinearity into the system dynamics to widen the output response. Some of the earliest reported nonlinear energy harvesting systems employed bistable mechanisms [7-9] and they were proven to be superior in terms of harvested power and frequency bandwidth in realistic multi-frequency stochastic vibration environments. Nonlinear monostable devices based on cantilever stretching effects were also studied [10, 11] and these configurations also demonstrated bandwidth widening effects. The present work aims to combine the effects of bistability and stretching based nonlinearity into a single novel device configuration in order to augment the overall performance of nonlinear VEH systems.

2. Design of the energy harvester

The device, consisting of FR4 device structure, NdFeB magnets, mild steel keepers and copper wire-wound coils is shown in figure 1. The device is based on an oscillator carved out of 0.15 mm thick FR4 sheets using computer controlled laser processing tools. The movable proof mass, consisting of four NdFeB magnets (8mm × 4mm × 2mm) and two mild steel plates (8mm × 8mm × 1.6mm) are bonded on either sides of a slot in the FR4 device structure using epoxy adhesive. The four magnets are arranged in oppositely polarized orientation in such a way that high flux density prevails in the slot between the magnet assemblies [3, 4]. The mild steel keepers are used on both sides of the four-pole magnet arrangement to concentrate the magnetic flux lines through the coil, as depicted in figure 1(b).

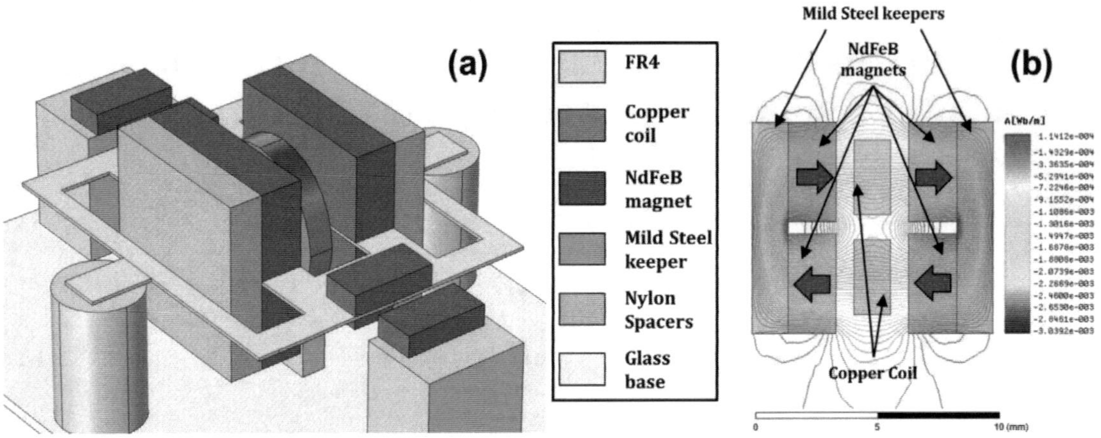

Figure 1: (a) Vibrational energy harvesting device structure. (b) Cross-section of the four-pole magnet configuration. Red arrows denote the polarity of the magnets.

When the mass vibrates in the vertical direction, the pair of centrally fixed and two-ends-guided cantilevers supporting the mass gets stretched in addition to their bending. This stretching introduces a nonlinear restoring force proportional to the third power of the proof mass displacement, into the system. Two more pairs of repulsively oriented magnets are arranged on both sides of the proof-mass. This magnetic repulsive force introduces bistable nonlinearity into the system which can be controlled by adjusting the gap between the repulsive magnets.

3. Electromechanical model and numerical analysis

The theory of operation of the device can be described as a spring-mass-damper system in conjunction with a nonlinear spring force, due to stretching of the spring arm. Another nonlinear force term is added due to the bistable potential well introduced by the magnetic repulsive force on the FR4 based oscillator.

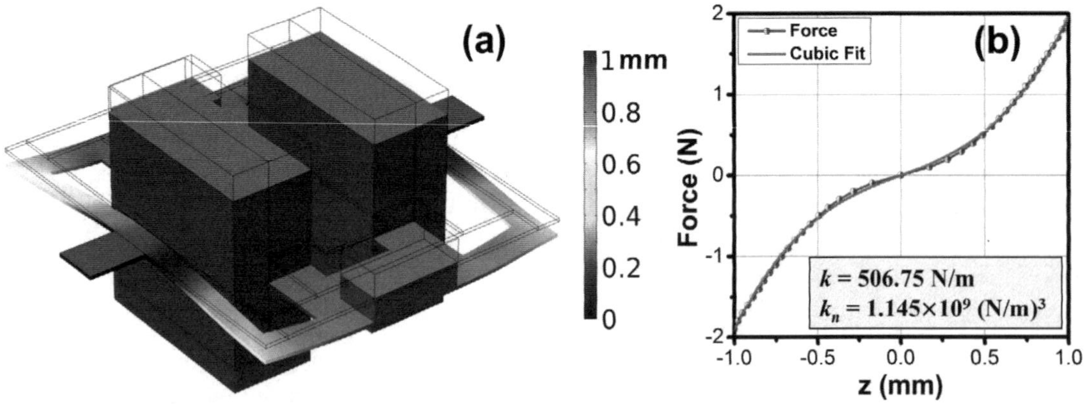

Figure 2: (a) Vertical deflection of the device under external force. (b) Applied force vs vertical deflection.

The equivalent mass, linear spring coefficient and total damping of the system is denoted by *m*, *k*, and *D* respectively. A periodic force *F* due to external vibration is applied in vertical (*Z*) direction. The resulting vertical displacement of the equivalent mass *m* from the initial equilibrium position is denoted by *z(t)* and the distance between the repulsively arranged pair of magnets become *r*. Since the cantilever arms supporting the equivalent mass are fixed at the middle and guided (i.e. allowed to move only in the vertical direction) on both ends, beams would undergo stretching as well as bending when vibrating with amplitude larger than the thickness of the beams [11]. The potential energy due to bending and stretching of the supporting beams is given by the Duffing potential function as,

$$U(z) = \frac{1}{2}kz^2 + \frac{1}{4}k_n z^4 \tag{1}$$

where k_n is the nonlinear spring coefficient due to stretching. The values of the linear and nonlinear spring constants are determined from stationary simulation using COMSOL Multiphysics [Figure 2]. The magnetic potential energy [12] due to the force of repulsion between the magnets is then given by

$$V_m(z) = -\frac{\mu_0}{4\pi r^3}\left[3m_1m_2\left(\frac{d}{r}\right)^2 - m_1m_2\right] = +\frac{\mu_0 m_1 m_2}{4\pi}\frac{\left(2d^2 - z^2\right)}{\left(z^2 + d^2\right)^{5/2}} = C\frac{\left(2d^2 - z^2\right)}{\left(z^2 + d^2\right)^{5/2}} \tag{2}$$

where μ_0 is the permeability of air and m_1, m_2 are the magnetic dipole moments due to the repulsively oriented magnets, and *C* is a constant pre-factor.

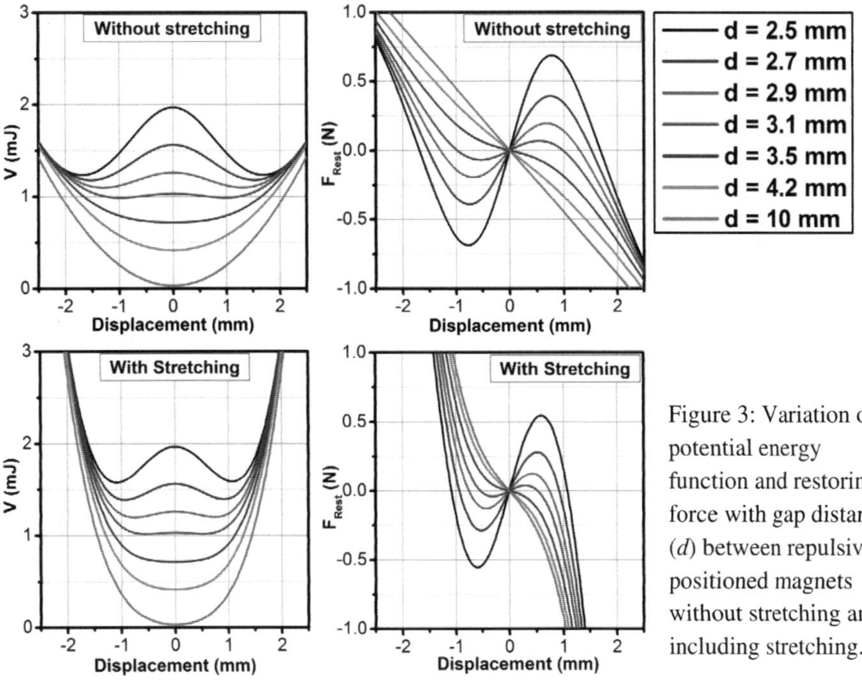

Figure 3: Variation of potential energy function and restoring force with gap distance (*d*) between repulsively positioned magnets without stretching and including stretching.

The total (mechanical and magnetic) potential energy and restoring force of the system is then given by,

$$V(z) = \frac{1}{2}kz^2 + \frac{1}{4}k_n z^4 + C\frac{\left(2d^2 - z^2\right)}{\left(z^2 + d^2\right)^{5/2}} \tag{3}$$

and,

$$F_{Rest}(z) = -\frac{\partial}{\partial z}V(z) = -kz - k_n z^3 - C\frac{3z(z^2-4d^2)}{(z^2+d^2)^{7/2}} \quad (4)$$

The variation of potential energy ($V(z)$) and restoring force ($F_{Rest}(z)$) with displacement (z) for the cases of the bistable VEH excluding stretching and including stretching at different gap distances (d) between the magnets are shown in figure 3. As d is gradually decreased from higher values, the potential energy profile and restoring force transform into nonlinear bistable from linear response. Subsequently, with inclusion of the stretching effects, the potential wells become narrower due to the hardening nonlinearity brought into effect by stretching of the cantilever spring arms. Figure 3 show that without stretching, the bistable restoring force follows almost a linear path while away from the equilibrium position. Then, as stretching effects are included, $F_{Rest}(z)$ follows a nonlinear cubic force path when deflected away from the equilibrium position. The dynamical equation of the system can be represented as follows,

$$m\ddot{z} + kz + k_n z^3 + C\frac{3z(z^2-4d^2)}{(z^2+d^2)^{7/2}} + D\dot{z} = F cos(\omega_0 t) \quad (5)$$

Total damping coefficient D is a combination of mechanical damping D_P and electromagnetic damping D_{EM}. The load power harvested across a load resistance (R_{Load}) is given by,

$$P_{Load} = \frac{R_{Load}}{R_{Coil} + R_{Load}} \frac{1}{T}\int_{t'}^{t'+T} D_{EM}(\dot{z})^2 dt \quad (6)$$

where R_{Coil} is the resistance of the copper coil. The analytical model was numerically simulated for different parameters and excitation conditions using MATLAB.

4. Simulation results and discussions

The numerical model of the proposed energy harvester was simulated under 1g (10 m/s^2) and 0.5g (5 m/s^2) acceleration in a frequency range from 10 Hz to 140 Hz. At 1g acceleration, when the device is linear, i.e. no effect due to stretching or bistability is included, the average power spectrum produces a sharp peak at the resonance frequency (58.7 Hz) and the peak power is 2.9 mW [figure 4(a)].

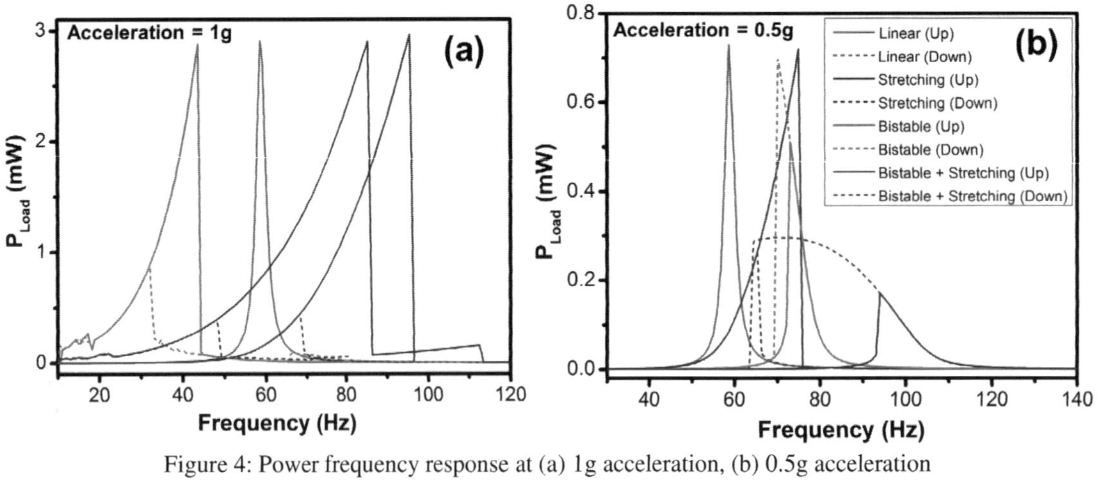

Figure 4: Power frequency response at (a) 1g acceleration, (b) 0.5g acceleration

As the repulsively oriented magnets are brought closer by reducing *d*, the system become bistable, and hysteresis occur in the frequency response. The oscillator jumps from one potential well to the other under the applied vibration (1g) producing peak power of 2.87 mW at 43.7 Hz frequency and the half-power bandwidth in upward frequency sweep is 6.62 Hz. When the system is made nonlinear based on stretching effects of the supporting beams and bistability is not included, peak power frequency shifts towards right (95.4 Hz), and the bandwidth becomes 12 Hz. However, combining both the effects of bistability and stretching, the bandwidth becomes even wider at 15 Hz, and the peak power frequency is shifted to 85.37 Hz, generating almost the same power as in all the other cases. The frequency response shows nonlinear spring hardening behaviour in all of these situations. Fig. 4(b) shows the power-frequency responses at 0.5g acceleration. When the system is made bistable in this case, the oscillator is confined within one potential well, producing spring softening behaviour in frequency response. The bandwidth in stretching only and bistable only cases are 6.7 Hz and 4.7 Hz respectively, with a peak power of about 0.7 mW. As both the effects of bistability and stretching are combined, the peak power comes down to 0.29 mW, while the bandwidth during downward frequency sweep becomes significantly wider at 31.5 Hz. Hence, in comparison to VEH devices employing only bistable or only stretching based nonlinearity, the bandwidth can be expanded significantly in a device incorporating the effects of bistability and stretching simultaneously. However, the effect of stretching on the frequency response become prominent in the bistable configuration when the oscillator is confined within a single potential well and deflected further away from the initial equilibrium, facilitating stretching dependent nonlinear force.

5. Conclusions
In this work the design and simulation of an electromagnetic vibration energy harvester, exploiting both bistability and stretching based nonlinearity was described. The expressions of total potential energy function and restoring force in the numerical model shows that while the effect of bistability is dominant for small displacements, i.e. near the equilibrium position, stretching based nonlinearity dominates when the displacement is larger. The simulation results show that simultaneous incorporation of bistability and stretching effects are able to widen the bandwidth significantly. However, if the oscillator is confined within one of the potential wells, the output power is lowered. Based on these results, suitable modifications and optimization of the device structure could achieve even wider frequency response.

Acknowledgements
This work is financially supported by Science Foundation Ireland (SFI) Principal Investigator (PI) project on 'Vibrational Energy Harvesting' grant no. SFI-11/PI/1201.

References

[1] Mitton N et al, *EURASIP Journal on Wireless Communications and Networking* 2012, **2012:247**.
[2] Roundy S, *Journal of Intelligent Material Systems And Structures*, **Vol. 16**, October 2005.
[3] Beeby S P et al, *Journal of Micromechanics and Microengineering*, **17** (2007) 1257–1265.
[4] Kulkarni S et al, *Journal of Applied Physics*, **99**, 08P511, 2006.
[5] Kulkarni S et al, *Sensors and Actuators A: Physical*, Volumes **145–146**, July–August 2008.
[6] Adhikari S et al, *Smart Materials and Structures*, **18** (2009) 115005.
[7] Cottone F et al., Nonlinear energy harvesting, *Physical Review Letter*, **102**, 080601 (2009).
[8] Ferrari M et al., *Sensors and Actuators A: Physical*, **162** (2010) 425–431.
[9] Stanton S C et al., *Physica D*, **239** (2010) 640-653.
[10] Marinkovic B and Koser H, *Applied Physics Letters*, **94**:103505 (2009).
[11] Mallick D et al., 2013 *J. Phys. Conf. Ser.*, 476 012088.
[12] Furlani E P, *Permanent Magnet and Electromechanical Devices: Materials, Analysis, and Applications*, (2001) Academic Press.

An Equivalent Circuit Model for Electrostatic Energy Harvester utilized Energy Harvesting System

K Minami, T Fujita, K Sonoda, N Miwatani, K Kanda, and K Maenaka

Graduate School of Engineering, University of Hyogo
2167 Shosha, Himeji 671-2280, Japan

E-mail: er14q043@steng.u-hyogo.ac.jp

Abstract. In this study, we report an equivalent circuit model of an electrostatic energy harvester for a SPICE circuit simulator. In order to simulate a harvesting system, the output power of the device is calculated in the simulator. The capacitance between the electrodes is obtained by FEM analysis by taking the fringing effect into account and the result is applied to a sub-circuit model for the simulator. Mechanical vibrations are converted into electricity by an equivalent circuit model of a mass-spring structure and an electrostatic energy harvester. The simulated output power and output waveform correspond with the measurement results of our electrostatic energy harvester. We also simulate the operation of a harvesting system connected with a power management IC.

1. Introduction

Recently, energy harvesters are attracted as a source of power for wireless sensor networks[1]. However, there are a few studies about harvesting system as electronic circuit components. We fabricated a electrostatic vibration energy harvester. Charge transfer on the counter electrode is caused by capacitive induction from an electret, which is a material with electric charge. The output power depends on the surface potential of the electret and the change in capacitance between counter electrodes and the base electrodes. It is possible to estimate the output power by applying those parameters to a SPICE circuit simulator. In a previous research, some circuit models of a PZT harvester[2] or an actuator[3] have been analyzed. We can also evaluate the harvesting system by combining a power management IC for vibration energy harvester.

In this study, the capacitance between the electrodes is obtained by FEM analysis and implemented in the equivalent circuit model of an electrostatic energy harvester. We also express a mass-spring system in a SPICE simulator. The simulation results are in good agreement with the measurement result. Then, the harvesting system is simulated by combining a power management IC.

2. Capacitance analysis

In our previous study, our electrostatic energy harvester consists of a Si counter electrode, an electret with base electrodes placed on a mass-spring structure, and a cover glass as shown in figure 1. The change in capacitance between the electrodes is caused by variation of the overlapping area due to mass vibration.

FEM analysis (ANSYS 14.0) is used to calculate the capacitance between the electrodes by taking the fringing effect into account. Figure 2 shows the FEM analysis model considered bipolar charged electret[5]. All dimensions are as same as our previous electrostatic energy harvester. A 2.5 μm thick CYTOP and 0.5 μm thick buried grid electrodes (BGE-N, BGE-P) are fabricated on a 0.5 μm thick SiO_2 isolation layer on a 500 μm thick Si substrate. The height of the slit-shaped Si counter electrode is 260 μm. The depth of all the elements is 10 mm. The change in capacitance for each displacement of the counter electrode is approximated as follows:

$$C(x) = C_{\mathrm{init}} + \Delta C\cos\left(\frac{1}{p}\pi x\right), \tag{1}$$

where C_{init}, ΔC, p, and x denote the initial capacitance, the change in capacitance, the pitch length of the base electrode, and the mass displacement, respectively. The change in capacitance between the counter electrodes and the BGE-N is anti-phase to that between the counter electrodes and the BGE-P.

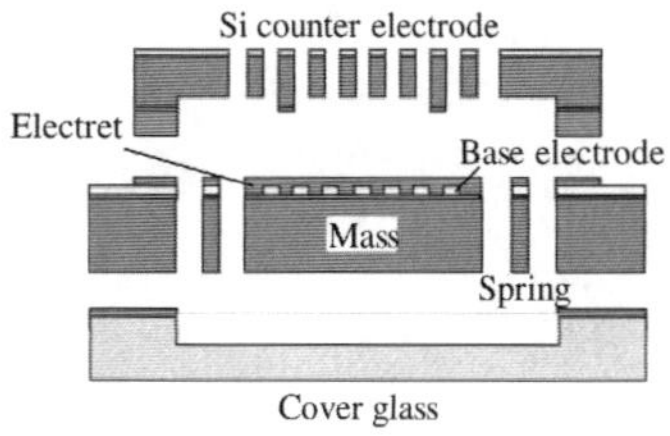

Figure 1. Electrostatic energy harvester[4].

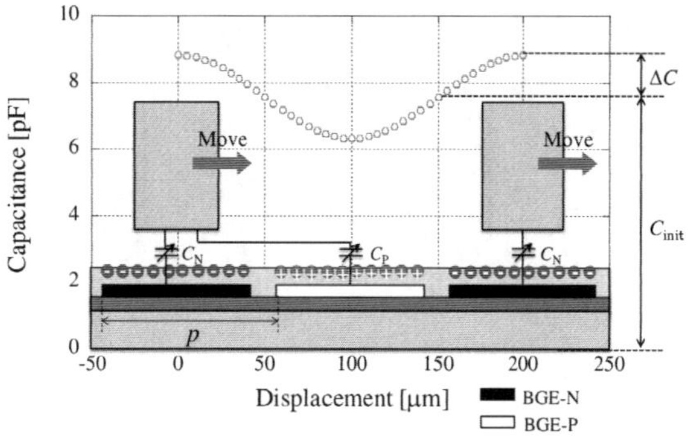

Figure 2. Analysis model and change in capacitance for displacement.

3. Equivalent circuit model

3.1. Variable Capacitance model

We developed an equivalent circuit model of the electrostatic energy harvester with a free SPICE circuit simulator LTSpice IV (Linear Technology Corp., USA). Figure 3 shows (a)the circuit symbol and (b)the internal circuit of a variable capacitance model for taking the time-function of capacitance $C(t)$. It converts controlling voltage into capacitance at the ratio of 1 pF/V. This model calculates the output current by following equation:

$$i = \frac{\mathrm{d}Q(t)}{\mathrm{d}t} = \frac{\mathrm{d}(C\times V)}{\mathrm{d}t} = \frac{\mathrm{d}(V_{\mathrm{con}}\times V_{\mathrm{in}})}{\mathrm{d}t}, \tag{2}$$

where V_{con} and V_{in} denote the capacitance controlling voltage and the voltage across the variable capacitance model, respectively.

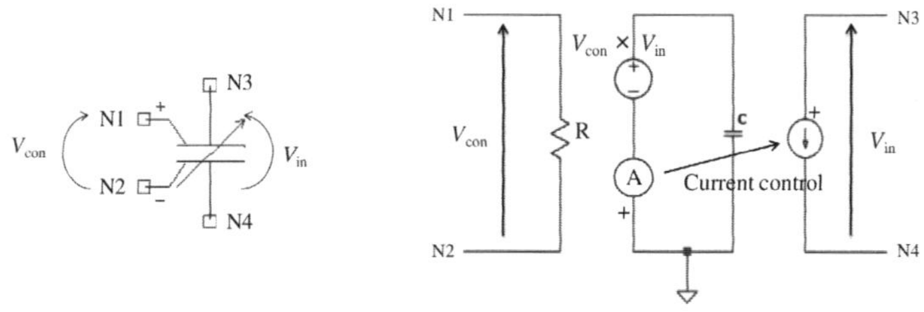

(a) Circuit symbol (b) Internal circuit

Figure 3. Variable capacitance model.

3.2. Equivalent circuit model of electrostatic energy harvester

The output transient voltage of the electrostatic energy harvester is calculated from the change in capacitance and the surface potential of the electret and time-function of the mass displacement as follows:

$$I(t) = \frac{\mathrm{d}Q(t)}{\mathrm{d}t} = V_{\mathrm{elect}} \frac{\mathrm{d}C(t)}{\mathrm{d}t}, \qquad V(t) = I(t)R_{\mathrm{L}}, \tag{3}$$

where V_{elect} and R_{L} denote the surface potential of the electret and the load resistance, respectively.

Figure 4 shows the equivalent circuit model of the electrostatic energy harvester connected with a load resistance. V_{p}, V_{n} denote the positive, negative potential surface voltage of the electret, respectively. Two variable capacitance models (C_{N}, C_{P}) connected in series with the bipolar voltage due to our bipolar charged electret. C_{N} expresses the capacitance between the counter electrode and BGE-N, C_{p} of the other expresses that between the counter electrodes and BGE-P. The voltage sources for controlling the variable capacitance models (V_{conN}, V_{conP}) are used to express the approximation equation of (1) that is obtained by FEM analysis. We consider the parasitic capacitance (C_{par}) and it is connected in parallel with the load resistance.

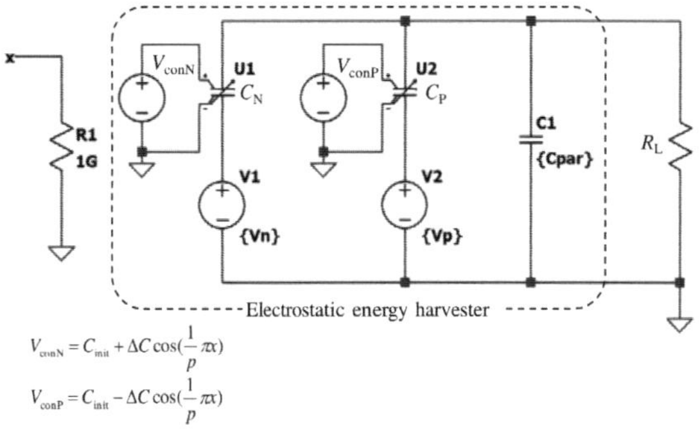

Figure 4. Equivalent circuit model of the electrostatic energy harvester connected with a load resistance.

3.3. Equivalent circuit model of mass-spring structure

It is possible to simulate the output power from human or machine motion by connecting harvester sub-circuit model with a mass-spring circuit model. A basic mechanical system consists of applying force (F_{m}), a proof mass (M), a damper (D), and a spring (K). The equation of motion is used to simulate the response of mechanical system and expressed as follows:

$$M\frac{d^2x(t)}{dt^2} + D\frac{dx(t)}{dt} + Kx(t) = F_{\mathrm{m}}(t). \tag{4}$$

The equation is solved by a RLC series circuit in a SPICE circuit simulator and the equation (4) is rewritten as follows:

$$L\frac{d^2q(t)}{dt^2} + R\frac{dq(t)}{dt} + \frac{1}{c}q(t) = V(t)\,. \tag{5}$$

Figure 5 shows the equation circuit model of a mass-spring structure with a limit displacement of the mass. We used a capacitor of 1 F as mathematic integrator to describe the mass displacement x (charge q in an electronic field) as voltage. There is a limit displacement of the mass $\pm x_{\mathrm{lim}}$ depending on the device size. When the mass contacts the frame(x=x_{lim}), the kinetic energy of the mass is reduced. At the same time, the mass receive a repulsive force from the frame. We used a resistor and a capacitor as the frame to implement with the fundamental damper and spring as shown figure 5. Here, D_{f} and K_{f} denote the damping coefficient and the spring constant of the frame, respectively; these parameters are determined from experimental results. The voltage controlled switches (SW) changes two closed-circuit in case that the mass displacement achieves its limit (x=x_{lim}).

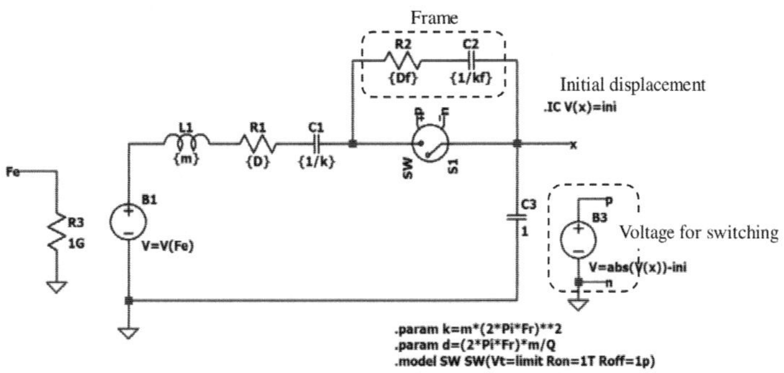

Figure 5. The equivalent circuit model of mass-spring structure with limit displacement of a mass.

4. Simulation and comparison with measurement

Our electrostatic energy harvester was attached to the shaker (ET-126; Labworks Inc., USA) and vibrated at 352 Hz; the resonant frequency of the mass-spring structure. The acceleration of 7 G is sufficiently large to ignore the electrostatic force, which is caused by charged electret. We simulated under the same condition in the LTSpice. Table 1 summarizes the simulation parameters. Figure 6 shows the measurement and the simulated maximum output power versus load resistance. From the result, the maximum output power is 9.8 μW with the optimal load resistance of 0.9 MΩ and it corresponds with simulated result. Figure 7 shows the output waveform connected with the load resistance of 0.9 MΩ. The response of the electrostatic energy harvester is simulated on the equivalent circuit model with our device. The output frequency of our electrostatic energy harvester is larger than the applied acceleration frequency because the amplitude of the vibration is larger than the electrode pitch.

We evaluate the harvesting system by combining with a power management IC with our harvester sub-circuit. We use the LTC3588-1 as a power management IC. Since the LTSpice and LTC3588-1 are released from the same company, we can easy to combine whole harvesting system in the LTSpice as shown in figure 8(a). From the simulation result, we found that our previous electrostatic energy harvester has too large parasitic capacitance of 70 pF that wastes the generated power to charge the battery by LTC3588-1. If the parasitic capacitance can be reduced to 50 pF, an appropriate power managed voltage of 1.8 V will be obtained at intervals of approximate 150 s as shown in figure 8(b).

Table 1. Simulation parameters

Parameter	Value
Air gap: g	43 μm
Electrode width: w	74 μm
Electrode pitch: p	100 μm
Initial displacement: x_{ini}	38 μm
Initial capacitance: C_{init}	7.57 pF
Change in capacitance: ΔC	1.24 pF
Surface potential of bipolar electret: V_p, V_n	± 250 V
Parasitic capacitance: C_{par}	70 pF
Mass weight: m	120 mg
Resonant frequency: F_r	352 Hz
Q factor: Q	100
Limit displacement of the mass: x_{lim}	200 μm
Damper of the frame: D_f	5
Spring of the frame: K_f	10 kN/m

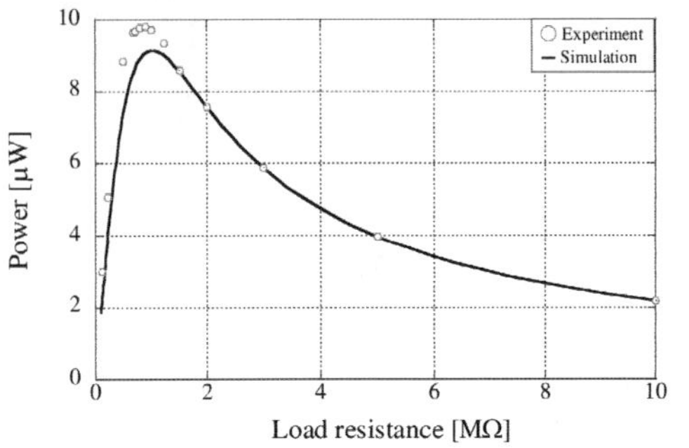

Figure 6. Output power versus load resistance.

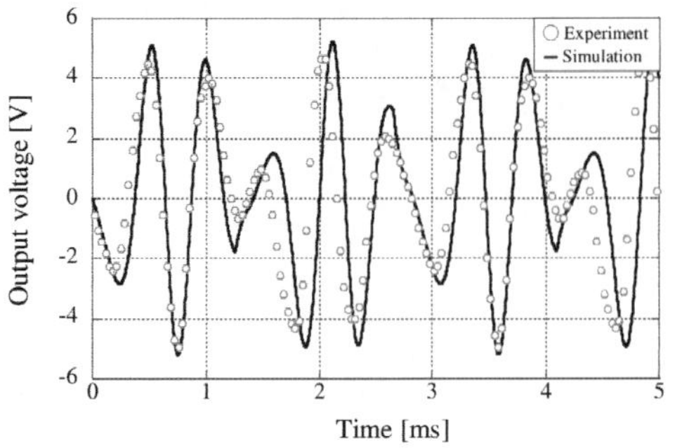

Figure 7. Output waveform connected with the load resistance of 0.9 MΩ

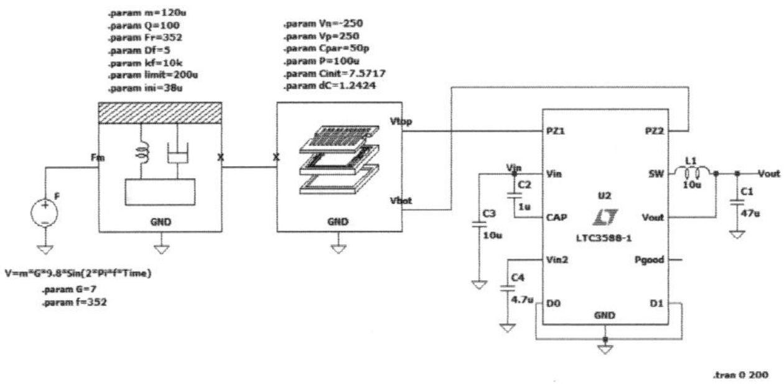

(a) SPICE circuit connected with LTC3588-1

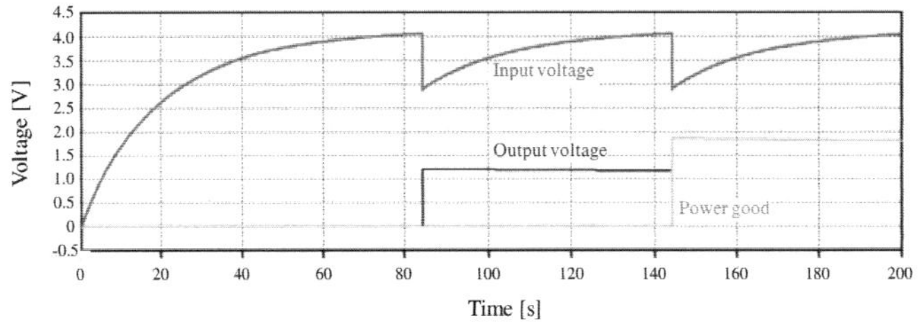

(b) Waveform connected with LTC3588-1

Figure 8. Harvesting system connected with LTC3588-1. The parasitic capacitance C_{par} is 50 pF.

5. Conclusion

In this study, an electro-mechanical model in a SPICE circuit simulator is investigated. The relationship between the capacitance and the counter electrode is analyzed by FEM analysis by taking the fringing effect into account. Applying the result to the sub-circuit model, the output power and output waveform can be calculated in the LTSpice. We evaluate harvesting system by combining a power management IC.

Reference

[1] Darwish A, Hassanien A E 2011 *Journal of Sensors* **11** 5561-5595
[2] Fain R M, Hopcroft M A, Wright P K, 2009 *Proc. PowerMEMS2009* **132** 1-9
[3] Maruyama S, Nakada M, Mita M, Takahashi T, Hiroyuki F, Toshiyoshi H 2012 *IEEJ Transactions on Sensors and Micromachines* **132** 1-9
[4] Fujita T, Fujii K, Onishi T, Kanda K, Higuchi K, Maenaka K 2011 *Proc. Eurosensors* **25** 733-736
[5] Fujita T, Onishi K, Fujii K, Katsuma K, Kanda K, Higuchi K and Maenaka K 2012 *Proc. PowerMEMS2012* 436-439

Simulation of a Novel Bridge MEMS-PZT Energy Harvester for Tire Pressure System

Edoardo Trabaldo[1,2], Elof Köhler[2], Henrik Staaf[2], Peter Enoksson[2] and Cristina Rusu[1]

[1] Acreo Swedish ICT AB, Sensor Systems dept., Gothenburg, Sweden
[2] Chalmers Univ. of Technology, dept. of Microtechnology and Nanoscience, Gothenburg, Sweden

E-mail: trabaldo@styudent.chalmers.se

Abstract. Self-powering is becoming an important issue for autonomous sensor systems. By having an on-the-go power source the life span increases in comparison to a limited battery source. In this paper, simulation of an innovative design for a piezoelectric energy harvester for Tire Pressure Measurement System (TPMS) is presented. The MEMS-based thin-film PZT harvester structure is in the form of a bridge with a big central seismic mass and multiple electrodes. This design takes the advantage of the S-profile bending and a short beam length to concentrate the piezoelectric effect in a small segment along the beam and maximize the power output for a given displacement. From simulation in Comsol Multiphysics, the 9mm x 5mm bridge, seismic mass of 8.7mg and resonance frequency of 615Hz, generates 1 μW by mechanical pulses excitation equivalent to driving at 60 km/h (roughly 180G).

1. Introduction

Energy harvesting is the key enabling approach for design of future small self-powered wireless sensor systems. Such devices require often very little power, but without energy-harvesting, they are nevertheless limited by the limited lifetime provided by conventional batteries. Energy harvesting for automotive applications has been of great interest in recent years to power sensors such as Tire Pressure Monitoring System (TPMS) [1], Vehicle Speed Sensor (VSS), and Strain Monitoring Sensor (SMS) [2]. The TPMS system used today is not suitable for heavy vehicles with many tires requiring more accurate control of the actual pressure in every tire and communicate that information remotely to the information center computer in the vehicle. Therefore fully autonomous TPMS systems based on accurate Micro Electro Mechanical Systems (MEMS) sensors are required with a complemented energy supply.

Vibration piezoelectric energy harvesters convert mechanical strain into electrical energy and have received much attention due to the simple configuration and high conversion efficiency as comparing to electro-static and electromagnetic harvesters [3]. There is an abundance of vibration energy available in car tires that can be converted into electrical energy using energy harvester methods [4] for MEMS devices. Utilization of MEMS technologies allow for a totally integrated system containing sensor, electronics, communication and energy source. However, the scaling remains an issue and the unfavorable scaling of power with miniaturization is still a challenge.

2. Specifications MEMS PZT harvester

Most of the articles agree that a *constant* power of about 3μW is required [5] for the TPMS. For applications of non-constant power harvesting (e.g. mechanical pulses) a rechargeable battery or a

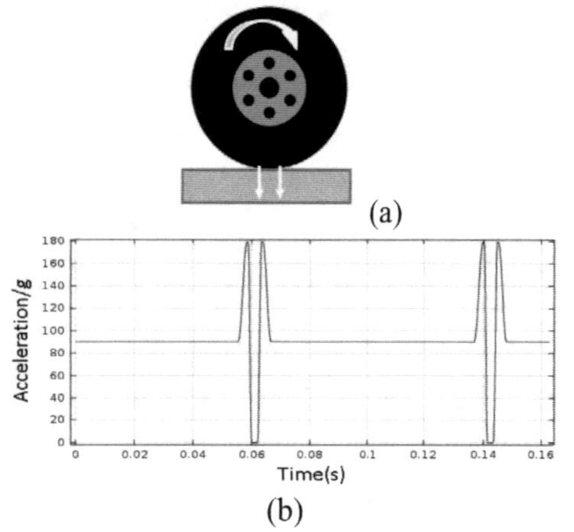

Figure 1. (a) Schematic of harvester placement at tire's inner liner, (b) Acceleration profile for 60km/h used in our simulations (calculated from [6]).

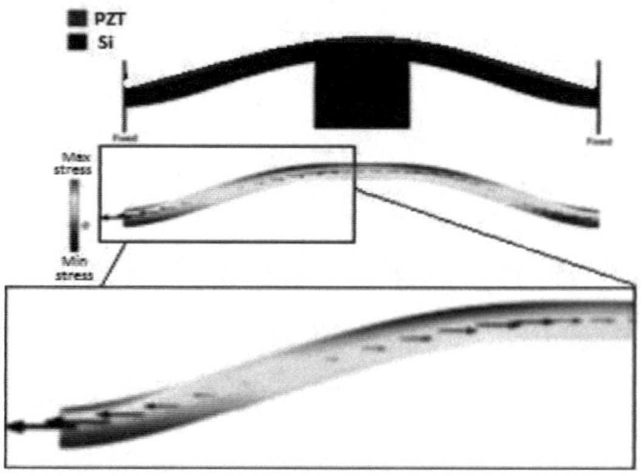

Figure 2. Schematic cross-section of bridge harvester structure showing the S-profile bending and strain (arrows) along the beam.

supercapacitor would be needed for collection of the converted energy. This is mainly for situations when the vehicle is stationary (traffic light, short parking, etc.) or at low speed whilst the tire pressure info is still needed. The current TPMS devices [6] have a weight of about 7g, a required lifetime of 8 years and it should withstand acceleration up to 2000g.

Two main vibration dynamics are applied on the harvester depending on its position on the tire. At the tire rim there are low amplitude and low frequency vibrations [7] while on the inner liner it consists of cyclic, high amplitude pulses / shocks [5]. The harvester structure (seismic mass) will respond to the mechanical pulses with decaying oscillations at its resonance frequency. During this relaxation, part of the mechanical energy is harvested.

Piezoelectric harvesters based on cantilever structure with rectangular or trapezoidal profile are the main design solutions proposed in the literature for TPMS. For these devices, d31 or d33 modes are employed [3, 7]. However, these designs cannot fulfil properly the application requirements as obtained from our simulations in Comsol Multiphysics (not shown here [8]). In order to find a suitable design, the power must be maximized while keeping the displacement of the vibrating structure within the MEMS cavity limits. We choose the inner liner as a suitable placing of harvester on the tire thus energy is harvested from mechanical pulse excitation. In Figure 1 the acceleration profile is presented (normalized g) at the inner liner of a standard tire diameter for a speed of 60km/h, calculated from [6] and used in the simulations giving a constant centripetal acceleration of 90g and a pulse width of 2.3ms.

The range of parameters utilized for the design which depend not only on TPMS requirements but also on the available MEMS fabrication technologies are shown in Table 1.

Table 1. Range of parameters utilized for our simulations.

Parameter	Range	Parameter	Range
l_{si}	2 to 10 mm	l_{mz}	≤400μm
w_{si}	2 to 10 mm	e_{31}	-15 C/m^2
t_{si}	8,10,40 and 100μm	$\varepsilon_{r\,PZT}$	1200
t_{PZT}	1μm	Frequency	100-600Hz
l_{PZT}	0 to l_{si}	Cavity pressure	10^{-3}mBar
Cavity heigth	450μm		

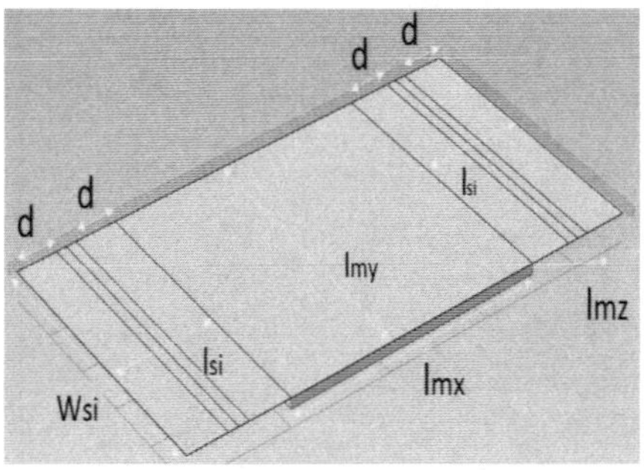

Variable	Value
l_{si}	2 mm
w_{si}	5 mm
t_{si}	8 µm
d	0.8 mm
l_{mx}	5 mm
l_{my}	5 mm
l_{mz}	0.15 mm
ε_{sq}	0.011

Figure 3. Bridge design with final parameter values.

Figure 4. Proposed harvester position inside MEMS cavity.

3. Comsol simulations for MEMS PZT harvester

These solutions were implemented for the new design:

- Short beam with big seismic mass giving high stress localized in a small sections along the beam.
- Bridge structure gives S-profile during vibrations for displacement reduction and stress maximization.

By placement at the inner liner, we found the optimal resonance frequency of the bridge structure of 600Hz (oscillation period 1.6ms). This is the first key factor for inner liner harvester design.

Due to the strain profile of the bridge structure, shown in Figure 2, special solution for harvesting the generated electric field is required. Due to opposite strain (opposite generated voltage), the piezoelectric material has to be polarized in opposite direction during the fabrication process. This gives an electric field uniformly aligned in the whole beam. Alternatively, different electrodes can be placed over the different strain zones along the beam. These electrodes can be connected in parallel or in series, depending on which values of current and voltage is needed. Nevertheless, the total power output is always the sum of the power of each single part. High acceleration amplitude, due to high speed, means that the beam needs high stiffness otherwise the displacement would be too high and the vibrations will be limited by the cavity height. The displacement is the second key factor in the design.

Quality factor, Q, is the third key factor that is mainly due to the mechanical damping, in particular the squeeze film damping. To increase the oscillation time after the initial pulse, Q needs to be maximized which corresponds to a minimum squeeze film damping. These three key factors are mostly influenced by the same structural parameters, but with different effects. Thus a compromise is needed for an optimal design.

Transient analyses for various shapes and sizes of the bridge structure have been performed using the acceleration profile from Figure 1. The bridge design with the best performance is shown in Figure 3 with the parameter values while the bridge harvester encapsulation is shown Figure 4.

The stationary displacement for a constant input acceleration of 90g is roughly 60µm. The mounting of the harvester will have the mid mass towards the centre of the tire such that the stationary displacement due to centripetal acceleration makes the distance between cavity wall and seismic mass larger (Figure 4). The quality factor, which mainly depends on the squeeze film damping ζ_{sq}, is calculated using the equations in [9] and [10]. For this design we obtain $Q= \frac{1}{2\zeta sq} = 45.8$ and a resonance frequency at 614Hz with optimal load resistance of 1000 Ohm.

The simulated displacement and power output for 60km/h (Figure 5) show that the oscillations are contained within the limits of the cavity. For this moderate speed, the average oscillations are 60µm and an average power output of 1µW is obtained; the maximum displacement is around 160µm, with power output of 46µW. The extra energy can be stored in a rechargeable battery or in case of intelligent wireless sensors it is better on a supercapacitor [11] due to a much higher amount of energy density over time.

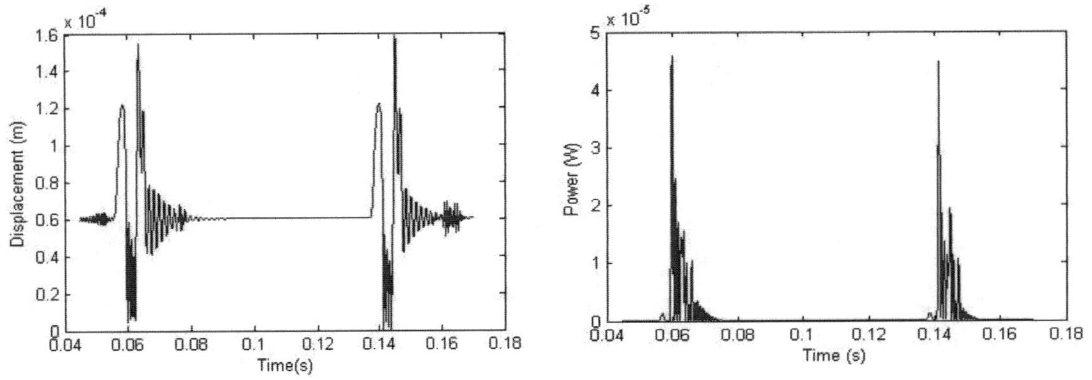

Figure 5. Simulated displacement and power for the design in Figure 3 at a vehicle speed of 60km/h.

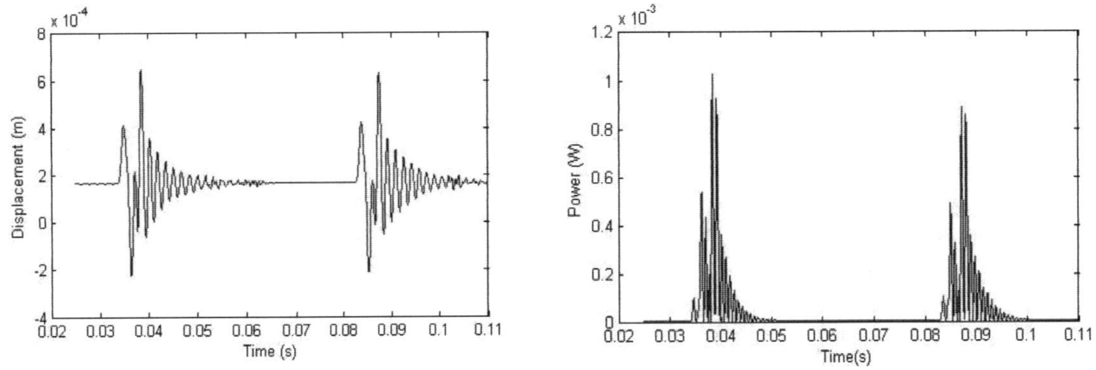

Figure 6. Simulated displacement and power for the design in Figure 3 at a vehicle speed of 96km/h.

In our study we performed test also for high vehicle speed (96km/h), in this case the acceleration profile is the same as the one reported in [6]. The simulated displacement and power output for 96km/h are shown in Figure 6, the oscillation average is around 300μm resulting in an average power of 40μW. The peaks at 610μm for displacement generates 1mW power output; these are not taken into account, since this displacement is beyond the cavity wall and for the study we did not use any contact model between the mass and the wall. This means the results for 96km/h must be considered only as an approximation of the harvester behaviour for high speed; from these we can have an idea about the magnitude of displacement and power output.

After transient analyses we performed a static shock test simulation for information on the maximum stress developed in the structure. For this, a constant load of 10^5g along the 3 coordinates was applied.

From [12] the fracture strength of Si is around 7GPa and we obtained maximum stress and displacement as shown in Table 2. The maximum displacement of the bridge allowed by the cavity is about 450μm and this would correspond to a static stress of 0.43GPa which is much lower than the Si fracture strength value. The same bridge structure presented in figure 3 was also studied for a continuous input vibration and compared with a similar in dimensions trapezoidal cantilever as in Figure 7. For 1m/s^2 a power of 3.9nW for the bridge and 8.4nW for the cantilever is obtained. In this case the power for the bridge is lower, but the displacement amplitude in the cantilever is 18μm while for the bridge it is only 0.65μm. The structure presented in this work (Figure 3) is optimized for mechanical pulses with high accelerations thus high stiffness. For 1m/s^2 and an optimized bridge design for continuous vibrations, the obtained power in the order of 150nW with displacement around 20μm. Thus the bridge structure has a performance 20 times higher than standard cantilever showing the advantages of this bridge design structure.

Table 2. Shock analysis for acceleration of 10000g

Direction	Stress [GPa]	Deflection
x	0.15	100μm
y	0.41	450μm
z	6.24	6.49mm

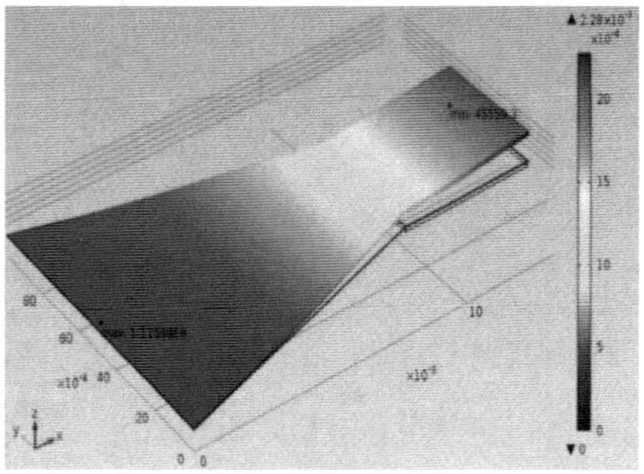

Variable	Value
l_{si}	10mm
w^l_{si}	10mm
w^r_{si}	4mm
t_{si}	40μm
l_{mx}	5mm
l_{my}	4mm
l_{mz}	0.15mm
ε_{sq}	0.024

Figure 7. Trapezoidal harvester simulated in comparison to bridge design.

4. Conclusion

The presented bridge design shows significant improvements in comparison to standard cantilever-based MEMS harvesters. All of the important requirements for a MEMS harvester for TPMS application are fulfilled in this design regarding MEMS-scale dimensions, power output above 1μW for medium / high car velocity, and withstanding accelerations of 10^5g. Further work is on MEMS harvester manufacturing and characterization for design validation.

Acknowledgements

This work is funded by VINNOVA - Swedish Governmental Agency for Innovation Systems under the project "Smart energy optimization via energy harvesting utilizing new Swedish piezo mems technology".

References

[1] Löhndorf M., KvisterØy T., Westby E. and Halvorsen E. *Proceedings of Power MEMS 2007; Freiburg, Germany.* 28–29 Nov. 2007; p. 331–334.

[2] D. Parthasarathy ,P. Enoksson and R. Johansson, *IEEE International Symposium on Robotic and Sensors Environments (ROSE)*, p.115 – 120, 2012.

[3] S. Roundy, P. K. Wright, J.M.'s Rabaey, Energy scavenging for wireless sensor networks with special focus on vibrations; *1st edition by Springer*, 2003.

[4] R. Brusarosco, F. Massimo, F. Mancosu and A. Calatroni, Method and system for generating electrical energy within a vehicle tyre, *US patent*, 2005.

[5] Frey, A.; Seidel, J.; Kuehne, I.; System Design of a Piezoelectric MEMS Energy Harvesting Module based on Pulsed Mechanical Excitation, *pages 29-32, Proceedings Power MEMS,* 2010.

[6] K.H. Mak, Vibration modelling and analysis of piezoelectric energy harvester, *PhD thesis; Univ. Nottingham;* 2011.

[7] S. Roundy. Energy harvesting for tire pressure monitoring systems: design considerations. *Proc. Power MEMS 2008 + microEMS 2008, Sendai, Japan*, Nov. 2008 .

[8] E. Trabaldo, *MSc thesis,* Design of a MEMS PZT energy harvester for the automotive industry, *Chalmers University of Technology,* June 2014.

[9] H. Hosaka, K. ITAO, S. Kuroda, Evaluation of energy dissipation mechanisms in vibrational microactuator; *IEEE 0-7803-1833-1/94;* 1994.

[10] K.Kadoya, N.Matsunaga and A.Nagashima; Viscosity and Thermal Conductivity of Dry Air in the Gaseous Phase; *J. Phys.Chem.Ref.Data, Vol.14, No.4,* 1985.

[11] L.G.H. Staaf, P. Lundgren, P. Enoksson, Present and future supercapacitor carbon electrode materials for improved energy storage used in intelligent wireless sensor systems , *Nano Energy (*2014*) 9, 128–141.*

[12] Petersen, K.E., Silicon as a mechanical material, *Proc. IEEE., Vol. 70, No. 5,* 1982.

A handy-motion driven, frequency up-converted hybrid vibration energy harvester using PZT bimorph and non-magnetic ball

M A Halim, H O Cho and J Y Park[1]
Micro/Nano Devices and Packaging Lab., Dept. of Electronic Engineering, Kwangwoon University, 447-1 Wolgye-dong, Nowon-gu, Seoul, 139-701 Korea.

E-mail: jaepark@kw.ac.kr

Abstract: We have presented a frequency up-converted hybrid type (Piezoelectric and Electromagnetic) vibration energy harvester that can be used in powering portable and wearable smart devices by handy motion. A transverse impact mechanism has been employed for frequency up-conversion. Use of two transduction mechanisms increases the output power as well as power density. The proposed device consists of a non-magnetic spherical ball (freely movable at handy motion frequency) to impact periodically on the parabolic top of a piezoelectric (PZT) cantilevered mass by sliding over it, allowing it to vibrate at its higher resonant frequency and generates voltage by virtue of piezoelectric effect. A magnet attached to the cantilever vibrates along with it at the same frequency and a relative motion between the magnet and a coil placed below it, induces emf voltage across the coil terminals as well. A macro-scale prototype of the harvester has been fabricated and tested by handy motion. With an optimum magnet-coil overlap, a maximum 0.98mW and 0.64mW peak powers have been obtained from the piezoelectric and the electromagnetic transducers of the proposed device while shaken, respectively. It offers 84.4μWcm^{-3} peak power density.

1. Introduction

Now a days, low power consuming microelectronic devices and sensors such as accelerometer, gyroscope, proximity sensor, heart-rate sensor, bluetooth, NFC, Wi-Fi, GPS tracker, and many more are being embedded within hand-held or wearable consumer electronics. These embedded devices and sensors consume power from a single power source of the consumer electronics. Use of the embedded sensors allows it to run out of power quickly. Generally, electro-chemical batteries have been using as power source to supply power to these portable and wearable electronic devices. The development of the power sources (batteries) are still slower than that of the device technologies, even though the devices require less power to be operated. Electro-chemical batteries have limited lifetime, needs to be replaced. They require periodic charging which is, sometimes, inconvenient or impossible. Since last few decades, a lots of approaches have been unveiled to sort out this problem. Many researchers around the world have been investigating to use environmental sources as the alternative power source to be supplied to the consumer electronics which is known as energy harvesting. Energy harvesting devices and systems capture physical energy from environmental sources and convert it into electrical energy by means of a number of energy conversion mechanisms depending on the type of energy

[1] To whom any correspondence should be addressed.

sources. In order to supply power to those electronic devices or recharge their power sources (batteries or super capacitors) in an emergency situation, energy harvesting form environmental sources e.g., mechanical vibration, wind, heat, light, radio wave etc. would have been an effective solution [1]. Mechanical vibration is one of the most attractive physical energy sources due to its abundance in nature and unlimited lifetime [2]. But most vibrations are of low frequencies and large amplitudes with various cyclic movements in different directions [3-5]. Piezoelectric, electromagnetic and electrostatic mechanisms are widely used techniques for harvesting energy from mechanical vibrations [2].

Generally, cantilevered spring-mass system with a specific resonant frequency is employed with these mechanisms. Harvested power is maximum when harvester's resonant frequency matches the applied vibration frequency. But employing cantilevered spring-mass system for low frequency (below 10 Hz) energy harvesting is quite challenging due to the size constraints for specific application and, of course, power flow decreases with decrease in resonant frequency [6]. Human motion (e.g., walking, running, shaking limbs etc.) also generates very low frequency (below 6 Hz) vibration which does not allow to employ the cantilever structure conveniently [7]. This is why, mechanical frequency up-conversion mechanism has been employed by a number of researchers [8-10]. Recently, we also proposed an impact based frequency up-converted electromagnetic energy harvester for hand-shaking vibration that employed an FR4 cantilever beam with parabolic-top proof-mass as high-frequency oscillator and a freely movable ball as a low frequency oscillator [11]. But, it was capable of generating a few hundred μW power, having low power density. In this work, we have approached to hybridize our previous work with a piezoelectric (PE) mechanism in addition to the existing electromagnetic (EM) mechanism without increasing the harvester volume. A PZT bimorph replaces the FR4 cantilever. It uses transverse impact mechanism by a non-magnetic ball that meets the reliability challenge from quick damage of the piezo-cantilever rather than direct impact. This hybrid approach generates significant power that can be supplied (after conditioning) to the portable and wearable consumer electronics from handy-motion vibration.

2. Structure of the hybrid energy harvester

Figure 1 shows the schematic drawing of the proposed impact based, frequency up-converted hybrid type handy motion driven energy harvester. It consists of a freely moveable non-magnetic ball within a rectangular channel with bottom opening at the centre. A poof-mass with parabolic-top is attached to a fixed-fixed piezoelectric bimorph cantilever beam that works as the spring-mass-damper system. The beam is aligned with the channel in such a way that the parabolic-top of the proof-mass can occupy the bottom opening of the rectangular channel so that the ball can slide over it during its back and forth motion in the channel. A cylinder magnet is attached to the piezo-cantilever, opposite to the proof-mass and a coil is also placed below the magnet, on the bottom cover of the device structure. When excitation is given by handy motion, the ball produces transverse impact on the proof-mass allowing the cantilever to vibrate. The piezo-beam vibrates with its resonant frequency in a direction perpendicular to the direction of ball movement and generates voltage due to strain experienced on its surfaces. Accordingly, the magnet attached to the beam also vibrates relative to the coil and induces emf voltage according to Faraday's law of electromagnetic induction. The vibration frequency (resonant frequency) of the cantilever beam is much higher than the applied handy-motion vibration.

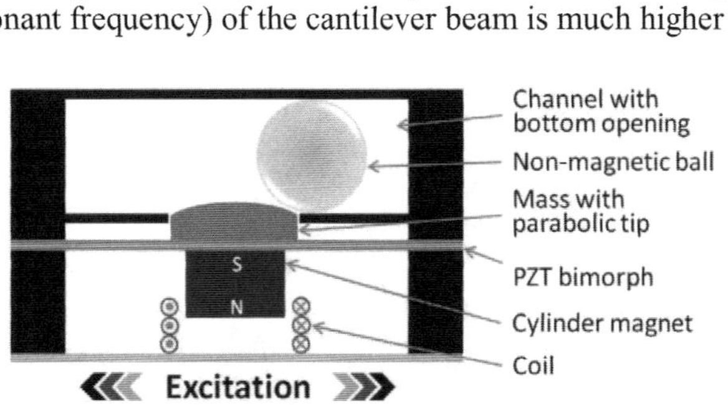

Figure 1. Schematic drawing of the proposed hybrid type handy motion driven vibration energy harvester.

3. Prototype fabrication

A macro-scale prototype was fabricated that consists of two transduction parts. The PE transduction part of the prototype comprises a Ø6×5 mm^2 NdFeB cylinder magnet and a cubic Fe mass of 6 mm length having a parabolic extension of 1 mm height at the top, both attached at the middle of either sides of a PZT bimorph (36×6×0.5 mm^3) cantilever beam. A suitable assembly of the magnet and a 1000-turn coil (0.1 mm copper wire) of 8 mm bobbin diameter assembled on a printed circuit board (PCB) constitutes the EM transduction part. Along with those, a Ø10.3 mm non-magnetic steel (type 316) ball that vibrates inside the rectangular channel (inner area 10.5×10.5mm^2) with opening (7×7mm^3) at the centre of one plane, is assembled within an aluminium structure to constitute the overall (40×30×16mm^3) prototype as shown in figure 2. The PZT beam was assembled in such a way that the channel opening is occupied by the parabolic-top of the cubic mass with a 0.4 mm overlap with the ball inside so that the ball can slide over it and produce a transverse impact while an excitation is given to the prototype by handy-motion.

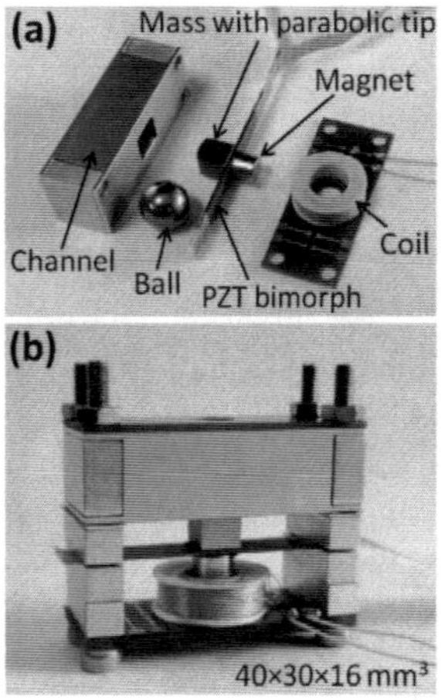

Figure 2. Photographs of the (a) components and (b) assembled prototype.

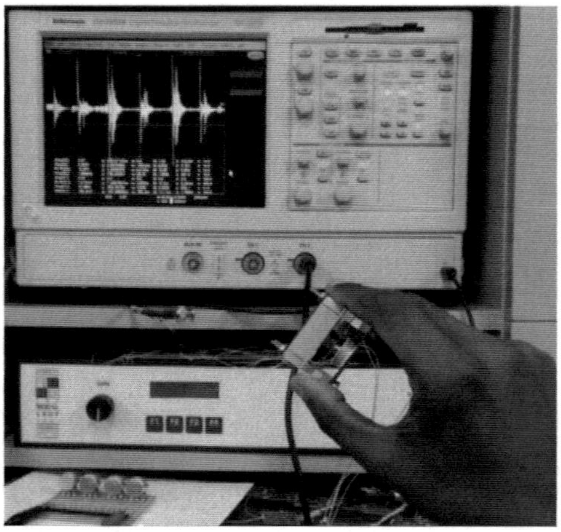

Figure 3. Photograph of the test setup while the fabricated harvester prototype was driven by handy-motion.

4. Experimental results and discussion

To date, most of the vibration energy harvesters have been characterized using electrodynamic shaker. In this case, our harvester prototype was tested by handy motion vibration, as we were intended to. Figure 3 shows the photograph of the test setup where the harvester outputs were connected to a digital storage oscilloscope (TDS 5052B) via continually adjustable load resistor (to measure load voltage and power) in order to observe (and record) the output response instantly.

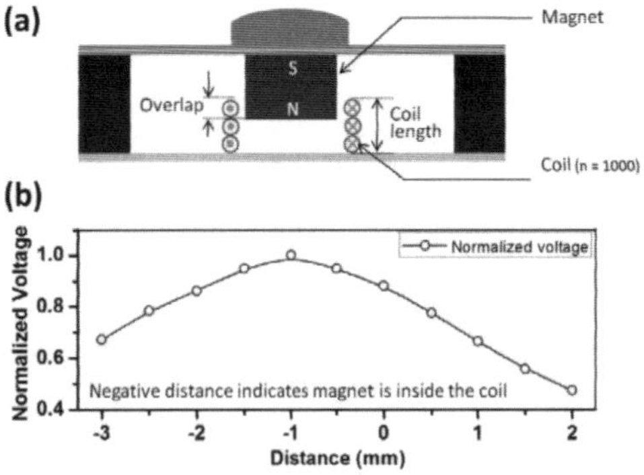

Figure 4. Optimization of magnet-coil overlap distance: (a) schematic structure, (b) Normalized voltage versus overlap distance.

In case of EM transduction part, the highest possible e.m.f. voltage is induced by electromagnetic induction when maximum flux linkage occurs between magnetic flux lines and coil turns. The optimal equilibrium position of the coil with respect to the magnet was measured in order to achieve maximum possible output. Results show that the EM transducer generates maximum open circuit voltage when the magnet (poled in its thickness direction) is immerged 1mm inside the hollow cylindrical coil, as shown in Fig. 4. The air gap between the magnet and coil was 1mm. Since the absolute values are primarily not of interest in determining the optimal equilibrium position of the magnet and coil, they were normalized to the maximum values.

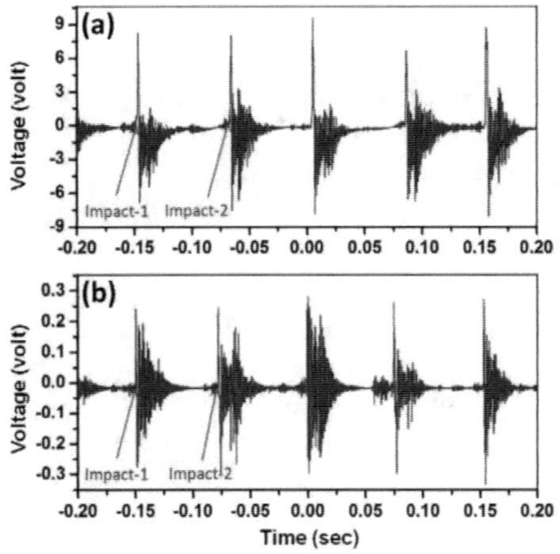

Figure 5. Instantaneous open circuit voltage waveforms generated by (a) PZT cantilever beam and (b) magnet-coil assembly of the fabricated prototype.

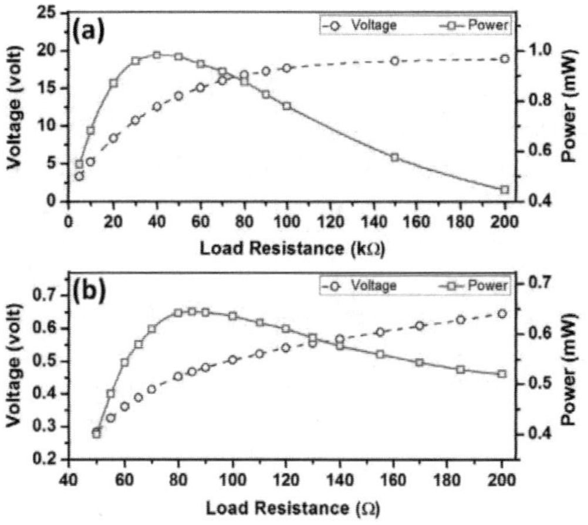

Figure 6. Load voltages (p-p) and powers (peak) versus load resistances generated by (a) PZT cantilever beam and (b) magnet-coil assembly of fabricated prototype.

Open circuit voltage waveforms generated by both PE and EM transduction parts as shown in figure 5, attenuate exponentially with time between two consecutive impacts which cause due to damping. Attenuation is not perfectly exponential due to process variation in assembling the components. Two consecutive maximum peaks were generated in one cycle of the applied vibration because the ball impacted on the parabolic top of the mass twice in one cycle of its back and forth movement. Figure 6 illustrates the peak-peak load voltage and peak power delivered to the load. A maximum 0.98mW and 0.64mW peak powers were generated while PE and EM transducers were connected to 40kΩ and 85Ω matched load resistances, respectively. Since peak voltage amplitudes reduces to almost zero as the time passes, before the next impact occurs, value of average powers reduce dramatically.

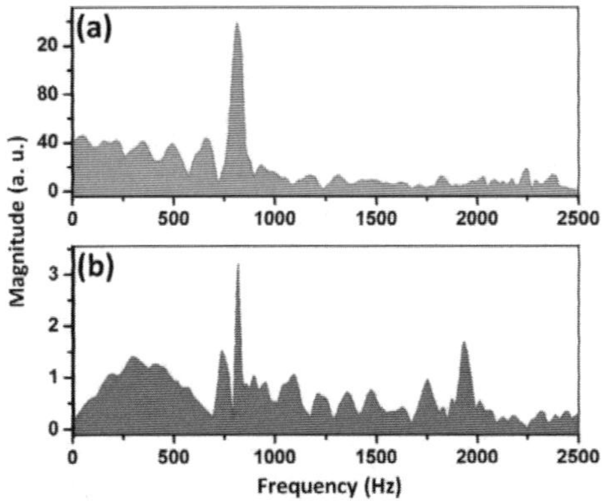

Figure 7. Frequency components of the voltage waveforms generated by (a) PZT cantilever beam and (b) magnet-coil assembly obtained by FFT.

By applying Fast Fourier Transform (FFT) analysis, frequency of the generated voltage waveforms for both PE and EM transduction parts were found to be 816Hz as shown in figure 7 whereas the frequency of handy-motion vibration was around 5Hz. As both PE and EM transduction parts generate power in this hybrid device, it offers much higher peak power density (84.4μWcm^{-3}) than its EM-only counterpart (21.5μWcm^{-3}). There are still much room for its performance improvement. In order to improve its performance, the damping of the cantilever beam vibration needs to be reduced by proper design choice. Beside this, the shape of the mass's top surface and the overlap area of the mass top and the ball needs to be optimized. Damping matching between mechanical and electrical damping would increase its performance as well.

5. Conclusions

We have proposed and demonstrated a handy-motion driven, frequency up-converted vibration energy harvester in which a piezoelectric transduction part and an electromagnetic transduction part have been hybridized. This hybrid approach increases the power density significantly. A macroscale prototype has been fabricated and tested by hand-shaking. The piezoelectric transducer generated 0.98mW peak power whereas the electromagnetic transducer generated 0.64mW peak power across their respective optimum load resistances. Frequencies of the generated voltage waveforms were up-converted by transverse impact mechanism which offers reliable operation of the harvester. The frequency up-conversion feature increased the power flow of the device. Since the output amplitudes degrade with time, average power of the device becomes significantly low which is not good enough to power up an electronic circuit, the test results show its ability to be implemented in powering portable hand-held and wearable consumer electronics from handy-motion with further optimization in design parameters (e.g., stiffness, damping, overlapped contact area). Future work will include more optimized design, miniaturization and implementation of the proposed handy-motion driven hybrid vibration energy harvester.

Acknowledgements

The authors are grateful to acknowledge the Basic Science Research Program (2013R1A1A2A 10064810) and the Pioneer Research Centre Program (2010-0019313) through the National Research Foundation of Korea (NRF) funded by the Ministry of Science, ICT and Future Planning, Korea.

References

[1] Yildiz F, 2009 *The J. Tech. Studies* **1** 40-48
[2] Beeby S P, Tudor M J and White N M, 2006 *Meas. Sci. Technol.* **17** R175-195
[3] Harne R L and Wang K W, 2013 *Smart Mater. Struct.* **22** 023001
[4] Mitcheson P D, Yeatman E M, Rao G K, Holmes A S and Green T C, 2008 *Proc. IEEE*, **96** 1457-1486
[5] Wang D -A and Ko H H, 2010 *J. Micromech. Microeng.* **20** 025019
[6] Williams C B and Yates R B, 1996 *Sens. Actuators A: Phys.* **52** 8-11
[7] Ju S, Chae S H, Choi Y, Lee S, Lee H W and Ji C -H, 2013 *Smart Mater. Struct.* **22** 115037
[8] Pozzi M, Aung M S H, Zhu M, Jones R K and Goulermas J Y, 2012 *Smart Mater. Struct.* **21** 075023
[9] Wei S, Hu H and He S, 2013 *Smart Mater. Struct.* **22** 105020
[10] Halim M A and Park J Y, 2014 *J. Appl. Phys.* **115** 094901
[11] Halim M A and Park J Y, 2013 *J. Phys. Conf. Series* **476** 012119

Storing Deformation Energy of ZnO fine wires in a Mechanical Battery

O G Súchil[1], G Abadal[1] and F Torres[1]

1 Dept. Enginyeria Electrònica, Universitat Autònoma de Barcelona, Bellaterra, 08193, SPAIN

E-mail: oscargilberto.suchil@e-campus.uab.cat

Abstract. The characterizations for ZnO arrays on different substrates have been done with the aim of fabricating an experimental battery that uses the strain energy as energy source. The storage principle is based on a common spring which is loaded by the force associated to the energy to be harvested. By following the same principle our approach is based on pressing an array of fine wires (fws) grown vertically on two different substrate surfaces. The ZnO has been chosen as a fabrication material due to its manufacturing which is cheap and simple using the hydrothermal method. A statistical study for quantify the ZnO fws physical dimensions as density, diameters and lengths was firstly done. Secondly a mechanical study showed the mechanical behaviour of the ZnO fine wires as a unit study where boundary conditions of free-fixed and pinned-fixed were used as possible behaviours. Subsequently an electrical IV characterization was done by using an experimental compressor that allows analyse of the current response of the ZnO array to different compression distances in order to seek for a relation between mechanical and electrical properties. The thermionic-emission was considered to relate the contacted area with the current response.

1. Introduction

As part of a continuous interest on powering nanodevices and self-powered systems previous works have been done on storage energy in the mechanical domain. For instance the study of the potential for storing energy by the elastic deformation on carbon nanotubes (CNTs) have been reported [1]. It was reported that the mechanical energy storage is limited due the strength of the material rather than an electrochemical reaction, making easier the energy storage control unlike in an electrochemical battery. Other approaches have been done by molecular dynamics simulation using body-centered cubic tungsten nanowire where it was revealed a theoretical actuation stress > 3GPa, actuation strain > 30% and finally an average efficiency storage > 90% [2, 3]. An important disadvantage that has to be noted is the high cost and complex fabrication of these carbon nanotubes, therefore the fabrication of ZnO nanowires was taken as an alternative material. Its production is easier and low cost through the hydrothermal method that has been successfully used and reported previously [4, 5]. Our mechanical storage approach is based on the same principle of a spring submitted to compression that suffers a deformation and in consequence stores a potential energy than can be released in a single discharge or multiple gradual discharges.

For our particular case the spring is given by a ZnO fine wires array grown on a substrate, our optimization process has been reported previously [6]. For this work we also present a new sample of

ZnO finewires grown in a different substrate by a ZnO deposition method called Greene method [9] where it was used a Silicon substrate rather than Si/Au. The main focus of this work is the electro-mechanical characterization of the ZnO fine wires array to evaluate its energy storage capability and by comparing both types of samples.

2. Methodology

From our previous experiments reported the best sample of a ZnO array growth on Si/Au and Si substrate were taken.

2.1. Statistical study

Each selected sample was submitted to a SEM image treatment, by taking a 90° and a 45° tilt view of a selected area to quantify diameters and lengths by the construction of a grid. This grid was divided in multiple 100 μm^2 cells. Each cell contains the measures of diameters, lengths and densities. Once the hyper matrix was built the standard deviation for each parameter was calculated and average dimensions were approached.

2.2. Theoretical model

Once the average dimensions for the fine wires were calculated a theoretical model can be performed by using the linear elastic theory taking the linear buckling as a first elastic region limit, limiting the stress and strain without reaching the yield stress on the material. To complete the study two boundary conditions are used as possible behaviors which are *free-fixed* and *pinned-fixed.* Finally all the theoretical computing is corroborated with a FEM model using the same conditions than in the theoretical one and a relative error is calculated to evaluate accuracy.

2.3. Experimental model

The general set-up for the compression is described in figures 1 and 2. The main aim of the compression test is to measure the IV characteristic of the ZnO arrays when the compression distance is varied. The compression test is done by using an experimental compressor conformed by a metal support that holds an angled compressor arm, which have fixed an electrode made of a 500 μm Si with a gold coating of 100nm with dimensions of 50 mm^2. In order to close the circuit the upper electrode has to be approached close enough until it makes contact with the ZnO array located in the samples support. Once the contact is ensured a slope voltage is introduced with the purpose of measuring the output current response. According to previous studies on characterization of ZnO nanowires it has been reported [8, 9] that the thermionic-emission is the main conduction mechanism and can be described by the next expression:

$$I = SA^{**}T^2\exp\left(-\frac{\varphi_s}{kT}\right)\exp\left(\frac{\left(q^7N_D(V+V_{bi}-kT/q)/(8\pi^2\varepsilon_s{}^3)\right)^{1/4}}{kT}\right) \quad (1)$$

Where S is the area of the source Schottky barrier, A^{**} is the effective Richardson constant, q is the electron charge, k is the Boltzmann constant, N_D is the donor impurity density, V_{bi} is the build in potential for the barrier and ε_s is the permittivity of ZnO. By following the literature the equation (1) can be approached to a linear equation by plotting ln I – $V^{1/4}$ where the behaviour shows a qualitative predominance of the thermionic emission at reversely biased Schottky barrier. In order to analyse the relation of the varying compression with the current response, the varying area can be defined as follow:

$$C = \ln[SA^{**}T^2] - \frac{\varphi_s}{kT} \quad (2)$$

Where C is the independent term of the approached linear equation ln I – $V^{1/4}$ and its change is related with the number of fine wires contacted.

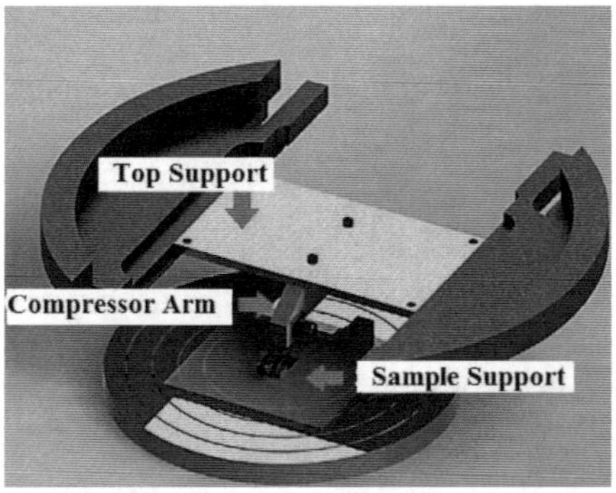

Figure 1. Experimental compressor

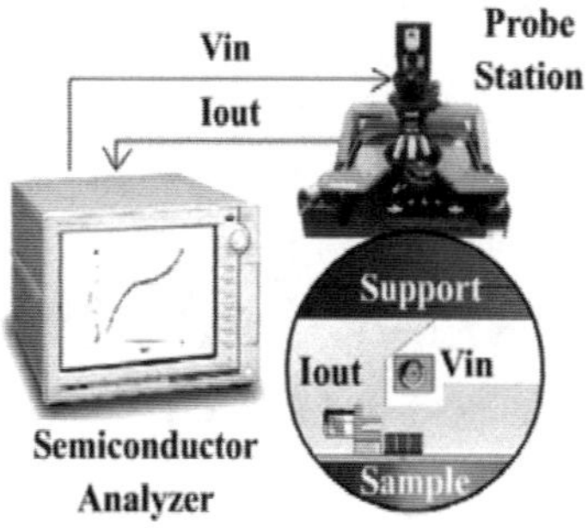

Figure 2. Compression set-up. The circle shows the Vin connector on the angled arm and the Iout connector on the samples support.

From the hydrothermal growth procedure the best samples were taken for two different substrates. HM_A is made of 500 μm of Si and it was coated with 100 nm Au. The second sample GM_B is made of 500 μm of Si doped type P with a seed deposition of $ZnNO_3$ by using the Greene Method [7] see figures 3 and 4.

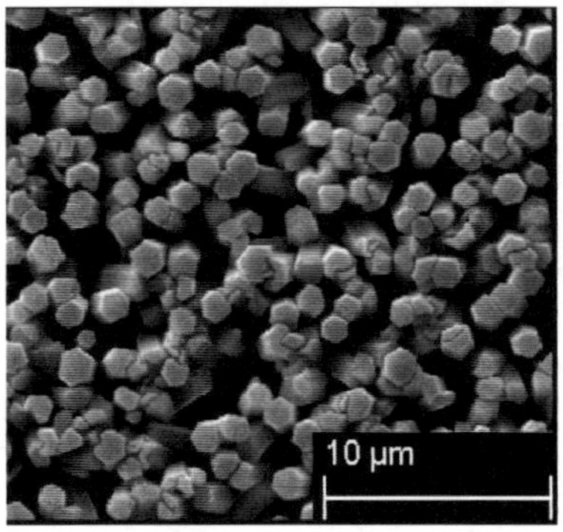

Figure 3. HM_A ZnO array growth on Si/Au substrate.

Figure 4. GM_B ZnO array growth on Si substrate by Greene Method.

3. Results

From the statistical analysis the general dimensions for the fine wires were obtained and plotted in figures 5 to 8.

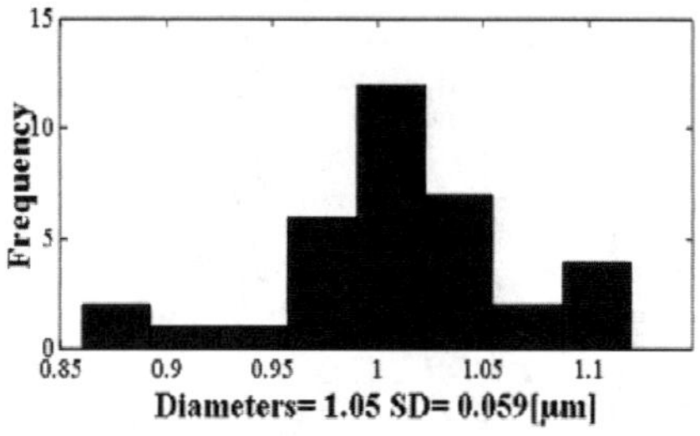

Figure 5. HM_A diameters histogram.

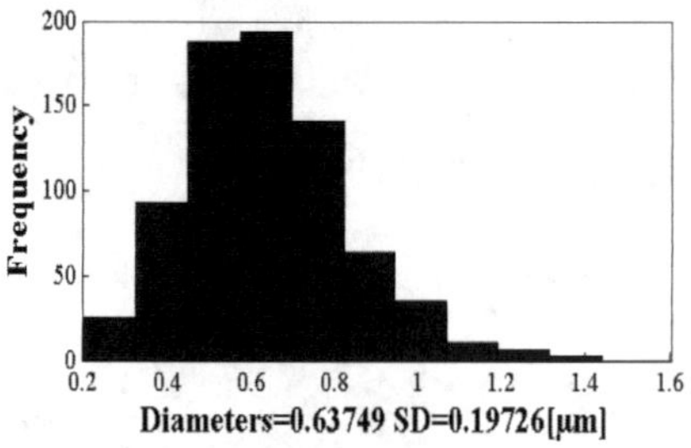

Figure 6. GM_B diameters histogram.

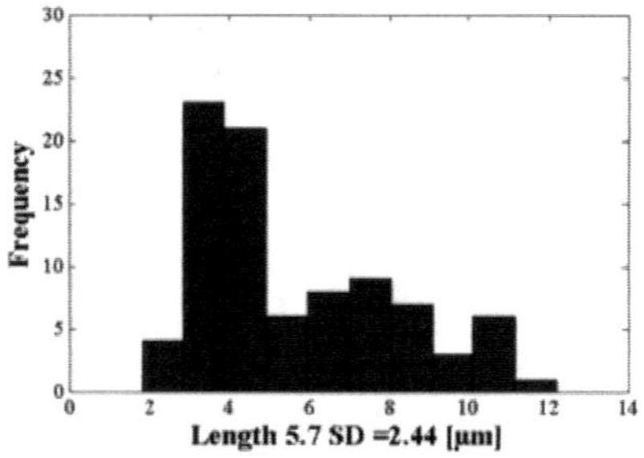

Figure 7. HM_A lengths histogram.

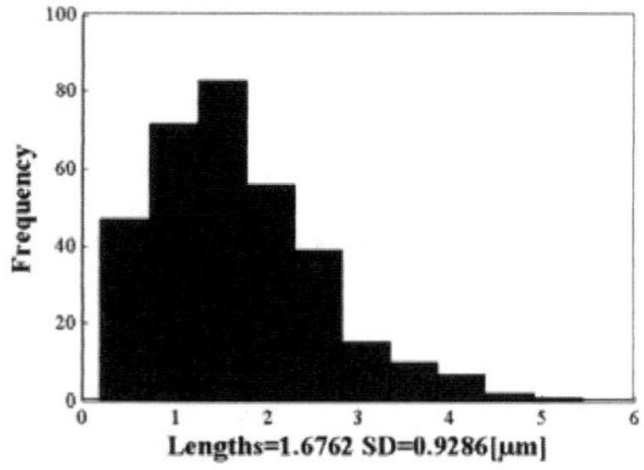

Figure 8. GM_B lengths histogram.

The corresponding densities of each sample were a 0.28 SD=0.043 fws/μm^2 for HM_A and 0.68 SD=0.432 fws/μm^2 for GM_B. The theoretical and FEM computing are reported in table 1.

Table 1. Theoretical model results.

Sample	Boundary Condition	Critical Stress σ_c(MPa)	Critical Strain ε_c(%)	Strain Energy U(pJ)	Strain Energy Density U/V(MJ/m^3)
HM_A	Free-fixed	88.03	0.54	1.19	0.239
GM_B	Free-fixed	401.28	2.48	2.93	9.94
HM_A	Pinned-fixed	718.63	4.43	79.7	15.93
GM_B	Pinned-fixed	3274.86	20.22	195.14	662.02

The FEM computing was compared with the theoretical calculations showing a good agreement with a general relative error < 10%. From the experimental model it was observed only a variation of current against different compression distances for the sample HM_A. For the case of GM_B the currents remained almost constant. Figure 9 shows the IV response for the sample HM_A.

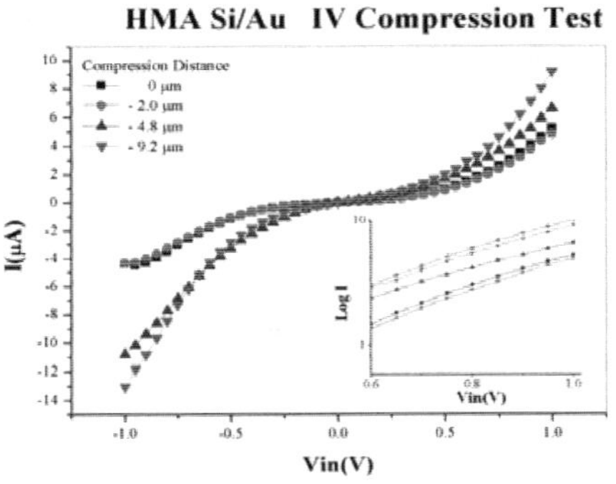

Figure 9. IV Characteristics for different compressions shows as larger is the compression larger is the current. Inset shows a clear increasing trend on the currents.

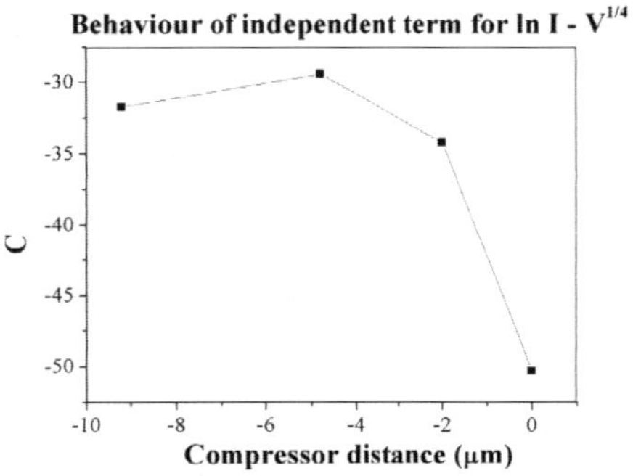

Figure 10. Independent term behaviour shows an increment with the increase of the compression distance.

4. Conclusions

It has to be noticed that the output currents showed a regular Schottky behaviour and an increasing trend of as larger the compression larger the output current and is consistent with the increase of the independent term observed in figure 10. It can be inferred that the contacted area *S* has a relation with the independent term as was shown in equation (2), this area increases when the compressor lowers towards the substrate and contacts a larger number of fine wires allowing to conduct more current. A further study has to be done in order to relate accurately the increase of the contacted area with the mechanical properties of the material and extract an experimental strain that allows to compute the strain energy that can be stored.

Acknowledgments

Funding from the MINECO under contract TEC2013-41767-R (PISCES) has been greatly acknowledged by the authors and with collaboration of CONACYT coordinated by the Mexican government.

References

[1] Hill F A, Havel T F, Hart A J and Livermore C 2009 Storing elastic energy in carbon nanotubes *J.Micromech.Microeng.* **19** 094015
[2] Li S, Ding X, Li J, Ren X, Sun J and Ma E 2010 High-efficiency mechanical energy storage and retrieval using interfaces in nanowires *NanoLett.* **10** 1774-1779
[3] Fthenakis Z G, Zhu Z, Teich D, Seifert G and Tománek D 2013 Limits of mechanical energy storage and structural changes in twisted carbon nanotubes ropes *Phys.Rev.B* **88** 245402
[4] Vayssieres L, Keis K, Hagfeldt A and Lindquist S 2001 Three-dimensional array of highly oriented crystalline ZnO microtubes *Chem. Mater.* **13** 12
[5] Yang R, Qin Y, Dai L and Wang Z L 2008 Power generation with laterally packaged piezoelectric fine wires *Nat. Nanotechnol.* **4**
[6] Súchil O G, Abadal G and Torres F 2013 Effect of temperature and pH level on the hydrothermal growth of ZnO finewires *Nanoenergy Letters* **5** 12-15
[7] Greene L E, Law M, Tan D H, Montano M and Goldberger J 2005 General route to vertical ZnO nanowire arrays using textured ZnO seeds *NanoLett.* **5** 1231-1236
[8] Wang Z L 2008 Piezoelectric Potential Controlled Polarity Reversible Schottky Diodes and Switches of ZnO wires *NanoLett.* **8** 3973-3977.
[9] Wang Z L 2008 Flexible Piezotronic Strain Sensor *NanoLett.* **8** 3035-3040.

Power Electronics Design of a Solar Powered In-car Wireless Tag for Asset Tracking and Parking Applications

D Zhu [1a], J Henaut [b] and S P Beeby [a]
[a]Electronics and Computer Science, University of Southampton, SO171BJ, UK
[b]STERELA, France

E-mail: dz@ecs.soton.ac.uk

Abstract. This paper reports the design and testing of a power conditioning circuit for a solar powered in-car wireless tag for asset tracking and parking application. Existing long range asset tracking is based on the GSM/GPRS network, which requires expensive subscriptions. The EU FP7 project CEWITT aims at developing a credit card sized autonomous wireless tag with GNSS geo-positioning capabilities to ensure the integrity and cost effectiveness for parking applications. It was found in previous research that solar cells are the most suitable energy sources for this application. This study focused on the power electronics design for the wireless tag. A suitable solar cell was chosen for its high power density. Charging circuit, hysteresis control circuit and LDO were designed and integrated to meet the system requirement. Test results showed that charging efficiency of 80 % had been achieved.

1. Introduction

Existing long range asset tracking systems transmit the position of the items being monitored between servers and end-users using GSM/GPRS network. Costs to use such communication channels are relatively high, which limits the applications of existing long range asset tracking systems. If the long range communication cost can be decreased, many more 'tracking' applications could be realised and benefit our daily life. The EU FP7 project CEWITT [1] aims at developing a credit card sized autonomous wireless tag with GNSS geo-positioning capabilities to ensure the integrity and cost effectiveness for parking applications. For this particular project, the Galileo system developed by the European Union and European Space Agency is used.

It was found in previous research that solar cells are the most suitable energy sources to power the CEWITT system [2]. This study focused on the power electronics design for the solar cell to power the wireless tag. The average power consumption of the system is around 200 mW with an operation voltage of 3 V. This means the average current consumption is 66 mA and peak current consumption can reach 100 mA. In addition, the satellite navigation chip used in this project is sensitive to ripple on its supply voltage. A maximum ripple of 30 mV is allowed. In this paper, system description is first introduced followed by the complete system design. Test results will then be presented.

2. System description

Figure 1 shows the system block diagram. Maximum output power of the solar cell is extracted by using the maximum power point tracking (MPPT) technique. A solar cell charges the battery through a

[1] To whom any correspondence should be addressed.

battery charger that can prevent the battery from overcharge. The battery voltage is regulated by a low dropout regulator (LDO) to a required 3 V_{DC}. A hysteresis control circuit cuts off the load from the battery when its voltage is below the low-end threshold and restart the system at a high-end threshold to ensure the stability of the system.

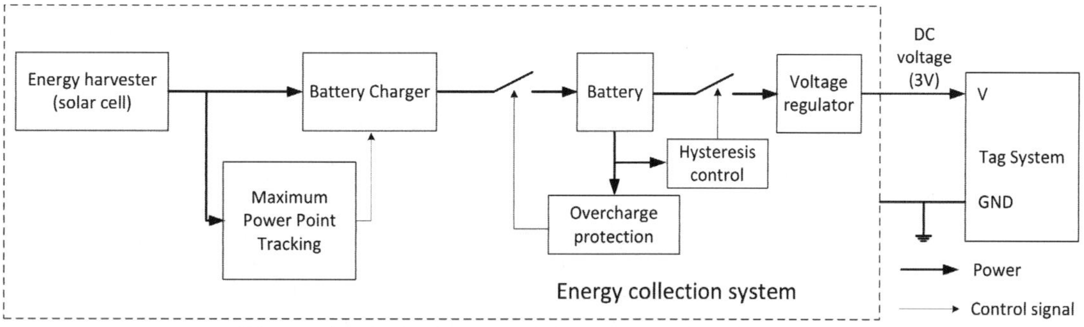

Figure 1. Block diagram of the energy collection system.

3. System design

3.1. Solar cell
A comprehensive market review has been done to look for the suitable solar cell for the CEWITT tag. Considering the size restriction and the power requirement of the CEWITT tag, Ixolar SLMD121H10 was chosen as for its relatively high power density.

3.2. Charging circuit
The high efficiency step-up DC/DC converter, LTC 3105, was chosen for maximum power point tracking. It can operate from input voltages as low as 225 mV and is integrated maximum power point controller (MPPC). A user programmable MPPC set point maximizes the energy that can be extracted from any power source.

The Li-Ion/Polymer shunt battery charger LTC 4071 was chosen as the battery charger. It allows simple charging of Li-Ion/Polymer batteries from very low current, intermittent or continuous charging sources and has a very low operating current of only 550 nA.

3.3. Hysteresis control circuit
A hysteresis control circuit is necessary to prevent the battery from over-discharging by disconnecting the load from the battery when its voltage is below the low-end threshold, V_L, and restart the system at a high-end threshold, V_H. Two designs have been made to realise this function.

3.3.1. Design 1. Figure 2(a) shows the Design 1 of the hysteresis control circuit. For Design 1, V_H is 3 V plus the voltage drops on the two diodes, D1 and D2. V_L is 3 V plus the voltage drop on D2 only. When the battery voltage is above V_H, the output of the voltage detector XC61C_3V is high, which enables the LDO and it has a 3 V DC output. The high level of the output of the voltage detector turns on the switch M1. This causes the gate of M2 to be low and the switch M2 is turned on. When M2 is on, D1 is shorted so that the output of the voltage detector will maintain the high level until the battery voltage drops below V_L when the output of the voltage detector becomes low and the LDO is disabled. In this case, M1 is off and so is M2 because its V_{GS} is 0. The LDO will remain disabled until the battery voltage exceeds V_H and the loop repeats. The advantage of this circuit is that its energy consumption is very low irrespective of the LDO being enabled or disabled. However, the two threshold voltages rely on the voltage drops of relevant diodes, which could be affected by the V/I relationship of individual diodes and the ambient temperature.

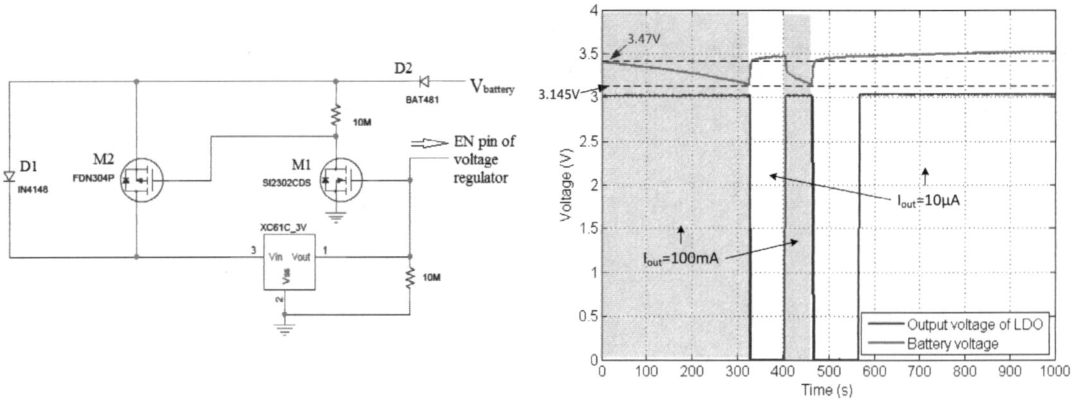

(a) circuit (b) test result

Figure 2. Hysteresis design 1

The test results of Design 1 are shown in Figure 2(b). The entire circuit was connected to a 30 Ω resistive load at the beginning of the test so that the output current of the LDO is 100 mA. The battery voltage dropped gradually as the load drew energy from it. The LDO was disconnected when the battery voltage dropped to 3.145 V (V_L). As the load is disconnected, the battery voltage recovered gradually towards its nominal voltage i.e. 3.7 V. When the battery voltage rose to 3.47 V (V_H), the LDO was enabled and the load started to draw energy from the battery again until the battery voltage dropped to V_L. The load resistance was then adjusted to 300 kΩ, which made the output current of the LDO to be 10 μA. With such low current consumption, the battery was able to rise beyond V_H even with the LDO enabled, due to its self-recovery and being charged by the external source. This design consumes less than 1 μA of current.

3.3.2. Design 2. Figure 3(a) shows the Design 2 of the hysteresis control circuit. This design is based on the hysteresis range enlargement capability of the voltage detector with N-channel output. For Design 2, V_L is 3 V plus the voltage drop on D3 and V_H is decided by the ratio of R1 and R2. A switch M1 is used to short R3 when the LDO is in operation to provide the accurate V_H and allow R3 to be connected in series with R2 in order to reduce the current consumption when the LDO is disabled. This design can provide more accurate threshold voltages but it consumes about 25 μA of current when the LDO is enabled. The test results of Design 2 are shown in Figure 3(b). It has similar performance as Design 1 but its threshold voltage is slightly different (V_L = 3.135 V and V_H = 3.452 V).

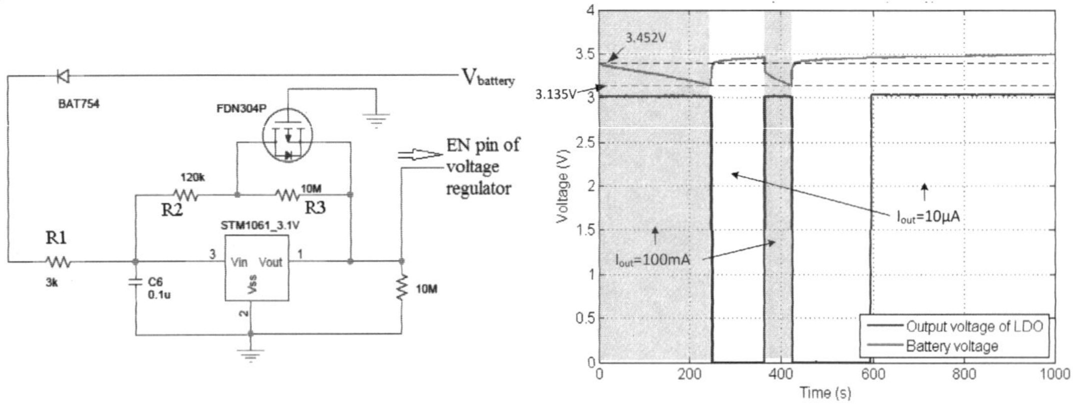

(a) circuit (b) test result

Figure 3. Hysteresis design 2

3.4. Low DropOut regulator (LDO)

Five LDO's have been selected after a comprehensive market survey. Their specifications from datasheets are listed in Table 1. TI TPS78001 was chosen because its low dropout voltage and low current consumption. TI TPS71730, TI TPS79930 and TI LP5900 were chosen because of their high ripple rejection. However, their current consumption is very high. The Torex XC6215 has been used in other satellite navigation systems and it was used in the test as a reference. It was found that LP5900 has the lowest ripple and TPS78001 has the lowest dropout voltage when their output current is 100 mA. Considering that the power consumption of LP5900 is considerably higher than TPS78001, TPS78001 was chosen as the LDO for this application.

Table 1. Comparison of various LDOs.

	I_L=10μA		I_L=100mA			
	Ripple (mV)	V_{out} of LDO (V)	Ripple (mV)	V_{out} of LDO (V)	Dropout Voltage (mV)	I_q (μA)
TPS78001	14.2	3.015	16.5	3.0106	-4.4	4
TPS71730	1.5	3.0101	1.9	2.9819	-28.2	100
TPS79930	1.3	3.0370	65.3	3.0671	+30.1	40
LP5900	1.1	3.0025	0.799	2.9943	-8.2	140
XC6215	2.2	2.9991	7.3	2.9702	-28.9	N/A

4. Test results

Figure 4 shows the final circuit design. Hysteresis design 1 was chosen as it consumes less energy. The battery used in the test was a 300mAh LiPo battery. The charging circuit was tested in a sunny spell day and a cloudy day.

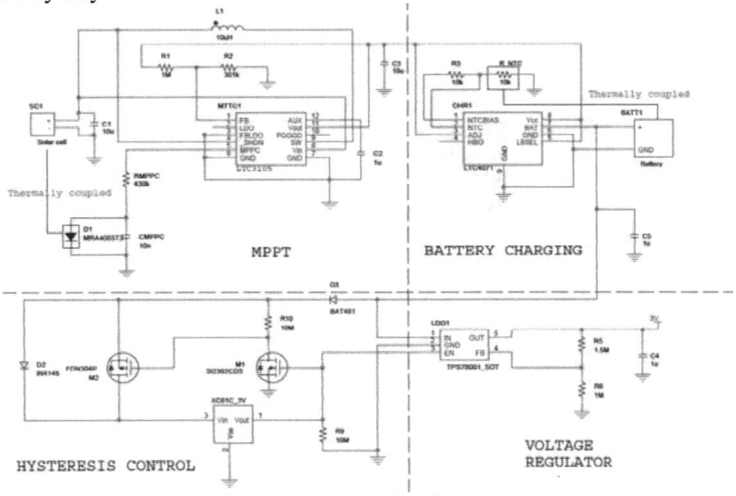

Figure 4. Final circuit design.

4.1. Sunny day

This test was carried out in a sunny spell day when the illumination was between 15 and 30 klux. Figure 5(a) shows the voltage of the solar cell and the battery and Figure 5(b) shows the input and charging power. During the test, the battery was discharged at 100 mA twice. For the first discharge, the duration was 1 min 35 sec and the battery voltage dropped from 3.287 V to 2.989 V. The battery voltage was recovered to 3.287 V after being charged for 16 min 6 sec. The average charging power during this period was 27.8 mW. For the second discharge, the duration was 5 min 5 sec. The battery voltage dropped from 3.373 V to 2.898 V. The battery voltage was recovered to 3.373 V after being charged for 47 min 51 sec. The average charging power during this period was 27.5 mW. Average charging efficiency was 80 %.

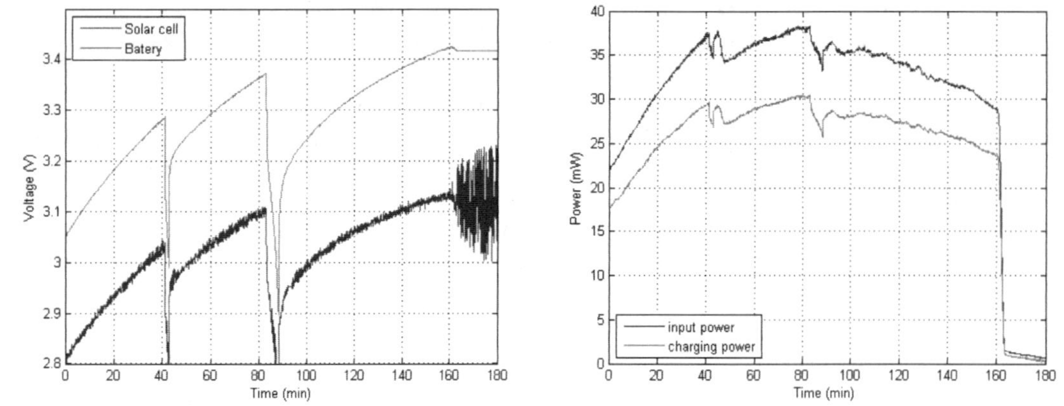

(a) voltage (b) power

Figure 5. Test results in a sunny day.

4.2. Cloudy day

This test was carried out in a cloudy day with a period of sunny spell lasting 20 minutes. The illumination was below 15 klux for most of the time. Figure 6(a) shows the voltage of the solar cell and the battery and Figure 6(b) shows the input and charging power. The average charging power in this case was 9.1 mW. Average charging efficiency was 75 %.

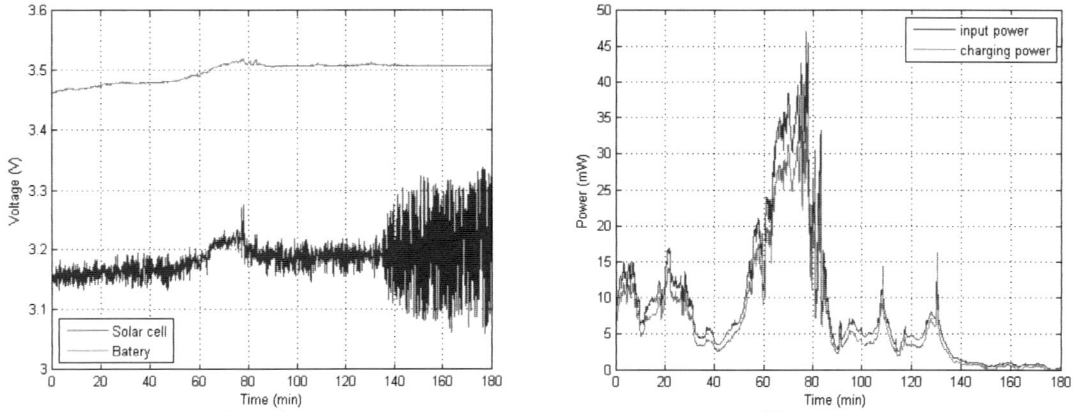

(a) voltage (b) power

Figure 6. Test results in a cloudy day.

5. Conclusions

This paper presents the design of a power conditioning circuit for a solar powered wireless tag. Ixolar SLMD121H10 was chosen as the solar cell for its suitable size and high power density. MPPT and charging circuit ICs were chosen to achieve maximum charging efficiency of 80 %. Hysteresis circuit was designed to protect the battery from over discharging. A suitable LDO was selected to minimise ripple of its output to meet the requirement of the satellite navigation IC.

6. References

[1] EU FP7 project CEWITT, http://www.cewitt.ecs.soton.ac.uk/

[2] Zhu D, Wang L, Henaut J and Beeby S 2014 *Proc. Eurosensors 2014.*

Acknowledgments

This work is supported by the European Union Seventh Framework Programme (FP7) research project CEWITT (http://www.cewitt.ecs.soton.ac.uk/).

A MICRO-STRUCTURED SI-BASED ELECTRODES FOR HIGH CAPACITY ELECTRICAL DOUBLE LAYER CAPACITORS

Valdas Krikscikas[1], Hiroyuki Oguchi[1], Hiroshi Yanazawa[2], Motoaki Hara[1], and Hiroki Kuwano[1]

[1] **Tohoku University, Japan,**
[2] **MEMS-CORE, Japan**

oguchi@nanosys.mech.tohoku.ac.jp
valdas@nanosys.mech.tohoku.ac.jp

Abstract. We challenged to make basis for Si electrodes of electric double layer capacitors (EDLC) used as a power source of micro-sensor nodes. Mcroelectromechanical systems (MEMS) processes were successfully introduced to fabricate micro-structured Si-based electrodes to obtain high surface area which leads to high capacity of EDLCs. Study of fundamental properties revealed that the micro-structured electrodes benefit from good wettability to electrolytes, but suffer from electric resistance. We found that this problem can be solved by metal-coating of the electrode surface. Finally we build an EDLC consisting of Au-coated micro-structured Si electrodes. This EDLC showed capacity of 14.3 mF/cm_2, which is about 530 times larger than that of an EDLC consisting of flat Au electrodes.

1. Introduction

Micro-sensor nodes have been developed to construct a sensor network to realize social monitoring system. One of the key components of the micro-sensor nodes is an energy source, and an electric double layer capacitor (EDLC) is anticipated to be a suitable device. If we can fabricate EDLCs on a single chip using microelectromechanical systems (MEMS) processes, manufacturing accuracy and degree of designing freedom should be increased. Thus we came up with fabricating EDLCs using MEMS compatible Si. Si has, however, never been studied as a material for electrodes of EDLC, and little is known for its behavior in EDLC applications.

Thus in this study, firstly we demonstrated fabrication of high surface area electrodes by MEMS processes, and investigated fundamental properties of the electrodes. Then we fabricated EDLCs consisting of Si electrodes, and evaluated device performance.

2. Experiment

To obtain high surface area Si electrodes by MEMS processes, we tested three ways: deep reactive ion etching (DRIE), anode oxidization, and glancing angle deposition GLAD. The anode oxidization is conducted for p-type Si(100) substrates with resistivity of 1 - 10 Ωcm. The surface of the substrate was etched in the acid solution containing 50% HF and ethanol by flowing current of 10 mA/cm^2 for 60 min. GLAD is a method to deposit a film by impinges atomic or molecular flux to the substrate from an oblique angle [1]. Figure 1 shows a schematic of the GLAD setup used in this study. The GLAD takes place on the side wall of the micro-pillars that makes an almost right angle with depositing atoms. A Cr adhesion layer and then an Au nano/micro-rods were deposited while rotating a sample stage at 20 rpm at room temperature.

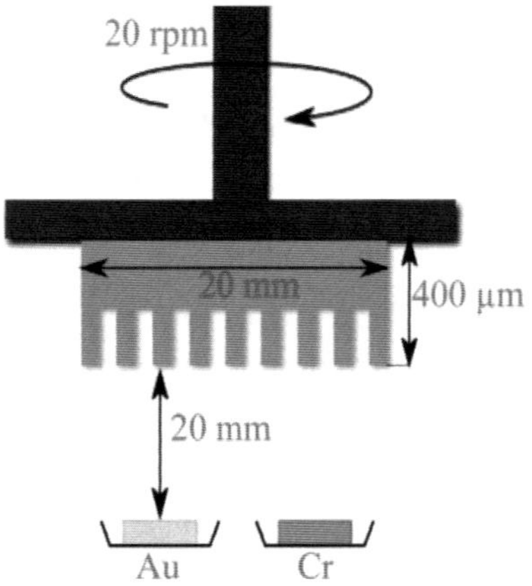

Figure 1: A schematic of the GLAD physical vapor deposition system including detailed information of deposition.

Wettability of the Si electrodes to electrolytes was studied by contact angle measurements. Micro structures were observed by scanning electron microscopy (SEM). Electrical resistivity of the Si electrodes was measured by an ac impedance method. Capacity of the EDLCs were evaluated by observing electrode potential during discharging.

3. Results and discussion

To demonstrate fabrication of high surface area Si electrodes by MEMS processes, we conducted microfabricaion on the Si(100) substrates. Figure 1(a) shows a scanning electron microscopic (SEM) image of the Si electrode obtained after DRIE. Densely and periodically aligned Si micro-pillars (50 μm width × 50 μm depth × 300 μm height, 50 μm gap in between) confirm high surface area. In addition, fine needle-like structures observed at the bottom of the pillars may further increase the surface area. Then, in pursuit of further increase in the surface area, we tried to construct more complicated micro structures using anode oxidization method in which electrochemical surface etching takes palce by flowing current of 1.2 mA/cm^2 for 60 min in a solution of HF : ethanol = 1 : 2. As a result, micro-porous of 1.5 μm-diameter and 15μm-height was obtained on the entire surface [Fig. 1(b) and (c)]. Calculation indicates that surface of the micro-porous Si electrodes has about 30 times higher surface area than that of flat Si electrodes.

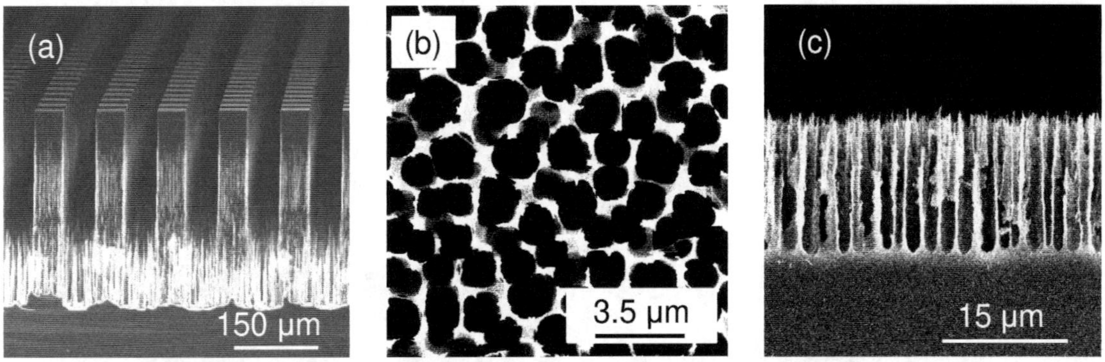

Figure 2. (a) A SEM image of a Si electrode obtained after DRIE. (b) Plain view and (c) cross-sectional SEM images of micro-porous structures.

To judge capability of Si as electrodes of EDLCs, we investigated resistance and wettability to electrolyte. The resistance was measured for the EDLC-type setup consisting of Si electrodes or Pt electrodes. $BMIMPF_6$ was used an ionic liquid electrolyte. Figure 3(a) shows impedance plots of Si and Pt electrodes. The plots intersect with real part of the impedance (horizontal axis) at 200 Ω and 0 Ω for Si and Pt, respectively. This indicates that Si is very resistive and not suitable as electrodes without modification. The wettability was evaluated by contact angle measurements for Si substrates without and with micro-porous on the surface. Figure 3 shows the side view images of water droplet on Si (b) without and (c) with micro-porous. Contact angle for (b) and (c) are 76.3° and 7.4°, respectively, clearly illustrates better wettability for the latter. Thus the micro-porous Si is good for electrodes because electrolytes will easily diffuse into the micro-porous during charging/discharging of EDLCs.

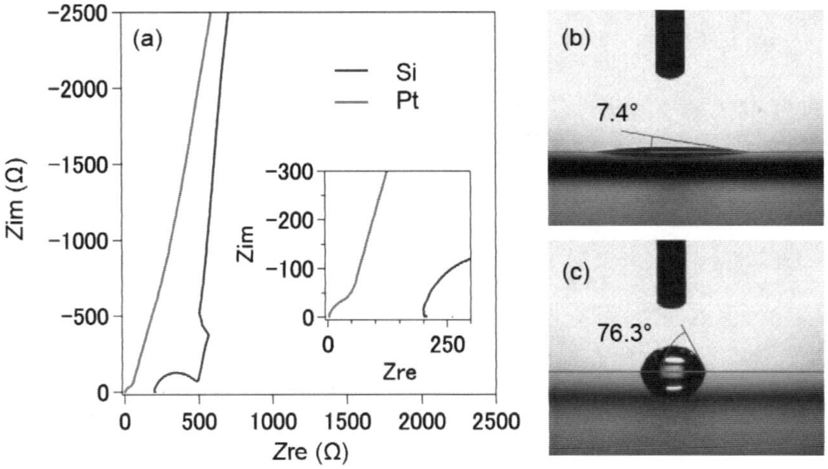

Figure 3. (a) Impedance plots of EDLCs consisting of Si and Pt electrodes. Side views of water droplet on (b) flat and (c) micro-porous Si electrodes.

From studies above, we learned that MEMS can fabricate high surface area Si electrodes with good wettability to electrolyte, but resistance need to be lowered by metal layer coating. These understandings motivated us to fabricate double micro-structured Si electrodes (DMSE) having miro-pillars covered by a micro-structured Au layer, which further increase surface area. Figure 4 shows (a) a plain view and (b) a cross-sectional view of Si rectangular micro-pillars of 100 μm depth and 15 μm width with a 10 μm gap in between. This pattern increases the surface area by 10.6 times than the flat substrate. Figure 4(c) shows a cross-sectional scanning microscopic image of Au nano/micro-rods deposited on the micro-pillars. Morphology of the nano/micro-rods changes along depth of the micro-pillars because of shadowing effect during Au deposition. As show in the inset, near substrate surface, the micro-rods with a diameter of several micrometers densely cover whole area of the side wall, indicating large increase in the surface area. Each rod is apparently separated by air gap aligned parallel to the axis of the rods, which may provide preferable diffusion path for electrolytes of EDLC during charging/discharging.

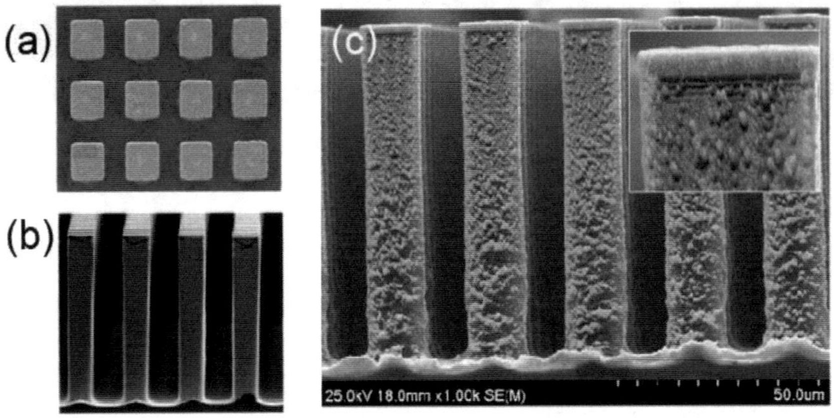

Figure 4: (a) Plain and (b) cross-sectional SEM views of silicon micro-pillars. (c) A cross-sectional SEM image of Au nano/micro-rods deposited on the micro-pillars.

Finally we implemented charge-discharge measurements by constructing EDLCs consisted of DMSE or Au electrodes. The $BMIMPF_6$ was used as an electrolyte. EDLCs were charged for more than 1 h by applying 1 V between electrodes, and then electrode potential was observed during discharging by flowing current of 0.1 μA. Fig. 5 shows discharging curve. Potential dropped rapidly for flat Au electrodes but very slowly (for more than 2 h) for the DMSE, clearly indicates significant increase in the capacity for the latter due to increased surface area. Actually calculated capacitance for DMSE is 14.3 mF/cm_2, which is about 530 times larger than that for flat Au.

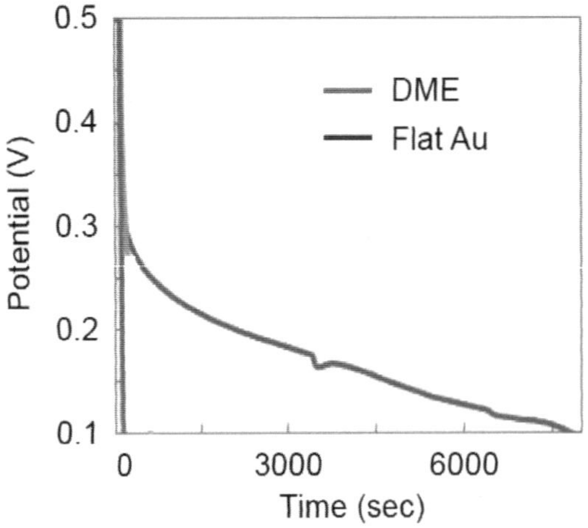

Figure 5: Discharging curves of EDLCs with DME and flat Au as electrodes.

4. Conclusion

This study challenged to make basis for the Si electrodes for EDLCs. We successfully demonstrated that MEMS processes can fabricate micro-structured high surface area Si electrodes. Furthermore we revealed that the micro-structured Si electrodes benefit from high wettability to the electrolytes, and electric resistivity can be lowered by surface metal coating. Based on these understandings, we fabricated the double micro-structured Si electrodes with micro-pillars covered by Au nano/micro-rods. The EDLC consisting of this electrode showed capacity of 14.3 mF/cm_2, which is about 530 times larger than that for flat Au. We believe that this study pave the way for Si-based EDLCs for the micro-sensor nodes.

References

[1] J. P. Norgaard, N. B. Lorentzen, R. Petersen and M. B. madson. *Growth of Nanostructures by Glancing Angle deposition (2007).*

Carbon nanotubes film preparation on 3D structured silicon substrates by spray coating technique for application in solar cells

Y Xiang, M Li, C Lin, P Liu and J Zhang

National Key Lab of Micro/ Nanometer Fabrication Technology, Institute of Microelectronics, Peking University, China

E-mail: zhangjinwen@pku.edu.cn

Abstract. This paper firstly reports the preparation of carbon nanotubes (CNTs) film on silicon substrate of three-dimensional (3D) inverted pyramid structure (IPS) by spray coating. The effect of different substrate temperatures, spraying times and opening sizes on CNTs sidewall covering properties were investigated. The results show that the CNTs covering ratio of sidewall is much lower than that of flat surface and gradually decrease with depth. 40μm×40μm opening obtained the best sidewall covering by CNTs suspension of 40μg/ml at 120°C after 30min spraying so that the CNTs can reach the bottom of IPS and cover about 68.9% sidewall area. At last, it is demonstrated that the output power of the CNTs film-Si solar cell can be enhanced 5.7 times by this method compared to that of the plane structure.

1. Introduction

Carbon nanotubes (CNTs) film is a two-dimensional network structure, formed by many CNTs arranged orderly or disorderly. Because of the low defect, high conductivity and unique physical and chemical properties, it has broad application in film transistors, sensors, batteries, and so on [1].

There are many CNTs film growth methods which can be divided into two categories of direct growth and solution deposition methods. The former one mainly includes graphite arc, chemical vapor deposition (CVD) and laser evaporation. As for the latter, the grown CNTs are dispersed into solution, then the suspension is transferred to the substrate, and finally the agent is removed to form CNTs film. Compared to the direct growth method, the solution deposition method absorbs more interest in industry and academia because of low temperature (<150°C), non-vacuum environment and low cost. Among the solution deposition methods, Langmuir-Blodgett (LB) and chemical self-assembly are used to control the orientation or positioning of CNTs [2, 3]. While, the drop coating and spray coating has the advantages of the simple operation, time-saving and the needlessness of modifying CNTs so that they are widely used in the preparation of random CNTs film [4, 5].

Usually, the output characteristics have positive correlation to the active area. Therefore, forming three-dimensional (3D) inverted pyramid structure (IPS) on silicon substrates is a conventional approach to enlarge surface area of micro-devices for improving their performances, especially in solar cells. Although various CNTs films deposition techniques have been studied in detail, there is little report on the preparation of CNTs film on 3D structured substrates. In this paper, we firstly reports the preparation of CNTs film on 3D inverted pyramid structured silicon substrate by spray coating, and the application in the CNTs film-Si solar cell was proved at last.

2. Design

For solar cells, the output characteristics get better along with the increase of the surface area. However, because electron and hole is bound to recombine in the transport process, the shorter of the distance before the collection of metal electrode, the lower is the probability of recombination. So the metal electrode is usually designed to be square arrays structure and the width is designed to be 20μm with the consideration of the technique limitation. In addition, 10μm×10μm, 40μm×40μm and 80μm×80μm three different opening size arrays are designed to study effects of different 3D structures on the sidewall coverage and optimize the spray coating technique parameters at last.

We choose the n-type Si (100) wafer (2-4Ωcm) as substrate. CNTs film is generally regarded as metallic, which forms a Schottky junction with the silicon substrate. Both front and back electrodes are made of Cr/Au and play roles of collecting and transporting carriers. The role of the SiO_2 layer was to isolate the front electrode and the Si surface to avoid short-circuit. Finally, the microcell was set up on a piece of Printed Circuit Board (PCB), and the front electrode of each microcell was conducted to the external circuit. When the CNTs film-Si heterojunction is illuminated, electron hole pairs are excited. Those generated in the depletion region can be swept across the junction by the built-in electric field into external circuit generating current. Holes are collected by the p-type SWNTs thin film, and electrons are collected by the n-type Si substrate.

3. Fabrication

The fabrication process started on n-type Si (100) wafer. 300nm SiO_2 were thermally grown on both sides of Si wafer, and the back side one was wiped off by buffered hydrofluoric acid (BHF). Then phosphorus ions were implanted to the back surface followed by annealing at 1000°C. 30nm/150nm Cr/Au were deposited on the front surface. The Cr/Au and SiO_2 of the front surface were patterned respectively into 10μm×10μm, 40μm×40μm and 80μm×80μm square opening arrays. After that the Si was wet-etched by KOH until self-stop, achieving 3D IPS. 30nm/150nm Cr/Au were deposited on the back side of Si wafer and was annealed at 320°C to form ohmic contact with Si. Finally, the wafer was diced into 5mm×6mm pieces on which we would deposit CNTs thin film. In order to contrast with the 3D IPS, we also designed plane structures. The fabrication process of plane structures is similar to the process introduced above except for the KOH wet-etched step.

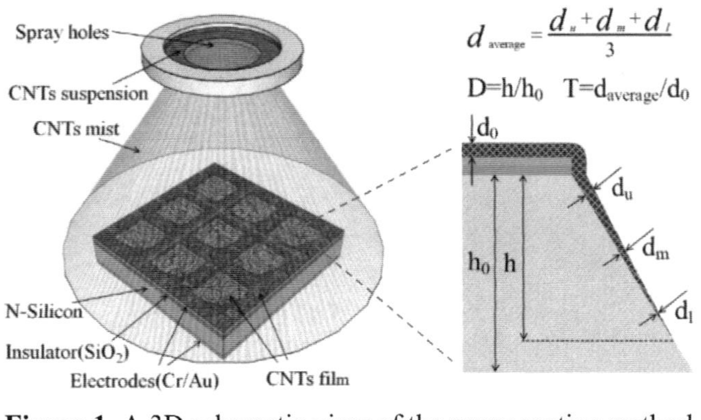

Figure 1. A 3D schematic view of the spray coating method (inset: measurement parameters of CNTs film).

Spray coating technique is a method to atomize the CNTs suspension which is sprayed to the hot substrate to get CNTs film (Fig. 1). At first, CNTs powders (purity>90%) were ultrasonically dispersed in dimethylformamide (DMF). Then CNTs suspension was converted into mist and sprayed onto baked

Si substrate by ultrasonic sprayer. At the same time, the Si substrate was heated to evaporate DMF rapidly and totally. Different spraying conditions were investigated experimentally, including the substrate temperature of 70°C, 100°C, 120°C 150°C and 200°C, spraying time of 15, 30 and 45min and CNTs suspensions concentration of 20 and 40μg/ml.

4. Test and results

The sidewall covering degrees are characterized by the depth ratio of IPS that CNTs film reached and the CNTs film thickness ratio along the sidewalls. For plane structures, the thickness of CNTs film can be measured directly. However, for 3D IPS, the thickness of CNTs film is measured by average method because the sidewall covering properties change with the depth of the groove. We define the thickness of CNTs film on plane structures as d_0, the CNTs film covering ratio of plane structures as C_0. As for the IPS, we define the depth of IPS as h_0, the depth CNTs reached as h, and the thickness of CNTs film in different depth on 3D structures sidewall as d_u, d_m, d_l (Fig. 1 insert). The average thickness on 3D structures sidewall is $d_{average}= (d_u+d_m+d_l)/3$. The depth ratio is $D=h/h_0$ and the thickness ratio is $T=d_{average}/d_0$. So the covering ratio can be expressed approximatively as $C_0\times d_{average}/ d_0$.

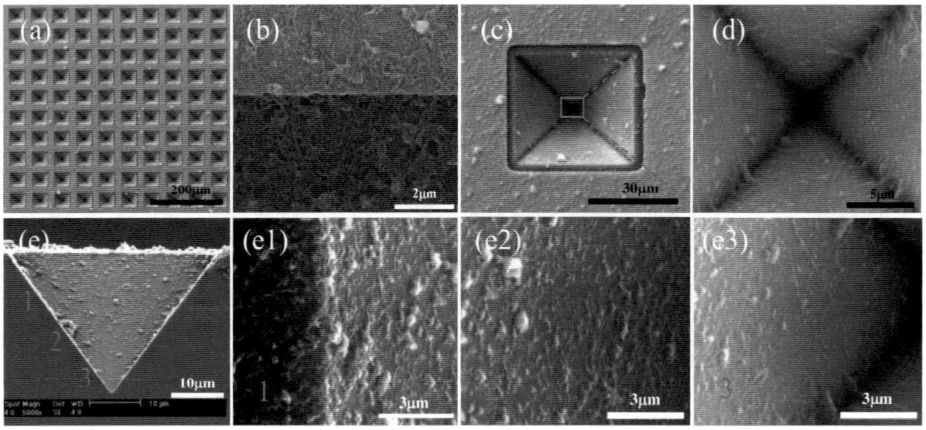

Figure 2. SEM images of CNTs film prepared by spraying 30min at 120°C on 3-D structured Si substrates with 40μm×40μm opening: (a) IPS array covering; (b) planar covering on Au and Si surface forming a network structure; (c) single IPS covering; (d) enlarged view of bottom area covering of IPS; (e) cross-section of IPS covering; (e1)-(e3)enlarge views of the IPS sidewall covering at points 1,2,3 of (e).

Fig. 2 is the SEM of the CNTs film prepared by spraying 30min at 120°C with 40μm×40μm opening. Spray coating technique can form uniform CNTs film in whole array (Fig. 2(a)) and the surface of Au and Si (Fig. 2(b)). This spraying condition obtained the best sidewall covering of D=1 and T=0.725, indicating that the CNTs can reach the bottom of IPS and cover about 68.9% (estimated C_0=95%) sidewall area (Fig. 2(c) and (d)). Fig. 2(e) gives the SEM of the whole sidewall covering. We enlarged 1, 2 and 3 position (Fig. 2(e1~e3)) and can find the CNTs covering ratio of sidewall is much lower than that of flat surface and gradually decreases with depth.

Fig. 3 is the curve of depth ratio versus spraying time and temperature of three different opening sizes. The CNTs film covering ratio increases along with the spraying time (Fig. 3(a)). What's worth noting is that the depth ratio of 40μm×40μm opening size is better than 10μm×10μm and 80μm×80μm opening sizes at 120°C after 30min spraying. With the increase of temperature, the CNTs film covering ratio shows a downward trend after the first increases (Fig. 3(b)). The 40μm×40μm and 80μm×80μm opening sizes get the best depth ratio at 120°C, while the 10μm×10μm opening size decreases above 100°C. And, it was found that concentration of the CNTs suspension has almost no influence on 3D covering properties.

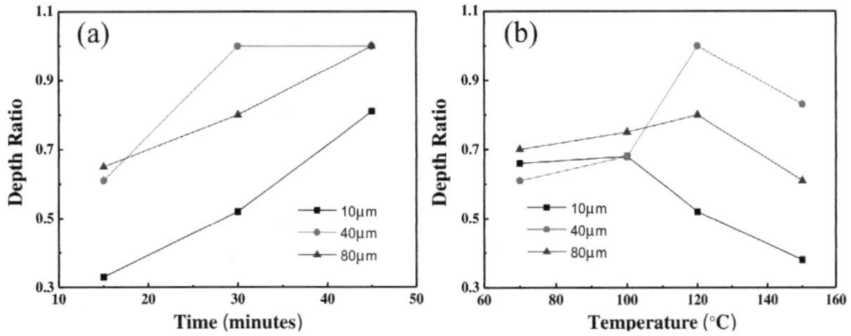

Figure 3. (a) The depth ratio versus time when temperature is 120°C and CNTs suspension is 40μg/ml; (b) The depth ratio versus temperature when spraying time is 30 minutes and CNTs suspension is 40μg/ml.

Fig. 4 is the Current-Voltage (I-V) curves and Power-Voltage (P-V) curves under light (AM 1.5G, 100mW/cm2 illumination) for IPS and plane structures. It can be seen that the short circuit current and the open circuit voltage of IPS structure are respectively I_{SC}^{IPS}=17.5μA and V_{OC}^{IPS}=0.32V, while I_{SC}^{plane}=3.7μA and V_{OC}^{plane}=0.26V for plane structure. So the output power of the CNTs film-Si solar cell can be enhanced 5.7 times by IPS structure.

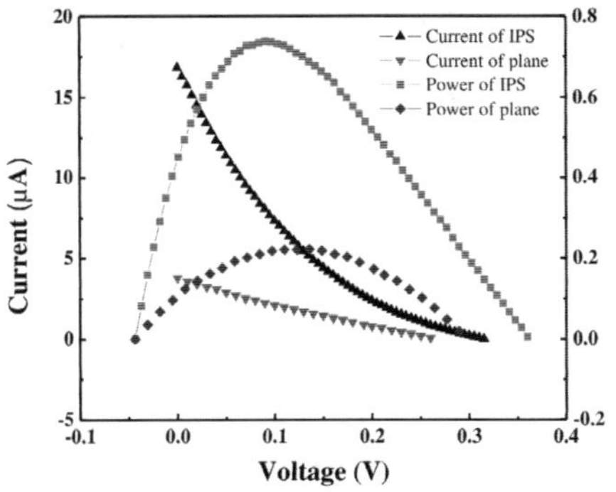

Figure 4. The Current-Voltage (I-V) curves and Power-Voltage (P-V) curves under light (AM 1.5G, 100mW/cm²[2] illumination) for IPS and plane devices, with I_{SC}^{IPS}=17.5μA , V_{OC}^{IPS}=0.32V and I_{SC}^{plane}=3.7μA, V_{OC}^{plane}=0.26V respectively.

5. Conclusion

This paper firstly reports the preparation of carbon nanotubes (CNTs) film on 3D inverted pyramid structured silicon substrate by spray coating. The effect of different substrate temperatures, spraying times and opening sizes on CNTs sidewall covering properties were investigated. The results show that the CNTs covering ratio of sidewall is much lower than that of plane surface and gradually decrease with depth. With the increase of spraying time, the CNTs film covering ratio increases. While covering ratio shows a downward trend after the first increases with the increase of temperature. 40μm×40μm opening obtained the best sidewall covering by CNTs suspension of 40μg/ml at 120°C after 30min spraying so that the CNTs can reach the bottom of IPS and cover about 68.9% sidewall area. Finally, it is proved that the output power of the CNTs film-Si solar cell can be enhanced 5.7 times by this method.

References

[1] Hu L B, Hecht D S, and Gruner G 2010 *Chem. Rev.* **110** 5790
[2] Yeji K, Minami N, Weihong Z, Kazaoui S, Azumi R and Matsumotu M 2003 *Jpn. J. Appl. Phys.* **42** 7629
[3] Yan D, Juhala J L, Griffith B W and Wei X 2013 *Microelectron. Eng.* **103** 18
[4] Small W R, Walton C D, Loos J and Panhuis MIH 2006 *J. Phys. Chem.* B **110** 13029
[5] Ramasamy E, Lee W J, Lee D Y and Song J S 2008 *Electrochem. Commun.* **10** 1087

AlN thin film growth using electron cyclotron resonance reactive sputtering

N. H. Hung, H. Oguchi, H. Kuwano
Department of Nanomechanics, Tohoku University, Japan
E-mail: hung@nanosys.mech.tohoku.ac.jp

Abstract. In this report, we investigated conditions to deposit stoichiometric aluminium nitride (AlN) thin films grown on (100)-oriented Si substrates under various Ar/N_2 gas flow rates at a wide range of temperature from room temperature (RT) to 350°C using electron cyclotron resonance (ECR) reactive sputtering. This study revealed that stoichiometric of thin film can be controlled by N_2/Ar flow rate and that stoichiometric N/Al = 1 was archived at N_2/Ar = 2. This study also revealed that crystallinity can be controlled by substrate temperature. From RT to 200°C, thin films were amorphous or poly-crystal, at 350°C however, thin film was mainly [110] and [100] AlN. Obtained thin films are densely packed and have very low root mean square (RMS) roughness of 0.41 nm which is much less than other sputtering methods.

1. Introduction

AlN is one of the III-nitride semiconductors having a large direct band gap of about 5.9–6.02 eV [1], high hardness, and good thermal conductivity [2]. In recent years, much attention has been paid on fabricating AlN thin films for applications in a micro-electronics field, especially as an important component of energy harvesters [3] and surface acoustic wave (SAW) [4] or bulk acoustic wave (BAW) [5] devices. It has been the most standard lead-free piezoelectric material for such application because of the high piezoelectric properties.

For AlN powder production, there are two major processes that were used [6]: direct nitridation of aluminium in presence of N_2 or NH_3 [7] and carbon thermal reduction of alumina [8]. For thin films, however, only the former approach likely to be available. For example, metal organic chemical vapour deposition [9], plasma-assisted molecular beam epitaxy [10], and chemical vapour deposition [11] have been used so far. In addition to these techniques, RF and DC reactive sputtering were often chosen since those techniques are suitable to prepare high quality AlN thin films [12]. Furthermore, these techniques allow us to control film stoichiometry and crystallinity by adjusting flow rate of gases during deposition.

ECR sputtering is another type of sputtering methods. This method can improve crystallinity [13] more effectively than other sputtering methods even at lower growth temperature due to activated plasma formed by reactive gases (in our case, nitrogen). It also improves density and flatness of the thin films that is favourable for micro device fabrication [3]. So far, however, ECR sputtering has been seldom reported for AlN films [13], [14].

In this study, we investigate the composition of film deposited by ECR sputtering by changing N_2/Ar gases flow rate, and crystallinity of thin film by changing substrate temperature. Furthermore, we evaluate density and flatness of the film

2. Experiments

Figure 1 shows an illustration of the ECR ion beam sputtering system used for AlN deposition. The system has two ion-guns that generate plasma to sputter targets or to N_2 gas to facilitate gas-film reaction. Nitrogen gas is introduced into the chamber directly or through the ion gun. The deposition rate of films grown on the (100) silicon substrates is 0.9 nm per minute averagely. Typical sputtering parameters were shown in Table 1. Prior to AlN film deposition, silicon substrates were cleaned by two standard cleaning processes of Radio Corporation of America (RCA1 and RCA2) in order. The Al-target was pre-sputtered before each deposition for 30 minutes in Ar and then 10 minutes in the mixture of Ar and N_2. Thickness of the films was measured by a surface profiler. Field emission scanning electron microscopy (FE-SEM, Hitachi SU-70) was used to perform morphological analysis. The composition was confirmed by using the EDX spectroscopy and the crystal structure of the samples was measured using an X-ray diffraction (XRD, Bruker D8). An atomic force microscopy (AFM, Nikon TE2000-U) was used to examine thin film's surface roughness.

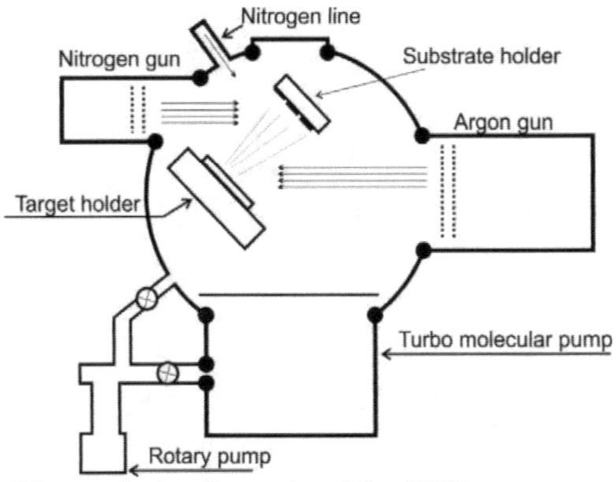

Figure 1. A schematic of the ECR apparatus

Table 1. Deposition conditions

Substrate temperature	RT – 350°C
Ar gas flow rate	0.6 sccm
N_2 gas flow rate	0 – 1.2 sccm
Ar ion beam current	12 mA/cm^2
N_2 ion beam current	0 and 4 mA/cm^2
ECR microwave power	100 W (both guns)
Ar accelerating voltage	2000 V
N_2 accelerating voltage	20 V
Back pressure	5x10^{-3} Pascal
Depositing pressure	0.33 Pascal

3. Results and discussions

To optimize the growth condition of Al nitridation, we investigated compositions of the thin films deposited at RT by changing N_2/Ar gas flow rate (Figure 2). Composition of the films increases almost linearly with N_2/Ar flow rate, and reaches to almost stoichiometric value (N : Al ~ 1 : 1) when the ratio of the N_2/Ar = 2.

Figure 3 shows XRD *2θ-θ* patterns of the stoichiometric films deposited at different temperatures. Thin films did not show any peak of AlN for growth temperature of up to 200°C, regardless of normal and activated nitrogen. However, at 350°C, a peak of AlN (110) and (100) diffraction planes appeared, indicating that crystallinity of the film was improved by increasing substrate temperature. Judging from the intensity ratio, a primary AlN domain has [110] orientation normal to the substrate.

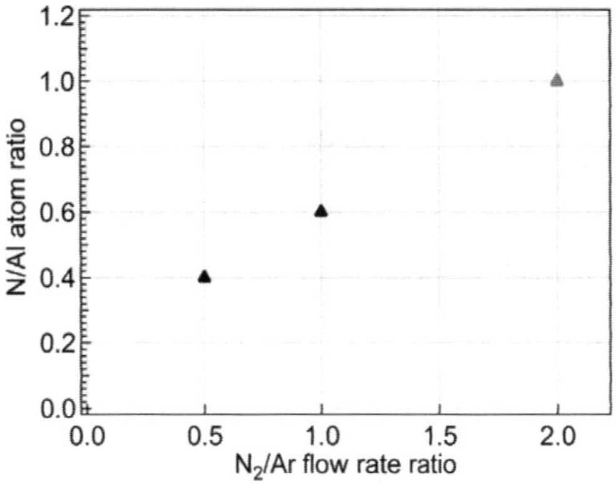

Figure 2. Dependence of Al and N ratio in the thin film deposited at room temperature as a function of Ar and N_2 flow rate.

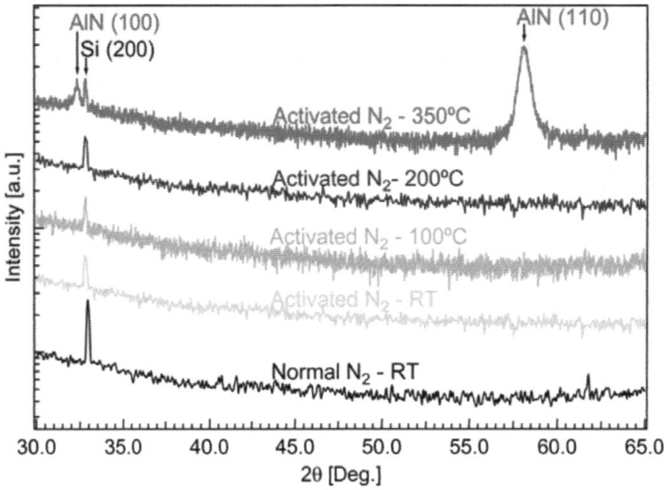

Figure 3. X-ray 2θ–θ scan of AlN deposited on Si at various substrate temperatures.

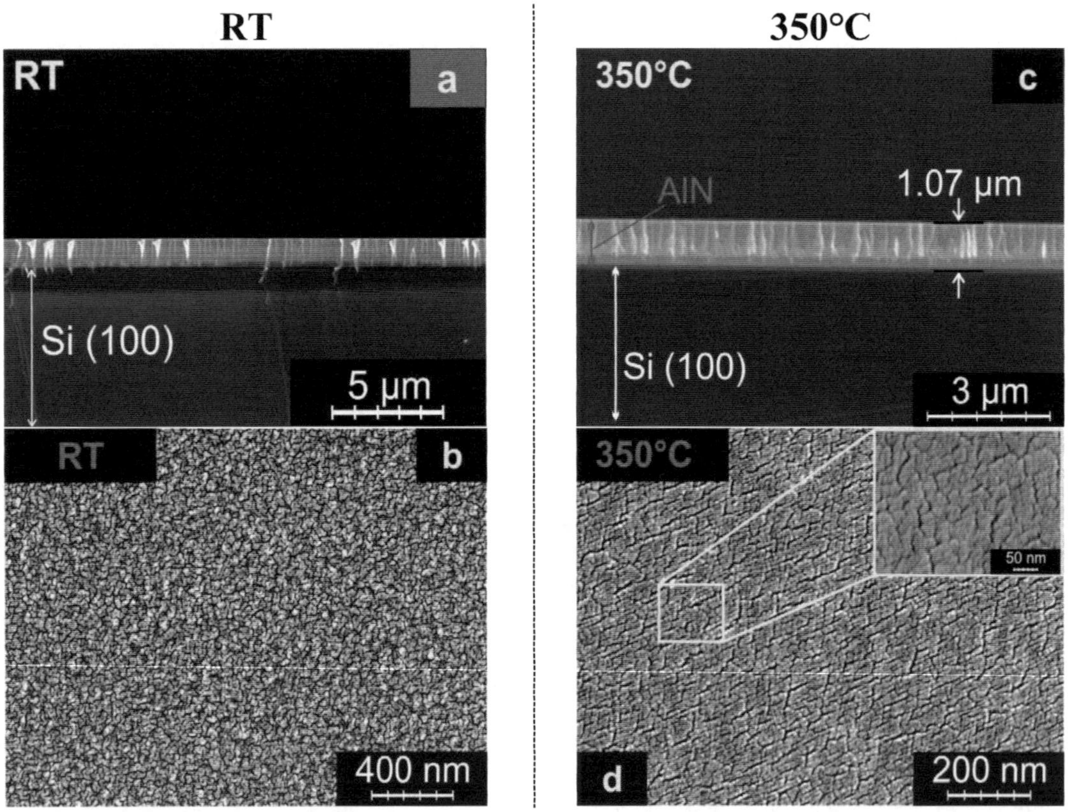

Figure 4. Cross-sectional FE-SEM and plane view images of AlN thin film deposited at RT and 350°C.

Typical cross-sectional and plane view FE-SEM images of the thin film deposited at RT and 350°C were shown in Figure 4. As shown in the figure 4(a) and 4(c), the cross-sectional images of the thin films deposited at RT and 350°C, respectively reveal that the both films are densely packed with high homogeneity and surface flatness. Surface morphologies of the films were shown in the figure 4(b) (deposited at RT) and figure 4(d) (deposited at 350 °C). One could be seen from the images, the surface of thin films kept almost identical for low and high temperature deposition. Cracks which were observed on the surface of thin films have widths in range of 7 – 10 nm. The cross-sectional image revealed that the cracks did not go through the thin film to the substrate (vertical lines on the images were created when sample had been broken). This will help avoiding short between top and bottom electrodes that is often a problem suffered by researchers using films grown by normal sputtering in micro device fabrications.

Figure 5 shows an AFM image (1 μm x 1 μm) of the thin film surface. Closely packed pebble-like nanostructure was observed. The measured RMS roughness and peak to valley value of the typical films are 0.41 nm and 3.8 nm, respectively. These values are clearly less than those of the films deposited by RF sputtering [15]-[17]. The flatness of our films is preferable in, for instance, SAW devices [15] which favour as low roughness as possible. It should be noted that roughness of thin films that was deposited by other sputtering methods often increase with substrate temperature while AlN prefers high temperature.

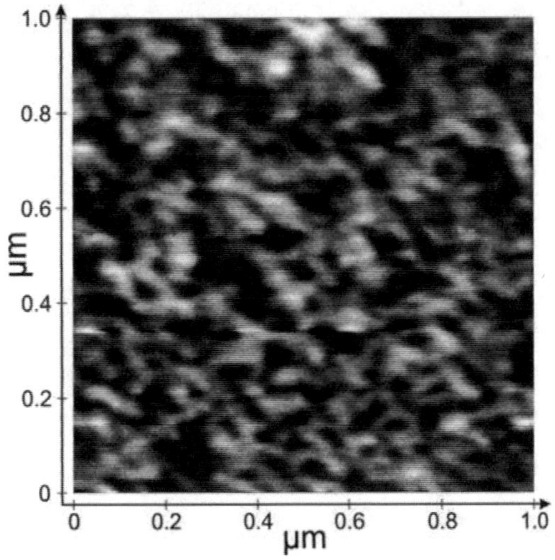

Figure 5. An AFM surface image of AlN thin film deposited at 350°C.

4. Conclusion

We investigated the optimum growth conditions to obtain stoichiometry and high crystallinity in the AlN thin films growth by ECR sputtering. The composition was controlled by changing N_2/Ar gas flow ratio during the film growth, and realized stoichiometry at N_2/Ar = 2. Crystallinity of the AlN was poor for the films deposited at the temperature range between RT and 200°C, but improved at 350°C. The AlN film deposited at 350°C are closely packed and consisted mainly of [110]-oriented domains. Its thickness is homogeneous and surface is flat as proved by RMS roughness of 0.41 nm. These results indicated that the AlN films grown by the ECR sputtering is suitable to fabricate functional micro-electromechanical devices.

References

[1] Shah A and Mahmood A 2012 *Phys. B Cond. Matter.* **407** 3987.
[2] Ait Aissa K, Achour A, Camus J, Le Brizoual L, Jouan P Y and Djouadi M A 2014 *Thin Sol. Fil.* **550** 264.
[3] Zhang J, Cao Z, and Kuwano H 2011 Jpn. J. Appl. Phys. **50** 09ND18.

[4] Chen S W, Lin H F, Sung T T, Wu J D, Kao H L and Chen J S 2003 Elec. Lett. **39** 1691.
[5] Hara M. and Kuwano H. 2012 Jpn. J. Appl. Phys. **51** 07GC11
[6] Sakurai T, Yamada O and Miyamoto Y 2006 *Mater. Sc. Eng.* A **415** 40.
[7] Weimar A W, Cochran G A, Eisman G A, Henley J P, Hook B D, Mills L K, Guiton T A, Knudsen A K, Nicholas N R, Volmering J E and Moor W G J. 1994 *Am. Ceram. Soc.* **77** 3.
[8] Lefort P and Billy M 1993 *J. Am. Ceram. Soc.* **76** 2295.
[9] Chen Y, Song H, Lia D, Suna X, Jianga H, Lia Z, Miaoa G, Zhanga Z and Zhoub Y 2014 *Matter. Let.* **114** 26.
[10] Yusoff M Z M, Mahyuddinc A, Hassanb Z, Yusofb Y, Ahmadb M A, Chinb C W, Hassanb H A and Abdullahb M J 2013 *Supper. Mic.* **60** 500.
[11] You Y, Ito A, Tu R and Goto T 2013 *J. Crys. Grow.* **365** 1.
[12] Schneider J M, Rohde S, Sproul W D and Matthews A 2000 *J. Phys. D. Appl. Phys.* **33** R173.
[13] Okano H, Tanaka T, Shibata K and Nakano S 1992 *Jpn. J. Appl. Phys.* **31** 3017.
[14] Ryunosuke H, Takeshi M, Masayoshi E and Shuji T 2010 *IEEJ Trans. Sens. Micr.* **30** 523.
[15] Kuang X, Zang H, Wang G, Cui L, Zhu C, Jin L, Sun R and Han J 2012 *App. Surf. Sc.* **263** 62.
[16] Vashaei Z, Aikawa T, Ohtsuka M, Kobatake H, Fukuyama H, Ikeda S and Takada K 2009 *J. Crys. Grow.* **311** 459.
[17] Kar J P, Boss G and Tuli S 2005 *Mater. Sc. Sem. Proc.* **8** 646.

A flexible micro biofuel cell utilizing hydrogel containing ascorbic acid

Hideaki Goto , Yudai Fukushi , and Yasushiro Nishioka

Department of Precision Machinery Engineering, College of Science and Technology, Nihon University, 7-24-1 Narashinodai, Funabashi, Chiba 274-8501, Japan

E-mail nishioka@eme.cst.nihon-u.ac.jp

Abstract. This paper reports on a biofuel cell with a dimension of 13x24 mm^2 fabricated on a flexible polyimide substrate. I its porous carbon-coated platinum (Pt) electrodes of 3 mm in width and 10 mm in length were fabricated using photolithography and screen printing techniques. Porous carbon was deposited by screen printing of carbon black ink on the Pt electrode surfaces in order to increase the effective electrode surface area and to absorb more enzymes on the electrode surfaces. It utilizes a solidified ascorbic acid (AA) aqueous solution in an agarose hydrogel to increase the portability. The maximum power and power density for the biofuel cell with the fuel unit containing 100 mM AA were 0.063 μW and 0.21 $\mu W/cm^2$ at 0.019 V, respectively.

1. Introduction

Microelectromechanical Enzymatic biofuel cells that use glucose in a human body to produce electricity for embedded microsystems have been of special interest. This is because they work under mild conditions of room temperature, neutral pH and atmospheric pressure, which make them amenable to the human body [1]-[14]. Nishizawa et al. reported a biofuel cell with a microchannel designed to continuously deliver fuel to the biofuel cell [15]. Fukushi et al. also reported a flexible glucose biofuel cell with a microchannel fabricated on a polyimide (PI) substrate [16]. However, these devices sometimes require complex biofuel delivering systems, which causes portability loss of the biofuel cells. We reported a portable simple-structured biofuel cell with planar two electrodes on a PI substrate that utilized agarose hydrogel containing glucose as a fuel [17].

Ascorbic acid (AA) also known as vitamin C, is an environmentally and biologically friendly compound. Thus, it is interesting to utilize an aqueous AA fuel absorbed in hydrogel. Typical hydrogels include collagen, polyacrylamide and agarose as polymers, and they are expected to store a AA aqueous solution. Collagen has been reported to be easily decomposed by ultraviolet rays in the sunlight [18], and it is not suitable for a portable hydrogel solid fuel. On the other hand, agarose has been reported to be harmless to human body, and the diffusions of AA and oxygen in agarose gels are almost the same as those in water [19]. Thus, we considered that agarose is suitable for a portable hydrogel solid fuel containing AA. In this paper, a AA biofuel cell with carbon electrodes modified with enzymes that utilizes an agarose hydrogel containing an aqueous solution of AA is reported.

Journal of Physics: Conference Series **557** (2014) 012048 doi:10.1088/1742-6596/557/1/012048

2. Experimental

2.1 Design of Ascorbic acid fuel cells

Figure 1 shows the schematic view of the fabricated AA biofuel cell. The anode with an area of 3x10 mm^2 is covered by carbon, the cathode with an area of 3x10 mm^2 is covered by carbon modified by bilirubin oxidase (BOD), and the reference electrode with an area of 3x7 mm^2 is covered by Ag|AgCl. The solid agarose gel as a portable fuel containing 100 mM AA solution was placed on these electrodes. The size of the agarose gels was 13x 20 mm^2 with the thickness ranging from 1 to 4 mm.

Figure 2 shows the power generation mechanism. On the anode, the AA is oxidized at the anode electrode surface and dehydro ascorbic acid is generated. As a result, electrons and protons were produced on the anode. The electrons were transferred from the anode to the cathode via an external circuit generating electricity. Oxygen (O_2) in the fuel solution is reduced by BOD immobilized on the cathode to water on reaction with electrons and protons.

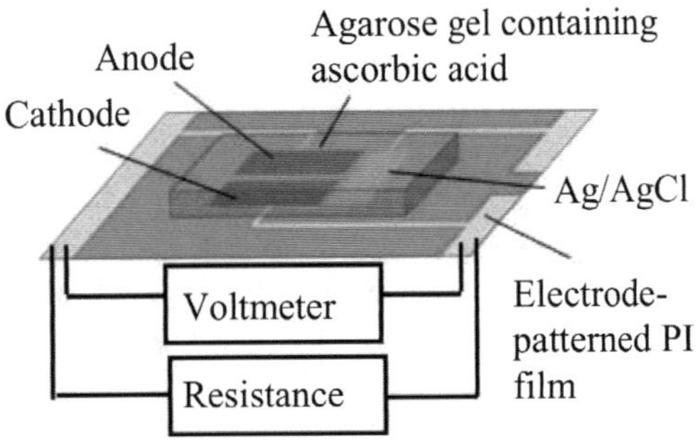

Figure 1. Schematic of the AA fuel cell with agarose gel containing ascorbic acid

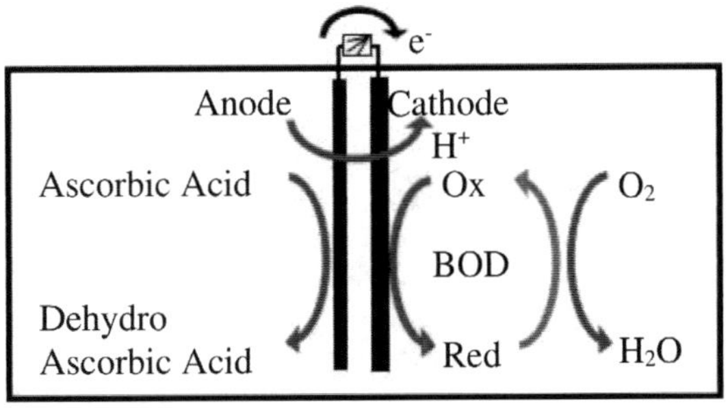

Figure 2. Possible mechanism of power generation.

2.2 Ascorbic acid fuel fabrication process

A mixture of ascorbic acid, agarose (Agarose Basic, TAKARA), and a phosphate buffer solution (PBS) (50 mM, ph7.0) was boiled to dissolve at 100 °C. The weight percentage (wt%) of agarose was varied for 0.5, 1.5, and 2.5 wt% to know the optimum condition for the agarose concentration.

2.3 Fabrication of enzyme-modified electrodes

The Pt wiring patterns were deposited on a 50 μm thick PI film (Kapton) using a mask in a radio-frequency sputtering equipment. The carbon-modified anode and cathode electrodes were patterned on the surfaces of the Pt film by screen printing using a carbon paste (MRX-713J-A, TAMURA) that was then cured at 150 °C for 15 min. Finally, the surfaces of the carbon paste film were coated by screen printing with Ketjenblack EC 300J (KB, Lion Inc.), and polyvinylidene fluoride (PVDF, KUREHA Inc) mixed ink was then cured at 80 °C for 20 min. Ag|AgCl reference electrodes were patterned on the surfaces of Pt film by screen printing with Ag|AgCl ink (BAS Inc.) that was then cured at 120 °C for 5min. The increased effective specific surface area increases the amount of enzyme adsorbed to the electrode surface was modified electrode surfaces in KB. The preparations of enzyme coated electrodes are described as follows. BOD with an amount of 1 mg was mixed with a 25 μl PBS (50 mM, pH 7.0). Then, a 2 μl portion of the resulting solutions was put onto the cathode, and left to dry for 20 min.

Figure 3 shows photographs of (a) a fabricated biofuel cell with a solidified AA aqueous solution on the electrodes and (b) an agarose AA fuel with agarose weight ratio 2.5 wt%. Exudation of water was not observed when holding the fuel with fingers, which suggests the portability of the fuel.

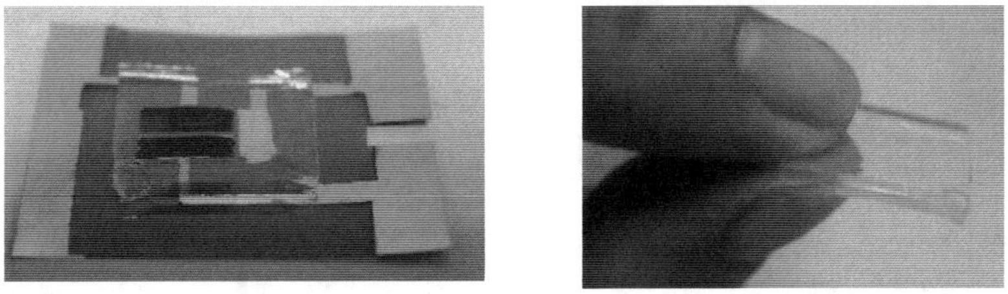

(a) (b)

Figure 3. Photographs of (a) the fabricated biofuel cell with an agarose fuel, and (b) an AA fuel unit containing 2.5% agarose.

2.3Measurements

Cyclic voltammogram (CV) measurements were performed with the agarose hydrogel containing AA 100 mM on the anode and the cathode at room temperature. As a reference, these electrodes were immersed in an aqueous solution containing 100 mM AA. When the CVs of the anode were measured, the surface of the cathode Pt area was not covered by carbon and enzymes, and the Pt area was used for the counter electrode. When the CVs of the cathode were measured, the surface of the anode Pt area was not covered by carbon and enzymes. The reference electrode was prepared by coating the reference electrode area with the Ag|AgCl ink. The electrochemical reactions within the biofuel were investigated by plotting the oxidation or reduction current densities J of the anode and the cathode. The oxidation and reduction current densities were analyzed by the Randles-Sevcik equation [20] below;

$$J = 2.99 \times 10^5 n(\alpha n)^{\frac{1}{2}} A D^{\frac{1}{2}} C v^{\frac{1}{2}} \tag{1}$$

where α is transfer coefficient, n is the number of reaction electrons, A is the electrode area, D is the diffusion coefficient of fuel or oxygen, C is the concentration of fuel or oxygen, v is the potential sweep rate.

The generated power was evaluated by measuring the cell voltage while varying the external load resistance between 0 and 2 MΩ. The power was derived using the following equation;

$$W = \frac{V^2}{R} \tag{2}$$

where R is the load resistance, and V is the generated voltage appearing between the terminals of the fuel cell (see Fig.1). The power density was calculated by dividing the 30 mm^2 of the anode area.

3. Results and discussion

Figure 4 shows the CV curves of the anodes using the solidified hydrogel containing 100 mM AA and agarose with different concentrations from 0.5 wt% to 2.5 wt%. The CV curve for the anode in the AA PBS solution in 50 ml of air-saturated 100 mM AA solution is also shown for comparison in Fig. 4. These measurements were done at the potential sweep rate of 20 mV/s. The CV curves for these 4 devices are very similar. Thus, it can be said that the oxidation of AA near the anode is very similar, and the agarose concentration has minimal impact on AA oxidation. Figure 5 shows the CV curves of the cathodes using the solidified hydrogel containing 100 mM AA and agarose with different concentrations from 0.5 wt% to 2.5 wt%. The CV curve measured in 50 ml of air-saturated 100 mM AA solution is also shown in Fig. 5. These measurements were done at the potential sweep rate of 20 mV/s.

The reduction peak current densities at -0.8 V for the biofuel cells with the agarose hydro gels were very similar. The agarose concentration has minimal impact on reduction of oxygen. However, the reduction current density of the cathode for the biofuel cell immersed in the 100 mM AA solution in a beaker showed much larger reduction current due to the larger amount of dissolved oxygen.

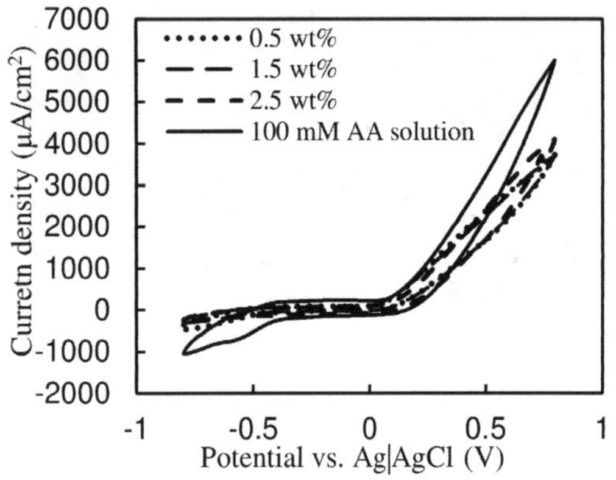

Figure 4. Cyclic voltammograms for the anode in 100 mM AA PBS and the anodes contacted with agarose gel fuel containing 100 mM AA with agarose concentration 0.5 wt%, 1.5 wt%, and 2.5 wt%.

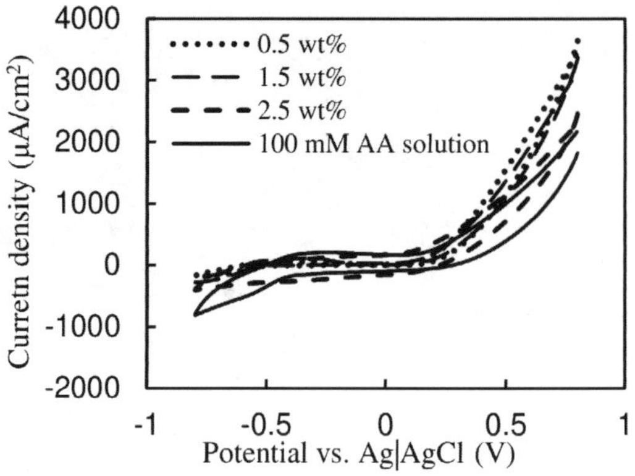

Figure 5. Cyclic voltammogram for cathode in 100 mM AA solution and agarose gel of agarose concentration 0.5 wt%, 1.5 wt%, and 2.5 wt%.

Figure 6 shows the relationships between the generated current density and the voltage across the load resistor for the biofuel cells with the biofuel cell units containing agarose with different concentrations. They were also compared with the biofuel cell in a 100 mM AA PBS in a 50 ml beaker. The generated currents suddenly dropped when the voltage increased, and stayed above 0.5 V. There was almost no difference within these biofuel cells in the early stage of the biofuel cell operation. It should be noted that the current density of the fuel cells during the power generation is different from those of the CV curves where the voltages are scanned from outside.

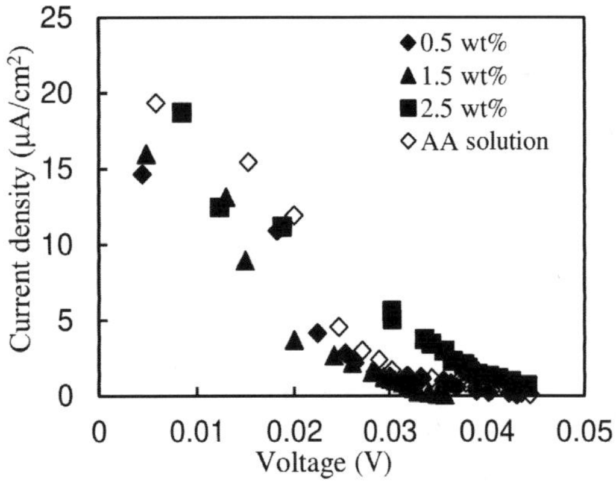

Figure 6. Current density-voltage relationships during fuel cell operation using the biofuel units containing different agarose concentration and the 100 mM AA PBS solution.

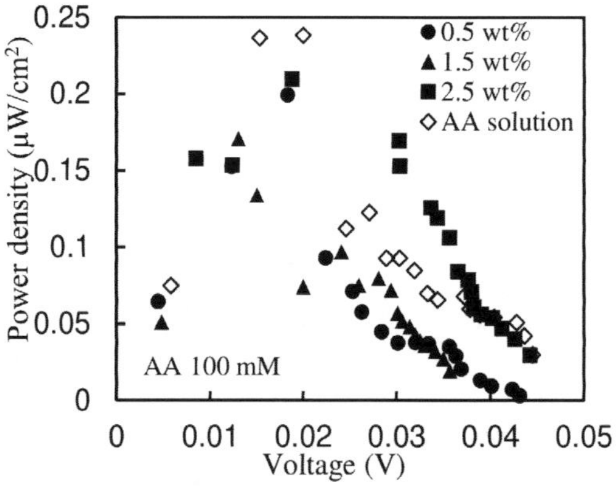

Figure 7. Output voltage dependences of power density for the biofuel cells with fuel units with different agarose concentration.

Figure 7 shows the relationships between the power density and the voltage. The maximum power and power density for the biofuel cell with the fuel unit containing 2.5 wt% agarose were 0.063 μW and 0.21 μW/cm^2 at 0.019 V, respectively. There are almost no differences on the power generation performances within the biofuel cells investigated here. Therefore, the optimal concentration of agarose was determined to be 2.5 wt% in order to keep the shape of the AA fuel units, which increased the portability of the fuel units.

4. Conclusion

A portable biofuel cell with a cathode modified with bilirubin oxidase was fabricated on a flexible polyimide substrate. The anode and the cathode with an area of 3x10 mm^2 were separated with a gap of 1 mm. The portable AA biofuel was prepared by solidifying a 50-200 mM AA solution containing agarose. The maximum power and power density were 0.063 μW and 0.21 μW/cm^2 at 0.019 V, respectively when the solidified hydrogel biofuel with a dimension of 13x24x4 mm^3 containing a 100 mM AA solution and 2.5 wt% agarose was used.

5. Acknowledgments

This research was conducted in supports of the staffs in the Micro Functional Device Research Center of the College of Science and Technology, Nihon University.

References

[1] N. Yuhashi, M. Tomiyama, J. Okuda, S. Igarashi, K. Ikebukuro and K. Sode. 2005 *Biosens. Bioelectron.* **20** 2145
[2] I. Willner and E. Katz. 2005 *Bioelectronics, Wiley-VCH, Weinheim.*
[3] G.T.R. Palmore, H. Bertschy, S.H. Bergens and G.M. Whitesides. 1998 *J. Electroanal. Chem.* **443** 155
[4] A. T. Yahiro, S. M. Lee, D. O. Kimble. 1964 *Biochim. Biophys.* **88** 375
[5] K. C. Swades and R. L. Derek. 2003 *Nature Biotechnology.* **21** 1229
[6] D. R. Bond and D. R. Lovley. 2003 *Appl. Environ. Microbiol.* **69** 1548
[7] M. J. Moehlenbrock and S. D. Minteer. 2008 *Chem. Soc. Rev.* **37** 1188
[8] K. G. Lim and G. T. R. Palmore. 2007 *Biosens. Bioelectron.* **22** 941
[9] Y. Kamitaka, S. Tsujimura, N. Setoyama, T. Kajino and K. Kano. 2007 *PCCP Phys. Chem. Chem. Phys.* **9** 1793
[10] S. Tsujimura, K. Kano and T. Ikeda. 2002 *Electrochemistry.* **70** 940
[11] S. Tsujimura, M. Fujita, H. Tatsumi, K. Kano and T. Ikeda. 2001 *PCCP Phys. Chem. Chem. Phys.* **3** 1331
[12] T. Chen, S. C. Barton, G. Binyamin, Z. Q. Gao, Y. C. Zhang, H. H. Kim and A. Heller. *J. Am. Chem. Soc.* **123** 8630
[13] F. Sato, M. Togo, M. K. Islam, T. Matsue, J. Kosuge, N. Fukasaku, S. Kurosawa and M. Nishizawa. 2005 *Electrochem. Commun.* **7** 643
[14] S. C. Barton, J. Gallaway and P. Atanassov. 2004 *Chem. Rev.* **104** 4867
[15] M. Nishizawa, M. Togo, A. Takamura and T. Abe. 2005 *Proceedings of the fifth international workshop on micro and nanotechnology for power generation and energy conversion applications*
[16] Y. Fukushi, S. Koide, R. Ikoma, W. Akatsuka, S. Tsujimura and Y. Nishioka. 2013 *J. Photo. Polym. Sci. Tech.* **26** 303
[17]H. Goto, Y. Fukushi, and Y. Nishioka. to be published in IEICE Trans. Eleectronics.
[18] S. Sakura, D. Fujimoto, K. Sakamoto, A. Mizuno and K. Motegi. 1982 *Can. J. Biochemistry.* **60** 525
[19] A. Polson and D. G. Parkyn. 1969 *Biopolymers.* **7** 107
[20] A.J. Bard and L.R. Falkner, 2001 *Hoboken: Wiley and Sons.* 226

Design, Fabrication and Prototype testing of a Chip Integrated Micro PEM Fuel Cell Accumulator combined On-Board Range Extender

A. Balakrishnan*, C. Mueller, H. Reinecke
Laboratory for Process Technology, IMTEK-Department of Microsystem Engineering, University of Freiburg, Georges-Koehler-Allee 103, 79110 Freiburg, Germany

E-mail: arvind.balakrishnan@imtek.uni-freiburg.de

Abstract. In this work we present the design, fabrication and prototype testing of Chip Integrated Micro PEM Fuel Cell Accumulator (CIμ-PFCA) combined On-Board Range Extender (O-BRE). CIμ-PFCA is silicon based micro-PEM fuel cell system with an integrated hydrogen storage feature (palladium metal hydride), the run time of CIμ-PFCA is dependent on the stored hydrogen, and in order to extend its run time an O-BRE is realized (catalytic hydrolysis of chemical hydride, $NaBH_4$). Combining the CIμ-PFCA and O-BRE on a system level have few important design requirements to be considered; hydrogen regulation, gas-liquid separator between the CIμ-PFCA and the O-RE. The usage of traditional techniques to regulate hydrogen (tubes), gas-liquid phase membranes (porous membrane separators) are less desirable in the micro domain, due to its space constraint. Our approach is to use a passive hydrogen regulation and gas-liquid phase separation concept; to use palladium membrane. Palladium regulates hydrogen by concentration diffusion, and its property to selectively adsorb only hydrogen is used as a passive gas-liquid phase separator. Proof of concept is shown by realizing a prototype system. The system is an assembly of CIμ-PFCA, palladium membrane and the O-BRE. The CIμ-PFCA consist of 2 individually processed silicon chips, copper supported palladium membrane realized by electroplating followed by high temperature annealing process under inter atmosphere and the O-BRE is realized out of a polymer substrate by micromilling process with platinum coated structures, which functions as a catalyst for the hydrolysis of $NaBH_4$. The functionality of the assembled prototype system is demonstrated by the measuring a unit cell (area 1 mm^2) when driven by the catalytic hydrolysis of chemical hydride ($NaBH_4$) and the prototype system shows run time more than 15 hours.

1. Introduction

The driving force for research in the field of micro power sources is due to the tremendous growth in portable electronics applications, for example cell phones, laptops, cameras and global positioning systems (GPS) [1]. But the biggest limiting factor for these portable electronics applications are the capacity and size of batteries utilized in them [2]. This ultimately leads to the developing new micro power sources with high energy density and reasonable power delivery. Miniaturized hydrogen driven fuel cell systems has be considered to be a good candidate for such requirements due to its high energy density and eco friendliness. Two major kinds of low temperature fuel cell which are under focus are alkaline fuel cell and acidic fuel cell. μ-PEMFC systems (type of acidic fuel cell) have been deeply investigated in the recent decades owing to its advantages over other fuel cell types, but there are also few technical drawbacks in terms of on board hydrogen storage. In large scale fuel cell systems hydrogen in stored either in gaseous form (high pressure required) or liquefied form (very low temperatures required). Whereas in micro scale systems they are stored in the form of hydrides, hydrides are further classified into metal hydrides and chemical hydride. Metal hydrides store hydrogen in their atomic lattice and their hydrogenation and dehydrogenation are reversible under specific operating conditions. Chemical hydrides have chemically bonded hydrogen molecules under normal conditions and release the hydrogen when it undergoes a catalytic hydrolysis reaction. Some major advantages of chemical hydride over metal hydride are that they have the ability to release hydrogen under ambient conditions and they are easy to handle and doesn't require sophisticated handling systems.

A novel concept of fuel cell integrated metal hydride (palladium on-board hydrogen storage) system on a silicon substrate was presented by Edler et al. [3], Chip Integrated Micro PEM Fuel Cell Accumulator (CIμ-PFCA). In this concept, the PEM fuel cell was directly screen printed on the metal hydride storage and the fuel cell has a self-breathing cathode, the capacity of the system was dependent on the thickness of the metal hydride layer. One important feature of the system is that it has CMOS electronics in built, shown in figure 1. The hydrogen storage capacity of these systems were increased by replacing bulk palladium by palladium-polymer composite mixture, an increase in capacity and an increase in the run time of the system was demonstrated [4].

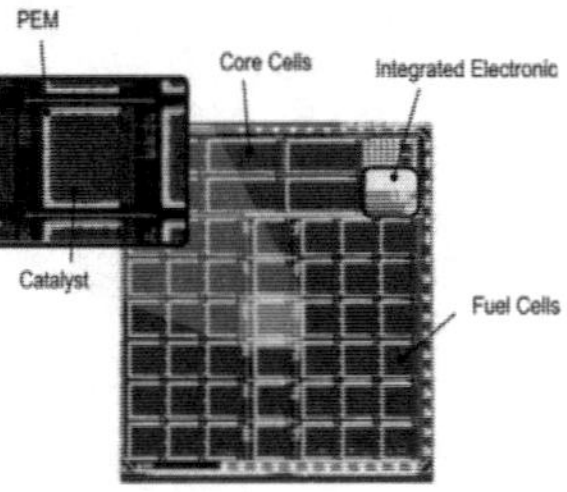

Figure 1. Chip integrated micro PEM Fuel Cell Accumulator (CIμ-PFC Chip integrated micro PEM Fuel Cell Accumulator (CIμ-PFCA); integrated on-board H_2 storage (palladium-polymer composite) and integrated CMOS electronics A); screen printed fuel cell, metal hydride storage and integrated (palladium on-board hydrogen storage), CMOS electronics

The storage capacity of the above mentioned system can be further increased by combining it with an On-Board Range Extender (O-BRE). O-BRE is a hydrogen production unit, where the hydrogen is produced from the catalytic hydrolysis of chemical hydride (sodium borohydride, $NaBH_4$). In this presented work $NaBH_4$ has been utilized due to its advantages of high gravimetric density, easy handling, non-toxic nature, longer shelf life and ambient operating conditions. The hydrolysis of $NaBH_4$ generates gaseous hydrogen, sodium borate and heat as reaction products (equation (1)) [5]. One mole of $NaBH_4$ reacts with two moles of water to delivers four moles of hydrogen under ideal conditions. The by-product $NaBO_2$ is non-toxic and can be regenerated into $NaBH_4$ by suitable processing techniques.

$$NaBH_4 + 2H_2O \rightarrow NaBO_2 + 4H_2 + (-300\ KJ/mol) \quad (1)$$

The self-hydrolysis of $NaBH_4$ is controlled by increasing the pH of the solution and then using an active catalyst to reinitiate the hydrolysis reaction.

1.1 System concept

In this work we present the design, fabrication and prototype testing of Chip Integrated Micro PEM Fuel Cell Accumulator (CIμ-PFCA) combined an On-Board Range Extender (O-BRE), but replacing the external hydrogen transport tubing and external gas-liquid phase separators by a palladium membrane, the schematic view of the system is shown in figure 2. There are 3 major parts that are marked (1) CIμ-PFCA, (2) palladium and palladium-alloy membrane, (3) O-BRE. Combining

CIμ-PFCA and O-BRE have few important design requirements to be considered; hydrogen regulation, gas-liquid separator between the CIμ-PFCA and the O-BRE. Traditional techniques to regulate hydrogen (tubes), separators (bulk porous hydrophobic modules) are less desirable in the micro domain. So in this work we present, our approach of using palladium membranes eliminating the use of above mentioned components. The property of palladium to selectively adsorb and transport hydrogen (concentration diffusion) is used as passive regulator and gas-liquid phase separator. The concept of using palladium and palladium-alloys to selectively filter, adsorb and transport hydrogen is a well-established technique used in small & large scale hydrocarbon reforming system (high temperature of operation > 350 °C) [6]. But in micro scale and meso scale system integration, using high temperatures is a drawback for the complete system in terms of system handling and external power requirement. So in this presented work the membrane is operated at ambient conditions. Some of the major advantages of such a system are; the system has high degree of freedom in terms of design (user defined) and the chemical hydride used and this could be transformed into an energy harvesting device with few design changes.

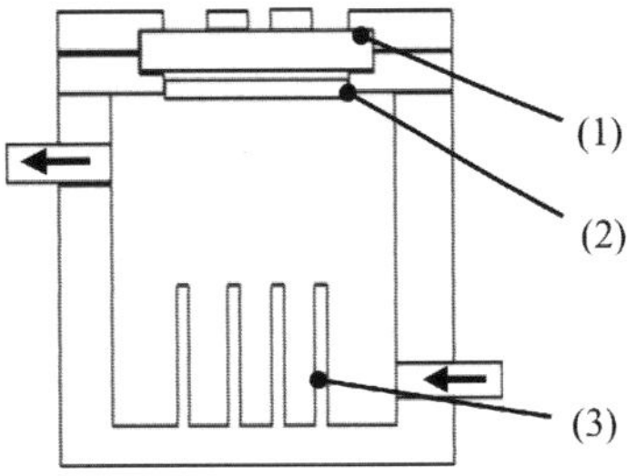

Figure 2. Schematic view of the complete system; (1) CIμ-PFCA: Screen printed fuel cell & hydrogen storage chip, (2) palladium and palladium-alloy membrane: for hydrogen regulation, (3) On-board range extender (O-RE): catalytic hydrolysis of $NaBH_4$ for on-board hydrogen production.

2. Experimental

The prototype system realization was done in 4 steps; (i) fabrication of CIμ-PFCA (ii) fabrication of palladium membrane (iii) fabrication of O-BRE and (iv) system assembly. Detailed description is presented in the following section.

2.1 *Fabrication*

Fabrication of CIμ-PFCA: this is performed as a two silicon chip assembly (chip 1&2). Chip 1 consists of screen printed cathode layer, PEM layer & sputtered anode current collector, palladium-polymer composite casted in the wet etched groove of the silicon chip from the back side. Chip 2 consists of a μ-tunnel for the hydrogen transport which is realized by a combination of wet and dry etching of bulk silicon. The schematic view of CIμ-PFCA is shown in figure 3 (cross section). The complete fabrication process chain is described elsewhere [4]. The assembly of the 2 individually processed silicon chips is performed by room temperature polymeric bonding process. The size of the chip is 1 cm² consisting of 42 unit cells each has a size of 1 mm².

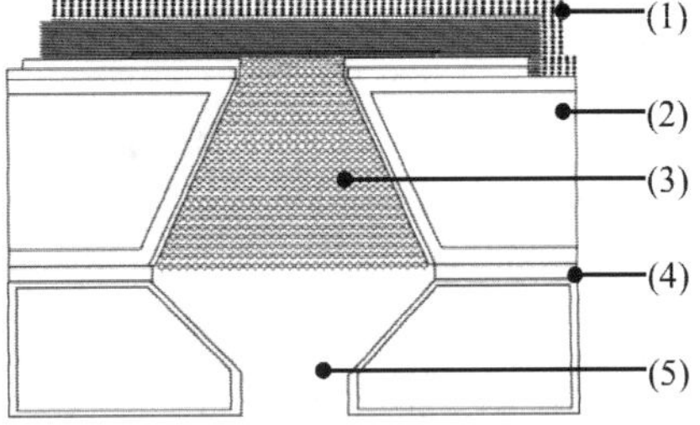

Figure 3. Cross section schematic view of CIμ-PFCA; (1) screen Printed PEM fuel cell, (2) silicon base chip with wet etch groove and sputtered anode current collector, (3) palladium-polymer composite (hydrogen storage), (4) low temperature polymeric bonding, (5) silicon chip with μ-tunnel for hydrogen transport.

Fabrication of palladium membrane: an electroplated sandwich stack consisting of 15 μm copper, 5 μm palladium, and 15 μm copper was performed. Copper act as mechanical support structures for the 5 μm palladium. The electroplated stack was selectively etched for copper only at the centre (diameter of ~ 5 mm) by protecting the other regions with photoresist. The fabricated palladium membrane with copper support structures was annealed at 500 °C for 20 min under argon atmosphere; to improve the mechanical stability of the membrane structures, for detailed electroplating parameters please refer [7].
Fabrication or O-BRE: the O-BRE was fabricated out of polymeric substrate. The miniaturization of O-BRE at step one would be a hurdle for us to study the system for better understanding. The O-BRE was fabricated with (Poly(Methyl MethAcrylate)-PMMA) substrate by micro milling process. Some of the reasons for using PMMA substrates are: low cost, fast to fabrication, alkaline stability and transparent. All these reasons give us an upper hand studying the system and are easily adaptability. The O-BRE has a total chamber volume of 28 cm³ (frustum shaped) it consists of catalyst (platinum) for the hydrolysis of chemical hydride, reactant inlet and outlet, nitrogen flush inlet and a temperature sensor.
The complete prototype system assembly was performed by fixing the CIμ-PFCA to a PCB board and wire bonded to larger contact pads for measurement convenience (figure 5 (a)), then the system is stacked as per the figure 2 and mechanically screwed together with O-Ring sealing's which are chemically stable. The completely assembled prototype system is shown in figure 5 (b).

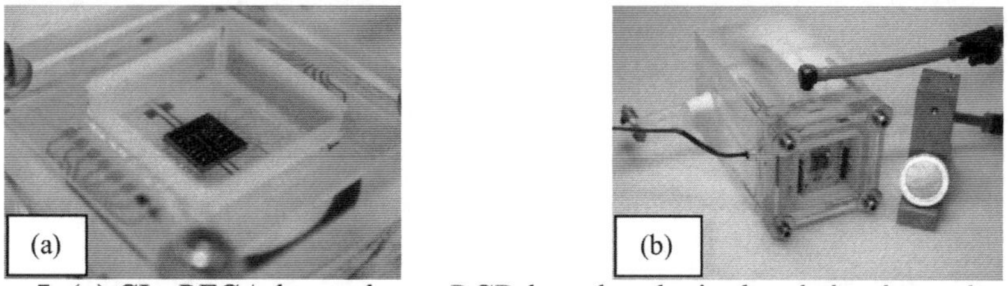

Figure 5. (a) CIμ-PFCA located on a PCB board and wire bonded to larger bond pads for measurement convenience, (b) completely assembled system: CIμ-PFCA, copper supported Pd membrane and O-BRE.

2.2 *Performance measurement*

The prototype system is tested for its performance using an experimental measurement platform consisting of syringe to insert the reactant and a computer controlled potentiostat to measure the system voltage. The prototype system is tested with pure hydrogen flow for measuring the I-V polarization characteristics of the CIμ-PFCA. A stock solution was prepared by dissolving $NaBH_4$ in 6M sodium hydroxide (NaOH) and a small amount of 99% pure sodium dodecyl sulphate (surfactant). The reason for using the surfactant is to reduce the surface energy of the platinum catalyst surface ultimately reducing the by-product clogging effect on the active sites. Before inserting the chemical hydride to the prototype system, it was constantly flushed with nitrogen gas, and then 5 ml of the chemical hydride stock solution was mixed with 15 ml of water and then inserted into the reactor. Open Circuit Voltage (OCV) and the voltage response for a constant electrical load of the unit cell which has total area 1 mm² and was measured under constant ambient conditions.

3. Result & discussion

The prototype system was tested for its functionality when driven by the hydrogen produced form the O-BRE. The chemical hydride was inserted and the open circuit voltage was measured till the unit cell reached a stable OCV value and then an electrical load of 10 MΩ was inserted and the voltage response of the unit cell was measured, the measurement results are shown in figure 6. From the figure one can observe that system reaches a stable OCV value of 600 mV about 250 min after the chemical hydride was inserted. The reason for this can be attributed to the time delay in the permeation and diffusion of hydrogen through the palladium bulk membrane (rate limiting factor). A similar effect was reported by [8] as "time lag", lower pressure difference causes increased time lag until a steady state permeation is achieved across the membrane. The time lag could be reduced by using thinner layers of palladium membrane. When the unit cell was subjected to a constant electrical load (10 MΩ) discharge for about 7 hours there is a drop in the voltage, depicted in figure 6. After the electrical load discharge the OCV of the system was measured till the system was completely depleted. The testing of the prototype system shows the feasibility of combining CIμ-PFCA and an O-BRE with a palladium membrane in-between.

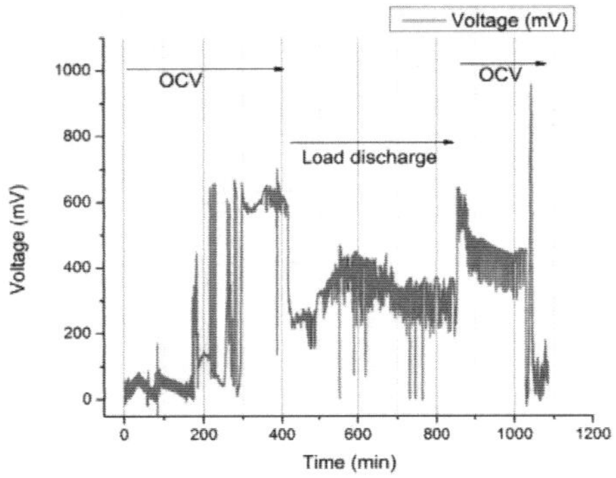

Figure 6. CIμ-PFCA integrated On-board range extender system response with palladium buffer layer

4. Conclusion & outlook

Design, fabrication and prototype system testing of CIμ-PFCA combined O-BRE are presented. The concept of using palladium membrane instead of tubes and gas-liquid phase separators is proved to be functional. The prototype system functionality is demonstrated by measuring the CIμ-PFCA (unit cell) performance when driven by the hydrogen produced from the O-BRE. Future works are to focus on; (i) in-situ measurement of the prototype system (ii) fabricating the palladium membrane on chip level with the CIμ-PFCA assembly, and (ii) to design a self-controlled hydrogen production mechanism form O-BRE (catalyst actuation mechanism). This work sets the platform to develop a system level adaptable CIμ-PFCA (ready to connect to any hydrogen source (shown in figure 7)). This paves the way for a considerable grown in the field of energy sufficient autonomous microsystem for various small scale sensing applications.

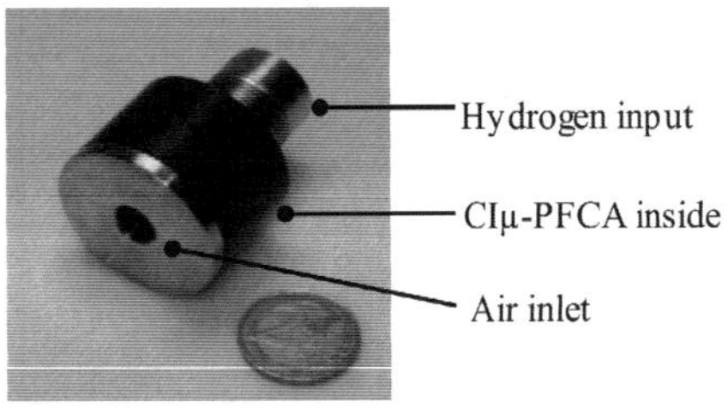

Figure 7. Prototype for system level adaptable CIμ-PFCA.

5. Reference

[1] A. Kundu et al., Journal of Power Sources, 170 (2007), pp. 67–78
[2] Y. Kojima et al., Journal of Power sources, 125 (2004), pp. 22-26
[3] G. Erdler et al., "Chip integrated fuel cell" Sensors and Actuators A 132 (2006) 331–336
[4] Arvind Balakrishnan at.al, Advanced Materials Research Vols. 608-609 (2013) pp 904-912
[5] Zhang Q, Smith G, Wu Y, Mohring, International journal of hydrogen energy 2006; 31; 961
[6] Deshpande, kishori et al, journal of microelectromechanical system, 19.2(2010): 402-409
[7] Arvind Balakrishnan, Journal of Physics: Conference Series 476 (2013) 012030
[8]Wei-Hsin Chen, I-Han Chiu, Applied Energy 87 (2010) 1023-1032.

Thermal management of power sources for mobile electronic devices based on micro-SOFC

S Murayama[1], F Iguchi[1], M Shimizu[1], H Yugami[1]

[1]Graduate School of Engineering, Tohoku University, Japan

E-mail: s_murayama@energy.mech.tohoku.ac.jp

Abstract. Small power sources based on micro-SOFC for mobile electronic devices required two conditions, *i,e*, thermally compatibility and thermally self-sustain, because of high operating temperature over 300 °C. Moreover, high energy efficiency was also required. It meant that this system should be designed considering thermal management. In this study, we developed micro-SOFC packages which have three functions, thermal insulation, thermal recovery, and self-heating. Heat conduction analysis based on finite element method, and thermochemical calculation revealed that vacuum thermal insulation was effective for size reduction and gas-liquid heat exchanger could reduce the temperature of outer surface. We fabricated the package with three functions for proof of concept and evaluated. As a result, it was suggested that developed package could satisfy both two requirements with high efficiency.

1. Introduction

Recently, high power and long usable power sources are strongly required as development of mobile electronic devices such as mobile phones and laptop computers. As new power sources, micro solid oxide fuel cells (micro-SOFCs) which utilize liquid fuels attract much attention for high power density [1]. Long operation is possible for micro-SOFCs by refueling and it is not restricted by battery charge. Hence, it is expected to be the power source of recent mobile electronic devices which consume a large amount of power. In previous, diverse types of micro-SOFCs [2, 3] and a maximum power density of 1 W/cm^2 at 500°C was reported [4].

As shown in the literature, high performance of micro-SOFCs was exhibited. However, they have many problems of durability and repeatability for practical use. Additionally, it is necessary to consider how to heat up micro-SOFC cells to operation temperature, compatibility with other electronic circuits and thermally self-sustained condition. We already solved the problem of heat up with internal micro-heater. In the system, power density of 0.14W/cm^2 at 300°C was achieved [5].

In this study, thermal design for micro-SOFC package was proposed in order to adopt the micro-SOFCs into a power source of compact mobile electric devices. We fabricated the micro-SOFC package with thermal insulation and experimentally confirmed the insulation performance.

2. Thermal design of micro-SOFC system

Applying micro-SOFC system as independent power sources, thermally self-sustained condition is required, *i.e.* micro-SOFC packages should keep operating temperature more than 300°C only by the generated heat during power generation. Micro-SOFC cells are heated up to operating temperature with internal micro heater using stored electricity and power generation is started. For the reaction of fuel cells as shown in Eq. (1), usable electric energy is produced from the difference of enthalpy in Eq. (1). Rest of the energy is released as heat which is usable to keep operating temperature. For thermally self-sustained condition, the amount of heat generated during power generation needs to exceed that of heat loss from surface of the package and heat transfer by exhaust gas. We calculated thermally self-sustained condition of a micro-SOFC system fueled by methanol and water that has reformer and heat exchanger from exhaust gas as shown in Fig.1. The result confirmed that heat loss with exhaust gas was dominated in this system. Hence, recovering the heat of exhaust gas with the gas-liquid heat exchanger was essential to achieve thermally self-sustained condition.

$$H_2(g) + \frac{1}{2} O_2(g) \rightarrow H_2O(g), \Delta H_{298} = -242 \text{ kJ mol}^{-1} \quad (1)$$

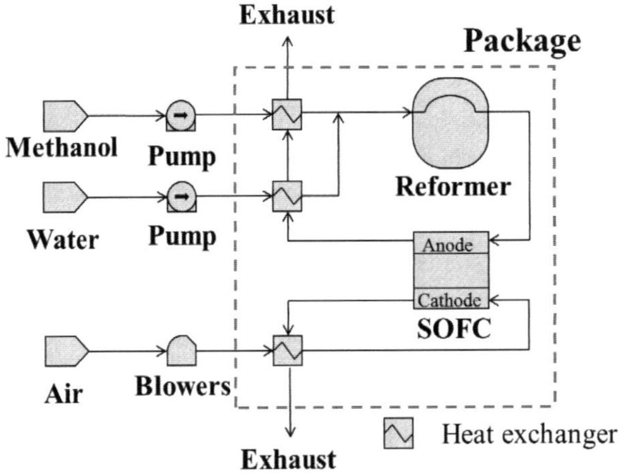

Fig.1 Schematic illustration of micro-SOFC system. This system consists of micro SOFC cell, reformer and heat exchanger.

Micro-SOFC packages also require thermal insulation to keep the surface temperature below about 85°C, which is maximum operating temperature of other electronic circuits. However, the insulation structure size is restricted to contain micro-SOFC package in compact mobile electric device. We applied vacuum insulation for the package. Vacuum insulation technique is widely used for thermal insulation in MEMS devices such as gas sensors.

Proposed design of the micro-SOFC package with vacuum insulation and the heat exchanger was shown in Fig.2. Silicon substrate with SOFC cell and micro heater was contacted by glass flow channels. Glass flow channels were surrounded by thermal insulator such as glass fibers and, the vacuum insulator was set on the outside. Glass flow channels were connected to the heat exchanger filled with liquid fuels. The heat of the exhaust gas was recovered by the liquid fuel through heat exchanger.

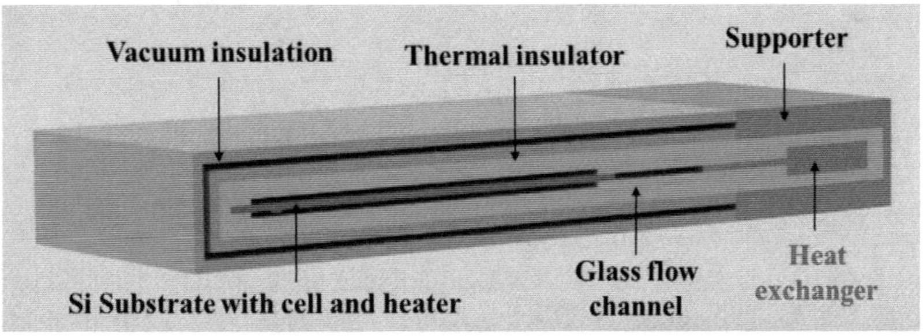

Fig.2 Designed micro-SOFC package with the vacuum insulator and the heat exchanger.

To estimate thermal insulation performance of the proposed micro-SOFC package, temperature distribution was calculated by the finite element method (MSC Marc 2010r2) as shown in Fig.3. Temperature of the cell is fixed 300°C and outer package was exposed external air (20°C). In the case of the package without heat exchanger, surface temperature of the package was kept from 53 to 88°C. It could meet the requirement when the cell temperature was 300°C by means of only vacuum insulation. In the case of the package with heat exchanger, surface temperature of the package was kept from 38 to 56°C. In this time, the heat exchanger was filled with 50 vol.% methanol, which boiling point was 73.5°C, as liquid fuel. As long as the liquid is evaporated, temperature of the heat exchanger is kept constant. Therefore, heat conduction from high temperature glass flow channels to vacuum insulator was prevented and low surface temperature was achieved. In accordance with the result, vacuum insulation could achieve low surface temperature as long as the cell temperature up to 300°C. Furthermore, the surface temperature can be reduced more with heat exchanger. Thus, the gas-liquid heat exchanger was effective for not only achieving thermally self-sustained condition but also reducing surface temperature of the package.

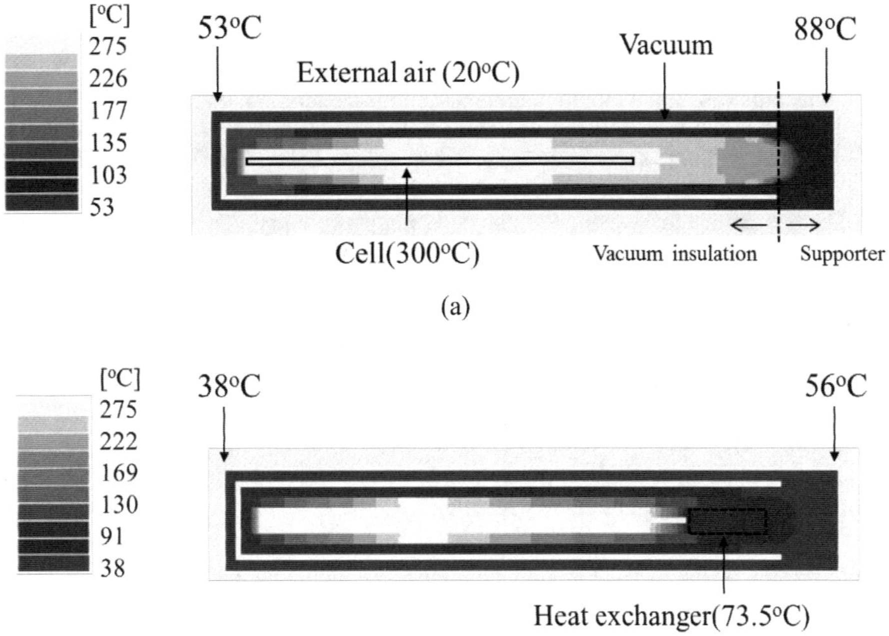

(b)

Fig.3 Cross sectional temperature distribution of the package (a) without heat exchanger and (b) with heat exchanger

3. Cooling performance of the insulation package

We actually fabricated the micro-SOFC package, as shown in Fig. 4, to evaluate cooling performance of the insulation package. Considering the convenience for evaluation test and fabrication process, dimension of package was decided to 46 mm×38 mm×18.5 mm. The vacuum package and the heat-exchanger were made of stainless-steel. Silicon substrate with cells and micro-heater was fabricated in accordance with our previous result [5]. Glass flow channel was made of Tempax glass. Thermal insulator (YUTAKA CORPORATION, ceramic paper, 0.1 W/(mK)) was installed between glass flow channels and vacuum package to prevent heat conduction. The base pressure in a vacuum zone was maintained under 0.2 Pa. Silicon substrate was heated up to 500°C by internal micro heater fabricated on the cell. Surface temperature of the package was measured by thermocouple. Cooling performance was evaluated at three conditions; heat exchanger (i)without liquid, (ii)with water, and (iii)with 50 vol% ethanol solution.

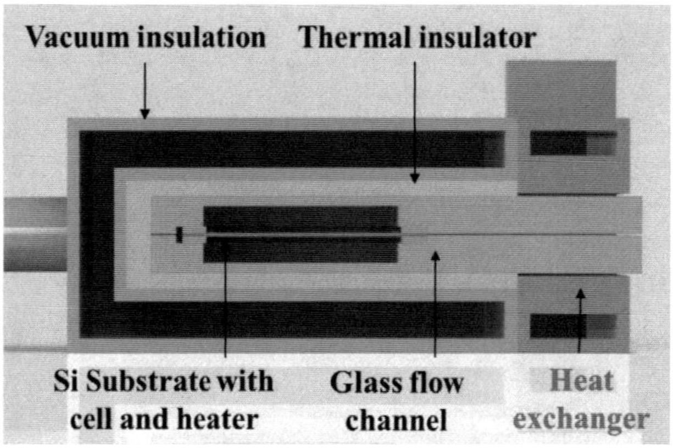

Fig.4 Cross sectional image of the fabricated micro-SOFC package

As shown in Fig. 5, micro heater was heated up to operating temperature of micro-SOFCs. In the case of (i), surface temperature became 152.8°C when micro heater temperature was 500°C. It was caused by heat conduction from glass part to surface of the package with the inside wall. In the case of (ii) and (iii), surface temperature was 96.7°C and 89.6°C, respectively, when micro heater temperature was 500°C. The heat exchanger was kept the boiling point of the liquid, i.e. 100°C for (ii), 89°C for (iii), as long as the liquid was evaporated. In accordance with the result, it was possible to keep surface temperature of the package sufficiently low with heat exchanger. Therefore, we can additionally reduce surface temperature using low boiling point liquid such as methanol because temperature of the heat exchanger depends on boiling point of the filled liquid. In consideration of the convenience for evaluation test and fabrication process, fabricated package became larger than the proposal design. However, it was considered that the cooling performance was not drastically changed because size of the vacuum zone was independent of the performance. Therefore, the vacuum package downsized to install into mobile electronics devices also keeps high insulation performance. Here we focused on the cooling performance of the heat exchanger to reduce surface temperature. In the future we will propose the micro SOFC package which can achieve thermally self-sustained condition and low surface temperature at the same time.

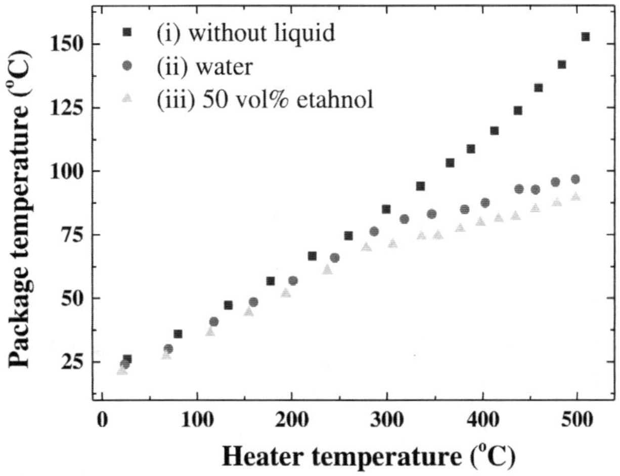

Figure.4 Evaluated cooling performance at three conditions; heat exchanger (i)without liquid, (ii)with water, and (iii)with 50 vol% ethanol solution.

4. Conclusion

We thermally designed the micro-SOFC package for applying micro-SOFCs to power source of mobile electronic devices. Thermally self-sustained condition is required for portable power source. Considering heat loss due to exhaust gas, recovering the heat of exhaust gas with the heat exchanger was adopted in the micro-SOFC system. We applied vacuum insulation for downsizing and the proposed package with vacuum insulation and heat exchanger. The proposed package could achieve thermally self-sustained condition and low surface temperature. We focused on cooling performance of the heat exchanger to reduce surface temperature and experimentally confirmed. As a result, surface temperature of the package was achieved below 90°C for electronic circuit. However the package has a problem with thermally self-sustained condition. We will continuously design the micro-SOFC package for compatibility with thermally self-sustained condition and low surface temperature.

Acknowledgements

This study was performed in R&D Center of Excellence of integrated Microsystems, Tohoku University under the program "Formation of Innovation Center for Fusion of Advanced Technologies" supported by Special Coordination Funds for Promoting Science and Technology.

References

[1] Evans A, Bieberle-Hutter A, Rupp J L M and Gauckler L J 2009 Review on microfabricated micro-solid oxide fuel cell membranes Journal of Power Sources 194 119-29

[2] Kerman K, Lai B K and Ramanathan S 2011 Pt/Y0.16Zr0.84O1.92/Pt thin film solid oxide fuel cells: Electrode microstructure and stability considerations Journal of Power Sources 196 2608-14

[3] Kerman K, Tallinen T, Ramanathan S and Mahadevan L 2013 Elastic configurations of self-supported oxide membranes for fuel cells Journal of Power Sources 222 359-66

[4] Shim J H, Park J S, An J, Gur T M, Kang S and Prinz F B 2009 Intermediate-Temperature Ceramic Fuel Cells with Thin Film Yttrium-Doped Barium Zirconate Electrolytes Chemistry of Materials 21 3290-6

[5] Iguchi F, Kubota K, Inagaki Y, Tanaka S, Sata N, Esashi M and Yugami H 2011 OPERATION OF MICRO-SOFC BY AN INTERNAL MICRO HEATER. Proceedings of PowerMEMS2011 411-414

Carbon-nanotube nanofluid thermophysical properties and heat transfer by natural convection

Y Li[1], S Suzuki[2], T Inagaki[1] and N Yamauchi[3]
[1] Ibaraki University, 4-12-1 Nakanarusawa, Hitachi, Ibaraki 316-8511, Japan
[2] Graduate School of Science and Engineering, Ibaraki University, Japan
[3] Fukushima National College of Technology, Japan

E-mail: yanrong@mx.ibaraki.ac.jp
Abstract. We measured the thermophysical properties of suspensions of carbon nanotubes in water as a type of nanofluid, and experimentally investigated their heat transfer characteristics in a horizontal, closed rectangular vessel. Using a previously constructed system for high-reliability measurement, we quantitatively determined their thermophysical properties and the temperature dependence of these properties. We also investigated the as yet unexplained mechanism of heat transport in carbon-nanotube nanofluids and their flow properties from a thermal perspective. The results indicated that these nanofluids are non-Newtonian fluids, whose high viscosity impedes convection and leads to a low heat transfer coefficient under natural convection, despite their high thermal conductivity.

1. Introduction
The present study was performed for quantitative evaluation of the previously undetermined properties of CNT nanofluids as a heat transport medium, from a materials engineering viewpoint. The CNT (MWNT; Meijo Nano Carbon Co., Ltd.) nanofluids (CNFs) were composed of 5 wt% and 1 wt% CNTs in pure water, under atmospheric pressure. The typical physical properties of CNFs are 1 - 5 μm in length, purity > 95%, content of carbon nanotube are 5 wt % and 1 wt %, and the solvent is water. The diameter of the CNTs as determined from SEM and TEM images was approximately 20-30 nm and their thickness was a few nanometers.

2. Experimental apparatus and measurement methods

2.1. CNF thermophysical properties
The method for measurement of density with a hydrometer, follows the JIS Z 8804 (JIS, 2001) and (JIS K 2249 JIS, 2001) standards. The viscosity was measured with a cone-and-plate viscometer, which is also used to determine whether the working CNF is a Newtonian fluid. The viscosity μ_R [Pa·s] of the CNF sample was obtained from the shear rate D [1/s] and the shear stress of the CNF sample placed between the rotating cone rotor and the cup plate, as follows.

$$D = 2\pi\omega r/60 \times 1/r\varphi = 2\pi\omega/60\varphi \text{ [1/s]} \quad s_c = 3M/2\pi R^3,\ M = F \times \theta/100 \text{ [Pa]} \quad \mu_R = s_c / D \text{ [mPa · s]} \qquad (1)$$

where ω [rpm] is the rotor's rate of rotation, θ [%] is the viscometer index, φ is the cone angle (1°34′), R [m] is the rotor radius (24 mm), r [m] is the rotor radial coordinate, M [N·m] is the viscous drag torque, and F [N·m] is the spring full-scale torque (67.4 μN·m). In this measurement system, the Newtonian fluid properties were thus obtained from the relationship between the shear rate and shear stress. The viscosity was also measured with a tuning-fork vibration viscometer. The adiabatic method was used to determine the specific heat. Its construction was based on a reported specific heat measurement method (Shimizu, 1994; Inagaki et al., 2011, 2012, 2013). The specific heat C_p [J/(kg·K)] of the CNF mass m [kg] in the insulated vessel was obtained from the measured temperature rise ΔT [K] and the heating time τ [s] while heating at a constant electric power Q_e [W] using

$$C_p = Q_e\tau / (m\Delta T) - W / m \text{ [J/(kg·K)]} \tag{2}$$

where W [J/K] is the heat capacity of the apparatus obtained from the known specific heat C_{pstd} (J/(kg·K)] of a standard heating medium using

$$W = Q_e\tau/\Delta T - C_{pstd}m \text{ [J/K]} \tag{3}$$

where $W = 3.796T + 25.10$ [J/K], which was the value used in Eq. (2).

The thermal conductivity was determined by a transient hot-wire method (JSTP, 1990; Inagaki et al., 2012, 2013). The thermal conductivity λ [W/(m·K)] of the CNF in the vessel was obtained from the measured temperature rise ΔT_w [K] at the surface of the resistance wire held taut in the CNF, between elapsed times t_1 and t_2 [s] from initiation of the uniform resistance wire heating, and the CNF thermal conductivity λ [W/(m·K)] was calculated using

$$\lambda = I^2 Re\ln(t_2/t_1) / (4\pi\Delta T_W) \text{ [W/(m·K)]} \tag{4}$$

where I [A] is the applied electric current and Re [Ω/m] is the electrical resistance per submerged length of the resistance wire.

The temperature dependence of each property was estimated by a least squares approximation, and the standard error of the estimate was defined (JSME, 1987) as

$$SEE = [\sum_{k=1}^{N}(X_{meas,k} - X_{caic,k})^2 /(N - C)]^{1/2} \text{ [variable]} \tag{5}$$

where $X_{meas,k}$ is the measured value at a given temperature, $X_{caic,k}$ is the approximated value at that temperature, N [-] is the total number of measurement points, and C [-] is the number of constants in the equation for the approximation curve. For all measured values relating to the thermophysical properties, outliers were discarded by the modified Thompson τ technique.

2.2. Heat transfer by natural convection

The heat transfer vessel (HT) was a horizontal, closed rectangular vessel having the previously reported configuration (Inagaki et al., 2011, 2012, 2013).

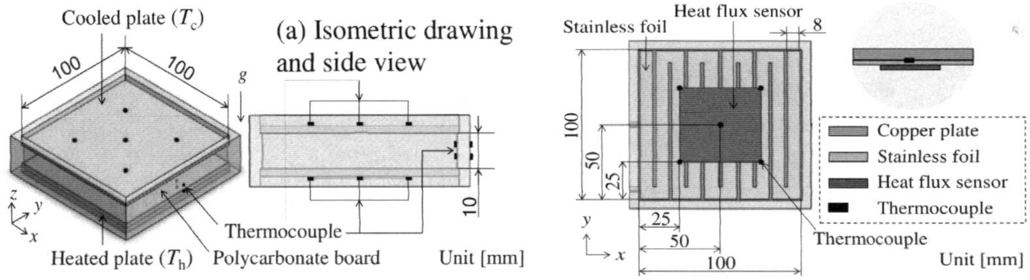

Fig. 1 Schematic illustration of experimental apparatus. (b) Detail of heating part (Right)

The HT vessel configuration is shown schematically in Figs. 1 (a) and 1 (b). The outside surface of the upper copper plate, was cooled naturally at room temperature. Although Joule heating was thus applied from the lower surface in this experimental configuration, the heat transfer surface was formed by contact of the fluid via a 5-mm-thick copper plate. To estimate the heat loss from the vessel bottom, a heat flux sensor was placed between the insulation layer and the heat transfer surface. The 50-mm-thick glass wool insulation on the bottom and sides of the vessel formed an adiabatic wall. The mean of the temperature outputs of the lower set of five was taken as the heated surface temperature T_h [K], and that of the upper set of five as the cooled surface temperature T_c [K].

In calculating the coefficient of heat transfer by natural convection, the net heat flux q' [W/m^2] was

$$q' = q_{in} - q_{loss} = VI/A - (q_b + q_s) \text{ [W/m}^2\text{]} \tag{6}$$

where q_{in} [W/m^2] is the input heat flux, q_{loss} [W/m^2] is the heat loss, V [V] is the input voltage, I [A] is the input current, A [m^2] is the wet surface area, q_b [W/m^2] is the heat loss from the vessel bottom as determined from the heat flux sensor, and q_s [W/m^2] is the heat loss from the vessel sides. In the present study, the ratio of the heat loss q_{loss} [W/m^2] to the thermal flux input q_{in} [W/m^2] was approximately 40%.

3. Results and discussion

3.1. Thermophysical properties of liquid phase state

Figure 2 (a) show the temperature dependence of the CNF density obtained with the hydrometer. As can be seen, at 300 K the density of CNF 5 wt% was ρ = 1032 kg/m^3, whereas the literature value for the water solvent was ρ = 997 kg/m^3. For CNF 1 wt% at the same temperature, the density was ρ = 1001 kg/m^3, which is 1.004 times that of water. The density for the CNF were shown to be higher than those for pure water but to approach the values for water at low concentrations. The least-squares approximation of the relationship between the CNF density as defined in this study and the temperature yielded the following results.

$$\rho = 973.73 + 0.749T - 1.848 \times 10^{-3}T^2 \ (293 - 343 \text{ K}) \ [\text{kg/m}^3] \ (\text{CNF: 5 wt\%}) \quad (7)$$

$$\rho = 753.85 + 1.940T - 3.722 \times 10^{-3}T^2 \ (293 - 343 \text{ K}) \ [\text{kg/m}^3] \ (\text{CNF: 1 wt\%}) \quad (8)$$

The standard error of the estimate (*SEE*) was 1.25 kg/m^3 for Eq. (7), and 0.737 kg/m^3 for Eq. (8).

Figure 2 (b) show the temperature dependence of the viscosity measured with the tuning-fork vibration viscometer. As can be seen, the viscosity for CNF 5 wt% at 300 K was μ_V = 49.82 mPa·s, and those for CNF 1 wt% was μ_V = 2.57 mPa·s. A least squares-approximation of the dependence of the CNF viscosity on temperature yielded the following results.

$$\mu_V = 591.47 - 2.940T + 3.782 \times 10^{-3}T^2 \ (293 - 343\text{K}) \ [\text{mPa·s}] \ (\text{CNF: 5 wt\%}) \quad (9)$$

$$\mu_V = 6.149 \times 10^4 - 9.507 \times 10^{-2}T + 5.877T^2 - 1.815 \times 10^{-2}T^3 + 2.802 \times 10^{-5}T^4 - 1.728 \times 10^{-8}T^5$$
$$(293 - 343\text{K}) \ [\text{mPa·s}] \ (\text{CNF: 1 wt\%}) \quad (10)$$

The *SEE* was 2.55 mPa·s for Eq. (9), and 0.125 mPa·s for Eq. (10). Thus, the viscosity for the CNF are larger than those for pure water but approach the values for water as the concentration decreases.

Figures 2 (c) and 2 (d) show the CNF viscosities as measured with the cone-and-plate viscometer. This method was employed because, unlike the tuning-fork vibration method, it permits resolution of the important question of whether the CNF is a Newtonian fluid. The measurements were performed at rotational speeds of N = 10, 20, 50, and 100 rpm.

Figures 2 (e) and 2 (f) show the relationship found between the shear rate D [1/s] and the shear stress s_c [Pa] for the CNFs at 303 K at different rotational speeds. The relationship is seen to be nonlinear in all cases, thus confirming that the present CNFs are non-Newtonian fluids.

Figure 2 (g) shows the temperature dependence of the specific heat as obtained by adiabatic specific heat measurement method. The specific heat for CNF 5 wt% at 300 K was found to be C_p = 3841 J/(kg·K). The value given in the literature for water at 330 K is C_p = 4179 J/(kg·K). For CNF 1 wt%, the specific heat was C_p = 4097 J/(kg·K), representing a ratio to that of water of 0.980. A least-squares approximation of the relationship between the CNF specific heat and the temperature yielded the following results.

$$C_p = 4672.3 - 3.859T + 3.631 \times 10^{-3}T^2 \ (293 - 343 \text{ K}) \ [\text{J/(kg·K)}] \ (\text{CNF: 5 wt\%}) \quad (11)$$

$$C_p = 5908.2 - 11.47T + 1.809 \times 10^{-2}T^2 \ (293 - 343 \text{ K}) \ [\text{J/(kg·K)}] \ (\text{CNF: 1 wt\%}) \quad (12)$$

The *SEE* for Eq. (11) and (12) was 46.5 and 5.00 J/(kg·K), respectively.

Figure 2 (h) shows the dependence of the thermal conductivity on temperature, as obtained by the transient hot-wire method. The results show that the thermal conductivity for CNF 5 wt% at 300 K is λ = 0.6773 W/(m·K). The literature value for water at 330 K is λ = 0.6104 W/(m·K). For CNF 1 wt%, the thermal conductivity at 300 K was found to be λ = 0.6242 W/(m·K), representing a ratio to that of water of 1.02. A least-squares approximation of the relationship between the CNF thermal conductivity and temperature yielded the following results.

$$\lambda = -1.220 + 0.01068T - 1.451 \times 10^{-5}T^2 \ (293 - 343 \text{ K}) \ [\text{W/(m·K)}] \ (\text{CNF: 5 wt\%}) \quad (13)$$

$$\lambda = -0.796 - 0.007709T + 0.9913 \times 10^{-5}T^2 \ (293 - 343 \text{ K}) \ [\text{W/(m·K)}] \ (\text{CNF: 1 wt\%}) \quad (14)$$

The *SEE* of the estimates for Eq. (13) and (14) was 9.25 x 10^{-5} and 1.21 x 10^{-2} W/(m·K), respectively.

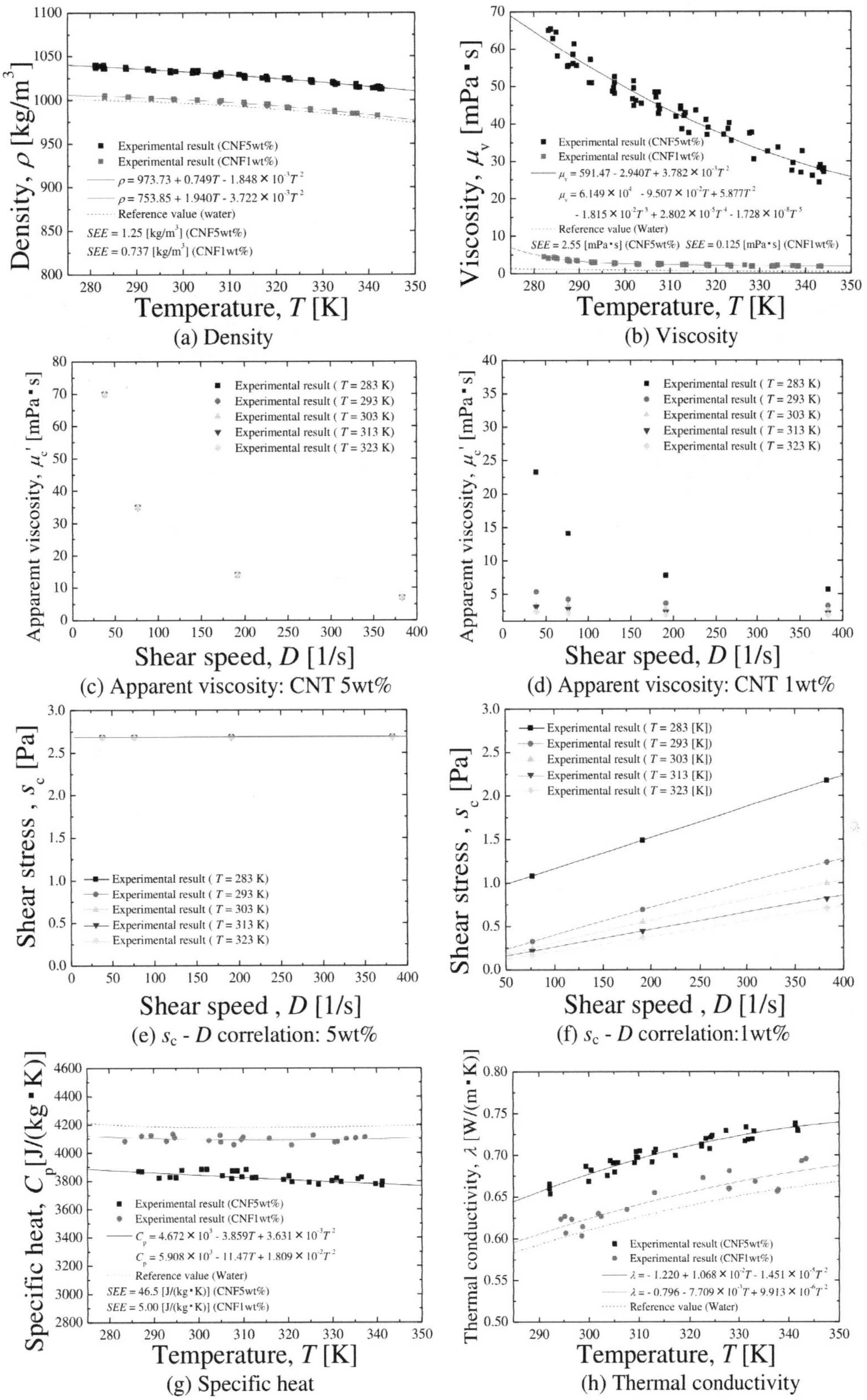

(a) Density (b) Viscosity

(c) Apparent viscosity: CNT 5wt% (d) Apparent viscosity: CNT 1wt%

(e) s_c - D correlation: 5wt% (f) s_c - D correlation:1wt%

(g) Specific heat (h) Thermal conductivity

Fig.2 Natural convection heat transfer ratio in a horizontal enclosed rectangular container

3.2. Heat transfer characteristics in natural convection

Figure 3 shows the heat transfer characteristics of the CNF 5 wt% thermal medium in a state of induced natural convection in the horizontal, closed rectangular vessel, with the mean Nusselt number $\overline{Nu}_l$ [-] as the vertical axis and the net heat flux q' [W/m^2] as the horizontal axis. Since a CNF is a non-Newtonian fluid, we investigated the heat transport mechanism of the CNF 5 wt% based on the heat transfer coefficient of the mean heat dissipation area rather than performing the nondimensionalization generally applied for Newtonian fluids. For reference, the figure also shows the heat transfer coefficient for the mean heat dissipation area with pure water as the working fluid, which was found to be in close agreement with the heat transfer correlation equation for natural convection in a fluid layer in a horizontal, closed rectangular vessel (JSME, 1986, 2009). As shown in this figure, the mean heat transfer coefficient with CNF 5 wt% as the working fluid is substantially smaller than that for pure water. In comparison with the thermophysical properties of pure water as the reference standard, the difference in the viscosity (by about 50 times) for CNF 5 wt% is far larger than the difference in the heat transfer coefficient (about 1.2 times), which can be attributed to a substantial suppression of buoyancy-induced convection. This is also clear from the results of the nondimensional analysis of the heat transfer coefficient shown in Fig. 3. In short, the Rayleigh number Ra_l [-] for the CNF is small because of its high viscosity, and as a consequence its heat transfer capacity is reduced.

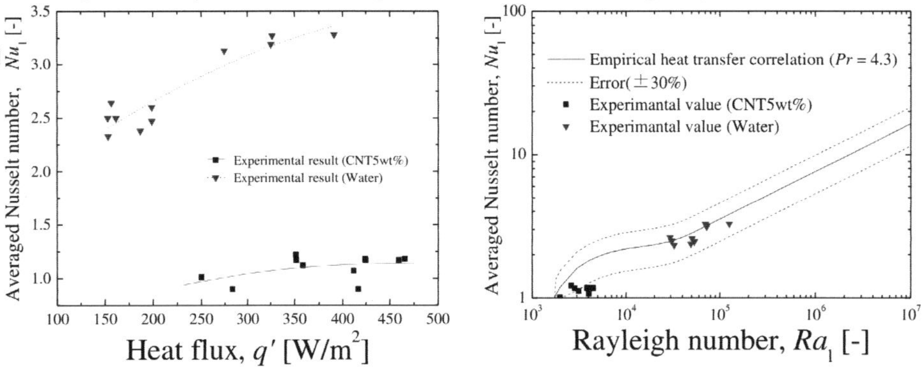

Fig.3 Natural convection heat transfer in a horizontal enclosed rectangular container

4. Conclusion

In the present study, we have investigated the thermophysical properties of CNFs, and in particular the temperature dependence of the thermal properties in CNF (1 and 5 wt%), and presented the results in the form of a heat transfer database that is essential for development of a heat transport process in a system configuration designed for actual use. The experimental results also confirm that the CNFs are non-Newtonian fluids. The effect of the 5 wt% CNF on viscosity outweighs that on thermal conduction, and suppression of buoyant convection clearly results in low heat transfer by natural convection.

References

Inagaki T and Kitazawa G 2011 *Kagakukogaku Ronbunshu* **37** 261
Inagaki T and Uezu T 2012 *Kagakukogaku Ronbunshu* **38** 263
Inagaki T and Isshiki T 2013 *Kagakukogaku Ronbunshu* **39** 33
JIS K 2249 2001 *Crude petroleum and petroleum products-determination of density and petroleum measurement tables based on a reference temperature*
JIS Z 8804 2001 *Methods of measuring specific gravity of liquid*
Meijo nano carbon 2010 *Carbon Nano-tube Fulid and its application* JP 4635103 B2
Shimizu Y 1994 *Univ. of Ryukyu, Department of Education, Bulletin* **45** 349
Japan Society of Mechanical Engineers 1986, 2009 *JSME data book: Heat Transfer 5th Edition* 56
Japan Society of Mechanical Engineers 1987 *Measurement uncertainty*
Japan Society of Thermo - Physical Properties 1990 *Thermo-physical Properties Handbook* 568

Self-Biased Inductor-less Interface Circuit for Electret-Free Electrostatic Energy Harvesters

E Lefeuvre[1], S Risquez, J Wei, M Woytasik and F Parrain

Univ Paris-Sud - CNRS, Institut d'Electronique Fondamentale, rue Ampère, bat. 220, 91405 Orsay Cedex, France

E-mail: elie.lefeuvre@u-psud.fr

Abstract. This paper presents a simple and efficient interface circuit for electrostatic energy harvesting devices, which uses neither controlled electronic switch nor inductor. A built-in voltage multiplier enables to get bias voltages higher than the circuit output voltage, resulting in an increased power output. This interface circuit was implemented using off-the-shelf components. Its operation was validated from 1V to 14V output voltages. Measured output power ranged from 10nW to 650nW using a capacitor varying between 45pF and 155pF at 15Hz. Measurements showed that it was possible to initiate the energy conversion cycles with start-up voltages as low as 100mV.

1. Introduction

Energy harvesting from mechanical vibration using electrostatic devices have focused significant research efforts over the last decade. This energy conversion principle is particularly interesting because the transducer is a simple mechanically-actuated variable capacitor, which can be easily manufactured and miniaturized using MEMS technologies. At micro-scale, tolerated electric fields are quite high, and theoretical power densities of electrostatic devices compete piezoelectric and electromagnetic devices ones.

Generating electrical energy using a variable capacitor without internal polarization (electret) requires to charge and discharge the capacitor in a synchronous way with the mechanical motion of its moving electrode. For this purpose, various power management electronic circuits have been proposed. However, practical implementation of such active interfaces remains difficult at microwatt scale, making a real obstacle for the development of electret-free electrostatic energy harvesting systems [1]. Indeed, at microwatt scale, the power losses of the electronic switches used in actives interfaces and the power consumption of their control circuit generally turn out to be more important than the electrical power generated by the variable capacitor itself. Yen et al proposed a simplified circuit using only one active switch without synchronization constraint [2], but energy-efficient implementation is also difficult to achieve at microwatt scale in this case [3].

Another drawback which is common to most of these interface circuits is the use of bulky magnetic components (inductors, transformers), that limits miniaturization. Moreover, in some specific applications such as medical implants, the use of magnetic parts is not possible because of compatibility requirements with Magnetic Resonance Imaging.

[1] To whom any correspondence should be addressed.

We are focusing here on interface circuits working without inductor and without active switches. In this domain, de Querioz et al have proposed a circuit based on Bennet's "doubler of electricity" [4]. Contrary to interface circuits using active switches, this circuit showed to be remarkably easy to implement and viable from the energy point of view down to the nanowatt scale. Dragunov et al proposed circuits with increased output current derived from this concept [5]. Based on the same principle, de Querioz recently presented a solution to increase the bias voltage and the output power [6], but this solution required using at least two variable capacitors.

The approach which led to the new circuit presented in this paper started from the analysis of another very simple circuit frequently used to assess the functionality of electrostatic energy harvesting devices [7]. Our objective beyond this work is to develop strongly integrated and highly reliable energy harvesting systems for active medical implant applications [8].

2. Approach and theoretical analysis of the proposed interface circuit

2.1. Interface circuits with pre-charged biasing capacitor

One of the simplest circuits which can be used to harvest energy with an electret-free variable capacitor is depicted in Fig. 1 (a). It is composed of a pre-charged capacitor C_{bias} connected in series with a load resistor R_{load} and the variable capacitor C_{var} [7]. Apart for testing electrostatic energy harvesting devices, this circuit has no real interest because of the alternating nature of the load voltage. From this remark comes the idea that the principle of the biasing capacitor could be useful if combined with a rectifier. This is for instance the case of the circuit depicted in Fig. 2 (b), in which the load resistor is replaced by a diode rectifier (*D1, D2*). The electrical energy generated by the variable capacitor is stored in the capacitor C_{store}.

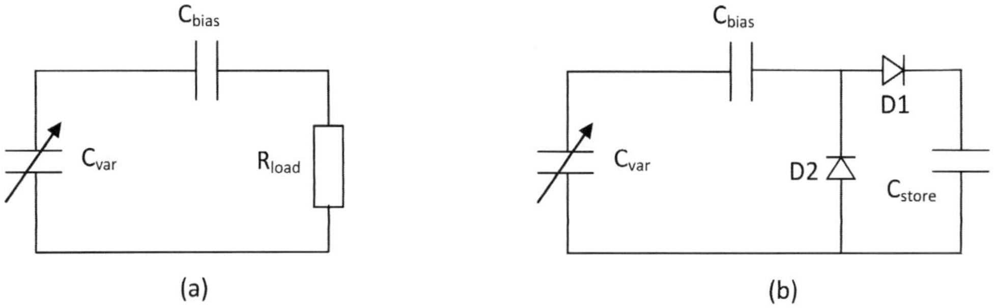

Figure 1. Schematic diagrams of (a) interface circuit with biasing capacitor and resistive load, (b) interface circuit with biasing capacitor, diode rectifier and energy storage capacitor.

Before presenting a solution to pre-charge the capacitor C_{bias} and keep it charged, operation of this circuit is first analysed with the help of the simulation waveforms shown in Fig. 2. In this simulation, C_{var} is varied at 100Hz between its maximum value C_{max}=300pF and its minimum value C_{min}=100pF. C_{bias} is pre-charged at 15V and C_{store} is pre-charged at 10V. The ON-state voltage drop across the diodes is noted V_D. Leakage currents are neglected in the following analysis.

According to Fig. 2, four time sequences can be identified over an energy conversion cycle:

- From t_0 to t_1, the diodes *D1* and *D2* are blocking, C_{var} is decreased at constant charge $Q_0=C_{max}\cdot(V_{Cbias}-V_D)$ and its voltage is increased from $V_{Cbias}-V_D$ to $V_{Cbias}+V_{Cstore}+V_D$.
- From t_1 to t_2, the diode *D1* is conducting, C_{var} voltage is equal to $V_{Cbias}+V_{Cstore}+V_D$ and the charge $Q_1=C_{max}\cdot(V_{Cbias}-V_D)-C_{min}\cdot(V_{Cbias}+V_{Cstore}+V_D)$ flows through C_{var}, C_{bias} and C_{store}.
- From t_2 to t_3, the diodes *D1* and *D2* are blocking, C_{var} is increased at constant charge $Q_2=C_{min}\cdot(V_{Cbias}+V_{Cstore}+V_D)$ and its voltage is decreased from $V_{Cbias}+V_{Cstore}+V_D$ to $V_{Cbias}-V_D$.
- From t_3 to t_4, the diode *D2* is conducting, C_{var} voltage is equal to $V_{Cbias}-V_D$ and the charge $Q_3=C_{min}\cdot(V_{Cbias}+V_{Cstore}+V_D)-C_{max}\cdot(V_{Cbias}-V_D)$ flows through C_{var} and C_{bias}.

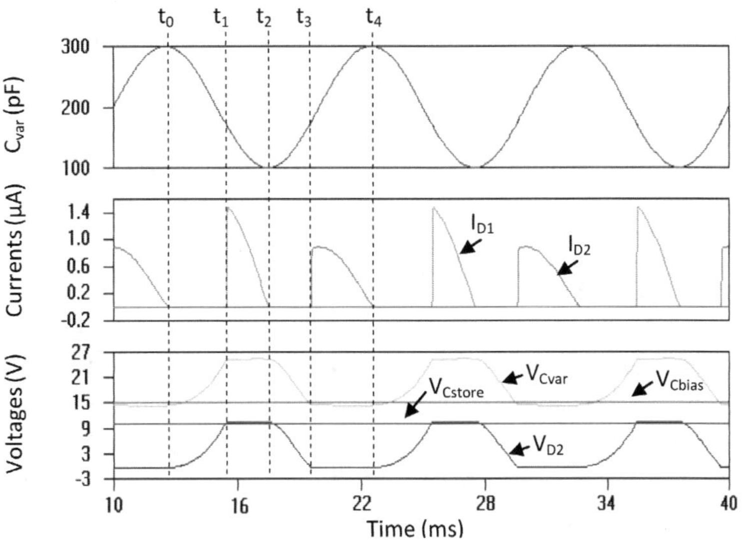

Figure 2. Simulation waveforms of the circuit with biasing capacitor, diode rectifier and energy storage capacitor (C_{bias}=10nF, C_{store}=10µF).

C_{bias} was chosen much larger than C_{var}, so the variations of V_{Cbias} are negligible over an energy conversion cycle. Similarly, C_{store} was chosen much larger than C_{bias} and the voltage V_{Cstore} can be considered as constant over several energy conversion cycles. By neglecting the voltage drop across the diodes, the energy harvested per cycle can be approximated by (1).

$$W_{Cycle} = \left(C_{\max} \cdot V_{Cbias} - C_{\min}\left(V_{Cbias} + V_{Cstore}\right)\right) \cdot V_{Cstore} \tag{1}$$

2.2. Interface including a charging circuit for the biasing capacitor

The previous the energy conversion cycle analysis showed the total charge flowing through C_{var} and C_{bias} over a period of operation (Q_1+Q_3) is null. Therefore, the average voltage across C_{bias} remains stable if one neglects the leakage current of the components. In practice, only a small fraction of the harvested energy should be sufficient to keep C_{bias} charged and compensate the effect of leakage currents.

A first observation of the circuit waveforms of Fig. 2 shows that it would be possible to recharge C_{bias} by connecting a diode *D3* between top electrodes of C_{store} and C_{var}: each time *D2* is conducting, C_{bias} would be recharged by C_{store} at the voltage V_{Cstore}.

Equation (1) indicates that the harvested energy increases with C_{bias} voltage, so it would be interesting to get higher voltages than V_{Cstore} across the biasing capacitor. A simple solution to increase voltages consists in using voltage multiplier circuits, but these circuits require variable input voltage. The circuit waveforms show that the voltages across *D1* and *D2* vary periodically between *0* and V_{Cstore}. Consequently, it should be possible to connect the input of a voltage multiplier circuit across *D1* or *D2* in order to recharge C_{bias} at a voltage which would be a multiple of V_{Cstore}. Starting from these remarks and observations, several implementations can be envisaged [9].

One of the simplest solutions is depicted in Fig. 3. In this new interface, the biasing capacitor is charged by a built-in voltage multiplier circuit, which may include from 1 to n cells. By neglecting the voltage drop across the diodes and the voltage ripple across C_{bias} and C_{store}, the voltage across C_{bias} can be expressed as (2).

$$V_{Cbias} = n \cdot V_{Cstore} \tag{2}$$

According to (1) and (2), the energy harvested per cycle using *n*-cells of the voltage multiplier can be expressed as (3). This equation is valid only if C_{max}/C_{min} if greater than the threshold $(n+1)/n$. Otherwise, the voltage variation across C_{var} is smaller than V_{Cstore} and *D1* and *D2* are always blocking so there is no energy conversion. It is interesting to note that this threshold becomes lower when the number of cells is increased, making possible the energy conversion at lower mechanical displacement magnitude of the variable capacitor electrodes.

$$W_{cycle} = n \cdot V_{Cstore}^2 \cdot C_{\min}\left(\frac{C_{\max}}{C_{\min}} - \frac{n+1}{n}\right) \tag{3}$$

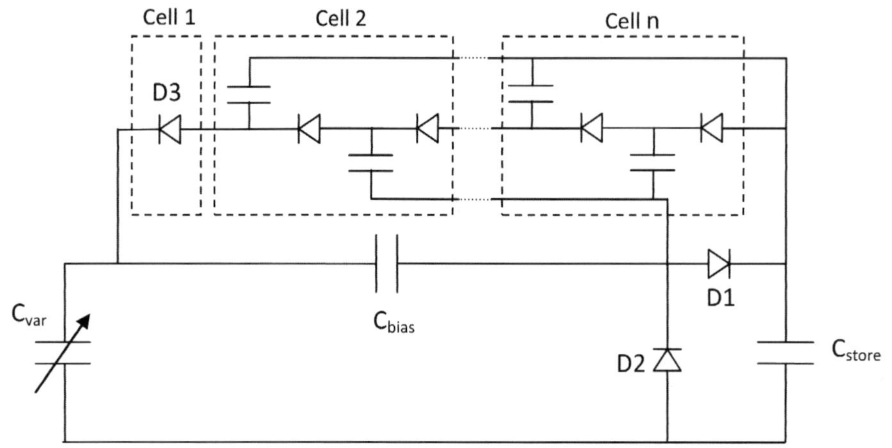

Figure 3. Schematic diagram of the self-biased interface circuit.

Fig. 4 (a) shows plots of theoretical energies harvested per cycle versus the capacitance variation ratio C_{max}/C_{min} for *n* between 1 and 4. As comparison, the maximum theoretical energy harvested per cycle using Yen's circuit [2] was also plotted in this figure. These results indicate for instance that for $C_{max}/C_{min}=2$, the self-biased circuits with 2, 3 and 4 cells theoretically harvest respectively 4, 12 and 16 times more energy than Yen's circuit for a given voltage V_{Cstore}. Fig. 4 (b) shows plots of the theoretical charge-voltage cycles for a number of cells *n* between 1 and 4.

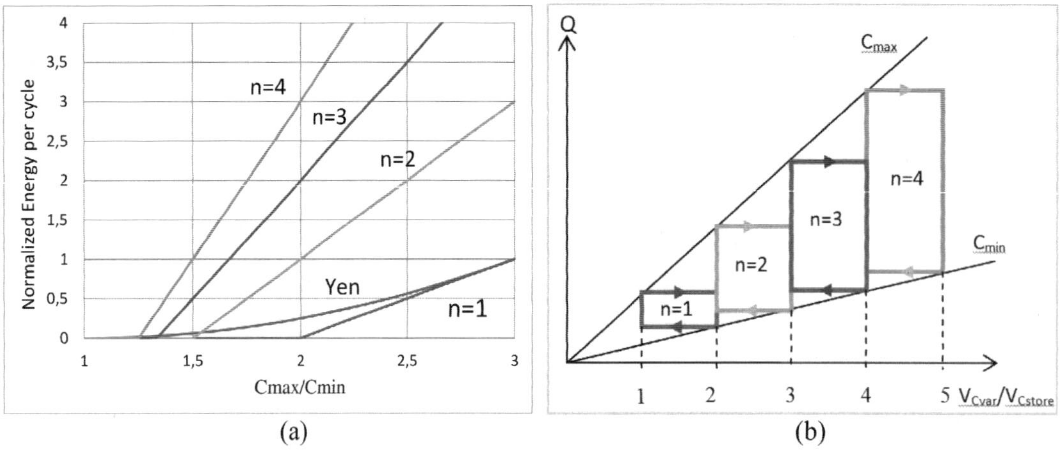

Figure 4. (a) Normalized energy converted per cycle vs capacitance ratio C_{max}/C_{min}, (b) charge-voltage diagrams of the variable capacitor

3. Experimental results

The circuit of Fig. 3 was implemented using a macro-scale rotating variable capacitor (C_{max}=45pF, C_{min}=155pF), low leakage PAD5 diodes, 560pF capacitors for the voltage multiplier, and C_{bias}=2.4nF. The frequency of operation was set to 15Hz. The voltages were measured using AD549JHZ high input impedance op-amp. Experimental and theoretical harvested powers plotted in Fig. 5 (a) exhibit similar shape as a function of V_{Cstore}. The differences which can be observed between theory and measurements indicate that the losses of the system increases with the number of cells. The analysis of the circuit practical limitations and efficiency will be deepened in further work. Fig. 5 (b) shows the start-up of the system, with C_{store} pre-charged at 1 volt, showing the ability of the circuit to self-start from very low voltages and self-adapt to variations of the output voltage. Our experiments showed that the circuit start-up was possible with initial output voltages as low as 100 mV.

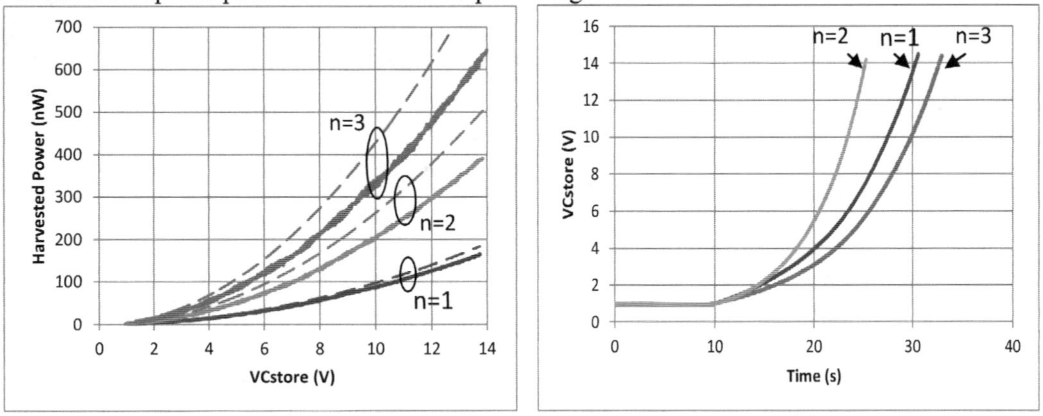

(a) (b)

Figure 5. (a) Experimental and theoretical harvested powers (resp. plain and dashed lines) vs voltage V_{Cstore}, (b) experimental start-up of the self-biased circuit (C_{store}=4.7nF).

4. Conclusion

This paper presents a new interface circuit for electrostatic energy harvesting devices, only composed of diodes and capacitors. As there is neither active switch nor control circuit, practical implementation is very simple and robust. Expected characteristics in terms of power output were confirmed by experimental results. Our future work will focus on the analysis of the effect of the components imperfections on the system performances.

5. References

[1] Mitcheson P D and Green T C 2012 *IEEE Trans. Circ. Syst.* **59** n°12 3098
[2] Yen B C and Lang J H 2005 *IEEE Trans. Circ. Syst.* **53**(2) 288
[3] Kempitiya A, Borca-Tasciuc D-A and Hella M M 2013 *IEEE Trans. Ind. Elect.* **60** n°12 5639
[4] de Queiroz A C M and Domingues M 2011 *IEEE Trans. Circ. Syst.* **58** n°12 797
[5] Dragunov V and Dorzhiev V 2013 *J. Phys.: Conf. Ser.* **476** 557
[6] de Queiroz A C M 2014 *In Proc. IEEE Circuits and Systems (LASCAS)* pp. 1-4
[7] Basset P, Galayko D, Cottone F, Guillemet R, Blokhina E, Marty F and Bourouina T 2014 *J. Micromech. Microeng.* **24** 035001
[8] Deterre M, Risquez S, Bouthaud B, Dal Molin R, Woytasik M and Lefeuvre E 2013 *J. Phys.: Conf. Ser.* **476** 012039
[9] Lefeuvre E, French Patent Application N°1456987, filed on July 18th 2014

Acknowledgments
The research leading to these results has received funding from the European Union Seventh Framework Programme (FP7/2007-2013) under grant agreement n°604360.

Position-insensitive long range inductive power transfer

Christopher H Kwan, James Lawson, David C Yates and Paul D Mitcheson

Department of Electrical and Electronic Engineering, Imperial College London, South Kensington Campus, London, SW7 2AZ, United Kingdom

E-mail: christopher.kwan09@imperial.ac.uk

Abstract. This paper presents results of an improved inductive wireless power transfer system for reliable long range powering of sensors with milliwatt-level consumption. An ultra-low power flyback impedance emulator operating in open loop is used to present the optimal load to the receiver's resonant tank. Transmitter power modulation is implemented in order to maintain constant receiver power and to prevent damage to the receiver electronics caused by excessive received voltage. Received power is steady up to 3 m at around 30 mW. The receiver electronics and feedback system consumes 3.1 mW and so with a transmitter input power of 163.3 W the receiver becomes power neutral at 4.75 m. Such an IPT system can provide a reliable alternative to energy harvesters for supplying power concurrently to multiple remote sensors.

1. Introduction

Far-field approaches to transferring power with electromagnetic fields over long distances (up to 10 m) suffer from many drawbacks that limit their potential to remotely power a network of sensors. Methods with dedicated transmitters that radiate power are generally quite inefficient. Employing antennas with narrow beam patterns can result in improvements in efficiency as the receivers are able to intercept a larger amount of the radiated energy. An example of these types of systems can be found in [1], where hundreds of µW could be transmitted over a distance of 1 m. However, the directivity of the antennas means that the freedom of movement of such receivers is severely restricted. Ambient RF energy harvesters are another candidate for far-field energy transfer; these do not require a separate transmitter as the energy is derived from mobile base stations, making them especially advantageous in urban environments. However, received power levels are limited to tens of µW as demonstrated in [2] with cm scale antennas.

Inductive power transfer (IPT) systems, which operate in the near field, are a particularly attractive alternative to existing far-field techniques. In a previous system [3], it was shown that more than 10 mW can be transferred to a load up to 6m from the transmitter. An issue with this system, though, is that different levels of load current draw in the sensor node detuned the magnetic link, resulting in poor efficiency and a significant drop in received power. Also moving the receiver too close to the transmitter could cause damage to the receiver electronics due to excessive received voltage. This paper proposes an improved long-range IPT system for reliable long-range powering of sensors with milliwatt-level consumption. It overcomes the problems of the previous system by using an ultra-low power switch-mode impedance emulator to present a constant impedance, and wireless feedback to modulate the transmitter power.

2. Receiver impedance emulation

For IPT systems, the amount of power transferred is maximised when an optimal load is connected to the receiver's resonant tank. In the case of a long-range system where coupling between the transmitter and the receiver is extremely weak, the optimal load Z_{load} can be easily determined as it is simply the conjugate match of the receiver's resonant tank at the operating frequency [3]. Thus,

$$Z_{\text{load}} = \left(\frac{1}{\frac{1}{R_{\text{RX}}+j\omega L_{\text{RX}}} + j\omega C} \right)^* \tag{1}$$

where R_{RX} is the parasitic RX coil resistance, L_{RX} is the RX coil inductance, C is the RX tuning capacitance for the resonant tank, ω is the angular operating frequency of the system, and $(x)^*$ denotes the complex conjugate of x.

Different levels of sensor node activity, resulting in variations in load current draw, mean that the load impedance changes over time, detuning the magnetic link. This leads to poor efficiency and a collapse in power available to the load. Therefore, a method of impedance emulation is required so that a constant optimal impedance is always presented to the output of the bridge rectifier in the receiver. Switch-mode DC-to-DC converters can be controlled in such a way that the desired impedance is presented. An open-loop converter control scheme is preferred as the power consumption requirements can be minimised. Both the buck-boost converter (Figure 1a) and the flyback converter (Figure 1b) are candidates for an open-loop impedance emulation circuit [4], which is to be inserted between the bridge rectifier and the sensor node.

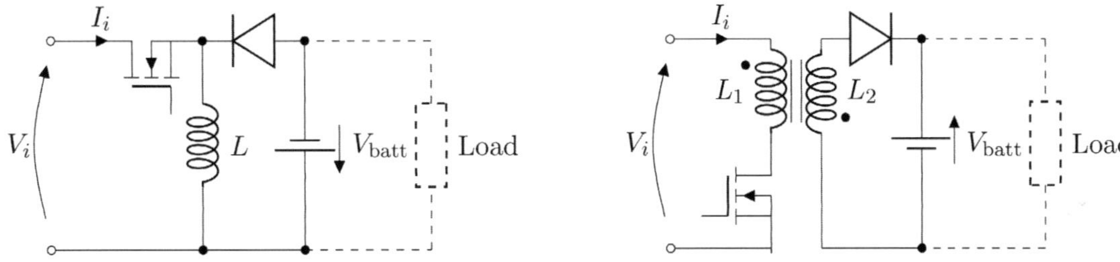

(a) Buck-boost (b) Flyback

Figure 1: Open-loop impedance emulator topologies

2.1. Buck-boost topology

A buck-boost converter operating in discontinuous conduction mode and with open-loop control (i.e. fixed duty cycle) can act as an impedance emulator. Buck-boost circuits were used in [5] as part of the power processing circuitry for MEMS energy harvester in order to present the correct electrical impedance to the generator. For the long-range IPT system, a Li-ion coin cell is placed across the output of the impedance emulator in order to fix the output voltage at $V_{\text{batt}} = 3.7\,\text{V}$. Any excess power not consumed by the sensor node is stored in the cell, whilst the cell can also be used to power the sensor node and impedance emulator if the receiver is out of the transmit power range.

To determine the impedance seen at the input of a buck-boost converter in discontinuous conduction mode, the average current into the converter $I_{i,average}$ when a voltage V_i is applied at the input is first calculated:

$$I_{i,peak} = \frac{V_i}{L} t_{on} = \frac{V_i}{L} \frac{\delta}{f} \Longrightarrow I_{i,average} = \frac{1}{2} I_{peak} \frac{t_{on}}{T} = \frac{1}{2} I_{peak} \delta = \frac{V_i \delta^2}{2Lf}$$

where $f = \frac{1}{T}$ is the switching frequency of the MOSFET, L is the value of the inductor, t_{on} is the time during which the MOSFET is switched on, and $\delta = \frac{t_{on}}{T}$ is the duty cycle of the switching signal.

Therefore, the input impedance of the buck-boost converter is:

$$R_{\text{emulated}} = \frac{V_i}{I_{i,average}} = \frac{2fL}{\delta^2} \tag{2}$$

The buck-boost circuit remains in discontinuous conduction mode only if the current in the inductor can return to zero during the MOSFET off time. From this we obtain an upper bound for the operating duty cycle:

$$\frac{V_i}{L}t_{on} \leq \frac{V_o}{L}(T - t_{on}) \Longrightarrow \delta \leq \frac{1}{1 + \frac{V_i}{V_o}} \tag{3}$$

In this system, V_o is fixed at 3.7 V by the battery, and the maximum expected V_i is 55 V (higher voltages could damage the receiver electronics). From Equation 3, the maximum duty cycle for which the converter remains in discontinuous mode is $\delta = 0.0630$. This would mean that the MOSFET would have to switch at unrealistic speeds since the duty cycle is very short as a result of the large step-down ratio. Another issue with the buck-boost impedance emulator is that the output voltage is negative.

2.2. Flyback topology

To overcome these limitations of the buck-boost impedance emulator, a flyback impedance emulator was designed. Flyback circuits are used in [6] for emulating an impedance for low-power RF and wind energy harvesters, and in [7] as part of a high-power-factor rectifier. The expression for emulated resistance is the same as Equation 2, but with the buck-boost inductor L replaced by the primary inductance of the flyback transformer, L_1:

$$R_{\text{emulated}} = \frac{2fL_1}{\delta^2} \tag{4}$$

The flyback circuit remains in discontinuous conduction mode only if the current in the secondary winding of the transformer can return to zero during the MOSFET off time:

$$\frac{V_i}{L_1}t_{on} \leq \frac{V_o}{nL_2}(T - t_{on}) \Longrightarrow \delta \leq \frac{1}{1 + \frac{nL_2V_i}{L_1V_o}} = \frac{1}{1 + \frac{V_i}{nV_o}} \tag{5}$$

where L_2 is the secondary inductance of the flyback transformer and $n = \sqrt{\frac{L_1}{L_2}}$ is the turns ratio.

With $V_o = 3.7$ V, $V_i = 55$ V and $n = 10$, the maximum duty cycle is $\delta = 0.402$, which means the MOSFET can be switched at more reasonable speeds. In addition, the output voltage of a flyback converter is positive. Like the buck-boost impedance emulator in Section 2.1, the flyback can be operated in discontinuous conduction mode and in open loop, keeping power consumption low.

The implemented flyback impedance emulator has the following parameters: $f = 19$ kHz, $L_1 = 140$ mH, $L_2 = 1.6$ mH, $\delta = 0.24$. This gives an emulated impedance of 92.3 kΩ, which is close to the optimal load presented to the rectifier. As $n = 9.35$ here, the maximum duty cycle is 0.386, which means that the flyback circuit will remain in discontinuous conduction mode.

Due to the extremely low magnetic coupling in this system, the optimal impedance presented to the rectifier is high (around 100 kΩ in this case), causing the received current to be very small (100s of µA). In reality the emulated impedance is not perfectly constant with variations in input voltage due to parasitic oscillations in discontinuous conduction mode (Figure 2a) but nevertheless, the emulator performance was sufficient, restricting impedance changes to within +20/-50% of the optimal value. Figure 2b shows the efficiency of the impedance emulator with input voltage.

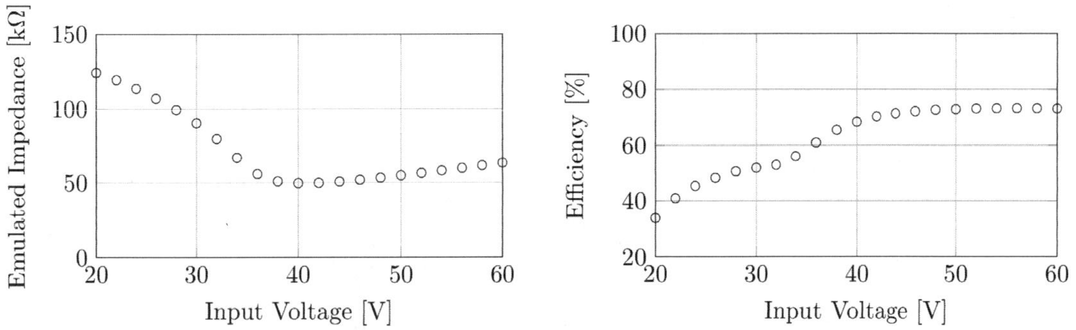

(a) Emulated impedance with input voltage

(b) Efficiency with input voltage

Figure 2: Performance of flyback impedance emulator

3. Transmitter power control

In order to maintain constant received power as the receiver position is changed and to prevent damage to the receiver electronics due to excessive received voltage, transmitter power modulation was implemented. At the receiver side, V_{RECT} (shown in Figure 3), the voltage at the output of the rectifier and at the input of the flyback impedance emulator, is measured by a microcontroller in the sensor node. This voltage measurement is then sent to a microcontroller at the power transmitter side via a 433 MHz FSK wireless data link.

The microcontroller at the power transmitter side reads the measured V_{RECT} value with an FSK data receiver, and this value is used to control the voltage of the DC power supply to the Class-E inverter, V_{CC}, in order to modulate the input power. The objective of the control algorithm is to keep V_{RECT} close to 50 V. Any voltage above 60 V is deemed excessive and could potentially cause damage to the receiver electronics.

The method of V_{CC} is as follows: if V_{RECT} is greater than 55 V, then V_{CC} is decreased by 10 V. If V_{RECT} is less than 45 V, then V_{CC} is increased by 2 V, but only if V_{CC} is less than a certain value called $V_{\mathrm{CC\_MAX}}$. $V_{\mathrm{CC\_MAX}}$ is an upper limit on V_{CC} that is set in order to keep V_{DS} of the MOSFET in the Class-E inverter less than 1000 V, which is its breakdown voltage. A hysteresis band for V_{RECT} (between 45 V and 55 V) is used to prevent frequency fluctuations in input power.

4. Results

Figure 3 shows a schematic diagram of the receiver coil and electronics, with the resonant tank on the left hand side, followed by the bridge rectifier, the impedance emulator and the sensor node with an RF data transmitter. Figure 4 shows post-emulator received power against distance between the receiver and the transmitter. Received power is constant at around 30 mW up to 3 m as the power is modulated, but beyond 3 m the transmit power saturates (limitations in the MOSFET voltage rating of the driver amplifier) and the received power decreases. The receiver electronics and feedback system consumes 3.1 mW and so with a transmitter input power of 163.3 W the receiver becomes power neutral at 4.75 m. At 5 m, received power falls to 2 mW.

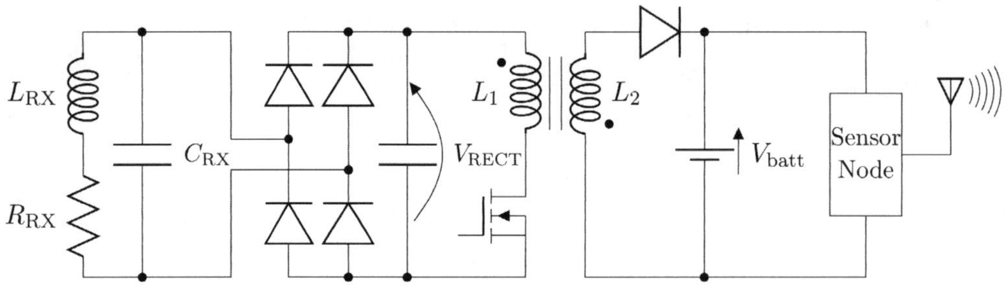

Figure 3: Receiver coil and electronics

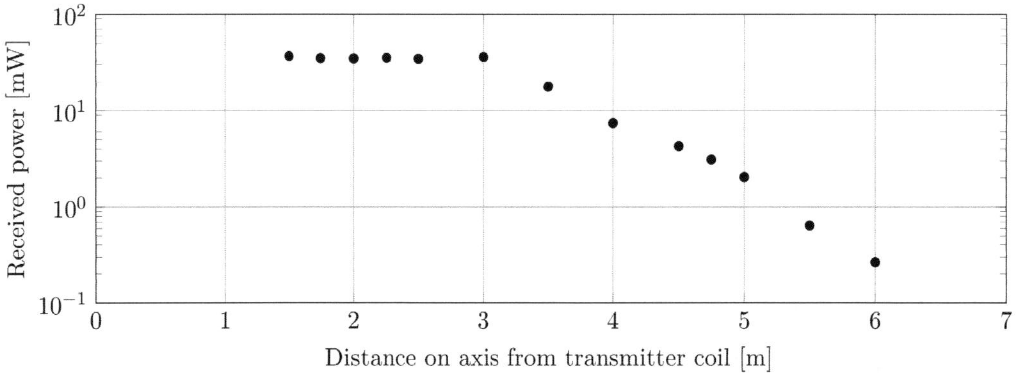

Figure 4: Received power with distance between transmitter and receiver coil

5. Conclusion

With the use of an ultra-low power flyback impedance emulator to provide constant loading and wireless feedback to maintain constant transmitter power, it has been demonstrated that inductive power transfer can be used to reliably power a sensor node. IPT systems are therefore a viable alternative to existing far-field methods such as RF energy harvesters for supplying power concurrently to multiple remote sensors.

Acknowledgments

Funding is gratefully acknowledged from Her Majesty's Government.

References

[1] Visser H 2012 *Microwave Conference (EuMC), 2012 42nd European* pp 317–320

[2] Pinuela M, Mitcheson P and Lucyszyn S 2013 *Microwave Theory and Techniques, IEEE Transactions on* **61** 2715–2726 ISSN 0018-9480

[3] Lawson J, Pinuela M, Yates D C, Lucyszyn S and Mitcheson P D 2013 *Journal of Physics: Conference Series* **476** 012005 URL `http://stacks.iop.org/1742-6596/476/i=1/a=012005`

[4] Erickson R and D M 2001 *Fundamentals of Power Electronics* (Springer) chap 18 2nd ed

[5] D'hulst R, Mitcheson P D and Driesen J 2006 *Proceedings of PowerMEMS 2006* pp 215–218

[6] Paing T and Zane R 2006 *Power Electronics Specialists Conference, 2006. PESC '06. 37th IEEE* pp 1–7 ISSN 0275-9306

[7] Erickson R, Madigan M and Singer S 1990 *Applied Power Electronics Conference and Exposition, 1990. APEC '90, Conference Proceedings 1990., Fifth Annual* pp 792–801

Micro hydrogel tubes for culture of oil production microbes in an open environment

K Mikami, H Onoe, and N Miki

Department of Mechanical Engineering, Keio University
3-14-1 Hiyoshi Kohoku-ku Yokohama Kanagawa 211-0033 Japan

E-mail: miki@mech.keio.ac.jp

Abstract. This paper proposes a microbe culture technology in an open environment using hydrogel microtubes. Oil production microbes, such as *Aurantiochytrium*, have been discovered and are currently studied to produce fuels instead of fossil fuels. Scale-up of the culture systems of such microbes is critical for practical applications. The microbe culture in a nutrition-rich open environment, such as lakes and sewage plants, is preferable in terms of scales and cost, however, contamination of other native microbes is often fatal to the target microbes. Hydrogel microtubes that can be manufactured by microfluidic devices can encapsulate the microbes and allow oxygen and nutrition to diffuse through the tube walls while preventing other microbes from intruding inside. Therefore, the microbes can be cultured in any environments and scale-up is also possible. In this paper, we demonstrate that the contamination of microbes was prevented and target microbes were successfully cultured inside the microtube.

1. Introduction

Oil production microbes, such as *Aurantiochytrium*, have been discovered and are currently studied to produce fuels instead of fossil fuels [1]. Carbon dioxide generated from such fuels is not ruled to be the one to contribute global warming since the circulation of carbon is much more rapid than that of fossil fuels that were produced millions or more years ago. In addition to discover or genetically engineer more efficient oil production microbes, scale-up of the culture systems of such microbes is critical for practical applications. The microbe culture in a nutrition-rich open environment, such as lakes and sewage plants, is preferable in terms of scales and cost, however, contamination of other native microbes is often fatal to the target microbes.

Hydrogel microtubes that can be manufactured by microfluidic devices were proposed to host cell cultures [2]. The size can be controlled by the flows and several materials can be used to produce the tubes. The tubes can encapsulate the cells while they allow oxygen and nutrition to diffuse through the tube walls and prevent other microbes from intruding inside. The tube enabled encapsulation of foreign cells without being attached by the immune system. We came up with the idea of utilizing the tubes to culture microbes inside. Since the tubes protect the microbes from other competitive microbes physically, the target microbes, even which have very little competitiveness against others, can be cultured under any environments. It is also to be noted that scale-up of the culture system is also possible because the microtubes can be mass produced by piling up the microfluidic systems. Figure 1 shows the concept of our study. In this paper, we demonstrate that the contamination of microbes was prevented and target microbes were successfully cultured inside the microtubes.

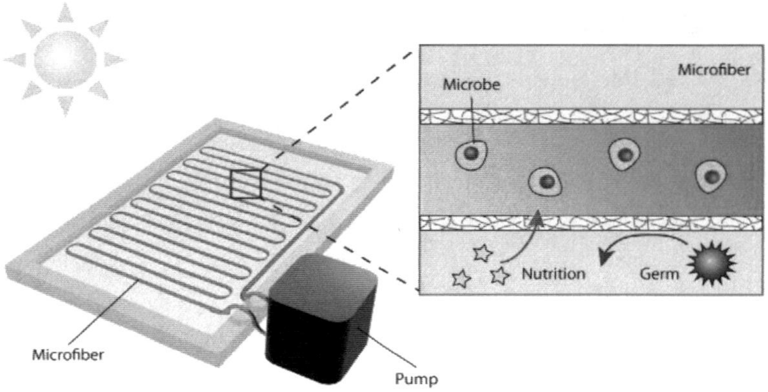

Figure 1 Culture of microbes in hydrogel microtubes under open environment.

2. Hydrogel tube formation

Figures 2 (a) and (b) show the microfluidic device to form hydrogel tubes and a photo of the production. The channel inside was a pulled glass tube and had a tip diameter of 100 μm. Calcium chloride solution was introduced to the inside channel and sodium alginate to the outside channel. When they are mixed, they form an alginate calcium gel tube. When microbes were mixed in the calcium chloride solution, the microbes were encapsulated inside the tube. The gel can allow molecules smaller than 60 kDa to permeate through, which is preferable for our purpose, i.e., oxygen and nutrition can diffuse into the tube while other microbes can be blocked. The tube thickness and inner diameter with respect to the ratio of the volumetric flow rate of sodium alginate to calcium chloride are shown in Figure 3. The inner diameter of the tube was found not to depend on the ratio while the membrane thickness increased with the ratio. The membrane thickness and inner tube diameter will determine diffusion of the oxygen, nutrition, and byproducts and thus, the microbe cultures. They will be further investigated in future work.

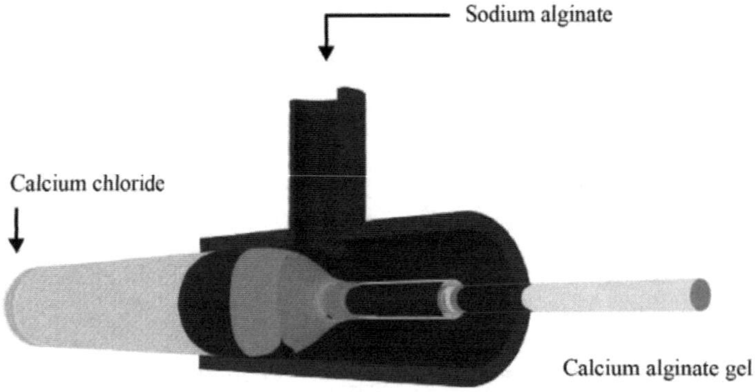

(a)

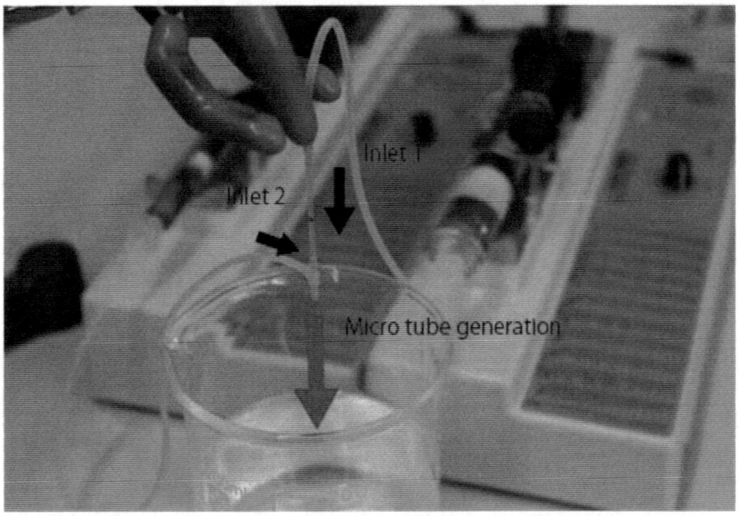

(b)

Figure 2. (a) Schematic view of the microfluidic device to produce alginate calcium hydrogel tube. (b) The produced gel can be collected easily.

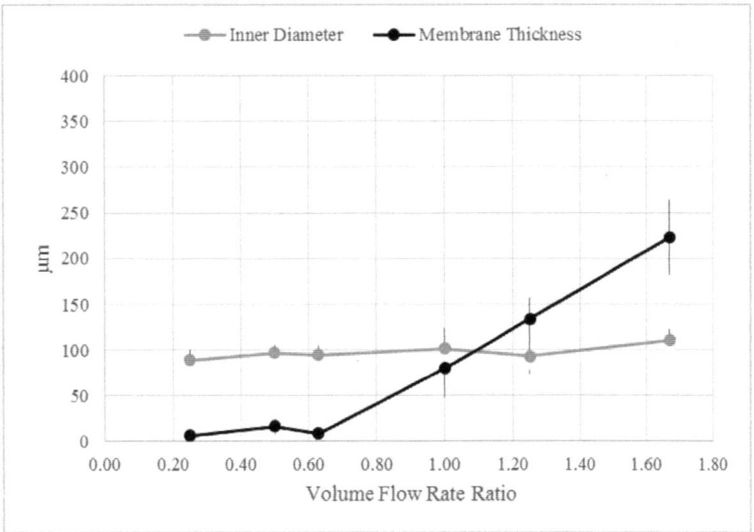

Figure 3. The inner diameter and tube wall thickness of the hydrogel tube with respect to the ratio of the volumetric flow rate of sodium alginate to calcium chloride.

3. Proof-of-concept experiments

We conducted proof-of-concept experiments using *Coryne Glutamicum* and *Vibrio alginotlytius*. Figure 4 shows the conceptual image of the experiments. *Coryne Glutamicum* is encapsulated in the hydrogel tubes and the tubes are immersed in culture media, which contain *Vibrio alginotlytius*. The cell densities of the microbes after 1 h of incubation without and with encapsulation of *Coryne Glutamicum* inside the tube are shown in Figures 5 (a) and (b). With the presence of *Vibrio alginotlytius*, *Coryne Glutamicum* were extincted, as shown in Figure 5 (a). However, when they are encapsulated in the fiber, they successfully increased the cell numbers. The results verified that the hydrogel tube can successfully culture cells inside while protecting them from contaminants outside.

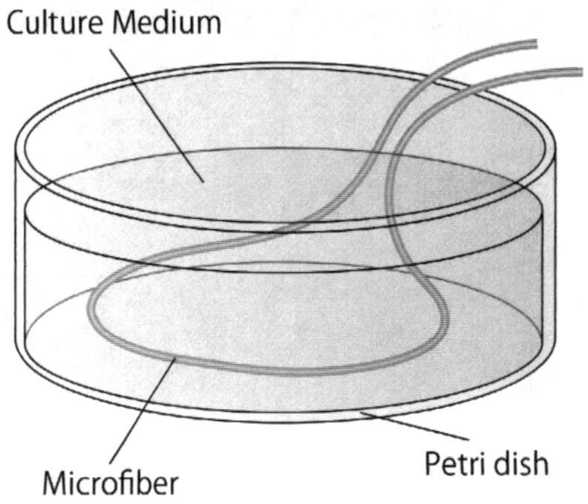

Figure 4. Proof of concept experiments. *Coryne glutamicum* is encapsulated in the microtube and *Vibrio alginolyticus* is contained in the culture media.

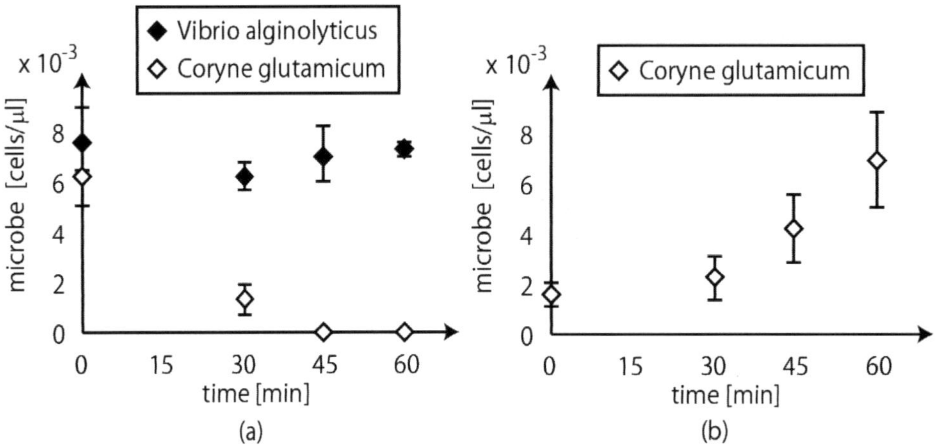

Figure 5. Cell densities of *Coryne Glutamicum* (a) when they are cultured together with *Vibrio alginolyticus* and (b) when they are cultured inside the microtube. *Coryne glutamicum* were extinct when they were mixed with *Vibrio alginolytics*, whereas the cell density successfully increased when they were culture in the tube.

4. Conclusions

Hydrogel fibers, which can be mass-producible using microfluidic technologies, were verified to be effective scaffold for culture of microbes even with the presence of competitive microbes. The culture technique proposed herein can be readily applicable to culture various microbes that produce useful materials, such as oil producing bacteria.

References

[1] Kaya K, Nakazawa A, Matsuura H, Honda D, Inouye I, Watanabe MM 2011 *Biosci. Biotechnol. Biochem.* **75** 2246

[2] Onoe H, Okitsu T, Itou A, Negishi K, Gojo R, Kiriya D, Sato K, Miura S, Iwanaga S, Kuribayashi-Shigetomi K, Matsunaga YT, Shimoyama Y, and Takeuchi S 2013 *Nat. Mat.* **12** 584

Advances on Ultrafast Silicon Field Emitter Array Photocathodes for Coherent Radiation Sources Based on Inverse Compton Scattering

C Dong[1], M Swanwick[2], P D Keathley[1], F X Kärtner[1,3] and L F Velásquez-García[2]
[1] Department of Electrical Engineering and Computer Science, Massachusetts Institute of Technology, 77 Massachusetts Avenue, Cambridge, MA 02139, USA
[2] Microsystems Technology Laboratories, Massachusetts Institute of Technology, 77 Massachusetts Avenue, Cambridge, MA 02139, USA
[3] Center for Free-Electron Laser Science, DESY and Department of Physics, University of Hamburg, Notkestralse 85, D-22607, Hamburg, Germany

E-mail: Velasquez@alum.mit.edu

Abstract. Low-cost, compact, and coherent X-rays sources would enable exciting applications such as biomedical imaging of soft tissue and real-time visualisation of molecules at a widespread scale. A promising approach to implement such an X-ray source is based on inverse Compton scattering of a series of nanostructured electron sheets accelerated to relativistic speeds. Photon-triggered field emission arrays can readily produce planar arrays of electron bunches with pC-level sheet charge at high repetition rates using intense laser pulses. In this article, the performance of single-crystal, ultrafast, photon-actuated silicon field emitter arrays is investigated for varying emitter height. Charge vs. incident photon pulse energy characteristics and quantum efficiency of the devices are reported.

1. Introduction

Conventional X-ray imaging is based on absorption, i.e., on registering the spatial change in the intensity of an X-ray beam after passing through the object being imaged [1]. A typical X-ray source for absorption imaging is composed of a thermionic cathode and a reflection anode inside a vacuum chamber with an X-ray transmission window. Electrons emitted by the cathode are accelerated toward the anode using a large bias voltage; deceleration of the electrons in the anode produces X-rays, a mix of bremsstrahlung radiation (broad, continuous spectrum) and fluorescence (emission at specific peaks corresponding to atomic shell transitions of the anode material). A fraction of the X-rays escape the vacuum chamber through a transmission window made of a low-absorption material, e.g., beryllium [2]. X-ray absorption imaging works best in structures made of dense, i.e., high-Z materials, but are not good to image samples made of light, i.e., low-Z materials. For imaging low-Z materials, far greater contrast would be attained if, instead of measuring the relative attenuation of the X-ray beam after passing through the sample, the change in phase of the photons is tracked [3]. Applications of coherent X-ray sources include biomedical imaging, cancer treatment, protein nanocrystallography, and real-time imaging of molecules. In order to be possible phase contrast imaging, a coherent X-ray source is needed. Current coherent X-ray sources are synchrotrons, very expensive and large particle accelerators; therefore, phase contrast imaging is limited to research and high-end applications.

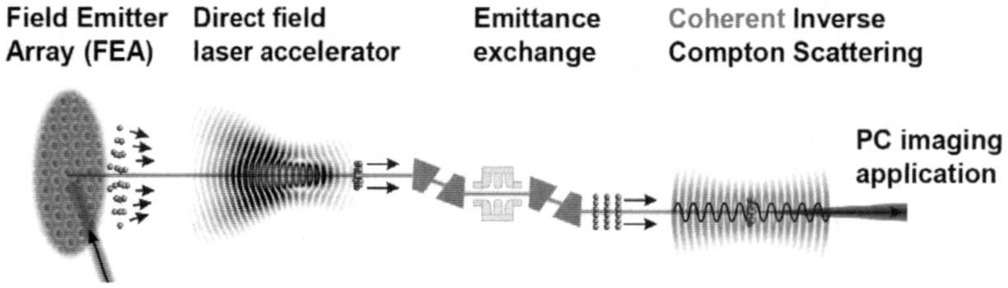

Figure 1. Schematic of the compact coherent X-ray source proposed in [4].

A promising implementation of a low-cost, compact, and coherent X-ray source based on inverse Compton scattering [4] is shown in figure 1. In this approach, a series of nanostructured charge sheets (a planar array of electron bunches) is produced by a photocathode; the sheets are then accelerated to relativistic speeds using an optical field. Then, the charge sheets experience an emittance exchange from the transverse phase space plane into the longitudinal plane (i.e., each sheet, originally advancing in the direction of its normal, now advances in the direction of one of its edges). Finally, a second optical field interacts with the electron bunches, decelerating them and causing coherent X-ray emission. The brilliance and flux of such compact coherent X-ray sources would be similar to those of a synchrotron facility, but with orders of magnitude smaller footprint, power assumption, and cost. A Coherent source based on a laser undulator has been previously proposed [5] but for very different sets of electron beam and laser parameters. In the approach shown in figure 1, the bunch charge and peak current are much lower, as is the electron beam emittance. In addition, the use of a structured electron beam and the exchange of the emittance of the beam results in visibly less number of gain lengths and power consumption, as well as less stringent focusing and pulse length requirements on the laser.

State-of-the-art ultrafast cathodes are flat surfaces that use highly reactive materials to lower the workfunction and increase the quantum efficiency of single-photon absorption of ultraviolet pulses. However, these devices have several disadvantages including (i) they need to be fabricated, stored and operated in ultra-high vacuum, and (ii) production of high current density pulses reduces their lifetime due to the rapid degradation of the low workfunction material [6]. Photon-triggered field emission cathodes are a promising technology to bypass these shortcomings. In these devices, electron tunneling occurs when the electric field of high-intensity optical pulses interacts with field enhancing structures [7]. Most work on ultrafast photo-triggered field emission cathodes has focused on single tips that are serially manufactured. We recently demonstrated batch fabricated photon-triggered field emission cathodes composed of massively multiplexed (> 100,000 tips, 4.6-million tips.cm^{-2}), uniform (<1 nm tip radius standard deviation) arrays of nano-sharp high-aspect-ratio silicon pillars, capable of pC-level emission with multi-kHz repetition [8]. These devices are made using standard CMOS batch fabrication processes, are stored at atmospheric pressure, and can be operated at lower vacuum levels compared to standard photocathodes with no degradation; in addition, the structure of the cathode shapes the emission as a series of planar arrays of electron bunches, which is essential for the coherent X-ray source approach we described. In a recent article we reported the effect of emitter pitch scaling on the performance of the field emitter arrays [9]; in this article we investigate the effect of varying the emitter height on the performance of the field emitter arrays; we provide charge vs. incident laser pulse energy characteristics as well as estimate the quantum efficiency of the cathodes.

2. Device fabrication

To fabricate the devices, an array of coin-like features is formed on a <100> n-Si wafer wafer using a 250 nm-thick chemical vapor deposited (LPCVD) silicon-rich silicon nitride film, projection lithography, and reactive-ion etching (RIE). Then, an array of coin-like features is created on top of the first array using a 500 nm-thick plasma enhanced chemical vapour deposited (PECVD) silicon dioxide film, projection lithography, and RIE; each oxide feature is concentric within 0.1 μm to the nitride feature right underneath. Third, the emitter body is etched using deep reactive-ion etching

(DRIE); the wafers are then cleaned and oxidized. The oxidation step thins the pillars and the nitride features act as a diffusion barrier to form the tips. The oxide and nitride films are removed using wet processing, leaving highly uniform arrays of single-crystal silicon pillars. Finally, the wafers are diced into 22-mm squares. For our experiments we made arrays of emitters with 10 µm emitter pitch, 2, 4, and 8.5 µm emitter height, and 6 nm tip radius. A fabricated device is shown in figure 2.

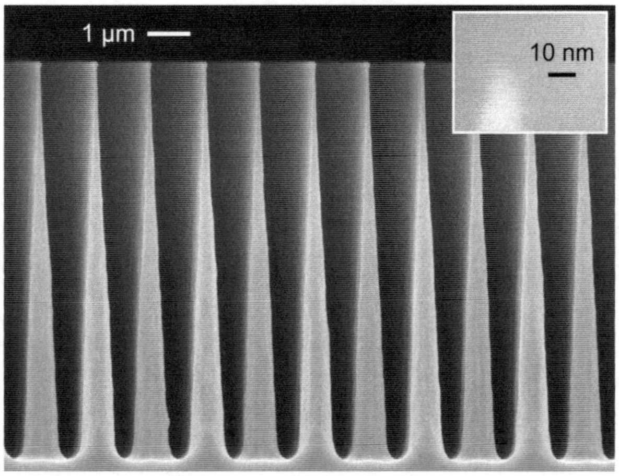

Figure 2. Close-up of an array of electron emitters; each emitter is 8.5 µm tall and the emitter pitch is 1.25 µm.

3. Experimental procedure

An schematic of the photocathode vacuum system is shown in figure 3. The measurement of the charge vs. incident laser energy characteristics takes place in a vacuum chamber with base pressure in the order of 10^{-7}-10^{-8} Torr. The cathode chip is dipped in 2% hydrofluoric acid to remove the native oxide right before characterization and is then connected to ground through a picoammeter (Keithley 6485). The anode electrode, a 0.25" diameter plate, is placed 3 mm above the tips and is connected with a source-measuring unit (SMU: Keithley 237 or Keithley 248), which provides various voltage biases between 10V and 5000 V. The cathode was excited with 35 fs 800nm laser pulses at 3 kHz repetition rate from a regeneratively amplified titanium sapphire oscillator seed. The laser pulses hit the samples at an 84° grazing incidence, creating a 765 µm by 80 µm ellipse. The intensity of the incident wave is controlled by changing the waveplate angle through a LabView script and switching different filter lenses. The data are collected using a LabView script.

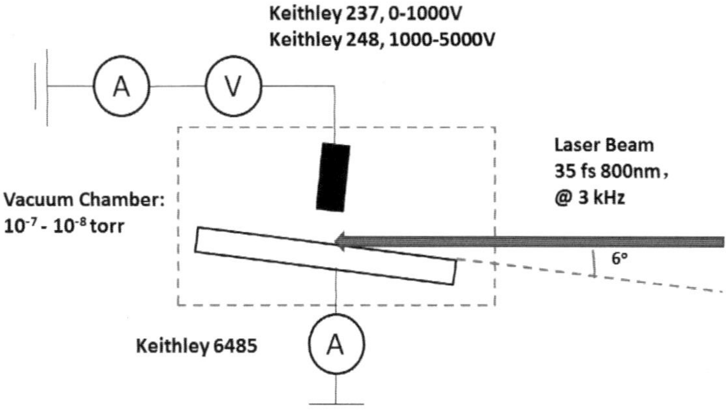

Figure 3. Schematic of the testing rig used to characterize the photocathodes.

4. Experimental results

Figure 4 shows the charge vs. incident laser pulse energy for devices with 10 μm emitter pitch and 2, 4, and 8.5 μm emitter height. At low intensities (<0.3 μJ), an electron absorbs three photons simultaneously in order to escape from the tip surface; this results from silicon's electron affinity (4.05 eV) and the energy of the incident photons (1.55 eV). The slope of the low-energy portion of the characteristics in figure 3 is about 3, consistent with the three-photon absorption process. For larger incident laser pulse intensities (>1 μJ), the optical field bends the barrier that traps the electrons within the tips and electron tunneling occurs. This regime is commonly referred to as the strong-field regime and it follows a time-averaged Fowler-Nordheim model [10]. In this regime, the expected slope of the charge-incident photon energy characteristic is ~1.2 if no space charge effects take place [8]. A slope of 1.06 is obtained in figure 3 for the strong-field regime, in close agreement with the model.

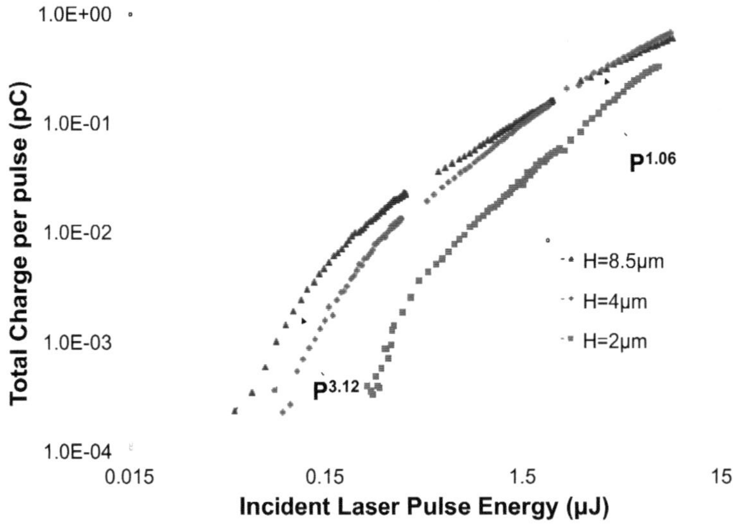

Figure 4. Total emitted charge vs. laser pulse energy for arrays of Si field emitters with 6 nm tip radius, 10 μm pitch and 2 - 8.5 μm height.

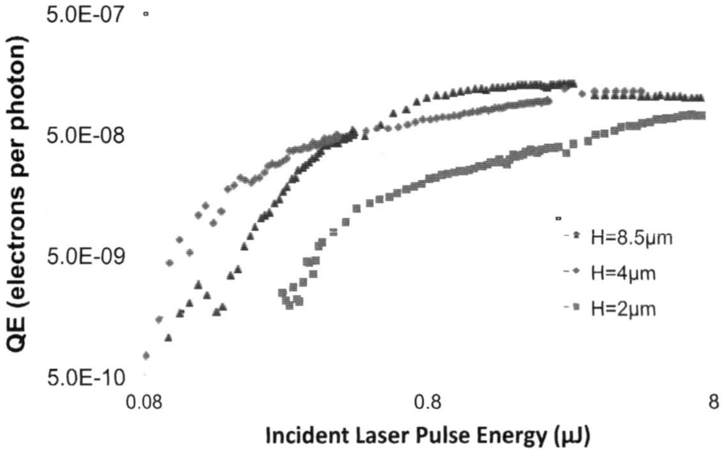

Figure 5. Quantum efficiency versus incident photon energy for arrays of Si field emitters with 6 nm tip radius, 10 μm pitch and 2 - 8.5 μm height.

In figure 4 the charge per incident photon pulse increases for larger the emitter height, as expected from the larger emitter field factor of taller emitters. However, for large enough intensity (>1.5 μJ), the characteristics converge and no further gains in electron emission are observed. There seems to be an optimum height for a fixed emitter pitch to maximize the total current emission. This is an important result given the greater difficulty for making longer emitters compared to making shorter emitters.

We estimated the quantum efficiency (QE) of our devices based on the data shown in figure 4 and the specifications of the laser; the result is shown in figure 5. We observe that in general larger emitter aspect-ratio results in higher QE, although for tall enough emitters and incident photon pulses of enough intensity the QE characteristics converge and become flat. The estimated maximum QE is $\sim 1.3\times10^{-7}$. The QE values are high taking into account that only a very small fraction of the area of the cathode emits electrons, as the emitter tip diameter is about 12 nm and the emitter pitch is 10 μm, resulting in only about a millionth of the area participating in the emission. Nonetheless, the emission is greatly enhanced as compared to that of on planar Si, which has a QE of 10^{-10} [8].

5. Conclusion

We investigated the performance of planar arrays of high-aspect-ratio single-crystal silicon nanosharp electron emitters for varying emitter height. We observed multi-photon emission for low-intensity incident photon pulses (<0.3 μJ), and strong-field emission for high-intensity incident photon pulses (>1 μJ), in agreement with the literature and the models. We observed that the charge emitted per incident photon pulse improves with increasing emitter height; however, for high-intensity incident photons, there is a limit characteristic and no further improvement is perceived. The estimated maximum quantum efficiency at about 1.3×10^{-7} –a high value considering that only about a millionth of the area is involved in the emission process. The quantum efficiency of the devices is better for arrays with taller emitters; however, for incident photon pulses of enough intensity the quantum efficiency characteristic converges and becomes flat.

Acknowledgments

The devices were made at the Microsystems Technology Laboratories of the Massachusetts Institute of Technology. This work was funded in part by the Defense Advanced Research Projects Agency / Microsystems Technology Office, grant N66001-11-1-4192 (Program Manager D Palmer). Any opinions, findings, and conclusions or recommendations expressed in this publication are those of the authors and do not necessarily reflect the views of the US Government and therefore, no official endorsement of the US Government should be inferred.

References

[1] Beckhoff B, Kanngießer B, Langhoff N, Wedell R, and Wolff H 2006 *Handbook of Practical Xray Fluorescence Analysis* (Berlin: Springer)
[2] Slack C M and Ehrke L F 1941 *J. Appl. Phys.* **12** 165
[3] Donath T, Pfeiffer F, Bunk O, Grünzweig C, Hempel E, Popescu S, Vock P, and David C 2010 *Invest Radiol.* **45** 445
[4] Graves W S, Kärtner F X, Moncton D E, and Piot P 2012 *Phys. Rev. Lett.* **108** 263904
[5] Bacci A, Ferrario M, Maroli C, Petrillo V, and Serafini L 2006 *Phys. Rev. ST Accel. Beams* **9** 060704
[6] Guo Q, Takahashi K, Saito K, Akiyama H, Tanaka T, and Nishio M 2103 *Appl. Phys. Lett.* **102** 092107
[7] Hommelhoff P, Sortais Y, Aghajani-Talesh A and Kasevich M A 2006 *Phys. Rev. Lett.*, **96** 077401
[8] Swanwick M E, Keathley P D, Fallahi A, Krogen P R, Laurent G, Moses J, Kärtner F X, and Velásquez-García L F 2014 *Nano Letters* **14** 5035
[9] Swanwick M E, Dong C D, Keathley P D, Fallahi A, Kärtner F X, and Velásquez-García L F 2014 *Tech. Digest of the 27th Int. Vac. Nanoelectronics Conf., Engelberg, Switzerland, July 6 – 10, 2014*, pp. 69 – 70
[10] Yalunin S V, Gulde M, and Ropers C 2011 *Phys. Rev. B* **84** 195426

Optimization and Experiment of Electromagnetic Energy Harvester by Using NdFeB Sputtered on High-aspect-ratio Corrugated Si

K Yamaguchi, Y Tanaka, T Fujita, N Takehira, K Sonoda
K Kanda, and K Maenaka
Department of EECS, Graduate School of Engineering, University of Hyogo
2167 Shosha, Himeji, Hyogo 671-2280, Japan

E-mail: er14a049@steng.u-hyogo.ac.jp

Abstract. This paper shows a design optimization of MEMS (Microelectromechanical Systems) electromagnetic energy harvester by using finite element method (FEM) analysis to obtain a maximum generated power. The electromagnetic energy harvester consists of the sputtered NdFeB magnet film on a high-aspect-ratio corrugated Si structure and a counter Au serpentine coil. The each dimension of the device is optimally designed by using FEM simulation. After the prototyping with MEMS fabrication process, the measurement results of the power generation are 2.65 mV of the induced electromotive force and 7.60 nW of the output power with optimum load resistance of 231 Ω at the vibration condition of 100 Hz, 294 $\mu m_{p\text{-}p}$.

1. Introduction

Recent years, wireless sensor networks have been attracted in many fields such as human healthcare and infrastructure monitoring systems. Vibration electromagnetic harvester is one of the expected power sources for these systems because it has (1) high current generation unlike to electrostatic or piezoelectric harvesters, (2) simple mass-spring structure with magnets and coils, and (3) no limit to reduce the gap between magnet and coil because there is no pull-in or sticking issue.

In our first generation of the electromagnetic harvester was fabricated on the trench etched Si structure and the NdFeB magnetic material was sputtered and polished to fabricate a magnetic stripe [1]. The harvester generated 760 nW at vibration condition of 100 Hz, 400 $\mu m_{p\text{-}p}$. To improve a generated power, we have established the optimization for device design e.g. air gap between magnet and coil, length of adjoining magnet areas etc. in the second generation device [2]. However, the further miniaturization of magnet causes poor quality of the NdFeB magnet because the sputtered magnet film is difficult to deposit in the narrow Si trench. In order to overcome the issue, we propose third generation electromagnetic harvester that consists of the sputtered magnet film not in the trench but on a high-aspect-ratio corrugated Si surface.

In this paper, we describe the optimal design of the electromagnetic harvester with NdFeB magnetic array on the corrugated Si microstructure and single layer serpentine-shaped coil. The optimum dimensions are calculated by finite element method (FEM) and numerical analysis. We also show the prototyping of novel harvester and its experimental results for power generation.

2. Electromagnetic harvester

The principle of power generation by the electromagnetic harvester is shown in figure 1. The harvester is composed with the magnet structure and the coil structure. Our magnet structure has the micro magnetic pole array that produces the micro distribution of the magnetic flux density on magnets surface. The coil structure is fabricated on the mass-spring structure. When the coil mass is excited by external vibration, the relative position of the coil and the magnet is changed, and then a magnetic flux density passing through the coil is also changed. The induced electromotive force is produced by this electromagnetic induction, and the induced current flows in the coil. The induced electromotive force, *V,* is the following equation:;

$$V = N\frac{d\Phi}{dx}\frac{dx}{dt} \tag{1}$$

where *N*, $d\Phi/dx$, and dx/dt denote the number of coil turns, magnetic flux changing ratio for position, and moving relative velocity between coil and magnet. From eq. (1), it is obvious that the much number of coil, fine pitch magnetic flux change and higher moving velocity are key issue for improving the electromotive force.

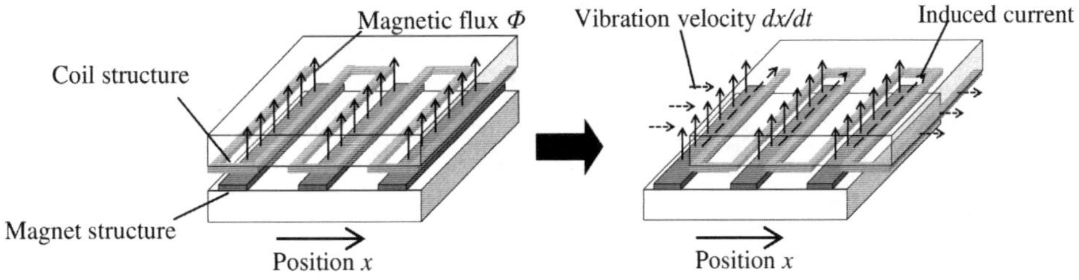

Figure 1. Principle of power generation

3. Optimal design of electromagnetic harvester

Figure 2 shows the structure of our electromagnetic harvester. The device size is $14 \times 14 \times 1$ mm^3, and the effective area for power generation is 9×9 mm^2. The magnet structure has the NdFeB magnet on a high-aspect-ratio corrugated Si surface (figure 2a). The coil structure has Au serpentine-shaped coil [3] with a single-layer coil wiring (figure 2b). Since there is no overlap area on the coil, the serpentine coil has very flat surface on the coil and has good capability to fabricate a fine patterned electrode. Thus, it is easy to decrease the air gap between the coil and the magnet array and increase the number of coil turns.

In order to maximize a generated power from the harvester, a structural optimization is required. We consider few device parameters for size optimization, i.e. corrugate-shaped Si trench pitch, *D*, line and space ration for the projection and the trench, *L:S*. By changing these parameters, we can estimate the total result for the harvesting energy that has a trade-off relationship on the (1) changing rate of the magnetic flux density of $d\Phi/dx$, (2) the impedance that proportional to the number of coil turns, *N,* and (3) the induced electromotive force, *V* that depends on *N* and $d\Phi/dx$. Therefore, it is possible to seek the optimal structure having a maximum generated power. The detailed optimal design flow has been described in references [2] and [3]. At first, we extracted the distribution of the magnetic flux density at the coil position that has a certain gap from the magnet surface (figure 2c) with magnet static field analysis on the FEM. To reduce the data number and to simplify the numerical calculation after the FEM analysis, we approximated the magnetic flux density distribution to a simple sinusoidal equation by using fast Fourier transform (FFT). Then, the induced electromotive force and the generated power for an arbitral external vibration are able to calculate with the approximation.
For the simulation condition, the interval of coil line, l_2, is a-half of the corrugate-shape trench pitch of *D*. The cross-sectional area of Au coil line is set to 6×20 μm^2 and Au resistivity is 2.21×10^{-8} Ωm. From the measurement results, the thicknesses of magnet films are 16 μm and 4 μm for top and bottom of the corrugated Si structure, respectively. And the coercive force of our NdFeB magnet is

0.9MA/m. The gap between coil and magnet array (figure 2c) is 30 μm as same as previous work's value. Figure 3 shows generated power for changing the trench pitch of *D* and the ratio of the projection and trench, *L:S*.

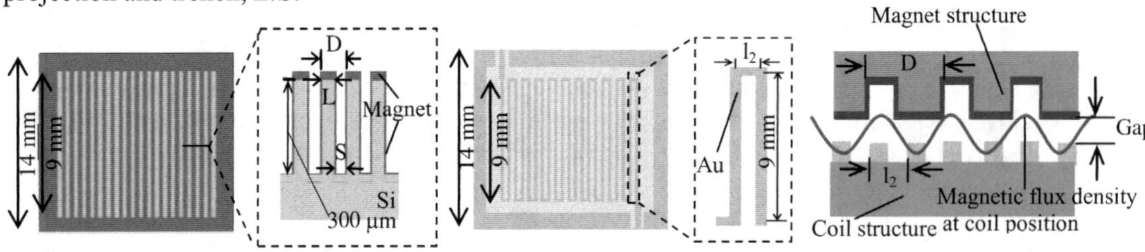

(a) Magnet structure (b) Coil structure (c) Cross-section of harvester

Figure 2. The detailed structure of the electromagnetic harvester. (a)Magnet structure, where *D* is trench pitch and *L:S* is ratio of line and space. (b) Structure of serpentine coil with folding lengths of l_2 and fixed length of 9 mm. (c) Cross sectional view of the harvester that has air gap.

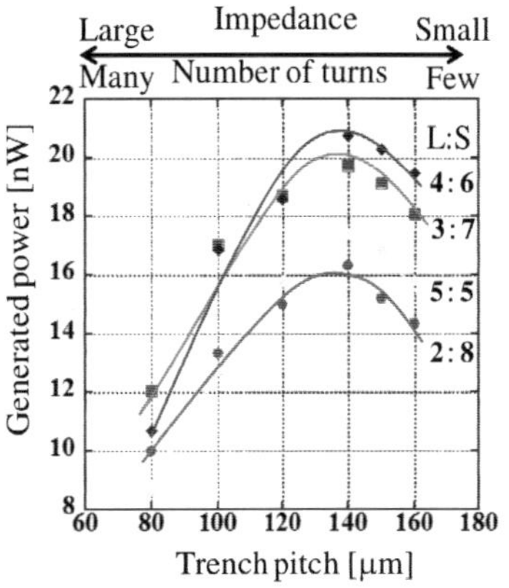

Figure 3. Simulation result with the 30 μm gap and the sinusoidal vibration (200 Hz, 200 $\mu m_{p\text{-}p}$). The optimal impedance and the optimal number of turns coil are varied with changing the trench pitch.

From the simulation results for 30 μm air gap, the optimal structure has 140 μm of trench pitch and 4:6 of *L:S* ratio. In that dimension, the number of coil turns is 64 and its impedance is 204 Ω. Finally, the generated power of 20.7 nW at the vibration condition of 200 Hz, 200 $\mu m_{p\text{-}p}$ are calculated. The result cannot compare with the previous work because of the different vibration condition, but we investigate the design optimization method.

In addition, we changed the gap between coil and magnet from 40 μm to 2 μm. The result of power estimation is shown in figure 5. All simulation results are after the optimization for each air gap. It is clearly to show that the narrower air gap and narrower trench pitch both are increasing the generated power.

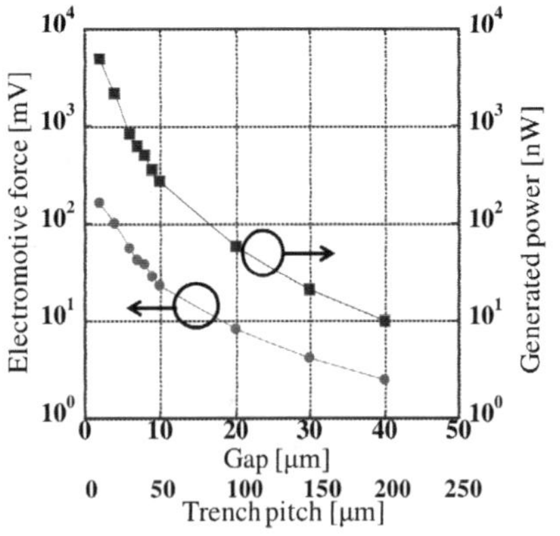

Figure 4. Simulation result for various air gap between the coil and the magnet. The vibration condition is fixed as the frequency of 200 Hz and the amplitude of $200\mu m_{p\text{-}p}$.

4. Fabrication of prototype and measurement

We fabricated the prototype that has no support spring but has magnet array and coil structure. We used optimal design dimension for 30 μm air gap, 140 μm trench width and 4:6 line and space ratio. Figure 5 shows the fabrication process flow of the prototype. Figure 5a shows fabrication of the magnet part. The micro corrugated shape on the Si wafer is fabricated by DeepRIE, and then sputtered the NdFeB/Ta multilayer film [4]. Then the magnet on the corrugated Si is magnetized with a pulse magnetization of 3.0 T magnetic field. As shown in the SEM image in figure 6, the NdFeB/Ta multilayer film is successfully covered on the top of the corrugated area. After the magnetization, the magnetic property is obtained as shown in figure 7. The magnet film shows 1.2 T magnetic flux density for the vertical axis.

The coil fabrication process is shown in figure 5b. The serpentine shaped spring was patterned with positive type photo resist (AZP-4620; AZ Electronic Materials plc, USA) on the Au/Cr seed layer for electroplating. The 20 μm width and 6 μm thickness Au coil is fabricated by the electroplating. Figure 8 shows the fabricated prototype of the coil and the magnet chip. The trench pitch of magnet and its ratio of L:S are 140 μm and 4:6, respectively. The coil impedance for the number of turns 64 is 231 Ω.

We performed vibration test. The coil chip was set above the fixed magnet chip with 30 μm constant air gap, and then vibrated by a shaker (ET-126B-1; Labworks Inc., USA). Since the coil mass is not supported by the spring but set on the shaker head, we can apply arbitrarily frequency and amplitude combination as the mass vibration. Figure 9 shows measurement apparatus to measure the electromotive force and applied acceleration of the mass, which measured by the accelerometer (MMA7361L; Freescale Semiconductor Inc., USA).

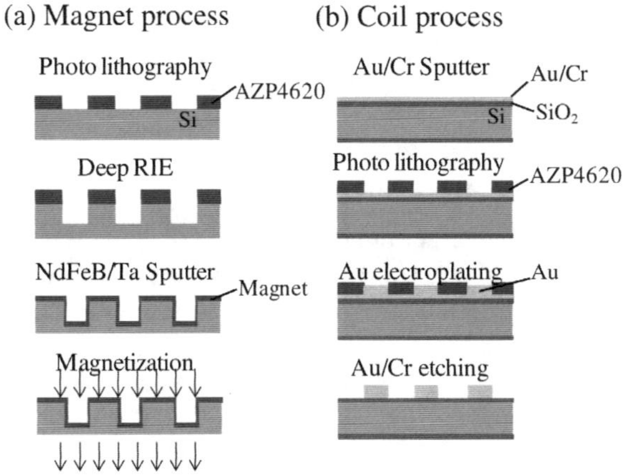

Figure 5. Process flow of prototype

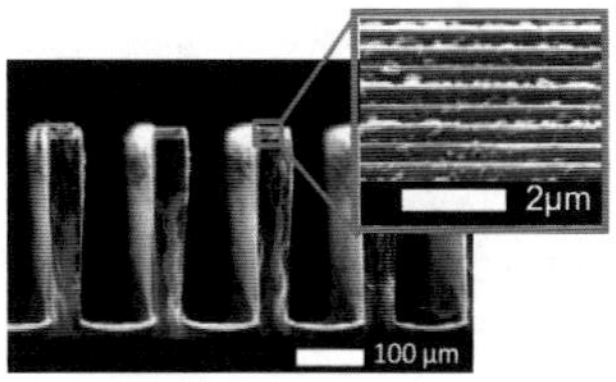

Figure 6. SEM image of the magnet on corrugated Si. Thickness of magnet is 16μm with 40 layers of NdFeB/Ta.

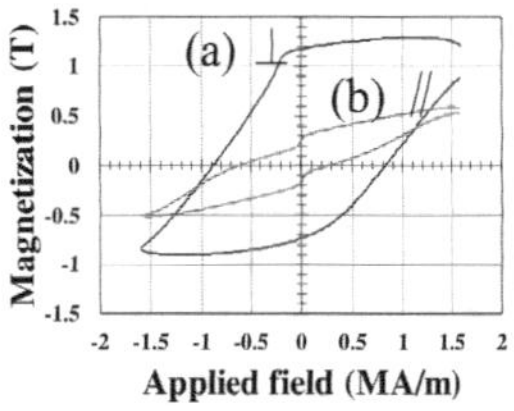

Figure 7. B-H hysteresis curve with VSM. (a) ⊥ is the vertical result to magnet surface. (b) // is the horizontal result to the magnet surface .

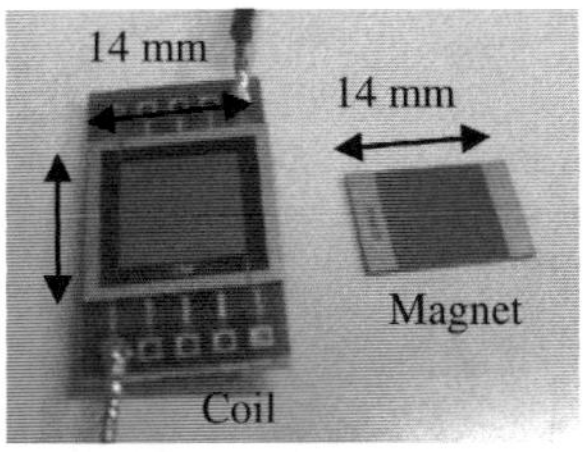

Figure 8. Prototype of the magnet chip and the coil chip

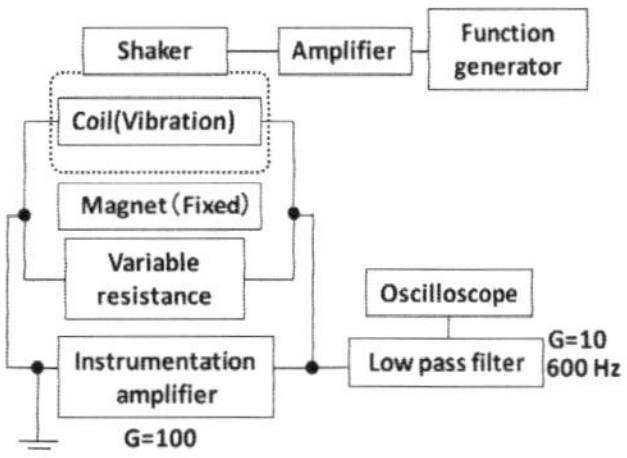

Figure 9. Measurement system for prototype harvester

In order to compare with our previous work, the coil chip was vibrated at the sinusoidal vibration of 100 Hz, 294 μm_{p-p}. Figure 10a shows the harvesting result of the measured electromotive force, power and applied acceleration to the mass, which measured by the accelerometer. The device generates 2.65mV electromotive force and 7.60 nW power with optimum load resistance of 231 Ω. On the same condition, we also simulated harvesting energy by using the FEM and numerical calculation method. Figure 10b shows the induced electromotive force of 2.99 mV and the generated power of 9.65 nW with same load resistance. These results from the measurement and simulation show the extremely high agreement. Therefore this optimization and design strategy could be useful for future design of high performance electromagnetic harvester.

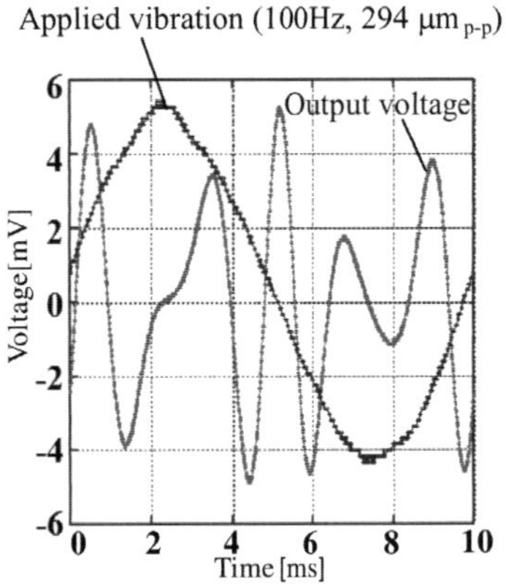

(a) Measurement

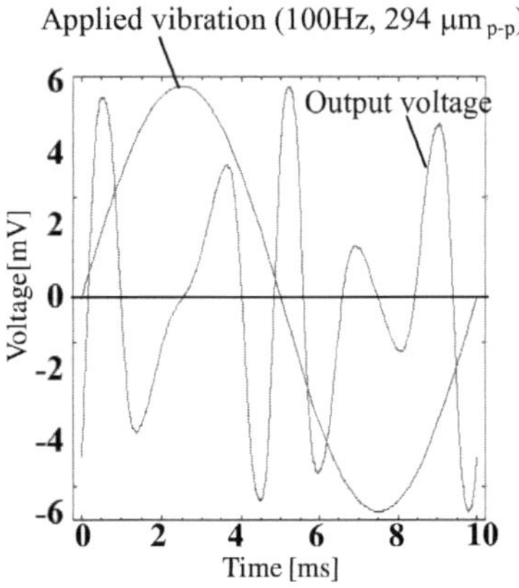

(b) Simulation

Figure 10. Result of power generation experiment. (a)measurement result and (b)simulation result at sine vibration (100 Hz, 294 μm_{p-p}). Regarding the waveform of power generation, the measurement extremely agree with the simulation.

5. Conclusion

We demonstrated the design optimization method for NdFeB electromagnetic vibration harvester with Si corrugated structure. The optimal structure for 30 μm fixed air gap has 140 μm trench width and 6:4 line and space ratio. From the measurement of the fabricated prototype, the harvested power of 7.60nW at the vibration of 100 Hz, 294 μm_{p-p} is obtained, which is about ten times large power from the previous work with same condition. We also investigated the numerical calculation method shows good agreement with the measurement data of the actual harvester. It will be a strong tool for the design optimization for the harvester.

References

[1] Miki S et al 2012 *Tech. Dig. Int. Conf. on MEMS2012 (Paris)* pp 1221-24
[2] Tanaka Y, et al 2013 *Proc. The 2nd Int. Workshop on e-Wise (Beijing)* pp 1656-58
[3] Tanaka Y et al 2012 *Tech. Dig. Int. Conf. on PowerMEMS2013(London)*
[4] Jiang Y G et al 2011 *Fabrication of a Vibration-driven Electromagnetic Energy Harvester With Integrated NdFeB/Ta Multilayered Micro-magnets* Journal of Micromechanics and Microengineering vol. 21 095014

Myocardial Cell Pattern on Piezoelectric Nanofiber Mats for Energy Harvesting

X Liu[1,2], X Wang[1,2], H Zhao[3], Y Du[3]
[1]Institute of Microelectronics, Tsinghua University, Beijing, China
[2]Tsinghua National Laboratory for Information Science and Technology, Beijing, China
[3]Department of Biomedical Engineering, Tsinghua University, Beijing, China

E-mail: wxh-ime@tsinghua.edu.cn

Abstract. The paper presents *in vitro* contractile myocardial cell pattern on piezoelectric nanofiber mats with applications in energy harvesting. The cell-based energy harvester consists of myocardial cell sheet and a PDMS substrate with a PVDF nanofiber mat on. Experimentally, cultured on specifically distributed nanofiber mats, neonatal rat ventricular cardiomyocytes are characterized with the related morphology and contraction. Previously, we have come up with the concept of energy harvesting from heart beating using piezoelectric material. A bio-hybrid energy harvester combined living cardiomyocytes, PDMS polymer substrate and piezoelectric PVDF film with the electrical output of peak current 87.5nA and peak voltage 92.3mV. However, the thickness of the cardiomyocyte cultured on a two-dimensional substrate is much less than that of the piezoelectric film. The Micro Contact Printing (μCP) method used in cell pattern on the PDMS thin film has tough requirement for the film surface. As such, in this paper we fabricated nanofiber-constructed PDMS thin film to realize cell pattern due to PVDF nanofibers with better piezoelectricity and microstructures of nanofiber mats guiding cell distribution. Living cardiomyocytes patterned on those distributed piezoelectric nanofibers with the result of the same distribution as the nanofiber pattern.

1. Introduction

Making good use of advanced smart materials in the field of biological and biomedical applications is promising to construct various biological robotics or "bio-bots" enabling sensing [1], drug delivery [2], tissue engineering [3] and energy transduction [4]. Combining biological entities such as DNA, cells or tissues with soft materials can yield soft bio-bots with the ability to dynamically sense and adapt to environmental cues or themselves cues with or without external stimuli triggered. Mechanical interactions are fundamental to cellular biology and physiology. Recent advances in cardiac muscle tissue engineering have yielded dense tissues that form a syncytium, with the coordinated propagation of electrical signals and synchronous contraction of engineered muscle [5]. This advantageous property has helped to produce machines that include self-assembling microelectromechanical-system-based cantilevers [6], 2D biohybrid "muscular thin films" [7], and "crab-like" robots [8]. These systems were powered by applied electric field stimulation or spontaneous contraction of engineered cardiac muscle, which have also been used as power sources for loco-motive machines or implanted microdevices.

Previously, we have come up with the concept of energy harvesting from heart beating using piezoelectric material [9]. Figure 1A illustrates the potential applications of cell-based energy harvesting which tends to adhere to the heart tissue to supply the pacemaker or other implanted microdevices. According to the filament sliding mechanism, the cardiomyocyte drives the piezoelectric fiber to bend periodically, as shown in the meshed of Figure 1A. A bio-hybrid energy harvester combined living cardiomyocytes, PDMS polymer substrate and piezoelectric PVDF film with the electrical output of peak current 87.5nA and peak voltage 92.3mV, as shown in Figure 1B. However, the bending angle of the bio-hybrid film just attains less than 20°, which is much lower than expected. There are several challenges existing in the design and fabrication of the device. The thickness of the cardiomyocyte cultured on a two-dimensional substrate is much less than that of the piezoelectric film. The Micro Contact Printing method used in cell pattern on the PDMS thin film has tough requirement for the film surface [7]. As such, we fabricated nanofiber-constructed PDMS thin film to realize cell pattern, due to PVDF nanofibers with better piezoelectricity and microstructures of nanofiber mats guiding cell distribution. The conceptual drawing of the cell-based energy harvester in Figure 1C consists of myocardial cell sheet and a PDMS substrate with a PVDF nanofiber mat on. The PVDF nanofibers were electrospun with specific collectors used.

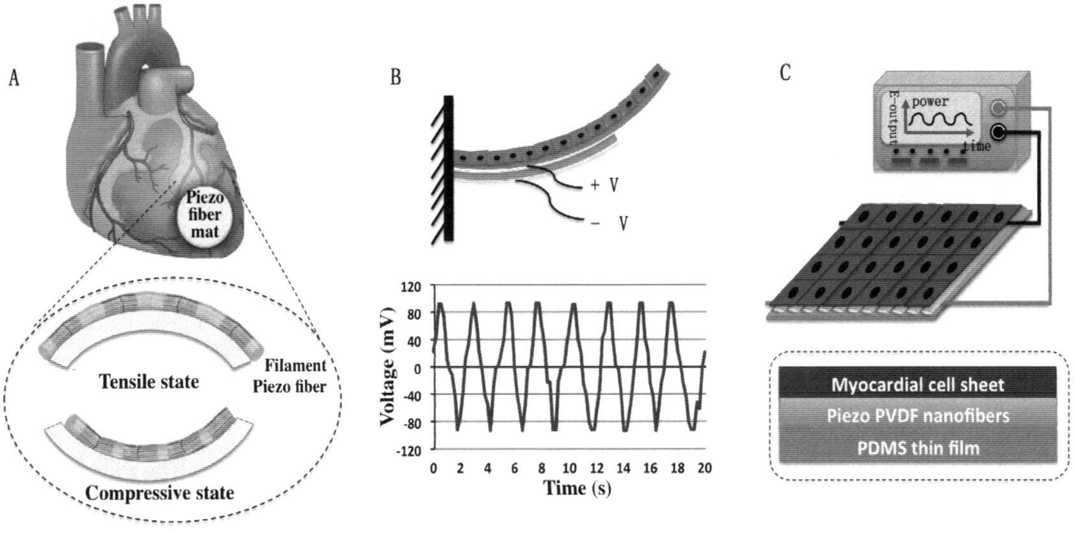

Figure 1. (A) Schematic of cell-based energy harvester with piezoelectric fiber mat used on heart tissue. The filament sliding mechanism causes the attached piezoelectric fiber bending; (B) previous work on bio-hybrid energy harvester based on piezoelectric PVDF film and its voltage output (C) schematic of device operating principle and the device structure.

2. Experimental

To construct the structure of the piezoelectric bio-bot, we applied far-field electrospinning technology to produce and used specific grounded gap collectors collecting electrodes to settle monolayer of PVDF piezoelectric nanofibers on a semi-cured PDMS thin film with two gold microwires partially imbedded in. The schematic of the device design is outlined in Figure 1C, and shows several key features. First, the piezoelectric nanofiber arrays are partially imbedded into the underlying PDMS substrate. Second, two gold microwires, as the device electrodes, are sandwiched between the nanofiber arrays and the PDMS film. We observed both nanofiber patterns for their ability to support cardiomyocytes and provide a compliant settlement for myocardial cell sheet contraction. In Figure 2A, a pair of parallel electrodes designed as the gap collector was grounded to collect uniaxially aligned nanofibers. When one end of the electrospun nanofiber lies across one of the two electrodes, the other end of the fiber is enforced to settle down across the other electrode under the electrostatic force between the electrospun nanofiber and the grounded electrodes. So the monolayer of PVDF nanofibers was formed and self-ordered on the PDMS film substrate with their ends across the collecting electrodes (Figure 2B). Scanning electron micrographs yield quantitative information on the degree of alignment, as shown in Fig. 2C. The nanofiber alignment is with up to 75% perpendicular between nanofibers and the collecting electrode. In contrast, randomly distributed straight nanofibers were realized by concentric ring electrode (Figure 2D) and the statistical result of the angle is uniform distributed (Figure 2E&2F).

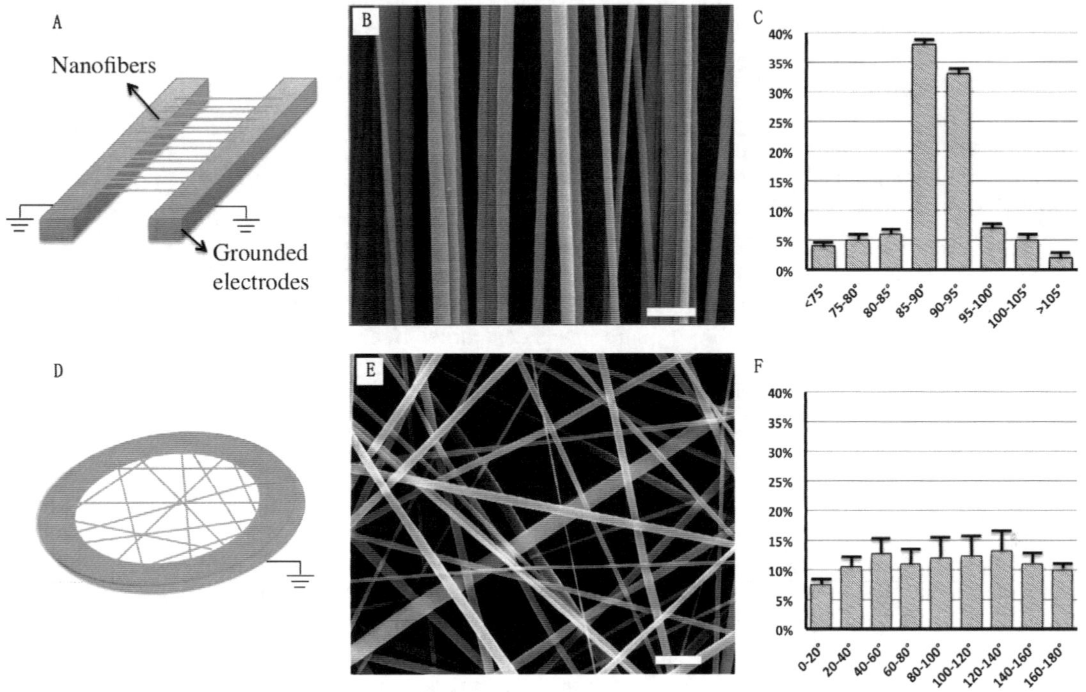

Figure 2. (A, D) Schematic illustrations of grounded electrodes used for collecting electrospun piezoelectric nanofibers were composed of parallel electrode pair (A) and concentric ring electrode (D). (B, E) SEM images of aligned (B) and random (E) nanofibers collected on PDMS films. (C, F) The statistical angles between nanofiber and the collecting electrode. **(Scale bar in B, E is 4μm.)**

To extract quantitative information about cardiomyocyte contractions, the piezoelectric response of PVDF nanofibers must be fully characterized. We analyzed crystalline structure and dipole orientation of the nanomaterial. X-ray diffraction (XRD) patterns provide information on the long-range order and the crystalline structure of electrospun PVDF nanofibers, shown in Figure 3. The results indicate that the overall crystallinity of the material in the aligned arrays is as the same as that of in the random networks. The nanofiber exhibits a prominent peak at $2\theta=20.6°$, corresponding to the (110) reflection of the β phase formation. A distinguishing feature of the electrospun nanofibers is that the 30 kV high voltage significantly enhances the fraction of the polar β-phase.

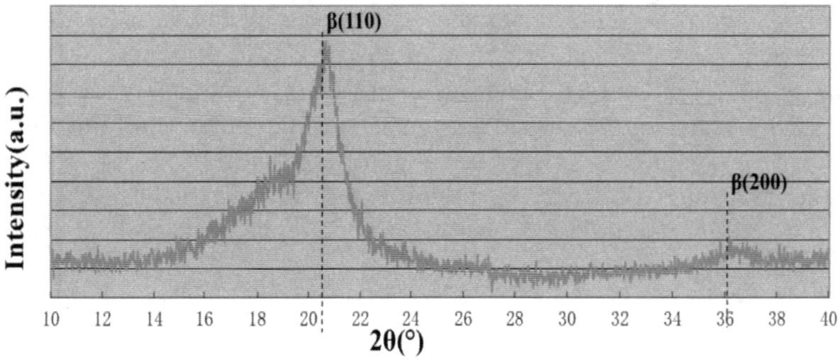

Figure 3. The XRD pattern of electrospun PVDF nanofiber mats which shows clearly the β(110) phase.

3. Results and discussion

We engineered two kinds of 2D myocardial tissue on fibronectin-coated monolayer PVDF nanofibers: anisotropic (Figure 4A) or isotropic (Figure 4B) cell sheets by passive seeding of dissociated ventricular cardiomyocytes. The loaded cells exerted traction forces on the nanofibers via integrin attachments to compact the cells into a 2D muscle strip over time. For both kinds of nanofiber pattern, the nanofibers served as physical cues for the inter- and intracellular organization of cardiomyocytes into a tissue and the uniaxial coupling of sarcomere ensembles over length scales from micronmeters to centimeters. Synchronized actuation of the array of contractile cardiomyocytes is critical for optimal functionality and can be achieved by their electrical coupling through gap junctions. From the 1st day to the 5th day of cell culture, the cardiomyocytes loaded on the aligned nanofibers are adhering to the nanofibers gradually and forms the uniaxial cell pattern eventually. Compared with Figure 4B, the 5th-day cell image in Figure 4A shows uniaxially aligned living cardiomyocytes on parallel nanofibers. The muscle strips therefore displayed functional behavior characteristic to physiological skeletal muscle. Anisotropic 2D myocardium had uniaxial alignment of cell bodies (Figure 4A). The 3D deflection of the bio-bot film depended on the direction of cell alignment relative to the nanofiber pattern. Instead of relying on external signal simulation, the contractions of the myocardial cell sheets can be controlled and paced via spontaneous muscle contraction. This bio-inspired design mimics the in vivo musculoskeletal arrangement in which force transmission occurs from a contracting muscle to bone through a connecting tendon.

Results presented here indicate that aligned PVDF nanofibers can be formed into flexible, settled monolayer, by use of electrospinning onto a specifically designed gap collector. The process yields alignment at both the level of the fibers and the cardiomyocytes, thereby enabling high piezoactive β-fraction without further processing and excellent contractile response, respectively. At one hand, soft robotic film can form a biocompatible interface with cells to act as sensitive extracellular probes to detect minute cellular deformations. At other hand, this can generate electricity to supply implanted devices, acting as biogenerator. As expected, the stiffer PDMS structures offered a greater resistance to bend; thus, films with higher elastic moduli exhibited a lower deflection in response to passive tension forces exerted by the muscle strips.

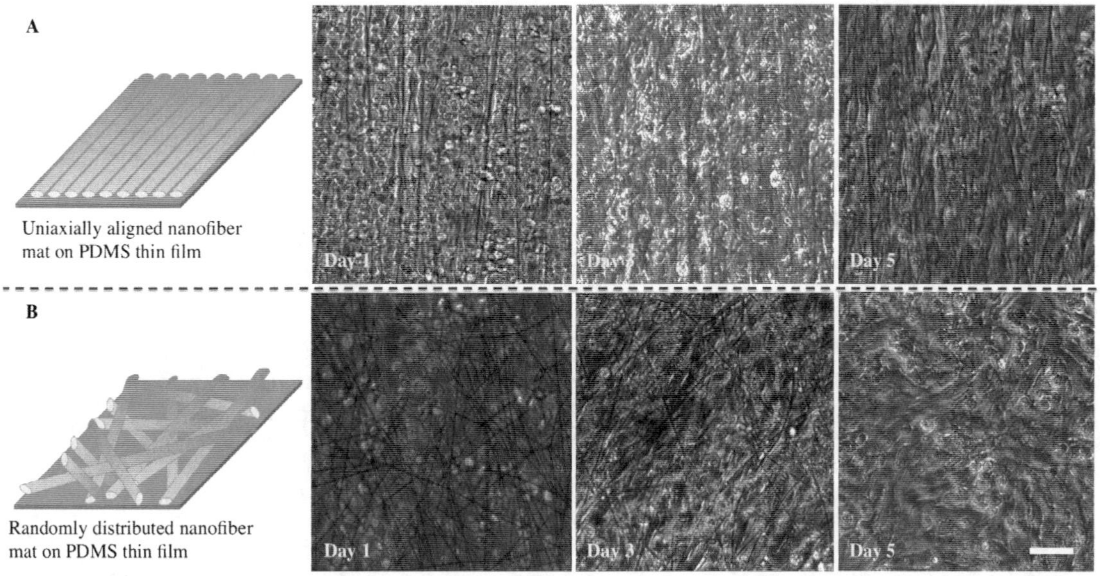

Figure 4. Microstructures of 2D cardiomyocyte sheets with piezoelectric PVDF nanofiber mats: with uniaxial alignment (A) and random pattern (B), respectively. (A) From the 1st day to the 5th day of cell culture, the cardiomyocytes loaded on the aligned nanofibers are adhering to the nanofibers gradually and forms the uniaxial cell pattern eventually. (B) The growing process of the cardiomyocytes is as the same way as (A), but causes the random cell pattern. **(Scale bar: 50um)**

4. Conclusion

To summarize, we demonstrate a cell pattern method on piezoelectric nanofiber mats with applications in energy harvesting. From sarcomeregenesis to the integration of the biochemical and electrical networks, engineered muscle remains an attractive method for building actuators and powering devices from the micro to macro scales. Our research shows a feasible approach to scavenge the biomechanical energy inside the body, such as heartbeat, blood flow, muscle stretching, or even irregular vibration. The new epidermal bioenergy paradigm thus holds considerable promise as a viable autonomous power source for implantable self-powered electronics.

5. Acknowledgements

This work is support by grants from Tsinghua National Laboratory for Information Science and Technology (No. 042003130).

6. References

[1] C. L. Schmidt and P. M. Skarstad 2001 *Journal of Power Sources* **97–98** 742
[2] L. R. Hochberg, M. D. Serruya, G. M. Friehs 2006 *Nature* **442** 164
[3] R. A. Roeder, G. C. Lantz, L. A. Geddes 2001 *Biomedical Instrumentation and Technology* **35** 110
[4] S. M. Kurtz et al. 2010 *Pacing and Clinical Electrophysiology* **33** 705
[5] Vunjak-Novakovic G, et al. 2010 *Tissue Eng Part B Rev* **16** 169
[6] Xi J, Schmidt J J, Montemagno C D 2005 *Nat Mater* 4(2): 180
[7] Feinberg A W, et al. 2007 *Science* **317** 1366
[8] Kim J, et al. 2007 *Lab Chip* **7** 1504
[9] X. Liu et al., *Proc. MEMS 2014*, pp. 159-162.

Coated Porous Si for High Performance On-Chip Supercapacitors

K. Grigoras, J. Keskinen, L. Grönberg, J. Ahopelto, and M. Prunnila

VTT Technical Research Centre of Finland, P. O. Box 1000, FI-02044 VTT Espoo, Finland

Email: mika.prunnila@vtt.fi

Abstract. High performance porous Si based supercapacitor electrodes are demonstrated. High power density and stability is provided by ultra-thin TiN coating of the porous Si matrix. The TiN layer is deposited by atomic layer deposition (ALD), which provides sufficient conformality to reach the bottom of the high aspect ratio pores. Our porous Si supercapacitor devices exhibit almost ideal double layer capacitor characteristic with electrode volumetric capacitance of 7.3 F/cm^3. Several orders of magnitude increase in power and energy density is obtained comparing to uncoated porous silicon electrodes. Good stability of devices is confirmed performing several thousands of charge/discharge cycles.

1. Introduction

Electrochemical double layer capacitors (EDLC), commonly called as supercapacitors, offer an option for short term high power storage [1, 2]. These energy storage elements could find many applications in the field of energy harvesting and MEMS if reliable integrated elements could be realized. High surface area carbon materials (as activated carbon) are typically used as supercapacitor electrodes, but those are challenging to integrate on chip with silicon devices. Supercapacitor elements based on Si nanostructures could be an optimal solution here, and there are numbers of reports dealing with silicon nanowires and similar structures [3], but these devices exhibit relatively low capacitance.

Porous silicon (PS) (see figures 1a and b) is an attractive candidate due to ease fabrication and integration possibility, but its resistance is relatively high and it suffers from high reactivity - features resulting in low power density and poor stability [4, 5]. Coating of PS has been recognized as a possible route to reduce the resistance and increase the stability [4-6], but even for coated electrodes the performance has still been several orders of magnitudes below that of carbon based supercapacitors. This is due to the fact that the power density is eventually limited by the resistance of the complex Si nanostructure and, therefore, in addition of the stability the coating has to provide high conductivity. In this work, we use atomic layer deposition (ALD) technique to get a conformal (see figure 1c), low resistance and stable coating on the complex porous silicon surface. Our results demonstrate that TiN coated PS is a promising material for micro supercapacitors.

Journal of Physics: Conference Series **557** (2014) 012058 doi:10.1088/1742-6596/557/1/012058

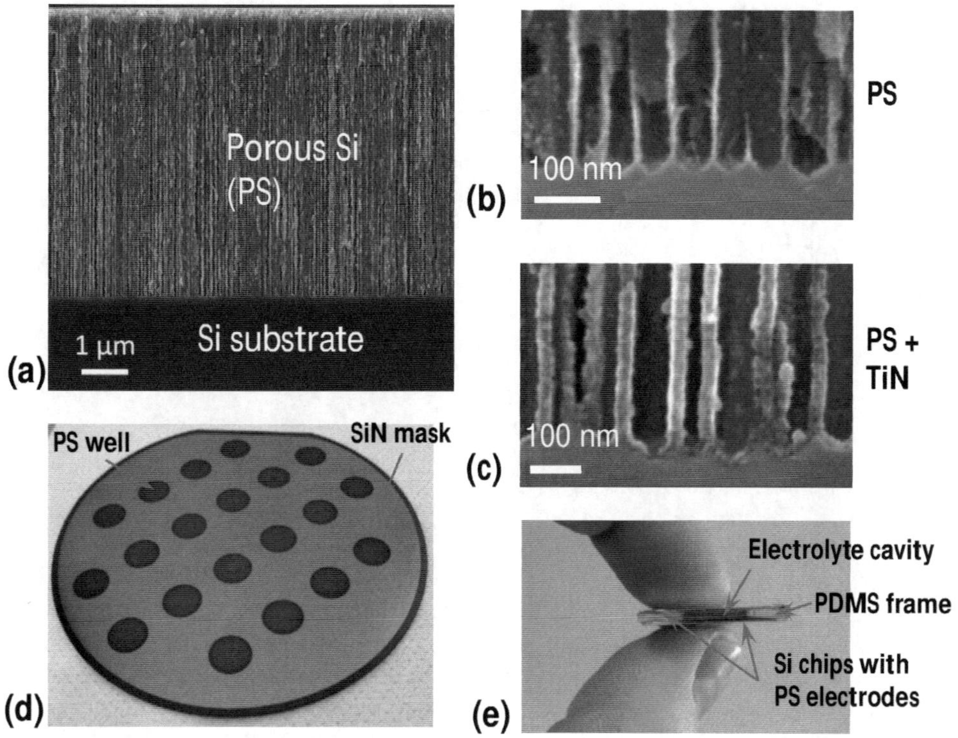

Figure 1. (a) Cross-sectional SEM image of 7 um thick PS layer. (b) High magnification SEM image of PS. (c) Optical image of a 150 mm Si wafer with PS wells for supercapacitor electrodes. (d) High magnification SEM image of PS matrix with a ~10 nm atomic layer deposited TiN layer. (e) Test bench device to characterise the properties of PS/TiN electrodes.

2. Experimental

Porous silicon was prepared by electrochemical etching of highly-boron-doped silicon wafers (1 - 4 mΩ-cm) in a 1:4 solution of 50 % aqueous-HF and ethanol. A whole 150 mm diameter wafer with a patterned silicon nitride mask was processed resulting in 20 identical porous areas 1.5 cm2 each (see figure 1 d). Approximately 6 μm thick porous layer was prepared during 30 minutes anodization step at 5.3 mA/cm2 current density. Porosity of the layer was about 88% as evaluated from gravimetrical measurements. Pore diameter was about 50 nm (see figure 1b). The TiN layer was deposited by ALD using TiCl and ammonia as precursors and nitrogen as carrier and purging gas. As can be observed from figure 1c, close to ideal conformality was achieved.

As a test bench for material study we have been using two PS/TiN electrodes with a PDMS frame that forms a cavity for the electrolyte, as shown in figure 1(e). The 2 mm thick PDMS frame served as a separator and as a reservoir for electrolyte, which was 0.5 M of tetraethyl-tetrafluoroborate in propylene carbonate (TEABF4/PC).

3. Results and discussion

Galvanostatic charge/discharge curves of coated porous Si supercapacitors at constant current are shown in figure 2. Almost symmetric triangular shape of voltage response indicates good performance as an EDL capacitor. The efficiency of the cell is about 82% and evaluated capacitance is 3.9 mF.

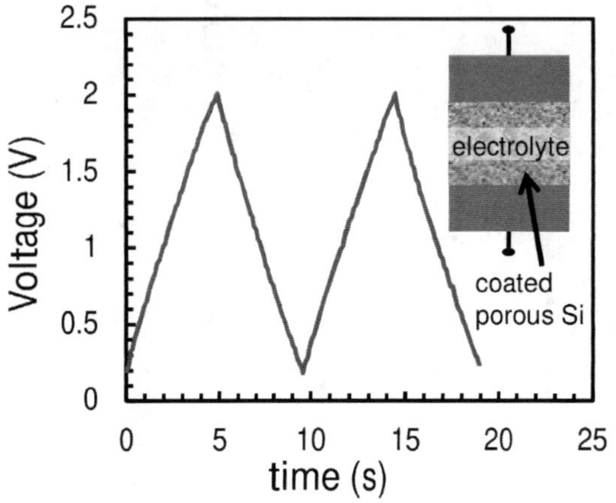

Figure 2. Galvanostatic charge/discharge of TiN coated porous Si supercapacitor structure at current of 1 mA. Current polarity changes at the instances where the time derivative of the measured voltage changes the polarity.

Electrochemical impedance spectroscopy (EIS) was used to examine the frequency characteristics and equivalent series resistance (ESR) of our devices. We found close to ideal capacitive behaviour and ESR value of 18 Ω, which is very close to the values obtained from galvanostatic measurements (17 Ω). The ESR mainly arises from the 2 mm thick electrolyte region in between the coated PS electrodes and only ~1 Ω or less comes from the electrodes themselves.

Good performance as EDL capacitor was confirmed also by cyclic voltammetry (CV) measurements. The CV curves of figure 3 exhibit almost ideal rectangular shape – a feature that has not been observed for porous silicon based devices before [4, 5, 6]. Such shape is an indicator of a very low ESR of our electrodes due to the conductive coating. From figure 3 we also can observe that the capacitance retention behavior is excellent. During the 5500 scans the CV characteristics change very little. In fact, if we inspect only the capacitance (at 1 V) after first drop/transient during the first 200-300 cycles the capacitance slowly approaches the initial value.

To compare our devices with other results, the volumetric values of specific energy and specific power were calculated, taking into account total porous volume of both electrodes. Our obtained energy value of 1 mWh/cm3 is comparable with the best results for on-chip capacitors and power value of 7.3 W/cm3 is at least order of magnitude larger than other results for silicon (and porous silicon) based supercapacitors.

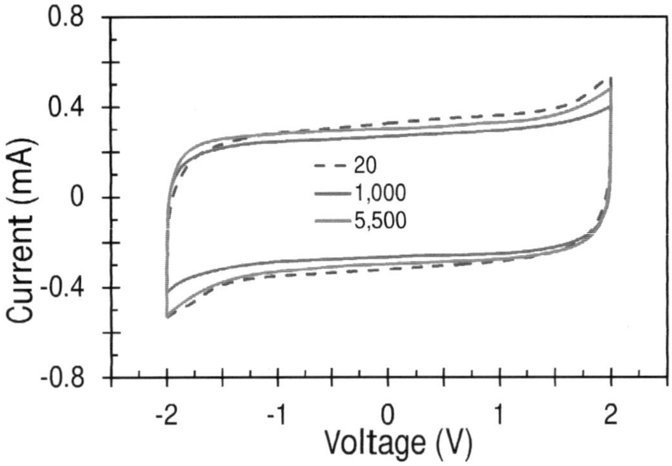

Figure 3. Cyclic voltammetry (CV) curves of TiN coated porous Si supercapacitor structure at 100 mV/s voltage scan rate. Few CV curves during 5500 cycles shown.

4. Conclusions

Un-coated pristine porous silicon cannot be used as electrode material for supercapacitors due too large reactivity of the surface and poor overall conductivity. Even in highly doped material the resistivity of the porous Si nanostructures significantly reduce the power extraction. TiN coating by ALD greatly improves the electrode performance by increasing the conductivity and passivating the surface. ALD is the best method for the coating as it provides the required conformality and tuneability for the coating layer parameters. Fabricated TiN coated porous Si supercapacitor electrodes show good figures of merit: electrode volumetric capacitance up to 7.3 F/cm3, power density of 32 W/cm3 and energy density of 1 mWh/cm3. Together with the demonstrated high stability these values pave the way for on-chip supercapacitor technologies that can be integrated with MEMS energy harvesters.

Acknowledgments

This work has been partially funded by the Academy of Finland through the Finnish Center of Excellence in Atomic Layer Deposition and Tekes - The Finnish Funding Agency for Innovation.

References

[1] P. Simon and Y. Gogotsi, “Materials for electrochemical capacitors”, *Nature Materials,* Vol. 7 (2008), pp. 845-854

[2] Y. Zhang et al, “Progress of electrochemical capacitor electrode material: A review”, *Int. J. Hydrogen Energy, Vol. 34* (2009), pp. 4889-4899

[3] J. Alper, M. Vincen, C. Carraro, R. Maboudian, “Silicon carbide coated silicon nanowires as robust electrode material for aqueous micro-supercapacitor”, *Appl. Phys. Lett.*, Vol. 100 (2012), pp. 163901-3

[4] S. Desplobain, G. Gautier, J. Semai, L. Ventura, and M. Roy, “Investigation on porous silicon as electrode material in electrochemical capacitor”, *Phys. Stat. Sol. C*, Vol. 4 (2007), pp. 2180-2184

[5] L. Oakes *et al.*, “Surface engineered porous silicon for stable, high performance electrochemical supercapacitors”, *Sci. Rep., Vol.* 3 (2013), pp. 3020-7

[6] S.E. Rowlands, R.J. Latham and W.S. Schlindwein, “Supercapacitor devices using porous silicon electrodes”, *Ionics Vol. 5* (1999) pp. 5 144-149.

A novel silicon/carbon nanocomposite anode for high performance lithium-ion micro-battery

X Z Li[1,2], X H Wang[1,2], S W Li[1,2], X P Qiu[3]

[1]Institute of Microelectronics, Tsinghua University, China
[2]Tsinghua National Laboratory for Information Science and Technology, China
[3]Department of Chemistry, Tsinghua University, China

Email: wxh-ime@mail.tsinghua.edu.cn

Abstract. This paper reports a novel nanocomposite anode for lithium-ion battery, with high initial specific capacity (1200mAh/g) and good capacity retention (850mAh/g remaining after 30 cycles). Sufficient silicon nanoparticles (SiNPs) at anode make a significant contribution to specific capacity increase. Moreover, nano void space between SiNPs and carbon scaffold provide enough space for expansion and contraction of SiNPs during the process of lithium ion intercalation and deintercalation to ensure a long cycle life. The porous carbon scaffold is obtained from Si/SiO_2-templated SU-8 photoresist. As such, this design and fabrication makes it possible to implement direct prototyping of three dimensional (3D) micro-battery on chip.

1. Introduction

Lithium-ion secondary batteries are widely considered as the most useful devices for electrochemical energy storage in the 21st century. Because of their high gravimetric and volumetric capacity, Li-ion batteries have a wide application range from portable electronics to electric vehicles and other renewable energy [1]. With the ever increasing power requirements for applications, the Li-ion batteries are in desperate need of anode materials that can yield a much higher specific capacity than the traditional graphite anode (theoretical capacity of 372 mAh/g). Silicon is one of the most attractive anode materials for next-generation Li-ion batteries because of its natural abundance, low discharge potential, and most importantly, high theoretical capacity (4200 mAh/g in a composite of $Li_{4.4}Si$), which is an order of magnitude beyond that of conventional graphite anodes [2]. Unfortunately, silicon undergoes large volume change (over 300%) during lithium-ion intercalation and deintercalation [3]. The volume change can lead to anode structure failure and electrical contact loss, and can also cause continuous consumption of the active material and Li-ion for reformation of solid electrolyte interphase (SEI) layers [4]. As a consequence, the electrode suffers rapid capacity decay.

Recent works have shown that silicon nanostructures can effectively solve this limitation. A number of silicon nanoscale morphologies have been investigated to reduce the structural degradation due to the volume change during cycling and therefore to overcome the mechanical instability in bulk Si. These nanoscale morphologies, including thin films [5, 6], nanowires/nanotubes [7, 8] and nanoparticles [9-11], facilitate relaxation of the stress associated to the intercalation/deintercalation process and thus demonstrating high capacities in initial cycles. However, after long cycles, most of these all-Si anodes still demonstrate obvious capacity fade due to silicon's separation from current collector. Many materials with good conductivity have been tested to stabilize Si-collector contact.

Among these efforts, anodes of nano-Si/C demonstrate the most outstanding performance. Carbon is effective not only in enhancing the electronic conductivity but also in stress relief due to its lightweight and ductile nature. A lot of promising results in terms of cycle stability have been obtained by designed nanocomposites of silicon and various carbon sources, such as graphene, PVDF, pitch, acetylene and polydopamine [12]. Some of these Si/C nanocomposite anodes also introduce suitable networks of interconnected pores in order to further buffer the volume change of the silicon. However, none of these techniques are compatible with standard microfabrication protocols.

In this work, a novel Si/void/C nanocomposite anode is designed and fabricated. Aiming at the direct prototyping of 3D micro-battery on chip, we use SU-8 photoresist as the carbon source to construct a scaffold to support SiNPs for the first time. The SiNPs make the main contribution to the high specific capacity, and the carbon scaffold from SU-8 builds an efficient conductive pathway. With rationally designed void space to accommodate silicon expansion and contraction during cycling, the anode maintains a stable structure and achieves a long cycle life.

2. Design and fabrication

The composite of silicon and carbon is gaining more and more attentions in the search of new high-performance Li-ion battery anodes. The limitations of carbon-based or all-silicon-based anodes could be addressed by designing and controlling the microstructure of Si/C composite anode, with the help of nanoscale fabrication techniques. The principle is based on the fact that graphitic carbon is able to release the stress generated by expansion and contraction of silicon, meanwhile the high conductivity of carbon can make up the electrical connection failure caused by silicon volume change.

In this work, we use SiNPs as the main active materials and SU-8 photoresists as the carbon source, which consist of epoxy resin SU-8, triaryl sulfonium salt and organic solvent. When exposed to ultraviolet light, the resin will form a highly structured cross-linked matrix. Moreover, SU-8 can retain micro patterns after high-temperature carbonization, which has been widely utilized in carbon microelectromechanical systems (C-MEMS) and micro power devices, such as micro supercapacitors [13] and three-dimensional micro batteries [14]. The good thermal stability and chemical resistance of SU-8 shown in these researches inspire a brand new way to fabricate on-chip batteries with Si/C composite anode, using a Si/SiO_2-templated synthetic procedure.

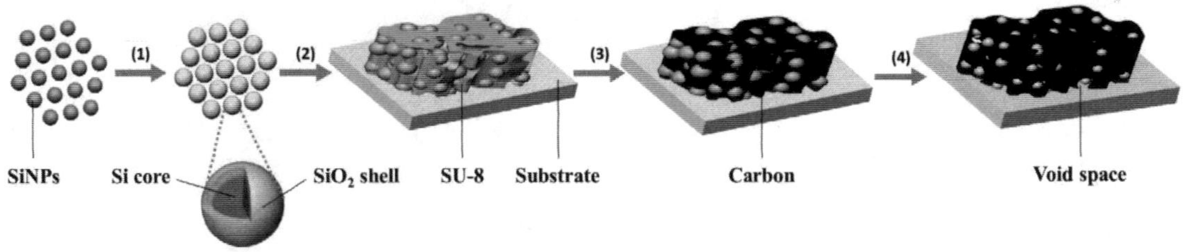

Figure 1: Schematic of the design and fabrication of the Si/void/C nanocomposite anode. 1) TEOS oxidation process to generate SiO_2 sacrificial layer; 2) Mixing Si/SiO_2 nanocomposite particles with SU-8 photoresist and spin-coating on Si substrate; 3) Pyrolysis process to carbonize SU-8; 4) Etching away SiO_2 layer, forming void space between carbon scaffold and SiNPs.

The scheme of the overall design and fabrication of the Si/void/C nanocomposite anode is shown in Fig.1. In the 1st step, SiNPs (30~50 nm in diameter) coated by SiO_2 layer are obtained by typical Stöber method, with hydrolysis and condensation of tetraethoxysilane (TEOS) in a water-alcohol-ammonia medium [15]. The thickness of the SiO_2 sacrificial layer and therefore the void space size can be easily controlled by controlling the coating time, PH, and TEOS concentration. In order to accommodate the maximum volume expansion (300%) of each SiNPs, the thickness of sacrificial coating layer should be over 10nm. In following experiments, the thickness of the SiO_2 layer is controlled to be about 10~15 nm, considering size variation. In step 2, the Si/SiO_2 particles are dispersed in SU-8 photoresist to form Si/SiO_2-templated SU-8 composite, where SU-8 serves as the precursor of subsequent carbon scaffold. Considering both specific capacity and stability of the electrode, the ratio of each component in this step is 20wt% for Si/SiO_2 particles and 80wt% for SU-8. After spin-coated on silicon substrate, the Si/SiO_2-templated SU-8 composite is exposed under ultraviolet light, forming highly structured cross-linked polymer matrix. In step 3, the polymer contained in the materials is pyrolyzed in nitrogen atmosphere at 900℃ for 2h. And the SU-8 is turned into carbon scaffold to provide a considerably strong mechanical support for the particles. At last, to remove the SiO_2 sacrificial layer, the $Si/SiO_2/C$ composite is etched in HF aqueous solution and the space once occupied by the nano SiO_2 layer becomes void space.

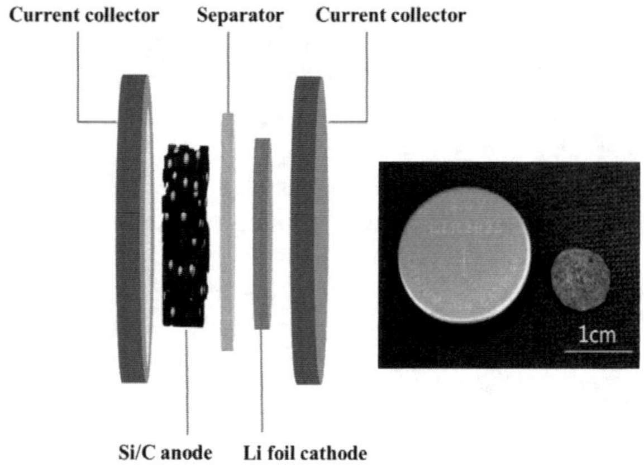

Figure 2. a) Packaging diagram of Li-ion battery: Si/void/C nanocomposite as the anode; Li foil as the cathode.; b) Photograph of Si/C composite anode (right) and packaged testing cell (left).

The working electrodes are fabricated using a typical aqueous slurry method. Si/void/C nanocomposite electrode is prepared by homogeneously mixing appropriate amount of Si/void/C nanocomposite, carbon black and carboxymethyl cellulose sodium (CMC) binder to form a slurry, with a mass ratio of 8:1:1. The mass loading of silicon in the mixture is kept at 35%. The viscous slurry is then cast onto a copper foil and dried at 80 ℃ under vacuum. For electrochemical tests, coin-type cells (CR2025) are fabricated inside a N2-filled glove box, using circular Si/void/C nanocomposite electrode as the working electrode and Li metal foil as the counter electrode. As shown in Fig.2 (a), two electrodes are separated by a Celgard 2400 separator, which is soaked in the electrolyte. 1 mol/L LiPF6 in a mixture of ethylene carbonate, diethyl carbonate and dimethyl carbonate (EC : DEC : DMC, 1: 1 : 1 by volume) is used as the electrolyte, and with 10% fluoroethylene carbonate (FEC) as the electrolyte additive to increase the cycling efficiency and improve cycling stability. Fig.2 (b) shows the composite anode on a copper foil and the packaged cell for testing.

The SEM image of the cross-section of the Si/void/C nanocomposite anode before cycling is shown in Fig.3. We can see that SiNPs are embedded in carbon scaffold with some nano void space between them clearly, which provides enough space for expansion of individual SiNPs during lithium-ion intercalation.

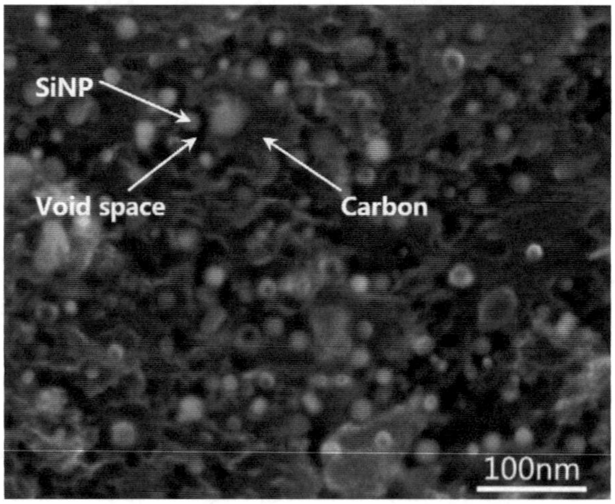

Figure 3: SEM image of the cross-section of Si/void/C nanocomposite anode before cycling.

3. Results and discussion

The galvanostatic charge-discharge tests are performed using a battery testing system (BTS-3000, Neware) at room temperature, with voltage cutoffs set at 0.01V and 1.2V vs. Li/Li^+. Specific capacity is calculated based on the total mass of Si/void/C composite. The current density of charge/discharge is 400 mA/g, corresponding to a rate of 0.1C.

The SiNPs account for about 35% of the anode composite by weight, thus the anode has a theoretical capacity of 1470mAh/g, on the premise that the capacity contributed by carbon is negligible when compared with silicon. From the discharge/charge voltage profiles for different cycles in Fig.4, the actual initial capacity is about 1200mAh/g. In the first few discharge processes, there is a long discharge plateau below 0.3V due to the SiNPs crystalline structure. The discharge/charge curves remain similar in shape from 1st to 10th cycle with a small decrease in the capacity, indicating a stable electrochemical behavior. This excellent cycling stability is attributed to the presence of void space and the structural stability of carbon scaffold from SU-8, therefore individual SiNPs can expand and contract inside the carbon scaffold without rupturing the outer shape of the structure. The capacity retention is 71% after 30 cycles. The irreversible capacity loss is a result of the irreversible insertion of Li into silicon and amorphous carbon. Besides, the continuous formation of SEI layers could make a significantly influence, because the carbon accounts for over 60% of the composite and directly contact the electrolyte, with a high surface area structure.

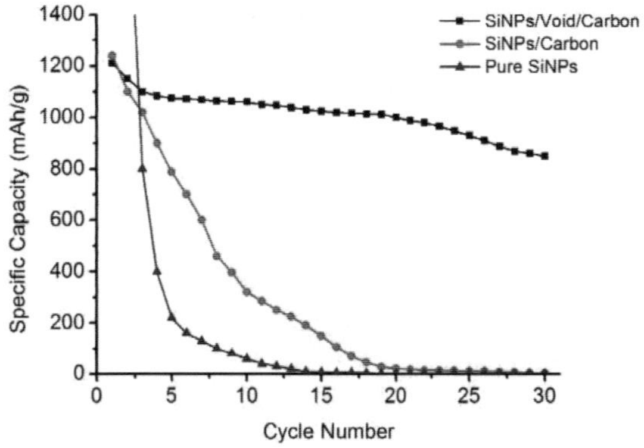

Figure 4: Galvanostatic discharge and charge profiles of the Si/Void/C nanocomposite anode for different cycles: 1st, 2nd, 5th, 10th, 20th and 30th.

The Si/void/C nanocomposite demonstrates much higher capacity and better capacity retention when compared to pure SiNPs and Si/C composite without void space, as shown in Fig.5. Our Si/void/C nanocomposite anode has an initial specific capacity of 1200mAh/g and a capacity of 850mAh/g remaining after 30 cycles. Under the same conditions, the pure SiNPs anode has a high initial capacity but decays quickly to almost zero after only several cycles, due to its structure failure, SEI formation and poor electrical conductivity. Si/C composite, which is without void space inside, shows a similar initial capacity to Si/void/C structure. However, poor capacity retention of less than 10% is observed after only 15 cycles. The above comparison has strongly proved that the Si/void/C composite anode shows outstanding performance, mainly because of the advantage of allowing the Si particles to expand freely, with a stable and conductive carbon scaffold to support silicon.

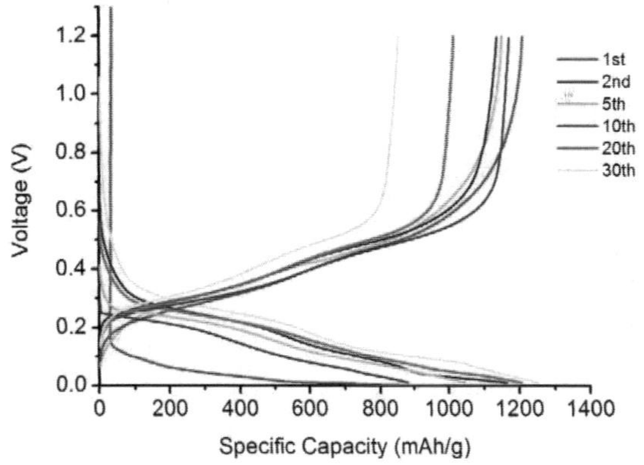

Figure 5: Galvanostatic cycling of different structures: a) pure SiNPs; b) Si/C composite (without void space) and c) Si/C composite (with void space).

4. Conclusions

In summary, we have designed and fabricated a Si/void/C nanocomposite anode from Si/SiO_2-templated and demonstrated its high capacity (1200mAh/g) and stable cycling behavior (71% capacity retention after 30 cycles). These results suggest that Si/void/C nanocomposite can improve the structural stability and cycling performance of the silicon-based anode, with rationally designed void space to accommodate volume expansion during cycling, and stable carbon scaffold to release the stress and provide an efficient conductive pathway. Besides, the SU-8 carbon source along with the fabrication technology opens up a new route for direct prototyping of Si-based on-chip 3D micro-battery in the future.

Acknowledgments
This work is supported by grants from Tsinghua National Laboratory for Information Science and Technology (No. 042003130). The authors also thank the members in the laboratory of Department of Chemistry, Tsinghua University, for providing necessary experimental equipments and helpful instructions.

References

[1] Tarascon J M and Armand M 2001 Issues and challenges facing rechargeable lithium batteries *Nature.* **414** 359-67
[2] Arico A S, Bruce P, Scrosati B, Tarascon J M and Van S W 2005 Nanostructured materials for advanced energy conversion and storage devices *Nature Materials.* **4** 366-77
[3] Boukamp B A, Lesh G C and Huggins R A 1981 All-Solid lithium electrodes with mixed-conductor matrix *J. Electrochem Soc.* **128** 725-9
[4] Lee Y M, Lee J Y, Shim H, Lee J K and Park J 2007 SEI layer formation on amorphous Si thin electrode during precycling *J. Electrochem Soc.* **154** A515-9.
[5] Ohara S, Suzuki J, Sekine K and Takamura T 2004 A thin film silicon anode for Li-ion batteries having a very large specific capacity and long cycle life *J. Power Sources.* **136** 303-6
[6] Arie A A, Song J O and Lee J K 2009 Structural and electrochemical properties of fullerene-coated silicon thin film as anode materials for lithium secondary batteries *Mater. Chem. Phys.* **113** 249-54
[7] Chan C K, Patel R N, O'Connell M J, Korgel B A and Cui Y 2010 Solution-grown silicon nanowires for lithium-ion battery anodes *ACS Nano.* **4** 1443-50
[8] Song T, et al 2010 Arrays of sealed silicon nanotubes as anodes for lithium ion batteries *Nano Lett.* **10** 1710-6
[9] Holzapfel M, Buqa H, Scheifele W, Novak P and Petrat F M 2005 A new type of nano-sized silicon/carbon composite electrode for reversible lithium insertion *Chem. Commun.*1566-8
[10] Ng S H, Wang J, Wexler D, Chew S Y and Liu H K 2007 Amorphous carbon-coated silicon nanocomposites: A low-temperature synthesis via spray pyrolysis and their application as high-capacity anodes for lithium-ion batteries *J. Phy. Chem. C.* **111** 11131-8
[11] Magasinski A, Dixon P, Hertzberg B, Kvit A, Ayala J and Yushin G 2010 High-performance lithium-ion anodes using a hierarchical bottom-up approach *Nature Materuaks.* **9** 353-8
[12] Terranova M L, Orlanducci S, Tamburri E, Guglielmotti V and Rossi M 2013 Si/C hybrid nanostructures for Li-ion anodes: An overview *J. Power Sources.* **246** 167-77
[13] Shen C W, Wang X H, Zhang W F and Kang F Y 2013 Direct prototyping of patterned nanoporous carbon: a route from materials to on-chip devices *Scientific Reports.* **3**(2294) 1-6
[14] Min H S, Park B Y, Taherabadi L, Wang C L, Yeh Y, Zaouk R, Madou M J and Dunn B 2008 Fabrication and properties of a carbon/polypyrrole three-dimensional microbattery *J. Power Sources.* **178** 795-800
[15] Stöber W, Fink A and Bohn E 1968 Controlled growth of monodisperse silica spheres in micron size range *J. Colloid Interf. Sci.* **26** 62

Regenerative braking systems with torsional springs made of carbon nanotube yarn

S Liu[1], C Martin[1], D Lashmore[2], M Schauer[3], and C Livermore[1]
[1] Mechanical and Industrial Engineering Department, Northeastern University, Boston, MA 02115 USA
[2] College of Engineering and Physical Sciences, University of New Hampshire, Durham, NH 03824-2222 USA
[3] Nanocomp Technologies Inc., Merrimack, NH 03054 USA

E-mail: liu.san@husky.neu.edu

Abstract: The demonstration of large stroke, high energy density and high power density torsional springs based on carbon nanotube (CNT) yarns is reported, as well as their application as an energy-storing actuator for regenerative braking systems. Originally untwisted CNT yarn is cyclically loaded and unloaded in torsion, with the maximum rotation angle increasing until failure. The maximum extractable energy density is measured to be as high as 6.13 kJ/kg. The tests also reveal structural reorganization and hysteresis in the torsional loading curves. A regenerative braking system is built to capture the kinetic energy of a wheel and store it as elastic energy in twisted CNT yarns. When the yarn's twist is released, the stored energy reaccelerates the wheel. The measured energy and mean power densities of the CNT yarns in the simple regenerative braking system are up to 4.69 kJ/kg and 1.21 kW/kg, respectively. A slightly lower energy density of up to 1.23 kJ/kg and a 0.29 kW/kg mean power density are measured for the CNT yarns in a more complex system that mimics a unidirectional rotating regenerative braking mechanism. The lower energy densities for CNT yarns in the regenerative braking systems as compared with the yarns themselves reflect the frictional losses of the regenerative systems.

1. Introduction

Many portable devices are limited by the lack of power supplies that offer both high energy density and high power density. Fuel cells and batteries offer high energy density but poor power density, whereas mechanical springs and capacitors have high power density but low energy density [1]. Fuel-burning devices can offer both high energy density and high power density, but their efficiency decreases at a smaller size scales, increasing their environmental cost. Carbon nanotubes (CNTs) offer an appealing alternative. Their low relative density (1.33-1.4 times of water) [2], high stiffness (1 TPa) [3], and high extendibility (up to greater than 13% elastic strain in [4]) offer exceptional performance as elastically-deformed, energy storing springs for small scale systems. Idealized continuum modelling of individual CNTs under load has estimated the upper bound on energy density at 5000 kJ/kg under the optimum achievable tensile loading. This energy density, if achievable, would be more than 1000 times the energy density of steel springs, and more than 5 times greater than that of lithium-ion batteries. In addition, CNT springs offer the potential for high power operation just as traditional mechanical springs do [1, 5], potentially enabling solutions for portable systems requiring both high energy density and high power density.

In practice, the energy densities of elastically-deformed CNT springs fall below ideal maximum values. For example, the CNT springs of [6] show an average energy density of 4.2 kJ/kg under incremental cyclic tensile test, along with demonstrated peak power densities of up to 170 kW/kg for quick-release sling-shot devices and mean power density of 2.5 kW/kg for slow-release, metered systems. Part of the challenge lies in scaling up from the single-CNT scale to the milli and macro scales. Using CNT yarns instead of individual CNTs increases the total energy stored at the expense of reduced energy density; the yarn's non-ideal inner structure permits non-uniform loading, slippage, and defects to reduce its performance under load. For systems like those of [6], a second challenge of effective energy release lies in the mismatch between the characteristics of the power supply (which offers high force and extremely small stroke) and the load (which typically requires lower force and higher stroke, except for applications that require the highest possible power density).

The present research employs CNT springs in torsion rather than in tension to address some of the challenges of CNT springs for driving mechanical loads. Loading CNT springs in torsion can increase the stroke (at the expense of slightly lower peak power) and offer better matching between the power supply and load characteristics. The driving torques produced by torsional loading couple readily into the rotating systems that comprise a great many common machines. Although torsional loading may in theory offer reduced energy density as compared with more spatially uniform axial loading, the reduction may be moderated by a reduction in slippage as the originally untwisted fibers become ever more tightly packed with twisting. In addition, torsion applied as in the present experiments (with fixed separation between the yarn's ends) can also result in more local axial stretching, thereby enhancing energy density for torsional actuation. Thus, torsional actuators can offer improved performance for applications that require longer stroke and are tolerant of slightly lower peak powers.

Previous studies of CNT assemblies under torsion have focused on systems in which the ends of the CNT assemblies are free to move closer together and further apart as the assembly twists and untwists. The research of [7, 8] examined several different hybrid, coiled CNT yarn structures, typically with hierarchical twist architectures, as contractile linear actuators that operate by the twisting and untwisting of CNT yarns. The largest contractile energy density in [7] was 1.36 kJ/kg at 5.6% linear contraction. In [8], electrolyte-filled twist-spun CNT yarns were demonstrated as torsional CNT artificial muscles that output a combined contractile and torsional power density of 0.92 kW/kg and 0.062 kW/kg, respectively. Additionally, a torsional experiment in which CNT fibers support a rotational pendulum at a height that is free to vary with rotation demonstrated up to 24% elastic shear strain [9]. In [10], CNT yarns that were overtwisted to form entanglements and released to rotate a cylinder demonstrated an optimum 8.3 kJ/kg energy density and a power density of less than 1 kW/kg. These tests suggest that CNT yarns under torsion may offer comparable energy density to CNT yarn under tensile loading, but the performance of twisted yarns subject to a length constraint has not been explored.

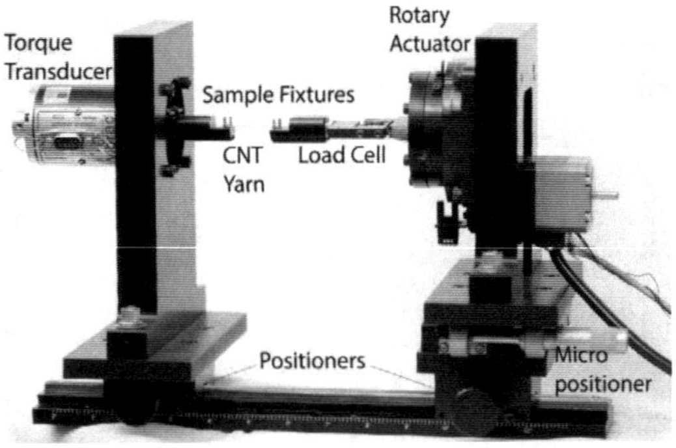

Figure 1. A custom-built torsional testing system with a torque transducer, an axial force transducer, a rotary actuator, two sample fixtures and three linear positioners.

Experiments were performed to better mimic the behaviour of real-world systems in which the ends of a torsional spring are tethered at a fixed separation; the fixed-separation approach may also increase energy density by maximizing energy stored in the stretching of CNTs that follow a helical path around the CNT yarn. To characterize energy storage under these conditions, a custom-built torsion tester (figure 1) twists one end of a yarn relative to the other while maintaining nearly constant separation between the yarn ends and measuring reaction torque and axial force.

2. Characterization of mechanical behavior under torsional loading

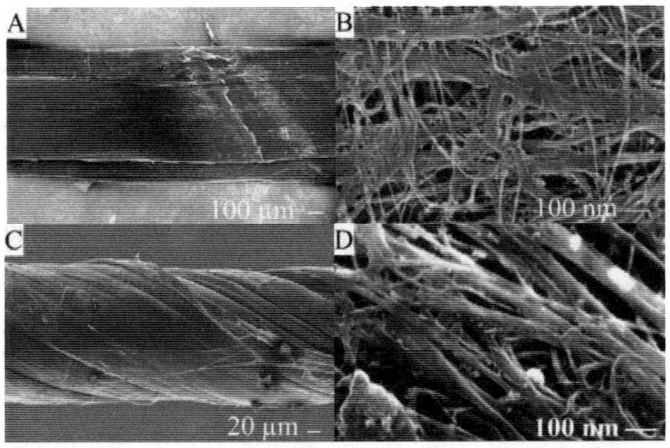

Figure 2. Scanning electron microscope (SEM) images of: (A) and (B) an untwisted CNT yarn; (C) and (D) a 2-ply CNT yarn after torsional test.

Continuous untwisted CNT yarns (figure 2A, 2B) are grown by a gas phase synthesis process that incorporates a free-flowing catalyst. They are comprised of both single-walled and multi-walled CNTs with diameters typically between 3 nm and 8 nm. The untwisted yarn has typical cross-sectional dimensions of 1 mm x 130 µm and has a mean linear mass density of 9.55 tex (where 1 tex = 1 mg/m). CNT yarn specimens with a typical gauge length of 25 mm are attached to sample fixtures using epoxy (Pacer Z-Poxy). The fixtures are then mounted on the torsion tester, and their axial spacing is adjusted using a micropositioner to remove slack in the yarn and achieve a small, starting axial preload of 0.2N. A rotary actuator (Oriental Motors DG85) twists the yarn at a constant rotational speed, while the torque transducer (Vibrac mini static 0.05 oz-in) and load transducer (SMD S250-5kg) detect the resulting torque and axial force. Figures 2C and 2D show the resulting yarn structure. The compliance of the transducers is much smaller than the compliance of the yarn, so that the supports may be approximated as rigid with fixed separation. The torque, force, and angle of rotation are collected by a NI USB-6221 BNC DAQ device via a Labview interface.

CNT yarns were tested under incrementally-increased torsional loading. The total rotation angle is increased from zero in increments of five turns until failure. The results are repeated five times per increment before increasing the number of rotations in the next increment. Figure 3 plots the results of a cyclic torsional loading of a 25.4 mm-long, 2-ply CNT yarn. The raw data include smaller torque variations with the periodicity of the rotation; these variations reflect a small axial misalignment of the motor and torque transducer. For clarity, the data of Figure 3A have been post-processed to show only the overall incremental loading performance. The rotational speed is 15 turns/min, and the yarn rotates to 30 turns before failing on the way to 35 turns. During loading, the yarn transitions from a non-circular cross-section to a "twisted ribbon" structure that approximates a circular cross-section (figure 2C). The maximum specific torque is measured at 540 N-m/kg. The maximum extractable energy density of 6.13 kJ/kg is calculated by integrating the area under the untwist curve of the fifth cycle in the largest increment loading group. Among the six specimens, the average angle of twist per unit length is more than 7400 rad/m, average input and output energy efficiency is about 53%, and average energy density is about 3.82 kJ/kg, comparable to the 4.2 kJ/kg energy density in tension [5].

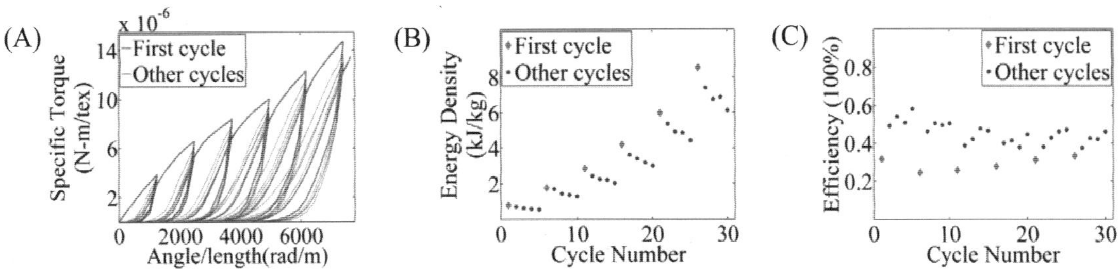

Figure 3. Torsion test data for a 25.4 mm long, two-ply CNT yarn cyclically twisted to failure with 5 cycles per increment (5 turns per increment) at the rate of 15 turns/min until breakage. (A) Plot of specific torque vs. rotation shear strain (angle per length); (B) plot of energy density of each twisting cycle; (C) plot of energy efficiency of twist and untwist of each cycle.

The high energy densities, comparable to those obtained for pure tension, might initially seem surprising. Because strains are much larger near the periphery of a torsionally-loaded shaft than they are near its center, twisting is not expected to load all of the material to its limits, resulting in a lower energy density than in axial loading. However, twisting of a filamentary structure with fixed separation of the ends results in stretching of the helically-wrapped filaments as described above and in [11]. In practice a given filament will sample both the inner and outer regions of the yarn's cross-section, and slippage between the filaments offers the opportunity for filaments to be loaded to a more spatially uniform average axial stress, potentially increasing energy storage as compared with torsional loading of a shaft. The twisted yarn's high performance may also reflect increased load transfer and reduced slippage due to radial compression of the yarn under large torsional loads. The yarn's structural reorganization under loading is apparent in figure 3A, in which the first loading cycle for each increment creates a new torque-angle loop, which is relatively stable over continuing load cycles at that level of deformation. Yarn reconditioning is therefore most apparent in the first load cycle at a given level, whereas a more stable hysteresis loop appears in the subsequent cycles. Figures 3B and 3C plot extractable energy density and energy efficiency as calculated from the hysteresis loops of figure 3A. Although energy density decreases over the load cycles at each increment, efficiency increases, demonstrating the structure's reorganization into a more stable configuration.

3. Regenerative systems driven by CNT yarn

Two model regenerative braking systems that capture energy in and deliver energy back from CNT torsional springs were demonstrated. In the first demonstration, the CNT yarn couples directly to a wheel on a shaft (figure 4A). Energy is stored in the actuator by twisting the disk quickly to an initial high speed to simulate braking and storing the kinetic energy in the twisted CNT yarn until the speed drops back to zero. The wheel is then allowed to accelerate freely under the yarn's actuating torque. Axial force is measured by a load cell as described in section 2. The rotation of the wheel is measured by an optical sensor (Philtec RC171) that detects passage of periodic features on the disk's surface, and the corresponding angular velocity and rotational acceleration are calculated from the measured rotation angle. The extracted energy density is calculated from the maximum measured kinetic energy of the wheel (determined from the measured angular speed and the rotational inertia of the wheel and the shaft) and the mass of the yarn. Because some of the extracted energy is lost to friction, this approach underestimates the energy density of the energy-storing yarn. The measured acceleration similarly underestimates the effects of the yarn's driving torque, because it represents the difference between acceleration of the wheel by the yarn and the deceleration of the wheel by friction. A lower bound on maximum power density may nonetheless be extracted from angular velocity, angular acceleration, and moment of inertia using the results shown in figure 4. The resulting values for the case of figure 4 are a maximum energy density of 4.69 kJ/kg and a mean power density of 1.21 kW/kg over a 3 s release time (figure 4B). The corresponding stroke was large, at 14 turns.

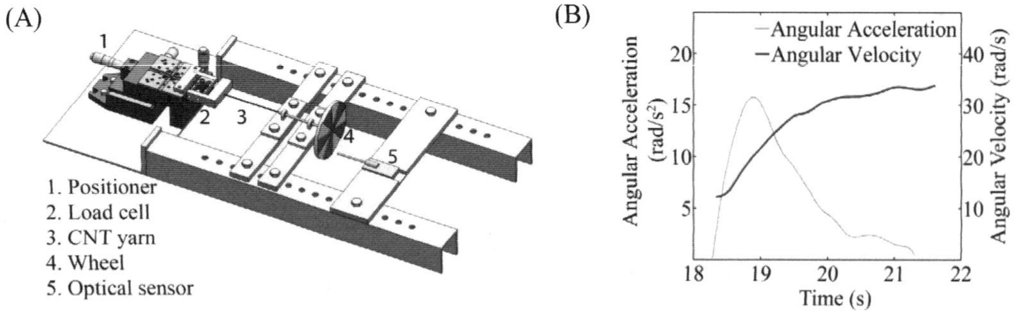

Figure 4. (A) Schematic diagram of the first regenerative braking test system in which the CNT yarn couples directly to a wheel on a shaft, and (B) plot of wheel angular velocity and power density vs. time. The data are determined from measured rotation angle vs. speed accelerating time data, when the wheel of figure 4 is driven by a 6 cm-long, 2 ply CNT yarn.

The second model regenerative braking system (figure 5A) also employs a single wheel. However, it is equipped with a ratchet unit that couples to the yarn by rotating the wheel to wind the spring, decouples it with the ratchet gears for transmission gear changeover, and enables the yarn to drive the wheel to release the energy in the same rotation direction as during energy capture. Figures 5B and 5C plot the kinematic variables over time from a test using a 6.2 cm, 4-ply CNT yarn. Figure 5B shows the initial angular velocity of 54.3 rad/s is decelerated to zero at time 6.7 s by storing energy into the CNT spring; then the CNT spring starts to reaccelerate the wheel at time 80 s. The maximum angular velocity is 18 rad/s, with a total acceleration time of 3.2 s. The 0.29 kW/kg mean power density is calculated from the system's moment of inertia, angular acceleration and angular velocity over the duration of energy release (80-83.2 s). The 1.23 kJ/kg energy density is calculated from the maximum velocity after energy release, the moment of inertia, and the mass of the wheel and shaft. Energy and power density are slightly lower than in the first demonstration due to larger frictional damping.

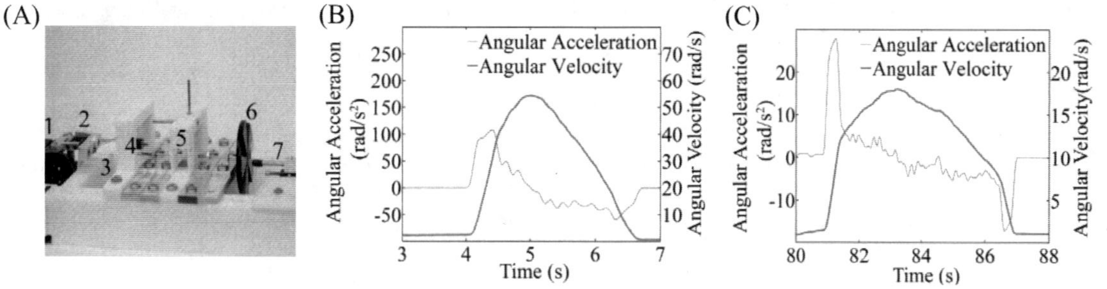

Figure 5. (A) Photograph of the second model regenerative braking system. A ratchet and two alternate sets of gears enable energy to be stored and released to continue travel in the wheel's original direction. (1. Positioner, 2. load cell, 3. CNT yarn, 4. ratchet and pawl, 5. changeover gears, 6. wheel, 7. optical sensor). (B) Measured angular acceleration and angular velocity plotted vs. time during energy capture. (C) Angular acceleration and angular velocity during energy release.

4. Conclusion

CNT yarn has been demonstrated as large stroke, high energy density, high power density torsional springs for regenerative braking systems with fixed separation between the spring ends. The maximum energy density measured in torsion tests is up to 6.13 kJ/kg. Demonstrations of regenerative braking using untwisted CNT yarn as the energy-storing medium obtain energy and mean power densities of the CNT yarns of up to 4.69 kJ/kg and 1.21 kW/kg, respectively. Advances in CNT yarn preparation and system design can offer the potential for still higher energy and power densities in the future.

Acknowledgements

The authors would like to thank Dr. Frances Hill for many useful discussions. This research was supported by the National Science Foundation under award #1434824.

References

[1] Christen T and Carlen M W 2010 *J. Power Sources* **91** 210-216
[2] Collins P and Avouris P G 2000 *Scientific American* **283** 62
[3] Qian D, Wagner G J and Liu W K 2002 *Appl. Mech. Rev.* **55** 495-533
[4] Chang C, Hsu I, Aykol M, Hung W, Chen C and Cronin S B 2010 *ACS Nano* **4** 5095–100
[5] Hill F A, Havel T F and Livermore C 2009 *Nanotechnology* **20** 255704
[6] Hill F A et al. 2014 *J. Energy* DOI: 10.1016/j.energy.2014.08.021
[7] Lima M D et al. 2012 *Science* **338** 928–932
[8] Foroughi J et al 2011 *Science* **334** 494–497
[9] Wu A S et al. 2012 *Appl. Phys. Lett.* **100** 201908
[10] Li Y et al. 2013 *ACS Nano* **7** 8128-35
[11] Grason G M and Bruss I R 2012 *PNAS* **109** 10781-86

Composite carbon-based ionic liquid supercapacitor for high-current micro devices

M Cowell[1], R Winslow[1], Q Zhang[2], J Ju[1], J Evans[2] and P Wright[1]

[1]Department of Mechanical Engineering, University of California at Berkeley, 6141 Etcheverry Hall, Berkeley CA, 94709 USA

[2]Department of Materials Science and Engineering, University of California at Berkeley, 210 Hearst Mining Building, Berkeley CA, 94720 USA

E-mail: martin.cowell@berkeley.edu

Abstract. Manufacture and performance of a composite carbon-based supercapacitor that employs a gel polymer ionic liquid electrolyte to achieve stable, long cycle life, high-current draw energy storage is discussed in this paper. This supercapacitor when cycled galvanostatically can achieve a discharge capacitance of 43.0 mF per square centimeter of substrate by leveraging the strengths of a composite electrode composition. The printed manufacturing process takes place in ambient conditions at room temperature enabling high-current, rechargeable energy storage to be built onto many substrates. Single-cell discharge power densities have reached 404 μW/cm^2, which could enable many technologies when paired with a MEMS energy harvester.

1. Introduction

Powering high-current micro devices with low power generation often found at the micro-scale [1] requires that energy be trickle-charged into a storage device capable of high-current discharge. While arrays of micro piezoelectric vibrational energy harvesters have been used to increase system output power [2], the power required to energize a wireless radio is still much higher [3]. Charge storage via an electric double layer, such as found in a supercapacitor, offers high-current discharging while maintaining long cycle-life performance. The thermal and chemical stability of the ionic liquid electrolyte 1-butyl-3-methylimidazolium tetrafluoroborate ([BMIM+][BF4-]) allows the thick film capacitor to be employed in harsh environments where aqueous electrolytes would otherwise evaporate [4], while the polymer binder allows for structural flexibility.

Experimental testing reached a maximum voltage of 1.0 V and achieved a discharge efficiency of 84.2%, discharge power of 404 μW/cm^2, and discharge energy of 5.03 μW-hr/cm^2. Long cycle life testing of up to 100,000 cycles showed no sign of precipitous cell degradation.

Previous work confirmed a proof of concept for the printability of carbon based ionic liquid supercapacitors but left composition optimization and manufacturing process control unfinished. The research presented in this paper improves upon the proof of concept with a composite carbon electrode to achieve higher capacitance, energy, and power density.

2. Cell Manufacturing

2.1. Layer Composition

Symmetrical capacitor geometry was employed with dry electrodes composed of activated carbon (AC), acetylene black (AB), graphite (GR), and poly(vinylidene fluoride-co-

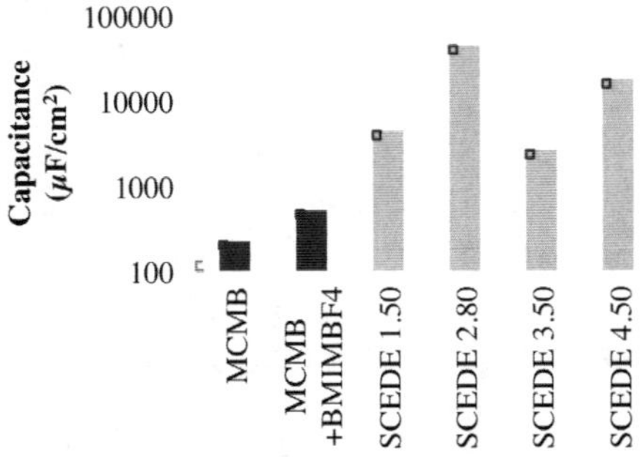

Figure 1. Capacitance performance improvements over previous work by leveraging AC, AB, and GR.

Table 1. Composite electrode variations. NMP was used to achieve rheology suitable for printing.

Composite Electrode Version	Electrode Composition AC:AB:GR:PVDF-HFP (Mass Ratio)
1.50	0:1:0:1
2.50	1:0:1:2
2.80	2:0:2:1
3.50	0:9:1:10
4.50	17:1:2:20

hexafluoropropylene) (PVDF-HFP) with mass ratios outlined in table 1. Previous work by our lab used mesocarbon microbeads (MCMB) and AB as active electrode materials [5], but these have been found to have an average capacitance of 0.50 mF/cm^2 as seen in figure 1. To improve capacitance performance, composite electrodes leverage the high surface area of AC with the electrically conductive microstructure of AB [6] and structural reinforcement of GR. The solvent n-methylpyrrolidone (NMP) was used to suspend the active material and solvate the polymer binder to achieve favorable rheological properties for printed manufacturing.

The gel polymer electrolyte (GPE) was composed of (50 wt%) PVDF-HFP and (50 wt%) [BMIM+][BF4-]. NMP was also used as a solvent to ensure a printable rheology. The gel polymer layer electrically separated the electrodes while contributing ion migration pathways to support the electric double layer characteristic of supercapacitors.

2.2. Layer Dispenser Printing and Assembly

The electrolyte was printed via CNC pneumatic dispenser printer, and electrodes were stencil cast (figure 2). They were dried in an oven at 80°C for 12 hours, driving off the NMP to leave only the active and binder materials. Once dry, electrodes were peeled off their stainless steel substrate without damaging their structural integrity. Dried gel polymer electrolyte layers were peeled off their glass substrate (figure 3). The dry layers were wetted with [BMIM+][BF4-] and stacked to create parallel-opposed cells (figure 4), which were housed in CR2023 coin cells for cycling and performance characterization.

3. Testing and Performance Results

Short cycle galvanostatic testing was performed on the Gamry Reference 600^{TM}. Constant current charge and discharge cycling was performed as outlined in supercapacitor testing standard IEC 62391-1:2006 [7]. During this cycling regime supercapacitors were looped ten times through steps outlined in table 2.

Energy was quantified via numerical integration of the voltage-current product vs. time curve during charging and discharging described by

$$W = \sum_{t=t_o}^{t_f} I_t V_t \Delta t \tag{1}$$

where I_t is the time series current in amperes, V_t is the time series voltage in volts, Δt is the time step between samples in seconds, t_f and t_o are the final time and initial time in seconds.

Figure 2. Stencil cast electrodes before drying.

Figure 3. Gel polymer ionic liquid electrolyte after oven drying.

Figure 4. SEM of stacked cross section of carbon electrode and GPE.

Table 2. Constant current charge and discharge regime.

Step Name	Limit/ Condition
Constant current charge	1.00 V
Rest	5.00 s
Constant current discharge	0.00 V
Rest	5.00 s

Power during charge and discharge was quantified by

$$P = W/(t_f - t_o) \tag{2}$$

Bulk capacitance was calculated using the time rate of change of voltage relation between current and capacitance, based on states at the beginning and end of charge and discharge:

$$C = I(t_f - t_o)/(V_f - V_o) \tag{3}$$

where I is the average current in amperes, V_f is the voltage at the end of the charge or discharge step, and V_o is the voltage at the start of the charge or discharge step.

Equivalent series resistance (ESR) was calculated using Ohm's law for voltage drop seen at the transition between charge, rest, and discharge steps:

$$ESR = (V_{o-1} - V_o)/I \tag{4}$$

where V_{o-1} represents the voltage immediately before charging or discharging begins. The voltage drop across the cell becomes substantial at higher discharge rates.

Coulombic efficiency was defined as

$$\eta_C = \sum_{t=t_o}^{t_f Discharge} I_t \Delta t / \sum_{t=t_o}^{t_f Charge} I_t \Delta t \tag{5}$$

where the number of coulombs transferred was determined through numerical integration of current-time series data.

Energy efficiency was defined as

$$\eta_E = W_d/W_c \tag{6}$$

where W_d represents the discharge energy in joules and W_c represents charge energy in joules.

Figures 5 and 6 show data taken from a cell whose electrodes (1 cm x 1 cm each) were composed of composition 2.80 and separated by a single layer of GPE 60 μm thick. The cell was subjected to constant current charge/discharge regimes using the same magnitude of current during charging and discharging steps (10 and 1000 μA/cm^2 for the two tests). The high voltage limit of 1.0 V was selected to avoid the possibility of side reactions associated with the electrolysis of trace water, but the cells were expected to be able to sustain voltages of 2.5 V to 4.0 V without breakdown of the ionic liquid electrolyte [8]. Results from the two tests are shown in table 3. Performance metrics were normalized by a footprint area of 1 cm^2.

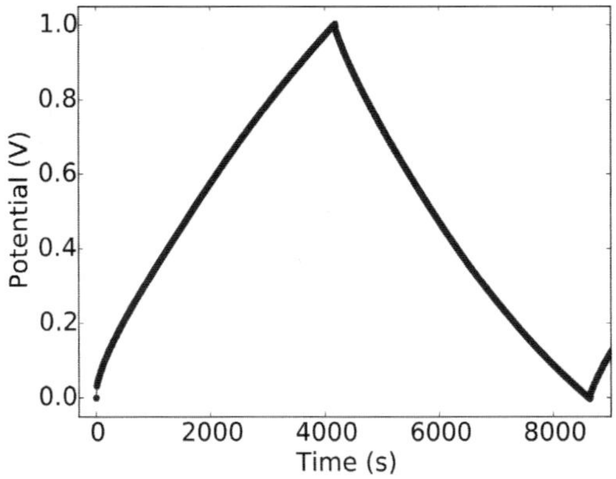

Figure 5. Constant current (10 μA/cm^2) charge/discharge voltage response of supercapacitor cell using electrode composition 2.80.

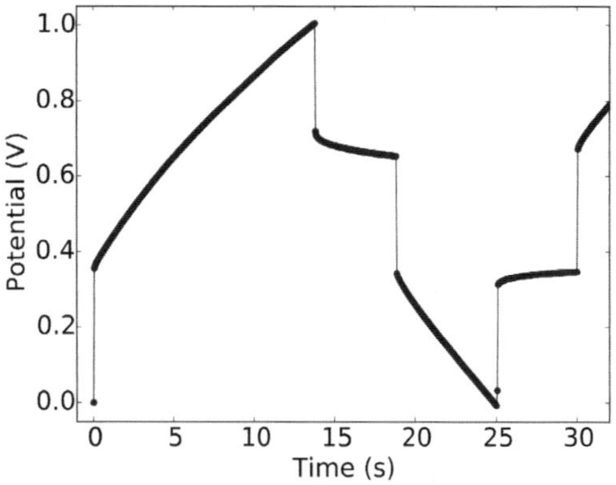

Figure 6. Constant current (1000 μA/cm^2) charge/discharge voltage response of supercapacitor cell using electrode composition 2.80.

Table 3. Performance metrics of supercapacitor cell containing electrode 2.80 tested at two different current densities.

	10 μA/cm^2		1000 μA/cm^2	
	Charge	Discharge	Charge	Discharge
Energy (μW-hr/cm^2)	6.75	5.03	1.61	0.300
Power (μW/cm^2)	5.46	4.26	855	164
Capacitance (mF/cm^2)	45.0	43.0	20.4	18.0
ESR (Ω)	1020	1150	325	307
Coulombic Efficiency (%)	95.4		97.5	
Energy Efficiency (%)	74.4		18.7	

As is evident in table 3, the testing regime heavily influenced the performance of the supercapacitor cell. The highest reported discharge energy density (5.03 μW-hr/cm^2) and discharge capacitance density (43.0 mF/cm^2) were attained by composition 2.80 at a discharge current of 10 $\mu A/cm^2$; discharge power density (404 $\mu W/cm^2$), composition 1.50 at 1000 $\mu A/cm^2$; lowest reported discharge ESR (79.0 ohms), composition 2.50 at 1000 $\mu A/cm^2$; highest reported coulombic efficiency (100%), composition 3.50 at 1000 $\mu A/cm^2$; and highest reported energy efficiency (84.2%), composition 1.50 at 50 $\mu A/cm^2$. (Please note the 100% coulombic efficiency is accurate to three significant figures, and does not represent perfect charge transfer.)

Long cycle life testing of 100,000 cycles was performed on the cell containing electrode composition 2.80. Figure 7 illustrates the cell's capacitance degradation versus cycle number (points plotted for every 1,000 cycles). The cell's capacitance degraded from 28.5 mF/cm^2 to 17.9 mF/cm^2 over the course of the test, a decrease of 37.0%. Note, however, that the largest drop (22.5% down to 22.1 mF/cm^2) occurred between cycle number 1 and 1,000, suggesting an initial "break-in" process occurred within the capacitor's first thousand cycles. Once the capacitor was "broken-in," the long cycle life capacitance only dropped by 14.5% over the course of the last 99,000 cycles.

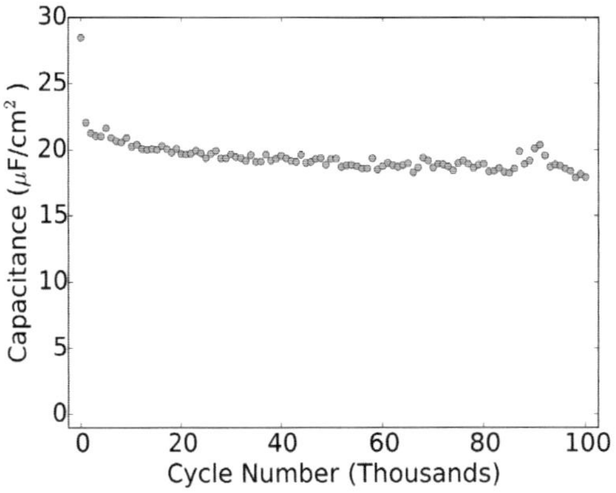

Figure 7. Long cycle life effects on cell capacitance

4. Conclusion

A high-current printable supercapacitor of composite carbon electrode composition employing gel polymer ionic liquid electrolyte has been shown to provide excellent discharge power to support trickle charging provided by micro-scale energy harvesters. This strong attribute promises to pair with micro-energy harvesting devices to meet the demands of high power devices, such as modern wireless radios, for applications in printed flexible wireless sensor nodes. The printing process employed can be rapidly scaled up for high-throughput manufacturing of printed energy storage to be integrated with MEMS device solutions.

5. Acknowledgments

The authors acknowledge the FlexTech Alliance for their sponsorship with award #RFP13-161.

References

[1] Blystad LCJ, Halvorsen E, Husa S. 2010 *IEEE* **57** 908-19
[2] Liu J-Q, Fang H-B, Xu Z-Y, Mao X-H, Shen X-C, et al. 2008 *Microelectronics J.* **39** 802-6
[3] Wong ACW, Dawkins M, Devita G, Kasparidis N, Katsiamis A, King O, et al. 2013 *IEEE* **48** 186-98
[4] Crosthwaite J, Muldoon M, Dixon J, et al. 2005 *J. Chem. Thermodynamics* **37** 559-68
[5] Ho CC, Steingart D, Evans J, Wright P. 2008 *ECS Transactions* **16** 35-47
[6] Zhang H, Zhang W, Cheng J, Cao G, Yang Y. 2008 *Solid State Ionics* **179** 1946-50
[7] International standard: fixed electric double layer capacitors. IEC 62391-1; 2006
[8] Armand M, Endres F, MacFarlane DR, Ohno H, Scrosati B. 2009 *Nature Materials* **8** 621-9

Development of a High-performance Fluorpolymer Electret Mixed with Nano-particles and Its Application to Vibration Energy Harvesting

M Suzuki[1], **T Takahashi**[1], **and S Aoyagi**[1]

Faculty of Engineering Science, Kansai University, 3-3-35 Yamate-cho, Suita, Osaka 564-8685, Japan

E-mail: m.suzuki@kansai-u.ac.jp

Abstract. We have been developing small power generation device of capacitance-type to be converted to electrical energy vibration energy using an electret. In this Study, dielectric nano-particles were mixed with an electret made of fluorocarbon polymer. As a result, implanted charge density of the electret was successfully enhanced thanks to the mixing of particles. A small sized vibration energy harvester (VEH) was fabricated using the fluorocarbon mixed with dielectric nano-particles. As a result of applying vibration (20 Hz, 0.65 G) to the fabricated VEH, The maximum generated power of approximately 50 μW was obtained.

1. Introduction

Recently, vibrating energy harvester (VEH) which generates electrical power from environmental vibration was widely developed [1-5], because it is prospective power supply for low-power electronic devices such as wireless sensor networks [1, 2]. Since frequency of environmental vibration is usually on the order of a few to several tens of hertz, electrical power generated by electrostatic VEH using an electret are generally higher than that generated by a electromagnetic VEH using a magnet [3]. Electret is dielectric material that keeps electric charge semi- permanently. Previously, many type of electrostatic VEH using electret were developed. We also developed a novel VEH consists of a fixed electret on bade electrode and a high dielectric constant plate such as barium titanate ($BaTiO_3$) which is suspended by vertically vibrated spring with counter electrode with, as shown in Fig. 1 [4, 5].
Surface charge density of electret is most impotent characteristic for VEHs because generated power from a VEH is theoretically proportional to square of it. Conventionally, fluorocarbon polymers [1-5] were frequently used as electret in VEHs because they keep higher concentration of electrical charge than other materials. CYTOP is famous fluorocarbon polymer as an electret, of which surface charge density and its time stability are better than other fluorocarbon polymer [6]. Additionally, it is reported that trapped charge density in a CYTOP can be enhanced when nano-clusters are added into the CYTOP because the nano-clusters are polarized by the electrical field applied during charge implantation, and they work as the charge trapping sites by their polarization [7]. Because of the above mentioned background, we propose a novel electret made of CYTOP in which many nano-dielectric-particles are added. It is expected that the charge density in CYTOP increases with dielectric constant of the nano-particles. Relationship between surface potential of CYTOP with nano-particles and consent ratio or dielectric constant of nano-particles are also evaluated.

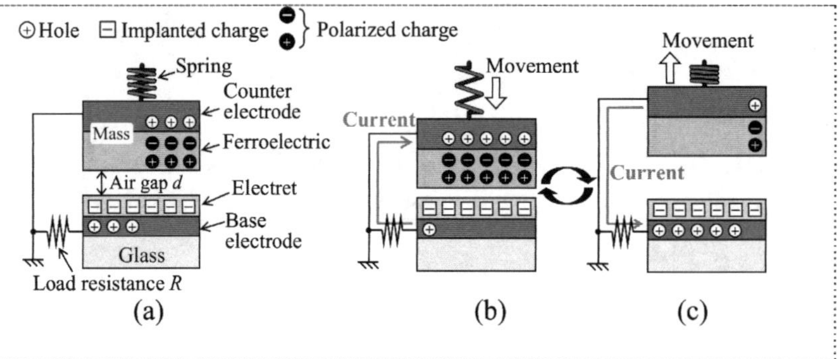

Figure 1. (a) Schematic diagram of developed Vibration energy harvester (VEH). (b) When the upper electret and dielectric plate supported approach the electret, dielectric plate was polarized by electric potential difference from the electret, and it attracts electric charge. (c) When the upper electret and dielectric plate move away from the electret, electric charge is attracted to the lower electrode and electret. The movement of the electric charge is changed to electric energy by the load resistance.

2. Experimental method

Fabrication process of the CYTOP with nano-particles is described as follows. At first, liquid CYTOP and nano-particles ($BaTiO_3$ or CaF_2) was mixed by a centrifugal mixer (THINKY CORPORATION, model:ARE310). Here, typical diameter of the $BaTiO_3$ and CaF_2 particles were 50 nm and 100 nm, respectively. Revolution speed of the mixer and mixing time were 2000 rpm and 60 sec, respectively. Next, mixed particles in CYTOP were dispersed using ultrasonic homogenizer (NIHONSEIKI KAISHA LTD., model:US-150T). Applied power and time of the ultrasonic were 150 W and 5 h. Then, the mixture was spin coated on a on the thermal oxidation (SiO_2) layer of a low-resistive silicon substrate, and dried by heating on an 80°C of hotplate. Then, it was baked using a clean oven. The baking temperature and baking time are 100°C and 1 h, respectively. Finally, negative electric charge was implanted to the fabricated electret by corona discharge method, as shown in Fig. 2.

The schematic diagram of the fabricated electret device is shown in Fig. 3. Thickness of CYTOP layer and SiO_2 layer are 5.4 μm and 1 μm, respectively. The low-resistive Si substrate acts as a lower electrode for the electret. Several kinds of samples were fabricated from which the mixture ratio differs, and these characteristic as an electret were compared.

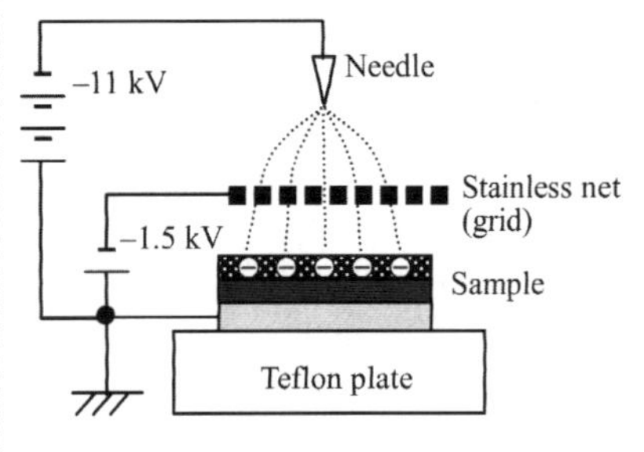

Figure 2. Schematic of setup for corona discharge to implant electron to the fluorpolymer film..

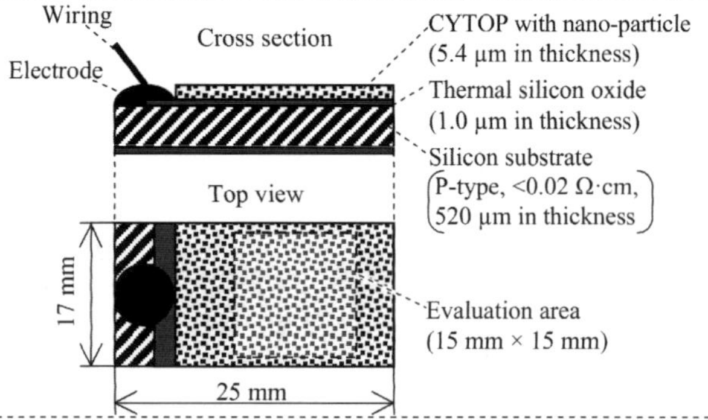

Figure 3. Schematic diagram of the fabricated electret device.

3. Results and discussion

3.1. Surface potential of CYTOP with nano-particles and its stability

Time degradation of surface potential of electrets made of mixture of CYTOP and $BaTiO_3$ nano-particles is shown in Fig. 4. As compared with pure CYTOP, the initial surface potential of CYTOP which mixed the $BaTiO_3$ nanoparticle was also high, and the reduction speed was also slow. Surface potential monotonically increased with the mixed rate of a nanoparticle, when the mixed rate is 10% or lower. However, the surface potential of the electret rapidly decreased with time progress, in which the mixture ratio is 20%. As a result of observation of the sample using scanning electron microscopy (SEM), it is confirmed that there is many aggregate of the nano-particle. Therefore, it is presumed that the implanted electric charge discharged through the aggregate of nano-particles, as shown in Fig. 5. Time degradation of surface potential of electrets made of mixture of CYTOP and CaF_2 nano-particles is shown in Fig. 6. This result indicates that the surface potential of CYTOP also increases by mixing CaF_2 nano-particles. However, the amount of increase in the surface potential of CYTOP/$BaTiO_3$ is better than that CYOP/CaF_2. The reason why this result was obtained is discussed below. When nano-particles are mixed in a CYTOP, the particles are polarized by electrical filed of corona discharge. Additionally, degree of electrical polarization probably increases with dielectric constant of the particles. Therefore, that capability of the $BaTiO_3$ to capture negative electric charge maybe larger than that of the CaF_2 because dielectric constant of $BaTiO_3$ is larger than that of CaF_2.

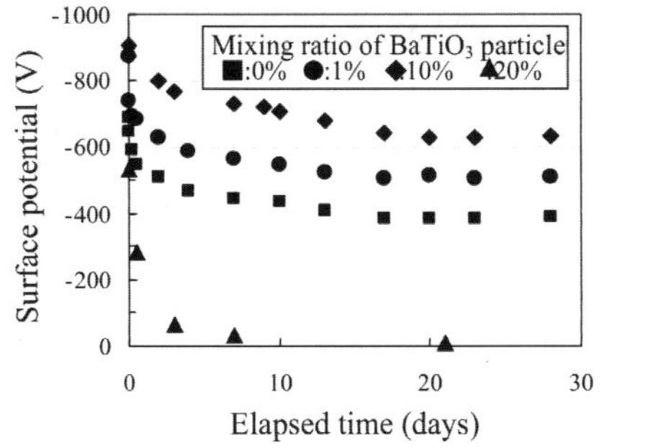

Figure 4. Time degradation of surface potential of electrets made of mixture of CYTOP and $BaTiO_3$ nano-particles.

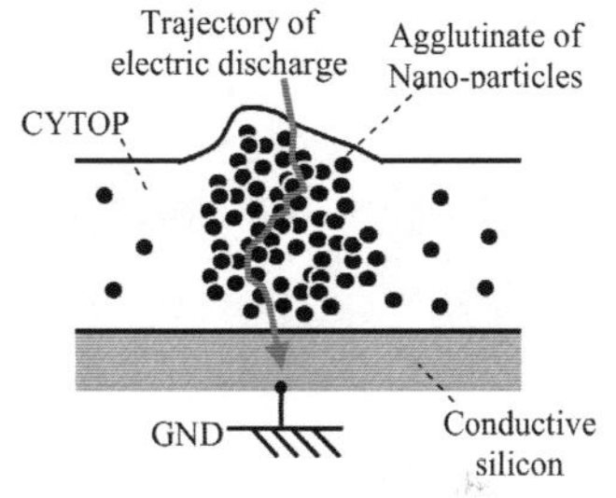

Figure 5. Schematic illustration describing how electric charge discharges through the aggregate of nano-particles.

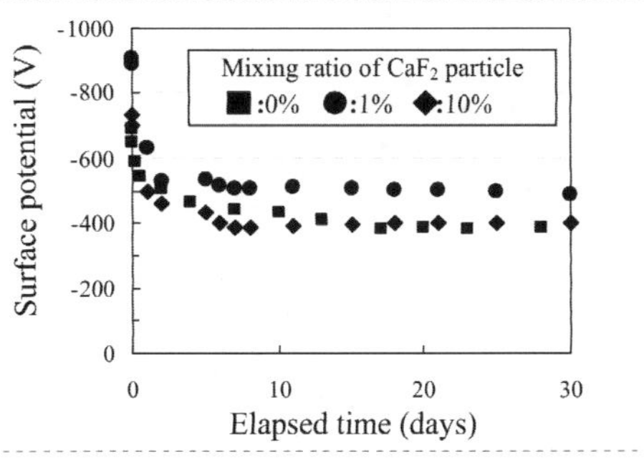

Figure 6. Time degradation of surface potential of electrets made of mixture of CYTOP and CaF_2 nano-particles.

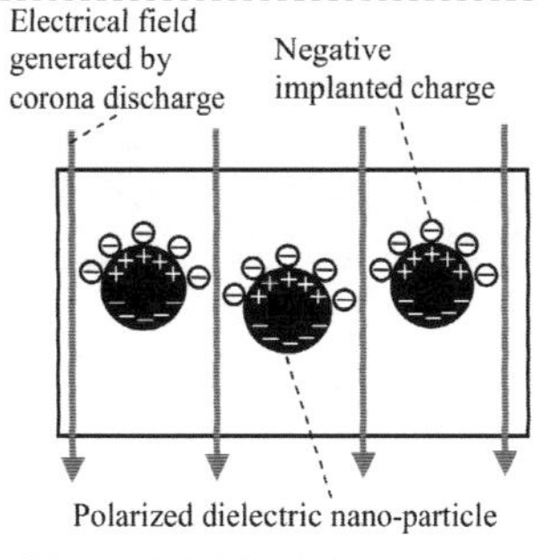

Figure 7. Schematic illustration describing how polarized nano-particle captures electric charge [4].

3.2. Fabrication of VEHs using CYTOP with nano-particles electret and their characterization

Five kinds of small VEHs were fabricated using CYTOP mixed with the particles. Schematic diagram and photographs of the fabricated VEH are shown in Fig. 8. Then electric power generated by the fabricated VEHs was characterized. The measurement system for the characterization is shown in Fig. 9. The result of characterization is shown in Fig, 10. This result shows that the maximum power of the approximately 50 μW was obtained when the VEH using CYTOP mixed with 1% of $BaTiO_3$ nano-particles as the electret. This generation power is about 10 times as large as the VHE using pure CYTOP. On the other hand, the generation power of VEH using CYTOP with 10% of $BaTiO_3$ becomes much smaller than that using pure CYTOP as the electret. This result indicates that the mixing of small amount (>10%) of dielectric particles into fluorpolymer electret is effective to increase the capability of power generation of VHE.

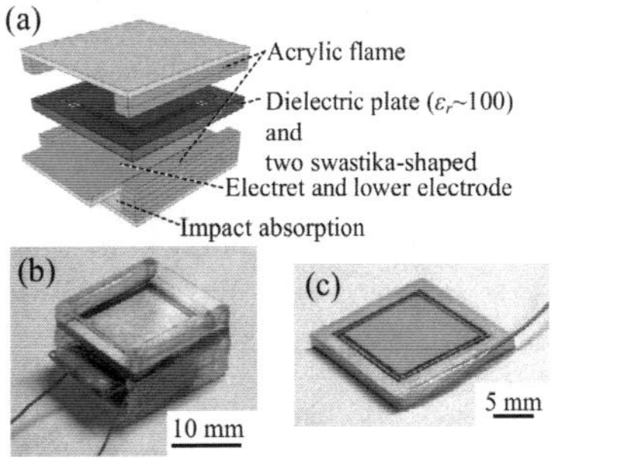

Figure 8. (a) Schematic illustration and (b) photograph of fabricated small vibration energy harvesting device, and (b)swastika-shaped flat spring used in the device.

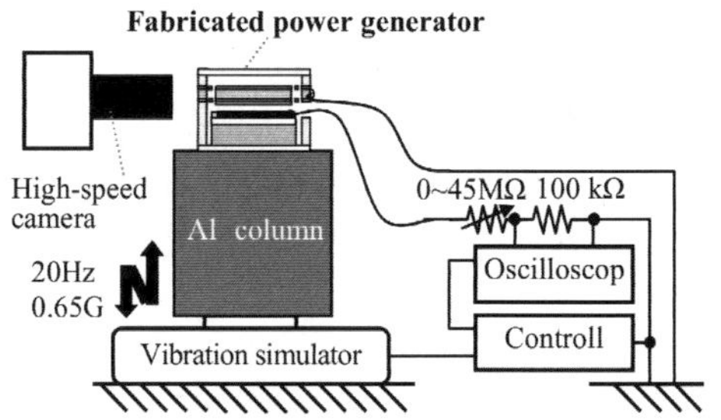

Figure 9. Setup for measuring generation power of the fabricated vibration energy harvesting device.

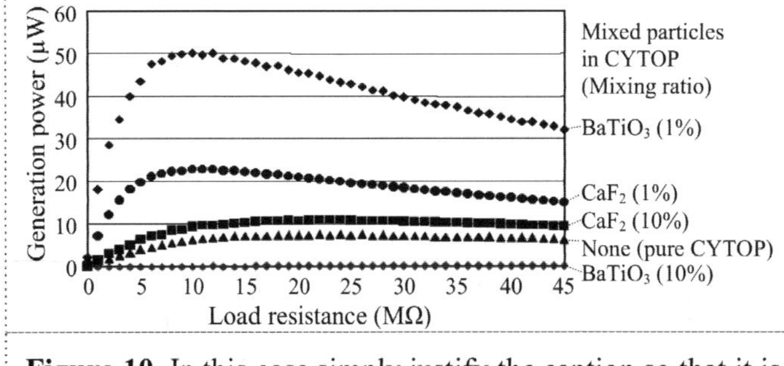

Figure 10. In this case simply justify the caption so that it is as the same width as the graphic.

4. Conclusion

In this study, we succeeded in raising the electric charge density of an electret made of one of fluorpolymer "CYTOP" by mixing dielectric particles. When the concentration of the particles in the CYTOP is smaller than several percent, the surface potential of the CYTOP with nano-particles after charge implantation monotonically increases with concentration of the nano-particles. However, the mixed nano-particles made an aggregate, and implanted charge discharged through it. This result indicates that the nano-particles should be mixed in the CYTOP with optimized concentration. We also developed an electrostatic vibration energy harvester (VEH) using CYTOP with nano-particles as an electret, which generated more electric power than the VEH using pure CYTOP.

5. References

[1] S. P. Beeby, M. J. Tudor, and N. M. White 2011 *Meas. Sci. Technol.*, **17**, 175-195
[2] K. Matsumoto, K. Saruwatari, Y. Suzuki 2011 *in Digest Tech. PowerMEMS'11*, 134-137
[3] T. Tsutsumino, Y. Suzuki, N. Kasagi, and Y. Sakane 2006 *in Digest Tech. IEEE Internat. Conf. MEMS'06* 98-101
[4] T. Takahashi, M. Suzuki, T. Nishida, Y. Yoshikawa, and S. Aoyagi, 2012 *in Digest Tech. MEMS'12* 1265-1268
[5] T. Takahashi, M. Suzuki, T. Nishida, Y. Yoshikawa, and S. Aoyagi 2013, *in Digest Tech. MEMS'13* 877-880
[6] Y. Sakane, Y. Suzuki, N. Kasagi *Journal of Micromechanics and Microengineering* 2008 **18** 104011(6 pp)
[7] K. Kashiwagi, K. Okano, N. Tanabe, Y. Sera, T. Miyajima, Y. Morizawa, Y. Sakane, Y. Hamatani, F. Nonaka, A. Asakawa, and Y. Suzuki 2011 *in Proc. IEEE ISE14*, 71-72

Acknowledgments

This work was supported in part by a grant of "Strategic Research Foundation Grant-aided Project for Private Universities": Matching Fund Subsidy MEXT (Ministry of Education, Culture, Sport, Science, and Technology, Japan), 2010-2014 (S1001048). This work was supported in part by JSPS (Japan Society for the Promotion of Science) KAKENHI (26870728). The CaF_2 nano-particles used in this study were crushed from micrometre scale to nanometre scale by the aid of THINKY CORPORATION.

Development of a pre-packaged MEMS electret energy harvester before charging

Seonwoo Kim[1], Qianyan Fu[1], Kei Hagiwara[1], and Yuji Suzuki[1]

[1]Department of Mechanical Engineering, The University of Tokyo,
7-3-1 Hongo, Bunkyo-ku, Tokyo 113-8656, Japan

Email : skim@mesl.t.u-tokyo.ac.jp

Abstract. A novel MEMS electret vibration energy harvester sealed in a package before charging is developed, which should improve the package reliability and long-term stability of the charges. With an early prototype in package, vertical electret on the comb drives has been successfully charged with soft X-ray photoionization, and up to 3.58 μW has been obtained at 570 Hz and 2.2 g oscillation, which corresponds to the effectiveness as high as 42%.

1. Introduction

Rapid progress of low-power-consumption electronics opens up the possibility of powering wireless sensor nodes by harvesting ambient energy from the environment, which will make significant contribution to the vision of Internet of Things (IOT) [1, 2, 3]. Especially, structural vibration is one of the most promising energy sources [2]. Among various energy conversion principles from kinetic energy to electricity, electrostatic/electret energy harvesters have advantage in efficient use of ambient vibration energy at low frequency vibration [3,4].

Hoffmann et al. [5] developed an in-plane electrostatic energy harvester with overlapping-area-change comb drives and achieved output power of 1.26 μW (RMS value) with 6.5 g acceleration at the resonant frequency of 1460 Hz. Guillemet et al [6] prototyped an electrostatic vibration energy harvester with gap-closing combs and obtained output power of 2.2 μW with acceleration of 1 g at 150 Hz. Whereas these electrostatic generators with comb drives can realize relatively-high effectiveness (the ratio of the electrical power output and the VDRG limit [7]), external power source is inevitable for the priming voltage.

In our previous studies, we developed in-plane electret MEMS energy harvesters [8, 9]. Fu & Suzuki [9] prototyped an electret energy harvester with vertical electrets on the comb drives, which are charged with soft X-ray irradiation [10]. By combining the overlapping-area-change and the gap-closing mechanisms, they achieved 1.6 μW with 2 g acceleration at 266 Hz. The effectiveness is as high as 57 %.

For long-term stability of electret, energy harvesters should be packaged to avoid unwanted effect of dust and moisture. However, in previous electret energy harvesters, electret is always charged before packaging, which limits the use of thermal packaging methods due to the charge decay of electret during the thermal cycle.

In the present study, an in-plane MEMS electret energy harvester with comb drives packaged before charging is developed, and the soft X-ray irradiation is used for charging vertical electrets on the comb drives.

2. MEMS electret vibration energy harvester with comb drives

Figure 1 shows fabrication process for an in-plane MEMS electret energy harvester with comb drives [9]. The Silicon-on-Insulator (SOI) MEMS technology is used to microfabricate the device. Process starts with standard lithography on a 4-inch SOI wafer with a 70 μm-thick device layer (Fig. 1a). The device layer is etched with DRIE for comb drives, springs, and etching holes (Fig. 1b). Then, vapor HF is used to remove the buried oxide layer through etching holes and release the structures (Fig. 1c). The device is bonded to a plastic package and the electrical connections are made with wire bonding (Fig. 1de). This is followed by parylene-C deposition on the whole structure, forming a 1.06 μm-thick electret layer on both sides of the interdigitated combs (Fig. 1f). Finally, the lid is bonded on the package in dry air (Fig. 1g), and the electret layer is charged with soft X-ray irradiation of 9.5 keV acceleration voltage (L9490, Hamamatsu) through the lid (Fig. 1h). The bias voltage is 100V.

Figure 2 shows the MEMS electret energy harvester thus fabricated. The seismic mass is 4.893 mg supported with four tilted folded-beam springs, which enable the maximum allowable traveling distance as large as 45 μm. The maximum capacitance change ΔC is 14.7 pF in the harvester with 600 finger pairs for the overlapping-area-change (OV) type and 452 finger pairs for the gap-closing (GC) one. The gap for OV is 8 μm, while the initial/minimum gaps for GC are 33/2 μm.

Figure 3 shows the frequency response before and after the parylene-C deposition measured with a stroboscopic video microscopy (MSA-500, Polytec). The resonant frequency is decreased from 755

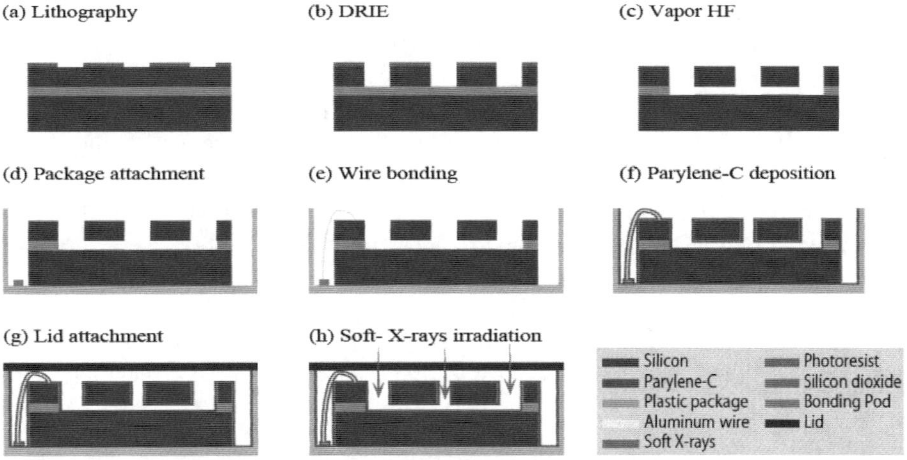

Figure 1. Process flow.

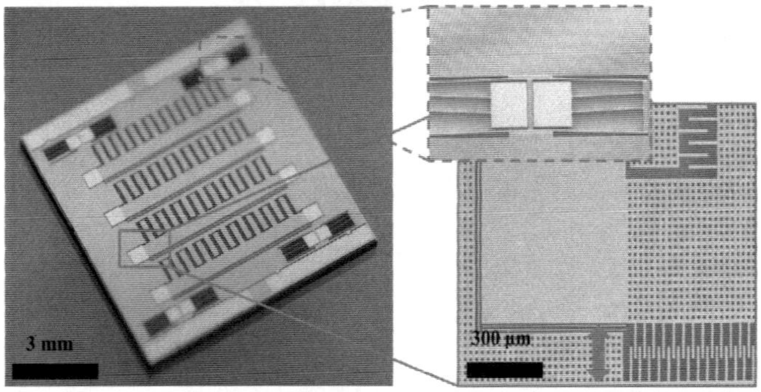

Figure 2. Overview of the MEMS electret energy harvester with comb drives [9].

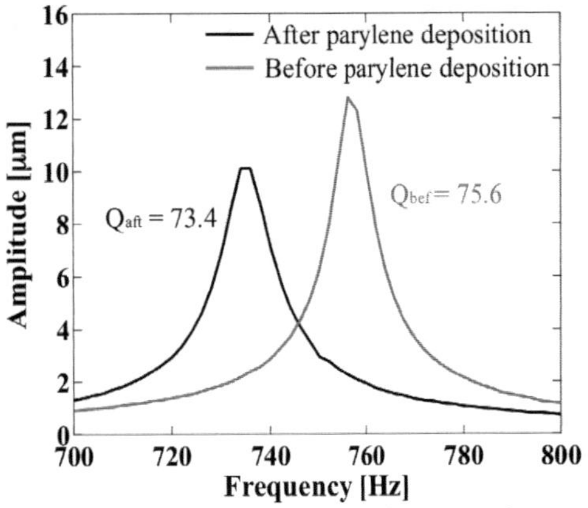

Figure 3. Frequency response before and after parylene-C deposition.

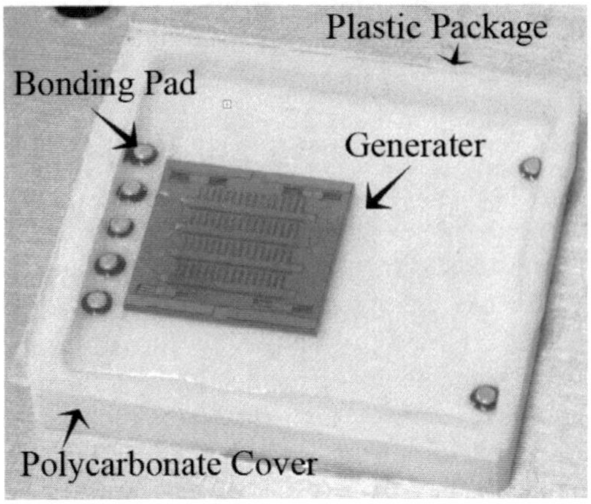

Figure 4. Prototype of a packaged MEMS electret vibration energy harvester with 0.5 mm-thick polycarbonate lid.

Hz to 735 Hz due to the added mass with the parylene-C deposition. The quality factor remains higher than 70 even after the deposition.

3. Charging experiment

In Fu and Suzuki [9], the charging time for 150 s is enough to get uniform surface potential in open air. In the present study, the penetration rates for different lid materials are measured to estimate proper charging time for devices in package. Figure 5 shows the setup to measure the penetration rate of the soft X-ray with 9.5 kV acceleration voltage. During charging, movable fingers of the comb drives are grounded, and fixed fingers are connected to a source measure unit (2600, Keithley) for applying the bias voltage and for measuring the ion current. The distance between the soft X-ray bulb to the surface of device is 5 mm, while the distance between the surface and the lid is 2.5 mm. The penetration rates of 500 μm-thick polycarbonate (PC), 300 μm-thick glass, and 100/200 μm-thick aluminum lids are measured. Table 1 summarizes the measured ion current and the penetration rate of different lid materials. The penetration rate for 500 μm-thick PC is 13 %, while 100 μm-thick aluminum is 7 %. The penetration rate can be increased when an X-rays source of higher energy is employed.

In order to confirm the charging performance of the pre-packaged device, a Kelvin force microscopy (KFM, SPM-9600, Shimadzu) is used to measure the surface potential of the vertical

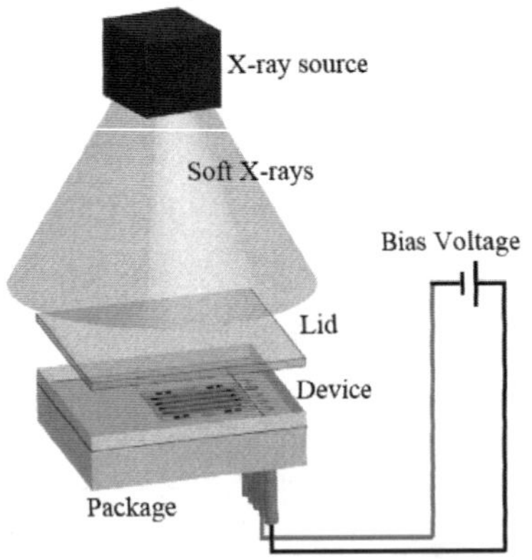

Figure 5. Setup for soft X-ray charging.

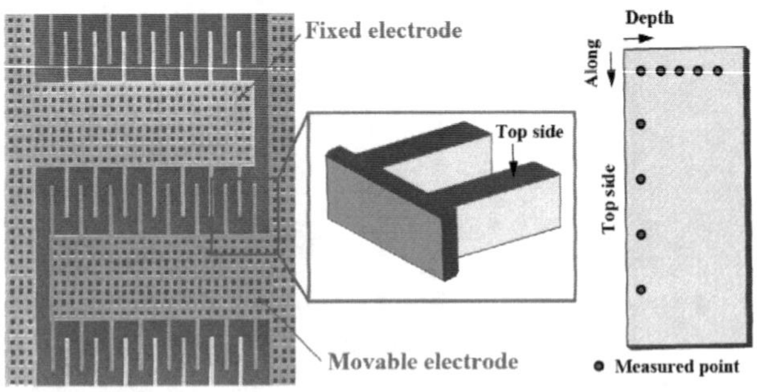

Figure 6. KFM measurement of surface potential for the vertical electrets.

Table 1. Ion current and penetration rate of different lid materials.

	w/o lid	PC 500 μm	Al 100 μm	Al 200 μm	Glass 300 μm
Ion Current [nA]	22.2	3.74	1.56	0.98	0.71
Penetration Rate [%]	-	13	7.0	4.4	1.2

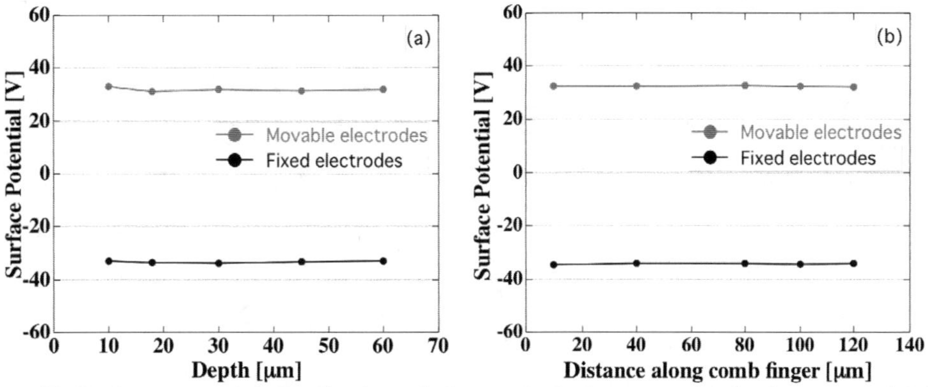

Figure 7. Surface potential distribution of the vertical electrets. (a) Surface potential in the depth direction, (b) Surface potential in the longitudinal direction.

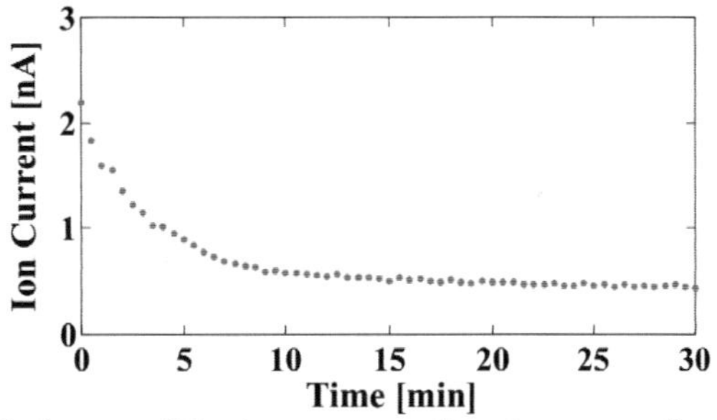

Figure 8. Temporal change of the ion current after the onset of soft X-ray irradiation.

electrets. After charging through a 100 μm-thick aluminum lid for 2143 seconds, the device is intentionally broken, and one of the comb fingers is fixed onto the KFM stage. Figure 5 shows the measured points.

Figure 6 shows the surface potential distribution of the vertical electrets. The result shows uniform surface potential distribution with ±32 V (surface potential difference of 64 V), which is somewhat lower than the applied bias voltage of 100 V. Longer charging time than estimates with the penetration rate seems necessary to fully charge the device.

4. Power generation experiment

A device in a plastic package with 500 μm-thick polycarbonate lid is charged in a setup shown in Fig. 5. The device is exposed to soft X-rays for 30 minutes with 100V bias voltage to ensure complete charging. Figure 8 shows the ion current during the charging process.

The device is fixed onto an electromagnetic shaker (ET-126B-1, Labworks), and in-plane oscillation is applied. The amplitude is measured with a laser displacement meter (LC-2430, Keyence).

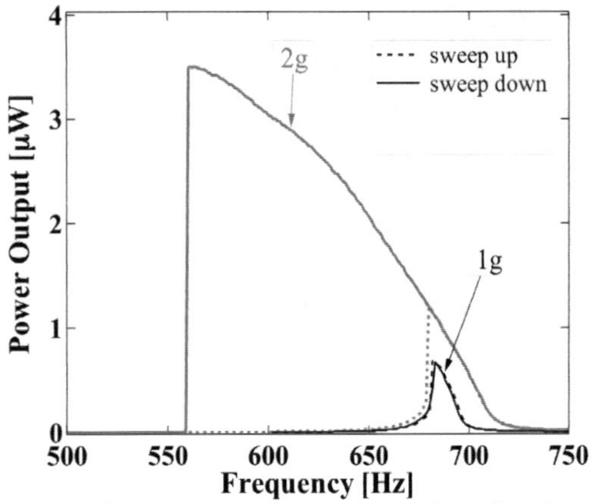

Figure 9. Power output versus the vibration frequency.

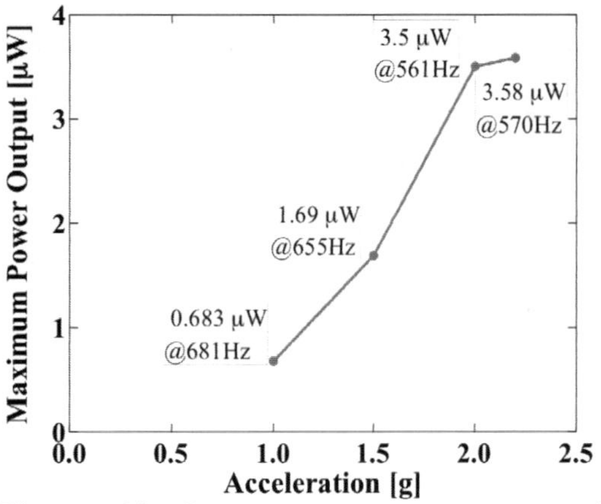

Figure 10. Power output versus external acceleration.

Figures 9 and 10 show the output power of the present device for external sinusoidal acceleration. The external load is chosen as 2.03 MΩ. At 1 g, 0.683 μW is obtained at 680 Hz with minimal hysteresis. At 2.2 g, up to 3.58 μW is obtained at 570 Hz. The effectiveness (the power output divided by the VDRG limit [9]) at 2.2 g is as high as 42%. With increasing acceleration, the power output exhibits a marked hysteresis. The nonlinear behavior becomes significant due to the softening mechanical springs at high acceleration.

5. Conclusion

We have developed a novel pre-packaged MEMS electret energy harvester and successfully charged with soft X-rays irradiation for the first time. The maximum output power of 3.58 μW has been obtained at 570 Hz and 2.2 g. The effectiveness at 2.2 g is as high as 42 %. Note that, with the present charging method after packaging, thermal packaging process operated at 300℃ (450℃ for parylene HT electret) can be used. Although plastic package and lid are used in the present early prototype, metal/ceramic package/lid can be used as far as sufficiently-large transmission rate of the soft X-ray is obtained.

This work is partially supported by NEXCO Group Companies' Support Fund to "Disaster Prevention Measures on Expressways". Photo-mask is made using the University of Tokyo VLSI Design and Education Center (VDEC)'s 8-inch EB writer F5112+VD01 donated by ADVANTEST Corporation. One of the authors (Seonwoo Kim) was supported through the Leading Graduates Schools Program, "Global Leader Program for Social Design and Management" by the Ministry of Education, Culture, Sports, Science and Technology.

References

[1] R.J.M. Vullers et al. Solid-State Electr. **53** 684–693 (2009)
[2] S. Roundy et al. Compt. Commun. **26** 1131–1144 (2003)
[3] Y. Suzuki, IEEJ Trans. Electr. Electr. Eng. **6** 101-111 (2011).
[4] Y. Sakane et al., J. Micromech. Microeng. **18**, 104011 (2008).
[5] D. Hoffmann et al., J. Micromech. Microeng. **19**, 094001 (2009)
[6] R. Guillement et al., IEEE MEMS 2013, 137-140 (2013)
[7] P. D. Mitcheson et al., *Proc. IEEE* **96**, 1457-1486 (2008).
[8] Y. Suzuki, et al., J. Micromech. Microeng. **20**, 104002 (2010)
[9] Q. Fu, and Y. Suzuki, IEEE MEMS'14, San Francisco, 409-412 (2014).
[10] K. Hagiwara, et al., IEEE Trans. Dielectr. Electr. Insul. **19**, 1291-1298 (2012).

Performance enhancement of an out-of-plane electret-based vibrational energy harvester with dual charged plates

K. Tao[1], S. W. Lye[1]*, J. M. Miao[1]* and Xiao Hu[2]
[1] School of Mechanical and Aerospace Engineering, Nanyang Technological University, 50 Nanyang Avenue, 639798 Singapore
[2] School of Materials Science and Engineering, Nanyang Technological University, 50 Nanyang Avenue, 639798 Singapore

E-mail: mswlye@ntu.edu.sg; jmiao@pmail.ntu.edu.sg

Abstract. This paper presents a novel out-of-plane electret-based vibrational power generator (EVPG) that has both negative and positive charged electret plates integrated into a single seismic mass system. Compared with the conventional single-charged two-plate configuration, the proposed device not only exhibits an enhanced output voltage magnitude but also has a wide operational bandwidth due to spring softening nonlinearity according to the experimental analysis. With the acceleration changes from 0.1g to 0.5g, the operating 3-dB bandwidth can be increased 2.6 times from 2.5 Hz to 6.5 Hz, which indicates higher energy conversion efficiency than a linear energy harvester in the practical scenario of broadband random vibrations.

1. Introduction

In recent years, the energy harvesting devices draw a lot of attention as a promising power source for wireless sensor networks (WSNs), which are attractive in many potential fields, such as healthcare, infrastructure and transportation intelligent monitoring[1, 2]. Such energy harvesting technology could offer viable solutions to the maintenance and replacement of battery in WSN to meet the vast and impending demand. Mechanical vibration sources are ubiquitous and readily available in the ambient environment. Several techniques have been proposed in the past decade to perform the electromechanical energy conversion such as piezoelectric, electromagnetic and electrostatic methods. Among these, the electrostatic/electret energy harvesting systems have the advantages both of compatibility with silicon CMOS processes and miniature size [3-5].

A conventional type EVPG generally has a laterally movable seismic mass integrated with parallel interdigital electrodes and stripe-patterned electret [4, 6]. The eletret films are normally designed with tens or hundreds of micros to be compatible with the small movement of the spring-mass system. They are first micro-patterned, etched and then charged. However, these processes may harm the performance and charge stability of the eletret films [7, 8]. The existence of strong fringe effect caused by small-dimension electret stripes also plays an important role in limiting the change of overall capacitance change[9]. Furthermore, the harvesters can only perform optimally when the two electrodes are precisely aligned with each other; slight misalignment may severely deteriorate the performance of the EVPG[10].

In this paper, a new EVPG with an out-of-plane gap closing scheme is proposed, where both negative and positive charged electret plates are integrated into a single spring-mass system. Since the capacitance change of the proposed EVPG mainly relies on the variation of air gap between the two electrodes, micro-patterning of the electret thin film and precise alignment are not required. Figure 1 shows the schematic diagram of the new concept with dual charged plates compared with the conventional out-of-plane gap closing EVPG. By introducing another positive charged electret plate, the increased electric field between the counter and base electrodes not only leads to an enhanced output voltage magnitude but also gives rise to a stronger nonlinear effect of the spring-mass system. These aspects will be further investigated in the ensuing sections

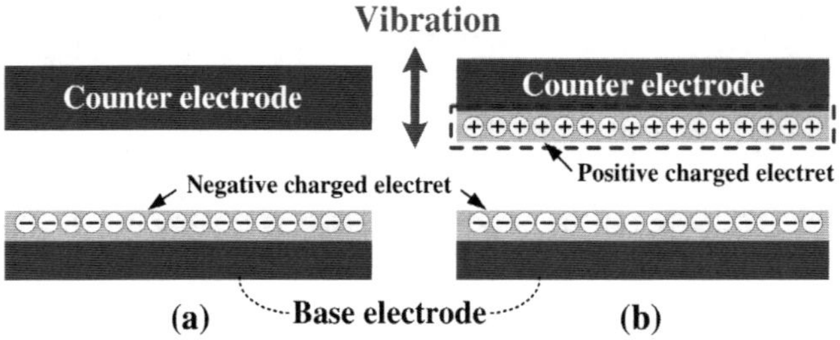

Figrue 1. Schematic diagram of the new concept compared with the conventional two-plate out-of-plane gap closing EVPG

2. Device design and fabrication

Figure 2 shows the schematic structure of the EVPG that seeks to harvest kinetic energy with an out-of-plane gap closing scheme. It consists mainly of two parallel-silicon plates. The top plate composes of a movable spring-mass structure with gold electrodes on it. The seismic mass with a height of 300 μm and a diameter of 6 mm is suspended by three sets of spiral springs. Each spiral beam is with a height of 300 μm and a width of 50 μm at a beam spacing of 250 μm. The electrets mounted on top and bottom plates can be charged and used as a positive/negative permanent voltage source. When the mass with top electrode is driven by an external vibration, the air gap between the electrode and electret would change, causing a variation of the capacitance and thus generating an alternating current in the external circuit.

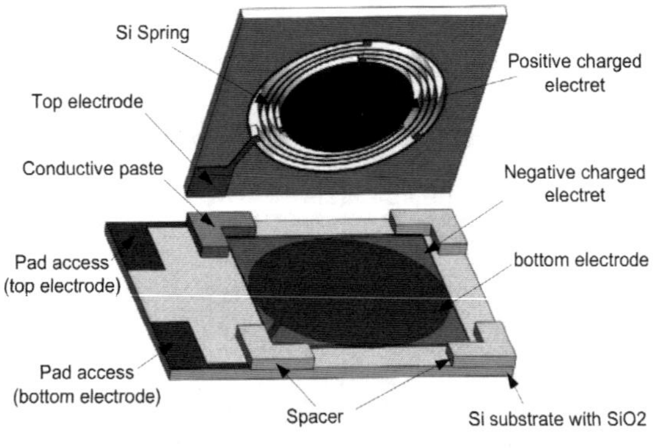

Figure 2. 3-D schematic view of the proposed out-of-plane EVPG with dual charged electret plates

The presented out-of-plane micro EVPG is fabricated using MEMS micromachining technique based on silicon substrate. The surface of the silicon is deposited with 1 μm SiO_2 as an insulation layer by plasma-enhanced chemical vapour deposition (PECVD). The fabrication of spring-mass structure adopts a two-mask process. The electrode is fabricated through a lift-off process with an adhesive film of reactive metal Cr of 50 nm followed by a 300 nm metallic film of non-reactive metal Au. An Al thin film with 600 nm thickness is sputtered on the backside of the top plate that facilitates heat dispassion. Then the wafer is bonded with a carrier wafer by a 5 μm thick thermal grease. Reactive-ion etching (RIE) and deep reactive-ion etching (DRIE) are used to remove the exposed SiO_2 and Si layer to define the geometry of the spring-mass structure, respectively. A commercial-grade film of LDPE (GOODFELLOW, USA)) thin film with a thickness of 15 μm is used as electrets in the prototype fabrication. The electret thin film is mounted on both the top and the bottom electrodes and then charged with a surface potential of 800V and -800V by the corona charging, respectively, which can serve as a constant voltage bias. After all the components are prepared, the EVPG prototype is finally constructed by assembling the top spring-mass plates and the bottom plates with charged electrets by a flip-chip package. Figure 3 shows optical photograph of an assemble out-of-plane EVPG prototype and a fabricated spring-mass structure. The overall size of the prototype is only about $10\times12\times1$ mm^3.

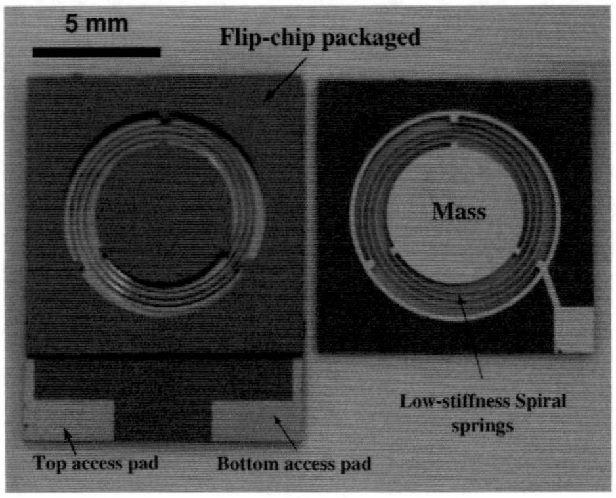

Figure 3. Optical photographs of an assemble out-of-plane EVPG prototype and a fabricated spring-mass structure

3. Power generation characterization

The fabricated EVPG prototype is characterized using the experimental setup as shown in Figure 4. It mainly consists of a shaker, an accelerometer, a power amplifier, a function generator, and a data acquisition system. The prototype is attached to a linear piezoelectric shaker that mimics the external ambient excitations. The harmonic excitation signal is created by the function generator, adjusted by the power amplifier and finally fed to the piezoelectric shaker. The motion of the central seismic mass is captured by a high speed camera. An acceleration sensor is fixed on the base of the holder to monitor the excitation acceleration. A data acquisition system (DAQ NI USB-6289 M series) is utilized to record the electrical output of the prototype.

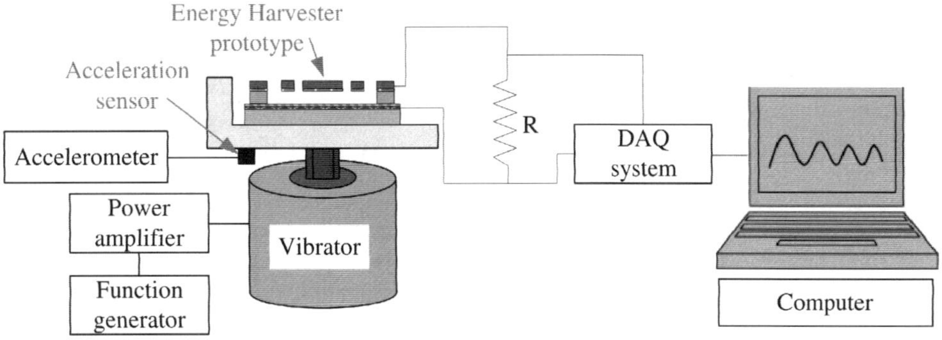

Figure 4. Block diagram of the experiment setup

During the experiment, the characterizations are carried out with excitation frequencies manually swept from 60 Hz to 90 Hz at different acceleration levels from 0.1g to 0.9 g having a load resistance of 50 MΩ. The testing is performed twice before and after the top electrode is positively charged. Figure 5 shows the comparison of output voltage responses with different excitation accelerations for dual-charged and single-charged electret plates with a load resistance of 50MΩ. It is found that the output voltage amplitudes increased from 0.35V to 1.92V for single-charged type and from 0.62V to 3.5V for dual-charged type when accelerations increase from 0.1g to 0.9g. The voltage magnitudes in both types are almost proportional to the excitation accelerations, while the increase rate of the dual-charged type is almost twice than that of single-charged plate and has a much larger magnitudes. This phenomenon also accompanies with a stronger spring softening phenomenon occurred in dual-charged type that the resonant frequencies shift from 77.9 Hz at the acceleration of 0.1g to 73.4 Hz when the acceleration increased to 0.8g in an frequency-up sweep process.

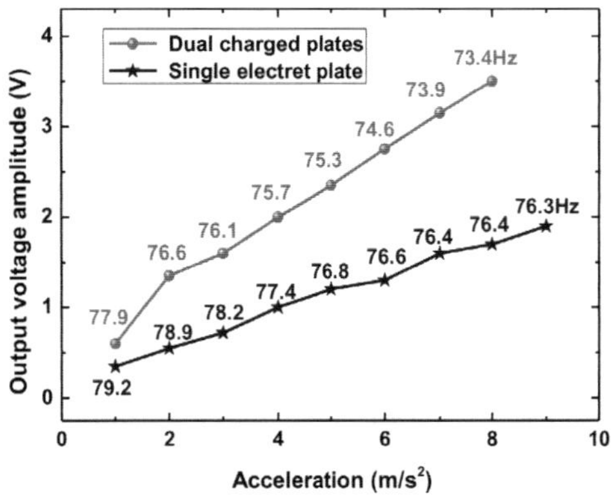

Figure 5. Comparisons of the output voltages with different excitation accelerations for dual-charged and single-charged electret plates in the frequency-up sweeps

This phenomenon is further investigated and illustrated in Figure 6, which shows the frequency responses with excitation accelerations increased from 0.3g to 0.5g. The frequency-up and frequency-down curves at the accelerations of 0.3g, 0.4g and 0.5g are recorded as shown in Figure 6(a), (b) and (c), respectively. The 3-dB bandwidth has been increased from 2.6 Hz to 6.5 Hz when the acceleration increases form 0.1g to 0.5g due to the spring nonlinearity caused by the electret/electrostatic force. This indicates that this type of EVPG can be further exploited for scavenging low-level ambient energy from broadband random vibrations.

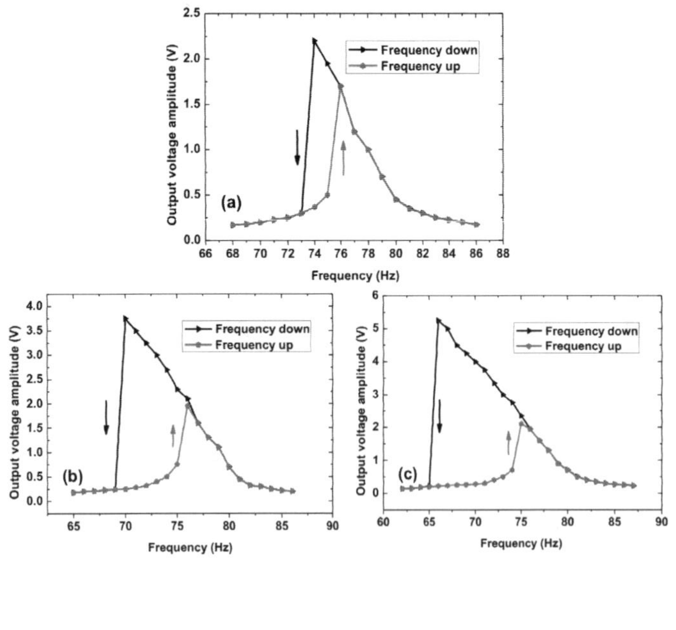

Figure 6. Measured frequency responses of the output voltage in up and down frequency sweeps at different excitation acceleration levels: (a) 0.3g; (b) 0.4g; (c) 0.5g

5. Conclusion

In this paper, a novel out-of-plane MEMS EVPG with dual-charged electret plates is designed, fabricated and tested. By the merit of this configuration, the proposed device not only exhibits an enhanced output voltage magnitude but also has a wide operational bandwidth due to spring softening nonlinearity induced by electret/electrostatic force. Spring softening nonlinearity and wide band frequency response were also investigated, where the spring softening effect was observed to be stronger with increase in the acceleration levels.

6. Acknowledgements

This research is supported by Energy Research Institute conducted at Nanyang Technological University, Singapore.

References

[1] Beeby S P, Tudor M J, and White N M, 2006 *Meas. Sci. Technol.* **17** 175-R195

[2] Mitcheson P D, Yeatman E M, Rao G K, Holmes A S and Green T C, 2008 *Proceedings of the IEEE* **96** 1457-1486

[3] Okamoto H, Suzuki T, Mori K, Cao Z, Onuki T and Kuwano H, 2009 *International Journal of Energy Research* **33** 1180-1190

[4] Suzuki Y, 2011, IEEJ Trans. *Electr. Electron. Eng.* **6** 101-111

[5] Tao K, Liu S, Lye S W, Miao J M and Hu X, 2014 *J. Micromech. and Microeng.* **24** 065022

[6] Boisseau S, Despesse G and Sylvestre A, 2010 *Smart Materials & Struct*, **19**, 105013

[7] Fujita T, Toyonaga T, Nakade K, Kanda K, Higuchi K and Maenaka K 2010 *Procedia Engineering* **5** 774-777

[8] Sonoda K, Onishi T, Minami K, Fujita T, Kanda K and Maenaka K, 2013 *J. of Phys.: Conf. Ser* **476** 012077

[9] Chiu Y and Lee Y C, 2014 *J. Micromech. and Microeng.* **23** 015012

[10] Minakawa Y, Chen R and Suzuki Y 2013 *17th Int. Conf. Solid-State Sensors, Actuators and Microsystems (TRANSDUCERS' 13)*, Barcelona pp 2241-4

A Design of Mechanical Frequency Converter Linear and Non-linear Spring Combination for Energy Harvesting

K Yamamoto[1, 2], T Fujita[1], A Badel[2], F Formosa[2], K Kanda[1] and K Maenaka[1]
[1] Graduate School of Engineering, University of Hyogo, Himeji, Japan
[2] SYMME Laboratory, University of Savoie, Annecy, France

E-mail: er14h053@steng.u-hyogo.ac.jp

Abstract In this study, the improvement of energy harvesting from wideband vibration with random change by using a combination of linear and nonlinear spring system is investigated. The system consists of curved beam spring for non-linear buckling, which supports the linear mass-spring resonator. Applying shock acceleration generates a snap through action to the buckling spring. From the FEM analysis, we showed that the snap through acceleration from the buckling action has no relationship with the applied shock amplitude and duration. We use this uniform acceleration as an impulse shock source for the linear resonator. It is easy to obtain the maximum shock response from the uniform snap through acceleration by using a shock response spectrum (SRS) analysis method. At first we investigated the relationship between the snap-through behaviour and an initial curved deflection. Then a time response result for non-linear springs with snap through and minimum force that makes a buckling behaviour were obtained by FEM analysis. By obtaining the optimum SRS frequency for linear resonator, we decided its resonant frequency with the MATLAB simulator.

1. Introduction

The Energy harvesting (EH) technologies have been studied for a wide research region [1]. We are focusing on the vibration type energy harvester (VEH) for human application [2]. In the majority of the previous works, the linear vibration mechanism has been used for large vibration amplitude at resonant situation. However, that kind of ideal resonance can hardly be obtained from an actual ambient vibration e.g. automotive or human movement. Thus, some group try to generate energy from the wideband or unstable vibration with nonlinear spring structures [3] [4].

In this study we propose a novel combination of linear and nonlinear spring structures for VEH fabricated by the MEMS (Micro Electro Mechanical Systems) technology. The nonlinear part that generates shock acceleration is realized by the snap through behaviour of a curved clamped-clamped beam spring. Then the generated shock acceleration is applied to another involved linear resonator system for harvesting energy. By using this combination, we can generate shock acceleration with uniform duration and a high mechanical Q factor linear resonator.

The Q factor for the linear dumped oscillation resonator directly impacts the harvested energy. We can design a high Q factor resonator separately without buckling mechanism. In our previous work, we successfully generated about 10 μW power at the resonance frequency of 433 Hz, and 200 μm amplitude from the electret type harvester with a Q factor of 130. High Q device shows large output however it has very narrow bandwidth to resonance. Thus we consider the combination of our linear electret harvester with additional shock generation mechanism with buckling spring. This method can be useful not only to electret type but also any vibration type harvester. On the other hand, the nonlinear spring system shows good performance for the random vibrations. A bistable type VEH can effectively collect mechanical power from the wideband frequency, however it mainly generates energy during snap through behaviour because of its low Q factor.

In next section, we explain a design method for nonlinear buckling spring and evaluate a time response of its buckling behaviour.

2 Designs of spring combination

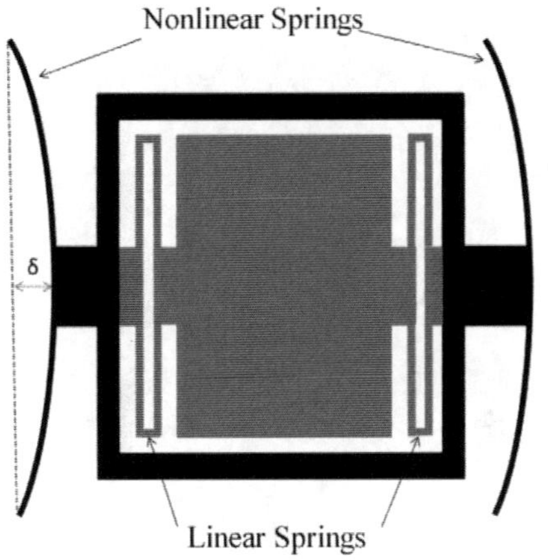

Figure 1. The proposal design of Combination springs

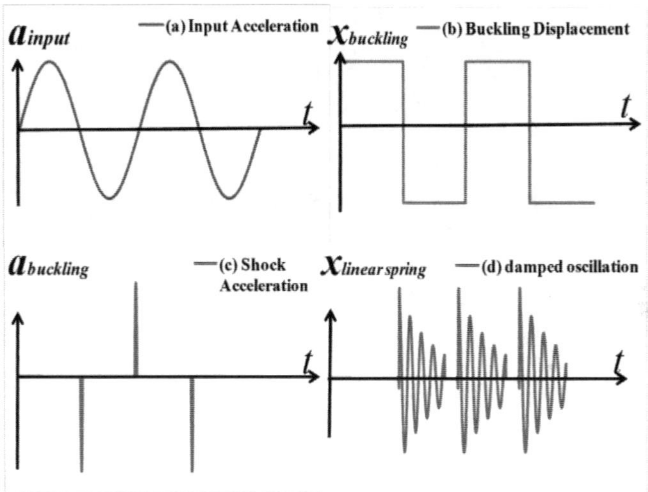

Figure 2. The ideal operation of mechanical frequency converter

The proposal design of our combination device is shown in figure 1. A pair of clamped-clamped curved beam-spring supports an inner frame. The inner frame contains the linear resonance harvester with high Q factor. By applying external acceleration to the harvester (Fig. 2a), the inner frame shows snap through movement (Fig. 2b). The snap through behaviour generates high and uniform acceleration as shown in Fig. 2(c). It means that the output frequency always constants against various external oscillation. Then from the maximum shock response by SRS (Shock response spectrum) analysis, the resonant frequency of the linear resonator can be obtained.

2.1 Buckling spring design

From the theoretical calculation, buckling beam displacement depends on the applying force and on its stiffness. However the initial shape of our buckling spring is not straight but curved. The beam stiffness and required force for the snap through action are then calculated by the FEM software (ANSYS 15). Figure 3 shows the simulation model, the beam length L is 10 mm, beam width is 60 μm, beam thickness is 525 and the initial deflection was varied from 50 μm to 200 μm. The proof mass size is 10 mm × 10 mm × 525 μm, 0.12 mg. We just considered the simplified mass instead of the inner frame resonator and two curved beam springs for preliminary calculation.

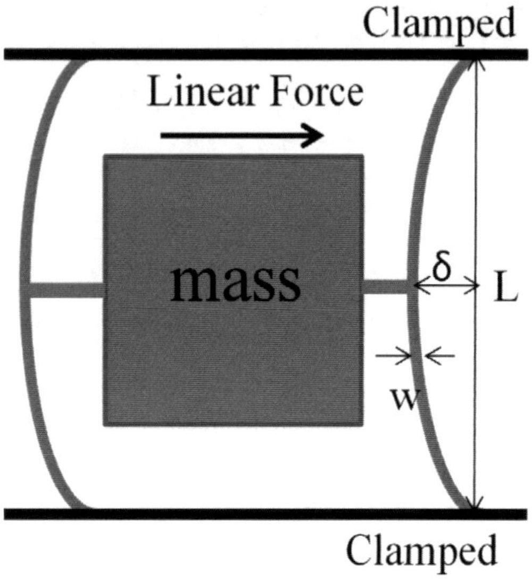

Figure 3. Simplified model for non-linear spring-mass system analysis

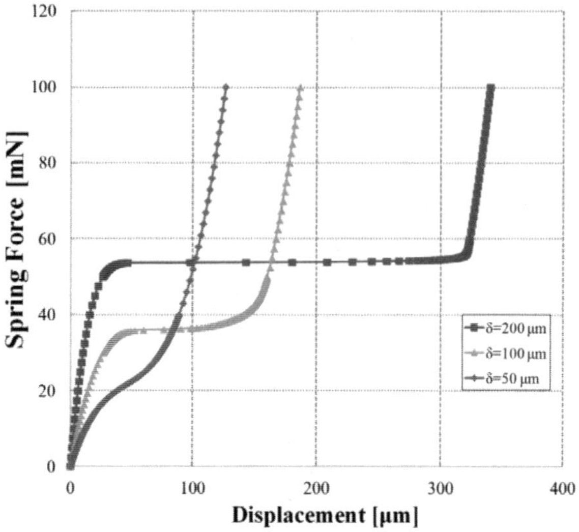

Figure 4. The F-X curve by FEM Simulation (ANSYS 15.0 Workbench)

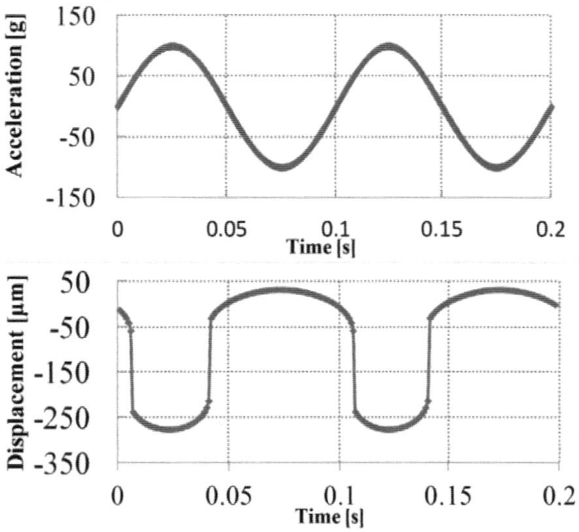

Figure 5. Time transient oscillation (ANSYS 15.0 Workbench)

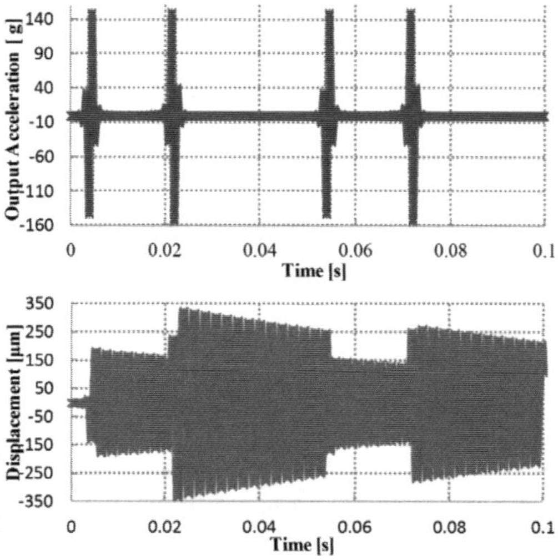

Figure 6. The shock acceleration and damped oscillation with inner linear springs

The Simulation result is shown in figure 4. Buckling displacement and required force for snap through were decreased by initial deflection. The initial deflection controls the nonlinear spring stiffness.

We also made the transient analysis of the curved spring model. Figure 5 shows the applied acceleration with frequency of 10 Hz, amplitude of 100 g and the simulated result of displacement for an initial deflection δ of 100 μm. Because of the lightweight mass and high spring-stiffness, large amplitude of acceleration was required for the snap-through to occur, however the result shows the snap-through behavior clearly as expected.

2.2 SRS optimized damped oscillation

From a shock acceleration of snap-through action as described above, the curved snap-through displacement and its duration time are extracted. From the response to the applied snap-through shock, we can evaluate the damped oscillation for the inner linear resonator. The vibration waveform was calculated by using the MATLAB software as shown in figure 6. There are two setting parameters of linear spring, the Q factor of 130 and the resonant frequency of 433 Hz. These parameters are decided from our previous work. In the result, the damped oscillation occurred at each snap-through shock acceleration. The response has however not been optimized yet.

3. The shock response method for inner linear spring.

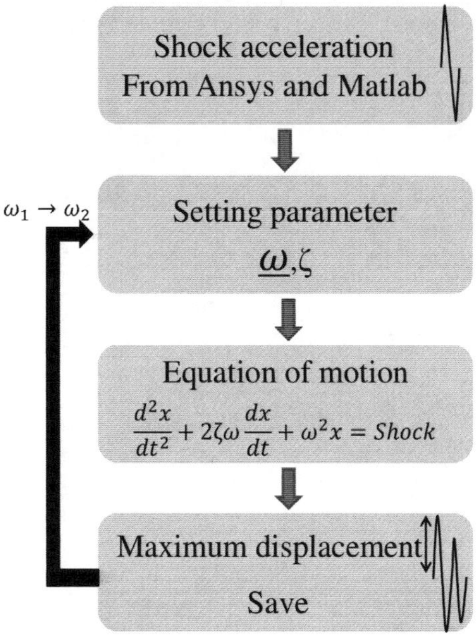

Figure 7. The flow graph of SRS

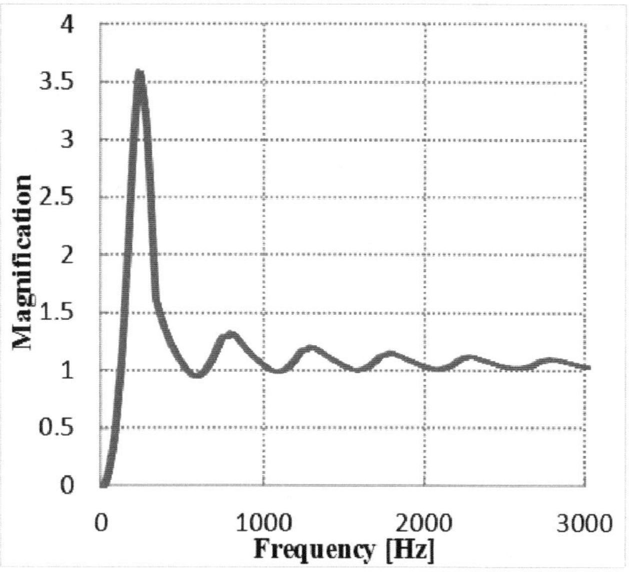

Figure 8. The Simulation result of the proposed SRS analysis

The SRS analysis is a well-known method to evaluate a response for shock. In order to optimize the spring parameter design, we introduce the SRS analysis as a spring combination tool. The simulation flow of SRS is shown in figure 7, which consists of four parts. The first step, the shock acceleration waveform from ANSYS is simulated, and then converted to the MATLAB analysis data. The second, the two parameters of resonant frequency ω and damping coefficient ξ are set. Next, the equation of motion is solved for increasing values of ω. The result shows damped oscillation that depends on the parameters ω and ξ. By recording the maximum displacement of the oscillation, the SRS result can be obtained as Figure 8. Considering the SRS result, it is shown that if we design the linear resonator with a 250 Hz resonant frequency, 3.6 times larger displacement will be applied to the linear resonator than on the nonlinear spring, which will also improve greatly the generated power.

4. Conclusion

We reported the optimum design for mechanical frequency converter by using linear and nonlinear spring combination. The spring stiffness and motion for snap-through behaviour of a nonlinear buckling spring are simulated. From the result, the SRS optimized resonant frequency design of the linear resonator is investigated. Matching the SRS maximum frequency, the linear resonant harvester in damped oscillation mode will be applied 3.6 times larger initial displacement from the wideband applied acceleration.

Reference

[1] Paradiso, J. A., & Starner, T. *'Energy Scavenging for Mobile and Wireless Electronics'. IEEE Pervasive Computing, 4(1), 18–27*

[2] Maenaka, K., Masaki, K., & Fujita, T. *Application of Multi-Environmental Sensing System in MEMS Technology-Monitoring of Human Activity. Networked Sensing Systems, 47–52.*

[3] W Q Liu et al. *'Novel piezoelectric bistable oscillator architecture for wideband vibration energy harvesting' Smart Mater. Struct. 22 (2013) 035013*

[4] Vocca, H., Neri, I., Travasso, F., & Gammaitoni, L. Kinetic '*Energy harvesting with bistable oscillators'. Applied Energy, 97, 771–776.*

IR Thermal Imaging Device using Photo-Patternable Temperature Sensitive Paint

T Tsukamoto, M Wang and S Tanaka
Department of Bioengineering and Robotics, Tohoku University
6-6-01, Aoba Aza Aramaki Aoba-ku, Sendai, Miyagi, 980-8579, Japan

E-mail: t_tsuka@mems.mech.tohoku.ac.jp

Abstract. This paper reports an infrared-to-visible transducer array made of temperature sensitive paint (TSP) for low-cost thermal imaging application. A novel fabrication process using a photo-patternable temperature sensitive paint (PTSP) combined with an SU-8 transfer method was developed. The developped process is simpler than before, and prevents the TSP structure from plasma-induced damage and sticking across a sacrificially-etched gap. The self-suspended structure as small as 100 μm was successfully fabricated with a large gap of 40 μm from the substrate. The heated object of 300°C was detected with a resolution of about 0.4 mm.

1. Introduction

An infrared (IR) sensor array is widely used for a non-contact thermal imaging such as disease survey in airports, failure analysis of electric components and scientific experiments. Nowadays, some consumer applications of the IR sensor array such as home security, human activity monitoring and precise fire detection have been proposed, and a low cost sensor array is required. Various types of IR sensors have been developed, including a bolometer [1], a thermopile [2], a pyroelectric [3] and a mechanical types [4]. However, the bolometer and pyroelectric sensor array require complicated structures made of non-CMOS-compatible materials on a read-out LSI, which increases the fabrication cost. On the other hand, the thermopile array is one of the promising low-cost devices, but it needs a large in-plane thermal isolation structure, which limits the spatial resolution. Anyway, a large array of thermopiles also need sensor fabrication on a read-out LSI.

Recently, an array of IR-to-visible transducers was proposed as the low-cost thermal imaging device [5, 6]. Figure 1 shows the working mechanism of the device. A thin film made of a temperature sensitive paint (TSP) and an IR absorber is suspended by a thermal isolation structure. The TSP, which has been used as a temperature sensor in a wind tunnel experiment [7] and a micro channel evaluation [8], emits the visible luminescence whose intensity decreases with temperature [9]. The thermally isolated TSP thin film is heated by IR light radiated from the observation object, thus the incident IR power can be converted to the intensity of visible light, which can be observed by a CCD/CMOS camera. The optical read-out method used here does not need any electric wirings, which promises high thermal isolation of each pixel and makes it scalable to large pixel numbers.

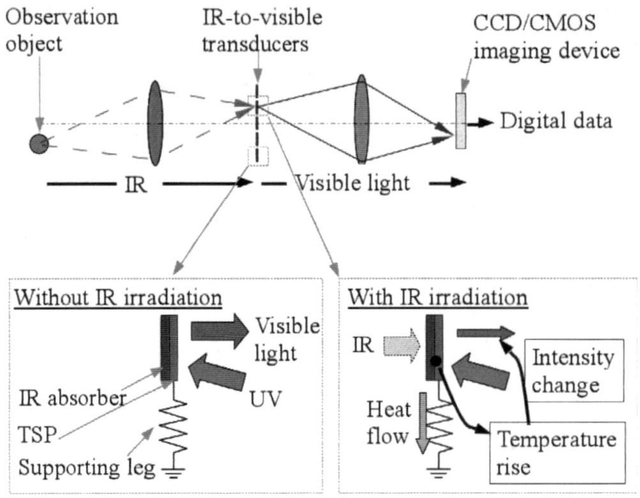

Figure 1. Principle of the thermal imaging using IR-to-visible transducer.

The previous reported IR-to-visible transducer was made by using reactive ion etching (RIE) for the micro-patterning of the TSP [6]. However, RIE sometimes caused plasma-induced damage in the TSP, which degrades the sensitivity. In addition, a thick sacrificial layer was required to prevent the self-suspended structure from sticking to a substrate, leading to severe control and long deposition time. To solve these problems, we propose a new fabrication process using a photo-patternable TSP (PTSP) combined with au SU-8 transfer method to make a large-gap self-suspended structure without a thick sacrificial layer.

2. IR-to-visible transducer using PTSP

2.1. Structure of the device

Figure 2 shows the structure of the device. The sensing part is made of the PTSP, a reflector and an IR absorber. The thermal isolation supporting structure consists of PTSP narrow beams and SU-8 long posts. The high-aspect-ratio SU-8 structure enables the self-suspended structure separated far away from the substrate, which prevents the structure from sticking. A glass was used as the substrate to transmit both excitation UV light and visible luminescence.

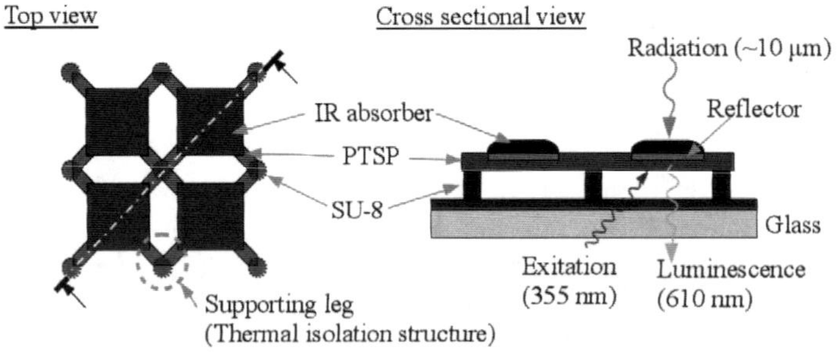

Figure 2. Structure of the IR-to-visible transducer using the PTSP.

2.2. Photo-patternable temperature sensitive paint

The conventional TSP consists of a luminescent material such as $Eu(TTA)_3$, a polymer binder such as PVB or PMMA and a solvent such as MEK or IPA. To add a photo-patternability to the TSP, we use SU-8 as the polymer binder and the solvent [10]. The composition of the PTSP was 0.03 g of

$Eu(TTA)_3$ and 1.0 mL of SU-8 5. The temperature sensitivity at the room temperature was about –0.3%/K.

2.3. Fabrication

Figure 3 shows the fabrication process. First of all, the Ge thin film, which was used as the sacrificial material, was deposited on a temporary glass substrate and partially etched to make a mold for the IR absorber. A black-body paint (JSC-3, Jpana sensor Co., Ltd., Japan) mixed with SU-8 5 was spread into the mold. After planarization, an Al thin film was deposited and patterned. The PTSP and SU-8 3050 were patterned by photolithography. Another glass substrate for a device was covered with SU-8 5 without UV exposure. Both temporary and device wafers were aligned and heated at 95°C for bonding using the unexposed SU-8 as a bonding material. Finally, the Ge sacrificial layer was etched by XeF_2 and the temporary glass substrate was separated.

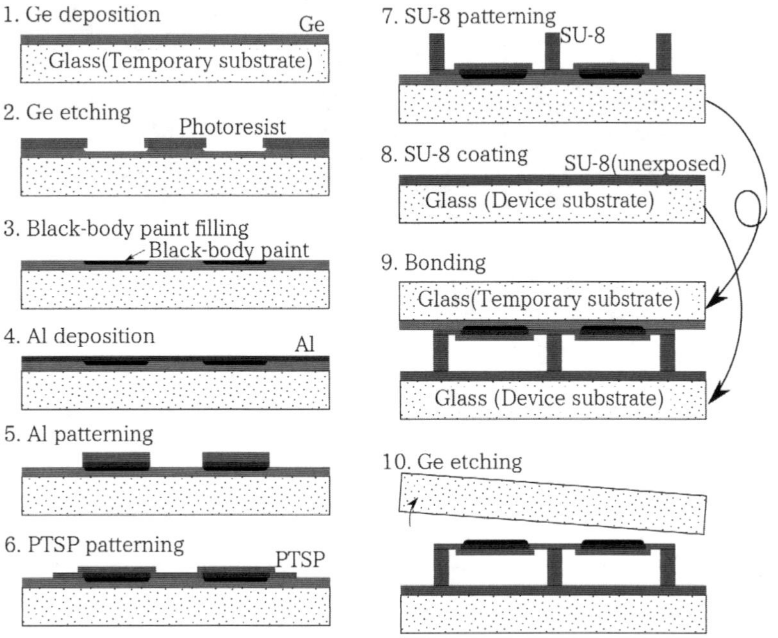

Figure 3. Fabrication process.

Figure 4 shows the scanning electron micrograph of the device. The size of each sensing element was 100 μm. The distance between the self-suspended structure and the substrate was as high as 40 μm.

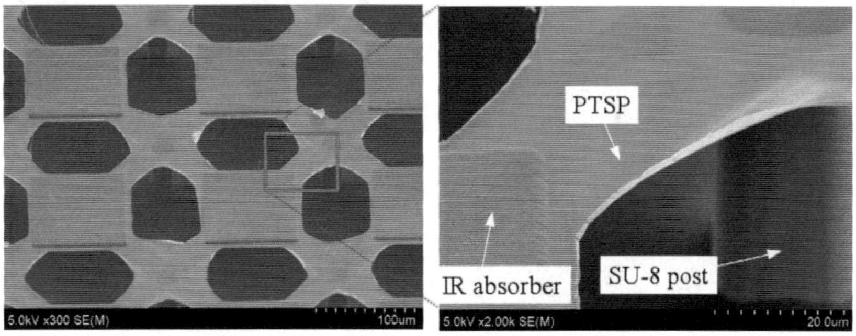

Figure 4. SEM of the self-suspended IR-to-visible transducers.

3. Thermal imaging

Figure 5 shows an experimental setup for the thermal imaging. The fabricated IR-to-visible transducer array was placed in a vacuum chamber to enhance the thermal isolation of each element. A Ge objective lens and a glass window were placed on both sides of the chamber. The PTSP was excited by UV-LEDs (NS355L-5RLO, Nitride Semiconductors Co., Ltd., Japan), and the luminescence was observed by the CCD camera (BU-50LN, Bitran Corp., Japan) through an optical band-pass filter. All the optical components were placed in a black box to eliminate the ambient light. A ceramic heater shown in Fig. 6 was used as an observation object. A K-type thermocouple was attached to the heater using ceramic glue to measure the real temperature.

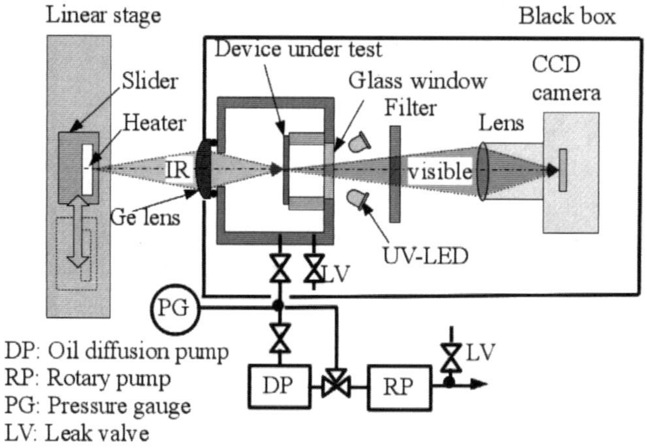

Figure 5. Experimental setup.

Figure 7 shows a captured luminescence. The temperature rise of the sensing element, $\Delta T(x, y)$, at the point (x, y) was calculated as

$$\Delta T(x,y) = S^{-1}\left(\frac{I(x,y) - I_0(x,y)}{I_0(x,y)}\right) , \quad (1)$$

where S, I and I_0 are the sensitivity of PTSP, the intensity of luminescence with and without IR irradiation, respectively [11].

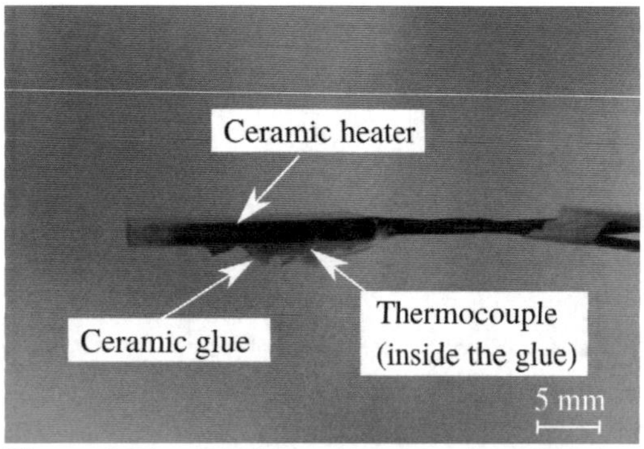

Figure 6. Ceramic heater (observation object).

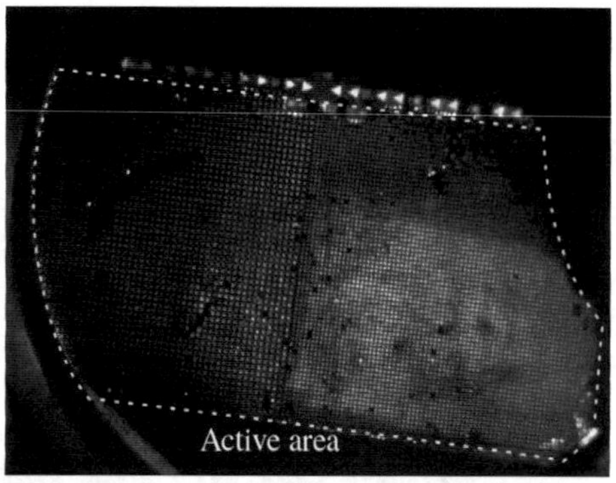

Figure 7. Luminescence from the device.

Figure 8 shows the obtained thermal images of the heater at different locations. The temperature of the heater was kept at 700°C. The value indicated in Fig. 8 was the temperature rise of the sensing element calculated by Eq. (1). The shape of the heater with a ceramic glue (Fig. 6) was clearly observed. The resolution of the thermal image was as small as about 0.4 mm. The thermal time constant of the sensing element was measured about 2 s. The minimum detectable temperature was as high as 300°C due to not-yet-optimized process and structure.

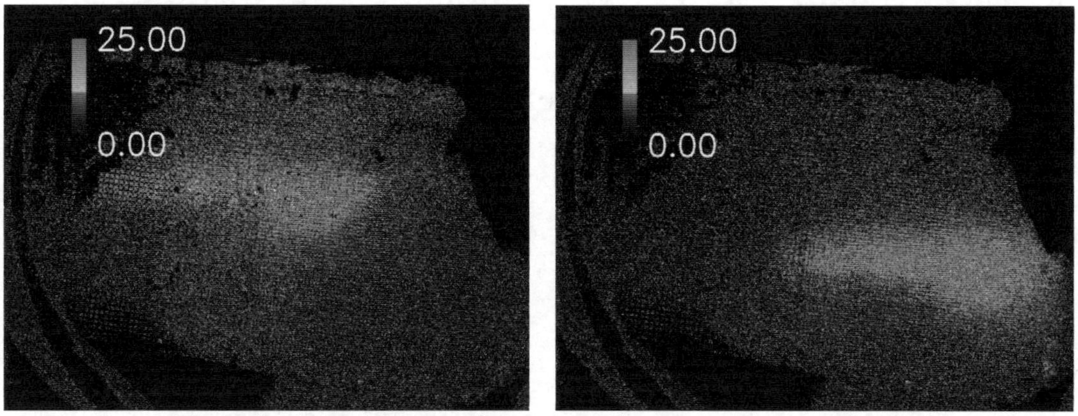

Figure 8. Thermal images of the ceramic heater at various locations.

4. Conclusion

An IR-to-visible transducer array using photo-patternable temperature sensitive paint was developed for low-cost IR imaging. A transfer method was employed to fabricate a large-gap self-suspended structure without an ultra-thick sacrificial layer. The self-suspended structure with 100 μm square pixel was fabricated with a gap as high as 40 μm. The thermal image was successfully observed. The obtained resolution, thermal response time and minimum detectable temperature were about 0.4 mm, 2 s and 300°C, respectively.

References

[1] Wood R A 1993 *Proc. IEDM1993* (Washington DC, USA) 175–8
[2] Kruse P W 2002 *SPIE Press* 49–56
[3] Teranishi N 1997 *Semiconduct. Semimet.* **47** 203–18
[4] Steffanson M, Gorovoy K, Kampmann R, Kleindienst R, Sinzinger S and Rangelow I W 2012 *Proc. MME2012* (Ilmenau, Germany) C11
[5] Tsukamoto T, Esashi M and Tanaka S 2013 *Proc. MEMS 2013* (Taipei, Taiwan) 421–4
[6] Wang M, Tsukamoto T and Tanaka S 2014 *Proc. APCOT 2014* (Daegu, Korea) 9–2
[7] Nagai H, Ohmi S, Asai K, Nakakita K 2008 *J. Thermophys. Heat Transfer* **22** 373–81
[8] Huang C-Y, Li C-A, Wang H-Y, Liou T-M 2013 *J. Micromech. Microeng.* **23** 037001
[9] Kolodner P and Tyson J A 1982 *Appl. Phys. Lett.* **40** 782–4
[10] Tsukamoto T and Tanaka S 2013 *J. Phys.: Conf. Ser.* **476** 012073
[11] Tsukamoto T, Esashi M and Tanaka S 2013 *J. Micromech. Microeng.* **23** 114015

Acknowledgement

This research was partly supported by Creation of Innovation Centers for Advanced Interdisciplinary Research Areas Program.

A tree-on-a-chip: design and analysis of MEMS-based superheated loop heat pipes exploiting nanoporous silicon membranes

I.-T. Chen[1], **and A. D. Stroock**[1,2]

[1]School of Chemical and Biomolecular Engineering
[2]Kavli Institute at Cornell for Nanoscale Science
Cornell University, Ithaca, NY, USA

E-mail: ic249@cornell.edu

Abstract. This paper reports the design, fabrication and analysis of a plant-inspired, MEMS-based superheated loop heat pipe (SHLHP) that would exploit nanoporous membranes to allow for operation with large capillary pressures and superheated liquid. The operating principles of SHLHPs differ from conventional designs in 1) the un-coupling of the working fluid from its saturation curve to eliminate limitations associated with temperature head and sub-cooling conditions and 2) the possibility of maintaining sub-saturation throughout the device to eliminate film condensation and improve the condenser thermal conductivity. Nanoporous silicon membranes integrated with DRIE channels are fabricated and characterized. The ability of the membrane to hold liquid under tension is tested by equilibrating water-filled device with various relative humidity and observing the cavitation events within individual voids underneath the membrane. Silicon membranes with desired functionality are further incorporated with patterned glass substrates to form prototype MEMS-based SHLHPs.

1. Introduction

Among many different technologies transferring heat, heat pipes (HPs) – closed-circuit, two-phase heat transfer devices in which a working fluid transfers its latent heat as it cycles between an evaporator and a condenser – have offered an attractive alternative in an array of applications[1], such as integrated circuits, systems for the storage of energy, vehicles and aircrafts, environmental controls in buildings, and energy efficiency of industrial processes. Unlike many other cooling technologies where external pumping power is required, HPs are passive systems that the central components required for HPs are wick structures that allow for the existence of a pressure difference, ΔP_{cap} [Pa] between the liquid and vapor phases of the working fluid based on capillarity, as described approximately by the Young-Laplace law:

$$\Delta P_{cap} = 2\gamma \cos\theta_c / r_p \tag{1}$$

where γ [Pa m] is the surface tension of the liquid, θ_c is the contact angle and r_p [m] is the radius of the pore. This capillary pressure can drive the working fluid around its cycle if a difference of temperature exists between the evaporator and the condenser.

For applications regarding heat transfer over long distance or with liquid motions against gravity /acceleration field, conventional HPs with large pores of the wicks can no longer provide sufficient capillary pressure head to overcome the total pressure losses along the fluid cycle. Loop heat pipes

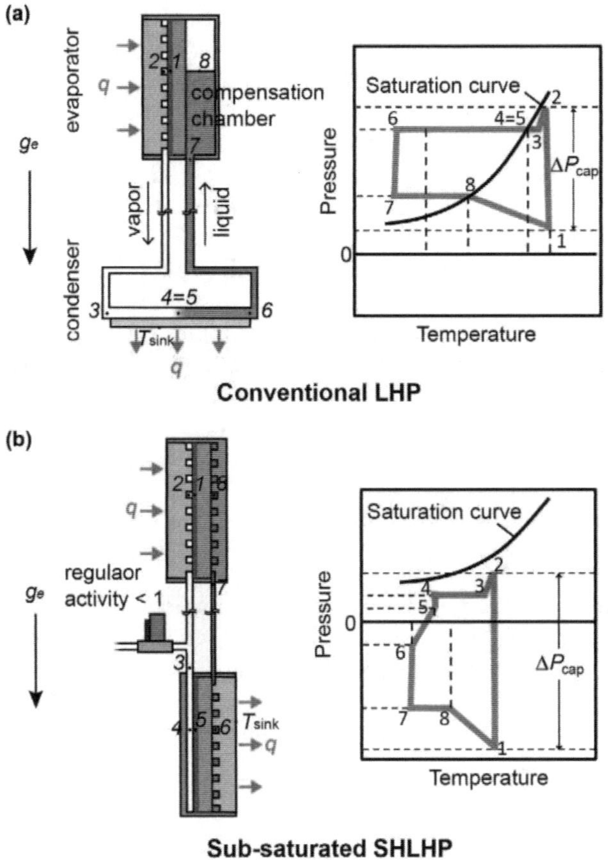

Figure 1. Schematic cross-sectional views and working cycles on *P-T* diagram for **(a)** a conventional LHP and **(b)** a sub-saturated SHLHP. The numbered points in the cycles correspond to state of the working fluid at the labelled points. An acceleration, g_e, acts along the pipe axis. Heat enters at the evaporator with a rate q. ΔP_{cap} represents the capillary pressure required to overcome the total pressure drop in the entire loop. In *P-T* diagrams, The red lines indicate the working fluid is in vapor phase (Path 2-4); the blue lines (Path 5-1) indicate liquid phase.

(LHPs), Fig. 1a, have been developed to extend the limits of conventional HPs. The main design rules of LHPs are[2]: 1) minimization of the distance that the liquid must move within the wick by placing a thin membrane in the evaporator, separating macroscopic conduits for the liquid and vapor, and 2) the introduction of a compensation chamber in the cycle to accommodate excess working fluid and allow the system to adapt to changing heat loads and temperatures. LHPs have been shown to provide robust operation and improved performance relative to conventional HPs, including higher heat load capacities and compatibility with a wider variety of architectures for adaptation to specific applications[3]. However, in conventional LHPs, vapor-liquid coexistence in the compensation chamber forces the thermal cycle to remain close to the saturation and leads to two constraints – the temperature head and the sub-cooling conditions. Additionally, the condensate film in the condenser adds conductive resistance and may cause undesired oscillations in response to heat load steps[4].

Motivated by vascular plants - the transpiration process of which is analogous to the operation of LHPs, we proposed, in the paper published previously[5], a new design of LHP - Superheated Loop Heat Pipe (SHLHP – Fig. 1b) with introductions of 1) nanoporous membranes allowing larger capillary pressures be maintained between the liquid and the vapor, 2) removal of the compensation chamber such that the entire liquid path can become superheated and hence decoupled from the saturation curve, and 3) a regulator maintaining sub-saturated state throughout the loop, eliminating liquid from the vapor path.

2. Operating principles and requirements

To gain a qualitative understanding of the differences between the conventional LHP and the SHLHP designs, one can compare their working cycles on P-T diagrams. Fig. 1 illustrates an idealized form of the working cycles of the conventional LHP and the sub-saturated SHLHP, with same pressure differences across the vapor path ($\Delta P_{2\text{-}3}$) and the liquid path ($\Delta P_{6\text{-}7}$) and the same condenser temperature (T_{sink}).

As shown in the working cycle for the conventional case, in the evaporator wick surface (from which evaporation occurs), the liquid (represented as Point 1) is near thermodynamic equilibrium with the vapor in the evaporator (Point 2, slightly sub-saturated due to curved menisci). Path 2-3 represents the adiabatic motion of the vapor in the vapor path. As the vapor enters the condenser, the temperature drops until condensation occurs at a macroscopic, vapor-liquid interface on the saturation curve (Point 4 is the vapor; Point 5, liquid). Path 5-6 represent sub-cooling of the liquid before it leaves the condenser. The latent heat released upon condensation is evacuated to a heat sink. Path 6-7 represents the adiabatic motion of the liquid in the liquid path. As the liquid enters the evaporator (Point 8), it is heated by the conduction through the wick membrane. Due to the presence of vapor in the compensation chamber, Points 8 is brought to saturation. Path 8-1 corresponds to the liquid motion through the wick membrane to the evaporating meniscus; the liquid confined within the wick membrane becomes superheated.

The above working cycle descriptions for the conventional design illustrate three important conditions for the operation of a conventional LHP: 1) the capillary pressure (ΔP_{cap}) developed by the porous wick must be able to overcome the total pressure drop in the entire loop:

$$\Delta P_{cap,\max} \geq \Delta P^{vap} + \Delta P^{liq} + \Delta P^{wick} \tag{2}$$

2) The motive temperature head condition - $\Delta P_{2\text{-}8}$ is responsible for driving the fluid motion through all components except the wick membrane. Given that these two points are saturated (or nearly so for Point 2), the temperature difference $\Delta T_{2\text{-}8}$ is determined by the shape of the saturation curve. This temperature difference must be established across the wick to create the corresponding pressure difference that drives fluid around.

$$\Delta P_{2-8} \cong \left.\frac{dp_s}{dT}\right|_{\bar{T}_e} \Delta T_{2-8} \tag{3}$$

where dp_s/dT is the slope of the saturation curve at the evaporator temperature. 3) The sub-cooling condition - The heat leaked through the evaporator wick membrane into the compensation chamber, q_{leak} [W], must be balanced by the sensible heat of returning cold working fluid:

$$q_{leak} = \frac{\Delta T_{2-8}}{\mathrm{R}_{\mathrm{wick,e}}} = QC_p^{liq}(\Delta T_{8-7})\,. \tag{4}$$

This balance implies that the sub-cooling of the liquid before it enters evaporator, $\Delta T_{8\text{-}7}$, grows with the increasing of $\Delta T_{2\text{-}8}$ and the decreasing of the thermal resistance of the evaporator wick, $\mathrm{R}_{\mathrm{wick,e}}$ [K W^{-1}]. Combined with Eq. 3, the required sub-cooling is coupled to $\Delta P_{2\text{-}8}$ via $\Delta T_{2\text{-}8}$.

The working cycle for SHLHP illustrates the effect of eliminating the compensation chamber from the conventional design. Without vapor on the liquid side of the evaporator, Point 8 is no longer constrained to be on the saturation curve. As the demand for pressure drop grows with additional heat load, acceleration or viscous drag, the pressures in the liquid at Points 7, 8, and 1 drop deeper into the superheated region. This use of reduced pressure or even tension (i.e. negative pressure, as presented in Fig. 1b) lessens the demand for elevated pressure in the vapor at Point 2, and the temperature head condition (Eq. 3) of the conventional design does not apply. Further, $\Delta T_{8\text{-}7}$ is independent of the pressure load (via Eq. 4 and as the lack of dependence of $\Delta T_{2\text{-}8}$ on load). With a condenser membrane and a regulator that pins the fluid activity < 1, the entire working cycle moves below the saturation line. Replacing the saturated condensation process with condensation of sub-saturated vapor directly on the nanoporous membrane could substantially reduce the temperature drop in the condenser.

3. MEMS-based SHLHP design

The global design of a sub-saturated SHLHP is similar to that of reversible loop heat pipes (RLHPs) in which the vapor and liquid paths are separate and there are wick membranes in both the evaporator

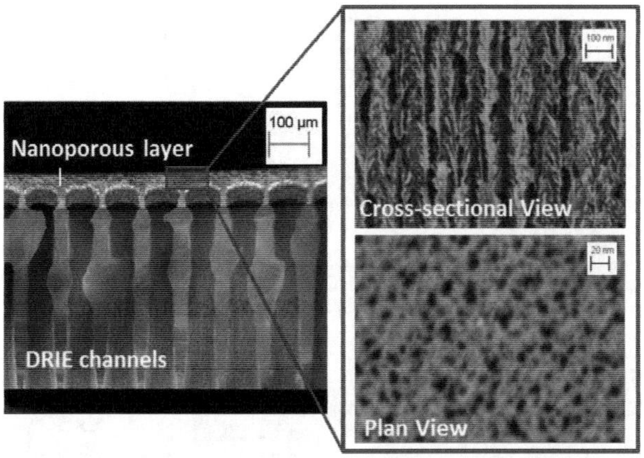

Figure 2. Electron micrographs of the two-layer membrane formed in silicon via electrochemical and deep reactive ion etches (DRIE): zoomed cross-sectional view (upper right) and plan view (lower right) showing sub-10 nm pores in the nanoporous layer.

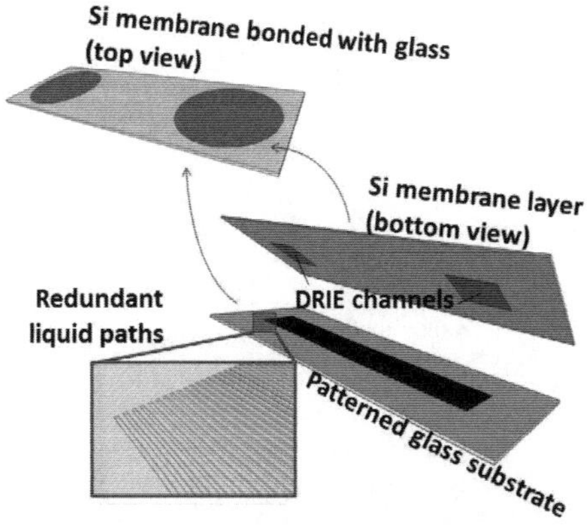

Figure 3. A design schematic of a prototype MEMS-based SHLHP. Silicon substrate with porous silicon membranes and DRIE channels is bonded to glass substrate with redundant liquid paths.

and condenser[5]. Two of the major challenges for the successful realization of SHLHPs are the potential for increased hydraulic resistance relative to conventional wicks and the increased proneness to dry-out due to boiling along the superheated liquid.

A desired membrane should meet the following criteria: 1) uniform, sub-micrometer to nanometer pores, 2) large permeability to avoid introducing excessive liquid tension, and 3) ability to sustain high pressure difference without fracture or collapse. We proposed a layer of nanoporous membrane be supported by microchannels (Fig. 2) that connect to the liquid paths. The nanoporous layer allows for the generation of large ΔP_{cap} between the liquid and vapor phases; the supporting microchannel layer presents a high permeability to flow and serves as a structural support of the thin nanoporous layer. Further, to ensure the continuing operation of a SHLHP after dry-out or boiling events, we propose hydraulically independent, redundant liquid paths (Fig. 3) to provide passive robustness to cavitation.

3.1. Materials and methods

For the prototype SHLHP, we proposed a MEMS-based, microscale SHLHP design, fabricated completely on silicon and glass wafers (Fig. 3). Nanoporous silicon membranes were prepared by electrochemical anodic etching of highly doped, p-type, <100> oriented silicon wafers with a resistivity of 0.003 Ω-cm. One silicon wafer was mounted between a Teflon cell and an Aluminium bottom electrode with one side of the silicon wafer in contact with the electrolyte, a 50:50 (v/v) solution of 49% hydrofluoric acid and 95% ethanol in the Teflon cell. Electrochemical etching was done under constant current density of 30 mA cm^{-2} for 20 minutes, resulting in a porous silicon layer of approximately 20 μm in thickness with a pore diameter of 5-10 nm (Fig. 2).

4. Results and discussion

4.1. Model analysis for steady-state operation

We presented in our previous paper[5] a unified model for steady state operation of both the conventional and superheated designs. By numerically solving the coupled heat and momentum balances and local thermodynamic equilibria (equations not shown), we can model the steady-state evaporator temperature at fixed heat flows and T_{sink}. Fig. 4 shows the operating curves for both conventional and SHLHP designs with geometry parameters identical to our MEMS-based device; evaporators are above the condensers (i.e. $g_e = 1g$). The evaporator temperatures of the SHLHP are more than 100°C lower than those of the conventional case.

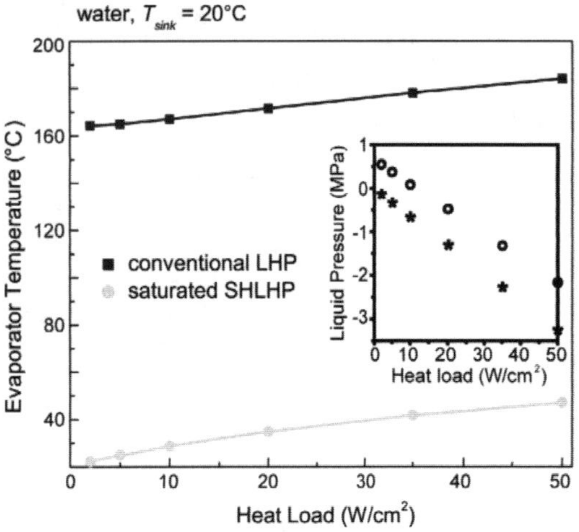

Figure 4. Operating curves for conventional and SHLHP designs with water as working fluid. Length, diameter and number of liquid paths: 6cm, 30μm and 100. Thickness and diameter of channels of the DRIE layer: 280μm and 50μm. Thickness and diameter of pores of the nanoporous layer: 20μm and 10nm.

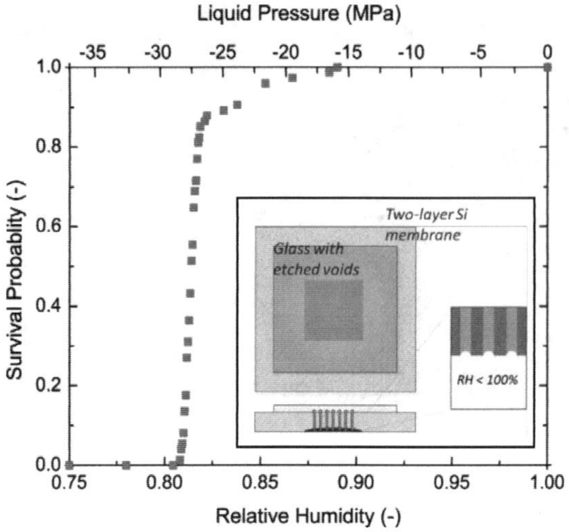

Figure 5. Survival probability for water in the device when silicon membranes were equilibrated with various water vapor activities (relative humidity) at 15°C. Corresponding negative pressures in the liquid within the glass cavities are given on the top axis. Insert: schematics illustrating the experiment set-up.

4.2. Stability limit of the nanoporous silicon membrane

The insert in Fig. 4 shows the predicted liquid pressure profile in the saturated MEMS SHLHP as a function of heat load; the tension introduced into the SHLHP can be more than -3 MPa. We estimated the ability of the silicon membrane to hold liquid under tension by equilibrating water-filled device with various relative humidity and observing the cavitation events within individual glass cavities underneath the membrane (Fig. 5). Our membrane sustains a cavitation pressure (probability reaches 1/2) of ~ -27MPa at 15°C; 100% of the voids survived tension up to -15MPa, much larger compared to the tensions required for the MEMS SHLHP operation.

5. Conclusions

This paper reports the design, fabrication, model analysis and membrane characterization of a MEMS-based SHLHP that exploits nanoporous silicon membranes to allow for operation with large tension in the liquid to provide better performance compared to conventional LHP. This MEMS-based SHLHP allows us to demonstrate for the first time the use of liquid at negative pressure in technologies. After scale-up, SHLHPs could be particularly valuable in applications such as cooling of avionics and energy management in buildings in which heat must be transferred over large distances and against gravity or acceleration.

References

[1] Faghri A 1995 *Heat Pipe Science and Technology* (New York: Taylor and Francis)
[2] Ku J 1999 *Proc. Intl. Conf. on Environmental System* (Denver, Colorado: Society of Automotive Engineers, Inc)
[3] Maydanik Y F 2005 *Appl. Therm. Eng.* **25** 635-657
[4] Pouzet E, Joly J L, Platel V, Grandpeix J Y and Butto C 2004 *Int. J. Heat Mass Transfer* **47(10-11)** 2293-2316
[5] Chen I T, Amit P and Abraham D S 2014 *AIChE J.* **60.2** 762-777
[6] Kim B H and Peterson G P 2005 *J. Thermophys. Heat Transfer* **19(4)** 519-526

Influence of geometric patterns of microstructured superhydrophobic surfaces on water harvesting performance via dewing

Donghyun Seo[1], Choongyeop Lee[2], and Youngsuk Nam[1]

[1]Department of Mechanical Engineering, Kyung Hee University, 1732 Deokyoungdaero, Giheung, Yongin, 446-701, Korea
[2]School of Aerospace and Mechanical Engineering, Korea Aerospace University, 76 Hanggongdaehakro, Deagyang, Goyang, 412-791, Korea

E-mail: ysnam1@khu.ac.kr

Abstract. We investigate how the wetting state of microstructured SHPo surfaces influences water harvesting performance via dewing by testing two different patterns including posts and grates with varying structural parameters. On grates, the observed Cassie wetting state during condensation is well described by the thermodynamic energy criteria, and small condensates can be efficiently detached from the surfaces due to the small contact line pinning force of Cassie droplets. Meanwhile, on posts, the observed wetting state is dominantly the Wenzel state regardless of the thermodynamic energy of each state, and the condensates are shed only after they grow to a sufficiently large size to overcome much larger pinning force of the Wenzel state. Based on mechanical force balance model and energy barrier consideration, we attribute the difference in the droplet shedding characteristics to the different dynamic pathway from the Wenzel state to the Cassie state between posts and grates. Overall, the faster droplet shedding helps enhance the water harvesting performance of the SHPo surfaces by facilitating the condensation on the droplet-free area, as evidenced by the best water harvesting performance of grates on the Cassie state amongst the tested surfaces.

1. Introduction

Recently, water harvesting from air-borne moisture has received the renewed interest as the promising way to address the global water shortage problem[1]. Many previous researches have studied the effects of surface wettability on the water harvesting performance employing the macroscopic properties such as apparent contact angle and contact angle hysteresis for the calculations[2]. But the effects of microscale geometries have not been well-understood. In this work, to clarify the influence of microscopic structural features on the dew harvest performance, we fabricated microposts and microgratings with varying structural parameters. And we investigated the condensates mobility and resulting water harvesting performances by used the microscopic geometric information in modeling.

2. Experimental method

Table 1 shows the investigated designs. Considering the equilibrium energy criteria E^*, we designed samples with Cassie ($E^*<1$) and Wenzel ($E^*>1$) wetting morphologies for posts and gratings[3] samples were fabricated with DRIE techniques followed by TFTS vapor deposition. The condensation experiments were conducted within the environmental chamber fixing the supersaturation level at 2.21 as shown in figure 1.

Table 1. Geometric parameters of test samples for experiments. The C and W indicate the Cassie and Wenzel states, respectively. The r is the roughness factor and φ_s is the solid fraction.

Surface		Grate width or post radius (μm)	Pitch (μm)	Height (μm)	r	φ_s	E^*
Grating	C I	20	60	60	3.00	0.33	0.76
	C II	20	80	60	2.50	0.25	0.91
	W I	20	60	20	1.67	0.33	1.36
	W II	20	80	20	1.50	0.25	1.51
Post	C I	20	60	60	3.09	0.35	0.74
	C II	22	80	60	2.30	0.24	0.99
	W I	20	60	20	1.70	0.35	1.34
	W II	22	80	20	1.43	0.24	1.59

3. Results and discussion

3.1. Sliding behaviors of condensed droplets

Figure 2 shows the behavior of water condensation on each test surface, captured using a CCD camera on the macro scale. Because of the hydrophobic nature, all surfaces exhibit high contact angles (> 130°) as well as dropwise condensation. The condensation behavior appears to be clear difference in the sizes of the pinned droplets which might be attributed to the different pinning force on each surface[2, 4]. On Grating C I, most of the surface area is clear of droplets, and the average size of pinned droplets is smaller than those on Grating W I since the mobility of droplets on the Cassie state is higher than the Wenzel state. Interestingly, we do not observe any difference in condensation behavior between Post C I and Post W I in contrast with gratings. Furthermore, the size of remaining droplets on posts is generally larger than those on gratings. To understand this difference, a microscopic force analysis is applied to calculate the pinning force at the contact line using the geometric details of surface patterns.

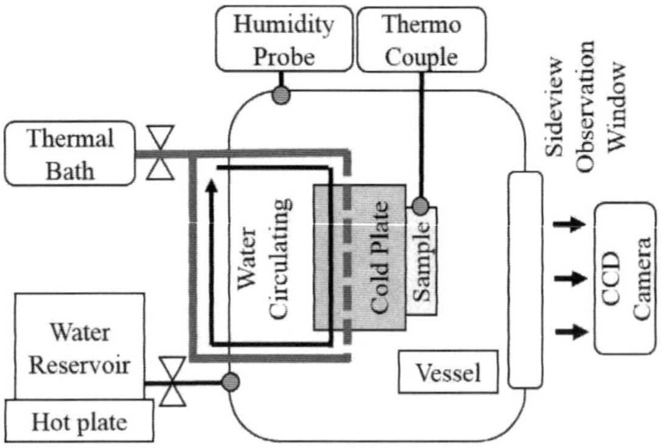

Figure 1. Experimental setup for harvesting water.

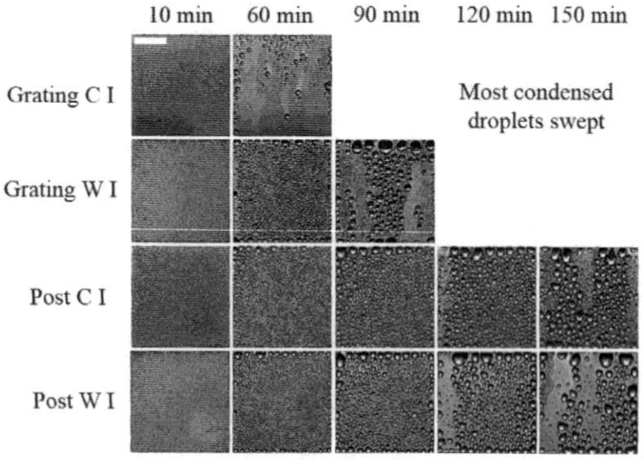

Figure 2. Time-lapse images of the water condensation until the most condensed droplets are removed. (scale bar: 10mm)

Figure 3(a-f) shows a schematic illustration of the contact line force when the droplet begins to slide on each surface. To slide on the surface, the droplet should overcome the maximum pinning force imparted by surface structures, which is dependent on the structure shape and wetting state, as detailed in the following. And for simplicity, we only consider the complete Cassie or Wenzel state.

$$F_{C.Grating} = w\gamma(\cos\theta_r - \cos\theta_a)/L, \quad (1)$$

$$F_{W.Grating} = \gamma(2H + L)(\cos\theta_r - \cos\theta_a)/L, \quad (2)$$

$$F_{C.Post} = \pi D\gamma \cos\theta_r / L, \quad (3)$$

$$F_{W.Post} = \{\pi D\gamma\cos\theta_r + 2H\gamma(1+\sin\theta_r) + L\gamma(\cos\theta_r - \cos\theta_a)\}/L, \quad (4)$$

where w is the width of the grates, γ is the surface tension of the liquid-air interface, L is the center to center distance of structures (i.e., defined as pitch), H is the height of the structures, and D is the diameter of the post. The diameter of a sliding droplet can be calculated by comparing pinning force and gravitational force, $\rho gV=bF_{Pinning}$ where ρ is the water density, g is the gravitational acceleration, V is the volume of a droplet, and b is the contact diameter of the droplet[5].

Figure 3(b) compares the predicted sliding droplet diameters by the above calculation with those actually observed in the experiments, which shows a good agreement between the prediction and the experiment for gratings. However, for posts, the actual sliding droplet size on Post C surfaces is much larger than the prediction and is comparable to that on Post W surfaces, which imply that the droplets are not on the Cassie state on Post C surfaces and the wetting criteria based on E^* might be inapplicable to posts.

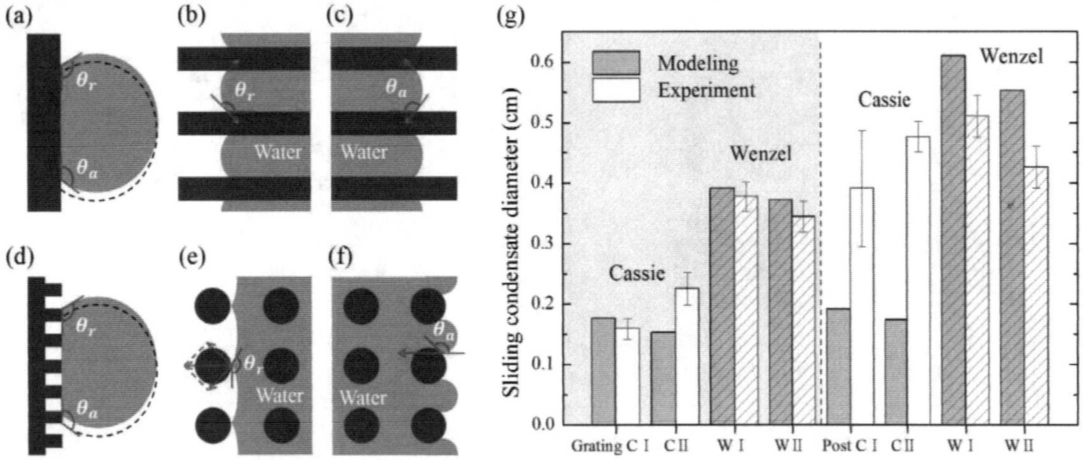

Figure 3. (a-f) Contact line force between a droplet and structures when the droplet begins to slide on the (a) Cassie grating, (b-c) Wenzel grating, (d) Cassie post, and (e-f) Wenzel post surfaces. (g) Droplet diameters calculated by modelling and observed by experiments.

3.2. Wetting states during condensation

To further understand the condensation behavior and the resultant wetting state on each surface, optical microscopy was employed to determine the actual wetting state on micro-scale by focusing on the top of surface structures [3]. As shown in figure 4(a), the droplets on most surfaces initially show satisfactory agreement with the expectation of the wetting state by the E^* criteria. However, the droplet on Post C I surface begins to irregularly spread out at 50 min, and become to take a similar morphology with that on Post W I surface after 70 min, indicating that the droplet is no longer on the Cassie state.

Figure 4(b) shows time-lapse images of droplets coalescence motion. When two large droplets merge on the surfaces, some parts of the droplet can penetrate into structures, as surface energy released during merging is converted to fluctuations or kinetic energy, which provide the sufficient energy for the transition from the Cassie state to the Wenzel state if the structures are not tall enough compared to the merged droplet[3]. On Grating C I surface, the liquid-solid interface elongates along the direction of the grates during merging, and the droplet vibrates along the same direction because of the difference in the energy barrier along and across gratings. Later, the pinned interface returns to the top of gratings and the droplet relaxes to a more spherical shape, implying occurrence of de-wetting transition from the Wenzel state to the Cassie state. Meanwhile, on the post C I surface, the liquid-solid interface does not return to the top of posts after penetrating into the structures with no sign of the transition from the Wenzel state to the Cassie state on posts, even when the Cassie state is energetically more stable state. This difference was attributed to the different dynamic pathway from the Wenzel state to the Cassie state between posts and grates. It has been suggested that a large energy barrier prevents the transition from the Wenzel to the Cassie state on posts, while it is absent or is negligibly small on grates.

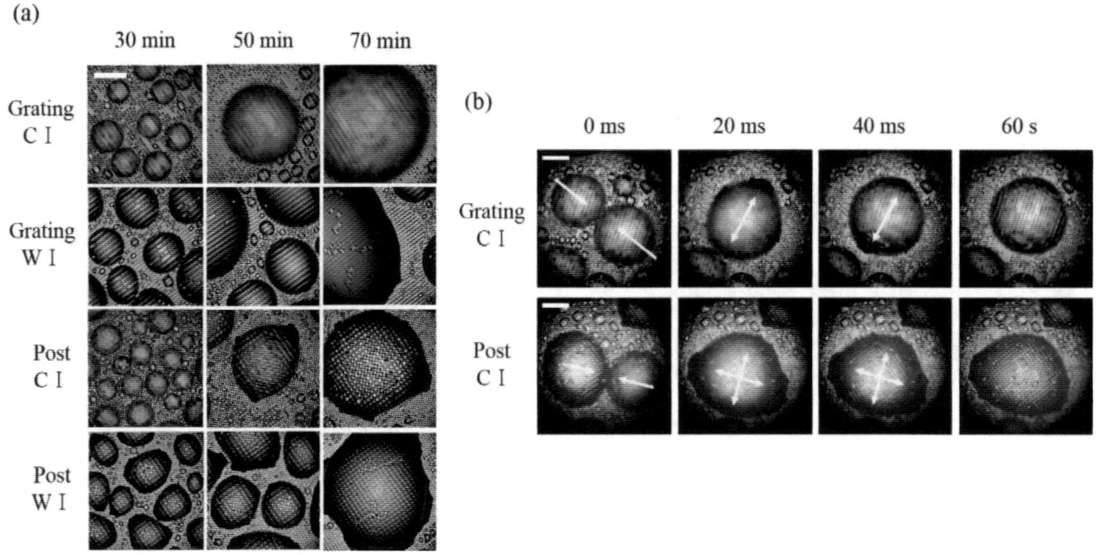

Figure 4. (a) Time-lapse images of water condensation motion in micro scale. (b) Time-lapse images of droplets coalescence motion. (scale bar: 0.6 mm)

3.3. Water harvesting performance

We investigate how the difference in droplet mobility and wetting state (i.e., the Cassie and Wenzel state) of SHPo surfaces affect the water harvesting performance. To compare the harvesting characteristics, the average mass and falling frequency of droplets detached from each surface were measured, as shown in figure 5(a). With the smaller pinning force of droplets on gratings than on posts, the harvesting falling frequency is noticeably higher on grates. In addition, the Cassie state grating surfaces exhibit higher falling frequency and smaller droplet mass ranges compared to the Wenzel state grating surfaces due to the smaller pinning forces on the Cassie state compared to the Wenzel state. On posts, all the tested samples similarly exhibit the low falling frequency and large mass ranges, as the condensed droplets are on the Wenzel state, which is associated with the strong pinning force.

The harvesting performance is calculated by multiplying the average droplet mass and falling frequency for a unit time. Figure 5(b) shows our main results about water harvesting performance of each SHPo surface. Despite the similar surface wettability and surface area, the performance of the grating surfaces is noticeably better than the post surfaces. Also, on grates, the Cassie state is better in water harvesting performance than the Wenzel state except for Grating C II surface, which exhibits a partial Wenzel state with E^* close to a unity. The reason that the Cassie state surface with the small pinning force and the high droplet mobility is associated with the better water harvesting performance can be explained by the consideration that the droplets resting on the surfaces act as the thermal resistance against heat transfer between the surface and vapor in air, thus delaying the water condensation. It agree with our experimental results that the water harvesting performance was the best on Cassie gratings where the droplet mobility was the highest or the critical falling mass of droplets was the smallest amongst the tested surfaces.

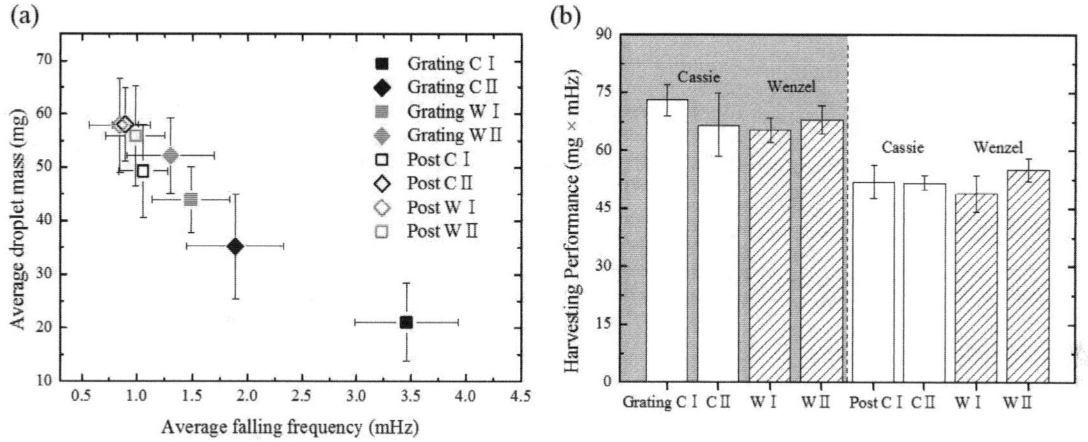

Figure 5. (a) Average mass and frequency of falling condensate on each test surface. (b) Water harvesting performance.

4. Conclusion

We experimentally studied the dew harvesting performance of SHPo surfaces with two different patterns (posts and gratings) and varying geometric parameters. It was found that thermodynamic energy criterion could be applied to determine the observed wetting state between the Cassie and Wenzel states on gratings during condensation, while it failed to predict the actual wetting state on posts. To characterize the dew harvesting performance, we measured both the mass and frequency of each fallen droplet in a temperature/humidity controlled environment to clarify the relationship between the droplet dynamics and water harvesting performance. Our results show that the different wetting state imparts the different pinning force against the droplet movement and thus affects average mass and falling frequency of collected droplets, which are the crucial factors in harvesting the water from dew. In the present study, the grates on the Cassie state with the highest droplet mobility exhibited the best harvesting performance amongst the tested surfaces, which was attributed to the enhanced heat transfer between the surface and vapor with the facilitated removal of smaller droplets from the surface.

References

[1] F T Malik, R M Clement, D T Gethin, W Krawszik and A R Parker 2014 *Bioinspiration & Biomimetics* **9** 031002
[2] Anna Lee, Myoung-Woon Moon, Hyuneui Lim, Wan-Doo Kim and Ho-Young Kim 2012 *Langmuir* **28** 10183–10191
[3] Ryan Enright, Nenad Miljkovic, Ahmed Al-Obeidi, Carl V. Thompson and N. Wang 2012 *Langmuir* **28** 14424–14432
[4] Sunwoo Kim and Kwang J. Kim 2011 *Journal of Heat Transfer* **133** 081502
[5] C. G. L. Furmidge 1962 *Journal of Colloid Science* **17** 309–324

A micro heat exchanger microfabricated from bulk aluminium

G. Scotti[1] and S. Franssila[2]

[1] Researcher, Aalto University, Department of Materials Science and Technology, Espoo, Finland
[2] Professor, Aalto University, Department of Materials Science and Technology, Espoo, Finland

E-mail: gianmario.scotti@gmail.com

Abstract. We report a micro heat exchanger microfabricated from a bulk aluminium alloy substrate. The device is comprised of two 18 cm long microchannels, one on each side of an aluminium chip, capped on both sides with a thin self-adhesive polymer film. In spite of a cheap and facile fabrication method, initial experiments with the device show promising results. Area densities from 25000 m^{-1} to 45000 m^{-1} have been achieved. Compared to our previous work on aluminium microfluidic devices produced with a similar technology but from a different, less pure alloy, in this study the etched surfaces are significantly smoother, and present less photoresist delamination.

1. Introduction

Micro heat exchangers are miniature heat exchanger devices produced with microfabrication techniques. The advantage of using microfabrication methods for heat exchangers is that the surface/volume ratio (area density) can be made to be very high [1], which is beneficial for heat exchange efficiency. Furthermore, micro heat exchangers can be integrated with other power MEMS devices (such as microreformers or micro fuel cells), with concentrated solar power systems [1], or with electronic components for cooling [2]. Micro heat exchangers require a smaller amount of material to fabricate compared to the traditional, macro-world counterparts, while having the same or better performance [2]. Additionally, the amount of working fluid is typically very small, as a consequence of the small total volume of the microchannels [2].

In [3] micro heat exchangers were made using traditional MEMS technologies and materials: a silicon wafer was etched with TMAH and capped with Pyrex glass by anodic bonding. However, due to robustness and high thermal conductivity, micro heat exchangers are often micromachined from metals: in [4] LIGA was used for micromachining a nickel micro heat exchanger, whereas in [5] microchannels were created in copper by a wire cutting machine. Due to its high thermal conductivity (237 W m^{-1} K^{-1}), aluminium is also an interesting material [6] [7] [8], but not many devices have been implemented at the microscale: none of the cited works implements microchannels with a hydraulic diameter below 1 mm.

In this work we present a micro heat exchanger made with a modification of the technology previously used for the micromachining of micro fuel cells from bulk aluminium [9] [10]. Unlike in [9] [10] where the alloy used was aluminium 6061, in this work we utilized aluminium 1050A, which contains much fewer alloying elements: min. 99.5 w% of aluminium, less than 0.25 w% Si, less than 0.4 w% Fe, and trace amounts of Cu, Mn, Mg and Zn.

2. Experimental

2.1. Micro heat exchanger structure

The micro heat exchanger discussed in this work is formed by two serpentine channels micromachined on each side of a 700 μm thick aluminium alloy 1050A chip. Structurally, this is similar to plate-fin heat exchangers [1]. The microchannels are capped on both sides with a thin polymer sticky film. The arrangement of the inlet holes (figure 1) allows also for the heat exchanger chips to be stacked, though this has not been tested yet. The smaller circles visible on the chip in figure 1(a) indicate the position of the fluidic inlets on the opposite side. Figure 1(b) shows a schematic of the arrangement of the channels and inlet holes on both sides of the chip.

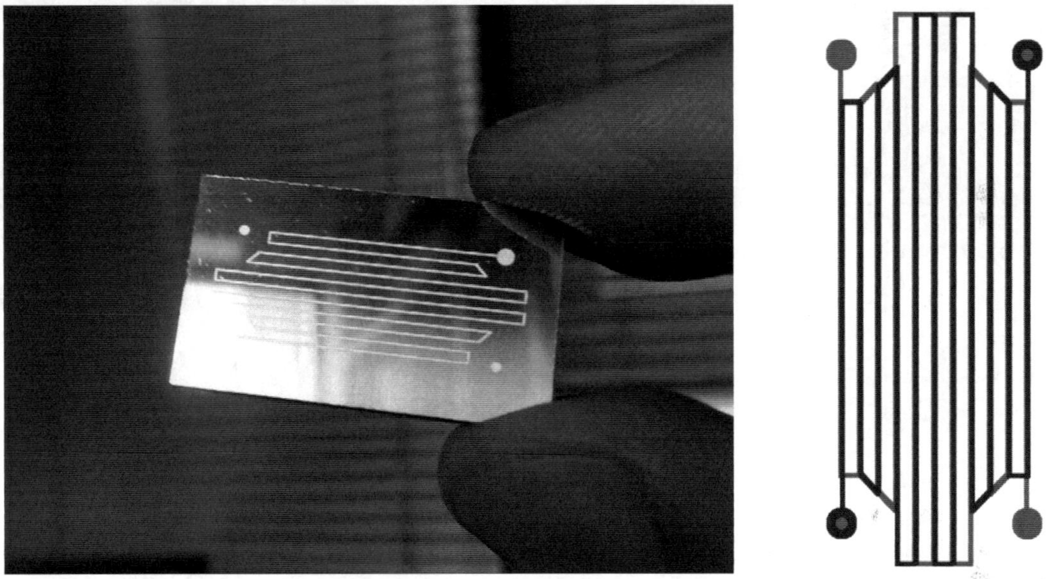

(a) (b)

Figure 1. A micro heat exchanger chip. (a) Photograph of chip (capping removed) with human fingers for size comparison. (b) Schematic drawing of the arrangement of the microchannels and inlet holes on both sides of the chip.

2.2. Microfabrication

Our microfabrication process started with priming of the aluminium sheet with HMDS and spinning of common DNQ/novolac resist AZ 5214 on both sides of a mirror polished, 4" x 4", 700 μm thick aluminium sheet (figure 2(a)). The spinning was done at 4000 RPM for a dried resist thickness of 1.4 μm. After each spinning the resist was soft-baked at 90°C in an oven. It is important that the resist is carefully soft-baked after the fist spinning so that the vacuum of the spinner chuck would not affect the already spinned resist layer. Lithography of the front side was done without alignment, while the back side required back-side alignment (figure 2(b)). After the UV exposure and development, the resist was hard-baked at 120°C in an oven for 20 minutes. The channels were etched in a commercial anisotropic aluminium etchant (an aqueous solution of 74% H_3PO_4 and 2.5% HNO_3) for 160 minutes at 50°C, resulting in an etch depth of ~75 μm (figure 2(c)). Both sides of the wafer were etched simultaneously. The aluminium sheet was placed between two dicing frames with a sticky tape stretched across (figure 2(d)). This tape acts as a simple capping layer (figure 3). Finally, the chips were diced and separated. Fluidic inlets were obtained by simply punching a hole at the inlet point with a needle or similar tool. Note that the arrangement of the inlets allows one to drill the chip through and have access to the microchannels from the opposite side of the chip (figure 1(b)).

Figure 3 shows a micropatterned aluminium sheet placed between two sticky tapes stretched on dicing frames/rings. After this capping step, one of the dicing frames must be removed before dicing can proceed, otherwise it will be in the way of the rotating blade.

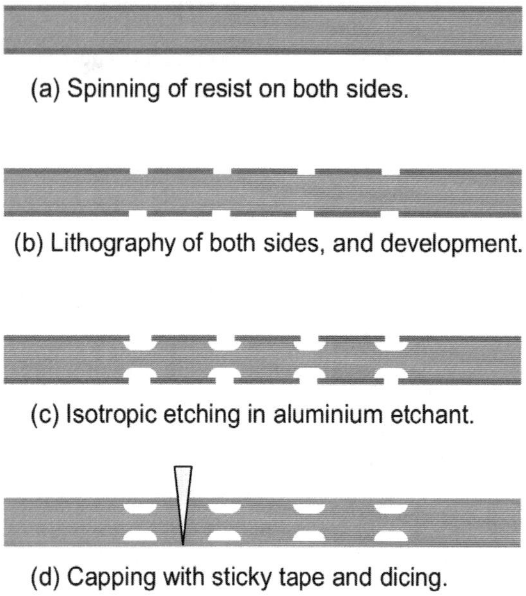

Figure 2. Process flow for the microfabrication of the bulk aluminium micro heat exchanger chips. (a) Spinning of photoresist. (b) Lithography (front and back). (c) Etching in isotropic aluminium etchant, to depth of ~75 μm. (d) Capping with sticky tape and dicing.

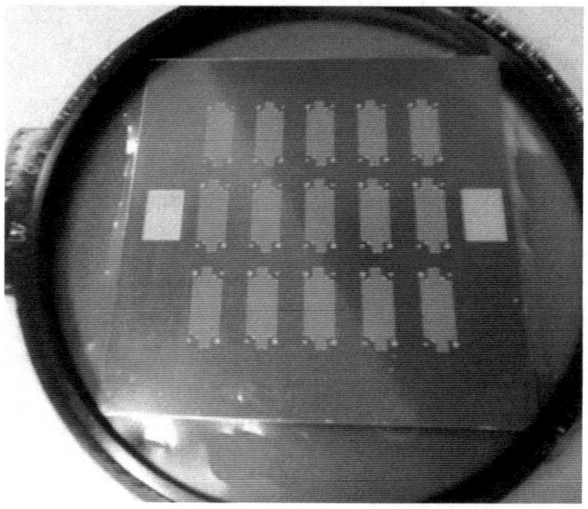

Figure 3. Micropatterned aluminium plate placed between two sticky tapes in dicing frames. After capping, one of the dicing frames must be removed before dicing.

2.3. Characterization

For the heat exchange characterization we placed the capped chips in a polycarbonate jig with fluidic fittings (figure 4).

Two peristaltic pumps pumped hot water through one side of the chips, and cold water through the other side. Both flows were set to 50 μL min^{-1}, which is close to the maximum attainable with the available peristaltic tubes. Temperatures were measured with an Amprobe® infrared thermometer, by temporarily removing the fittings and letting the water (first at the inlet, then at the outlet) leak onto a patch of cleanroom wipe. When the wipe was sufficiently wet, the infrared thermometer was pointed at it and the temperature was recorded. The ambient temperature was 20°C. The cold water was obtained by having water and ice mixed in a beaker, while hot water was taken from a beaker placed on a hotplate at 80°C.

Figure 4. Micro heat exchanger chip with capping inserted in measurement jig. Fluidic fittings are threaded into the polycarbonate blocks.

3. Results and discussion

The measured temperatures of the water on the "hot" side were 21°C at inlet and 18°C at outlet. On the "cold" side, the temperatures were 16°C at the inlet and 18°C at the outlet. Due to the method used to measure the temperature of the water, it is certain that some thermalisation of the water with ambient air took place, and that the temperature differences of the water going in vs. flowing out of the micro heat exchanger must have been larger. While the temperature measurement methodology is far from optimal, the experiment demonstrates qualitatively the validity of the design.

A separate set of results of the experiment relate to the microfabrication itself:

- The etched surfaces resulted in much smoother etched surfaces compared to the ones obtained in previous work [9] [10], with RMS roughness ~550 nm vs. ~960 nm
- The etch rate is about half that found in [9] [10]: ~0.45 $\mu m\ min^{-1}$ vs ~0.85 $\mu m\ min^{-1}$
- The photoresist adhesion appears to be significantly improved; the angle between the etched surface and the horizontal plane of the aluminium sheet is very close to 90°, and it is fairly sharp compared to what was obtained in [9] [10]. It can be concluded that no resist delamination occurred.

The first two effects can be easily attributed to the choice of aluminium alloy — 1050A in this work vs. 6061 in our previous works. It is known that aluminium etching is affected, and in fact catalyzed, by impurities [11]. Furthermore, the fact that 1050A has much fewer insoluble alloying elements, causes a much lighter (in fact, invisible/not noticed) patina to be deposited on the etched surface. As a consequence, roughness caused by the non-etchable patina should be smaller than in the case of aluminium 6061.

We do not know, at this point, what is the cause for the better photoresist adhesion. This important aspect will require further research.

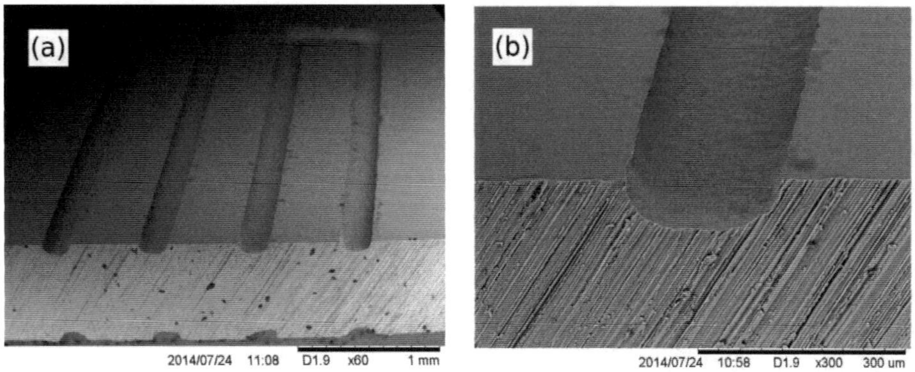

Figure 5. Scanning electron microscope images of microchannels. (a) Microchannel on one side, and channel profiles on both sides, are imaged. (b) Close-up of one channel near the dicing point. The undercut is equal to the etch depth, ~75 μm

4. Conclusions

The micro heat exchanger presented in this work is still a work in progress, but in spite of a less-than-ideal characterization method, it shows promising heat exchange characteristics. The very cheap and facile fabrication process opens many new possibilities for power MEMS applications, especially considering the improved results obtained with aluminium 1050A compared to our previous work.

The device can be improved by using other capping methods; attempt to thermocompressively bond aluminium chips to form a capped stack is a worthwhile research. Other channel topologies, e.g. parallel, should be better suited for heat exchange, and experimenting with narrower channels to further increase surface density, should be attempted. Improving the characterization method is very important as well. Finally, it would be of great value to discover the reason for better photoresist adhesion to aluminium 1050A.

References

[1] Li Q, Flamant G, Yuan X, Neveu P, and Luo L 2011 Compact heat exchangers: A review and future applications for a new generation of high temperature solar receivers *Renew. Sust. Energ. Rev.* **15** 4855– 4875.

[2] Kew P A and Reay D A 2011 Compact/micro-heat exchangers e Their role in heat pumping equipment *Appl. Therm. Eng.* **31** 594e601.

[3] Munkejord S T, Mæhlum H S, Zakeri G R, Nekså P, and Pettersen J 2002 Micro technology in heat pumping systems *Int. J. Refrig.* **25** 471–478.

[4] Harris C, Despa M, and Kelly K 2000 Design and Fabrication of a Cross Flow Micro Heat Exchanger *J. Microelectromech. S.* **9** 502-508.

[5] Jiang P-X, Fan M-H, Si G-S, and Ren Z-P 2001 Thermal-hydraulic performance of small scale micro-channel and porous-media heat-exchangers *Int. J. Heat Mass Tran.* **44** 1039-1051.

[6] Antohe B V, Lage J L, Price D C, and Weber R M 1996 Numerical characterization of micro heat exchangers using experimentally tested porous aluminum layers *Int. J. Heat and Fluid Flow* **17** 594-603.

[7] Omri M, Barrau J, Riera S, and Fréchette L G 2012 A novel hybrid microchannel heat exchanger configuration aiming for uniform wall temperature *Proc. PowerMEMS 2012* (Atlanta, USA) pp. 407-411

[8] Fernando P, Palm B, Ameel T, Lundqvist P, and Granryd E 2008 A minichannel aluminium tube heat exchanger – Part I *Int. J. Refrig.* **31** 669–680.

[9] Scotti G, Kanninen P, Kallio T, and Franssila S 2012 A high-power micro fuel cell microfabricated from aluminum *Proc. PowerMEMS 2012* (Atlanta, GA, USA) pp. 375-378

[10] Scotti G, Kanninen P, Kallio T, and Franssila S 2014 Bulk-Aluminum Microfabrication for Micro Fuel Cells *J. Microelectromech. S.* **23** 372-379.

[11] Catotti A J, Grad P P, and Cohn H, "Process of etching aluminum foil for electrolytic capacitor," US Patent 2,853,445, 1958.

PowerMEMS 2014 IOP Publishing
Journal of Physics: Conference Series **557** (2014) 012070 doi:10.1088/1742-6596/557/1/012070

Can Sensors Solve the Deterioration Problems of Public Infrastructure?

Chitoshi Miki,
Tokyo City University, Japan

Abstract Various deteriorations are detected in public infrastructures, such as bridges, viaducts, piers and tunnels and caused fatal accidents in some cases. The possibility of the applications of health monitoring by using sensors is the issues of this lecture. The inspection and diagnosis are essential in the maintenance works which include appropriate rehabilitations and replacements. The introduction of monitoring system may improve accuracy and efficiency of inspection and diagnosis. This seems to be innovation of maintenance, old structures may change smart structures by the installation of nerve network and brain, specifically. Cost-benefit viewpoint is also important point, because of public infrastructures. .

The modes of deterioration are fatigue, corrosion, and delayed fracture in steel , and carbonization and alkali aggregate reaction in concrete. These are like adult disease in human bodies. The developments of Infrastructures in Japan were concentrated in the 1960th and 1970th. These ages are approaching 50 and deterioration due to aging has been progress gradually. The attacks of earthquakes are also a major issue. Actually, these infrastructures have been supporting economic and social activities in Japan and the deterioration of public infrastructure has become social problems. How to secure the same level of safety and security for all public infrastructures is the challenge we face now.

The targets of monitoring are external disturbances such as traffic loads, earthquakes, winds, temperature, responses against external disturbances, and the changes of performances. In the monitoring of infrastructures, 3W1H(WHAT, WHERE, WHEN and HOW) are essential, that is what kind of data are necessary, where sensors place, when data are collected, and how to collect and process data. The required performances of sensors are accuracy, stability for long time. In the case of long term monitoring, the durability of systems needs more than five years, because the interval of regular bridge inspection works are five years. The supply of powers sometimes becomes serious problem. Some case studies will be presented here.

About the Speaker

Professor Chitoshi Miki is the Vice President, Professor of Advanced Research Laboratories of Tokyo City University, and Professor Emeritus of Tokyo Institute of Technology. He received his Master of Engineering (1972) and Doctor of Engineering (1979) degrees from Tokyo Institute of Technology, Japan. In 2014 he received Honorary Doctor degree from Thammasat University, Thailand. He worked on Research Associate of Tokyo Institute of Technology (1972-1979), Assistant Professor (1979-1980) and Associate Professor (1980-1981) of University of Tokyo, Post-Doctoral Researcher of Lehigh University, USA (1991), Associate Professor of Tokyo Institute of Technology (1981-1990), Visiting Scientist of Lehigh University, USA (1984-1985), Professor (1991-2012) and University Senator (1999-2002) of Tokyo institute of Technology, Dean School of Engineering (2003-2005) and Executive Vice President (2005-2007) of Tokyo Institute of Technology, Japan.

His research area includes fatigue and fracture of steel structures, fracture mechanics, corrosion, non-destructive evaluation, bridge design and bridge maintenance. He has published 13 books and written more than 350 technical papers. From 2011 and 2013, Prof. Miki is appointed as Chief adviser of AUN SEED Net, and Executive director (Acting) of AUN SEED Net, respectively. He received 10 Awards and Prizes; Japan Society of Civil engineers (JSCE) Award, Tanaka Prize (Excellence in Research Paper) (1979, 1995, 2002, 2007), JSCE Award, Best Paper Prize (1992), Japan Society of Welding, Best Achievement Award (1999), Ministry of Industry and Trading, Outstanding Contribution Award (2004), Japan Society of Steel Construction, Best Achievement Award (2010), JSCE Award, Book of the Year (2012), Ernst Gaßner Award, Fraunhofer LBF (Germany) (2013).

Improved mechanical reliability of MEMS electret based vibration energy harvesters for automotive applications

M Renaud[1], T Fujita[2], M Goedbloed[1] , C de Nooijer[1] and R van Schaijk[1]
[1]Holst Centre-Imec, Eindhoven, the Netherlands
[2]University of Hyogo, Himeji, Japan

E-mail: Michael.Renaud@imec-nl.nl

Abstract. Current commercial wireless tire pressure monitoring systems (TPMS) require a battery as electrical power source. The battery limits the lifetime of the TPMS. This limit can be circumvented by replacing the battery by a vibration energy harvester. Autonomous wireless TPMS powered by MEMS electret based vibration energy harvester have been demonstrated. A remaining technical challenge to attain the grade of commercial product with these autonomous TPMS is the mechanical reliability of the MEMS harvester. It should survive the harsh conditions imposed by the tire environment, particularly in terms of mechanical shocks. As shown in this article, our first generation of harvesters has a shock resilience of 400 g, which is far from being sufficient for the targeted application. In order to improve this aspect, several types of shock absorbing structures are investigated. With the best proposed solution, the shock resilience of the harvesters is brought above 2500 g.

1. Introduction

With the purpose of enhancing driving safety, tire pressure monitoring systems (TPMS) are legally mandatory for all newly produced passenger cars in the US. Similar legislation will soon pop-up in Europe and Asia. Actual TPMS solutions are mounted on the valve of the tire. Future TPMS may be embedded in the inner liner of the tire. This would represent a first step towards the concept of intelligent tire. Additional features of intelligent tire may include monitoring of the forces between the road and the tire for future active safety systems [1]. Sensors placed in the inner liner of a tire require compact and long lasting electrical power sources that do not need to be replaced during the lifetime of the tire. While electrochemical batteries remain an option, their limited lifetime can be an issue. These considerations combined with the abundance of mechanical vibrations in the tire makes vibration energy harvesting a promising alternative.

Our group previously demonstrated fully autonomous TPMS powered by MEMS fabricated electret based vibration harvesters [2]. The harvester is described in Figure 1a. It consists of a stack of three bonded wafers. The central wafer contains a mechanical resonator made of a proof mass and springs etched in silicon. An illustration of the design for the silicon mass spring system is given in Figure 1b. The dimensions of the mass are $10\times10\times0.65$ mm^3. The mass is connected to the frame by folded silicon beams. The springs on both sides of the mass are linked by a truss for increased stability of the oscillating structure. The maximum displacement of the mass is defined by the gap between the anchors of the springs. It is equal to ±100 μm. The proof mass supports a corrugated electret on its bottom side, obtained by Corona charging of a SiO_2/Si_3N_4 stack. The bottom wafer is made of glass and supports a set of two metallic electrodes that are connected to a load circuit. The top glass wafer is used as capping to protect the device and to allow vacuum encapsulation. With external vibrations, the mechanical resonator oscillates and an electric current is induced between the electrodes on the bottom wafer.

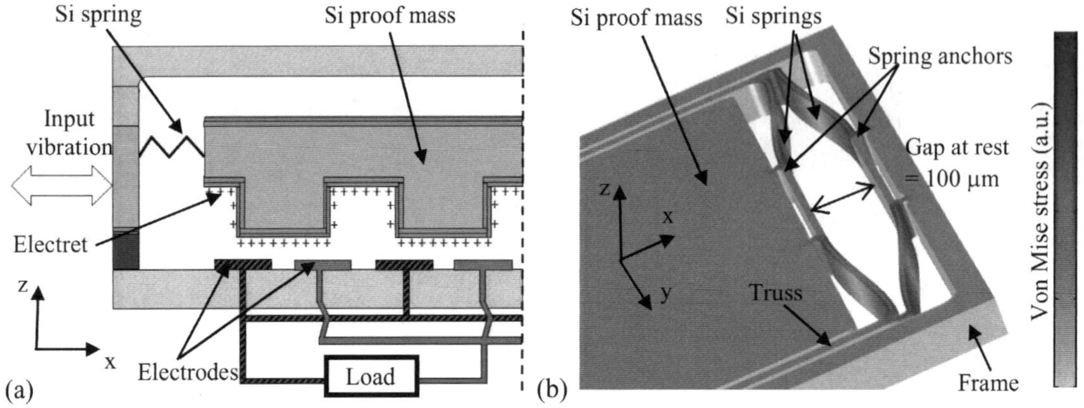

Figure 1. (a) Side-view schematic of the electret based harvester. (b) Top view schematic of the mass spring system.

While the functionalities of the harvester have been demonstrated, its mechanical reliability, particularly in terms of shock resilience, remains a challenge. Our objective is to make harvesters that survive half sine shocks with amplitude in the range of 2500 g and duration of 0.6 ms, which would be within safety margins of the standard requirements of the automotive industry [3]. As discussed in [4], our first generation of devices do not meet this requirements so that solutions for improving the shock resilience needs to be found. Inspired by the work of Yoon [5], solutions are found in the form of mechanical stoppers. Different types of stoppers, illustrated in Figure 2, are investigated.

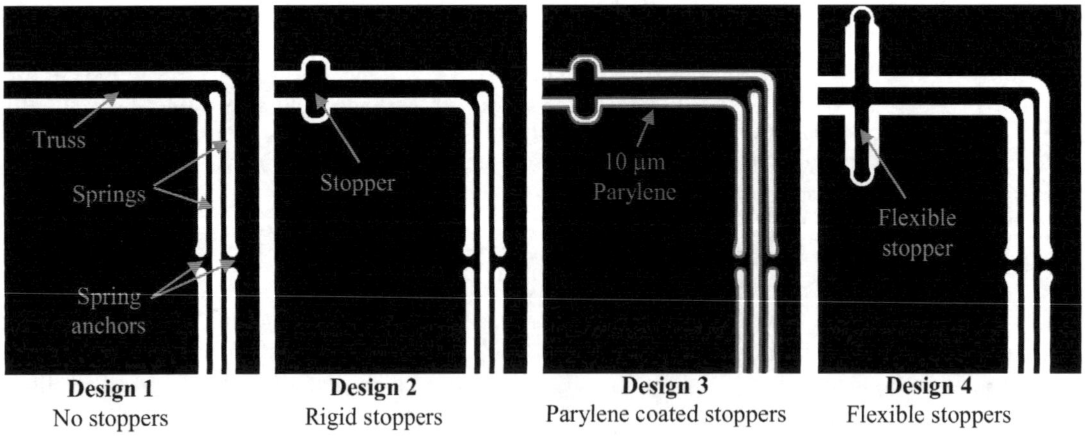

Figure 2. Illustration of the different mechanical stoppers investigated in this article.

2. Stoppers free structures (design 1)

As detailed in [4], shock tests are performed using a drop machine. For each shock amplitude, 3 successive shocks are applied. The status of the devices is checked after each shock. A first series of experiments suggested that the x direction in Figure 1a is the most critical. Therefore, we focused on investigating the reliability of the harvester with shocks applied along the x direction. In order to limit the manufacturing efforts necessary to provide a sufficient quantity of test samples, the subsequent shock tests were not performed on fully functional harvesters but on simplified test devices. The test devices are made of FR-4 boards on which the central wafer in Figure 1, i.e. the mass spring system of the harvester, is glued. The distribution of the shock amplitude leading to failure for structures with design 1 is given in Figure 3a. Failure of the samples is observed in the form of fracture of the silicon springs. This results from the spring anchors impacting each other. A close up picture is given in Figure 3b. It can be seen that more than 60% of the 27 tested samples fail for 400 g shocks and that 100% of the samples are broken at 700 g. This result is clearly not satisfying.

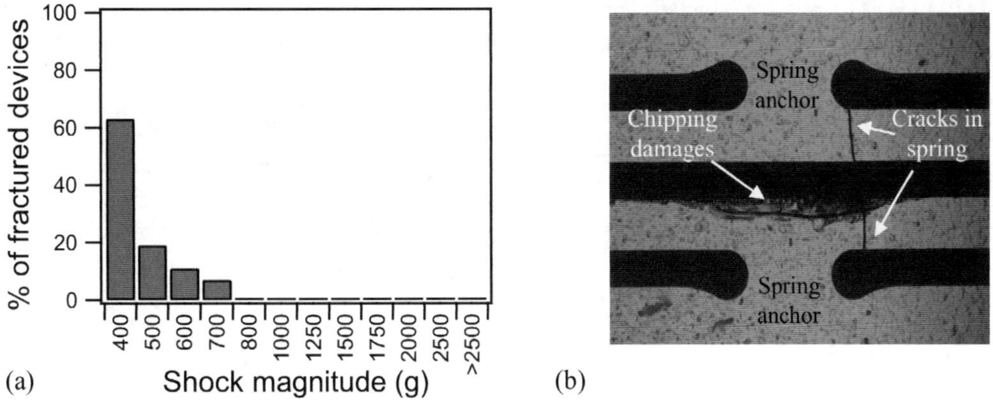

(a) (b)

Figure 3. (a) Distribution of the shock resilience for the stoppers free devices (27 samples). (b) Picture of the broken area.

3. Rigid stoppers structures (design 2)

In order to avoid contact between the anchors of the springs, rigid stoppers are introduced in design 2. The stoppers are designed in such a way that, when they contact, there is still some distance between the anchors of the springs. As shown in Figure 4a, the shock resilience is improved compared to the stoppers free design. However, the median shock survival rate is about 700 g which remains unsatisfying. Failure of the springs occurs when the stoppers themselves are fractured by the impact. In this case, contact between the anchors of the springs is not avoided and the springs break.

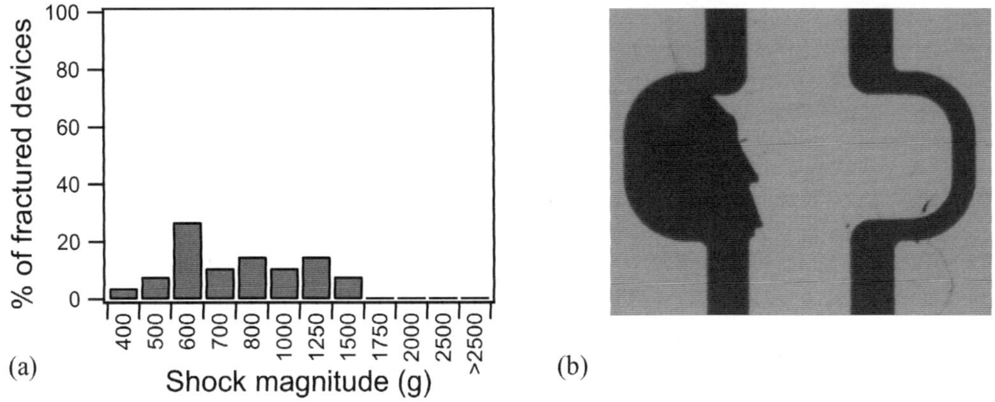

(a) (b)

Figure 4. (a) Distribution of the shock resilience for the rigid stoppers devices (26 samples). (b) Picture of a broken stopper.

4. Parylene coated stoppers (design 3)

In order to avoid damaging the stoppers and, ultimately, to avoid contact between the spring anchors for ensuring the survival of the springs, the amplitude of the contact force during impact should be reduced. This can be done by increasing the compliance of the contact surfaces. It is possible to do so by coating the rigid stoppers of the last section with a layer of compliant material. Parylene is not only a compliant material but is also conformal and non-brittle. It is therefore a good candidate. Note that, for manufacturing simplicity, the whole test samples are covered with 10 μm Parylene. The layout of the devices is adjusted so that the maximum displacement of the mass in the coated devices remains 100 μm.

The distribution of the shock amplitude leading to fracture of the springs is given in Figure 5a. Only one of the 28 tested samples has damaged springs, showing that the Parylene coating is a very good solution to improve the shock resilience of the harvesters. As shown in Figure 5b, signs of impact are visible on the bumpers of the tested devices. The Parylene is locally delaminated by the impact force. To check if this could become an issue after a large quantity of repeated impacts, some samples were tested under a series of 30 shocks with 2500 g amplitude. None of them showed signs of failure.

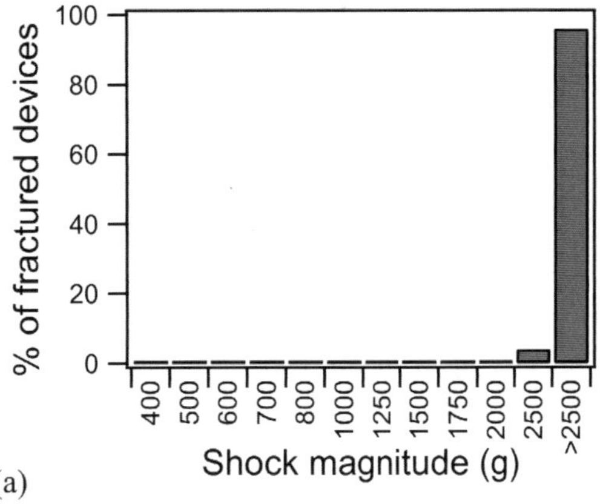

(a)

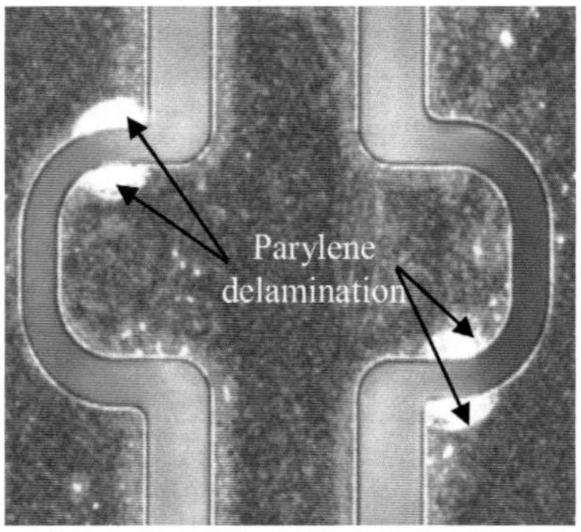

(b)

Figure 5. (a) Distribution of the shock resilience for the devices with rigid bumpers coated with Parylene (28 samples). (b) Picture of a stopper with delamination of Parylene resulting from impact.

While the Parylene coating is a very effective solution to enhance the shock resilience of the test devices, it is not straightforward to integrate it in the manufacturing process of the complete harvester. Therefore, a silicon only solution to improve the shock resilience is considered in the next section.

5. Flexible stoppers (design 4)

In the previous section, the impact force on the stoppers was reduced by increasing the compliance of the contact surfaces with a layer of Parylene. Another solution to reduce the impact force is to make the bumpers structurally flexible [5]. Several variations of flexible stoppers with different stiffness are investigated. Theoretical considerations for the various considered dimensions are given in [4]. We observed that the more flexible are the stoppers, the higher is the median shock resilience. The distribution of the shock amplitude leading to fracture of the springs for the most flexible tested stoppers is given in Figure 6a. The flexible stoppers concept is convincing as the median shock resilience of the test samples is brought up to 2500 g. This result can be enhanced further by making the stoppers even more flexible. One drawback of this approach is that the area occupied by the stoppers increases as they are made more flexible. This is detrimental for the total footprint of the harvester.

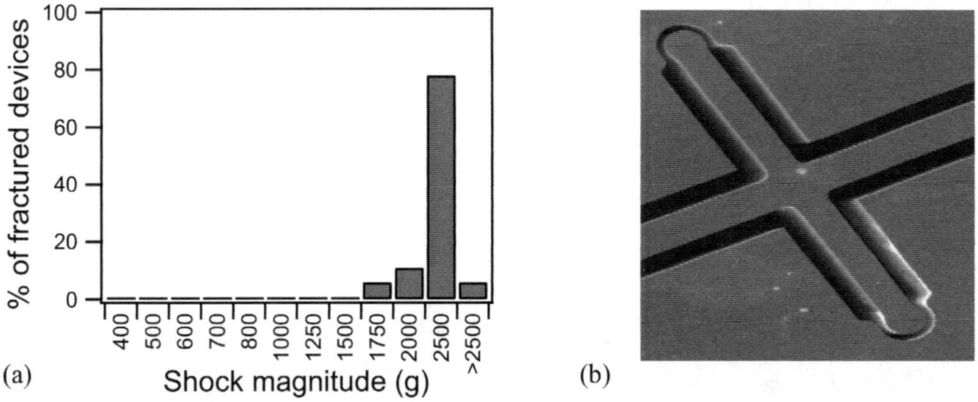

Figure 10. (a) Distribution of the shock resilience for the devices with flexible stoppers (27 samples). (b) SEM pictures of a pair of flexible stoppers.

6. Conclusion

Wireless TPMS powered solely by MEMS fabricated electret based vibration energy harvesters have been demonstrated by our group in previous articles. To attain the grade of commercial product with the devices, the mechanical reliability of the harvester has to be improved. Indeed, the tire environment is harsh, particularly in terms of mechanical shocks. Devices designed for this environment should survive shocks with amplitudes exceeding 2500g. It is shown in this article that our first generation of harvesters had a shock resilience of 400 g which was clearly not sufficient. For improving this aspect, several types of mechanical stoppers are considered. Rigid stoppers are first studied. They allow increasing the median shock resilience to 700 g which remains unsatisfying. Parylene covered stoppers is the second considered option. It gives very good results. 96% of the tested devices were not broken with 2500 g shocks. However, the integration of Parylene in the process flow of the harvesters is not straightforward. Therefore, a silicon only solution would be preferred and the concept of flexible stoppers is investigated. The concept is convincing, as we could bring the shock resilience of the tested samples to 2500 g. It is possible to enhance further this number by increasing the flexibility of the stoppers.

Acknowledgements

This work is supported by FP7-NMP-2013-SMALL-7 (604169), SiNERGY (Silicon Friendly Materials and Device Solutions for Microenergy Applications), and by NanoNextNL, a micro and nanotechnology programme of the Dutch Government and 130 partners.

References

[1] Ergen S C, Sangiovanni-Vincentelli A L, Sun X, Tebano R, Alalusi S, Audisio G and Sabatini M 2009 The tire as an intelligent sensor *IEEE Trans. on CAD of Integrated Circuits and Systems* **28** 941-55

[2] Elfrink R, van Acht V, Stanzione S, Bembnowicz P, Tutelaers M and van Schaijk R 2014 Fully autonomous Tire Pressure Monitoring System (TPMS) powered by a vibrational electrostatic energy harvester *Proc. Smart Systems Integration 2014* 69-76

[3] AEC-Q100-Rev-G-Failure Mechanisms Based Stress Test Qualification for Integrated Circuits.

[4] Fujita T, Renaud M, Goedbloed M, de Nooijer C and van Schaijk R 2014 Reliability Improvement of Vibration Energy Harvester with Shock Absorbing Structures *Proc. of Eurosensors 2014*

[5] Yoon S W, Lee S, Perkins N C and Najafi K 2011 Shock-protection improvement using integrated novel shock-protection technologies *J. Microelectromech. Sys.* **20** 1016-31

Flexible Electret Energy Harvester with Copper Mesh Electrodes

Yi Chiu[1], Ming Hsuan Lee[1], Wei-Hung Hsu[2]
[1] Department of Electrical and Computer Engineering, National Chiao Tung University, Hsin Chu, Taiwan, R.O.C.
[2] Green Energy and Environment Research Laboratories, ITRI, Hsin Chu, Taiwan
E-mail: yichiu@mail.nctu.edu.tw

Abstract. Flexible energy harvesters are desired in biomedical applications since human motion is often complicated and aperiodic. However, most demonstrated flexible energy harvesters employ piezoelectric materials which are not biocompatible. Therefore we propose a PDMS-based flexible energy harvester with Parylene-C electret suitable for biomedical applications. To address the reliability issues of sputtered metal electrodes, we use copper mesh electrodes to improve the reliability. The proposed flexible harvester was fabricated and characterized. The measured power of the proposed harvester was 3.33 μW in the compression tests at 20 Hz and 8.5 nW in the finger bending tests at 2 Hz.

1. Introduction

When energy harvesters are used in biomedical applications such as physiological monitoring or human gesture detection [1], flexible harvesters are advantageous since human motion is often complicated and aperiodic. Flexible energy harvesters have been demonstrated by using piezoelectric materials [2, 3]; however, piezoelectric materials are not IC-process compatible nor biocompatible. In [4], we have demonstrated a flexible electret energy harvester based on biocompatible PDMS and Parylene. However, the sputter-deposited metal electrodes were prone to cracks upon the bending and stretching of the flexible substrate. To improve the reliability of the electrodes, we propose and demonstrate a flexible electret energy harvester with embedded copper mesh as the electrodes in this paper. The electrode reliability and the output power of the fabricated harvester are presented.

2. Principle and design

Figure 1(a) shows the schematic view of the proposed harvester composed of two PDMS electrode layers and a spacer layer. The device is a gap-closing capacitive harvester biased by Parylene electret. The gap-closing type is featured with robust structure, simple fabrication processes and thus low cost. Copper mesh is embedded in the electrode layers for electric connection and to improve electrode reliability. The copper mesh is made of copper wires with 50-μm diameter and has 200 meshes per inch, as shown in Figure 1(b). When an external force is applied to the harvester, the gap between the electrodes is changed and thus an output current is induced. Since the deformation of the PDMS structure can be large for large applied force, the nonlinear force-displacement relationship of the device was measured, as shown in Figure 1(c), and used in the electromechanical simulation of power generation. The capacitance area of the proposed harvester is 1 cm × 1 cm and the initial gap is 1 mm.

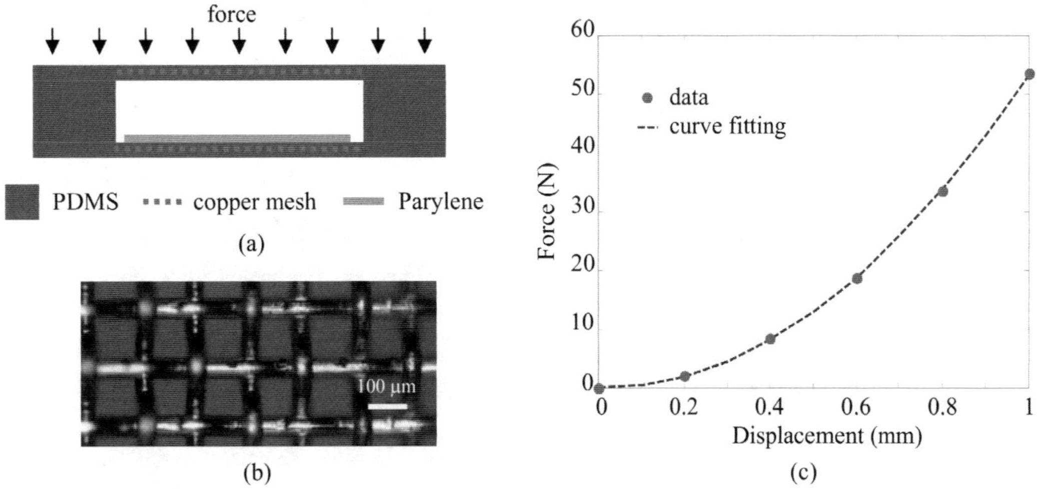

Figure 1. (a) Schematic of proposed flexible harvester, (b) photograph of copper mesh, (c) force-displacement relationship of fabricated device.

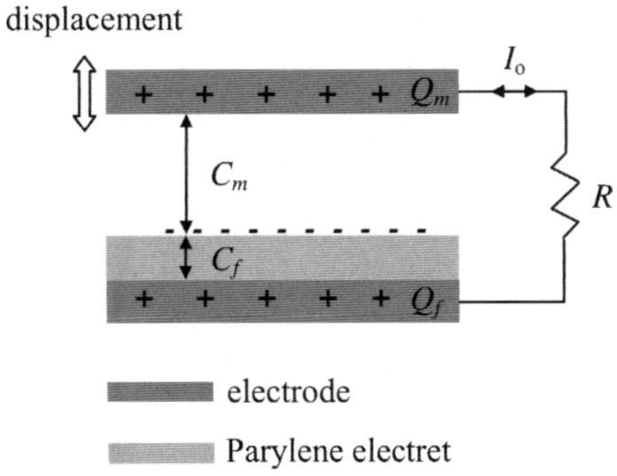

Figure 2. Electrical model of harvester.

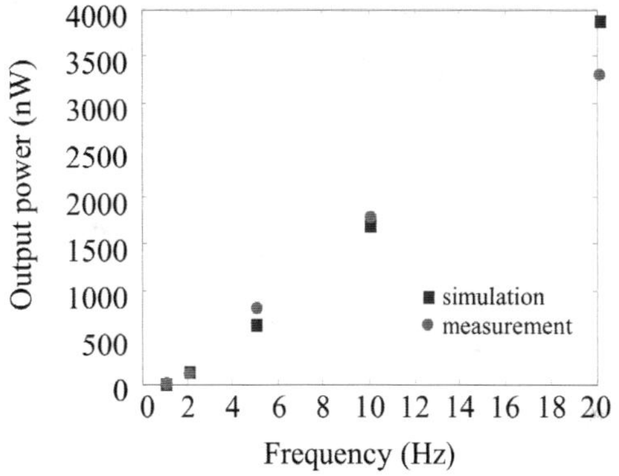

Figure 3. Simulated and measured power with 1000-MΩ load.

The electrical model of the harvester is shown in Figure 2. The output current induced by the time-varying capacitance is given by

$$I_0 = -\frac{dQ_m(t)}{dt} = -\frac{V_{suf}}{R} + \frac{Q_m(t)}{RC_{eq}(t)},$$

where V_{suf} is the surface potential of the electret, C_{eq} is the total capacitance of C_m and C_f connected in series, and Q_m is the charge on the movable electrode. Since the equation is a complex electro-mechanically coupled differential equation, the output current and power were estimated by Simulink simulation, as shown in Figure 3.

3. Device fabrication

The fabrication processes of the proposed energy harvester are shown in Figure 4. First, flexible electrodes were fabricated by casting PDMS between two glass substrates with the copper mesh sandwiched in between. After curing, the copper mesh was embedded in the PDMS flexible electrodes. (Figure 4(a)). The total thickness of the electrodes was controlled by a feeler gauge during the casting to be 250 µm. A 5-µm-thick Parylene-C was then coated on the cured electrodes as the electret layer (Figure 4(b)). A 1-mm-thick PDMS spacer was then bonded to the electrode (Figure 4(b)). The electret was charged by corona discharging at 100 °C with -4 kV needle voltage and -500 V grid voltage (Figure 4(c)). Finally, another electrode was bonded to the top of the spacer (Figure 4(d)). Figure 5 is the photograph of a fabricated harvester with soldered leads. The long term charge stability of Parylene-C electret on PDMS substrate in this process is shown in Figure 6. After 724 days of charging, the surface potential decayed to about 52% of the initial value, comparable to the data in [5].

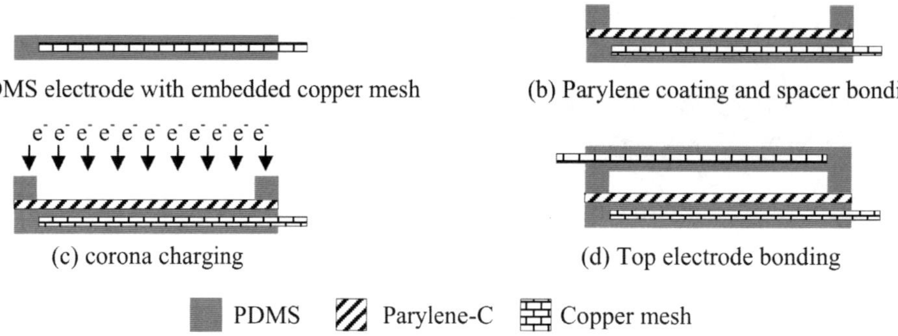

(a) PDMS electrode with embedded copper mesh

(b) Parylene coating and spacer bonding

(c) corona charging

(d) Top electrode bonding

Figure 4. Fabrication processes.

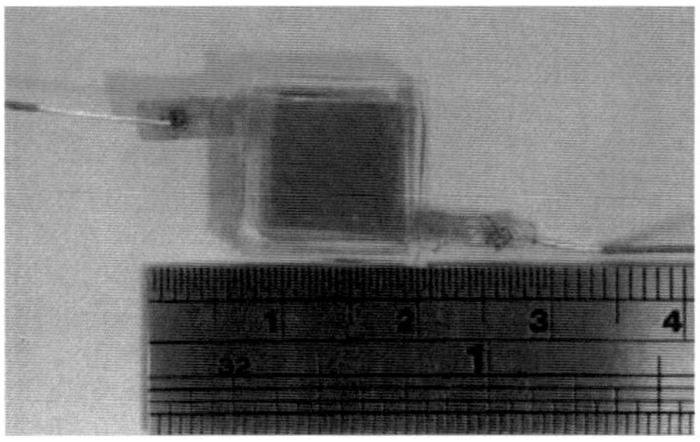

Figure 5. Photograph of a fabricated harvester.

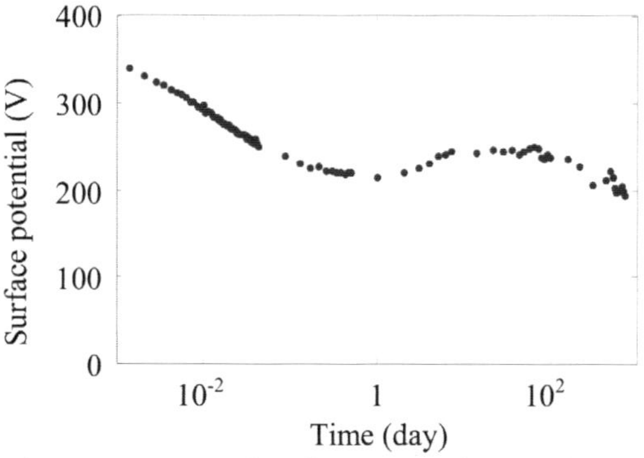

Figure 6. Measured surface potential of Parylene-C electret on PDMS

4. Measurement and discussion

4.1. Electrode reliability

Two types of electrodes were prepared for reliability test. Figure 7(a) shows a 1 cm × 1cm PDMS electrode with embedded copper mesh; Figure 7(b) shows a 1 cm × 1cm 100-μm-thick PDMS electrode with 0.5-μm-thick sputtered aluminium on the surface. Both types of electrodes were attached to another 1 cm × 2 cm 1-mm-thickness PDMS substrates and then placed on a C-shaped acrylic test fixture (Figure 7(c)). The test fixture was loaded in a Bose Electroforce 3200 dynamic mechanical analyzer (DMA) which applied a 5 Hz and 3-mm displacement at the center of the test samples. In this pure bending test, the metal electrode experienced a tension force. Figure 8(a) shows the circuit used to characterize the electric conduction of the electrode. If electric conduction is good, the 1-V DC bias can be observed on the oscilloscope; if the conduction deteriorates, the observed voltage will decrease. For the copper-mesh electrode, the observed voltage remains at the 1-V bias level during a 60-min test, indicating no degradation of conduction. For the sputtered-Al electrode, the recorded voltage started at the bias level but started to show fluctuation after 10 min of test. Figure 8(b) is the voltage after 12 min of test, showing significant electrode degradation. After 16 min of test, the sputtered-Al electrode became open-circuited and the oscilloscope showed a flat zero voltage signal.

Significant cracking of the Al electrode was observed on the surface after the 16-min test. The reliability of the copper mesh as an electrode material was thus confirmed.

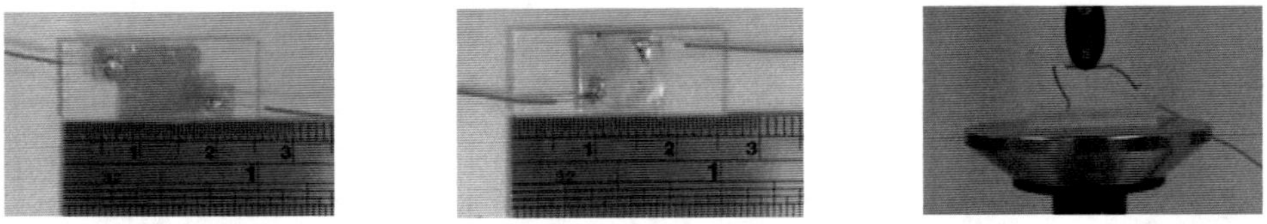

(a) Copper mesh electrode (b) Sputtered-Al electrode (c) Bending test setup

Figure 7. Electrode reliability test samples and setup.

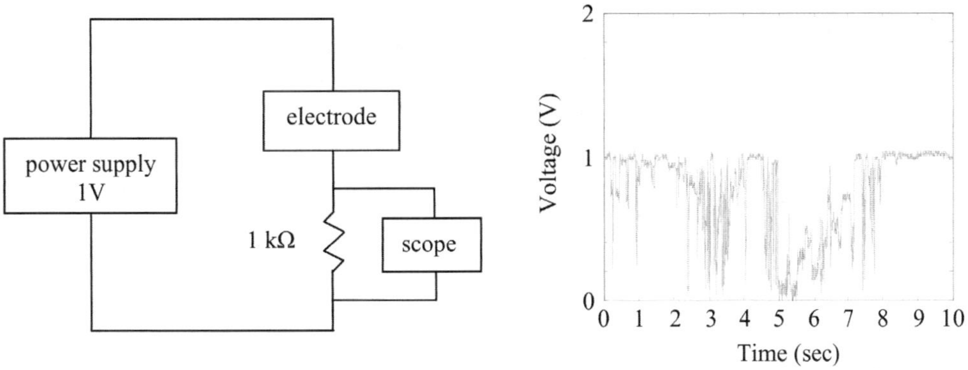

(a) (b)

Figure 8. (a) Electrode reliability test circuit, (b) observed signal for sputter-Al electrode after 12 min of test.

4.2. Power generation

Two power generation experiments were conducted. The DMA was first used to apply a uniform sinusoidal displacement on the device to mimic scavenging energy from periodic human activity such as walking or running. The measured output with 1.0-mm stroke on a 1000-MΩ load agreed with the simulation well, as shown in Figure 3. The output power at 2 Hz is 0.13 μW; the power increases with frequency and reaches 3.33 μW at 20 Hz. The maximum output power density was estimated to be 15.4 $\mu W/cm^{-3}$. The linear increase of power with the frequency implies that the harvester can be modeled as a constant energy source per cycle at low output levels. Figure 9 shows the simulated and measured output power vs. load at 2 Hz. At such low frequency, the output capacitive impedance of the harvester and thus the optimum load are very large, as shown in Figure 9. Thus, proper power conditioning circuits are needed to match the load to the harvester to improve the power transfer efficiency.

In the second experiment, the harvester was attached to a human finger knuckle to scavenging energy from the finger motion. Figure 10 shows the experimental setup and measured output voltage. The angle of finger bending was about 90°. With a 1000-MΩ load, the output power was 8.36 nW for a harvester with electret surface potential of 350 V. The experiment demonstrated that the proposed harvester could scavenge energy from human motions or when attach to deformable surfaces. The low power output was due in part to the low actuation frequency and in part to the low effective area utilized in the bending deformation as compared to compression.

5. Conclusion

This paper presents a flexible electret energy harvester with robust copper mesh electrodes. The measured power was 3.33 μW in compression tests at 20 Hz and 8.36 nW in human finger tests at about 2 Hz. With further optimization of the device design and fabrication processes, improved output power levels can be expected.

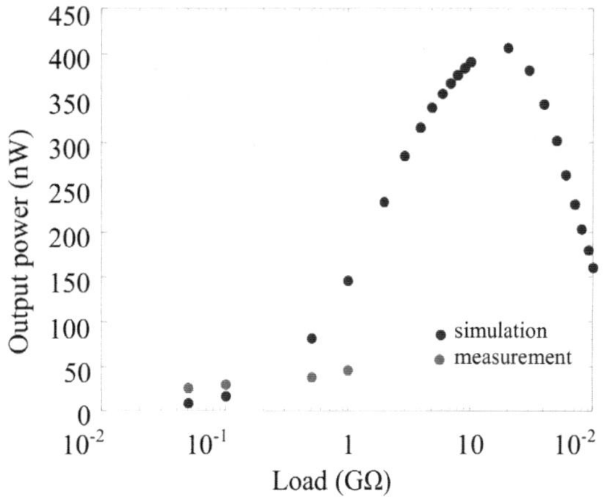

Figure 9. Simulated and measured output power with various loads at 2 Hz.

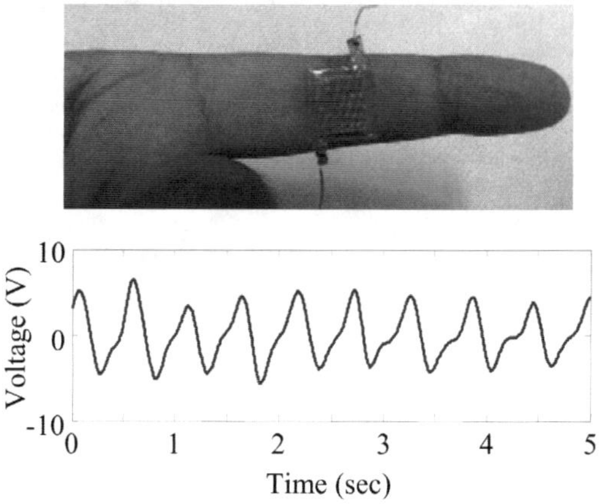

Figure 10. Testing of harvester attached to a human finger.

Acknowledgments
The project was supported in part by the Ministry of Science and Technology, Taiwan, R.O.C., (NSC 102-2221-E-009-191 and MOST 103-2221-E-009 -218) and the Bureau of Energy, Taiwan, R.O.C., (103-E0210). The authors were grateful to the use of facilities at the Chip Implementation Center, the National Center for High-performance Computing, the National Nano Device Laboratory, and the National Chiao Tung University Nano Facility Center, Taiwan, R.O.C.. The authors appreciate the assistance of Mr. Sheng-Tsai Wu of Industrial Technology Research Institute, Taiwan, R.O.C., in the DMA measurement.

References

[1] T. Starner and J. Paradiso, "Human-generated power for mobile electronics," *Low-Power Electronics Design*, CRC Press, 1-35, 2004
[2] M. Lee, C.-Y. Chen, S. Wang, W. S. Cha, Y. J. Park, J. M. Kim, L. J. Chou, and Z. L. Wang, "A hybrid piezoelectric structure for wearable nanogenerators," *Adv. Mater.*, 24 1759-1764, 2012
[3] X. Chen, S. Xu, N. Yao, and Y. Shi, "1.6 V nanogenerator for mechanical energy harvesting using PZT nanofibers," *Nano Lett.*, 10, 2133-2137, 2010.
[4] Y. Chiu and S.-H. Wu, "Flexible electret energy harvesters with parylene electret on PDMS substrates," *J. Phys.: Conf. Ser.* 476, 012037, 2013
[5] S. Genter and O. Paul, "Parylene-C as an electret material for micro energy harvesting," *Proc. Power MEMS 2012*, 317-320, 2012

Journal of Physics: Conference Series **557** (2014) 012073 doi:10.1088/1742-6596/557/1/012073

DEVELOPMENT OF POWER SUPPLY PACKAGE FOR ELECTRET VIBRATION GENERATOR

Yutaka Hattori
THHINK WIRELESS TECHNOLOGIES JAPAN LTD. Kanagawa.Japan

y.hattori@thhink-jp.com

Abstract. Development of power supply package for electret vibration generator. To use the electret vibration generator widely in the market the power supply unit of the package where the circuit of making to direct current had been integrated with the control circuit was developed. And the application was examined.

1. Introduction

The vibration generator and two electric circuits were designed as one unit.

Because two or more pieces were able to be combined according to the usage, the package design of the hope output of the user was enabled.

Module is vibration generator combined 2 pieces to suppress the difference of the performance of the vibration generator and the electric circuit, and leveled the performance when mass-producing it.

(**Figure 1**.Vibration Generator: OMRON)

2. Design of module

The device adopted the diode and the low loss capacitor to make the output of the Polymer electret type vibration generator of high impedance direct current efficiently.

Output of the Polymer electret type vibration dynamo is a high voltage of $\pm$120V or more of impedance and $\pm$ output that about 20MΩ is high, and the diode that

Rectifies this exchange output should have enough directions of easy flow resisting pressure and an appropriate recovery characteristic.

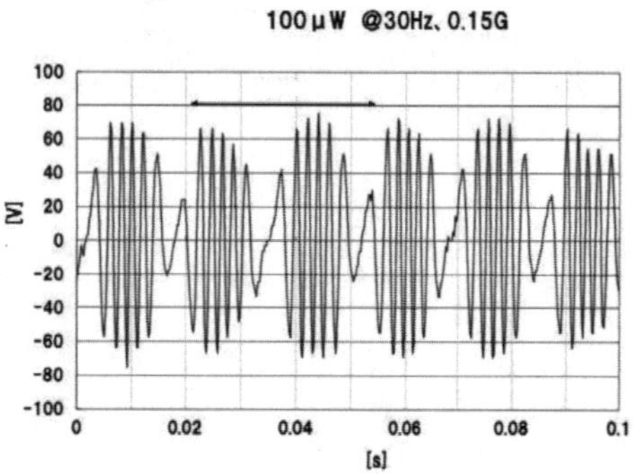

(**Figure2**.Vibration Generator output: ±75V)

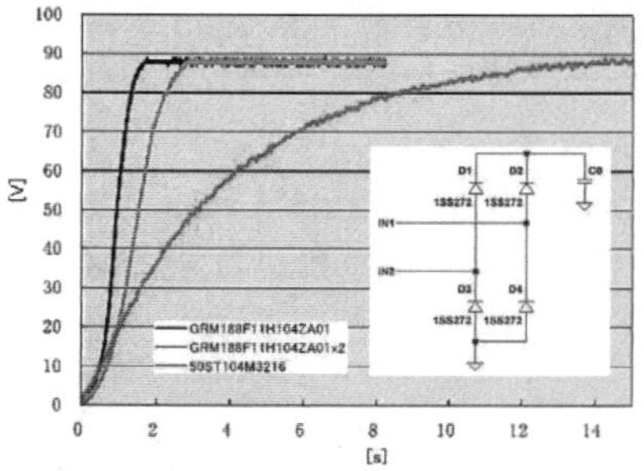

(**Figure3**.Charge characteristic)

Moreover, the capacitor where direct current was temporarily saved had the Chemical capacitor and the film type, etc. ceramic type of a low loss was especially adopted here.
This adopted the converter circuit in the circuit that controlled the direct-current output after it had rectified it with stability.
The direct-current output saved in the capacitor by the continuing vibration rises up to about 90V at time.

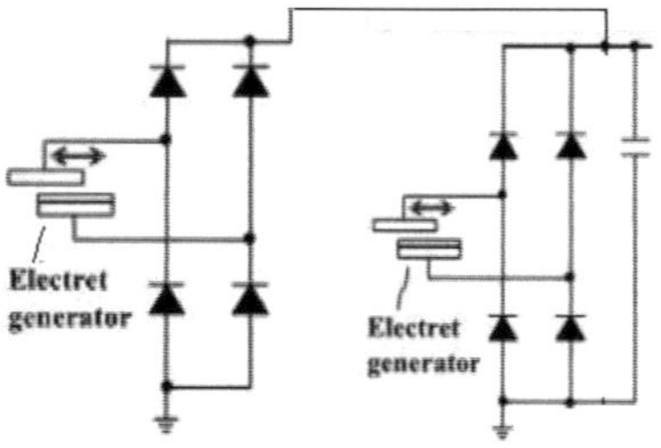

(**Figure4**.Vibration generator mixer)

(**Figure5.**Generator mounted circuit board)

It uses the DC-DC converter circuit of the fixed voltage drive for the control of this voltage, and conversion into the output of 5V.
The need the composition that is appropriate for the vibration type generator with a little amount of power generation (100μW level) to compose this converter.
Adopted not the bipolar transistor of the current control.
Type but MOS-FET of the voltage control type in the switch device that controlled the converter, and aimed at the decrease of the control current.

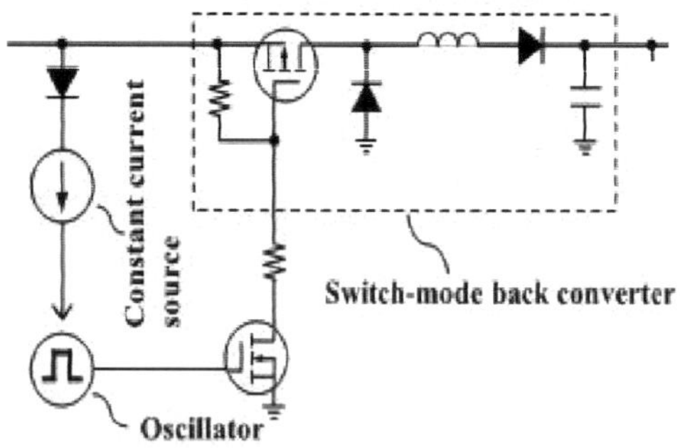

(**Figure6.**Switching converter circuit)

Adopt a resonance coil of a low loss and a Large-scale capacitor of 100μF in the output circuit from which the stability of the output voltage is requested.

Moreover, to prevent the output voltage from Rising, Zener-diode is inserted in the final steps.

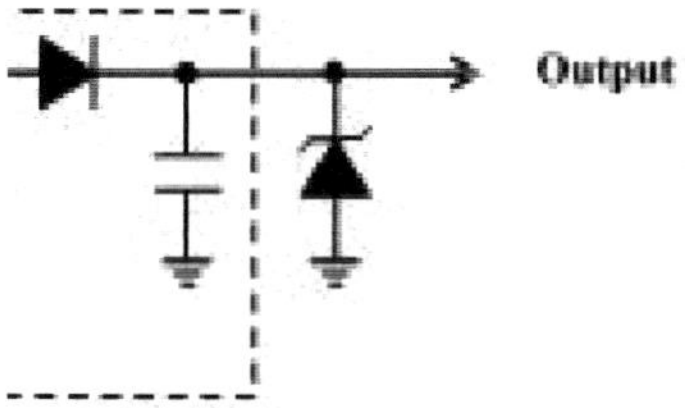

(Figure7. Output Zener diode)

Obtain the addition impedance and obtain the power supply of 90μW or more as for the output at this about 5V impedance at 300KΩ.

The diode for the armature voltage control has been added.

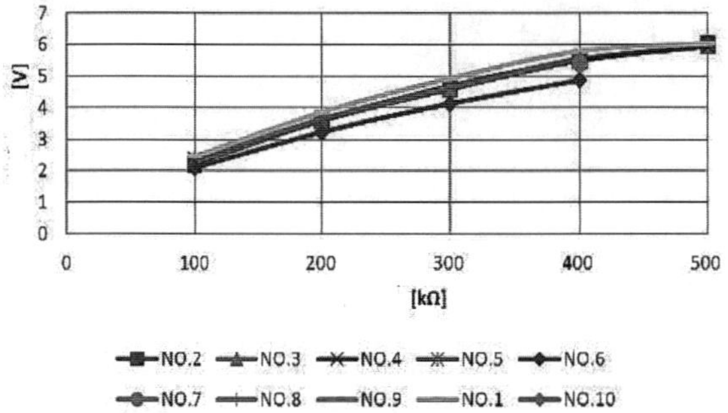

(Figure8.Output impedance / voltage)

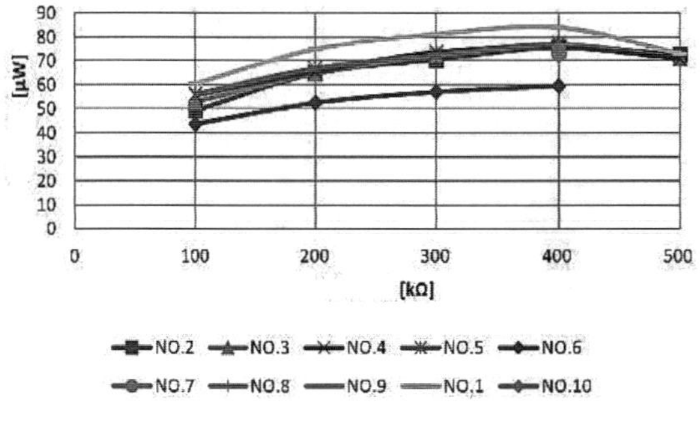

(Figure9.Output impedance/power-mW)

The power outlet becomes and when the output becomes "zero" because of the vibration stop etc., the reset signal might become necessary as a condition of connected equipment drive such as digital wireless machines.

For model BP-06, the reset signal output terminal has been added.

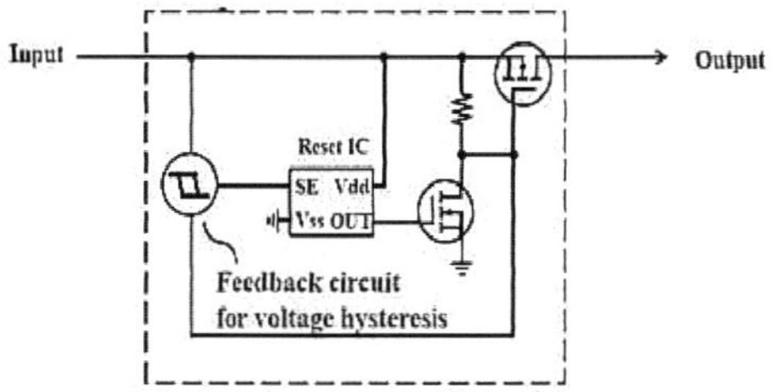

(Figure10.For Reset signal module)

3. Battery Pack Specification

MODEL: BP-02
3.2V / 0.10mW (38×54×10 ㎜ 20g)
30Hz / 0.15G

MODEL: BP-06
3.2V / 0.30mW (55×70×10 ㎜ 35g)
30Hz / 0.15G
With Reset Signal Output

(**Figure11**.BP－02 & USB Type Output)

(**Figure12**.BP－06)

4. Conclusions

At developed a small, light vibration generator module so that the development of the environmental generator system might improve in the speed by using the Polymer electret type vibration generator.

The equipment of the temperature hygrometer with a liquid crystal type clock on the market was made to work without trouble by 5V output model. &Wireless Module.

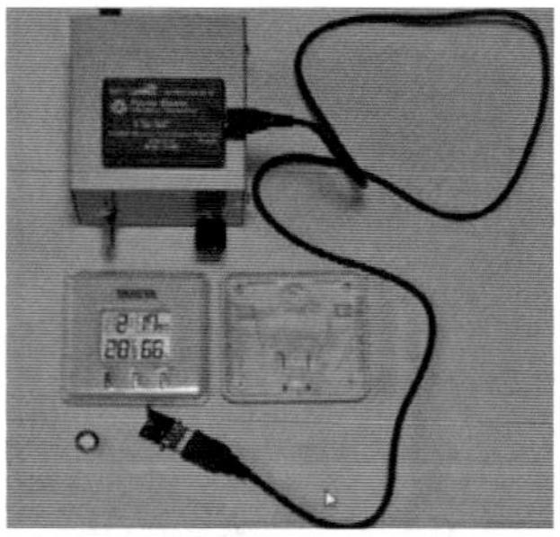

(**Figure13**.Liquid crystal panel)

(**Figure14**.Wireless Module)

(**Figure15**.Wireless transmitter Demonstration)

Journal of Physics: Conference Series **557** (2014) 012073 doi:10.1088/1742-6596/557/1/012073

5. References

[1] Yuji Suzuki "Development of a MEMS energy harvester with high-performance Polymer electrets". PowerMEMS2010 47-52

[2] Tatsuakira Masaki, Kenji Sskurai, Toru Yokoyama, Masayo Ikuta, Masasi Doi, Tomomori Seki, Masatosi Oba "Power Output enhancement of vibration-drivn Electret generator using concave Electrodes" PowerMEMS 2010 57-60

[3] Makoto Ueta, Yuji Suzuki "Evaluation of RF Sensor Module with Vibration Electret Energy Harvester" 5th-Micro Nano symposium2013.11 5-7

[4] Tomomori Seki "Electret Vibration Generator" TECHNO FRONTIER 2014 July 16-18, 2014

[5] Yu Jia ,Yan.Kenich Soga and Ashwin A Seshia "Parametrically excited MEMS vibration energy harvesters with design approaches to overcome the initiation threshold amplitude" 25 Octoberstacs.iop.org/JM/23/114007

[6] Tom Krupenkin & J, Ashley Taylor"Reverse electrowetting as a new approach to high-power energy harvesting" Nature Communications 2, Article number.448 dol: 10.1038/ncomms 1454 07 February 2012

[7] Guang Zhu, Jun Chen, Tiejun Zhang, Qingshen Jing, Zhong Lin Wang. "Radial-arrayed rotary electrification for high Performance triboelectric generator." Nature communications DOI: 10 1038/ncomms4426,4 Mar 2014

[8] Yuji Suzuki ,Masato Edamoto,Nobuhide Kasagi, Kimiaki Kashiwagi, Yoshitomi Morizawa,Toru Yokoyama ,Tomonori Seki , Msamoto Oba. "Micro Electret Energy Harvesting Device with Analogue Impedance Conversion Circuit."PowerMEMS2008-microEMS 2008,Sedai,Japan.November 9-12 2008

Multi-frequency vibration-driven electret generator for wireless sensor applications

T Masaki, N Yoshitake, S Kamiyama, M Nabeto, T Seki, M Oba and D Uchida

OMRON Corporation, 686-1 Ichimiyake, Yasu, Shiga 520-2362, Japan

E-mail:tatsuakira_masaki@omron.co.jp

Abstract. We had reported the power output enhancement of an electret harvester and the operation of wireless sensor modules only on electricity from our electret harvester[1]. Since vibration is random and irregular in actual fields, there is the issue that power output of vibration harvesters can't be obtained outside the range of frequencies designed to resonate the inner mass. As a way to solve, we fabricated four electret harvesters tuned to different resonant frequencies only by assembling different springs and weights, and developed voltage conversion circuits with 73% efficiency from 70V AC to 3.6V DC. We demonstrated to obtain stable DC power output at a broad range of frequencies from 27.6Hz to 30.6Hz by combining the four electret harvesters and voltage conversion circuits.

1. Introduction

It is expected that Internet of Things (IoT) described as a network of devices to sense and collect data around physical objects expands further. In order to expand IoT, almost IoT devices need to connect wirelessly to form networks. It depends on IoT devices with wireless interfaces capable of operating for years without battery replacement. Unfortunately, it is difficult to deploy many more IoT devices for a reason of power source. Replacing the batteries in each IoT devices is very difficult to implement with the expansion of these devices. For sustainability of wireless network operations, energy harvesting techniques have obtained much attention recently.

For example, IoT applications include the structural health monitoring for bridges and railway. There is a growing need to enhance safety and useful life-span of these society's infrastructures. In the actual fields of bridges or railway, vibration is random and irregular. Many naturally occurring vibrations don't have fixed frequency spectrums and vibrate over a broad range of frequencies. It is important for vibration energy harvesters to generate electricity from random vibrations. In this paper, we present mainly following two aspects.

1. High efficiency voltage converting circuits techniques
2. Multi-frequency electret harvester combined four harvesters tuned to different vibration frequencies respectively

2. Electret harvester

There are three kinds of vibration-driven microgenerators: electromagnetic, piezoelectric and electrostatic/electret types have been reported. Compared to electromagnetic or piezoelectric generators, electrostatics/electret is thought to have advantages in the viewpoint of compatibility to microfabrication, adaptability to low frequency vibration. In addition, vibrating stress doesn't apply to the electricity-generating parts in electrostatics/electret generators.

Suzuki found out CYTOP (Asahi Glass Co., Ltd) as an electret material[2] and his group proposed MEMS electret harvesters fabricated everything by MEMS process[3]. However, power output of fully-MEMS harvester using electret was under 10μW despite high frequency (>100Hz) vibrations in some previous reports[4][5][6]. For fully-MEMS harvesters, reduced size necessarily means reduced mass, meaning reduced output power in an inertially-driven device. Furthermore, the smaller harvesters have smaller space available for proof mass motion. We thought that power output need to enhance more as a generator for IoT devices and developed partially using high precision processing technology[7][8].

As a basic design of our electret harvester, there are mainly three points. To obtain the large power output from vibration, 1) giving high surface potential (-700V) to electret electrodes using CYTOP, 2) amplifying the amplitude ($2mm_{0\text{-}P}$) of inner mass by coil springs, and 3) narrowing the separation gap (70μm) between electret and collector electrodes, make advantage. As ambient vibrations are generally low in amplitude, the use of a mass-spring system generates a resonance phenomenon and amplifies the relative movement amplitude of the inner mass compared to the vibrations amplitude to enhance the harvested power. Thanks to coil springs, it is easy to obtain large amplitude and to design the resonance frequency of inner mass as shown in Fig.1. Our electret harvester is a very compact and lightweight generator. As shown in Fig.1, its size and weight are respectively $20\times20\times4$ mm^3 and 3.7g, and similar in size to coin battery. This is aimed at extending the range of applications.

Fig.2 shows the output voltage trace at 30Hz and 1.47ms^{-2} (0.15G) acceleration. We prepared the electret harvester tuned to this vibration frequency and acceleration as an example of ambient vibration. AC outputs with peak-to-peak voltage over 70V were obtained. Fig.3 shows the load voltage and power output to different loads at resonant frequency of 30Hz. As we expected, output voltage increased with the increase in load R. Maximum power output was obtained at R=15MΩ and its value was over 100μW. We designed the power output over 100μW when the amplitude of inner mass is $2mm_{0\text{-}P}$ at resonant frequency of 30Hz.

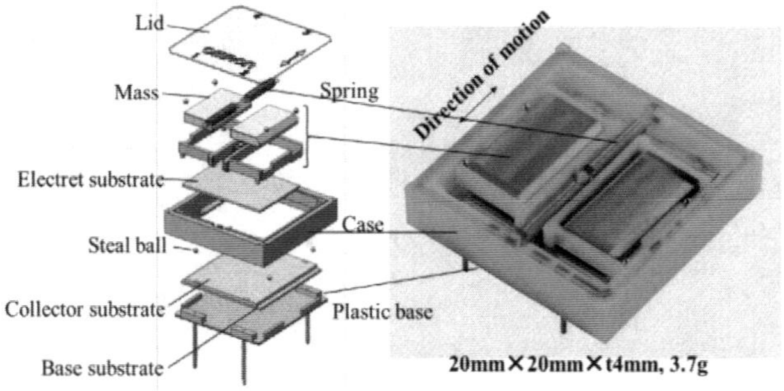

Figure 1. Schematic and photo of the electret harvester. Electret harvester in picture is assembled from a glass lid to view the inside.

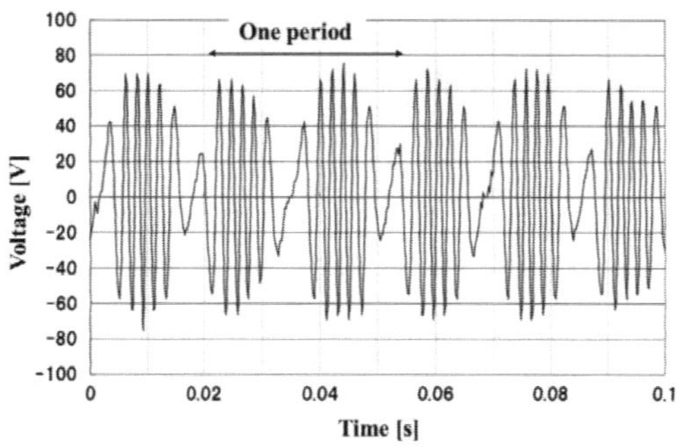

Figure 2. Time trace of the output voltage at 30Hz and 0.15G.

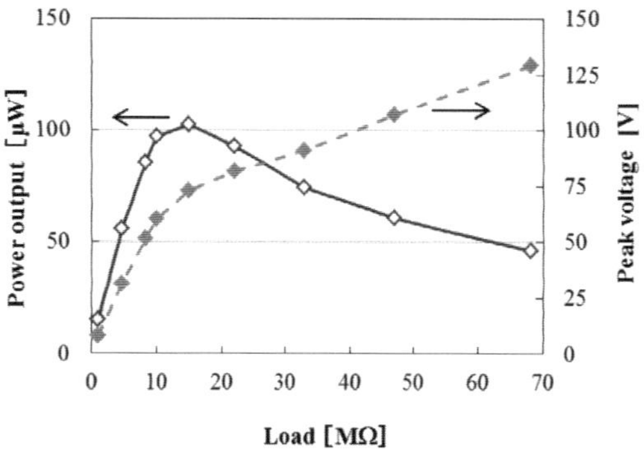

Figure 3. Power output and load voltage versus loads at 30Hz and 0.15G.

3. Experiment

3.1. High efficiency voltage conversion circuits

Since an output impedance of the electret harvester is very large, it is necessary to step down the output voltage from 70V AC. A required voltage after converting depends on the driving voltage of IoT devices and it is usually between the voltage of about 1.2V and 3.6V. There is a gap between the voltage of electret harvester and the required voltage. In addition, power output of 100μW is not enough. It is important to reflect these characteristics into the circuits.

In this time, we present voltage conversion techniques. Fig.4 shows the schematic of our circuits. First capacitors C_1 are connected in series. This C_1 group divides high voltage of the electret harvester to close the voltage gap. AC output voltage of the electret harvester charges once multiple capacitors C_1 connected in series and simultaneously supplies to a load circuit connected in parallel to each second capacitor C_2. Charging efficiency E_{C1} between the electret harvester and the C_1 group is represented by the following equation. Charging efficiency E_{C1} is a big part of conversion efficiency of our circuits.

$$E_{C1} = \frac{\text{amount of energy accumulated in C1 group per unit time}}{\text{amount of electric power generated by electret harvester with match resistance}}$$

Fig.5 shows theoretical correlations between the charging efficiency E_{C1} defined as described above and the voltage of C_1 group. Maximum efficiency is obtained when the voltage of C_1 group is equal to half of the output voltage of electret harvester. In the actual circuits, the number of C_1 we designed is ten. We achieved to convert with high efficiency of 73% as shown in Table 1.

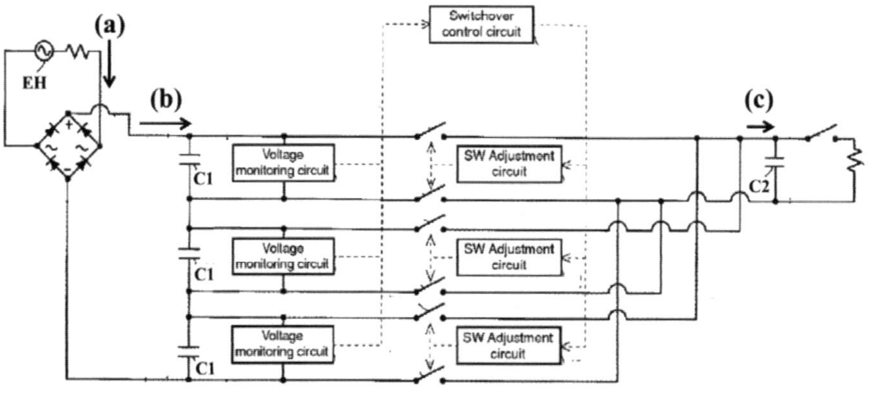

Figure 4. Schematic of our voltage conversion circuits.

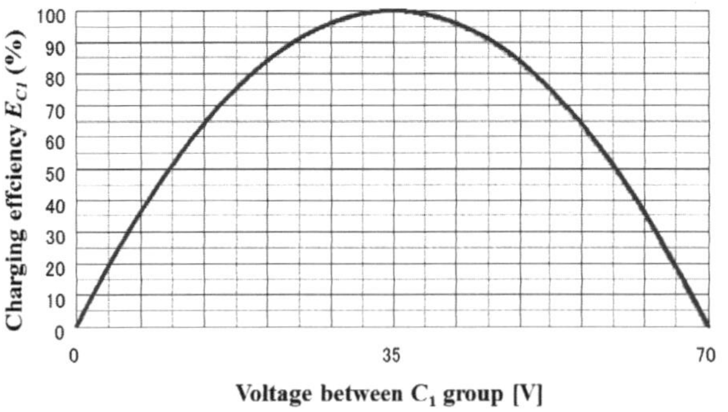

Figure 5. Theoretical Correlations between the charging efficiency and the voltage of C_1 group.

Table 1. Experimental results on our circuits. Measuring points at (a), (b) and (c) are plotted respectively on Fig.2.

	(a) Generator 70V	(b) Rectfied & Smoothed	(c) Down Convert 3.6V
Power output	71 μW AC	60 μW DC	52 μW DC
Efficiency	-	85%	73%

3.2. Multi-frequency electret harvester

There are three methods to widen the range of frequency adaptable to random vibrations in the actual fields. 1) Multi-frequency[9][10], 2) nonlinear springs[11][12], and 3) dual mass-spring systems[13][14] were reported. From the aspects of design and structure, we thought that the multi-frequency method has many advantage over another methods.

As a demonstration, we fabricated four electret harvesters tuned to different resonant frequency by assembling springs and weight as a specification of Table 2. Spring constant can be changed easily by coil designs such as wire diameter, outside dimension and coil tunes. As shown in Fig.6 and Fig.7, we demonstrated to obtain stable DC power outputs by combining four harvesters and high efficiency AC/DC conversion circuit described above.

Table 2. Resonant frequencies designed by spring constant and weight

Electret harvester	Design resonant frquency (Hz)	Spring constant (N/mm)	Weights (g)
# EH 1	27.6	0.0261	1.73
# EH 2	28.4	0.0276	1.73
# EH 3	29.7	0.0302	1.73
# EH 4	30.6	0.0320	1.73

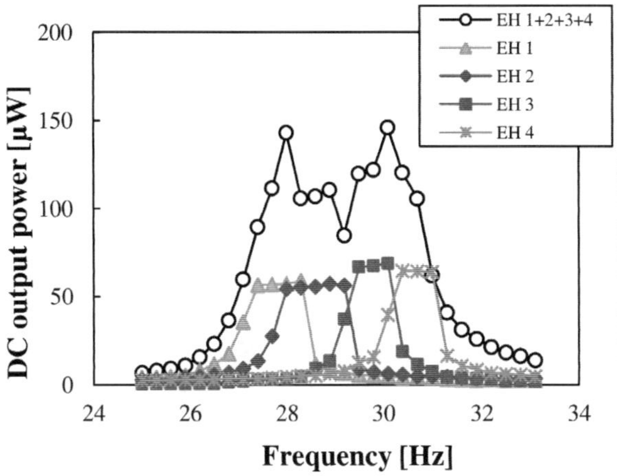

Figure 6. Experimental results of DC power outputs of four electret harvesters tuned to the frequency shown in Table 2.

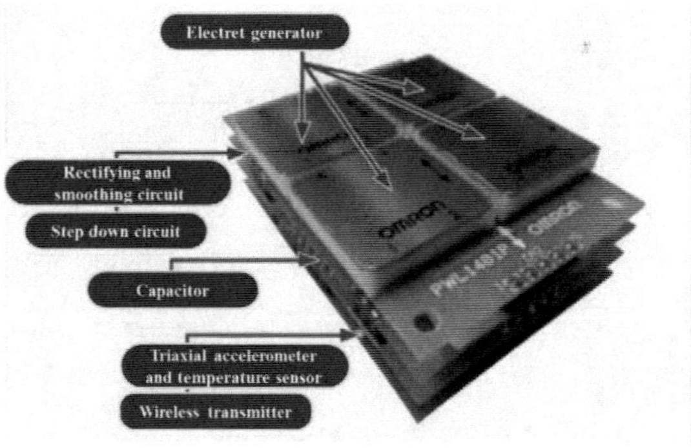

Figure 7. Photo of our wireless sensor module. External size is 65×45×t25mm^3.

4. Conclusion

We developed high efficiency voltage conversion circuits and achieved to convert the voltage over 70V AC into that of 3.6V DC with high efficiency of 73%.

We fabricated four electret harvesters tuned to different resonant frequencies only by assembling springs and weights. We demonstrated to obtain stable DC power output by combining four electret harvesters and AC/DC conversion circuits at a broad range of frequencies from 27.6Hz to 30.6Hz.

References

[1] Masaki T, Sakurai K, Yokoyama T, Ikuta M, Sameshima H, Doi M, Seki T and Oba M 2011 *J. Micromech. Microeng.* **21** 104004
[2] Arakawa Y, Suzuki Y and Kasagi N 2004 *Proc. PowerMEMS2004 (Kyoto)* p 183
[3] Suzuki Y, Miki D, Edamoto M and Honzumi M 2010 *J. Micromech. Microeng.* **20** 104002
[4] Roundy S, Wright P K and Rabaey J M 2003 *Energy Scavenging for Wireless Sensor Networks with special focus on vibrations* (Springer)
[5] Sterken T, Fiorini P, Altena G, Van Hoof C and Puers R 2007 *Proc. Transducers2007 (Lyon)* p 129
[6] Hoffmann D, Folkmer B and Manoli Y 2008 *Proc. PowerMEMS2008 (Sendai)* p 15
[7] http://www.omron.co.jp/press/2008/11/c1111.html# (accessd 13 Oct. 2014)
[8] http://techon.nikkeibp.co.jp/english/NEWS_EN/20081117/161303/?P=1 (accessd 13 Oct. 2014)
[9] Sari I, Balkan T and Kulah H 2008 *Sensors and Actuators A: Physical* **145-146** 405
[10] Ferrari M., Ferrari V, Guizzetti M, Marioli D and Taroni A 2008 *Sensors and Actuators A: Physical* **142** 329
[11] Erturk A, Hoffmann J and Inman D J 2009 *Applied Physics Letters* **94** 254102
[12] Stanton S C, McGhee C C and Mann B P 2009 *Applied Physics Letters* **95** 174103
[13] Tang X and Zuo L 2011 *J. Sound and Vibration* **330** 5199
[14] http://www.takenaka.co.jp/news/2013/11/04/index.html (accessd 13 Oct. 2014)

OH-LIF measurement of $H_2/O_2/N_2$ flames in a micro flow reactor with a controlled temperature profile

T Shimizu[1], **H Nakamura**[1], **T Tezuka**[1], **S Hasegawa**[1] **and K Maruta**[1,2].

[1]Institute of Fluid Science, Tohoku University, 2-1-1, Aoba-ku, Sendai, Miyagi, 980-8577, Japan

[2]Far Eastern Federal University, 8 Suhanova St., Vladivostokok 690950, Russia

E-mail: shimizu@edyn.ifs.tohoku.ac.jp

Abstract. This paper presents combustion and ignition characteristic of $H_2/O_2/N_2$ flames in a micro flow reactor with a controlled temperature profile. OH-LIF measurement was conducted to capture flame images. Flame responses were investigated for variable inlet flow velocity, U, and equivalence ratio, ϕ. Three kinds of flame responses were experimentally observed for the inlet flow velocities: stable flat flames (normal flames) in the high inlet flow velocity regime; unstable flames called *Flames with Repetitive Extinction and Ignition* (FREI) in the intermediate flow velocity regime; and stable weak flames in the low flow velocity regime, at ϕ = 0.6, 1.0 and 1.2. On the other hand, weak flame was not observed at ϕ = 3.0 by OH-LIF measurement. Computational OH mole fractions showed lower level at the rich conditions than those at stoichiometric and lean conditions. To examine this response of OH signal to equivalence ratio, rate of production analysis was conducted and four kinds of major contributed reaction for OH production: R3($O + H_2 \Leftrightarrow H + OH$); R38($H + O_2 \Leftrightarrow O + OH$); R46($H + HO_2 \Leftrightarrow 2OH$); and R86($2OH \Leftrightarrow O + H_2O$), were found. Three reactions among them, R3, R38 and R46, did not showed significant difference in rate of OH production for different equivalence ratios. On the other hand, rate of OH production from R86 at ϕ = 3.0 was extremely lower than those at ϕ = 0.6 and 1.0. Therefore, R86 was considered to be a key reaction for the reduction of the OH production at ϕ = 3.0.

1. Introduction

Hydrogen produced by renewable energies attracts attentions as one of the alternative fuels for petroleum fuels. Hydrogen has huge energy per unit mass and its combustion never exhausts carbon dioxide because it contains no carbon atom. However, it is important to investigate combustion characteristics to develop practical combustion devices because the present knowledge on hydrogen combustion is not sufficient.

This study employed a micro flow reactor with a controlled temperature profile (MFR)[1] to investigate hydrogen combustion. This reactor has a narrow channel, whose diameter is smaller than the ordinary quenching diameter, and the channel is heated by an external heat source to give a stationary temperature profile along the inner surface of the reactor. This reactor has been applied for various hydrocarbons in our past studies [1, 2]. They showed three kinds of flame responses for variable inlet flow velocities: stable flat flames (normal flames) in the high inlet flow velocity regime; unstable flames called *Flames with Repetitive Extinction and Ignition* (FREI) in the intermediate flow velocity regime; and stable weak flames in the low flow velocity regime. Especially, weak flames were often studied because the weak flame branch was regarded as a part of the ignition branch of

Fendell curve [3] in the theoretical analysis [4] and unsteady multi-stage oxidation of hydrocarbons was examined by stable multiple weak flames [2].

In our previous study for $H_2/O_2/N_2$ mixture using a MFR, two kinds of flame responses were experimentally observed for variable inlet flow velocities, *U*, by measurements of OH chemiluminescence using an UV camera [5]. One is normal flame in $U > 70$ cm/s and another is FREI in $U = 11$ - 70 cm/s. Gas sampling and analysis showed the consumption of H_2 and 1-D computational results indicated the existence of weak flame. However, weak flame was not observed in the low inlet flow velocity regime experimentally. Mantzaras et al. conducted 3-D numerical simulation of hydrogen flames in a heated micro channel [6-8] and they showed axisymmetric combustion modes for $U = 0.5 – 500$ cm/s in tube diameter of $d = 1.0$ mm [8]. Steady mild combustion in the low flow velocity, oscillatory ignition / extinction in the intermediate flow velocity and steady flames in the high flow velocity were obtained at $\phi = 0.5$.

Thus, the existence of weak flame of $H_2/O_2/N_2$ mixture is still an open question. Therefore, the existence of weak flame of $H_2/O_2/N_2$ mixture was investigated by OH-LIF measurement at various ϕ (0.6, 1.0, 1.2 and 3.0) in this study.

2. Experiment and computation

2.1. Experimental setup and conditions

A schematic of experimental setup is shown in Figure 1. A quartz tube with an inner diameter of 2 mm was used as a reactor channel and was heated by a H_2/air flat-flame burner to obtain a stationary temperature profile from 300 K to 1300 K along the inner surface of the reactor. A $H_2/O_2/N_2$ mixture was supplied to the reactor. The mixture was diluted in $H_2/O_2/N_2 = 2:1:9$. Flame responses to variable inlet flow velocity were observed by OH-LIF measurement. Wavelength of 283 nm by DYE-Laser (FINE ADJUSTMENT, 0312P043) which was converted from 532 nm of Nd:YAG-Laser (LOTIS TII, LS-2137/3) was chosen to excite OH. A laser beam was run through a pinhole with diameter of 2 mm after concentrated by convex and concave lens. The laser was introduced from the downstream side of the reactor. An image-intensified CCD camera (ANDOR, DH334T-18-E3) with UV lens, an OH band-pass filter (center wavelength: 313 nm; half value width: 10 nm; and transmissivity > 60%) and two 300 nm high-pass filters were used to capture flame images. Background subtraction image processing was done for captured images. All experiments were conducted at atmospheric pressure.

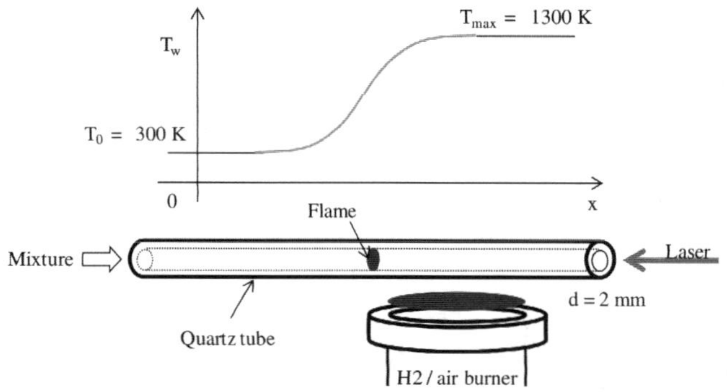

Figure 1. Schematic of experimental setup and provided wall-temperature profile.

2.2. Definition of flame locations

The peak position in the luminosity profile of OH-LIF images was selected to define the normal flame position because 1-D computation showed that the peak position in the heat release rate (HRR) profile corresponded to that in the OH mole fraction profile. On the other hand, the position of the half maximum in OH profile rise was used for that of weak flame because it agrees with the position of the HRR peak for weak flames in the computation. For FREIs, wall temperatures at the extinction and ignition positions were identified from the upstream and downstream edge of more than 50 instantaneous OH-LIF images.

2.3. Computational conditions

One-dimensional steady-state flame code based on PREMIX was used [1]. The energy equation includes convective heat transfer between gas and wall. Four kinds of mechanism (GRI 3.0 [9], San Diego Mech. [10], Princeton H_2-O_2 Model [11] and Aramco Mech. 1.3 [12].) were used to examine their performances in the present reactor. Computations were conducted for the same conditions as those of the present experiments.

3. Results and discussions

3.1. Flame responses to equivalence ratio and inlet flow velocity

Figure 2 shows observed flame responses to the inlet flow velocity at $\phi = 1.0$. Three kinds of flame responses to the inlet flow velocity were observed: normal flame in the high flow velocity regime; FREI in the intermediate flow velocity regime; and weak flame in the low flow velocity regime. These three kinds of flame responses were observed not only at $\phi = 1.0$ but also at $\phi = 0.6$ and 1.2. At $\phi = 3.0$, however, weak flames were not observed while normal flames and FREI were observed.

The measured flame positions were shown in Figure 3. Normal flames shift to the lower temperature side with the decrease of the inlet flow velocity for all equivalence ratios studied. Normal flames at $\phi = 1.2$ and 3.0 are located at the lower temperature region than those at $\phi = 0.6$ and 1.0. Weak flames shift to the higher temperature side with the decrease of the inlet flow velocity. Weak flames at $\phi = 1.2$ are located at the lower temperature region than those at $\phi = 0.6$ and 1.0.

Figure 4 shows computational HRR peak positions at various inlet flow velocities and equivalence ratios. Normal flames shift to the lower temperature side with the decrease of the inlet flow velocity for all equivalence ratio studied and normal flames at $\phi = 1.2$ and 3.0 are located at the lower temperature region than those at $\phi = 0.6$ and 1.0 which are the same tendencies in the experiment. Steady solutions were not obtained in the intermediate flow velocity region where unsteady solutions of FREI were expected. Weak flames were not observed in the experiment at $\phi = 3.0$ but weak flame solutions in the low flow velocity region were obtained in the computation.

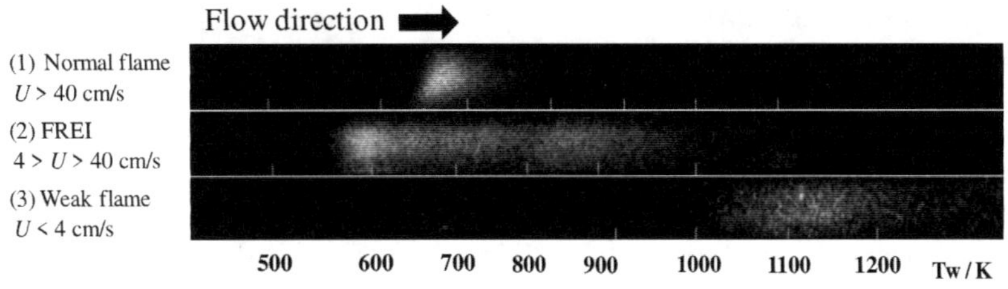

Figure 2. Flame responses captured by OH-LIF measurements at ϕ=1.0.

3.2. OH concentration for variable equivalence ratios

A reason why OH-LIF was not able to obtain weak flames at $\phi = 3.0$ was investigated by computations. Figure 5 shows that the maximum OH mole fractions for variable equivalence ratios. The maximum OH mole fraction was extremely decreased in the rich side. This low level of OH mole fraction made the OH-LIF measurement difficult to observe weak flames at $\phi = 3.0$. Rate of production analysis was conducted for OH to investigate the significant decrease of OH mole fraction in the rich side. GRI 3.0 was used for this analysis because the computational weak flame position with GRI 3.0 was in the best agreement with the experimental flame position compared with the other mechanisms.

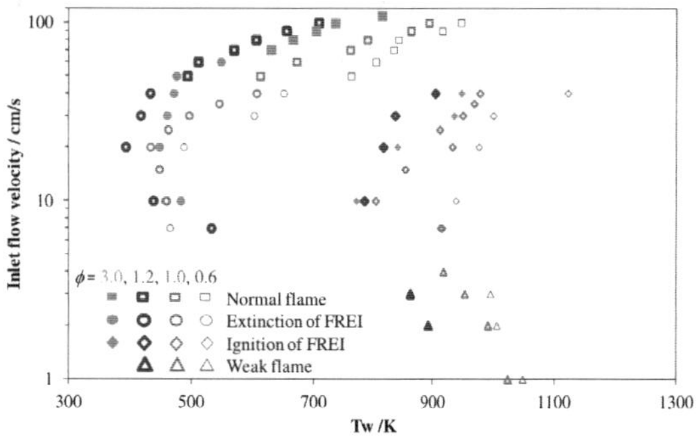

Figure 3. Measured flame position and extinction and ignition points of FREI in various flow velocities at $\phi = 0.6$, 1.0, 1.2 and 3.0.

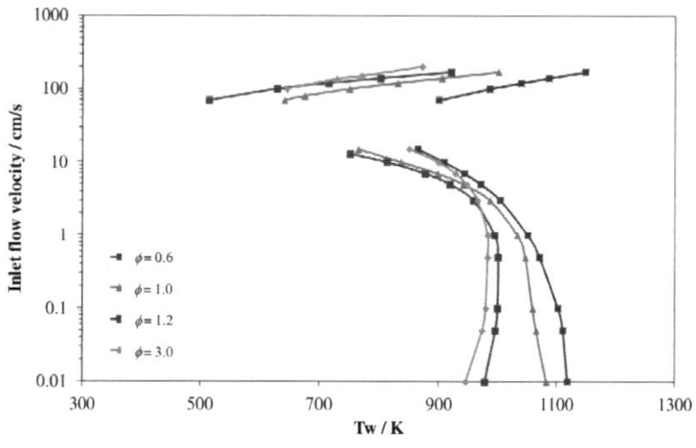

Figure 4. Computational HRR peak positions in various flow velocities at $\phi = 0.6$, 1.0, 1.2 and 3.0.

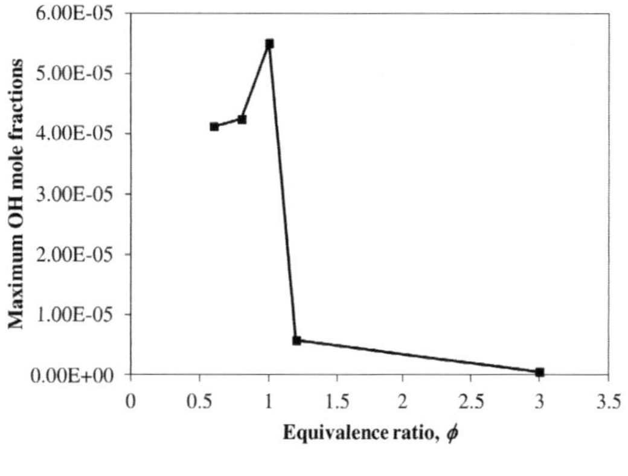

Figure 5. Maximum OH mole fraction for variable equivalence ratios in computation.

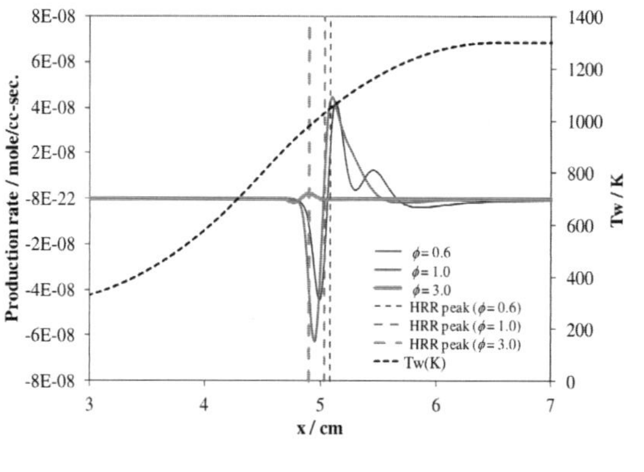

Figure 6. Computational net rate of OH production at $\phi = 0.6$, 1.0 and 3.0.

Net rate of OH production at $\phi = 0.6$, 1.0 and 3.0 were shown in Figure 6. Net rate of OH production at $\phi = 3.0$ was much lower than those at $\phi = 0.6$ and 1.0. Major contributed reactions for OH production were investigated at the HRR peak position. As a result, four major contributed reactions for OH production were found: R3($O + H_2$ <=> $H + OH$); R38($H + O_2$ <=> $O + OH$); R46($H + HO_2$ <=> $2OH$); and R86($2OH$ <=> $O + H_2O$). Three of them, R3, R38 and R46, did not show big difference between $\phi = 3.0$ and $\phi = 0.6$ and 1.0. However, R86, which forms OH radicals through the backward reaction, became significantly small at $\phi = 3.0$. Therefore, it is considered that the low level of OH mole fraction in the rich side would be due to the equivalence ratio dependence of R86.

4. Conclusion

Ignition and combustion characteristics of $H_2/O_2/N_2$ flames were investigated using a micro flow reactor with a controlled temperature profile at various equivalence ratios and inlet flow velocities. OH-LIF measurement was applied for capturing flame images. As a result, three kinds of flame responses: normal flame in the high inlet flow velocity regime; FREI in intermediate flow velocity regime; and weak flame in the low flow velocity regime, were experimentally observed at $\phi = 0.6$, 1.0 and 1.2. On the other hand, only two kinds of flame responses: normal flame and FREI, were observed at ϕ = 3.0. Computational OH mole fraction was remarkably decreased in the rich conditions. Therefore it is considered that weak flames were not observed in experiment at ϕ = 3.0. Rate of production analysis was conducted to find the difference in reaction process at $\phi = 0.6$, 1.0 and 3.0. Four major reactions for OH production: R3(O + H_2 <=> H + OH); R38(H + O_2 <=> O + OH); R46(H + HO_2 <=> 2OH); and R86(2OH <=> O + H_2O), were identified. The reaction rates of R3, R38, R46 were not changed so much when the equivalence ratio was varied but the reaction rate of R86 became significantly small at $\phi = 3.0$.

Acknowledgement

The part of the study was supported financially by the Ministry of education and science of Russian Federation (project 14. Y26.31.0003).

References

[1] Maruta K, Kataoka T, Kim I. N, Minaev S. and Fursenko R, 2005 *Proc. Combust. Inst.* **30** 2429-2436.

[2] Yamamoto A, Oshibe H, Nakamura H, Tezuka T, Hasegawa S. and Maruta K, 2011 *Proc. Combust. Inst.* 2 **33** 3259-3266.

[3] Nakamura H, Yamamoto A, Hori M, Tezuka T, Hasegawa S and Maruta K, 2013 *Proc. Combust. Inst.* **34** 3435-3443.

[4] Minaev S, Maruta K and Fursenko R, 2007 *Combust. Theor. Model* 2 **11** 187-203.

[5] Saruwatari K, Nakamura H, Tezuka T, Hasegawa S and Maruta K, 2012 *ICFD* OS11.

[6] Pizza G, Frouzakis E. C, Mantzaras J, Tomboulides A. and Boulouchos K. 2008 *Comb. Flame* **152** 433-450.

[7] Pizza G, Frouzakis E. C, Mantzaras J, Tomboulides A. and Boulouchos K. 2008 *Comb. Flame* **155** 2-20.

[8] Pizza G, Frouzakis E. C, Mantzaras J, Tomboulides A. and Boulouchos K. 2010 *J. Fluid Mech.* **658** 463-491.

[9] Smith G. P, Golden D. M, Frenklach M, Moriarty N. W, Eiteneer B, Goldenberg M, Bowman C. T, Hanson R. K, Song S, Gardiner W. C. Jr. Lissianski V. V, Qin Z. GRI-Mech 3.0. Available at http://www.me.berkeley.edu/gri mech/.

[10] Prince J. C, Williams F. A, 2011 *Proc. 7th U. S. National Tech. Meeting Combust. Inst*, 20-23 Available at http://web.eng.ucsd.edu/mae/groups/combustion/mechanism.html.

[11] Burke M. P, Chaos M, Ju Y, Dryer F. L and Klippenstein S. J, 2004 *Int. J. Chem. Kinet.* **36** 566-575 Available at http://www.princeton.edu/mae/people/faculty/dryer/homepage/kinetic_models/h2-o2-model-update/.

[12] Metcalfe W. K, Burke S. M, Ahmed S. S, Curran H. J, 2013 *Int. J. Chem. Kinet.* 10 **45** 638–675 Available at http://www.nuigalway.ie/c3/Mechanism_release/frontmatter.html.

An experimental and numerical study of flames in narrow channels with electric fields

DC Murphy[1], M Sánchez–Sanz[2], C Fernandez-Pello[1]
[1]Department of Mechanical Engineering, University of California Berkeley
[2]Dept. Ingeniería Térmica y de Fluidos, Universidad Carlos III de Madrid
E-mail: dmurphy0516@gmail.com

Abstract. The advancement of microscale combustion has been limited by quenching effects as flames cease to be much smaller than combustors. The long studied sensitivity of flames to electrical effects may provide means to overcome this issue. Here we experimentally and numerically investigate the potential of electric field effects to enhance combustion. The results demonstrate that, under specific conditions, externally electric fields will sustain combustion in structures smaller than the quenching distance. The analysis proposes a reduced mechanism to model this result and provides a study of the governing parameters. We find good qualitative agreement between the model and experiments. Specifically, the model is found to successfully capture the capacity to increase and decrease flame speed according to electric field magnitude and direction. Further, in both experiments and computations the sensitivity to electrical enhancement increases for more energetic mixtures. We do find that the model underpredicts the maximum achievable speed enhancement observed, suggesting that additional phenomena should be included to expand the range of conditions that can be studied.

1. Introduction

In this paper we examine the use of electrical effects as a means of improving the performance of micro-combustion systems. One of the primary technical challenges in miniaturizing combustors is the phenomenon of quenching, wherein flames are extinguished as heat losses to closely spaced combustor walls become so great that they overcome the rate of chemical release and the combustion reaction can no longer sustain itself [1]. To date, progress has been made in the field of micro-combustion by reducing thermal losses from the flame [2] and by increasing the pressure and reactivity of the fuel-air mixtures [3] to overcome quenching. By introducing new physical processes it may be possible to further reduce the scale of combustion.

As early as 1899, it was known that electric fields could be utilized to modify flame geometry [4]. Since that time, experimental studies have found that electric fields can improve the speed of flames [5, 6]. In the context of quenching, this increase in flame speed corresponds to an increased rate of chemical heat release, and a reduction in quenching. We demonstrate below that an applied electric field will increase the propagation speed of flames at the limit of quenching and that this increases prevents the occurrence of quenching. Further, we describe a basic model to understand the enhancement of laminar flames.

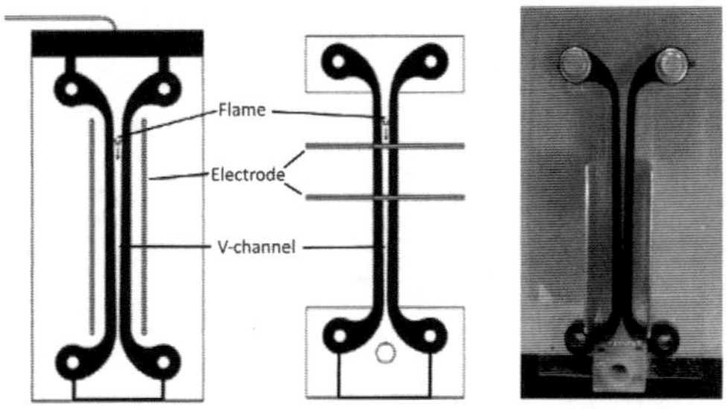

Figure 1: Diagrams of the V-channel for perpendicular (left) and parallel (center) electric fields. Photograph of the V-channel for perpendicular fields (right)

2. Experimental

2.1. Methodology

An experiment was developed to observe the propagation speed and quenching behavior of premixed methane-air flames in a narrow channel while subject to externally applied electric fields. The apparatus, referred to here as the V-channel and shown schematically in Figure 1, consisted of two Polymethylmethacrylate (PMMA) blocks which formed the walls of the V-Channel. Oriented vertically, these formed a gap which decreased from at 5 mm the the entrance, to 2 mm at the bottom over a 100 mm length. The channel walls were 12 mm thick and held between clear PMMA sheets to enclose the channel while providing optical access. The channel was operated in two configurations. One established the electric field perpendicular to the channel walls, the other established a fields parallel to the walls. For the perpendicular field configuration, the electrodes were flat aluminum plates, 107 mm square, held symmetrically around the channel walls at a distance of 25 mm from one another. For the parallel configuration, the electrodes were 90 mm square plates with centrally located 15 mm square holes, to permit the V-channel to pass through. The parallel plates were spaced either 25 mm or 75 mm apart to achieve higher and lower electric field strengths for a given voltage. When spaced 25 mm apart, the plates were positioned along the V-channel such that quenching would occur in the region between them without an applied filed.

Prior to each test the channel and chambers were purged with the desired methane-air mixture until no fewer than sixty full changes of gas had occurred. Results were recorded using a high speed video camera at 2000 fps and a Z-type schlieren system and subsequently processed with an automated cross-correlation algorithm to measure flame propagation speeds. Spark ignition of the gas mixture occurred at the top of the channel. Flames propagated downward while exhaust gases exited from the top of the test volume. In the perpendicular field configuration, experiments were performed at 5 equivalence ratios from 0.85 to 1.2. In the parallel field configuration, four equivalence ratios from .95 to 1.2 were tested. To minimize the impact of environmental factors, each nominal equivalence ratio and configuration was tested multiple times, alternatively applying electric field and allowing the flame to propagate freely, providing a reference to which the electrically altered flames could be compared. In all cases, 6 tests were conducted with electric field and 6 without. For a 1.0 equivalence ratio with perpendicular field an additional 6 tests with and without field we conducted.

Two high voltage power supplies were used to energize the electrode plates. The first produced an alternating voltage (AC) of 20 kV at 60Hz and was used on both the perpendicular and parallel configurations. In the parallel configuration, this resulted in an electric field of 800 kV/m with 25mm plate spacing and 266 kv/m for 75mm spacing. The second power supply operated at a constant (DC) 4.5 kV, producing 180 kV/m or 60 kV/m, depending on plate spacing. The DC supply was tested under both polarities, such that the electric field lines lay

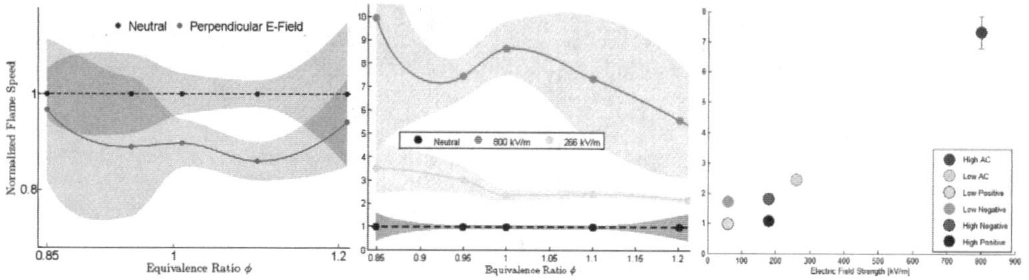

Figure 2: Variaiton in the flame propagation with applied electric field and stoichiometry. **Left:** 800 kV/m AC with perpendicular orientation, **Center:** 800 and 266 kV/m AC in parallel orientation, **Right:** Stoichiometric flames in AC and DC fields with parallel orientation. 'high/low' indicates 800/266 kV/m for AC and 180/60 kV/m DC

.

along or against the direction of propagation (referred to as positive and negative, respectively in the left plot of Figure 2). Flame propagation occurred under quiescent conditions.

2.2. Results

The changes in propagation speed with electric field and stoichiometry are presented in Figure 2. Primarily, we observe large increases in propagation speed due to the AC and negative DC fields in the parallel field configuration. The positive DC field showed minimal impact on the propagation speed and the perpendicular AC field produced a small but measurable reduction in speed. For the AC experiments, the degree of enhancement or impairment increased with the electric field magnitude. For those conditions where the fuel concentration was varied, the electric field was most effective at lean and stoichiometric conditions, although large variability in the measured speeds makes it difficult to detect any consistent trend as flames became increasingly lean. In all cases where the propagation speed was enhanced, the flames did not quench until they reached the farthest extent of the electric field, beyond the location at which quenching would have otherwise occurred.

3. Numerical

3.1. Formulation

In this analysis, we consider a lean planar premixed flame propagating with a constant velocity and subject to an electric field parallel to the direction of propagation. The equations describing the structure of this flame in the presence of an electric field begin with the mass and species conservation equations for two-step reaction [7]. Additionally, the model needs to consider the effect that the electric field exerts on the charged species. For the weakly ionized conditions in a flame, these effects may be accounted for by the addition of an electrically induced drift velocity for charged species as well as accounting for variations in the electric field caused by those species. For the full development of this model and a discussion of the assumptions, the reader may consult [8]. For a lean flame which reacts primarily by a two-step reaction with a neutrally charged intermediate, but also includes reactions for ionization and recombination, the non-dimensional governing equations may be presented as:

$$\frac{d\theta}{dx} = \frac{d^2\theta}{dx^2} + \mu Q\left(\omega_{II} + \omega_{IV}\frac{q_{IV}}{q_{II}}\right) \tag{1}$$

$$\frac{dY_F}{dx} = \frac{1}{Le_F}\frac{d^2Y_F}{dx^2} - \mu\omega_I \tag{2}$$

$$\frac{dY_Z}{dx} = \frac{1}{Le_Z}\frac{d^2Y_Z}{dx^2} + \mu\left[\omega_I - \omega_{II} - \omega_{III}\right] \tag{3}$$

$$\frac{dY_{Z^+}}{dx} = -\mu^{1/2}\frac{d\left(EY_{Z^+}\right)}{dx} + \frac{1}{Le_{Z^+}}\frac{d^2Y_{Z^+}}{dx^2} + \mu\left[\omega_{III} - \omega_{IV}\right] \tag{4}$$

$$\frac{dY_{e^-}}{dx} = \mu^{1/2}\left(\frac{m^+}{m^-}\right)^{1/2}\frac{d(EY_{e^-})}{dx} + \\ + \left(\frac{m^+}{m^-}\right)^{1/2}\frac{1}{Le_{Z^+}}\frac{d^2Y_{e^-}}{dx^2} + \mu\left[\omega_{III} - \omega_{IV}\right] \tag{5}$$

with boundary conditions $\theta = Y_Z = Y_{Z^+} = Y_{e^-} = Y_F - 1 = 0$ at $x \to -\infty$ and $\theta' = Y_F' = Y_Z' = Y_{Z^+}' = Y_{e^-}' = 0$ at $x \to \infty$. Where Le, m, θ, Y_F, Y_Z, Y_{Z^+}, and Y_{e^-} are the Lewis number, mass of a charged species, temperature, and mass fractions of fuel, neutral radical, ionized radical and electrons. The remaining variable μ is the dimensionless eigenvalue of the problem which relates to the laminar flame speed S_L by the proportionality $\mu \propto S_L^{-2}$. The non-dimensional reaction rates are written as

$$\omega_I = \beta^2 Y_Z Y_F \exp\left\{\beta\frac{\theta - 1}{1 + \gamma(\theta - 1)}\right\} \tag{6}$$

$$\omega_{II} = Y_Z \tag{7}$$

$$\omega_{III} = \beta^2 \mathcal{A} Y_Z \exp\left\{(\beta + \Delta\beta)\frac{\theta - 1}{1 + \gamma(\theta - 1)}\right\} \tag{8}$$

$$\omega_{IV} = \mathcal{B} Y_{Z^+} Y_{e^-} \tag{9}$$

with $\mathcal{A} = \frac{A_{III}}{A_I}\frac{W_{Z^+}W_F}{W_Z\bar{W}}\frac{e^{-\Delta\beta/\gamma}}{Y_{F_0}}$, $\mathcal{B} = \frac{A_{IV}}{A_{II}}\frac{\bar{W}W_Z}{W_{Z^+}W_F}Y_{F_0}$ and $\Delta\beta = \gamma\frac{E_{III} - E_I}{R_g T_c}$ representing the effect of a differential activation energy between the chain-branching and the chemi-ionization steps.

In the above equations, the electric field E is the result of both the externally imposed field E_0 and the two charged species. The spatial variation due the charged species is $dE/dx = \mu^{1/2}/\epsilon(Y_{Z^+} - Y_{e^-})$. This my be integrated, subject to the condition $E(x \to -\infty) = E_0$ to complete the formulation.

3.2. Results

The problem defined by Eqs. (1)- (5), with the corresponding boundary conditions, was solved numerically to compute the eigenvalue μ and the profiles of temperature and species. Figure 3 shows a representative flame structure for neutral conditions as well as for a positive and negative external fields. Fundamentally, the addition of the ionization/recombination reactions serves to delay the heat release, reducing the transfer of heat to the unburnt gases and lowering flame speed. The external electric field is therefore able modify S_L by shifting the relative location at which recombination occurs and heat is generated.

For comparison to our experimental results, it is most illustrative to examine the effects of both E_0 and the heat release parameter Q (which relates to ϕ in the experiments), as shown in the

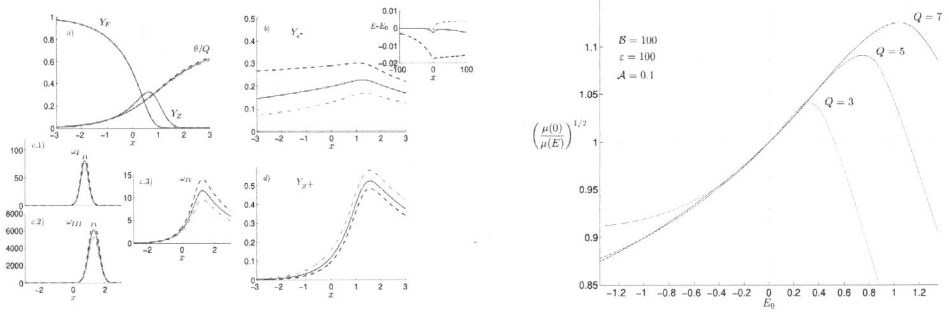

Flame Structure Flame Speed

Figure 3: **(left)** a) Profiles of fuel Y_F, radical Y_Z and temperature θ/Q, b) and d) Electron and proton mass fraction profiles, c.1) chain-branching reaction rate ω_I, c.2) proton-electron production rate ω_{III} and c.3) proton-electron recombination reaction rate. ω_{IV} for $E_0 = 0$ (blue, solid line), $E_0 = 0.65$ (black, dashed line) and $E = -0.65$ (red dot-dashed lines) with $\beta = 10, \Delta\beta = 1, \mathcal{A} = 0.1, \mathcal{B} = 100$, $\varepsilon = 100$, $m^+/m^- = 100$ and $Q = 5$. The inset of figure 3b) represents the auto-induced electric field created by the charge displacement near the flame. **(right)** The variation of flame speed, $\mu^{-1/2}$ with E_0 and Q.

right of Figure 3. Here we find that flame speed increases with E_0 until a critical value is reached and S_L begins to decrease. The observed critical point occurs when E_0 becomes large enough that it overcomes the auto-induced field, causing separation of the charged species that reduces recombination.

4. Conclusion

The presence of an externally electric field can, under appropriate conditions, enhance millimeter-scale combustion. This can increase the rate of combustion and prevent quenching. Although the full extent to which quenching is prevented has not been determined, the large observed increases in flame speed indicate that significant gains can be realized over conventional combustors. The model demonstrates partial agreement with experiments. Chiefly, it captures the increase in flame speed with positive parallel electric fields and that the enhancement increases with the available energy in the mixture. The predicted decrease in speed for negative parallel fields was not observed in the experiments. Small decreases were found for perpendicular fields, showing that ion transport can be detrimental to the flame, but this is beyond the scope of the one dimensional model. Also the maximum degree of enhancement observed far exceeded predictions. Future efforts will progressively expand this model to include more complex features while remaining elementary enough to provide a clear understanding of the process.

References

[1] Fernandez-Pello, A. C. *Proc. of the Comb. Inst.* **29(1)**, (2002) 883-899.
[2] Ronney, P. D. (2003) *Cmb. & Flm.*, **135(4)**, 421
[3] Tsuji, Y., Sprague, B., Walther, D. C., Pisano, A. P., Fernandez-Pello, C., & Fernandez-Pello, C. *43 rd AIAA Aerospace Sci. Mtng. & Exh.* (2005).
[4] Chattock, A. P., *The London, Edinburgh, and Dublin Phil. Mag. and J. of Sci.* **48.294** (1899): 401-420.
[5] Jaggers, H. C., and A. Von Engel. *Combustion and Flame* **16(3)** (1971): 275-285.
[6] Marcum, S. D., & Ganguly, B. N. (2005). *Cmb. & Flm.*, **143(1)**, 27-36.
[7] Dold, J. W., Thatcher, R. W., Omon-Arancibia, A., & Redman, J. *Proc. of the Comb. Inst.* **29(2)**, (2002) 1519-1526.
[8] Sánchez-Sanz, Mario, Daniel C. Murphy, and C. Fernandez-Pello. *Proc. of the Comb. Inst.* In press. (2014).

MEMS wireless temperature sensor for combustion studies

Minhyeok Lee[1], Yoshihiro Kawahara[2,3], Kenichi Morimoto[1], and Yuji Suzuki[1]

[1] Department of Mechanical Engineering, The University of Tokyo, Japan
[2] Department of Information and Communication Engineering, The University of Tokyo, Japan
[3] JST, Presto, Japan

E-mail: mlee@mesl.t.u-tokyo.ac.jp

Abstract. A MEMS wireless wall temperature sensor for combustion studies is proposed. Electrical resistance change in a LCR circuit is used to measure the temperature through inductive coupling the sensor coil and the read-out coil. Equivalent circuit model and 3-D electromagnetic simulation are employed to design sensor configuration. The resonant frequency is increased with increasing the resistance due to the temperature increase. The prototype sensor was successfully fabricated with MEMS technologies. The impedance phase angle shows a sharp dip at the resonant frequency, which is in good accordance with the equivalent circuit model. The measured temperature sensitivity is found to be as high as 6 kHz/K, when the distance between the read-out and the sensor coils is 0.71 mm.

1. Introduction

Precise wall temperature measurement is crucial in various combustion studies, such as wall-flame interaction [1], catalytic combustion [2], and the thermal/chemical wall quenching phenomenon [3]. Contact thermometry using thermocouple or RTD (Resistance Temperature Detector) has been widely used for measuring the wall temperature in combustion fields. However, physical contact with the target easily introduces disturbances to the temperature field, which makes the precise measurement difficult. On the other hand, non-contact thermometry based on infrared radiation or temperature-dependent fluorescent material can only be applied when the optical access is allowed [4].

We previously proposed a wireless temperature sensor using a LCR circuit, in which the resonant frequency is changed by the capacitance change [5]. We employ Al_2O_3 for dielectric material of the capacitor in the LCR circuit, by which the temperature coefficient of permittivity (TCP) is around 0.3%/K. However, the impedance phase did not exhibit a sharp dip because the resistance of the sensor coil was around 10 times higher than the designed value. Moreover, considering the temperature coefficient of resistance (TCR) of the electrodes, the resistance change will cancel out the capacitance change on the resonant frequency. For a wide temperature range used in combustion studies, the effect of resistance change is much larger than that of the capacitance change. In the present study, a MEMS wireless wall temperature sensor based on TCR of electrodes is proposed, and performance of a prototype sensor is evaluated for a non-contact wall temperature measurement.

[1] Correspondence to: Minhyeok Lee. E-mail: mlee@mesl.t.u-tokyo.ac.jp
Department of Mechanical Engineering, The University of Tokyo
7-3-1 Hongo, Bunkyo-ku, Tokyo, 113-8656, Japan

2. Design of the sensor

Figure 1 shows the concept of non-contact wall temperature measurement using the MEMS wireless temperature sensor. The sensor consists of a thin-film coil and a capacitor and is attached on the target wall surface. A read-out coil is placed on the other side of the wall. The wall temperature change results in the resonant frequency change of the LCR circuit, which is measured by inductive coupling between the sensor and the read-out coils [5-7].

The present sensor is essentially a LCR resonant circuit. The sensor coil is inductively coupled with the read-out coil with a coupling coefficient k as described later. The read-out coil is connected to a network analyzer (NA), and the impedance phase dip that appears at the resonant frequency of the LCR circuit is measured. Based on the resonant frequency, the resistance of the sensor and thus the wall temperature are estimated. The equivalent circuit model for the read-out coil and the sensor is shown in figure 2. R, L, C, M, and v are the resistance, the self-inductance, the capacitance, the mutual-inductance and the voltage, respectively. Subscript e and s denote the values for the read-out coil and the sensor, respectively. The circuit equations can be written as follows:

$$\begin{cases} v_e(t) = R_e i_e + L_e \dfrac{di_e}{dt} + M \dfrac{di_s}{dt} \\ 0 = R_s i_s + L_s \dfrac{di_s}{dt} + \dfrac{1}{C_s} \int i_s \, dt + M \dfrac{di_e}{dt} \end{cases} \tag{1}$$

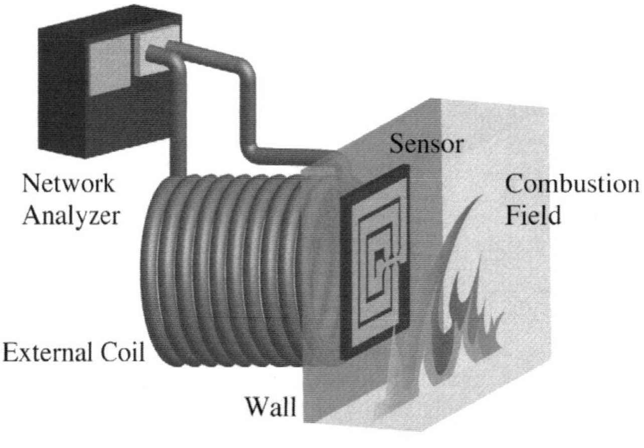

Figure 1. Principle of MEMS wireless wall temperature sensor. Wall surface temperature can be measured without contact from the backside of the wall.

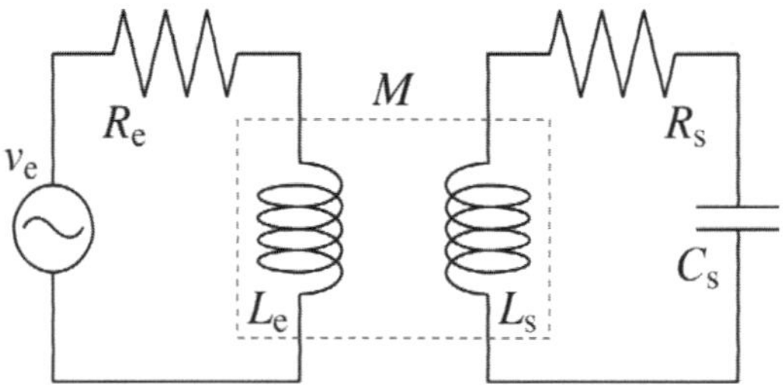

Figure 2. Equivalent circuit model for the read-out coil (left) and the sensor (right).

By solving these simultaneous equations, the ratio of the external voltage to the external circuit current, i.e., the impedance of the external circuit $Z_e(\omega)$ is calculated. R_e is much smaller than R_s, and can be neglected. The impedance phase angle can be found as follows:

$$\angle Z_e(\omega) \cong \arctan\left\{\frac{\left(1-k^2\right)L_s^2C_s^2\omega^4+\left\{R_s^2C_s^2+\left(k^2-2\right)L_sC_s\right\}\omega^2+1}{k^2R_sL_sC_s^2\omega^3}\right\} \quad (2)$$

We can define the resonant frequency by getting the extreme value of equation (2):

$$\omega_0 \cong \sqrt{\frac{\left\{R_s^2L_s^{-1}C_s+\left(k^2-2\right)\right\}+\sqrt{\left\{R_s^2L_s^{-1}C_s+\left(k^2-2\right)\right\}^2+12\left(1-k^2\right)}}{2\left(1-k^2\right)}}\cdot\frac{1}{\sqrt{L_sC_s}} \quad (3)$$

Note that k is the coupling coefficient representing intensity of the coupling between two coils [6], which strongly depends on the distance between the coils:

$$k=\frac{M}{\sqrt{L_eL_s}} \quad (4)$$

From equation (3), we can notice that the resonant frequency ω_0 is increased when the sensor resistance R_s is increased. In the present analysis, the coupling coefficient is assumed as $k = 0.3$, which is reasonable for near-distance inductive coupling. Figure 3a shows the analytical results based on the equivalent circuit model when resistance is in range of 10 ~ 20 Ω. The resonant frequency is increased with the resistance, while the dip is decreased. 3-D electromagnetic simulation using ANSYS HFSS (High Frequency Structure Simulator) is also made for comparison. The sensor is modelled as a closed LCR circuit as designed and both ends of the read-out coil are connected to a lumped port, which functions as a NA. Metal layers on the sensor are assumed as a 2-D plate to reduce computational loads. To change the sensor resistance value, a lumped resistor element is connected to the sensor circuit in series. The frequency is swept from 1 MHz to 100 MHz with a step size of 1 MHz. The results are plotted in figure 3b. Although the 3-D simulation results are somewhat different from the analytical solution, both results show similar resonant frequency shifts due to the resistance change.

3. MEMS fabrication of the sensor

The microfabrication process starts with sputtering of a 500 nm-thick Au layer on the glass substrate with a 50 nm-thick Cr or Ti film as the adhesion layer. The Au layer was then patterned into the bottom electrode using standard lithography. Then, a 100 nm-thick SiO_2 layer was deposited as a

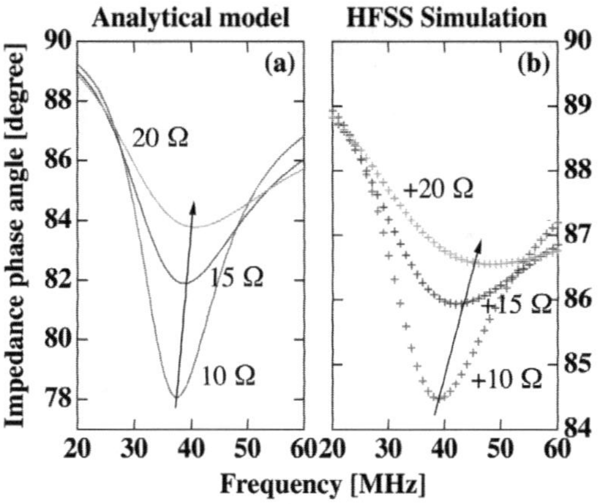

Figure 3. Analytical (a) and HFSS simulation (b) results for L_s=0.1 μH and C_s=0.2 nF when R_s changes in range of 10 ~ 20 c.

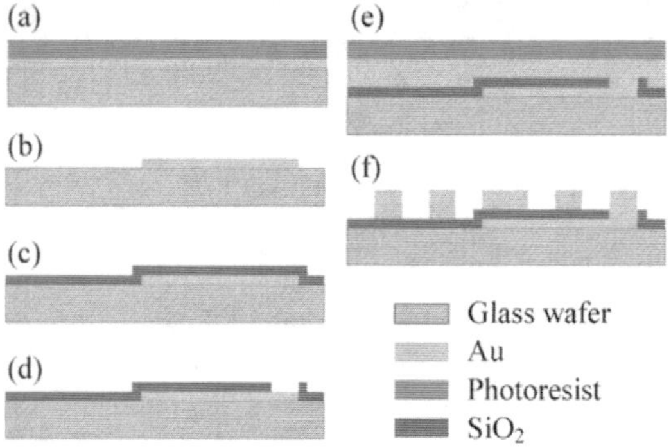

Figure 4. MEMS Fabrication process; (a) Sputtering and resist-coating, (b) Bottom metal layer patterning, (c) SiO_2 layer deposition, (d) SiO_2 patterning, (e) Sputtering and resist-coating, (f) Upper metal layer patterning.

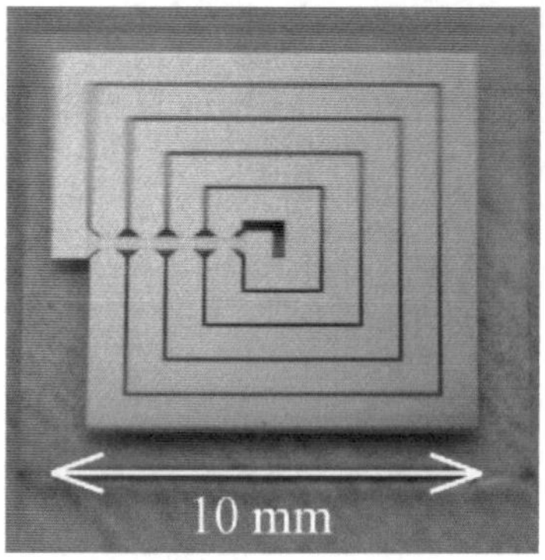

Figure 5. Prototype of the wireless temperature sensor. Dimensions are 10 mm × 10 mm.

dielectric layer by plasma CVD with TEOS (tetraethyl orthosilicate) as a precursor. A contact hole for connecting two metal layers was etched using buffered HF. Finally, another 1.5 μm-thick Au layer was sputtered and patterned to form a counter electrode and a 5-turn spiral coil. Figure 5 shows a photograph of the completed sensor with a size of 10 mm squared.

4. Results and discussion

The read-out coil was made by winding Cu wire (diameter: 1.12 mm) in a spiral shape. The prototype sensor and the read-out coil were attached to each side of a resin film. The distance between two coils is 0.68 mm. The read-out coil is connected to a NA (ZVL6, Rohde & Schwarz), and the resonance profile was measured. Figure 7 shows the impedance phase angle changes measured in the frequency range up to 100 MHz. A sharp impedance phase dip over 35 degrees is clearly seen near the resonant frequency. The measured results are well fitted with the equivalent circuit model (equation (2)), proving the effectiveness of the present circuit model analysis. The fitting value of the coupling coefficient k is approximately 0.6 when the distance between coils is 0.68 mm, while it decreases below 0.1 when the distance is beyond 5.6 mm. With decreasing coupling coefficient, the sensitivity is also decreased because the impedance phase dip becomes dull. Improvement of the sensitivity for longer distance can be achieved by changing the coil shape or by improving the measurement

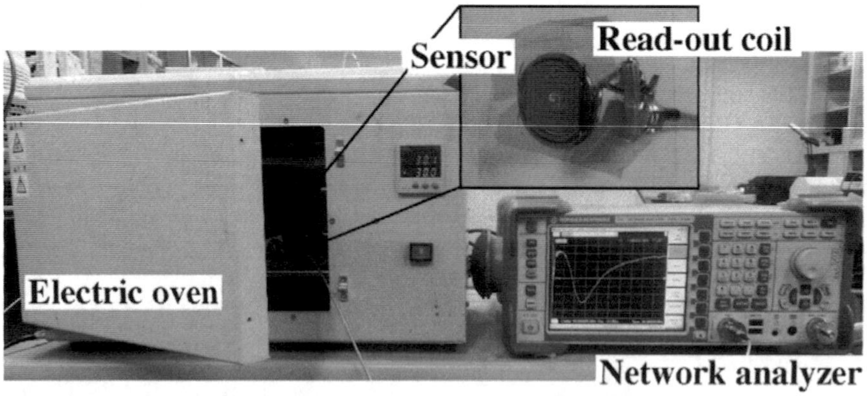

Figure 6. Experimental set-up to evaluate sensor performance. The read-out coil is coupled with the sensor and its impedance phase angle is measured by NA.

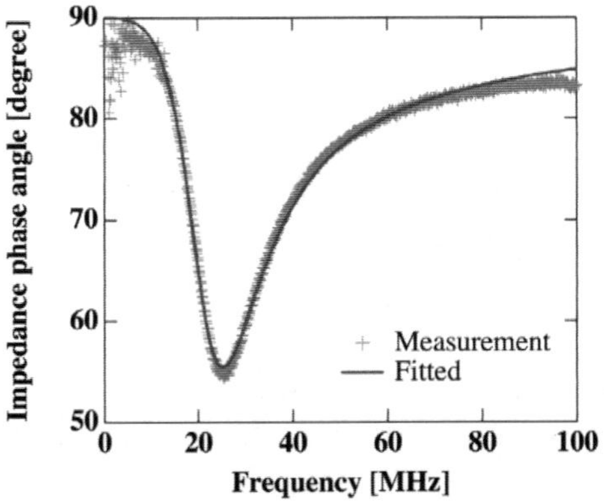

Figure 7. Impedance phase angle versus frequency. Solid line is a curve-fitted circuit model with R_s=22.4 Ω, L_s=0.25 μH, and C_s =0.23 nF.

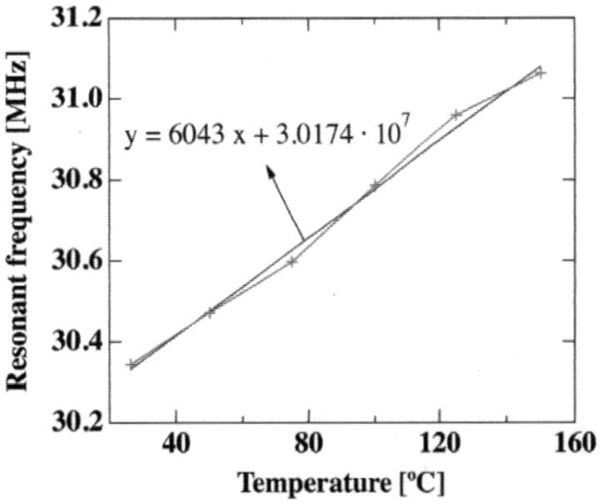

Figure 8. Resonant frequency versus temperature. A measured sensitivity against temperature changes is about 6 kHz/K.

algorithm.

Figure 6 shows the experimental set-up to evaluate the performance as a wireless temperature sensor. The sensor and the read-out coil were fixed on a 0.15 mm-thick slide glass, which corresponds to the distance between coils of 0.71 mm. The device was put into an electric oven for the measurement at different temperatures from 26.5 to 150 °C. The resonant frequency at each temperature is plotted in figure 8. The frequency sensitivity is found to be around 6 kHz/K. When the measurement error in determining the resonant frequency is assumed to be ~ 50 kHz, the temperature accuracy is ±10 K, which is high enough for most combustion studies.

5. Conclusion

A novel MEMS wireless temperature sensor for combustion studies is proposed. Resonant frequency change due to the resistance change is employed. A prototype sensor is successfully microfabricated using MEMS technologies. A sharp impedance phase dip in a good accordance with the equivalent circuit model has been observed. Frequency sensitivity around 6 kHz/K is achieved, which corresponds to the temperature accuracy of ±10 K.

Acknowledgement

This work is supported by JSPS Grant-in-Aid for Exploratory Research (No. 26630060). Photo-mask is made using the University of Tokyo VLSI Design and Education Center (VDEC)'s 8-inch EB writer F5112+VD01 donated by ADVANTEST Corporation. CVD of SiO_2 was supported by NIMS Nanofabrication Platform in Nanotechnology Platform Project sponsored by the Ministry of Education, Culture, Sports, Science and Technology (MEXT), Japan.

References

[1] Poinsot T, Haworth D and Bruneaus G 1993 *Combust. Flame* **95** 118
[2] Prasad R, Kennedy L A and Ruckenstein, E 1984 *Catal. Rev. Sci. Eng.* **26** 1
[3] Saiki Y and Suzuki Y 2013 *Proc. Combust. Inst.* **34** 3395
[4] Childs P R N, Greenwood J R and Long C A 2000 *Rev. Sci. Instrum.* **71** 2959
[5] Mitarai R, Morimoto K and Suzuki Y 2013 *Proc. 50th National Heat Trans. Symp., Sendai,* 568
[6] Akar O, Akin T and Najafi K 2001 *Sensors Actuators* A **95** 29
[7] Lee M and Suzuki Y 2014 *35th Int. Symp. Combust., San Francisco, Work-in-Progress-Poster* W4P036

Investigation of the storage and release of oxygen in a Cu-Pt element of a high-temperature microcombustor

Z Khaji[1], P Sturesson[1,2,3], K Hjort[1], L Klintberg[1], and G Thornell[1,2]
[1]Uppsala University, Division of Microsystems Technology, Ångström Laboratory, Box 534, SE-7521, Uppsala, Sweden
[2]Uppsala University, Ångström Space Technology Centre, Box 534, SE-75121, Uppsala, Sweden
[3]Swedish National Defence College, Division of Military Technology, Box 27805, SE-11593, Stockholm, Sweden

Email:zahra.khaji@angstrom.uu.se

Abstract. A miniature combustor for converting organic samples into CO_2 with application in carbon isotopic measurements has been manufactured and evaluated. The combustor was made of High-Temperature Co-fired Ceramic (HTCC) alumina green tapes. The device has a built-in screen printed heater and a temperature sensor made of platinum, co-sintered with the ceramic. A copper oxide oxygen supply was added to the combustor after sintering by *in-situ* electroplating of copper on the heater pattern followed by thermal oxidation. Scanning Electron Microscopy (SEM), Energy Dispersive Spectroscopy (EDS) and Thermal Gravimetric Analysis (TGA) were used to study electroplating, oxidation and the oxide reduction processes. The temperature sensor was calibrated by use of a thermocouple. It demonstrates a temperature coefficient resistance of 4.66×10^{-3} /°C between 32 and 660 °C. The heat characterization was done up to 1000 °C by using IR thermography, and the results were compared with the data from the temperature sensor. Combustion of starch confirmed the feasibility of using copper oxide as the source of oxygen of combustion.

1. Introduction

Isotope ratio measurements of carbon have multiple applications in a wide range of disciplines, *e.g.*, earth sciences, forensic science, biomedicine, and archeology. For isotopic analysis of the ratio of ^{13}C to ^{12}C, the sample is usually converted to CO_2 [1]. Therefore, combustion plays an important role in sample preparation.

In, *e.g.*, biomedicine, the demand for isotopic analysis of ultra-small samples is continuously increasing [2]. One way to reduce the analyte amount necessary is to miniaturize the sample analysis and preparation system, and perhaps even integrate the two.

Using lithography techniques and wet etching, H. J. Tobias *et al.* [1] fabricated a high temperature microreactor ($78\times30\times2$ mm^3) for on-line conversion of organic gas compounds in high purity fused silica. Cu and Ni wires were inserted in the reactor after manufacturing, and were thermally oxidized *in situ* to be used as the source of oxygen during the combustion. Heating was done externally with a high-temperature furnace. The combustion took place at 950 °C, and the reactor capillaries were gas tight up to 1000 °C.

Using high-precision ceramic tape casting of nanoporous alumina, T. Okamasa *et al.* [3] developed a catalytic microcombustor ($25\times25\times5.3$ mm^3) with the potential to sustain 800 °C as governed by the brazing temperature of the ceramic. However, it was evaluated only up to 390 °C. The oxygen required for combustion was fed in to the combustor through an air inlet. The combustor had an embedded tungsten igniter and was submerged in a bath to keep the temperature constant.

D. A. Merritt *et al.* [4] made a tubular microvolume reactor (200×5 mm^2) with nonporous alumina packed with metal wires oxidized after manufacturing (CuO/NiO) to be used for on-line combustion of gases. CuO and NiO could be used as the only source of oxygen at temperatures above 850 °C and 1150 °C, respectively.

This paper presents the fabrication and evaluation of a miniature combustor based on High-Temperature Co-fired Ceramic (HTCC) technology. The combustor has a built-in heater and temperature sensor. After sintering, the heater is Cu electroplated *in situ* and thermally oxidized in order to provide for an integrated oxygen storage.

2. Experimentals

2.1. *Design, material and fabrication*

The combustor, designed to be operated at up to 850 °C, consists of four ceramic layers: inlet and outlet layers (I), conductor layer, which includes the heater pattern and the temperature sensor (III & IV), and combustor wall layer (II), figure 1.

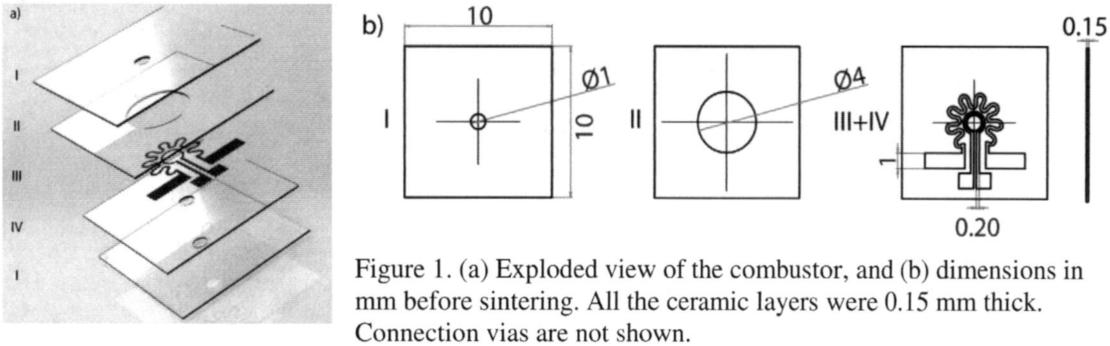

Figure 1. (a) Exploded view of the combustor, and (b) dimensions in mm before sintering. All the ceramic layers were 0.15 mm thick. Connection vias are not shown.

The connection pads for the temperature sensor and the heater are accessible through vias made between the outlet and the combustion layers (I & II) during manufacturing.

The flower shape of the heater pattern and its dimensions were chosen to be able to heat up the chamber uniformly. The temperature sensor is circular and centered inside the heater pattern.

The device was fabricated by milling and lamination of alumina green tapes (ESL 44007-150 μm, Electro Science Laboratories, USA). Conductive patterns were screen printed using platinum paste (ESL 5571, ESL Electro Science Laboratories, USA). A detailed description can be found in [5].

Copper, which is not compatible with the HTCC process, was deposited after sintering by electroplating (J-PLATE Cu 90H, J-KEM International, Germany) on the platinum heater inside the cavity, according to the manufacturer's data sheet. Electroplating durations of 7 and 46 minutes, providing two thicknesses of Cu film, were chosen.

The copper was thermally oxidized in an oven (Logotherm L5/S17, Nabertherm, Germany) at 500 °C for 10 hours, since the ratio of CuO to Cu_2O is greatest at around 500 °C, and CuO should be completely reduced at about 850 °C [6].

2.2. *Characterization*

The combustor sample was investigated with X-ray microscopy (XT V 130, Nikon, Japan).

In order to investigate electroplating inside the cavity, additional samples were prepared and analyzed using Scanning Electron Microscopy (SEM) and Energy Dispersive Spectroscopy (EDS). The following material compositions were analyzed: (a) platinum printed on alumina, (b) Cu electroplated on top of the platinum pattern, (c) oxidized Cu electroplated pattern, and (d) Cu achieved after reduction of the oxide. The thickness of the interfaces and the abundance of aluminum, oxygen, platinum and copper were measured from sample cross sections prepared by cutting, epoxy mounting, grinding and polishing.

The reduction of copper oxide was done in a furnace (T5HT, A.D.A.M.E.L, France) at 900 °C for an hour in presence of argon to avoid oxidation during cooling.

The study of copper oxidation and oxide reduction was complemented with Thermogravimetric Analysis (TGA) (Q500, TA Instruments, USA).

The integrated temperature sensor was calibrated by heating the combustor slowly to 660 °C in an oven (FB1410M-33, Thermo Scientific, USA), while the ambient temperature was measured by a thermocouple (K type) and the resistance of the temperature sensor was measured using 4-wire sensing technique and digital multimeter (34401A, HP, USA).

The heat distribution over the combustor floor (layers I, III, IV) during resistive heating was studied using IR imaging (Thermovision A40, FLIR Systems, Sweden), and its correlation with the temperature senor signal was investigated.

For demonstration, corn starch was distributed over a CuO layer on a bulk alumina sample placed on a heat stage inside a soda-lime glass tube. The tube was evacuated to about 2.5 torr and surrounded by nitrogen, and the heat stage was resistively heated up to above 850 °C.

3. Results and discussion

The dimensions of the combustor after sintering were measured to be 8.3×8.3×0.4 mm^3.

The lateral and vertical shrinkage of the chip after sintering was 16% and 21%, respectively. The integrity of the realized combustor, and especially its electrical connectivity were revealed by X-ray imaging, figure 2.

3.1. SEM and EDS results

The SEM images of the Cu electroplated cross section samples before oxidation as well as after reduction are shown in figure 3. The EDS analysis confirms that the copper oxide had been completely converted to copper after reduction, figure 4(left).

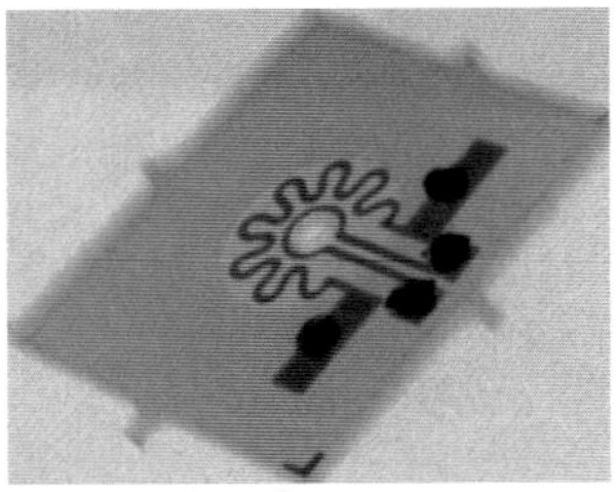

Figure 2. X-ray image of the combustor, showing the connection vias between different layers.

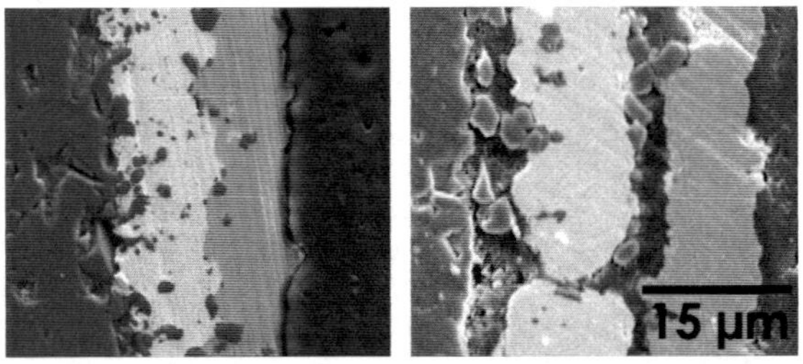

Figure 3. SEM images of the cross section of a thin layer of Cu electroplated on Pt (left) before oxidation, and (right) after oxide reduction. From left to right: alumina substrate, Pt film, Cu film and epoxy material used for cross-section preparation.

For some samples the oxidation step had caused delamination of the copper oxide and the platinum film. This was found to remain after the reduction step, figure 3(right). No delamination was observed directly after the electroplating, however.

For some of the samples, the EDS analysis shows two oxide layers with different proportions of oxygen and copper, figure 4 (center), for other samples just one, figure 4 (right).

Furthermore, migration of copper into the platinum film is noticed after the oxidation step, figure 4.

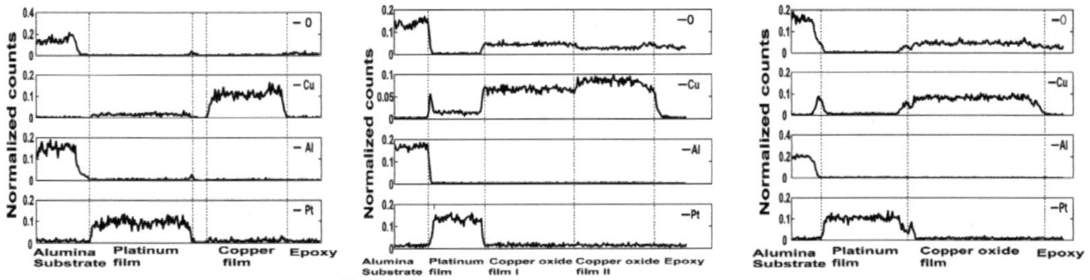

Figure 4. Composition of cross-sections. Sample with thin layer of Cu after reduction (left). Sample with thick layer of Cu after oxidation (center), and sample with thin layer of Cu after oxidation (right).

3.2. TGA results

TGA analysis results are presented in figure 5. The weight of the sample is almost constant until 500 °C, where air is introduced and oxidation is believed to start. The weight of the sample increases by more than 0.2%. At around 750 °C, the weight begins to decrease. This decrease increases at around 850 °C, which corresponds with the decomposition temperature of the oxide, and continues until the sample weight reaches its initial value.

Assuming that the weight increase of the sample at 500 °C is due to oxidation, it was calculated that 8.5 μmoles of oxygen had been absorbed, which is much less than the amount of copper(around 38.5 μmoles) in the sample. This shows that all copper had not been oxidized. However, almost all the oxide gained during oxidation had been decomposed.

3.3. Temperature sensor calibration results

Over the calibration range of 32 to 660 °C, the temperature sensor is linear (R^2 of 0.999 for the line fitted) and exhibits a temperature coefficient resistance of 4.66×10^{-3} /°C.

3.4. Heat characterization results

Figure 6 illustrates the resistive heating of a combustor floor sample with probes to 1000 °C. The heat distribution is asymmetric in the beginning of heating, (right), but becomes more symmetric at 1000 °C, (left). The power consumption at 1000 °C was around 10 W and the device was operated at that temperature for 10 minutes.

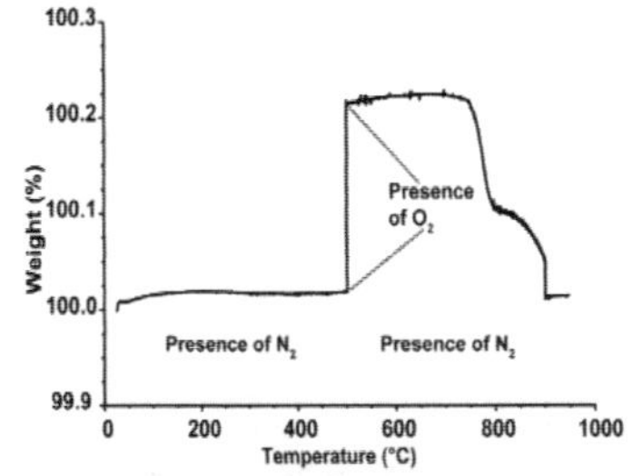

Figure 5. Results of TGA analysis of the Cu electroplated sample. (The step at 500 °C corresponds to a dwell time of six hours.)

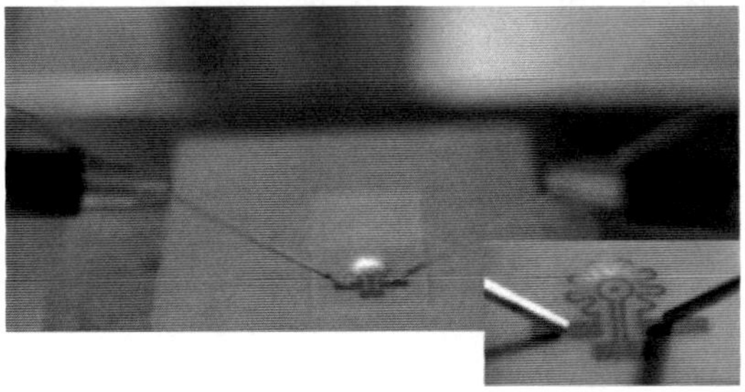

Figure 6. Resistive heating of the heater pattern using probes.

Figure 7. IR image of the heater pattern driven with 10 W.

At 1000 °C, IR imaging revealed a heat gradient resulting from cooling by the probes, figure 7. From the IR images, the average temperature of the area between the sensor and heater was extracted during three phases of heating and starch combustion in normal atmosphere. Referring to the sensor, the thermograph gave 98, 244, and 384 at 105, 243, and 316 °C, respectively, *i.e.*, a deviation increasing with temperature.

3.5. Corn starch combustion results

At temperatures below the reduction temperature of CuO (around 725 °C according to EkviCalc calculations) no combustion was observed, figure 8 (left). While heating and before the reduction, an intermediate phase of gelatinization of starch was observed, figure 8 (center). As soon as the reduction started, the combustion took place. Almost no starch was left after the combustion, figure 8 (right). Red spots appeared after the reduction, figure 8 (right), indicated that the reduction of copper had occurred. If significant leakage of oxygen had occurred, further oxidation would have happened during cooling, and the red spots would have disappeared.

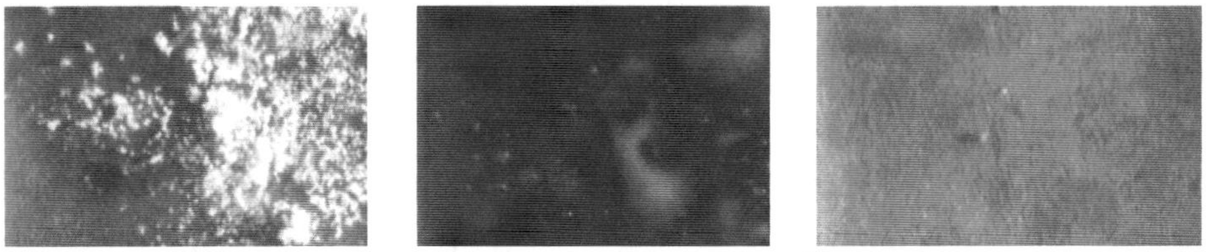

Figure 9. Left: Corn starch on CuO layer before heating, Center: Intermediate gelatinization of starch, and Right: Partially reduced CuO after combustion and cooling.

4. Conclusion and future work

A miniature high-temperature combustor has been operated to 1000 °C, although the integrity of the thick platinum and copper films suffered from the chemical and thermal cycling. EDS and TGA analysis, as well as the partial combustion of starch, verified that the *in situ* electroplating of copper and the use of copper oxide as the source of oxygen were successful.

In the future, the oxidation and decomposition recipes should be chosen more carefully in order to achieve robust and homogenous surfaces with better adhesion between the different materials. Alternatively, separate elements can be used for heating and supplying oxygen. Furthermore, the combustion should be fully studied with a residual gas analyzer in order to thoroughly measure the amount of the gases produced during the reduction and combustion, and the temperature sensor should be improved.

Acknowledgments

The Knut and Alice Wallenberg foundation is acknowledged for funding the laboratory facilities. The Centre for Natural Disasters Science (CNDS) and the Swedish National Space Board are acknowledged for funding. Jan Bohlin and Anders Lund at the Department of Chemistry at Uppsala University, Candor Sweden AB, and the Swedish distributor of J-KEM International are acknowledged for contributing with hardware and knowledge. Martin Berglund at the Department of Engineering Sciences at Uppsala University is acknowledged for his contribution with data.

References

[1] Tobias H J and Brenna J T 2010 *J. Microfluid. Nanofluid.* **9** 461-70.
[2] Salehpour M, Håkansson K and Possnert G 2013, *J. Nucl. Instrum. Meth.* B **294** 97-103
[3] Okamasa T, Lee G, Suzuki Y, Kasagi N and Matsuda S 2006 *J. Micromech. Microeng.* **16** S198-205
[4] Merritt D A, Freeman K H, Ricci M P, Studley S A and Hayes J M 1995 *J. Anal. Chem.* **67** 2461-73
[5] Sturesson P, Khaji Z A, Knaust S, Sundqvist J, Klintberg L and Thornell G 2013 *J. Phys.: Conf. Ser.* **476** 012055
[6] Zhu Y, Mimura K, Lim J, Isshiki M and Jiang Q 2006 *J. Metal. Mater. Trans.* A **37** 1231-7

Interaction of Two Micro-slot Flames: Heat Release Rate and Flame Shape

K Kuwana[1,4], S Kato[2], A Kosugi[1], T Hirasawa[2] and Y Nakamura[3]

[1] Department of Chemistry and Chemical Engineering, Yamagata University, 4-3-16 Jonan, Yonezawa, Yamagata 992-8510, Japan
[2] Department of Mechanical Engineering, Chubu University, 1200 Matsumoto-cho, Kasugai, Aichi 487-8501, Japan
[3] Department of Mechanical Engineering, Toyohashi University of Technology, 1-1 Hibarigaoka, Tempaku, Toyohashi, Aichi 441-8580, Japan

E-mail: kuwana@yz.yamagata-u.ac.jp

Abstract. This paper studies the interaction between two identical micro-slot diffusion flames. Here, we define a micro-slot flame as a slot flame of which the slot width is less than about 1 mm. Because of its smallness, a micro-slot flame has a high heating density and can be used as a small heat source. However, the heat release rate of a single micro-slot flame is limited, and therefore, multiple micro-slot flames may be used to increase total heat release rate. As a first step, this paper considers a situation in which two micro-slot flames are used with certain burner spacing. When two diffusion flames are placed closely, flame shape changes from that of an isolated flame. Studying such flame shape change and resultant change in total heat release rate is the topic of this paper. Experiment is conducted and total heat release rate is measured by integrating CH* chemiluminescence recorded using a CCD camera and an optical filter of the wavelength of 430 nm. Two different burner materials, copper and glass, are tested to study the effect of heat loss to burners. An analytical model is applied to predict flame shape. In addition to the classical Burke-Schumann assumptions, two slot flames are modeled as line sources with zero width, enabling a simple analytical solution for the critical burner spacing at which two flames touch each other. The critical burner spacing is a key parameter that characterizes the interaction between two micro-slot flames. Computational fluid dynamics (CFD) simulations are then conducted to test the validity of the present theory. CFD results are favorably compared with the theoretical prediction.

1. Introduction

A jet diffusion microflame can be used as a small heat source [1-7]. In some applications, multiple microflames may be simultaneously used to enhance heating performance [8-10]. When multiple microflames are simultaneously used, the flames should not interfere with each other [10-12].

This paper studies the interaction of two identical micro-slot flames. A micro-slot flame is a flame on a slot burner whose width is less than about 1 mm, and an array of parallel micro-slot flames is capable of more uniform heating than that of circular-port microflames. Further, heat transfer behavior through the burner wall to the fresh fuel for a slot burner is different from a circular-port burner because of geometrical effects [5, 13-15]. Another aspect of interacting two slot flames is that

[4] To whom any correspondence should be addressed.

they can be easily handled by a computational fluid dynamics (CFD) simulation because of their two dimensional nature, whereas the interaction between two circular-port flames is a three-dimensional phenomenon.

This paper focuses on the effects of burner spacing on flame shape and heat release rate. Experimental observations are first presented, followed by a simple model that predicts the flame shapes; the model is validated by comparing the model prediction with CFD results.

2. Experimental method

Figure 1 shows the experimental setup. Methane-air diffusion flames are established on two identical 1-mm-wide slots, each having 1-mm-thick burner walls. The distance between burner rims, is varied from 0 to 8 mm. The average exit velocity of methane is fixed at 40 mm/s.

Images of CH* chemiluminescence (hereinafter called CH* emission) are taken using a CCD camera with an optical filter of the wavelength of 430 nm. The intensity of CH emission is considered a good indicator of heat release rate for both premixed and diffusion flames.

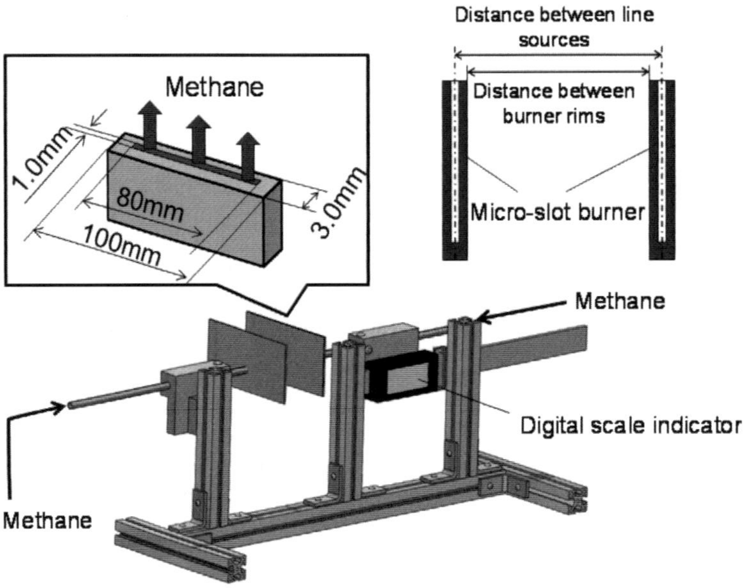

Figure 1. Experimental setup.

3. Experimental observations

Figure 2 shows integrated CH* emission (equivalently, total heat release rate). When the distance between burner rims is greater than about 5 mm, the total heat release rate is nearly constant, suggesting that each flame behaves similarly to an isolated flame as evidenced by the flame shapes shown in figure 3(a).

The total heat release rate tends to increase when the distance between burner rims decreases. This is mainly because with a decrease in the distance between burner rims, flame height increases (flame is established farther away from burners), reducing the heat loss to burners. Figure 3 shows that flame height increases with a decrease in the distance between burner rims. When the distance between burner rims is 0 mm and the two slot burners are in contact, the total heat release rate is increased by about 30% from the reference value. Under this condition, however, two flames merge into a single flame as shown in figure 3(d), losing characteristics as small-scale flames.

The local minimum of total heat release rate observed in figure 2 can be explained by heat loss to burners and local extinction. As shown in figure 3(c), the flames under a condition near the local minimum are merged, and an inner flame and an outer flame are separately formed. The inner flame is much closer to the burners than the outer flame, being significantly influenced by heat loss. In addition, the tip of the outer flame is locally extinguished because of the inner flame's blocking effect of oxygen, further lowering total heat release rate. The influence of heat loss is also confirmed by the observation that the total heat release rate of glass burners is greater than that of copper burners; the latter material has a higher thermal conductivity and takes more heat from flame. From the experimental results thus far, predicting flame shape appears to be important when designing a device using multiple micro-slot flames.

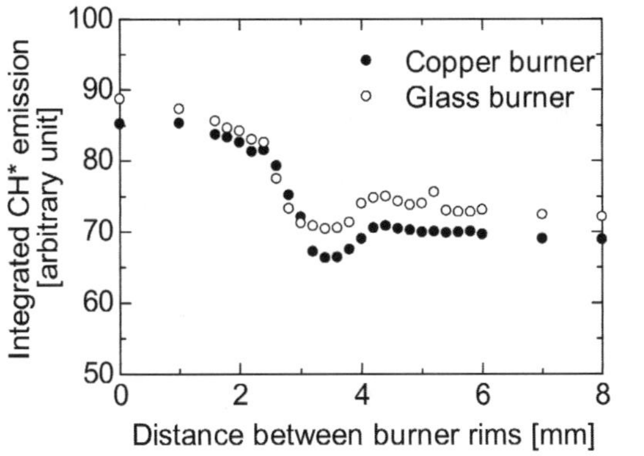

Figure 2. Integrated CH* emission and distance between burner rims.

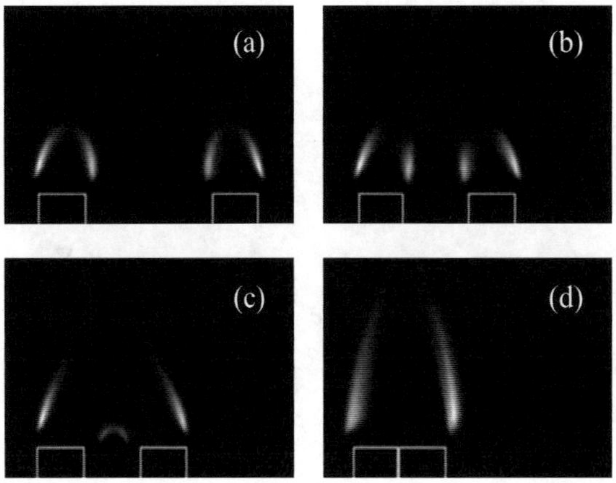

Figure 3. Flame shapes for different distances between burner rims: (a) 8 mm, (b) 4.4 mm, (c) 3.4 mm, and (d) 0 mm.

4. Line-source model

One of the simplest methods to predict the shape of a jet diffusion flame is to assume that the flow is uniform everywhere as in the pioneering study by Burke and Schumann [16]. They also assumed that thermophysical properties are constant, Lewis numbers for all species are unity, and the molecular transport terms in the flow direction is negligible. Analytical solutions for flame shapes were then derived; which have been confirmed to be qualitatively correct under various conditions. In this study, following Ref. [20], two micro-slot flames are modeled as diffusion flames in the uniform flow of velocity U.

When a typical combustible gas burns in air, $Z_{st} \ll 1$, where Z_{st} is the stoichiometric mixture fraction. Then, the exit of a slot burner may be regarded as a line with zero width, and the following solution for the mixture fraction equation is obtained:

$$Z = \frac{Pe^{1/2}}{2\pi^{1/2}\bar{z}^{1/2}}\left[\mathrm{e}^{-Pe(2\bar{x}-\bar{a})^2/16\bar{z}} + \mathrm{e}^{-Pe(2\bar{x}+\bar{a})^2/16\bar{z}}\right] \tag{1}$$

where $Pe = UL/D$ is the Péclet number (L: slot width, D: diffusivity), and a is the distance between line sources. Overbar indicates that the variable is dimensionless in units of L.

Figure 4 shows flame shapes predicted by equation (1) for varied distance between line sources. It is confirmed that the present simple model can qualitatively reproduce experimentally observed flame shape shown in figure 3.. From equation (1), the critical distance between line sources at which two flames touch each other is given by

$$a = \left(\frac{8}{\mathrm{e}\pi}\right)^{1/2} \frac{L}{Z_{st}} \tag{2}$$

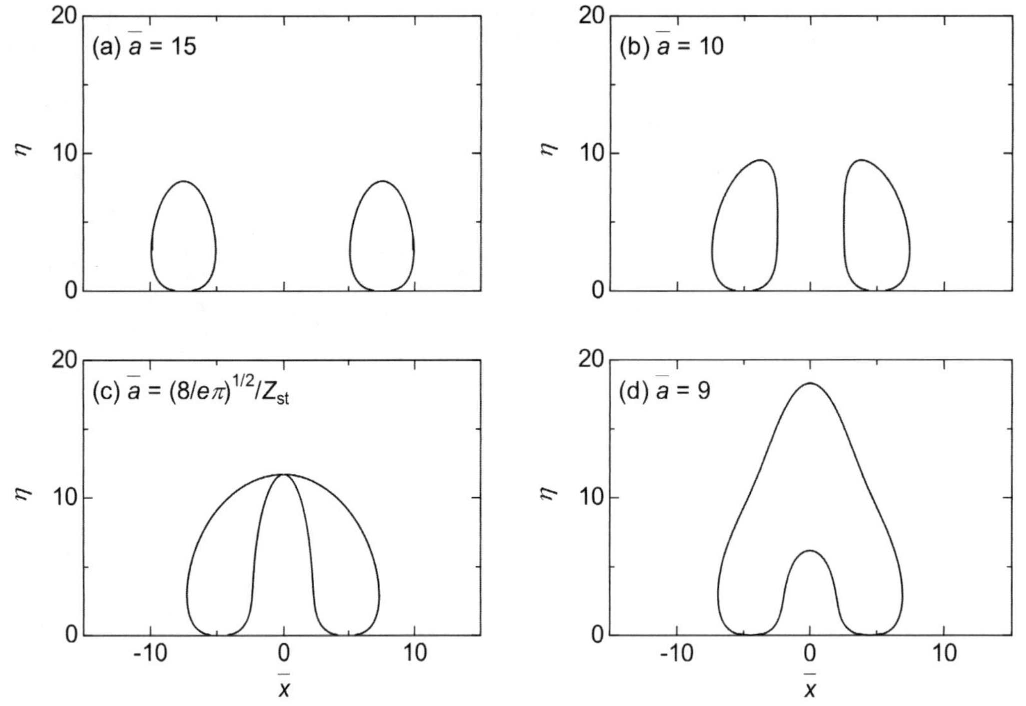

Figure 4. Theoretically predicted flame shapes, where $\eta = \bar{z}/Pe$.

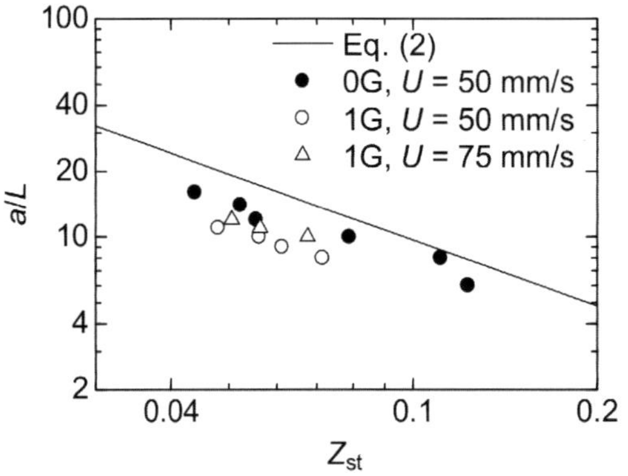

Figure 5. Validation of theoretical model comparing with numerical solutions.

5. CFD validation

This section tests the validity of model prediction, equation (2). Considering the difficulty in experimentally varying Z_{st}, CFD simulations are conducted. In CFD, Z_{st} can be easily varied by, for example, varying the oxygen mass fraction in the air. At the bottom of computational domain, inlet boundary conditions of methane and air with the same inlet velocity are adopted, similarly to the analytical model.

Figure 5 compares equation (2) with CFD predictions under various conditions. Overall, equation (2) predicts the dependence of critical distance between line sources on Z_{st} reasonably well. Since a uniform flow is assumed in the model, equation (2) predicts the same critical burner pitch for different gravity levels or fuel exit velocities, whereas the CFD result depends on these parameters. However, the dependence of critical distance between line sources on these parameters is rather weak, and equation (2) can be used for a first estimate.

Equation (2) tends to overestimate the critical burner pitch. A major reason for the error is because the model does not consider buoyant effects. Flow acceleration due to buoyancy tends to reduce the flame size, decreasing the critical burner pitch. This explains why the model works relatively well for 0G CFD results. Therefore, a remedy for the model is to consider acceleration due to buoyancy. Then, the model accuracy for predicting the influence of gravity level and possibly of fuel exit velocity will be improved; such a model will be studied in a later work.

6. Conclusions

The interaction between two identical micro-slot flames is studied. Experimental observations show that the total heat release rate basically increases with a decrease in burner spacing, but it has a local minimum. These trends are attributed to the shapes of the flames, indicating the importance of predicting flame shape. In particular, the critical distance between burner rims at which two flames touch each other is identified as a representative parameter to characterize the flame interaction. Experiments of different burner materials (copper and glass) further confirm heat loss effects on total heat release rate.

A simple analytical model is then developed to predict the flame shape and the critical burner spacing. The model assumes a uniform flow, similarly to the original Burke-Schumann model. In addition, two slot burners are modeled as two line sources to enable a simple analytical solution, from which the critical distance between line sources is obtained as a function of the stoichiometric mixture fraction. Considering the difficulty in experimentally changing the stoichiometric mixture fraction, CFD simulations are conducted to validate the analysis. The CFD results agree reasonably well with the analysis.

References

[1] Ban H, Venkatesh S and Saito K 1994 *J. Heat Transfer* **116** 954
[2] Matta L M, Neumeier Y, Lemon B and Zinn B T 2002 *Proc. Combust. Inst.* **29** 933
[3] Cheng T S, Chao Y C, Wu C Y, Li Y H, Nakamura Y, Lee K Y, Yuan T and Leu T S 2005 *Proc. Combust. Inst.* **30** 2489
[4] Cheng T S, Chen C P, Chen C S, Li Y H, Wu C Y and Chao Y C 2006 *Combust. Theory Modell.* **10** 861
[5] Nakamura Y, Yamashita H and Saito K 2006 *Combust. Theory Modell.* **10** 927
[6] Chen C P, Chao Y C, Cheng T S, Chen G B and Wu C Y 2007 *Proc. Combust. Inst.* **31** 3301
[7] Kuwana K, Tagami N, Mizuno S and Ida T 2009 *Proc. Combust. Inst.* **32** 3115
[8] Hirasawa T and Nakamura Y 2009 Japanese Pat. No. 5358198
[9] Nakamura Y and Hirasawa T 2010 *Denshizairyo* supplementary volume **7** 155 (in Japanese)
[10] Kuwana K, Hirasawa T and Nakamura Y 2012 *J. Combust. Soc. Jpn.* **54** 279 (in Japanese)
[11] Hirasawa T, Gotanda K, Masuda H and Nakamura Y 2012 *Combust. Sci. Tech.* **184** 1651
[12] Hirasawa T, Gotanda K, Kamata Y and Nakamura Y 2012 *Visualization of Mechanical Processes* **2** DOI: 10.1615/VisMechProc.2013003813
[13] Hirasawa T, Sumi M and Nakamura Y 2013 *J. Japan Soc. Exper. Mech. (2013 Special Issue)* **13** s75
[14] Fujiwara K and Nakamura Y 2013 *Combust. Flame* **160** 1373
[15] Hossain A and Nakamura Y 2015 *Proc. Combust. Inst.* **35** (in press)
[16] Burke S P and Schumann T E W 1928 *Ind. Eng. Chem.* **20** 998

Effect of surface reactions on ignition delay of methanol/air mixture

J P Huo, H L Yang, L Q Jiang, X H Wang and D Q Zhao

Guangzhou Institute of Energy Conversion, Chinese Academy of Sciences, China

E-mail: huojp@ms.giec.ac.cn

Abstract. A surface reaction kinetic model for the combustion of methanol/air mixture was developed in order to investigate the ignition inhibitory mechanism of wall on the premixed gas in a micro closed volume. In this model, except for H, O, OH and CH_3 radicals, the absorption of hydrogen peroxide and hydroperoxyl on the surface were also considered. By applying CHEMKIN-Pro software, the model was integrated into the calculation of homogeneous combustion process of gas mixture. Surface reactions were found resulting in the increase of ignition delay time. The sensitivity analysis showed that the loss of hydrogen peroxide on the wall was the main reason, due to the direct suppression effect on the generation and accumulation of OH in the radical pool. However, the loss of hydroperoxyl would take the place of hydrogen peroxide as the main inhibitory factor when the sticking coefficient became as large as the order of 10^{-3}. In addition, the ignition delay time increased with sticking coefficient or surface-area-to-volume ratio. Enhancing the initial temperature of premixed gas was able to reduce the inhibitory effect of surface reactions.

1. Introduction

Micro-scale internal combustion engine is considered as an ideal power source for unmanned vehicles and portable electronic instruments, owing to large energy density and long lifetimes. The fabrication and assembly of micro-engines have got fast development due to the major improvement of MEMS technology. The physical length scale of micro-engine is usually below 1 mm. Therefore, the surface-area-to-volume ratio become so large that great heat loss and radical loss on the internal wall show non-ignorable or even significant inhibitory effect on the ignition and flame propagation of gas mixture. Fundamental studies on micro-combustion inside a small closed vessel were carried out by researchers worldwide [1].

Norrish and Reagh [2] measured the slow oxidation reaction rate of several hydrocarbons in some cylindrical silica reaction vessel varying different diameters. It was found that the reaction rate dropped abruptly as the diameter of the vessel was sufficient reduced. The major reason was that more percentage of radicals was destroyed by internal surface with decreasing scale of the vessel. A number of similar oxidation experiments at low temperature were described in reference [3], and their results indicated that the surface had inhibitory influence on gas phase oxidation, especially strong when the internal diameter was less than 5 mm. Bowman and Wilk [4] investigated methanol auto-ignition including surface effect with three spherical glass reaction vessels of different sizes. The experimental results showed that heterogeneous surface effect would retard the ignition of the system and the ignition delay time increased with surface-area-to-volume ratio. Besides silica and quartz, other different materials like stainless steel, alumina and zirconia ceramics, were adopted as wall material to study the surface effect on combustion process in a narrow channel [5, 6]. The results implied that stainless steel has the largest radical removal ability in the quenching process.

Metallic material, e.g. stainless steel, is favored as the manufacturing material for micro-combustor. The inhibitory effect of radical loss at wall on the ignition of hydrocarbons may not be neglected as micro-combustor has large surface-area-to-volume ratio. However, the mechanism of radical loss at wall is still unclear. In this study, a surface reaction kinetic model for the combustion of methanol/air mixture is developed and the heterogeneous surface effects on ignition delay are examined.

2. Model

2.1. Model setup

The surface kinetic mechanism consists of absorption reactions of radical species by the surface and desorption reactions of stable species which are formed by recombination reactions of absorbed species [7]. However, there is no consensus among researchers on which radical species should be considered as they are capable of being absorbed and which can be neglected. Aghalayam *et al* [7] first built a surface reaction model for premixed hydrogen-air impinging on a flat surface. The model consisted of five radical species including H, O, OH, HO_2 and H_2O_2. A similar surface reaction model for methane-air was setup by Raimondeau *et al* [8], in which only CH_3, H, O, OH was considered, exclusive of HO_2 and H_2O_2. However, Held and Dryer [9] indicated that the low temperature oxidation rate of methanol/oxygen in a closed vessel was determined by H_2O_2 (+wall) $\rightarrow H_2O+1/2O_2$. But the loss of HO_2 at wall has a very small impact on the reaction rate.

Table 1. Surface kinetic mechanism of methanol/air

Absorbed species	Reactions [a]	Sticking Coefficient γ or Pre-Exponential (s^{-1})
H	$H + * \rightarrow H*$	$0 \sim 1$
	$2H* \rightarrow H_2 + 2*$	10^{13}
O	$O + * \rightarrow O*$	$0 \sim 1$
	$2O* \rightarrow O_2 + 2*$	10^{13}
OH	$OH + * \rightarrow OH*$	$0 \sim 1$
	$2OH* \rightarrow H_2O + O* + *$	10^{13}
	$2O* \rightarrow O_2 + 2*$	10^{13}
HO_2	$HO_2 + 2* \rightarrow OH* + O*$	$0 \sim 1$
	$2OH* \rightarrow H_2O + O* + *$	10^{13}
	$2O* \rightarrow O_2 + 2*$	10^{13}
H_2O_2	$H_2O_2 + 2* \rightarrow 2OH*$	$0 \sim 1$
	$2OH* \rightarrow H_2O + O* + *$	10^{13}
	$2O* \rightarrow O_2 + 2*$	10^{13}
CH_3	$CH_3 + * \rightarrow CH_3*$	$0 \sim 1$
	$CH_3* + H* \rightarrow CH_4 + 2*$	10^{13}
	$2CH_3* \rightarrow C_2H_6 + 2*$	10^{13}

a) * denotes reactive surface site and for example H* means an adsorbed H radical.

Base on the fundamental research of referred groups, a new surface reaction model for methanol/air mixture is established in this study. Table 1 tabulates the model, which has several basic assumptions: (1) Six radical species can be destroyed at wall, including H, O, OH, HO_2, H_2O_2 and CH_3. (2) The sticking coefficient γ of each radical species is identical. (3) The surface recombination reactions are taken to have zero activation energy and their pre-exponential are estimated at 10^{13} s^{-1} [7].

2.2. Model validation

The effect of surface reaction on the ignition and the oxidation process were study numerically using CHEMKIN-Pro software [10]. The detail gas phase kinetic mechanism of methanol/air has been developed by Li *et al* [11]. The surface kinetic mechanism was integrated into the calculation of homogeneous combustion process of gas mixture in order to investigate the inhibitory effect of surface.

Experimental verifications for the heterogeneous surface effect on the ignition of mixtures in an adiabatic system are quite rare, but various static reactor experiments have been conducted to study the surface effect. The calculated results have fine agreement with both experimental data [12, 13] for $4.2\times10^{-6}\sim8\times10^{-6}$ of γ, shown in figure 1 and figure 2.

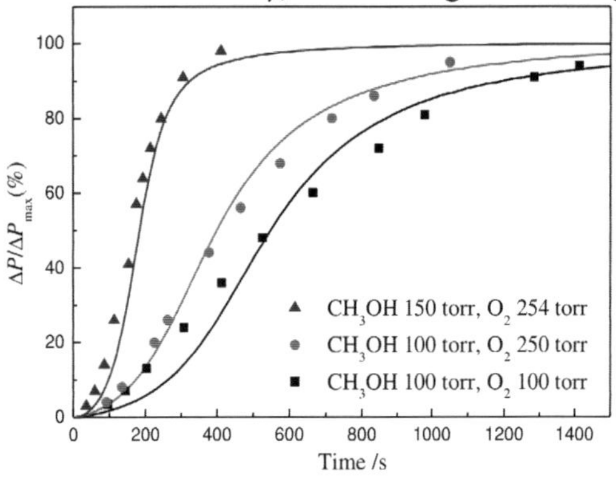

Figure 1. Experimental (scatters) and calculated pressure profile (lines) of reference [12]. The γ are 6×10^{-6}, 8×10^{-6} and 8×10^{-6} respectively for 404 torr, 350 torr and 200 torr cases under 710 K.

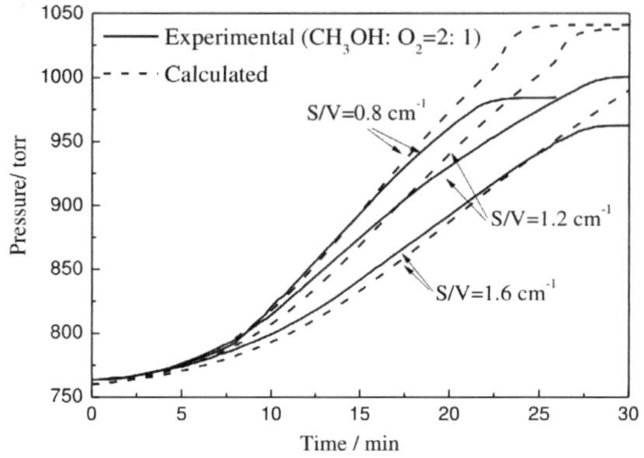

Figure 2. Measured and calculated pressure profile of reference [13]. The γ are 6.0×10^{-6}, 4.5×10^{-6} and 4.2×10^{-6} respectively for 0.8 cm^{-1}, 1.2 cm^{-1} and 1.6 cm^{-1} cases under 663 K.

The developed surface reaction model for methanol oxidation is demonstrated to be applied to predict the influence of surface on gas phase reaction rate. The sticking coefficients γ is a crucial parameter that determines the absorption intensity of the internal surface. However, it is difficult to measure the value of γ as it has correlation with surface material, microscopic surface texture and physical property of the surface, etc. Therefore, a certain discrepancy of γ may occur in the estimation for reaction vessels that are even made of the same material. The value of γ is at the order of 10^{-6} for quartz. For stainless steel, γ may be of more than 100 times larger than quartz [14].

3. Numerical method

In order to investigate the effect of surface reaction on the ignition delay of methanol/air mixture in an adiabatic system, surface reaction mechanism of methanol in this study has been integrated into the calculation of homogeneous reactions using CHEMKIN-Pro software. The ignition delay time τ_{ign} was defined by the maximum molar concentration of OH. The calculations were performed for initial pressure of 3 atm and a stoichiometric equivalence ratio. In this study, the main interest is to find out the dominant surface reaction affecting the ignition of mixture at initial temperature T_0 below 1000 K. In addition, the sticking coefficient γ and surface-area-to-volume-ratio *SV* are two important parameters for surface reactions, and their effects on ignition delay are also analyzed.

4. Results and discussions

The gas phase kinetic mechanism of methanol/air has got comprehensive study by Held and Dryer [9]. At low temperature, OH is mainly produced by the decomposition of H_2O_2 during the induction time.

In the numerical simulations, the cases of inert surface and reactive surface were compared. For inert surface, the system proceeds homogeneous combustion and no radical is absorbed. For reactive one, the surface can be regard as the sink of radicals, reducing the radical concentration in the gas phase. In figure 3, the growth rate of H_2O_2 mole fraction in the gas phase is lower for reactive-surface case, resulting in slower production of OH as the reaction rate of R6 is lower. Since the absorbed H_2O_2 by the surface converges to H_2O and O_2 eventually, less OH is produced by the decomposition H_2O_2. This is the main reason why the reactive-surface case has longer ignition delay time.

Table 2. Low temperature kinetic mechanism of methanol/air [9]

No.	Reactions	Reaction type
R1	$CH_3OH + O_2 = CH_2OH + HO_2$	Chain initiation
R2	$CH_3OH + HO_2 = CH_2OH + H_2O_2$	Chain propagation
R3	$CH_2OH + O_2 = CH_2O + HO_2$	Chain propagation
R4	$CH_2O + HO_2 = HCO + H_2O_2$	Chain propagation
R5	$HCO + O_2 = CO + HO_2$	Chain propagation
R6	$H_2O_2(+M) = 2OH(+M)$	Chain branching
R7	$CH_3OH + OH = CH_2OH + H_2O$	Chain propagation
R8	$CH_2O + OH = HCO + H_2O$	Chain propagation

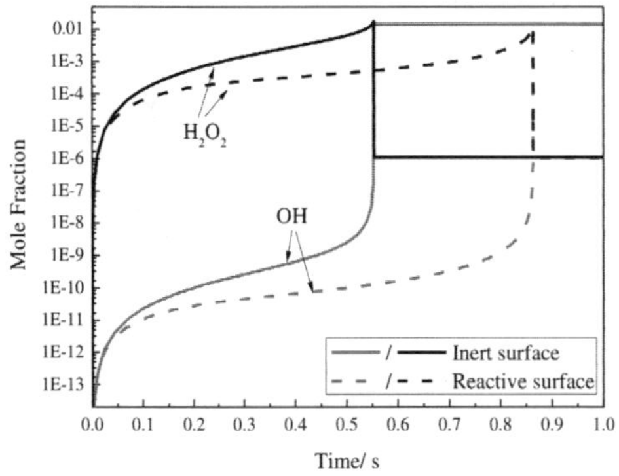

Figure 3. Mole fraction profiles of OH and H_2O_2 radicals for inert-surface and reactive-surface. (SV=10 cm^{-1}, γ=7×10^{-4}).

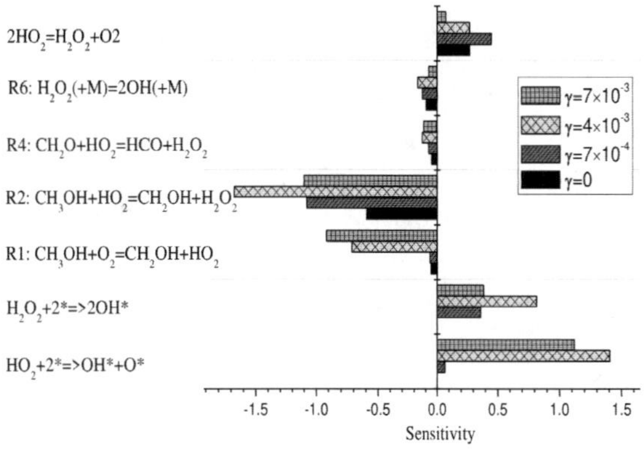

Figure 4. Normalized sensitivity of ignition delay time with respect to reaction rate (T_0=900 K, SV=10 cm^{-1})

A sensitivity analysis was performed to study the reaction sensitivity to ignition delay time as a function of γ. The normalized sensitivity was computed using the formula as Zhang *et al* [15]. It was found that the sensitivity of absorption reactions for H, O, OH and CH_3 were so small that can be neglected. Therefore, they are not shown in figure 4, which gives out seven largest sensitivities among gas phase reactions and surface reactions. When γ=7.0×10^{-4}, the absolute values of the sensitivity for the absorption of HO_2 and H_2O_2 were much smaller than gas phase reactions. It means that the ignition delay time is mainly determined by gas phase reactions when γ is small. But the inhibitory effect of surface reaction is still significant. The main cause of this inhibitory effect is the absorption of H_2O_2 and the absorption of HO_2 can be ignored, the same conclusion deducted in reference [9]. However, the sensitivity of R1 increases abruptly when γ increases to 4.0×10^{-3} and the sensitivity for the absorption of HO_2 surpasses the one of H_2O_2. Since HO_2 is an important chain carrier, it comes out through R1 and is removed rapidly by R2 and R4, and absorption reaction of HO_2. When γ is quite large and the effect of absorption of HO_2 cannot be ignored, less HO_2 is consumed by R2, which produces most of the H_2O_2, resulting in lower concentration of H_2O_2, shown in figure 3. Therefore, the surface effect is determined by the absorption of H_2O_2 when γ is small. But when γ is as large as the order of 10^{-3}, the absorption of HO_2 should be considered.

The effects of γ and SV on ignition delay were also explored. As figure 5 shows, for T_0=800 K, the ignition delay time has slightly change when γ<1.0×10^{-4}, but increasing rapidly when γ>1.0×10^{-4} because of the surface effect. When T_0 is up to 900 K, the rapid increase just emerged after γ>1.0×10^{-3}. And for T_0=1000 K, the ignition delay time increases quite slightly. It illustrates that the enlargement of T_0 greatly help weaken the inhibitory effect caused by surface reaction because the reaction rate of the gas-phase chain reactions increase quickly and more radicals are produced, compensating the radical loss on surface. Figure 6 illustrates that the ignition delay time increases with SV, which associates with the shape and the internal volume of reaction vessel. The larger of the SV, the more possibility of radicals impinge to the surface and get absorbed. In addition, as *SV* becomes larger, the growth rate of ignition delay time is faster when γ is large. Therefore, besides γ, *SV* also is one of important surface parameters that decribe the intensity of inhibitory effect of surface on ignition delay.

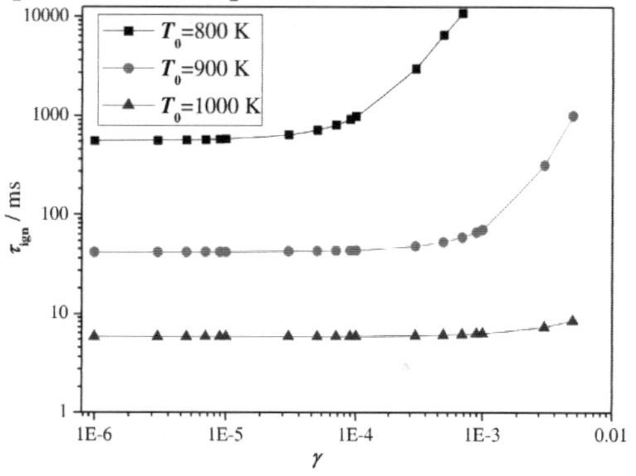

Figure 5. Comparison of ignition delay times for different temperatures versus sticking coefficients (SV=10 cm^{-1})

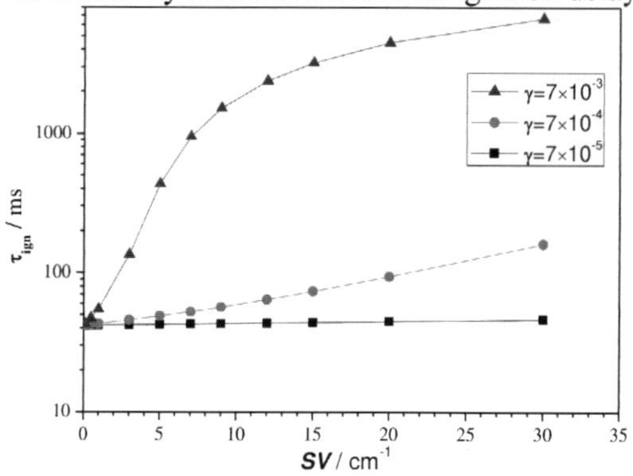

Figure 6. Comparison of ignition delay times for different sticking coefficients versus surface-area-to-volume ratios (T_0=900 K)

5. Conclusions

In this paper, a surface kinetic reaction mechanism for methanol oxidation in a small closed volume was developed. Using the surface model, the inhibitory effect of surface reaction on ignition delay was also investigated. It was found that the absorption of H_2O_2 on the surface was the main reason, due to the direct suppression effect on the concentration of OH in the radical pool. However, the absorption of HO_2 would take the place of H_2O_2 as the main inhibitory factor when the sticking coefficient became as large as the order of 10^{-3}. Both the sticking coefficient and surface-area-to-volume ratio were two important parameters that determined the intensity of surface effect. And the ignition delay time increased with sticking coefficient or surface-area-to-volume ratio. In addition, enhancing the initial temperature of the mixture was able to reduce the inhibitory effect of surface reactions.

Acknowledgement

The current work was supported by National Science Foundation of China (No.51336010) and National Basic Research Program of China (No. 2014CB239600).

References

[1] Ju Y G and Maruta K 2011 *Prog. Energy Combust. Sci.* **37** 670
[2] Norrish R G W and Reagh J D 1940 *Proc. R. Soc. Lond.* A **176** 432
[3] Lewis B and Von Elbe G 1987 *Combustion, flames and explosions of gases* (Orlando: Academic Press) pp 25-76
[4] Bowman M J and Wilk R D 1996 *SAE paper*
[5] Kim K T, Lee D H and Kwon S 2006 *Combust. Flame* **146** 21
[6] Yang H L, Feng Y X, Wang X H, Jiang L Q and Zhao D Q 2013 *Proc. Combust. Inst.* **34** 3380
[7] Aghalayam P, Bui P A and Vlachos D G 1998 *Combust. Theory Model.* **2** 519
[8] Raimondeau S, Norton D and Vlachos D G 2002 *Proc. Combust. Inst.* **29** 906
[9] Held T J and Dryer F L 1998 *Int. J. Chem. Kinet.* **30** 811
[10] *CHEMKIN-Pro* 2010 (San Diego: Reaction Design, Inc.)
[11] Li J, Zhao Z W, Kazakov A, Chaos M, Dryer F L and Scire J J 2007 *Int. J. Chem. Kinet.* **39** 109
[12] Fort R and Hinshelwood C N 1930 *Proc. R. Soc. Lond.* A **129** 285
[13] Bone W A and Gardner J B 1936 *Proc. R. Soc. Lond.* A **154** 298
[14] Saiki Y and Suzuki Y 2012 *Proc. Combust. Inst.* **34** 3401
[15] Zhang Y J, Huang Z H, Wei L J, Zhang J X and Law, C. K. 2012 *Combust. Flame* **159** 926

A Bistable Vibration Energy Harvester with Closed Magnetic Circuit

Takahiro Sato[1], **Takeshi Sugisawa, Hajime Igarashi**
Graduate School of information science and technology, Hokkaido University,
Kita 14, Nishi 9, Kita-ku, Sapporo, 060-0814, JAPAN.

E-mail: tsato@em-si.eng.hokudai.ac.jp

Abstract. In this work, to increase magnetic flux passing through the electric coil in a bistable vibration energy harvester, the magnetic circuit is made closed by introducing two coil systems which have magnetic core in their axis holes. The magnetic resistance of the magnetic circuit, composed of silicon steel and thin air gaps, is supressed to be small. The double well potential is realized from the spring force and nonlinear magnetic force between the magnets and the magnetic core. Two harvesters with opened and closed magnetic circuits are manufactured for comparison. It is also shown that the closed magnetic circuit can effectively improve the output power.

1. Introduction

Energy harvesting devices have been receiving much attentions as voltage source for low-powered wireless sensor nodes with autonomous operations [1]. The electromagnetic vibration energy harvester (VEH) transforms vibration energy into electrical energy through magnetic induction [2],[3]. Conventional VEHs produce electric power through linear spring-damper oscillation. Thus, maximum power is obtained at the resonant frequency and the output power drops as operating frequency goes far away from the resonance. In order to make effective energy harvesting from real-world vibrations, the operational bandwidth of VEHs should be broaden.

For the wideband VEHs, various approaches have been proposed [4]. The use of the bistable structure is one of the attractive approaches [5]. Bistable VEHs are characterized by a double-well potential. As for the bistable VEHs, the inertial mass of VEH transits between two equilibrium positions if the oscillator can overcome the potential barrier by the vibrations regardless of frequency. Hence, the bistable VEHs have a possibility to realize broadband operation. In fact, it has been shown that bistable VEHs can harvest electrical power under noise excitations [6]. Moreover, the magnetic flux passing through the coil in VEHs should be increased to improve power generation efficiency.

In this work, to increase magnetic flux in a bistable VEH, we make the magnetic circuit closed by introducing two coil systems which have magnetic core in their axis holes. The magnetic resistance of the magnetic circuit, composed of silicon steel sheets and thin air gaps, is supressed to be small. The double well potential is realized from the spring force and nonlinear magnetic force between the magnets and the magnetic core. Two harvesters with opened and closed magnetic circuits are manufactured for comparison. It is also shown that the closed magnetic circuit can effectively improve the output power.

---

[1] Corresponding author: Takahiro Sato

2. Bistable harvester with closed magnetic circuit

It is necessary to improve the operational bandwidth and the output power of VEH. The electromotive force induced in the coil is equal to the time derivative of the magnetic flux across the coil. The magnetic flux should be, therefore, increased to improve the output. Introduction of magnetic materials in the magnetic circuit allows us to increase the flux linkage with the coil. When introducing magnetic materials, attraction magnetic force, which is nonlinear with respect to displacement, is generated between magnets and magnetic materials. This nonlinearity would broaden the frequency bandwidth of the VEH. It is thus expected that appropriate magnetic circuit design leads to the improvements of both the bandwidth and the output power.

On the basis of the above mentioned insight, we have been developed a VEH harvester model using magnetic material [7, 8]. In this work, a harvester which has H-shaped magnetic core shown in figure 1 is considered. This model is here called H1-harvester. For the magnetic core in the axis hole of the coil bobbin, we employ the silicon steel sheets whose relative permeability is much higher than the space permeability. The magnetic flux is expected to be much larger in this VEH in comparison with the VEH whose magnetic core is made of Ferrite [7, 8]. It is therefore expected that the H1-harvester can improve the output power. However, the magnetic resistance is still large because the magnetic flux outside the coil has to pass the air region.

For this reason, another harvester with a closed magnetic circuit is presented, as shown in figure 2, called here H2-harvester. In this model, two H-shaped magnetic cores are introduced in two coils so as to form closed magnetic circuit, as shown in figure 2(b). It is thus expected that the flux linkage with the coils can effectively be increased in comparison with the H1-harvester.

In these two models, nonlinear attraction magnetic force is generated between the magnets and the magnetic cores. When carefully tuning the magnetic force and spring constant, the bistable potential is realized, as will be mentioned below. Thus, it is expected that introduction of the magnetic cores not only improve the efficiency but also broaden the bandwidth of the harvester.

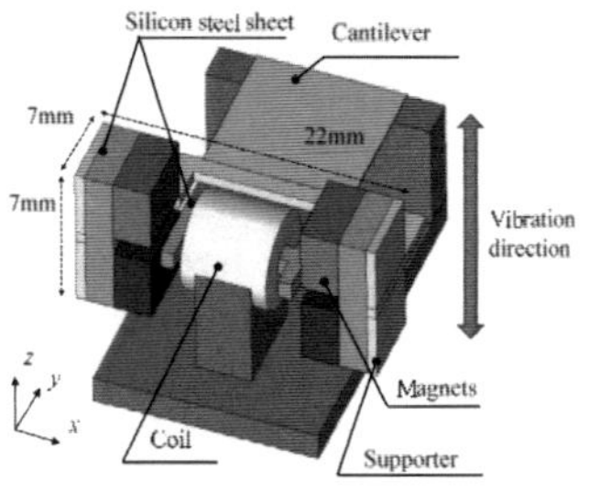

Figure 1. H1-harvester model with opened magnetic circuit.

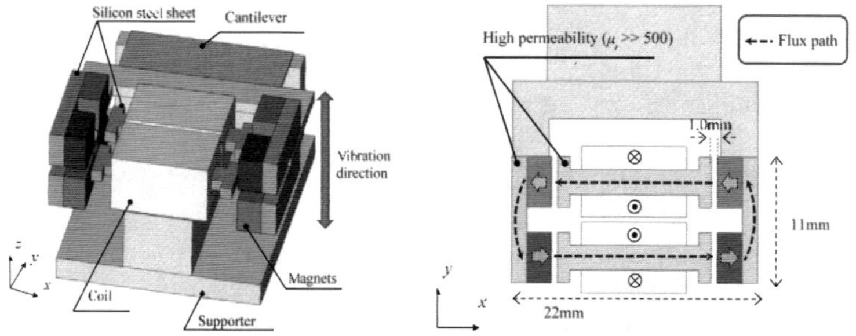

(a): 3D overview (b): Closed magnetic circuit

Figure 2. H2-harvester model with closed magnetic circuit.

We now consider the potential energy in the proposed harvesters. The potential energy, E, in the VEH is written by

$$E(z) = E_{\mathrm{mag}}(z) + \frac{1}{2}kz^2, \tag{1}$$

where k is spring constant, and z is the relative displacement of the coil and magnets. Moreover, E_{mag} is magnetic energy which can be computed by the finite element method as follows:

$$E_{\mathrm{mag}}(z) = \iiint_V E_m(z)dV, \tag{2}$$

where

$$E_m(z) = \begin{cases} \dfrac{\|\boldsymbol{B} - \boldsymbol{B}_r\|^2}{2\mu} & \text{magnet,} \\ \displaystyle\int_0^{\|\boldsymbol{B}\|} \|\boldsymbol{H}(b)\| \mathrm{d}b & \text{otherwise,} \end{cases} \quad (3)$$

and $\boldsymbol{B}_r$ is remnant flux density in magnet. Note that $E_m(z)$ is a function of z because flux distribution, $\boldsymbol{B}$, varies depending on z.

Figures 3 and 4 show the potential energy in the H1- and H2- harvesters, respectively, for three different values of k. It can be found that bistable potential structure can be realized when setting k to proper values. Moreover, when carefully tuning k, the potential barrier gets low, which makes it easy to transit the oscillator between two equilibriums. For example, in the H2-harvester, the appropriate value of k is about 1100N/m.

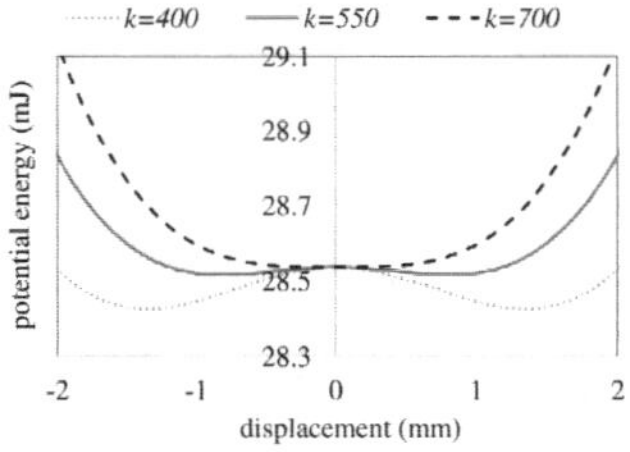

Figure 3. Potential energy in H1-harvester.

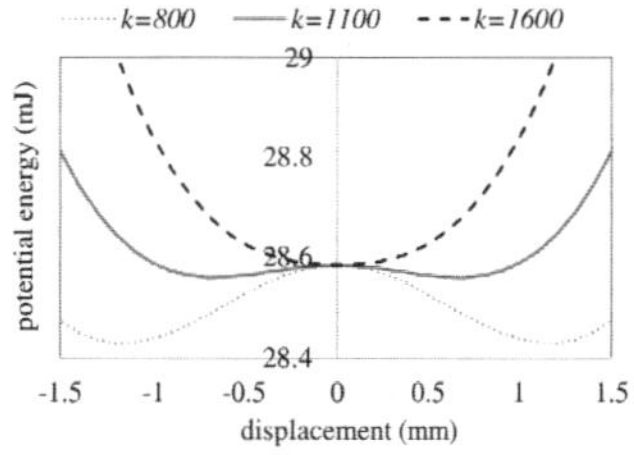

Figure 4. Potential energy in H2-harvester.

3. Experiments

The proposed harvesters was manufactured. Figure 5 shows the manufactured H2-harvester. The total mass of the oscillator (the magnets, core, and their keepers), effective spring constant of the cantilever, k, and the number of the turn in the coils are summarized in Table 1. To evaluate the performance with high potential barrier, k for the H1-harvester is set to relatively small.

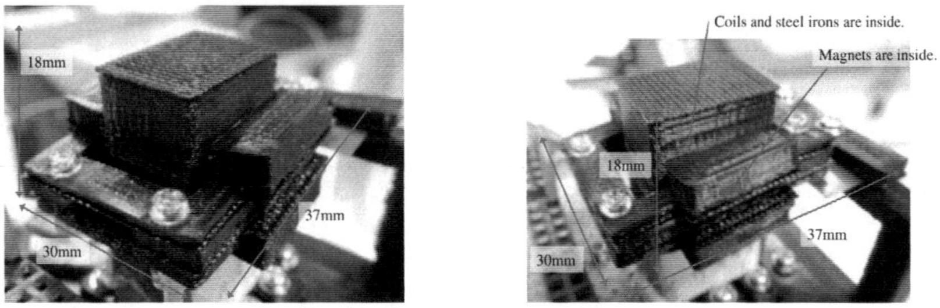

Figure 5. Manufactured H2-harvester.

Table 1. Mechanical parameters in manufactured harvesters

	mass of oscillator (g)	effective spring constant of cantilever (N/m)	Number of turns in coil
H1-harvester	9.1	392	700
H2-harvester	8.0	1150	840

3.1. Measurements

The output power against frequency for different frequencies was measured. Sinusoidal vibration was applied to the base of the proposed harvesters, and the acceleration level was fixed to about 1.0G at all the frequencies. A resistive load of 75Ω was connected to the coils, and the load voltage was measured by an oscilloscope.

Figures 6 and 7 show the output power and load voltage, respectively, plotted against the input frequency in the H1-harvester. These figures show that the maximum is located at about 25Hz, and

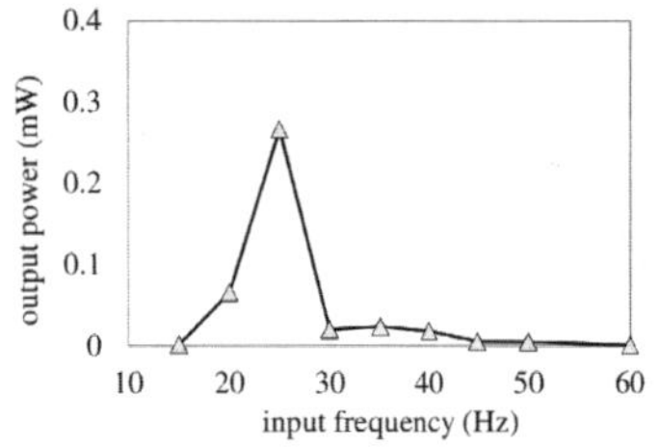

Figure 6. Output power in H1-harvester.

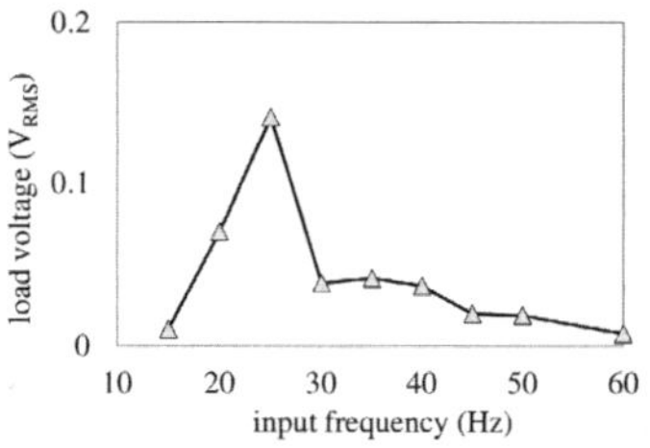

Figure 7. Load voltage in H1-harvester.

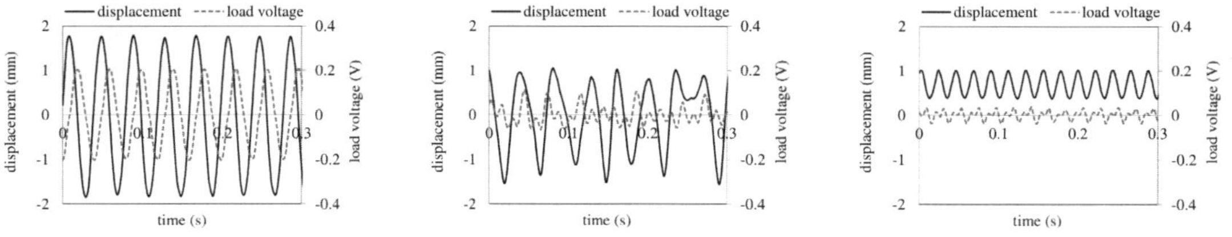

(a): 25Hz (b): 35Hz (c): 45Hz

Figure 8. Time-variation of displacement and load voltage in H1-harvester.

around which the outputs drop. The time-variation of displacement and load voltage are shown in figure 8, from which it is found that H1-harvester has three distinct behaviour modes. These modes well correspond to the behaviour modes of bistable VEH: interwell, chaotic, and intrawell oscillations [5]. Thus, this result strongly suggests that the manufactured H1-harvester has a bistable property. Nevertheless, the bandwidth is not broaden. This would be due to the fact that the potential barrier in the manufactured H1-harvester is high when k is about 400, as shown in figure 3. High potential barrier makes it difficult to produce the interwell oscillation, as a result, the bandwidth of the bistable VEH is narrow.

The output power and load voltage in the H2-harvester are shown in figures 9 and 10, respectively, which show that the output of the H2-harvester is much greater than that of the H1-harvester. Although efficiency of two harvesters is not simply compared because the numbers of turns in them are different, this result clearly indicates superiority of the superiority in the closed magnetic circuit. However, the H2-harvester again has a relatively narrow peak at about 40Hz. The time-variation of displacement and load voltage are shown in figure 11. At 35Hz, the amplitude of the oscillation is large. On the other hand, at 50Hz, the oscillation of the displacement is biased and its amplitude is

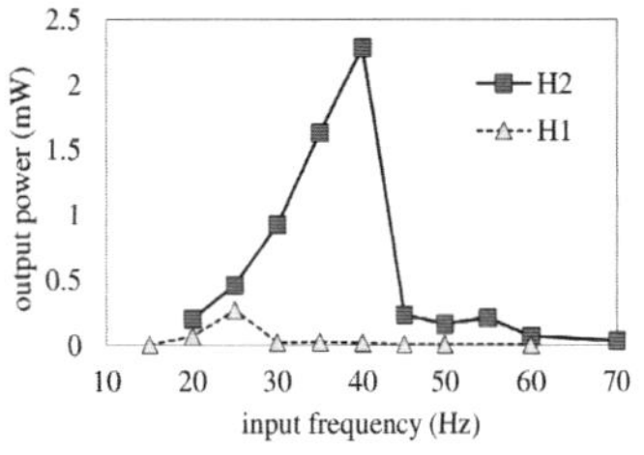

Figure 9. Output power in H2-harvester.

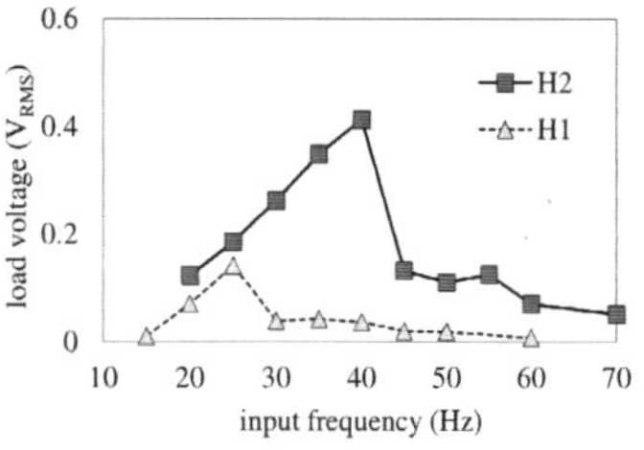

Figure 10. Load voltage in H2-harvester.

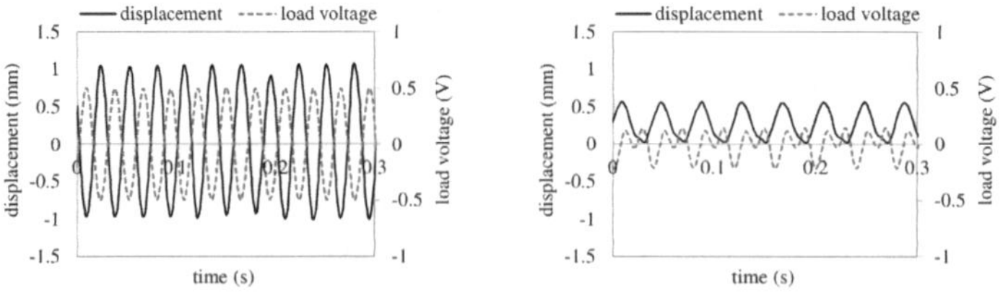

(a): 35Hz (b): 50Hz

Figure 11. Time-variation of displacement and load voltage in H2-harvester.

much smaller than that at 35Hz. From these results, it can be said that the H2-harvester has two distinct behaviour modes depending on the input frequency. This fact suggests that the H2-harvester does not have a bistable property.

3.2. Discussions

Here, the reason why the H2-harvester does not have the bistable property is discussed. One of the considerable reason is due to the manufacturing error. As for bistable VEHs, the coil axis is aligned to the center of the magnet at the neutral position. The potential energy profiles for the misaligned coil are shown in figure 12, from which it is found that the H2-harvester does not have a double-well potential whereas H1-harvester has. Because the potential barrier in the H1-harvester is relatively high, bistable structure is maintained against the small misalignment. On the other hand, in the H2-harvester, the manufacturing results in disappearance of the double-well structure because of the low potential barrier. From these discussions, it can be concluded that bistable VEHs with low potential barrier can have broad bandwidth although it can easily loose the double-well potential due to manufacturing errors. In summary, the broadband bistable VEH is not robust against the manufacturing errors.

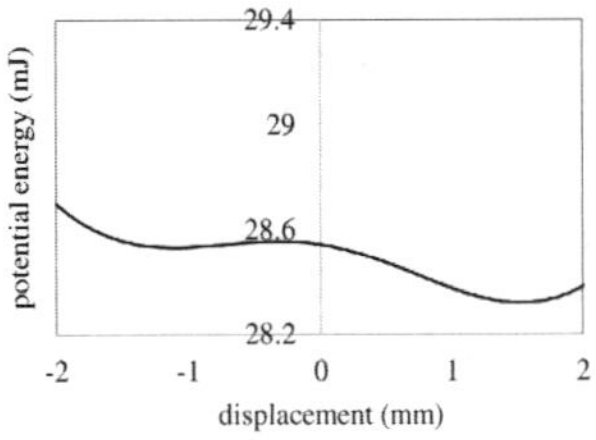

(a): H1-harvester

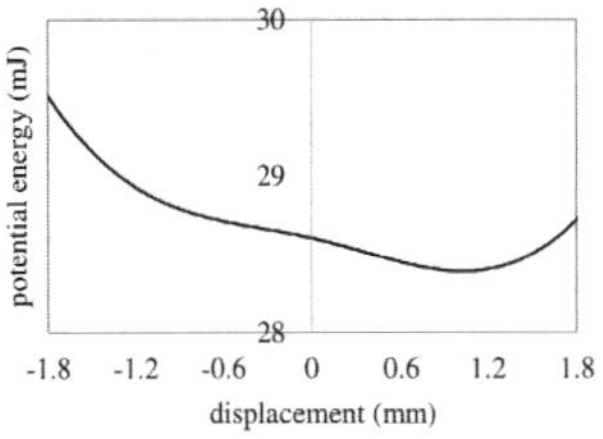

(b): H2-harvester

Figure 12. Potential energy when misaligned (0.2mm).

4. Conclusions

This paper has presented the bistable harvester which has the closed magnetic circuit. The experimental results have suggested that the manufactured H1-harvester is bistable while H2-harvesters is not. It has also shown that the output power in the H2-harvester is much greater than that in the H1-harvester. The reason why the H2-harvester does not have double-well potential has been discussed.

For the future work, the bistable H2-type harvester will be developed by improving the manufacturing accuracy.

5. Acknowledgment

This study was partly supported by Grant-in-Aid for JSPS Fellows and JSPS KAKENHI Grant (B) Number 24310117.

6. References

[1] E. Sazonov, H. Li, D. Curry, P. Pillay, 2009, *IEEE Sensors Journal*, vol. 9, no. 11, pp. 1422-1429..

[2] S. P. Beeby, R. N. Torah, M. J. Tudor, P. G. Jones, T. O'Donnell, C. R. Saha, S. Roy, 2007, *J. Micromech. Microeng.*, Vol. 17, no. 7 pp. 1257-1265.

[3] D. Zhu, S. Roberts, T. Mouille, M. J. Tudor, S. P. Beeby, 2012, *Smart. Mater. Struct.*, vol. 21, no. 10, 105039.

[4] L. Tang, Y. Yang, C. K. Soh, 2010, *Journal of Intelligent Material Sytems and Structures*, vol. 21 no. 18, pp. 1867-1897.

[5] R. L. Harne, K. W. Wang, 2013, *Smart. Mater. Struct.*, vol. 22, no. 2, 023001.

[6] S. P. Beeby, el. al., 2012, *Smart. Mater. Struct.*, vol. 22, no. 7, 075022.

[7] T. Sato, K. Watanabe, H. Igarashi, 2014, *IEEE Trans. Magn.*, vol. 50, no. 2, 7007604.

[8] T. Sato, H. Igarashi, 2013, *Journal of Physics: Conference Series*, vol. 476, pp. 622-626.

PowerMEMS 2014 IOP Publishing
Journal of Physics: Conference Series **557** (2014) 012082 doi:10.1088/1742-6596/557/1/012082

Fabrication and performance analysis of MEMS-based Variable Emissivity Radiator for Space Applications

Changwook Lee[1], Hyung-Ung Oh[1] and Taegyu Kim[1]
Depertment of Aerospace Engineering, College of Engineering, Chosun University
309 Pilmun-daero, Dong-gu, Gwangju 501-759, Republic of Korea

E-mail: -changwook-@hanmail.net

Abstract. All Louver was typically representative as the thermal control device. The louver was not suitable to be applied to small satellite, because it has the disadvantage of increase in weight and volume. So MEMS-based variable radiator was developed to support the disadvantage of the louver MEMS-based variable emissivity radiator was designed for satellite thermal control. Because of its immediate response and low power consumption. Also MEMS-based variable emissivity radiator has been made smaller by using MEMS process, it could be solved the problem of the increase in weight and volume, and it has a high reliability and immediate response by using electrical control. In this study, operation validation of the MEMS radiator had been carried out, resulting that emissivity could be controlled. Numerical model was also designed to predict the thermal control performance of MEMS-based variable emissivity radiator.

1. Introduction

MEMS based sensor and actuator technology have come to stage of commercialization. Active efforts have been made to apply to aerospace, JAXA, NASA and ESA of advanced aerospace countries attempt to graft MEMS technology onto the aerospace application[1-6]. If MEMS technology applies to the satellite, which would be having the ability such as weigh lighting, miniaturization, low power consumption by high integration. The satellite operated additional thermal control systems to maintain allowable temperature of electrical device. Because the satellite carried out mission at space environment To control temperature of satellite[7], Louvers are usually used. Louvers have good thermal control ability but they have a disadvantage of becoming weight heavier[8-9]. So MEMS based variable radiator was developed to support the disadvantage of the louver. Table 1 shows a characteristic of louver and variable emissivity radiator. Microscopic device with MEMS technology could be produced in bulks. So it could be developed at low cost. Also MEMS technology applied to aerospace, it expected to have leading technology and intellectual property. Therefore this study carried out thermal performance analysis using a numerical method, calculation based EM was fabricated, temperature of EM radiator was measured by 00000 camera. CFD model was verified by comparing experiment value. Based on verified CFD model, the performance characteristics of variable emissivity radiator and fixed emissivity were examined by orbit analysis commercial tool. When variable emissivity radiator was applied, temperature could be controlled without additional

[1] To whom any correspondence should be addressed.

Table 1. Comparison of the variable emissivity radiator with the louver

	Results	characteristics
Louver	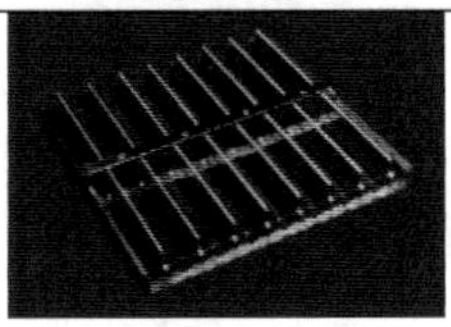	· Weakness by launch enviroment · Increase in weight and volume · Low operation reliability · High thermal control performance ·
MEMS radiator		· High strength by environment · Light weight and miniaturization · Good response · High operation reliablity

heater and power consumption. Even if variable emissivity radiator used a half area than fixed emissivity radiator, efficiency of variable emissivity radiator was superior to fixed emissivity radiator.

2. CFD model design and experiment

Operation principle of variable emissivity radiator was that emissivity was controlled as bead with low emissivity and low thermal conductivity moved between radiation plates by electrical discharge. CFD model was proposed to analyse performance of radiator as shape and material properties. Also Ansys fluent was used to analyse heat transfer which conductivity and radiation was occurred simultaneously. EM variable emissivity radiator with identical CFD model properties was fabricated for verification of analysis results. Variable emissivity radiator consists of the radiation plates with high emissivity and low emissivity, Electrodes for discharging bead and cover. Figure 2 shows schematic diagram of test device for thermal control test. EM radiator was placed on the heating plate for copying the generation of heat of box temperature. In the temperature range of 30°C ~ 70°C, Cover temperature was measured as bead location. When low temperature range is a copy, EM radiator model was placed in the low temperature chamber, temperature changes about both materials were measured by IR camera.

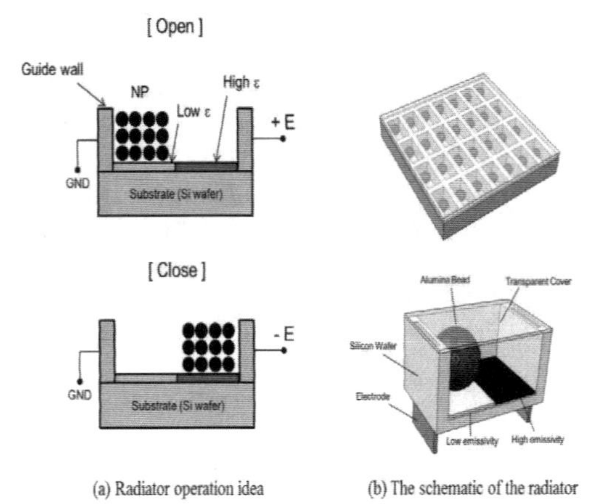

Figure 1. Variable emissivity radiator

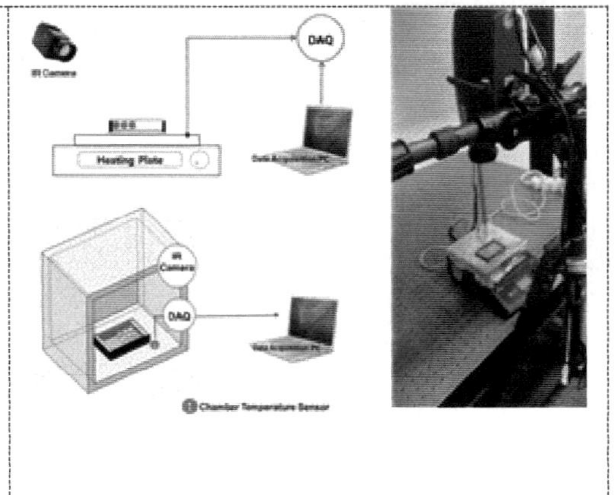

Figure 2. Experiment setup for thermal test

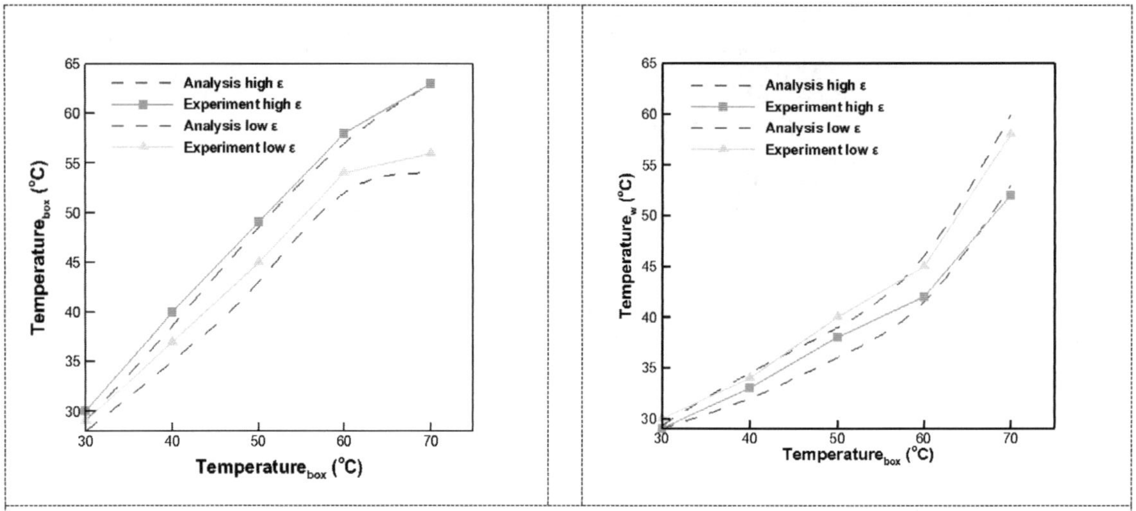

Figure 3. Comparison of analysis result with the experiment result of radiation plate

3. Analysis result verification
In the low-temperature range, after placing the EM radiator in a low-temperature chamber, an IR camera was used to measure the temperature change for two materials with time. Figure 4-5 shows the radiator's heat data measured through the experiment and its analysis result. The reason why the thermal diffusion velocity was increased in the open state is that a large amount of radiant heat was transferred directly toward the outer side to increase the thermal diffusion velocity. On the other hand, for the closed state, the bead is on the high-emissivity radiation plate, so for the heat transfer on the high-emissivity radiation plate, the heat is complexly transferred in the order of radiation-conduction-radiation, and the temperature of the outer side was measured about 3℃ lower due to the low heat transfer characteristic of the bead. For the low-emissivity radiation plate, the temperature rose by about 2℃, and it is considered that the temperature rose because the heat is transferred only by thermal radiation for the outer side. The temperature change in the open and closed states obtained through the analysis and experiment has the same tendency, so it could be considered that the result of the analysis is consistent with that of the experiment.
Based on the CFD model result analysed earlier, the heat control performance was verified in the deep space environment. To carry out the analysis, the SINDA, which is a commercial tool, was used to conduct the orbit thermal analysis of the fixed and variable emissivity radiators. As a result of the thermal analysis, the fixed emissivity radiator requires the electric power of more than 10 W for driving the heater because it reduces the equipment's temperature due to its high heat emission performance even in the low temperature condition, but the variable emissivity radiator performs heat emission like the fixed emissivity radiator in high temperature and changes into low emissivity in low temperature, so it could keep the equipment's allowable temperature without using a heater.should *not* be inserted using Word's footnote command because this will place the reference in the wrong place—at the bottom of the page (or end of the document) rather than next to the address. Ensure that any numbered superscripts used to link author names and addresses start at 1 and continue on to the number of affiliations. Do not add any footnotes until all the author names are linked to the addresses. For example, to format

4. Results
The MEMS variable emissivity radiator is a heat control device using electrical discharge. Because there is no attempt to apply the MEMS technology into the space in Korea, this study was carried out

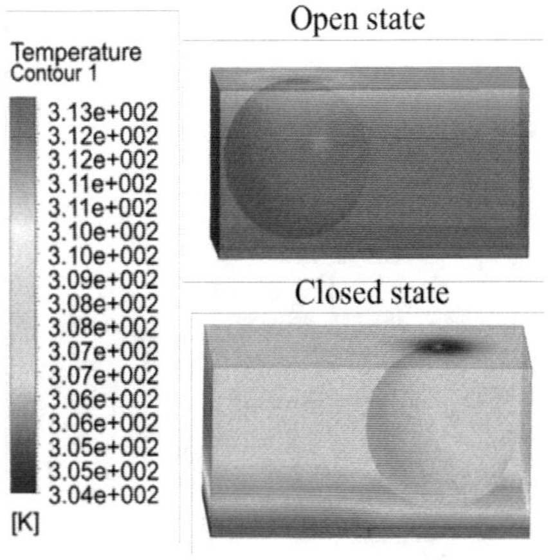

Figure 4. Temperature contour

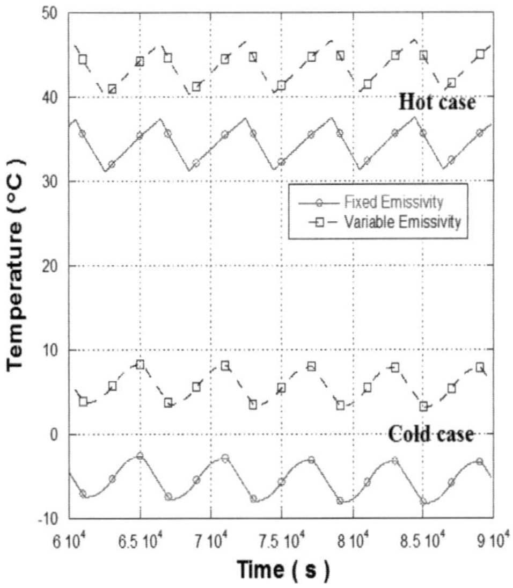

Figure 5. Temperature profiles of fixed and variable emissivity radiator

to apply the MEMS technology. The CFD analysis was carried out based on a mathematical model, and an EM of the variable emissivity radiator was made to verify the validity of the analysis model. As a result of analyzing the verified CFD model in the space environment, it was found that the heat could be controlled more efficiently than the existing heat control device.

(1) It was found that the EM variable emissivity radiator could control the emissivity to radiate and cut off heat in the temperature range (-25 ~ 70 ℃).

(2) For the existing studied radiator, its variable emissivity is 0.74, and for the variable emissivity radiator in this study, its emissivity could be changed up to 0.84.

(3) It was found that the variable emissivity radiator could keep the allowable temperature without using a heater in the deep space environment by using only the 50 % area compared to the fixed emissivity radiator.

The variable emissivity radiator made in this study is a space heat control system applying MEMS, which would be a basic study for the development of heat control devices applying MEMS in Korea, and if a sufficient space verification and heat control test is completed in the future, it could be expected to use it as a small satellite component for a satellite heat control system.

[1] A. Darrin, R. Osiande and J. Champion . 2000 AIP Conference Proceedings Vol. 504. 803
[2] Abeysinhe, D.C. 2001 IEEE Photonics technology. Vol. 13,. No. 9. Pp. 993-995
[3] W.J Tak, J. U. Jo, M. S. Lee, B.K. Kim 2010 The Korean society for aeronautical & space science. Vol .38. No. 8 pp. 767-773
[4] De Aragron, A.M.J. 1998 Micromechanics and Microengineeing. Vol. 8, No. 2, pp. 54-56
[5] W.C. Tang.1998 IEEE Aerospace Conf. MAR 21-28 , Snowmass. Colorado
[6] W. Biter. S. Oh, S. Hess. 2002 American Institute of Physics, 2002 AIP Conference Proceedings. Vol. 608.

[7] H. Bang, H. park 2000 The Korean society for Aeronautical & Space sciences. Vol. 28. No. 5. pp 153-160
[8] M. Koedinger, t. Brissonnaud. 1997 6th European Symposium on space Environmental Cotrol systems. Noordwijk
[9] R. Osiander, S.L Firebaugh, J.L champion, D.Farrar, M.A.G Darrin 2004 IEEE Sensors Journal. Vol. 4. Pp525-531

Improving the dielectric and piezoelectric properties of screen-printed Low temperature PZT/polymer composite using cold isostatic pressing

A Almusallam, K Yang, Z Cao, D Zhu, J Tudor, S P Beeby
Electronics and Computer Science (ECS), University of Southampton, SO17 1BJ, UK

E-mail: asa1g09@ecs.soton.ac.uk

Abstract. This paper reports an improvement in dielectric and piezoelectric properties of screen-printed PZT/polymer films for flexible electronics applications using Cold Isostatic Pressing (CIP). The investigation involved half and fully cured PZT/polymer composite pastes with weight ratio of 12:1 to investigate the effect of the CIP process on the piezoelectric and dielectric properties. It was observed that the highest dielectric and piezoelectric properties are achieved at pressures of 5 and 10 MPa for half and fully cured films respectively. The relative dielectric constants were 300 and 245 measured at 1 kHz for the half and fully cured samples. Using unoptimised poling conditions, the initial d_{33} values were 30 and 35 pC/N for the half and fully cured films, respectively. The fully cured sample was then poled using optimized conditions and demonstrated a d_{33} of approximately 44 pC/N which is an increase of 7% compared with non-CIP processed materials.

1. Introduction

Many researchers exploit the piezoelectric effect in energy harvesting, sensing and actuating applications. Piezoelectric materials can be used in energy harvesting and sensing by applying a mechanical force (e.g. squashing or compression) to the material which produces a charge and voltage across the material (i.e. the direct effect). It can be also exploited inversely in actuating systems by applying a voltage across the material which results in a change in its geometry (i.e. the inverse effect).

Piezoelectric materials can be found in four forms: single crystal materials, ceramics, polymers and composites. Ceramic piezoelectric materials (e.g. lead zirconate titanate PZT, Barium titanate $BeTiO_3$) are brittle but offer high d_{33} (350-560 pC/N) and high dielectric constants (2900). Piezoelectric polymers such as PVDF provide relatively low d_{33} (20-30 pC/N) [1], low dielectric constants (~ 8 at 1kHz) [2] but they have good mechanical flexibility. Relatively high d_{33}, ε_r and mechanical flexibility can be provided by piezoelectric composite materials which are a mixture of piezoelectric ceramic powders and a polymer matrix. The polymer used can itself be piezoelectric [3] or alternatively standard non-piezoelectric materials can be employed [4, 5]. Using piezoelectric polymers such as PVDF can be problematic since it has positive piezoelectric coefficients and if mixed with PZT (which has negative coefficients) can reduce the overall d_{33} of the composite material [3]. Using piezoelectric composites instead of piezoelectric polymers gives the advantage of producing a material with a higher piezoelectric coefficient d_{33} and dielectric constant ε_r values [4, 5], improved electromechanical coupling coefficient [6] and greater opportunity for use in mass production because of the lower price of its ingredients compared to PVDF and its copolymers counterparts. However, piezoelectric composites suffer from porosity and air voids that affect its dielectric and piezoelectric properties. This porosity increases when increasing the weight percentages of the ceramic phase above 50% [7]. It occurs during the curing and evaporation of the solvent phase in the composite after printing as the solvent is only dissolved in the polymer phase. During evaporation, the solvent leaves pores in the polymer which then become filled with air.

Thick film deposition such as screen-printing can produce films with typical thickness between 1-100 μm [8]. However, some other reviews state that this range can be shifted to higher thicknesses 10-200 μm [9]. The use of screen-printing can offer mass production, simplicity and a high range of ceramic loadings in the composite (e.g. < 90%).

Applying high pressure to the printed film can densify the material and reduce the number and size of the air voids. *Cold Isostatic Pressing* (CIP) is a technique that applies a homogenous and continuous (i.e. depending on the holding time) force across the surface of the material at room temperature. CIP can also be used to process complicated shapes and structures without deformation [10].

This paper presents an evaluation of the effect of CIP on screen-printed composite PZT/polymer films by measuring the effect it has on the mechanical, dielectric and piezoelectric properties of the printed material.

2. Experimental

2.1. Paste Formulation and Screen-printing

The 0-3 type PZT/polymer composite material consists of three constituents, PZT-5H powder, thermoplastic polymer and solvent. The PZT powder was a mixture of 2 μm (Pz29, Ferroperm Piezoceramics) and 0.8 μm (Pzt-S-55, Sunnytec) particles with a weight ratio of 4:1. The thermoplastic polymer was dissolved in a solvent to convert it from a solid to liquid phase. Then, the PZT and the polymer were blended together with a weight ratio of 12:1. A triple roll mill was employed to disperse the PZT particles inside the polymer producing a homogeneous paste. This provides a screen-printable paste without lumps that has consistent and repeatable mechanical piezoelectric and dielectric properties. The material was screen-printed using a DEK 248 screen-printer between two electrodes on Kapton polyimide 300 HN substrate (75 μm thickness). This printing formed the capacitive structure shown in Figure 1.

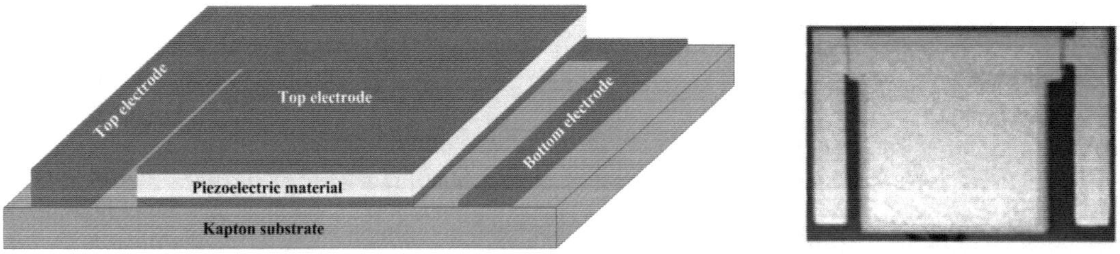

(a) (b)

Figure 1: Printed capacitive structure. (a) Schematic of the capacitive structure. (b) Top view of the screen-printed printed device

2.2. Curing printed materials

A silver-polymer paste (DuPont 5000) was used for printing the bottom and top electrodes. The curing conditions of the composite films were governed by the curing requirements of the polymer matrix. The screen-printed materials were cured in a box oven. Bottom electrode was cured at 120 °C for 10 min. The PZT/polymer composite and top electrode films were cured at 90 °C for 8 min. Although both the top and bottom electrode are formed from the same material, the top electrode was cured at lower curing temperature so that it does not over-cure and affect the piezoelectric film, which might lead to a variation in the properties of the printed composite.

2.3. Initial Dielectric and Piezoelectric Properties before CIP

Poling ferroelectric materials such as PZT is an important phase to activate its piezoelectric behaviour. The poling process was conducted by applying an external electric field concurrently with elevated temperatures for a specific time. A variety of poling methods can activate the piezoelectric properties of the material. Direct contact poling method was used in this investigation. Electrodes offer a basic setup to ensure an even distributed electric filed across piezoelectric material. The applied electric field is given by $E = V/d$, where V, d and E are the applied voltage, the thickness of the piezoelectric material and external electric field applied. It was essential to take reference d_{33} and ε_r measurements in order to identify if CIP leads to any improvement in these properties. The d_{33} and dielectric constant measurements were performed using piezometer (PM35, PiezoTest) and Precision Impedance Analyser (WAYNE KERR, UK) respectively. The capacitances of the devices were obtained from the impedance analyser. Then, the dielectric constants were calculated with the aid of the following equation $c = (A.\varepsilon_r.\varepsilon_0)/d$, where A is the area of the electrodes, ε_r is the relative permittivity or the dielectric constant of the material, ε_0 is the permittivity of free space and d is the average thickness of the piezoelectric film.

2.4. Cold Isostatic Pressing (CIP) Process

The CIP process was performed using Cold Isostatic Pressing Machine (CIP-20TA, MTI Corporation) shown in Figure 2. The sample is placed in the cylindrical chamber that is filled with pressing medium as shown schematically in Figure 2a. The die is pushed by an external force through the hole from the top of the cylindrical chamber pressurizing the hydraulic fluid which then applies a homogenous pressure across the sample. The compression can be applied continuously for the duration of the holding time.

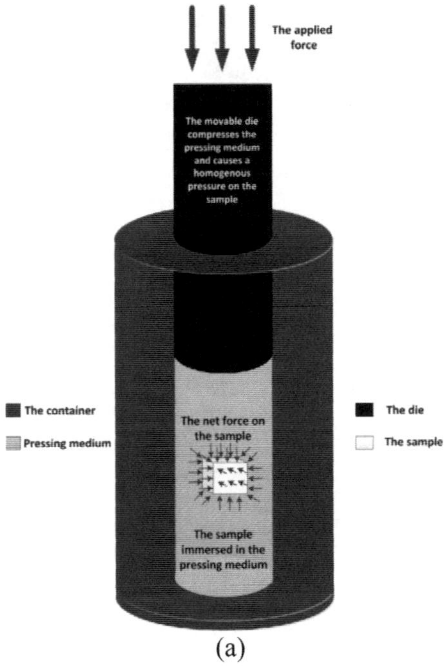

(a)

(b)

Figure 2: A schematic shows the CIP process (a). Actual CIP machine (b)

3. Results and Discussion

3.1. Dielectric Constants after Applying CIP

After applying CIP on two devices at different pressures, 5 dielectric constant readings were taken for each device giving a total of 10 measurements at every pressure point. The overall dielectric constant of the half cured sample is greater than the fully cured ones because of incomplete evaporation of the solvent. Once the solvent is fully evaporated and voids are left (i.e. because of the poor cross linking of the polymer), these voids are filled with air and the dielectric constant is reduced. Figure 3 shows that the dielectric constant increases to 300 and 245 with applied CIP pressures up to 5 and 10 MPa for the half and fully cured samples respectively. After 5 MPa, further increases in CIP pressure actually reduce the dielectric constant of the half cured sample whilst the fully cured sample remains fairly constant after 10 MPa.

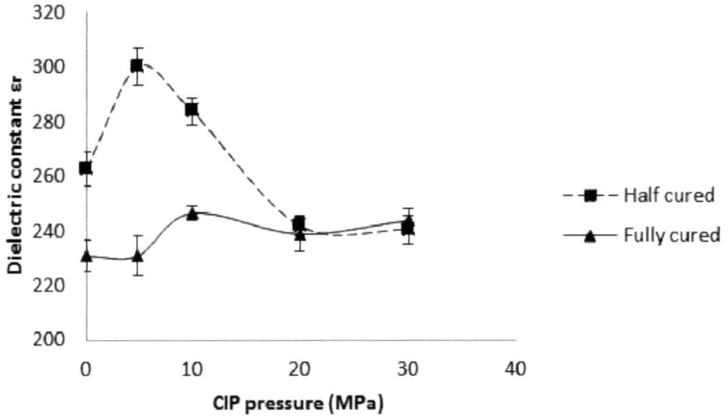

Figure 3: Effect of the applied pressure on the dielectric constant (at 1 kHz) for half and fully cured samples

3.2. Piezoelectric Coefficient d_{33} after Applying CIP

Two devices were used for every pressure point and five d_{33} measurements were taken for every device giving 10 measurements in total. Half cured samples were put in the oven for complete curing before the poling process. Similarly, the d_{33} values were improved by increasing the CIP on the samples reaching their maximum values 30 and 35 pC/N at pressures 5 and 10 MPa for the half cured and cured samples, respectively. Above these pressures, the CIP negatively affected the piezoelectric properties of the film for both the half cured and fully cured materials.

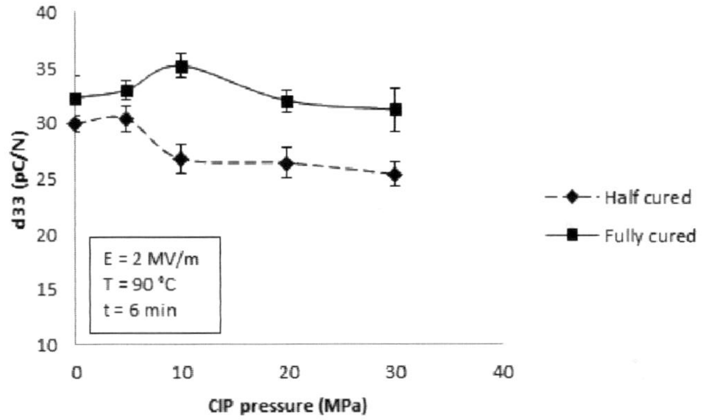

Figure 4: Effect of the applied pressure on the d_{33} values for half and fully cured samples

3.3. d_{33} Values at Optimum Poling Conditions

CIP improved the piezoelectric properties of the cured samples by 6 and 7% for samples poled at with normal and optimum poling condition, respectively. Figure 5 shows the maximum d_{33} achieved is 44 pC/N obtained at optimum poling conditions ($E = 3.7\ MV/m$, $T = 90\ °C$ *and* $t = 6\ min$).

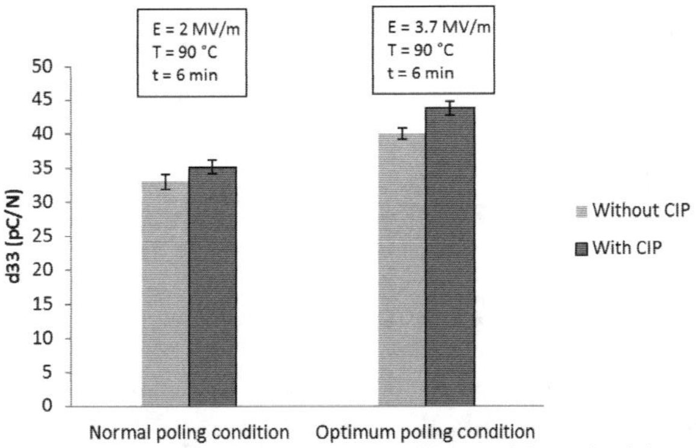

Figure 5: d_{33} values comparisons between PZT/polymer films with and without 10 MPa CIP at normal and optimum poling conditions

4. Conclusion

CIP process does improve the dielectric and piezoelectric properties of PZT/polymer films. This paper has confirmed this improvement but there is a limit to the pressure that should be applied (i.e. there is an optimum pressure that can be applied in order to maximise the properties of the material). After applying a CIP pressure of 10 MPa, the PZT/polymer film showed a maximum d_{33} of 44 pC/N with an increase of 7% compared to films without CIP.

5. Acknowledgment

The authors thank EPSRC for supporting this research with grant reference EP/IO05323/1 and King Abdullah Saudi Scholarship Programme for their sponsorship of Ahmed Almusallam.

6. References

[1] Ueberschlag P 2001 PVDF piezoelectric polymer *Sensor Review,* vol. 21, pp. 118-126

[2] Varadan V V Roh Y R Varadan V K and Tancrell R H 1989 Measurement of all the elastic and dielectric constants of poled PVDF films *IEEE 1989 ULTRASONICS SYMPOSIUM,* pp. 727-730

[3] Dietze M Es-Souni M 2008 Structural and functional properties of screen-printed PZT-PVDF-TrFE composites *Sensors and Actuators* **143**, pp. 329-334, 2008.

[4] Almusallam A Torah R N Yang K Tufor J and Beeby S P 2012 Flexible Low Temperature Piezoelectric Films for Harvesting from Textiles *PowerMEMS 2010, December 3-6, Atalanta, USA, Atalanta, USA,*

[5] Almusallam A Torah R N Zhu D Tudor M J and Beeby S P 2013 Screen-printed piezoelectric shoe-insole energy harvester using an improved flexible PZT-polymer composites *PowerMEMS 2013, London*

[6] Madhusudhana Rao C V Prasad G 2009 Charactrization of piezoelectric polymer composites for MEMS devices *Bulletin of Materials Science* **35**, pp. 579-584

[7] Hossain M E Liu S Y O'Brien S Li J 2014 Modelling of high-k dielectric nanocomposites *Acta Mechanica,* vol. **225**, pp. 1197–1209

[8] Priya S Ryu J Park C Oliver J Choi J and Park D 2009 Piezoelectric and magnetoelectric thick films for fabricating power sources in wireless sensor nodes *Sensors,* vol. 9, pp. 6361-6384

[9] Torah R N 2004 Optimisation of the piezoelectric properties of thick-film piezoceramic devices *PhD, Electronics and Computer Sience, University of Southampton*

[10] Weerasinghe H C Prasad M S George P S and Yi-Bing C 2012 Cold isostatic pressing technique for producing highly efficient flexible dye-sensitised solar cells on plastic substrates," *Progress in Photovoltaics: Research and Applications,* vol. 20, pp. 321–332.

PowerMEMS 2014 IOP Publishing
Journal of Physics: Conference Series **557** (2014) 012084 doi:10.1088/1742-6596/557/1/012084

Thermal Energy Harvesting from Wildlife

P. Woias, F. Schüle, E. Bäumke, P. Mehne, M. Kroener
Laboratory for Design of Microsystems, IMTEK, University of Freiburg, Georges-Koehler-Allee 102, 79110 Freiburg, GERMANY

E-mail: woias@imtek.de

Abstract. In this paper we present the measurement of temperature differences between the ambient air and the body temperature of a sheep (Heidschnucke) and its applicability for thermoelectric energy harvesting from livestock, demonstrated via the test of a specially tailored TEG system in a real-life experiment. In three measurement campaigns average temperature differences were found between 2.5 K and 3.5 K. Analytical models and FEM simulations were carried out to determine the actual thermal resistance of the sheep's fur from comparisons with the temperature measurements. With these data a thermoelectric (TEG) generator was built in a thermally optimized housing with adapted heats sink. The whole TEG system was mounted to a collar, including a data logger for recording temperature and TEG voltage. First measurements at the neck of a sheep were accomplished, with a calculated maximal average power output of 173 µW at the TEG. Taking the necessity of a low-voltage step-up converter into account, an electric output power of 54 µW is available which comes close to the power consumption of a low-power VHF tracking system.

1. Introduction

Today, wildlife monitoring is a subject of increasing interest from several reasons: Climate change as well as human intrusion into natural habitats and stopover sites takes a growing influence on migration paths, times and routines of migratory species. Some theories predict even a disappearance of animal migration [1] with drastic effects on ecology and human environment. On the other hand we see a growing intrusion of wildlife into densely populated areas, and, with that, the general necessity to combine human and wildlife environments in a suitable way (see e.g. [2]). Endangered species have to be tracked from obvious reasons. Also, livestock farming over large areas is essentially based on a wildlife-like migration, i.e. migration paths and actual positions of herds have to be monitored with a minimal amount of cost. Additional interest is on the retrieval of relevant physiological data and behaviour of livestock, as an indicator of e.g. grazing behaviour, medical condition and fertility [3,5]. This is also the scope of recent studies on an environmentally triggered behaviour of livestock, to be used as an indicator for impending natural disasters [4] or environmental catastrophes.

In all these scenarios one main challenge is to develop suitable methods and equipment for long-term observations of animals. While several more or less indirect methods (e.g. regular hunting, wildlife observation and counting) are available, position sensing with GPS transmitters - typically fitted on collars - is the most used method hereof [5]. As these systems are battery-powered today, their operational lifetime is limited and, in addition, determined by their functionality. Exhaustion of the battery requires a replacement of any collar. For that purpose the collared animal has to be found, trapped and, if required, sedated, with a considerable amount of time and cost. From ethical reasons a collar should be removed from the animal after its time of service. This can be accomplished via automatic release buckles that leave, however, the collar in the natural environment and require a cost-intensive recovery. Other concepts try to save on battery power via a drastic reduction of all wireless communication. Instead, acquired position data are stored in the collar's internal memory and accessed after removing the collar from the animal or after recovery from its release position.

Having all drawbacks of battery-based systems in mind, energy harvesting can be considered a reasonable alternative or addition to supply a wearable tracking system continuously with electric power. Among all conceivable concepts of energy harvesting (solar, motion, force,...) thermoelectric energy harvesting turns out to be advantageous from several reasons: (1) In almost every situation we may find small temperature differences between the body of an animal and ambient air, water, snow or ground. We have already shown in a previous study that harvesting from temperature gradients of only 1 K allows operating a wireless sensor node (WSN) with a low power RF interface [6]. (2) Mammals, frequently the target for wildlife tracking, regulate their body temperature thus providing a reliable heat source or sink (3) Motion as well as solar energy harvesting do heavily depend on animal activity and nesting behaviour which may be irregular and non-reliable for energy harvesting.

While thermoelectric energy harvesting for human applications is the focus of several publications [7,8] animal-based TEG systems are not found in literature. Therefore we have conducted a study with a small generator system to get deeper insight into these specific requirements and the potential of thermoelectric energy harvesting on smaller wildlife. To mimic the conditions of wildlife as close as possible all experiments were performed on a freely grazing sheep ("Heidschnucke").

2. Temperature measurements: Experimental set-up, equipment and results

In a first series of experiments the temperature difference between the animal and ambient air was determined. The corresponding measurement system uses two coin-sized temperature data loggers (DS1922L, Maxim). The data loggers were calibrated before use with a Pt 100 temperature sensor system as a reference. Two data loggers with almost identical calibration curves (deviation between 0,054 K and 0,069 K) were mounted onto both faces of a 20 mm wide nylon collar. The measurement interval and temperature resolution of both data loggers was set to 300 s and 0.0625 K, respectively. Fig. 1 shows, as a typical result, the temperatures at the inner and outer side of the collar over a period of 24 hours in May 2014, starting at midnight. In addition, temperature data of local weather observation are added to the graph. In this measurement graph the average temperature difference over the collar is 3.79 K. The slight fall of temperature during the day is reflected in the temperature data obtained from the sensor at the collar outside. In a first measurement run an average temperature difference of 2.5 +/- 0.39 K was found over a period of 14 days. As detrimental effect a certain heat feed-through from the inner side of the collar to the outer side could not be neglected. Moreover, it was observed, that the collar moved freely around the sheep's neck, which lead to variations in temperature. Therefore a bell was added to the collar in a second and third 2-week measurement period to keep the data loggers constantly in the same orientation and in close contact to the sheep's neck. As a result higher temperature differences of 3.53 +/- 0.33 K and 2.99 K ± 0,353 K, respectively, could be achieved in these two campaigns. The data shown in Fig. 1 are taken from the second measurement campaign with the highest gradients.

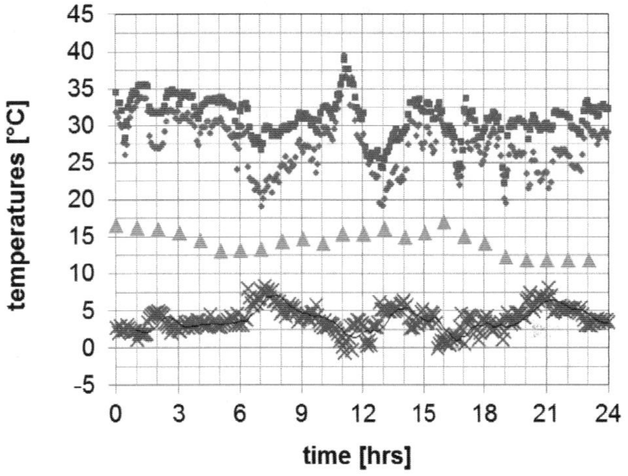

Fig. 1: 24-hour measurement of temperatures at the collar's inside (◆) and outside (■), calculated temperature difference (✖) over the collar, local weather data (▲)

3. TEG system design and heat flux model

As a result of the temperature measurements described above a design for a TEG system was set up according to Fig. 2. This system uses a small TEG with a footprint of 18 x 21 mm² (TEG-083-230-07, Thermalforce) in a circular polymer housing, with a heat sink as TEG-air interface (ICK S R 50x20, Fischer Elektronik) on its top and an aluminium thermal connector to the fur at its bottom. One of the pins of the heat sink was replaced by an alumina can with an embedded PT 1000 temperature sensor, to measure the air temperature at the heat sink. A second PT 1000 temperature sensor was embedded in the thermal connector to measure the temperature at the fur. In general, the fur of a mammal reduces the heat flux between skin and ambient and will therefore be a problem for thermoelectric energy harvesting. For that purpose the thermal connector to the fur was designed as protruding element to flatten the fur and to increase, by that, the thermal heat flux through the TEG. TEG and temperature sensors are electrically accessed via a common waterproof connector. A miniaturized 3-channel datalogger (uLOG, SparkFun) with three micro-power instrumentation amplifiers was built into a separate housing at the collar to record all data.

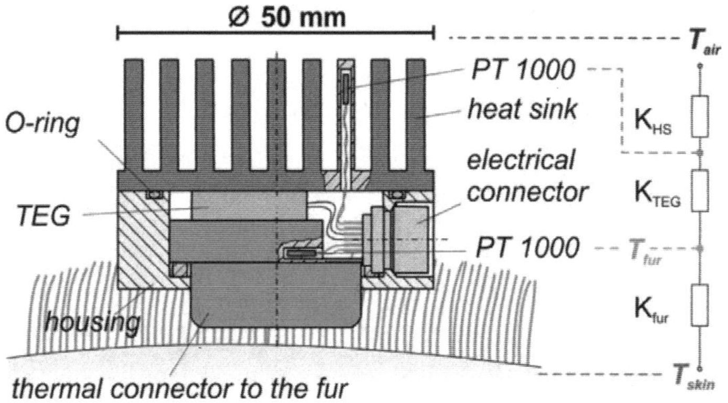

Fig. 2: Schematic of the TEG system with its thermal resistive divider (left), photograph (right)

In a simplified model, the electrical output power P of a TEG can be described with equ. (1) to

$$P = \underbrace{\left(\frac{K_{\text{TEG}}}{K_{\text{fur}} + K_{\text{TEG}} + K_{\text{HS}}}\right)^2}_{\textit{thermal interface}} \cdot \underbrace{\left[\alpha \cdot (T_{\text{skin}} - T_{\text{air}})\right]^2}_{\textit{Seebeck effect}^2} \cdot \underbrace{\frac{R_{\text{L}}}{(R_{\text{G}} + R_{\text{L}})^2}}_{\textit{electrical interface}} \qquad (1)$$

Here, α denotes the TEG's Seebeck coefficient, R_L and R_G the load resistance and the TEG's internal resistance, respectively. T_{skin} and T_{air} are the temperatures of skin and air (see Fig. 2). K_{fur}, K_{TEG} and K_{HS} form a thermal temperature divider with the thermal resistances of the fur, the TEG and the heat sink between TEG and ambient air, respectively, as shown in Fig. 2. While reliable data are available on K_{TEG} and K_{HS} , data for K_{fur} are lacking. In general, fur shows a highly variable thermal behavior depending on several physiological and environmental influences. Realistic values for K_{fur} were taken from [9,10,11] with a worst case and best case scenario (see table 1). These data show that the thermal resistance of the thermal connector (see Fig. 2) is always much smaller than K_{fur} and negligible.
With a temperature difference of T_{skin} - T_{air} = 10 K and for optimal load matching ($R_L = R_G$) equ. (1) would give an output power of 158 μW and 401 μW with an open circuit voltage of 48 mV and 76 mV,

Table 1. Model parameters for equ. (1)

parameter	**Value**
K_{HS}	5 K/W
K_{TEG}	20 K/W
K_{fur}	46 (best) and 88 (worst) K/W
α	27 mV/K
R_G	3.6 Ohm

depending on the value of K_{fur} . The temperature difference between the upper side of the fur, i.e. the bottom side of the TEG, and the ambient air would be 2.1 K and 3.5 K, respectively. Therefore, the measured average temperatures T_{fur} - T_{air} between 2.5 K and 3.5 K are approximately matching with an assumed total temperature difference T_{skin} - T_{air} of 10 K, with all uncertainties concerning K_{fur} in mind.

4. TEG system test under real-life conditions, system considerations

In another measurement campaign the TEG system was mounted to a collar and applied at the animal for a real-life test. Fig. 3 shows exemplary data obtained during a 48-hour period in October 2014, starting at 9:30 A.M. The open-circuit voltage of the TEG was calculated from the acquired temperature data to eliminate the influence of electronic noise that unfortunately had been present during the measurement. Also, air temperature data from a local weather station are included in the graph. The elevation of the TEG voltage between hour 17 and 24 presumably reflects the influence of the cold night, whereas the negative voltages in hour 25 show even a reversal of the temperature gradient during the warm morning. Influence of the colder, second night is not found. Obviously the animal has taken shelter against the lower temperature.

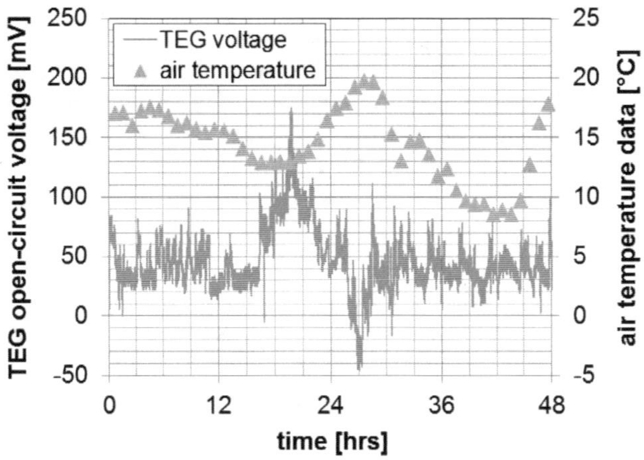

Fig. 3: calculated TEG open-circuit voltage during a 48-hour measurement (left), photograph of the sheep with mounted TEG and data logger (right)

These data have to be interpreted in a system context to evaluate the usefulness of the tested device: It has been shown in a previous publication that low output voltages of a TEG can be up-converted to higher, usable system voltages with specially tailored step-up converters [12] that run at an input voltage of 10 mV already and with a power conversion efficiency of appr. 33%. The system described in [12] exhibits a typical input resistance of 1.8 Ohm which acts as effective load resistance R_L for the TEG with an internal resistance of 3.6 Ohm. Therefore according to equation (1) a no-load input voltage of appr. 30 mV is required to start step-up conversion, i.e. all voltages below 30 mV and also all negative voltages will not be valid for a realistic power calculation. With that the output voltage in Fig. 3 would, under ideally load-matched conditions, yield an average output power of 173.9 μW. With the non-ideal load matching present here and after neglecting all voltages below +30 mV, the useful average output power of the TEG is 164.8 μW. The step-up converter with a power conversion efficiency of approximately 33 % [12] would deliver 54 μW of average electrical power to a wireless system. In total, the available electrical energy is approximately 10.2 Ws, harvested over 48 hrs.

The usefulness of this amount of energy has to be discussed with a view on potential applications:

A low-power and low range VHF tracker, the smallest unit used for animal tracking, will typically send out RF bursts with a lowest repetition rate of 30 pulses per minute [13]. A realistic power figure for such a transmitter would be 10 mW over a minimal burst time of 15 ms [13]. The average power consumption of the transmitter would then be 75 μW, which is already in the range of the small TEG system tested in this study.

5. Summary and conclusion

This paper presents the results of a feasibility study of energy-autonomous wildlife tracking, as far as the authors know for the first time. A German sheep ("Heidschnucke") has been used as a test animal due to its good availability and access, and also due to the similarity of its physiology to other, more relevant wild animals, e.g. deer. To mimic the conditions of wildlife as close as possible the sheep was kept freely grazing. In a first measurement series the temperatures at the inside and outside of the animal's collar were recorded. Typical temperature differences of 2 to 3 K were found in three measurement campaigns, each lasting over 14 days. In a second step a small commercial TEG was used as core element for a TEG system, made of a polymer housing, a temperature connector to the animal's fur and a heat sink as thermal interface to ambient air. Temperature sensors were integrated into the system for a determination of the TEG's hot side and cold side temperature. With this system an electric output power of 54 µW was obtained, sufficient for a low range VHF tracking system.

On a first sight the obtained electrical power seems to be too low to justify the effort for a TEG system. However, in this study a small TEG with a footprint of only 18 x 21 mm² has been used, to comply in any case with the requirement that the mass of a tracking system should not exceed 2 to 5 % of the animal's weight [13]. The majority of animals of interest, e.g. deer, reindeer, moose or bear, would allow for much larger devices with a much higher output power. Therefore, TEG systems come into reach that can either prolong the lifetime of a battery-operated wildlife tracker or take the role of the main power supply with a rechargeable battery for auxiliary power. In both cases the operational lifetime of the wildlife tracking system would be significantly prolonged. Also, a larger range of functionality could be conceived with, e.g., tracked animals used as mobile environmental sensors.

6. References

[1] Wilcove D S and Wikelski M 2008 Going, Going, Gone: Is Animal Migration Disappearing? *PLoS Biology* **6** (7) 1361-1364

[2] Schadt S, Knauer F, Kaczensky P, Revilla E, Wiegand T and Trepl L 2002 Rule-based assessment of suitable habitat and patch connectivity for the Eurasian lynx *Ecological Applications* **12** (5) 1469-1483

[3] Frost A R, Schofield C P, Beaulah S A, Mottram T T, Lines J A and Wathes C M 1997 A review of livestock monitoring and the need for integrated systems *Computers and Electronics in Agriculture* **17** 139-159

[4] Maier E 2014 A Four-Legged Early-Warning System *MaxPlanckResearch* **2** 58 - 63, retrieved from http://www.mpg.de/8252362/W004_Environment-Climate_058-063.pdf, October 2014

[5] Ungar E D, Henkin Z, Gutman M, Dolev A, Genizi A and Ganskopp D 2005 Inference of Animal Activity From GPS Collar Data on Free-Ranging Cattle *Rang. Ecol. Man.* **58** 256-266

[6] Moser A 2012 Thermoelectric Energy Harvesting from Transient Ambient Temperature Gradients *Doctoral thesis*, University of Freiburg, Germany

[7] Leonov V Energy Harvesting for Self-Powered Wearable Devices in: Bonfiglio A, De Rossi D (eds.) Wearable Monitoring Systems, Springer, pp. 27-42

[8] Vullers R J, Van Schaijk R, Visser H J, Penders J and Van Hoof C 2010 Energy Harvesting for Autonomous Wireless Sensor Networks *IEEE Solid-State Circuits Magazine* **2** (2) 29-38

[9] Cena K and Clark J A 1974 Heat Balance and Thermal Resistances of Sheep's Fleece *Physics in Medicine and Biology* **19**, 51-65

[10] Cena K and Monteith J L 1975 Transfer processes in animal coats II. Conduction and convection *Proceedings of the Royal Society of London. Series B, Biol. Sciences* **188** 395-411

[11] Keren E N and Olson B E 2006 Thermal balance of cattle grazing winter range: Model development *Journal of Thermal Biology* **31**, 371-377

[12] Woias P, Islam M, Heller S and Roth R 2013 A low-voltage boost converter using a forward converter with integrated Meissner oscillator *IOP Journal of Physics: Conf. Series* **476** 012081

[13] Markham A 2008 On a Wildlife Tracking and Telemetry System: A Wireless Network Approach *Doctoral thesis*, University of Cape Town, South Africa

Analytic modeling of a high temperature thermoelectric module for wireless sensors

J.E. Köhler, L.G.H. Staaf, A.E.C. Palmqvist and P. Enoksson

Chalmers University of Technology, 412 96 Göteborg, Sweden

E-mail: elof@chalmers.se

Abstract. A novel high temperature thermoelectric module with thermoelectric materials never before combined in a module is currently researched. The module placement in the cooling channels of a jet engine where the cold side will be cooled by high flow cooling air (550°C) and the hot side will be at the wall (800°C). The aim of the project is to drastically reduce the length of the wires by replacing wired sensors with wireless sensors and power these (3-10mW) with thermoelectric harvesters. To optimize the design for the temperature range and the environment an analytic model was constructed. Using known models for this purpose was not possible for this project, as many of the models have too many assumptions, e.g. that the temperature gradient is relatively low, that thick electrodes with very low resistance can be used, that the heat transfer through the base plates are perfect or that the aim of the design is to maximize the efficiency. The analytical model in this paper is a combination of several known models with the aim to examine what materials to use in this specific environment to achieve the highest possible specific power (mW/g).

1. Introduction

Measuring pressure, temperature, stress and acceleration is essential for safe operation of jet engines. For test engines in development this means thousands of sensors and kilometers of wire connecting the sensors. The aim of this project is to place a thermoelectric module in the middle and the back of a test jet engine where it is complicated to place sensors due to both high temperature and the need of extensive cable wiring. Replacing the wired sensors with less wired or wireless sensors, coupled to an energy harvester as power source, can drastically reduce the cable wiring and the weight in total by not using cables. A high temperature thermoelectric module with thermoelectric materials never before combined in a module is currently researched specifically for this purpose [1].

A thermoelectric energy harvester is a device that converts a heat gradient into electricity. This is usually done by connecting alternating p-type and n-type semiconductors with high Seebeck coefficients in parallel thermally and in series electrically. The Seebeck coefficient (S) is the generated voltage difference per degree temperature difference of a material. It is material- and temperature dependent with negative values for the n-type and positive for p-type semiconductors. Electrical conductivity (σ) and thermal conductivity (κ) are also important when considering materials for thermoelectric energy harvesting to reduce Joule heating and maintaining a high temperature gradient. These properties are combined into the materials figure-of-merit, zT, value (zT=$\sigma S^2 T/\kappa$) that gives a measure of the efficiency of the thermoelectric material.

In the hot parts of a jet engine the walls are cooled down with air through cooling channels. Placing thermoelectric energy harvesters inside the cooling channels would give a hot side of 800-900°C, and with cooling air of 500-600°C on the cold side this means that a thermoelectric module may achieve a temperature gradient of roughly 200°C, if the heat transfer through the module is low enough.

To optimize the design for the temperature range and the environment an analytic model was constructed. Using known models for this purpose was not possible for this project, as many of the models have too many assumptions, e.g. that the temperature gradient is relatively low, that thick electrodes with very low resistance can be used, that the heat transfer through the base plates are perfect or that the aim of the design is to maximize the efficiency. In this project the aim of the design is to maximize the specific power output (mW/g).

2. Analytical Model

The thermally conductive parts of the device are the legs, contacts, base plates and the surrounding material. As a first analysis, the thermal conductance and electrical resistance in the thermoelectric legs are described. Each leg has a thermal conductance

$$K_i = \frac{\kappa_i A_i}{\ell_i} \tag{1}$$

Where κ_i is the average thermal conductivity in the temperature range, A_i the area of the leg and ℓ_i the length of the leg, giving a total thermal conductance of n couples of legs and surrounding thermally parasitic material

$$K = n(\frac{\kappa_N A_N}{\ell_N} + \frac{\kappa_P A_P}{\ell_P}) + \frac{\kappa_{Pa} A_{Pa}}{\ell_{Pa}} \tag{2}$$

where index i is N for n-type and P for p-type semiconductor and Pa the surrounding thermally parasitic material. For the electrical resistance the legs are in series and with similar derivation as for thermal conductance this gives the total resistance

$$R = n((\frac{\sigma_N A_N}{\ell_N})^{-1} + (\frac{\sigma_P A_P}{\ell_P})^{-1}) \tag{3}$$

with σ the average electrical resistivity in the temperature range. The thermal conductance and electrical resistance can then be used to formulate an energy transport diagram [2]. In Figure 1 the energy transport to the leg-electrode junctions can be seen. For the hot junction this gives following energy balance

$$Q_H = P_{TE} + Q_{Cond} - Q_{Joule} = nS_{avg}T_H I + K\Delta T - \frac{I^2 R}{2} \tag{4}$$

where Q_H is the heat input, $Q_{Cond} = K\Delta T$ the conduction away from the hot junction, $Q_{Joule} = I^2R/2$ the resistive heating and $P_{TE} = nS_{avg}T_H I$ the electrical energy from the thermoelectric effect, where T_H is the temperature and S_{avg} the average Seebeck value for n-type and p-type material in the temperature interval, a simplification that is known to introduce only minimal errors [3]. For one couple the Seebeck effect induces an open circuit voltage $V = S_{avg}\Delta T$ which from Ohm's law gives the short circuit current $I = V_{oc}/R$. For n couples this gives

$$I_{sc} = nV_{oc}/nR = n \cdot S_{avg} \cdot \Delta T/R \tag{5}$$

However, to extract power from the module there has to be a load R_L attached. For maximum power and efficiency this load should be resistance matched with the module resistance, $R_L = R$. By introducing $\mu = R_L/R$ the power output for an n-coupled module is

$$P = (n \cdot S_{avg} \cdot \Delta T)^2 \mu / R(\mu + 1)^2 \tag{6}$$

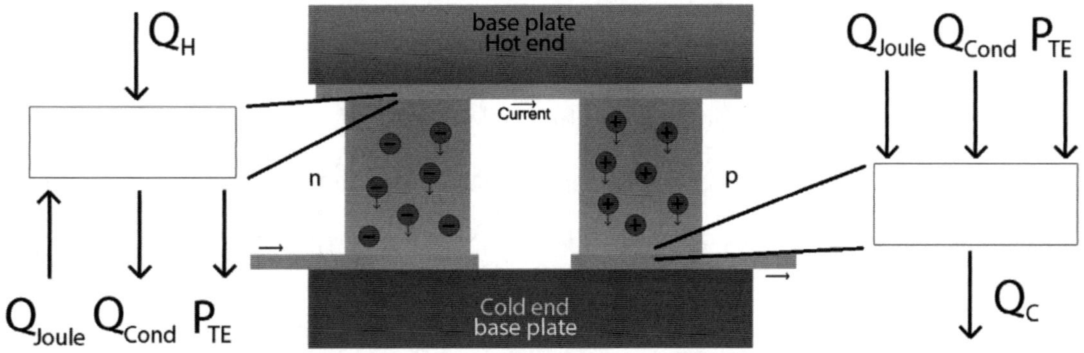

Figure 1. *Energy balance transport of thermoelectric module for hot junction (left) and cold junction (right).*

The simplification of perfect heat transfer through the base plates will now be corrected. Adding the two base plates to the calculations gives the total thermal resistance of the device.

$$R_{th} = \frac{1}{K} + \frac{2\ell_C}{\kappa_C A_C} \tag{7}$$

with the thermal conductivity of the plate material κ_C, the area A_C and the thickness of one base plate ℓ_C. From this the temperature on the legs can be calculated. This gives a new temperature difference, T_{diff}, between the hot- and cold side of the legs to be used in the calculations. One more aspect to consider are the electrodes and the contact resistance $R_{contact}$ [4]. Introducing $\delta = R_{contact}/R$ that we can add to the resistance in Equation 5 gives the device output power approximation

$$P = (nS_{avg}T_{diff})^2\mu/R(\mu + (1+\delta))^2 \tag{8}$$

Resistance of cables and the voltage amplification to 3.3V with a DC-DC step up booster [5] was also implemented in the model to match the requirements of the wireless sensors.

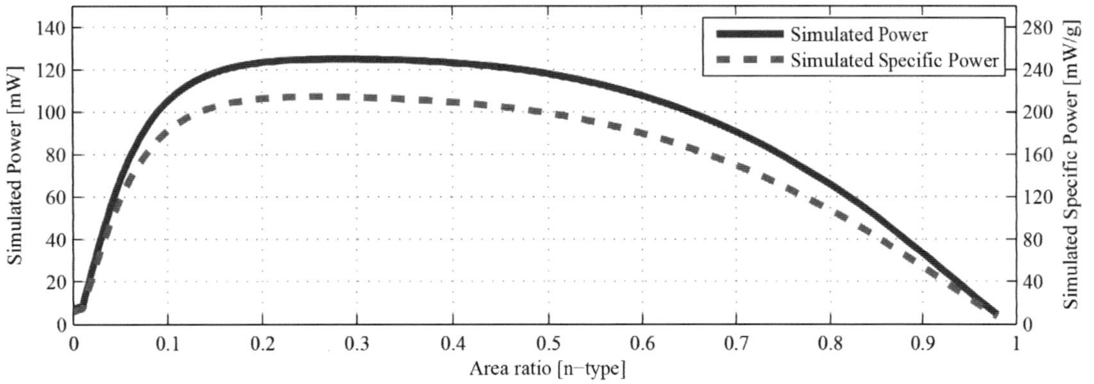

Figure 2. Estimated power output (blue, solid line) and specific power output (green, dashed line) of a 17-couple module as a function of the area ratio of the n-type:p-type material with a temperature gradient of 200°C. A DC-DC converter circuit with an efficiency of approximately 30% and 2m long wires with a resistance of 0.16Ω/m was included to the model.

3. Results

Some of the important variables examined by the model were the number of couples, area of legs, temperature gradient of environment, thermal conductance of base plates, area ratio between n-type and p-type material, thickness of electrodes, load resistance, parasitic heat conductance and several different materials for base plates, thermoelectric legs and electrodes, with the temperature dependence of all material properties included in the model. The simulation was done in temperature intervals set to 0.1°C and integrated over the whole temperature range (600-800°C). The fixed values was the module size of $1cm^2$ with the leg height set to 1mm, as this is the minimum height that we could handle with the current fabrication techniques.

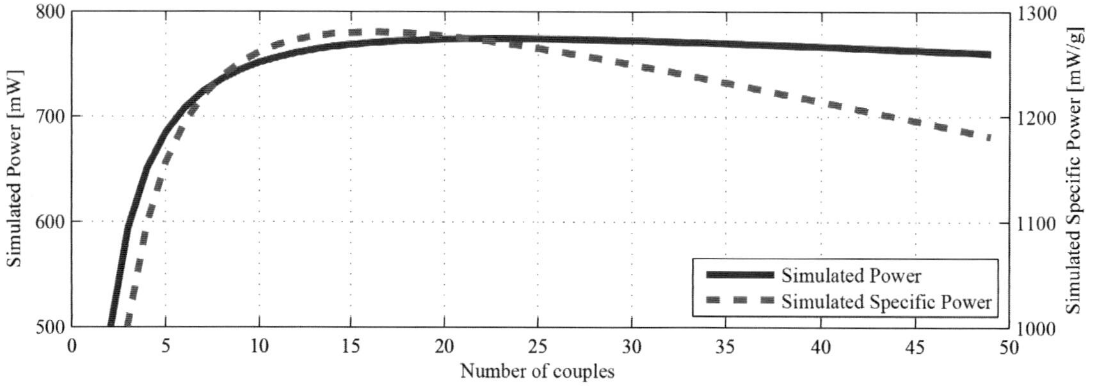

Figure 3. Simulated power output (blue, solid line) and specific power output (green, dashed line) of a thermoelectric module as a function of the number of couples with a temperature gradient of 200°C. No DC-DC converter circuit was included in this simulation and the load resistance was matching the module.

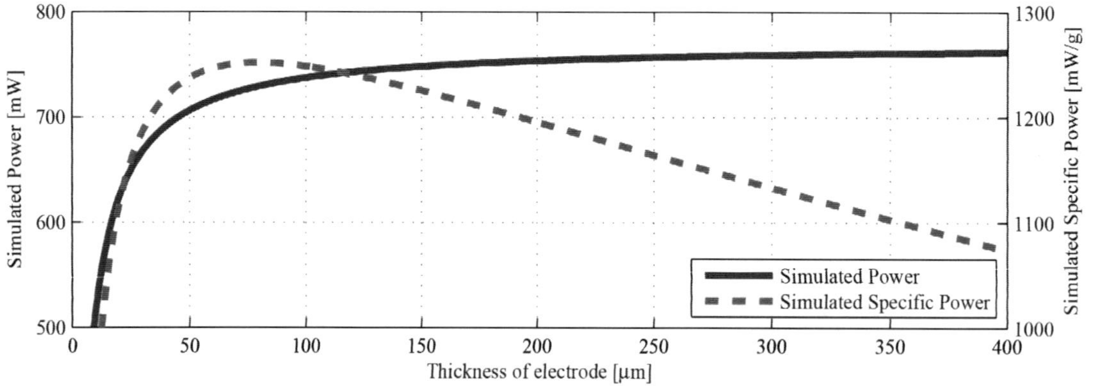

Figure 4. Estimated power output (blue, solid line) and specific power output (green, dashed line) of a 17-couple module as a function of the thickness of the Mo electrode with a temperature gradient of 200°C. No DC-DC converter circuit was included in this simulation and the load resistance was matching the module.

The highest specific power output after DC-DC conversion and cable losses was given with an n-type:p-type area ratio of 1:3.7 (Figure 2), 17 couples (Figure 3), 65m thick Mo electrodes

(Figure 4) with thermoelectric materials n-type $Ba_8Ga_{16}Ge_{30}$ and p-type La-doped $Yb_{14}MnSb_{11}$, alumina base plates (0.25mm thick) and the structure material (stability and sublimation repression) Cotronics Thermeez 7020. This gives, with an environmental temperature gradient of 200°C, a simulated specific power output of 215mW/g or 125mW/cm^2. When the load resistance is matched for a case without DC-DC conversion it gives a corresponding 1275mW/g or 771mW/cm^2.

A test module with one couple was fabricated and measured giving an open circuit voltage approximately 70% of simulated value, see Figure 5. The low voltage could be explained by the fact that the module, to reduce stress, was not attached to the heater or cold sink with any conductive paste.

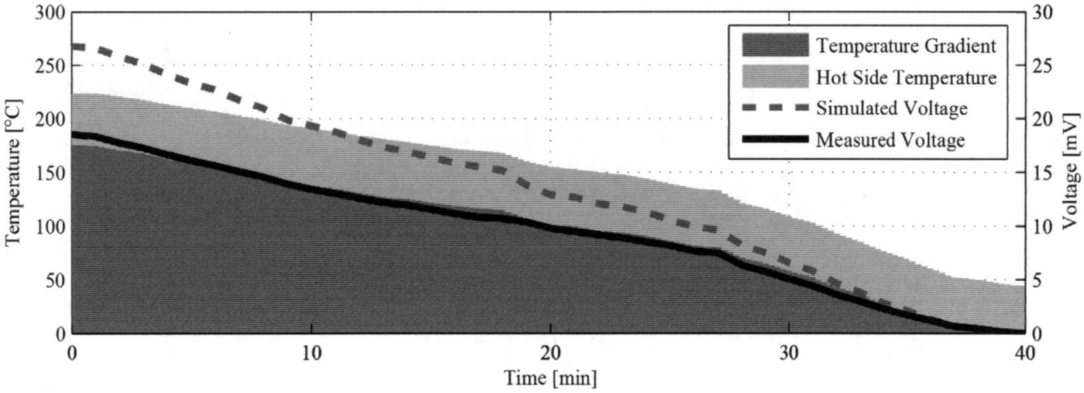

Figure 5. To verify the model, a 1-couple module was realized and measured. The figure shows the measurement and the simulated value with the measured value approximately 70% of the simulated value. Conductive paste was not used during the measurement which could explain the lower voltage.

4. Conclusion

The analytical model in this paper is a combination of several known models with the addition of DC-DC conversion and cable losses directly in the model with the aim to find the highest specific power instead of highest efficiency. For applications similar to the jet engine application mentioned in this paper where highest power-to-weight is wanted, this analytical model is very useful and give more suitable results than other known models.

5. Acknowledgement

The FP7 project Stargate is acknowledged for funding this project.

References

[1] J. E. Köhler, R. Heijl, L. G. H. Staaf, S. Zenkic, E. Svenman, A. Lindblom, A. E. C. Palmqvist and P. Enoksson *Smart Materials and Structures* **23** (2014) 095042

[2] A. Bitschi *Modelling of Thermoelectric Devices for Electric Power Generation* (Zurich: Swiss Federal Institute of Technology Zurich) (2009) ch 2

[3] S. A. Omer and D. G. Infield *Solar Energy Materials and Solar Cells* **53** (1998) 67-82

[4] D. M. Rowe *Thermoelectrics handbook:macro to nano* (Boca Raton: Taylor & Francis group) (2006) ch 39 p 1-15

[5] Linear Technology 2010 *LTC3108, Ultralow Voltage Step-Up Converter and Power Manager* LT 0813 Rev C Datasheet

Comparison of Five Topologies of Cantilever-based MEMS Piezoelectric Vibration Energy Harvesters

Y Jia and A A Seshia

Nanoscience Centre, University Cambridge, Cambridge, CB3 0FF, UK

E-mail: yj252@cam.ac.uk; aas41@cam.ac.uk

Abstract. In the realm of MEMS piezoelectric vibration energy harvesters, cantilever-based designs are by far the most popular. Despite being deceptively simple, the active piezoelectric area near the clamped end is able to accumulate maximum strain-generated-electrical-charge, while the free end is able to accommodate a proof mass without compromising the effective area of the piezoelectric generator since it experiences minimal strain anyway. While other contending designs do exist, this paper investigates five micro-cantilever (MC) topologies, namely: a plain MC, a tapered MC, a lined MC, a holed MC and a coupled MC, in order to assess their relative performance as an energy harvester. Although a classical straight and plain MC offers the largest active piezoelectric area, alternative MC designs can potentially offer higher average mechanical strain distribution for a given mechanical loading. Numerical simulation and experimental comparison of these 5 MCs (0.5 μ AlN on 10 μm Si) with the same practical dimensions of 500 μm and 2000 μm, suggest a cantilever with a coupled subsidiary cantilever yield the best power performance, closely followed by the classical plain topology.

1. Introduction

Cantilever-based topologies are the incumbent design of choice within the field of piezoelectric vibration energy harvesting (VEH), especially for MEMS harvesters [1]. This is primarily because the active piezoelectric area near the clamped end is able to accumulate strain energy, while the free end can house a stiff proof mass without significantly compromising the effective area of the piezoelectric generator since it experiences minimal strain anyway.

While a number of different examples of micro-cantilever (MC) designs for MEMS piezoelectric vibration energy harvesters have been previously explored [1], the most popular design is based on the classical rectangular plain MC topology [2] with variations in design usually driven by process constraints rather than design objectives. Nonetheless, alternative designs such as tapered cantilevers [3] have been explored in an attempt to achieve equal distribution of strain along the cantilever length. This paper demonstrates that systematic modifications in the specific topology of piezoelectric MEMS cantilever energy harvesters can result in significant differences in the output power response. Five different cantilever-based topologies are numerically and experimentally compared in order to rank the effectiveness and power performance of each design for the purpose of VEH.

2. Apparatus

Five different topologies implemented using an aluminium nitride(AlN)-on-silicon micromachining process with the same overall practical dimensions (2 mm by 0.5 mm) were investigated, as shown in figure 1, in an attempt to compare their relative power performance for vibration energy harvesting. Apart from the plain and the tapered designs, the other topologies included: a primary MC coupled with a subsidiary MC based on [4], a MC with etched holes throughout the surface as a means of stiffness reduction inspired by [5], as well as a MC formed by multiple parallel thin beams strongly coupled at the end.

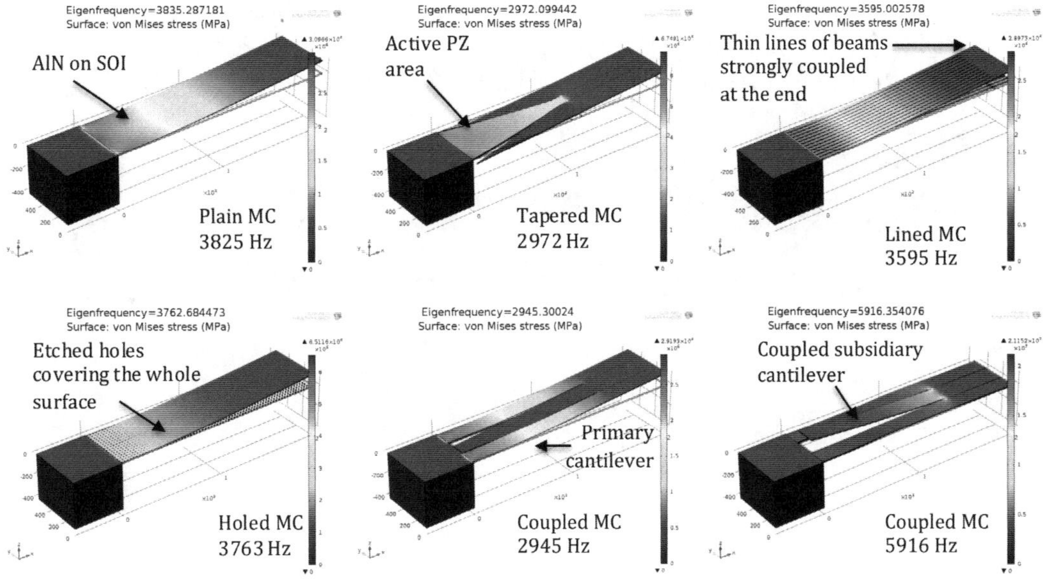

Figure 1. Designs of the five topologies of micro-cantilevers (MC), with the relevant resonant frequency information. Overall dimensions are the same: 2000 μm length and 500 μm width.

A summary of the 5 topologies and the employed acronyms are listed as the following,

- MCI: plain micro-cantilever (MC) with 2000 μm length and 500 μm width
- MCT: tapered MC with trapezium of 500 μm and 100 μm sides, and 1000 μm height.
- MCL: lined MC with 30 μm wide lines separated by 10 μm gap.
- MCH: holed MC with 5 μm square holes separated by 25 μm gap in both length and width.
- MCC: coupled MC with 1375 μm long and 180 μm wide subsidiary cantilever coupled onto a primary cantilever with 150 μm wide side beams.

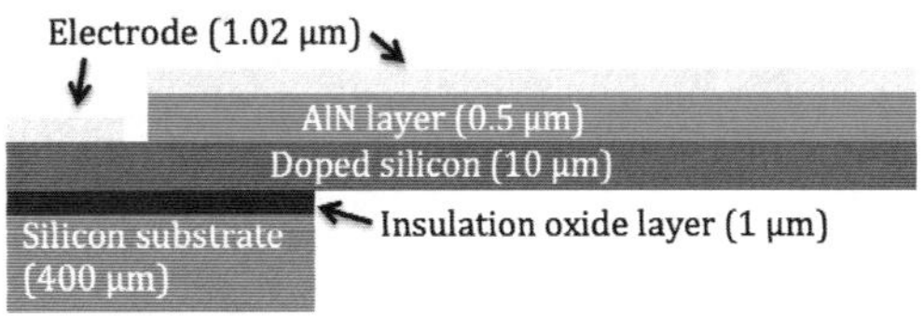

Figure 2. Stack of material used in the fabrication of the MEMS piezoelectric micro-cantilevers.

The experimental prototypes were manufactured using the material stack shown in figure 2 and the micrographs of the fabricated devices are presented in figure 3. For topologies with fine design features such as the MCL and the MCH, piezoelectric area is further reduced due to the need to provide clearance between the silicon, piezoelectric and the metal layers as well as over-etch of some of the small features during fabrication. Therefore, the actual sandwiched piezoelectric area is noticeably smaller than the silicon area for these designs.

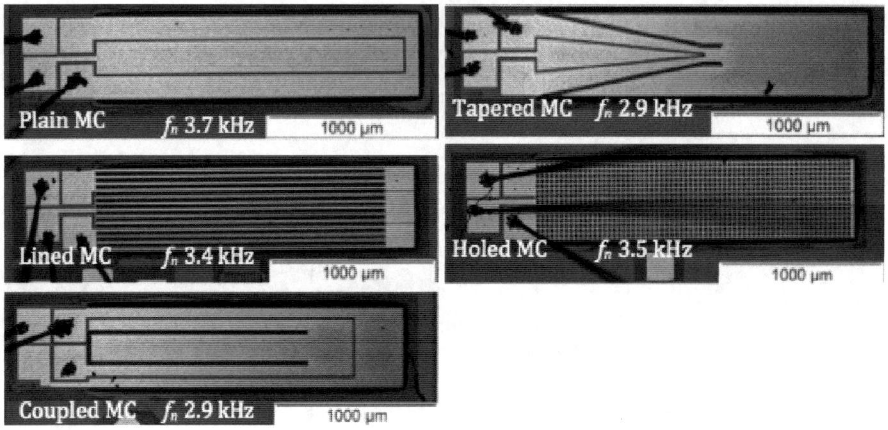

Figure 3. Micrographs of the 5 micro-cantilevers

3. Result

3.1. Simulation and calculation

COMSOL finite element analysis was used to evaluate the strain distribution of the 5 designs for a given loading. Figure 4 illustrates the strain distribution across the length of the active region of the MCs. The average strain of this distribution is shown in figure 5, demonstrating that the plain MCI fares the worst in terms of accumulating mechanical strain energy. It can be noted that MCT is the only one to have approximately constant strain distribution long the active piezoelectric area, while all others have a steady decreasing strain along the cantilever length.

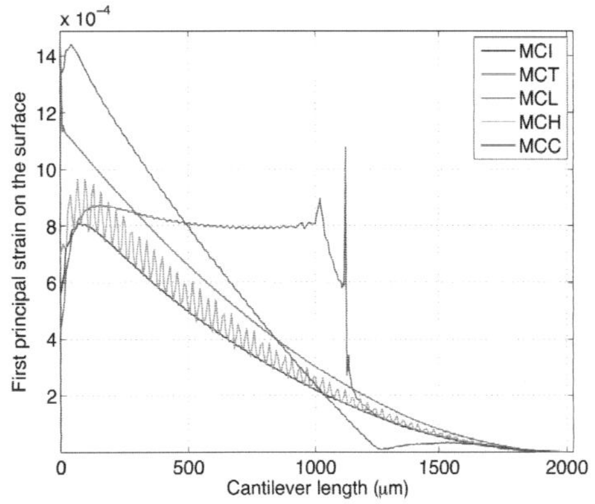

Figure 4. Strain distribution across the length of the 5 micro-cantilevers when subjected to a static loading of 1.5E-3 N. The origin of the x-axis is the clamped end.

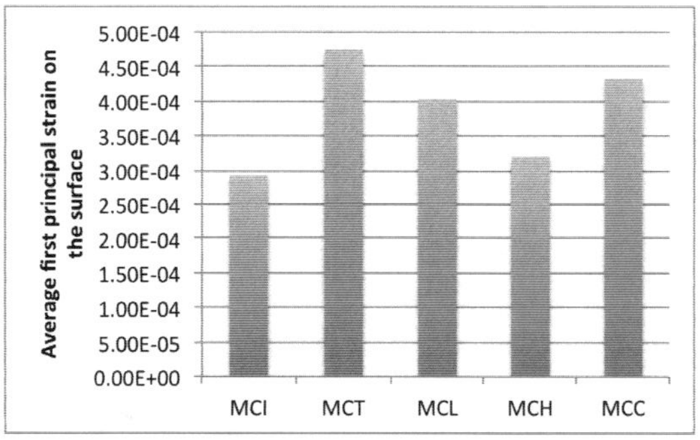

Figure 5. Average strain across the active transducer area of the cantilever length for the 5 micro-cantilevers when subjected to a static loading of 1.5E-3 N. While MCI ranks the lowest on this metric, it has the largest active piezoelectric transduction area.

While mechanical strain maximisation across the cantilever surface for a given input forcing is an important design metric, there is a trade-off involved in sacrificing the active piezoelectric area to accommodate the additional design complexity. Although all of the 4 alternative MC topologies are able to experience a higher mechanical strain for a given mechanical loading, the plain MC have by far the largest active piezoelectric area as shown in figure 6.

Along with the strain information extracted from COMSOL, electrical charge generated and peak power output can be estimated by equations 1 and 2 respectively.

$$q = d_{31}\varepsilon_{av}Ea_{pz} \tag{1}$$

where, q is the charge generated, d_{31} is the piezoelectric charge constant in the 31 mode, ε_{av} is the average strain experienced by the piezoelectric transducer, E is the elastic modulus of the piezoelectric material and a_{pz} is the active piezoelectric area.

$$P = \frac{\omega q^2}{C} \tag{2}$$

where, P is the peak power generated, ω is the frequency and C is the capacitance of the piezoelectric layer. Figure 7 shows the power predications for the 5 MCs when subjected to the same loading. MCH is predicted to perform the worst despite having a relatively high average strain, while MCC is estimated to perform the best and is closely followed by the classical MCI design. Therefore, the systematic optimisation of the power performance is primarily a compromise between the average mechanical strain and the active piezoelectric area.

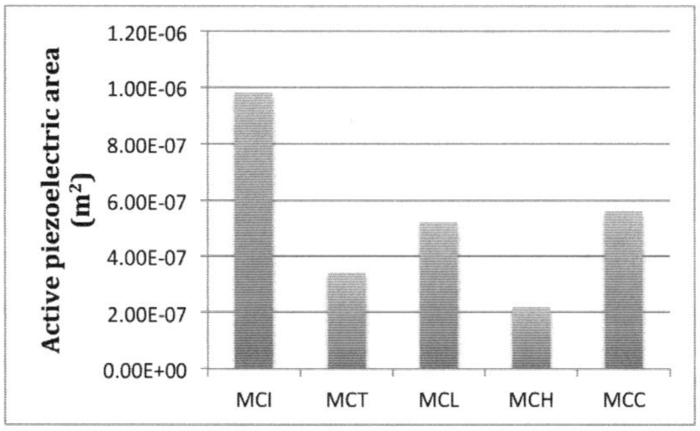

Figure 6. Calculated active piezoelectric area for the 5 micro-cantilevers when subjected to a static loading of 1.5E-3 N.

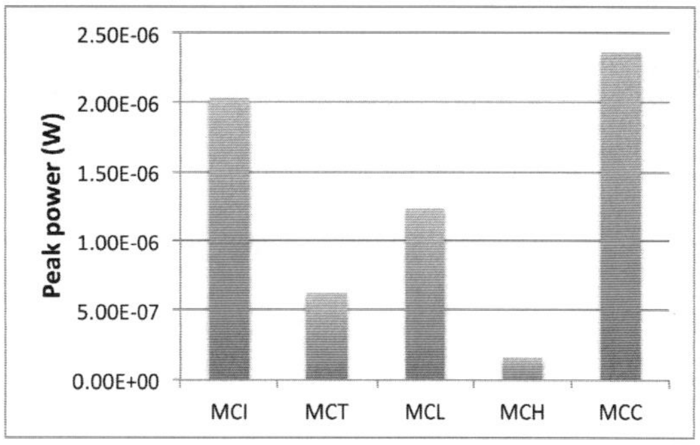

Figure 7. Calculated peak power for the 5 micro-cantilevers when subjected to a static loading of 1.5E-3 N.

3.2. Experiment

Experimental verification was carried out by electrically driving the piezoelectric transducers and measuring the mechanical motion by a laser doppler vibrometer (figure 8) to characterise the electrical-to-mechanical responsiveness, as well as exciting the prototypes on a mechanical shaker and measuring the voltage output across matched resistive loads at their resonant frequencies (figure 9) to determine the mechanical-to-electrical responsiveness.

The experimentally measured power output at 3 g of input acceleration for the 5 devices are: 1.01 nW for MCI, 0.45 nW for MCT, 0.7 nW for MCL, 0.03 nW for MCH and 1.35 nW for MCC. From figure 9, power values for MCH driven at lower acceleration levels were experimentally not measurable. The matched load resistance and natural frequencies for the various prototypes are: 0.3 MΩ at 3688 Hz for MCI, 1 MΩ at 2938 Hz for MCT, 0.6 MΩ at 3375 Hz for MCL, 1.2 MΩ at 3531 Hz for MCH and 0.6 MΩ at 2891 Hz for MCC.

The experimental results agree with the simulated predication of a relatively superior responsiveness from the MCC, closed followed by the MCI. MCH fares the worst, with MCL and MCT taking the 3rd and 4th positions respectively. The precise order of this power performance ranking is in agreement with the simulated results. Additionally, the MCC further benefits from

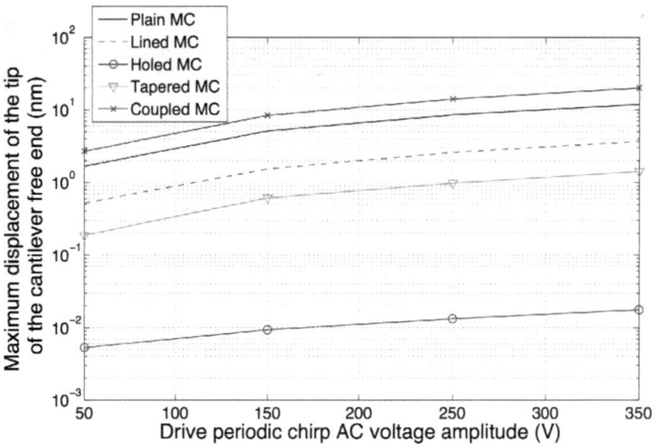

Figure 8. Laser vibrometer measured electrical-to- mechanical response (displacement). The coupled MC yielded the largest response.

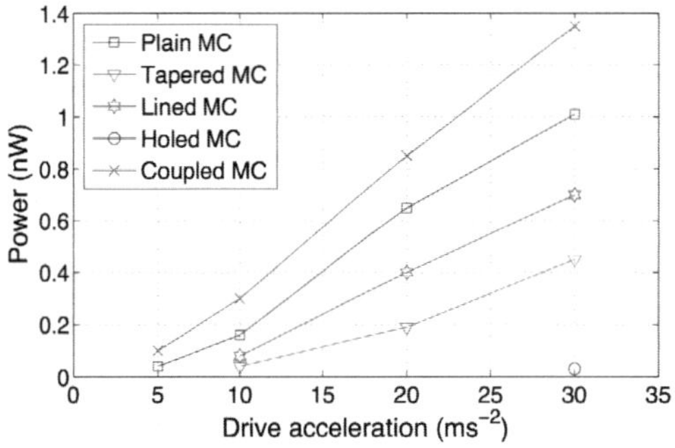

Figure 9. Experimentally measured power response of the mechanical-to-electrical response of the 5 MC. MCC produced the highest power responsiveness, with MCH being the worst.

dual frequency band sensitivity (from both its primary and subsidiary cantilevers), whereas the other 4 MCs only possess a single fundamental transverse mode.

Conclusion

Five cantilever-based topologies have been investigated for the purpose of vibration energy harvesting. Coupled micro-cantilevers ranked the highest in terms of power performance and was closely followed by the classical plain cantilever topology. While the tapered, lined and holed cantilevers all experienced higher average mechanical strain than the plain cantilever for a given loading, they performed worse due to the smaller active piezoelectric area. Therefore, a compromise is required while optimising designs between the maximisation of mechanical strain energy accumulation and the total active piezoelectric area.

References

[1] S. Priya and D. J. Inman, *Energy Harvesting Technologies*, Springer (2009).
[2] S. Roundy and P. Wright, *Smart Mater. Struct.*, 13 (2004), pp. 1131-1142.
[3] P. Glynne-Jones, S. Beeby and N. White, *IEE Proc. Sci. Meas. Technol.*, 148 (2001), pp. 68-72.
[4] Q. Tang and X. Li, *IEEE Transducers*, (2013), pp. 697-700.
[5] G. Sevilla, J. Rojas, S. Ahmed and A. Hussain, *IEEE Transducers*, (2013), pp. 2636-2639.

Characterization of piezoelectric material for micro thermal harvesters

E Trioux[1,2,3], S Monfray[1], T Skotnicki[1], S Basrour[2,3] and P Muralt[4]
[1] STMicroelectronics, Crolles, France
[2] Univ. Grenoble Alpes, TIMA Laboratory, F-38031 Grenoble, France
[3] CNRS, TIMA Laboratory, F-38031 Grenoble, France
[4] Ceramics Laboratory, EPFL, Lausanne, Switzerland

E-mail: emilie.trioux@st.com

Abstract. This paper presents the first realizations of a novel concept for thermal energy harvesting at micro scale. The devices proposed here are based on a two-step transduction combining thermo-mechanical and piezoelectric conversion. In this contribution, we present for the first time results on micro fabricated structures with integrated piezoelectric layers, focusing mainly on the characterization of the piezoelectric material. The process flow to get a bilayered bistable structure is briefly described, highlighting the way how to control the initial deflection. The characterization of the piezoelectric thin film is presented then. The $e_{31,f}$ coefficient is measured in both sensor and actuator mode and is found to be equal to -0,91 C.m^{-2} in both configurations. This value, close to the state-of-the-art, is very promising for the future thermal harvesting applications. Finally the buckling of a structure actuated by a voltage was observed and the corresponding displacement measured by laser interferometry.

1. Introduction

These last years have seen the development of many harvesting systems using all kind of surrounding energies to create electrical energy in order to power autonomously wireless sensor networks. The piezoelectric transduction is extensively used to harvest mechanical energy [1, 2]. Regarding thermal energy, a lot of efforts have been done in the conception and optimization of thermoelectric generators based on the Seebeck effect. In this paper are presented the first tiny devices combining both piezoelectric transduction and thermal harvesting. They represent an alternative to the use of thermoelectric materials to harvest thermal energy at microscopic scale using only CMOS standard processes for their fabrication. Indeed the thermoelectric generators need a disproportioned heat sink to maintain a temperature gradient across the material, which is limiting the downscaling of the devices, and the thermoelectric materials with the best efficiency are not CMOS compatible.

The working principle of the structures presented here has already been presented [3] at macroscopic scale. A bilayer bridge, initially curved downward, is put in contact with a hot source. The two layers of the structure have very different thermal expansion coefficients (CTE). As a consequence, huge thermal stresses are created when it is heated up. To release these stresses the structure buckles and becomes curved upward. Not anymore in contact with the hot source, the bridge is cooling down, and for the same reasons huge thermal stresses are generated, that lead to the inverse buckling of the structure. A thermal cycle with regular oscillations is then installed. The innovative idea of these microscopic devices is that the piezoelectric element, instead of being an external commercial buzzer that is impacted at every buckling by the structure, is directly integrated in the device. Indeed one of the two layers composing the bridge is a piezoelectric material. In this way, the piezoelectric transduction is converting the whole displacements of the structure in electrical charges and no more mechanical energy is lost during the impact on a buzzer. In this contribution the piezoelectric film will be fully characterized in sensor mode as well as in actuator mode on cantilever test structures, and the first buckling movement observed on the bridge structures is presented.

2. Description of the structures

The structures as shown in Figure 1 were fabricated. A first layer of 200nm of aluminium nitride (AlN) was deposited by reactive, pulsed, direct current magnetron sputtering on a silicon wafer with 1µm thick thermal silicon oxide on both sides. Then 100nm of platinum (Pt) was sputtered and patterned by dry etching to create the bottom electrodes of the devices (Figure 2). The active piezoelectric layer of AlN is sputtered on top of the patterned Pt electrodes, followed by one layer of aluminium (Al) which has a huge CTE (24 10^{-6} K^{-1}) compared to the CTE of AlN (4,5 10^{-6} K^{-1}).

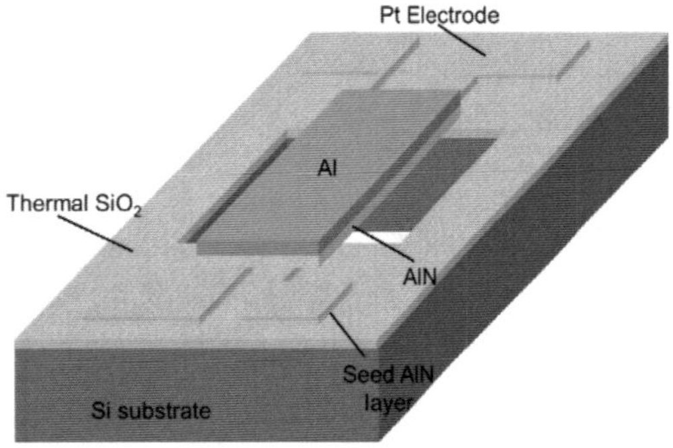

Figure 1. Scheme of a rectangular structure

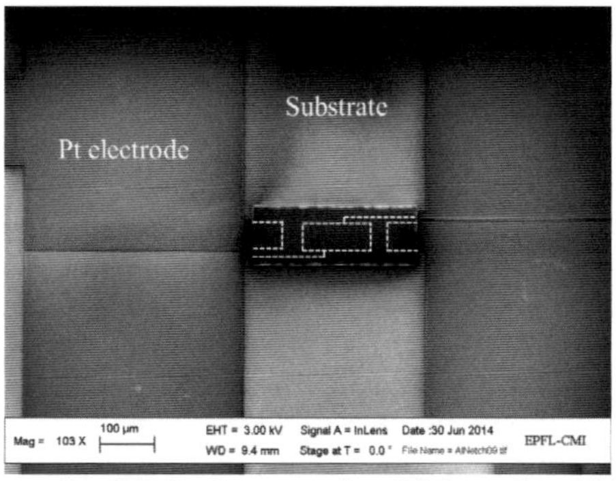

Figure 2. SEM picture of a 300µm by 100µm plate (dashed line: bottom Pt electrode)

Rectangular structures with different length to width ratio were fabricated, for instance 400µm by 400µm or 2000µm by 200µm in order to see the influence of this ratio on the buckling temperature. The design of the electrodes is realized in such a way to have two bottom Pt electrodes and one top, floating electrode in Al. In this way, the piezoelectric charges are collected both at the anchors where the stresses are higher and at the centre of the plate where the displacements are higher. The charges should not be compensated between the two electrodes, because the plate will always have a different sign of curvature at its centre and its anchors.

The initial curvature of the structures is obtained by tuning the deposition parameters of the different layers to adjust the residual stress. During the sputtering of the AlN layers, as explained in [4], it is possible to adjust the power density applied on the substrate to modify the number of collision and by consequence the stress induced in the thin film. The major drawback of tuning this parameter is that when the power density is reduced to decrease the stress in the film, the (100) crystalline orientation of the film is reduced. As a high quality for the AlN bottom layer is not required, the stress in this seed layer can be tuned. By adjusting the deposition temperature of the Al, the stress in the active AlN layer and in the Al layer can compensate. The Figure 3 illustrates a 400µm by 400µm device, with a maximum curvature at the centre of 4,5µm.

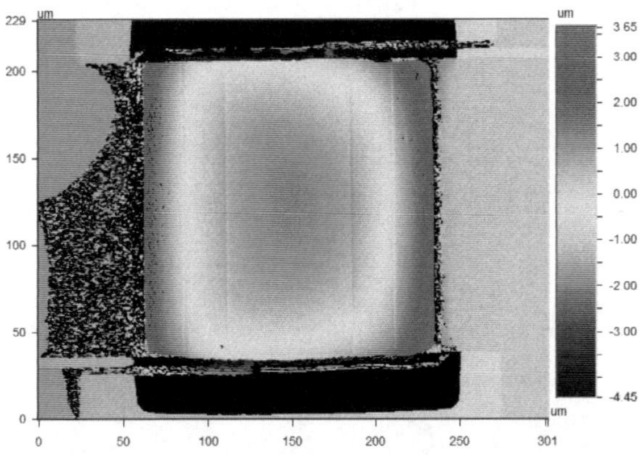

Figure 3. Picture of a 400μm by 400μm structure on the optical profilometer showing the final curvature after the back side opening

The curvature is an important parameter to tune the buckling temperature of the device, which increased directly with the deflection. Nevertheless this temperature is limited by the low Al melting point.

3. Characterization of the piezoelectric film

Before any experiment of harvesting, one has to be sure that the piezoelectric film is of good quality in order to have a high charge yield. Two different methods of characterization have been tested on test cantilever structures described in [5]. The cantilever has the full wafer thickness and the same stack as the rectangular structures presented in the former paragraph. The cantilever has standard dimensions (15mm by 1,5mm) as well as the top electrode, which defines the actuation/sensor area (3mm by 1mm).

3.1. Actuator mode

The test cantilever is first tested in actuator mode. The structure is clamped mechanically and a laser interferometer is focused on its free end to measure the displacements. A voltage is applied between the bottom and the top electrode. From the displacements measured at the free end, the stress induced in the piezoelectric film can be calculated as follow:

$$\sigma_1 = \frac{1}{3}\frac{Y_{sub}}{(1-v_{sub})b_f}\left[b_f + (1-b_f)(1-v_{sub})\right]\frac{w(x_2)}{x_1(2x_2-x_1)}\frac{t_{sub}^2}{t_p}$$

where Y_{sub} is the Young modulus of the substrate, υ_{sub} its Poisson ratio, b_f is the ratio between the cantilever width and the electrode width, x_1 is the length of the electrode and x_2 the measurement point of the interferometer, t_{sub} the thickness of the substrate and t_p the thickness of the piezoelectric layer, $w(x_2)$ the displacement along z-direction of the point x_2.

The stresses σ_1 generated in the x-direction by the actuation voltage are directly proportional to the piezoelectric coefficient e_{31f} according to this equation:

$$\sigma_1 = -e_{31,f}E_3$$

The Figure 4 shows the e_{31f} coefficient obtained for two different cantilevers, one being oriented along the (110) direction of the Si substrate, the other one along the (100) direction. The value for the Young modulus and the Poisson ratio were taken in [6]. The e_{31f} coefficients in both cases, respectively -1,042 C.m^{-2} and - 0,919 C.m^{-2} are comparable, they present only 10% difference.

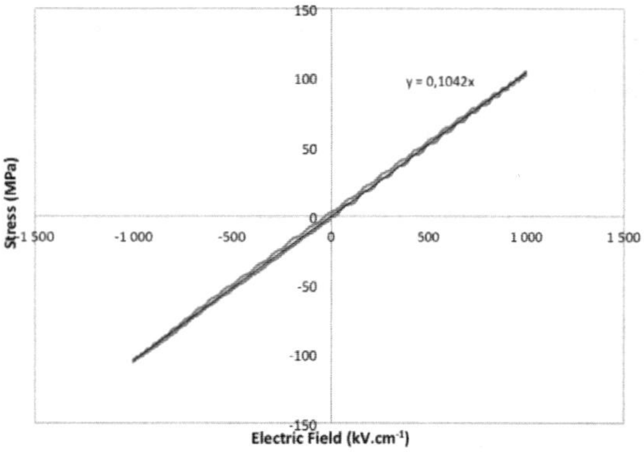

Figure 4. a) Evolution of the stress with the electric field across a 1µm thick AlN layer for a cantilever, oriented along the (110) direction of the Si substrate

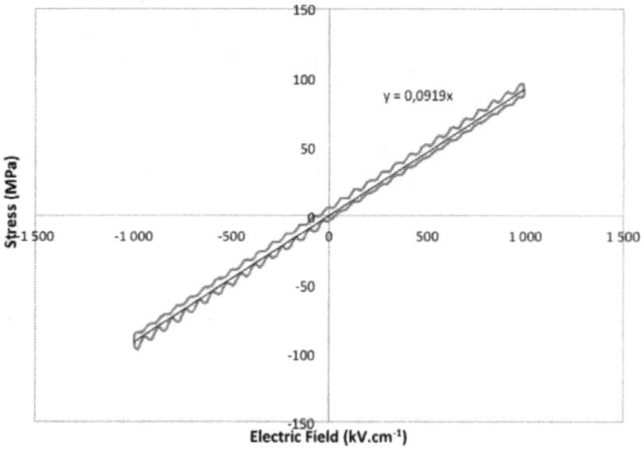

Figure 4. b) Evolution of the stress with the electric field across a 1µm thick AlN layer for a cantilever, oriented along the (100) direction of the Si substrate

These results are the first indication that the film is of high piezoelectric quality. Indeed the state of the art for the $e_{31,f}$ coefficient obtained with the same geometry of cantilever is -1,05 C.m-2, with a top electrode that was not patterned by dry etching but deposited by lift-off technique. The underneath piezoelectric layer is not damaged by the dry etching plasma; its quality is higher then.

3.2. Sensor mode

The cantilever test structures are then tested in sensor mode. The sample is put in the same conditions as before: it is clamped and two probes are contacting the top and bottom electrodes. An alumina tip is put in contact with the sample at a distance x_2 from the clamp. The tip displacement is controlled by a piezoelectric actuator to furnish exactly 4,5µm of displacement at a frequency of 110Hz.

The charges are collected by a charge amplifier and registered on an oscilloscope. Using the same technique as in [3], the strain corresponding to this displacement was found to be $2{,}4.10^{-5}$, and the $e_{31,f}$ value was -0,89 $C.m^{-2}$ for a sample oriented along the (110) direction. By changing the charge amplifier by a instrumentation amplifier, one can measure directly the voltage generated by the displacement of the free end of the cantilever, and get the $h_{31,f}$ coefficient, which was $-4{,}1.10^{9}$ $V.m^{-1}$. For energy harvesting systems one can use the following figure of merit (FOM) as described in [4]. For these samples the FOM was of 3,2 $GJ.m^{-3}$, which is not as good as non-patterned AlN film (around 11 $GJ.m^{-3}$) and PZT thin films (18 $GJ.m^{-3}$).

This difference could one more time be explained by the micromachining of the top electrode, which can damage the underlying piezoelectric layer.

4. First buckling in actuator mode

In this section, the buckling of a rectangular device of 1400µm long by 700µm width is presented. A voltage is applied between the two Pt bottom electrodes to actuate the structure. The displacement is measured with the same laser interferometer setup used to characterize the piezoelectric film on the cantilever.

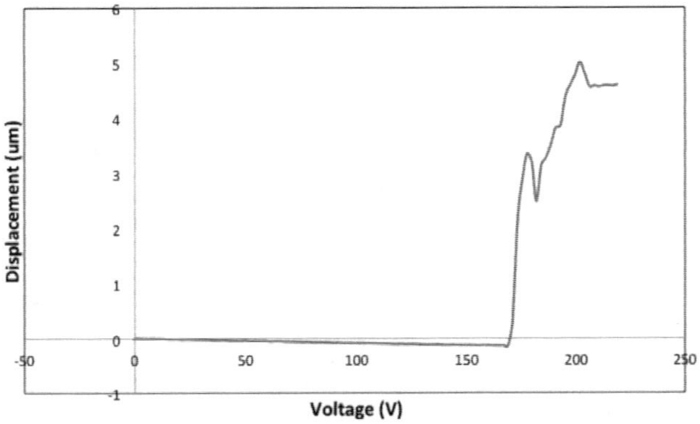

Figure 5. Displacement and buckling of a 1400µm by 700µm bridge when applying an actuation voltage

The device presented here is 1400µm long by 700µm wide, and had an initial deflection of 6µm downward. A slight movement in the direction of the curvature can be observed at low voltage, but suddenly at 170V, the plate change curvature and a huge displacement of 5µm is measured. The applied voltage necessary to cause buckling is high due to non-optimized structures. Indeed devices with higher initial curvature are stiffer and would present a higher buckling voltage. Also, as the device has a high output impedance, an impedance matching circuit can be designed in order to decrease considerably the applied voltage.

5. Conclusion

In this paper the first structures presenting a novel concept of thermal energy harvesting at microscale are presented. After a brief description of the working principle as well as the fabrication process, the characterization of the piezoelectric film is detailed. Two kinds of test were performed in order to measure the performance of the thin film both in sensor and actuator mode. The film quality was found to be good, the e_{31f} coefficients were between -1,04 C.m^{-2} in actuator mode and -0,92 C.m^{-2} in sensor mode, which hits well the literature values for AlN thin films. On the other hand, its quality is not optimal because of the fabrication process of the top electrode, which is dry etched. This process and especially the plasma used to etch Al (Cl_2 and BCl_3) is damaging the AlN layer underneath. Finally an actuation voltage is applied to one rectangular structure until it buckles. A displacement of 5µm is achieved. The challenge now is to cause the buckling thermally, by local heating. One way to do it could be to apply a current through the continuous platinum path and heat the device by Joule effect, or to use a powerful laser and to focus it on the Al layer.

6. References

[1] Marzencki M, Defosseux M and Basrour S 2009 *J. Microelectromech. Syst.* **18** 1444
[2] Isarakorn D et al. 2001 *Smart Mater. Struct.* **20** 25015
[3] Puscasu O et al., 2012 *IEEE Int. Electron Devices Meeting* San Francisco USA 291
[4] Dubois M A and Muralt P, 2001 *J. Appl. Phys.* **89** 11
[5] Mazzalai A, Balma D, Chidambaram N, Matloub R and Muralt P, 2014 *J. Microelectromech Syst.* **99**
[6] Hopcroft M A, Nix W D and Kenny T W, 2010 *J. Microelectromech. Syst.* **19** 2

New nonlinear vibration energy harvesters based on PVDF hybrid fluid diaphragm

F.Huet, F.Formosa and A.Badel
Univ. Savoie. SYMME, F-74000 Annecy, FRANCE
E-mail: Florian.Huet@univ-savoie.fr

Résumé. A low resonance frequency piezoelectric energy harvesting using a hybrid fluid diaphragm (HFD) is presented. This paper describes the design, fabrication and measurement of such device for harvesting energy from environmental vibrations. The HFD consists in an incompressible fluid confined between two thin piezoelectric membranes. The output voltage and power of the PVDF HFD are studied based on experimental and simulation results. Compared with conventional vibration harvester, this proposed solution is very simple and suitable for miniaturization and integration.

1 Introduction

An increasing interest in wireless sensors networks has been growing up over recent years. The implementation of communicating sensors networks in industrial application, transport or buildings offers possible improvements : productivity gains, reliability and energy performance are feasible by the exploitation of the extensive collected information (temperature, humidity, electric consumption ...). With the rapid advancement of low-power wireless sensor nodes and MEMS technology, microscale energy harvesting has attracted worldwide research interests for powering [1]. Vibration-based energy harvesting has been exploited using transduction mechanisms including piezoelectric, electromagnetic and electrostatic [2, 3]. In order to extract the maximum power from the environment, the resonant frequency of a linear harvesting device has to match the dominant frequency of the excitation. Indeed frequencies of ambient available vibration sources are relatively low (normally less than 200 Hz) [4]. As a result, many vibration-based piezoelectric energy harvesters of centimeter scale have been demonstrated to operate at low frequencies [5]. This paper reports the modelling, fabrication and testing of a new piezoelectric fluid-membrane structure for vibration energy harvesting. Compared to classical cantilever architectures, this approach takes advantage of the mechanical stretch of circular diaphragms. However, when miniaturized, they present high resonance frequency [6] and it hinders their direct use. The concept of hybrid fluid diaphragm (HFD) consists in an incompressible fluid confined between two thin piezoelectric membranes, it allows the realization of low resonance frequency membranes. Moreover, Pressure fluctuation harvesting [7] can be especially targeted.

2 Piezoelectric HFD concept

We propose here, an innovative architecture for vibration energy harvesters. Fig.1 describes the geometry as well as the main geometrical parameters of the structure. The incompressible fluid is used as an inertial mass which allows a drastic reduction of the resonance frequency to effectively harvest ambient vibrations. The height of the structure can be easily tunned to provide more or less inertia effect. In a previously published work, the mechanical behavior of a HFD was presented [8]. As an example of the frequency reduction, the use of glycerin as the inertial fluid allowed the resonance frequency of Kapton® (R = 10 mm, H = 10 mm and h = 100 μm) membranes to decrease from 1200 to 180 Hz.

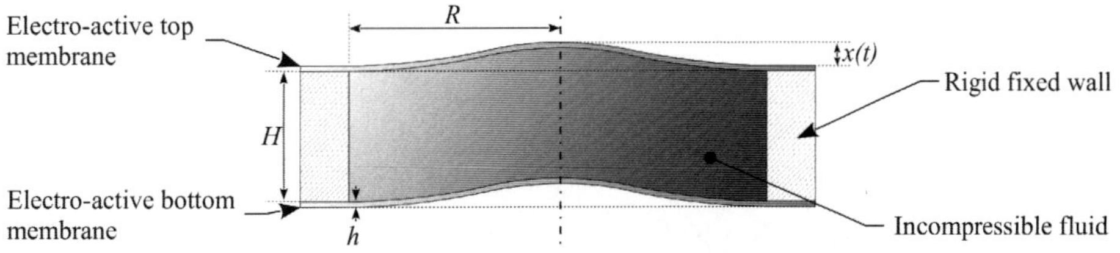

FIGURE 1. Architecture of the piezoelectric HFD

2.1 Theoretical model

A comprehensive theoretical model was proposed in [8]. Based on the Ritz method, approximate solutions for the fluid and the mechanicals diaphragms behaviours are set. This prior model is modified and extended here. Because, the final investigated dynamic behaviour of the HFD in the vicinity of its first resonance frequency is notably lower than the first natural frequency of a single diaphragm, it is thought that a simple static behaviour of the latter can be used for approximating the solution. By doing this, the model is simplified. Moreover, the electromechanical nature of the piezoelectric membranes is added in the model through the usual linear piezoelectric constitutive relations. The modelling strategy is developed using the following assumptions :

- Ideal geometry and homogeneous material properties
- Incompressible and inviscid fluid
- Elastic cylindrical diaphragms are considered
- The prestress effect of the diaphragms is taken into account
- Large amplitude deformation
- The analysis is restricted to axially symmetrical motions
- The analysis is restricted to in-phase motion of the top and bottom diaphragms

As a result, a simple lumped model can be derived presented by the schematic representation in Fig.2 and Eq.1-2. This system is constituted by a moving fluid mass (m_f) tied to a vertical spring ($k + n$) standing for the flexural diaphragm stiffness and two springs (b) standing for the non-linear stretching associated to large amplitude motions. The piezoelectric conversion (α) is associated with the latter (Fig.2a). The diaphragm deformationsinduce a voltage between the diaphragm electrodes. The input force is $m_i\gamma$ with γ the frame acceleration. The damping coefficient (c) embodies all the mechanical losses (the viscous losses within the membrane and the fluid dissipation). The piezoelectric generator behaves as a voltage source whose value is proportional to the product of the diaphragm bending and velocity (Fig.2b).

$$m_f\ddot{x} + c\dot{x} + (k+n)x + 2bx^3 + \alpha\frac{x}{h}V = m_i\gamma \tag{1}$$

$$\alpha\frac{x}{h}\dot{x} - C\dot{V} = I \tag{2}$$

If the displacement is small, the nonlinear behavior effects are minor. Because a single layer of PVDF is used, during a flexural motion, the same amount of PVDF film is compressed and

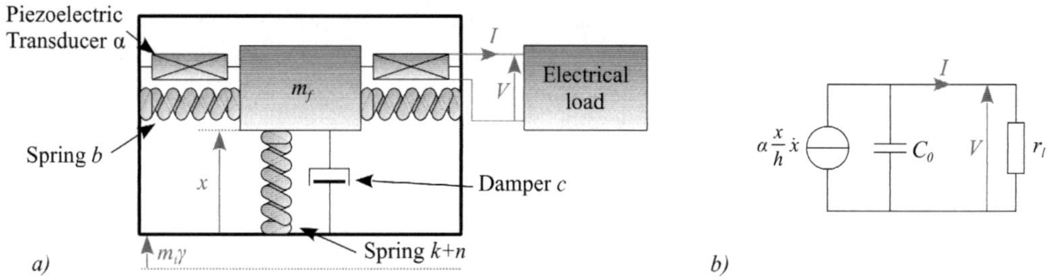

FIGURE 2. Non-linear inertial generator a) mechanical model and b) electrical model

stretched across the thickness so the global generated voltage is zero. From the previous equations, the voltage generation due to large deformation of the membranes is obvious.Is is related to the stretching of the whole section. Table 1 details the lumped parameters, their units and relationship to the geometry and material properties.

TABLE 1. Model Parameters

Parameter	Symbol	Value	Unit
Fluid inertial mass	m_f	See [8] $f(\rho, H, R)$	kg
Fluid excitation mass	m_i	$\frac{1}{3}H\pi\rho R^2$	kg
Flexural stiffness	k	$\frac{-16Eh^3\pi}{9R^2(1-\nu^2)}$	N/m
Large deformation stiffening coefficient	b	$\frac{Eh\pi(-7505-4250\nu+2791\nu^2)}{19845R^2(1-\nu^2)}$	N/m^3
Electromechanical coupling coefficient	α	$\frac{2d_{31}Eh\pi}{3(1-\nu)}$	N/V
Diaphragm capacitance	C	$\varepsilon_{33}\varepsilon_0\frac{\pi R^2}{h}$	F
Prestress flexural stiffness	n	Measured	N/m
Damping	c	Measured	N.m/s^2

2.2 Prototype

A HFD prototype has been realized aiming at the validation of the proposed model. Fig.3a introduces a section of the HFD design. Fig.3b shows the prototype realized in PVC. The first step for the assembly procedure is the clamping of the cylinders to obtain two equal halves of the final HFD. This operation is realized without fluid. When tightening, the membranes are stretched over a low shoulder to get a flat surface producing mechanical prestress (coefficient n in Eq.1). The next step is to put the two half-HFDs together plunged in a fluid to get a fluid-filled HFD without bubbles to ensure incompressibility of the cavity. The tightness between the half-HFDs is assured by a seal that can be seen in Fig.3a. It is also possible to adjust the internal pressure, and consequently the flatness of the diaphragms, with a small tuning screw on one half-body (Fig.3b). The film used to make the diaphragms was home-made from a solution of Poly(VinyliDeneFluoride-co-TriFluoroEthylen) also noted as P(VDF-TrFE). It is a copolymer with 70 % of VDF and 30 % of TrFE with piezoelectric property. The mechanical and piezoelectric properties of the fabricated films are similar to industrial ones. The main advantage of this choice is the tailor-made structures. The thickness h of the films is 80 μm with variations of some microns. For the fluid, we used glycerin for its high density.

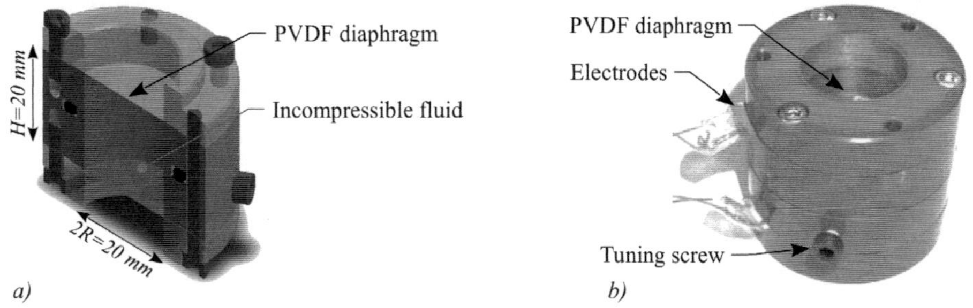

FIGURE 3. HFD a) design and b) prototype

3 Experimental and theoretical results

The piezoelectric HFD was tested on an electro-dynamic shaker driven with a slow varying increasing frequency sweep from 80 Hz up to 140 Hz. The first resonance frequency of the system is 119.6 Hz. Using an accelerometer and a dedicated control unit, the shaker is driven using a closed-loop control to ensure constant acceleration amplitude throughout the frequency range. The displacement and the velocity of the top diaphragm center are measured using a differential laser vibrometer. In parallel, the control unit drives a set of electrical load resistances (r_l). The generated power can be rigorously assessed varying the load between 0.8 MΩ and 14 MΩ. Several tests were performed for different acceleration levels (around 40 m/s^2) allowing the nonlinear behavior to be triggered.

3.1 Temporal results

The model highlights a frequency doubling voltage, which is clearly visible comparing Fig.4a (displacement of the diaphragm center at 119.6 Hz) and Fig.4b (voltage). Two dominant harmonics are shown performing the Fourier transform of the voltage signal (Fig.4c). The first is at the excitation frequency and the second is about the double (224.2 Hz). This first frequency shows the voltage produced by the membrane bending and the second from the nonlinear behavior. This observation highlight a non-homogeneity in the membranes thickness,which was not taken into account in the model.

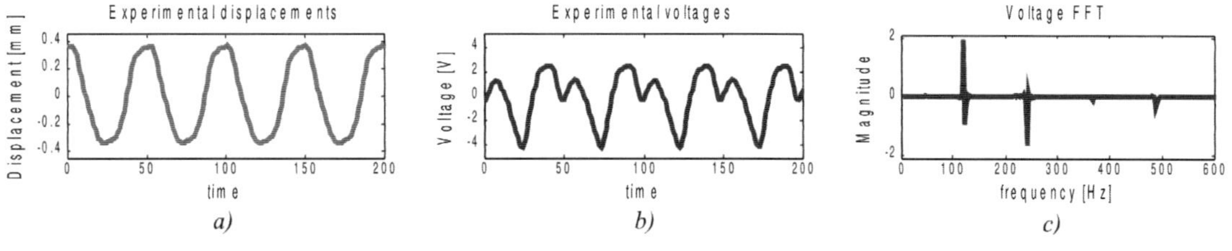

FIGURE 4. Experimental results a) displacements, b) voltages during the time for 119 Hz at 40 m/s^2 and c) voltage Fourier transform of the voltage signal

3.2 Experimental results and simulation comparison

Experiments using frequency sweep and various load resistances have been performed and the results (red grid in Fig.5) are compared to simulation results (color scale). Fig.5a shows the displacements with respect to the excitation frequencies and load resistances. The nonlinear effects associated with large displacement are very marked. The electric conversion has a negligible impact on the displacement, reflecting a weak electromechanical coupling.

The displacement amplitude at resonance frequency is 376 μm. Fig.5b shows the power, whose maximum is about 4.18 μW at the resonance frequency and for a load resistance of 2.88 MΩ. However the mechanical stiffening effect allows to have a high bandwidth, about 41.58 Hz. The simulations show a good correlation with the experimental results near the peak of power, however for other results furthest from the peak, we find the voltage error induced by the suspected membrane inhomogeneity.

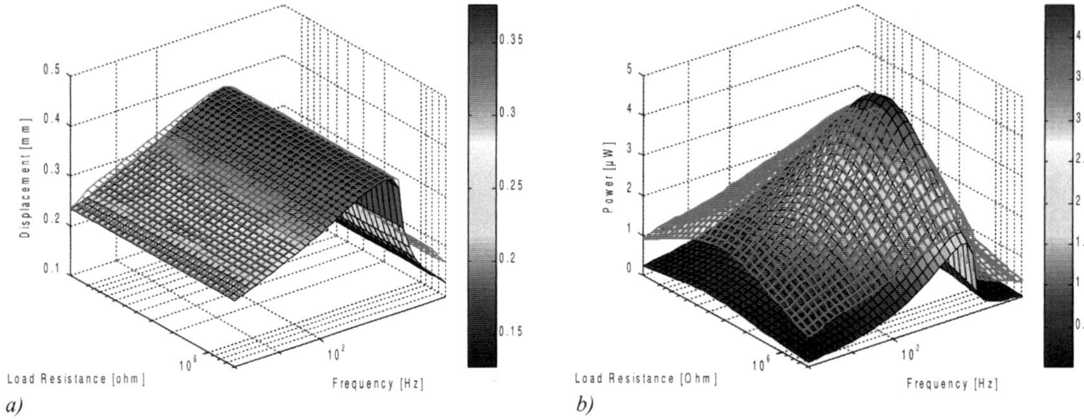

FIGURE 5. Experimental and simulation results comparison, a) displacements and b) power

4 Conclusion

A complete model of a piezoelectric inertial generator has been developed and a prototype realized. Simulation results showed that the voltage from the flexural mechanical effect is not negligible when the PVDF film is not perfect, but in the immediate area of the maximum power, the model is rather predictive and it correlates well. The generator performances are low but it has a very wide bandwidth and the first resonance frequency is about 100 Hz. Now the model can be used as a design tools to aim at a set requirements from a given applications. A future approach is to double the improved PVDF films. By reversing the polarity, this will fully exploit the voltage from the diaphragms flexion allowing performance gain.

References

[1] J. A. Paradiso, "Energy Scavenging for," *IEEE Pervasive Comput*, vol. 4, no. 1, pp. 18–27, 2005.

[2] S. P. Beeby, M. J. Tudor, and N. M. White, "Energy harvesting vibration sources for microsystems applications," *Measurement Science and Technology*, vol. 17, pp. 175–195, Dec. 2006.

[3] B. P. D. Mitcheson, M. Ieee, E. M. Yeatman, S. M. Ieee, G. K. Rao, S. M. Ieee, A. S. Holmes, M. Ieee, T. C. Green, and S. M. Ieee, "Human and Machine Motion for Wireless Electronic Devices," *Proc. IEEE*, vol. 96, no. 9, pp. 1457–1486, 2008.

[4] S. Roundy, P. K. Wright, and J. Rabaey, "A study of low level vibrations as a power source for wireless sensor nodes," *Comput. Commun.,*, vol. 26, no. 11, pp. 1131–1144, 2003.

[5] J.-Q. Liu, H.-B. Fang, Z.-Y. Xu, X.-H. Mao, X.-C. Shen, D. Chen, H. Liao, and B.-C. Cai, "A MEMS-based piezoelectric power generator array for vibration energy harvesting," *Microelectronics Journal*, vol. 39, pp. 802–806, May 2008.

[6] X.-r. Chen, T.-q. Yang, W. Wang, and X. Yao, "Vibration energy harvesting with a clamped piezoelectric circular diaphragm," *Ceramics International*, vol. 38, pp. 271–274, Jan. 2012.

[7] C. Mo, L. J. Radziemski, and W. W. Clark, "Experimental validation of energy harvesting performance for pressure-loaded piezoelectric circular diaphragms," *Smart Materials and Structures*, vol. 19, p. 075010, July 2010.

[8] F. Formosa, A. Badel, and H. Favrelière, "Development of low frequency, insulating thick diaphragms for power MEMS applications," *Sensors and Actuators A : Physical*, vol. 189, pp. 370–379, Jan. 2013.

Modeling and experimental characterization of a fluttering windbelt for energy harvesting

E Arroyo, S Foong and K L Wood

Singapore University of Technology and Design, 20 Dover Drive, 138682 Singapore

E-mail: e_arroyo@sutd.edu.sg

Abstract. Wind energy harvesters based on fluttering offer a valuable and efficient alternative to the traditional wind turbines. A longer life expectancy and cheaper fabrication is attained through the absence of gears or bearings. This article presents the theoretical and experimental study of a novel windbelt-based energy harvester, designed to harvest from continuously changing low-speed winds. A theoretical model is derived to explore the scaling effect on the critical flutter frequency, and experimental results validate the theoretical predictions.

1. Introduction

Aeroelastic flutter is a self-feeding vibration of an elastic structure in a fluid flow. It occurs from a flutter limit wind speed, when there is a positive feedback between the structure natural vibration and the aerodynamic forces exerted by the fluid flow, leading to a large amplitude of vibration only limited by the mechanical and aerodynamic damping. Although fluttering is usually considered as a destructive phenomenon, it has been recently demonstrated that the self-feeding vibration of a membrane or a cantilever can be used with a piezoelectric [1, 2] or an electromagnetic device to harvest energy from wind. A few micro and macro-scales prototypes have been demonstrated [3, 4], and the company Humdinger Wind Energy [5] commercializes a small device harvesting 2.2 mW AC at a wind of 6 ms^{-1}. An experimental characterization of a Humdinger windbelt of 50 cm length has been performed by Pimentel et al. [6] to study the influence of the tension and the wind angle of attack on the harvested power. The latter is the only published work focusing on the windbelt optimization. The present paper aims at extending this study by providing tools for the optimization of a flutter-based energy harvester. A theoretical model is derived from Theodorsen flutter theory [7] and allows the simulation of the flutter frequency and critical wind speed evolution as a function of the ribbon parameters. An experimental analysis completes the results.

2. Prototype and optimization objectives

The proposed prototype is an omnidirectional wind energy harvester suitable for the low-speed and omni-directional winds of Singapore [8]. It consists of four ribbons separated by 45 degrees accelerating walls (figure 1). Each ribbon exploits the flutter phenomenon to convert the wind energy into mechanical vibrations. This mechanical energy is then in turn transformed into electrical energy with an electromagnetic generator: when a light copper coil attached to the fluttering ribbon moves within the static electromagnetic field generated by a fixed magnetic circuit with permanent magnets, a voltage is generated across the coil ends according

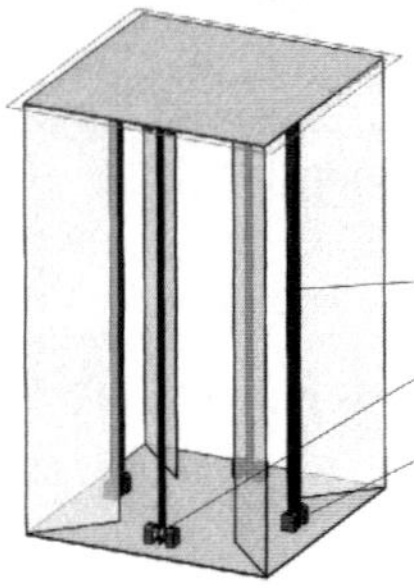

Figure 1. Schematic of the prototype.

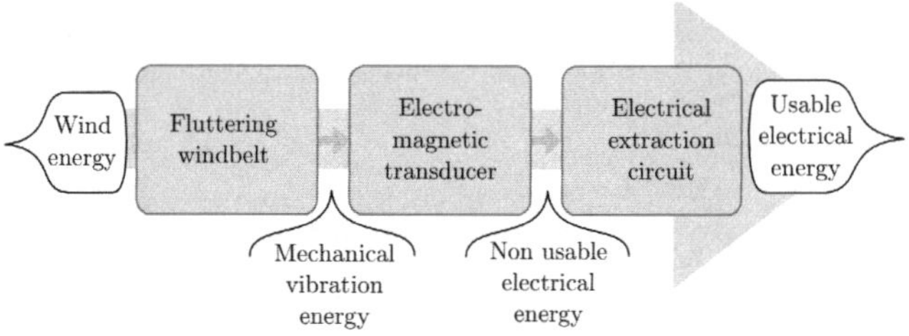

Figure 2. Structure of the energy harvester.

to Faraday's law of induction (figure 2). Finally, an energy extraction circuit is needed to rectify and adjust the generated voltages in order to supply a load.

Optimization of this harvester can be approached in three directions:

- **Maximizing the harvested power**. Previous studies have shown that maximizing the power generated by an electromagnetic generator can be reached by improving the electromechanical coupling coefficient and reducing the electromagnetic generator losses coefficient (reducing the coil resistance for instance). Another more obvious way is to increase the input vibration acceleration [9]. Thus from an aerodynamic viewpoint only, optimizing the fluttering ribbon will aim at a getting the maximum vibration acceleration. This optimization has been experimentally studied in a previous work [8].
- **Maximizing the power density**. This can also be seen as minimizing the size. In our case, reducing the harvester size mainly means reducing the ribbon length which is much bulkier than the electromagnetic device itself. The effect of scaling the ribbon is theoretically studied in section .
- **Enlarging the range of wind speeds at which it is harvesting energy**. This aim can be translated as reducing the minimal wind speed at which the flutter starts. In this objective, the critical flutter wind speed will be studied as a function of several design parameters.

3. Theoretical results

3.1. Model of the fluttering ribbon

We consider a thin and long ribbon clamped at both ends, and pre-strained by a tension force T. The ribbon is subjected to an incompressible fluid flow of speed U. When this velocity reaches the flutter critical velocity U_c, a self-sustained dynamic oscillation of large amplitude and stable frequency f_c occurs (figure 3). The model comprises a finite elements modal analysis performed with the ANSYS software, which determines the first uncoupled heaving and torsion modes frequencies. The second part of the program is a two-dimensional, two degrees of freedom model based on Theodorsen flutter theory [7, 10]. The modal resolution of the coupled heaving and torsion mode gives an estimation of the coupled flutter frequency f_c as well as the critical wind speed U_c (figure 4). The mass repartition is supposed symmetrical, and this model does not take into account the effect of concentrated masses.

3.2. Theoretical results

The ribbon dimensions and characteristics are summed up in table 1. The simulation results show that for these parameters the flutter frequency varies between 10 and 70 Hz, depending on the tension T. The limit wind speed varies between 1 and 6.4 m/s. At a tension of 0.5 N, the

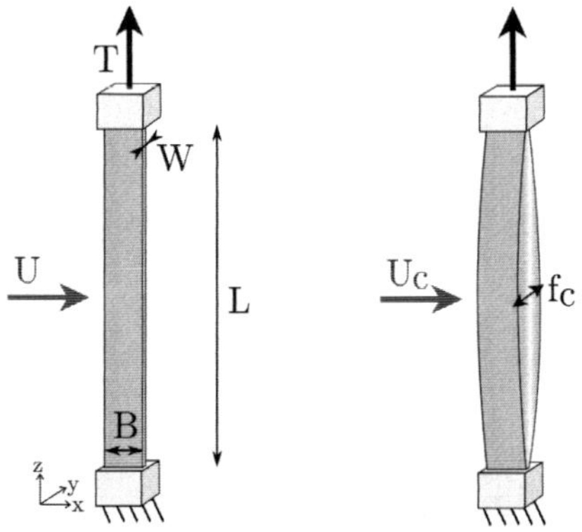

Figure 3. Windbelt and notations, before and after the critical wind speed.

	Value
B	0.013 m
L	0.450 m
W	15 10^{-6}m
T	0.5 N
d	1779 kg/m^3
E	2.51 10^9Pa
Poi	0.23
m	0.35 10^{-3}kg/m

Table 1. Windbelt parameters.

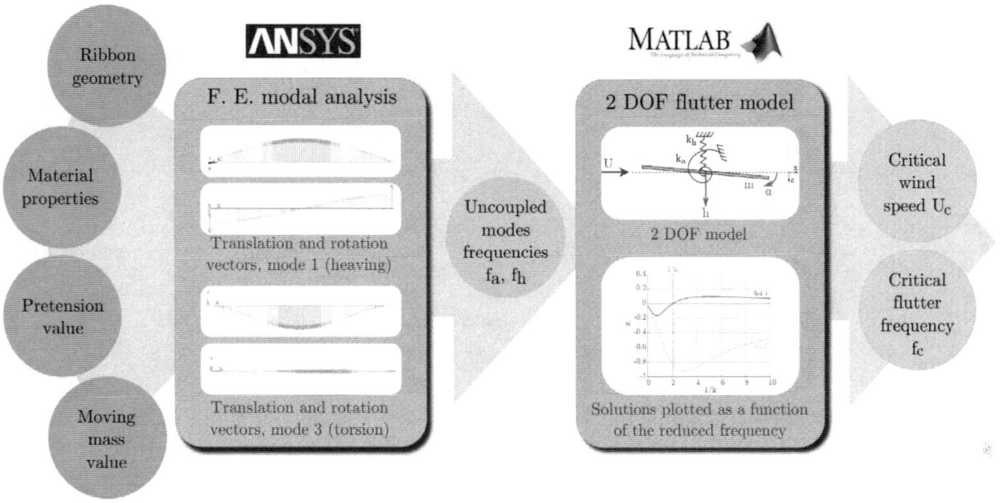

Figure 4. Schematic of the fluttering model.

calculated frequency is 31 Hz and the critical wind speed is 2.5 m/s, which seems suitable for our application targeting low wind speeds.

When the ribbon tension is varied from 0.1 to 3 N, the figure 5 shows the evolution of U_c and f_c as a function of T, for several lengths L. The frequency is, as expected, an increasing function of the tension, as is the flutter limit speed. Both increase as well when the ribbon length is decreased.

If all the ribbon dimensions are scaled by a factor s the model predicts an increase of the critical wind speed together with an increase of the flutter frequency with the miniaturization, as shown in figure 6. Note that U_c is increasing at a faster rate when the length only is reduced. So when miniaturizing such a system, one should reduce the width and thickness at least at the same rate at the length.

4. Experimental measurements

A PET ribbon of same dimensions and parameters as the simulation is vertically tensed between one force sensor on the top and a stepper motor on the bottom (figure 7). The force sensor retrieves the tension of the ribbon and this information is used to control the stepper motor and

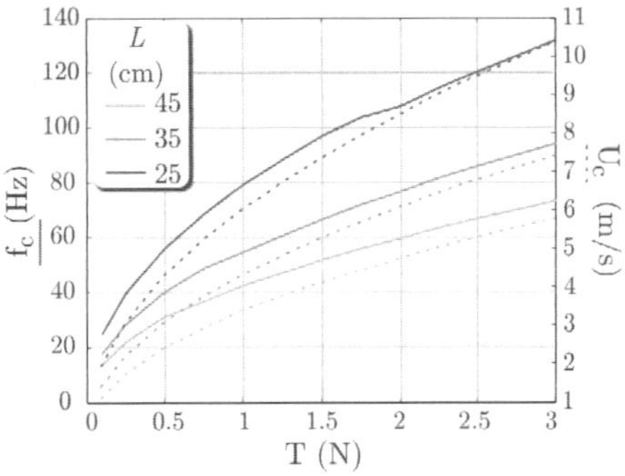

Figure 5. Flutter frequency and critical wind speed as a function of the ribbon tension T, for several ribbon lengths L.

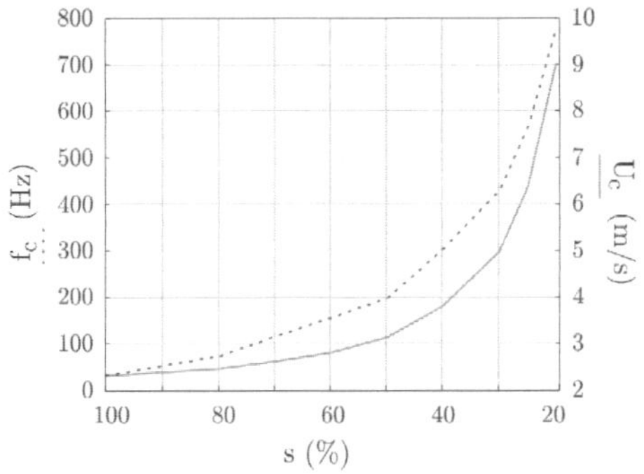

Figure 6. Flutter frequency and critical wind speed as a function of the scaling factor s.

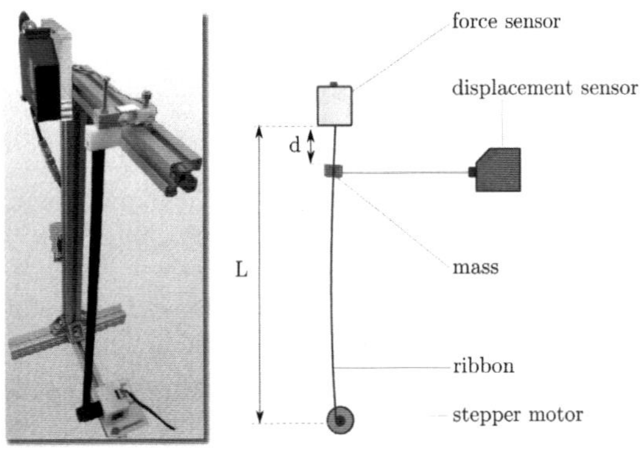

Figure 7. Schema and photo of the experimental setup.

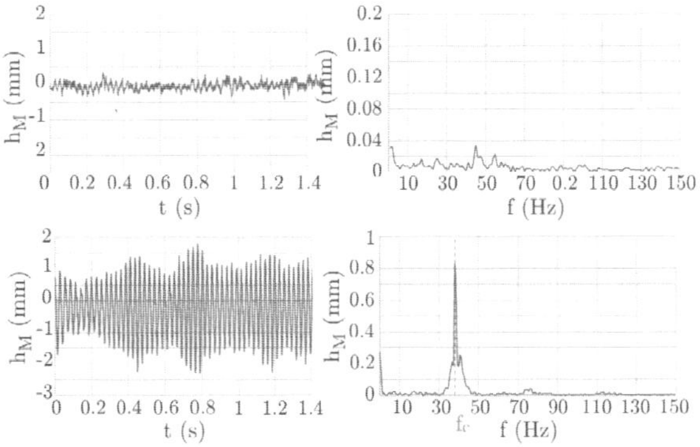

Figure 8. Flutter amplitude h_M and frequency f as a function of time t and FFT, before (upper) and after (lower) the wind critical speed.

adjust the tension to the desired value. This setup allows to get a fixed and reproducible tension of the membrane. A mass, in this case two symmetric wound coils, is fixed on the ribbon close to the top edge. A laser displacement sensor pointing towards the coil is used to measure its vibration amplitude and frequency.

The first measurements are performed with two 0.5 g coils attached symmetrically on top of the ribbon. As expected, there is a minimal wind speed at which the flutter starts. As an example, figure 8 shows the vibration amplitude respectively before and after the flutter limit in one random case when the tension is low. The critical wind speed at 0.5 N is measured between 1 and 2 m/s, and the recorded flutter frequency is 30 Hz at low wind speeds. These values correspond well to the theoretical values calculated in section 2.

The figure 9 shows the flutter frequency as a function of the wind speed when the tension is varied from 0.5 to 3 N. Contrary to the theoretical results, the flutter frequency slightly increases with the wind speed. It also increases with the ribbon tension, accordingly with the theoretical results previously shown in figure 5. Another point is that the more tensioned the ribbon, the higher the wind speed necessary to induce flutter. This too corresponds to the theoretical results.

When the ribbon length is decreased from 45 cm to 25 cm, the critical wind speed increases

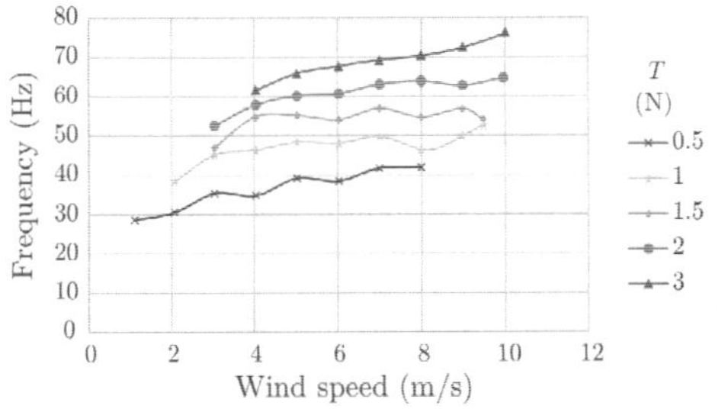

Figure 9. Flutter frequency as a function of the wind speed, for several tensions.

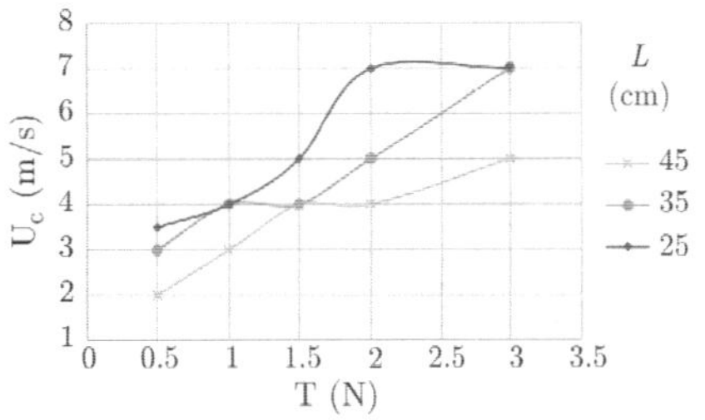

Figure 10. Critical wind speed U_c as a function of the ribbon tension for several lengths.

as shown by figure 10, accordingly with the theoretical results presented in section 3. The discrepancies between the theoretical and experimental results might be due to the difficulty to measure U_c in some cases, probably because of the not perfect balance of the two hand-wound coils.

5. Conclusion

This work presents two tools developed to help designing a windbelt-based energy harvester made of a fluttering ribbon and an electromagnetic transducer: a theoretical model and an experimental setup are used to give an estimation of the two characteristic parameters of the aerodynamic subsytem – the critical flutter frequency and the critical wind speed – as a function of the ribbon dimensions and the material properties. Simulation and experimental results notably show that the critical speed increases when the dimensions are scaled down. Thus in order to target low-speed winds, this increase has to be compensated by slightly reducing the ribbon tension for instance. Further works will focus on including the electromechanical subsystem and on studying the influence of the electromechanical coupling on the critical flutter frequency, in order to develop a global model of the harvester.

References

[1] Kwon S D 2010 *Applied Physics Letters* **97** 164102 ISSN 00036951
[2] Li S, Xian S, Yuan J and Lipson H 2011 *Journal of Applied Physics* **109** 026104 – 026104–3
[3] Kim S H, Ji C H, Galle P, Herrault F, Wu X, Lee J H, Choi C A and Allen M G 2009 *Journal of Micromechanics and Microengineering* **19** 094010 ISSN 0960-1317
[4] Fei F, Mai J D and Li W J 2012 *Sensors and Actuators A: Physical* **173** 163–171 ISSN 09244247
[5] Frayne S Humdinger Wind Energy URL `http://www.humdingerwind.com`
[6] Pimentel D, Musilek P, Knight a and Heckenbergerova J 2010 *2010 9th International Conference on Environment and Electrical Engineering* 81–84
[7] Theodorsen T 1935 General Theory of Aerodynamic instability and the mechanism of flutter Tech. Rep. NACA TR 496
[8] Arroyo E, Foong S, Marechal L and Wood K L 2014 *Proceedings of the ASME 2014 Dynamic asystems and control conference* pp 2–6
[9] Arroyo E, Badel a and Formosa F 2013 *Journal of Intelligent Material Systems and Structures* **24** 2023–2035 ISSN 1045-389X
[10] Turnock D L 1985 Two degree of freedom flutter solution for a personal computer Tech. Rep. NASA Technical Memorandom 86381

Analysis of Upper Bound Power Output for a Wrist-Worn Rotational Energy Harvester from Real-World Measured Inputs

T Xue[1], X Ma[2], C Rahn[2] and S Roundy[1]

[1] University of Utah, Department of Mechanical Engineering, 50 S. Central Campus Drive, Salt Lake City, UT, USA
[2] Pennsylvania State University, Department of Mechanical and Nuclear Engineering, University Park, PA, USA

E-mail: tiancheng.xue@utah.edu

Abstract. Energy harvesting from human motion addresses the growing need for battery-free health and wellness sensors in wearable applications. The major obstacles to harvesting energy in such applications are low and random frequencies due to the nature of human motion. This paper presents a generalized rotational harvester model in 3 dimensions to determine the upper bound of power output from real world measured data. Simulation results indicate much space for improvement on power generation comparing to existing devices. We have developed a rotational energy harvester for human motion that attempts to close the gap between theoretical possibility and demonstrated devices. Like previous work, it makes use of magnetically plucked piezoelectric beams. However, it more fully utilizes the space available and has many degrees of freedom available for optimization. Finally we present a prototype harvester based on the coupled harvester model with preliminary experimental validation.

1. Introduction

Energy harvesting has been a promising solution as the alternative to traditional systems powered by batteries when energy independence is required. As the power consumption becomes a major obstacle to the emerging market for wearable technologies, kinetic energy contained in human daily activities naturally appeals to the researchers in the field of energy harvesting. Due to the low and random frequencies of human motion, which usually occurs around 1 Hz [1], resonant harvesters cannot benefit from the peak dynamic magnification. The well-known Seiko Kinetic watch converts the kinetic energy of a rotor to electricity using an electromagnetic generator [2]. Such rotational designs with non-resonant configuration are the common approaches that researchers have adopted to overcome the limitation of conventional linear harvesters. While the Seiko Kinetic watch boosts its energy density by a sophisticated high-ratio gear train, another strategy, which is often referred to as frequency up-conversion, is to excite a higher-frequency resonance in a transducer via a slower-frequency motion. Prior reported non-resonant harvesters include a piezoelectric harvester with in-plane magnetic plucking based on a planar rotor model [3] and a magnetic spherical harvester [4].

The goal of this paper is to provide an estimate of the maximum power output from a rotational harvester regardless of the energy conversion mechanism using real-world measured data such as walking and jogging as inputs. We derived a generalized 3D rotor model to fulfil the task. Simulation results show that the upper bound on power generation from an eccentric rotor during normal activities is approximately an order of magnitude greater than what has been demonstrated [3], which motivates us to develop an optimized non-resonant piezoelectric harvester with different magnet-beam configurations from previous designs in the literature. We use multiple piezoelectric beams and magnets in alternate configurations that allow us to optimize the system electromechanical coupling and provide more design variables to tune the system dynamics.

2. Generalized Rotational Harvester Modelling

2.1. Generalized Rotor Model in 3 Dimensions

In order to provide an estimate of the maximum possible power generation from real-world multidimensional inputs, a full rotor model in 3 dimensions is required. The planar rotor model described in [5] only accounted for one rotational or linear excitation separately. We extended this model to 3 dimensions with 6 effective inputs simultaneously as illustrated in Figure 1. An electrical and a mechanical rotational damper are included in the model to represent extracted and lost energy respectively. The electrical damper represents any kind of ideal energy transducer while the mechanical damper represents a combined pathway for all kinds of energy losses (e.g. friction). As in many previous published studies [6, 7], this model makes the assumption that the power dissipated through an optimal viscous damper that represents the electromechanical transducer is the maximum electrical power that can be extracted from the system. The rotation of the rotor is constrained in the XY plane and the governing equations of the rotor motion are given by:

$$m\begin{bmatrix} -l\cos\phi_z\cdot\dot{\phi}_z^2 - l\sin\phi_z\cdot\ddot{\phi}_z \\ -l\sin\phi_z\cdot\dot{\phi}_z^2 + l\cos\phi_z\cdot\ddot{\phi}_z \end{bmatrix} = -m\begin{bmatrix} \ddot{X} \\ \ddot{Y} \end{bmatrix} + \begin{bmatrix} F_x \\ F_y \end{bmatrix} + m\begin{bmatrix} g_x \\ g_y \end{bmatrix} \tag{1}$$

$$I_G\left(\ddot{\theta}_z + \ddot{\phi}_z\right) = -\left(D_e + D_m\right)\dot{\phi}_z + F_x l\sin\phi_z - F_y l\cos\phi_z \tag{2}$$

Where m, I_G, and l are the mass, the moment of inertia and the eccentric distance of the rotor respectively. $\ddot{X}$, $\ddot{Y}$ are acceleration inputs to the system. F_x, F_y and g_x, g_y are the forces from the housing and gravity acting on the rotor respectively in each local coordinate. ϕ_z denotes the relative angular displacement between the rotor and the housing, which determines the power output as:

$$P = D_e\dot{\phi}_z^2 \tag{3}$$

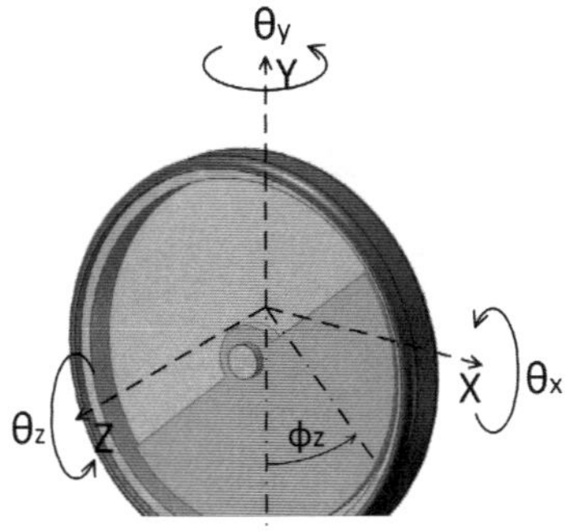

Figure 1. Dynamic model of eccentric rotor. θ_x, θ_y and θ_z denote rotational inputs to the housing while there are linear accelerations in X, Y and Z coordinates as well

2.2. Simulation with Real-World Measured Data

The input data was measured using the TEMPO Inertial BSN Platform [8], which captures motion in full six degrees of freedom. One subject wearing the platform on his wrists was instructed to conduct a series of activities involving wrist motion including walking, jogging, running, typing, and reading. Gravitational force in X-Y plane is determined by the orientation calculated using the gyroscope data. Given certain real-world input, the power output was found to be a function of the electrical damping coefficient and rotor size when the mechanical damping coefficient was held constant. Simulation results, given in Figure 2, show that a generalized harvester placed on the wrist of a walking subject can provide an average power of 110 μW for a brass rotor of 16mm in radius and 3mm in thickness with reasonable friction. An optimal level of electrical damping is required to achieve the maximum power output. It is also obvious that the system generates more power with a lower mechanical damping level.

Figure 3 details the upper bound on power output from the generalized harvester in common daily activities. Typing produces very low power as it involves little wrist movement whereas there is a large amount of potential power in intense activities such as running and jogging. Reading and walking provide similar amounts of power as they are both mild activities involving some back-and-forth wrist motion. This generalized model provides an estimate of maximum power generation in an ideal scenario in which the electromechanical coupling can be closely approximated by an optimal linear viscous damper and parasitic electrical losses are neglected. In more specified cases, certain constraints will lower the maximum power output.

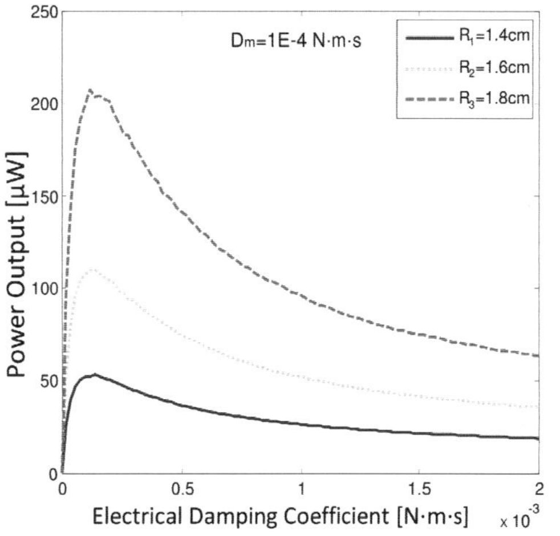

Figure 2. Power output vs. Electrical damping coefficient from the generalized rotational harvester during walking

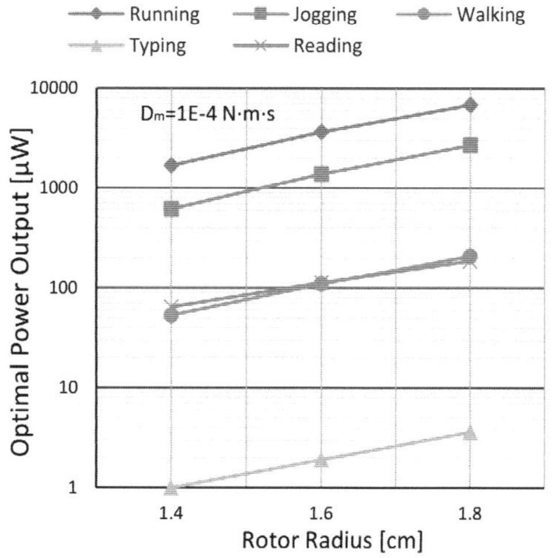

Figure 3. Upper bound on power generation from the generalized rotational harvester in different activities

3. Piezoelectric Rotational Harvester Design

We adopted the out-of-plane beam deflection configuration (i.e. the beam deflects out of the X-Y plane defined in Figure 1 rather than in the X-Y plane) in the design due to its potential for multiple beams and the simplicity of manufacture. As illustrated in Figure 4, the harvester uses repulsive magnets to achieve the actuation with magnets embedded in the rotor and on the tip of the beams. A 6-cantilever-beam array is attached on the lid to fully utilize the space within the device. Energy conversion is achieved by piezoelectric beams as three of them vibrate synchronously while the semi-circular rotor swings. Figure 5 shows a photo of the prototype harvester made with off-the-shelf piezoelectric beams and magnets. The 15-mm-radius rotor is built with slots for variable magnet spacing. We attached only one 0.38-mm-thick bimorph beam, manufactured by Piezo Systems, on the plate for the initial test. Its active length was shortened to 10 mm to fit in the device. With an NdFeB magnet (1/16 inch^3, grade N52) on its tip, the resonant frequency of the beam is around 1000 Hz. The effective volume of the prototype inside the housing is 6 cm^3 while the total device weights 27 g with only one assembled beam and one pair of magnets attached.

In order to model this system, the equations in (1) and (2) above need to be augmented with a piezoelectric bimorph beam model and a magnetic interaction model. Many researchers have developed different approaches to formulate mathematical models for the piezoelectric harvesters. The beam model adopted in this work is based on the distributed parameter model presented in [9]. A similar modelling procedure is described in [10]. The model is altered as the excitation is the magnetic force on the tip instead of the base of the beam. The magnetic interaction force is approximated for simplicity using a magnetic dipole model, which predicts a steeper force gradient than the more accurate finite element model. We derived such a coupled rotor-magnet-beam model with multiple design variables for future optimization.

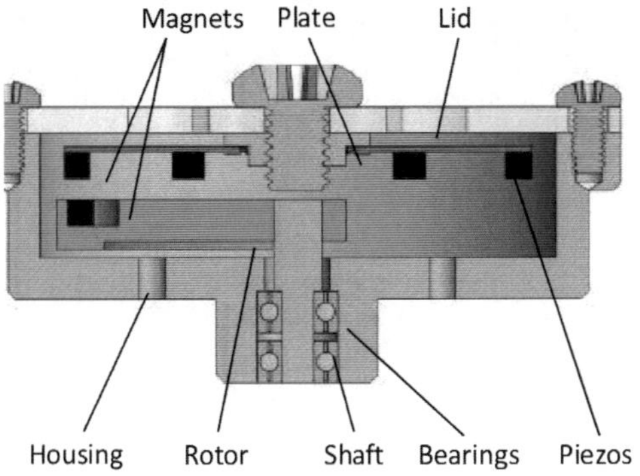

Figure 4. Piezoelectric rotational harvester CAD model (with 6 beams) in section view

Figure 5. Photo of the piezoelectric rotational harvester prototype (with one beam)

4. Experiment

4.1. Experimental Set Up

We developed the experimental set up shown in Figure 6 to test the harvester in a pseudo wrist-worn situation. The current configuration is to test motions in the vertical plane, however the servo can be mounted with different configurations. The one-beam prototype harvester is attached to the swing arm controlled by the servo. Since the definition of certain human motion is vague and subjective, by varying the swing profile (angle, speed and harvester location on the swing arm), we created a series of pseudo human motion from gentle to relatively intense walking for qualitative comparison with simulation results of the coupled harvester model. Voltage output was measured from a resistive load of 100kΩ as a match to simulation.

4.2. Experimental Results

Figure 7 shows the average power harvested from one piezoelectric beam in the test while simulation results indicate a power output ranging from 0.8 to 4 μW with high degrees of dependency on initial conditions. Simulation with a very gentle real-world walking motion also demonstrates power output above 1 μW. Preliminary test results show that power generation benefits from higher swing radius and higher frequency. (Note, that for a given angular velocity, a 60 degree rotation will have a higher frequency than a 120 degree rotation.) The one-beam harvester can generate an average power output of 3.8 μW in a relatively intense pseudo walking motion. With the possibility of 6 beams embedded, the simulation model suggests that the total power output will be at least 3 times the power generated by one beam. Due to the high resonant frequency of the shortened off-the-shelf beam, the ring down phenomena of the plucking is not clearly demonstrated in the walking motion. Performance improvement can be achieved with a softer custom made piezoelectric beam.

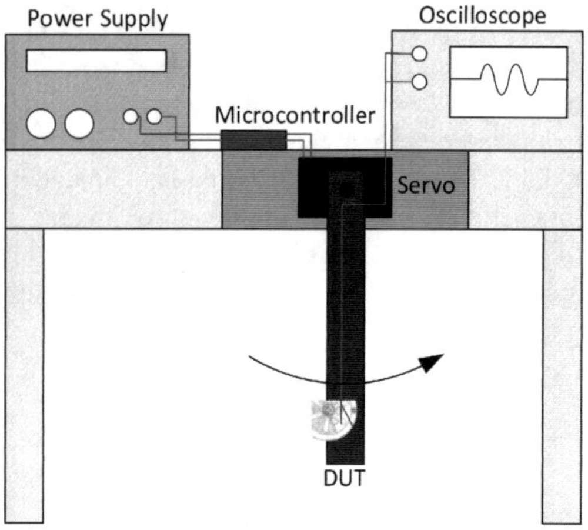

Figure 6. Experimental set-up for testing the one-beam prototype in the swing motion

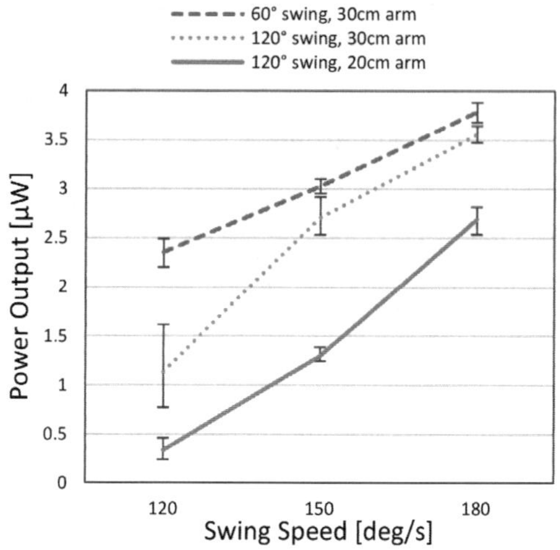

Figure 7. Average power output from pseudo walking motions

5. Conclusions

This paper presents a generalized 3D rotational harvester model to provide an estimate on maximum power output from wrist motion in different activities. Simulation results indicate that there is space for performance improvement over reports of existing devices. We have developed a coupled piezoelectric rotational harvester model with multiple design variables for future optimization. Preliminary experimental results of the one-beam prototype demonstrate an average power output of 3.8 μW in pseudo walking motion while improvement can be made by finishing the prototype with the full 6-beam assembly.

Future work will include the experimental validation of the full 6-beam prototype harvester, further optimization based on all the design variables of the coupled harvester model, and the incorporation of custom designed and fabricated piezoelectric elements.

6. Acknowledgement

The authors would like to thank the NSF Nanosystems Engineering Research Center (NERC) for Advanced Self-Powered Systems of Integrated Sensors and Technologies (ASSIST) for financial support as well as the UVA Center for Wireless Health for providing the wrist motion data.

References

[1] Yun J, Patel S N, Reynolds M S and Abowd G D 2011 *IEEE Trans. Mob. Comput.* **10** 669–83
[2] http://www.seikodigitalwatches.com/5XXX/5M6263A.pdf, Accessed October 2014.
[3] Pillatsch P, Yeatman E M and Holmes A S 2013 *J. Phys. Conf. Ser.* **476** 012010
[4] Rao Y, McEachern K M and Arnold D P 2013 *J. Phys. Conf. Ser.* **476** 012011
[5] Yeatman E M 2008 *Proc. Inst. Mech. Eng. Part C J. Mech. Eng. Sci.* **222** 27–36
[6] Mitcheson P D *et al.* 2008 *Proceedings of the IEEE* **96** 1457–86
[7] Halvorsen E Le C P Mitcheson P D and Yeatman E M 2013 *J. Phys. Conf. Ser.* **476** 012026
[8] Barth A T, Hanson M A, Powell H C and Lach J 2009 *The 6th International Workshop on Wearable and Implantable Body Sensor Networks* 71–76
[9] Erturk A and Inman D J 2008 *Journal of Vibration and Acoustics* **130** 1–15
[10] Pillatsch P, Yeatman E M and Holmes A S 2014 *Smart Materials and Structures* **23** 025009

SPICE modelling of a coupled piezoelectric-bimetal heat engine for autonomous Wireless Sensor Nodes (WSN) power supply

J. Boughaleb[1,2,4], S. Monfray[1], G. Vine[1], P.J. Cottinet[4], A. Arnaud[1], S. Boisseau[3], A.B. Duret[3], S. Quenard[2], O. Puscasu[1], C. Maitre[1], S. Trochut[1], F. Hasbani[1], T. Di Gilio[1], V. Heinrich[1], P. Urard[1], J.C. Grasset[1], F. Boeuf[1], D. Guyomar[4], T. Skotnicki[1]

[1]STMicroelectronics, 850 rue Jean Monet – 38920 Crolles, France
[2]CEA-LITEN, 17 rue des Martyrs – 38054 Grenoble Cedex 9, France
[3]CEA-LETI, 17 rue des Martyrs – 38054 Grenoble Cedex 9, France
[4]LGEF, INSA Lyon, 8 rue de la Physique – 69621 Villeurbanne Cedex, France

E-mail: jihane.boughaleb@st.com

Abstract. This paper deals with an electrical modelling and optimization of a thermal energy harvester dedicated to power autonomous systems. Such devices based on bimetal strips and piezoceramics turn thermal gradients into electricity by a two-step conversion mechanism. This work focuses first on a demonstration of a ST-WSN (GreenNet demonstration platform) supplied by the harvester to validate, for the first time, the harvesters viability. That demonstration focuses attention on the need for an optimized power management circuit for piezoelectric generators able to reach output voltages up to 20V. The work deals then with the proposal of an equivalent lumped element model of the piezoelectric transducer with its SPICE implementation to enable the optimization of a dedicated power management circuit based on the **P**ulsed **S**ynchronous **C**harge **E**xtractor (PSCE). Simulations using the SPICE model and the power management circuit lead to an increased extracted power by 144%.

1. Introduction

The field of energy harvesting has become a predominant research area, especially, with the development of wireless sensors and communication node networks. Industrial and health monitoring, smart buildings, internet of things, these are some domains where node networks can be widely used. However, the impractical aspect of wired systems and the problems related to the power supply by short-lifetime batteries are raised for those applications. This brought attention on ambient energy harvesters and explains the great interest in energy scavenging devices over the last years. In this paper, a thermal energy harvester previously presented in [1,2,3] is studied. The harvester is based on coupling a bimetal to a piezoelectric transducer where a two-step conversion mechanism occurs. The heat flowing through the bimetal strip is converted into mechanical oscillations insuring the thermo-mechanical conversion; the second step consists in the conversion of the bimetallic strip impacts into electricity by a piezoelectric transducer. A bimetal strip is characterized by two equilibrium positions as shown in Fig. 1(a). Due to the bimetal thermo-mechanical bi-stability explained in [4], the bimetal snaps up and hits the piezoelectric membrane once its temperature raises and reaches a threshold temperature called the snap-up temperature. At that point, the bimetal is at its upper-position and releases a part of its energy to the piezoelectric. While being in contact with the piezoelectric acting as a cold surface, the temperature of the bimetal decreases until it reaches a certain snapping-back temperature that makes it switch-back to its initial position and the piezoelectric to be freed from any strain. Thanks to the structure design, a significant thermal gradient is maintained across the harvester [5] allowing the bimetal thermo-mechanical oscillations between a hot source and a piezoelectric membrane acting as a cold surface.

2. Electrical characteristics and demonstration of a ST-WSN (GreenNet)

Prototypes are fabricated out of a Peek substrate to take advantage of its high glass transition temperature around 143°C, as well as its high elastic modulus (3,8GPa) that enhances the piezoceramic clamping and reduces the mechanical energy losses. The heat engine presented here uses a shell-like bimetal made in Invar (Fe-Ni 36%) acting as the low thermal expansion coefficient (CTE) layer and NC4 (Fe-Ni 22%-Cr 3%) acting as the high CTE layer. The bimetal has a three-degree thermal hysteretic behaviour, its dimensions are 1,8cm x 3,6cm x 300µm, and it applies a 1,9N force on the piezoelectric transducer every snap-up.

In this configuration, the quality factor is increased three times in comparison with the previous devices made of Teflon and it reaches 30 with Peek substrate. A usable output power of 2,56µW is measured across a simple diode bridge and a buffer capacitor (C=10µF) for a 5mm-thick prototype on a hot source between 83°C and 105°C and cooled down by ambient air at 25°C without any heat sink.

Thanks to the low power consumption of electronic circuits and the requirements of a single WSN node that can be as low as 1µW in a sleep state [6], a demonstration of a ST-WSN (GreenNet) supplied by this harvester is realized in asynchronous mode (unidirectional data transmission). As the harvester AC output signal is not compatible with standard electronic circuits, a voltage doubler and a 57µF storage capacitor are connected to the device, followed by a level detection circuit. The harvested energy is first stored in the capacitor until a threshold voltage value is reached and detected by a low power comparator. It is then released to supply the WSN. The GreenNet emission circuit transmits 64 bits of data with IEEE 802-15-4 protocol toward the GreenNet reception module every 29s using two devices and every 15s using three cells. Every data reception is illustrated by a switching led as shown In Fig. 1.c. The architecture of this WSN are shown in Fig 1.b and the evolution of the storage capacitor voltage is shown in Fig 1.c. This experiment validates definitively the ability of these harvesters to power WSN and highlights the necessity of a dedicated power management to increase the amount of harvested power in comparison with the standard circuits currently used (rectifier + accumulator), which will be studied in the next part of this paper.

(a) (b)

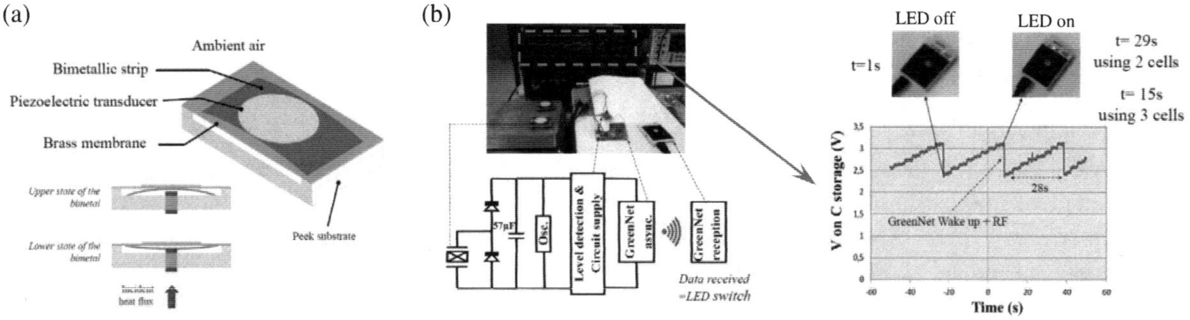

Figure 1. (a) Piezoelectric and bimetal heat engine, (b) Test of a ST-WSN (GreenNet) in an asynchronous mode powered by the thermal energy harvesters and state of the reception node

3. SPICE modelling

An equivalent electrical model of the harvester is needed to design and simulate a power management circuit optimized for piezoelectric generators with output voltages as high as 20V. A SPICE of the structure model is established based on the lumped element parameters of a piezoelectric transducer. That model is inspired from Mason's representation to describe the piezoelectric oscillators [7]. The considered generator, excited near its resonance frequency and undergoing little displacements can be modelled, in the mechanical domain, as a vibrating mass m taking into account the effective beam mass associated with the deformation mode of the system, a spring K_m representing the stiffness of the device, a damper D_m and the capacitance of the piezoceramic C_p (Fig. 2.a). In the electrical representation (Fig. 2.b), the piezoelectric transducer is modelled as a transformer whose primary is the mechanical part of the system and whose secondary is its electrical part. The mechanical elements are consequently, replaced by their electrical analogous quantities: an inductance for the inertial term, a capacitance for the stiffness term and a resistance for the mechanical losses. Moreover, the piezoelectric losses are supposed negligible [8], and the coupling factor between the mechanical and the electrical parts is the force factor α. This parameter extraction is realized using a vibrometer and the experimental setup is illustrated in Fig. 2.c. Tables 1 and 2 show respectively, the experimental parameters to extract and the calculation of the components of the electro-mechanical model of the piezoelectric transducer.

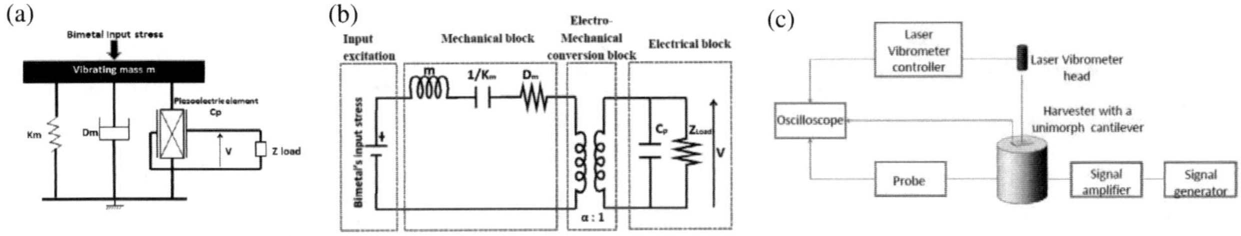

Figure 2. Equivalent Lumped Element Model of the piezoelectric system in the neighbourhood of resonance in (a) mechanical and , (b) electrical representations, (c) experimental setup used to extract the piezoelectric force factor

Table 1. Measured parameters of the system

f_1	Short circuit resonance frequency
f_2	Open circuit resonance frequency
γ	Piezoceramic open circuit voltage to the beam displacement ratio
C_0	Clamped capacitance of the piezoceramic
Q_m	Mechanical quality factor

Table 2. Extracted parameters for the model

$$\alpha = \gamma C_0$$

$$K_m = 1/C_m = \alpha \gamma \frac{f_1^2}{f_1^2 - f_2^2}$$

$$L_m = \frac{K_m}{4\pi^2 f_1^2}$$

$$R_m = \frac{2\pi L_m f_2}{Q_m}$$

Assuming that the mechanical parameters are frequency dependent and that their values change significantly with the boundary conditions of the piezoceramic [9], two motional branches are employed to simulate the piezoelectric response when the bimetal snaps up and vibrates with the piezoceramic or when it snaps down allowing it to vibrate freely (Fig. 3). We show that a succession of electrical pulses models accurately the impact of the bimetal on the piezoelectric membrane. The duration of these pulses as well as their amplitude allows obtaining an electrical representation of these impacts whether it concerns a snap up or a snap down (Fig. 4.c). Finally, voltage controlled switches are used to establish the connection of the voltage sources to one of the two motional branches depending on the bimetal state. The final SPICE model is shown in Fig. 3.

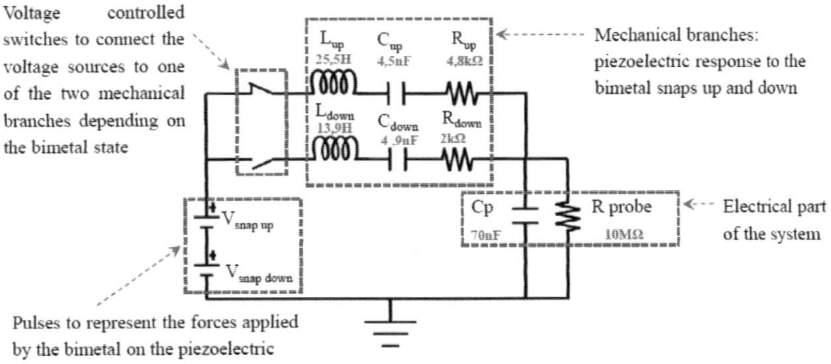

Figure 3. Equivalent SPICE model of the piezoelectric transducer taking into account the bimetal impacts

Experimental measurements and simulations using this SPICE implementation are performed and the results of the voltage evolution across the piezoelectric capacitor, showing a good agreement, are exposed on Fig. 4 (a 10MΩ impedance probe is used for simulation and experiments). Thanks to that SPICE model, a standard circuit (voltage doubler + 10μF storage capacitor) is also simulated, and 2,56μW output power are obtained (Fig. 5.b), showing as well, a good agreement with the experimental measurements carried out previously. These steps confirm the validity of this modelling.

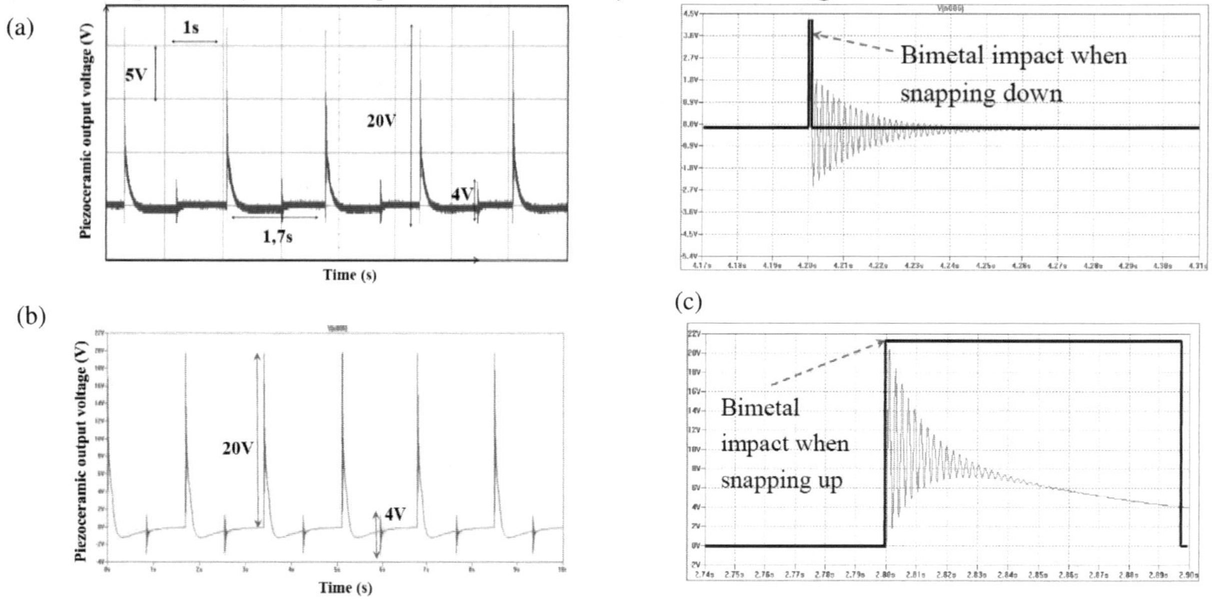

Figure 4. (a) Experimental piezoelectric signal measured with a 10MΩ probe, (b) SPICE simulation results of the piezoceramic output signal with with a 10MΩ probe, and (c) bimetal impact modelling

4. Power management circuit PSCE

To optimize the power transfer from the piezoelectric generator to a storage capacitor, thanks to the SPICE model, the enhanced power performance of the harvester is predicted using PSCE (Pulsed Synchronous Charge Extractor) nonlinear technique studied by T. Hehn [10]. This circuit is shown in Fig. 5.a. It includes a rectifier, an inductance acting as temporary energy storage and three switches S1, S2, S3. Its working principle is explained in Fig. 6, revealing three phases as shown by the timing diagram and the control unit generating the switches controlling signals. During the first state, all the switches are opened and the piezoelectric capacitor charge rises until the voltage reaches a maximum. Once the peak detector detects that maximum voltage across the piezoceramic, phase B starts: the switch S1 is closed and an LC oscillator is then created. This circuit permits the discharging of the capacitor into the inductance and the functioning mode ends when the voltage across the piezo-capacitor reaches the value 0V detected by a zero crossing detector. At this point the energy is completely transferred into the inductor and phase C is initiated to enable the discharge of the inductance into a storage capacitor C_{out} during a last phase 3. The switch S1 is opened, S2 and S3 are closed and another LC oscillating circuit is constituted. This phase is terminated when a comparator detects a zero current in the circuit.

The switches control circuit is supplied by the storage capacitor C_{out}. At the startup, when this capacitor is discharged, the PSCE circuit is equivalent to a standard circuit constituted of a rectifier and a storage capacitor because only the switch S2 is closed to charge the storage capacitor. The complete circuit is designed and simulated using CADENCE software. The consumption of the controlling circuit is estimated to be lower than 600nW. The simulation of charging a 10μF capacitor leads to an improvement by 144% of harvested power improvement, reaching 6,25μW per cell and up to 19μW with 3 cells (Fig. 5.b). The physical implementation of the simulated PM circuit would allow operating in synchronous mode (bidirectional data) for future Internet of Things applications.

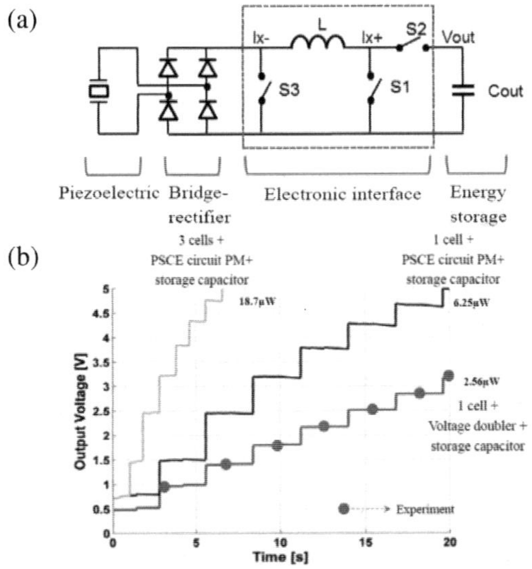

Figure 5. (a) Interface circuit of PSCE technique, (b) Simulations of power efficiency improvements using PSCE circuit and the SPICE model (C_{out}=10µF)

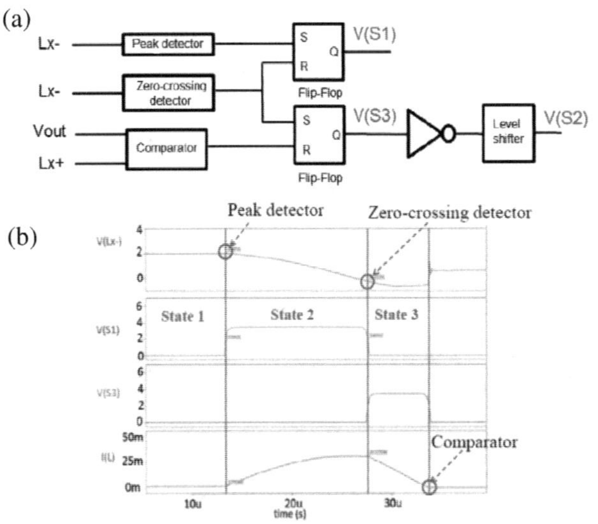

Figure 6. (a) Block diagram of the switches controlling unit, (b) Timing diagram of the switches controlling signals

5. Conclusion

This paper presents the improvements and optimizations performed on a thermal energy harvester based on bimetals and piezoelectric materials. A first demonstration of supplying a ST-WSN GreenNet in asynchronous mode with these harvesters is realized confirming their efficiency and viability without any heat sink and cooled down by natural convection only. An electrical model of the harvester is then established based on a lumped element model of a piezoelectric system and a SPICE implementation is proposed for the different working phase of the device: when the bimetal snaps up and when it snaps down. This model is even more necessary to design a dedicated power management inspired from PSCE technique. The circuit's simulations results reveal a power gain of 144% in comparison with the standard circuits which is promising for the future implementation of autonomous WSN for Internet of Things.

Acknowledgments

This work was funded by the Fonds Unique Interministériel (FUI), through the HEATec project.

References

[1] T. Skotnicki 2009 Dispositifs de conversion d'énergie thermique en électricité In: France Patent FR2951873 (A1).

[2] O. Puscasu et al 2014 Sensor Actuat a-phys A 214 pp. 7-14.

[3] A. Arnaud et al 2013 Proc. PowerMEMS London pp. 295-299.

[4] A. Arnaud et al Thermodynamic modeling of the thermal instability of composite beams applied to thermal energy harvesting, submitted to international journal of thermal sciences.

[5] J. Boughaleb et al Theoretical modeling and design optimization of a thermally-matched energy harvester submitted to Sensors and actuators

[6] Merett G W et al 2010, Wireless devices and sensor networks In: Beeby S and White N (Ed.), Norwood Atech House Chapter 2 p. 9.

[7] W.P. Mason 1935 Proc. of the Institute of Radio Engineers IEEE 23 pp. 1252-1263.

[8] L. Pinna et al 2009 Proc. ICECS Yasmine Hammamet IEEE pp. 587-590.

[9] S. Priya and D.J. Inman 2009 Energy harvesting technologies chapter 4 pp 107-128.

[10] T. Hehn et al PowerMEMS 2008, pp. 9-12.

Evaluation of Electrostatic Force on Bipolar Charged Electret

K Sonoda, K Minami, N Miwatani, T Fujita, K Kanda, and K Maenaka
Department of EECS, Graduate School of Engineering, University of Hyogo
2167 Shosha, Himeji, Hyogo 671-2280, Japan

E-mail: eu10x003@steng.u-hyogo.ac.jp

Abstract This paper presents an evaluation of an electrostatic vibration energy harvester with the bipolar charged electret. The energy harvester with the size of 13 × 12 × 1.2 mm^3 was fabricated. The output power of the bipolar charged with ±250 V harvester was 9 μW when the acceleration was 1.4 *g* at 352 Hz with 0.9 MΩ load resistance. The effectiveness against the velocity-damped resonant-generator (VDRG) limit was 2.5%. The electrostatic forces of the actual device with DC bias, which simulates charged electret with monopolar and bipolar were experimentally and numerically verified. We estimated the electrostatic force by measuring the vibration amplitude versus applied acceleration of the electret mass. As a result, we investigated the bipolar charged device can reduce the effect of electrostatic force as low as no bias condition. The numerical model of the energy harvester considering the electrostatic force by FEM static analysis was also established. The comparison between the numerical model and the measurement results showed a similar inclination.

1. Introduction

We study an electrostatic energy harvester for an ambient vibration. Electrostatic energy harvester generally uses electret, which holds electric charges for a long year. Charging electret with high surface potential and decreasing a parasitic capacitance[1] is effective to improve output power, although surface charge density and potential are limited by thickness of the electret film[2]. High surface potential thick film electret is fabricated by multi-layer electret film[3]. However, surface potential also increases electrostatic force, which restrains mass movement[4]. In order to overcome this issue, a reduction method of the electrostatic force for horizontal direction with a dual-phase arrangement and for vertical direction by using a repulsive force, which electrets are charged with monopolar charge have been reported[5]. We proposed a bipolar charging method by selective charging method[6] in order to form an electret, which has not only negative charge but also positive charge. From the analytical and numerical estimation, we investigated the reduction for electrostatic force on our bipolar charged electret device.

In this paper, we demonstrate that the bipolar charged electret is effective to decrease the electrostatic force because the potential difference between the electrodes can be reduced in comparison with the monopolar charged electret for the same surface potential difference of the electret e.g. monopolar charged with -200 V, bipolar charged with ±100 V. From the generated power of the bipolar charged device, we also confirmed the reduction of electrostatic force of bipolar biased electret by comparison between the measurement and the numerical analysis. In order to formulate a numerical model considering an effect of electrostatic force, the electrostatic force of static condition is estimated by FEM analysis.

2. Bipolar Charged Energy Harvester

The surface potential after charging the CYTOP (CTL-809M; Asahi glass Co., Ltd., Japan) with the thickness of 2.5 μm was examined. In conventional electrostatic energy harvesters, electrets are charged mainly with negative charge. The grid potential during the charging process was applied from 0 V to 1 kV with negative and positive potentials. The surface potential is measured by an electrostatic voltmeter (Model 279; Monroe Electronics Inc., USA) and it is shown in figure 1. The surface potential of the CYTOP is saturated around 600 V because of the material dielectric breakdown voltage for 2.5 μm thickness. The characteristic of the positive charging indicated as similar as the negative charging one.

The energy harvester with the size of 13 × 12 × 1.2 mm^3 was fabricated and charged with ±250 V bipolar potential as shown in figure 2. The diagram of bipolar charged electret on the device are shown in figure 3. The actual device consists of 50 pieces of analysis model. We already reported that the test specimen of the bipolar charged electret, which is charged with the pitch of 1 mm[7]. The energy harvester, which consists of the electrode pitch of 100 μm, was charged with bipolar charge and the

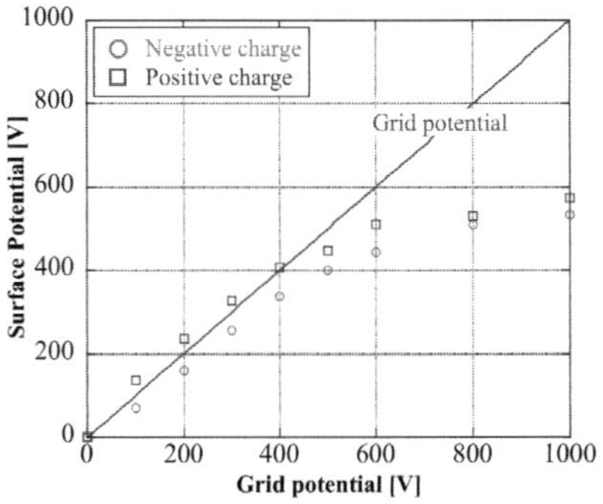

Figure 1. Characteristic of surface potential saturation for 2.5 μm thickness CYTOP film.

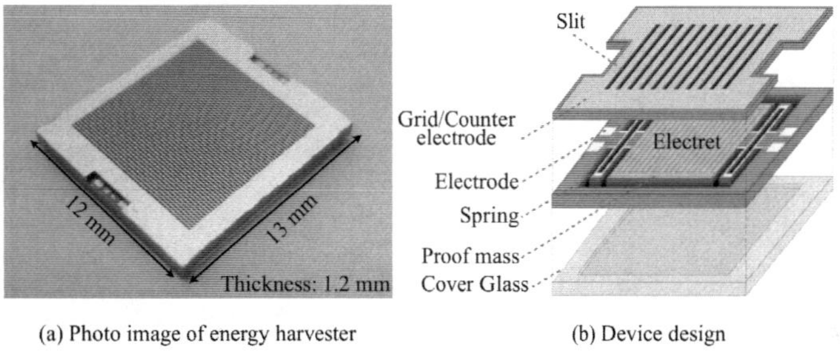

Figure 2. Photo image of energy harvester and schematic diagram of device design.

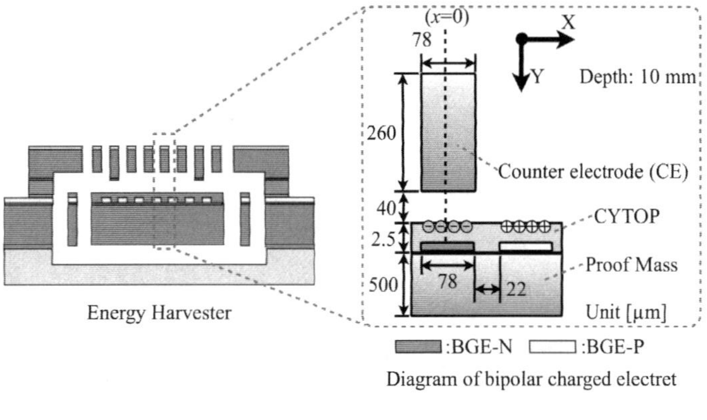

Figure 3. Device design of energy harvester and diagram of bipolar charged electret.

output power was investigated. The oscillation amplitude and the weight of the mass are 200 μm and 120 mg, respectively. Figure 4 shows the output power versus external load when applied acceleration is 7 *g* at 352 Hz. The maximum output power appeared at 0.9 MΩ optimum load resistance. The optimum resistance for maximum output power also can be calculated by the following equation: $R_{OPT} = (2\pi fC)^{-1}$, where *f* is the peak frequency of the output waveform by FFT analysis and *C* is the parasitic capacitance of the device. The optimum resistance is calculated as approximately 1.1 MΩ load resistance. The parasitic capacitance of the device was approximately 80 pF. The output waveform is shown in figure 5. The peak frequency of the output waveform was 1756 Hz, which is 5 times of 352 Hz because it is an initial position of the actual device that the counter electrode has an offset of +50 μm due to a process error from arrangement in figure 3. The calculated optimum resistance was according to the measurement resistance at the maximum output power. The maximum effectiveness against the velocity-damped resonant-generator (VDRG) limit is 2.5% when the harvester outputs 9 μW by applying acceleration of 1.4 *g* at 352 Hz with 0.9 MΩ load resistance.

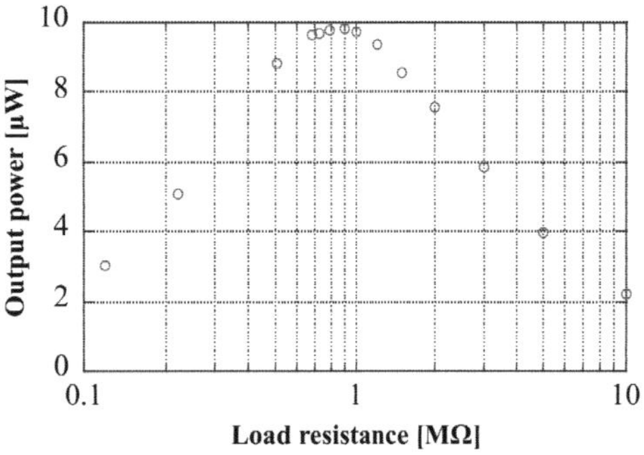

Figure 4. Output power vs. load resistance. Optimum load resistance is 0.9 MΩ with applied acceleration of 7 *g* at 325 Hz.

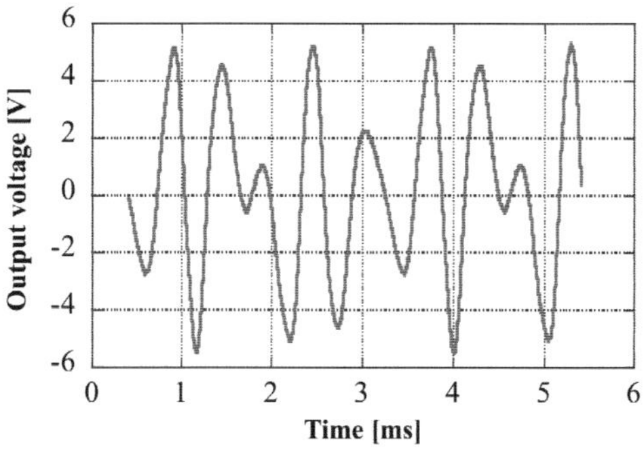

Figure 5. Output waveform of bipolar charged device.

3. Numerical Model of Energy Harvester

The electrostatic forces under static condition of the monopolar and bipolar charged harvester were obtained by FEM analysis. The electrostatic force versus the displacement of the proof mass changes sinusoidally. The electrostatic forces along to vertical and horizontal directions were respectively reduced to 15% and 20% from the same potential difference for monopolar charged electret[7]. The formula of the electrostatic force is calculated by curve fitting and applied to the numerical model. The numerical model from equation of motion of the energy harvester considering the electrostatic force as shown in figure 6 is given by

$$m\ddot{x} + D\dot{x} + kx + F_e(x) = F(t) \quad (1)$$

where x, m, D, k, F_e, and F are the expression for the displacement of the proof mass, the weight of the proof mass, the damping coefficient, the spring constant, the electrostatic force of horizontal axis

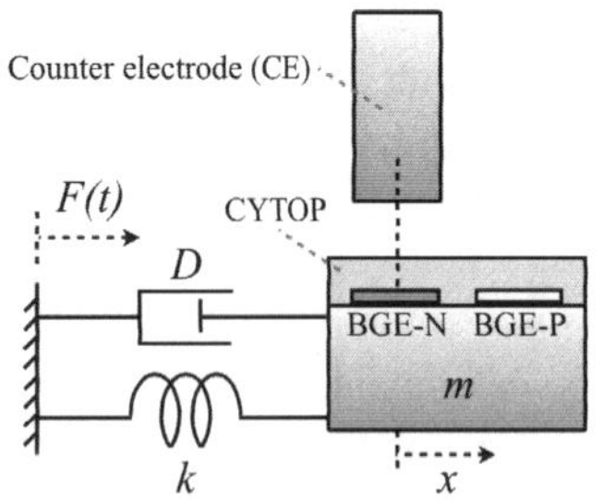

Figure 6. Numerical model of the electrostatic energy harvester.

under static condition, and the applying force, respectively. The mechanical quality factor of the proof mass is approximately 150. The numerical model is evaluated by the comparison with the characteristics of actual device.

4. Evaluation of Electrostatic Force of Monopolar and Bipolar Charging

We investigated the electrostatic force of the device, which didn't use fixed charged electret, by measuring the vibration amplitude of the proof mass versus applied acceleration as shown in figure 7. During the measurement and the simulation, the DC bias voltage is applied to the buried grid electrodes (BGEs) and the counter electrode (CE) is grounded as similar as actual harvesting operation. The DC bias voltage simulates charged electret. The applied bias voltage conditions and its resonant frequency were measured as shown in Table 1. Because the electrical spring stiffness is changed with different bias voltage, all measurement and simulation were carried out with different resonant frequency. The experiment results and the numerical analysis value are shown in figure 8. The mass movement of no bias condition showed only mechanical spring stiffness. When -200 V bias voltage is applied to the BGE-N (blue marker in figure 8), the mass movement showed smaller amplitude i.e. higher electrostatic force was occurred between the CE and the BGE-N. On the other hand, applying -200 V bias voltage to the BGE-P, the result showed lower electrostatic force because the CE was pulled laterally by BGE-P (green marker in figure 8). When ±100 V bipolar bias was applied, the mass movement showed as similar as the no bias condition. As a result, the bipolar charged electret was able to reduce the effect of electrostatic force as low as no bias condition. We confirmed that the numerical model with the electrostatic force from FEM static analysis accorded with the measurement results. In future, the evaluation of output power versus applied acceleration with load resistance for measurement and simulation is required to reproduce an actual harvesting operation.

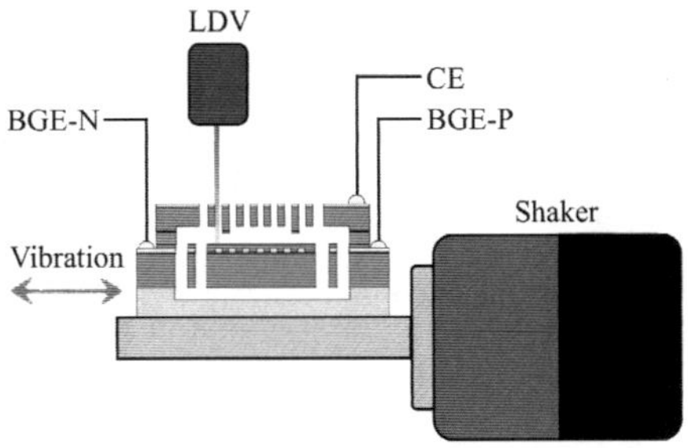

Figure 7. Experiment setup for evaluating the electrostatic force with DC bias between CE and BGEs. Laser Doppler Vibrometer (LDV) measures the amplitude of mass during vibrating. Force measurement of fabricated device was performed with controlled bias voltage and variable acceleration on shaker.

Table 1. Experiment potential on BGEs and resonant frequency f_r. Applied bias voltage conditions and its each resonant frequency were measured. CE is grounded under all conditions.

	BGE-N [V]	BGE-P [V]	f_r [Hz]
No bias	-	-	433.0
Monopolar	-200	0	427.2
	0	-200	425.7
Bipolar	-100	+100	434.7

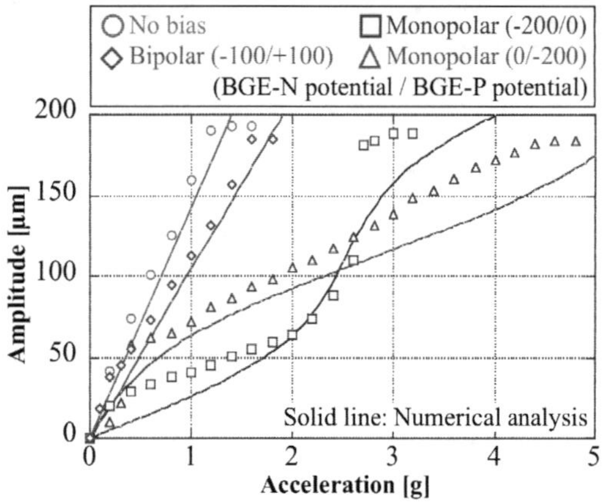

Figure 8. Vibration amplitude of proof mass versus applied acceleration. Comparison of measurement result and numerical analysis was performed with each resonant frequencies that depend on the electrostatic force. CE is grounded under all conditions.

5. Conclusion

The energy harvester with bipolar charged electret with the size of 13 × 12 × 1.2 mm³ was fabricated. The output power of the bipolar charged with ±250 V harvester was 9 μW when the acceleration was 1.4 g at 352 Hz with 0.9 MΩ optimum load resistance. The effectiveness against the VDRG limit was 2.5%. The numerical model of the energy harvester considering the electrostatic force obtained by FEM static analysis was also established. The characteristic of vibration amplitude versus applied acceleration was investigated and the results indicated that the bipolar charged device can reduce the effect of electrostatic force as low as no bias condition. The comparison between the numerical model and the measurement results showed a similar inclination.

References

[1] Masaki T et al., *J. Micromech. Microeng.*, 21, 104004, pp.1-5 (2011)
[2] Suzuki Y, *IEEJ Trans. 2011*, 6, pp.101-111 (2011)
[3] Wada Y et al., *Proc. of The 4th Symposium on Micro-Nano Engineering*, OS4-1-4 (2012) (in Japanese)
[4] Altena G et al., *Proc. of Transducers'11*, pp.739-742 (2011)
[5] Suzuki Y et al., *J. Micromech. Microeng.*, 20, 104002, pp.1-8 (2010)
[6] Toyonaga T et al., *The Technical Digest of APCOT 2010*, SSASN5, p.290 (2010)
[7] Sonoda K et al., *Proc. of PowerMEMS 2013* (London, UK, Dec. 3–6), pp. 366–370 (2013)

Performance enhancement of a low frequency vibration driven 2-DOF piezoelectric energy harvester by mechanical impact

M A Halim and J Y Park[1]
Micro/Nano Devices and Packaging Lab., Dept. of Electronic Engineering, Kwangwoon University, 447-1 Wolgye-dong, Nowon-gu, Seoul, 139-701 Korea.

E-mail: jaepark@kw.ac.kr

Abstract. We present a low frequency vibration driven 2-DOF piezoelectric energy harvester with increased performance, in terms of both bandwidth and output power, by mechanical impact. It consists of two series spring-mass systems (positioned in a parallel manner) one of which responds to low frequency vibration, engages with the harvester base stopper periodically by piecewise linear impact, and transfers a secondary shock to the second spring-mass system comprising of power generating element. It introduces a non-linear frequency up-conversion mechanism which, in turn, generates increased output power within a wide range of applied frequency. A 2-DOF prototype harvester without stopper shows two narrow resonant peaks and delivers maximum 2.11μW peak power to its matched load resistance at 17Hz frequency and 0.5g acceleration. On the other hand, it offers a -3dB bandwidth of 15Hz (9Hz-24Hz) and delivers maximum 202.4μW peak power to its matched load resistance at the same operating condition when a stopper is placed below the primary mass at 0.5mm distance. Generated power increases up to 449μW as the acceleration increases to 1g.

1. Introduction

During last few decades, A lots of efforts have been deployed to develop piezoelectric energy harvesters to scavenge power effectively from ambient vibration for powering low-power-consumption small wireless electronics [1,2] so that these harvesters can be an alternative to eliminate the need for recharging or replacing their external power sources such as batteries. Generally, a vibration energy harvester generates maximum power when its resonant frequency matches ambient vibration frequency and power flow decreases with the decrease in resonant frequency; the ambient vibrations, unfortunately, are of low frequencies with eccentric nature which may drift over time [3,4]. A conventional piezoelectric vibration energy harvester is designed as a single-degree-of-freedom (SDOF) model in the form of a single mass-loaded cantilever beam (made of either a piezoelectric bimorph or a piezoelectric layer attached to a non-piezoelectric layer) which is efficient at its 1st resonance mode; 2nd and higher resonance modes with comparatively low response levels occur in excessively high frequencies and are generally ignored. This instinctive behaviour results in a narrow operating bandwidth making itself inefficient for power harvesting from a wide range of low frequency vibrations. In order to overcome the narrow bandwidth problem, a number of researchers have demonstrated broadband harvester by using a multi-resonant piezoelectric beam, a second

[1] To whom any correspondence should be addressed.

piezoelectric beam as stopper, two-mass cantilever (piezoelectric) beam constituting a 2-degree-of-freedon (2-DOF) system, and many more [5-7].

A conventional 2-DOF piezoelectric vibration energy harvester comprising two spring-mass systems generates two close resonant peaks that increase its effective bandwidth than a 1-DOF energy harvester [8]. It also increases the output power by increasing the strain produced in the piezoelectric element because the spring-mass system with non-piezoelectric element works as a dynamic magnifier [9]. However, its power generation capability is still limited in low frequency vibration (below 30Hz). Also, such 2-DOF harvesters cannot be considered as true wideband because the 2nd resonance peak is still considerably smaller and quite far away from the 1st resonance peak, as well as contain a deep valley between two peaks. In order to overcome these constraints, we have proposed a mechanical impact based 2-DOF piezoelectric energy harvester that generates much higher output power from wideband low frequency vibration as compared to its without-stopper counterpart. It uses a flexible dynamic magnifier that not only decreases the resonant frequency but also increases the displacement limit in order to enhance dynamic magnification. The dynamic magnifier impacts on the harvester base stopper by piecewise linear motion during its vibration, even at a wide range of low frequencies; delivers secondary shock and introduces non-linear frequency up-conversion mechanism to the power generating element which, in turn, increases the output power significantly by increasing the strain and decreasing the source resistance within it.

2. Harvester design

Figure 1 shows the schematic structure of the proposed harvester. It is consisted of two parts: a primary spring-mass system comprising a flexible primary beam (32×11×1 mm^3 polycarbonate) with a primary mass (Aluminium, 6.1 gram) attached to one end and, a relatively smaller and rigid secondary beam (28×6×0.6 mm^3 PZT bimorph) with a small secondary mass (NdFeB cylinder, 1.05 gram) attached to one end, other end being clamped on the primary mass in a parallel manner over the primary beam, constituting secondary spring-mass system. The primary spring-mass system acts as a dynamic magnifier. In order to enhance its performance, a base stopper is placed at the bottom and the overall system is clamped to the anchor (Aluminium) in such a position that a small gap (1 mm) exists between the fender (lower part) of the primary mass and the harvester base as shown in figure 1(b). The overall spring-mass system as in figure 1(a) constitutes a 2-DOF system. While low frequency periodic excitation with sufficiently large acceleration is applied, the overall system starts vibrating as a 2-DOF system. On the other hand, while the primary mass engages the stopper, it results in a retardation of vibration motion and frequency response diverges from normal behaviour into non-linearity, broadening operating frequency bandwidth.

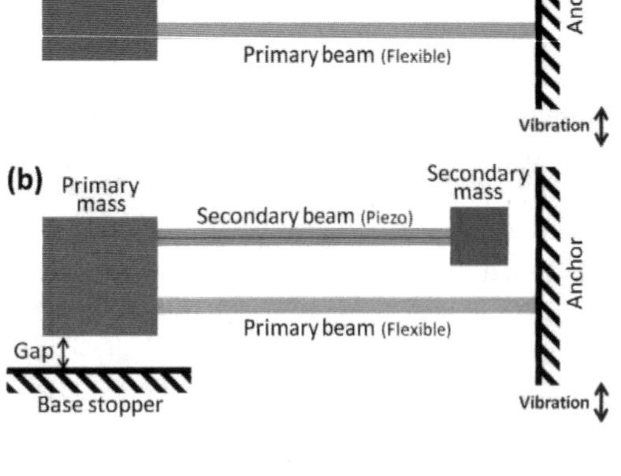

Figure 1. Schematics of the proposed 2-DOF energy harvester (a) without stopper and (b) with stopper.

The dynamic equation of motion for the conventional 2-DOF system (figure 1(a)) can then, be expressed as [7]

$$
\begin{aligned}
&m_2\frac{d^2x_2}{dt^2}+k_2(x_2-x_1)+c_m\left(\frac{dx_2}{dt}-\frac{dx_1}{dt}\right)=0\\
&m_1\frac{d^2x_1}{dt^2}+k_1(x_1-x_0)=k_2(x_2-x_1)+c_m\left(\frac{dx_2}{dt}-\frac{dx_1}{dt}\right)
\end{aligned}
\tag{1}
$$

in which m_1, k_1 and x_1 are the mass, stiffness and displacement of the primary spring-mass system, respectively; m_2, k_2 and x_2 are the mass, stiffness and displacement of the secondary spring-mass system, respectively; x_0 is the displacement of the vibrating base; and c_m is the mechanical damping. On the other hand, in the stopper-engaged 2-DOF system (figure 1(b)), the impact mechanism induces a secondary shock to the secondary spring-mass system. As a result, the secondary mass vibrates at its own natural frequency (higher than that of the primary spring-mass system) with exponential decay. The average stress on the piezoelectric layer is increased due to larger displacement of the secondary mass resulting in increased output voltage or power as compared to the earlier one. In the time interval between consecutive impacts, the dynamic equations of motion can be given as

$$
\begin{aligned}
&m_1\frac{d^2x_1}{dt^2}+k_1x_1+k_2(x_1-x_2)+c_m\left(\frac{dx_1}{dt}-\frac{dx_2}{dt}\right)=m_1a\sin\omega_0t\\
&m_2\frac{d^2x_2}{dt^2}+k_2(x_2-x_1)+c_m\left(\frac{dx_1}{dt}-\frac{dx_2}{dt}\right)=\delta_2(t-\tau)
\end{aligned}
\tag{2}
$$

in which ω_0 and a are the applied angular frequency and acceleration, respectively. $\delta_2(t\text{-}\tau)$ is the impulsive force generated by the secondary shock (while impact occurs) at time $(t\text{-}\tau)$, $t \geq \tau$. Then the response relative displacement of the secondary mass can be expressed by Duhamel's integral as [8]

$$x_2=\frac{1}{m_2\omega_d}\int_0^t p(\tau)e^{-\xi_2\omega_2(t-\tau)}\sin[\,\omega_d(t-\tau)]dt \tag{3}$$

where ω_d is the damped natural frequency, $p(\tau)$ is the arbitrarily varying excitation force at time τ, and ζ_2 is the damping ratio of the secondary beam.

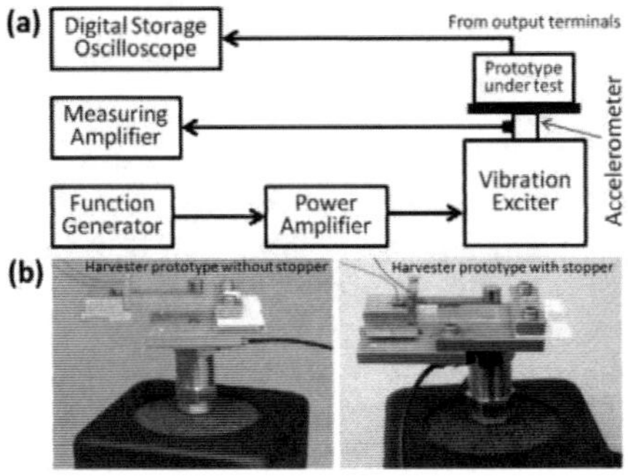

Figure 2. (a) Block diagram of the experimental setup and (b) photographs of the prototype without stopper and with stopper under test.

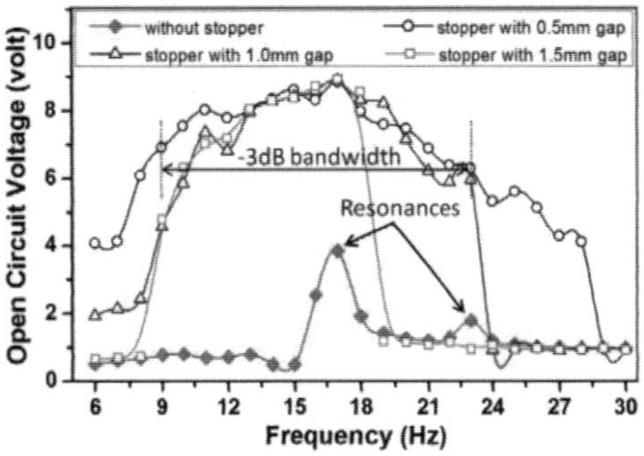

Figure 3. Frequency response of the 2-DOF energy harvester without stopper and with stopper under various stopper gap at 0.5g acceleration .

3. Experimental results and discussion

Figure 2 illustrates the schematic diagram of the complete experimental setup along with the photograph of the harvester prototype without stopper and with stopper under test. The frequency response of the peak-to-peak open circuit voltage of the prototype with stopper at various gap distances along with the prototype without stopper is shown in figure 3. The prototype without stopper shows two narrow resonant peaks at 17Hz and 23Hz, and that with stopper shows non-linear, wideband frequency response. It shows a maximum 15Hz -3dB bandwidth that reduces as the gap distance increases. When the primary mass impacts on the stopper, the effective stiffness of the primary beam changes, allowing the primary mass to deviate from its normal motion and enables the resonance to extend over a wider range. Besides, the peak-to-peak open circuit voltage from the prototype with stopper is found to be more than double than that of the prototype without stopper at the frequency of its 1st resonance. An impulsive force during impact increases the stress on the PZT beam that allows to increase the voltage.

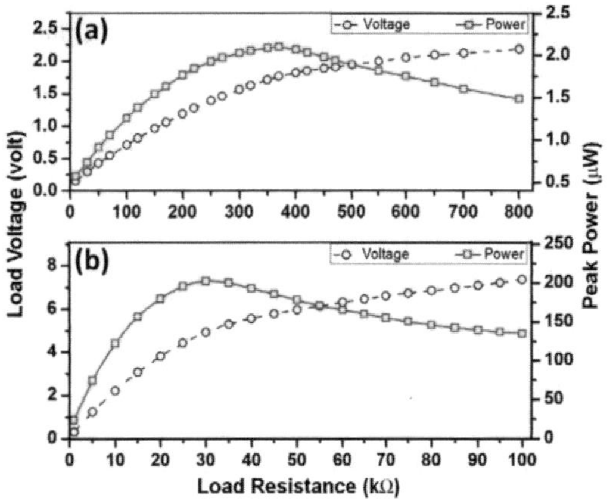

Figure 4. Load voltage and power of the proposed 2-DOF energy harvester (a) without stopper and (b) stopper with 0.5mm gap at 17Hz and 0.5g.

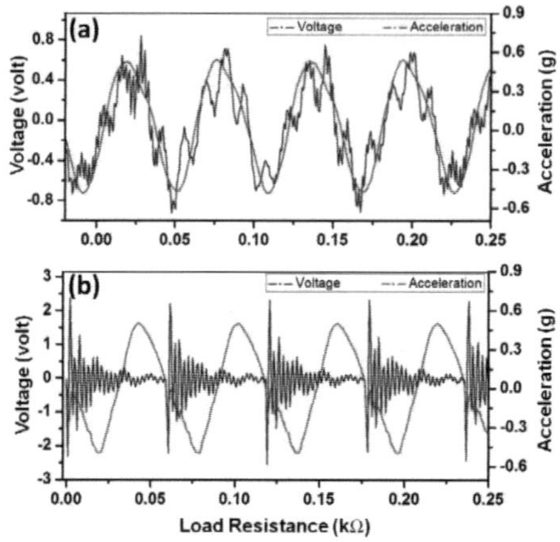

Figure 5. Instantaneous voltage waveforms of the proposed 2-DOF energy harvester (a) without stopper and (b) stopper with 0.5mm gap at 17Hz and 0.5g.

As shown in figure 4, the load voltages across various load resistances have been measured to determine the maximum power delivering capability of the proposed device. It was done at the 1st resonant frequency (17Hz) of the prototype without stopper for convenient comparison. The generated power is experimentally equal to $V_{p-p}^2/4R_l$; where V_{p-p} is the peak-peak load voltage across the load resistance R_l. Results show that the prototype with stopper delivers maximum 202.4μW peak power to a 30kΩ load resistance at 0.5g acceleration when the stopper was placed below at 0.5mm gap distance. On the other hand, the prototype without stopper delivers 2.51μW peak power to its optimal load resistance, 370kΩ at the same operating condition. Mathematically, the optimal R_l can be determined by [9]

$$R_l = R_s = 1/\omega_n C_p \tag{4}$$

where R_s, ω_n, and C_p are the source resistance, natural frequency, and capacitance of the piezoelectric beam, respectively. The frequency of the secondary (piezoelectric) beam is up-converted after the impact which, in turn, reduces the source resistance.

Figure 5 shows the instantaneously generated voltage waveform across the corresponding optimal load resistances of the harvester prototype without stopper and prototype with stopper at 17Hz frequency and 0.5g acceleration. The peak-to-peak voltage of the second one right after an impact is quite high, as compared to the first one, but decays exponentially with time due to damping and becomes almost zero before the next impact occurs. The up-converted frequency of the decaying waveform is 375Hz, obtained by Fast Fourier Transform (FFT) analysis. As seen, both the waveforms contain subharmonics, resulting in waving of the corresponding outputs. It happens because the flexible primary spring-mass system vibrates at low frequency with large displacement along with the secondary spring-mass system (power generating element). However, generated voltage and power increases in both cases as the acceleration increases as shown in figure 6. A maximum power of 449μW and 11.2μW was delivered to corresponding optimal load resistances of the prototype with and without stopper, respectively at an acceleration of 1g. All experimental results reveals that the proposed impact based 2-DOF piezoelectric vibration energy harvester performs better, in terms of both bandwidth and power, than the conventional one.

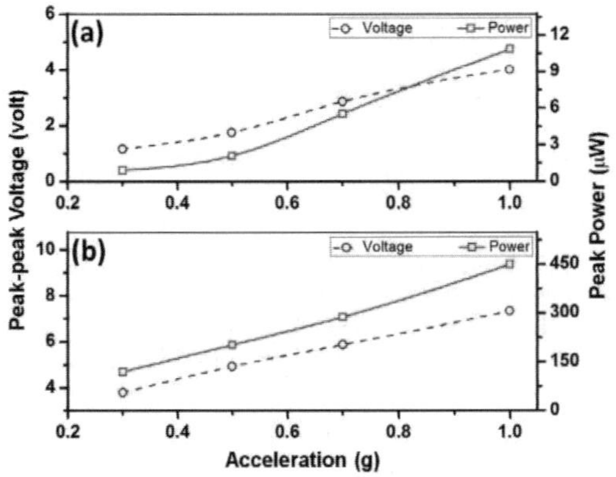

Figure 6. Output voltage and power across the corresponding optimal load of the proposed 2-DOF energy harvester (a) without stopper and (b) stopper with 0.5mm gap at 17Hz and various base excitations.

4. Conclusions

An impact based low frequency driven 2-DOF piezoelectric vibration energy harvester has been proposed and demonstrated that exhibits significantly improved performance in terms of both output power and wideband operation in low frequency vibration energy harvesting, compared to its conventional counterpart. It offers maximum 15Hz -3dB bandwidth and 202.4μW peak power at 0.5g which is increased up to 449μW at 1g acceleration. Piecewise linear impact of the primary mass on the mechanical stopper transforms an impulsive force to the secondary power generating element, introducing a frequency up-conversion mechanism and non-linear motion of the primary mass as well. Non-linearity results in wideband operation and frequency up-conversion allow the power to increase by increasing stress and decreasing source resistance in the generating element. Experimentally obtained results show that it can be used in automobile applications. Use of flexible beam material and mechanical stopper offers its reliable operation even when subjected to shock vibration. Optimization through further analysis and fabrication through MEMS process can further improve its performance.

Acknowledgements

The authors are grateful to acknowledge the Basic Science Research Program (2013R1A1A2A 10064810) and the Pioneer Research Centre Program (2010-0019313) through the National Research Foundation of Korea (NRF) funded by the Ministry of Science, ICT and Future Planning, Korea.

References

[1] Anton S R and Sodano H A, 2007 *Smart Mater. Struct.* **16** R1–R21
[2] Beeby S P, Tudor M J and White N M, 2006 *Meas. Sci. Technol.* **17** R175-195
[3] Mathers A, Moon K S and Yi J, 2009 *IEEE Sens. J.* **9** 731–739
[4] Arrieta A F, Hagedorn P, Erturk A and Inman D J, 2010 *Appl. Phys. Lett.* **97** 104102
[5] Qi S, Shuttleworth R, Oyadiji S O and Wright J, 2010 *Smart Mater. Struct.* **19** 094009
[6] Liu H, Lee C, Kobayashi T, Tay C J and Quan C, 2012 *Smart Mater. Struct.* **21** 035005
[7] Tang X and Zuo L, 2011 *J. Sound Vib.* **330** 5199–5209
[8] Lalanne C 2009 *Mechanical Vibration and Shock Analysis–volume 2* (New Jersey: Wiley) p 51
[9] Gu L, 2011 *Microelectr. J.* **42** 277–282

Equivalent circuit model of an impact-based piezoelectric energy harvester

S H Kim, S Ju, C H Ji and S J Lee
Department of Electronics Engineering, Ewha Womans University, 52, Ewhayeodae-gil, Seodaemun-gu, Seoul, Korea

E-mail: slee@ewha.ac.kr

Abstract. This paper presents an equivalent circuit model of an impact-based piezoelectric energy harvester. The harvester alternately generates two different shapes of damped-oscillating signals due to the asymmetric structure of the device. Those signals have strong high frequency components through mechanical impact from low frequency vibration. Equivalent circuit model with dual voltage source is proposed to precisely describe the behaviour of the harvester. The simulation results of the harvester with typical conversion circuits were compared with measurement data to verify the correctness of proposed model.

1. Introduction

Research on energy harvesting devices has been widely conducted on demand for sustainable and remote energy power supplies in applications such as implantable devices and wireless sensor networks [1, 2]. Ambient energy sources are, for example, sunlight, temperature difference, and vibration. Among harvesters using these sources, vibrational-to-electrical energy converters based on piezoelectric materials are drawing much attention because piezoelectric materials have large power densities and ease of use [1]. The energy generated by a piezoelectric harvester, however, is not adequate to be directly applied to loads. So a conversion circuit between a harvester and a load circuit is required to extract the maximum power from the harvester and transfer it to the load circuit efficiently.

To predict the operation of the interface circuit, an equivalent circuit model of the harvester is needed. Piezoelectric harvesters are often represented as a current source model with a resistor in parallel [3-5]. Although this model is simple, its usefulness is limited to the case of a single resistive load [1]. More accurate and analytical models, which reflect mechanical and structural characteristics of harvesters, have been also presented [1, 2]. These models generally assume a cantilevered beam type harvester excited by movements at its base, and cannot express the behaviour of an impact-based piezoelectric energy harvester reported by the authors precisely. The targeted harvester in [6] generates an asymmetric damped-oscillating signal based on indirect impact using a movable ball. Therefore a new circuit model is needed to account for specific voltage waveforms of the harvester.

This paper presents an equivalent circuit model of the impact-based piezoelectric energy harvester reported by the authors. Next section describes the structure of the harvester and characteristics of electrical signal generated from it. The proposed circuit model is presented and evaluated with respect to the measurement results. The final section draws conclusions.

2. Modeling for the impact-based piezoelectric energy harvester

2.1. Description of the harvester

The impact-based piezoelectric energy harvester [6], as shown in figure 1, consists of a freely movable ball within an aluminum housing and a cantilever-type beam. The beam is covered with piezoelectric material and has a proof mass at one end. When the harvester is shaken, the ball moves and impacts each sidewall of the case to enable the beam to vibrate. It generates higher frequency signal than the applied force, which could significantly increase the energy conversion efficiency by generating high frequency signal through mechanical impact from low frequency vibration.

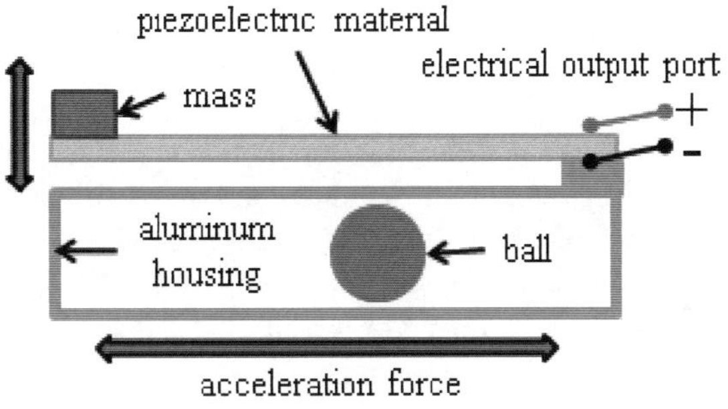

Figure 1. A schematic of the targeted energy harvester using a ball and a cantilever-type beam.

To observe electrical responses of the harvester, a shaker exciting an attached device vertically was used. Open-circuit output voltage of the harvester was measured at 20 Hz excitation with 3g acceleration, which was known as the optimum condition in this harvester [6]. Figure 2(a) shows the open-circuit voltage was generated twice in one period of the applied low frequency vibration because the ball moved upwards and downwards in one cycle of its movement. Because the force of a collision

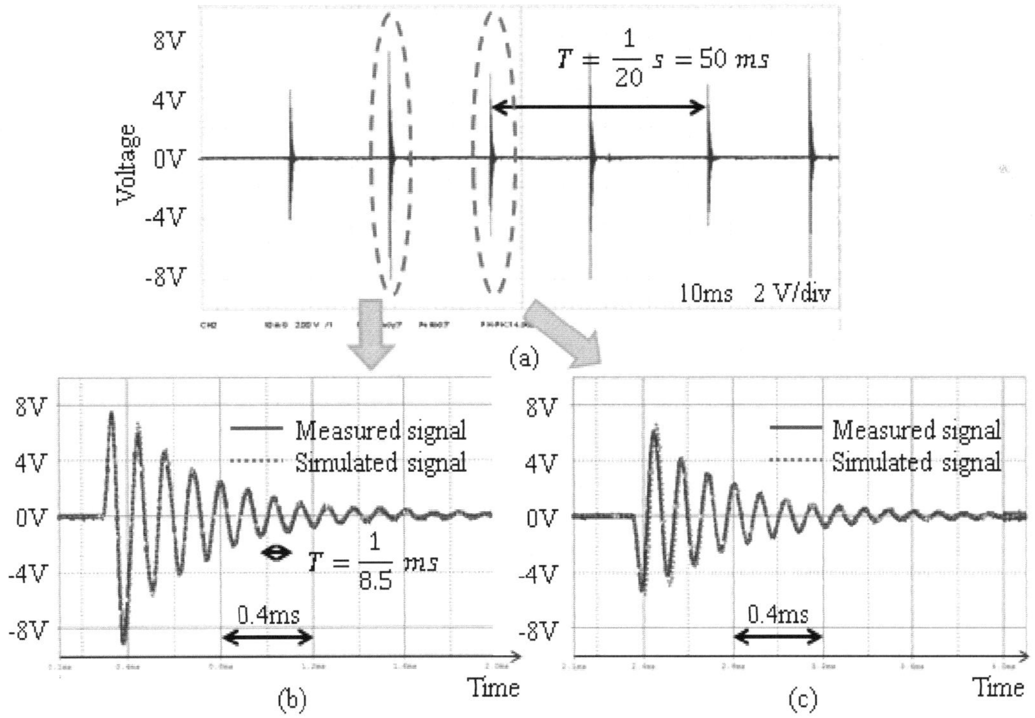

Figure 2. **(a)** A measured signal at open-circuit during 150 ms, **(b)** a measured vs. simulated positive-first voltage at open-circuit during 2 ms and **(c)** a measured vs. simulated negative-first voltage at open-circuit during 2 ms

at the bottom was not equal to that at the upper end, signals were induced in two different waveforms alternately. In addition, the second peak in the waveform was larger than the first peak due to the asymmetric structure of the device as shown in figure 2(b-c). The oscillating frequency of the generated signal through impacting was around 8.5 KHz. The signal gets attenuated almost exponentially after positive-negative peaks and returns to zero.

2.2. Equivalent circuit model

To depict the observed behavior, an equivalent circuit was built as shown in figure 3(a). It is composed of a mechanical portion and an electrical portion as was proposed by [7]. The mechanical system includes an inductor, L_m, a capacitor, C_k, and a resistor, R_{d1}, representing the mass, mechanical stiffness, and mechanical damping respectively. The electrical part consists of a capacitor, C_p, and a resistor, R_{d2}, to represent internal capacitance of the piezoelectric material and loss respectively. A transformer is used to model electromechanical coupling between the mechanical and electrical parts. Due to two asymmetric damped-oscillating signals of different amplitude and phase, however, two independent voltage sources, V_p and V_n, were used in this model instead of one sine wave source. As shown in figure 3(b), each voltage source generated two signals with opposite sign at intervals of an applied vibration.

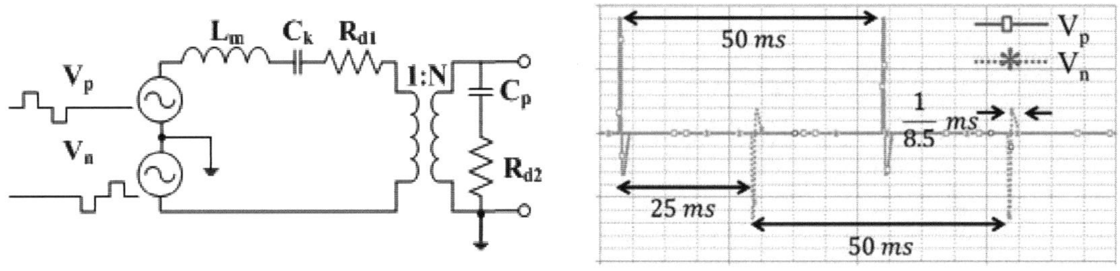

(a) (b)

Figure 3. (a) An equivalent circuit model of the targeted energy harvester **(b)** each signal of two voltage sources

Table 1. Measured peak-to-peak voltages across various loads: resistance, capacitance and inductance.

R(Ω)	Vp-p	C(nF)	Vp-p	L(mH)	Vp-p
500	6.6	1.1	15.2	1	1.6
600	7.5	2.2	14.2	5	12.8
700	8.0	5.0	12.4	10	26.8
800	8.8	10.0	9.9	15	42.0
900	9.6	20.0	7.1	20	42.3
1000	10.1	47.0	4.0	27	45.0
5000	15.4	94.0	2.3	43	38.7
∞(open)	16.6	235.0	1.0	100	25.4

Model parameters were determined by comparing results of Pspice simulation with experimental data, which were conducted by changing loads: inductors, capacitors and resistors. Table 1 lists the measured peak-to-peak voltages across various loads. Figure 4 shows that simulated voltages in the case of capacitive and inductive load conditions are very well-matched with the measurement data.

3. Experimental results

A voltage doubler and a full-bridge converter in figure 5(a,b) were tested to verify the model. SB140 diodes from VISHAY were used as rectifiers. The output capacitor value was 22 uF, and a 27 mH

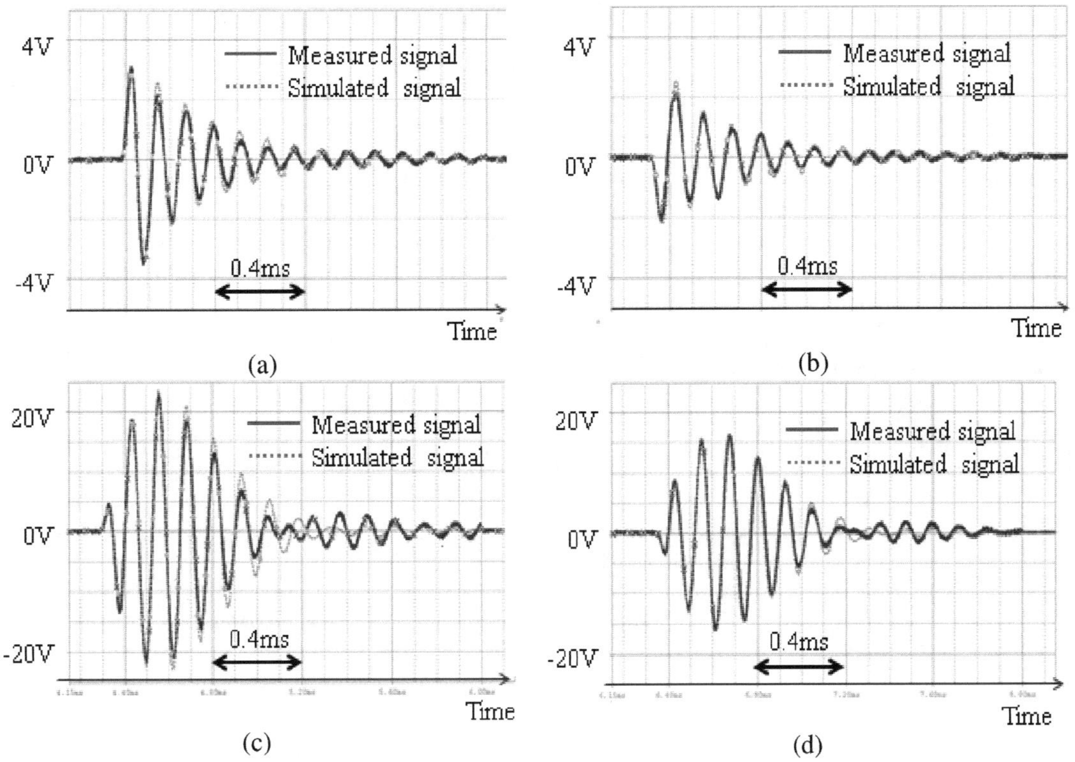

Figure 4. Measured vs. simulated signals during 1.5ms **(a)** a positive-first voltage across a load capacitor C = 20 nF, **(b)** a negative-first voltage across a load capacitor C = 20 nF **(c)** a positive-first voltage across a load inductor L = 27 mH, and **(d)** a negative-first voltage across a load inductor L = 27 mH

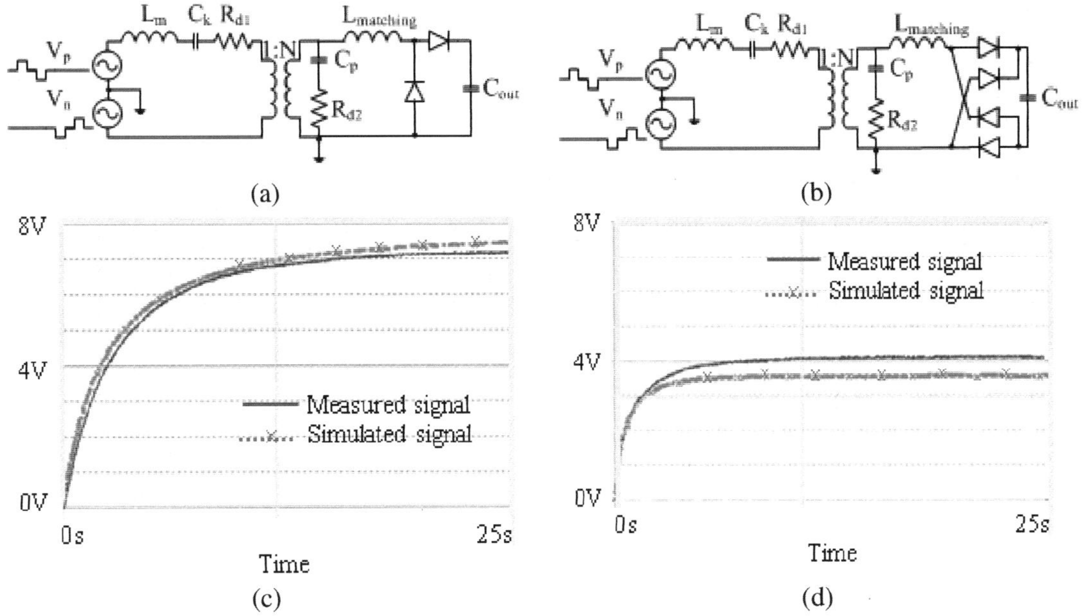

Figure 5. **(a)** a voltage doubler, **(b)** a full-bridge rectifier, **(c)** measured vs. simulated output signals of a voltage doubler, and **(d)** measured vs. simulated output signals of a full-bridge rectifier

inductor was used as a matching circuit. Measured and simulated voltages across the capacitor are shown in figure 5(c, d). The output of a voltage doubler increased more slowly and produced higher steady-state voltage than that of full-bridge converter in both cases. Figure 5(c, d) shows good agreement between measured data and simulation results.

4. Conclusion

This paper suggests an equivalent circuit model that is tailored to the behavior of an impact-based piezoelectric energy harvester. The harvester alternately generates two asymmetric damped-oscillating signals of different amplitude and phase at intervals of a vibration. Proposed circuit model includes two independent voltage sources as well as mechanical and electrical equivalent components to describe the behavior. This model is validated through PSpice simulations and proven to be quite accurate compared with experimental data.

Acknowledgments

This work was supported by the Ewha Global Top 5 Grant 2013 of Ewha Womans University.

References

[1] Erturk A 2009 Electromechanical modeling of piezoelectric energy harvesters. In: *Dept. Engineering Mechanics* Virginia Polytechnic Institute and State University, Blacksburg, VA

[2] Yang Y and Tang L 2009 Equivalent circuit modeling of piezoelectric energy harvesters *Journal of Intelligent Material Systems and Structures* **20** 2223-35

[3] Elliott A D T, Trice M R and Mitcheson P D 2012 Implementation of single supply pre-biasing with sub-35μw control overhead for piezoelectric energy harvesting *Proc. PowerMEMS 2012*

[4] Dicken J, Mitcheson P D, Stoianov I and Yeatman E M 2012 Power-extraction circuits for piezoelectric energy harvesters in miniature and low-power applications *Power Electronics, IEEE Transactions on* **27** 4514-29

[5] D'Hulst R, Sterken T, Puers R, Deconinck G and Driesen J 2010 Power processing circuits for piezoelectric vibration-based energy harvesters *Industrial Electronics, IEEE Transactions on* **57** 4170-7

[6] Ju S, Chae S H, Choi Y, Jun S, Park S M, Lee S, Lee H W and Ji C-H 2014 *Proc. APCOT 2014*, pp 13-3

[7] Roundy S and Wright P K 2004 A piezoelectric vibration based generator for wireless electronics *Smart Materials and structures* **13** 1131

Inductive energy harvesting from variable frequency and amplitude aircraft power lines

T T Toh[1], S W Wright[1], M E Kiziroglou[1], J Mueller[2], M Sessinghaus[3], E M Yeatman[1] and P D Mitcheson[1]

[1] Department of Electrical and Electronic Engineering, Imperial College London, UK
[2] Wireless Competence Centre, Airbus Operations GmbH, Hamburg, Germany
[3] SILVER ATENA Electronic Systems Engineering GmbH, Hamburg, Germany

E-mail: tzern.toh@imperial.ac.uk

Abstract. This paper presents a non-contact method of harvesting energy from an aircraft power line that has an AC current of variable amplitude and a frequency range of 360-800 Hz. The current and frequency characteristics of the aircraft power line are dependent on the rotation speed of the electrical generators and will therefore change during a flight. The harvester consists of an inductive coil with a ferrite core, which is interfaced to a rectifier, step-down regulator and supercapacitor. A prototype system was constructed to demonstrate reliable output voltage regulation across a supercapacitor that will supply a peak power of 100 mW under duty cycled load conditions. The system could fully charge a 40 mF supercapacitor to 3.3 V in 78 s from a power line current of 1.5 A_{rms} at 650 Hz.

1. Introduction

Energy harvesting for aircraft-based applications has gained considerable interest as a means of realising a localised power source, thus reducing the weight and complexity of aircraft wiring looms. Kinetic harvesters have had little impact in this environment, however, partly because of reliability issues and also the difficulty of operating with the low vibration levels typical of modern aircraft. Contrastingly, thermoelectric generators used with phase-change materials have been demonstrated for powering sensors outside the heated cabin environment [1].

The research presented here focuses on another non-kinetic energy harvester for aircraft applications, coupling the magnetic field from a variable amplitude and frequency aircraft power supply line into an inductive coil. In this case, the metal frame of the aircraft forms the ground return path. The target application is a wireless sensor node that typically requires tens of mW and the harvesting system should be easily retrofitted onto an aircraft power line cable.

2. Inductive energy harvester design and characterisation

Figure 1(a) shows a schematic of the split-ring cores, inductive coil and power line cable, and figure 1(b) shows an equivalent circuit model of the inductive energy harvester [2]. It consists of a current source I to represent the power line current, N is the number of turns of the inductive coil, the winding and leakage inductances are L_w and L_{leak} respectively, a winding resistance R_w and the losses in the ferrite core are modelled by a resistor R_i. A tuning capacitor C_t is chosen to resonate with L_w at a specific frequency. Finally, a load resistor R_L completes the circuit. All the reactive and resistive components have been referred to the secondary side. For the frequencies of interest, 360-800 Hz, the loss in the core is negligible and hence R_i is only present in figure 1(b) for completeness.

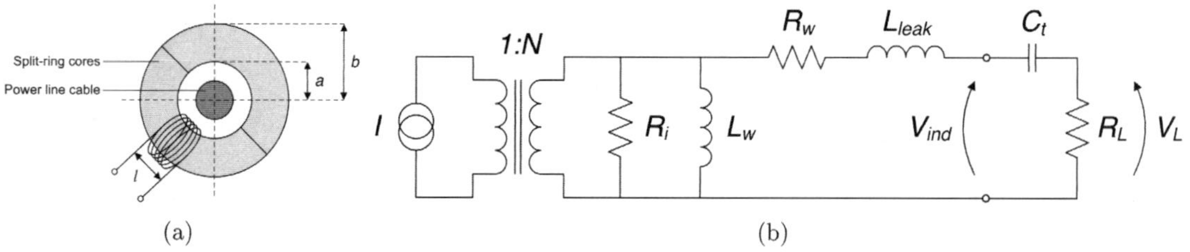

Figure 1. (a) Schematic of the inductive coil energy harvester and (b) equivalent circuit of the power line cable and inductive coil connected to a tuning capacitor and load.

2.1. Number of coil turns, N

The induced open-circuit voltage of the coil may be deduced by use of Ampere's Law. In (1), f and I are the power line frequency and current respectively, μ_r is the relative permeability of the ferrite core and a, b and l represent the inner and outer core radii and the coil length respectively, as shown in figure 1(a).

$$V_{ind} = 2\pi f N I \mu_0 \mu_r \ln\left(\frac{b}{a}\right) l \tag{1}$$

There is freedom when designing the coil to change the number of turns N and the coil wire diameter, which determines the coil resistance. Intuitively, it seems desirable to maximise N because the induced voltage across the coil is proportional to N. However, in a practical application there is a volume constraint on the total device size and for a given volume, increasing the number of turns increases the coil resistance since thinner wire is necessary to accomodate the additional turns. Consequently, the design strategy is to select N such that the induced coil voltage is above the minimum input voltage of the power electronics circuit and the winding resistance is comparable to the expected input impedance of a switch-mode interface circuit.

2.2. Tuning capacitor, C_t

The combination of coil inductance and tuning capacitor determines the resonant peak of the harvesting system according to (2) [3]:

$$f_{res} = \frac{1}{2\pi\sqrt{L_w C_t}} \tag{2}$$

In an aircraft application, C_t should be chosen such that resonance occurs at the dominant frequency of the flight cycle during the cruising phase. If, however, a dominant frequency does not exist, it will usually be better to have resonance occur in the lower frequency range. This is because the output power increases with frequency and hence, more power is harvested at the higher frequencies compared to the lower frequencies.

Figure 2(a) plots the power provided to a resistive load that is matched to the coil resistance for various values of C_t and also with no tuning capacitor. Figure 2(b) shows the variation in bandwidth for different tuned frequencies. The measurements in figure 2 were taken from a prototype harvester with 1100 turns hand wound onto a split ferrite core. The characteristics of the coil are summarised in table 1.

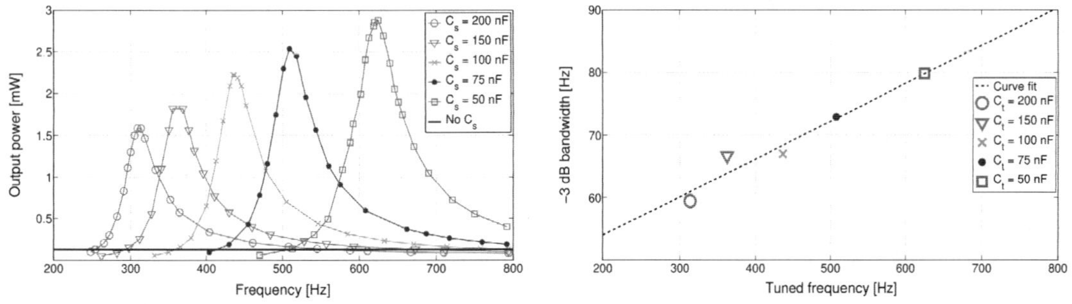

(a) (b)

Figure 2. Measurements of (a) output power dissipated into a matched load resistor against power line frequency and (b) -3dB bandwidth for different tuned frequencies, from a power line current of 0.9 A_{rms}.

Table 1. Measured inductive coil and ferrite core characteristics for $N = 1100$.

Parameter	Coil			Ferrite core (Würth Elektronik: 7427021)			
	Inductance	Resistance	Wire diameter	Inner diameter	Outer diameter	Length	μ_r
Value	1.4 H @ 500 Hz	100.9 Ω	100 μm	8.2 mm	16.3 mm	20 mm	≈600

3. Power management electronics

A block diagram of the interface electronics topology is shown in figure 3. A signal generator and an audio amplifier were used to simulate the aircraft power line current for test purposes.

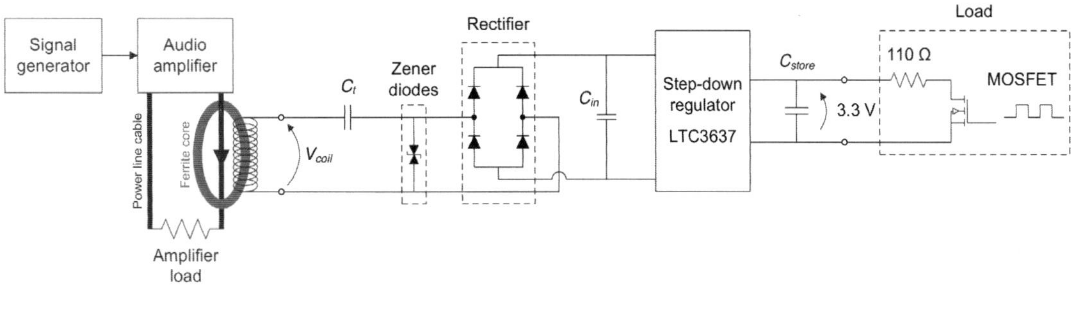

Figure 3. Power processing stages for the power line energy harvesting system.

A bidirectional Zener diode is used to protect the power processing electronics from excessive induced voltages due to surges in the power line. The induced coil voltage needs to be rectified. The simplest implementation is a full-wave bridge rectifier using four BAT46W Schottky diodes with N chosen to provide a sufficiently high voltage to overcome the diode voltage drops. Supercapacitors were chosen for energy storage because of their larger number of charge and discharge cycles before degradation and higher specific power rating compared to batteries.

The LTC3637 step-down regulator has an input voltage range of 4-76 V and it will be used to regulate the voltage across the supercapacitors at 3.3 V. Since the LTC3637 requires a minimum input voltage of 4 V, the number of coil turns N should be sufficiently large to exceed this threshold. Therefore, a 1100 turn coil was chosen because this could be reliably hand wound using 100 μm enamelled wire, given the size of the split-ring ferrite core.

3.1. Supercapacitor charging time

The choice of supercapacitors depends on the duty cycle and power consumption of the sensor node. Higher value capacitors will require a longer charging time compared to lower values ones but they can store more energy. So, there is a trade-off between the energy requirement of the load and the permissible charging time.

The system was constructed as shown in figure 3, using a 40 mF Cellergy CLG04P04L17 supercapacitor as the storage element. This was fully discharged prior to each measurement. The energy harvesting capability was tested for various power line currents and the process of charging the supercapacitor, without a tuning capacitor, is shown in figure 4(a). Figure 4(b) summarises the the charging times for $C_t = 100\ nF$. The induced voltage peaks at the tuned frequency of 421 Hz, resulting in a shorter charging time for a given current.

The charging time can be shortened by looping a slackened length of power line cable multiple times around one half core, as shown in figure 4(c), and subsequently aligning it with the second half core using a clamp. The thinner enamelled wire on the left half core is the inductive coil. Figure 4(d) shows the effect of doubling or tripling the loops of power line cable.

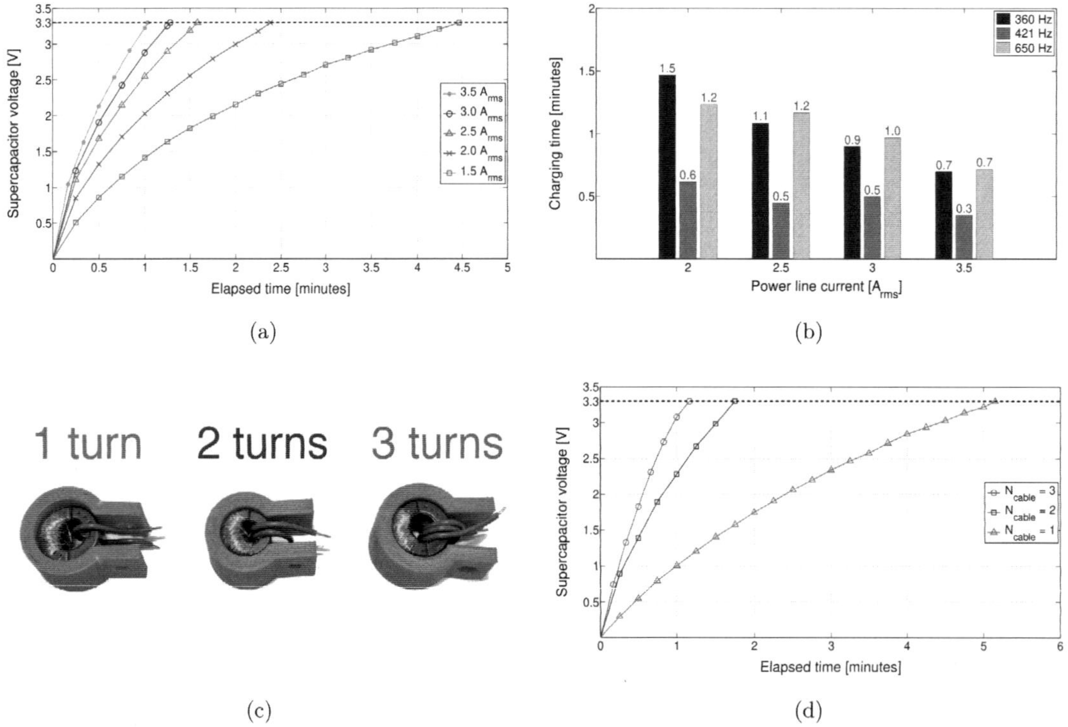

Figure 4. (a) Measured charging times of a 40 mF supercapacitor at 650 Hz and a single turn of power line cable carrying 1.5 A_{rms} without C_t, (b) charging time variations for $C_t = 100\ nF$ with a single turn power line cable, (c) photograph showing multiple turns of a power line cable, and (d) charging times of a 40 mF supercapacitor using multiple turns of a cable carrying 1.5 A_{rms}.

The prototype could fully charge a 40 mF supercapacitor to 3.3 V in 78 s with a power line current of 1.5 A_{rms} and three turns of the power line cable, *i.e.* delivering 0.22 J to the supercapacitors in 78 s.

4. System characterisation

A typical application for this system would be a wireless sensor node, requiring about 100 mW at a 1% duty cycle [1]. To emulate such a load, a 110 Ω resistor in series with a MOSFET switch was connected in parallel with the supercapacitor. When the MOSFET is turned on by the appropriate gate drive, the resistor draws ≈100 mW from the system.

Initially, the load is disconnected and the supercapacitor was charged using the inductive coil. During the charging phase, the LTC3637 input voltage is held at the under-voltage lockout threshold, which ranges between 3.5-3.65 V. In this mode, all the harvested energy is used to charge the supercapacitor. Once the supercapacitor is charged to 3.3 V, the LTC3637 input voltage will change according to the induced coil voltage.

Figure 5 shows measurements of the rectified coil voltage and supercapacitor voltage for power line currents of 1.5 A_{rms} and 3 A_{rms} at 650 Hz. In this example, the 40 mF supercapacitor can sufficiently supply 100 mW once per second, for four consecutive gate drive pulses. Whenever the 110 Ω was connected, the supercapacitor voltage remains at 3.3 V whilst the rectified coil voltage decreases because the harvested energy is being used to recharge the supercapacitor. Once the supercapacitor is fully charged, the rectified coil voltage will rise accordingly. In the 3 A_{rms} case, the rectified coil voltage never decreases to the lockout threshold because the supercapacitor is being recharged at a faster rate. This agrees with the charging times observed in figure 4.

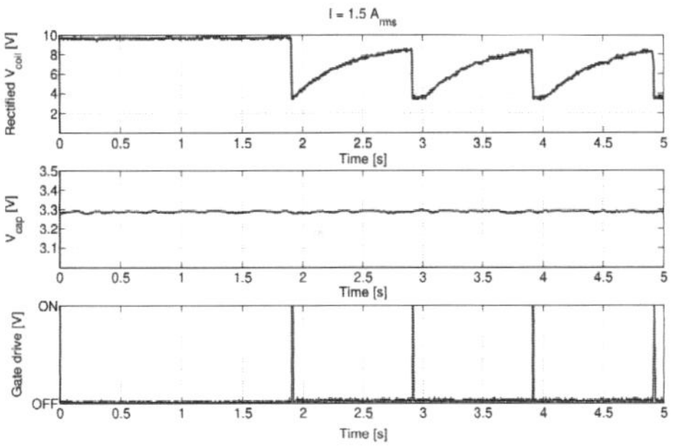

(a)

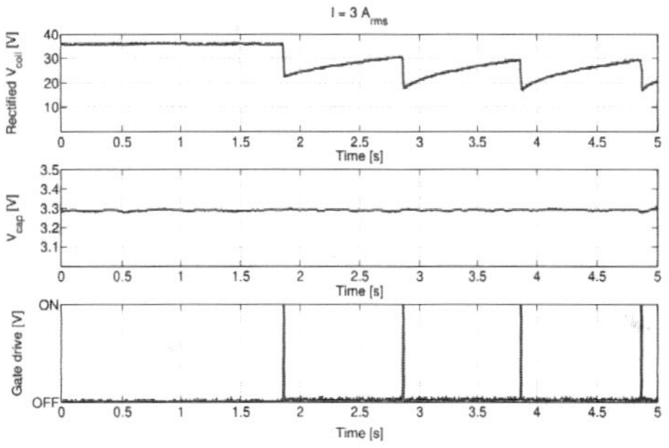

(b)

Figure 5. Experimental measurements using a continuous power line current of (a) 1.5 A_{rms} and (b) 3 A_{rms}, with a switched load to draw 30 mA once per second at 1% duty cycle.

5. Conclusions

A prototype power line energy harvesting system has been developed for aircraft applications. A tuning capacitor can be used to resonate out the coil inductance at specific frequencies. The system successfully charged a 40 mF supercapacitor to 3.3 V and supplied 100 mW to a duty cycled load. The weight and volume of the prototype are 80 g and 103 cm^3 respectively and this equates to an energy density of 2.7 mJ/g or 2.1 mJ/cm^3. The energy density of the system can be increased because the prototype was designed for laboratory tests.

References

[1] Toh T T *et al.* 2014 *Sensors and Actuators* A: Physical **211** 38–44
[2] Moon J *et al.* 2013 *Applied Power Electronics Conference* (Long Beach, USA) pp 3151–3159
[3] Roscoe N M and Judd M D 2013 *IEEE Sensors* **13**(6) 2263–2270

Reliability of vibration energy harvesters of metal-based PZT thin films

Y Tsujiura, E Suwa, F Kurokawa, H Hida and I Kanno

Department of Mechanical Engineering, Kobe University, Rokkodai-cho, 1-1 Nada-ku, Kobe, Japan

E-mail: tsujiura@stu.kobe-u.ac.jp

Abstract. This paper describes the reliability of piezoelectric vibration energy harvesters (PVEHs) of $Pb(Zr,Ti)O_3$ (PZT) thin films on metal foil cantilevers. The PZT thin films were directly deposited onto the Pt-coated stainless-steel (SS430) cantilevers by rf-magnetron sputtering, and we observed their aging behavior of power generation characteristics under the resonance vibration condition for three days. During the aging measurement, there was neither fatigue failure nor degradation of dielectric properties in our PVEHs (length: 13 mm, width: 5.0 mm, thickness: 104 μm) even under a large excitation acceleration of 25 m/s^2. However, we observed clear degradation of the generated electric voltage depending on excitation acceleration. The decay rate of the output voltage was 5% from the start of the measurement at 25 m/s^2. The transverse piezoelectric coefficient ($e_{31,f}$) also degraded with almost the same decay rate as that of the output voltage; this indicates that the degradation of output voltage was mainly caused by that of piezoelectric properties. From the decay curves, the output powers are estimated to degrade 7% at 15 m/s^2 and 36% at 25 m/s^2 if we continue to excite the PVEHs for 30 years.

1. Introduction

In the research of piezoelectric micro-electromechanical systems (MEMS), piezoelectric vibration energy harvesters (PVEHs) using $Pb(Zr,Ti)O_3$ (PZT) thin films have been widely investigated because of their high piezoelectric properties [1–6]. Having good compatibility with conventional semiconductor microfabrication techniques, Si substrates are used for most of the piezoelectric MEMS devices including PVEHs. However, brittle Si is not the best material for PVEHs which incur continuous large vibration. One of the solutions for this problem is to use flexible materials such as organic or metal materials as a base substrate [7–10]. Organic materials have excellent flexibility more than metal, however, their melting points are usually less than 200°C; this indicates that it is impossible to deposit the PZT thin films directly on organic substrates using conventional deposition methods because of their high deposition temperature more than 500°C.

On the other hand, the PZT thin films can be deposited directly on metal substrates since melting points of metals are higher than deposition temperature of piezoelectric thin films, indicating that we can use the same fabrication process with Si-MEMS devices. In recent years, PVEHs of metal cantilevers have been actively studied because they can generate the larger displacement at the resonance without break down unlike brittle Si cantilevers [11–13]. However, there have been no detailed study on the long-term reliability for the power generation characteristics of the PVEHs of PZT thin films on metal cantilevers.

In this study, we fabricated the PZT thin-film PVEHs of stainless-steel (SS430) foil cantilevers using rf-magnetron sputtering. To clarify the relationship between aging behavior of the output power or the piezoelectric properties and excitation magnitudes, we measured the output voltage of our PVEHs for three days (4320 minutes) at the various vibration conditions. Based on the short-range aging characteristics, we discuss the long-term reliability of the PZT thin films on SS430 substrates.

2. Device fabrication

Figure 1 shows a schematic illustration of the PVEH of PZT thin films deposited on SS430 foil cantilever. After deposition of Pt/Ti bottom electrodes, 4.0-μm-thick PZT thin films were then directly deposited onto the Pt-coated SS430 substrate (length: 20 mm, width: 5.0 mm, thickness: 100 μm) by rf-magnetron sputtering [14]. The PZT thin films on SS430 substrate were annealed in atmosphere at 600°C for 1.5 hours. Subsequently, the Pt top electrode was prepared through a shadow mask. We fixed one end in the longitudinal direction of the PZT/SS430 substrate by a clamping jig, and the PVEH of PZT/SS430 unimorph cantilever (length: 13 mm, width: 5.0 mm, thickness: 104 μm) was fabricated.

From X-ray diffraction (XRD) measurement, PZT thin films on Pt-coated SS430 substrate have the perovskite structure with a random orientation (figure 2). We also evaluated the dielectric properties of the PZT thin films using an LCR meter (NF ZM2353), and the relative dielectric constant (ε_r) and the dielectric loss (tan δ) were measured to be 425 and 0.018, respectively. The transverse piezoelectric properties of the PZT thin films were assessed from the direct piezoelectric effect for the PVEHs. After the application of a negative unipolar electric voltage on the top electrode as a poling treatment (1000 Hz, −20 $V_{p\text{-}p}$), we applied periodical displacement to the tip of the cantilever. The generated electric voltage between the top and bottom electrodes was measured using a lock-in amplifier (SIGNAL RECOVERY 7265). From the relationship between the measured electric voltage and the applied displacement, the transverse piezoelectric coefficient $e_{31,f}$ was calculated using equation (1), [15]

$$e_{31,f} = -\frac{4l^3 C_p}{3wh_s(1-\nu_s)(x_1-x_0)(2l-x_1-x_0)} \cdot \frac{V}{\delta} \quad (1)$$

where h_s and ν_s are the thickness and Poisson's ratio of a base substrate; w and x_0 (x_1) are the width and the extremities of the top electrode; l, δ, V, and C_p are the length of the cantilever, applied displacement to the cantilever, output electric voltage, and capacitance, respectively. When the frequency of applied displacement was 30 Hz, the transverse piezoelectric coefficient $e_{31,f}$ was calculated to be −3.5 C/m^2 from equation (1).

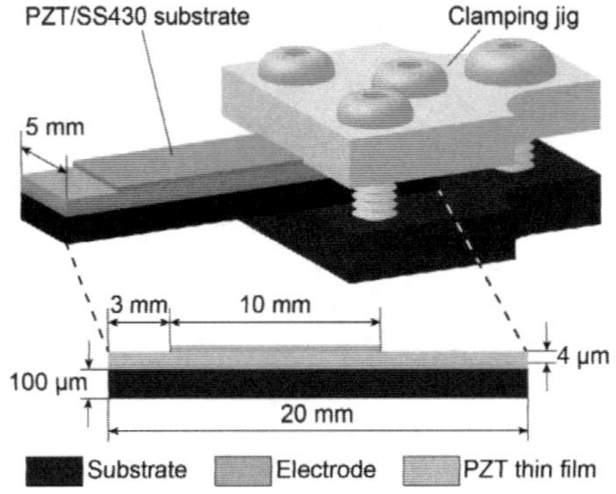

Figure 1. Schematic illustration of the PVEH of PZT thin films on SS430 foil cantilever.

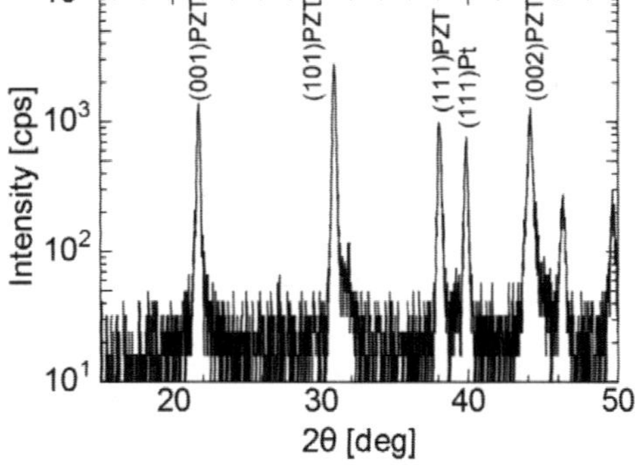

Figure 2. XRD patterns of PZT thin films on Pt-coated SS430 substrate.

3. Power generation reliability of PVEHs

3.1. Basic characteristics of power generation

Figure 3 shows a schematic illustration and a photograph of the measurement setup for PVEHs. The PVEH was mounted on the vibration exciter (EMIC 512-A). The acceleration pickup (EMIC 731-B) was attached on the base of the cantilever for monitoring the acceleration amplitude. The top electrode of the PZT thin films was connected to a gold wire using silver paste. The generated electric voltage across an external load resistance was measured using a frequency response analyzer (FRA; NF FRA5095). Average output power P of the PVEHs was calculated from the root-mean-square (rms) voltage V_{rms} in a load resistance R using

$$P = \frac{(V_{rms})^2}{R} \tag{2}$$

First, we measured both the resonance frequency and the optimal load resistance which maximize the output power of the PVEHs. Figure 4 shows the frequency response of the output voltage and the cantilever displacement (at 7 mm from the fixed end) of the PVEHs. We swept the excitation frequency, and the output voltage and the cantilever displacement were measured using an FRA and a laser Doppler vibrometer, respectively. As shown in figure 4, clear peaks in both the output voltage and the cantilever displacement appeared at the same frequency of 323.7 Hz at the acceleration of 5 m/s^2. Furthermore, when the acceleration was raised until 25 m/s^2, the resonance frequency increased

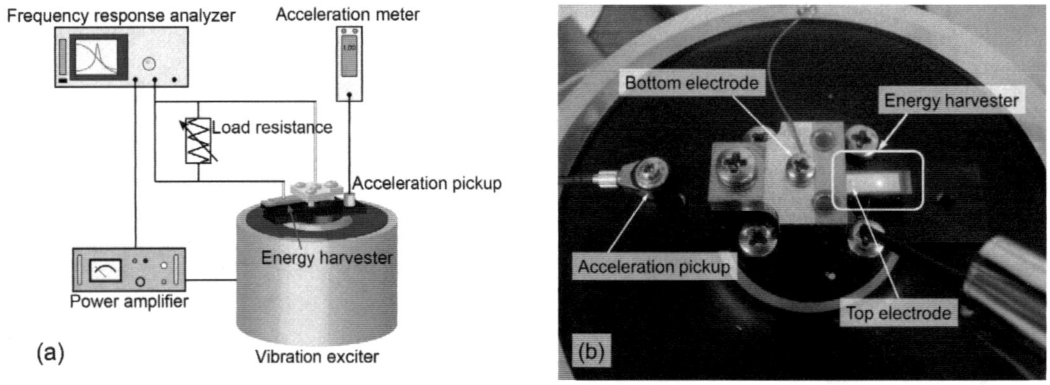

Figure 3. Measurement setup for power generation performance of PVEHs: (a) a schematic illustration and (b) a photograph.

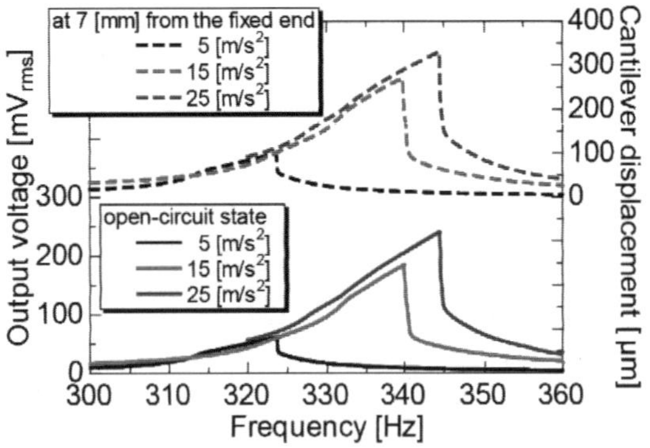

Figure 4. Output voltage in an open-circuit state and cantilever displacement as functions of frequency.

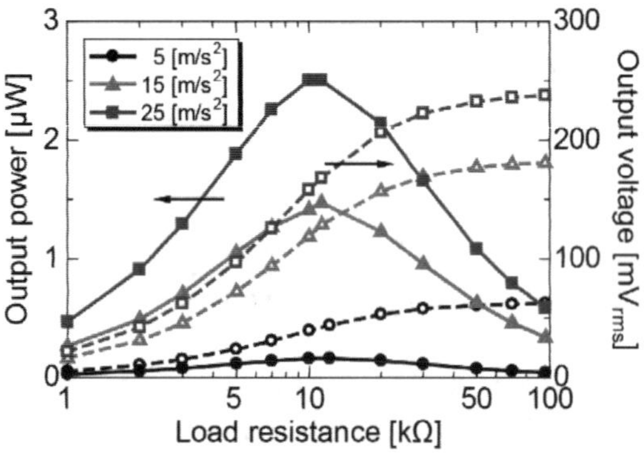

Figure 5. Output power and output voltage as functions of load resistance.

to 344.4 Hz because of nonlinear resonance of hardening spring effect. Figure 5 shows the output power and output voltage as functions of the load resistance at the resonance frequency, and the optimal load resistance was measured to be approximately 12 kΩ. In the aging measurement, the PVEHs were excited under the conditions of resonance frequency, optimal load resistance, and acceleration obtained in this subsection. Just before the aging measurement, we applied a negative unipolar electric voltage (1000 Hz, −20 $V_{p\text{-}p}$) on the top electrode for one minute as a poling treatment.

3.2. Aging behavior

Measuring the aging behavior of power generation characteristics, we excited the PVEHs at their resonance frequency measured in the previous subsection 3.1 and adjusted the cantilever displacement to be constant.

During the aging measurement for three days, neither fatigue failure nor degradation of dielectric properties (ε_r and tan δ) was observed in our PVEHs under each acceleration of 5, 15 and 25 m/s^2. Figure 6 shows the output voltage in an open-circuit state as a function of vibration time. At the acceleration of 5 m/s^2, the output voltages were almost constant during the three-day measurement. On the other hand, when the acceleration was increased up to 15 and 25 m/s^2, the output voltages monotonically reduced 2.5 mV_{rms} (1.4%) and 11.7 mV_{rms} (4.8%), respectively.

Figure 7 shows the normalized output power as a function of vibration time. Under accelerations of 15 and 25 m/s^2, the output powers reduced 4 and 11% during the measurement period, respectively. The output power at 15 m/s^2 decreased linearly throughout the measurement, whereas the output power at 25 m/s^2 fell rapidly from the times that exceed 200 minutes. The equations in figure 7 represent the decay curves at the accelerations of 15 and 25 m/s^2 calculated using the least square method; x and y mean the vibration time and the normalized output power, respectively. The decay rates of the output powers were 0.93% at 15 m/s^2 and 6.4% at 25 m/s^2 by every decade on a logarithmic time scale. Assuming that the decay curves are appropriate secularly, the output powers of the PVEHs are estimated to decay 7% at 15 m/s^2 and 36% at 25 m/s^2 if we continue to excite the PVEHs for 30 years.

Furthermore, we evaluated the reliability of the PVEH without vibration excitation by measuring the piezoelectric properties. Figure 8 shows the aging characteristics of the transverse piezoelectric coefficient $e_{31,f}$. No clear degradation of the $e_{31,f}$ was observed at the accelerations of 0 and 5 m/s^2. On the other hand, when the accelerations were 15 and 25 m/s^2, the $e_{31,f}$ respectively reduced 1 and 4%, which are almost the same decay rates with those of the output voltages. This indicates that the degradation of power generation performance was mainly caused by that of piezoelectric properties.

From the aging measurement, the PZT thin films on SS430 foil cantilevers are expected to possess good long-term reliability of power generation performance at even large excitation acceleration of 15 m/s^2.

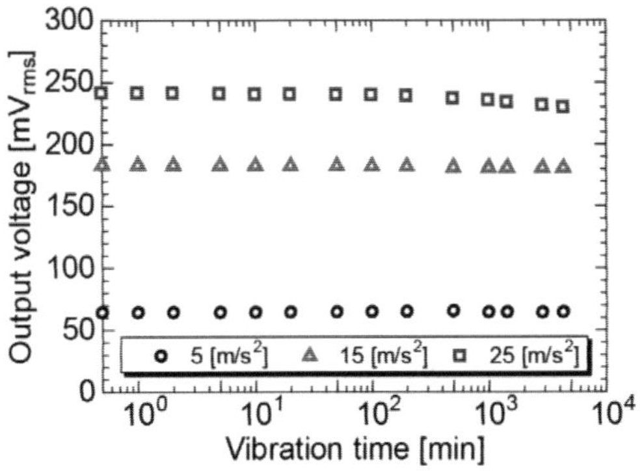

Figure 6. Output voltage in an open-circuit state at the resonance frequency as a function of vibration time.

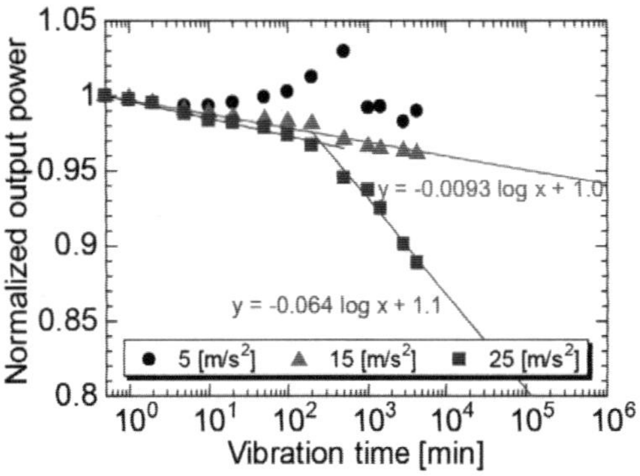

Figure 7. Normalized maximum output power as a function of vibration time.

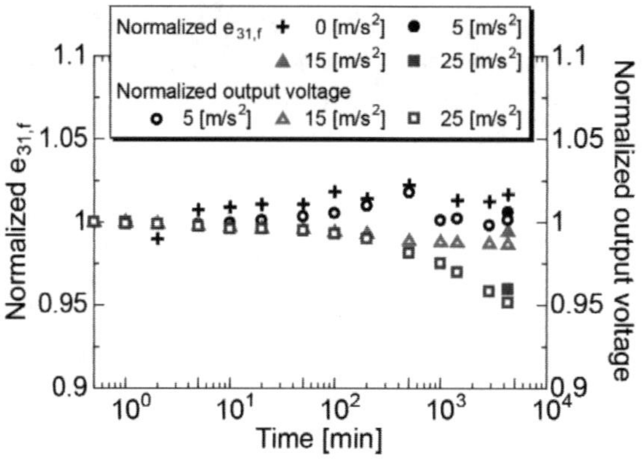

Figure 8. Aging behavior of the transverse piezoelectric coefficient $e_{31,f}$.

4. Conclusions

In this study, we evaluated the reliability of the PZT thin-film PVEHs of SS430 foil cantilevers. During the aging measurement for three days, there was neither fatigue failure nor degradation of dielectric properties in our PVEHs even under a large excitation acceleration of 25 m/s^2. However, we observed clear degradation of the output voltage depending on excitation acceleration, and the decay rate of the output voltage was 5% from the start of the mesurement at 25 m/s^2. The transverse piezoelectric coefficient ($e_{31,f}$) also degraded at almost the same decay rate with that of the output voltage; this indicates that the degradation of the output voltage was mainly caused by that of piezoelectric properties. From the decay curves, the output power at 15 m/s^2 is estimated to decay only 7% if we continue to excite the PVEH for 30 years, whereas the output power at 25 m/s^2 fell rapidly from the times that exceed 200 minutes. Therefore, the PZT thin films on SS430 foil cantilevers are expected to possess good long-term reliability of power generation performance at even large acceleration of 15 m/s^2.

References

[1] Fang H -B, Liu J -Q, Xu Z -Y, Dong L, Wang L, Chen D, Cai B -C and Liu Y 2006 *Microelectronics J.* **37** 1280–4
[2] Muralt P, Marzencki M, Belgacem B, Calame F and Basrour S 2009 *Procedia Chem.* **1** 1191–4
[3] Lee B S, Lin S C, Wu W J, Wang X Y, Chang P Z and Lee C K 2009 *J. Micromechanics Microengineering* **19** 065014
[4] Shen D, Park J -H, Noh J H, Choe S -Y, Kim S -H, Wikle H C and Kim D -J 2009 *Sens. Actuators A.* **154** 103–8
[5] Isarakorn D, *et al.* 2011 *Smart Mater. Struct.* **20** 025015
[6] Xu R, Lei A, Dahl-Petersen C, Hansen K, Guizzetti M, Birkelund K, Thomsen E V and Hansen O 2012 *Sens. Actuators A* **188** 383–8
[7] Hwang G -T, *et al.* 2014 *Adv. Mater.* **26** 4880–7
[8] Oh S R, Wong T C, Tan C Y, Yao K and Tay F E -H 2014 *Smart Mater. Struct.* **23** 015013
[9] Sivanandan K, Achuthan A T, Kumar V and Kanno I 2008 *Sens. Actuators A* **148** 134–7
[10] Matsushita S, Kanno I, Adachi K, Yokokawa R and Kotera H 2012 *Microsyst. Technol.* **18** 765–71
[11] Morimoto K, Kanno I, Wasa K and Kotera H 2010 *Sens. Actuators A* **163** 428–32
[12] Kuwano H, Wang Q and Cao Z P 2012 *Micro Nano Lett.* **7** 1158–61
[13] Lin S -C and Wu W -J 2013 *Smart Mater. Struct.* **22** 045016
[14] Tsujiura Y, Adachi K and Kanno I 2012 *Proc. PowerMEMS (Atlanta, USA, 2–5 December 2012)* pp 500–3
[15] Dubois M and Muralt P 1999 *Sens. Actuators A* **77** 106–12

Feasibility of energy harvesting from a rotating tire based on the theory of stochastic resonance

Y. Zhang[1], R. Zheng[2], K. Nakano[3]

Institute of Industrial Science, The University of Tokyo, 4-6-1, Komaba, Meguro, Tokyo, 153-8505, Japan

Email: zysgluck@iis.u-tokyo.ac.jp, topzrc@iis.u-tokyo.ac.jp, knakano@iis.u-tokyo.ac.jp

Abstract. Recently the use of nonlinear bi-stable micro-electro mechanical systems (MEMS) to achieve automobile tire vibration power generation has made some progress. However, the theory of stochastic resonance has not been successfully applied to automobile tires, which can produce a larger vibrational response than for a typical resonance while inputting a weak periodic force and noise excitation into a nonlinear bi-stable system. Hence, in this paper, in view of the principle of stochastic resonance, a new model is derived by positioning a magnetic end mass attached to a cantilever beam and another permanent magnet with the same polarity on the frame. Due to the road noise excitation along with the periodic force inputted to the mechanism, whether the phenomenon of stochastic resonance can happen will be discussed. Meanwhile, on the basis of Kramers rate and duffing equations the preliminary experimental device is also designed.

1. Introduction

The ability to harvest or scavenge energy from a vehicle operating environment can realize many sensing applications in which regularly replacing or recharging batteries is either difficult or expensive. Many potential sources of energy have been studied, including vibrations, temperature gradients, and human motion. Harvesting energy from a vehicle tire rotating environment may be considered a beneficial method for a tire pressure monitoring system (TPMS) [1].

Most vibration energy harvesting systems operate effectively only at resonance [2]. Thus, for rotational environments to enlarge the narrow operating bandwidth is the key. In order to improve the bandwidth three general approaches are as follows: tuning the resonance frequency of the harvester during operation, implementing multi-mode oscillators, and employing non-linear oscillators with bistable systems to widen the bandwidth. These methods are equally applicable to rotational systems. Among that the use of nonlinear bi-stable MEMS to achieve automobile tire vibration power generation has made some progress. However, the theory of stochastic resonance has not been successfully applied to automobile tires, which can produce a larger vibrational response than for a typical resonance while inputting a weak periodic force and a noise excitation into a nonlinear bi-stable system [3].

Therefore, in view of the principle of stochastic resonance, a new model is derived to demonstrate that tire rotation can make the phenomenon of stochastic resonance occur. A bi-stable mechanism is designed by positioning a magnetic end mass attached to a cantilever beam and another permanent magnet with the same polarity on the frame shown in Figure 1. The cantilever beam is excited by noise excitation and there is an interaction between elastic force and magnetic force of the beam, generating a nonlinear response by adjusting the distance *d* between the two magnets, then the system can show bi-stability. Due to the configuration of two magnets, a repulsive magnetic force will act between the two magnets. When the mechanism is mounted on a rotational tire shown in Figure 2, the road noise excitation coupled with the periodic force caused by the tire rotation and gravity of the magnetic end mass provides the external condition of the occurrence of stochastic resonance.

Different from the previous study [4], [5] where a weak external periodic force was manually added, this system can be self-provided from the gravity and rotation of the tire, thereby without consuming extra energy. In this paper, a simulation study was processed to validate this proposal, and by theoretical calculations the stochastic resonance of this system can be generated referring to Kramers rate equation. It indicates that the phenomenon of stochastic resonance occurs when the tire rotates at the frequency of 98 rad/s with the speed of 103 km/h. Based on the mathematic analysis and simulation results the design of a small-scale experimental device is proposed simultaneously.

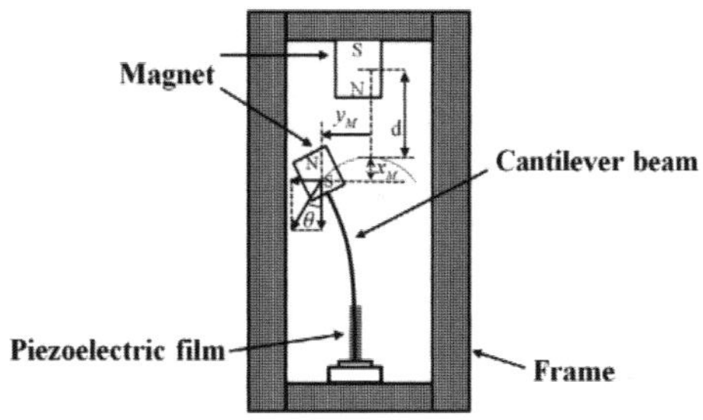

Figure 1. Structure of the power generation device using stochastic resonance.

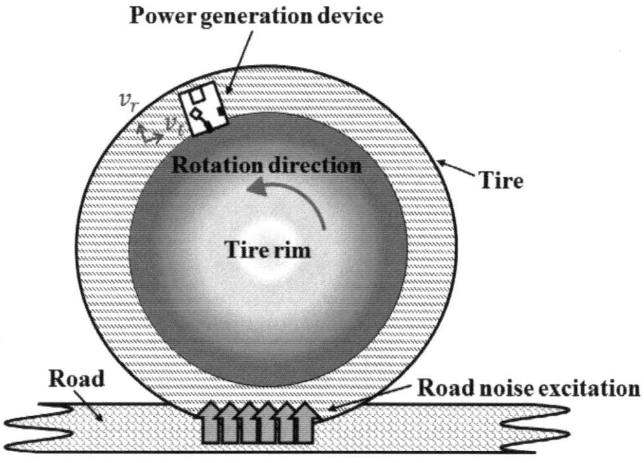

Figure 2. Illustration of the power generation device mounted to a rim.

2. Mathematic model and simulation results

If the magnetic force and gravity of the magnet end mass are taken into consideration, which may cause the angle of the beam to be θ and the tangential displacement of the end mass to be y_M, the centripetal displacement to be x_M. when the tire rotates at a constant speed. By adjusting the distance d between two magnets, a repulsive magnetic force F_M and a tangential component F_H are also changing to make the mechanism vibrate between two stable states. F_H can be written in terms of F_M and expanded into a Taylor series, calculated around $y_M = 0$ and truncated as follows,

$$F_H = F_M \sin\theta = \frac{F_M}{d}\frac{y_M}{\left[1+\left(y_M^{\,2}/d^2\right)\right]^{1/2}} \approx \frac{F_M}{d}y_M - \frac{F_M}{2d^3}y_M^{\,3} \tag{1}$$

When considering the tire rotation, due to the centripetal force, it is necessary to take into account the effect of gravity. Therefore, the total force of the tangential direction f_H is expressed as

$$f_H = \frac{F_M}{d}y_M - \frac{F_M}{2d^3}y_M^{\,3} + G\sin\omega t \tag{2}$$

When the tire with a radius of r rotates at the frequency of ω with the speed of v_S, the total kinetic energy of the end mass will be defined by E_K as

$$E_K = \frac{1}{2} m \left[(\dot{y}_M + v_S)^2 + \dot{x}_M{}^2 \right] \tag{3}$$

Here, according to the thin beam theory, the potential energy of the beam with the Young's modulus and the moment of inertia of E and I, respectively, is given as follows

$$U = \frac{1}{2} EI \int_0^l (w'')^2 dx \tag{4}$$

where w is the transverse displacement of the beam and $w'' = d^2w/dx^2$. Assuming $y_M(t) = w(l,t)$, where l is the length of the beam, the following equation can be derived as

$$w(x,t) = v(x)w(l,t) = \left(\frac{3x^2}{2l^2} - \frac{x^3}{2l^3} \right) y_M(t) \tag{5}$$

where $v(x)$ is a static linear shape function of the cantilever beam. Assuming the transverse deflection is tiny, x_M is given as

$$x_M = \int_0^l \left[\sqrt{1 + \left(\frac{dw}{dx} \right)^2} - 1 \right] dx = \int_0^l \frac{1}{2} \left(\frac{dw}{dx} \right)^2 dx = \frac{1}{2} \left[\int_0^l (v')^2 \right] y_M{}^2 \tag{6}$$

According to the following generalized Lagrange's equation and equations (2) to (6), it gives

$$\frac{d}{dt} \left(\frac{\partial E_K}{\partial \dot{y}_M} \right) - \frac{\partial E_K}{\partial y_M} + \frac{\partial U}{y_M} = f_H \tag{7}$$

The motion equation is derived as follows

$$m\ddot{y}_M + c\dot{y}_M + \left[EI \int_o^l (v'')^2 dx - \frac{F_M}{d} \right] y_M + \frac{F_M}{2d^3} y_M{}^3 = N(t) \sin \omega t + G \sin \omega t \tag{8}$$

where c is the viscous damping coefficient, $N(t) = \sqrt{2D}\xi(t)$ is the noise excitation from road, G as the periodic force from gravity of the end magnet. Since the rotation of tire, the magnitude and direction of these two parts will become to be $N(t)\sin\omega t + G\sin\omega t$. $EI\int_o^l (v'')^2 dx$ is the spring coefficient, $\frac{F_M}{d}$ is the linear coefficient of the two magnets, $\frac{F_M}{2d^3}$ is the nonlinear coefficient. When $EI\int_o^l (v'')^2 dx - \frac{F_M}{d} = -a$, $\frac{F_M}{2d^3} = b$, the equation (8) will be transformed into:

$$m\ddot{y}_M + c\dot{y}_M - ay_M + by_M{}^3 = N(t)\sin\omega t + G\sin\omega t \tag{9}$$

Base on this mathematic model, the numerical simulation is implemented, where $a = 157$, $b = 5\times10^7$, $D = 0.001$, $c = 0.27N/m/s$, $G = 0.1N$.

On the basis of Kramers rate equation, the stochastic resonance frequency can be calculated by the following equation as

$$f = \frac{a}{\sqrt{2}} \exp\left(-\frac{a^2}{4bD} \right) \tag{10}$$

where D is the intensity of the noise excitation. By calculating the frequency of stochastic resonance according to equation (10), it can be indicated that stochastic resonance can occur at the frequency of 98 rad/s.

In order to confirm whether the stochastic resonance can occur and distinguish the stochastic resonance and normal resonance, numerical simulations are carried out under three conditions, i.e. only the noise excitation, the periodic force and both of them are applied on the device, respectively. In the case of only the noise excitation is applied or the condition that the frequency of the periodic force is 98 rad/s, the displacement responses show that the inter-well dynamics cannot be activated as shown in figure 3.

Journal of Physics: Conference Series **557** (2014) 012097 doi:10.1088/1742-6596/557/1/012097

From figure 4, it can be validated that the cantilever beam can be brokenly excited to vibrate between two potential wells when the noise and periodic excitations are applied at the same time. And as predicted by the calculated parametric excitation frequency of 98 rad/s, the cantilever beam can keep the inter-well motion all the time, which indicates the phenomenon of the stochastic resonance.

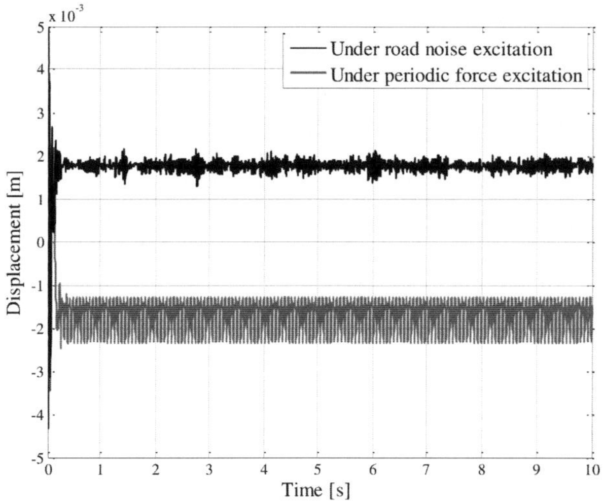

Figure 3. Responses under road noise excitation or periodic force excitation: the black line fluctuating at 1.8×10^{-3} m and the grey line fluctuating at -1.8×10^{-3} m represent the responses under road noise and periodic force excitation respectively.

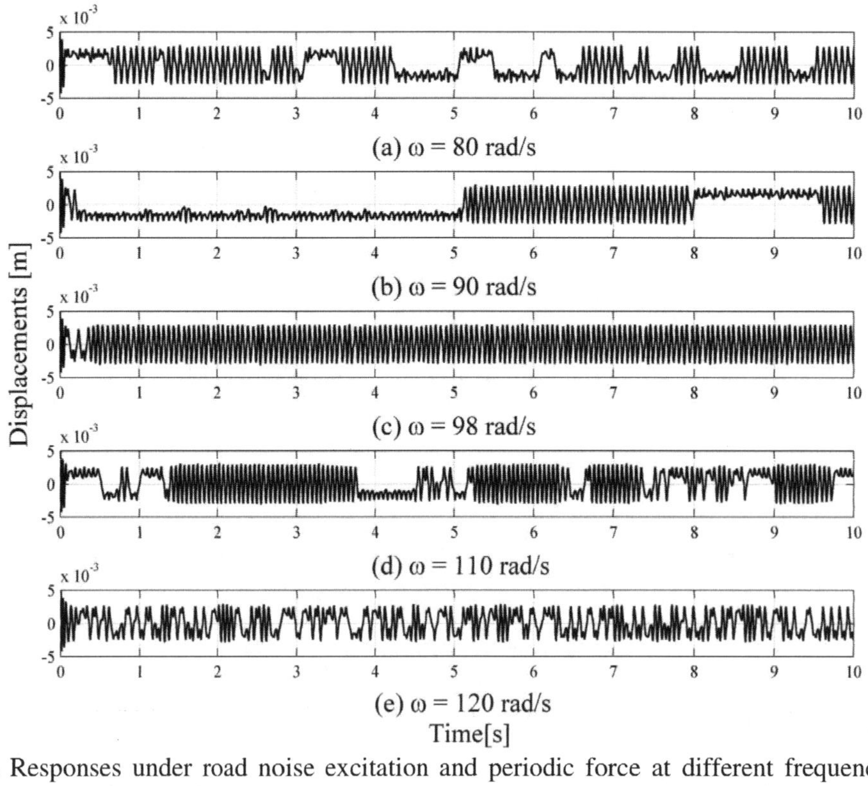

Figure 4. Responses under road noise excitation and periodic force at different frequencies of tire rotation: (a) 80, (b) 90, (c) 98, (d) 110, and (e) 120 rad/s.

3. Design for experimental device

For the mathematic mode, the potential energy can be expressed as:

$$U(y_M,t) = -\frac{1}{2}ay_M{}^2 + \frac{1}{4}by_M{}^4 \tag{11}$$

Differentiating equation (11) with respect to time lead to:

$$F_M{}' = -ay_M + by_M{}^3 \tag{12}$$

where $F_M{}'$ is the nonlinear spring force. After differentiating $F_M{}'$ the nonlinear spring coefficient k' can be given as follows

$$k' = -a + 3by_M{}^2 \tag{13}$$

Also, the equilibrium points of two potential wells can be expressed by

$$y_0 = \pm\sqrt{\frac{a}{b}} \tag{14}$$

Substituting equation (13) into (14), the nonlinear spring coefficient k' can be obtained. Therefore, the size of the brassiness cantilever beam is designed to be $5cm \times 2cm \times 0.5mm$. The conceived experimental setup is shown as figure 5. Finally, the photo of the fabricated power generation device is shown in figure 6 and it is currently under testing.

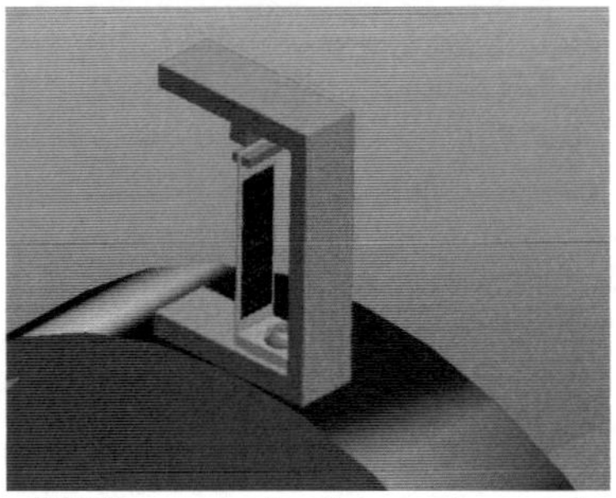

Figure 5. The conceived experimental setup

Figure 6. The generation power device

4. Conclusion

In this study, the feasibility of energy harvesting from a rotating tire based on the theory of stochastic resonance was discussed. From the mathematic mode and numerical simulation results, it can be shown that the phenomenon of stochastic resonance occurs when the tire rotating at the frequency of 98 rad/s. Meanwhile, using the assigned and fabricated device the experiment is conducting well now.

References

[1] S. J. Roundy, and J. Tola, "An Energy Harvester for Rotating Environments Using Offset Pendulum Dynamics," *Proceeding of The 17th International Conference on Transducers & Enrosensors 2013*, pp. 689 – 692.

[2] S Hashimoto, Y Zhang, N Nagai, Y Fujikara, J Takahashi, S Kumagai, M Kasai, K Suto, H Okada,W Jiang, "Multi-Mode and Multi-Axis Vibration Power Generation Effective for Vehicles," *Proceedings of the IEEE-ISIE2013,* pp. 1 – 6.

[3] L. Gammaitoni, P. Hanggi, P. Jung, and F. Marchesoni, "Stochastic Resonance," *Reviews of ModernPhysics*, 70 (1998), pp. 223 – 287.

[4] K. Nakano, M. P. Cartmell, H. Hu, and R. Zheng, "Feasibility of Energy Harvesting Using Stochastic Resonance Caused by Axial Periodic Force," *Strojniški vestnik - Journal of Mechanical Engineering*, 60 (2014), pp. 314 – 320.

[5] R. Zheng, K. Nakano, H. Hu, D. Su, M. P. Cartmell, "An application of stochastic resonance for energy harvesting in a bistable vibration system," *Journal of Sound and Vibration,* 333 (2014),pp. 2568 – 2587.

Modelling and experimental verification of more efficient power harvesting by coupled piezoelectric cantilevers

L G H Staaf[1], E Köhler[1], D Parthasarathy[2], P Lundgren[1] and P Enoksson[1]

[1]Department of Microtechnology and Nanoscience, Chalmers University of Technology, 412 96 Gothenburg, Sweden

[2]Volvo Technology AB, CTP, 412 58 Gothenburg, Sweden

v96staaf@chalmers.se

Abstract. A new piezoelectric energy harvester design is proposed in order to achieve a wider bandwidth without compromising energy conversion efficiency. By coupling two cantilevers where the tip of the bottom one is attached to the base of the upper one, the simulated harvester will have a wider bandwidth and higher power output compared with two simulated single tuned single cantilevers. This is a compact design, using only half the area compared to two parallel single cantilevers at the price of a small increase in height. The measured coupled harvester has approximately 1.7 times higher energy output than the combination of two measured tuned single cantilevers achieved by a coupling with less mechanical damping. With an improved coupling the power output is increased to 2.3 times higher than two single tuned cantilevers.

1. Introduction

Harvesting energy with piezoelectric cantilevers utilizing ambient vibrations is environmental friendly, since it replaces batteries. Battery-free operation enables new applications to be incorporated in harsh environments such as gas turbines [1].

A single piezoelectric cantilever has a narrow range of resonance frequencies where it effectively can convert vibrations to electricity. To widen this range of effective harvestable resonance frequencies, without losing power output, is a very important challenge when developing new designs [2]. Over the recent years of research many new approaches have been reported on how to widen the bandwidth with little or no loss of power output. Li et al (2014) proposed an array of structures with a different resonant frequency for each structure [3]. Soliman et al (2008) used an amplitude limiter to gain broader bandwidth [4]. Petropoulos et al (2004) used coupled oscillators [5]; Spreemann et al (2006) used magnets to achieve a nonlinear harvester resulting in broader bandwidth [6]. Ramlan et al (2012) tested bi-stable structures [7], and Zhu et al (2010) used a large inertial mass (large device size) with a high degree of damping [2]. Another way to increase the power output and the bandwidth has been suggested by Zhou et al (2011) where a primary beam was used as a multimode magnifier [8]. In this design the secondary beam multi-mode energy harvester is attached to the primary beam multimode dynamic magnifier and this design has a broader bandwidth and the harvested energy is greatly increased. To further enhance this effect Wu et al (2012) describe a two degrees of freedom

(2DOF) piezoelectric energy harvester [9], where the secondary beam is cut inside the main beam. This makes the design less bulky, results in a broader bandwidth with two closely spaced peaks, and gives a high power output from both beams.

The presented and developed harvester in this paper uses commercial components and is a 2DOF where one cantilever acts like a magnifier and by the design is also magnified by the other cantilever. This design enhances the power output and gives a wider bandwidth than an array of two tuned single cantilevers. The design gives a small increase in volume but utilizes the bottom beam more efficiently for higher energy output, much due to the improved and extended strain profile of this beam.

2. Design

The design attaches a MIDE v22b cantilever on top of a MIDE v21b cantilever (both commercially available from MIDE) with the v22b cantilever folded backwards (Figure 1) over the bottom MIDE v21b that is clamped between two alumina bars. By using this design the bottom cantilever will act as if attached on both ends and the bottom cantilever will magnify the top cantilever movement. Hence the harvester will deliver a higher power output.

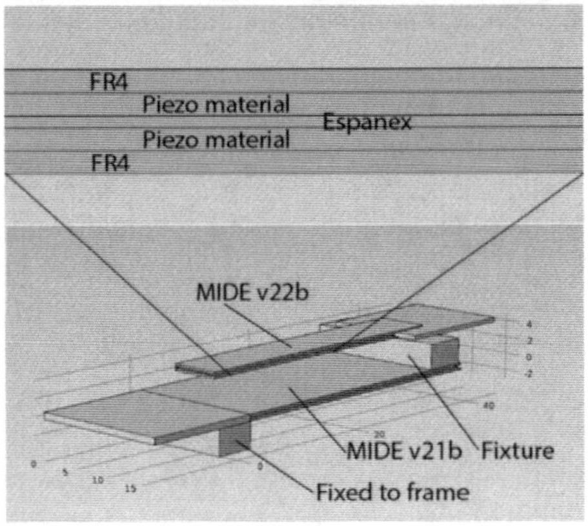

Figure 1. The coupled model in COMSOL Multiphysics.

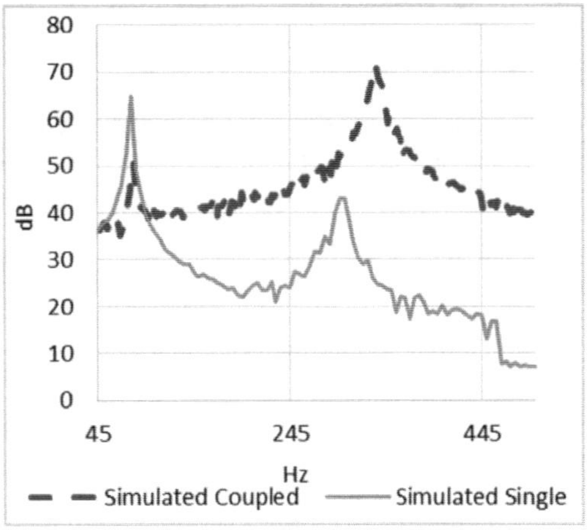

Figure 2. Red dotted line is the simulated couple harvester vs the green line is the two single cantilevers, compared by power output by voltage square and the simulated couple harvester has 10.5 times higher energy output.

3. Simulation

An extension to the single cantilever theory has to be applied in order to describe two connected beams as a couple for enhanced energy output. W. Zhou et al (2011) [8] extend the single cantilever theory to cover two coupled beams with a multi-mode design. This design is proved to yield a higher energy output and to broaden the bandwidth.

Our modification of their design uses two piezoelectric cantilevers folded over each other (Figure 1) instead of extending in the same plane. This design will be able to harvest from both cantilevers with an extra voltage output for the bottom cantilever originating from the top cantilever where the coupling will act more like an attachment for the bottom cantilever. This design is realized with industrial components and will consume only half the area compared to two single cantilevers. The height is slightly increased but in total the design provides an appropriately manageable harvester.

A model of a single MIDE v21b cantilever [10] is built within COMSOL Multiphysics. The real MIDE v21b is a biomorph piezoelectric cantilever with five layers (figure 1): FR4, Piezo material (PZT), Espanex, Piezo material and FR4 [11]. The length and width for the layers are measured with calipers and the thicknesses of the layers are measured with a microscope (table 1) as these were not available in the specification. The length, width and thickness are carefully calibrated so that the cantilever has the resonance frequency 274 Hz in COMSOL, which is the specified resonance frequency without tuning weight given by MIDE [10]. With the tuned single cantilever (v21b) as starting point another top cantilever (v22b) is attached at the free moving tip of the first (bottom) one, so that the top one extends out over the bottom one (figure 1). For comparison, two single harvesters were tuned so that each one showed resonance at one of the two resonance frequencies of the coupled harvester. The voltage responses as a function of frequency for the two differently tuned cantilevers were then added together.

Material properties	Young's modulus (E_p, Pa)	Piezoelectric constant (d_{31}, m V^{-1})	Dielectric constant (ε_{33}, F m^{-1})	Density (ρ, K m^{-3})	Poisson's ratio	L × w × t (mm × mm × mm)
PZT	63×10^9	-190×10^{-12}	830×10^{-10}	7800	0.3	36 × 15.2 × 0.18
FR4	22×10^9	-	-	1920	0.28	36.5 × 17 × 0.18
Espanex	3.2×10^9	-	-	1300	0.34	36 × 17 × 0.09
Damping material	3.5×10^5	-	-	-	0.28	5 × 17 × 0.01

Table 1, materials and constants used in the simulation for the MIDE v21b cantilever.

4. Result

A single v21b cantilever and a single v22b cantilever are simulated from 40-500 Hz. The voltage outputs are added together. The coupled model is also simulated from 40-500 Hz. The voltage output is converted to power by voltage square and the simulation result is compared (Figure 2) and the couple harvester shows a 10.5 times higher power output integrated over frequency band 40 - 500 Hz.

The prototype I is assembled by a bottom v21b cantilever, attached by two aluminum clamps. The coupling to the top v21b is also of aluminum and has a weight of 3.5 g (Figure 3). The protecting plastic and contact plastic protections are removed to reduce weight from the top v22b. The measured power output for the double single harvester is compared with the measured energy output for the coupled harvester (Figure 4), where the measured coupled harvester has 1.7 times higher energy output. To gain a higher power output, as predicted by the simulations, the coupling fixture between v21b and v22b has to be reduced in weight.

Figure 3. Built prototype I with MIDE v21b at the bottom, the coupling fixture and the MIDE v22b at the top.

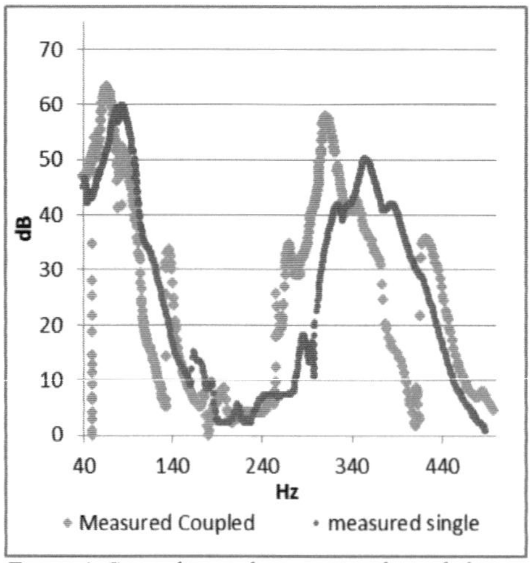

Figure 4. Green diamonds is measured coupled harvester I power output, the blue circles is measured single power output. The coupled power output is 1.7 times higher than the single power output.

Prototype II has a coupling of two screws with sleeves to hold the beam at the correct height, and a screw holder of FR4 with two holes for the screws that hold it together around the top cantilever were tested. The coupling weighed 1.8 g. A weight of 0.5 g is added to the top cantilever to make the two resonance frequencies closer for a broader bandwidth harvester. With this coupling the power output to the coupled harvester was 2.3 times higher compared with two single cantilevers (Figure 5). In figure 5 a simulation of the new coupling and the added weight is added and a comparison with two single cantilevers, one v21b and one v22b.

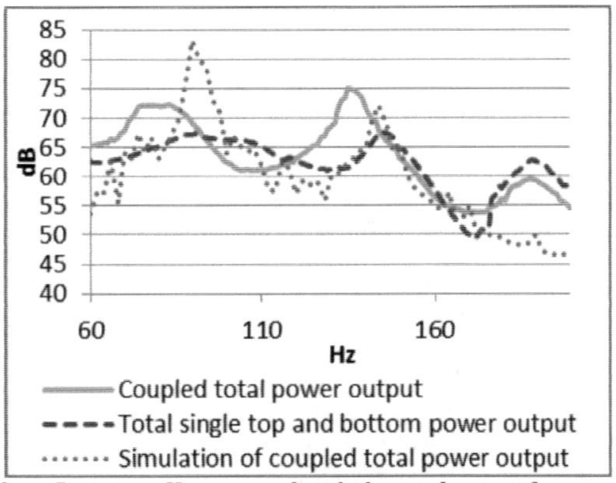

Figure 5. Measured results from Prototype II compared with the total output from two single cantilevers and compared with simulation of the improved coupling. The power output for the prototype II is 2.3 times higher than for two single cantilevers.

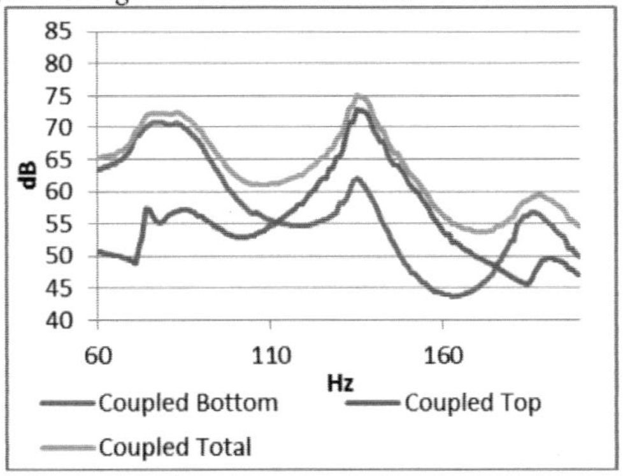

Figure 6. Coupled Prototype II where the green line is output power in total and the blue is the bottom cantilever and the red is the top cantilever.

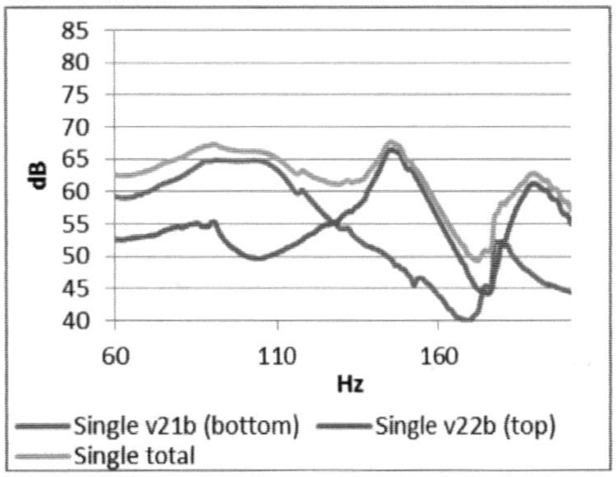

Figure 7. The two single cantilevers with the v21b as blue and v22b as red and the green line is the power output in total from both.

In Figure 6 the power output for the bottom and top cantilever contribution can be seen that adds to the total power output. In Figure7 the two measured single cantilevers are shown and put together to a total output from the two cantilevers. For the coupled harvester in Figure 6 the coupling effect on the bottom cantilever from the top cantilever is seen on the middle power peak, which is absent for the single v21b cantilever seen in Figure 7. In Figure 8 the coupled bottom cantilever power output is compared with the single v21b cantilever with a weight of 3.79 g as the attachment and the v22b weight in total. The primary and secondary resonance frequency for the single cantilever has a lower power output compared with the coupled cantilever which has three peaks where the left is primary, the rightmost one is secondary and the one in the middle is due to the magnification of the top cantilever. In Figure 9 the top coupled power output is compared with the single v22b power output. The coupled top has a higher power output at the primary resonance frequency but a lower output at the secondary resonance frequency.

5. Conclusion

By combining two piezoelectric cantilevers with a coupling, the measured power output is increased by 1.7 times compared to using the same cantilevers uncoupled; also the bandwidth is increased. With a lighter coupling of only 1.8 g and an added weight on the top cantilever the power output for the coupled harvester is increased to 2.3 times higher than for the two single cantilevers, and the power output peaks are brought much closer indicating that a broader bandwidth is reachable if the peaks are managed to merge together.

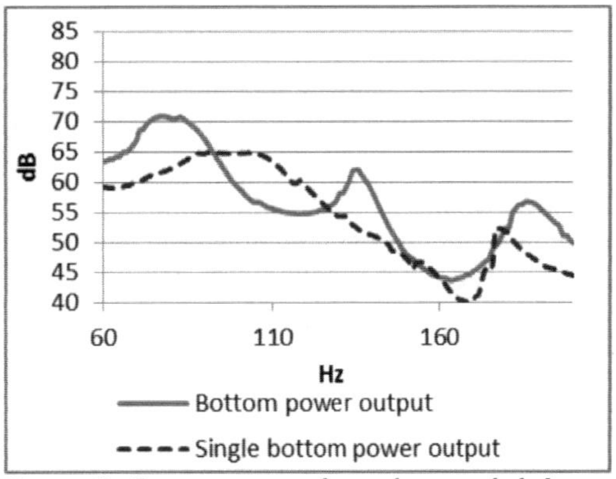

Figure 8. Power output from the coupled bottom cantilever compared with the same single cantilever. The middle peak on the coupled bottom line is the contribution from the coupling - working as an attachment and enhancing the strain on the bottom cantilever.

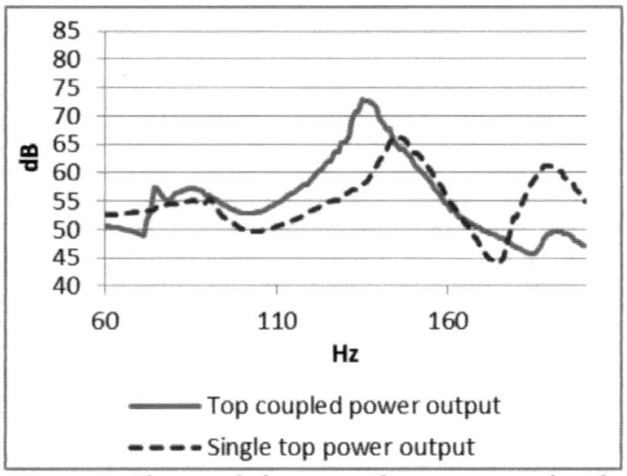

Figure 9. The coupled top cantilever compared with a single v22b. The coupled top cantilever is magnified by the bottom cantilever hence has a higher power output.

Acknowledgement

The FP7 project Stargate and the Swedish Science Council are greatly acknowledge for funding this project.

References

[1] Liang J, Liao W, (2012) Impedance Modeling and Analysis for Piezoelectric Energy Harvesting Systems, ASME TRANSACTIONS ON MECHATRONICS, VOL. 17, NO. 6

[2] Zhu D, Tudor MJ and Beeby SP (2010), Strategies for increasing the operating frequency range of vibration energy harvesters: a review, Meas. Sci. Technol. 21

[3] Li, Y. (2014). Modeling and tuning of energy harvesting device using piezoelectric cantilever array PhD Thesis, West Virginia University, USA

[4] Soliman M S M, Abdel-Rahman E M, El-Saadany E E and Mansour R R (2008) A wideband vibration-based energy harvester J. Micromech. Microeng. 18 115021

[5] Petropoulos T, Yeatman E M and Mitcheson P D 2004 MEMS coupled resonators for power generation and sensing Micromechanics Europe (5–7 September (2004), Leuven, Belgium)

[6] Spreemann D, Folkmer B, Maurath D and Manoli Y (2006) Tunable transducer for low frequency vibrational energy scavenging Proc. Eurosensors (Göteborg, Sweden)

[7] Ramlan R, Brennan M J, Mace B R and Kovacic I (2012) On the performance of a dual-mode non-linear vibration energy harvesting device, Journal of intelligent material systems and structures: VOL: 23 ISSUE: 13 PAGE: 1423-

[8] Zhou W, Reddy Penamalli G and Zuo L (2012) An efficient vibration energy harvester with a multi-mode dynamic magnifier, Smart Mater. Struct. 21

[9] Wu H, Tang LH, Yang YW, et al. (2013) A novel two degrees-of-freedom piezoelectric energy harvester. Journal of Intelligent Material Systems and Structures 24: 357–368.

[10] MIDE Engineering Smart Technology Products, Product datasheet available at: www.mide.com/pdfs/Volture_Datasheet_001.pdf

[11] MIDE Engineering Smart Technology Products, Material datasheet available at: www.mide.com/pdfs/volture_specs_piezo_properties.pdf

Tunable Vibration Energy Harvester for Condition Monitoring of Maritime Gearboxes

D Hoffmann[1], A Willmann[1], B Folkmer[1] and Y Manoli[1,2]

[1] HSG-IMIT – Institute of Micromachining and Information Technology, Wilhelm-Schickard-Str.10, 78052 Villingen-Schwenningen, Germany
[2] Fritz Huettinger Chair of Microelectronics, Department of Microsystems Engineering – IMTEK, University of Freiburg, Germany

daniel.hoffmann@hsg-imit.de

Abstract. This paper reports on a new tuning concept, which enables the operation of a vibration generator for energy autonomous condition monitoring of maritime gearboxes. The tuning concept incorporates a circular tuning magnet, which interacts with a coupling magnet attached to the active transducer element. The tuning range can be tailored to the application by careful design of the gap between tuning magnet and coupling magnet. A total rotation angle of only 180° is required for the tuning magnet in order to obtain the full frequency bandwidth. The tuning concept is successfully demonstrated by charging a 0.6 F capacitor on the basis of physical vibration profiles taken from a gearbox.

1. Introduction

The application of conventional vibration transducers (single spring-mass system) on variable-speed machinery is not practical since the bandwidth of these devices is very limited. In order to effectively generate energy from variable vibration frequencies, harvesters should be inherently broadband or should be tunable to the dominant frequency. To date a variety of tunable vibration harvesters have been reported in literature [1]. Varying the total spring stiffness is a favored tuning technique, which can be realized by magnetic, piezoelectric and electrostatic principles. The magnetic principle is a well-known tuning technique, which utilizes the magnetic stiffness between two magnets. This technique is well applicable to cantilever-based energy harvesters by attaching a coupling magnet to the transducer structure. An axial magnetic force is then generated if a second magnet (tuning magnet) is placed opposite to the coupling magnet [2]. A common approach for modifying the total spring stiffness and thus the Eigen-frequency of the energy harvester, is to adjust the distance between the coupling magnet and the tuning magnet by translational motion of the tuning magnet [3]. For this concept we identified two drawbacks: First, the coupling magnet and the tuning magnet can be configured either in attractive or repulsive mode. However, using both modes in combination would enhance the feasible tuning bandwidth. Second, this concept requires a linear actuator, which is more complex than a rotary actuator. Indeed, a rotary actuator can also be deployed, however, only in combination with a gearing mechanism.
In this paper we propose a magnetic tuning concept, which is based on the rotational motion of a circular tuning magnet. In this manner, both coupling modes (attractive and repulsive) can be utilized for tuning the Eigen-frequency of the energy harvester.

Figure 1. Gearbox (Reintjes) for work boats and ferries

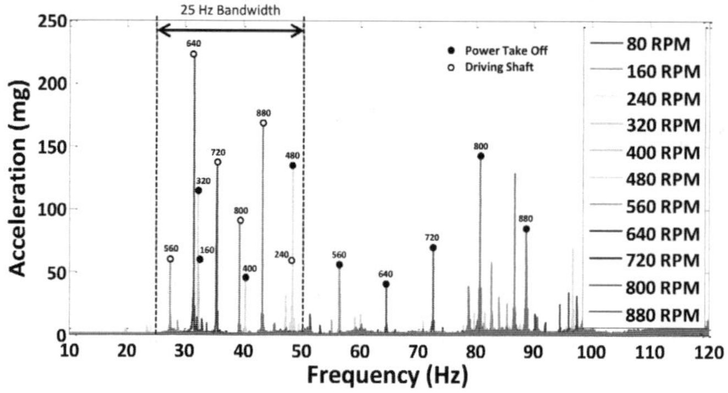

Figure 2. Frequency spectra of a gearbox (sensor position SP6, z-axis) operating at different revolution speeds

Moreover, the tuning magnet can be attached directly to a rotary actuator, making a gearing mechanism with its accompanying mechanical losses dispensable.

2. Vibration Analysis

The movable parts of a gearbox as shown in Figure 1 operate at variable speed depending on the operation mode and the travel speed of the vessel. The revolution speed at the output shaft ranges from 80 RPM to 880 RPM when the driving shaft is driven at speeds from 236 RPM to 2600 RPM. The revolution speed in this paper is always denoted with respect to the output shaft.

Vibration profiles were recorded at 6 different positions using triaxial accelerometers. The revolution speed at the output shaft was varied from 80 RPM to 880 RPM in steps of 80 RPM. A frequency spectrum was calculated for each position, axis and revolution speed. The frequency spectra of a particular position and axis were then combined into a single diagram as shown in Figure 2. The dominant frequencies are directly correlated to the revolution speed of the two main shafts of the gearbox (driving shaft, power take off). The fundamental oscillation modes of the driving shaft are located in a frequency band between 4 Hz and 45 Hz. In general, the amplitude of the dominant frequencies varies with the revolution speed and strongly depends on the measurement position. However, dominant frequencies below 25 Hz (i.e. < 500 RPM) always have amplitudes below 10 mg. In consideration of both main shafts a frequency band can be identified (see Figure 2), which comprises dominant frequencies of all revolution speeds with usable amplitudes except 80 RPM. Since smaller vessels such as work boats do not operate at constant speed, the revolution speed of the gearbox also varies over time. In this respect, a potential tunable energy harvester must be able to adjust its Eigen-frequency within a frequency band of 25 Hz to 50 Hz.

3. Concept and Setup

The concept of the tunable energy harvester is shown in Figure 3(a) and comprises a coupling magnet attached to a cantilever-based electromagnetic transducer and a circular tuning magnet. The tunable frequency band can be defined by adjusting the gap between the tuning magnet and the coupling magnet whereas the center frequency is determined by the stiffness of the leaf spring. By rotation of the tuning magnet between 0° and 180° the Eigen-frequency can be varied between a minimum value (repulsive coupling) and a maximum value (attractive coupling).

The electromagnetic transducer incorporates a closed magnetic circuit suspended on a cantilever and a fixed coil. The thickness of the leaf spring is 0.7 mm. The architecture of the electromagnetic transducer and significant parameters are described in more detail in [4].

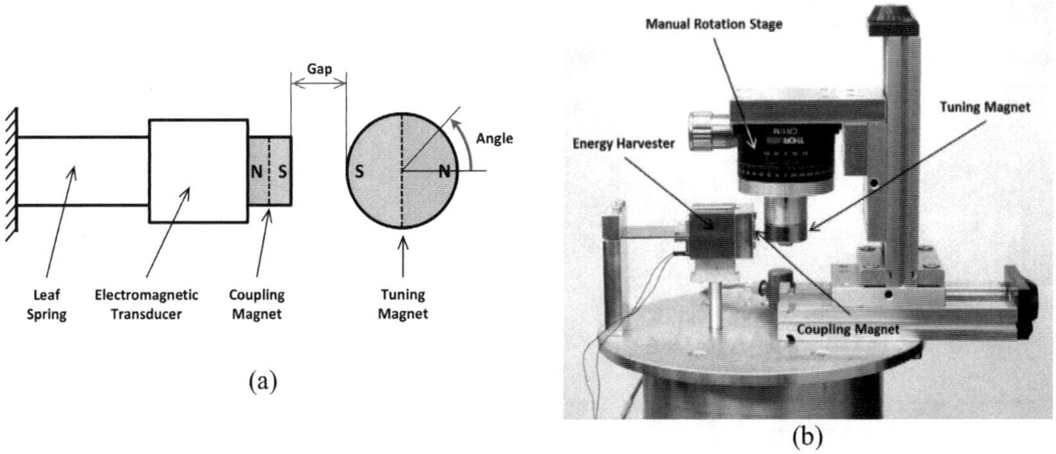

Figure 3. (a) Schematic (top view) showing the concept of the tunable vibration energy harvester. (b) Proof-of-concept implementation of the tunable energy harvester with manual rotation stage

A proof-of-concept implementation of the tunable energy harvester is shown in Figure 3(b) utilizing a manual rotation stage for rotational manipulation of the tuning magnet. The horizontal and vertical position of the tuning magnet can be modified using conventional translation stages. The frequency response (output voltage) of the energy harvester is shown in Figure 4(a) for different angles between 0° and 180°. The gap between coupling magnet and tuning magnet is 6 mm. The resulting frequency bandwidth is about 10 Hz ranging from 34.5 Hz to 44.5 Hz. By varying the gap between the two magnets the tuning bandwidth can be tailored to the application. The smaller the gap the larger is the bandwidth of the tunable Eigen-frequency of the energy harvester (Figure 4(b)). A gap of 3.5 mm is required in order to achieve a tunable frequency band of 25 Hz to 50 Hz. Hereby, the center frequency is around 40 Hz, which is determined by the stiffness of the leaf spring and the mass of the magnetic circuit. By varying the thickness of the spring element and thus its stiffness the center frequency can also be tailored to the application.

4. Experimental Characterisation

For characterization of the tunable energy harvester physical vibration data (sensor position 6, z-axis) from the gearbox was replicated on a shaker system. First, a load resistance of 800 Ohm was attached to the coil and the instantaneous power output was measured for each vibration profile over time. The total play time for each vibration profile was 1 min. The energy harvester was manually tuned to the

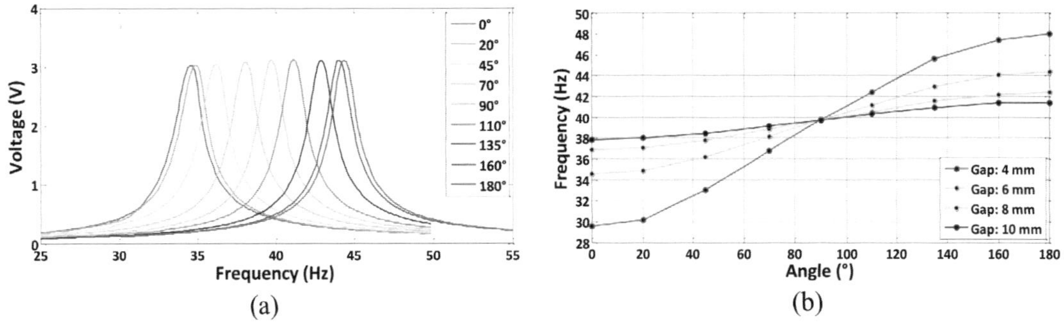

Figure 4. (a) Frequency response of the tunable vibration energy harvester at a gap of 6 mm for different angle adjustments of the tuning magnet. (b) Eigen-frequency as a function of the angle position for different gaps between the coupling magnet and the tuning magnet

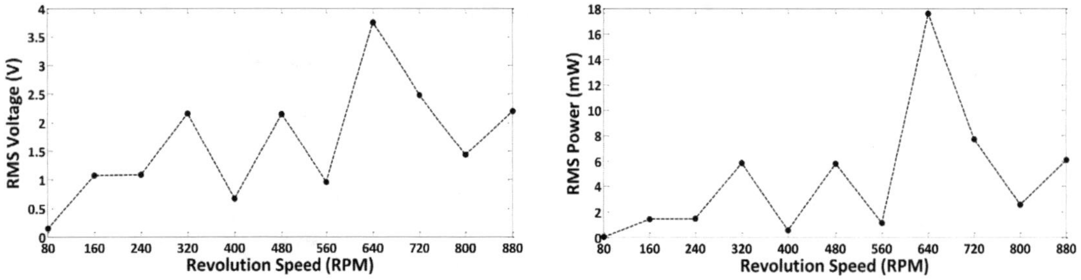

(a) (b)

Figure 5. Output vs. revolution speed at 800 Ohm load resistor: (a) RMS voltage. (b) RMS power

respective dominant frequency of each vibration profile. The average output power was calculated and plotted as a function of revolution speed (Figure 5(b)). At the lowest possible revolution speed (80 RPM) the average output power is almost zero (some μW). In this case the dominant frequency resides below 25 Hz and thus the energy harvester cannot be tuned. In general, the excitation amplitudes at 80 RPM are very low. For the remaining revolution speeds the power output is equal to or greater than 1 mW except for 400 RPM (0.5 mW). The highest power output (about 18 mW) was achieved at the critical revolution speed of 640 RPM. At this particular revolution speed the dominant frequency is located at 31.5 Hz exhibiting an amplitude of 225 mg (Figure 2).

The targeted condition monitoring system requires a current of 30 mA for a time period of 10 s in order to perform one measurement task. Assuming an acceptable voltage drop of 1 V at the energy storage, a capacitance of 0.6 F is required for performing the measurement. The activation cycle is one hour. A power management unit was designed and implemented for charging the capacitor and to power the condition monitoring system at a regulated output voltage of 3.0 V. The upper and lower threshold value of the integrated hysteresis is 2.9 V and 1.9 V, respectively. If the upper hysteresis threshold is reached (voltage at capacitor = 2.9 V), the voltage-regulated output port is enabled. If the voltage at the capacitor falls below 1.9 V, the output port is disabled. From this point of view there are two charging scenarios: start-up charging (cold start, the initial capacitor voltage is equal to zero) and operation charging (warm start, the capacitor voltage is 1.9 V). In case of a cold-start the upper hysteresis level is 3.3 V.

Since the activation cycle of the condition monitoring system is 1 hour, a charging time of 59 min and 50 sec is available to charge the capacitor. The charging time for the cold-start scenario is shown in Figure 6(a). For most revolution speeds the time required to charge the 0.6 F capacitor to a voltage

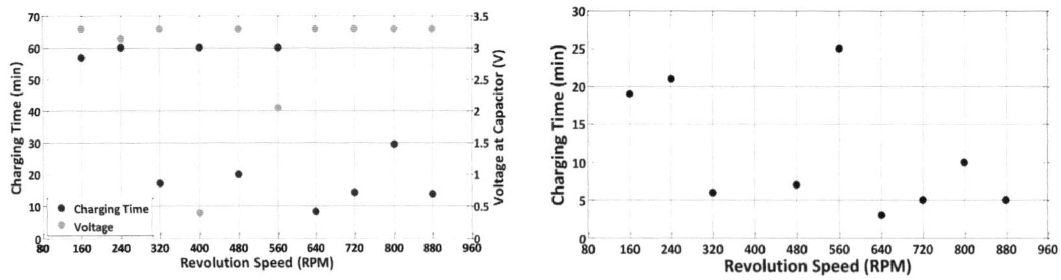

(a) (b)

Figure 6. (a) Start up charging time and voltage level at the capacitor vs. revolution speed. (b) Charging time during operation vs. revolution speed level of 3.3 V is less than 60 min. For example, at a revolution speed of 160 RPM a charging time of 57 min is necessary. At a revolution speed of 560 RPM a voltage of 2 V is gained after a time period of 1 hour. Figure 6(b) shows the charging time for the warm-start scenario (the voltage level at the capacitor is 1.9 V). For all revolution speeds except 80 RPM and 400 RPM the upper hysteresis level of 2.9 V at the capacitor is achieved in less than 30 min. Once the capacitor is pre-charged to a level of 1.9 V an activation cycle of 1 hour appears possible with respect to the worst case (revolution speed of 560 RPM). In this case, there are 35 min left for harvesting energy for the tuning mechanism. Within 35 min the capacitor can be recharged at least one more time from the lower hysteresis level (1.9 V) to the upper hysteresis level (2.9 V). Consequently, an energy packet of 1.44 J is available for the tuning mechanism. In other words, a current of 0.3 A is available for 1s.

5. Conclusion

In this work we presented a new concept for tuning the Eigen-frequency of a cantilever-based electromagnetic energy harvester. The functionality and feasibility of the concept was demonstrated on the basis of a proof-of-concept implementation, which included a manual tuning mechanism.

The demonstrated tuning concept is based on a magnetic principle and incorporates a circular tuning magnet, which interacts with a coupling magnet attached to the active transducer element. The circular tuning magnet merely requires a rotational motion with angle adjustments ranging from 0° to 180°. In this manner the tuning magnet can be attached directly to a rotary actuator without the need of a gearing mechanism. Only a half turn of the actuator is required for utilizing the entire tuning range.

The tuning range can be designed by careful adjustment of the gap between tuning magnet and coupling magnet. The smaller the gap the larger is the tuning bandwidth. The center frequency of the tuning bandwidth is adjusted by the stiffness of the cantilever in consideration of the mass of the magnetic circuit. In this work the tuning bandwidth and center frequency was tailored to a specific condition monitoring application targeting the monitoring of maritime gear boxes.

The new concept proposed in this paper will facilitate a simple and energy efficient implementation of a self-powered and self-tunable energy harvesting system, which is part of the ongoing work.

Acknowledgements

This work was financially supported by the Federal Ministry of Education and Research, Germany (BMBF: 03SX350C).

References

[1] D. Zhu, M. J. Tudor, and S. P. Beeby, "Strategies for increasing the operating frequency range of vibration energy harvesters: a review," *Meas. Sci. Technol,* vol. 21, no. 2, p. 22001, 2010.

[2] D. Zhu, S. Roberts, M. J. Tudor, and S. P. Beeby, "Closed Loop Frequency Tuning of a Vibration-Based Micro-Generator," in *PowerMEMS*, 2012.

[3] A. S. Weddell, D. Zhu, G. V. Merrett, S. P. Beeby, and B. M. Al-Hashimi, "A Practical Self-Powered Sensor System With a Tunable Vibration Energy Harvester," in *PowerMEMS*, 2012.

[4] D. Hoffmann, B. Folkmer, and Y. Manoli, "Comparative Study of Concepts for Increasing the Bandwidth of Vibration Based Energy Harvesters," in *PowerMEMS*, 2012.

Self power generating piezoelectric elements applied to switching circuits

T.Imai[1], **S.Fujimoto**[1], **M.Ichiki**[2]
[1]Tokyo City University, Japan
[2]National Institute of Advanced Industrial Science and Technology (AIST), Japan
1-28-1 Tamazutumi, Setagayaku, Tokyo, Japan

E-mail: g1481004@tcu.ac.jp

Abstract. In this study, we focused on lead zirconate titanate (PZT) as a power generating piezoelectric element. Niobium was added to each of the PZT elements to improve their power generation characteristics. The purpose of the study was to develop a high-efficiency PZT generator element that utilizes the vibration loads in the support members of a structure. We have previously reported the power generation characteristics of laminated PZT elements under vibration loads. Effect of vibration load, vibration frequency and number of PZT layers on generation characteristics of PZT elements was evaluated in the vibration test. We evaluate the power generation of laminated PZT elements and present the results of an experiment using a switching circuit as a load circuit in order to confirm the suitability of the laminated PZT element as a power source.

Keywords : Piezoelectric element, PZT, Laminating

1. Introduction

The necessity for the development of micro power supply technologies for mobile phones and portable electronics has increased in recent years. Methods of self power generation using the vibration loads of structures and human motion have been receiving significant attention. To this end, the generation of electricity by harnessing vibration forces using piezoelectric elements has been attempted.[1] However, piezoelectric elements only generate power on the order of several microwatts, so improving their power generation capacity is of the utmost importance. The use of lead zirconate titanate (PZT) with high piezoelectric properties and PZT doped with additives is known to improve the electrical properties of piezoelectric elements[2]. Fujimoto et al. have reported the effects of vibration loads on the power generation characteristics of a PZT element doped with niobium (Nb)[3] and have proposed the use of laminated PZT elements to significantly improve the power generation characteristics of PZT elements[4].

2. Electrical Properties and Structure Laminated PZT elements

The structure of the 11-layer PZT element and the appearance of the PZT element and the laminated PZT elements consisting of 3, 5, 7, and 11 layers are shown in Figure 1. Niobium at 1.0 mol% was added to each of the PZT elements to improve their power generation characteristics. Copper electrode plates with a diameter of 16 mm and a thickness of 0.2 mm were inserted between the PZT elements. In addition, a brass plate with a diameter of 16 mm and a thickness of 2 mm was affixed at both ends. Each electrode was divided into a positive electrode and a negative electrode and connected by a wire. The laminated PZT element was covered with a vinyl cover about 1 mm in thickness on the side for insulation. All of the PZT elements were electrically connected in parallel by these structural electrodes. When a vibration load was exerted axially on the laminated PZT elements, all of the PZT elements emitted an electric charge corresponding to the vibration load. This shows that the laminated (folded) PZT elements are efficient power generation elements. Measurements of the electrical properties of the produced laminated PZT elements are shown in Table 1. The piezoelectric constant was measured by a *d_{33} meter* (MODELZJ-6B, made by Academia Sinica), and the capacitance was measured by an *LRC meter* (ZM2353, made by NF Corporation). It was confirmed for the 11-layer PZT element that the values of the piezoelectric constant and the capacitance are related to the number of folds (layers).

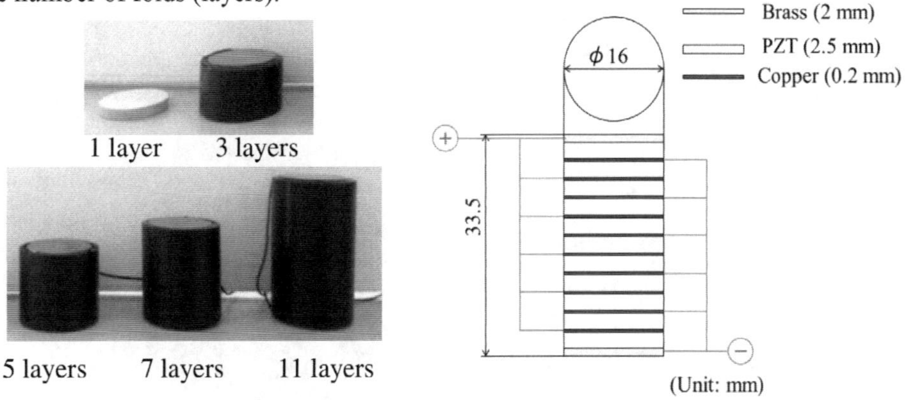

Fig. 1. Photographic view and structure of laminated PZT elements.

Table 1. Specifications of laminated PZT

Number of layer	Piezoelectric constant d_{33} (pC/N)	Capacitance C (pF)
1	320	1053
3	952	2843
5	1582	4790
7	2170	6726
11	3550	11560

3. Power generation characteristics of laminated PZT elements

A schematic of the power generation experiment using a vibration load is shown in Figure 2. This experimental device consists of a connection device and a rubber block that transmit the vibration load of the vibration table to the piezoelectric element and a device that holds the piezoelectric element in place. The voltage generated by the PZT element is measured by a voltmeter (Omniace RA2300, made by NEC). By actuating the vibration table and maintaining the initial compressive load acting on the PZT elements, a vibration test using a sinusoidal vibration load was performed with this device. In this experiment, the initial compression load given was 1000 N. Vibration experiments were conducted at six frequencies (5, 10, 15, 20, 25, and 30 Hz) and eight vibration loads (50, 100, 150, 200, 250, 300, 350, and 400 N). The fluctuating loads and voltage generated from the PZT elements at that time were measured as time history data. As an example of the result of the vibration experiments, the time histories of the load and vibration-generated voltage acting on the 7-layer PZT elements when the amplitude of the vibration load was 400 N with a vibration frequency of 30 Hz are shown in Figure 3. Because it has electric noise, the exciting frequency (30 Hz) has been removed using a band-pass filter. An initial compressive load of 1000 N was applied as the static load, but it is not included in the time histories of the load. In the experiments, it was found that the vibration load and generated voltage had a phase of 30°, and the maximum voltage was about 100 V.

The generated voltage and electric power as functions of the vibration load are shown in Figure 4 for each laminated PZT element. The generated voltage is increased 2.9-fold for the 3-layer PZT element, 4.1-fold for the 5-layer PZT element, and 4.8-fold for the 7-layer PZT element. In addition, although a 5.1-fold voltage increase was noted in the 11-layer PZT element compared with the single-layer PZT element, the generated voltage and electric power tended to become saturated with an increase in the number of laminated layers. The generated voltage of all of the laminated PZT elements was found to increase proportionally under load, and the generated electric power was found to increase proportionally to the square of the number of layers. In the case of the 11-layer PZT element, this corresponded to about 5 mW when the vibration load was 400 N, and the generated power increased about 25-fold compared to the single-layer PZT element.

The effects of the vibration frequency on the generated voltage and electric power for each of the laminated PZT elements when vibrated at 400 N are shown in Figure 5. The generated voltage and electric power for all of the laminated PZT elements increased almost proportionally to the load. However, the capacitance of the PZT elements was proportional to the number of layers, the resistance of the voltmeter was much larger, and the power generation characteristics of each of the laminated PZT elements became saturated.

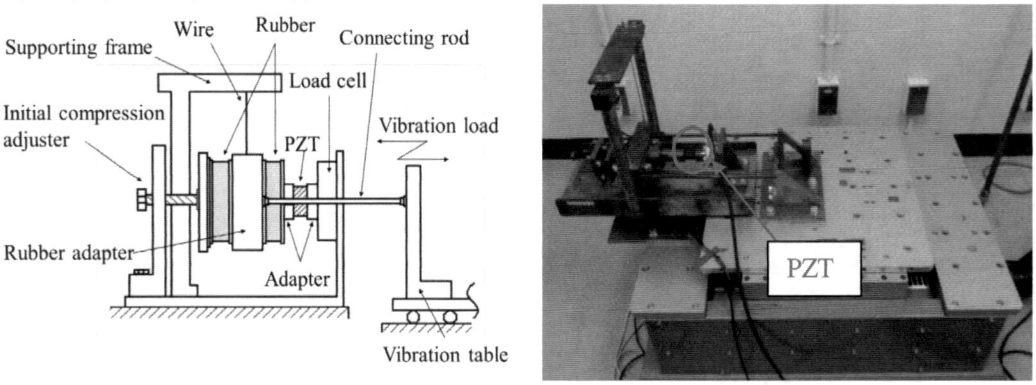

Fig. 2. Experimental setup for self power generation using a vibration load.

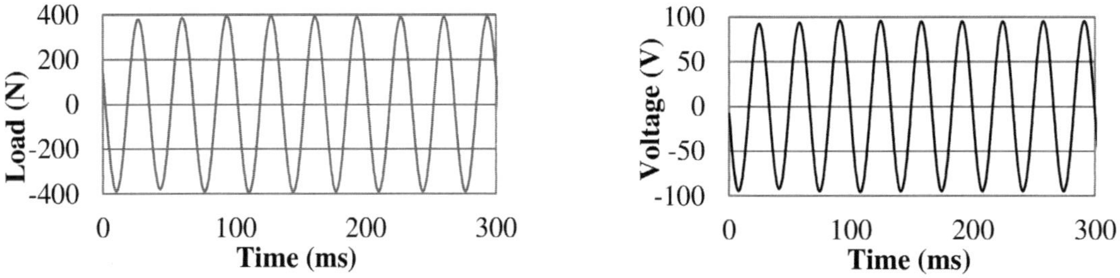

Fig. 3. Time histories of generated voltage and load for 11-layer PZT (30 Hz, 400 N).

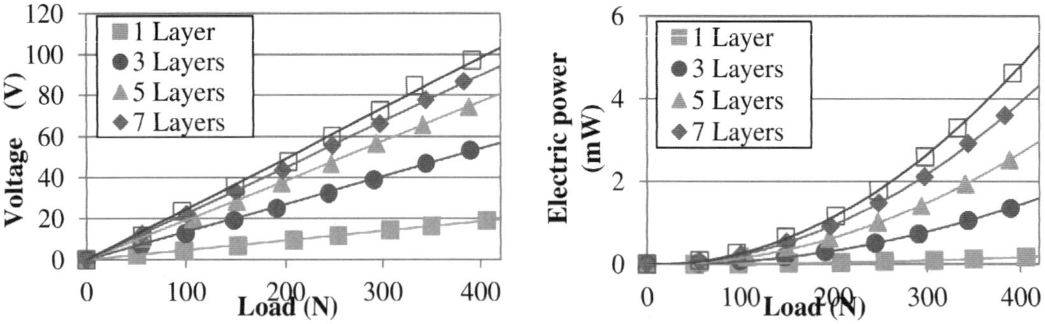

Fig. 4. Generated voltage and electric power functions of the vibration load for each PZT element (vibrated at 30 Hz).

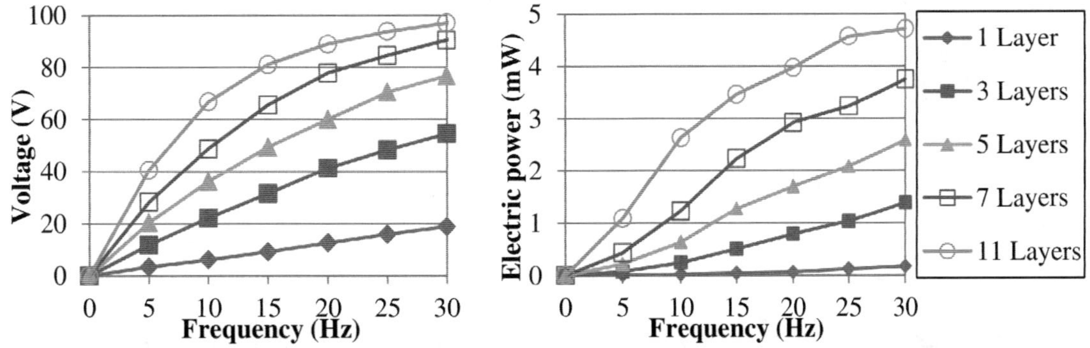

Fig. 5. Effect of frequency on generated voltage and electric power for each of the laminated PZT elements (vibrated at 400 N).

4. Experiment of laminated PZT elements applied to switching circuits

In an experiment, we applied the 11-layer laminated PZT elements to switching circuits (Kyowa Electric Instruments) that were turned on when they reached the driving voltage. The generated voltage for the 11-layer PZT elements was converted to 3.4 V by a constant voltage module (LTC3588, Strawberry Linux Corporation). The circuit diagram for the laminated PZT element and the switching circuit with an LED lamp are shown in Figure 6. The voltage generated by the laminated PZT element using the experimental setup (Figure 2) was rectified and smoothed and reached the LED lamp through the switching circuit. The capacity of the condenser used was 470 μF. In this experiment, we measured the on-off frequency of the LED lamp using the switching circuits set at 2.5 V. This experiment was conducted at five frequencies (10, 15, 20, 25, and 30 Hz) and four vibration loads (50, 100, 150, and 200 N). The time history of the voltage of the laminated PZT with the switching circuit at a vibration load of 200 N vibrated at 30 Hz is shown in Figure 7. The generated voltage of the laminated PZT elements was about 3.4 V. It is assumed that the generated voltage was affected by the constant voltage module. The time history of the LED lamp voltage using the 2.5 V switching circuit (30 Hz, 200 N) is shown in Figure 8. It shows that the voltage fluctuated and that the LED lamp was lit when the voltage was at or above 2.5 V. The LED lamp was turned on and off at a frequency of 1.4 Hz.

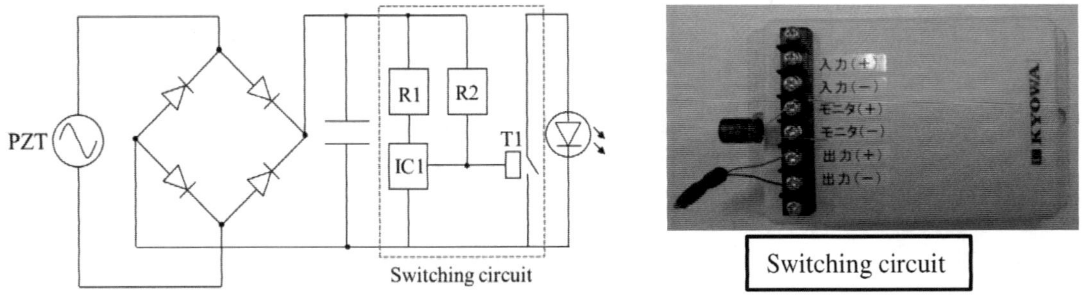

Fig. 6. Circuit diagram for a laminated PZT element and a switching circuit with an LED lamp.

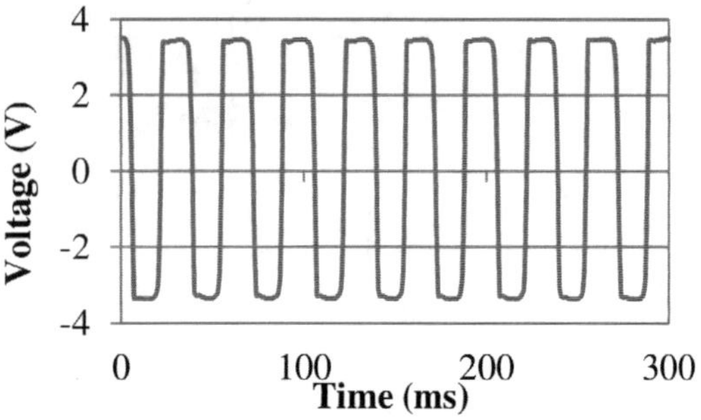

Fig. 7. Time history of the voltage of the laminated PZT with a switching circuit (vibrated at 30 Hz, 200 N).

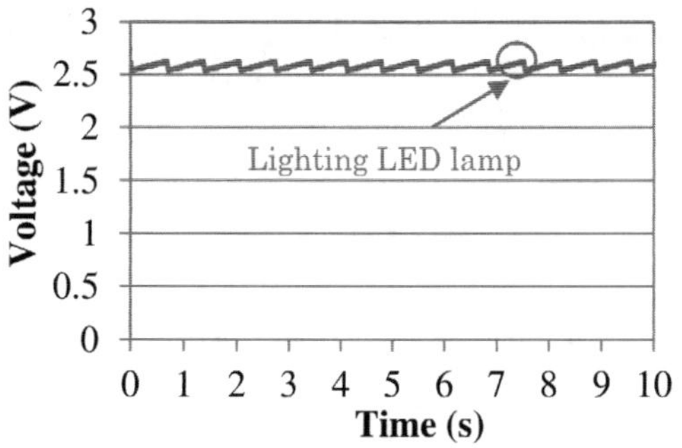

Fig. 8. Time history of the voltage of the capacitor with a switching circuit (vibrated at 30 Hz, 200 N).

5. Conclusions

The following conclusions were made from this study.

- The power generation characteristics were greatly improved by increasing the number of layers. The generated voltage of all of the laminated PZT elements was found to increase proportionally under load, and the generated electric power was found to increase proportionally to the square of the number of layers. In the case of the 11-layer PZT element, this was about 5 mW when the frequency was 30 Hz and the vibration load was 200 N, and the generated power increased about 25-fold compared to a single-layer PZT element.
- The generated voltage and electric power of all of these test pieces increased almost proportionally to the load. However, it is assumed that the capacitance of the PZT elements is proportional to the number of layers.
- The LED lamp was turned on and off at a frequency of 1.4 Hz when the frequency was 30 Hz and the vibration load was 200 N.

Based on these results, the laminated PZT elements can be applied to monitoring sensors and other applications.

References

[1] Arms, S.W., Townsend, C.P., Churchill, D.L., Galbreath, J.H., and Mundell, S.W., "Power Management for Energy Harvesting Wireless Sensor," Smart Structures and Materials 2005: Smart Electronics, MEMS, BioMEMS and Nanotechnology, edited by Varadan, V.K., Proceedings of SPIE, 5763 (2005), p. 267.

[2] Sekine, N., Kitahara, T., Ichiki, M., and Fujimoto, S., "Study on Power Generation Method using Structural Vibration (Power Generation Characteristics of Piezoelectric Element)," Proceedings of the Annual Conference of the Japan Society of Mechanical Engineers, No. 080-1, Vol. 6 (2008), p. 385.

[3] Fujimoto, S., Ichiki, M., Yano, S., and Kitahara, T., "Study on Method of Vibration Power Generation using Piezoelectric Element (1st Report, Generation Characteristics of Piezoelectric Element Doped with Additive)," Japan Journal of Design Engineering, Design Engineering, Vol. 47, No. 12 (2012), p. 54.

[4] Yamagishi, S., Fujimoto, S., Ichiki, M., and Kitahara, T., "Study on Power Generation Method using Structure Vibration (Power Generation Characteristics of Laminated Piezoelectric Elements)," Japan Society of Mechanical Engineers Kanto Branch 19th Annual Meeting Conference Proceedings, No. 130-1 (2013), pp. 355–356.

Output power of piezoelectric MEMS vibration energy harvesters under random oscillations

K Kariya[1], T Yoshimura[1], S Murakami[2] and N Fujimura[1]

[1]Osaka Prefecture University,
1-1 Gakuen-cho, Naka-ku, Sakai, Osaka 599-8531, Japan
[2]Technology Research Institute of Osaka Prefecture,
2-7-1 Ayumino, Izumi, Osaka 594-1157, Japan
E-mail: tyoshi@pe.osakafu-u.ac.jp

Abstract. Environmental vibrations include random oscillations of different frequencies and amplitudes. Energy harvesters recover the energy associated with these vibrations. Properties of the vibrations and output power are characterized for cantilever-type piezoelectric vibration energy harvesters using (100)-orientated $BiFeO_3$ films subject to both ideal and random oscillations. The displacement and output power under random oscillations were smaller than those under ideal oscillations. This decrease originates with the decreasing acceleration of the fundamental wave with the spurious component having little influence on the resonance response.

1. Introduction

Recently, the concepts "Internet of Things" and the "trillion sensor universe" have attracted considerable attention. The underlying motivation is that energy harvesting from light, heat, microwaves, and vibrations has potential as power sources for sensor nodes. Among them, vibration energy harvesters (VEHs) are promising because they can provide relatively constant electric power.[1,2] Three energy transduction mechanisms for VEHs are available: electromagnetic induction, electrostatic, and piezoelectric effect. VEHs using the piezoelectric effect have advantages of high energy density and potential for miniaturization by micro-electro-mechanical systems (MEMS).[2–5] The figure of merit (FOM) of the piezoelectric thin films for VEHs is given by $(e_{31,f})^2/\varepsilon_0\varepsilon_{33}$, where $e_{31,f}$ is the effective transverse piezoelectric coefficient, ε_0 the permittivity in free space, and ε_{33} the relative permittivity.[6,7] FOM higher than 20 GPa have been reported for Pb-based ferroelectric films such as $Pb(Zr,Ti)O_3$, because of a high $e_{31,f}$ (~−20 C/m^2).[8,9] AlN thin films also have relatively large FOM (11 GPa) because of a low ε_0 (~11).[9] We have focused on $BiFeO_3$ because it has high spontaneous polarization (~100 μC/cm^2) and low ε_{33} (~100), which suggests that $BiFeO_3$ potentially has large FOM.[10,11] We have reported that (100)-epitaxial and (100)-orientated $BiFeO_3$ films have FOM of 11 and 14 GPa, respectively, and also demonstrated MEMS VEHs using $BiFeO_3$ films.[12–14] VEHs using (100)-orientated $BiFeO_3$ films have an output power of 10.5 $\mu W\cdot mm^{-3}\cdot G^{-2}$ (G=9.8 m/s^2), which is approximately four times larger than that using a polycrystalline film.[15,16] The output power is comparable to those of the best-performing VEHs using $Pb(Zr,Ti)O_3$ and AlN films.[2]

The results described above were obtained using monochromatic vibrational waves. However, environmental vibrations have random oscillations comprising various frequencies and amplitudes. For instance, the power spectrum of an oil-sealed rotary pump (Fig. 1) shows spurious components:

the harmonic and sub-harmonic waves. In this study, to discuss the influence of the spurious components on VEH performance, the output power is characterized using vibrations composed the fundamental, harmonic, and sub-harmonic waves.

2. Experiment

$LaNiO_3$ films, which are used as bottom electrodes and seed layers for the (100)-oriented growth of $BiFeO_3$ films, was deposited on Pt/Ti/SOI diaphragms using the rf-magnetron sputtering method. Sputtering was performed at an rf power of 50 W and a working pressure of 10^{-2} Torr. The 200-nm-thick films were deposited at 300 °C under a gas mixture of O_2 (30%) and Ar (70%) and then annealed at 600 °C under O_2 atmosphere by rapid thermal annealing (RTA) to reduce oxygen defects in the films. The resistivity of the films was around 5×10^{-4} Ω·cm. $BiFeO_3$ films were prepared on the $LaNiO_3$ film using the sol-gel method. The films were crystalized at 500 °C by RTA. The film thickness was 250 nm. The 2θ-ω XRD profile of the film (not shown here) indicated that the $BiFeO_3$ and $LaNiO_3$ have (100) preferential orientation and no secondary phase and misoriented grains.[13,14] The $e_{31,f}$ coefficient of the film before the MEMS process was −2.8 C/m^2. The unimorph-type cantilever (Fig. 2) was fabricated using a conventional MEMS process. The details are described in Ref. 16. The width, length, and thickness of the cantilever were 3.20, 0.400, and 0.006 mm, respectively. The top electrode was 0.380 × 1.00 mm × 3.50 μm. A 0.4-mg Cu mass of dimension 0.380 × 0.470 × 0.250 mm was attached to the free end of the cantilever to match the frequency of the environmental vibrations (60–200 Hz).

Vibrations were applied to the VEHs toward the vertical direction using a shaker. We applied several waves to the shaker to characterize the output power of the VEHs under ideal and random vibrations. Figure 3 depicts profiles of these vibrational waves, which have the following specifications: wave 1 (fundamental wave of 92.3 Hz), wave 2 (fundamental wave + sub-harmonic wave of 46.15 Hz), wave 3 (fundamental wave + harmonic wave of 184.6 Hz), and wave 4 (fundamental wave + harmonic wave + sub-harmonic wave). Each were prepared by a function generator (HP 33120A, Hewlett Packard). The amplitude of each vibration was the same. The applied acceleration was measured using an accelerometer (352A24, PCB Piezotronics, inc.). The displacement of the cantilever was determined using a laser head LK-020 and laser displacement sensor (LK-G5000, Keyence Corporation). To characterize the generated electrical power of the VEH, a load resistance was connected to the $BiFeO_3$ film. The output voltage across the load resistor was measured using a lock-in amplifier (LI5640, NF Corporation).

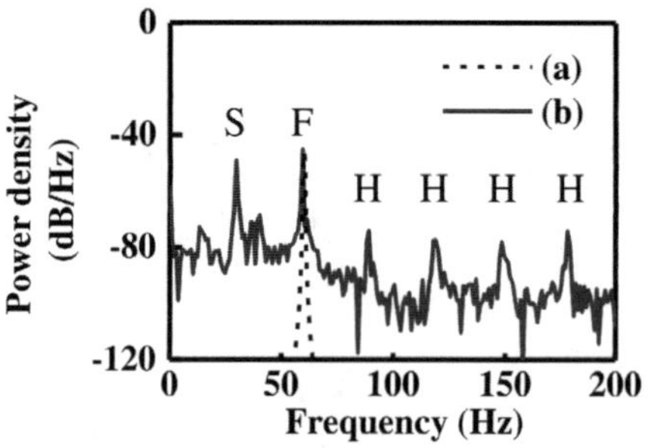

Figure 1. Power density of the vibrations of an oil-sealed-rotary pump composed of (a) ideal vibration, (b) real vibration, for which F, H, and S indicate the fundamental, harmonic, and sub-harmonic waves, respectively.

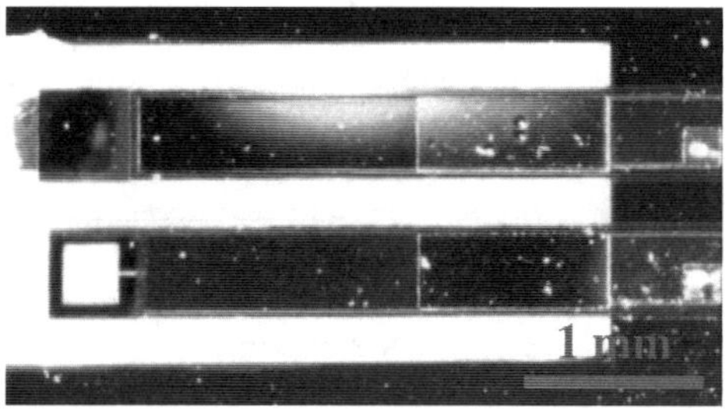

Figure 2. Piezoelectric VEHs using (100)-orientated $BiFeO_3$ films.

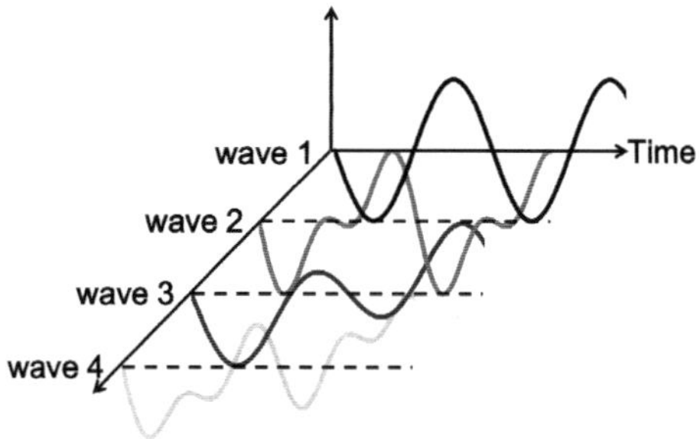

Figure 3. Prepared random oscillation.

3. Results and discussion

3.1. Output power under ideal vibration

Figure 4 shows the open-circuit output voltage at different accelerations under wave 1, the ideal vibration. The accelerations of 0.009, 0.017, 0.035, and 0.052 G_{rms} were applied. The output voltage is normalised by the acceleration because the induced charge via the piezoelectric effect of the $BiFeO_3$ thin film is proportional to the applied strain. The hysteresis behaviour, which is the difference of the power responses between the frequency sweep-up and sweep-down, was observed. With increasing acceleration, the curves of the output voltage near resonance frequency become asymmetric. Simultaneously, the resonance frequency gradually shifts to higher frequencies with increasing acceleration. This nonlinear resonance is caused by the increase in stiffness of the cantilever at large deformation. The resonance frequency, *Q*-factor, and maximum normalised output voltage are 92.3–93.9 Hz, 280, 2.8 $V{\cdot}G^{-1}$, respectively. Figure 5 shows the normalised output power as a function of load resistance. The maximum output power is obtained at 1.0 MΩ, which is consistent with the calculated resistance (8.6 kΩ) to obtain impedance matching. The maxim output power is 2.6 $\mu W{\cdot}mm^{-2}{\cdot}G^{-2}$. Static output power P_{Rmax} and generalized electro-mechanical coupling coefficient K^2 are given by:

$$P_{R\max} = \frac{MQ^2A^2}{\omega_r}\frac{9d}{4l^3u}\left(\frac{1}{4}x^3 - lx^2 + l^2x\right)K^2 \ , \tag{1}$$

$$K^2 = \frac{(1-v)^2}{E}\frac{e_{31,f}^2}{\varepsilon_0\varepsilon_{33}} \ , \tag{2}$$

where M is the mass of the proof mass, Q the Q-factor, A the acceleration of the applied vibration, ω_r the angular velocity at resonance frequency, and d, x, v, E, l, and u are the thickness of the piezoelectric film, length of the top electrode, Poisson's ratio, Young's modulus, length, and thickness of the cantilever, respectively. Using values for the $e_{31,f}$ coefficient and ε_r of the (100)-orientated $BiFeO_3$ thin film on (100) $LaNiO_3/Pt/Ti/SiO_2/Si$, which are −2.8 C/m^2 and 123, respectively, K^2 expected from Eq. (2) is 2.3%, which is approximately five times larger than that of a polycrystalline film.[15] The $e_{31,f}$ coefficient and K_r^2 calculated using Eqs. (1) and (2) are −3.0 C/m^2 and 2.5 %, which are consistent with those of the film before the MEMS process.

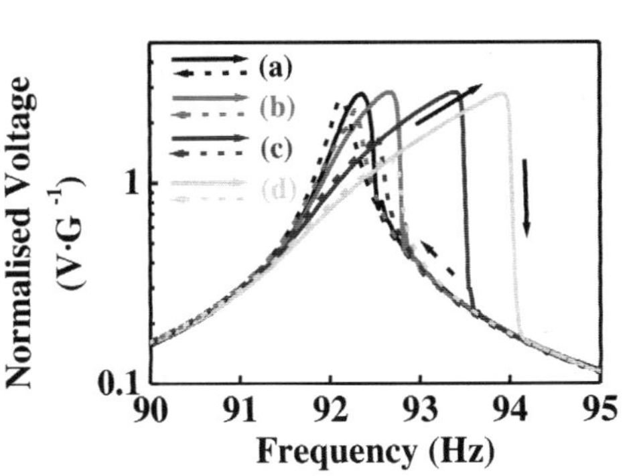

Figure 4. Frequency dependence of the normalised voltage at (a) 0.009, (b) 0.017, (c) 0.035 G_{rms}, and (d) 0.052G_{rms}.

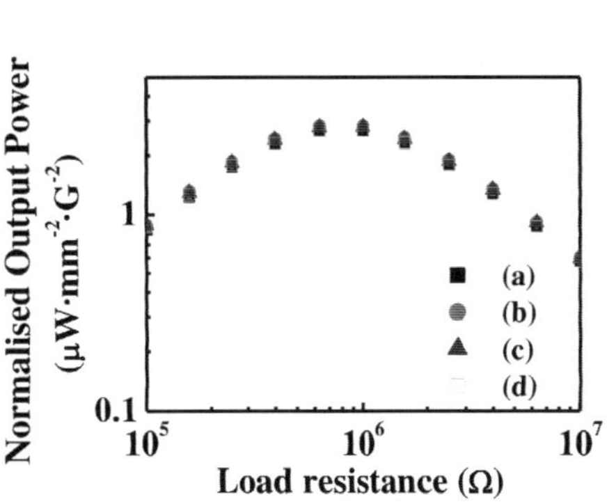

Figure 5. Normalised output power as a function of load resistance at (a) 0.009, (b) 0.017, (c) 0.035 G_{rms}, and (d) 0.052 G_{rms}.

3-2. Vibration and output power under random vibration

The relative displacement and output power of the VEH under random vibrations were characterized. Figure 6 shows the waveforms of the input vibrational acceleration and displacement of the tip of the cantilever under waves 1–4. Each displacement waveform is different from that for acceleration under random vibrations and all the waveforms look roughly sinusoidal. This indicates that only a single frequency component is included in each wave. The results of fast Fourier transform analysis indicate that the VEH responds only to the resonance vibration of 92.3 Hz, which is the resonant frequency of the VEHs. Nevertheless, the displacement amplitude under random vibrations is smaller than that under an ideal vibration of the fundamental; the maximum accelerations of waves 1–4 are similar.

The waveforms of acceleration were also Fourier analysed. Figure 7 shows that the ratios of the maximum acceleration of waves 2–4 with respect to wave 1. The displacement calculated using the acceleration ratio and the maximum displacement at wave 1 (0.232 mm) are 0.148, 0.167, and 0.128 mm for waves 2, 3, and 4, respectively, which are smaller than the experimental results (0.188, 0.192, and 0.172 mm, respectively). Similarly, the output power of each of the waves 1–4 also depends on the acceleration ratio. The maximum output power normalised by the area of the cantilever was 1.2, 0.73, 0.77, and 0.62 $nW \cdot mm^{-2}$ for waves 1, 2, 3, and 4, respectively. The ratios of the output power for waves 2–4 do not correspond to the acceleration ratio as shown in Fig. 7.

To investigate the effect of spurious component on output power in detail, the component of the output power at 92.3 Hz under wave 1-4 were characterized by the lock in amplifier. The results are shown in Fig. 8. The output power is independent of the wave. This indicates that the spurious component has no influence on the resonance response. The curves shown Fig. 8 have two regimes. For lower accelerations, the output power is proportional to the square of the acceleration. For higher accelerations, above 0.01 G, the output power is saturated because of nonlinear resonance. The incline of the curve at saturation gradually decreases with increasing

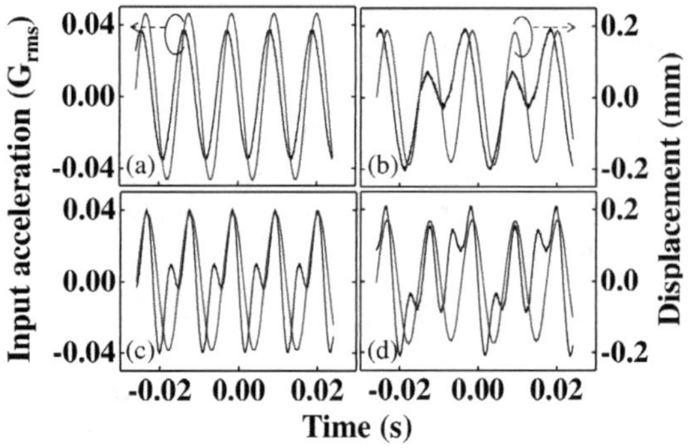

Figure 6. Waveform of input acceleration and displacement under (a) wave 1, (b) wave 2, (c) wave 3, and (d) wave 4.

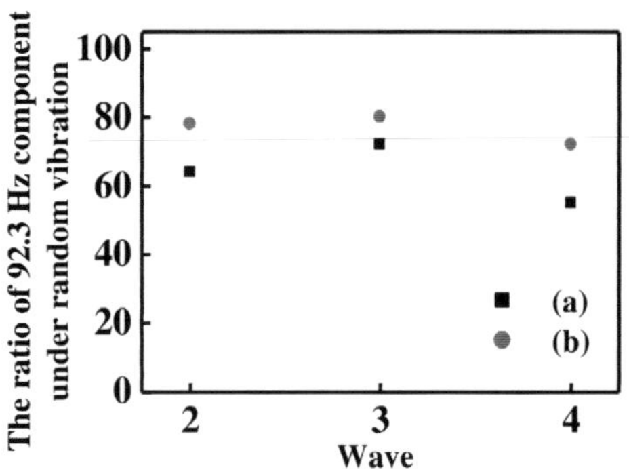

Figure 7. The ratios of 92.3 Hz component on (a) acceleration and (b) output power under random vibration.

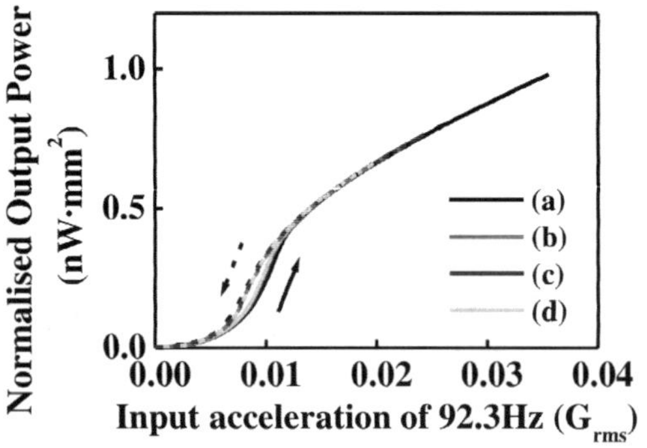

Figure 8. Normalised output power as a function of acceleration of 92.3 Hz under (a) wave 1, (b) wave 2, (c) wave 3, and (d) wave 4.

acceleration. It appears that this nonlinearity causes a difference in the acceleration and power or displacement ratio shown in Fig. 7.

4. Conclusion

A piezoelectric MEMS VEH using (100)-orientated $BiFeO_3$ film was fabricated and the output power of a cantilever-type VEH under ideal and random vibrations were characterized. The maximum output voltage and power at the ideal resonance vibration were 2.8 $V\cdot G^{-1}$ and 2.6 $\mu W\cdot mm^{-2}\cdot G^{-2}$, which is comparable to the best-performing VEHs using $Pb(Zr,Ti)O_3$ and AlN. The K^2 of the VEH using (100)-orientated $BiFeO_3$ film is 2.3 %, which is approximately five times larger than that using a polycrystalline film. To characterize the vibration and output power properties of the VEH, several waves composed of the fundamental (92.3 Hz), harmonic (184.6 Hz), and sub-harmonic (46.15 Hz) were applied to the VEH. The waveforms of displacement of the mass were independent of those of acceleration under ideal and random vibration and were only composed of a single frequency component corresponding to the resonance fundamental. In contrast, the displacement under random vibrations was smaller than that under ideal vibrations. To uncover its origin, the output power properties as a function of acceleration of the resonance frequency under ideal and random vibrations were characterized. The results were independent of the form of the vibration. This indicated that the decrease in displacement stems from the nonlinear resonance of the VEH and that the resonance response depends only on the fundamental component of the wave.

Acknowledgement
This work was supported by the Industrial Technology Research Grant Program in 2011 from the New Energy and Industrial Technology Development Organization (NEDO) of Japan.

References

[1] Roundy S, Wright P K, and Rabaey J 2013 *Comput. Commun.* **26**, 1131
[2] Kim S G, Priya S, and Kanno I 2012 *MRS Bull.* **37**, 1039
[3] Roundy S and Wright P K 2004 *Smart Mater. Struct.* **13**, 1131
[4] Mitcheson P D, Reilly E K, Toh T, Wright P K, and Yeatman E M 2007 *J. Micromech. Microeng.* **17**, S211
[5] Priya S 2007 *J. Electroceram.* **19**, 167
[6] Miyabuchi H, Yoshimura T, Murakami S, and Fujimura N 2011 *Jpn. J. Appl. Phys.* **50** 09ND1
[7] Trolier-Mckinstry S and Muralt P 2004 *J. Electroceram.* **12** 7
[8] Calame F and Muralt P 2007 *Appl. Phys. Lett.* **90** 062907
[9] Wan X, Houwman E P, Steenwelle R, van Schaijk R, Nguyen M D, Dekkers M, and Rijnders G 2014 *Appl. Phys. Lett.* **104**, 092902
[10] Wang J, Neaton J B, Zheng H, Nagarajan V, Ogale S B, Liu B, Viehland B, Vaithyanathan V, Schlom D G, Waghmare U V, Spaldin N A, Rabe K M, Wuttig M, and Ramesh R 2003 *Science* **299** 1719
[11] Yun K Y, Ricinschi D, Kanashima T, Noda M, and Okuyama M 2004 *Jpn. J. Appl. Phys.* **43** L647
[12] Ujimoto K, Yoshimura T, Murakami S, and Fujimura N 2012 *Appl. Phys. Lett.* **100** 102901
[13] Kariya K, Yoshimura T, Murakami S, and Fujimura N 2014 *Jpn. J. Appl. Phys.* **53**, 08NB02
[14] Kariya K, Yoshimura T, Murakami S, and Fujimura N 2014 *Jpn. J. Appl. Phys.* **53**, 09PA14
[15] Yoshimura T, Murakami S, Wakazono K, Kariya K, and Fujimura N 2013 *Appl. Phys. Express* **6** 051501
[16] Murakami S, Yoshimura T, Satoh K, Wakazono K, Kariya K and Fujimura N 2013 *J. Phys.: Conf. ser.* **476** 012007

Influence of Various Conducting and Inorganic Dielectric Drift Barriers on the Charge Decay in Parylene-C Electret Layers

S Genter and O Paul

Department of Microsystems Engineering (IMTEK), University of Freiburg, Germany

E-mail: simon.genter@imtek.uni-freiburg.de

Abstract. This paper reports on an attempt to improve the charge carrier life time in electret layers made of Parylene-C by implementing a patterned drift barrier close to the surface. Using various, differently structured metals and inorganic materials as a drift barrier, the initial surface potential directly after charging was increased by about 11.5% compared to electret layers made of pure Parylene-C. Furthermore, a higher charging efficiency was achieved. The charge carrier distribution has a higher homogeneity and the usually observed surface potential reduction at the rim of the electret layer was less pronounced. Against expectations, the long-term stability did not improve among the chosen materials and patterns. Regarding side effects observed with unstructured drift barriers, the electret layers with a structured drift barrier do not suffer from the horizontal discharging.

1. Introduction

Recently the development of micro energy harvesting devices transducing ambient energy into the electrical domain has received much attention [1–5]. These devices need no costly maintenance and thus are suitable for deployment in wireless sensor nodes and in medical implants such as cardiac pacemakers. They can contribute to extending battery lifetime and in some applications may even completely replace batteries.

Vibrational energy can be harvested by electrostatic transduction relying on charged electret materials [6]. An electret material is able to store charge carriers quasi-permanently [7]. The stability of implanted ionic charge carriers is mandatory to guarantee the long-term functionality of such a device. Among other polymers like Cytop and Teflon, Parylene-C has been considered as a promising electret material [8–10]. Previously the charge decay in simple Parylene-C layers with thicknesses up to 23 µm has been investigated [11]. The highest charge carrier stability was achieved using 11-µm-thick samples charged at 100°C, which showed a decay of 20% after 140 days. An attempt to reduce the charge decay using a SiN drift barrier close to the electret surface revealed the effect of an additional discharging path along the electret-barrier interface [12]. In this work, an improved design using a variety of conducting and inorganic barrier materials to eliminate this effect and increase the long-term stability is presented.

2. Theory

2.1. Charging of Electrets

Among different charging techniques the corona discharge method was chosen [7]. Such a setup comprises three electrodes as shown in Fig. 1. First, the ground potential is connected to the base electrode under the electret layer. Second, a potential U_{Needle} of several kV is connected to a needle-shaped electrode. The local electrical field near the tip is sufficiently high to ionize the surrounding medium. At negative potentials, anions are accelerated towards the ground potential and thus implanted into the electret layer. To control the resulting surface potential U_{S} a third, mesh-shaped grid electrode set to the potential U_{Grid} is placed between the needle tip and the electret. Since U_{S} increases while the charging proceeds, the potential difference between grid and electret tends asymptotically to zero so that U_{S} is limited by U_{Grid}. In practice, U_{S} is found to reach about 70% of U_{Grid}.

2.2. Discharging

In polymer-based electret materials, discharging mechanisms such as resistive loss, ion diffusion, compensation by charges from the environment and ionic drift are present. The latter has been identified to be the predominant discharging effect in Parylene-C over longer times [7]. The implanted charge carriers drift through the electret layer with the drift velocity $v_{\mathrm{D}} = \mu_{\mathrm{I}} E(x,t)$ proportional to the self-induced electrical field $E(x,t)$ depending on the momentary charge carrier distribution $\rho_{\mathrm{I}}(x,t)$. Here, μ_{I}, x and t denote the charge mobility, the coordinate perpendicular to the electret surface and time, respectively. Thus, implementing a drift barrier layer close to the surface is conjectured to impede the charge carrier movement through the layer. Materials with low ionic mobilities and a high dielectric constant are promising candidates.

3. Methods

3.1. Test Chip Fabrication

The geometry of test chips is schematically shown in Fig. 2. The test chips are based on Pyrex substrates with a Cr/Au/Cr layer sandwich serving as the base electrode. A 7.5-µm-thick Parylene-C layer is deposited using the *Labcoter 2* from *SCS Coatings* at a pressure of 22 mTorr. Using an image reversal resist, the 200-nm-thick drift barrier layer made of materials such as Ti, Au, Al, Pd, Pt, Cr, SiC, SiN and SiO_x is deposited onto the Parylene-C layer by plasma enhanced chemical vapor deposition (PECVD) or evaporation. This barrier layer is structured into a continuous layer with dimensions of $7 \times 7\,\mathrm{mm}^2$ or as an array of smaller squares with

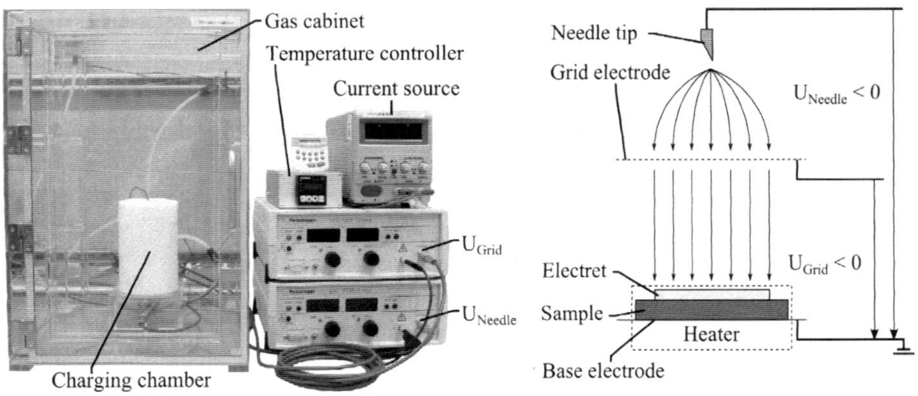

Figure 1. Photograph of the custom-made corona discharge setup (left) and schematic view of the charging chamber (right).

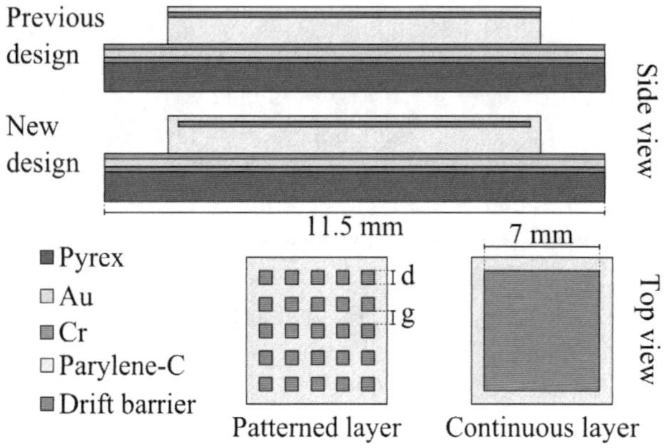

Figure 2. Schematic view of the test chips. The cross-section (top) shows the layer stack and the top view (bottom) the pattern of the drift barrier.

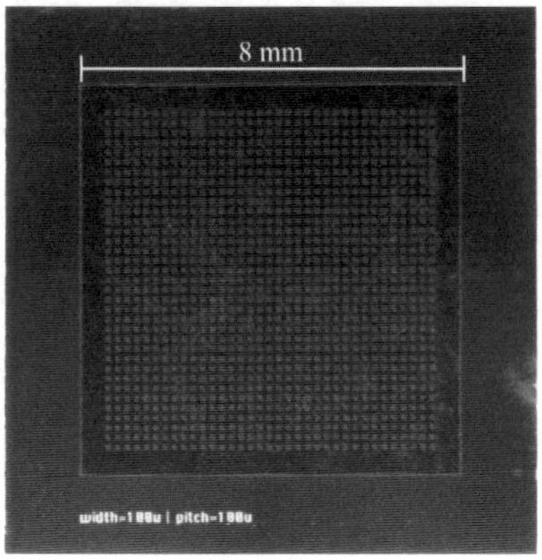

Figure 3. Photograph of one fabricated test chip with a $g = d = 100\,\mu$m pattern.

a side length d and gap g of 100 or 200 µm. A second, 0.5-µm-thick, Parylene-C layer is then deposited to bury the drift barrier. Finally, the entire layer stack is structured by reactive ion etching into $8 \times 8\,\text{mm}^2$ squares and the wafer is diced into chips. One fabricated test chip is shown in Fig. 3. Each fabricated wafer uses only one of the listed drift barrier materials and comprises test chips without drift barrier. The previous design is shown in Fig. 2. In this design, the drift barrier reached the rim of the layer stack. That design suffered from charge carrier drift along the interface between the drift barrier and Parylene-C [12]. Besides the structured drift barrier layer, the main difference to the previous samples is the fully encapsulated drift barrier which is hypothesized to suppress this horizontal loss channel.

3.2. Charging of the Test Chips
To charge the electret layers, the custom-made corona discharge setup shown in Fig. 1 is used. The high-voltage sources *LNC-10000 neg* and *LNC-1200 neg* from *Heinzinger*, Germany, supply the potentials for the needle tip and the mesh electrode. For the experiments, the potentials $U_{\text{Needle}} = -8\,\text{kV}$ and $U_{\text{Grid}} = -1\,\text{kV}$ are applied to the needle tip and grid electrodes, respectively. Charging is performed at room temperature for a duration of 3 min using air as the charging medium.

3.3. Surface Potential Measurement
The surface potential $U_{\text{S,initial}}$ was measured immediately after charging. For this purpose the electrostatic voltmeter *Isoprobe 279* from *Monroe Electronics*, USA was used to measure surface potentials of $-3\,\text{kV} < U_{\text{S}} < 3\,\text{kV}$ with an accuracy of 0.1% of the full scale. As shown in Fig. 4 the voltmeter probe is mounted above the chuck with the test chips. The xyθ-stage enables the system to measure the surface potential at all positions in the plane of the charged electret.

4. Experimental
4.1. Initial Surface Potential
The initial maximum surface potential $U_{\text{S,initial}}$ and the mean value over the width of the electret with a continuous barrier were measured and are compared among the investigated materials in Fig. 5. The $U_{\text{S,initial}}$ values for samples with incorporated Al, SiN and SiO_{x} barriers are lower than those of samples with a pure Parylene-C layer. Other materials show an increase of up to 11.5% with respect to pure Parylene-C samples in the maximum and mean $U_{\text{S,initial}}$.

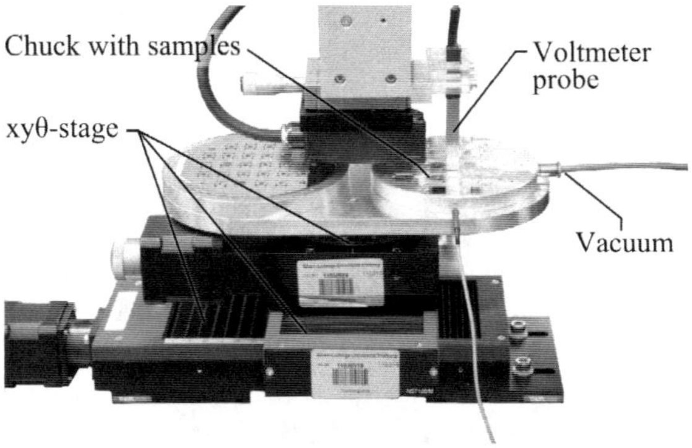

Figure 4. Photograph of the setup for the investigation of the surface potential U_S.

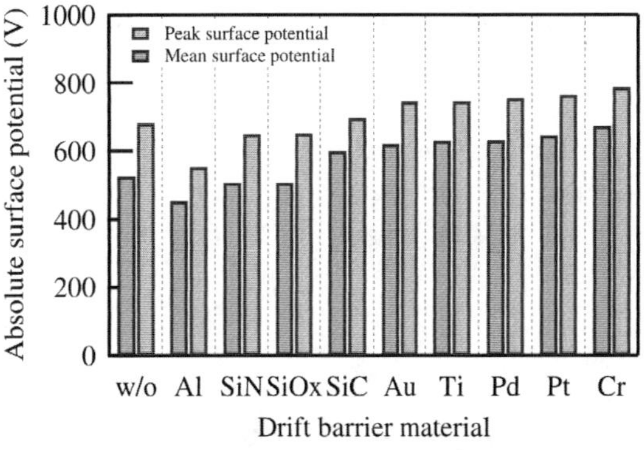

Figure 5. $U_{S,\text{initial}}$ for the variety of materials using a continuous drift barrier.

Furthermore, drift barriers made of metal materials are superior to the investigated dielectric materials.

Moreover, we investigated the initial U_S profiles line scans with 200 µm step size, as shown in Fig. 6. The red dashed line indicates the ideally measurable profile shape assuming a sharp transition from zero potential beside the electret to a fixed, constant potential on the electret. The slope in the transition area is due to the 500-µm-wide measurement spot of the voltmeter probe. Using SiN and SiO_x as a drift barrier, the measured profile shape is approximately congruent with the pure Parylene-C layer. Except for Al, all other materials result in a profile shape closer to the ideal case. Due to the electrical conductivity the continuous metal layers assume a homogeneous potential. Thus, the potential difference between the electret surface and the grid electrode is equal for all locations which significantly improves the charging process. Increased $U_{S,\text{initial}}$ values were not observed using the patterned drift barrier designs. Due to the Parylene-C isolation between the individual square elements of the structured drift barriers, the individual squares of the arrays do not necessarily assume the same potential.

In comparison to the results using the previous design, undesired discharging effects introduced by the drift barrier are suppressed by the new design. This is illustrated in Fig. 7. After a time span of 15 days the profile has flatten as expected. In comparison to unstructured

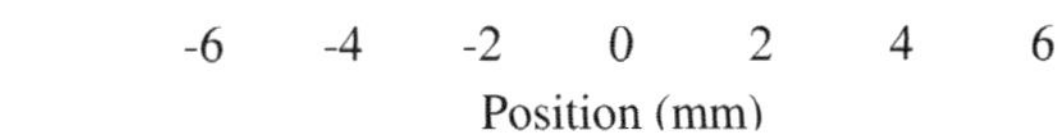

Figure 6. Profile shape of $U_{S,\text{initial}}$ using a continuous drift barrier and the ideally measurable profile shape.

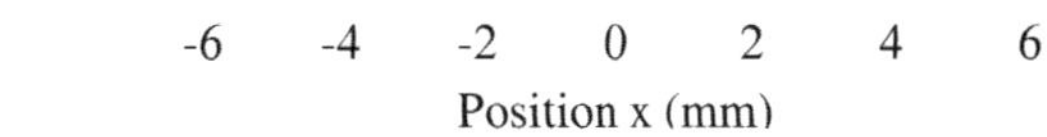

Figure 7. Profile shapes of U_S (t) measured at test chips with a continuous drift barrier made of SiN after several time spans.

and unencapsulated SiN drift barriers, the profile width did not decrease which indicates the elimination of the discharging at the edge of the electret-barrier interface.

4.2. Decay of the surface potential

To further investigate the charge stability, the surface potentials were monitored over a time span of 100 days. In case of metals such as Au, Cr, Pd, Ti and Pt, no significant difference between the sample without drift barrier and the various designs was observed. After 100 days the relative decay is in the range of 40% to 50%. Considering the absolute U_S values, the initial potential difference between the different patterns vanishes after a time span of about 10 days.

Different observations were made for the other investigated materials. In agreement to the lower $U_{S,initial}$ value of the sample with an Al barrier, the long term stability was inferior compared to other materials, as well. This sample lost its charge carriers rapidly within minutes. Furthermore, samples using SiC, SiN and SiO_x deposited by PECVD resulted in a highly accelerated charge decay rate, which is also true for samples where the barrier layer was again removed completely from the Parylene-C layer. Apparently, the PECVD process alters the polymeric structure of the previously deposited Parylene-C layer and makes it less able to store charge carriers.

5. Conclusion

Parylene-C based electret layers including a structured drift barrier out of various metals or inorganic materials have been fabricated and successfully charged. Metal barriers increased the efficiency of the charging process. Higher initial surface potentials and a higher charge carrier homogeneity were achieved. Patterned drift barriers with 100 and 200 μm width and spacing resulted in no improvement compared to pure Parylene-C layers. Furthermore, test chips with a continuous drift barrier showed no horizontal discharging in contrast with the previous design with unencapsulated barrier layers. However, the investigated drift barriers were not able to improve the long term stability of ionic charge carriers in Parylene-C.

References

[1] Matsumoto K et al Vibration-Powered Batteryless Sensor Node using MEMS Electret Generator 2011 *Tech. Dig. PowerMEMS* (Seoul, South Korea, 15-18 Nov 2011) 134-7

[2] Woias P Micro Energy Harvesting as Core Technology for Energy-Autonomous Embedded Systems 2011 *Tech. Dig. PowerMEMS* (Seoul, South Korea, 15-18 Nov 2011) 98

[3] Bartsch U et al Low-Frequency Two-Dimensional Resonators for Vibrational Micro Energy Harvesting 2010 *J. Micromech. Microeng.* **20** (3) 035016

[4] Beeby S P et al Energy Harvesting Vibration Sources for Microsystems Applications 2006 *Meas. Sci. Technol.* **17** R175-R195

[5] Crovetto A et al MEMS Fabricated Energy Harvesting Device With 2D Resonant Structure 2012 *Tech. Dig. PowerMEMS* (Atlanta, USA, 2-5 Dec 2012) 125-8

[6] Bartsch U Electret-Based Resonant Micro Energy Harvesting in Two Dimensions 2010 (Der Andere Verlag, Germany) *MEMS Technology and Engineering Series*, **15** 66

[7] Sessler G M 1998 *Electrets - Third Edition* (Laplacian Press)

[8] Wada Y et al Aging Characteristics of Electret Used in a Vibration-based Electrostatic Induction Energy Harvester 2011 *Dig. Tech. Papers Transducers* (Bejing, China, 5-9 June 2011) 2626-9

[9] Yamashita Ket al Vibration-driven MEMS Energy Harvester with Vertical Electrets 2010 *Tech. Dig. PowerMEMS* (Leuven, Belgium, 30 Nov - 3 Dec 2010) 165-8

[10] Lo H et al Parylene-based electret power generators 2008 *J. Micromech. Microeng.* **18** (10), 104006

[11] Genter S, Paul O Parylene-C As An Electret Material For Micro Energy Harvesting 2012 *Tech. Dig. PowerMEMS* (Atlanta, USA, 2-5 Dec 2012) 129-132

[12] Genter S, Paul O Influence of a SiN Drift Barrier on the Charge Decay in Parylene-C Electret Layers 2013 *Tech. Dig. PowerMEMS* (London, UK, 3-6 Dec 2013) 401-5

Pt@TiO2- Au nanoCORRUGATed PLASMONIC STUCTURE for visible-light active photocatalysis

Yu-Hsuan Peng [1], Chun-Sheng Chen [1], Chun-ting Lin[2], Fan-Gang Tseng [1,3*]
1.Department of Engineering and System Science, National Tsing Hua University, TAIWAN
2. Instrument Technology Research Center, National Applied Research Laboratories, Hsinchu 300, TAIWAN
3.Division of Mechanics, Reschanics, Researchnter for Applied Science, Academia Sinica, TAIWAN

E-mail : kobe800819@hotmail.com

Abstract. In this paper, we design and fabricate a NanoFinger Plasmonic Structure (NFPS) utilized in water splitting for hydrogrn production by visible light irradiation. This structure can generate numerous hot electrons from localized surface plasmon resonance (LSPR) during visible light irradiation. Through this hot electrons transferred via titanium dioxide to platinum nanoparticles, hydrogen ions in water can be efficiently reducing into hydrogen, while hot holes left in gold oxidizing into oxygen. Experimental results demonstrated the photoelectrochemical signal can be enhanced by 1.96~2.09-fold in on-off test under visible light, indicating the possibility for high efficient hydrogen generation.

1. Introduction

Sunlight is a source of clean energy which is non-polluting, inexpensive, and endlessly renewable. However, it can hardly be applied freely in demand. As a result, to store solar energy into another energy forms, such as hydrogen, is a more practical way for real applications [1]. There were various methods developed in the past decade to transfer solar energy into hydrogen, including electrolysis of water using a solar cell, reforming of biomass, and photocatalytic [2]. Among them, photocatalytic reactions based on electron-hole pair production in semiconductors can generate hydrogen without any CO2 or other pollution produced. However, most of light spectra employed were in UV region, posing low efficiency on the utilization of the whole sun-light energy [3]. Recently, it was reported that hot electrons transferred from plasmonic structure to neighboring photocatalyst could enhance the photo response for solar water splitting, especially in visible light region [4]. In this paper, we propose a simple yet high efficient plasmonic structure [5] incorporated with Pt-TiO2-Au sandwich nano fingers to provide numerous hot spot regions for harvesting photons over entire solar spectrum. As shown in Fig. 1, the hot electrons generated from the surface plasmons in the gold nanocorrugation structures can be transferred via neighboring titanium dioxide into the platinum nanoparticles, thus help on generating hydrogen in higher efficiency.

2. Experimental

2-1 PS ball array

i. The PS ball is diluted with water and 99% alcohol and then place it in ultrasonic agitation for 15 minutes.

ii. The wafer was deal with hydrophilic treatment by O2 plasma.

iii. PS ball self-assembled underwater 6-12 hours.

iv Removed the water. The PS ball array is finished when it gets dry naturally.

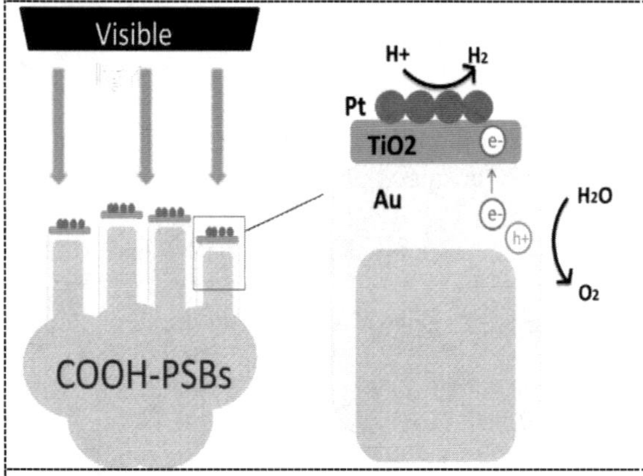

Figure 1. Schematic of hot electrons transfer among nano-structure to product hydrogen, and hot holes product oxygen at Au.

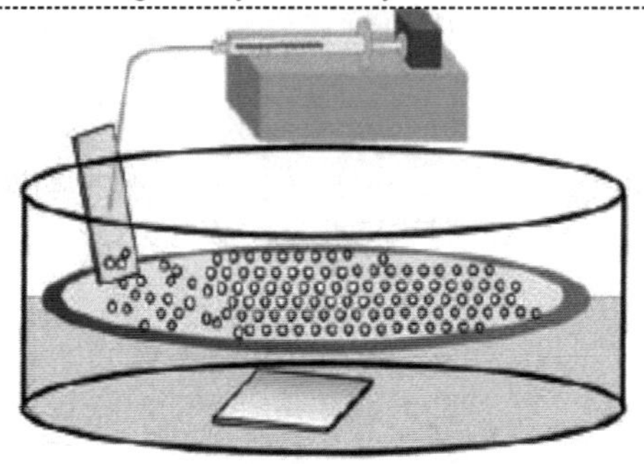

Figure 2. Schematic of PS ball self-assembled equipment

2-2 NanoFinger structure fabrication process

The first step is to deposit PS ball array on silicon wafer or ITO, and then take treatment O2 plasma to NFPS A (10 sccm, 75 mtorr, 200W) and NFPS B (5 sccm, 60 mtorr, 55 W) which has been structured by Reaction Ion Etching (RIE). Then we load Au, TiO2 and Pt on substrate to make NanoFinger structure step by step[6].

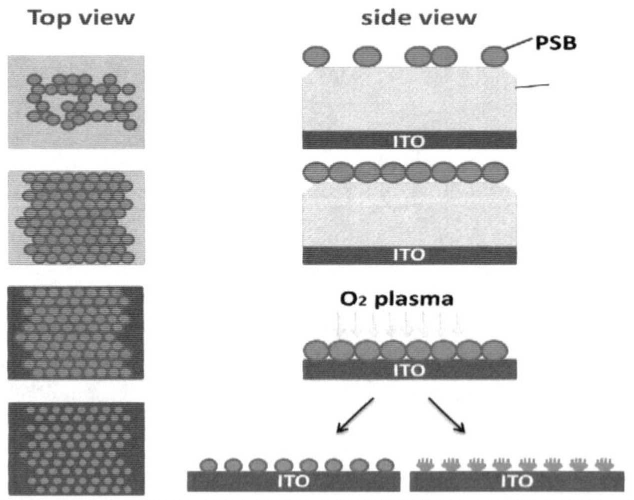

Figure 3. Fabrication process of NanoFinger photocatalyst.

3. Results

The structure of our investigation using SEM is shown in Fig. 4. The NFPS B aspect ratio is higher than the NFPS A. As the result, we confirm that NFPS A is better than the NFPS B at surface plasmonic resonance.

Photocurrents generated by the NFPS immersed in 0.5M H2SO4 electrolye were measured in three-electode photoelectrochemical cell with platinum mesh as counter electrode and standard calomel electrode as the reference electrode.

Visible light and infrared illumination were obtained from xenon lamp fitted with an AM 1.5 filter illuminated passing wavelengths ranging from 400 nm to 1000 nm. As shown in fig.2 on-off test was conducted from 100s to 2000s, with 0 V applied voltage . Even without semiconductor, LSPR caused the Au-TiO2 NFPS performed 1.19-fold photocatalytical enhancement under visible light irradiation, and Au-TiO2-Pt NFPS can perform 1.96~2.09-fold enhancement even better than Au-TiO2 NCPS.

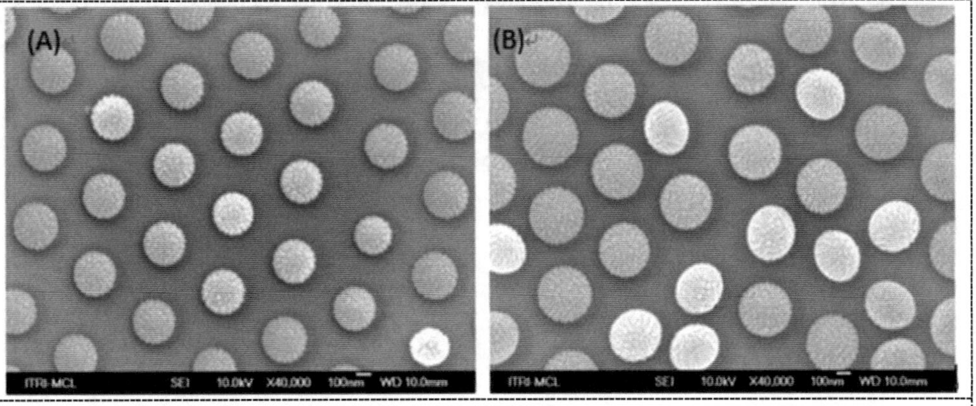

Figure 4. SEM images of (A) NFPS A (B) NFPS B

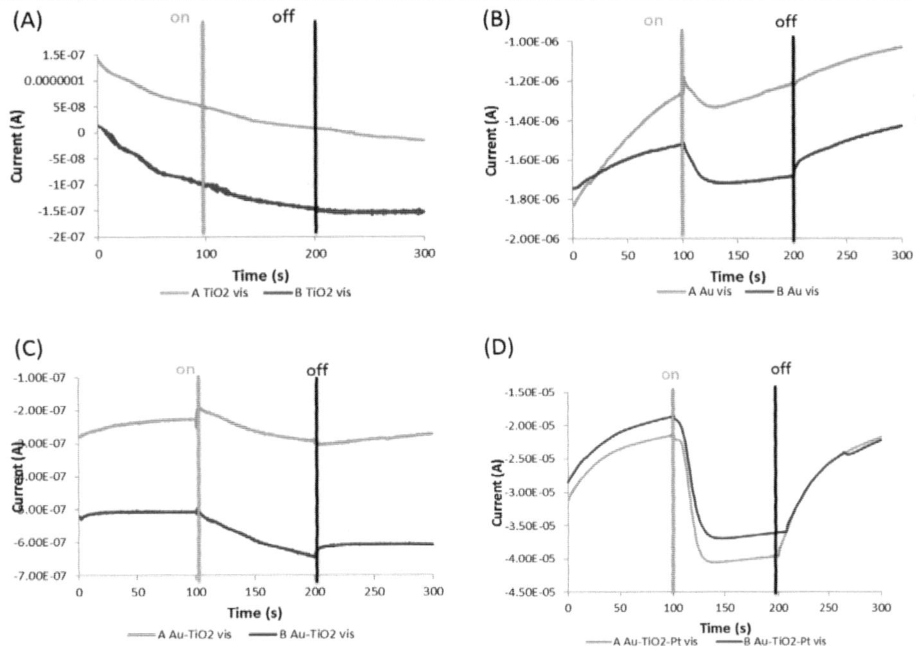

Figure 5.Photoelectrochmical on-off test results of (A)only TiO2 (B)only Au (C)Au-TiO2 (D)Au-TiO2-Pt, at 0 V applied voltage and under visible light irradiation.

4. Conclusion

As the result, we have developed Au-TiO2-Pt NanoFinger structure which is periodic. This structure can generate numerous hot electrons from localized surface plasmon resonance (LSPR) during visible light irradiation. Experimental results demonstrated the photoelectrochemical signal can be enhanced by 1.96~2.09-fold in on-off test under visible light, indicating the possibility for high efficient hydrogen generation.

References

[1] Chen, H.M., et al., Nano-architecture and material designs for water splitting photoelectrodes. *Chem Soc Rev*, 2012. 41(17): p. 5654-71.7

[2] Kudo, A. and Y. Miseki, et al.,Heterogeneous photocatalyst materials for water splitting. *Chem Soc Rev*, 2009. 38(1): p. 253-78.

[3] G.P. Anderson et al., MODTRAN2: Evolution and Applications. *SPIE Proceedings,* 1994, pp. 790-799.

[4] Mubeen, S., et al., An autonomous photosynthetic device in which all charge carriers derive from surface plasmons. *Nat Nanotechnol*, 2013. 8(4): p. 247-51

[5] Hsieh, H.-Y., et al., Au-coated polystyrene nanoparticles with high-aspect-ratio nanocorrugations via surface-carboxylation-shielded anisotropic etching for significant SERS signal enhancement. *The Journal of Physical Chemistry C*, 2011. 115(33): p. 16258-16267.

[6] Ho, C.-C., et al., Fabrication of monolayer of polymer/nanospheres hybrid at a water-air interface. *Acs Applied Materials & Interfaces*, 2011. 3(2): p. 204-208.

An investigation of PDMS structures for optimized ferroelectret performance

J Shi, D Zhu and S P Beeby
Electronics and Electrical Engineering Group, Electronics and Computer Science University of Southampton, Southampton, SO17 1BJ, UK

E-mail: js4g11@soton.ac.uk

Abstract: This paper reports the ANSYS simulation and fabrication processes for optimising PDMS ferroelectret performance. The proposed model extends the previously published analytical models and compares this with simulation of individual void geometry. The ferroelectret material is fabricated from PDMS using 3D-printed plastic moulds. The analytical model and Ansys simulation results predict the variation in performance of the PDMS ferroelectret with the different void geometry and surface charge density. The theoretical maximum piezoelectric coefficient d_{33} achieved was about 220 pC/N. The experimental maximum d_{33} obtained was 172 pC/N.

1. Introduction

A ferroelectret is a type of polymer electret that displays piezoelectric like properties, which is a polymer film with cellular structure. The thickness of ferroelectret films typically ranges from a few dozen to a hundred microns. The geometry of internal voids typically a lens shape with a thickness and width from a few to dozen micrometers. The separated positive and negative charges which are trapped on the upper and lower gas-polymer interfaces in the voids cause the piezoelectric like property of ferroelectrets, because the separated charged form macroscopic dipoles in the voids. The macroscopic dipole moment can be determined by the quantity of the charges and the distance between the separated charges. When the ferroelectret is compressed by an external mechanical load, its volume decreases and compensating charges are generated on the external surface. The resulting macroscopic behaviour of ferroelectret is very similar with the well-known piezoelectric materials; but its microscopic mechanism of charge generation is entirely different from the latter. For ferroelectret, the interior voids not only reduce the overall mass and stiffness, but also generate great dipoles when the separated positive and negative charges are trapped on the upper and lower surface of the voids. Due to the combined effects of the internal dipole moment and the anisotropic cellular polymer structure, the majority of volume change takes place in the voids. Therefore when loaded the distance between positive and negative charges decreases, the dipole moment decreases. The amount of compensating charge decreases and a flow of charge is generated, which presents a strong piezoelectric like effect for ferroelectret [1]. Hence, ferroelectrets are utilized as functional materials in electromechanical sensors and actuators, also as electrically charged layers in electrostatic transducers [2, 3].

The piezoelectric like properties of ferroelectrets has already been theoretically analysed [4-7]. However, most ferroelectrets are formed by a modified film blow and extrusion process for producing the desired cellular structures. The results of this fabrication method are usually ill-controlled in both individual void geometry and overall cellular structure, and the existing modelling methods to approximate the piezoelectric like properties of these materials and structures are simplified models that do not consider individual void geometry and void distribution. Therefore, in this paper we report a design, fabrication and testing methodology for optimising PDMS ferroelctret performance based on polymer MEMS fabrication technology.

2. Model

Piezoelectricity of ferroelectrets has been theoretically analyzed with a simplified model based on charged parallel polymeric and gaseous layers, which includes the finite-element calculations [4-7]. The ANSYS simulation work of this research modelled a cellular PDMS structures (Figure 1) containing a layer of rectangular voids. The ANSYS results predicted deformation of the voids when the structure is subject to an external mechanical load. Assuming the charge densities of the voids surface was uniform, the variation of the charge on the external surface can be calculated by the average of the deformations. Then the piezoelectric coefficient d_{33} can be estimated as:

$$d_{33} = 2\varepsilon_1\sigma(x_2\Delta x_1 + x_1\Delta x_2)(2x_1 + \varepsilon_1 x_2)^{-2}F^{-1} \quad (1)$$

Where ε_1 is the dielectric constant of air and the relative dielectric constant of PDMS; σ is the charge density on void surface; x_1 and x_2 are the thickness of solid and void layers, respectively; Δx_1 and Δx_2 are the variations of the thickness and F is the external force.

For analytical model, the piezoelectric coefficient d_{33} can be directly obtained from:

$$d_{33} = 2\varepsilon_1\sigma tr_1(1 - sr_1)^2 sr_1{}^{-1}[2 + \varepsilon_1 tr_1]^{-2}c_{33}{}^{-1} \quad (2)$$

Where sr_1 is the ratio of the effective area of the void layers to the effective area of the solid layers; tr_1 is the ratio of the thickness of void layers to the thickness of solid layers; σ is the charge density on void surface; ε_1 is the dielectric constant of air and the relative dielectric constant of PDMS; c_{33} is the elastic modulus of PDMS.

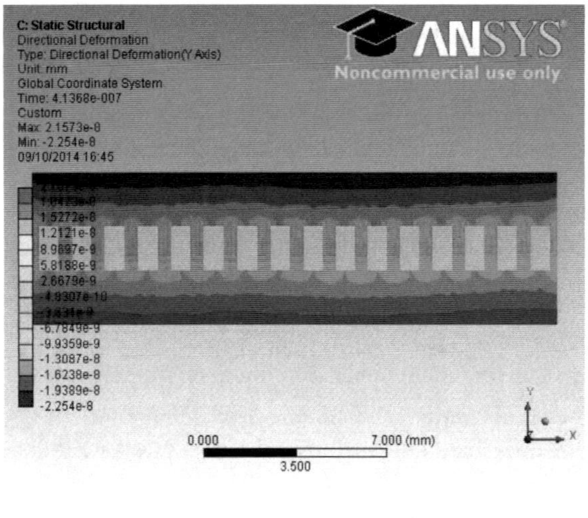

Figure 1. Cellular structures with a layer of rectangular voids

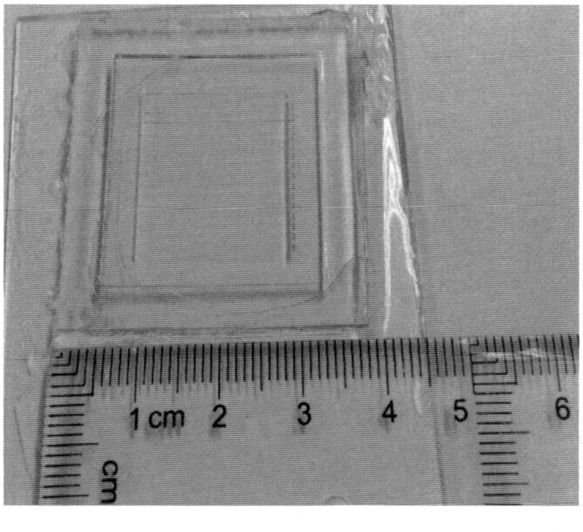

Figure 2. The image of a prepared 3D-printed mould

3. Fabrication and Measurement

3.1. 3D-printed mould preparation

A 3D CAD model of the mould structure was constructed with Solidworks 2012 (Dassault Systems, MA, USA), exported as an STL-format file, and printed on a Connex350™ 3D printing system (Stratasys, MN, USA). The used moulds can be made from VeroWhitePlusTM, VeroGery™ or VeroClear™, generated in 'glossy' printer mode on a bed of FullCure®705 support material. Because the 3DP material inhibits PDMS polymerization and the printed structure tends to warp some extent when it is removed from the printer, the resulting 3D-printed structure is not directly suitable as a mould for PDMS casting. The moulds should be baked in oven at 80 °C overnight. In this bake step, the moulds were placed and glued to a glass slide by a drop of degassed non-polymerized PDMS, with its base in contact with the PDMS film. After this procedure, the glass-backed moulds were exposed to a silane vapour for 1 hour to coat it with a thin layer of trichloro(1H,1H,2H,2Hperfluorooctyl)silane (Sigma Aldrich, MO, USA). An example of a finished mould of VeroClear™ material, with outer dimensions of 38 × 38 mm, is shown in Figure 2. Depending on the maximum precision of the 3D printer and the restrictions of fabrication process, the thickness of the solid layer and void layer was selected as 2 mm and 1.6 mm, respectively; the width of the voids and the gap between the voids was selected as 0.75 mm and 0.5 mm, respectively.

3.2. PDMS casting and Bonding
Liquid PDMS and curing agent (Sylgard 184 from Dow Corning, MI, USA) were mixed at a 10:1 weight ratio and the mixture was degassed in a vacuum desiccator. The degassed PDMS was poured into the glass-backed moulds which were again degassed and then baked at 80 °C for 1 hour. After detaching the polymerized PDMS from the moulds, an oxygen plasma treatment was applied on the patterned surface of PDMS layers (Femto Asher, Diener, Germany, 30 S at 35-40W) and then bonded to the other PDMS layers. To optimise the bonding effect, the samples were assembled with bonding times of 30 s, 60 s, 90 s and 120 s, separately. The bonded PDMS was then baked in oven at 80 °C for 1 hour to improve the bonding effect, and also recover the original hydrophobic PDMS surface chemistry. The schematic of PDMS casting and bonding processes, and the cross-section view of the bonded sample are illustrated in Figure 3 and Figure 4, separately.

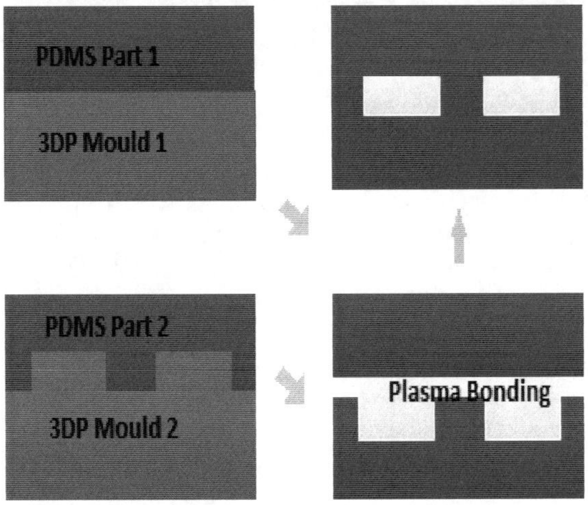

Figure 3. Schematic of fabrication processes

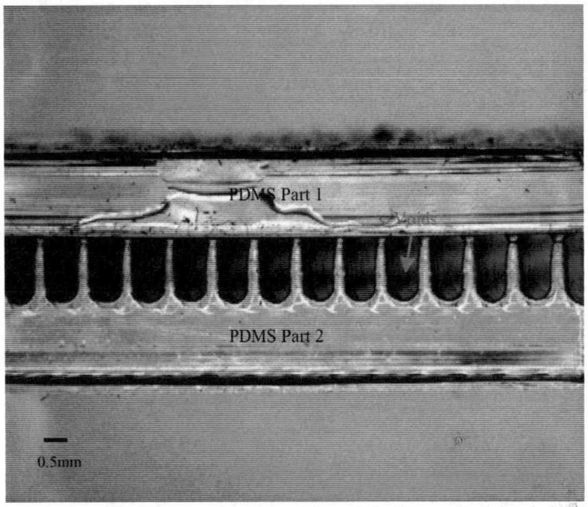

Figure 4. Image of cross-section view of single layer ferrelectret foam

3.3. Corona charging and Measurement
The samples were polarized with a point-to-plane corona discharge in air under ambient conditions. A corona-tip voltage of -25 kV and a charging time of 30 s were employed. Its piezoelectric charge constant d_{33} was measured using a PiezoMeter (PM300, Piezotest Ltd).

4. Results and Discussion
An analytical result is provided in Figure 5, which numerically illustrates the impact of the different void geometry. While the ratio of the effective areas of the solid and void layers is increased, the piezoelectricity of PDMS ferrelectret was reduced. In other word, the piezoelectric properties of the PDMS ferreoelectret improved with increasing void area. In contrast, when the ratio of the thickness of the solid and void layers is increased, the piezoelectric properties of the PDMS ferroelectret was improved and the optimised point has been demonstrated at around 0.8, with a peak piezoelectricity of 220 pC/N. However, considering the feasibility of the fabrication processes, the optimized structure which can be achieved with $sr_1=0.4$ and tr_1=0.8. Based on these ratios, the geometry of void structure is designed as 0.75 mm in width and 1.6 mm in height, and the distance between the voids and the thickness of the whole film is selected as 0.5 mm and 5.6 mm, respectively. For this structure, the theoretical d_{33} values is 62.5 pC/N at surface charge density = 100 μC/m^2.

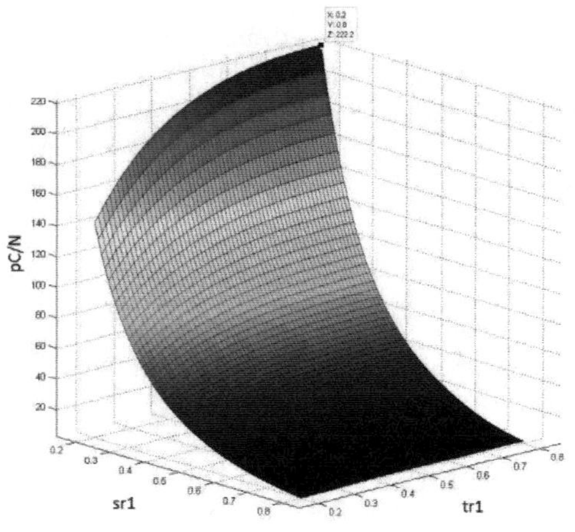

Figure 5. Analytical rectangle model results varying with the size of voids

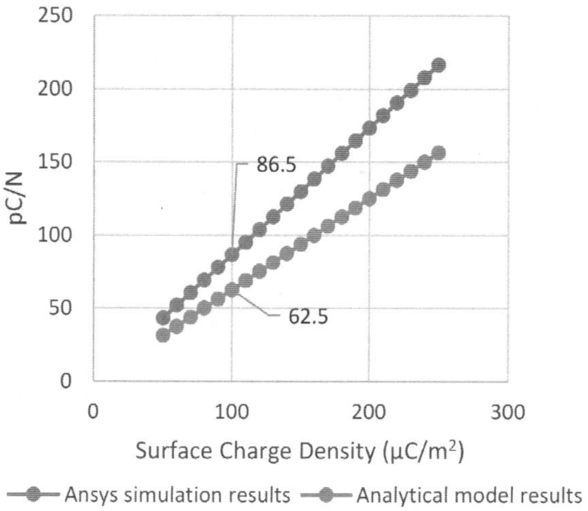

Figure 6. Analytical model results and Ansys simulation d_{33} results for rectangle voids

Figure 6 shows a comparison of ANSYS simulation results and analytical model results to illustrate the influence of the surface charge density based on the designed geometry. There was an obvious deviation between the ANSYS simulation results and analytical model results because the analytical model was based on the principle of a rigid body. In other word, the bending effects that happened on the upper and lower gas-polymer interfaces of the voids was also not considered.

The realized geometry of the void inside the PDMS ferroelectret is illustrated in Figure 7. The realized geometry of void is different to the model due to the precision of fabrication process. The d_{33} value of the PDMS ferroelectret was experimentally measured to be around 70 pC/N. The performance of the PDMS ferroelectret can be influenced by plasma treatment. The impact of plasma treatment time is illustrated at Figure 8. The performance of the PDMS ferroelectret was substantially grown during the plasma treatment duration increasing. The experimental maximum d_{33} obtained is about 170 pC/N at 120 s plasma treatment condition. The possible reason of this phenomenon is the surface charge density of PDMS can be significantly improved by increase in plasma treatment time.

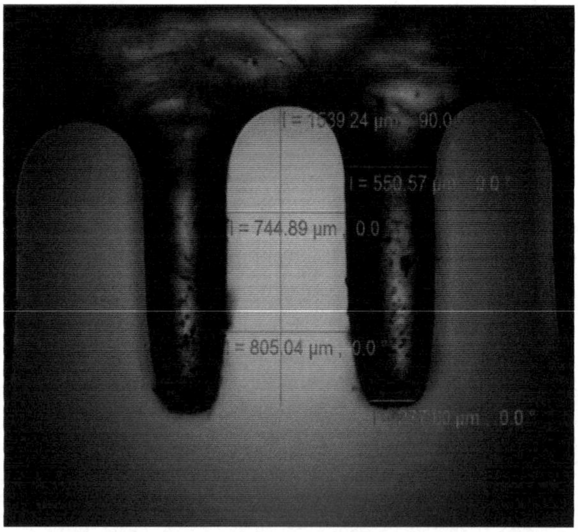

Figure 7. The image of cross section view of Realized voids

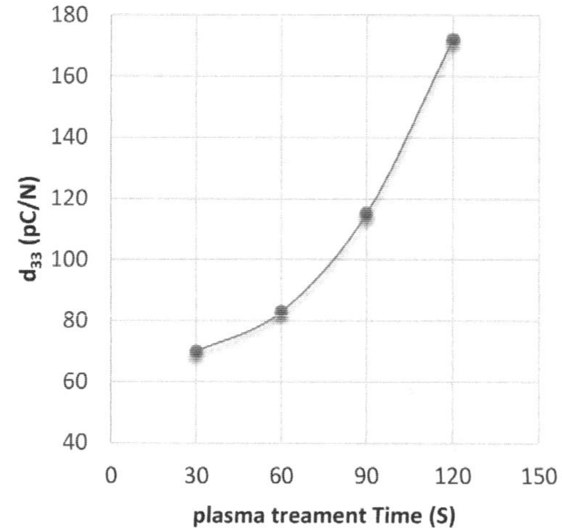

Figure 8. The piezoelectric performance varying with duration of plasma treatment

5. Conclusion and future work

In this work, the simulation and testing of PDMS ferroelectret with a variety of void geometries and surface charge density has been investigated. The piezoelectric properties of PDMS ferroelectret are improved with an enhancement of the void area and the optimized structure was achieved with the ratio tr_1=0.8 for a single void layer of PDMS ferroelectret material. In addition, the performance of PDMS ferroelectret can be significantly affected by the plasma treatment. In the future, PDMS ferroelectret will be optimised for different shapes, multilayer void structure and scaling down based on a MEMS fabricated mould.

6. Acknowledgement

This work was performed under the SPHERE IRC funded by the UK Engineering and Physical Sciences Research Council (EPSRC), Grant EP/K031910/1.

7. References

[1] M. Paajanen, "The cellular polypropylene electret material-Electromechanical properties", Dissertation, VTT Publications 436, Technical Research Centre of Finland, Tampere University of Technology, 2001.

[2] J. Backman. "Audio applications of electrothermomechanical film (ETMF)", *J. Aud. Eng. Soc.*, vol. 38 (5), pp. **364-371**, 1990.

[3] M. Paajanen, J. Lekkala, and K. Kirjavainen, "Electromechanical film (EMFi) - a new multipurpose electret material", *Sens. Actu. A: Phys.*, vol. 84 (1-2), pp. **95-102**, 2000.

[4] J. Hillenbrand and G. M. Sessler, "Piezoelectricity in cellular electret films", *IEEE Trans. Dielect. Electr. Insul.*, vol. 7 (4), pp. **537-542**, 2000.

[5] M. Paajanen, J. Lekkala, and H. Välimäki, "Electromechanical modeling and properties of the electret film EMFi", *IEEE Trans. Dielect. Electr. Insul.*, vol. 8 (4), pp. **629-636**, 2001.

[6] M. Paajanen, H. Välimäki, and J. Lekkala, "Modelling the sensor and actuator operations of the ElectroMechanical Film EMFi", Proceedings, 10th International Symposium on Electrets, 22-24 September 1999, Delphi, Greece (IEEE Service Centre, Piscataway, NJ, USA 1999), pp. **735-738**.

[7] G. M. Sessler and J. Hillenbrand, "Electromechanical response equations of cellular electret films", Proceedings, 10th International Symposium on Electrets, 22-24 September 1999, Delphi, Greece (IEEE Service Centre, Piscataway, NJ, USA 1999), pp. **261-264**.

Porous Ge based Electric Double Layer Capacitors with High Compatibility for Low Threshold Voltage Diode Rectifiers

A. Sekiguchi, M. Hara, H. Oguchi and H. Kuwano

Department of Nano-Mechanics, Tohoku University, 6-6-01, Aoba, Aramaki, Aza-Aoba, Sendai, 980-8579, Japan

E-mail: sekiguchi@nanosys.mech.tohoku.ac.jp

Abstract. The final goal of this study is to develop all-Ge-based energy harvesting modules consisting of mechanical generators, rectifiers, capacitors and power managing digital circuits. To make basis for the development of modules, we focused on development of the Ge electrodes for electric double layer capacitors (EDLC). Firstly, to establish a method to make high surface area Ge electrodes, we studied surface etching of the Ge(100) substrates in the acid solution. We found that the substrates with higher dopant concentration forms smaller size micro-pores with higher area density that leads to higher surface area. Secondly we studied stability of Ge in the ionic liquid (IL) electrolyte by ac impedance measurements. The impedance plots verified that Ge is reactive in IL electrolyte, thus not appropriate for electrodes without surface modification. Finally we oxidized Ge surface and tested the stability again. The impedance plot of the surface-oxidized Ge showed little reaction, proving that we succeeded to improve the surface stability.

1. Introduction

The energy harvester modules have been collecting attention as an on-site power generator for micro devices. Generally components of the energy harvester modules such as mechanical generators, rectifiers, capacitors and power managing digital circuits are fabricated separately and integrated at last. If the components are made of the same material, we can fabricate those on a single substrate simultaneously. Then manufacturing cost and size of the modules can be reduced. Also, its parasitic capacitance and leak current will be decreased.

Ge should be a promising material because it is known for the established MEMS fabrication processes and lower threshold voltage than that of Si, which enables to make low power consumption diode circuit. Moreover, micro porous can be made on Ge by etching in acid solution. This enables us to prepare large surface area electrodes for making large capacity EDLCs. However, so far the etching conditions to make large surface area were not understood. In addition there has been no report of Ge as electrodes of EDLCs.

In this study, we challenged to make basis for the development of Ge electrodes for EDLCs and investigated two electrochemical properties of Ge. At first, we investigated the relationship between the porous structure and the doping level of a Ge(100) substrate in acid solution to find out the etching conditions to realize large surface area. Then we evaluated the stability of Ge electrodes in an ionic liquid BMIM-AC electrolyte by observing chemical reaction at the electrode/electrolyte interface.

2. Experiment

In this study, the ionic liquid is adopted as an electrolyte because it has larger chemical window (2~4 V) than conventional electrolytes such as H_2SO_4 (1 V) [1].

2.1. porous structure formation

We prepared three Sb-doped one-side polished Ge(100) substrates with different doping levels 3.0×10^{14}, 3.5×10^{15}, and 1.0×10^{16} atm/cm^3. The substrates were n-type because it suppresses hole-mediated chemical reaction and limits number of reaction sites and avoids incorporation of micro-pores. The size is 2 x 2 cm^2 square with thickness of 500 μm. The Ge substrate was etched by HCl solution [HCl : $CH_3CH(OH)CH_3$: H_2O = 1 : 1 : 6] in a double cell etching setup (Fig. 1) [2,3]. Etching was conducted for 1 h by flowing current of 2.5 mA/cm^2 between Pt anode and cathode. A Ge substrate on the current flow path undertakes electrochemical reactions [4].

$$Ge + H_3O^+ + e^- \rightarrow GeH + H_2O \qquad (1)$$
$$H_3O^+ + GeH + e^- \rightarrow Ge + H_2 + H_2O \qquad (2)$$

on the anode side, and

$$Ge + nh^+ + 3H_2O \rightarrow H_2GeO_3 + 4H^+ + (4\text{-}n)e^- \qquad (3)$$

on the cathode side, respectively. H_2GeO_3 solves into the solution and pore is formed on the cathode side of the Ge surface. [3,4]

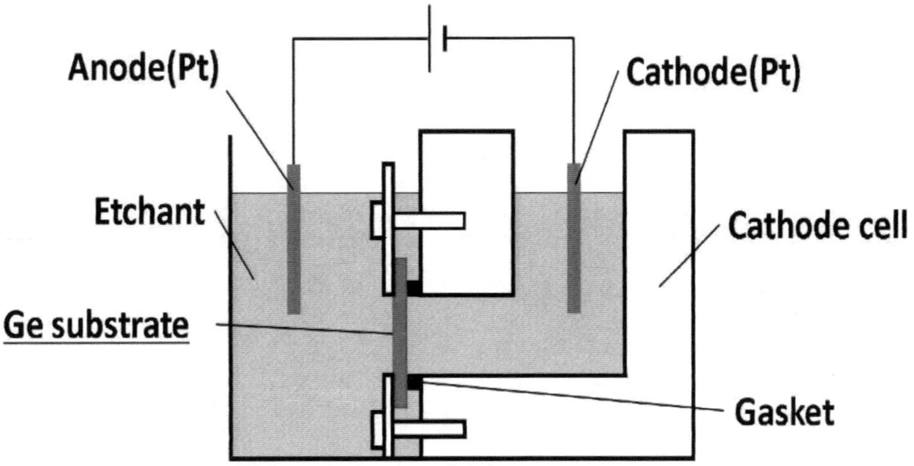

Figure 1. Setup to fabricate porous structure

2.2. Stability in IL

In order to investigate chemical stability of the Ge substrate in IL, we constructed the EDLC consisting of Ge electrodes, and conducted ac impedance measurements. We used 1-Butyl-3-methylimidazolium acetate (BMIM-AC) as an IL electrolyte because that does not contain reactive fluorine. Additionally, comparison was made for Ge with oxidized surface layer prepared by thermal oxidization at 550 ℃ under O_2 pressure of 0.1 MPa for 1 h [5].

3. Results and discussion

3.1. Porous structure formation

Figure 2 shows plan [(a) – (c)] and cross-sectional [(d)-(f)] views of scanning electron microscopy (SEM) images of surface of Ge(100) substrates with different dopant levels after the etching. From these observations, it was verified that porous structure strongly depends on the dopant levels. As shown in Figure 2 (a), for the lowest dopant level (3.0×10^{14} atm/cm^3), etching uniformly takes place on the surface and fails to form porous structure. In cross-sectional view [Fig.2 (d)], we could not see porous structure. With increase in the dopant levels, etching becomes less uniform and anisotropic (faster etching along [100] direction), leading to formation of the micro-porous structure. Interestingly, size and distribution of the micro-porous structure depends on the dopant level. As shown in Figure 2 (b), large cross-shaped macro-pores of about 70 × 70 μm^2 and small micro-pores of about 10 × 10 μm^2 appeared for 3.5×10^{15} atm/cm^3. As shown in Figure 2 (e), the pore has cone-shape, indicating that etching occurred selectively along depth. In contrast small micro-pore of less 5 × 5 μm^2 appeared densely for 1.0×10^{16} atm/cm^3 together with large micro-pore of about 20 × 20 μm^2. In Figure 2 (f), we can see that many micro-pores were created vertically. This fine and dense porous structure is adequate to prepare large surface area Ge electrodes for EDLCs. Note that the formation of the cross-shaped pore may be ascribed to the intrinsic crystal anisotropy and consequential anisotropic career density that causes a shortage of the doped career partly [3].

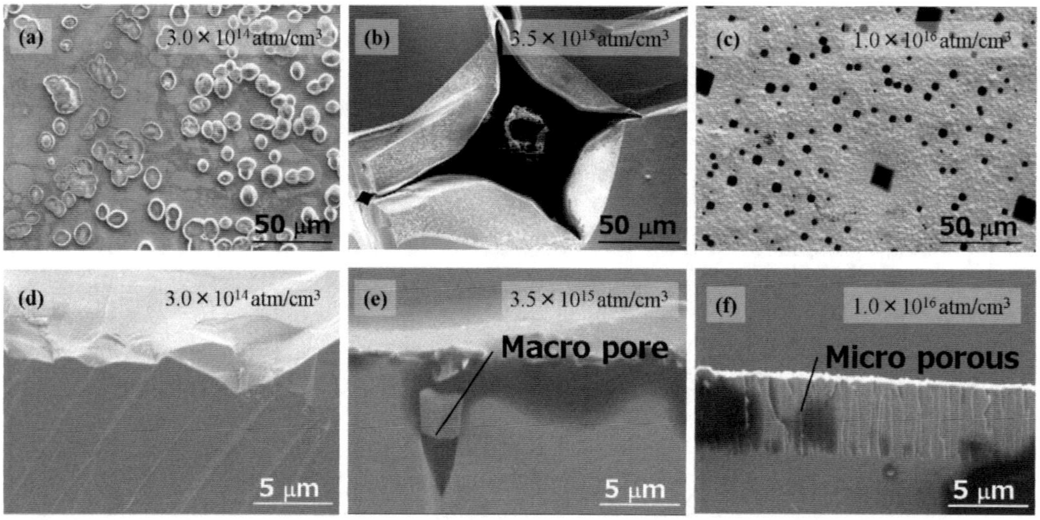

Figure 2. SEM images of pores on the n-type Ge substrate with different Sb doping amount: (a-c) plain views, and (d-e) cross-sectional views

3.2. Stability in IL

To evaluate adequacy of Ge as an electrode of the EDLC, we implemented ac impedance measurements (0.1 Hz ~ 1.0×10^6 Hz) of the Ge(100) substrates (dopant density ≧ 1.0×10^{16}) with and without surface oxide layer in BMIM-AC. Figure 3 shows obtained impedance plots. Typical impedance plot of the electrode/electrolyte interface in the EDLC in the inset indicates that the plot shows vertical straight line at the lower frequency region (on the right) if the electrical double layer is formed at the interface [6]. As shown in Figure 3 (a), in contrast, a plot of pure Ge bended downward with decrease in the frequency, indicating that chemical reaction took place at the interface. Thus pure Ge is not suitable as an electrode of EDLCs if IL is used. To solve this problem, we challenged to stabilize Ge surface by growing GeO_2 layer by considering that generally oxides are stable compounds due to their large formation energy. Actually the impedance plot of the surface-oxidized Ge showed almost straight line for entire frequency range. Thus significant improvement in the stability against IL by surface oxidization is verified. Looking at the plots closely, we noticed that the plots intersect with horizontal axis at Z_{re} of between 500 and 1000 Ω [Fig. 3(b)]. This indicated that even heavily doped Ge (0.029 ~ 0.054 Ω·cm) still suffers from high resistance. We believe that ongoing experiments will establish innovative method to reduce resistance.

In this study, porous structure formation and oxidizing Ge were conducted separately. In next study, we fabricate low resistive Ge electrode included porous structures and an oxidation layer. Then we will investigate the characteristics of the EDLCs.

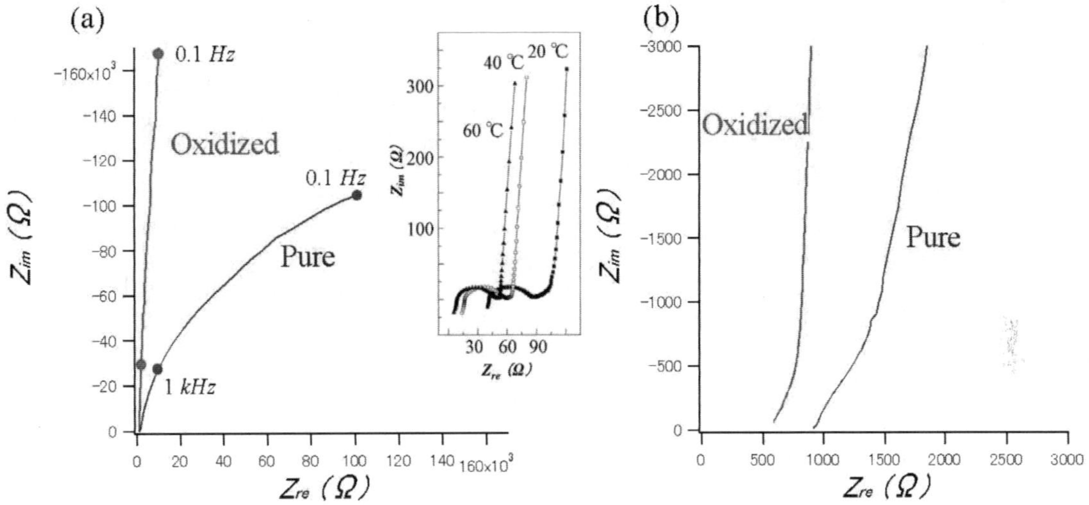

Figure 3. Impedance characteristic of the Ge and Ge oxide in BMIM-AC.

4. Conclusion

We investigated the relationship between porous structures and the doping levels. We also investigated stability of Ge for IL. The doping levels strongly affects porous structure formation in acid solution. At the high doping level (1.0×10^{16} atom/cm^3), we got micro-pores densely. So high doping condition is suitable for preparing large surface area on Ge substrate. Ge oxide layer showed stability for IL. By comparing pure Ge, we could see the improvement of stability in impedance plots. These results gave research direction to MEMS production of Ge based EDLCs. This study will finally open the way for all-Ge-based energy harvesting modules.

References

[1] Alan B. McEwen, Helen L. Mgo, Karen LeCompte, and jay L. Goldman 1999 *Jounal of the Electrochemical* Society **146** (5) 1686-1695
[2] S. Langa, J. Carstensen, M. Christophersen, H. F öll, I. M. Tiginyanu 2001 *Appl. Phys. Lett* **78** 1074-1076
[3] C. Fang, H. Föll, J Carstensen 2006 *Journal of Electroanalytical Chemistry* **589** 259-228
[4] P.J. Holmes (Ed.) 1962 *The electrochemistry os Semiconducteor, Academic press,* New York
[5] S. K. Sahari, A. Ohta, M. Matsui, K. Mishima, H. Murakami, S. Higashi, and S. Miyazaki 2013 *Journal of Physics* : Conference Series **417** 012014
[6] Bin. Xu, Feng. Wu, Renjie. Chen, Gaoping. Cao, Shi. Chen, Guoqing. Wang, Yusheng. Yang 2006 *Journal of Power Sources* **158** 773-778

Improvement on electrochemical performance by partial replacement of Ru@Pt core-shell nanocatalyst by temperature modification

Chih-Juei Chang[1], Liang-You Lin[1]and Fan-Gang Tseng[1,2]

[1]National Tsing Hua University and [2]Academia Sinica

Abstract. In this paper, the homemade open-loop reduction system (OLRS), and redox transmetalation method were utilized to produce the core-shell Ru (ruthenium)/Pt (platinum) catalysts on the carbon cloth (CC) for direct methanol fuel cell (DMFC) application. By adjusting pH value and heating to proper temperature of the ionized reduction environment, Pt^{4+} can be first converted into Pt^{2+} to allow partial Ru replacement with Pt by redox transmetalation and produce Ru@Pt core-shell nanostructures[1]. And we change the reduction temperature to see how it affects the efficiency of the DMFC.
The scanning electron microscopic (SEM) top-view micrographs showing that the apparent Ru@Pt nanoparticles successfully deposited on both the inner and outer surfaces of the hydrophilically-treated CC. At high SEM magnification, the small size and high-density distribution of the Ru@Pt nanoparticles were clearly observed on the hydrophilically-treated CC , and much more Pt@Ru catalyst deposit on the CC surface with the sample of 80℃. The electrosorption charges of hydrogen ion (Q_H) and the peak current density (I_P) of the samples in the cyclic voltammetry (CV) curves. The magnitude of peak current density is positive correlation to the temperature. However, the CO tolerance, indicated that the better CO tolerance contributed to the less Pt replace on Ru cluster, which allow the Ru oxidizing CO to CO_2 efficiently, is negative correlation¬¬ to the temperature. The sample of 50℃ shows the better combination catalyst efficiency between the CO tolerance and the electrochemical performance.

1. Introduction

With the rising price of the fossil, the alternative energy sources have become a significant issue for scientific and technological developments. Direct methanol fuel cells (DMFCs) utilize methanol as the anode fuel, mainly because it offers a high energy density, low pollution, fast refueling, and a low operating temperature [2-4]. Compared to proton exchange membrane fuel cell (PEMFC) using hydrogen gas as the anode fuel, DMFC has some advantages such as the safety of fuel storage, operation at room temperature, and so on. Therefore, DMFCs have potential applications for portable electronics. To reduce the amount of noble metal catalyst and achieve uniform dispersion on the carbon support for the achievement of better utilization and activity of electrocatalysts. And reduce the effect of CO poisoning during the catalytic reaction which is used to produce energy. We choose Ru@Pt core-shell nanoparticle structure, the characteristics of Ru is to oxidize CO become CO_2, and suppress the effect of CO poisoning. To enhance Ru@Pt catalyst efficiency, small size and uniformly distributed nanoparticles are highly desired. Conventionally, there are two common methods applied for Ru nanoparticle preparation, including chemical reduction [4, 6] and electro-deposition [7]. Chemical reduction provides the advantages of easier preparation, direct reaction, and uniform size distribution of the nanocatalysts. However, it usually takes much longer time (24 h) for catalyst preparation due to slow nucleation and growth process at low temperature (60℃). As a result, recent developed reflux technology[8]called Open-Loop reflux system(OLRS), can react at higher reaction temperature (160℃), substantially shorten the time required for catalyst preparation (~1.5 hr) and better electrochemical performance with Cyclic Voltammetry testing[5]. For preparing uniform Ru@Pt nanoparticles as catalyst, redox transmetalation reaction for Pt catalyst[11] provided better CO tolerance comparing to the high temperature OLRS. By adjusting the pH value, Pt^{4+} can be first converted into Pt^{2+} to allow partial Ru replacement with Pt and produce Ru@Pt core-shell nanostructures. To more control the Pt shell thickness, Patil et al[12]suggested the theredox transmetalation reaction at lower temperature. However, the temperature effect of Ru@Pt core-shell structure was not discussed clearly in the earlier documents. To find out exact effect of the electrocatalytic propertiesof Ru@Pt catalysts, the effect of the variable reduction temperature would be discussed in the paper.

2. Experiment

2.1 Ethanol Immersion Pretreatment & Hydrophilic Treatment of Carbon Supports

We used a 6 M sulfuric acid (H_2SO_4) solution as the oxidant for hydrophilic treatment of the CC supports. In order to remove air bubbles from the surface of the carbon supports to effectively immerse the whole surfaces of the supports in a H_2SO_4 solution for better Pt catalyst adhesion, vacuum means was added in the hydrophilic treatment process [2, 9]. However, it is still difficult to thoroughly immerse the three-dimensional, interwoven structures of carbon cloth fibers in the acidic solution. In the present work, the EtOH immersion pretreatment for the CC carbon supports was utilized prior to the above-mentioned hydrophilic modification of the carbon supports.

2.2 Open-Loop Reduction System for Ruthenium Catalyst

According to a homemade open-loop reduction system (OLRS) in our previous study [2], as shown in *Fig.1(a)*, we used 0.0222 g $RuCl_3 \cdot xH_2O$ and 0.1 mM ethylene glycol as a Ru catalyst precursor and a reducing agent, respectively. Heating to 160℃ and then naturally cooled to room temperature. Compared to the traditional reflux system, as shown in *Fig.1 (b)*, the reduction time of OLRS is only about 1.5 hours. The distribution of Ru is also more uniform with the OLRS contrast to the traditional reflux system.

2.3 Redox transmetalation reaction for Pt Catalyst

We used 0.0432 g $H_2PtCl_6 \cdot 6H_2O$ and 0.1 mM ethylene glycol as a Pt catalyst precursor and a reducing agent, respectively. Adjust the pH value of the solution with sodium hydroxide (NaOH). Then heating the solution to 80℃ to make Pt^{4+} in the solution convert to Pt^{2+}. Immersed the above-mentioned CC which is deposited Ru into the solution, control working temperature as constant (80℃, 50℃, 30℃) for 3 hours. The redox transmetalation is an efficient way to produce Ru@Pt core-shell nanocatalyst.

2.4 Electrochemical Measurements

Cyclic voltammetry (CV) was carried out to identify the electrochemical properties of the electrocatalysts. The working principle is to impose a variable voltage at the working electrode (WE) and analyze the current signal received with time to determine the oxidation-reduction (redox) status in the electrochemical reaction. In this case, we imposed time-varying triangular wave of potential in the WE, and then observed the relationship between potential (E, V_{SCE}) and current density (I, mA/cm^2) to realize the potential for the redox reaction, electrochemical mass activity (MA, A/g_{Pt}) and reaction rate [2, 5]. In the CV curves, we can also obtain the active electrochemical surface area (ESA) of Pt catalysts from the hydrogen ion electrosorption reactions (charge transfer density of H^+, QH, mC/cm^2) in a H_2SO_4 aqueous solution. The three-electrode cell was setup for all of the electrochemical tests. Saturated calomel electrode (SCE, 0.241 V vs. standard hydrogen electrode, SHE) and Pt-coated Ti mesh are used as the reference electrode (RE) and the counter electrode (CE), respectively. All of the electrochemical measurements were purged with nitrogen (N_2) gas to remove oxygen for extra oxidation reaction.

3. Results and discussion

Both the Q_H and the peak current density (I_P) of the different reaction temperature but the same reaction time electrodes with the EtOH immersion pretreatments obtained in the CV curves, as shown in *Fig. 2*and *Fig. 3*. Apparently, the electrode which reaction temperature is 80℃, has better Q_H and I_P

than the other two electrodes. *Table 1* lists the electrocatalytic properties of the three electrodes with different reaction temperature. The efficiency of the 80℃ electrode has better electrochemical performance than the others. For Q_H, the magnitude of 80℃ electrode is 3.28 times higher than the 50 ℃ electrode, and 22 times higher than the 30℃ electrode; and for I_P, the magnitude of 80℃ electrode is 3.39 times higher than the 50℃ electrode, and 5.06 times higher than the 30℃ electrode. However, the performance of reducing CO poisoning, the 80℃ electrode is surprisingly the lowest of the three, and the 30℃ electrode is the best. I_f/I_b in *Table 1* shows the ability to reduce CO poisoning, the higher magnitude the better ability to reduce CO poisoning.
The result data of the above EDS graph is shown in *Table1*. Indicated that there is a big difference of Pt content between the two electrodes, 80℃and 30℃.The atomic percent of 80℃ electrode is 18.94%, much more than which of 30℃ electrode, 2.17%. Which lead to better CO tolerance of the 30℃ electrode.
*Table 2*shows the microscopic picture of three electrodes under the different magnifications of SEM. At low magnification, we can see high density, small particle nanocatalyst attached around the CC fiber. As the higher the reduction temperature, the distribution of the nanoparticle become denser. At high magnification, the Pt particles of the 80℃ electrode seriously agglomerated. The particle size of the 80℃ electrode is also the largest, about 50 nm. In the contrary, the particle size of the 30℃ electrode is only about 16 nm. According to the above-mentioned, we inferred that the reduction temperature can affect the reaction rate while the redox transmetalation reaction. Therefore, the 80℃ electrode loads much more Pt than others, its electrocatalytic properties is the best. However, if the temperature was too high, it may cause the Pt particles cluster,which leads to the limitation of the mass activity, because of the too fast reaction rate of Redox transmetalation reaction for Pt Catalyst. For further exact analysis for electroactivity per unit mass, the Inductivity Coupled Plasma-Mass Spectrometer (ICP-MS) would be utilized in future work.

4. Conclusion

By redox transmetalation reaction, different redction temperatures effect obviously on the efficiency of electrocatalytic properties. For Q_H and I_P, we can find that 80℃ electrode is better than the 30℃ one (Q_H : 12.46 vs. 0.585 mC/cm^2, I_P : 130 vs. 25.7 mA/cm^2). However, for CO tolerance, the 30℃ electrode is yet better than the 80℃one (I_f/I_b : 1.15 vs. 1.627). Speculated that is because as the reaction temperature of Pt rises, it fastens the replacement reaction rate, make more Pt deposit on Ru and improves its electrocatalytic properties. However, the more Pt covered on Ru particle also means less Ru surface disposed, and reduce the CO tolerance of the electrode. Results from the SEM measurements show that 80℃ electrode has occurred the Ru@Pt cluster, and it is less clutch for the 50℃ electrode.

Acknowledgement

This research was financially supported by the Ministry of Science and Technology (MOST), Taiwan under grant MOST 103-3113-E-007-006. The authors also thank the Instrument Center at National Tsing Hua University for FE-SEM, EDS and ICP-MS examinations.

References

[1] B. J. Hwang et al.,"Controlled Synthesis and Characterization of Ru core -Pt shell Bimetallic Nanoparticles, " J. Phys. Chem. C 112, pp. 19922–19929, 2008.
[2] Y. S. Wu et al., "Highly efficient platinum nanocatalysts synthesized by an open-loop reduction system with a controlled temperature loop," *ElectrochimicaActa*64 (2012) 162–170.
[3] C. H. Wang et al., "High performance of low electrocatalysts loading on CNT directly grown on carbon cloth for DMFC," *Journal of Power Sources*, vol. 171, pp. 55-62, 2007.
[4] M. C. Tsai et al., "A catalytic gas diffusion layer for improving the efficiency of a direct methanol fuel cell," *Electrochemistry Communications*, vol. 9, pp. 2299-2303, 2007.
[5] N. Jha et al., "Pt–Ru/multi-walled carbon nanotubes as electrocatalysts for direct methanol fuel cell," *International*

 doi:10.1088/1742-6596/557/1/012106

Journal of Hydrogen Energy, vol. 33, pp. 427-433, 2008.
[6] M. C. Tsai et al., "An improved electrodeposition technique for preparing platinum and platinum–ruthenium nanoparticles on carbon nanotubes directly grown on carbon cloth for methanol oxidation," *Electrochemistry Communications,* vol. 8, pp. 1445–1452, 2006.
[7] M. Watanabe et al., "Preparation of highly dispersed Pt-Ru alloy clusters and the activity for the electrooxidation of methanol," *Journal of Electroanalytic Chemistry*, vol. 229, pp. 395-406, 1987.
[8] C. Bock et al., "Size-selected synthesis of PtRunano-catalysts: reaction and size control mechanism," *Journal of the American Chemical Society,* vol. 126, pp. 8028-8037, 2004.
[9] Y. S. Wu et al., "Characteristics of controlled Nafion® coating inside nano structured anode for three-phase zone arrangement in micro DMFC", *Proceedings of the 2010 5th IEEE International Conference on Nano/Micro Engineered and Molecular Systems*, Xiamen, China, pp. 698-701, January 20-23, 2010.
[10]L.Y Lin et al, " High Performance Nanocatalysts Supported on Micro/Nano Carbon Structures Using Ethanol Immersion Pretreatment for Micro DMFCs," *The 13th International Conference on Micro and Nanotechnology for Power Generation and Energy Conversion Applications(PowerMEMS),* The Royal Society, London, UK, 3-6 December, 2013
[11]N.Muthuswamyet. al,"Ru@Ptcoreeshellnanoparticles for methanol fuelcell catalyst: Controland effects of shellcomposition,"*International journal of hydrogen energy*, vol. 38, pp. 16631-16641, 2013
[12]P.Patil et al, "Sensitized solar cell from type-II CdTe/CdSe core/shell nanocrystals synthesized without seed purification at low temperature,"*Journal of Alloys and Compounds*, vol. 607, pp. 630-637, 2014

Figures& Tables

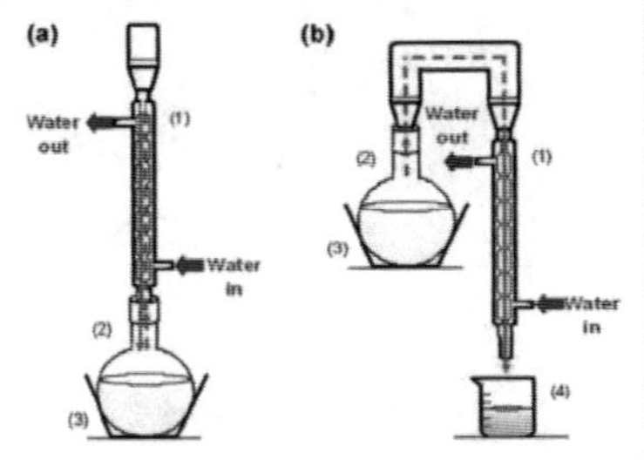

Figure 1. Schematic setup of the(a) traditional reduction system and (b)open-loop reduction system (OLRS).

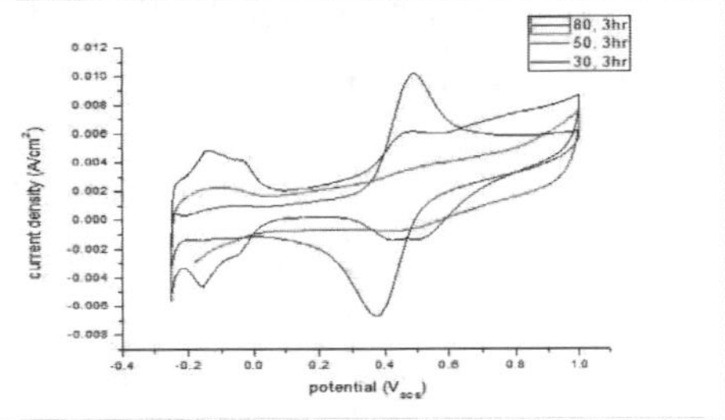

Figure 2. Cyclic voltammograms of the three different electrodes. Electrolyte: 0.5 M H_2SO_4 aqueous solution. Scan rate: 50 mV/s. Temperature: 25℃.

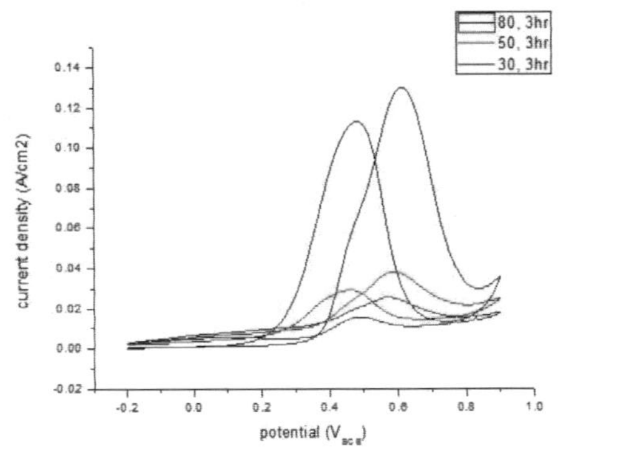

Figure 3. Cyclic voltammograms of the three different electrodes. Electrolytes: 1 M CH_3OH + 0.5 M H_2SO_4 aqueous solutions. Scan rate: 20 mV/s. Temperature: 25℃.

Table 1. The electrocatalyst property of different Pt replacement reaction temperature.

Electrodes	Q_H (mC/cm^2)	I_P (mA/cm^2)	I_f/I_b	Ru wt%	Pt wt%	Ru atom%	Pt atom%
80℃, 3hr	12.46	130	1.15	0.68 %	78.37 %	0.32 %	18.94 %
50℃, 3hr	3.803	38.3	1.316	11.85 %	50.37 %	3.36 %	7.41 %
30℃, 3hr	0.585	25.7	1.627	43.85 %	15.97 %	11.51 %	2.17 %

Table 2. Low- and high-magnification (1K &200K) SEM images of different electrodes.

Electrodes	Low-magnification (1K)	High-magnification (200K)
80℃, 3hr		
50℃, 3hr		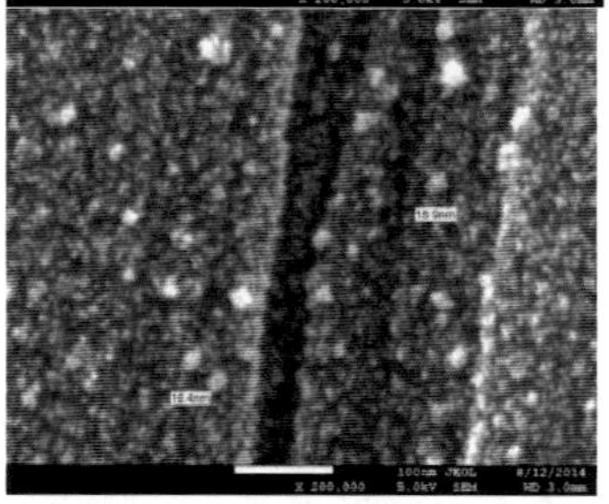
30℃, 3hr	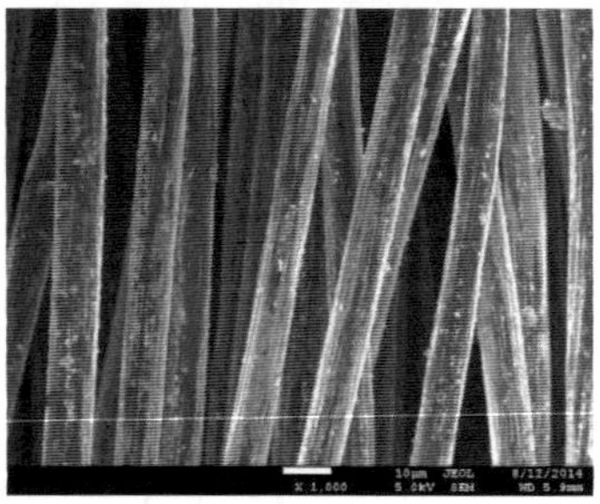	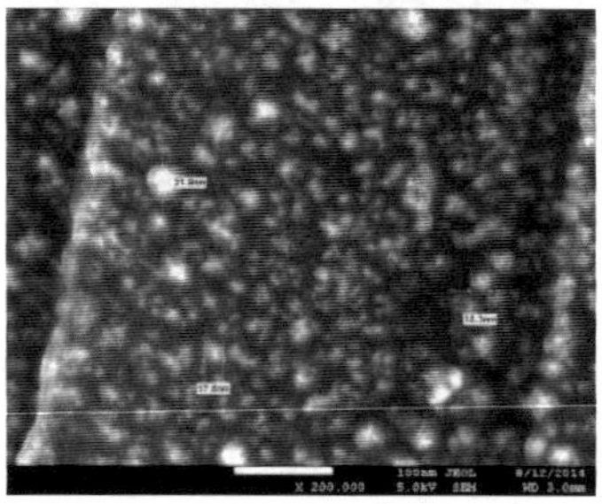

GLASS MICROPOROUS FIBER/NANOPOROUS POLYTETRAFLUOROETHENE COMPOSITE MEMBRANES FOR HIGH EFFICIENT PHOSPHORIC ACID FUEL CELL

Chia-Lien Lu[1] , Wei-Jia Lee[1] , Cheng-Ping Chang [2] and Fan-Gang Tseng[1,3]
1.Department of Engineering and System Science, National Tsing Hua University, TAIWAN
2.Interdisciplinary Program of Nuclear Science, National Tsing Hua University, TAIWAN
3.Division of Mechanics, Reschanics, Researchnter for Applied Science, Academia Sinica, TAIWAN

E-mail : heromac3@hotmail.com

Abstract. This paper reports a high efficient phosphoric acid fuel cell by employing a micro/nano composite proton exchange membrane incorporating glass microfiber (GMF) sealed by polytetrafluoroethylene (PTFE) nano-porous film. This multilayer membrane not only possesses both thermal and chemical stability at phosphoric acid fuel cell working temperature at 150~220℃ but also is cost effective. As a result, the inclusion of the high porosity and proton conductivity from glass microfiber and the prevention of phosphoric acid leakage from PTEF nano film can be achieved at the same time.The composite membrane maximum proton conductivity achieves 0.71 S/cm at 150 °C from AC impedance analysis, much higher than common phosphoric acid porous membranes For single cell test , The GMF fuel cell provides a 63.6mW/cm2 power density at 200mA/cm2 current density while GMF plus methanol treated PTFE (GMF+mPTFE) provides 59.2mW/cm2 power density at 160mA/cm2 current density for hydrogen and oxygen supply at 150 °C. When we change the electrodes that are more suited for phosphoric acid fuel cell, the GMF+mPTFE single cell gets higher performance which achieve 296mW/cm2 power density at 900mA/cm2 current density for hydrogen and oxygen supply at 150 °C.

1. Introduction

Phosphoric acid fuel cell (PAFC) belongs to medium temperature fuel cell. PAFC has many advantages, such as the power usage is enhance and supply often higher power density than PEMFC. PAFC operates at 150~220℃, so that the water produced in cathode vaporize rapidly, which simplify the water management design. Phosphoric acid has low vapor contents and good thermal stability at this temperature range (150~220℃). Liquid phosphoric acid play an important role in transporting protons which produced by hydrogen oxidation reaction from anode to cathode. Among many components in PAFC, the proton conductivity of the membrane is the most important part. Traditional Nafion will lose its moisture above 80℃, and the proton conductivity will decay. In the past decades,

Journal of Physics: Conference Series **557** (2014) 012107 doi:10.1088/1742-6596/557/1/012107

silicon carbide powders (SiC) were employed popularly as the matrix for containing phosphoric acid, for its high temperature endurance, mechanical integrity and resistant to phosphoric acid [1, 2]. However, the porosity is limited to 79~89% due to the space occupation of the power structure [3]. Therefore, this research introduce a way for improving the issue by using a composite proton exchange membrane. The micron pores (1-1.5 μm) of glass microfiber can provide higher (93%) porosity and better phosphoric acid possessing ability than those of SiC powder. However, the large opening of the membrane cannot effectively prevent the leakage of phosphoric acid to the reaction chamber [4]. Therefore, in this study, a 25μm thick PTFE thin film with nano pores (50-400 nm) with 70% porosity is employed to attach on the surface of glass microfiber in order to prevent phosphoric acid leakage, shown in Figure 1. (a). This composite proton exchange membrane not only possesses both thermal and chemical stability at phosphoric acid fuel cell's working temperature at 150~220℃ but also is cost effective. As a result, the inclusion of the high porosity and proton conductivity from glass microfiber and the prevention of phosphoric acid leakage from PTEF nano film can be achieved at the same time. And morphology as the schematic shown in Figure 1. (b).

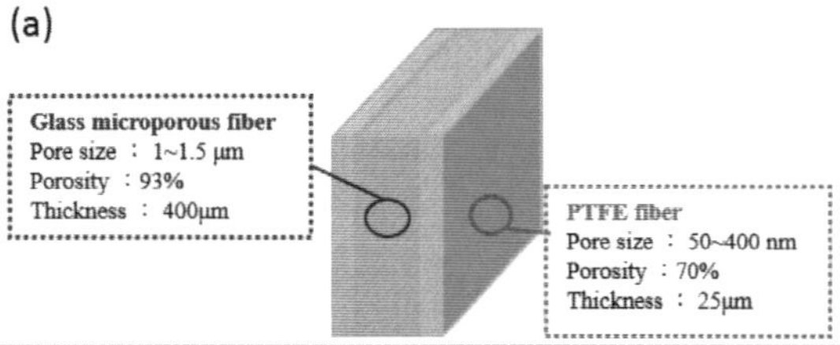

Figure 1. (a) GMF and PTFE composite membrane

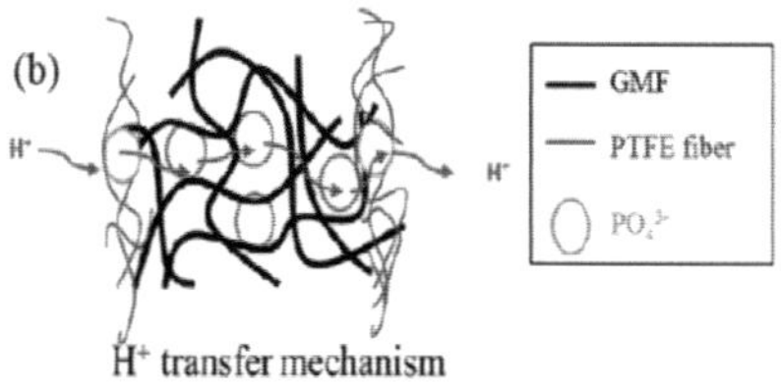

Figure 1. (b) composite membrane's proton transfer mechanism

2. Experimental

2.1 Preparation of composite membrane

Before assembling the two films together, we put PTFE in methanol or ethanol for 15 minutes with ultrasonic vibration as hydrophilic treatment [5], so that PTFE fiber can be fully loaded with phosphoric acid. After the treatment, put the methanol treated PTFE (mPTFE) into 85% phosphoric acid immediately for 20 minutes in room temperature to absorb phosphoric acid. Meanwhile, glass microfiber is soaks in 150℃ 85% phosphoric acid for 20 min to load phosphoric acid fully. After GMF and PTFE preparation, attach the two kinds of thin films together to form the three stacks composite film.

2.2 Preparation of the electrodes

Carbon clothes (CC) is used as the gas diffusion layer of phosphoric acid fuel cell. To prepare the electrodes, catalyst ink containing 20% Pt/XC72 carbon with 35 wt% PTFE as binder was sprayed on carbon cloth. Until the Pt loading reaches 1mg/cm^2, the electrodes was heated at 120℃ for 1 hour to evaporate solvent and sintered to melt the binder. Before the single cell test, the electrodes was soaked in 150℃ 85% phosphoric acid for 24 hours to enhance proton conductivity. Moreover, a PDMS ring is put around the membrane electrode assembly (MEA) to prevent phosphoric acid leakage, Shown in Figure 2.

2.3Analyses

AC impedance measurement is applied to measure the proton conductivity of the composite membrane. Also, a long term test is done to observe the durability of the membrane. Single cell test is applied to measure the performance of the MEA. The MEA consists of anode, cathode and GMF sealed with mPTFE which is filled with phosphoric acid. The cell was assembled at 5kgf·cm by cold press. Anode was supplied by H_2 at 100 sccm and cathode by O_2 at 100 sccm.

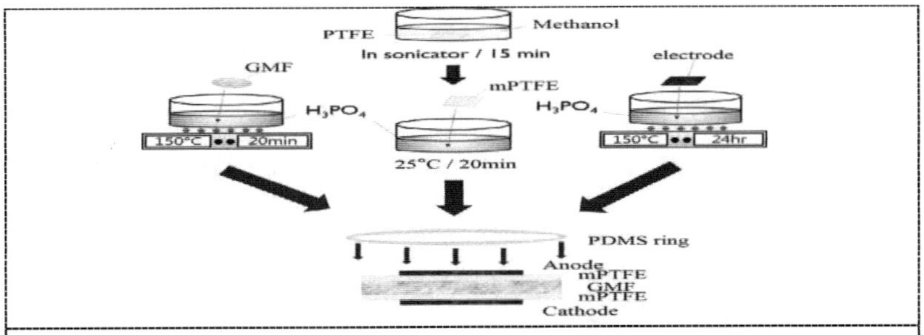

Figure 2. The preparation process illustrate of PTFE/GMF/electrodes

2. Results and discussion

3.1 AC impedance analysis

AC impedance test after assembling the three films together, we employed through-plane test to measure membrane proton conductivity in different temperature condition. Compare three data the red line is GMF, the blue line is GMF plus PTFE after methanol treatment (GMF+mPTFE), and the black line is Nafion117[R] with x-raid is Temperature and y-raid is proton conductivity , Nafion117[R] is a commercial proton exchange membrane , when test condition at 85 °C and 98% relative humidity that has maximum proton conductivity 6.5×10^{-2} S/cm, GMF has maximum proton conductivity 7.3×10^{-1} S/cm at 150 °C . The proton conductivity decreases with mPTFE + GMF Composite membranes, because of that structure increased a lot of resistance, shown in Figure 3.Although the proton conductivity decreases, The maximum proton conductivity of the composite membrane achieves high level 7.15×10^{-1} S/cm at 150 °C, which is much higher than those of common phosphoric acid porous membranes ($\sim10^{-2}$ S/cm at 150 °C)[6]. GMFB and GMFB+mPTFE composite proton conductivity compare, the experiment lasted for one hour, we can see only GMFB proton conductivity decreases 73%, while GMFB+PTFE composite membranes proton conductivity only decreases 21%.The results confirm that mPTFE nano porous is useful for preventing the phosphoric acid leakage, shown in Figure 4

*3.2 S*ingle cell test

The GMF fuel cell provides a 63.6mW/cm^2 power density at 200mA/cm^2 current density while GMF plus methanol treated PTFE (GMF+mPTFE) provides 59.2mW/cm^2 power density at 160mA/cm^2 current density for hydrogen and oxygen supply at 150 °C(Fig.5). From comparison with OCV, the nanoporous PTFE membrane is effectively preventing the fuel crossover from performance of the 60% higher OCV at 0.8 volt with similar power density (~59.2 mW/cm^2). When we change the electrodes that are more suited for phosphoric acid fuel cell, the GMF+mPTFE single cell gets higher performance which achieve 296mW/cm^2 power density at 900mA/cm^2 current density for hydrogen and oxygen supply at 150 °C(Fig.6).

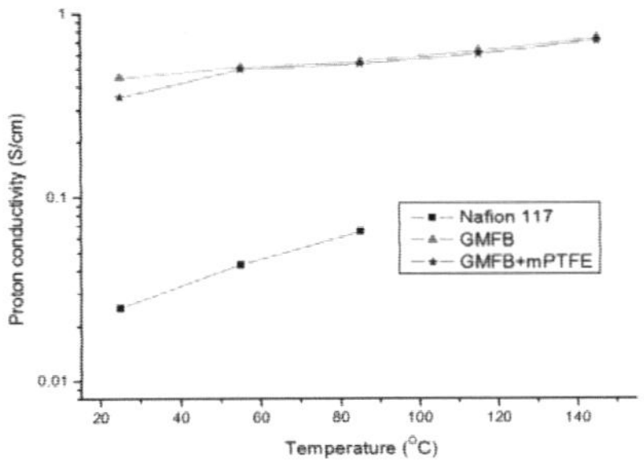

Figure 3.
Proton conductivity comparison among Nafion117,GMFB, and GMFB+mPTFE membranes at working temperatures

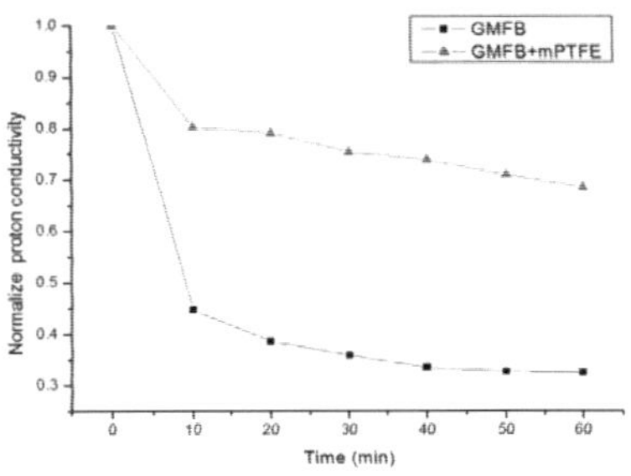

Figure 4.
Long-time test of Proton conductivity for GMFB and GMFB+mPTFE. (■)GMF and (▲) GMFB+mPTFE.

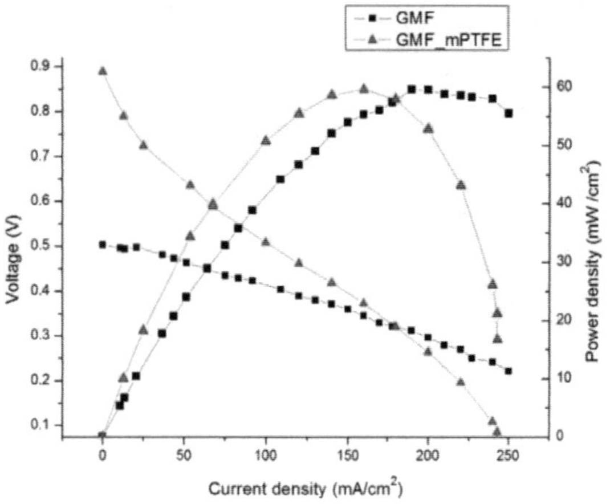

Figure 5.
Current density and power density of a single phosphoric acid fuel cell prepared (■) GMF and (▲) GMF+mPTFE.

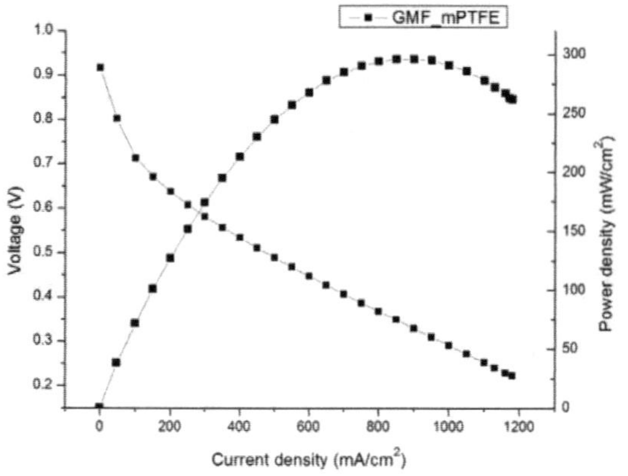

Figure 6.
Current density and power density of a single phosphoric acid fuel cell prepared by GMF+mPTFE

4. Conclusion

As a result, mPTFE+GMFB composite membranes can be achieved high porosity and proton conductivity from glass microfiber and PTEF nano film can be prevention of phosphoric acid leakage and 60% higher open circuit voltage (OCV) was obtained by employing this composite membrane. By AC impedance of the composite membrane, the proton conductivity of the composite membrane achieves 0.71 S/cm at 150 ºC. By the long term test, mPTFE is useful to prevent phosphoric acid leakage. By single cell test, the best performance of GMF+mPTFE composite membrane with proper electrodes, supplied with pure hydrogen and oxygen, achieve 296mW/cm^2 power density at 900mA/cm^2 current density at 150ºC.

References

[1] A. K. Shukla, M. Neergat, "A high-performance phosphoric acid fuel cell," *Journal of Power Sources,* vol. 102, pp. 317–321, 2001.

[2] S. Dheenadayalan, R. H. Song, D. R. Shin, "Characterization and performance analysis of silicon carbide electrolyte matrix of phosphoric acid fuel cell prepared by ball-milling method," *Journal of Power Sources,* vol. 107, pp. 98-102, 2002

[3] B. D. Yang. K. H. Yoon, "Preparation and characterization of matrix retainingelectrolyte for a phosphoric acid fuel cell by non-volatile solvent, NMP.pdf," *Journalof Power Sources,* vol. 124, pp. 47–51, 2003.

[4] N. Sammes, R. Bove, and K. Stahl, "Phosphoric acid fuel cells: Fundamentals and applications," *Current Opinion in Solid State and Materials Science,* vol. 8, pp.372-378, 2004.

[5] Y. S. Wu, L. Y. Lin, C. Chang and F. G. Tseng, "High Performance Nanocatalysts Supported on Micro/Nano Carbon Structures Using Ethanol Immersion Pretreatment for Micro DMFCs" (PowerMEMS), (2013).

[6] Q. Tang, K. Huang, G. Qian, and B. C. Benicewicz, "Phosphoric acid-imbibedthree-dimensional polyacrylamide/poly(vinyl alcohol) hydrogel as a new class of high-temperature proton exchange membrane," *Journal of Power Sources,* vol. 229, pp. 36-41, 5/1/ 2013.

Miniature fuel cell with monolithically fabricated Si electrodes - Uniformity of Catalyst Layer Thickness -

Itsuki Morisawa[1], Takahiro Suzuki[1], Noboru Katayama[2], Kiyoshi Dowaki[3], and Masanori Hayase[1]

[1]Department of Mechanical Engineering
[2]Department of Electrical Engineering
[3]Department of Industrial Administration
Faculty of Science and Technology
Tokyo University of Science
2641 Yamazaki, Noda, Chiba 278-8510, Japan

E-mail: mhayase@rs.noda.tus.ac.jp

Abstract. Uniformity of catalyst layer thickness was improved to obtained higher output with our miniature fuel cells. Though the miniature fuel cells demonstrated high power density about 500mW/cm^2 with 1 mm^2 reaction area, it was difficult to maintain the power density with larger reaction area and careful observation of the prototypes revealed that the porous Si layer, which should be etched away completely, remained on the channel bottoms. Non-uniform porous Pt catalyst layer thickness was suspicious for the residual porous Si, and rotation of the plating vessel in the Pt deposition process was performed to mitigate variation of the catalyst layer thickness. Though the power density was still low with larger reaction area, the larger output was successfully obtained with the vessel rotation.

1. Introduction

A lot of attention has been paid to miniature fuel cells for portable power sources and some products had been sold. But they did not achieve commercial success yet, and further miniaturization and cost reduction are needed. MEMS fabrication technology is an important tool to reduce the fuel cell structure to micrometer scales and is advantageous for mass production. Therefore, various studies using Si substrate and MEMS techniques have been performed [1-3]. Among those studies, Prinz et al demonstrated high output by their solid oxide fuel cell, and several research groups are now studying the concept, though there are difficulties in high temperature treatment and catalyst durability. Novel electrolyte concepts have also been explored and chemically modified porous Si shows promising results [4-5].

We also have proposed a miniature polymer electrolyte fuel cell (PEFC) design using monolithically fabricated Si electrodes. We found that porous Pt layer can be obtained on a Si substrate by immersing high-porosity porous Si into a Pt plating bath containing HF, and the porous Pt layer is used as a catalyst layer for the PEFC. Applying plasma etching on the backside of the porous Pt layer, fuel channels are opened on the Si substrate and monolithic Si electrodes are made. Porous metal layer is resistive to the plasma and the plasma etching is automatically stopped when the etching reaches to the porous metal layer.

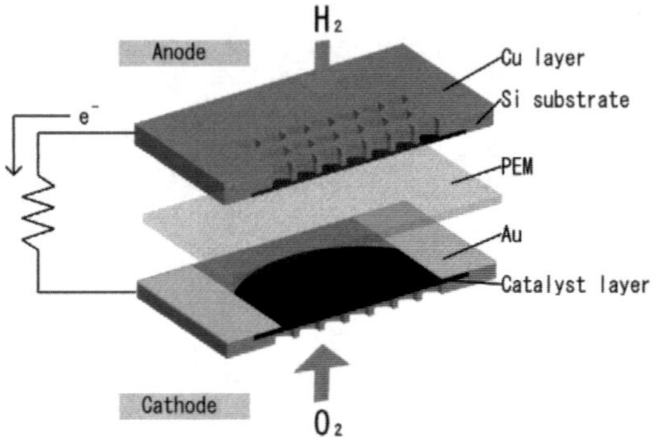

Figure 1. Schematic of the miniature fuel cell with monolithically fabricated Si electrodes.

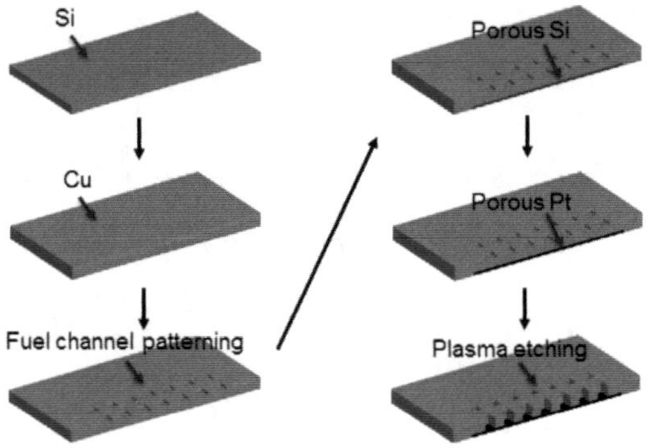

Figure 2. Fabrication process of the Si electrode.

Table 1 Condition for the catalyst layer formation.

Property of Si substrate	
Crystal Orientation	(100)
Type	N
Thickness of Si substrate	100μm
Resistivity [Ωcm]	0.001-0.003
Anodization condition	
Composition of solution	Water:HF(46%):Ethanol=5:3:2
Pt immesion plating condition	
Composition of solution	20mM H_2PtCl_6 + 1M H_2SO_4 + 450mM HF
Plating time [min]	15
Temperature [K]	283

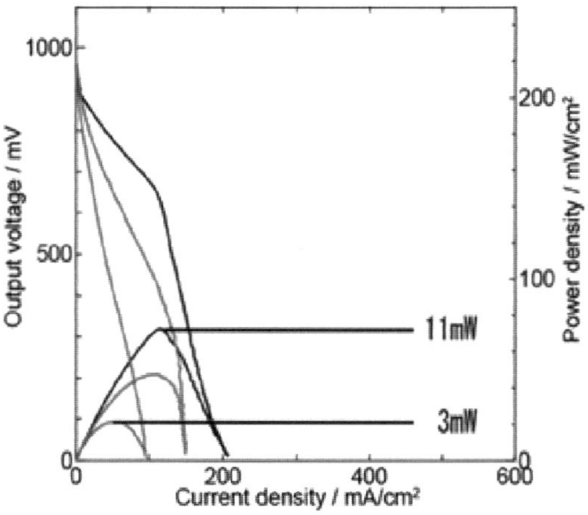

Figure 3. Polarization curves of cells with 4×4 mm^2 reaction area.

Recently, we found that the porous Si sometimes left on the porous Pt layer after the plasma etching, and the residual porous Si lowered the fuel cell performance by blocking reactant gas supply. Then, the residual porous Si was reduced by optimizing formation process of porous Si. The best performance was demonstrated by cells with the least residual porous Si [6] and the peak power density reached 500mW/cm^2 with 1×1 mm^2 reactive area. However, we still have difficulty in making the Si electrodes with no residual porous Si especially with larger reaction area.

In this study, we carefully observed the prototype cells, and attempted to reduce the residual porous Si to obtain higher output with larger reaction area.

2. Miniature fuel cell with Si electrodes

Figure 1 illustrates our fuel cell structure. The Si electrode chip has quite simple structure, in which a catalyst layer and fuel channels are monolithically fabricated on a Si wafer. A polymer electrolyte membrane (PEM) sheet is combined with the two Si electrodes using a Nafion® solution as an adhesive, and the thickness of the cell is about 230μm.

Figure 2 illustrates the fabrication process of the Si electrode. A highly As-doped Si wafer (resistivity 0.001-0.003 Ω cm) was used as a Si substrate. Due to its low resistivity, the Si substrate is expected to work as current paths. The wafer was cut into 15×15 mm chips by a dicing saw. After removing surface oxide films by a HF solution, copper thin film was deposited on the Si wafer by sputtering.

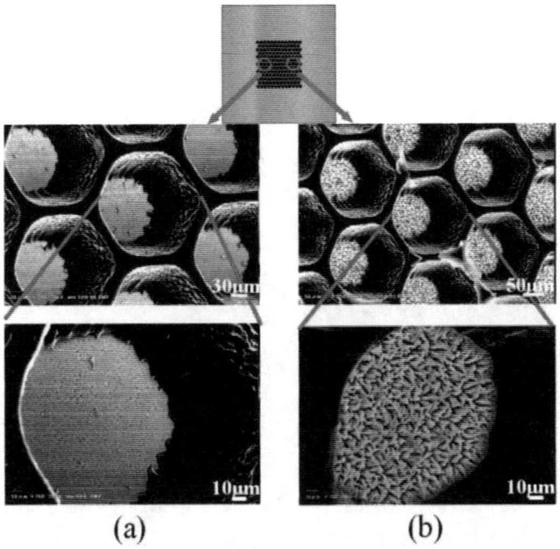

Figure 4. SEM images around channel bottoms on a Si electrode with 4×4 mm^2 reaction area. Almost no residual porous Si was observed in (a), while obvious residual porous Si was observed on some channels, as shown in (b).

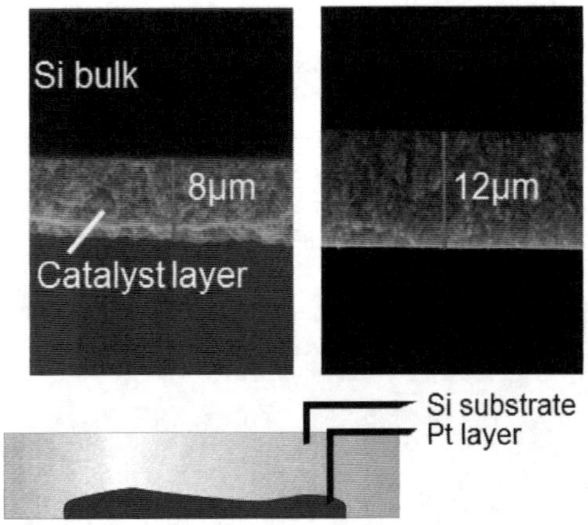

Figure 5. Cross sectional observation and schematic of the non-uniform thick porous Pt layer.

Copper thin film was patterned by conventional photolithography and wet etching, and a mask for the plasma etching was made. Porous Si layer was formed on the opposite side of the Si chip by anodization in an electrolyte containing HF. The porous Si layer was subsequently immersed in a Pt plating bath, and porous Pt layer was obtained. Detailed formation process is described in ref. [7]. Fuel channels were opened by applying plasma etching on the backside of the porous Pt layer with the copper thin film mask. Conventional parallel plate reactive ion etching system (Samco RIE-10N, Japan) was used and 18 sccm of SF_6 and 4 sccm of O_2 gases were supplied for the etching. In this plasma etching process, porous Pt layer was supposed to work as a stopping layer to the plasma etching, because the etching rate is low at the porous metal layer, and through-chip porous Pt layer can be relatively easily fabricated. In this way, monolithic Si based electrodes were prepared.

3. Non-uniform Catalyst Layer Thickness

Several prototype fuel cells were built with optimized porous Si formation conditions based on ref.[6]. Table 1 shows the conditions for the catalyst layer formation. To obtain larger output, reaction area was enlarged to 4×4 mm^2 from 1×4 mm^2, and power generation tests were performed. The result is shown in figure 3. Only 11 mW (70 mW/cm^2) was observed and the power density was far lower than the expected value (500 mW/cm^2). The performance also varied from cell to cell. In order to find out the possible factors for the deterioration, the Si electrode chips were observed carefully.

Figure 4 shows example SEM images around channel bottoms on a Si electrode with 4×4 mm^2 reaction area. As shown in the figure, on some channels there were no residual porous Si as we expected, while obvious residual porous Si was also observed on some channels. The residual porous Si may cause the performance deterioration.

With the optimized porous Si formation process, almost uniform thick porous Si layer is expected to be formed. However, non-uniform partial residual porous Si was observed. Cross-section of the Si electrode chips was observed widely, and significant non-uniformity in the porous Pt layer thickness was found in some chips. Even within a Si chip, thickness of the porous Pt layer varied from 8 – 12μm as shown in figure 5. Such variation in thickness and pore morphology of porous Si layer are hardly expected, and the wet Pt plating process was suspicious. Figure 6 shows a schematic of the non-uniform catalyst layer.

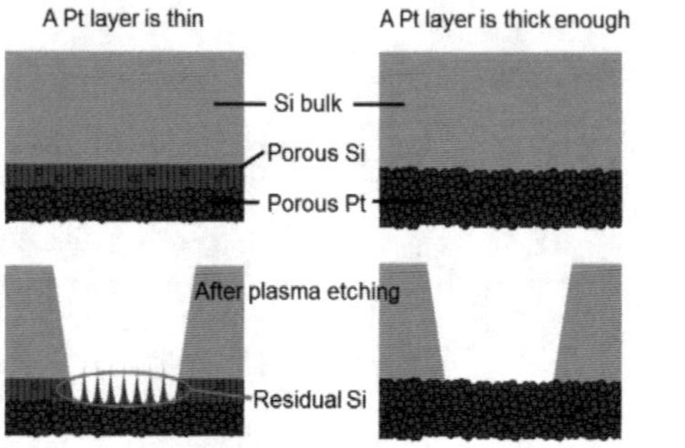

Figure6. Schematic of the non-uniform thick catalyst layer.

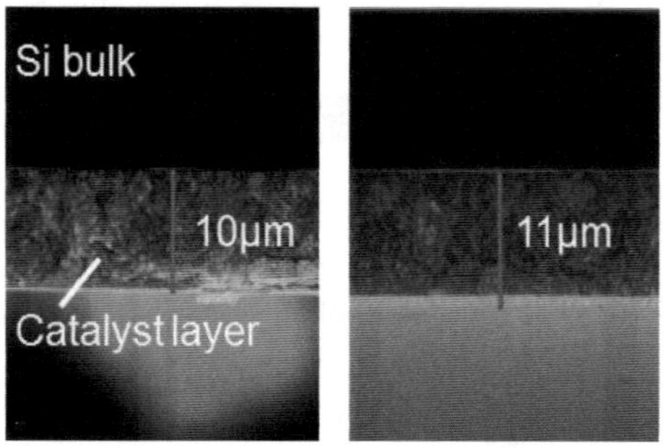

Figure 8. Cross-sectional SEM imaged of the porous Pt layer levelled by vessel rotation.

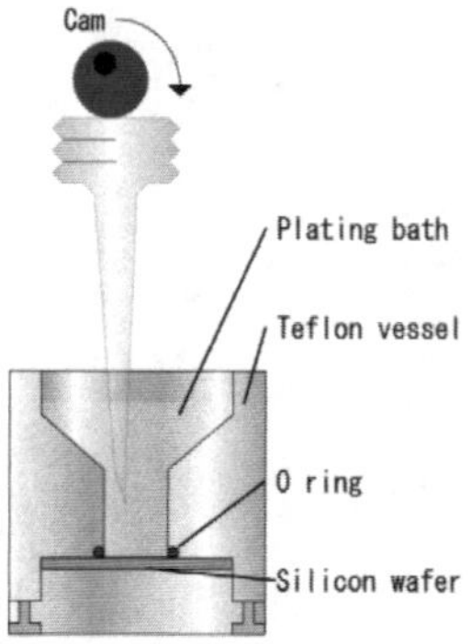

Figure 7. Plating vessel for the porous Pt formation.

In the wet plating process, Si is oxidized while Pt ion is reduced and deposited on the porous Si. Oxidized Si is removed by HF. Basically, this replacement reaction progresses from the top surface to the deeper region. However, small amount of Pt deposition in deep area was observed frequently even in an initial period of the wet plating process, and the porous Si could not be etched away due to the slight Pt deposition and the residual Si might be formed. Anyway, the thickness of the porous Pt rich region (bright region in SEM images) must be levelled.

4. Leveling Catalyst Layer Thickness

4.1 Vessel Rotation during the Pt Plating

Empirically, we found that strong agitation is necessary to obtain porous Pt layer in the wet plating process. Without agitation, deposition sometimes masks the top of the porous layer and significant uneven Pt deposition was observed. Therefore, syringe pumping has been used for agitation in the wet plating process as shown in figure 7. But uniform agitation cannot be expected with the syringe system. Agitation may be sensitive to the syringe tip location in the vessel.

To mitigate the agitation non-uniformity, we decided to rotate the plating vessel during the plating process. The vessel was rotated 3 times during the plating. With each time, 90 degree rotation was applied manually.

Figure 8 shows the cross section of the porous Pt layer formed with the vessel rotation in the plating process. Maximum and minimum thicknesses of the porous Pt layer in a whole cross section were 11 and 10μm, respectively. Though there is still variation, thickness of the porous Pt layer was successfully levelled. Using the vessel rotation technique, the Si electrode chips were fabricated. Figure 9 shows the typical channel bottoms. Magnified view of channel bottoms around each corner of $4 \times 4 mm^2$ reaction area showed almost no-residual porous Si, and the residual porous Si was successfully reduced by levelling the porous Pt layer thickness.

4.2 Power generation tests

Several Si electrode chips were made, and prototype fuel cells with $4 \times 4\ mm^2$ reaction area were constructed. To test the performance of the cells, 10 sccm of hydrogen gas and 5 sccm of oxygen gas were fed into the cells in a in a temperature controllable chamber at 318K. In order to obtain good ion

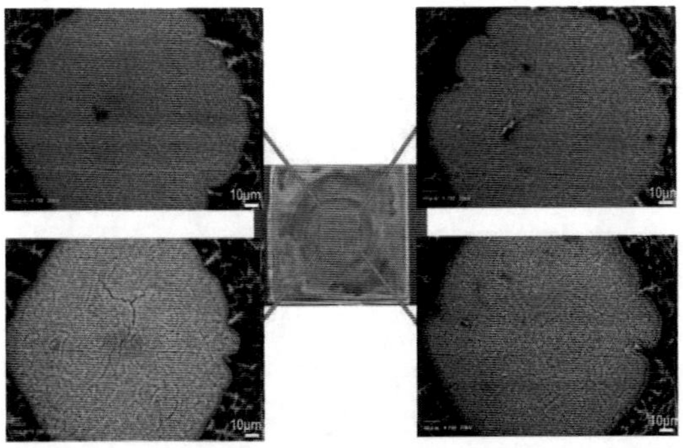

Figure 9. Channel bottoms on a Si electrode fabricated with the vessel rotation.

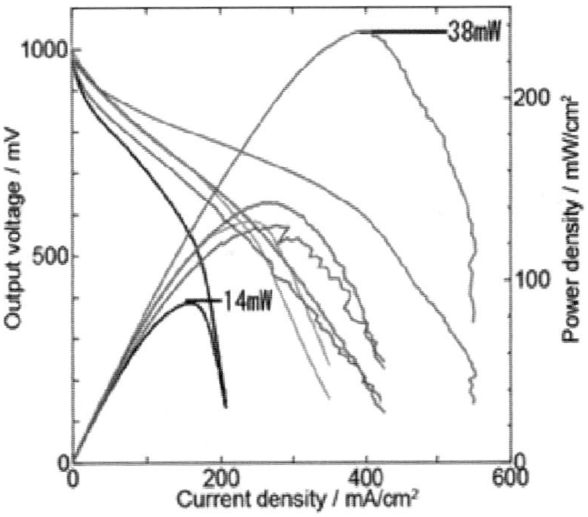

Figure 10. (a) Power generation tests with vessel rotation

conductivity of PEM, hydrogen gas was humidified by bubbling through a water containing tank which was heated to the testing temperature.

Figure 10 shows the result of power generation test. Without the vessel rotation, the output power varied from 3-11 mW as shown in figure 3, while larger output of 14-38 mW were obtained with the vessel rotation. It is found that non-uniform catalyst layer thickness was the major cause of the residual porous Si layer, and the residual porous Si lowered the fuel cell performance. However, the output still varied from 14-38mW, and the maximum power density was 240 mW/cm^2 and was still lower than expected. Therefore, different factors for the performance deterioration must exist. During our study, prototype cells with small reaction area sometimes showed very poor performance. Those cells were not well considered, and they were just thought as broken cells. Contact between catalyst and PEM might be poor in those cells. If the poor contact happened frequently, with larger reaction area, incursion of the poor contact area may be inevitable, and we think that the poor contact is most suspicious factor for the performance deterioration.

5. Conclusions

Residual porous Si, which blocks the fuel supply and deteriorates the cell performance, was found when the reaction area was enlarged to obtain higher output by our miniature fuel cells. The residual porous Si was caused by non-uniform thick porous Pt layer, and the porous Pt thickness was levelled by vessel rotation during the wet Pt plating process. By this improvement, residual porous Si was reduced. Prototype fuel cells were constructed with the least residual porous Si chips, and higher power generation performances were successfully obtained. However, the power density was still lower than that expected from small reaction area test results, and the peak outputs also varied from cell to cell. We think that poor contact between the catalyst and PEM is most suspicious, and further study especially about the contact will be performed.

References

[1] C Weinmueller, G. Tautschnig, N. Hotz, D. Poulikakos, 2010 *Journal of Power Sources,* **195**, 3849-3857
[2] N. Kuriyama, et al., 2008 *Sensors and Actuator*, **145-146**,354-362
[3] J. G. A. Brito-Neto S. Araki, M. Hayase, 2006 *J.Electrochem. Soc.,* **153**, C741-C746
[4] S. Moghaddam, et al., 2010 *Nature Nanotechnology*,**5**, 230 - 236
[5] R. Ogawa, S. Aoki, M. Hayase, 2012 *Intl. workshop on PowerMEMS,* pp. 391-394.
[6] M. Kobayashi, T. Suzuki, M. Hayase, 2014 *Journal of Power Sources*, **267**, 622-628
[7] T. Fujii, J. G. A. Brito-Neto, M. Hayase, 2011 *J. Electrochem. Soc.*,**158**, B355-B359

Condensation Heat Transfer Performance of Nano-Engineered Cu Surfaces

Hyunsik Kim and Youngsuk Nam*

Department of Mechanical Engineering, Kyung Hee University, 1732 Yongin, 446-701, Korea

*E-mail: ysnam1@khu.ac.kr

Abstract. We investigated condensate mobility and resulting heat transfer performance on Cu based water repellent surfaces including hydrophobic, superhydrophobic and oil-infused surfaces. We observed the transient microscale condensation behaviours up to 3 hours with controlling the supersaturation level at 1.64. We experimentally characterized the nucleation density, droplet size distribution and growth rate, and then incorporated them into the developed condensation heat transfer model to compare the condensation heat transfer performance of each surface. Due to the spontaneous coalescence induced jumping, super-hydrophobic surface can maintain the high heat transfer performance while other surfaces show a gradual decrease in heat transfer performance due to the increase in the thermal resistance across the growing droplets. We also quantified each thermal resistance values from the vapor to the surface through the droplets to find out the relative importance of each thermal resistance term.

1. Introduction

Condensation behavior has been studied actively for decades to improve heat and mass transfer performance. Condensation heat transfer performance can improve the efficiencies in water harvesting, thermal management facilities, power plants, as well as in desalination and air conditioning.

Copper is a useful material in many industrial applications due to its high thermal conductivity, easy manufacturing, and relatively low price In many systems such as power plants, copper has been applied to efficiently reject the heat to the surrounding via condensation.

To enhance condensation heat transfer performance, condensed droplets must be rapidly removed from the surface to minimize the thermal barrier. Recently, researchers showed that super-hydrophobic surfaces provides a higher mobility of condensates, which may enhance the heat transfer performance[1-3].

Here, we investigated the condensate mobility and resulting heat transfer performance on nano-engineered Cu surfaces. We introduced scalable Cu nano surfaces with unique wetting characteristics and observed the transient microscale condensation behaviors using a high-speed microscopy and a temperature and humidity controlled stage. Unlike previous heat transfer model studies based on the nucleation density correlations[2,4], we experimentally characterized the nucleation density, droplet size distribution and growth rate, and then incorporated them directly into the developed condensation heat transfer model to compare the condensation heat transfer performance of each Cu nano condenser surface more precisely.

2. Experimental method

2.1. Surface fabrication
We fabricate hydrophobic copper (HPo), super-hydrophobic copper oxide (SHPo), and oil infused super-hydrophobic copper oxide (Oil+SHPo) surfaces to observe the condensation phenomena on each surface. The CuO nanostructured copper surfaces are formed with a chemical oxidation process using hot alkali solution[1], and modified by TFTS (Sigma Aldrich) vapor deposition process. The Oil+SHPo is fabricated by infusing Krytox oil (DuPont) on the SHPo surface.

Figure 1 shows the field emission scanning electron microscopy (FE-SEM) images of CuO nanostructures which have uniform height approximately 1 μm. Table 1 shows contact angles (CA) on each surface, and we can predict the droplet mobility via the contact angle hysteresis. The HPo surface has almost 30° CA hysteresis, so the condensed droplets cannot be removed easily. The SHPo and Oil+SHPo surfaces have the CA hysteresis less than 5 degrees, so the droplets have high mobility.

Figure 1. FE-SEM image of nano-engineered copper oxide surfaces.

Table 1. Contact angle on each surface.

	Advancing(θ_a)	Receding(θ_r)	Static(θ_s)
HPo	122.6°±2.8	81.9°±8.8	118.7°±2.4
SHPo	165.1°±1.6	159.8°±3.3	161.2°±1.6
Oil+SHPo	120.1°±1.1	118.7°±1.5	119.3°±0.6

2.2. Condensation experiment
Figure 2 shows the experimental set-up for observing condensed droplet growth. Nitrogen gas is supplied the chamber through the DI water. Then measured temperature and humidity were 18.34±1.23℃ and 70.85±4.12%. The sample is cooled down to 5.93±0.11℃ by cold plate which is cooled by thermal bath. Droplets were condensed on the horizontal surface, and were observed via an optical microscope (BX51M, Olympus) and we take the images of droplets on three to five points of each surfaces.

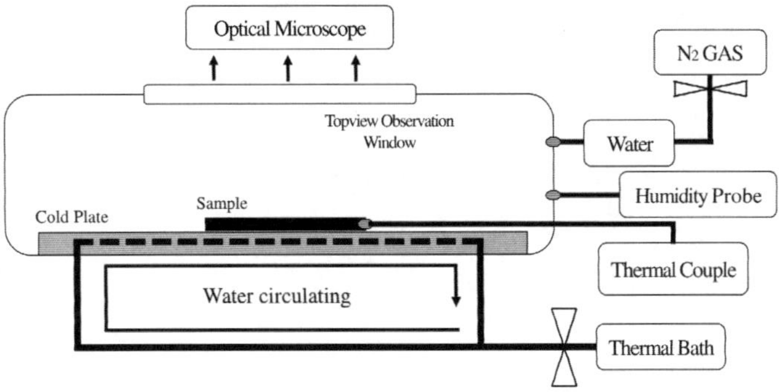

Figure 2. Schematic image of the condensation experiment setup.

To quantify the condensate rate, we used a super-saturation level (s)[5], then we can compare the condensation performance at the same conditions by making s=1.64.

3. Results and discussion

3.1. Droplet growth rate and surface coverage
Figure 3 (a) shows the condensed droplets on each surface up to 3 hours. Droplets on HPo continuously grow on their initial position due to high CA hysteresis, however droplets on Oil+SHPo

are merged actively with the neighbor droplets because they can move easily by low CA hysteresis. On both surfaces, most area are covered by droplets after ~180 min although some small droplets are present between the large droplets, because when the large droplets return to circular shape after two or more droplets merged, then the condensate nucleation start on the appeared refreshed area. The SHPo shows different behavior with other surfaces. Even after condensation time passed, the size are almost same compared with the initial stage due to the droplet jumping phenomena[3] i.e., the phenomena that droplets jump away from the surface without extra energy due to the released surface energy during the coalescence. Then the size of the droplets on SHPo seems to be near constant due to the repeated droplet disappearance and regeneration.

By measuring the observed images, we obtained the average individual droplet diameter (2R) and sum of droplet-surface contact area ($\pi(R\sin\theta)^2$) as a function of condensation time (Figure 3 (b) and (c)). The Oil+SHPo has high growth rate, then the average droplet size is higher than HPo at the initial ~30 min. Because the droplets on the Oil+SHPo have high droplet mobility by low CA hysteresis, then the droplets are merged easily. While the droplets on HPo grow up alone, on Oil+SHPo are become bigger by droplet merging phenomena. After 30 min, however, growth rate on Oil+SHPo decrease because no more merging happened, then the droplets on HPo become bigger after 120 min.

Surface coverage shows almost same tendency on both surfaces, although the growth rate was different. While the Oil+SHPo has large droplets by merging the HPo has high droplet density. Therefore surface coverage on HPo becomes similar with Oil+SHPo. The rate of surface coverage decreases on both surfaces after 30 min, because uncovered bare area is exposed by retuning circular shape after merging.

The SHPo maintains a small droplets size and low contact area even after long time. The average diameter and surface coverage are maintained near constant, under 15 μm and 0.05 due to the active droplet jumping behaviors.

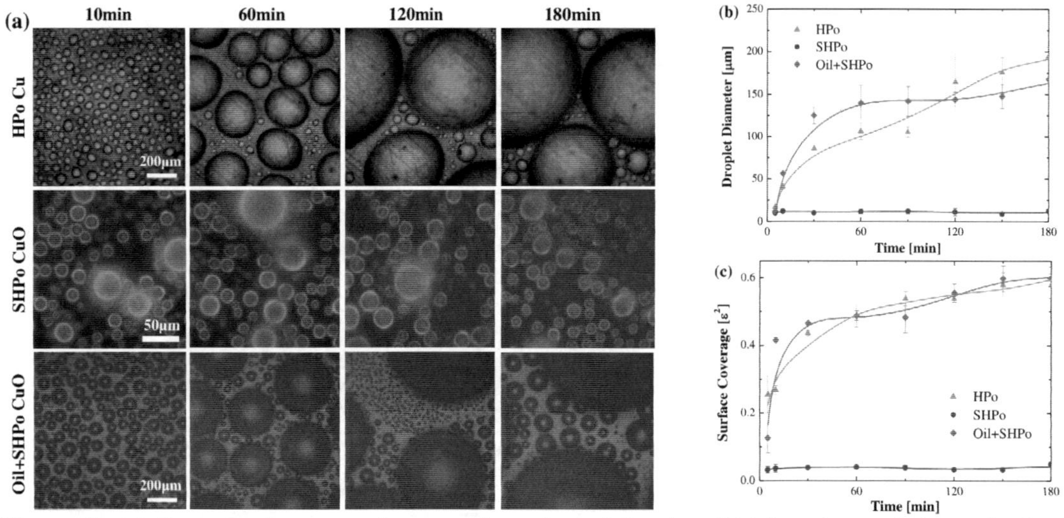

Figure 3. (a) Optical microscopy images of condensed droplet. (b) The observed dynamic droplet growth behaviour and (c) the normalized droplet-surface contact area by total area.

3.2. Heat transfer performance

The condensation heat transfer performance is calculated using the thermal network model for HPo, SHPo and Oil+SHPo (as in Figure 4 (a), (b) and (c)) including various thermal resistances at liquid vapor interface, inside droplet, structures and coatings. The HPo only has a surface coating while the SHPo has nanostructures and partially impregnated water above the Cu_2O layer which is formed when the process of change copper to copper oxide.

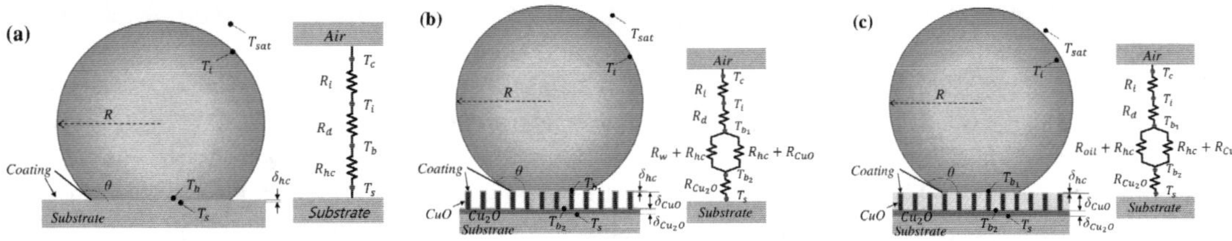

Figure 4. Schematic images of individual droplet heat transfer model and thermal resistance diagrams. (a) HPo, (b) SHPo and (c) Oil+SHPo surfaces.

Using the droplet growth rate and the nucleation densities obtained from the microscopic experiments, we calculate the heat transfer rate through the each individual droplet, measured via experiment.

$$q_{SHPo} = \frac{\pi R^2 \left(\Delta T - \dfrac{2T_{sat}\sigma}{Rh_{fg}\rho_w} \right)}{\dfrac{1}{2h_i(1-\cos\theta)} + \dfrac{R\theta}{4k_w \sin\theta} + \dfrac{1}{k_{hc}\sin^2\theta}\left[\dfrac{k_{CuO}\varphi}{\delta_{hc}k_{CuO} + hk_{hc}} + \dfrac{k_w(1-\varphi)}{\delta_{hc}k_w + hk_{hc}} \right]^{-1} + \dfrac{\delta_{Cu_2O}}{\sin^2\theta k_{Cu_2O}}} \quad \text{Eq. (1)}$$

Here, R is droplet radius and σ is water surface tension. h_{fg} and h_i are the latent heat of vaporization and condensation interfacial heat transfer coefficient. k_w, k_{hc}, k_a, k_{CuO} and k_{Cu2O} are the thermal conductivity of water, functional coating, air, nanostructures and Cu_2O layer, respectively. Finally, δ is the thickness of each parameter.

The overall heat flux was calculated based on Eq. (1) by aggregating the heat transfer through the individual droplets and expending the surface to 1 m^2 (Figure 5). The HPo and Oil+SHPo have high heat flux during the initial ~5 min, and the HPo has higher heat flux than the Oil+SHPo because the nanostructures and infused oil work as thermal resistance. However the heat flux on both surfaces are rapidly decrease during the initial ~30 min because droplets growth, while the SHPo is maintained near constant. Therefore, SHPo performs better than Oil+SHPo and HPo from approximately 30 min and 45 min, respectively. The SHPo has higher heat flux approximately 141% and 86% than Oil+SHPo and HPo, at 180 min, respectively.

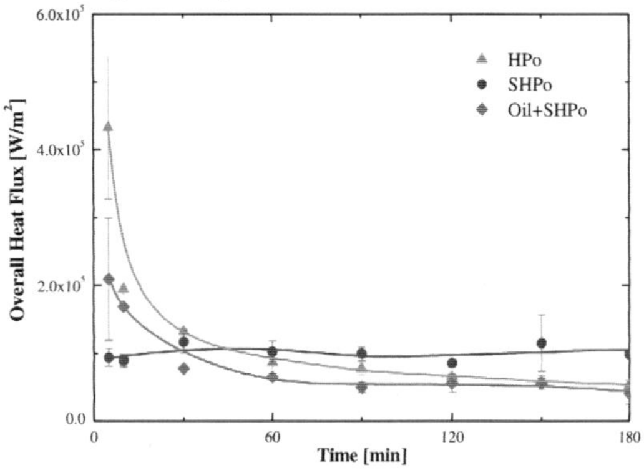

Figure 5. Overall heat flux through the surfaces.

Figure 6 shows the average thermal resistance in individual droplets by stacking each thermal resistance term at a specific time. The nanostructures & liquid resistance coloured blue indicate the resistance from nanostructures and permeated water and oil for the SHPo and Oil+SHPo, respectively.

Total thermal resistance on SHPo is higher than other surfaces at 5 min due to high droplet and nanostructures conduction resistance and the total resistance are near constant due to uniform droplet size. And the thermal resistance due to the nanostructures was significant during the entire condensation event due to the near constant droplet size, which implies that the reduction of the thermal resistance of nanostructures may be crucial to further increase the condensation heat transfer performance of SHPo.

The HPo and Oil+SHPo have lower thermal resistance, but droplets are continuously grow during condensation, then the total thermal resistance become much bigger than SHPo. For the Oil+SHPo, the thermal resistance due to the nanostructures and infused oil are significant at the beginning but the relative importance decreases as the droplet size and resulting conduction resistance through droplets get bigger. For the HPo, the droplet conduction resistance is the only influence term to heat flux.

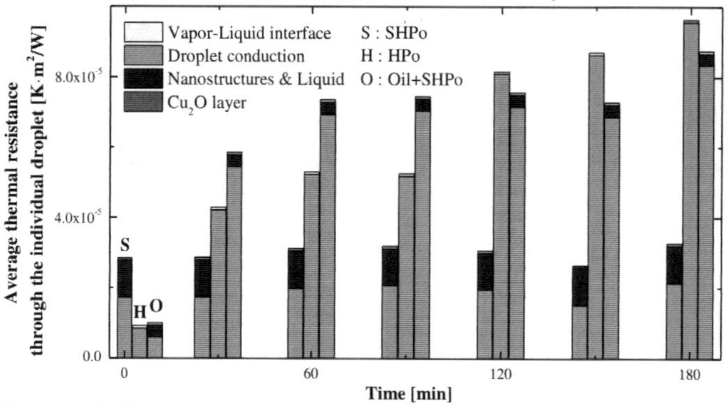

Figure 6. Thermal resistance through the individual droplet.

4. Conclusion

In this study, we made a stable surface, which has uniform nanostructures of 1 μm, by using a hot alkali solution. Then, we made three surfaces which have different characteristics including HPo, SHPo and Oil+SHPo. By 3 hours condensation experiments, droplet growth rate and surface coverage on SHPo are maintained near constant due to high condensate mobility named droplet jumping, while on other surfaces are continuously grow due to high CA hysteresis and merging phenomena. The overall heat flux calculated by using the droplet growth rate and the nucleation densities.

The HPo and Oil+SHPo had the higher total heat flux and overall heat transfer coefficient at the initial part of condensation time. However, as the droplets on HPo and Oil+SHPo grew continuously, resulting in decreased heat transfer performance. Then the heat transfer performance on the SHPo had higher heat flux approximately 141% and 86% than Oil+SHPo and HPo at 180 min., due to the SHPo maintained by droplet regeneration.

We showed the average thermal resistance through the individual droplet. Therefore, the conduction resistance through the droplet inside was significant on HPo and Oil+SHPo during the entire time. The droplet and nanostructure, however, were both significant on SHPo due to near constant droplet size. This work demonstrates scalable Cu-based nano surfaces with investigating the condense mobility and the resulting heat transfer performance on such surfaces to help develop advanced condensers.

References

[1] Y. Nam and Y.S. Ju 2013 *Journal of Adhesion Science and Technology* 27 2163-2176
[2] N. Miljkovic, R. Enright, and E.N. Wang 2012 *Acs Nano* **6** 1776-1785
[3] Y. Nam, H. Kim, and S. Shin 2013 *Applied Physics Letters* **103** 161601
[4] R. Enright, N. Miljkovic, N. Dou, Y. Nam and E.N. Wang 2013 *Journal of Heat Transfer* **135** 091304
[5] N. Miljkovic, R. Enright, Y. Nam, K. Lopez, N. Dou, J. Sack and E.N. Wang 2012 *Nano Letters* **13** 179-187

Power Generation Characteristics of Single Electrode Output Circuit in Electret Energy Harvester

L Bu[1], H Y Xu[1], B J Xu[1], L Song[2]

[1] School of Information Engineering, China University of Geosciences, Beijing, China
[2] State Grid Jibei Electric Power Co., Ltd., Beijing, China

Email: lingbu@cugb.edu.cn

Abstract. This paper reports the new output circuit using a single electrode in electret energy harvester, and proves that the single electrode is able to generate output power on grounded load. 3D numerical model of gap-closing type electret energy harvester is presented, and power generation characteristics are analysed and verified. Results show that the two electrodes are actually two independent current sources. Single electrode output circuit has two merits: when only one electrode is connected, it reduces wiring difficulty; when both electrodes are connected to grounded load respectively, it doubles output power compared with traditional output circuit. Using proposed circuit, maximal total power of 30mm×20mm prototype reaches 154.5μW@10Hz, 1.8mm sinusoidal vibration, and an LED has been successfully lighted up.

1. Introduction

Electret energy harvesters have been extensively explored in recent years due to good compatibility with planar processes and relatively higher output power in the low frequency range [1-2]. Latest research concentrates on proposing novel structures with improved output power [3], lowered working frequency [4], or broadened bandwidth [5]. These results prove the potential of electret energy harvesters in applications such as wireless sensor networks, implantable medical devices, and etc.

However, so far in literature most electret energy harvesters utilize the same simple output circuit, which takes the form of load resistors forming a loop with the inherent capacitor of electret energy harvester, as is shown in Fig.1(a). It is typically assumed that current flows from one electrode to another, requiring both electrodes connected to load resistors. Yet this assumption is contradictory to experimental results in [6], as the electret energy harvester with one wiring generates 200μW@20Hz vibration. In fact, little has been investigated about power generation characteristics of output circuit topology, and the mechanism of how the two electrodes function as current source remains unclear.

This paper presents two forms of single electrode output circuit with no loop. Power generation mechanism is theoretically explained and experimentally verified. The two electrodes are tested to be two independent current sources, and novel equivalent circuit is also presented.

2. Single Electrode Output Circuit

In the proposed circuit, "single electrode" can be base electrode (BE) or counter electrode (CE), i.e. only BE or CE is connected to output circuit, as is shown in Fig.1(b) and (c) respectively. Although there is no loop in either topology, power is obtained on grounded load resistances R_1 and R_2.

The principle of single electrode output circuit is in accordance with fundamental electrostatics. In electrostatic equilibrium, induced charges distribute on both BE and CE, and the total amount of induced charges is fixed due to the fixed amount of trapped charges in electrets. This means that increase of induced charges on BE results in decrease of that on CE, and vise versa.

In Fig.1(b), although the distance between trapped charges in electrets and BE remains constant, the electric field between them strengthens (i.e. increased induced charges on BE) as CE vibrates upward (i.e. decreased induced charges on CE), and weakens (i.e. decreased induced charges on BE) as CE vibrates downward (i.e. increased induced charges on CE). As long as load resistors are grounded, changing electric field leads to changing electrostatic equilibrium state, and the flow of induced charges between BE and ground forms the alternating current on load resistors, regardless of whether CE is connected or not. Similar principle also applies for Fig.1(c).

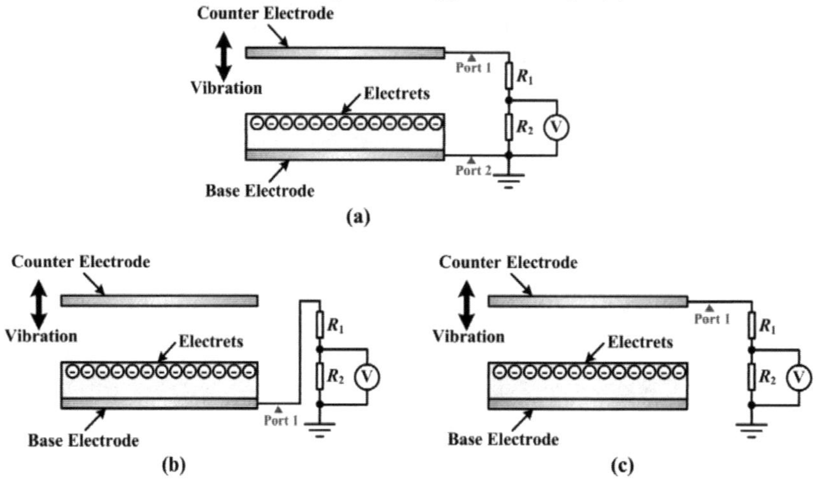

Fig.1 Output circuit topology: (a) traditional; (b) this work, with only Base Electrode (BE) connected; (c) this work, with only Counter Electrode (CE) connected.

3. Theory

A 3D model is adopted to numerically calculate the amount of induced charges on both BE and CE. As is shown in Fig.2(a), assume that electrode width w, electrode length l, position of CE g(t), electret thickness d are comparable in dimension. Adopting method of images, trapped charges and image charges are axially symmetrical with respect to electrodes, which is $\sigma_{img,CE}$ located at z=2g(t)-d for CE,

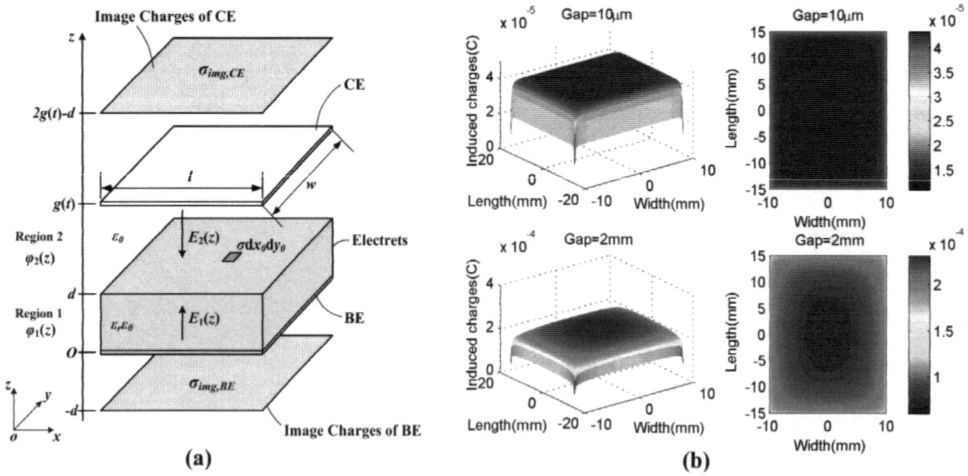

Fig.2 (a) 3D model of gap-closing type electret energy harvester
(b) Induced charge density on BE when g(t)-d=10μm (upper half) and g(t)-d=2mm (lower half).

and $\sigma_{img,BE}$ located at z=-d for BE. For simplicity, assume that $\sigma_{img,CE}$ and $\sigma_{img,BE}$ are uniform, and neglect electrode thickness and fringing electric field. σ represents trapped charge density of electrets, ε_r relative dielectric constant of electrets, and ε_0 vacuum permittivity.

Based on fundamental electrostatics, induced charge density of BE σ_{BE}, and that of CE σ_{CE} are:

$$\sigma_{BE}(z)=\varepsilon_r\varepsilon_0 E_1(z)=-\varepsilon_r\varepsilon_0\frac{\partial\varphi_1(z)}{\partial z},\quad \sigma_{CE}(z)=\varepsilon_0 E_2(z)=-\varepsilon_0\frac{\partial\varphi_2(z)}{\partial z} \tag{1-1, 1-2}$$

where $\varphi_1(z)$ and $\varphi_2(z)$ are potential functions in Region 1 and Region 2. Due to dielectric discontinuity between two electrodes, $\varphi_1(z)$ and $\varphi_2(z)$ are solved separately, both using superposition theorem:

$$\varphi_1(z)=\frac{1}{\varepsilon_r}\left(\varphi_{elct}(z)+\varphi_{img,BE,1}(z)+\varphi_{img,CE,1}(z)\right),\quad \varphi_2(z)=\varphi_{elct}(z)+\varphi_{img,BE,2}(z)+\varphi_{img,CE,2}(z) \tag{2-1, 2-2}$$

In Eq.(2-1)~(2-2), potential function caused by electrets $\varphi_{elct}(z)$, by image charges on BE $\varphi_{img,BE,i}(z)$, and by image charges on CE $\varphi_{img,BE,i}(z)$are derived by integrating potential from charge per area:

$$\varphi_{elct}(z)=\int_{-\frac{l}{2}}^{\frac{l}{2}}\int_{-\frac{w}{2}}^{\frac{w}{2}}\frac{-\sigma \mathrm{d}x_0\mathrm{d}y_0}{4\pi\varepsilon_0\sqrt{(x-x_0)^2+(y-y_0)^2+(z-d)^2}} \tag{3}$$

$$\varphi_{img,BE,i}(z)=\int_{-\frac{l}{2}}^{\frac{l}{2}}\int_{-\frac{w}{2}}^{\frac{w}{2}}\frac{\sigma_{BE,i}\mathrm{d}x_0\mathrm{d}y_0}{4\pi\varepsilon_0\sqrt{(x-x_0)^2+(y-y_0)^2+(z+d)^2}} \tag{4}$$

$$\varphi_{img,CE,i}(z)=\int_{-\frac{l}{2}}^{\frac{l}{2}}\int_{-\frac{w}{2}}^{\frac{w}{2}}\frac{\sigma_{CE,i}\mathrm{d}x_0\mathrm{d}y_0}{4\pi\varepsilon_0\sqrt{(x-x_0)^2+(y-y_0)^2+(z-2g(t)+d)^2}} \tag{5}$$

where i=1 or 2, $\sigma_{BE,i}$ is $\sigma_{img,BE}$ for solution in Region i, and $\sigma_{CE,i}$ is $\sigma_{img,CE}$ for solution in Region i.

Substituting potential functions $\varphi_1(z)$ and $\varphi_2(z)$ to interface continuity equations and zero potential boundary conditions solves $\sigma_{BE,i}$ and $\sigma_{CE,i}$, and further substituting $\sigma_{BE,i}$ and $\sigma_{CE,i}$ to Eq.(1-1)~(1~2) renders the numerical results for σ_{BE} and σ_{CE}. Fig.2(b) shows the numerically obtained σ_{BE}, which is in accordance with previous analysis that induced charges on BE increase as gap between BE and CE increases, and vice versa. Similar numerical results can be obtained for σ_{CE}.

Based on these, BE and CE are modeled as two separate current sources independently generating current, as is shown in Fig.3. i_{BE} and i_{CE} are total current of BE and CE, which is proportional to v_{vib}, vibration velocity of CE, as is shown in Eq.(6)~(7). C_{BEG} and C_{CEG} are ground capacitance parallel to grounded load. The two current sources can work individually or simultaneously.

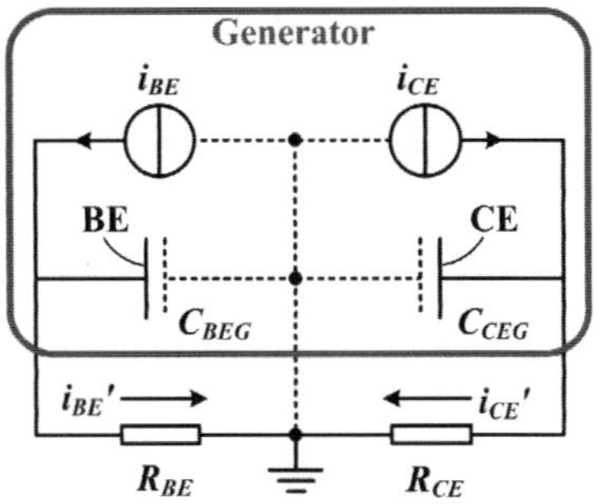

Fig.3 Equivalent circuit of gap-closing type electret energy harvester.

$$i_{BE}(t)=-\frac{\mathrm{d}Q_{BE}(t)}{\mathrm{d}t}=-\frac{\sigma_{BE}(g(t))lw}{\mathrm{d}(g(t))}\cdot\frac{\mathrm{d}(g(t))}{\mathrm{d}t}=-\frac{\sigma_{BE}(g(t))lw}{\mathrm{d}(g(t))}\cdot v_{vib} \tag{6}$$

$$i_{CE}(t)=-\frac{\mathrm{d}Q_{CE}(t)}{\mathrm{d}t}=-\frac{\sigma_{CE}(g(t))lw}{\mathrm{d}(g(t))}\cdot\frac{\mathrm{d}(g(t))}{\mathrm{d}t}=-\frac{\sigma_{CE}(g(t))lw}{\mathrm{d}(g(t))}\cdot v_{vib} \tag{7}$$

Power on load resistors is as follows:

$$P_{BE}=\frac{i_{BE}(t)^2 R_{BE}}{\left(1+j\omega C_{BEG}R_{BE}\right)^2},\quad P_{CE}=\frac{i_{CE}(t)^2 R_{CE}}{\left(1+j\omega C_{CEG}R_{CE}\right)^2} \tag{8-1, 8-2}$$

4. Experimental Verification

Fig.4 shows the experimental setup for verifying the proposed circuit. The upper slider of the linear ball guide is fixed, while the lower slider is driven by electromagnetic shaker. PMMA fixtures are fixed on the two sliders for clamping BE and CE. BE with electrets (area: 30mm×20mm, surface potential: -700V) is fixed on the upper slider, and CE vibrates up and down with the lower slider, thus forming a gap-closing type electret energy harvester. The displacement of CE is recorded by an eddy current sensor. During test, R_1 is variable from 10MΩ to 500MΩ, while R_2 is fixed 1MΩ to minimize the influence of probe resistance. The entire setup is effectively grounded to minimize EMI effects.

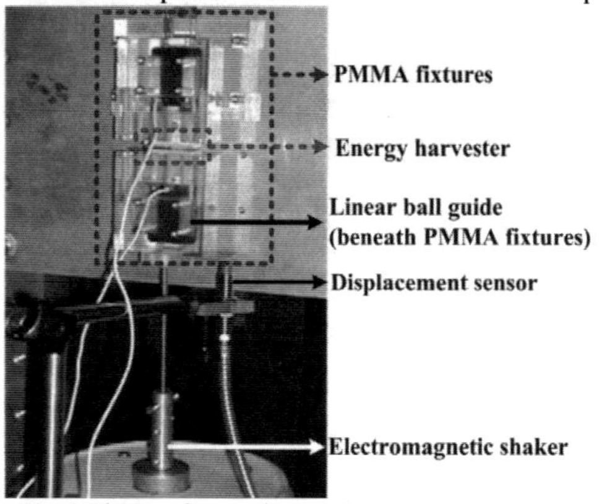

Fig.4 Photo of experimental setup.

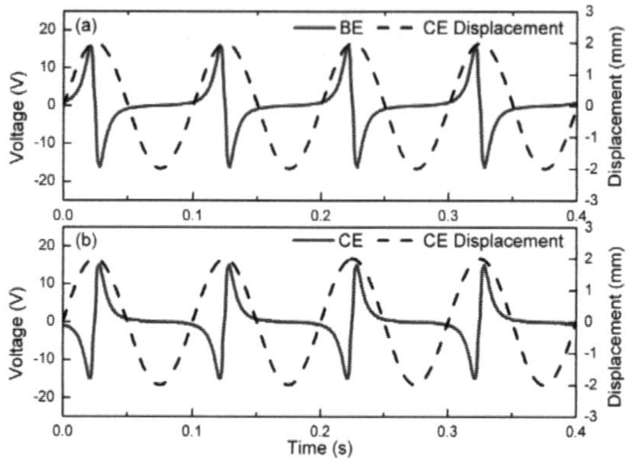

Fig.5 Measured voltage: (a) BE; (b) CE. R_1=10MΩ.

When only BE/CE is connected, measured voltage is shown in Fig.5(a) and (b), respectively. Voltage waveforms of BE and CE are in opposite phase. This is in accordance with electrostatic induction principle, since charge increases on BE implies that charge decreases on CE. Maximum BE/CE voltage occurs slightly before/after maximum CE displacement due to small gap and high vibration velocity. These results prove the power generation capability using single electrode output circuit, and exclude the possibility of displacement current.

Fig.6 displays the characterization results of output voltage. In Fig.6(a), BE/CE voltage decreases as gap between BE and CE increases. In Fig.6(b), BE/CE voltage increases as vibration frequency increases. These results agree well with theoretical analysis of gap-closing type electret energy harvester, thus proving measured voltage and power is indeed generated by electret energy harvester. Since voltage of BE or CE is almost the same, in practice connecting either electrode generates almost equal power. This greatly reduces wiring difficulty, for instance the vibrating CE can be floated while power is generated from the fixed BE.

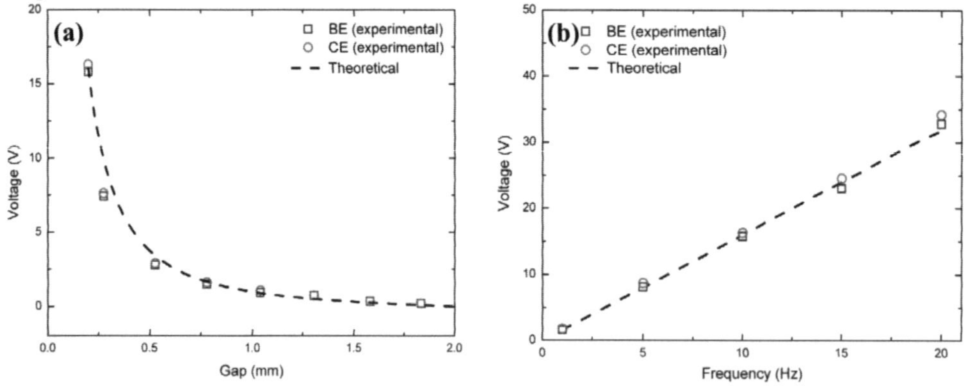

Fig.6 Relation between: (a) voltage and gap; (b) voltage and frequency. In both (a) and (b) R_1=10MΩ.

When BE and CE are respectively connected to grounded load, the total power doubles that of the traditional output circuit in Fig.1(a). As is shown in Fig.7, as BE and CE are respectively connected to

110MΩ grounded load resistance (i.e. total resistance is 220MΩ), maximal total power reaches 154.5μW@10Hz, 1.8mm sinusoidal vibration. This is two times of the maximal power using traditional output circuit, which is 74.9μW on 110MΩ under the same vibration condition. Therefore, single electrode output circuit also helps to double output power by utilizing generation capability of both electrodes.

Fig.8 shows photos of using single electrode output circuit to power a light emitting diode (LED). BE (or CE) is solely connected to Schottky rectifier bridge to convert AC voltage of harvester into DC voltage, and generated power is stored in 47μF capacitor. The LED is dim when there is no vibration, and is light under 100Hz, 0.5mm sinusoidal vibration, as is shown in Fig.8(a) and (b), respectively. These results further solidify the applicability of proposed single electrode output circuit.

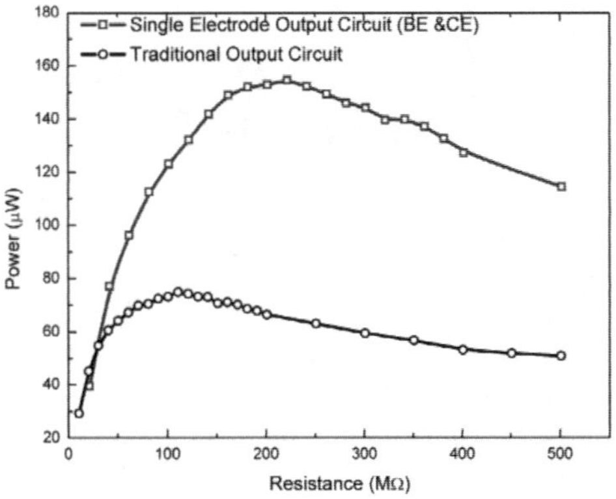

Fig.7 Comparison of power on various load.

Fig.8 Photos of LED application using only BE/CE for output: (a) dim LED under no vibrations; (b) light LED under 100Hz, 0.5mm sinusoidal vibration.

Conclusion

This paper reports the new output circuit using a single electrode in electret energy harvester, and proves that the single electrode is able to generate output power on grounded load. 3D numerical model of gap-closing type electret energy harvester is presented, and novel equivalent circuit is presented. The two electrodes are tested to be two independent current sources, and the experimental power generation characteristics agree well with theoretical analysis. Single electrode output circuit reduces wiring difficulty when only one electrode is used, and doubles output power compared with traditional circuit when two electrodes are respectively connected to grounded load. Using the proposed circuit, for the 30mmx20mm prototype, maximal total power reaches 154.5μW@10Hz, 1.8mm sinusoidal vibration, and an LED has been successfully lighted up.

Acknowledgements

This work is supported by Fundamental Research Funds for Central Universities (Grant No. 2-9-2014-010) of China.

References

[1] Y. Suzuki. *IEEJ T. Electr. Electr.*, 2011, 6(2), pp.101-111.
[2] L. Tang, et al. *J. Intel. Mat. Syst. Str.*, 2010, 21, pp.1867-1897.
[3] G. Altena, et al. *J. Phys.: Conf. Ser.*, 2013, 476 012078 (5pp).
[4] T. Galchev, et al. *Proc. IEEE MEMS*, 2014, pp.350-353.
[5] L. Bu, et al. *Micro Nano Lett.*, 2013, 8(3), pp.135-137.
[6] T. Takahashi, et al. *Proc. IEEE MEMS*, 2012, pp.1265-1268.

Analysis and Implementation of Real Number Extended Sliding Surface with Integral Compensation for DC-AC Converters

E C Chang[1], R C Wu and K Y Liao

Department of Electrical Engineering, I-Shou University, Kaohsiung City 84001, Taiwan, R.O.C.

E-mail: enchihchang@isu.edu.tw

Abstract. This paper proposes a real number extended sliding surface (RNESS) with integral compensation (IC) for the application of DC-AC converters. Classic sliding surface (CSS) is insensitive to system uncertainties, but in sliding action its system dynamics becomes a reduced-order dimension and thus lost the partial system dynamics. For recovering incomplete system response, the RNESS is designed and can retrieve incomplete system response of the CSS. However, steady-state errors still exist in system dynamics of the RNESS and cause high converter voltage harmonics. To overcome steady-state errors, a modified RNESS by the addition of IC is proposed. With the proposed method, the system yields a DC-AC converter with high-quality AC output voltage. Experiments are performed in support of the proposed method.

1. Introduction

DC-AC converters have been extensively used in many AC power conditioning systems. The DC-AC converters need to be demanded high-quality AC output voltage of low THD and fast dynamic response, and these can be obtained by employing feedback control. The control of the DC-AC converters is popularly designed through sliding surface control; however, the CSS is adopted, and thus incurs incomplete dynamics response [1-3]. To recover lost system dynamics, the RNESS is applied to DC-AC converters. However, steady-state errors still exist in DC-AC converters, and cause serious voltage harmonics [4], [5]. Thus, by combining a RNESS with IC (RNESS-IC), steady-state errors can be overcome and a closed-loop DC-AC converter will yield good performance under various loading. The proposed method is implemented digitally using a CPLD development board. Experiments are given to verify the effectiveness of the proposed method.

2. Mathematical Modeling of DC-AC Converter

Figure 1 shows the block diagram of a DC-AC converter, and the system dynamics can be written in the following.

---

[1] Corresponding Author: E C Chang is with the Department of Electrical Engineering, I-Shou University, No.1, Sec. 1, Syuecheng Rd., Dashu District, Kaohsiung City 84001, Taiwan, R.O.C. This work was supported by the Ministry of Science and Technology of Taiwan, R.O.C., under contract number MOST103-2221-E-214-027.

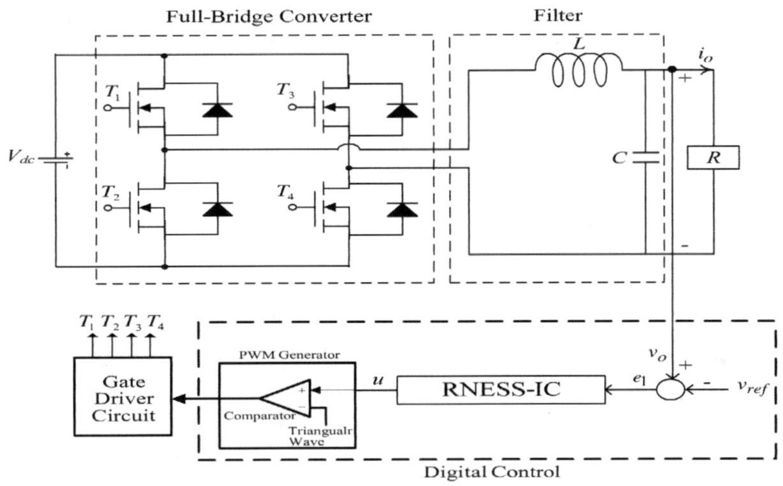

Figure 1. Block diagram of DC-AC converter.

$$\begin{bmatrix}\dot{e}_1\\ \dot{e}_2\end{bmatrix}=\underbrace{\begin{bmatrix}0 & 1\\ -\frac{1}{LC} & -\frac{1}{RC}\end{bmatrix}}_{\text{system matrix}}\begin{bmatrix}e_1\\ e_2\end{bmatrix}+\underbrace{\begin{bmatrix}0\\ \frac{1}{LC}\end{bmatrix}}_{\text{control matrix}}u+\underbrace{\begin{bmatrix}0\\ -\frac{1}{LC}v_{ref}-\frac{1}{RC}\dot{v}_{ref}-\ddot{v}_{ref}\end{bmatrix}}_{\text{external interferece}} \tag{1}$$

Thus, once the control signal u is designed well, the converter's output will remain the same as what is desired v_{ref}.

3. RNESS-IC Control Design

Generally, a CSS is constructed as

$$\sigma = \ker(C) \subset \Re \tag{2}$$

where $C \in \Re^{m\times n}$, and $rank(C) = m$.

The resulting closed-loop dynamics yields

$$\begin{bmatrix}\dot{x}\\ \dot{y}\end{bmatrix}=\begin{bmatrix}R_e A_e & 0\\ 0 & R_e A_e\end{bmatrix}\begin{bmatrix}x\\ y\end{bmatrix}+\begin{bmatrix}0\\ B\end{bmatrix}u;\quad A_e = system\ \text{matrix},\ B = control\ matrix \tag{3}$$

From (2) and (3), the system dynamics is $(n-m)$ reduced-order dimension, and deteriorates system performance. To recover the incomplete system dynamics, the RNESS will be designed as follows.

Firstly, defining the u and x with the pairs of complex conjugate ($v, \bar{v}$ and $z, \bar{z}$) as

$$u = \frac{1}{2}(v+\bar{v}) \tag{4}$$

$$x = \frac{1}{2}(z+\bar{z}) \tag{5}$$

where $v, \bar{v} \in C^m$ and $z, \bar{z} \in C^n$.

Substituting (4) and (5) into (3), the real system (3) yields the sum of the two complex system

$$\dot{z} = Az + Bv \tag{6}$$

$$\dot{\bar{z}} = A\bar{z} + Bv \tag{7}$$

Then, a real number extended sliding surface is designed as

$$\gamma = \Psi z = 0 \tag{8}$$

$$z = \begin{bmatrix}x_1 & \dots & x_m & \dot{x}_1 & \dots & \dot{x}_m\end{bmatrix}^T \tag{9}$$

where $\Psi \in C^{m\times n}$.

The sliding function matrix can be partitioned compatibly as

$$\Psi = \begin{bmatrix} \Psi_1 & \Psi_2 \end{bmatrix} = \begin{pmatrix} M_{11}+iN_{11} & & & 1 & & \\ & \ddots & & & \ddots & \\ & & M_{mm}+iN_{mm} & & & 1 \end{pmatrix} \tag{10}$$

where $\Psi_1 \in C^{m\times(n-m)}$ and $\Psi_2 \in C^{m\times m}$.

The system matrix thus becomes

$$A_z = (I - B(\Psi B)^{-1}\Psi)A = \begin{bmatrix} A_{11} - A_{12}\Psi_2^{-1}\Psi_1 & A_{12} \\ 0 & 0 \end{bmatrix} \tag{11}$$

Letting $z = x + iy$, we obtain

$$\begin{aligned} \bar{\dot{z}} + \dot{z} &= A_z(x+iy) + \bar{A}_z(x-iy) \\ &= 2(\operatorname{Re} A_z x - \mathrm{I_m} A_z y) \end{aligned} \tag{12}$$

Contrariwise,

$$\begin{aligned} \bar{\dot{z}} - \dot{z} &= A_z(x+iy) + \bar{A}_z(x-iy) \\ &= 2(\mathrm{I_m} A_z x + \operatorname{Re} A_z y) \end{aligned} \tag{13}$$

Therefore, the $\Re^{2n}$ dimensional system exists

$$\begin{bmatrix} \dot{x} \\ \dot{y} \end{bmatrix} = \begin{bmatrix} \mathrm{R_e} A_z & -\mathrm{I_m} A_z \\ \mathrm{I_m} A_z & \mathrm{R_e} A_z \end{bmatrix} \begin{bmatrix} x \\ y \end{bmatrix} \tag{14}$$

The classic situation can be verified as $\Psi = \sigma$, here Ψ is a real linear map, then the (14) has

$$\begin{aligned} \begin{bmatrix} \dot{x} \\ \dot{y} \end{bmatrix} &= \begin{bmatrix} \mathrm{R_e} A_z & 0 \\ 0 & \mathrm{R_e} A_z \end{bmatrix} \begin{bmatrix} x \\ y \end{bmatrix} \\ &= \begin{bmatrix} (I - B(CB^{-1}C)A & 0 \\ 0 & (I - B(CB^{-1}C)A \end{bmatrix} \begin{bmatrix} x \\ y \end{bmatrix} \end{aligned} \tag{15}$$

The control signal is assumed as the following form

$$\begin{aligned} v = & \underbrace{-(\Psi B)^{-1}\Psi A z}_{u_{ez}} \\ & \underbrace{-(\Psi B)^{-1}\Psi A z\left[F\mathrm{sign}(\mathrm{R_e}(\gamma) + iG\mathrm{sign}(\mathrm{I_m}(\gamma))\right]}_{u_{sz}} \end{aligned} \tag{16}$$

where u_{ez} is the equivalent control and u_{sz} represents a discontinuous switching control component.

Define $v = u + iw$, (4) and (16) lead to

$$\begin{aligned} u = & -\mathrm{R_e}(\Psi \mathrm{B})^{-1}\Psi A x + \mathrm{I_m}(\Psi \mathrm{B})^{-1}\Psi A y \\ & - \mathrm{R_e}(\Psi \mathrm{B})^{-1} F\mathrm{sign}(\mathrm{R_e}\gamma) + \mathrm{I_m}(\Psi \mathrm{B})^{-1} G\mathrm{sign}(\mathrm{I_m}\gamma) \end{aligned} \tag{17}$$

and

$$\begin{aligned} w = & -\mathrm{I_m}(\Psi \mathrm{B})^{-1}\Psi A x + \mathrm{R_e}(\Psi \mathrm{B})^{-1}\Psi A y \\ & - \mathrm{I_m}(\Psi \mathrm{B})^{-1} F\mathrm{sign}(\mathrm{R_e}\gamma) - \mathrm{R_e}(\Psi \mathrm{B})^{-1} G\mathrm{sign}(\mathrm{I_m}\gamma) \end{aligned} \tag{18}$$

Suppose $\Psi = M + iN$, the (17) and (18) is rewritten as

$$\begin{aligned} \begin{bmatrix} u \\ w \end{bmatrix} = & -\begin{bmatrix} \mathrm{R_e}(\Psi B)^{-1}\Psi A & \mathrm{I_m}(\Psi B)^{-1}\Psi A \\ \mathrm{I_m}(\Psi B)^{-1}\Psi A & \mathrm{R_e}(\Psi B)^{-1}\Psi A \end{bmatrix} \begin{bmatrix} x \\ y \end{bmatrix} \\ & + \begin{bmatrix} \mathrm{R_e}(\Psi B)^{-1} F & \mathrm{I_m}(\Psi B)^{-1} G \\ \mathrm{I_m}(\Psi B)^{-1} F & \mathrm{R_e}(\Psi B)^{-1} G \end{bmatrix} \\ & \cdot \mathrm{sign}\left(\begin{bmatrix} M & -N \\ N & M \end{bmatrix} \begin{bmatrix} x \\ y \end{bmatrix}\right) \end{aligned} \tag{19}$$

Employing $y = 1/2(z - \bar{z})$, the differential equation can be obtained as

$$\dot{y} = Ay + Bw \tag{20}$$

Thus, the resulting system is shown in (21) into $\Re^{2n}$ dimension, and thus recovers all the dynamics lost in the CSS design process as follows:

$$\begin{bmatrix} \dot{x} \\ \dot{y} \end{bmatrix} = \begin{bmatrix} \mathrm{R_e}A_z & -\mathrm{I_m}A_z \\ \mathrm{I_m}A_z & \mathrm{R_e}A_z \end{bmatrix} \begin{bmatrix} x \\ y \end{bmatrix} + \begin{bmatrix} B & 0 \\ 0 & B \end{bmatrix} \begin{bmatrix} u_{swz} \\ w_{swz} \end{bmatrix}, \quad \begin{matrix} u_{ez} = equivalent\ \ control \\ w_{sz} = sliding\ \ control \end{matrix} \tag{21}$$

Notice that the (1) with external interferences cause steady-state errors, and thus an IC $u_{ic} = \int e_1 dt$ is introduced to overcome steady-state errors. Finally, the proposed control law is $u = u_{ic} + u_{ez} + iw_{sz}$.

4. Experimental Results

The proposed system parameters are as follows: V_{dc}=200 V; v_o=110 V_{rms}, f =60 Hz; L=1 mH; C=30 μF; Switching Frequency f_s =12 kHz; R_{rated} =10 Ω. For no load to full load, the results with the RNESS-IC and CSS are shown in Figure 2 and Figure 3, respectively. The output voltage with the RNESS-IC only dips lightly, but the large voltage dip occurs in the CSS. For rectifier load ($C_d = 200\,\mu\text{F}$, $R_d = 45\,\Omega$) shown in Figure 4, the output voltage with the RNESS-IC is close to sinusoidal waveform (%THD=2.01%), but that with the CSS shown in Figure 5 has a high %THD of 6.23%.

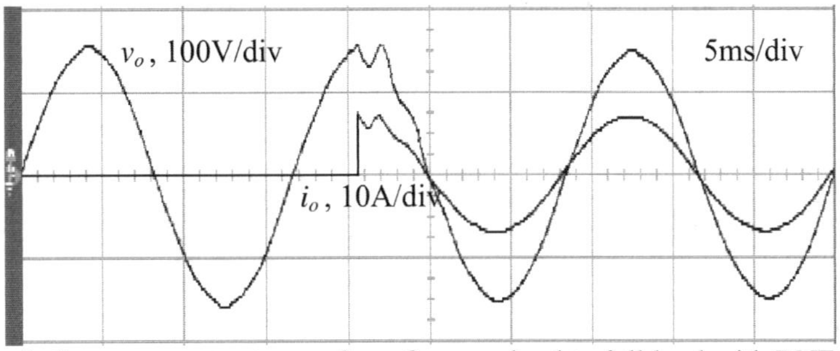

Figure 2. Converter output waveform from no load to full load with RNESS-IC.

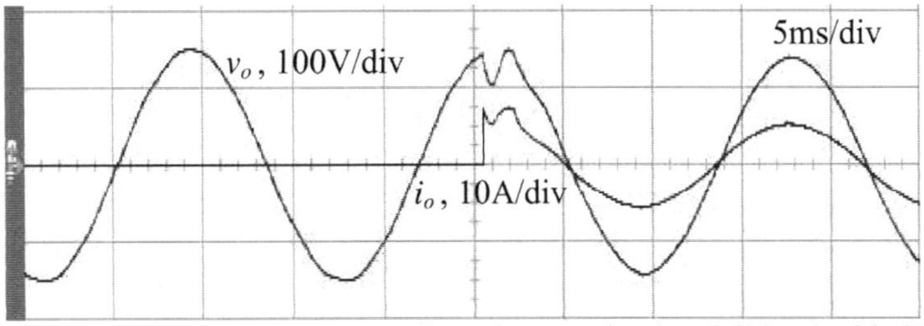

Figure 3. Converter output waveform from no load to full load with CSS.

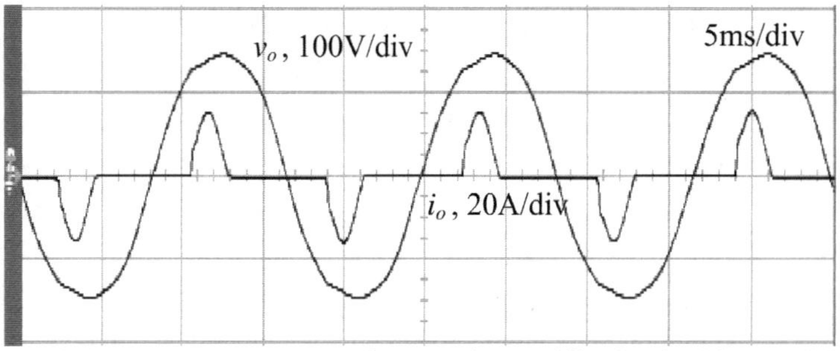

Figure 4. Converter output waveform under rectifier load with RNESS-IC.

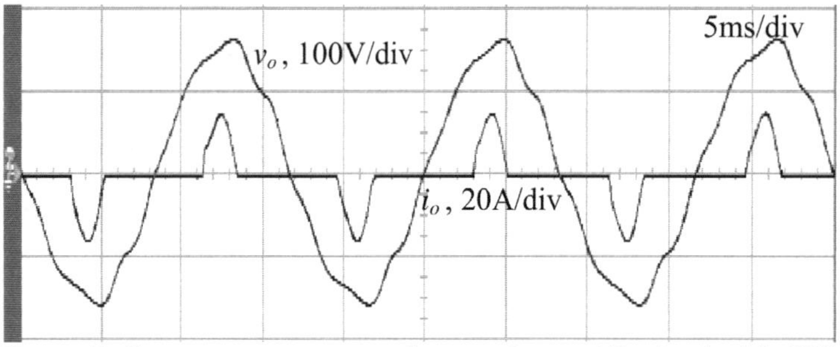

Figure 5. Converter output waveform under rectifier load with CSS.

5. Conclusions
The proposed converter recovers lost system dynamics and overcomes steady-state errors. Experiments show that the proposed system surpasses the results achieved under the classic system. Therefore, it is clear that the proposed method is robust in transient loading and steady-state loading.

6. Acknowledgment
This work was supported by the Ministry of Science and Technology of Taiwan, R.O.C., under contract number MOST103-2221-E-214-027.

7. References
[1] Tan S C, Lai Y M and Tse C K 2012 *Sliding Mode Ccontrol of Switching Power Converters: Techniques and Implementation* (*Boca Raton*: CRC Press)

[2] Abrishamifar A, Ahmad A and Mohamadian M 2012 Fixed switching frequency sliding mode control for single-phase unipolar inverters *IEEE Trans. Power Electronics*, vol 27, no 5, pp 2507-2514

[3] Yang X., Liu T, Huang L and Chen W J 2013 A sliding-mode controller with multiresonant sliding surface for single-phase grid-connected VSI with an LCL filter *IEEE Trans. Power Electronics*, vol 28, no 5, pp 2259-2268

[4] Shepit B M and Pieper J K 2001 Sliding mode control design for complex valued sliding manifold *Proc. Conf. IEEE Decision and Control* vol 1 pp 920-921

[5] Shepit B M, and Pieper J K 2003 Sliding-mode control design for a complex valued sliding manifold *IEEE Trans. Automatic Control* vol 48 no 1 pp 122-127

Journal of Physics: Conference Series **557** (2014) 012112 doi:10.1088/1742-6596/557/1/012112

Energy Conversion at Micro and Nanoscale

Luca Gammaitoni,
NiPS Laboratory, Università di Perugia, Italy

Abstract Energy management is considered a task of strategic importance in contemporary society. It is a common fact that the most successful economies of the planet are the economies that can transform and use large quantities of energy. In this talk we will discuss the role of energy with specific attention to the processes that happens at micro and nanoscale. The description of energy conversion processes at these scales requires approaches that go way beyond the standard equilibrium termodynamics of macroscopic systems. In this talk we will address from a fundamental point of view the physics of the dissipation of energy and will focus our attention to the energy transformation processes that take place in the modern micro and nano information and communication devices.

About the Speaker

Luca Gammaitoni, is Professor of Experimental Physics at the University of Perugia, in Italy and the director of the Noise in Physical Systems (NiPS) Laboratory. He is also the founder of Wisepower srl a university spin-off company. He obtained the PhD in Physics from the University of Pisa in 1990. Since then he has developed a wide international experience with collaborations both in Europe, Japan and the USA. His scientific interests span from noise phenomena in physical systems to non-equilibrium thermodynamics and energy transformations at micro and nanoscale.
The Noise in Physical Systems (NiPS) Laboratory is a research facility within the Physics Department of the University of Perugia. The laboratory (www.nipslab.org) has a long-standing tradition in studying physical systems in the presence of noise. Scientific interest ranges from stochastic nonlinear dynamics modelling to thermal noise measurements. Wisepower srl (www.wisepower.it) is a university spin-off company focused on the design and prototyping of micro-energy generators. The company holds patents on vibration-to-electricity nonlinear conversion technology.
He authored over 200 papers on top-level scientific journals and few books. He is also the author of 10 patents. His papers have been cited more than 12.000 times. H factor=48.

Tunable nonlinear piezoelectric vibration harvester

S Neiss, F Goldschmidtboeing, M Kroener and P Woias

Laboratory for Design of Microsystems, IMTEK, University of Freiburg,
Georges-Koehler-Allee 102, 79110 Freiburg, GERMANY

E-mail: sebastian.neiss@imtek.uni-freiburg.de

Abstract. Nonlinear piezoelectric energy harvesting generators can provide a large bandwidth combined with a good resonant power output. However, the frequency response is characterized by a strong hysteresis making a technical use difficult if the hysteresis cannot be compensated. We propose a tuning mechanism that allows both, a compensation of the hysteresis as well as maintaining the optimal work point. The compensation algorithm can reduce the hysteresis to a minimum of only 1.5 Hz and maintain a high energy oscillation in a large frequency window between 53.3 Hz and 74.5 Hz.

1. Introduction

Nonlinear mass-spring systems have found growing interest in recent research for vibrational energy harvesting [1, 2, 3, 4]. In contrast to linear vibrational generators, a nonlinear stiffness is intentionally introduced in the system. By this, such systems inherently comprise a large bandwidth combined with a good resonant power output [1, 2]. This distinguishes them from linear devices which show a contrary behavior of bandwidth and conversion efficiency [3]. As a consequence, generators with a nonlinear restoring force are better suited for energy harvesting from vibrations with a varying dominant frequency compared to a device with a linear stiffness.

2. Theory

Figure 1(a) shows an exemplary frequency response of a nonlinear piezoelectric harvester with softening characteristic. While power output can be high in a large frequency band, the response is characterized by a frequency hysteresis. If the excitation frequency is located in the hysteresis range, the oscillation will always start in the low power regime.

To overcome this limitation, a control mechanism is implemented which is schematically shown in figure 1. Starting from a low power oscillation with a phase shift between piezo voltage and excitation around $0\,^\circ$, the system's stiffness is first reduced to shift the jump-up point to a lower frequency than the excitation frequency. The jump is detected by measuring the change of the phase shift which is now larger than $-90\,^\circ$. Next, the stiffness is adjusted to have the harvester work with a higher power output at a phase shift around $-135\,^\circ$. When the working point is reached, the control keeps the oscillation in the attractor's high power orbit by adjusting the stiffness to maintain the phase shift between $-130\,^\circ$ and $-140\,^\circ$. In contrast to a linear system, the working point is not set to a phase shift of $-90\,^\circ$. In nonlinear systems, the jump-down is located at this point. Consequently, the working point is set to a point with a phase shift less than $-90\,^\circ$ to obtain a certain safety margin during operation.

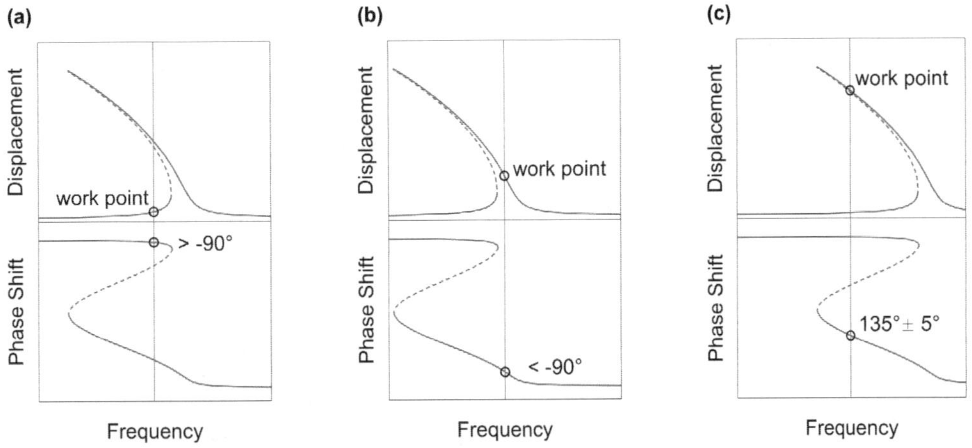

Figure 1. Schematic of the tuning and hysteresis compensation mechanism visualized using the frequency (upper graphs) and phase response (lower graphs). Starting from a low power oscillation (a), the stiffness is reduced until the jump-up occurs (b) and subsequently increased again to set the work point (c).

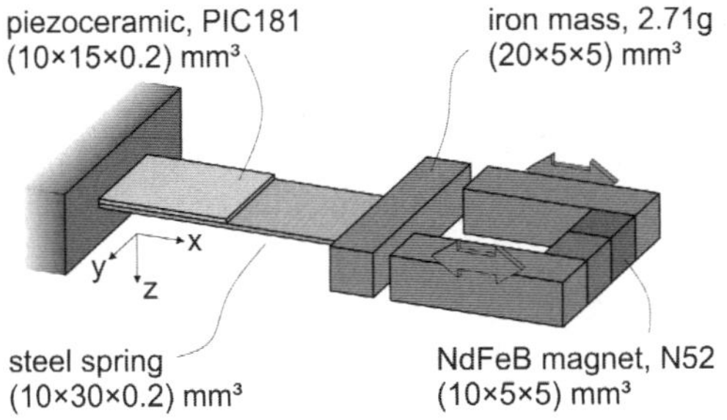

Figure 2. Schematic of the nonlinear harvester. The movable magnetic circuit allows for a change of the resonance frequency.

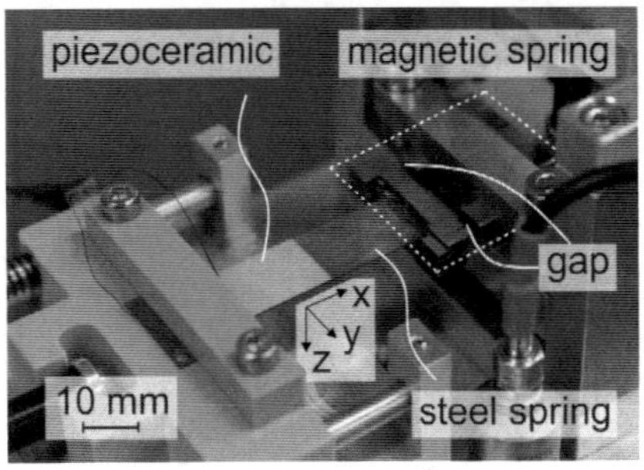

Figure 3. Actual harvester mounted on a motorized base allowing for a change of the gap size between proof mass and magnetic spring.

The proposed mechanism uses a harvester with adjustable stiffness. Thus, we developed the harvester shown in figure 2 and 3. The transducer comprises a steel spring with a piezoceramic sheet attached to its base. In addition, an iron proof mass (2.7 g) is mounted at the tip. A magnets flux is guided through this iron mass using a magnetic circuit which generates a nonlinear softening restoring force on the latter [5]. By changing the gap size between magnetic circuit and proof mass, the strength of the magnetic spring is altered thus changing the systems linear and nonlinear spring constant [6].

3. Dynamic characterization

Figure 4 shows the influence of the excitation amplitude on the open-loop frequency and phase response of the harvester. Stronger excitation leads to an increased non-linearity of the system resulting in an increased hysteresis window. This gives further rise to the need of a control in a nonlinear harvester. On the one hand, a drop of the excitation amplitude shifts the jump-down frequency, and consequently also the optimal working point, to higher frequencies. This can lead to a drastic drop in output power if the jump-down point exceeds the excitation frequency.

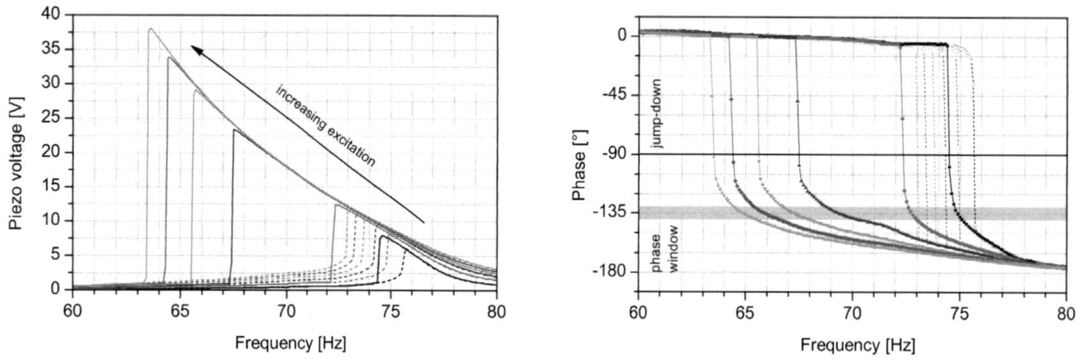

Figure 4. Frequency response (left) and phase response (right) of the harvester at different excitation amplitudes ($0.5 - 3\,\mathrm{m/s^2}$) and a gap size of 0.5 mm. Up-sweeps are indicated by dashed lines, down-sweeps by solid lines.

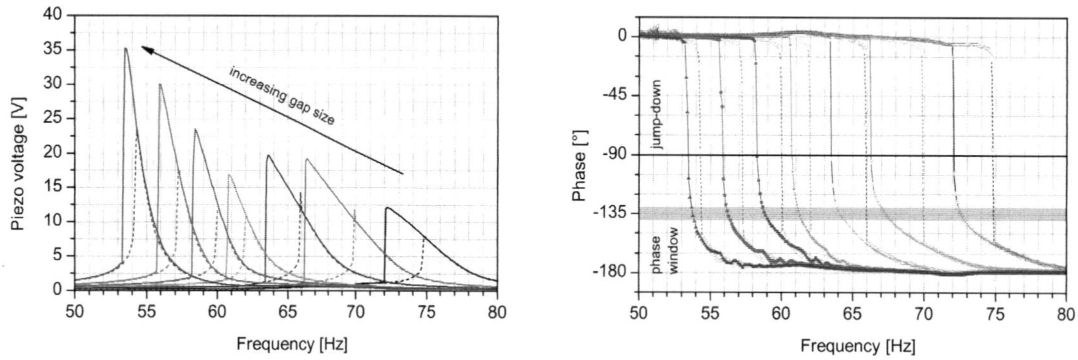

Figure 5. Frequency response (left) and phase response (right) of the harvester at different gap sizes (0.5 – 1.5 mm) and an excitation amplitude of $1\,\mathrm{m/s^2}$. Up-sweeps are indicated by dashed lines, down-sweeps by solid lines.

On the other hand, an increase of the excitation frequency shifts the jump-down point to lower frequencies. Again, the work point will not be at the optimal point any longer.

In a similar sense, figure 5 shows the influence of the gap size on both, frequency and phase response. The measurement proves that the whole frequency response can be shifted along the frequency axis as needed for the implementation of the tuning mechanism by changing the gap size. The measurement also shows that small gap sizes produce a larger hysteresis window thus indicating a stronger non-linearity.

For all measurement conditions, the jump-down point is located at a phase shift of $-90\,°$. An increase of the non-linearity caused by either a change of excitation amplitude or change of gap size only alters the slope of the phase response curve close to the jump-point. Therefore, this parameter is perfectly suited to determine the work point of the system.

4. Tuning and hysteresis control

The tuning mechanism explained above is implemented by a simple algorithm. If the phase shift is larger than $-130\,°$, the frequency response has to be shifted to lower frequencies which is achieved by increasing the gap size. If the phase shift is lower than $-140\,°$, the frequency response is shifted to higher frequencies by decreasing the gap size. For the automatic adjustment of the

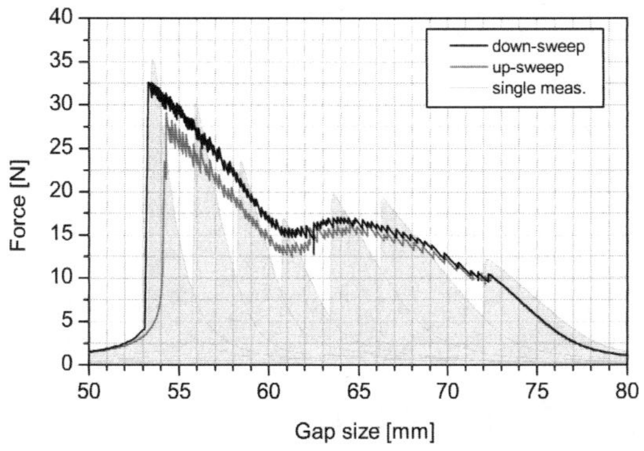

Figure 6. Frequency response of the harvester with activated hysteresis compensation at an excitation amplitude of $1\,{}^{\mathrm{m}}/_{\mathrm{s}^2}$. The control changes the gap size from $0.5-1.5$ mm to maintain a phase shift between $-130\,°$ and $-140\,°$. For comparison, the down-sweep frequency responses from figure 5 are given as well.

gap size, the harvester is mounted on a motorized linear table.

The algorithm in *LabVIEW* to test the usability of this approach. The algorithm itself is known in a similar implementation from linear devices [7]. However, the phase response shows a varying slope around $-135\,°$ depending on the excitation frequency and gap size. As a consequence, safety margin between operating point and jump-down frequency becomes small if either the gap or the excitation amplitude are small. This might be a problem since the oscillation becomes unstable the closer the frequency is located at the jump-down frequency. This gives rise to basically two questions which have to be answered to prove the usability of this method for nonlinear devices as well. First, a sudden change of the spring constant could destabilize the oscillation and lead to a sudden jump-down as soon as the motor is active. In a similar sense, the vibration that is induced by the motor itself could lead to a destabilization as well. This problem can be compensated if the safety margin is increased and the work point is set to a larger phase shift. However, this leads to an unwanted drop in power output.

Figure 6 shows the frequency response of the harvester with activated control at an excitation amplitude of $1\,{}^{\mathrm{m}}/_{\mathrm{s}^2}$. The hysteresis is reduced to a minimum of 1.5 Hz and the high energy oscillation is maintained between 53.3 Hz and 74.5 Hz. The noise in the frequency response is caused by the control itself. If the gap size is changed, the piezo voltage shows a step caused by the change of the working point. However, it is not possible to reduce the safety margin between operating point and jump-down frequency any further. During tests with a work point located at $-125\,° \pm 5$ degree, the activation of the motor occasionally lead to a destabilization of the oscillation. The measurement also shows another advantage of using a nonlinear device if a tuning mechanism is included. The configurations with strong non-linearity need less tuning activity compared to systems with weaker non-linearity.

In the single measurements without control presented above, the power dissipated in a resistive load with $300\,\mathrm{k}\Omega$ ranged from $\approx 130\,\mu\mathrm{W}$ at a gap of 0.5 mm to $\approx 1020\,\mu\mathrm{W}$ at a gap of 1.5 mm. With activated control, the power drops to $\approx 90\,\mu\mathrm{W}$ at a gap of 0.5 mm to $\approx 880\,\mu\mathrm{W}$ at a gap of 1.5 mm. This corresponds to a drop of power of about $15-35\,\%$. It is obvious that the drop is larger for systems with a smaller gap. Theses systems show a stronger non-linearity and consequently, the safety margin between operating point and jump-down point is larger resulting in an emphasized drop of the voltage amplitude.

Whenever a tuning mechanism is included in an energy harvesting device, the tuning energy consumption is a major figure that has to be evaluated and compared to the power delivered by the harvester itself. However, the presented system was built to test the usability of this approach and uses a *LabVIEW* based control in combination with a linear motor. Therefore, we limit the comparison at this point to the evaluation of the energy that is needed to move the harvester against the pulling force of the magnetic spring. Figure 7 shows the measured

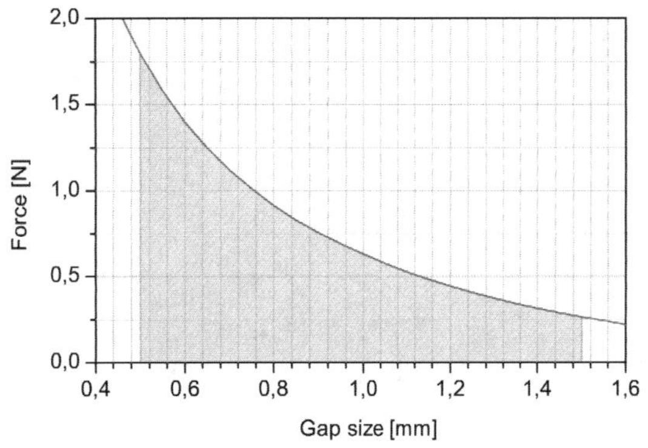

Figure 7. Force acting on the mass in beam direction, i.e. the tuning direction. The energy needed for one tuning cycle from 0.5 mm to 1.5 mm is given by the filled area.

force acting on the mass in beam direction, i.e. the force against which the harvester has to be moved. From this measurement, the energy that is needed to move the harvester from 0.5 mm to 1.5 mm is calculated to 780 µJ. However, the energy that is needed in a real system will be higher due to the limited efficiency of the actuation mechanism. With a reasonable efficiency of around 20 %, the energy would be in the range of a few Millijoule. Compared to the output power of the system, the energy for a full tuning cycle would be harvested in 40 s at maximum. We emphasize at this point that this is a theoretical estimation. Nevertheless, it proves the suitability of the magnetic springs in combination with this tuning mechanism to build a tunable nonlinear energy harvester.

5. Conclusion

The results of the measurements prove the proposed tuning and compensation mechanism to be working. The frequency hysteresis is nearly completely compensated and the harvester works always at an optimal working point. In addition the chosen implementation achieves both, a compensation of the hysteresis as well as maintaining the work point at a phase shift of around $-135\,^\circ$. In addition, the tuning energy is in the same order as the power delivered by the harvester. This allows for an implementation of an autonomous control if a suitable actuation mechanism replaces the current linear motor.

Acknowledgments

We gratefully acknowledge the financial support from the German Research Association (DFG) within the Research Training Group (GRK) 1322 "Micro Energy Harvesting".

References

[1] Ramlan R 2009 *Potential benefits of a non-linear stiffness in an energy harvesting device* (University of Southampton: PhD thesis).
[2] Mann B, Barton D and Owens B 2012 Uncertainty in performance for linear and nonlinear energy harvesting strategies *J. Intell. Mater. Syst. Struct.* **23** pp 1451-60.
[3] Zhu D, Tudor M and Beeby S 2010 Strategies for increasing the operating frequency range of vibration energy harvesters: a review *Meas. Sci. Technol.* **21** pp 022001.
[4] Ferrari M, Ferrari V, Guizzetti M, And B, Baglio S and Trigona C 2010 Improved energy harvesting from wideband vibrations by nonlinear piezoelectric converters *Sensor. Actuat. A Phys.* **162** pp 422-31.
[5] Furlani E P 2001 *Permanent Manget and Electromechanical Devices* (London: Academic Press).
[6] Neiss S, Kleber J, Woias P and Kroener M 2012 Reluctance Springs for Nonlinear Energy Harvesting Generators *Proc. PowerMEMS (Atlanta, GA)* Dec 02-05 2012 pp 153-156.
[7] Eichhorn C, Tchagsim R, Wilhelm N and Woias P 2011 A smart and self-sufficient frequency tunable vibration energy harvester *J. Mircomech. Microeng.* **21** pp 104003.

Experimental Validation of Damping Model for a MEMS Bistable Electrostatic Energy Harvester

C H Nguyen[1], D S Nguyen[2] and E Halvorsen[1]

[1] Department of Micro- and Nanosystem Technology, Buskerud and Vestfold University College, Raveien 215, N-3184 Borre, Norway

[2] Department of Mechanical Engineering, University of California, Berkeley, USA

E-mail: Einar.Halvorsen@hbv.no

Abstract. This paper validates lumped models of an asymmetric bistable MEMS electrostatic energy harvester against measurements. A conventional model of constant damping coefficient turns out to be ineffective in predicting or reproducing the device response. This shortcoming is demonstrated by the effective damping coefficient obtained from fits to experimental results being far from constant under varying operating conditions. Therefore, two different nonlinear models of the damping force in polynomial form are introduced and investigated. We find that the experimental results are well reproduced over the entire range of measured acceleration amplitudes by modeling with a phenomenological nonlinear damping force which is a high-order polynomial in the velocity.

1. Introduction

In designing a vibration energy harvester, it is challenging to make the generator adapt to the wide range of ambient vibration spectra. Nonlinearity of suspensions has been exploited to extend the working frequency range of the energy harvesters. In this approach, advantage is taken of the nonlinear dynamical behavior resulting from softening spring, hardening spring or bistable character of the system stiffness [1–3]. For example, an electrostatic MEMS device with an asymmetric bistable suspension showed an experimental bandwidth up to 715 Hz centered at 700 Hz [4]. In [5], 2.5 times wider bandwidth was achieved in a bistable electromagnetic generator than in a monostable one. It can often be difficult to obtain close agreement with experimental responses when modeling these bistable devices.

Damping is important to the bandwidth because it strongly affects the jump frequencies in frequency sweeps [6]. Therefore, understanding the behaviour of damping against a variety of working conditions is needed for device performance prediction. In this work, a bistable electrostatic energy harvester is characterized and modelled. The common model with a linear damping force [7] misses essential physics in the device. Motivated by this, we introduce instead nonlinear damping-force models in the form of polynomials in velocity and displacement. We have investigated to which degree these different models of parasitic mechanical damping result in simulated responses that agree with experiments on an asymmetric, bistable MEMS electrostatic energy harvester.

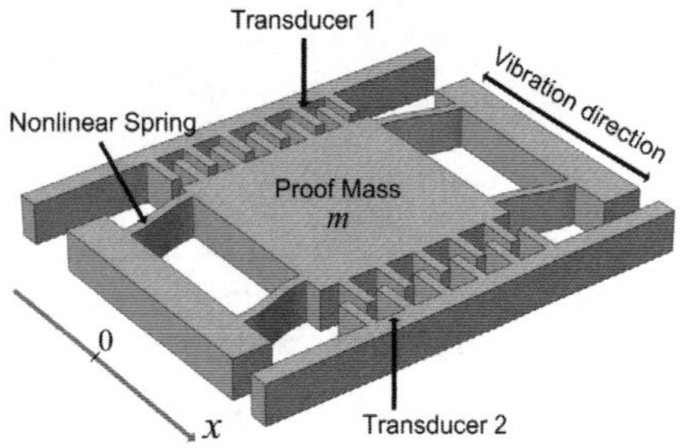

Figure 1. Schematic drawing of the bistable electrostatic energy harvester. The two capacitive transducers have different initial overlap areas.

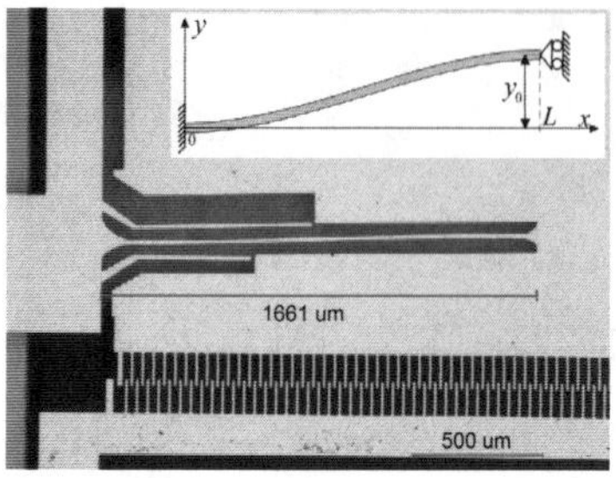

Figure 2. A corner of the device with a nonlinear spring and a part of transducer 2. Inset: drawing of the nonlinear spring $y_0 = 42\ \mu m$ and length $L = 1660\ \mu m$

2. Device Description and Characterisation

2.1. Bistable design of an electrostatic energy harvester

The harvester has a proof mass suspended by four springs as shown in figure 1. Taking advantage of the available space, two transducers, one on each side of the proof mass, are included in the generator. To obtain the bistable properties, the springs are designed to have a curved shape as shown in figure 2. The S-shaped springs have an initial displacement y_0 at the guided ends. The spring force vs. displacement curve, obtained by fitting the modeling to measurements, is shown in figure 3. For small displacements, the springs behave linearly around the 1st stable position at 0. For further displacement in the positive direction, the spring is stiffer (or hardening) which is mainly due to tension. In the reverse direction, the springs are partly in compression resulting in a stiffness reduction or softening effect over part of the displacement range when some of the elastic energy is released. Decreasing the displacement from zero, the stiffness continues to decrease and reaches the instability point at the displacement of -30 μm and then the stable point at the displacement of -60 μm.

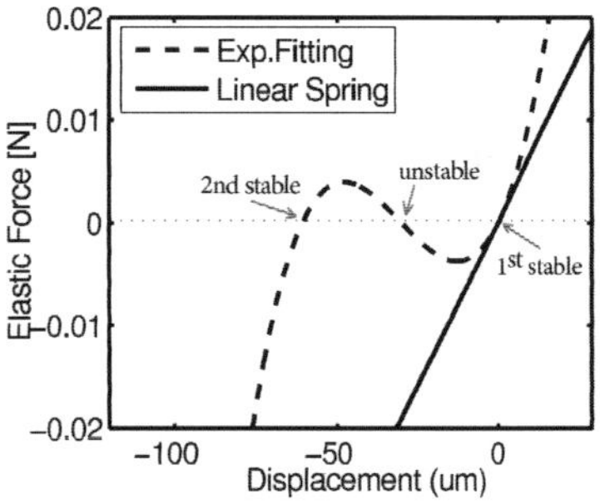

Figure 3. Nonlinear spring characteristic obtained from parameter fit. Linear spring is plotted for comparison

2.2. Characterisation method and results

For characterisation purposes, an external DC voltage is used for biasing of the device, a chirp excitation is taken as the input vibration source while the outputs of the generator are directly connected to resistive loads R_{L1}, R_{L2} at transducer 1 and transducer 2 respectively. The device is first characterized at several low acceleration amplitudes around 0.01 g to yield linear responses. Then, parameters such as the linear stiffness k_1, damping coefficient b, parasitic capacitance C_p and load capacitance C_L are determined from output voltage information, i.e. the resonant

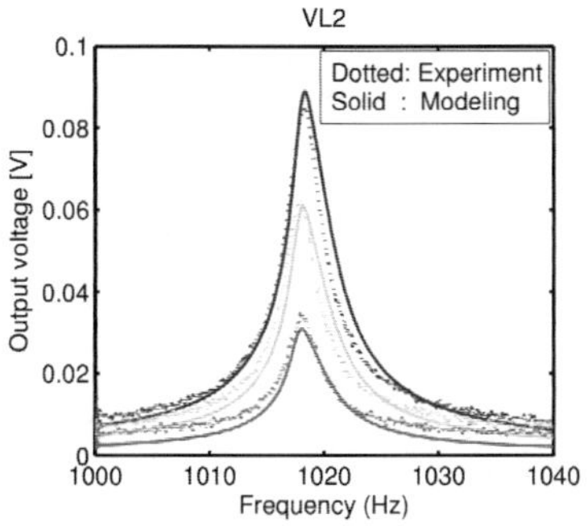

Figure 4. Output voltage at transducer 2 for acceleration amplitude of 0.01 g (linear regime) and bias voltage of 30 V, 20 V and 10 V (top to bottom)

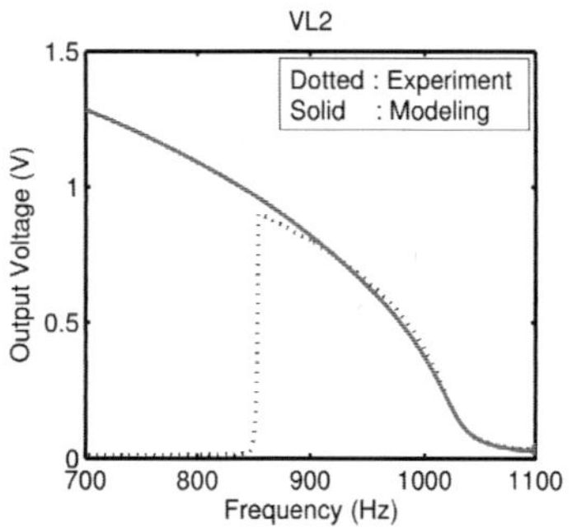

Figure 5. Output voltage at transducer 2 for acceleration amplitude of 0.28 g (nonlinear regime) and bias voltage of 20V.

frequency $f_0 = \frac{\omega_0}{2\pi} = \frac{1}{2\pi}\sqrt{\frac{k_1}{m}}$ and the quality factor $Q = \frac{f_0}{\Delta f} = \frac{m\omega_0}{b}$, where Δf is the bandwidth of the response. Next, in a range of higher acceleration amplitudes from 0.1 g to 0.3 g causing strong nonlinear effects, the higher-order stiffness parameters are obtained by fitting the output response from the device model with results from measurement. In this procedure, the previously established linear parameters are unchanged. The mass m and the dominant capacitor C_0 are calculated using the layout dimensions.

The parameters of the device are shown in Table 1. The linear behaviour of the generator is presented in figure 4 while the nonlinear one is shown in figure 5. As can be seen in the linear regime, the modeling results and experimental data are close with a small deviation far from the resonant frequency. The small deviation is due to noise in the measurement system. However, in the nonlinear regime, the amplitude of the voltage curve can only be fitted over a limited frequency range. The jump-down frequency is far from the observed value in the experiment. The same phenomenon was also observed for nonlinear piezoelectric energy harvester in [6].

The reason for this mismatch is that the linear damping is retained in our model for the nonlinear regime while in reality the situation is more complex. Experimentally this explanation is supported by the observation that the coefficient of the linear damping model needs to be adapted to the operation condition, i.e. acceleration amplitude and bias voltage, in order to reproduce the experiment results. This will be discussed in the next section.

3. Nonlinear Damping and Modeling

Figure 6 shows the effective linear damping value fitted to reproduce the experimental results for different conditions of the acceleration amplitude and the bias voltage. The effective linear damping is approximately linearly increasing when increasing the acceleration amplitude from 0 to 0.16 g and relatively constant beyond 0.16 g. We note that the effective damping curve depends on bias voltage, and that in particular the 30-V curve distinguishes itself from the 10-V and 20-V curves. We interpret this voltage dependence as a signature of inaccuracies in the representation of electrical damping which have little effect at 10-V and 20-V bias. We therefore conduct the remaining analysis at 20-V bias.

To describe the damping force that can give rise to effective damping coefficients as shown in figure 6, we introduce two models of the nonlinear damping force. One is a high-order polynomial in velocity and one is a third order form which also has displacement dependence:

Table 1. Bistable Electrostatic Energy Harvester Parameters

Notation	Description	Value	Unit
b	Linear damping coefficient	2.7x10^{-4}	Ns/m
m	Mass	15.18	mg
k_1	Linear stiffness	622	N/m
$k_{i=2:7}$	Nonlinear stiffness	2.96x10^7 , 2.8x10^{11} , -4x10^{14} 1.06x10^{19}, 1.36x10^{23}, 4.2x10^{26}	Nm^{-i}
C_0^1, C_0^2	Transducer dominant capacitance	7.07, 1.75	pF
C_p^1, C_p^2	Transducer parasitic capacitance	14.5, 12.5	pF
C_L^1, C_L^2	Parasitic load capacitance	23, 8	pF
R_{L1}, R_{L2}	Load resistance	20.5, 20.5	MΩ

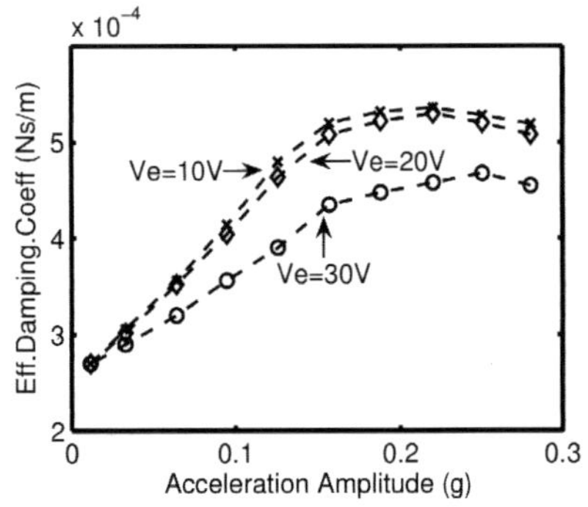

Figure 6. Effective damping coefficient extracted from measurements

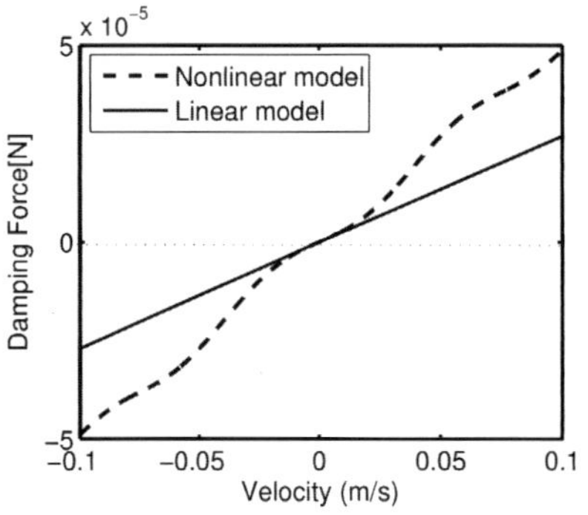

Figure 7. Nonlinear damping force obtained from parameter fit, compared to linear damper.

$$F_{nl\_damp}(\dot{x}) = b\dot{x} + \sum_{n=1}^{N} b_{2n+1}(\dot{x})^{2n+1} \tag{1}$$

$$F_{nl\_damp}(x, \dot{x}) = b\dot{x} + b_{30}(\dot{x})^3 + b_{32}x^2\dot{x} \tag{2}$$

where $b_3 = 0.234\mathrm{N(s/m)}^3$, $b_5 = -69.26\mathrm{N(s/m)}^5$, $b_7 = 8.73 \times 10^3\mathrm{N(s/m)}^7$, $b_9 = -4.98 \times 10^5\mathrm{N(s/m)}^9$, $b_{11} = 1.06 \times 10^7\mathrm{N(s/m)}^{11}$ and $b_{30} = 0.274\mathrm{N(s/m)}^3$, and $b_{32} = -1.24 \times 10^7\mathrm{Ns/m}^3$.

The form in (2) has been adopted from the analytic work on modeling of the nonlinear damping in a micromechanical oscillator in [8]. The high-order polynomial for the purely velocity-dependent nonlinear damping in (1) is mathematically motivated by the need capture a wide range of conceivable function forms. The damping force resulting from a fit to (1) is presented in figure 7 and is very different from the simple linear model $F = b\dot{x}$. The slope change in the force-velocity relation is made possible by using a 11th order of polynomial. This form is able to reproduce the experimental results over a wide range of acceleration amplitudes as shown by the agreement with the experimental results in figure 8. The form (2) gives significantly worse fits as seen from the mismatch in the responses already at moderate acceleration amplitudes, i.e. above 0.19 g, in figure 9.

It is not clear what is the mechanism behind the nonlinear damping. However, this device has a low linear damping, i.e. $Q \approx 350$, due to the large gap of 500 μm between the mass and the substrate, and due to the even larger distance to the lid. Thus, gas flow in the finger

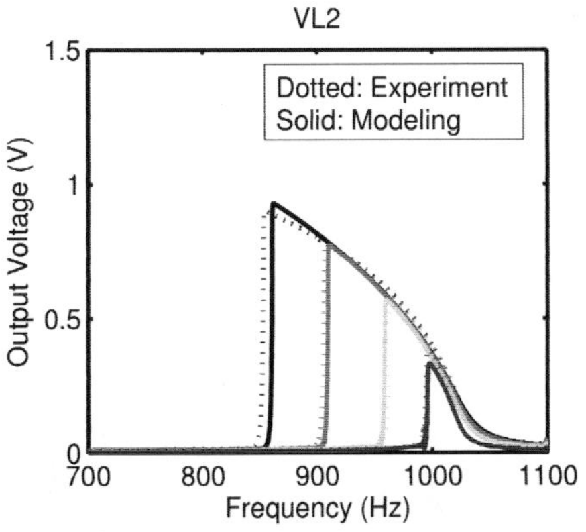

Figure 8. Output of transducer 2 with 11th order polynomial damping force for acceleration amplitudes of 0.28 g, 0.25 g, 0.19 g, 0.1 g (left to right), bias voltage of 20 V.

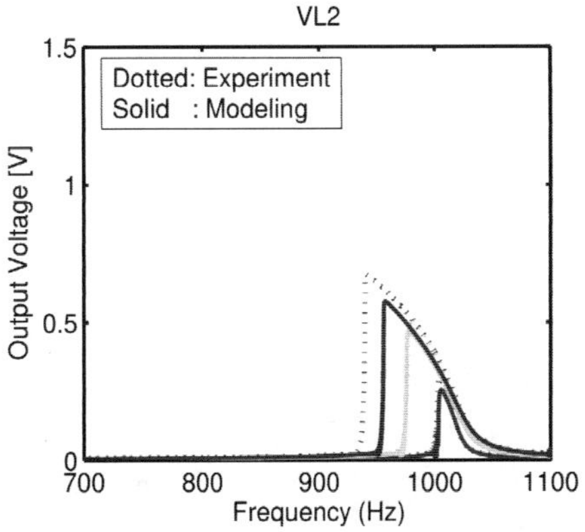

Figure 9. Output of transducer 2 with third-order velocity- and displacement-dependent damping force for acceleration amplitudes of 0.22 g, 0.19 g, 0.1 g (left to right), and bias voltage of 20 V.

structure is important for the damping, and could potentially include nonlinear effects at large displacements, which cause the gas volume within the finger structure to change significantly.

4. Conclusion

This paper investigated the behaviour of a bistable electrostatic energy harvester. It was demonstrated that the model of linear damping is not enough to reproduce the frequency response of this kind of device in the nonlinear regime. A systematic experiment was set up and revealed the dependency of the effective damping coefficient on excitation level and bias voltage. The bias-voltage dependence was used to indicate at which bias voltages one could expect inaccuracies in the electrical damping model to be insignificant for the mechanical damping estimates. Two different nonlinear damping models were fit to the experiment. A high-order expansion in velocity could successfully reproduce the experimental results. While this purely mechanical nonlinear damping can explain the device behavior as a function of acceleration amplitude, a physical explanation of the nonlinear damping is yet to be established and is a matter for further investigation.

Acknowledgments

This work was sponsored in part by the California Energy Commission under Grant no. 500-10-044, and in part by the Research Council of Norway under Grant. no. 229716/E20.

References

[1] A. Erturk, D. J. Inman, 2011 *J. Sound. Vib.* **330** pp 2339-2353
[2] S. C. Stanton, C. C. McGehee, B. P. Mann, 2010 *Physica D* **239** pp 640-653
[3] F. Cottone, L. Gammaitoni, H. Vocca, M. Ferrari, V. Ferrari 2012 *Smart Mater. Struc.* **21** 035021
[4] S. D. Nguyen, E. Halvorsen, I. Paprotny 2013 *Appl. Phys. Lett.* **102** 023904
[5] F. Cottone, P. Basset, H. Vocca, L. Gammaitoni, T. Bourouina 2014 *J.Intell. Mater Syst. Struct.* **25**
[6] G. Sebald, H. Kuwano, D. Guyomar, B. Ducharne 2011 *Smart. Mater. Struc.* **20** 10200
[7] S. D. Nguyen, E. Halvorsen 2009 *Proc. PowerMEMS* pp 411-414
[8] S. Zaitsev, O. Shtempluck, E. Buks, O. Gottlieb 2012 *Nonlinear. Dyn* **67** pp 859-883

Wideband Piezoelectric Energy Harvester Tuned Through its Electronic Interface Circuit

Adrien Badel[1,3], Elie Lefeuvre[2]

[1]Univ Savoie, SYMME, 74000 Annecy, France
[2]Univ Paris-Sud - CNRS, Institut d'Electronique Fondamentale, 91405 Orsay, France

adrien.badel@univ-savoie.fr

Abstract. This paper reports a novel technique enabling to tune the resonant frequency of linear inertial piezoelectric energy harvesters through the control strategy of the associated electronic interface circuit. When associated with piezoelectric devices exhibiting high electromechanical coupling, it is expected to enable the variation of the resonant frequency in large proportions.

1. Introduction

During the last decade, devices that can generate electrical power by exploiting ambient vibration energy have been developed as possible solution to power remote wireless sensors [1]. A large part of these developments use linear mechanical oscillators associated with piezoelectric transducers. Most of these linear piezoelectric energy harvester (PEH) exhibit squared coupling coefficient lower than 10%. They can efficiently operate in a narrow frequency bandwidth tuned to match the excitation frequency.

However, environmental excitations have broadband or time-dependent characteristics in which the energy is distributed over a spread spectrum of frequencies. Two strategies have been investigated: Developing nonlinear wideband oscillators and developing linear oscillator with resonance frequency tuning mechanisms. Nonlinear oscillators (hardening, softening or bistable) are well suited for broadband vibration spectrum [2], [3], whereas resonance frequency tuning is more appropriated for narrow band but time-dependent vibrations (which can be found on a motor whose rotation speed is varying). Frequency tuning is usually done through an additional component that passively or actively changes the stiffness or inertia of a linear mechanical oscillator [4].

This paper reports a new approach to tune the resonant frequency of linear inertial piezoelectric energy harvesters through the control strategy of the associated electronic interface circuit. When associated with piezoelectric devices exhibiting high electromechanical coupling, it enables to vary the resonant frequency in large proportions without any additional component.

Seddik et al previously proposed a principle of control through the electronic interface circuit [5]. It consisted in connecting shunt capacitors to the piezoelectric device. Consequently, the electromechanical structure stiffness was varied, and the resonant frequency was changed accordingly. They showed that it was possible to significantly vary the resonant frequency of highly coupled piezoelectric energy harvesters. However, because the shunt capacitance was varied step by step, the

[3] To whom any correspondence should be addressed.

resonant frequency could not be continuously tuned. With the proposed technique, it is expected that a continuous tuning is achievable on a noticeably larger frequency range.

2. High coupling coefficient piezoelectric energy harvester

A very high electromechanical coupling PEH was designed and manufactured for experimental validation of the proposed technique (Figure 1). The deformable part is composed of two 10x5x0.5mm^3 plates of PZN-PT single crystals bonded on both sides of a 100μm thick brass shim. The inertial mass is composed of two 40x5x2.5mm^3 steel plates bonded on both sides of the shim. The PZN-PT plates were manufactured by Microfine Company (Singapore). They are poled along the [011] axis and they exhibit a very high transverse coupling coefficient k_{31} of 0.92 along the [100] axis.

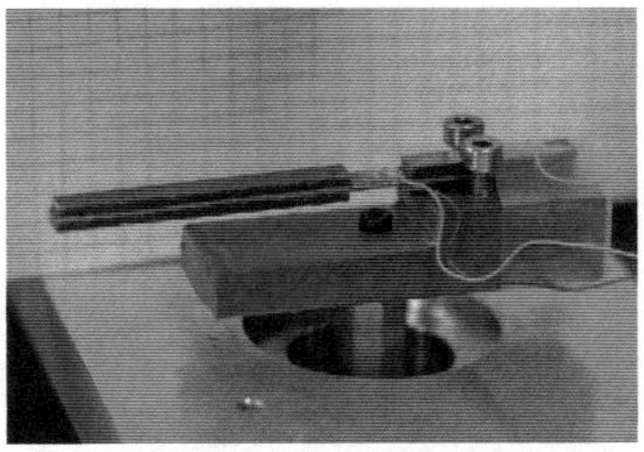

Figure 1. Photograph of the piezoelectric device with high electromechanical coupling

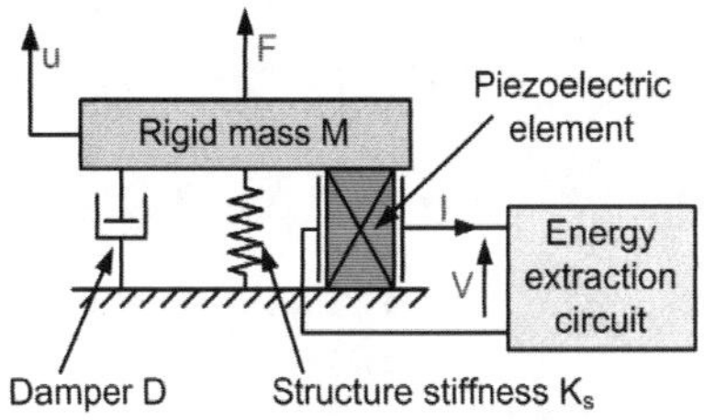

Figure 2. Model of the PEH

For a classical PEH working near to its resonance frequency, a system with a single degree of freedom [mass M + spring K_S + damper D + piezoelectric element] provides a relevant description (Figure 2). The corresponding governing equations are given by equations (1) and (2).

$$M\ddot{u} = F - K_{SC}u - D\dot{u} - \alpha V \tag{1}$$

$$I = \alpha\dot{u} - C_P\dot{V} - VR_P^{-1} \tag{2}$$

where F, u, V, I refer to the driving force, the displacement, the piezoelectric voltage and the current respectively. In the case of a vibration driven PEH, $F=Ma$, where a is the ambient acceleration. α stands for the piezoelectric force-voltage coefficient; C_P and R_P refer to the clamped capacitance and the dielectric losses respectively. K_{SC} is the global equivalent stiffness of the system encompassing the host structure K_S and the piezoelectric element stiffness.

If the PEH is not subjected to any external solicitation, the theoretical complex admittance can be obtained from (1) and (2):

$$\underline{Y}(\omega) = -\frac{\underline{I}}{\underline{V}} = \mathrm{i}C_P\omega\left(\frac{\frac{k^2}{1-k^2}}{1-\frac{\omega^2}{\omega_0^2}+\mathrm{i}\frac{\omega}{Q_m\omega_0}} + 1 + \mathrm{i}\xi_e\frac{\omega_0}{\omega}\right) \tag{3}$$

where ω_0 is the characteristic angular frequency of the short circuited PEH, k^2 is the squared coupling coefficient, Q_m is the mechanical quality factor of the structure and ξ_e is the dielectric losses coefficient. Experimental admittance diagrams of the device are represented on Figure 3, and compared with simulated ones obtained from an identified linear model. The expression of ω_0, k^2, Q_m and ξ_e, as well as there identified values, are given in Table 1.

The very good agreement between experimental and simulated admittance confirm the validity of the linear model, at least in the 40Hz – 90 Hz frequency range.

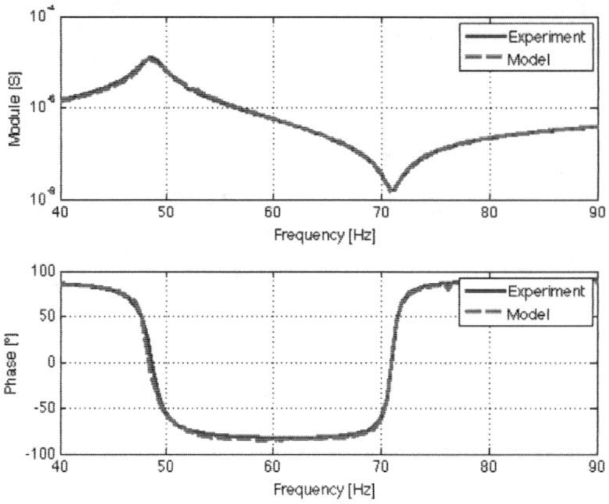

Figure 3. Module and phase of the piezoelectric device admittance (Model and experimental results)

Table 1. Expression and identified values of the PEH parameters

Parameter	Expression	Identified value
C_P	–	1.29nF
ω_0	$K_{SC}^{1/2}M^{-1/2}$	305 rad/s
k^2	$\alpha^2\left(\alpha^2+K_{SC}C_P\right)^{-1}$	53%
Q_m	$\left(MK_{SC}\right)^{1/2}D^{-1}$	28
ξ_e	$\left(R_PC_P\omega_0\right)^{-1}$	0.02

3. A new wideband energy extraction circuit dedicated to high coupling coefficient PEH

The proposed approach is derived from the previously developed SECE (Synchronized Electrical Charge Extraction) approach [6]. It is called FTSECE for Frequency Tuning SECE. It consists in letting the PEH in open circuit condition most of the time, and to extract the generated electrical charges two times a period of vibration. In contrast to the SECE technique, the FTSECE circuit does not extract energy at the piezoelectric voltage extremum, but with a phase shift ϕ. Moreover, whereas the piezoelectric voltage is null after each energy extraction phase in the SECE approach (all the electrical charges are extracted), it can be tuned with the FTSECE technique.

A possible electronic interface circuit for implementing the approach is schematically depicted in Figure 4. This circuit includes an electronic switch (SW), whose control is synchronized with the piezoelectric voltage. Two parameters of SW are varied through the control circuit: the ON state time duration (t_{ON}) and the lag time duration (t_{lag}) between the voltage extrema and the instant where SW is turned ON. The phase shift $\phi = \omega t_{lag}$ can be positive (SW is closed after the voltage extremum) or negative (SW is closed before the voltage extremum). If V_M is the piezoelectric voltage at the moment where SW is turned on, varying t_{ON} will allow to tune the voltage after the energy extraction phase between V_M ($t_{ON} = 0$) and $-\gamma V_M$ ($t_{ON} \simeq \pi\,(LC_P)^{1/2}$), where γ is the inversion coefficient of the LC_P electrical oscillating circuit.

The load resistance value has no influence on the harvested power provided that the DC load voltage is larger than the piezoelectric voltage amplitude. In this way, energy transfers from the inductor to the rectifier only occur right after SW is turned OFF. Varying t_{ON} enables to tune the amount of electrical energy extracted from the piezoelectric elements, and, if needed, to reverse the polarity of the piezoelectric voltage. Varying t_{lag} enables to modify the electromechanical structure stiffness by tuning the phase shift between the piezoelectric voltage and the strain in the piezoelectric element.

Typical waveforms for the FTSECE approach are given in Figure 5. The voltage on the piezoelectric elements can be expressed by:

$$V = \frac{\alpha}{C_P}(u+h) \tag{4}$$

where u and h are respectively given by :

$$\begin{cases} u = U_M \cos(\omega t) \\ h = -\dfrac{1-\beta}{1+\beta} U_M \cos(\phi) \mathrm{sqwave}(\omega t - \phi) \end{cases} \tag{5}$$

In (5), sqwave(t) is a square wave function whose amplitude is 1 and which as the same sign as sin(t). β is a coefficient that can be varied between 1 and $-\gamma$ tuning t_{ON}. Since the PEH is a resonant linear system, a first harmonic analysis is carried out to determine the amplitude U_M of the inertial mass. The first harmonic of h is given by:

$$\hat{h} = -\frac{4}{\pi}\frac{1-\beta}{1+\beta} U_M \cos(\phi)\sin(\omega t - \phi) \tag{6}$$

The expression of $\hat{h}$ in the Fourier domain can then be expressed as:

$$\underline{\hat{h}} = \frac{2}{\pi}\frac{1-\beta}{1+\beta}\sin(2\phi)\underline{u} + \mathrm{i}\frac{4}{\pi}\frac{1-\beta}{1+\beta}\cos^2(\phi)\underline{u} \tag{7}$$

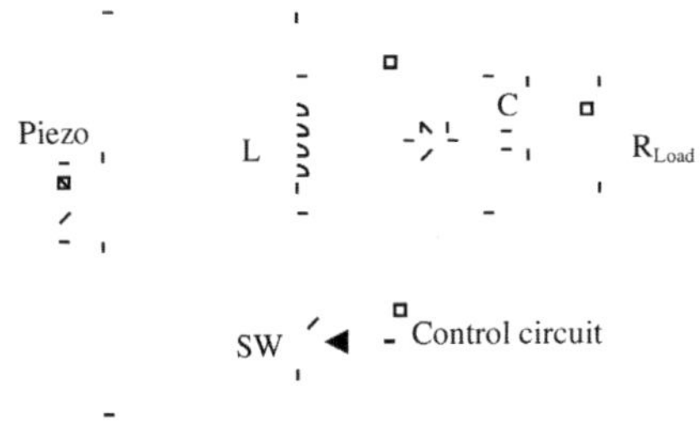

Figure 4. Schematic representation of the FTSECE electronic interface circuit

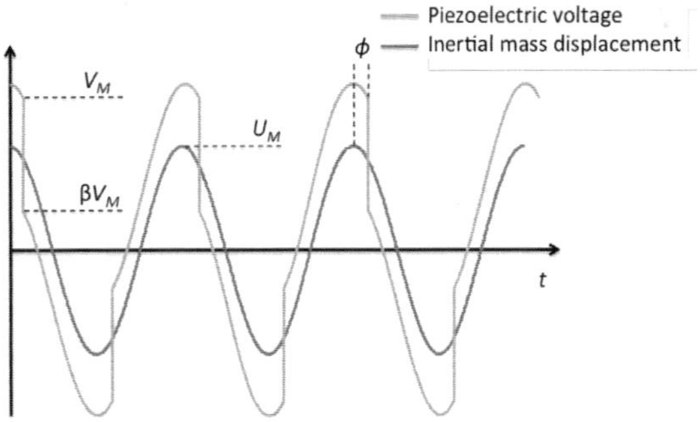

Figure 5. Typical waveforms for the FTSECE approach

From equation (1) written in the Fourier domain, substituting the voltage V by its expression (4) and using the expression (7) of $\hat{h}$ leads to the expression of the displacement of the inertial mass as a function of the ambient acceleration a:

$$\underline{u} = \frac{M\underline{a}}{K - M\omega^2 + \dfrac{\alpha^2}{C_P}\left(1 + \dfrac{2}{\pi}\dfrac{1-\beta}{1+\beta}\sin(2\phi)\right) + \mathrm{i}\left(D\omega + \dfrac{4}{\pi}\dfrac{1-\beta}{1+\beta}\dfrac{\alpha^2}{C_P}\cos^2(\phi)\right)} \tag{8}$$

It can be seen in (8) that ϕ and β (t_{lag} and t_{ON}) affect both the resonance frequency of the system and its damping. The change in the damping is related to the energy conversion. The power extracted from the PEH is calculated from the energy extracted two times a period of the ambient vibration:

$$P_{EX} = \frac{\omega}{2\pi} C_P V_M^2 \left(1-\beta^2\right) \tag{9}$$

The expression of V_M is given by (10) where U_M can be obtained from (8).

$$V_M = \frac{\alpha}{C_P} U_M \cos(\phi)\frac{2}{1+\beta} \tag{10}$$

Taking into account the losses in the inductor L the harvested power can be approximated by:

$$P_H = P_{EX}\frac{\gamma+\beta}{1+\beta} \tag{11}$$

The simulated harvested power using the proposed approach is given in Figure 6 in the case of a 0.3G sinusoidal ambient acceleration. For comparison purpose, the harvested power on a matched resistance is also given. A maximal power of 480μW is obtained over a 32Hz bandwidth. It can be seen that the maximal harvested power is slightly lower than the one obtained on a matched resistance. This is due to the losses in the inductor. Figure 7 shows the evolution of ϕ and β as a function of the frequency. Around 71Hz, which corresponds to the open-circuit resonance frequency of the PEH, only a small amount of the generated electrical charges are extracted and the extraction times coincide with the voltage extrema. When the vibration frequency is slightly shifted away from 71Hz, the switching phase shift ϕ is adjusted to tune the resonance frequency of the PEH. This phase shift induces a decrease of the generated electrical charges, which is compensated by extracting a larger percentage of them. If the vibration frequency is lower than 66Hz or larger than 75Hz, β becomes negative, which means that a portion of the generated electrical charges is injected back to the PEH with a reversed polarity. This effect increases the piezoelectric voltage amplitude and enhances the energy extraction.

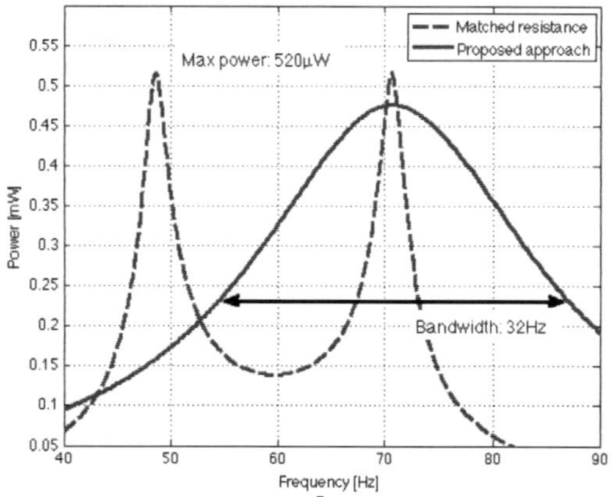

Figure 6. Simulated harvested power versus the vibration frequency at 0.3G RMS (plain line: proposed approach, dotted line: power dissipated in a simple resistor whose resistance is matched to the excitation frequency)

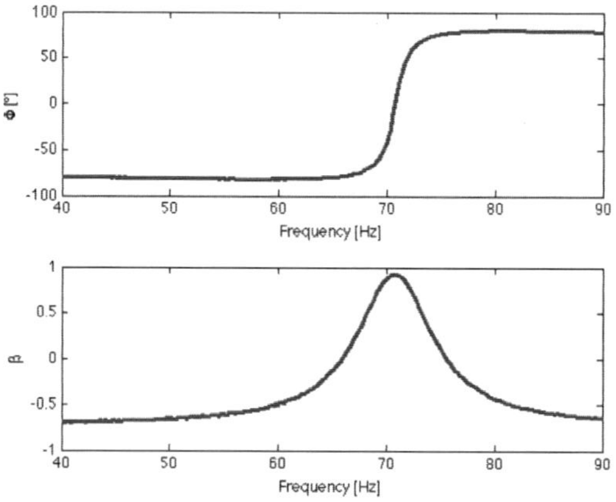

Figure 7. Evolution of the optimal value of ϕ and β as a function of the frequency

4. Conclusion

This paper presents the FTSECE technique. Combined with a high coupling coefficient linear PEH, it is expected to drastically enhance its bandwidth. This approach has not been experimentally validated yet, but the simulated results are supported by the very good agreement between the measured and modeled admittance.

References

[1] J. A. Paradiso and T. Starner, "Energy Scavenging for Mobile and Wireless Electronics," *IEEE Pervasive Computing*, vol. 4, no. 1, pp. 18–27, 2005.

[2] L. Gammaitoni, I. Neri, and H. Vocca, "Nonlinear oscillators for vibration energy harvesting," *Appl. Phys. Lett.*, vol. 94, no. 16, p. 164102, 2009.

[3] W. Q. Liu, A. Badel, F. Formosa, Y. P. Wu, and A. Agbossou, "Novel piezoelectric bistable oscillator architecture for wideband vibration energy harvesting," *Smart Mater. Struct.*, vol. 22, no. 3, p. 035013, 2013.

[4] C. Eichhorn, R. Tchagsim, N. Wilhelm, and P. Woias, "A smart and self-sufficient frequency tunable vibration energy harvester," *J. Micromech. Microeng.*, vol. 21, no. 10, p. 104003, 2011.

[5] B. Ahmed Seddik, G. Despesse, and E. Defay, "Autonomous wideband mechanical energy harvester," presented at the Industrial Electronics ISIE, IEEE International Symposium on, Hangzhou, China, 2012.

[6] E. Lefeuvre, A. Badel, C. Richard, and D. Guyomar, "Piezoelectric Energy Harvesting Device Optimization by Synchronous Electric Charge Extraction," *JIMSS*, vol. 16, no. 10, pp. 865–876, 2005.

A Miniature Coupled Bistable Vibration Energy Harvester

D Zhu[1]**, D C Arthur and S P Beeby**
Electronics and Computer Science, University of Southampton, Southampton, SO171BJ, UK

E-mail: dz@ecs.soton.ac.uk

Abstract. This paper reports the design and test of a miniature coupled bistable vibration energy harvester. Operation of a bistable structure largely depends on vibration amplitude rather than frequency, which makes it very promising for wideband vibration energy harvesting applications. A coupled bistable structure consists of a pair of mobile magnets that create two potential wells and thus the bistable phenomenon. It requires lower excitation to trigger bistable operation compared to conventional bistable structures. Based on previous research, this work focused on miniaturisation of the coupled bistable structure for energy harvesting application. The proposed bistable energy harvester is a combination of a Duffing's nonlinear structure and a linear assisting resonator. Experimental results show that the output spectrum of the miniature coupled bistable vibration energy harvester was the superposition of several spectra. It had a higher maximum output power and a much greater bandwidth compared to simply the Duffing's structure without the assisting resonator.

1. Introduction

Vibration energy harvesters collect ambient vibrations in the environment around the device and convert them into electrical energy [1]. The main hurdle preventing linear vibration energy harvesters from widespread use is that they only harvest energy efficiently when the device's resonant frequency matches the frequency of the vibration. Significant research has been done to increase the operational bandwidth of vibration energy harvesters and multiple approaches have been created [2]. One of the most promising ways to widen the bandwidth is to use a bistable structure [3].

A bistable structure is a type of nonlinear structure that consists of two potential wells. These two potential wells enable the inertial mass to jump between the two equilibrium positions under external mechanical vibrations regardless of their frequencies. It can be realised with a pre-stressed structure or by applying an external nonlinear force to a linear structure. Conventional bistable structures have been found to have broader operating frequency ranges than linear structures and harvest more energy than linear structures under white noise excitation [4]. Zhu and Beeby [5] proposed a coupled bistable structure consisting of two cantilevers with a repelling force created by permanent magnets between them. An energy harvester [6] based on the coupled bistable structure was later realised and proved that such harvester require lower excitation to trigger bistable operation and to operate significantly better than conventional bistable harvesters and linear energy harvesters under wideband vibrations.

This paper presents miniaturisation of a coupled bistable vibration energy harvester. The operational principle is first introduced followed by the design and experiment of a prototype device. Test results are then presented and discussed.

[1] To whom any correspondence should be addressed.

2. Proposed miniature coupled bistable energy harvester

Figure 1 shows the cross-section view of the proposed cylinder miniature coupled bistable energy harvester. It consists of two parts, i.e. a Duffing's nonlinear structure and an assisting resonator. A plastic tube is located in the middle of the energy harvester. Magnets are fixed to both ends of the tube. A shuttle magnet in the tube can move freely vertically. These three magnets are oriented so that the two end magnets repel the shuttle magnet. This forms a typical Duffing's nonlinear structure. The assisting resonator consists of a ring spring with a pair of magnets attached. The two end magnets and those on the assisting resonator generate two potential wells and the shuttle magnet is always trapped in one of them. It is also used to reduce the threshold acceleration that triggers the bistable operation. Two coils are placed above and beneath the assisting resonator. When the shuttle magnet moves vertically, voltage can be induced in both coils. Figure 2 shows the integration concept of the harvester. Please note that magnets on the assisting resonator are not shown.

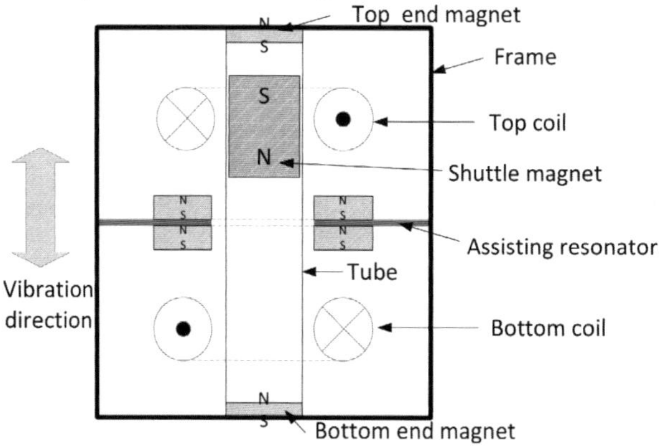

Figure 1. Structure of the miniature coupled bistable energy harvester.

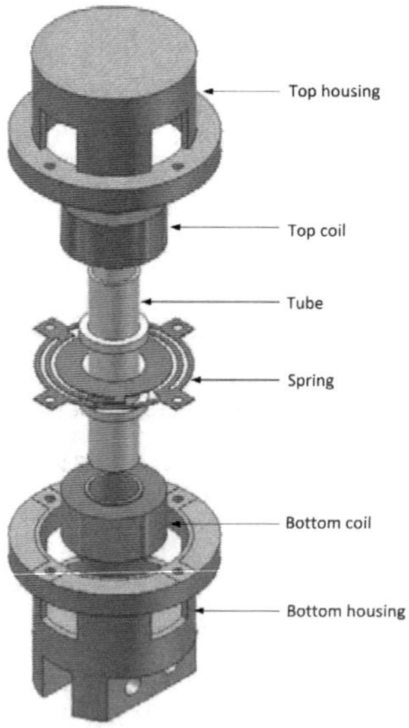

Figure 2. Integration concept of the harvester.

3. Experiment

3.1. Test setup

The proposed energy harvester is a cylinder device. It has a diameter of 40 mm and a total length of 56 mm including the section for mounting the harvester on the shaker. It has a similar size to a D battery as shown in Figure 3(a). The plastic tube is 44 mm long and has outer and inner diameters of 8.5 mm and 9.6 mm respectively. The two end magnets are both 1 mm thick and 8 mm in diameter. The shuttle magnet has a diameter of 8 mm and thickness of 4 mm. All magnets used in this device are NdFeB. The ring spring as shown in Figure 4(a) is 100 μm thick and made of BeCu. Two sets of magnets are attached to the spring. Apart from magnets, twelve tungsten pieces are also attached to the spring to decrease the resonant frequency of the assisting resonator. The total mass on the spring is 2.66 g. Each magnet has an area of 2×2 mm^2 and is 1 mm thick. Two coils as shown in Figure 4(b) were hand wound around plastic holders. Each coil has 500 turns of 254 μm thick copper wire. Resistances of the two coils were measured as 12.4 Ω and 12.6 Ω respectively. All plastic components in this device are 3D printed ABS plastic. The assembled harvester was attached to a Data Physics V55 shaker for testing as shown in Figure 3(b).

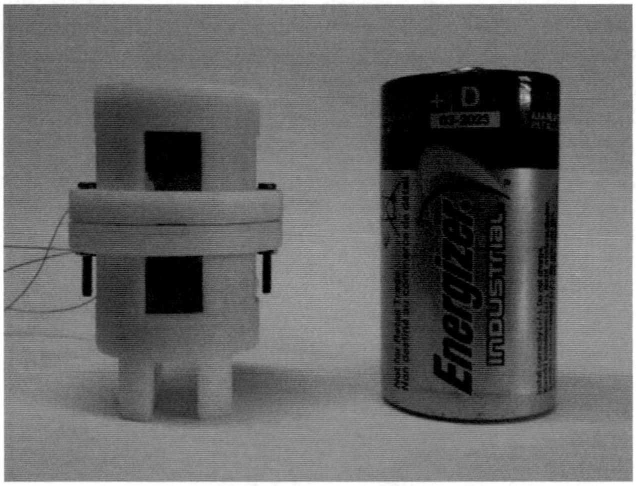

(a) Assembled energy harvester

(b) Harvester mounted on the shaker

Figure 3. The energy harvester.

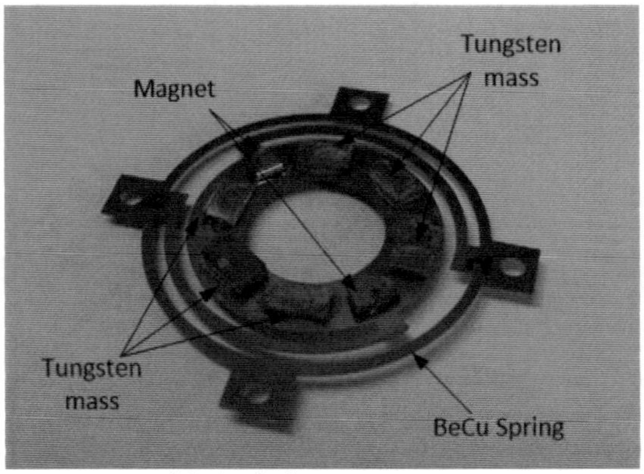

(a) Assisting resonator

(b) Coil

Figure 4. Components.

3.2. Results and discussion

The energy harvester was tested under sinusoidal excitations with various frequencies (10 to 40 Hz) and accelerations (0.5 to 0.8 g, 1 g = 9.81 m·s^{-2}). Two sets of tests were carried out. In the first test, the assisting resonator was removed so that the energy harvester is purely a Duffing's nonlinear structure. In the second test, the assisting resonator was in place, which formed a coupled bistable energy harvester. Both up-sweep and down-sweep tests were carried out and results were found identical in both tests.

Figure 5 shows the open circuit voltage of the two coils under various excitations. It was found that both maximum output voltage in the bottom coil and bandwidth were higher in the case with the assisting resonator, i.e. the coupled bistable harvester, compared to the case without the assisting resonator, i.e. Duffing's nonlinear energy harvester. However, the maximum output voltage in the top coil is lower in the coupled bistable harvester although its bandwidth is still larger. Furthermore, it was found that there are three peaks in the spectrum of the coupled bistable energy harvester. Peak 1 was the result of the resonance of the Duffing's nonlinear structure at 19 Hz, i.e. magnetic spring formed by the two end magnets. In this case, the shuttle magnet only moved within one of the potential wells. Therefore, the output voltage was low compared to other two peaks. Peak 2 was caused by the resonance of another Duffing's nonlinear structure at 25 Hz which was formed by one of the end magnet and magnets on the assisting resonator. During this resonance, bistable operation was triggered. Thus the output voltage was higher. Higher excitation generated higher force that enabled bistable operation to occur more frequently. Therefore, amplitude of this peak increased more than other two peaks with increasing accelerations. Peak 3 occurred at the resonant frequency of the assisting resonator at 35 Hz. Bistable operation was also triggered in this case. It is worth mentioning that due to the gravity, the shuttle magnet preferred to stay in the bottom potential well. Therefore, output voltage in the top coil was much lower than that in the bottom coil and the advantage of the coupled bistable harvester was not obvious.

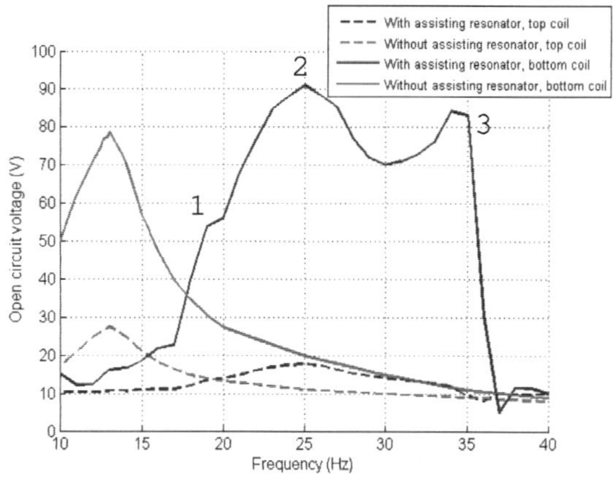

(a) 0.5 ***g***

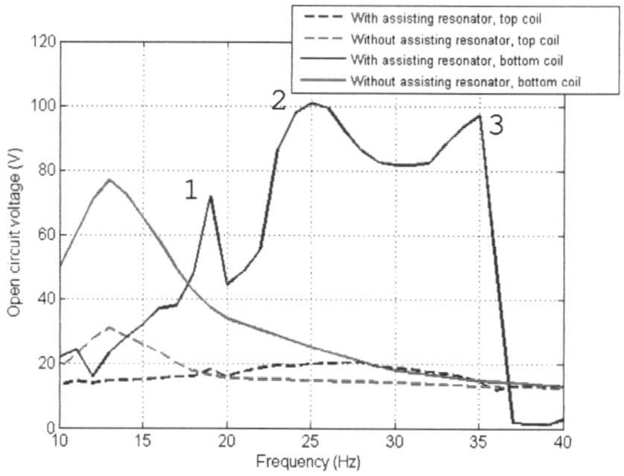

(b) 0.6 ***g***

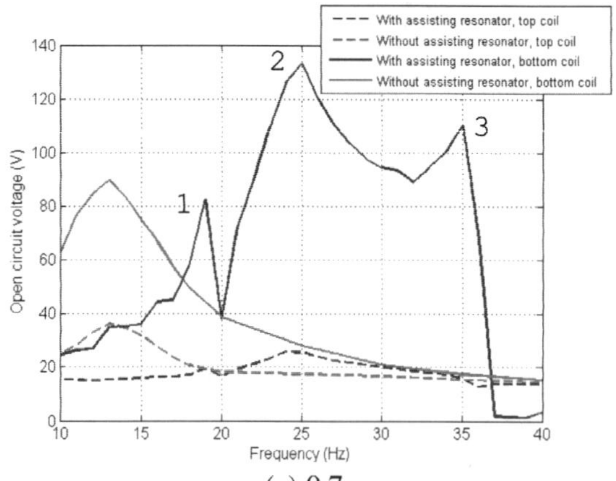

(c) 0.7 ***g***

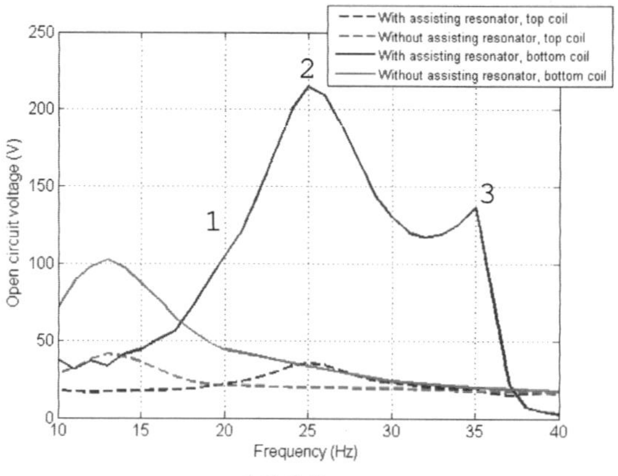

(d) 0.8 ***g***

Figure 5. Comparisons of open circuit voltage under various accelerations.

Figure 6 compares maximum output power in the bottom coil of the Duffing's nonlinear harvester and the coupled bistable harvester when connected to an optimal resistive load of 13 Ω. It can be found that maximum output power of the coupled bistable harvester increased more significantly than that of the Duffing's nonlinear harvester. This is because the shuttle magnet travelled more frequently between two potential wells when the excitation amplitude increased.

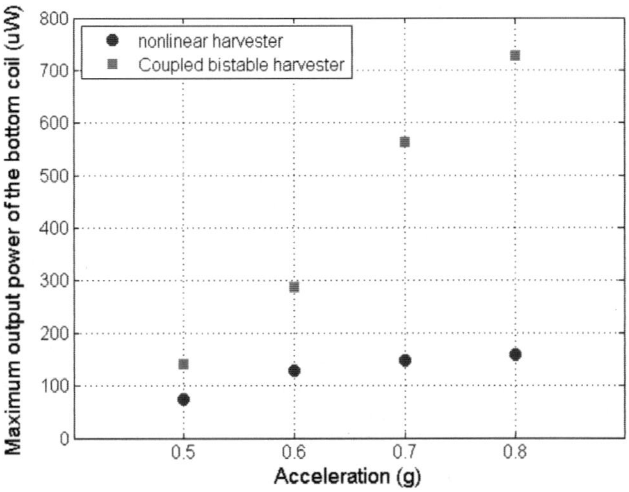

Figure 6. Comparisons of output power under various accelerations.

4. Conclusion and future work

This paper reports a miniature coupled bistable energy harvester that has a similar size of a D battery. Experimental results showed that the output spectrum of the miniature coupled bistable vibration energy harvester was the superposition of several spectra and the harvester had a higher maximum output power and much greater bandwidth compared to the Duffing's structure without the assisting resonator. The next step forward will be to investigate the performance of this harvester under wideband excitations such as white noise.

5. Reference

[1] Beeby S P, Tudor M J and White N M 2006 Energy harvesting vibration sources for microsystems applications *Meas. Sci. Technol.* **17** pp 175-195

[2] Zhu D, Tudor M J and Beeby S P 2010 Strategies for increasing the operating frequency range of vibration energy harvesters: a review *Meas. Sci. Technol.* **21(2)** 022001

[3] Harne R L and Wang K W 2013 A review of the recent research on vibration energy harvesting via bistable systems *Smart Mater. Struct.* **22** 023001 (12pp)

[4] Ferrari M, Ferrari V, Guizzetti M, Andò B, Baglio S and Trigona C 2010 Improved Energy Harvesting from Wideband Vibrations by Nonlinear Piezoelectric Converters *Sens. Actua.* A **162** pp. 425-431

[5] Zhu D and Beeby S P 2013 A coupled bistable structure for broadband vibration energy harvesting *Transducers 2013 & Eurosensors XXVII, Barcelona, Spain, 16 - 20 Jun 2013*

[6] Zhu D and Beeby S P 2013 A Broadband Electromagnetic Energy Harvester with a Coupled Bistable Structure *PowerMEMS 2013, London, UK, 3 - 6 December 2013*

Acknowledgments

Authors would like to thank Perpetuum Ltd for providing experiment equipment.

Development of a portable power system with meso-scale vortex combustor and thermo-electric device

D. Shimokuri [1], **T. Hara** [1] **and S. Ishizuka** [1]

Hiroshima University, 1-4-1 Kagamiyama, Higashi-hiroshima, Hiroshima, Japan

Email: cri@hiroshima-u.ac.jp

Abstract. In this study, a small scale power generation system with a meso-scale vortex combustor has been developed. The system was consisted of a couple of thermo-electric device and a heat medium. The medium was made of duralumin, 40 × 40 × 20 mm and 52 g weight, and the vortex combustion chamber of 7 mm inner diameter was embedded in it. It was found that a stable flame could be established in the narrow 7 mm channel even the mean axial velocity reached 1.2 m/s. And furthermore, the vortex flow significantly enhanced the heat transfer from the burned gas to combustion chamber, and as a result, the medium was heated to 300°C quickly (within 5 minutes) by the combustion of propane / air mixture for 145W input energy. The system could successfully generate 1.98 W (4.3 V and 0.46 A), which corresponded to the energy conversion rate of 0.7 % per unit thermo-electric device.

1. Introduction

Due to strong demands for small-scale power sources, meso- and micro- scale combustors have received considerable attention [1,2]. On such small scale combustors, the thermal and chemical quenching effects of the wall become significant due to its large surface to volume, hence, large heat loss to heat release, ratio. Flame quenching is an unavoidable issue for small-scale combustors.

One solution for the flame stability issue on the micro combustor is using the "vortex flow". Recently, we found that the flame could be easily stabilized in 3.6 mm tube for methane/air and propane/air mixture if the flow was rotating. In figure 1, appearance of flames in 3.6 mm tube for non-rotating flow (A) and rotating flow (B) are shown. In figure 1B, it can be seen that a large surface flame can be established in a vortex flow although the tube diameter (3.6mm) is close to the so-called "quenching diameter" of the mixture (3.5 mm). Even with vortex flow, the flame could not be stabilized in "μm" scale channels for those mixtures, however, it was found that the stable combustion range in a few millimeter scale channel was considerably extended. For example, in 3.6-mm tube, the stable combustion range of methane/ air mixture was $\Phi = 0.75$ to 1.2 , even for the high mixture flow rate 8.0 L/min (corresponding mean axial flow velocity is 13 m/s). According to these results, it can be said that the vortex combustor be the powerful energy source for the portable devices. And due to these high stability nature of the vortex

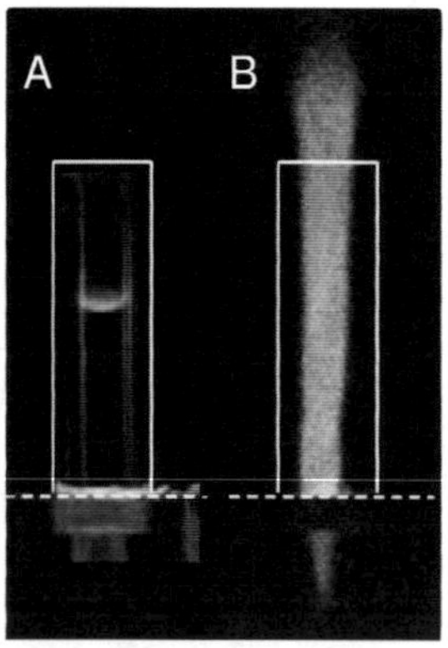

Fig. 1 Appearance of CH_4 / air flames in (A) non-rotating and (B) rotating flow (A: mixture flow rate Q_{mix} = 0.22, mean axial velocity V_x = 0.35 m/s. B: Q_{mix} = 2.0 L/min, V_x =3.3 m/s).

combustor, meso-scale vortex combustors have been successfully developed not only for gaseous fuels [3,4] but also for the liquid fuels [5-7].

However, the power system with the vortex combustor has never been developed. Then, in this study, a powerful portable power system with a meso-scale vortex combustor has been developed. On the system, a couple of thermo-electric devices (TED) and the vortex combustor are used. It is known that TED has poor conversion efficiency from heat to the electric power, however, it is promising device to generate the electric power from the heat of combustion. In this study, the temperatures of the gas and the heat medium as well as the system output are examined. The degree of the heat transfer from the burned gas to the combustor wall is also evaluated based on the obtained surface / gas temperatures.

2. Experimental

Figure 2 shows the appearance of the power system designed for 100 ~ 200 W input based on the heat of combustion of fuel (propane). Its weight is less than 400g and scale is comparable to a conventional fuel cell as shown in figure 2A. The schematic of the system is shown in figure 3. The system consists of a heat medium (figure 3A), thermo-electric device (figure 3B) and cooling plate and fan (figure 3C). The heat medium is 40 × 40 × 20mm and 52g weight, of which appearance is shown in Fig. 2B. The heat medium is directly heated up by the vortex combustion chamber (7mm i. d.) embedded in itself. The electric power can be obtained by a couple of thermo-electric devices placed on the surface of the heat medium. Note that, in this study, the cooling fan is powered by an outer power source to examine the maximum output of the system. The variable resistance (0 ~ 20 Ω) was installed in the circuit, and the electric current and voltage on the resistance were measured (see, lower part of figure 3). The obtained electric power can be a measure of the electric power that can be used by the actual electronic devices.

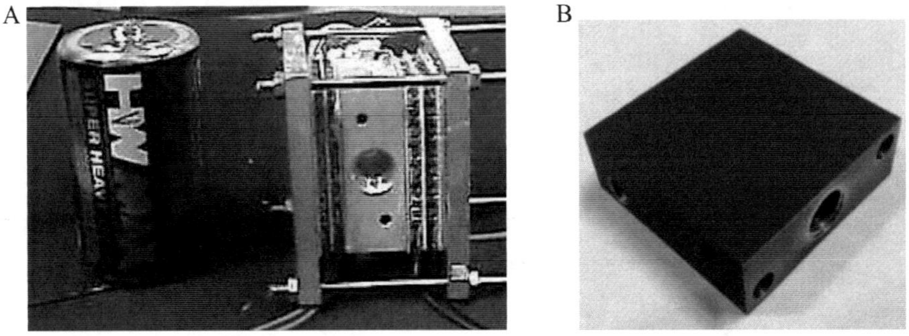

Figure 2. Appearance of the power generation system (A: the system, B: the heat medium).

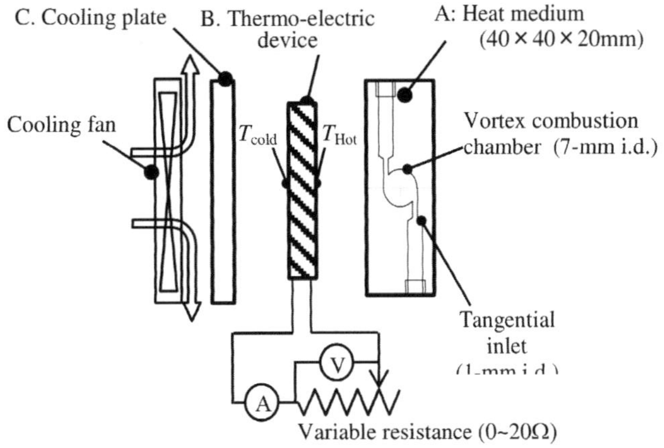

Figure 3. Schematic of the power generation system with a vortex combustor.

In figure 4A, the channel inside of the heat medium is shown. A homogeneous propane / air mixture enters into the medium from the two inlets. The channel is connected to the gas line (1 mm i. d.) in right-angle at the Z – Z' position, and then, the unburned gas is ejected tangentially into the combustion chamber (see figure 4B). In the experiment, the mixture is ignited by a torch at the opened "ignition port" shown in figure 4A. The flame appearance obtained with opened ignition port is shown in figure 4B. After the ignition, the port is closed by a stainless steel plug, and thus, the hot burned gas flows in one direction. The combustion chamber is 7 mm in inner diameter and 27 mm in length. As the flame length is estimated to be less than 10 mm in the experimental range of this study (input energy < 200W, mixture flow rate < 3 L/min), the half of the combustion chamber is expected to be occupied hot burned gas. The gas outlet channels are equipped to be close to the unburned mixture line to enhance the flame stability by the preheating of the unburned mixture.

During the power generation experiments, temperatures both on the heating side and cooling side of TED were measured by 250 μm K-type thermocouple. The measurement positions are shown in figure 3. The gas temperature was also measured by a series of R-type silica coated thermocouple installed in the gas channel. The measurement positions are shown in figure 4A, and each temperature is denoted as (1) T_f : flame temperature, (2) T_b: burned gas temperature and (3) T_{out} : exhaust gas temperature. Separately to the power generation experiment, the surface temperature profile on the heat medium was measured by a thermo graphic camera (Nihon avio, TSV-200EX) without TED.

In the experiment, the mixture was ignited under the stoichiometric condition of the air flow rate Q_a = 3.0 L/min which corresponds to input power of 200 W. When the medium surface temperature reaches the upper limiting temperature of the TED (260°C), the mixture condition was changed to Φ= 0.8 of Q_a = 2.7 L/min (145 W) to achieve the steady state.

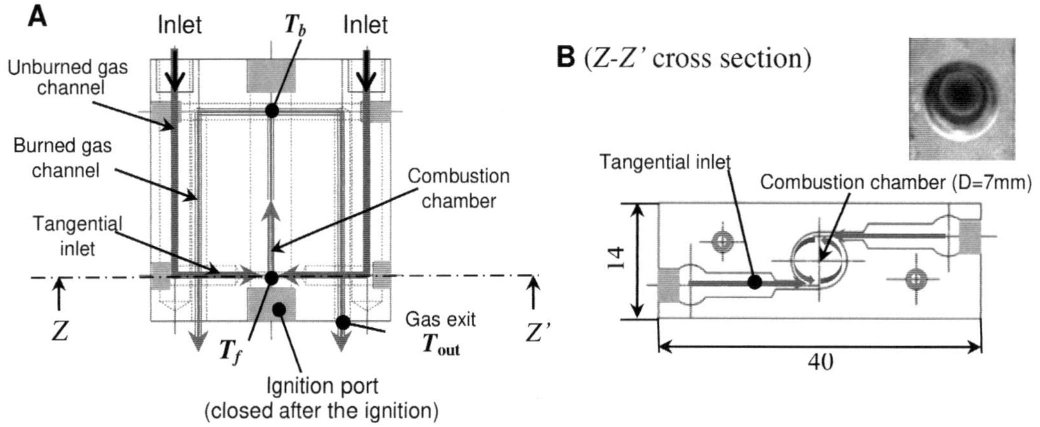

Figure 4. Schematic of the heat medium in which the vortex combustor is embedded.

3. Results and discussion

Figure 5A shows the variations of the heating side (T_{Hot}) and cooling side (T_{Cold}) temperature of TED as well as the output voltage obtained with the system. After the ignition (t = 0 min), the hot side of the TED was heated up to 200°C within 5 minutes. At t = 7 min., the input was decreased to 145 W from 200W because the limiting temperature of the TED (260°C) was reached. After that, the T_{Hot} takes almost constant of 250°C which indicates that the input energy is balanced with TED output and the heat loss to the surroundings. The cooling side temperature also increases with T_{Hot} and takes almost constant of 100°C after t = 7 min.

As for the output voltage, the maximum of 4.7 V was reached at t = 7min., and the steady output of 4.3 V could be obtained. As the output of the TED dependents on the temperature difference between the hot side and cold side of TED, the temperature difference $T_{Hot} - T_{Cold}$ is also shown in figure 5A by (+). It is clear that the output voltage traces the tendency of the temperature difference.

Next, the medium surface temperature was measured by thermography. In the experiment, the wall temperature was measured directly without TED. The obtained two dimensional temperature profile is shown in figure 6A, and the temperature distribution along the *A – A'* line (in figure 6A) is shown in figure 6B. In figure 6A, the gas flow line is also indicated by dashed line. Results show that the difference between the maximum temperature (T_{max} =260°C) and the minimum temperature (T_{min} = 246°C) is 14°C, which is within ± 3 % of average temperature of T = 253°C. It is also seen that the position of T_{max} is not centered but exists around the tangential injection.

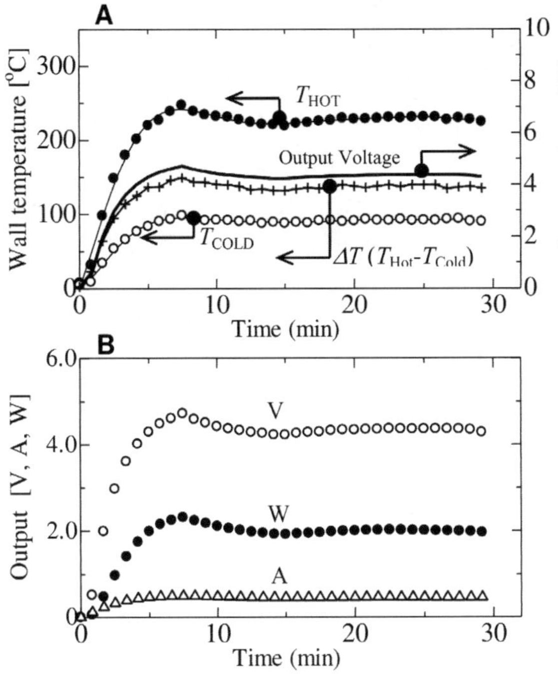

Fig. 5 Variations of wall temperatures (A) and output voltage, current and power (C) with time.

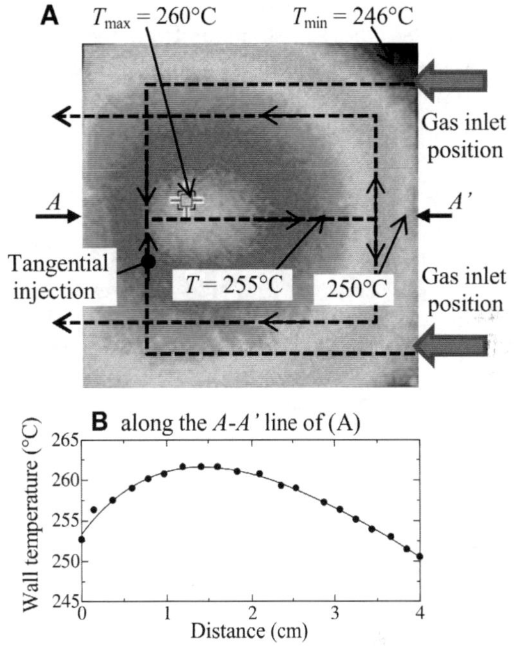

Fig. 6 Temperature profile (A: two dimensional profile, B: temperature distribution along *A-A'* line of (A). The results were obtained when T_{max} reached at 260°C).

Figure 5B shows the variations of the output voltage, electric current and the power of the system. The result shows that the maximum output of 2.3 W (4.7V × 0.49A) could be tentatively obtained at t = 7 min., and 1.98 W (4.3V × 0.46A) could be obtained continuously over 20 minutes.

The system output was examined for various internal resistances under the steady output state (after t = 15 min. in figure 5B). Results are shown in figure 7. It can be seen that the voltage monotonically increases whereas the electric current monotonically decreases with resistance. The output power takes its maximum of 1.98 W when the resistance is 10Ω. These results confirmed that the vortex combustor power system could successfully generate 1.98 W continuously. The conversion rate of input heat of

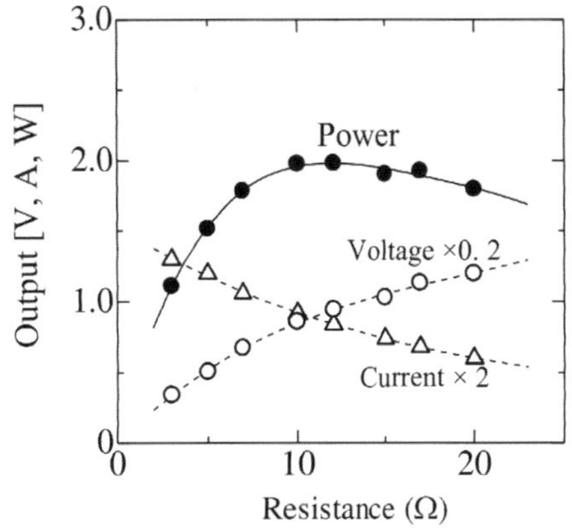

Fig. 7 Variations of output with internal resistance.

combustion to the electric output per unit TED, defined as,

$$\frac{\text{Output[W] / number of TED}}{\text{Input[W]}} \times 100\,[\%], \qquad (1)$$

was calculated to be 0.68%. This low conversion rate comes from the inefficient cooling. As shown in figure 5A, T_{Cold} reaches 100°C continuously. The optimization of the cooling side of TED will lead to the drastic increase of the output of the system.

Finally, the heat transfer from the burned gas to the combustor wall is discussed. The variations of the flame temperature T_f, burned gas temperature T_b and the exhaust gas temperature T_{out} are shown in figure 8. It can be seen that the temperature shows complicated behavior. The points are;

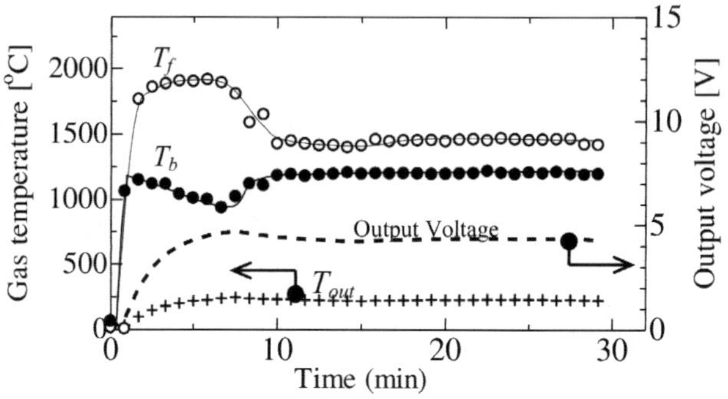

Fig. 8 Variations of the gas temperatures (measurement positions are shown in figure 4A).

(1) The exhaust gas temperature 250C° is enough lower than T_f, and furthermore, it is almost same as the combustor wall temperature (T_{Hot} in figure 5A). This indicates that the heat of the burned gas was effectively transferred to the combustor wall.
(2) The distance between the measurement position of T_f and T_b is only 27 mm and the residence time of the burned gas is estimated to be less than 20 ms, however, the maximum temperature difference between T_f and T_b is about 1000 K. This indicates the heat of the burned gas was rapidly transferred to the wall. Then, the Nusselt number was calculated for the 27 mm combustion chamber at t = 7 min using conventional heat transfer model in a cylindrical tube [10]. Resulting Nusselt number is 7, which is enough larger than that for the developed laminar flow (N_u = 3.7), although the Reynolds number on the experimental condition is about 100 (based on the mean axial velocity). Thus, it can be said that the heat transfer from the burned gas to the combustor wall is drastically enhanced by the vortex flow.

References

[1] Fernandez-Pello C A 2002 *Proc. Combust. Inst.* **29** 883.
[2] Dunn-Rankin D, Leal E M and Walther D C 2005 *Prog. Energ. Combust. Sci.* **31** 422.
[3] Wu M, Wang Y, Yang V, Yetter R A 2007 *Proc. Combust. Inst.* **31** 3235.
[4] Li Y H, Li H Y, Dunn-Rankin D, Chao Y, 2009 *Proc. 7th AsPaCC* ID: 10098.
[5] Pham T K, Dunn-Rankin D, Sirignano W A 2007 *Proc. Combust. Inst.* **31** 3269.
[6] Mattioli R, Pham T K, Dunn-Rankin D 2009 *Proc. Combust. Inst.* **32** 3091.
[7] Sirignano W A, Pham T K, Dunn-Rankin D 2002 *Proc. Combust. Inst.* **29** 925.
[8] Shimokuri D, Honda Y, Ishizuka S 2011 *Proc. Combust. Inst.* **33** 3251.
[9] Shimokuri D, Kratsu Y, Ishizuka S 2013 *Proc. Combust. Inst.* **34** 3403.
[10] Incropera F P 2002 *Fundamentals of Heat and Mass Transfer* 355, 482.

Development of Micro Cogeneration System with a Porous Catalyst Microcombustor

S Takahashi[1], M Tanaka[1] N. Ieda[1] and T Ihara[1]

[1] Department of Mechanical Engineering, Faculty of Engineering, Gifu University, 1-1 Yanagido, Gifu-shi, Gifu 501-1193, Japan

E-mail: shuhei@gifu-u.ac.jp

Abstract. The self-standing micro cogeneration system by coupling a microcombustor, thermoelectric modules and an air supply device was developed. The microcombustor has a porous monolithic Pt catalyst layer and the combustion efficiency of 90% was attained. A micro-blower was used to supply air to the combustor, and it was driven by a part of the electricity from the Bi-Te TE modules through a DC-DC converter. We investigated the optimal point where the output became maximal and the system stood by itself. At the optimal point, the input fuel enthalpy was 13.2W and the electricity of 403mW was generated from the TE modules. The micro blower used 280mW and the net electricity was 123mW. Therefore the final thermal efficiency was 0.93%. The efficiency was the same magnitude of the world smallest model plane engine TeeDee01 (COX Co. Ltd.) although the thermal input was less than its 1/20.

1. Introduction

Recent increase of portable electric devices and/or development of UMVs, such as micro drones, demands higher energy density power sources. Hydrocarbon fuels have large energy density compared with those of conventional batteries, which are attractive for large power/weight ratio generators or heat engines [1]. However, in order to use the combustion heat of hydrocarbon fuels in a small volume, it is necessary to overcome the huge heat loss due to large surface/volume ratio to sustain combustion.

In our previous research, we developed a micro-combustor, which has porous Pt catalyst layer inside a narrow ceramics tube whose inner diameter was 0.8mm [2]. The catalyst layer, which was built up by sintering technique, had a monolithic mesh structure and had a large surface area, and could sustain combustion with the aid of the surface reaction. We coupled our microcombustor with Bi-Te thermoelectric (TE) modules and succeeded in generating electricity with final efficiency of 3.5% [3, 4]. The corresponding energy density was about 420Wh/kg, which was 2.4 times that of typical Li-ion battery. The developed generation system was also able to provide heat of 5W, therefore it was expected to use as an emergency personal energy resource used in catastrophic situation right after natural disaster, for example.

However, the efficiency above did not take account of fuel/air supply system. Some of hydrocarbon fuels, such as butane, can be condensed by compression. Therefore, the supply of fuel to the combustor is not so difficult. On the other hand, the supply of sufficient air is critical issue to solve because the amount of air is more than that of fuel and the entrainment force of fuel is very small due to its small flow rate. In the present article, we drove air supply system with using the electricity from TE modules and discuss the feasibility of self-standing cogeneration system.

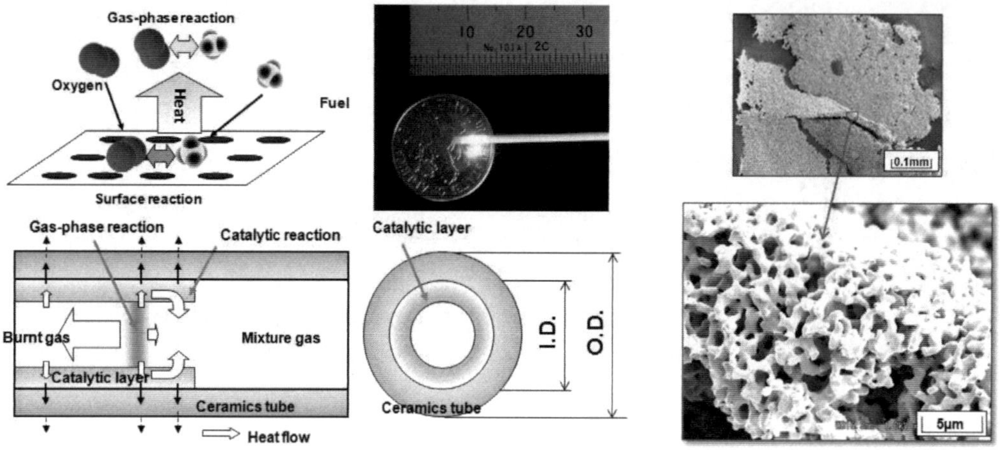

Figure 1. Concept of microcombustor and SEM image of the catalyst layer.

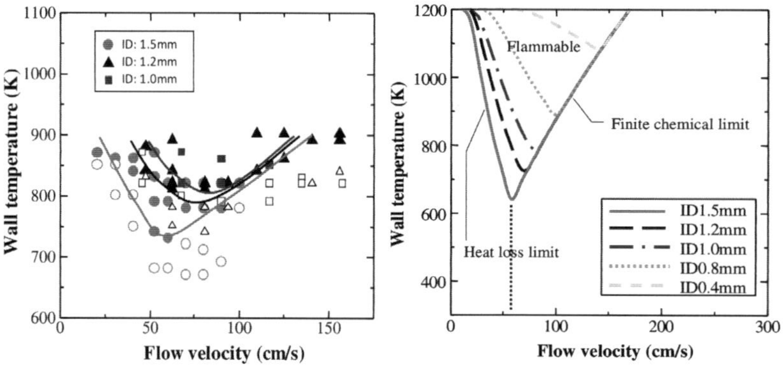

Figure 2. Limiting wall temperature for methane-air mixture in a narrow tube (left: exp., right: prediction).

2. Experimental setup

2.1 Microcombustor

The microcombustor used in the article is shown in Fig. 1. The combustor was a mullite ceramics tube and had a porous Pt catalyst layer on the inner wall. The length of the catalyst layer was 5mm, and the thickness was less than 50μm. The concept of sustaining combustion even in a narrow tube whose diameter is smaller than the quenching diameter is as follows. A part of the supplied mixture reacts on the catalyst later and increases the wall temperature. The rest mixture reacts in the core of the combustor with the aid of heated wall. The combustion limit in a narrow tube is determined by the blow-off limit due to high velocity, and quenching limit due to heat loss to the wall. In our previous scale analysis and experiments [5], the limiting wall temperatures are expressed as follows.

$$T_{w,q} = T_i - \frac{\eta_c T^*_{f,ad}}{(1+\theta)(1+2\Omega^*)} \quad \text{where } \theta = \frac{2h\alpha}{\rho c_p V_g{}^2 r} = 4Nu\,\mathrm{Re}^{-2}\,\mathrm{Pr}^{-2} = 4NuPe^{-2},\ \Omega^* = \sqrt{Nu}\,\frac{\delta}{r}$$

$$T_{w,b} = 300\left(\frac{V_g}{C_4}\right)^{\frac{1}{1.68}} \quad \text{where } C_4\text{=0.33}$$

The η_c, r, δ, V_g are the combustion efficiency, the radius of the tube, flame thickness, and flow velocity in the tube, respectively. Figure 2 shows the experimental results and analytical prediction of the combustion limit in a heated narrow tube. The same mullite ceramics tubes were used. It is found that both too high and too low flow velocity cannot sustain combustion.

The typical combustion efficiency with the inner diameter of 0.8mm is shown in Fig. 3. The fuel was methane and butane, respectively. It is found that the combustion efficiency of 90% was achieved

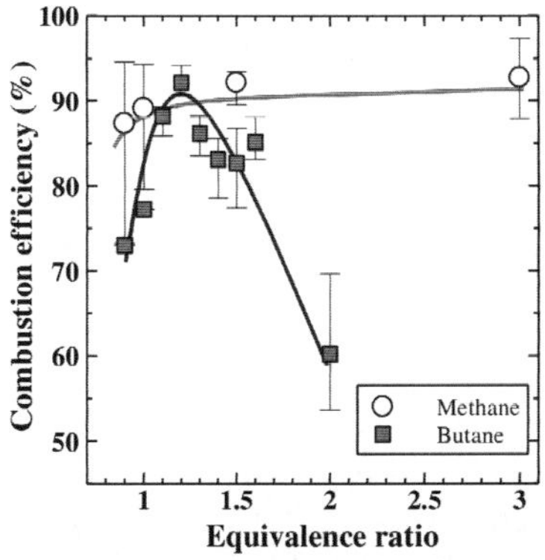

Figure 3. Combustion efficiency in the microcombustor.

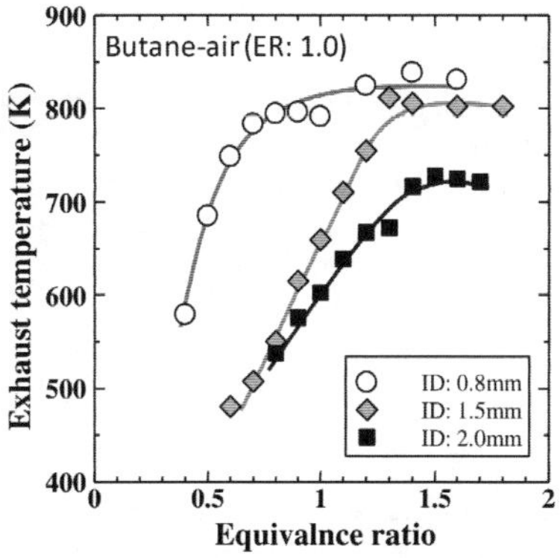

Figure 4. Exhaust temperature with varying the inner diameter.

Figure 5. Micro-blower.

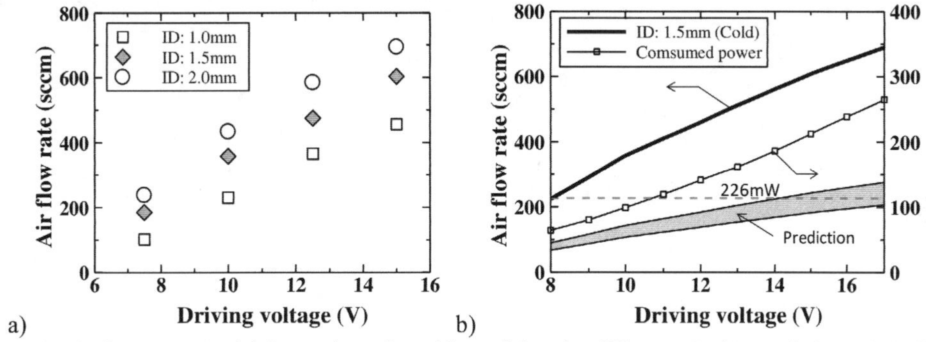

Figure 6. a) Air flow rate vs. driving voltage in cold condition for different ID, b) prediction of air flow rate during combustion for ID of 1.5mm.

near stoichiometric condition. Figure 4 shows the effect of the inner diameter on the exhaust gas temperature from the combustor when the total mixture flow rate is fixed at 70sccm (at 293K). The fuel was n-butane and the equivalence ratio was 1.0. The temperature was measured by a thermocouple whose diameter was 0.1mm. It is found that the larger the inner diameter is, the lower the exhaust temperature is. This means that the combustion efficiency becomes lower when the inner diameter is large. As shown in Fig. 2, the higher wall temperature was required to sustain combustion if the flow velocity was low. The result implied that mixture in the core of the tube did not react when the diameter was large. Therefore, it is found that relatively high velocity is required in order to sustain high combustion efficiency.

2.2 Setup of air supply system

In the present article, we set up air supply system by using a micro blower (Murata Manufacturing Co., Ltd.) shown in Fig. 5. The necessary air flow rate depends on the total mixture flow rate, which determines the combustion heat. Therefore, if the final conversion efficiency is the same, the larger the mixture flow rate is, the larger the generated electricity becomes. On the other hand, the required electricity becomes larger if the air flow rate is large. Hence, we first investigated the relation between flow rate and the driving voltage in the ceramics tube of different diameter with a differential pressure sensor (no catalyst, no burning). The result is shown in Fig. 6a. It is found that the micro blower consumes 100-250mW, and the air flow rate range significantly depends on the tube diameter. In Fig. 6a, even in the tube whose inner diameter is 1.0mm, the air flow rate of 450sccm was achieved with the voltage of 15V. This air flow rate corresponded to the combustion heat of 26W if the stoichiometric mixture was supplied. However, in actual, the flow rate reduced if catalyst later was set up. Additionally, if the mixture reacted in the tube, it increased the viscosity, and finally the flow rate was expected to be reduced in 30-40% of that in cold condition. Therefore, in order to avoid too much

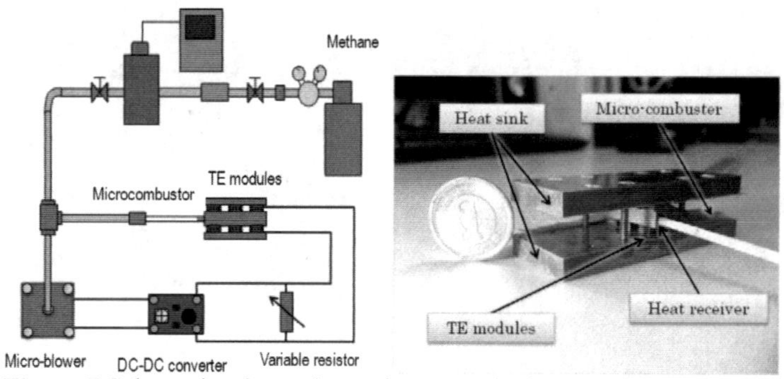

Figure 7 Schematic of experimental setup and micro-cogenerator.

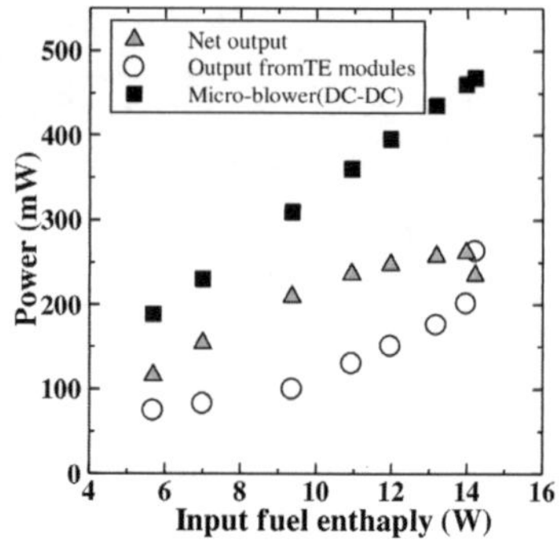

Figure 8. Output from TE modules vs. input fuel enthalpy.

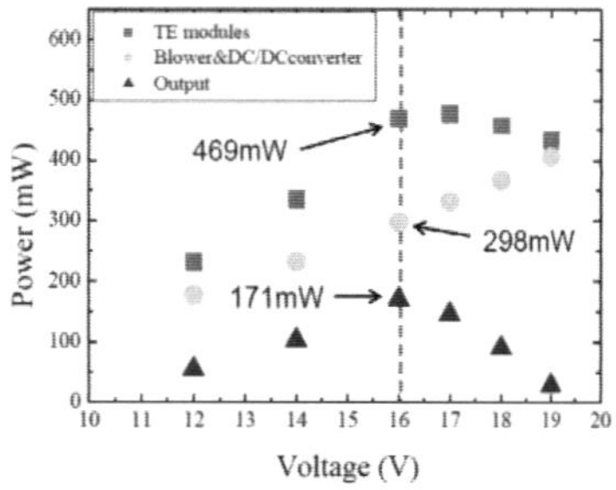

Figure 9. Net output vs. driving voltage of micro-blower.

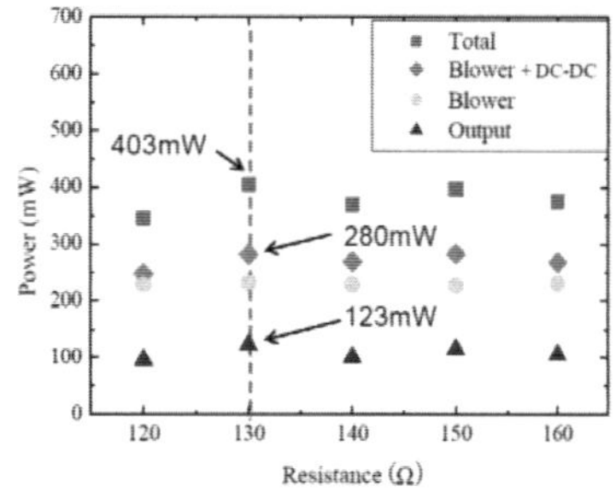

Figure10. Net output of self-standing system with varying load resistance.

Table 1. Specifications at design point.

Input fuel enthalpy	13.2W
Inner/outer diameters	1.5/2.5mm
Methane/air flow rate	23.8/226sccm
Size and number of TE modules (17 pairs)	4mm x 4mm 6 (in series)
Matching load resistance	130Ω
Output voltage	4.0V
Gross electricity from TE modules	403mW
Efficiency of TE modules	3.35%
Consumed power at micro-blower	247mW
Consumed power at DC-DC converter	33mW
Net output electricity	123mW
Final thermal efficiency	0.93%

pressure loss and to keep high combustion efficiency, we chose the inner diameter of 1.5mm in the present article. The expected flow rate vs. driving voltage is shown in Fig. 6b.

The schematic of the self-standing generation system is shown in Fig. 7. In order to generate the electricity from the heat of the exhaust gas, Bi-Te TE modules (1MD04-017-1, RMT Ltd.) were attached on the copper hot block with a hole, in which the microcombustor was inserted. The number of TE modules was changed from 2 to 6 according to the magnitude of the output heat from the combustor not to heat the copper block above 550K. The cold side of the TE modules were copper slabs, heat-sink type coolers. A part of generated electricity was supplied to DC-DC converter (LM2735: Strawberry Linux Co., Ltd.), whose conversion efficiency was 90%, and the output voltage was adjusted to drive the micro blower. The rest electricity was supplied to the load, which was variable resistor. The fuel was methane, which was supplied by a mass flow meter. The supplied air and fuel were mixed at the inlet of the combustor.

3. Results and Discussion

We first discussed the optimal input enthalpy to maximize the output electricity. Figure 8 shows the expected generated electricity, consumed electricity for the blower including DC-DC converter, and the net output electricity with varying the mixture flow rate, namely the input enthalpy of the fuel. The necessary electricity for the blower was predicted from the result in Fig. 6b. In Fig, 8, the generated electricity from TE modules was calculated assuming that the average final conversion efficiency was 3.3% according to the previous research [3]. The generated electricity increased with the input enthalpy, but the consumed electricity significantly increased when the input enthalpy was more than 14W. Hence we chose 13W as the input enthalpy for the experiment. According to the chosen input enthalpy, the number of TE modules attached on the copper block was selected as 6, and they were connected in series. The generated electricity was changed by the load resistor, and the matching load to maximize the output was 30Ω. The champion output was 533mW for 13.2W input enthalpy, which corresponded to the conversion efficiency of 4.0%.

Then, we investigated the optimal driving voltage for the blower by experiments. First, the micro-blower was driven by the separate power source not from the TE modules so that we could measure the consumed power and stabilize the system. The fuel flow rate was fixed at 23.8sccm, which corresponded to the enthalpy of 13.2W. In order to make stoichiometric mixture, the air flow rate of 226sccm was required. In Fig. 6b, the required air flow rate can be supplied in the range of voltage from 14V to 17V. Figure 9 shows the experimental result varying the driving voltage of the blower when the fuel flow rate was fixed. The output electricity from the TE modules was measured by setting the matching load as 30Ω. The result shows that the electricity has a peak of the output from the TE modules at 16V. It means that the stoichiometric mixture was supplied at the driving voltage of 16V. Therefore, we chose 16V as the driving voltage of the blower for 13W input condition. As the prediction, the net electricity was expected 170mW, which corresponded to the conversion efficiency of 1.3%

Finally, we found the optimal point for the self-standing generation system. The fuel flow rate was fixed at 23.8sccm, which corresponded to the input enthalpy of 13.2W. The driving voltage of the blower was set at 16V. We first drive the micro blower with a separate power source. After the micro combustor attained the steady state, we switched the power source to the output from the DC-DC converter. Then we adjust the load resistor and found the condition where the net output became maximal. Figure 10 shows the measured electricity with varying the load resistor. When the load resistor was 130Ω, the maximal net output 123mW was measured. The specifications at the optimal point are listed in Table 1. The final conversion was about 0.93%. The obtained efficiency was less than the expected one, but it was due to the lower thermal output caused by the individual difference among the developed combustors. If the output from TE modules were 500mW, the net output had increased to 200mW. The model plane engine, Tee Dee01 by OCX is known as the world smallest reciprocal engine [6]. Its displacement volume is 0.163cc and the input fuel enthalpy is about 300W. The net output is 5W, which corresponds to the thermal efficiency of 1.6%. Generally, the thermal efficiency is drastically reduced if the size of the engine becomes small. The developed system achieved the efficiency of the same magnitude whereas the thermal input was less than 1/20 of that of TeeDee01.

4. Summary

We developed a small self-stand generation system in which the air supply system was driven by a part of the generated electricity. The thermal input was 13.2W and the final output was 123mW; thus the final thermal efficiency was 0.93%. The magnitude of the efficiency was close to the world smallest conventional reciprocal engine TeeDee01although the power size was less than its 1/20.

References

[1] Fernandez-Pello A C 2002 *Proc. Combust Institute* **29** 883
[2] Takahashi S and Wakai K 2009 *45th AIAA/ ASME/ SAE/ ASEE Joint Propulsion Conference & Exhibit* AIAA 2009-5246
[3] Takahashi S, Yamada N and Wakai K 2009 *Proc. 7th ASPACC*, ID:10082
[4] Sugie K, Takahashi S, Ihara T and Wakai K 2010 *Proc. 8th ASPACC* 1087
[5] Imoto Y, Takahashi S, Ihara T and Wakai K 2004 *42nd Symposium (japanese) on Combustion* 181 (in Japanese)
[6] Balmer R T 2011 *Modern Engineering Thermodynamics* Elsevier 507

Power generation properties of Direct Flame Fuel Cell (DFFC)

S Endo[1] and Y Nakamura[2]

[1] Division of Mechanical and Space Engineering, Hokkaido University, N13 W8, Kita-ku, Sapporo, Hokkaido, 060-8628, Japan
[2] Department of Mechanical Engineering, Toyohashi University of Technology, 1-1 Hibarigaoka, Tenpaku, Toyohashi, Aichi, 441-8580, Japan

[1] E-mail: endo-sho@mech-hm.eng.hokudai.ac.jp
[2] E-mail: yuji@me.tut.ac.jp

Abstract. This paper investigated the effect of cell temperature and product species concentration induced by small-jet flame on the power generation performance of Direct Flame Fuel Cell (DFFC). The cell is placed above the small flame and heated product gas is impinged toward it and this system is the simplest and smallest unit of the power generation device to be developed. Equivalence ratio (ϕ) and the distance between the cell and the burner surface (d) are considered as main experimental parameters. It turns out that open circuit voltage (OCV) increases linearly with the increase of temperature in wide range of equivalence ratios. However, it increases drastically at which the equivalence ratio became small ($\phi \leq 2.0$) showing inner flame clearly. This result suggests that OCV depends on not only cell temperature but also the species concentration exposed to the cell. It is suggested that Nernst equation might work satisfactory to predict OCV of DFFC.

1. Introduction

Direct Flame Fuel Cell (DFFC), which is known as the one of solid oxide fuel cells (SOFCs; promising future power source [1]), can generate electricity by adopting flame directly to it. One of the potential features of DFFC is to be a good candidate of the miniaturized thermo-electro system in comparison with those of SOFCs. For SOFC-based power generation system, gas supplying system becomes rather massive [2], moreover, a heater is required to preheat the cell to work well. DFFC does not need such facility (i.e. the gas chamber and heater) because the fuel (gas) and heat are supplied by the flame. However, since the generated power might strongly be related to the thermal and species status exposed to the cell, there must be most suitable combination of cell and the flame for DFFC-based power generation system. Because the main target is to develop the miniaturizing power generation system based on DFFC, the applied flame must be also small and stable. Here, we propose to use the miniaturized jet flame (small-scale jet flame) [3] and see how the combination between DFFC and micro-jet flame work as the unique small power generation device.

In this work, we investigated the effect of cell temperature and product gas species concentration at the vicinity of the cell on the power generation performance of DFFC, especially the open circuit voltage (OCV), by using of fuel-rich premixed micro-jet flame.

2. Experimental

Figure 1 shows the schematic drawing of experimental setup. Methane and oxygen were used as a fuel and oxidizer, respectively. Although each gas flow rate changed, total flow rate was fixed at 1.0 m/s. The flow rates were controlled by the needle valve and the accuracy is well validated with soap-film flowmeter. 1.0 mm inner diameter burner was used to form the small-scale jet flame. Flame was partially premixed flame whose equivalence ratio was too rich to propagate itself to prevent any flashback. DFFC cell used in this work was developed/supplied by Kikusui Chamical Indstries Co., Ltd. and placed horizontally above the flames. The positions of burner and cell were adjusted with optical stage. The distance between the cell and the micro-burner (d) was varied from 5.0 mm to 3.0 mm to modify their heat and species interaction. Equivalence ratio (ϕ) was varied from 1.9 to 5.0 because DFFC cannot produce electricity under $\phi \leq 1.0$ [4]. K-type thermocouple was used to measure the cell temperature and discharge voltage (OCV) was measured by digital sensor (GRAPHTEC GL-220). The current value was obtained from the measured voltage and resistance. The change of flame shape was observed with DV camera. A gas chromatography was used to identify the product gas exposed onto the cell surface.

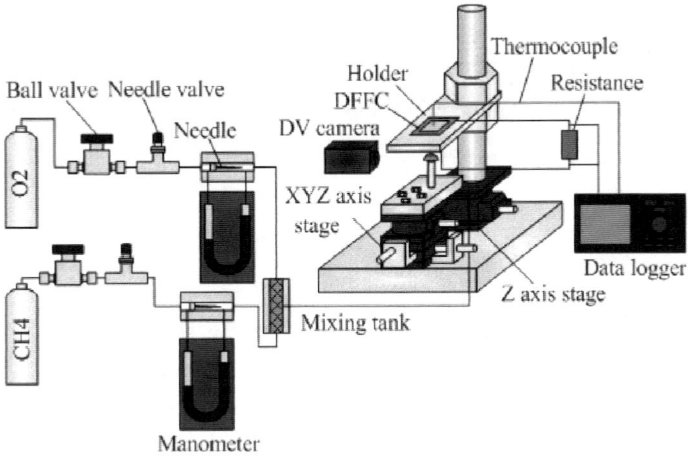

Figure 1. Schematic drawing of experimental setup.

3. Result and discussion

Figure 2 shows an example of the change of flame shape by the change of equivalence ratio ϕ and distance d. From this figure, it is obvious that the flame becomes smaller with the decrease of equivalence ratio and the flame shape is largely distorted by the distance d (especially $\phi \leq 2.0$, d=3mm). Under $\phi \leq 2.0$, except for d=3.0 mm, the inner flame was formed. Figure 3 shows the relationship between the cell temperature and OCV in various experimental conditions. It is understood that the OCV tends to linearly increase with the increase of cell temperature, however, it is also found that there was a singular points which OCV drastically increases in spite of the small change in temperature. Since the singular points in Fig.3 corresponds to the ones obtained at $\phi \leq 2.0$, it can be considered that gas species concentration changed with the change of flame shape in comparison with $\phi \geq 2.5$. From this result, it is expected that the power generation performance of DFFC depends on not only cell temperature but also gas species concentration within the flame. It is important to investigate how this singular point is given in detail in order to understand the controllability of power generation performance of DFFC. Next, let us look into deeply for the relation between the change of flame shape and OCV.

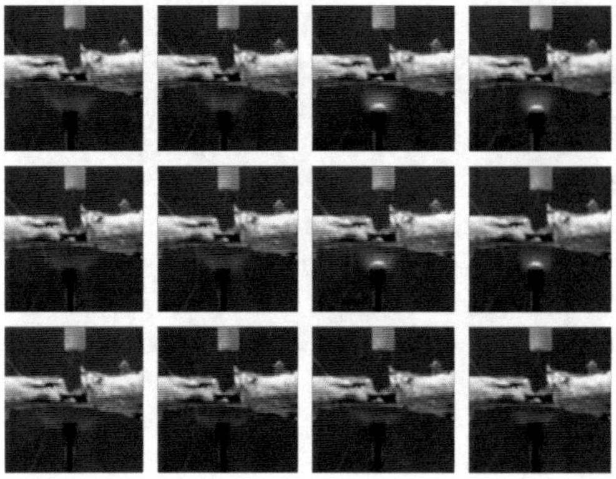

Figure 2. Change of flame shape by the change of equivalence ratio and distance between burner and DFFC cell.

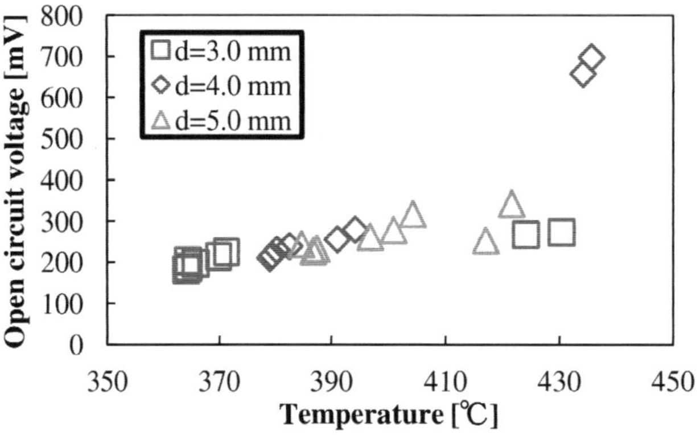

Figure 3. Relationship of cell temperature and open circuit voltage in various equivalence ratios.

Figure 4 shows the change of flame shape by the change of distance from d=3.5 mm to 3.0 mm under ϕ=2.0. From this figure, it is found that the inner flame disappeared at d=3.2 mm and such disappearance seems abrupt (not continuous) event. Figure 5 shows the relationship of OCV and cell temperature at each position. It is obvious that the closer the cell gets to the inner flame, the larger the OCV becomes and the voltage sharply decreased after the inner flame disappeared. More importantly, OCV increases even though the cell temperature hardly changed. These results indicate that the OCV performance changes by the change of distance d and this should correspond to the change of gas species concentration exposed onto the cell surface. Figure 6 shows the gas analysis by GC at each position under ϕ =2.0. It is understood that oxygen remains relatively when the inner flame disappeared and hydrogen, on the other hand, carbon monoxide were hardly formed. This change would affect the obtained OCV as found in previously.

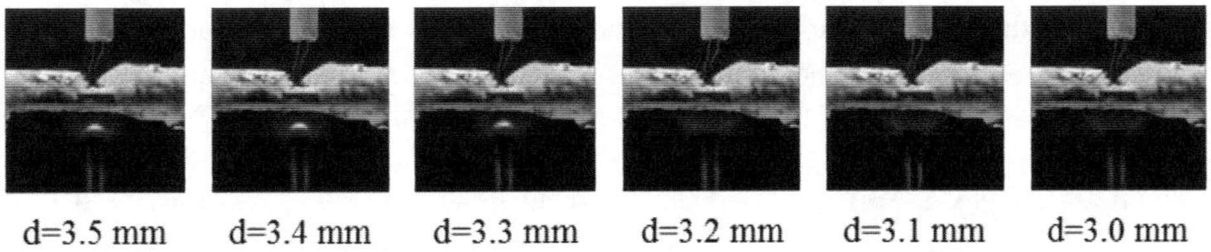

Figure 4. Change of flame shape by the change of distance between burner and DFFC cell under ϕ=2.0.

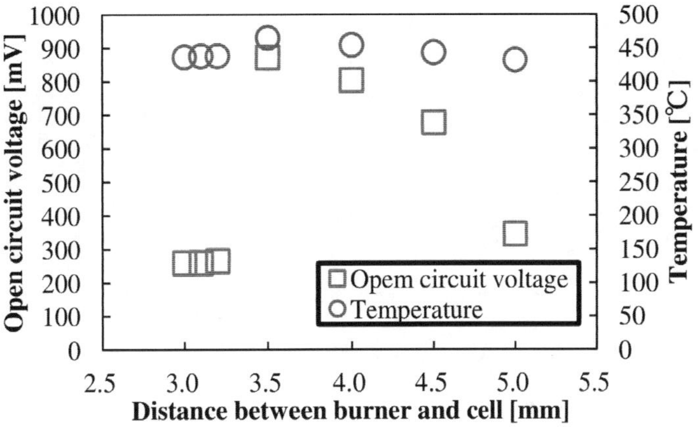

Figure 5. Relationship of open circuit voltage and cell temperature at various distances between burner and cell.

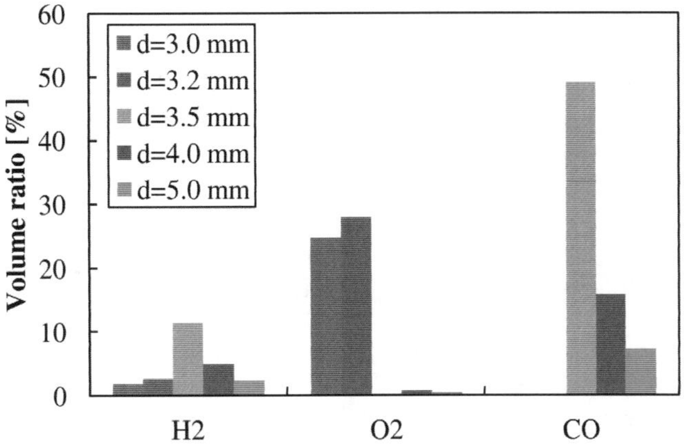

Figure 6. Analysis result of gas species concentration within the flame at each position under ϕ=2.0.

The difference of OCV performance under same condition can be considered to correspond to the local gas concentration at the vicinity of the cell and there is a theoretical equation called Nernst equation to evaluate the OCV [5]. Nernst equation is written as

$$E_{\mathrm{rev}} = \left|v_{\mathrm{O_2}}\right| \frac{RT}{n_{\mathrm{e}}F} \ln\left(\frac{p_{\mathrm{O_2,c}}}{p_{\mathrm{O_2,a}}}\right)$$

where E_{rev} is the Nernst potential (OCV), $v_{\mathrm{O_2}}$ is the absolute value of stoichiometric coefficient for oxygen, n_{e} is the number of transferred electrons, F is the Faraday constant (96 485.3365 s A/mol), R

is the universal constant (8.314 J /(mol K)), T is the cell temperature (K), $p_{O_2,c}$ and $p_{O_2,a}$ are the partial pressures of oxygen at the cathode and anode (Pa), respectively. According to Nernst equation, the OCV becomes large when the oxygen pressure of anode side is extremely small even the temperature is constant. Figure 7 shows the predicted OCV using Nernst equation with measured oxygen concentration detected by GC for $p_{O_2,a}$, while ambient value is used for $p_{O_2,c}$. Although there is a quantitative difference between the measured and predicted OCVs, an abrupt decrease of OCV is clearly captured as shown in dashed arrow in the figure.

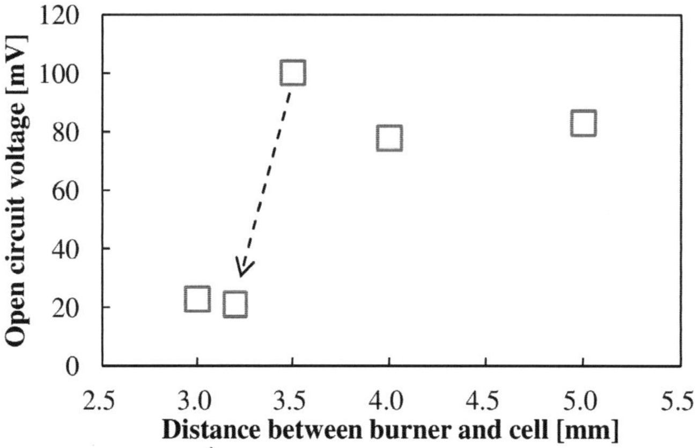

Figure 7. Comparison of the measured open circuit voltage and the calculation result of Nernst equation

4. Conclusion

In this work, we investigated the power generation performance of DFFC with micro-jet flame. The open circuit voltage (OCV) increases linearly with the increase of cell temperature when the inner flame does not appear clearly. On the other hand, when the inner flame is formed, the OCV increases drastically. These results imply that the OCV depends on not only the cell temperature but also the gas species concentration generated by the flame. Nernst equation might work to predict such drastic behaviour.

Acknowledgments

The authors are grateful for support provided by Kikusui Chamical Indstries Co., Ltd.

References

[1] You H, Gao H, Chen G, Abudula A and Ding X 2011 *Journal of Power Sources* **196** 2779-84
[2] Wang K, Ran R, Hao Y, Shao Z, Jin W and Xu N 2008 *Journal of Power Sources* **177** 33-39
[3] Lee K H and Kwon O C 2007 *Chemical Engineering Science* **62** 3710-19
[4] Kronemayer H, Barzan D, Horiuchi M, Suganuma S, Tokutake Y, Schulz C and Bessler W G 2007 *Journal of Power Sources* **166** 120-126
[5] Hanna J, Lee W Y, Shi Y and Ghoniem A F 2014 *Progress in Energy and Combustion Science* **40** 74-111
[6] You H Abuliti A, Ding X and Zhou Y 2007 *Journal of Power Sources* **165** 722-727

A Study on Energy Conversion Efficiency of Direct Flame Fuel Cell Supported by Clustered Diffusion Microflames

T Hirasawa[1] and S Kato[1]
[1] Department of Mechanical Engineering, Chubu University, 1200 Matsumoto-cho, Kasugai, 487-8501, Japan

E-mail: txh@isc.chubu.ac.jp

Abstract. The micro power generator based on clustered diffusion microflames and a direct flame fuel cell (DFFC) is proposed and its energy conversion efficiency is evaluated to investigate the optimum state of combustion. The clustered diffusion microflames are established on the 2.5 mm pitch 3-by-3 array of fuel jets with a diameter of 0.07 mm and a velocity of 24 m/s at the outlet by the interaction with other microflames around. The clustered diffusion microflames is more suitable for micro power generator than microflame array because of suppression of soot formation without air supply system. Although the conversion efficiency of the system with clustered microflames is quite sensitive to the separation distance between the burner and the cell, compared with the microflame array, it has attained the same level of conversion efficiency with microflame array, i.e. 0.45 %. The results of cell temperature suggest that the rapid decrease of conversion efficiency with increasing separation distance is caused by the decrease of species supply such as CO and H_2 rather than the decrease of heat supply. The observation of the secondary flame suggests that the reason of high sensitivity to the separation distance is because the cell position changes the flowrate of entrained ambient air and hence the flame equivalence ratio.

1. Introduction

One of potential micro power source that replaces conventional secondary batteries is a power generation system that uses a micro combustor as a heat source. The research and development on this kind of power generation system have been conducted in resent years [1-4]. The interest of this kind of power generation systems exists in the high thermal energy density (kJ/kg) of hydrocarbon. Since its thermal energy density is several ten times larger than the electric energy density of secondary batteries, its electric energy density is expected to overtake that of conventional batteries like Li-ion battery even at low energy conversion efficiency. In the power generation system that uses a micro combustor, there are a few kinds of approach, for example, the system based on thermal engines miniaturized by MEMS technology [1] or thermoelectric transducer [2, 3].

The present approach is different from them, that is, the system based on direct flame fuel cell (DFFC)[5], which uses solid oxide cell operated by a flame [6-8]. The way of power generation by DFFC is generally considered as follows; ideal fuels like CO or H_2 are supplied to the cell as intermediate products from hydrocarbon flame while the cell temperature is kept high enough by heat from the flame. If thermal efficiency attains more than 3 %, it will overtake the electric energy density of Lithium-polymer battery, assuming that the weight of device is a half of fuel weight and that fuel is dimethyl ether (DME)[5].

There are a few previous studies on DFFC [6-8], where they report the power density of cell but no energy conversion efficiency. However, evaluation and improvement of the conversion efficiency is necessary in the development of DFFC to make it a valuable micro power source. Therefore, the present study evaluates its energy conversion efficiency and aims to improve it with the improvement of combustion by use of microflames.

In order to improve conversion efficiency, decreasing the fuel consumption rate are important as well as increasing the electric power output. Thus, microflame array (MFA), which establish each flame with low fuel flowrate with high heat release density, is selected as one of potential flames to operate a solid oxide cell. Clustered diffusion microflames (CMF) established on the burners with tens of micrometers [9], is another potential flame to operate a solid oxide cell. CMF are established by the interaction with other microflames around because the diameter of each burner is too small to establish a single flame. CMF is suitable for micro power generator because it is able to suppress soot formation without air supply system. In the present study, therefore, the optimum condition to improve the conversion efficiencies of DFFC supported by MFA and CMF are investigated.

2. Experimental Setup

The experimental setup is shown in figure 1. Two types of burners are used for the experiments; one is for MFA and the other is for CMF. The MFA burner has an array of 2-by-2 coaxial double tubes, which issue fuel from the inner tube and air from the outer tube. The CMF burner has an array of 3-by-3 square poles with a 0.07 diameter fuel outlet. There is no air outlet in the CMF burner. Methane is used as fuel for all cases in the present study. The fuel flowrate for MFA burner is fixed at 60 sccm and air flowrate is adjusted to meet the equivalence ratio of mixture of fuel and air jet becomes 1.27. The fuel flowrate for CMF burner is fixed at 50 sccm.

Temperature of the cell surface (thin silver film) has been measured by using a type R thermocouple with wire diameter of 25 μm and thermography (FLIR, T390). The emissivity of thin silver film is determined to be 0.06 based on those results. After this calibration, temperature of cell cathode surface has been measured by using the thermography.

Two solid oxide cells with collector electrode areas of 0.66 cm^2 and 0.67 cm^2 are used for experiments, which are afterward called #6 and #7, respectively. It has been verified in advance that the difference of performance between those cell are negligibly-small.

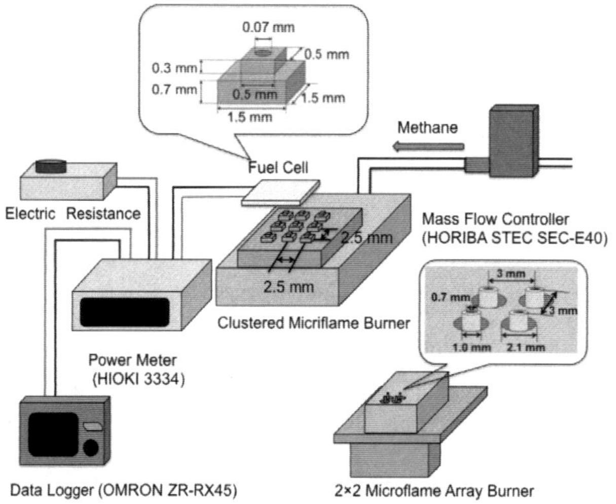

Figure 1. Schematic drawing of experimental setup

3. Results and Discussion

3.1. Influence of cell position on power generation

The power densities (W/cm^2) at the solid oxide cell above MFA and CMF are measured at various distances between cell and burner tip, h, and those value as well as evaluated conversion efficiencies based on a lower heating value are shown in figure 2. During those measurements, the load resistance is fixed at 2 Ω. No measurement result of power are obtained at h = 4 mm and below because flame goes around to cathode of the cell.

For burners of MFA and CMF, both the power density and the conversion efficiency shows maximum at h = 5 mm. At this distance, while there are no large difference between MFA and CMF, a little higher power density is attained at MFA and a little higher conversion efficiency is attained at CMF. The reason why CMF shows higher conversion efficiency at even at lower power density is simply because lower fuel flowrate supplied to the CMF burner. CMF shows a rapid decrease of power density and conversion efficiency at 6 mm and above, which is different from MFA. The direct photographs of MFA and CMF at various cell positions are shown respectively in figures 3 (a) to (d) and figures 3 (e) to (h) to investigate what causes this rapid decrease.

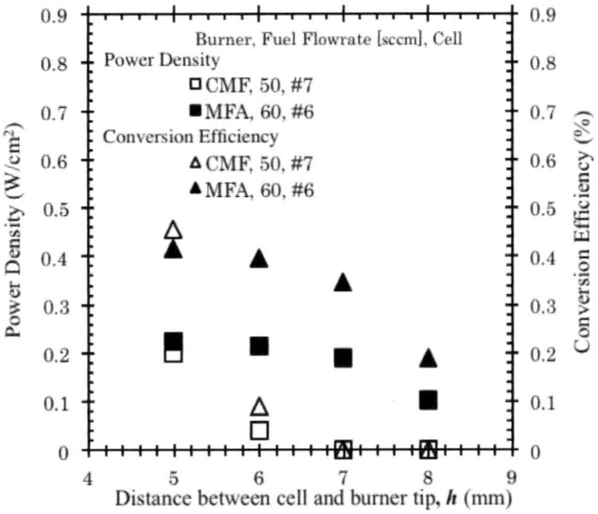

Figure 2. Effect of distance between cell and burner tip, h

The position of MFA flame-base does not change with cell position. This is because the mean velocity of fuel at the burner exit is 0.29 m/s, which is slow enough for diffusion microflames to be held regardless of cell position. Therefore, the influence of cell position on combustion state of MFA is smaller, which results in smaller change in conversion efficiency.

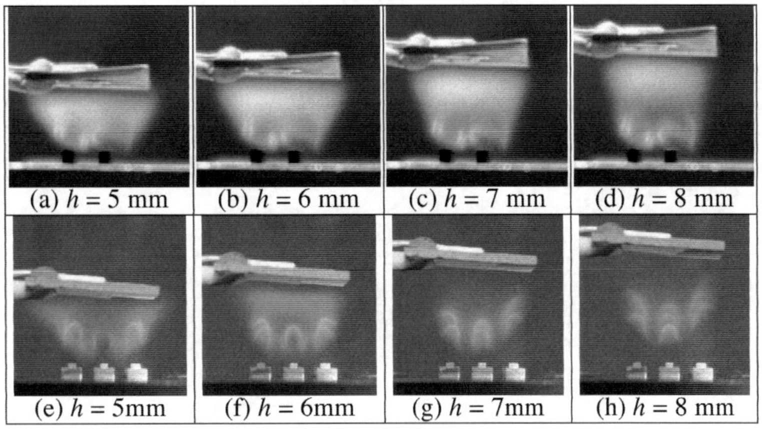

Figure 3. Direct photograph of MFA and CMF at various distance between burners and cell, h (MFA:(a)~(d), CMF:(e)~(h))

On the other hand, CMF is lifted a little at h = 8mm, and flame base position decreases with decreasing cell position from h = 8 to 6 mm. These CMF behavior and its mean velocity of fuel at the burner exit as high as 24 m/s suggests that flames are held in a stagnation flow where the cell works as a stagnation plate.

Thus, the decrease of cell position slows down the main flow velocity with divergence of the flow, which results in the decrease of flowrate of entrained ambient air, and hence attains the fuel rich condition. Consequently, the secondary flame is brightly observed at h = 5 mm. The maximum conversion efficiency has been attained when the secondary flame is brightly observed, which suggests that this combustion state is suitable for cell operation.

3.2. Influence of cell position on cell temperature

The surface temperature of cell is measured to specify what causes the rapid decrease of power density and conversion efficiency at *h* = 6 mm and above, specifically, whether the decrease of species supply such as CO and H_2 or the decrease of heat supply. The results are shown in figure 4 together with results on the conversion efficiency shown in figure 2. The comparison of the result of CMF between *h* = 5 mm and 6 mm shows that the change of cell temperature is negligibly small but the change of conversion efficiency is quite large. In addition, although the cell temperature decreases gradually in the range of *h* = 6 to 8 mm, the cell temperature is still larger than that of MFA. Therefore, it must be the decrease of species supply such as CO and H_2 that causes the power density and the conversion efficiency reduced largely at *h* = 6 mm and above.

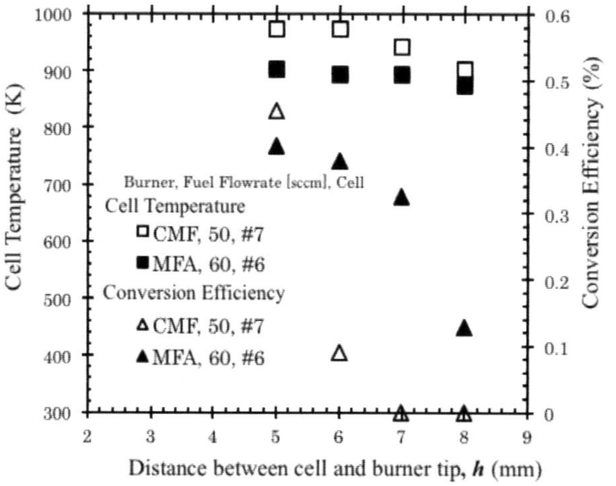

Figure 4. Effect of cell position on cell temperature

3.3. Influence of load resistance on power generation

The load resistance connected to the solid oxide cell above CMF is varied at *h* = 5 mm, and its influence is evaluated. The influences on cell voltage and current are shown in figure 5 (a) and influences on power density and conversion efficiency are shown in figure 5(b). The dashed line in figure 5(a) indicates electromotive force.

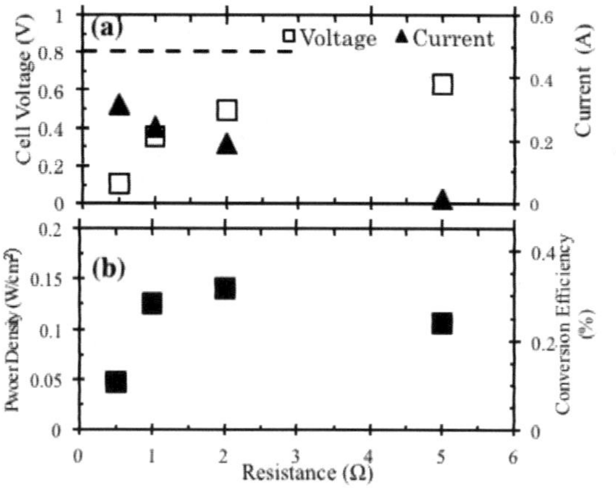

Figure 5. Effect of load resistance

As the load resistance decreases, the cell voltage and hence voltage efficiency are reduced but the current increases. Since the fuel in DFFC is supplied and consumed at constant rate regardless of the electric power output, the increase of current means increase of fuel utilization. Therefore, the reduction of load resistance directly leads to improvement of conversion efficiency unless the voltage efficiency becomes too low.

The conversion efficiency decreases rapidly with decreasing from 2 Ω and decreases gradually with increasing from 2 Ω. It is found that the conversion efficiency attains the maximum around 2 Ω, which means that it is the optimum value of load resistance on the present solid oxide cell.

4. Concluding Remarks

It is demonstrated that the clustered diffusion microflames properly supports the power generation of DFFC without air supply system. Although the conversion efficiency of the system with clustered microflames is quite sensitive to the separation distance between the burner and the cell, compared with the microflame array, it has attained the same level of conversion efficiency with microflame array, i.e. 0.45% at a separation distance of 5 mm and load resistance of 2 Ω. The results of cell temperature suggests that the rapid decrease of conversion efficiency owing to the change of the separation distance is caused by the decrease of species supply such as CO and H_2 rather than the decrease of heat supply. The observation of the secondary flame suggests that the reason of high sensitivity to the separation distance is because the cell position changes the flowrate of entrained ambient air and hence the flame equivalence ratio. Since the burner of clustered diffusion microflames works as an injector as well, designing of fuel jet will be a key to control the combustion state to be suitable for cell operation.

Acknowledgments

Authors wishing to acknowledge the micromachining of micro burners provided by NS Tool Co., Ltd and the grant supports provided by Chubu University Research Grant (A).

References

[1] Cho J H, Lin C S, Richards C D, Richards R F, Ahn J, Ronney P D 2009 *Proc. Combust. Instit.* **32** 3099
[2] Takahashi S and Wakai K 2009 45th AIAA/ASME/SAE/ASEE Joint Propulsion Conference & Exhibit, AIAA 2009-5246
[3] Weinberg F J, Rowe D M, Min G, Ronney P D 2003 *Proc. Combust. Instit.* **29** 941.
[4] Kim N I, Aizumi S, Yokomori T, Kato S, Fujimori T, Maruta K 2007 *Proc. Combust. Instit.* **31** 3243
[5] Hirasawa T, Sugiura H 2013 *Proc. Thermal Engineering Con. '13* 271 (in Japanese).
[6] Kronemayer H, Barzan D, Horiuchi M, et al. 2007 *J. Power Sources*, **166** 120
[7] Wang, K., Zeng, P. and Ahn, J. 2011 Proc. Combust. Instit. **33** 3431
[8] Zhu X, Wei B, Lü Z., et al. 2012 *J. Hydrogen Energy*, **37** 8621
[9] Hirasawa T, Sato N, Nakamura Y 2014 *J. Japan. Soc. Exper. Mech.* (2014 Special Issue) **14** s67

Wideband Piezomagnetoelastic Vibration Energy Harvesting

Anders Lei and Erik V. Thomsen

Department of Micro- and Nanotechnology, Technical University of Denmark, DTU Nanotech, Building 345B, DK-2800 Kongens Lyngby, Denmark

E-mail: `anders.lei@nanotech.dtu.dk`

Abstract. This work presents a small-scale wideband piezomagnetoelastic vibration energy harvester (VEH) aimed for operation at frequencies of a few hundred Hz. The VEH consists of a tape-casted PZT cantilever with thin sheets of iron foil attached on each side of the free tip. The wideband operation is achieved by placing the cantilever in a magnetic field induced by either one or two magnets located oppositely of the cantilever. The attraction force created by the magnetic field and iron foils introduces a mechanical force in opposite direction of the cantilevers restoring force causing a spring softening effect. In linear operation (without magnets) the harvester generates a RMS power of $141\,\mu W/g^2$ at 588 Hz with a relative bandwidth of 3.8% over a $100\,k\Omega$ load resistor. When operated with one magnet ideally positioned opposite the cantilever, a RMS power of $265\,\mu W/g^2$ is generated at 270 Hz with a relative bandwidth of 25%.

1. Introduction

Small-scale vibration energy harvesters (VEH) have attracted significant attention as power-source for autonomous wireless systems. A development highly stimulated by the significant advances in low power electronics and especially sensor units such as gyroscopes and accelerometers. One of the main challenges for VEHs is that a high quality factor provides high peak power but low frequency bandwidth and vice versa. High output power is essential, but low bandwidth is an issue if the source vibration frequency is drifting over time. Another challenge for small-scale VEHs is to keep the resonant frequency sufficiently low. This is usually achieved by vibrating structures with a very high aspect ratio between length and thickness which often results in an inherently fragile harvester. One approach to address these two key challenges is to introduce an external force in addition to the vibrating structure's internal spring restoring force. The result is a non-linear VEH that depending on the external force can be either multi-stable or monostable. A multi-stable harvester will have two or more stable points separated by a potential energy barrier. The harvester can therefore achieve high deflection at low frequencies due to switching between the stable points [1]. A monostable harvester will depending on the external force experience an either increasing of decreasing effective spring constant. The latter is desirable for energy harvesting as the effective resonant frequency is decreased while the harvester deflection is increased. High generated power can therefore be obtained with a high relative bandwidth.

The external force can be induced using both active or passive techniques. Since the objective of a VEH is to deliver power, passive techniques such as permanent magnets are naturally of preference. A setup with magnets usually consist of one or more external magnets and either

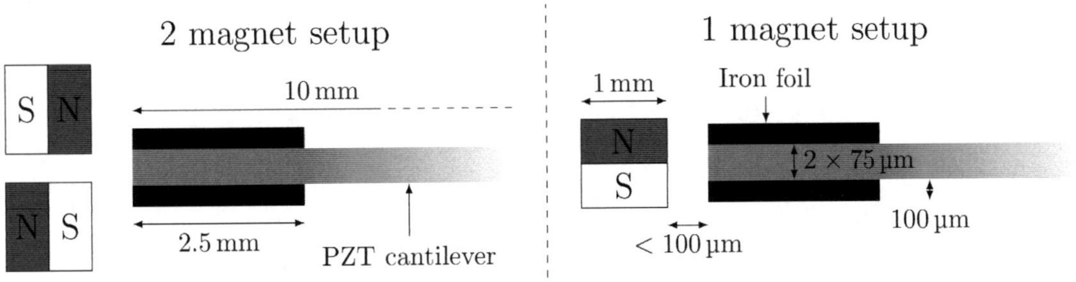

Figure 1. Sketch of the two magnetic setup investigated.

a permanent magnet or a ferromagnetic material on the harvester. Repelling forces can be achieved with a permanent magnet on the harvester but the realization of this is especially difficult on a small-scale. The majority of magnetic enhanced VEHs presented in literature are therefore proof-of-concept systems on a macroscopic scale [2–4] while only very few small-scale devices are reported [5]. The VEH presented in this work is comprised of a PZT based cantilever with dimensions suitable for a realistic packaging and implementation into a wireless sensor system. The magnetic setup utilizes ferromagnetic foils on the cantilever permanent magnets positioned oppositely the cantilever. Two different magnet setups with either one or two magnets are investigated, see figure 1. Two magnets creates the strongest attraction force but in an packaging assembly one magnet might be easier to implement.

2. Experimental setup

The VEH setup is sketched in figure 1 and is comprised of a 6 mm wide and 10 mm long cantilever diced from a tape-casted two-layered sheet of PZT. Each of the two PZT layers has a thickness of 75 µm with two outer and one middle conducting layers covering the full area. The cantilever is anchored to a PCB through soldering. The two PZT layers are poled in opposite directions, hence only electrical contacts to the two outer conducting layers are required. A 100 µm thick iron foil of 5.5 mm in width and 2.5 mm in length is attached on each side of the cantilevers free tip using glue. The magnets are attached in a fixture which is mounted on a XYZ 500 micropositioner stage from Quater Research and Development. The magnets are NdFeB with a size of 1 mm × 1 mm × 5.5 mm and a magnetization of 750 A m^{-1} from Webcraft GmbH. The PCB with cantilever and the stage with magnets are mounted on an aluminum fixture which then is attached to a TIRA TV51110 shaker system. The shaker is driven by a Agilent 33250A function generator and the input acceleration measured using a B&K Piezoelectric Accelerometer 8305 in connection with a B&K Type 2692-A-0I2 Charge Conditioner. The position of the magnet setup are measured using two laser displacement sensors from Micro-Epsilon. An optoNCDT 2300-10 for the vertical position and an optoNCDT 1401-5 for the horizontal position.

3. Measurements

The shaker is excited with a continuous sinusoidal frequency sweep from low to high value with a scan rate of 1 Hz s^{-1}. The harvester is connected to a 1040 resistance box from Time Electronics and the generated voltage is measured using a NI USB-6210 DAQ. The RMS voltage (V_{RMS}) frequency response is then extracted using a synchronization signal from the function generator, and the generated RMS power is calculated as $P = V_{RMS}/R$ where the load R is 100 kΩ for all measurements. The measured acceleration is stated in gravitational unit $g = 9.82\,\mathrm{m\,s^{-2}}$.

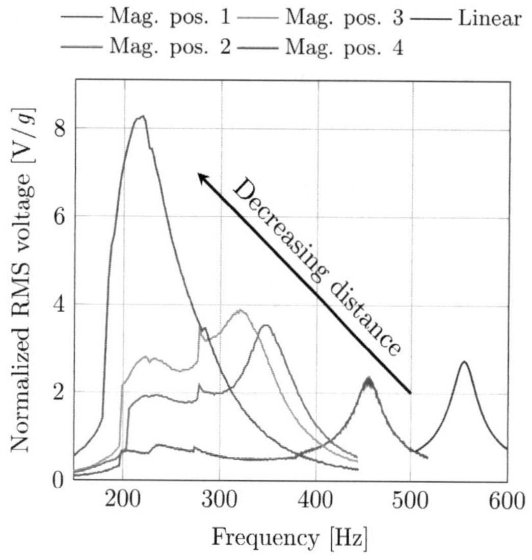

Figure 2. Harvester voltage response with no magnets (linear) and with the two magnet setup at four different distances.

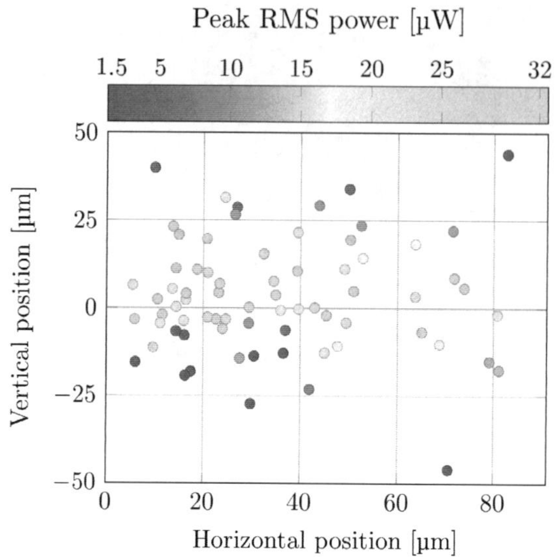

Figure 3. Peak RMS power as function of vertical and horizontal position of the two magnets relative to the cantilever tip.

3.1. Position study with 2 magnets

The induced magnetic force is naturally highly dependent on the strength of the magnetic field and hence the distance from cantilever tip with iron foils to the magnets. This distance dependency is examined in figures 2-3 using an excitation acceleration of $0.25g$. Figure 2 shows the voltage response for the linear case with no magnets present and for four different magnet positions where the distance is gradually decreased. As the magnets are moved closer to the cantilever the magnetic field around the iron foil increases in strength. The induced magnetic force decreases the linear spring force, the results is both a significant decrease in peak voltage frequency and an increase in generated voltage as the cantilever deflection increases due to the lower spring stiffness.

A comprehensive study of the magnet position dependency is seen in figure 3 as a scatter-plot with vertical and horizontal positions of the magnets relative to the cantilever tip. The beneficial non-linear effects of the two magnets vanished rapidly at horizontal distances of more than 100 µm from the cantilever tip. Ideally the magnets should be positioned less than 50 µm from the cantilever tip. The vertical axis is even more sensitive as only a position within a range of approx. 30 µm is of interest. The asymmetry around the cantilever axis in figure 3 is caused by a slight tilt between magnets and cantilever which is not possible to adjust with the XYZ 500 micropositioner stage. Despite using two permanent magnets with a high magnetization, a high precision of the magnet positions are required relative to the millimeter scale of the cantilever and magnet dimensions.

The non-linear force from the magnets not only increases the generated power, it also decreases the peak power frequency as seen in figure 2. This is naturally equally desirable as most ambient sources of vibrations are at a maximum of a few hundred hertz. Figure 4 shows the peak power measurements in figure 3 as function of peak power frequency with the magnitude of the scatter-plot being the absolute distance between magnets and cantilever tip. The correlation between lower peak power frequencies resulting in higher generated power is evident as is the required precision of the magnets position.

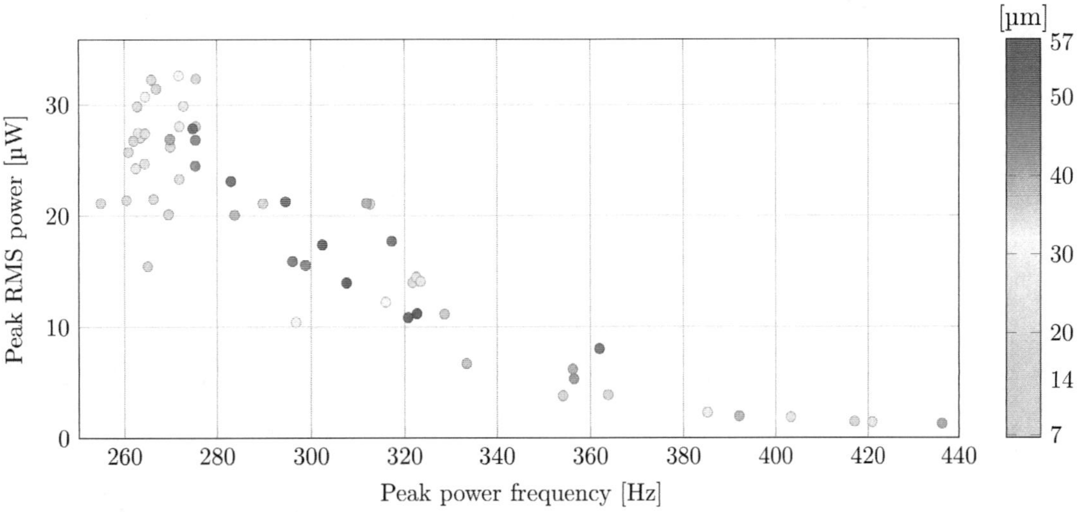

Figure 4. Peak RMS power as function of peak power frequency. Scatter magnitude is the absolute distance between magnets and cantilever tip.

3.2. Acceleration study for 1 and 2 magnet setup

The measurements in figures 2-4 are all performed with an input acceleration of $0.25g$. Figures 5-6 show the voltage response for different input accelerations with the 2 magnet and 1 magnet setup at a fixed optimal position. The ripples on the voltage response in the frequency range from 250 Hz to 350 Hz are caused by vibrations of the micro-stage. These stage vibrations results in asymmetric voltage responses at low accelerations for both magnet setups. A consequence of this voltage asymmetry is that the evaluation of peak voltage frequency and bandwidth are resulting in a high variation.

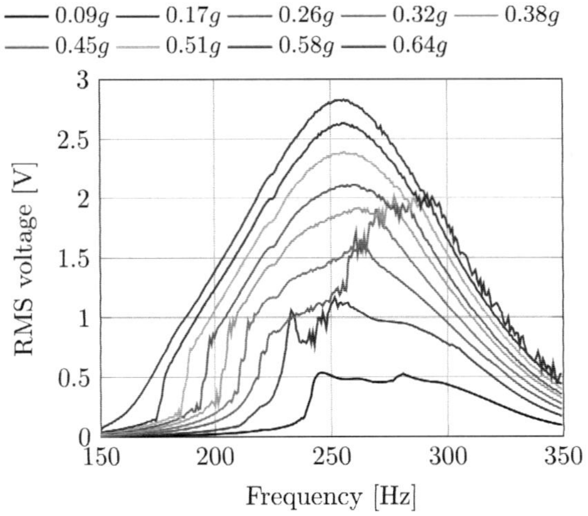

Figure 5. Voltage response with 2 magnets for different input accelerations.

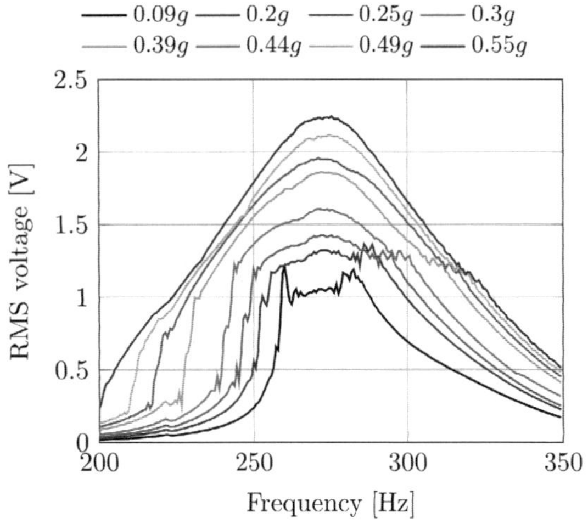

Figure 6. Voltage response with 1 magnet for different input accelerations.

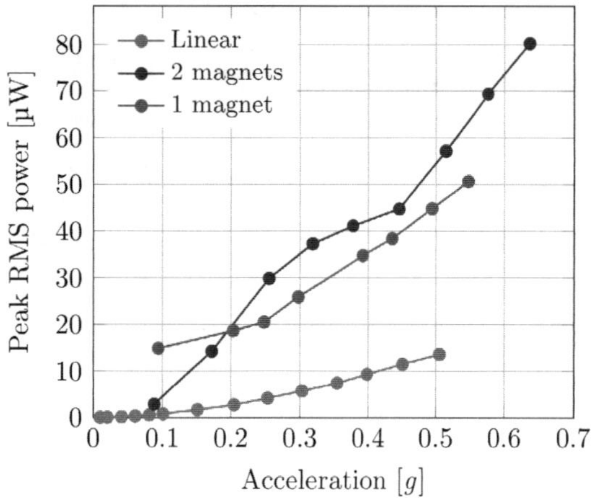

Figure 7. Peak RMS power as function of acceleration for the two magnet setups and without magnets.

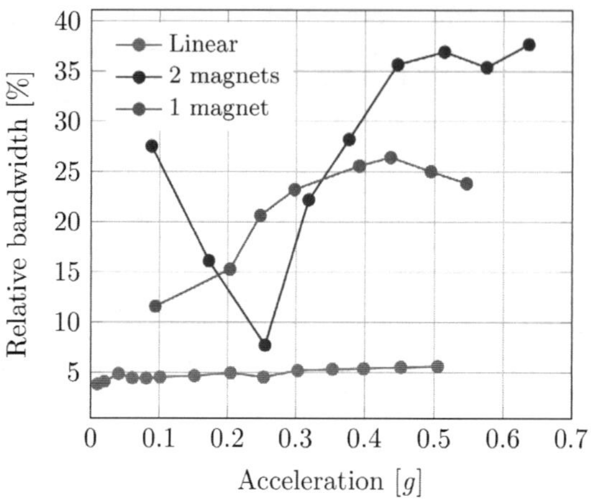

Figure 8. Relative bandwidth as function of acceleration for the two magnet setups and without magnets.

The summarized peak RMS power for both magnet setups are plotted in figure 7 together with the results for the harvester under linear operation without magnets. Using either of the two magnetic setups results in a significant increase in generated power compared to the linear harvester. While two magnets results in higher power than one magnet, the additional packaging challenge of precise positioning of two magnets should be taking into consideration. The relative bandwidth for the measurements in figure 7 are plotted in figure 8. Both magnet setup results in a considerable increase of the relative bandwidth. The fluctuations for the 2 magnet setup is caused by the asymmetric voltage response seen in figure 5. For the 1 magnet setup the magnetic field is perpendicular to the iron foil, the vertical vibrations of the stage is therefore considerable less influencing.

4. Conclusion

Wideband piezoelectric vibration energy harvesting has been demonstrated on a 6 mm wide and 10 mm long cantilever with external magnetic forces. The internal spring restoring force of the cantilever is compensated by attraction forces between thin iron foils at the free cantilever tip and magnets placed oppositely. The result is a decrease in effective resonant frequency, increase in generated power and a considerable enhancement of the relative harvesting bandwidth. In linear operation (without magnets) the harvester generates a RMS power of 141 $\mu W/g^2$ at 588 Hz with a relative bandwidth of 3.8% over a 100 kΩ load resistor. When operated with two magnets ideally positioned opposite the cantilever, a RMS power of 502 $\mu W/g^2$ is generated at 217 Hz with a relative bandwidth of 25%. With only one magnet, a RMS power of 265 $\mu W/g^2$ is generated at 270 Hz with a similar relative bandwidth of 25%.

References

[1] Stanton S C, Owens B A and Mann B P 2012 *Journal of Sound and Vibration* **331** 3617–3627 ISSN 0022-460X
[2] Erturk A, Hoffmann J and Inman D J 2009 *Applied Physics Letters* **94** 254102–254102–3 ISSN 00036951
[3] Ferrari M, Ferrari V, Guizzetti M, Ando B, Baglio S and Trigona C 2009 *Procedia Chemistry* **1** 1203–1206 ISSN 1876-6196
[4] Vocca H, Neri I, Travasso F and Gammaitoni L 2012 *Applied Energy* **97** 771–776 ISSN 0306-2619
[5] Ando B, Baglio S, Trigona C, Dumas N, Latorre L and Nouet P 2010 *Journal of Micromechanics and Microengineering* **20** 125020 ISSN 0960-1317

Vibration energy harvesting via parametrically-induced bistability

G Abadal[1], M López[2], W J Venstra[3], G Murillo[4] and F Torres[1]

[1]Departament d'Enginyeria Electrònica, Universitat Autònoma de Barcelona, 08193 Bellaterra, Barcelona, Spain, [2]NiPS Laboratory, Dipartimento di Fisica - Università di Perugia, I-06123 Perugia, Italy, [3]Kavli Institute of Nanoscience, Delft University of Technology, Delft 2628CJ, The Netherlands, [4]Department of Micro and Nanosystems, Institut de Microelectrònica de Barcelona IMB-CNM (CSIC), Bellaterra, 08193, Spain.

E-mail: gabriel.abadal@uab.cat

Abstract. The dynamic response to white Gaussian noise of a bistable non-linear vibration energy harvester based on the repulsive electrostatic interaction between a microcantilever and an electrode has been theoretically studied. The cantilever-electrode system can be brought from a linear regime characterized by a quadratic potential, when cantilever is far from the electrode, to a non-linear bistable regime characterized by a quartic potential, when both elements are close enough. This distance parameter, which is commonly used to tune bistability, is unusually used here also to inject the energy to the system in the form of displacement noise. Thus, the widening and shifting to the low-frequency region of the response spectrum as well as the enhancement of the *rms* out-of-plane vibration of the cantilever are both demonstrated through this parametrically-induced bistability.

1. Introduction

Energy Harvesting (EH) activity in the scientific community experienced a fast growth since approximately ten years ago. During this period, several transduction strategies have been successfully tested to convert ambient energy in the form of mechanical vibrations [1], temperature gradients [2] or electromagnetic radiations [3], [4] into useful electrical energy. However, Vibration Energy Harvesting (VEH) is playing with the advantage that the energy source, mechanical vibrations, are almost permanently available in a wide variety of situations such as industrial environments with machinery, transport vehicles, domestic environments with appliances and civil infrastructures.

On the other hand, MEMS technology has played a key role on keeping unaltered, or even enhancing, the VEH performance in terms of harvested power density, when volume of the VEH is unavoidably reduced following the miniaturizing trends imposed by ICTs evolution, since in many cases harvested power scales more slowly than volume.

However, when VEH devices are scaled down to micrometer dimensions, their efficiency decreases due to a mismatch in the frequency of the harvester response and the environmental vibrations, and due to the narrow bandwidth of the harvester's spectral response. To address this problem, the use of the chip substrate as inertial mass to reduce the resonance frequency of the harvester was tested in a CMOS-MEMS electrostatic implementation [5] and by using the AlN piezoelectric technology [6]. Alternatively, non-linear techniques have been also proposed, which widen the spectral bandwidth and improve the low-frequency response [7], [8], [9], [10]. A common approach is to non-linearize an initially linear mechanical element by making use of magnetic [7], electrostatic [10] or strain [11] interactions, which results in a bistable system with an energy barrier that can be tuned by the design. The device is then typically evaluated by measuring the broadband response to a white Gaussian displacement noise, which directly excites the transducer.

In this work we theoretically demonstrate that a wideband low-frequency response can also be obtained by injecting displacement noise through the bistability tuning parameter. Thus, the energy enters the system through a parametric excitation that tunes the system stochastically from a monostable to a bistable regime, rather than applying the noise to a bistable system in which the parameters are constant.

2. The Non-linear Energy Harvesting Device description

To demonstrate this concept, we use a previously developed electrostatic nonlinear VEH [10], which is based on a microcantilever with a trapped charge, as is shown schematically in Figure 1(left).

2.1. Geometry

A commercial Si_3N_4 triangular cantilever and a counter electrode (CE), both of which contain a permanent charge, q, of several fC, are placed at a controllable distance d_0 in the in-plane y direction. Tip ends of the cantilever and the CE and their trapped charges are precisely aligned in all three directions (in-plane, x,y and out-of plane, z).

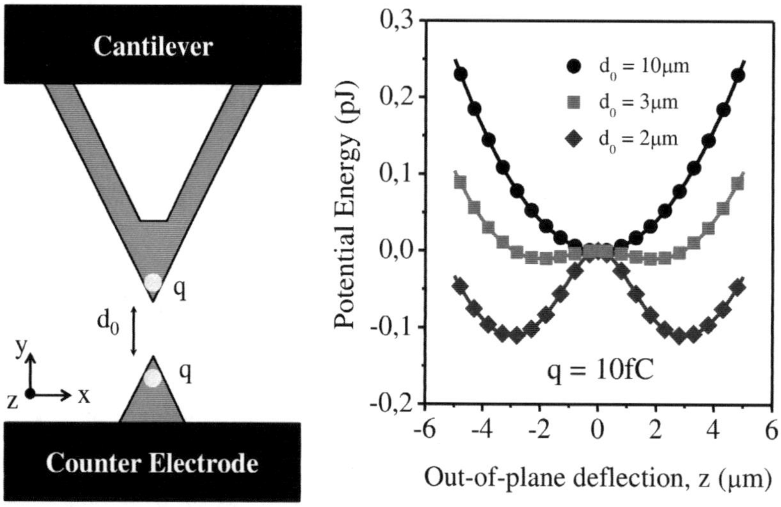

Figure 1. Schematic of the cantilever-based NLVEH (left) and potential energy (right) as calculated for three cantilever-electrode distances. For long distances (d_0=10μm) the potential is quadratic and monostable, while for short distances (d_0<3μm) it is quartic and bistable [10].

2.2. Potential energy

The strength of the repelling Coulomb force mutually exerted between the trapped charges of same sign is adjustable by varying d_0, which tunes the potential energy landscape of the cantilever from a quadratic potential (for large d_0) to a quartic potential (small d_0). In the latter case the system is bistable, with two symmetric potential wells that are separated by a potential barrier, as is shown in Figure 1(right). For the particular case of our previous work [10], an as measured implanted charge of 10fC is inducing a potential barrier of approximately 100fJ between two wells separated by a distance of around 3μm.

3. Dynamic response to noise

In our previous work [10], we demonstrated the existence of an optimum fixed distance, d_{opt}, for which, at a given input noise intensity, the rate of the random transitions maximizes and an optimum in the harvester's energy output is achieved [7,10]. Here we demonstrate that a similar optimum can be achieved when displacement noise is applied to the distance parameter, d_0. Figure 2 shows a simulation of the time evolution of the cantilever deflection when white Gaussian displacement noise (100 nm standard deviation) is applied around the working point defined by the static distance d_0.

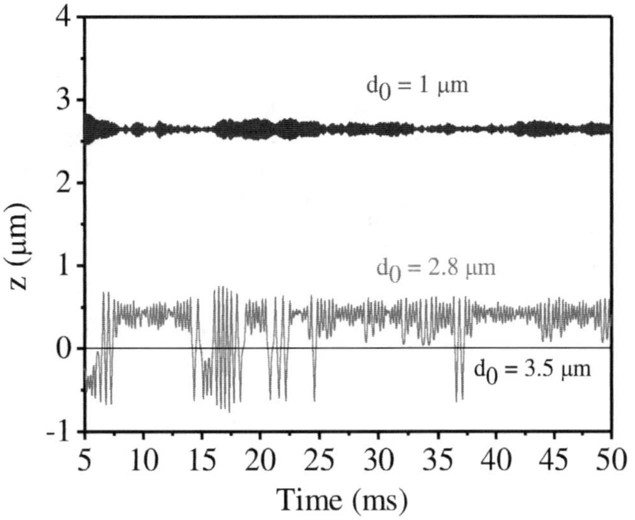

Figure 2. Time evolution of the cantilever out-of-plane deflection, z, when the cantilever-electrode distance is modulated by a 100 nm (standard deviation) white Gaussian displacement noise around a static distance d_0=3.5µm (black), d_0=2.8µm (red) and d_0=1µm (blue).

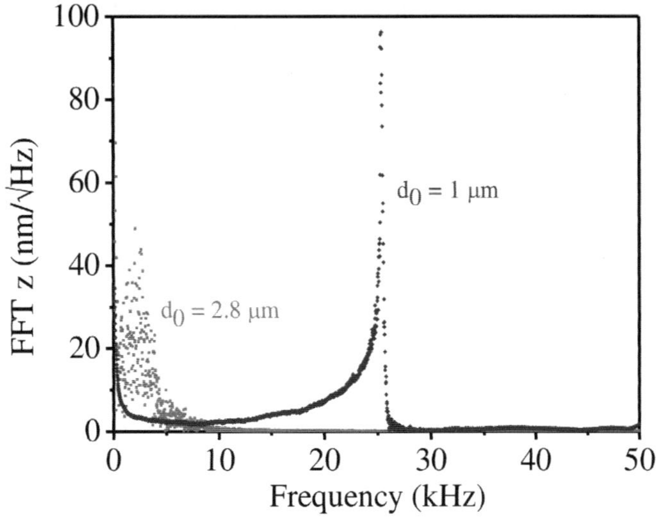

Figure 3. Frequency response of the cantilever as obtained from the FFT of the time series of figure 2 corresponding to the optimum distance (red curve, d_0=2.8µm) and the well trapping distance (blue curve, d_0=1µm).

Then, the working point d_0 is varied from a monostable regime with d_0=3.5 μm (black curve) to a deep in the bistable regime at d_0=1 μm (blue curve). In between, an optimum distance is found at d_0=2.8 μm (red curve), in which inter-well transitions occur.
In Figure 3, it is shown that an appropriate tuning of d_0, which corresponds to the optimum working point (d_0=2.8μm) results in a broadband response at a lower frequency, when compared for instance to the case of short working point distance (d_0=1μm), where the cantilever is trapped into on the potential wells.
To demonstrate the resulting enhancement on the harvested energy, the *rms* value of the cantilever tip deflection is plotted as a function of d_0 in Figure 4. A peak observed at d_0=2.8 μm, which marks a local optimum working point for the parametrically excited energy harvester, corresponds to the maximized inter-well transition regime (red curves in Figures 2 and 3). However, it is worth to notice that, additionally to this local maximum at d_0=2.8 μm, the vibration energy of the cantilever is also maximized in the deep bistable regime, as it is reflected by the increase of the out-of-plane *rms* deflection value produced when the working point is approaching d_0=0.5 μm. This second local optimum, which in this case becomes an absolute maximum, demonstrates that a tuning range for the energy enhancement wider than the first traditional one (d_0=2.8 μm) is also available.

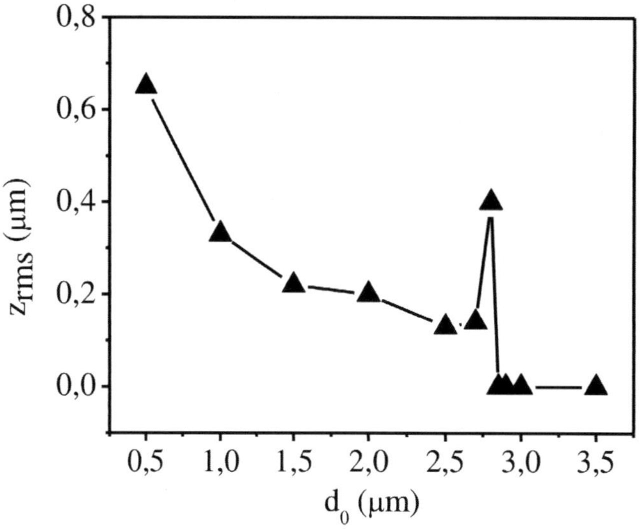

Figure 4. The *rms*-deflection of the cantilever tip measured at different static working point distances d_0, when white Gaussian displacement noise (100 nm standard deviation) is applied.

4. Conclusions

A theoretical analysis of the dynamic response to white Gaussian noise of a microcantilever based bistable non-linear vibration energy harvester has been presented. An unusual mechanism to introduce the energy to the system based on the injection of displacement noise through the bistability tuning parameter has been proposed. The widening and shifting to the low frequency band of the harvester spectral response and the enhancement of the *rms* vibration amplitude has been demonstrated. Additionally to the conventional optimum tuning value that produce an enhancement of the harvested energy, a new and wider optimum enhancement range is also found at the deep bistable regime.

Acknowledgments

Funding from the MINECO under contract TEC2013-41767-R (PISCES) has been greatly acknowledged by the authors.

References

[1] Roundy S, Wright P K and Rabaey J 2003 A study of low level vibrations as a power source for wireless sensor nodes *Computer Communications* **26** 1131-44

[2] Stevens J W 1999 Heat transfer and thermoelectric design considerations for a ground-source thermo generator. In: *Thermoelectrics, 1999. Eighteenth International Conference on,* pp 68-71

[3] Hagerty J A, Helmbrecht F B, McCalpin W H, Zane R and Popovic Z B 2004 Recycling ambient microwave energy with broad-band rectenna arrays *Microwave Theory and Techniques, IEEE Transactions on* **52** 1014-24

[4] Leland E S, Wright P K and White R M 2009 A MEMS AC current sensor for residential and commercial electricity end-use monitoring *Journal of Micromechanics and Microengineering* **19** 497-500

[5] Murillo G, Abadal G, Torres F, Lopez J L, Giner J, Uranga A and Barniol N 2009 Harvester-on-chip: Design of a proof of concept prototype *Microelectronic Engineering* **86** 1183-6

[6] Murillo G, Campanella H, Esteve J, Abadal G and Iop 2013 Novel optimized design of a piezoelectric energy harvester in a package for low amplitude vibrations *13th International Conference on Micro and Nanotechnology for Power Generation and Energy Conversion Applications (Powermems 2013)* **476** 012042

[7] Cottone F, Vocca H and Gammaitoni L 2009 Nonlinear Energy Harvesting *Physical Review Letters* **102** 080601

[8] Halvorsen E 2013 Fundamental issues in nonlinear wideband-vibration energy harvesting *Physical Review E* **87** 042129

[9] Khovanova N A and Khovanov I A 2011 The role of excitations statistic and nonlinearity in energy harvesting from random impulsive excitations *Applied Physics Letters* **99** 144101

[10] Lopez-Suarez M, Agusti J, Torres F, Rurali R and Abadal G 2013 Inducing bistability with local electret technology in a microcantilever based non-linear vibration energy harvester *Applied Physics Letters* **102** 153901

[11] Cottone F, Gammaitoni L, Vocca H, Ferrari M and Ferrari V 2012 Piezoelectric buckled beams for random vibration energy harvesting *Smart Materials and Structures* **21** 035021

Passively Self-Tuning Piezoelectric Energy Harvesting System

C G Gregg, P Pillatsch, P K Wright
University of California, Berkeley, Department of Mechanical Engineering,
Advanced Manufacturing for Energy, 2111 Etcheverry Hall, Berkeley CA, 94709
USA

E-mail: cgregg@berkeley.edu

Abstract. Real world systems that are candidates for vibrational energy harvesting rarely vibrate at a single frequency, nor are these frequencies constant over time. This necessitates that vibration harvesters operate over a wide bandwidth or tune their resonance. Most tunable devices require additional energy or active control to achieve resonance over various frequencies. This work presents a passively self-tuning energy harvester that autonomously adapts its resonant frequency to the input without requiring additional energy. The system consists of a clamped-clamped beam, a movable proof mass, and a piezoelectric patch bonded to the underside of the beam. It demonstrated an open-circuit voltage output of 668 mVrms at 160Hz, 0.65g input excitation. Discrepancies between displacement and voltage magnification factors upon tuning at higher frequencies are discussed, as well as instabilities of the system and sensitivity to proof mass characteristics.

1. Introduction

Energy harvesting is an important topic for the realization of ubiquitous wireless sensing and condition monitoring. Vibrational energy harvesting is attractive due to the abundance of ambient vibrations from machinery in manufacturing environments, but practical realization has been limited by the lack of broadband solutions. Dominant frequencies of real-world machinery are rarely known in advance and can shift with time, rendering traditional resonant harvesters impractical. Multiple approaches have been suggested for the active adjustment of resonant tuners [1-2], but most solutions require that some of the power generated be used for tuning. Ideal solutions should adjust their resonant frequency without additional energy input (passively) and without user intervention (autonomously).

Various self-tuning mechanisms are discussed in literature [3-6], including a string loaded with a moveable bead [4-6], a mass in a tube [6], and the beam with moveable mass system used in this work [3]. Previously, [3] demonstrated the ability of a fixed-fixed beam and movable proof mass system to exhibit self-tuning behavior. This work represents the first time the system in [3] has been utilized to make an operational energy harvesting device through the integration of piezoelectric materials. It also discusses the practical challenges of the system that have inhibited detailed system characterization and need to be overcome for real-world implementation.

2. Experimental Setup

A clamped-clamped beryllium-copper beam was fitted with a free-sliding proof mass (Figure 1-2) and mounted to a shaker. The proof mass was a 3D printed ABS bead with a screw that was used to adjust the size of the gap between the underside of the beam and the bead (Figure 3). A 3mm by 6mm single layer PZT piezoelectric element was soldered onto the underside of the beam near a clamping point (Figure 4) to exploit the maximum bending point of the beam and to avoid interference with the movement of the proof mass. The capacitance of this element was measured at 2.1 nF. A shaker was used to excite the system with a given input frequency, causing the mass to move (autonomously) along the beam until the system reached resonance. The displacement of the beam center and the input acceleration were recorded with a laser displacement sensor and accelerometer respectively. The open-circuit voltage output of the piezoelectric patch was also recorded.

Investigations into the stability of the system consisted of the same clamped-clamped beryllium-copper beam with proof mass, but without the piezoelectric element.

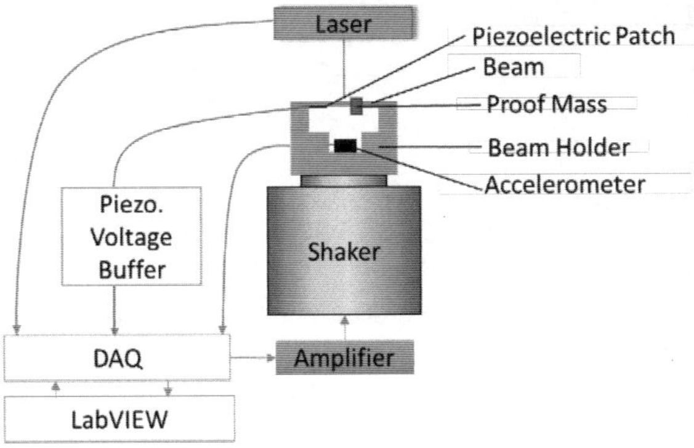

Figure 1. Schematic of the experimental setup.

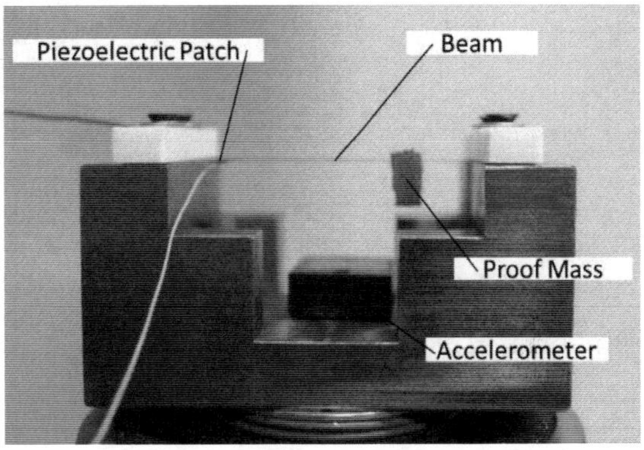

Figure 2. Photograph of the beam-mass system.

Figure 3. Illustration showing the proof mass 'gap' between the tip of the screw and the top of the 3D printed bead.

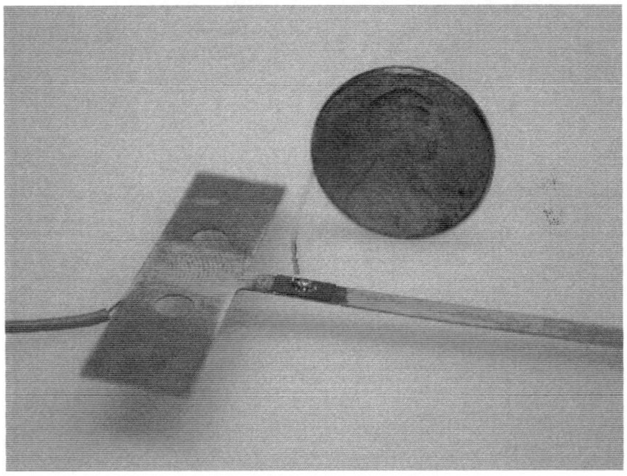

Figure 4. Close-up photograph of the piezoelectric patch bonded to the underside of the beam, with an American penny for scale.

3. Results

Figure 5 shows an example of self-tuning with an input frequency of 160 Hz and 0.65 g RMS acceleration. The proof mass was originally positioned near the center of the beam. When excitation was applied from the shaker, it moved outward (autonomously) from the center until the system achieved resonance. The self-tuning phenomenon is clearly identifiable by the simultaneous and sudden increase in the level of both mid-point beam displacement and open-circuit voltage output of the piezoelectric patch once resonance was achieved. Voltage output at 160 Hz was 668 mVrms. At the same excitation, power output was 0.33 μW when connected to a 470 kΩ impedance-matched resistive load.

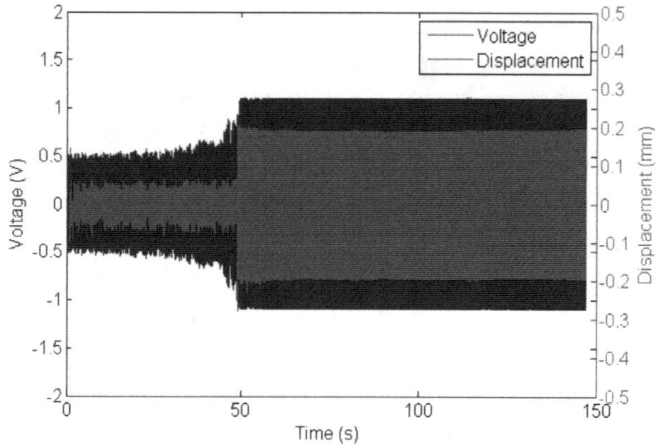

Figure 5. Self-tuning at 160 Hz on a beryllium-copper beam at 0.65 g RMS acceleration.

Figure 6 shows the relationship between open-circuit voltage and displacement magnification factors and frequency. Figure 7 shows the RMS voltage and displacement vs. frequency. All measurements were made at constant input excitation. It is hypothesized that the discrepancies between voltage and displacement magnification factors in Figure 6 at higher frequencies are due to the changing resonant positions of the proof mass along the beam at different frequencies. The different stable positions of the proof mass affect the shape of the first vibration mode and thus the beam deflection shapes and bending moments at the clamp points. Thus, it is thought that the stresses on the piezoelectric material (and output) are altered, while center beam displacement is less affected. Experimental verification of this hypothesis is necessary. At this stage, it is not fully understood why the voltage and displacement experience a drop at 164 Hz (Figure 7) and it will be cause for further investigation.

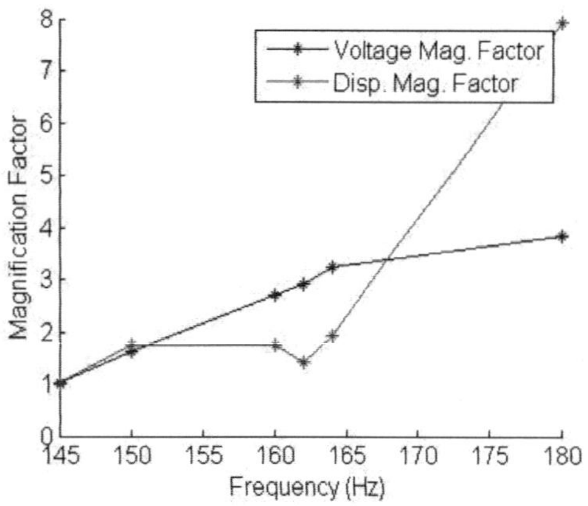

Figure 6. Magnification factor vs. frequency for both the piezoelectric voltage output and the beam center displacement.

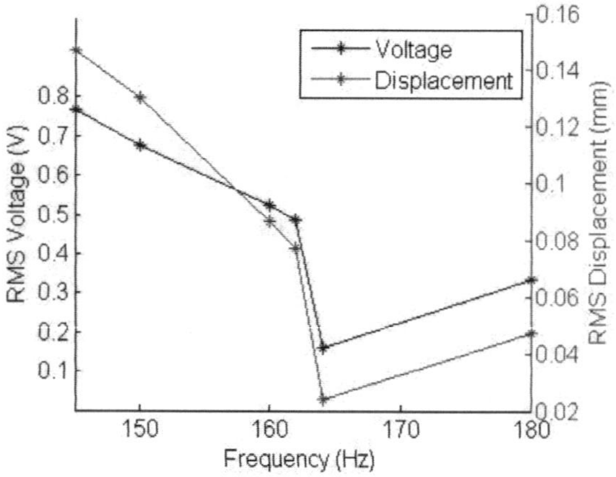

Figure 7. Tuned RMS voltage and RMS displacement vs. frequency.

Characterization of the effects of system parameters on voltage output and beam displacement revealed inconsistencies and instabilities of the system. The system is very sensitive to the characteristics of the proof mass, especially the size of the gap between the tip of the adjustment screw and the underside of the beam and mass. Adjustments on the order of fractions of a millimeter can change the behavior of the system. Such dependence on gap size was observed in similar self-tuning systems in literature [4]. Additionally, different gap sizes seem to be optimal only for certain combinations of mass and input acceleration.

Five distinct cases of behavior were observed: stable tuning, quasi-stable tuning, unstable tuning, damping, and inhibited motion. As shown before, Figure 5 is an example of stable tuning, where the mass moved to cause the system to enter and stay in a state of stable resonance. Figure 8 illustrates the quasi-stable tuning case, where though tuning behavior was observed, the mass would oscillate in and out of resonance about the tuned point. An example of unstable tuning is illustrated in Figure 9. If the gap size was not small enough, the mass would not remain tuned but rattle through its resonant point to the end of the beam. Alternatively, mass with a gap size that was too large could also exhibit self-damping behavior, as seen in Figure 10. Figure 11 shows the frequency dependence of the magnification factor for the same proof mass properties that yielded the case seen in Figure 10. Magnification factors below 1 confirm damping behavior and demonstrate damping over a large frequency range. Which of these behaviors, unstable tuning or self-damping, would occur when the gap was too loose is suspected to be a function of the mass, though more characterization is necessary to confirm. If the gap was too small, the motion of the mass was inhibited and no tuning behavior was observed. Further characterization would be aided by a redesign of the proof mass to include better control of gap size for better reproducibility.

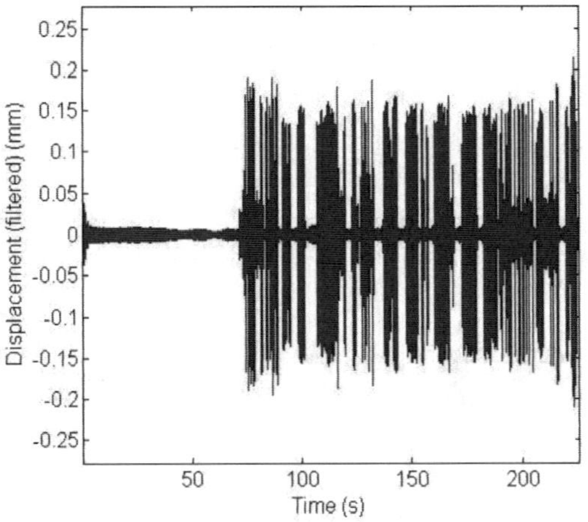

Figure 8. Example displacement plot of a case when the system displayed self-tuning behavior, yet the tuned position of the mass was unstable, oscillating in and out of resonance.

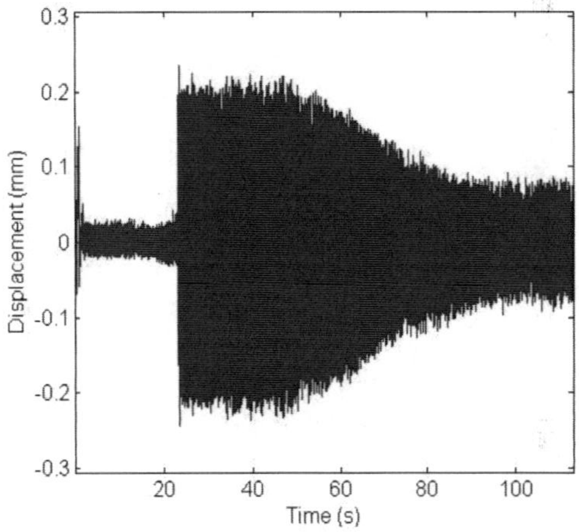

Figure 9. Example displacement plot of a case when the proof mass initially achieved resonance, but continued to move through the resonant point to the end of the beam, failing to achieve stability.

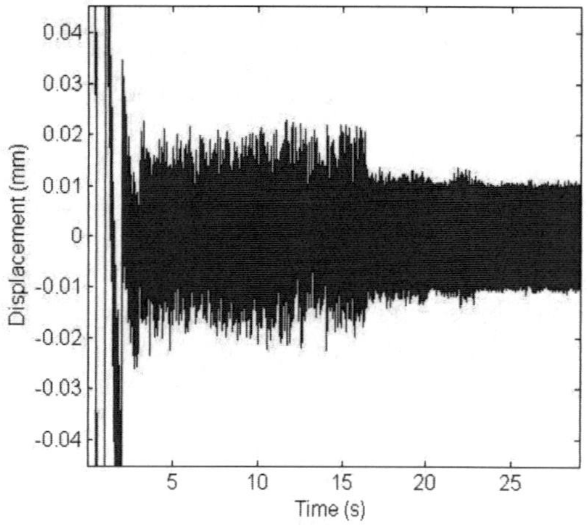

Figure 10. Example displacement plot of a case when the proof mass was 'loose' (large gap size) and exhibited self-damping behavior. The difference between damped and un-damped displacement is less in this case due to a lighter proof mass.

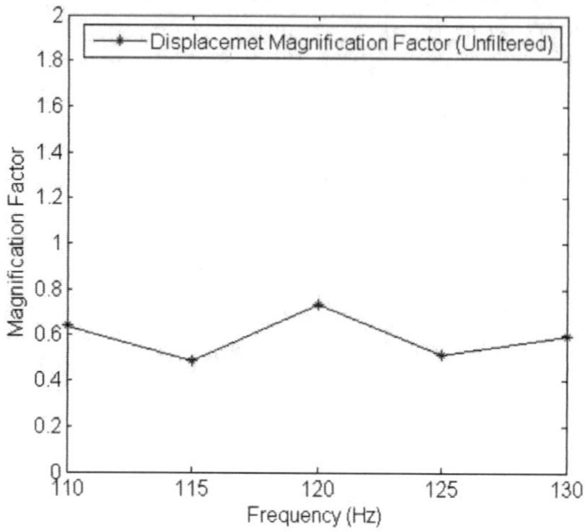

Figure 11. Magnification factor vs. frequency plot demonstrating the self-damping phenomenon across multiple frequencies.

4. Conclusions

This paper presents a passively self-tuning energy harvesting system. The integration of a small piezoelectric patch on the clamped-clamped beam and mass system displayed self-tuning behavior and yielded an open-circuit voltage of 668 mVrms at 160 Hz, 0.65 g excitation. Discrepancies between the magnification factors for the voltage and displacement were observed at higher frequencies (corresponding to proof mass tuning positions further from the center). These are suspected to result from the changing of the first vibration mode shape. The moving position of the proof mass alters the bending moments at the clamp points. However, experimental verification is needed.

This paper also discusses the practical challenges and instabilities associated with the system. The system behavior was strongly dependent on the characteristics (mass, gap size) of the moveable bead. Five distinct behavior cases were observed depending on the proof mass gap size: stable tuning, quasi-stable tuning, unstable tuning, self-damping, and inhibited mass movement. Previous publications alluded to such sensitivities [4-6], but rarely specified mass gap size as a sensitive parameter. Further characterization will focus on understanding the effect of gap size and proof mass on the system.

Acknowledgements

This research was supported by California Energy Commission award number 500-11-021.

References

[1] Ayala I N, Zhu D, Tudor M J, and Beeby S P 2009 *PowerMEMS.* 49-52.
[2] Roundy S, Zhang Y 2005 *Proceedings of SPIE.* 5649.
[3] Miller L M, Pillatsch P, Halvorsen E, Wright P K, Yeatman E M, and Holmes A S 2013 *Journal of Sound and Vibration.* **322** 7142-7152.
[4] Boudaoud A, Couder Y, Amar M B 1999 *The European Physics Journal B.* **9** 159-165.
[5] Babitsky V, Veprik A 1993 *Journal of Sound and Vibration.* **166** 77-85.
[6] Kozinsky I 2009 *PowerMEMS.* 388-391.

Electrostatic Force-induced Broadband Effect in Electret-based Vertical Vibration Energy Harvesters using Fine-grained Stainless Steel Oscillator

H Asanuma, M Hara, H Oguchi and H Kuwano

Department of Nano-Mechanics, Tohoku University, 6-6-01, Aoba, Aramaki, Aza-Aoba, Sendai, 980-8579, Japan

E-mail: haruhiko.asanuma@nanosys.mech.tohoku.ac.jp

Abstract. We propose a fine-grained stainless-steel as a promising material for a robust oscillator and investigate the dependence of frequency band width, resonance frequency, and output power on initial air gaps in electret-based vertical vibration energy harvesters. Beams of the oscillator showed a shallow side-etched depth less than 10 μm, as well as smooth edges. The oscillator succeeded in travelling over 1-mm displacement without fracture. Also, we found that broader frequency band, as well as lower resonance frequency, can be achieved with reducing the initial air gap, whereas the output power exhibited a peak value at an optimal initial air gap. The results may be attributed to the soft spring effect induced by the stronger electrostatic force. Maximum output power density and FWHM of frequency band width of our harvester are 4.7 $\mu W/cm^3$ and 14 Hz at initial air gap 0.3 mm and acceleration 4.9 m/s^2.

1. Introduction

Recently, considerable attention has been devoted to a miniature vibration energy harvester (VEH) as a power source for building an autonomous wireless sensor network [1-3]. An electrostatic VEH using a charged dielectric called "electret" is considered to be one of the promising power sources [4-10]. Electret-based VEHs are divided into two groups in terms of vibrational direction, vertical (or out-of-plane) and horizontal (or in-plane) one. An electret-based vertical VEH [6-10] has the advantage of low fabrication cost compared to a horizontal one because it can exclude the process of patterning electret, whereas it has problem with the development of a robust oscillator withstanding large displacement. Stainless steel with 83 times higher fracture toughness than that of silicon would be a promising material, however, an oscillator formed by wet etching bulk stainless steel involves in deeply side-etched beams and thus malfunction of vibration. Recently, a fine-grained stainless steel was developed for micro fabrication and robust springs [11], and it would meet our demands. Also, an initial air gap between the oscillator and the electret plays a critical role in the performance (frequency band width, resonance frequency, and output power) of electret-based vertical VEHs. Our intention in this study is to develop a side-etch free oscillator from a fine-grained stainless steel and to investigate the dependence of output power on initial air gaps.

2. Fabrication

We used a fine-grained stainless steel (Nippon Steel & Sumitomo Metal, Japan, SUS301L-SE1, grain size: 2 μm) and a typical stainless steel (Nilaco, Japan, SUS 304, grain size: 20-30 μm) with the same thickness of 0.08 mm to fabricate oscillators, and compared the resulting beam shape of them. Oscillators were fabricated through photolithography followed by immersed wet-etching with ferric chloride solution (38 wt%, 45 °C). Figure 1 shows images of a beam observed by laser scanning microscope and scanning electron microscope. The fine-grained SUS oscillator showed a shallower side-etched depth less than 10 μm and smooth edges, whereas the typical SUS one showed an over 60-μm side-etched depth and coarse edges, resulting in malfunction of vibration. Therefore, the fine-grained SUS is a promising material for a robust oscillator in electret-based vertical VEHs.

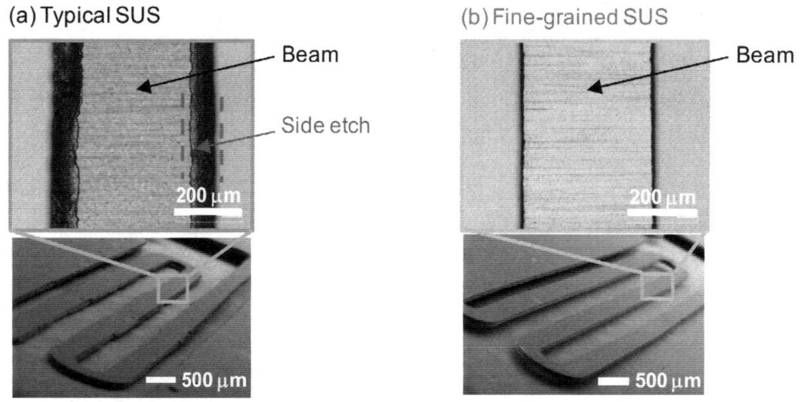

Figure 1. Images of oscillator's beam of (a) typical SUS and (b) fine-grained SUS observed by laser scanning microscope (upper) and scanning electron microscope (lower).

3. Experimental

We theoretically and experimentally investigate the dependence of output power (P_{out}) and peak-peak amplitude ($Amp_{p\text{-}p}$) of the oscillator on initial air gaps. Figure 2 shows a schematic of assembling of our harvester, photographs of the oscillator and harvester set on the shaker, and an experimental set-up for evaluation of output power. We used a laser Doppler vibrometer to measure amplitude of the oscillator during vibration. Applied acceleration and electrical load are 4.9 m/s^2 and 50.5 MΩ, respectively. The oscillator has the whole size of 2×2 cm^2, and the center square has the size of 1×1 cm^2. ABS resin was employed to form frames defining air gap distance by machined cutting. An electret was prepared by implanting charges onto a 10-μm-thick CYTOP polymer film (Asahi Glass, CTL-809M) spin-coated on a 1-mm-thick rigid copper plate using the point-to-grid corona discharging method.

We evaluated P_{out} and $Amp_{p\text{-}p}$ by numerically solving the following coupled differential equations, *equation of motion* (1) and *Kirchhoff's voltage law* (2) [7, 8]. Q, x, and y are the induced charges on the oscillator, the relative displacement of the oscillator, and the displacement forced by the shaker, respectively. Dot denotes the time derivation. First, we computed the air capacitance C_{air} and the spring constant k using FEM simulator (COMSOL Multiphysics), and substitute them into the coupled differential equation. The parameters are summarized in Table 1.

$$\begin{cases} m\ddot{x} + b\dot{x} + kx - \dfrac{d}{dx}\left(\dfrac{Q^2}{2C_{tot}}\right) - mg = -m\ddot{y}, \quad \text{where} \quad \dfrac{1}{C_{tot}(x)} = \dfrac{1}{C_{electret}} + \dfrac{1}{C_{air}(x)} & (1) \\ R\left(-\dot{Q}\right) = V_s + \dfrac{Q}{C_{tot}(x)} & (2) \end{cases}$$

Table 1. Parameters in numerical simulation.

Parameters		Value	Unit
Acceleration	a	4.9	m/s^2
Load	R	50.5	MΩ
Inertia mass	m	0.078	gram
Spring constant	k	37	N/m
Viscous damping coefficient	b	1.3×10^{-3}	Ns/m
Surface potential	V_s	-430	V
Relative permittivity	ε_r	2.1	-
Electret thickness	t	10	μm

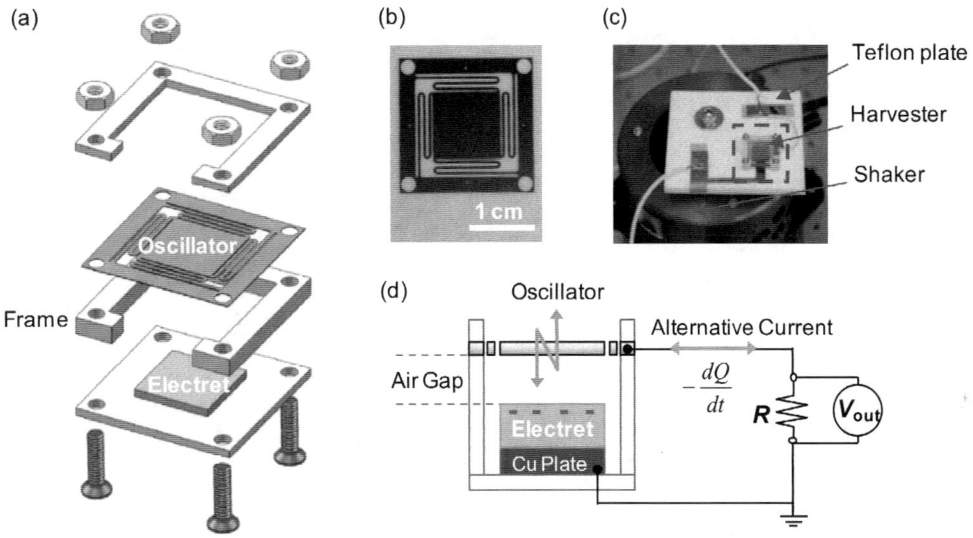

Figure 2. (a) Schematic of assembling, photographs of (b) oscillator and (c) harvester set on shaker, (d) an experimental set-up for evaluation of output power.

4. Results and discussion

Figure 3 shows simulated results of $Amp_{\text{p-p}}$ and P_{out} versus frequency with three different initial air gaps (0.4, 0.5, 1.0 mm). In simulation, a pull-in that the oscillator sticks to the electret was observed with initial air gaps less than 0.4 mm. Interestingly, although it decreased maximum P_{out} from the value obtained at the initial air gap of 0.5 mm, the reduced initial air gap of 0.4 mm exhibited broader frequency band and lower resonance frequency. These results may be attributed to the soft spring effect induced by the stronger electrostatic force.

Subsequently, we employed three different initial air gaps (0.3, 0.48, 1.05 mm) to investigate the theoretical prediction shown in Figure 3. Figure 4 shows experimental results corresponding to Figure 3. The oscillator fabricated from the fine-grained SUS succeeded in travelling over 1-mm displacement without fracture. Similarly, broader frequency band and lower resonance frequency have been observed by reducing initial air gaps. At 0.3-mm initial air gap, maximum P_{out} and FWHM are

2.8 μW (4.7 μW/cm^3 normalized with the volume) and 14 Hz, respectively. The discrepancies of output power between the simulation and the experiment may result from parasitic capacitance and/or inclination of the oscillator installed on machine-cut frames with design errors. Therefore, we found that a reduced initial air gap is effective in broader frequency band and lower resonance frequency in electret-based vertical VEHs.

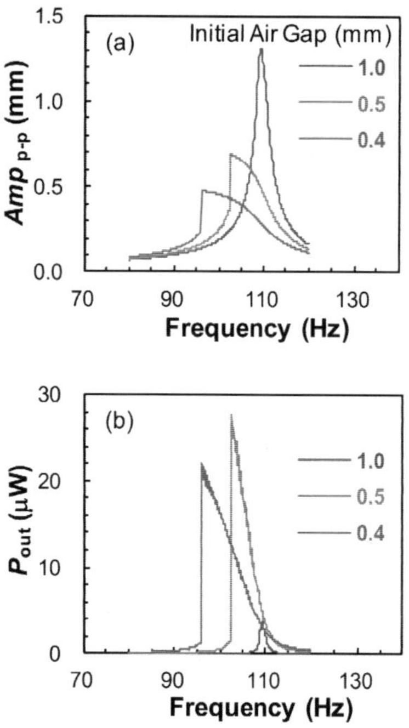

Figure 3. Simulated results of (a) peak-peak amplitude vs frequency and (b) output power vs frequency with three different initial air gaps.

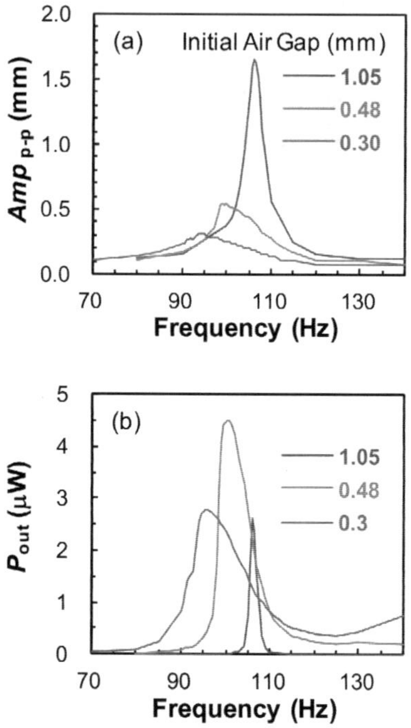

Figure 4. Experimental results of (a) peak-peak amplitude vs frequency and (b) output power vs frequency at three different initial air gaps.

5. Conclusion

In this study, we proposed a fine-grained stainless steel as a promising material for a robust oscillator and investigated the dependence of output power on initial air gaps in electret-based vertical vibration energy harvesters. Beams of the oscillator showed a hardly side-etched depth and smooth edges, and the oscillator succeeded in travelling over 1-mm displacement without fracture. Also, we found that reduced initial air gaps yields broader frequency band and lower resonance frequency, whereas it decreased output power from the value obtained at the optimum.

Acknowledgements

We would like to thank the Japan Society for the Promotion of Science (JSPS). This work was supported by Grant-in-Aid for JSPS fellows (No.H4254346).

References

[1] Kuwano H 1996 *Proc. of IEEE Seventh International Symposium on Micro Machine and Human Science* (Nagoya, Japan, 2-4 October 1996) pp 21-8
[2] Roundy S, Wright P and Rabaey J 2003 *Computer Communications.* **26** 1131
[3] Boisseau S, Despesse G and Seddik B 2012 *Electrostatic Conversion for Vibration Energy Harvesting, Small-Scale Energy Harvesting, Edited by Dr. Mickaël Lallart* (Rijeka: InTech) p 91
[4] Masaki T, Sakurai K, Yokoyama T, Ikuta M, Sameshima H, Doi M, Seki T and Oba M 2011 *J. Micromech. Microeng.* **21** 104004
[5] Crovetto A, Wang F and Hansen O 2014 *Journal of Micromechanical Systems.* **23** pp 1141-5
[6] Wang F and Hansen O 2014 *Sensors and Actuators A.* **211** pp 131-7
[7] Boisseau S, Despesse G, Ricart T, Defay E and Sylvestre A 2011 *Smart Materials and Structures.* **20** 105013
[8] Chiu Y and Lee Y C 2013 *J. Micromech. Microeng.* **23** 015012
[9] Asanuma H, Oguchi H, Hara M, Yoshida R and Kuwano H 2013 *Applied Physics Letters.* **103** 162901
[10] Asanuma H, Oguchi H, Hara M and Kuwano H 2013 *Proc. of PowerMEMS 2013* (London, UK, 3-6 December 2013) pp 192-6
[11] Sawada M, Adachi K and Maeda T 2011 *ISIJ International.* **51** pp 991–998

Piezoelectric energy harvester interface with real-time MPPT

A D T Elliott and P D Mitcheson

Department of Electrical and Electronic Engineering, Imperial College London, South Kensington, London, SW6 2BT, UK

E-mail: alwyn.elliott07@imperial.ac.uk

Abstract. Power of resonant piezoelectric harvesters can be severely limited if the damping force cannot be dynamically altered as the mechanical excitation level changes. The single-supply pre-biasing (SSPB) technique enables the Coulomb damping force to be set by a single voltage and so by varying that voltage, real-time adaptation to variations in the mechanical force can be implemented. Similarly the conduction angle of a diode bridge rectifier circuit can be altered by changing the biasing voltage applied. This paper presents a method of achieving this by altering the amount of energy transferred from the pre-biasing capacitor used in SSPB and the diode bridge rectifier to a storage battery via a buck converter. The control system was implemented on a FPGA and consumed 50 μW.

1. Introduction

The proliferation of wireless sensors has seen a dramatic increase with the development of ultra low power electronics and energy scavenging devices. However the capability of these sensors is limited by the variability in the applied mechanical excitation source from which energy is being harvested. This paper presents a complete system, which maximises the power extracted and transfers this to typical coin-cell battery, whilst adapting in real-time to variations in the mechanical excitation source.

Mechanical vibrations can be converted to electrical energy by mechanically stressing piezoelectric material. If we assume the mechanical stress takes a sinusoidal excitation form, the piezoelectric material will induce a sinusoidal output voltage. This voltage must be rectified before transferring to an energy storage device (e.g. battery [1]).

The simplest method of achieving this is through use of bridge rectifier circuit (Figure 1(a)). Maximum power extraction for the bridge rectifier technique (1) is achieved by setting the storage device voltage to an optimal value [2].

$$P_{diode_{max}} = f_0 C_p \left(V_{po} - 2V_D\right)^2 \text{ when } V_{opt_{BR}} = \frac{1}{2}\left(V_{po} - 2V_D\right) \quad (1)$$

where f_0 is the mechanical excitation frequency, C_p is the piezoelectric capacitance, V_{po} is the open-circuit voltage induced across the piezoelectric element with no electrical damping connected, and V_D is the voltage drop across the diode [3].

However more power can be extracted by modifying the charge on the piezoelectric material [3, 4]. The greatest improvement can be achieved using the single-supply pre-biasing (SSPB)

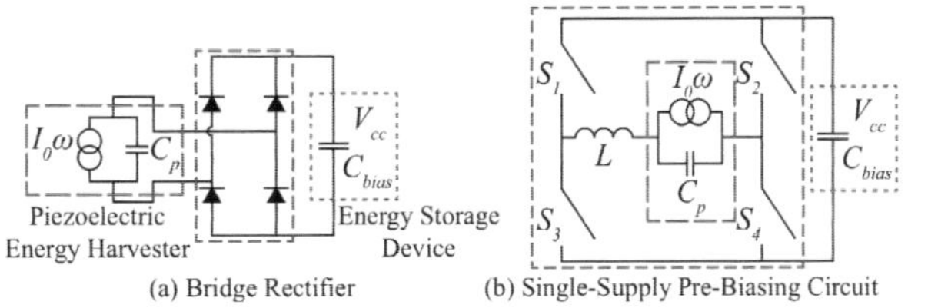

Figure 1. Piezoelectric energy harvester (where $I_0 = V_{po}\omega C_p$) with (a) bridge rectifier and (b) SSPB circuit.

technique (Figure 1(b)) [5], which operates as follows. Every half mechanical cycle, charge is placed on the piezoelectric material which will induce a force opposing the motion of the piezoelectric material. This acts to dampen the beam's motion, causing an increase in the power extracted. SSPB changes the circuit from a velocity-damped resonant generator (VDRG) to a coulomb-damped resonant generator (CDRG) [6], enabling the level of damping applied to be readily set by the pre-bias voltage, V_{cc}, applied. At the piezoelectric beam's extreme points of travel, charge from the pre-biasing capacitor, C_{bias}, is transferred on to the beam. This generates a Coulomb force to oppose the beam's motion until it reaches the opposite extreme position. The energy on the beam is then discharged back into C_{bias}, and the process repeats. Similarly to the bridge rectifier, the damping force can be optimally set by setting the voltage on the storage device in order to achieve maximum power extraction (2).

$$P_{SSPB_{max}} = {V_{po}}^2 f_0 C_p \left(\frac{8Q}{\pi}\right) \text{ when } V_{opt_{SSPB}} = 2V_{po}\frac{\gamma}{1-\gamma^2} \quad (2)$$

where Q is the Q-factor of the resonant current discharge path through the inductor and γ is the fraction of the magnitude of the voltage conserved on the capacitor of an RLC oscillator with Q-factor Q after a half-cycle of the oscillator [3].

Both techniques face an issue when the mechanical excitation force varies as the optimal voltage changes requiring the bias voltage to be changed in real time in order to maintain maximum power extraction. In this paper we present a system capable of detecting a change in the piezoelectric induced voltage (due to variance in the mechanical excitation force) and adjusting the damping force applied automatically by varying the amount of energy transferred from an intermediate energy storage element and a battery.

2. Maximum Power Point Tracking

Maximum Power Point Tracking (MPPT) for piezoelectric energy harvesters requires a system capable of monitoring the harvester and adjusting the conduction angle (rectifier case) or damping force (SSPB case) applied. The system must detect a change in the piezoelectric induced voltage, quantify the magnitude of the change, and apply an appropriate response. Since both the bridge rectifier and SSPB techniques use a voltage on a bias capacitor to apply the optimal conduction angle and damping force respectively, adding a buck converter with battery enables the bias capacitor voltage to be independently set. The MPPT scheme operates by adjusting the off-time of the buck converter causing the energy on the bias capacitor to be more or less frequently transferred, resulting in a decrease or increase in voltage respectively.

Figure 2 shows the suggested topologies to adjust the voltage applied for both (a) the bridge rectifier and (b) the SSPB circuit. The voltage on the biasing capacitor, C_{bias}, is controlled by varying the power transferred through the buck converter. If the time between energy transfers is increased, C_{bias} voltage will rise and a greater damping force will be applied to the piezoelectric material.

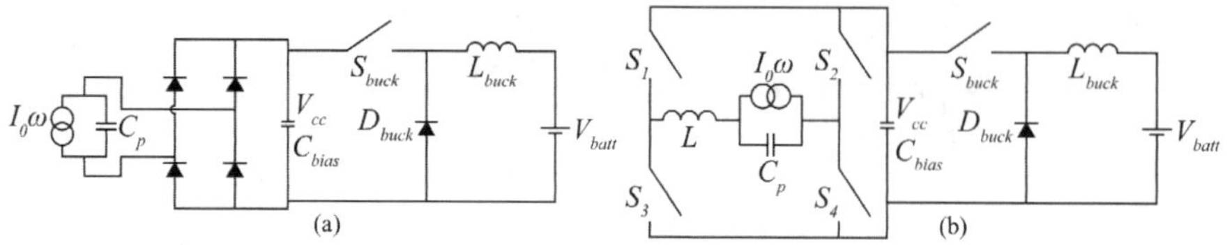

Figure 2. (a) Bridge rectifier and (b) SSPB circuit with buck converters and batteries.

A PSPICE simulation of the effect on power generation as energy is transferred from the pre-biasing capacitor to the storage battery is shown in Figure 3. The induced current was increased representing an increase in mechanical excitation. The power per cycle was measured over four energy transfer periods for two different buck converter off-times. The results show that the system with the longer buck converter off-time generated more power on average.

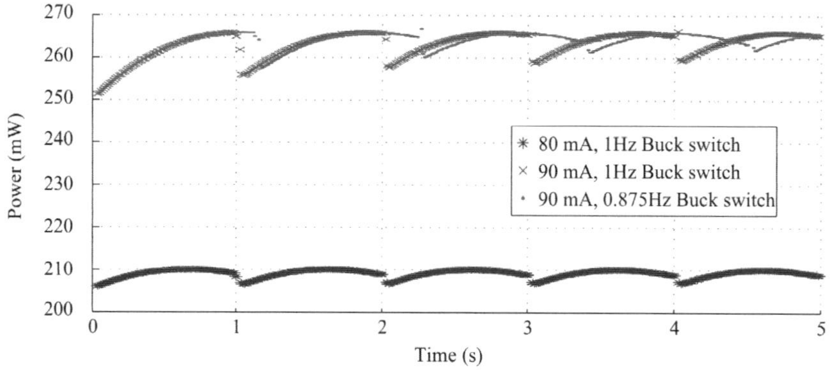

Figure 3. PSPICE simulation of the effect of adjusting the off-time of the buck converter on the power generated by a SSPB circuit for two different mechanical excitation forces.

3. Implementation

Implementing the proposed system requires three elements, a circuit capable of measuring the power generated, a controller to decide whether the power has increased or decreased since the previous measurement and a buck converter to transfer the energy. In designing all three circuits, power consumption must be minimised to ensure a high efficiency. Figure 4 shows the implementation used in this paper to demonstrate the MPPT technique.

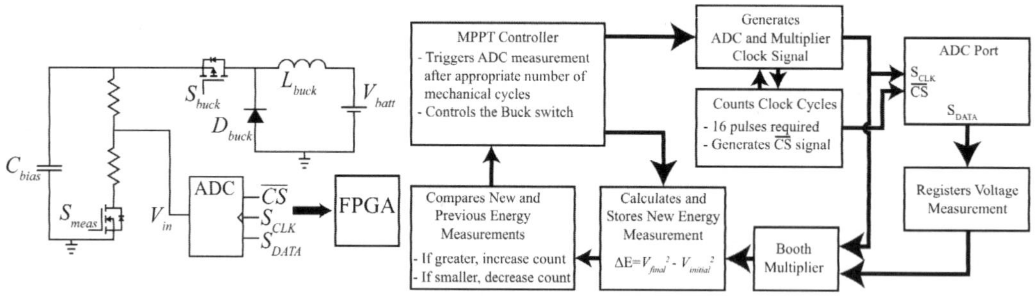

Figure 4. MPPT implementation and controller for either the bridge rectifier or SSPB circuit.

3.1. Change detection

The system needs to identify when the damping force needs to be adjusted due to a change in the mechanical excitation force to maintain maximum power extraction. A change can be detected by periodically measuring the energy transferred to the bias capacitor, C_{bias}. The energy being generated is found by comparing the square of C_{bias} voltage at the start, V_{init}, and end, V_{end}, over a set number of cycles (3). This result can then be compared to a previous measurement taken many cycles ago so that the trend for increasing or decreasing power extraction can be determined.

$$\Delta E = \frac{1}{2} C_{Bias} \left({V_{end}}^2 - {V_{init}}^2 \right) \tag{3}$$

Measuring the voltage on the capacitor with sufficient accuracy presents several difficulties which need to be overcome. Any power consumption used to measure the voltage detracts from the efficiency of the energy harvester. The measurement and control circuitry therefore should operate at the lowest voltage possible, however the voltage on the biasing capacitor maybe several times larger than this. The solution proposed in this paper is to use a potential divider which is switched in across the biasing capacitor during a measurement (Figure 4). The switch is an n-type MOSFET (BSS138) which can be driven by a low power FPGA (Igloo Nano), which can also be used to implement the SSPB scheme [7]. A low power Analog-to-Digital Converter (ADC AD7468) can then be used to measure the voltage and a Booth Multiplier algorithm [8] on the FPGA can square the value.

3.2. Controller

The controller shown in Figure 4 was implemented on the low power FPGA for MPPT. The controller works as follows: initially the buck converter transfers energy every 80 cycles, therefore at a count of 38 and 42 cycles the voltage is measured as V_{init} and V_{end} respectively. Both values are squared and V_{init} is subtracted from V_{end}. The result is compared against the previous energy measurement, 80 cycles previous. If the result is greater than the former, the count is increased allowing the voltage to rise and increasing the damping force. If the result is less, the counter is decreased, reducing the voltage the bias capacitor will attain, decreasing the damping force. However if it is the same, the counter stays the same and the damping force is not altered. The cycle counter signal for the SSPB signal is naturally generated by the switching pulses used to implement the scheme. The bridge rectifier however requires a secondary sense piezoelectric beam to be mechanically coupled to the generating beam and a zero crossing detector with an over-voltage protection MOSFET to generate the same signal [9].

The ADC and Booth Multiplier algorithm both need a clock signal in order to operate. The simplest solution is to use a crystal oscillator [10]. However the start-up time and power for a clock signal that may only be operating for a few microseconds, once a second renders this solution undesirable. Instead it is more efficient to generate the clock signal using a RC circuit. This technique is also used in SSPB circuit to generate the on-time pulses for the switches [7].

4. Results

The MPPT circuit was tested with a bridge rectifier circuit connected to a buck converter with a 1.2 V power supply representing the battery. The on-time for the buck switch was set to 118 μs and the off-time was varied by an Igloo Nano FPGA. A 50 nF piezoelectric loudspeaker was used to represent the harvester, which was mechanically excited at 200 Hz with different accelerations causing different induced open-circuit voltages, Vpo. Power generation on the biasing capacitor was measured using a Yokogawa WT210 power meter at 5 second intervals. The results were compared with the theoretical limit given in (1) demonstrating the circuit's functionality (Figure 5). It can be seen at high degrees of acceleration (corresponding to large

V_{po}), the system achieves close to the theoretical limit. However at lower excitation levels, performance is degraded due to the number of cycles between transfers and the number of cycles between measurements converging. This issue could be alleviated by increasing the resolution of the ADC so fewer cycles are required between measurements. The power consumption of the controller was measured as 50 μW when the piezo was inducing an open-circuit voltage of 6.75 V.

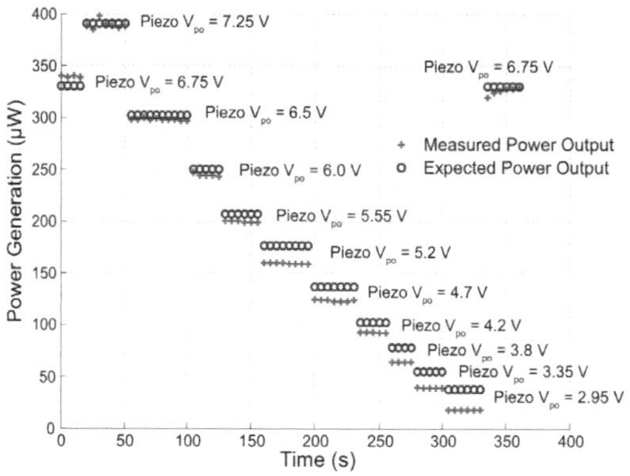

Figure 5. Comparison between measured power generation using MPPT implementation and bridge rectifier theoretical maximum power limit for several different excitation forces.

5. Conclusion

An algorithm and implementation have been presented which enable a piezoelectric energy harvester to automatically adapt to variation in mechanical excitation. The system is designed to track the maximum power point by adjusting either the conduction angle or the damping force for the bridge rectifier and SSPB circuit respectively. Future work will be focussed on reducing the power consumption of the controller and combining it with a SSPB circuit.

Acknowledgements

The authors would like to thank ABB and EPSRC for their sponsorship of this project.

References

[1] Gerasopoulous K, McCarthy M, Royston E, Culver J N and Ghodssi R 2008 *Journal of Micromechanics and Microengineering* **18** 1–8
[2] Roundy S, Wright P K and M R J 2003 *Energy Scavenging for Wireless Sensor Networks* (Boston MA: Kluwer Academic Publishers) ISBN 1-4020-7663-0
[3] Dicken J, Mitcheson P D, Stoianov I and Yeatman E M 2012 *IEEE Transactions on Power Electronics* **27** 4514–4529
[4] Lallart M and Guyomar D 2008 *Smart Materials and Structures* **17**
[5] Dicken J, Mitcheson P D, Elliott A D T and Yeatman E M 2011 *PowerMEMS* (Seoul, South Korea) pp 46–49
[6] Miller L M, Mitcheson P D, Halvorsen E and Wright P K 2012 *Applied Physics Letters* **100**
[7] Elliott A D T and Mitcheson P D 2012 *IEEE Sensors 2012* (Taipei, Taiwan)
[8] Booth A D 1951 *Quart. Journ. Mech. and Applied Math.* **4**
[9] Elliott A D T, Trice M R and Mitcheson P D 2012 *PowerMEMS* (Atlanta, Georgia) pp 347–350
[10] Integrated M 2003 Microcontroller clock-crystal, resonator, rc oscillator, or silicon oscillator? Application note 2154 Maxim Integrated

MEMS electrostatic vibration energy harvester without switches and inductive elements

V Dorzhiev[1], A Karami[2], P Basset[3], V Dragunov[1] and D Galayko[2]

[1] Department of Semiconductor Devices and Microelectronics, Novosibirsk State Technical University, Novosibirsk, Russia
[2] LIP6-UPMC – Sorbonne Universités, France
[3] Université Paris-Est, ESYCOM, ESIEE Paris, France

E-mail: dorzhiev@corp.nstu.ru

Abstract. The paper is devoted to a novel study of monophase MEMS electrostatic Vibration Energy Harvester (e-VEH) with conditioning circuit based on Bennet's doubler. Unlike the majority of conditioning circuits that charge a power supply, the circuit based on Bennet's doubler is characterized by the absence of switches requiring additional control electronics, and is free from hardly compatible with batch fabrication process inductive elements. Our experiment with a 0.042 cm^3 batch fabricated MEMS e-VEH shows that a pre-charged capacitor as a power supply causes a voltage increase, followed by a saturation which was not reported before. This saturation is due to the nonlinear dynamics of the system and the electromechanical damping that is typical for MEMS. It has been found that because of that coupled behavior there exists an optimal power supply voltage at which output power is maximum. At 187 Hz / 4 g external vibrations the system is shown to charge a 12 V supply with a output power of 1.8 μW.

1. Introduction

The rapid progress in wireless technologies is stimulating the development of new types of autonomous power supplies for microelectronic devices, instead of traditional batteries requiring periodic replacement or recharging. Power supplies converting energy of ambient energy sources such as vibrations, solar, thermal, wind, wave and RF energies into electrical energy (energy harvesters) are becoming promising substitutes of batteries. However, it is not possible to get rid of using traditional batteries up to now, partly, because of inconstant availability of ambient energy sources (for example, at sunset). Therefore, the best option for autonomous power supplies is the coupling of a rechargeable battery and an energy harvester.

Among the available ambient energy sources mentioned, the mechanical vibration energy source is of specific interest due to its wide availability. There are several types of vibration energy harvesters such as electromagnetic, piezoelectric, electrostatic, etc. Electrostatic vibration energy harvesters (e-VEHs) appear to be the most suitable for miniaturization. They are not as bulky as electromagnetic harvesters, and they do not suffer from fatigue or depolarization like piezoelectric devices.

One of the main challenges of e-VEH implementation is the design of conditioning and interface electronics. Conditioning circuits that use the energy of vibration for increasing the charge of the power supply are of particular interest. The majority of such conditioning circuits contain inductive elements, which are hardly compatible with batch fabrication process, and switches, which require additional power-consuming control electronics for the correct sequencing of the different phases of energy conversion [1–4]. Recently proposed in [5], a conditioning circuit based on Bennet's doubler operates without any external controllable switches or inductors. To date, its operation has only been proven with macroscale variable differential capacitors (a couple of capacitors varying synchronously with opposite phases), whose variation was induced by an external motor. However, practical e-VEHs are implemented in MEMS technologies. The use of a realistic MEMS device (a microresonator + a capacitive transducer) induces three consequences: (i) limited transducer capacitance variation magnitude, (ii) strong electromechanical coupling impacting the dynamics of the system, (iii) use of monophase (single) variable capacitor, rather than a differential one made of 2 capacitors. These consequences could adversely affect the operation of the circuit.

The paper presents the results of theoretical and experimental study of a monophase (single-capacitor) MEMS e-VEH with a conditioning circuit based on Bennet's doubler. This novel experimental study of the circuit with real MEMS device shows results that differ significantly from those obtained in the previous experimental study with a macroscale device.

2. Theory

The e-VEH conditioning circuit based on Bennet's doubler with a single variable capacitor is shown in figure 1. It consists of a power supply (V_0), three diodes ($D_1 - D_3$), a fixed capacitor (C_{store}) and a variable capacitor (C_{var}) representing the transducer, associated with a mechanical resonator. The load (R_{load}) may be connected in parallel to the power supply. We will analyze the operation of the circuit without losses, assuming that diodes are ideal (no reverse current and threshold voltage).

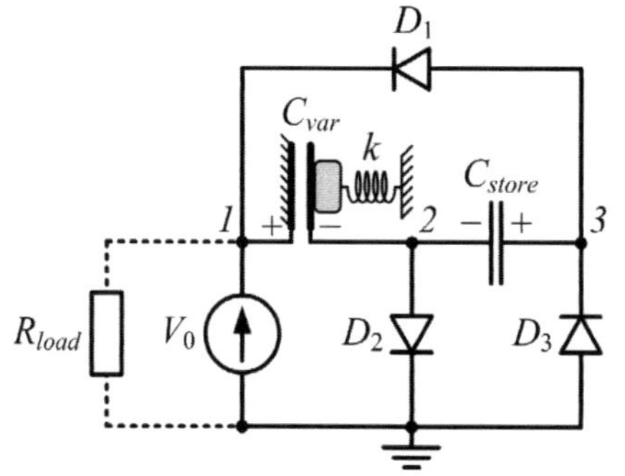

Figure 1. e-VEH conditioning circuit based on Bennet's doubler with a single variable capacitor.

The element called "power supply" may have different practical implementations: it can be a chemical accumulator or a large reservoir capacitor. As we will show, its voltage determines the energy conversion rate of the network. Hence, if the operation conditions of the circuit vary in time and if optimal circuit operation is maintained, this voltage may be variable (adjustable). However, the analysis of the circuit assumes a slow variation of power supply voltage in steady state: at the time scale of a mechanical vibration period, this voltage is constant.

2.1. Start-up stage

Suppose that C_{var} and C_{store} are initially discharged, C_{var} is maximum and the power supply is disconnected from the circuit. When the power supply is connected, the diode D_2 becomes ON (D_1 and D_3 are OFF) and C_{var} charges to V_0, whereas C_{store} is still discharged. So as C_{var} decreases because of the vibrations, its voltage increases, D_3 becomes ON (D_1 and D_2 are OFF) and a portion of the C_{var} charges flows into C_{store}. When C_{var} increases again, its voltage decreases, D_1 becomes ON (D_2 and D_3 are OFF) and a portion of the C_{store} charges flows back into C_{var}, after that D_2 becomes ON (D_1 and D_3 are OFF) and C_{var} is charged from the power supply till voltage V_0. Then, the capacitance decreases again and the cycle repeats. After several cycles, the system goes into steady state in which there is no more power supply discharge (D_2 is always OFF).

2.2. Steady state

In steady state, the total charge on the capacitors C_{var} and C_{store} is constant. During the variation of C_{var}, this total charge is continuously redistributed between C_{store} and C_{var}. The charging of the power supply occurs when charges flow from C_{var} into C_{store}.

Figure 2 presents the relationship between the variable capacitor charge Q and its voltage V in a steady-state operation. Point A corresponds to the state at which C_{var} is maximum; C corresponds to the state at which C_{var} is minimal; B and D correspond to the states at which diodes D_3 and D_1 respectively become ON. Voltages at which the diodes D_3 and D_1 opens (see figure 2): $V_1 = V_0\left(1+\left(C_{max}-2C_{min}\right)/\left(C_{store}+C_{min}\right)\right)$, $V_2 = V_0\left(2+\left(C_{max}-2C_{min}\right)/\left(C_{store}+C_{min}\right)\right)$, where C_{max} and C_{min} are respectively the maximum and minimum capacitances of C_{var} on the cycle.

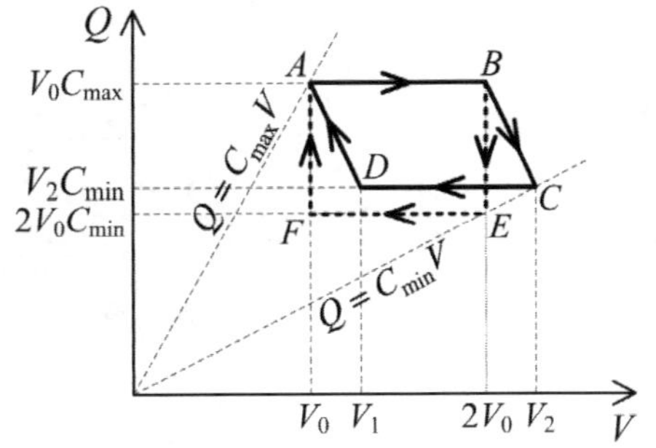

Figure 2. QV-diagram of the harvesting cycle in a steady state.

The cycle area $ABCD$ is equal to the energy converted by the transducer into electricity (harvested energy) during one operation cycle, which is defined by one C_{var} variation period. This energy can be estimated by the expression:

$$E_{cycle} = V_0^2 C_{store} \frac{1-2/\eta}{C_{store}/C_{max}+1/\eta}, \tag{1}$$

where $\eta = C_{max}/C_{min}$. If there are no losses in the diodes, this energy is totally injected into the power supply. Note that (1) is only positive if the requirement $\eta > 2$ is fulfilled. Otherwise, the voltage variation on the variable capacitor is not large enough to modify the states of the diodes, and the circuit does not operate in the energy conversion mode. The requirement of $\eta > 2$ is strong, and is often difficult to fulfill with MEMS transducers. However, it can be shown that by modulating the total charge on the capacitors C_{var} and C_{store}, energy conversion is possible for lower η, however, with different circuit topology.

Analysis shows that the larger C_{store} is, compared to C_{max}, the more energy is harvested and the more rectangular $ABCD$ is. When $C_{store} >> C_{max}$, the variable capacitor voltage ranges from V_0 to $2V_0$ and the energy is equal to the rectangular $ABEF$ area (figure 2):

$$E_{cycle\_rect} = V_0^2 C_{max}\left(1-2/\eta\right). \tag{2}$$

From (1) and (2), it follows that the harvested energy can be increased by increasing the power supply voltage or increasing the potential of the node 1 through circuit modification [6], since the supply voltage is determined by the load. However, increasing this voltage leads to strong electromechanical coupling, which brings restrictions on the operation of e-VEHs. For example, if the variable capacitor has parallel electrodes with out-of-plane gap-closing construction (figure 1) the power supply voltage cannot be more then $\sqrt{8kd_0^3\left(27\varepsilon S\right)^{-1}}$ (where k – linear spring stiffness, d_0 – gap between the electrodes in quiescent state, ε – electric permittivity, S – electrodes area), otherwise electrodes collapse (pull-in effect occurs) and the system stops operating [7]. By inserting this maximum allowed voltage in (1) one can estimate maximum harvested power taking into account design parameters. However, real devices have more complex constructions with more complicated electromechanical interactions and the optimal voltage at which the harvested energy is maximum can only be determined experimentally or by mathematical analysis.

3. Experiment

Figure 3 shows the schematic diagram of the used experimental equipment. The experiment was carried out using a 0.042 cm^3 batch fabricated MEMS e-VEH with in-plane cap-closing interdigitated comb geometry (figure 4). The e-VEH main characteristics and its fabrication technique are very similar to the one described in [8]. The measured unbiased capacitance variation ratio of the used e-VEH induced by harmonic vibrations at 187 Hz with acceleration amplitude 4 g, was $\eta = 3.25$ ($C_{\max} = 175$ pF).

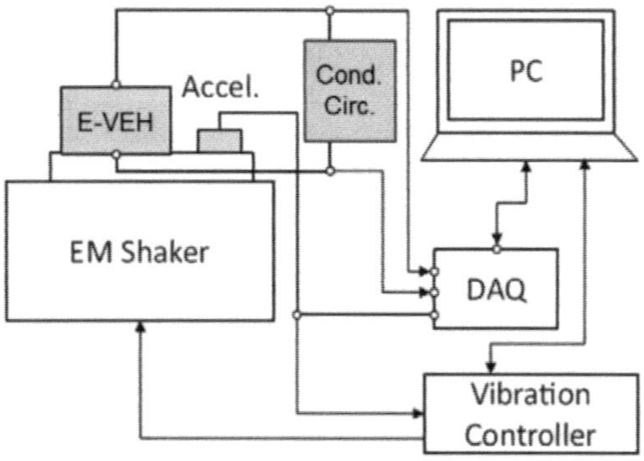

Figure 3. Schematic diagram of the used experimental equipment.

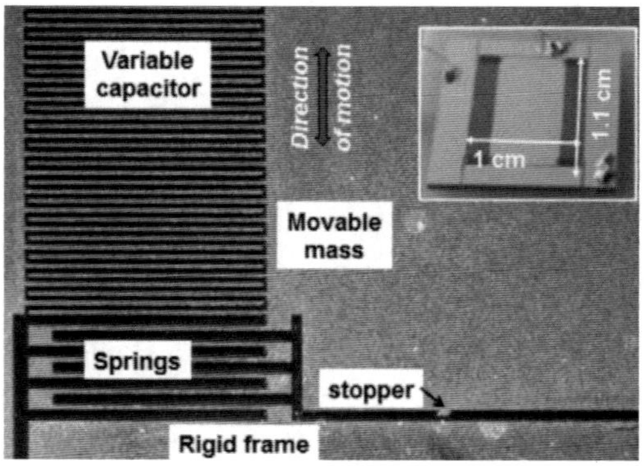

Figure 4. Photograph of the using MEMS e-VEH with interdigitated comb geometry.

To find the optimal power supply voltage, an initially pre-charged high-capacitance reservoir capacitor $C_{res} = 1$ μF was used. During the operation of the e-VEH, the voltage on C_{res} increased. Figure 5 shows the evolution of the voltage on C_{res} at 187 Hz / 4 g external vibrations and with $C_{store} = 2.8$ nF. At time $t = 0$, C_{res} is charged to 5 V. An increase of the voltage on C_{res} corresponds to an increase of the accumulated energy. It can be seen that voltage and accumulated energy (figure 6) initially increases, with an increasing rate (according to (1), the energy accumulation rate is proportional to V_0^2). Then, after several cycles of the harvester operation, the increasing rate of the voltage decreases, eventually leading to saturation at which voltage and energy on C_{res} stop increasing.

This saturation phenomenon was not observed in the experiment presented in [5], where a macroscale capacitor was used. The saturation is due to the electromechanical coupling and not to the reverse current of the diodes which becomes significant only above 30 V: (i) the resonance frequency of the harvester decreases because of its biasing by the conditioning circuit and the input frequency is not anymore in the mechanical system bandwidth for large V_0; (ii) the electromechanical damping increases with V_0, reducing the amplitude of vibrations, hence the parameter η. These phenomena enforce the fundamental law: only a limited amount of energy can be converted from a mechanical resonator at given external vibrations [9].

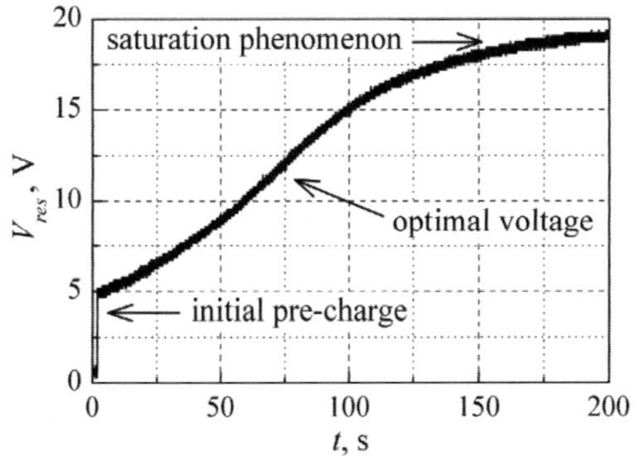

Figure 5. Measurement of C_{res} voltage during harvester operation, excited by 187 Hz / 4 g external vibrations.

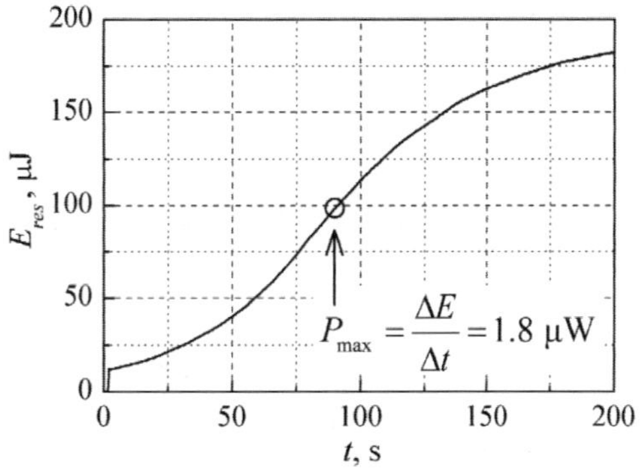

Figure 6. Evolution of the energy accumulated on C_{res}.

From figure 6, the available output power can be estimated as the time derivative of the energy stored on C_{res}. The maximum measured output power $P_{max} = 1.8\ \mu W$, obtained at $V_0 \approx 12$ V. Expression (1) predicts approximately the same result. Thus, it is seen that there is an optimum power supply voltage at which the e-VEH output power is maximum.

4. Conclusion

In this paper a novel experimental study of monophase (single-capacitor) MEMS e-VEH with conditioning circuit based on Bennet's doubler has been presented. This circuit allows charging a power supply without inductive elements and additional power-consuming control electronics.

The experiment was carried out using a 0.042 cm^3 batch fabricated MEMS e-VEH with in-plane cap-closing interdigitated comb geometry. The existence of an optimum power supply voltage at which e-VEH output power is maximum has been shown by the observation of the self-increasing voltage and energy on a reservoir capacitor used in place of the power supply. The presence of this optimum voltage is due to the electromechanical coupling in MEMS and had not been previously observed in studies with macroscale devices.

At 187 Hz external vibrations with 4 g acceleration amplitude and at optimal voltage of 12 V, the measured output power was 1.8 µW.

References

[1] Yen B C and Lang J H 2005 *IEEE Trans. Circuits Syst. I* **53** 288–95
[2] Meninger S, Mur-Miranda J O, Amirtharajah R, Chandrakasan A P and Lang J H 2001 *IEEE Trans. Very Large Scale Integr. (VLSI) Syst.* **9** 64–76
[3] Dudka A, Basset P, Cottone F, Blokhina E and Galayko D 2013 *J. Phys.: Conf. Ser.* **476** 012127
[4] Kempitiya A, Borca-Tasciuc D A and Hella M M 2013 *IEEE Trans. Power Electronics* **28** 609–14
[5] de Queiroz A C M 2010 *53rd IEEE Internat. Midwest Symp. Circuits Syst. (Seattle, WA, 1-4 August 2010)* pp 404–7
[6] Dragunov V and Dorzhiev V 2013 *J. Phys.: Conf. Ser.* **476** 012115
[7] Nathanson H C, Newell W E, Wickstrom R A and Davis J R 1967 *IEEE Trans. Electron Dev.* **14** 117–33
[8] Basset P, Galayko D, Cottone F, Guillemet R, Blokhina E, Marty F and Bourouina T 2014 *J. Micromech. Microeng.* **24** 035001
[9] Mitcheson P D, Green T C, Yeatman E M and Holmes A S 2004 *J. Microelectromech. Syst.* **13** 429–40

Low-power Methods of Power Sensing and Frequency Detection for Wideband Vibration Energy Harvesting

P Proynov, B H Stark, and N McNeill
Department of Electrical and Electronic Engineering, University of Bristol, UK.

E-mail: p.proynov@bristol.ac.uk

Abstract. Power maximisation techniques in wideband vibration energy harvesting typically require the periodic sensing of input power or excitation frequency. This paper presents low-power circuits and sensing methods to obtain this information. First, an excitation frequency measurement circuit is presented that permits a reduced timer run-time compared to reported methods. Second, a power sensing method is presented, which extends the measurement range of reported techniques by adapting to the levels of the available power. Experimental results for the frequency measurement circuit tested in the range 35-51 Hz show a power consumption of 3.7 μW. The power-sensing technique is experimentally validated over a power range of 370-690 μW, and its power consumption is 7.5 μW.

1. Introduction

Maximum power point tracking and electrical tuning for wideband energy harvesting require the sensing of the extracted power or of the excitation frequency [1] [2]. Harvested power levels can span orders of magnitude, and in particular, at excitation frequencies away from the harvester resonance, the available power is low. Therefore the quiescent power consumption of the control and sensing circuitry should be as low as possible. This paper presents two low-power sensing methods: one for frequency sensing, and the other for power sensing.

Frequency can be determined from acceleration, velocity, or displacement. Direct acceleration sensing requires accelerometers which add to the total volume of the system and the power overheads. Displacement sensing can be implemented by means of a piezoelectric strip attached to the beam of the harvester as demonstrated in [1], which generates a signal without the need for a power rail. This alternating signal, however, requires adaptive filtering again with associated quiescent consumption. Phase shifting of the piezoelectric signal to align with the harvester voltage allows a reduction in power consumption of the measurement circuit because the same signal can be used for polarity detection and gate driving of the power conditioning circuit and the operation of the control circuit. This work uses a sensorless method that deduces frequency directly from the harvester output voltage. Since the mechanical structure of an electromagnetic harvester filters out excitation frequency components except those in the vicinity of the resonant frequency, the time interval between zero-crossings of the generated voltage can be used to determine the excitation frequency that is relevant for tuneable harvesters.

Power can be inferred from output current measurement in maximum-power-point-tracked energy harvesting systems, if the energy storage element is large enough to maintain a near output constant voltage between perturb-and-observe cycles. E.g. in [3], the authors present a power sensing method using a switched capacitor, where the output current of the power converter is determined by measuring the voltage rise on a small sensing capacitor during a short period during which the main storage and load are disconnected. A limitation of this method is that if the measurement range is large, as would be expected for a wideband excitation, then at the lower end of the range, the relative measurement resolution is poor. Here, a low-power system is demonstrated which adapts the resolution to the power level to obtain high accuracy over the whole measurement range.

2. Method

The energy harvesting system including the low-power methods for power sensing and frequency detection is presented in Figure 1. The system consists of an electromagnetic energy harvester, power converter, storage element, and measuring and control circuitry. The signals generated by the frequency measurement and the power sensing circuits are read by a low-power microcontroller MSP430. The microcontroller is enabled by an interrupt at the zero-crossing of the input voltage and is active for a short period during which all measurements and control actions are performed. The microcontroller then remains in sleep mode for the rest of a harvester half-cycle.

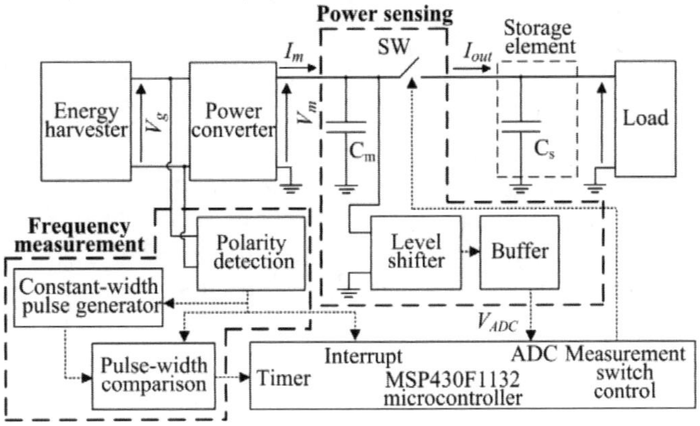

Figure 1. Low-power frequency detection and power sensing system.

2.1. Frequency detection

In digitally controlled systems, time intervals between two events are usually measured by running a timer on a microcontroller for the duration of the interval. Thus by timing the half-cycle period of the input voltage the excitation frequency can be measured [2]. However, the continuous operation of a timer, even on the low-power MSP430, requires 28 μW to run continuously at 85 kHz clock frequency. Here we demonstrate a method where the timer runs for only part of the harvester half-cycle. Figure 2 illustrates the principle: A polarity detector determines the beginning of the harvester cycle, upon which a pulse is generated, whose fixed width T_{rc} equals the half-cycle period of the excitation at the upper limit of the harvester frequency band. At the end of the fixed-width pulse, the timer is started in order to measure the remainder of the half-cycle. Therefore timer's on-time T_m approaches zero at the upper limit of the frequency band and is highest at the lower limit. The derived harvester frequency can be expressed as

$$f_h = \frac{1}{\frac{1}{f_{max}} + 2T_m}, \tag{1}$$

where $f_{max} = 1/2T_{rc}$ is constant and equals the highest frequency in the measurement range and T_m is the timer value.

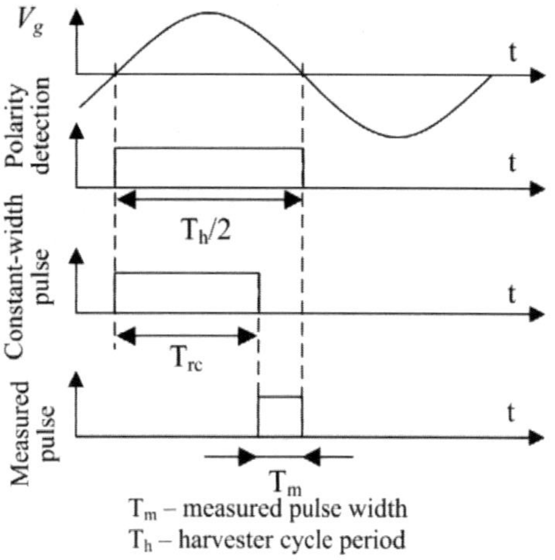

Figure 2. Waveform of the main pulses in the frequency measurement circuit.

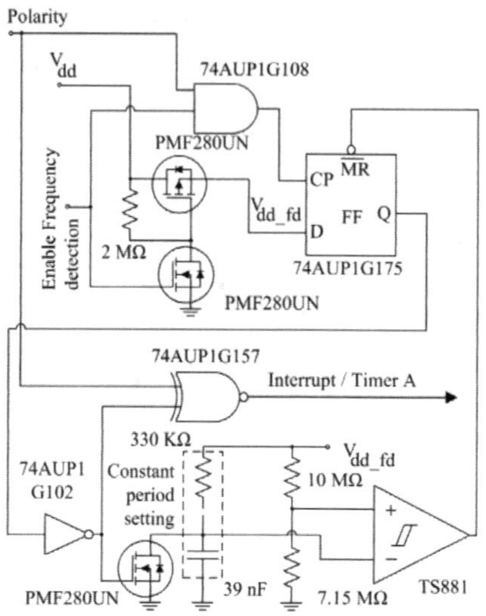

Figure 3. Low-power implementation of the frequency detection circuit.

The fixed-width pulse is generated by an RC circuit, a comparator, and a reset circuit (Figure 3). At the beginning of the positive half-cycle the voltage across the capacitor starts to increase from zero. When a pre-set threshold voltage is reached, the comparator generates a reset pulse for the flip-flop, the MOSFET switch is closed, and the capacitor is discharged to zero and remains zero until the beginning of the next positive half-cycle. The falling edge of the constant-width pulse sets an interrupt flag in the microcontroller and activates a 16-bit timer. The timer runs until a second interrupt is generated by the falling edge of the polarity detection circuit. The measured period T_m can be calculated as

$$T_m = \frac{K + K_{init}}{F_{clk}} \quad (2)$$

where K is the final timer value, K_{init} is a constant denoting the number of clock cycles between the interrupt and the start of the timer, and F_{clk} is the clock frequency of the oscillator used by the timer. In order to reduce the power consumption during the operation of the timer, the clock frequency is reduced from 5 MHz nominal to 85 kHz, and the CPU. Additionally, the low clock frequency ensures that the register that stores the timer value will not overflow. Nevertheless, decreasing the clock frequency reduces the resolution of the measurement. At 85 kHz the frequency resolution varies between 14 mHz at the lower limit of the bandwidth and 36 mHz at the upper limit.

The power consumption of the proposed frequency detection circuit is evaluated by comparison with the case where the microcontroller timer runs for the whole half-cycle period. The power measurements are performed at the resonant frequency of the harvester. The reduced timer operation has reduced the total average power consumption from 4.3 μW to 3.7 μW, 1.05 μW of which is consumed by the microcontroller and the remaining 2.65 μW by the components in the frequency detection circuit. It should be noted that the improvement the frequency detection circuit offers increases with the increase of the excitation frequency.

2.2. Power sensing

A small change in environmental factors such as excitation frequency and amplitude can result in significant change in the amount of extracted power. For example, for the harvester used in this work, a decrease of the frequency by 1.5 Hz from the resonance, 44.3 Hz, reduces the maximum extractable power by 70 %. The method proposed here aims to enable accurate operation of MPPT algorithms at a wide range of the excitation frequency and low quiescent overhead. In [3], a switched capacitor current sensing technique is deployed, as described in Section 1. The voltage rise ΔV_{C_m} on the measurement capacitor C_m is proportional to the output power:

$$\Delta V_{C_m} = \frac{\overline{P_{out}}_{\Delta t}\,\Delta t}{C_m \cdot V_{out}} \,. \qquad (3)$$

Since the output power P_{out} averaged over Δt is a function of extracted power, large variations in the extracted power result in large variations of the measured voltage. As the range and the resolution of the ADC are fixed, these variations affect the overall sensitivity and range of the measurement.

The improved power sensing technique described here increases the measurement sensitivity and range by adaptively changing the duration of the measurement period Δt according to the output power (Figure 5). The method relies on the measurement being synchronised with the harvester half-cycle, and the operational cycle of the MPPT control algorithm being much longer than the measurement period M. This allows the measurement period to be increased in steps equal to the period of the harvester cycle without affecting the voltage range across the storage element significantly. An algorithm for implementation of the method on a low-power MSP430 microcontroller is presented in Figure 4.

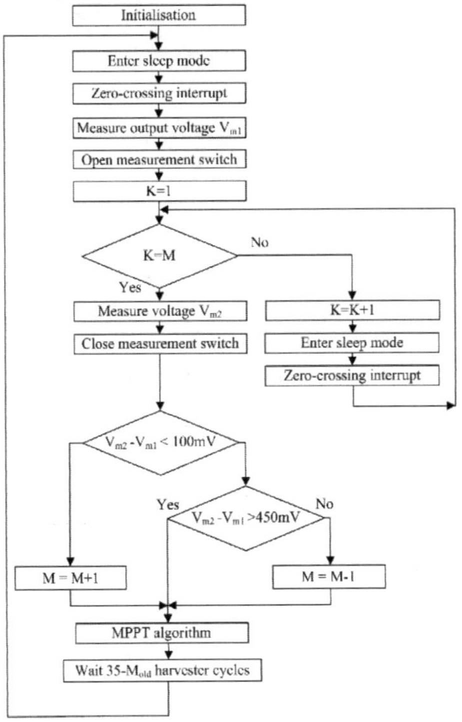

Figure 4. Algorithm for adaptive power sensing; M – number of harvester cycles during which the measurement switch is open; Vm1, Vm2 – measurements of the converter's output voltage at the beginning and at the end of the measurement period.

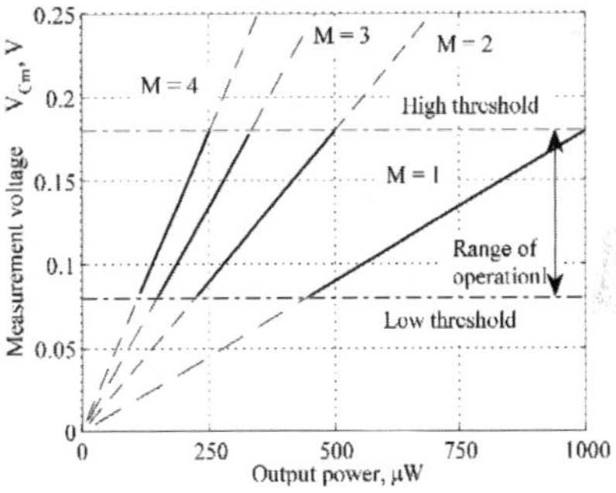

Figure 5. Theoretical measurement voltage ΔV_{Cm} as a function of the output power. If one of the threshold is reached, the duration of the measurement period M is changed. The measurement capacitor is 39 µF and the output voltage is 2.5 V.

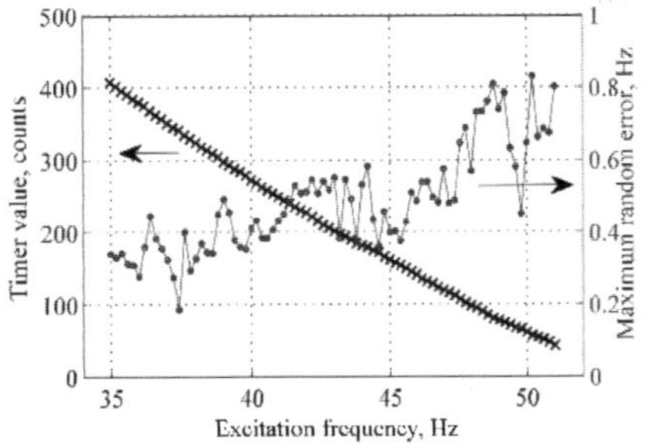

Figure 6. Measured timer value as a function of the excitation frequency and calculated maximum random error.

3. Experimental results

3.1. Frequency detection

To test the performance of the frequency detection method experimentally, the excitation frequency is swept between 35 Hz and 51 Hz at step size of 200 mHz. The length of the fixed-width pulse is T_{rc} = 9.09 ms, which corresponds to an upper limit of the harvester frequency of 55 Hz.

3.2. Power sensing

Figure 5 demonstrates how the measurement period is changes with output power. The excitation frequency and amplitude are held constant and the power is changed by sweeping the duty ratio of the switching converter. The measurement voltage reaches a threshold where the measurement period changes in steps of one harvester half-cycle, for example from 2 cycles (*M*=2) to 2½ cycles (*M*=2.5).

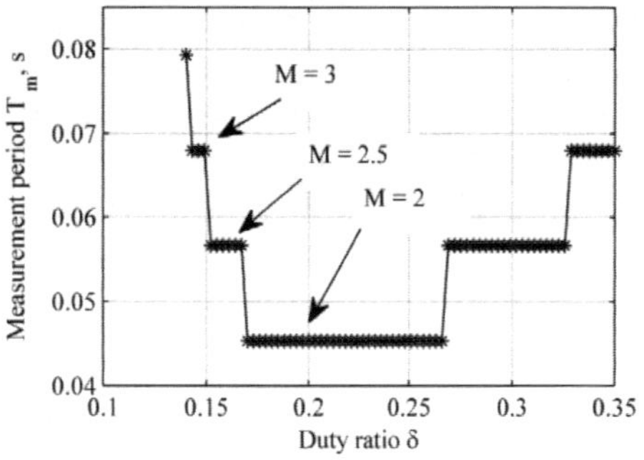

Figure 5. Measurement period against duty ratio.

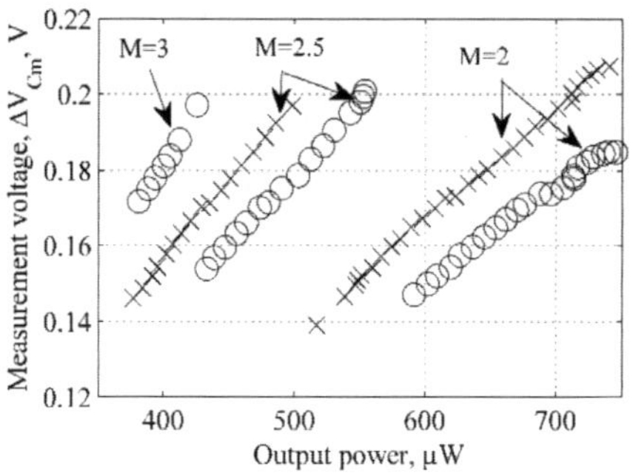

Figure 6. Measurement voltage and measurement period against power. (Output voltage: ○ - 4 V, × - 3.5 V).

The relationship between the measurement voltage ΔV_{C_m} and the output power at output voltages 4 V and 3.5 V is shown in Figure 6. When a decrease in the power causes the measurement voltage to fall below a threshold of 140 mV, the measurement period is increased by a half-cycle. Since the relationship between the value of the analogue-to-digital converter and the measurement voltage is constant, this results in increased measurement resolution, which is necessary at low power levels. If however the measurement voltage reaches the upper threshold at 210 mV the measurement period is reduced to prevent saturation of the ADC. The measurement voltage is thus kept between these two threshold limits, resulting in M being varied between 2, 2.5, and 3. At lower output voltage levels, due to higher output currents, the same range of measured power can be achieved with fewer steps. For example, at $V_{out=}$3.5 V *M* takes on only two values: 2.5 and 2. The results show that if the measurement period were kept constant at *M* = 2, then the sensitivity of the measurement circuit would be 3.75 mW/V whereas the adaptive system reaches sensitivity of 1.88 mW/V. The measured standard deviation for the power range between 370 and 750 μW at 3.5 V output voltage is 0.6 mV. The average power consumption of a discrete component implementation of the proposed method is 7.5 μW in a system which performs measurement once every 35 harvester cycles.

4. Conclusions

In this paper, two low-power methods for frequency detection and power sensing are demonstrated. The frequency detection method uses the harvester output directly, and has a total power consumption of only 3.7 μW. The method is compared with the case where a microcontroller timer is used to measure the frequency of the excitation, and reduction in the power consumption by 14 % is demonstrated. The power sensing method uses 7.5 μW, and is validated over a power range of 370 μW to 750 μW extending the operational range by 26 %. The results show that the proposed methods are suitable for application in MPPT for wide-band sub-milliwatt energy harvesting.

Acknowledgement: This work was performed under the SPHERE IRC funded by the UK Engineering and Physical Sciences Research Council (EPSRC), Grant EP/K031910/1.

References

[1] Szarka G, Burrow S G, Proynov P, Stark B H 2013 *IEEE Trans. Power. Elec.* **99** 201
[2] Weddell A, Merrett G, Barrow S and Al-Hashimi B 2012 *IET Wireless Sensor Syst.* 100
[3] Proynov P, Szarka G, McNeill N and Stark B 2013 *Electron. Lett.* **49** 151

Resonant frequency tuning of an industrial vibration energy harvester

T T Toh, S W Wright and P D Mitcheson

Department of Electrical and Electronic Engineering, Imperial College London, UK

E-mail: tzern.toh@imperial.ac.uk

Abstract. This paper presents preliminary results of tuning the resonant frequency of two industrial vibration energy harvesters. The VEH-450 from Ferro Solutions and the PMG17-50 from Perpetuum were tested using discrete reactive electrical loads. The former could be tuned to +0.5 Hz and -2 Hz from its natural resonant frequency of 50.5 Hz at 0.1g. The latter, however, has a broadband output power spectrum that spans ±10 Hz and its output voltage saturates at 7 V_{rms}, thereby rendering it un-tunable using the method presented here. A comparison of output power between a tuned VEH-450 and an un-tuned PMG17-50, normalised by harvester weight, shows that the former outperforms the latter only at a tuned frequency of 49.8 Hz. A discussion of a resonant frequency tuning circuit that can be fitted to an existing harvester without making modifications to the harvester is presented.

1. Introduction

Energy harvesting from industrial machinery typically requires a resonant device that is tuned to the dominant frequency. However, the host structure usually exhibits a vibration spectrum that is often broadband and/or time-varying. This broad frequency range represents an often untapped energy source that could increase the output power density of the harvesting system if the harvester was broadband capable and if there is no built-in rectifier [1]. Recently, considerable attention has been devoted to using mechanical [2] and electronic [3] methods of tuning a resonant harvester. These approaches have only been demonstrated on proof-of-concept harvester designs. This work, however, demonstrates resonant frequency tuning on a vibration harvester that was specifically manufactured for industrial applications. The system presented here will be suitable for retrofitting to a range of electromechanical energy harvesters, as long as direct access can be made to the coil terminals.

2. Industrial vibration energy harvesters

In this work, a VEH-450 from Ferro Solutions and a PMG17-50 from Perpetuum were chosen because they both provide access to the transducer coil terminals, which is necessary for resonant frequency tuning. Figure 1 shows a photograph of the VEH-450 and the PMG17-50 vibration energy harvesters and their physical dimensions are listed in table 1.

When excited with large accelerations, the output voltage of the PMG17-50 device was observed to clip as it approached a peak-to-peak value of 19.8 V. This suggests that Zener diodes are connected across the device terminals as a protection measure. To verify this, a

1 kHz signal was applied across the harvester terminals and as the peak-to-peak amplitude of the signal was varied, the transducer current and voltage were measured.

Figure 1. Photograph of a VEH-450 (left) and a PMG17-50 (right) energy harvester.

Table 1. Dimensions and electrical characteristics of the VEH-450 and the PMG17-50.

Parameter	Base diameter	Height	Weight	Resonant frequency	Coil inductance	Coil resistance
VEH-450	60 mm	68 mm	430 g	50.5 Hz	≈ 0.4 H	2.2 kΩ
PMG17-50	55 mm	55 mm	655 g	50 Hz	≈ 3.9 H	3.8 kΩ

Figure 2(a) shows the output voltage waveform of the PMG17-50 at a source excitation frequency of 47 Hz and at 50 Hz whereby the waveform is clipped. Figure 2(b) shows that the measured transducer current is proportional to the voltage up to about 7 V_{rms}. Beyond that, the current increases at a higher rate than the voltage, which suggests that the Zener voltage has been reached. Adding a load resistor in parallel to the transducer will dampen its response and lower its output voltage.

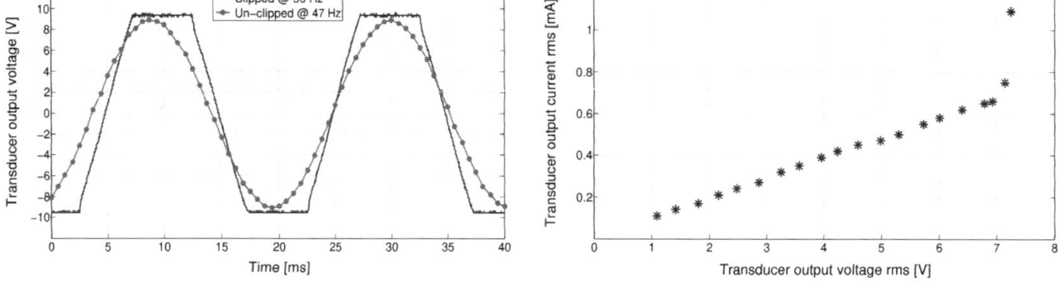

(a) (b)

Figure 2. PMG17-50: (a) transducer output voltage waveform at 47 Hz and at the resonant frequency of 50 Hz whereby the waveform is clipped and (b) measured I-V performance.

3. Resonant frequency tuning

In order to compare the output power levels of the VEH-450 and PMG17-50, they were both subjected to the same excitation of $0.1g$ at a range of frequencies and connected to an optimal load resistor to quantitatively analyse their power generation capabilities. The optimal resistive load for both harvesters is chosen based on the maximum power point at resonance at $0.1g$. This is approximately 28 kΩ and 7 kΩ for the VEH-450 and PMG17-50 respectively. The PMG17-50, as shown in figure 2, has a Zener diode connected across its terminals and the 7 kΩ load is sufficient to prevent the output voltage from clipping.

3.1. Discrete components

Both harvesters were mounted on a vibration platform and an accelerometer provides closed-loop feedback to a vibration controller. Discrete capacitors of several values and a single inductor were connected in parallel with the output terminals of the harvesters, in addition to the aforementioned optimal load resistances.

In order to make a fair comparison, the output power would have to be normalised. For a given excitation frequency and amplitude, the maximum output power of a resonant vibration harvester is proportional to the weight and displacement limit of the proof mass [4]. Since neither of these are easily measurable, the output power was normalised to the harvester weight, which is listed in table 1.

Figure 3(a) shows a plot of the output power normalised to the device mass against frequency for the two harvesters, for various reactive tuning component values at a fixed acceleration of $0.1g$. Figure 3(b) shows the measured susceptance of each reactive tuning component plotted against their tuned resonant frequency for the VEH-450 harvester.

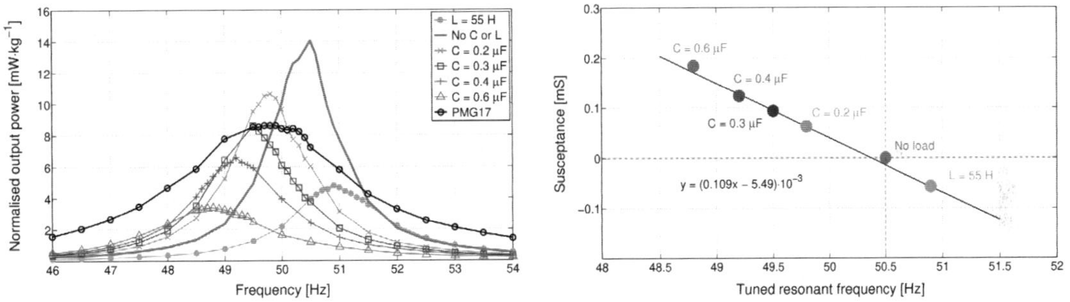

(a) (b)

Figure 3. Measured results from the VEH-450 and the PMG17-50 harvesters excited at $0.1g$: (a) output power against frequency for different reactive components in parallel with an optimum load resistance and (b) susceptance against the tuned resonant frequencies for the VEH-450.

For the VEH-450, its resonant frequency was distinctly shifted from 50.5 Hz with an output power at the tuned lower frequencies that is more than in the un-tuned case. With the inductor, the output power was lower than in its equivalent capacitive case because of the significant losses in the off-the-shelf inductor.

As can be seen, the PMG17-50 is inherently broadband and this was corroborated with the data sheet specifications. The PMG17-50 data presented in figure 3(a) is with a 50 nF tuning capacitor and the performance was similar to the case with just a 7 kΩ load resistor. The VEH-450 outperforms the PMG17-50 on two occasions: firstly, at their respective un-tuned resonant frequencies and secondly, when the VEH-450 was tuned to 49.8 Hz. Apart from those instances, the PMG17-50 generates higher output power at frequencies away from the un-tuned cases.

3.2. Synthesised components

In the previous section, resonant frequency tuning was demonstrated using fixed-valued reactive components for one specific frequency in each case. In a practical implementation, where the excitation frequency varies [1], the use of discrete components is inconvenient. Alternatively, synthesised and adaptable reactive loads have been shown to present an optimal tuning impedance under varying excitation conditions [3, 5, 6]. For each resonant frequency, an inductor could be replaced by a synthesised negative capacitor and vice-versa if the tuning was synthesised electronically [3]. Following on from [3, 5], which used an H-bridge to synthesise an optimum impedance, an improved implementation that will be suitable for retrofitting to an existing harvester is presented here.

A characteristic feature of impedance synthesis is that it requires, as inputs, measures of the transducer current and voltage. The former can easily be achieved using Hall-effect current sensors but, until now the latter would have required a separate winding on the transducer. This is because it is not possible to directly measure the source voltage, E_g in figure 4, when a current is flowing because of the voltage drop across the transducer's internal impedance [3].

The requirement of a separate winding would, in most cases, require intrusive modifications to the harvester and the host structure. In addition, this will reduce the energy density of the harvesting system because some of the space through which the magnetic flux passes will be occupied by the sense winding. Consequently, that space is not available for the transducer coil. Voltage sensing, in this case, will be achieved by a differential amplifier, removing the need for a separate sense winding.

Figure 4 shows the block diagram of the proposed system. The electromagnetic transducer is modelled by an AC voltage source, an inductance L_c and a resistance R_c. This is connected across the four MOSFET H-bridge with a battery connected across the other two terminals. By controlling the conduction of the relevant MOSFETs, power can flow from the transducer to recharge the battery, or to provide reactive power from the battery to the transducer.

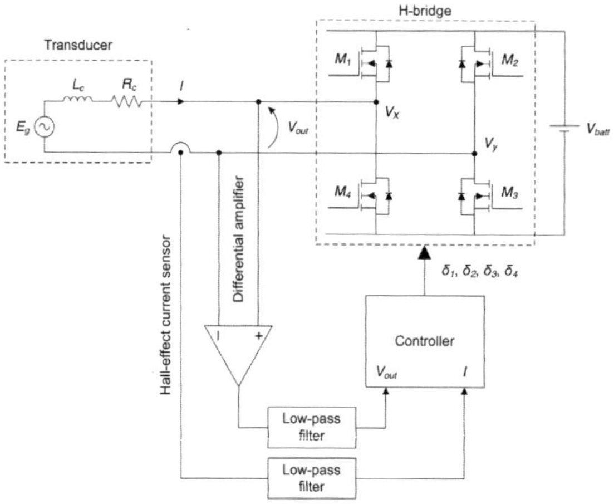

Figure 4. Schematic of an adaptive tuning circuit topology.

The controller calculates the desired current I_d based on the transducer voltage V_{out} and the values of an RLC synthesised load using (1).

$$I_d = \frac{V_{out}}{R} + C\frac{\mathrm{d}V_{out}}{\mathrm{d}t} + \frac{1}{L}\int_0^t V_{out}\,\mathrm{d}t \tag{1}$$

The controller adjusts the conduction of the MOSFETs by applying appropriate pulse-width modulated (PWM) signals to each gate. This controls the flow of power between the transducer and the battery. A PID-controller closes the control loop to maintain the optimum load across the transducer thereby maximising the power transferred, using the methods described in [1].

4. Harvester characteristics at typical PWM frequencies

The transducer is connected across the H-bridge and therefore its properties at typical PWM frequencies must be considered because it is now being used as part of a switching circuit. Figure 5 shows that the VEH-450 behaves very much like an inductor in series with a resistor throughout the 20 kHz range. Contrastingly, the PMG17-50 coil is capacitive at frequencies above 6.5 kHz and only has significant inductance at 2 kHz, suggesting an upper limit on the converter switching frequency for the PMG17-50 of approximately 2 kHz.

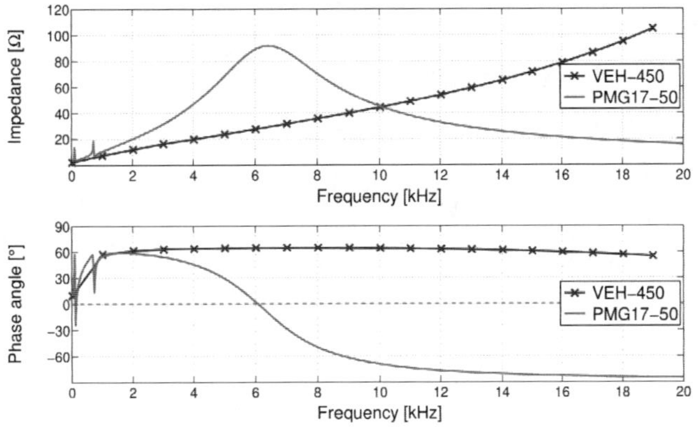

Figure 5. Impedance and phase variations of the VEH-450 and PMG17-50

5. Conclusion

The VEH-450 harvester was shown to be tunable using discrete reactive components. However, the PMG17-50 is intrinsically broadband and on its own, when compared to the reactively-tuned VEH-450, is capable of generating higher output power levels over a broader range of frequencies. If volume and weight are unimportant, the PMG17-50 will be a better choice. Nevertheless, the PMG17-50 is about 50% heavier but it occupies 32% less volume compared to the VEH-450. The tuning methods described here have potential in expanding the applicability and output power density of resonant vibration harvesters operating under broadband excitations. It also has the desirable feature of sensor-less operation and more importantly, the ability to be retrofitted onto existing harvesters and host structures.

Acknowledgments

The authors would like to thank ABB Switzerland AG for their support.

References

[1] Bowden J *et al.* 2012 *ASME IDETC/CIE* (Chicago, Illinois, US) pp 1263–1268
[2] Wu H *et al.* 2013 *J. Intell. Mater. Syst. Struct.* **24**(3) 357–368
[3] Mitcheson P D *et al.* 2011 *IEEE Trans. Circuits Syst. II* **58**(12) 792–796
[4] Mitcheson P D *et al.* 2007 *Journal of Micromechanics and Microengineering* **17**(9) S211
[5] Kaphengst N F *et al.* 2012 *PowerMEMS* (Atlanta, GA, US) pp 420–423
[6] Luo C *et al.* 2009 *International Journal of Electronics* **96** 1249–1264

Degradation of Piezoelectric Materials for Energy Harvesting Applications

P Pillatsch[1], N Shashoua[1], A S Holmes[2], E M Yeatman[2], P K Wright[1]
[1]University of California – Berkeley, Advanced Manufacturing for Energy, Etcheverry Hall, 94720, Berkeley, California, USA
[2]Imperial College London, Optical and Semiconductor Devices Group, Electrical and Electronic Engineering, SW7 2AZ, London, UK

E-mail: ppillatsch@berkeley.edu

Abstract.
The purpose of energy harvesting is to provide long term alternatives to replaceable batteries across a number of applications. Piezoelectric vibration harvesting provides advantages over other transduction methods due to the ability to generate large voltages even on a small scale. However, the operation in energy harvesting is different from typical sensors or actuators. The applied stress is often at the material limit in order to generate the maximum power output. Under these conditions, the degradation of the materials becomes an important factor for long term deployment. In this work bimorph piezoelectric beams were subjected to lifetime testing through electromagnetic tip actuation for a large number of cycles. The results of two measurement series at different amplitudes are discussed. The dominant effect observed was a shift in mechanical resonance frequencies of the beams which could be very detrimental to resonant harvesters.

1. Introduction

In the field of energy harvesting, i.e. generating electricity locally from surrounding sources, piezoelectric materials convert mechanical into electrical energy. The advantage of piezoelectric transducers compared to electromagnetic systems is that they provide sufficiently high voltages for efficient power processing, even at a small device scale [1]. At the same time, they do not need high priming voltages or small, i.e. hard to fabricate, gap sizes as do electrostatic transducers.

Piezoelectric materials can be used in two ways. They either generate a voltage when a stress is applied (sensor or generator), or, vice-versa, they deform when a voltage is applied (actuator). In sensor applications, very small stresses are usually sufficient in order to create a readable voltage, e.g. in pressure gauges. Research has been published on fatigue of PZT [2, 3], AlN [4] and the lifetime of multilayer stack actuators [5, 6]. However, most of these articles look at the problem from the electrical side rather than investigating the effects of mechanical loading. Even in papers that address mechanical degradation, as in [7, 8], the focus is not on energy harvesting from bending beams where the goal is to extract the maximum possible power. Nevertheless, [8] suggests that the mechanical degradation behaviour of piezoelectric materials could be much worse than the electrical one.

The present paper therefore investigates degradation behaviour of piezoelectric bimorph bending beams as they are commonly used for energy harvesting: namely at high stress levels.

 doi:10.1088/1742-6596/557/1/012129

2. Prototype And Experimental Set-Up

Figures 1 and 2 show an experimental set-up comprising a series connected bimorph piezoelectric bending beam (19.5 × 1 × 0.37 mm) that is actuated via an electromagnet and a permanent magnet at its tip. In contrast to a previous set-up with base excitation, this method allows large tip deflections even when operated "off–resonance". An Agilent E4980A precision LCR meter is used to record impedance spectra and material parameters such as capacitance and Q-factor. The system is controlled via a Labview program. Furthermore, a laser displacement sensor monitors tip deflection and the measured output voltage of the piezoelectric beam can be recorded after running it through a voltage attenuation op-amp circuit.

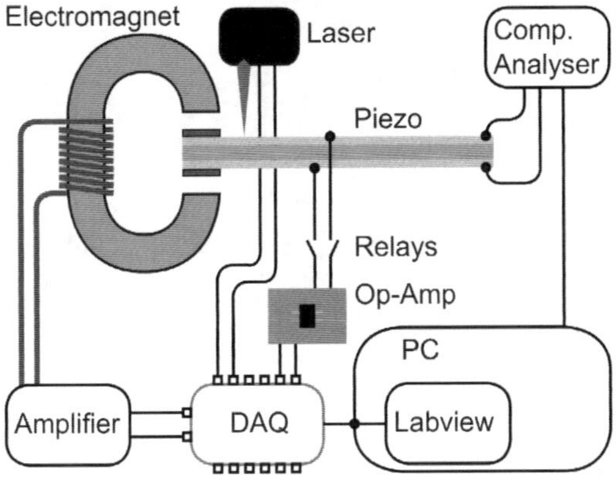

Figure 1: Layout of the measurement set-up

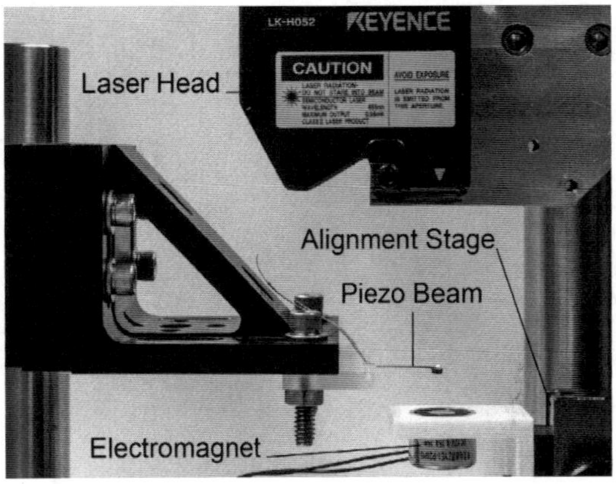

Figure 2: Photograph of the laser head, electromagnet and piezoelectric beam

In order to achieve reliable impedance values, no additional circuitry can be connected to the piezoelectric element during this measurement. This is why the measurement procedure follows three steps. First, reed relays are used to disconnect the op-amp circuit and connect the LCR meter to the piezoelectric beam for the measurement of electrical parameters. Second, the op-amp circuit is connected, the LCR meter disconnected and a low amplitude chirp signal is applied to the electromagnet. The resulting voltage and displacement recordings are used to determine the resonance frequency of the beams at this stage by applying an FFT to the signals. Finally, a fixed number of sinusoidal oscillations, e.g. 50000, with a set tip deflection is applied to the beam at a driving frequency of 100 Hz (below resonance), before repeating the entire procedure for a number of iterations.

3. Results

The above measurement procedure was applied to two beams separately. The only difference between the two measurements was the peak to peak tip deflection, set to 1 mm (yielding a maximum tensile beam strain of 0.073%) for beam 1 and set to 0.85 mm (maximum tensile beam strain of 0.062%) for beam 2. Beam 1 snapped to the electromagnet at 1.2 million cycles due to the close proximity of the electromagnet to the beam tip required to achieve 1 mm tip deflection, which ended the experiment. The tip deflection for beam 2 was consequently lowered to prevent this from happening. In addition, the resolution for the resonance frequencies and impedance spectra was improved for the measurements on beam 2. This beam was tested for 20 million cycles.

Figures 3 and 4 show the impedance magnitude spectra for both beams at the start and at the end of the measurements. The 1.2 million cycles impedance spectrum for beam 1 was recorded just before it snapped. Interestingly, the first beam shows a much larger shift in the curve with a dramatic difference in electrical resonance frequency even after only 1.2 million cycles. This is reflected in the corresponding impedance phase spectra in figures 5 and 6.

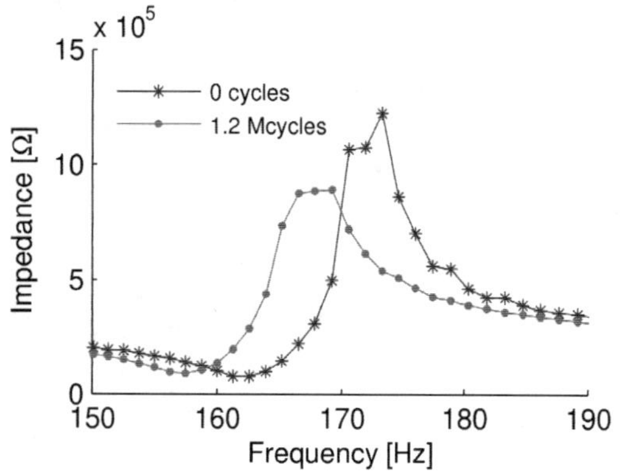

Figure 3: Impedance magnitude spectra (beam 1)

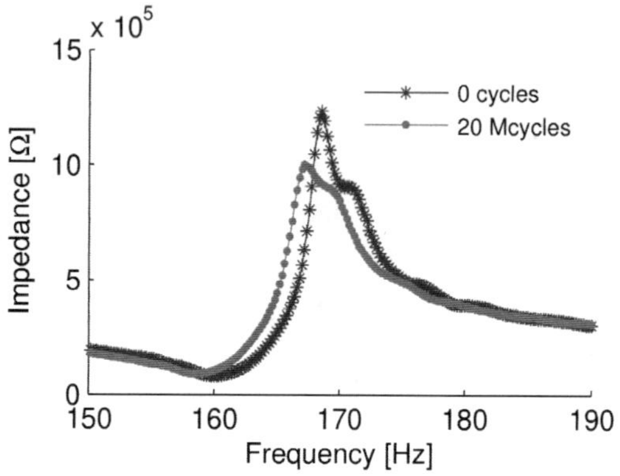

Figure 4: Impedance magnitude spectra (beam 2)

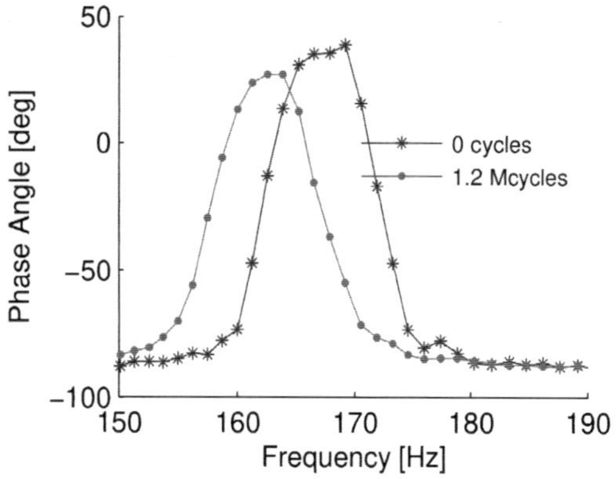

Figure 5: Impedance phase spectra (beam 1)

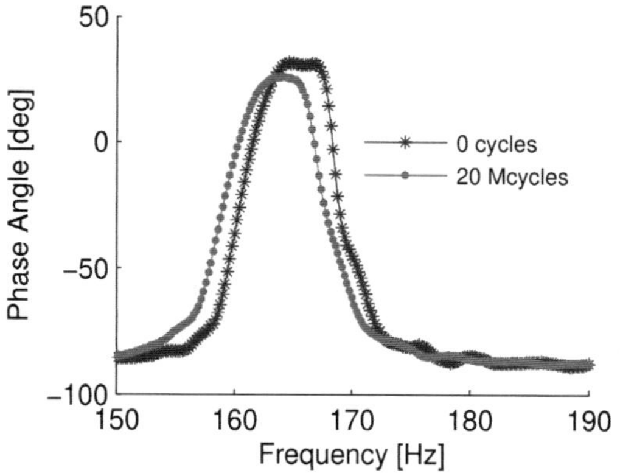

Figure 6: Impedance phase spectra (beam 2)

These results suggest that the small difference in peak to peak tip deflection can dramatically affect the cycle life of these piezoelectric beams.

Figures 7 and 8 compare the mechanical resonance frequencies as determined by the laser signal and the piezoelectric voltage, the electrical resonance frequencies from the impedance spectra and the frequencies of the impedance maxima for both beams up to 1.5 million cycles. After 1.5 million cycles, the frequencies for beam 2 started stabilising and showed little change over the following 18.5 million cycles. Congruent with the previous findings for the impedance spectra, the initial change in frequencies is much less pronounced for beam 2 than it is for beam 1, indicating a faster degradation in beam 1 due to the higher tip deflection. Nevertheless, in both cases a drop in mechanical resonance frequencies of 3 to 4 Hz can be observed. This is a dramatic shift, given that the bandwidth, based on Q-factors and a starting frequency around 165 Hz, is less than 4 Hz. This would be highly detrimental for a resonant harvester designed to operate around that frequency as the operation would be pushed beyond the half-power point and the entire system might fail. It is also important to note that the most significant shifts happen within the first million cycles, which, at a driving frequency of 100 Hz, equates to less than three hours of continuous run time.

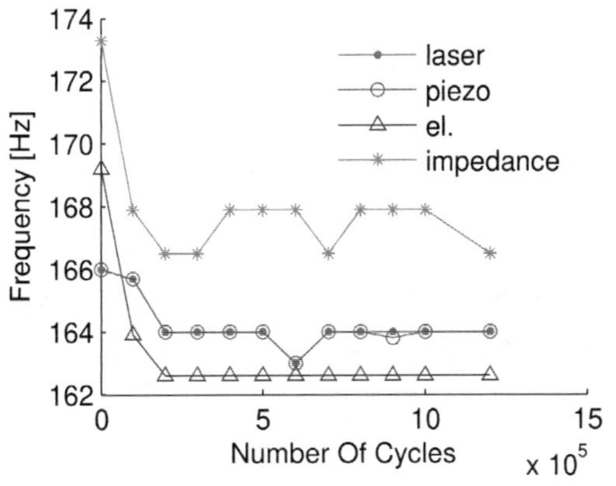

Figure 7: Resonance frequencies and frequencies of the impedance maxima vs. number of cycles (beam 1)

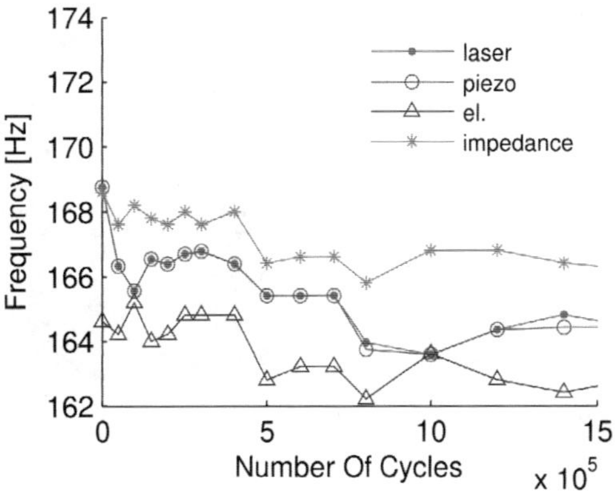

Figure 8: Resonance frequencies and frequencies of the impedance maxima vs. number of cycles (beam 2)

The variations in capacitance and Q-factor for beam 2 over the entire 20 million cycles are shown in figures 9 and 10 respectively. At this stage it is not entirely clear what causes the various spikes in both graphs. It is suggested this might be due to difficulties experienced when connecting to the LCR meter through LabView and will be investigated further. Nevertheless, a slight reduction in both capacitance and Q-factors appears to be the trend. In combination with the previous graphs, shifts in mechanical resonance frequencies are the clearest indicator of degradation.

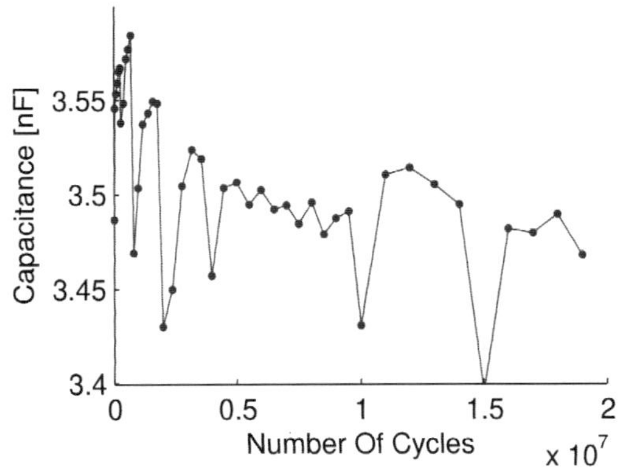

Figure 9: Capacitance vs. number of cycles (beam 2)

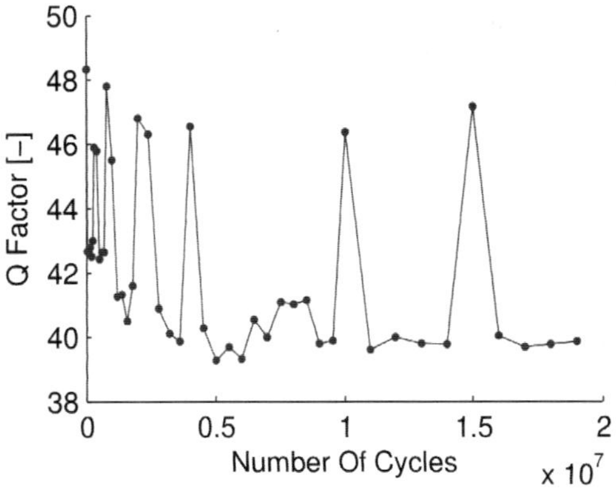

Figure 10: Q-factor vs. number of cycles (beam 2)

These shifts can be caused by micro-cracking of the piezoelectric layers as observed under magnification on a damaged sample of a piezoelectric beam, see figure 11. Such fractures provide an explanation for a stiffness reduction that causes the resonance frequency to decrease.

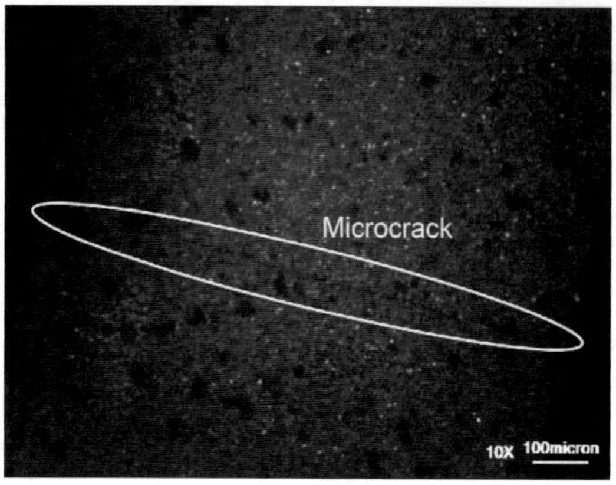

Figure 11: 10 times magnified image of a hairline crack on a damaged piezoelectric beam

4. Conclusions

This paper presents experiments on the degradation of piezoelectric bending beams with the particular application of energy harvesting in mind where the materials are operated close to their limits. At the same time, such systems are to be deployed in the field for many years and their lifetime needs to be guaranteed in order to inspire confidence and widespread adoption.

Experiments on two beams with different peak to peak tip deflections suggest that small changes in operating conditions can have a large effect on lifetime. The most pronounced effect was a shift in mechanical resonance frequency which could be especially detrimental for resonant harvester designs that might lose a significant amount of their power output even after short run times.

Ultimately, there is a broad scope for future research on this topic. Experiments will be performed on a larger number of piezoelectric beams at different tip deflections, different material compositions, larger cycle counts, different electric loads, etc. Further improvements on the experimental set-up are necessary and an analysis relating tip deflection and material stress will be helpful in order to make recommendations for safe operation of these materials in energy harvesting.

Acknowledgments

The authors would like to thank the Institution of Engineering and Technology (IET) for financial support in the form of a travel grant for conference attendance. This research was funded in part under California Energy Commission (CEC) grant number 500-11-021.

References

[1] Kim S G, Priya S and Kanno I 2012 *MRS Bulletin* **37** 1039–1050 ISSN 0883-7694
[2] Zhang Q M and Zhao J 1999 *IEEE transactions on ultrasonics, ferroelectrics, and frequency control* **46** 1518–26 ISSN 0885-3010
[3] Bhattacharyya M and Arockiarajan A 2013 *Smart Materials and Structures* **22** 085032 (12pp) ISSN 0964-1726
[4] Krupa K 2009 *Opto-numerical analysis of AlN piezoelectric thin film operating as an actuation layer in MEMS cantilevers* Ph.D. thesis
[5] Sherrit S, Jones C M, Aldrich J B, Blodget C J, Moore J D, Carson J W, Goullioud R and Jau B 2011 *2011 Aerospace Conference* 1–8
[6] Laurent M, Bödinger H, Steinkopff T, Lubitz K and Schuh C 2004 *IEEE International Ultrasonics, Ferroelectrics, and Frequency Joint 50th Anniversary Conference* pp 3–6 ISBN 0780384105
[7] Jiang N, Liu J, Tao T and Han L 2007 *Micro-NanoMechatronics* 255–259
[8] Cain M G, Stewart M and Gee M G 1999 Degradation of Piezoelectric Materials Tech. rep. NPL

Activation of piezoelectric property of PZT thin films by pulse poling

T. Kobayashi[1], Y. Suzuki[1,2], N. Makimoto[1], H. Funakubo[3], and R. Maeda[1]
[1]National Institute of Advanced Industrial Science and Technology (AIST), Tsukuba, Ibaraki 305-8564, Japan
[2]Ibaraki University, Hitachi, Ibaraki 316-8511, Japan
[3]Tokyo Institute of Technology, Yokohama, Kanagawa 226-8502, Japan

E-mail: takeshi-kobayashi@aist.go.jp

Abstract. We have demonstrated the pulse poling technique to activate the piezoelectric property of PZT thin films within 1 second. We have fabricated piezoelectric microcantilevers with Si proof mass using $Pb(Zr_{0.52},Ti_{0.48})O_3$ (MPB-PZT) thin films. By applying 1 kHz and 100 V of unipolar triangle pulse voltage, the positive piezoelectric constant $|d_{31}|$ of PZT thin film has been enhanced as high as 95 pm/V, which is larger than dc poled PZT thin films (67 pm/V).

1. Introduction

Microelectromechanical systems (MEMS) devices using lead zirconate titanate (PZT) thin films (PZT-MEMS) are utilized to gyros [1], inkjet heads [1], optical microscanners [2], ultrasonic probes [3] and so on. We are also developing new applications of PZT-MEMS including electrostatic field sensors [4] ,trigger sensor to activate wireless sensors [5,6] and energy harvesters [7,8].

In order to activate the piezoelectric property of PZT thin films, dc voltage of 10-100 kV/cm is usually applied for at least several minutes as dc poling [9,10]. Such a several minutes of procedure, however, affects productivity at mass production stage. One of the solution for this problem is to use very expensive probe card, which is available only for fixed electrode pattern.

The other possible solution is to reduce poling time by using pulse poling, where pulse voltage is applied [11-13]. Recently, we have found that bipolar pulse poling, where intermittent ac triangle wave larger than coercive field of the PZT thin films is applied, enhances the piezoelectric property of tetragonal composition (Zr/Ti = 30/70) PZT (tetra-PZT) thin films [14]. It is well known that the morphotropic phase boundary (MPB, Zr/Ti = 52/48) composition PZT (MPB-PZT) thin films have the largest piezoelectric constant among all compositions [15]. Thus, MPB-PZT thin films are most important for the practical application. Bipolar pulse poling can be effective for MPB-PZT thin films as well as tetra-PZT thin films. Moreover, unipolar pulse poling, where positive or negative pulse voltages are applied, can be also effective to enhance the piezoelectric property. Then, in the present study, we have investigated the influence of bipolar and unipolar pulse poling on the piezoelectric property of MPB-PZT thin films on MEMS based piezoelectric microcantilevers.

2. Device fabrication

2.1. PZT deposition and MEMS microfabrication

The 1.9-μm-thick PZT thin films were deposited by sol-gel method on thermally oxidized silicon-on-insulator (SOI) wafers with Pt (100 nm)/Ti (10 nm) bottom electrodes [16]. The SOI wafers have 10 μm structural Si and 1 μm buried oxide (BOX). The Pb/Ti/Zr composition of the precursor solution was 120/52/48. After PZT deposition, Pt (100 nm)/Ti (10 nm) top electrodes were deposited as top electrode. The deposited multilayers were fabricated into piezoelectric microcantilevers with Si proof mass as shown in Fig. 1(a,b). We have also fabricated piezoelectric microcantilevers without Si proof mass to estimate inverse piezoelectric constant d_{31} for comparison.

2.2. Pulse poling

After packaging, dc and triangle pulse voltage as illustrated in Fig. 1(c1-c4) was applied. The top and bottom electrodes were connected to signal and grand lines, respectively. Figure 1(a) is a conventional dc poling. When positive voltage is applied, the polarization direction becomes downward as shown in fig. 1(c1). The present conditions for dc poling is +20V and -20V for 5min. Figure 1(c2) represents bipolar pulse poling reported in our previous study [14]. Intermittent triangle wave of 30-100 V and 1 kHz was applied 10 times at 0.1 sec intervals. Since the last triangle wave is negative voltage, polarization direction is upward. Figures 1(c3) and 1(c4) illustrate unipolar pulse poling, which we newly employed in the present study. Positive or negative intermittent triangle wave of 30-100 V and 1 kHz was applied 10 times. When positive or negative triangle waves are applied, the polarization directions are downward and upward, respectively. Note that the respective coercive voltage of the present tetra- and MPB-PZT thin films are about 20 V and 7V, which means that the dc and pulse poling voltages are larger than the coercive voltage of PZT thin films.

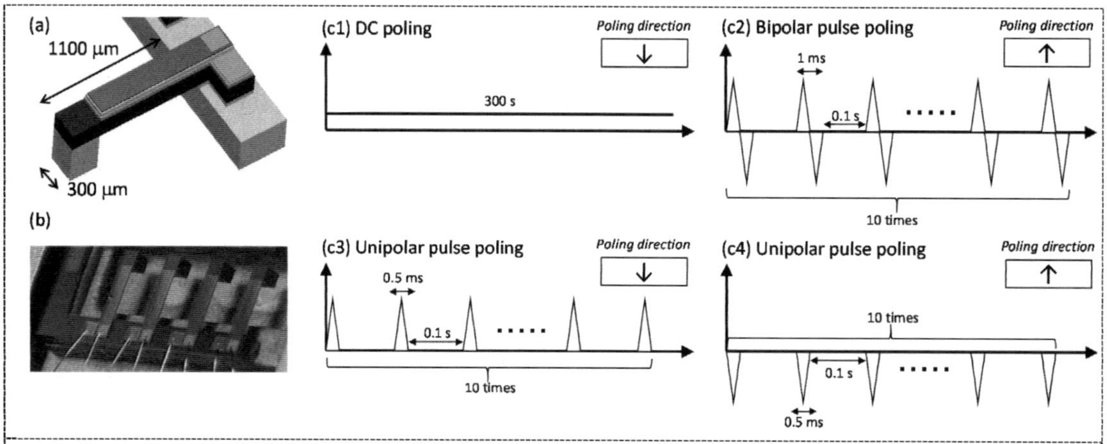

Figure 1. (a) Schematic of the piezoelectric microcantilevers with Si proof mass. (b) Packaged piezoelectric microcantilevers with proof mass. Waveforms of poling voltage for (c1) conventional DC poling, (c2) bipolar pulse poling, (c3) downward unipolar pulse poling, and (c4) upward unipolar pulse poling.

3. Test and results.

3.1. Piezoelectric property

The packaged piezoelectric microcantilevers were mounted on a shaker and vibrated to characterize the positive piezoelectric constant d_{31}, open circuit voltage (OCV), and ac power generation. The data are summarized as a function of pulse-poling voltage in the following. Figure 2(a) shows absolute values of positive piezoelectric constant d_{31} estimated from the output charge Q and acceleration a using the equations,

$$d_{31} = KQ/3L^2 s_{11}^{Si} t_{Si}(t_{Si} + t_{PZT})ma \quad (1),$$

$$K = \left(s_{11}^{Si}\right)^2 t_{PZT}^4 + 4 s_{11}^{Si} s_{11}^{PZT} t_{Si} t_{PZT}^3 + 6 s_{11}^{Si} s_{11}^{PZT} t_{Si}^2 t_{PZT}^2 + 4 s_{11}^{Si} s_{11}^{PZT} t_{Si}^3 t_{PZT} + (s_{11}^{PZT})^2 t_{Si}^4 \quad (2),$$

where m, L, s_{11}, and t are mass, microcantilever length, compliance and thickness of Si and PZT. Although the $|d_{31}|$ of dc poled MPB-PZT thin films in only 67 pm/V, that of unipolar pulse poled MPB-PZT thin films increases with voltage and reaches 95 pm/V at 100 V. Figure 2(b) also shows inverse piezoelectric constant d_{31}, estimated from the equations,

$$d_{31} = K\delta/3ABL^2V \quad (3),$$

$$A = s_{11}^{Si} s_{11}^{PZT}\left(s_{11}^{Si} t_{PZT} + s_{11}^{PZT} t_{Si}\right) \quad (4),$$

$$B = t_{Si}(t_{Si} + t_{PZT})/\left(s_{11}^{Si} t_{PZT} + s_{11}^{PZT} t_{Si}\right) \quad (5),$$

where δ is tip displacement of the piezoelectric microcantilevers actuated at the voltage V. The inverse piezoelectric constant shows similar tendency with that of the positive piezoelectric constant. The inverse $|d_{31}|$ reaches 105 pm/V at the pulse poling voltage of 100V, which is a little bit larger than positive $|d_{31}|$.

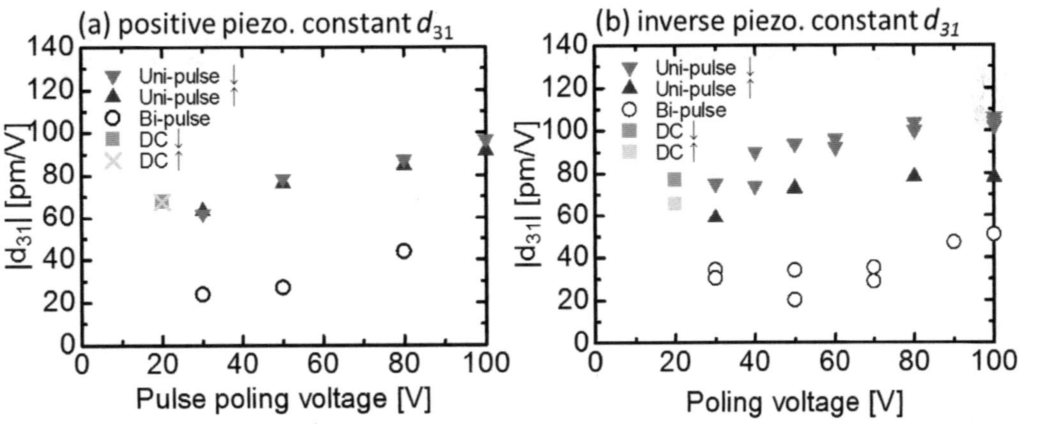

Figure 2. Absolute values of (a) positive and (b) negative piezoelectric constant d_{31} as a function of pulse poling voltage.

3.2. XRD characterization

The increase in $|d_{31}|$ of MPB-PZT thin films with pulse poling voltage can be derived from alignment of polarization direction by disappearance of 180° domain wall and increase in c-axis orientation. Then, we have investigated the variation of c-axis orientation with unipolar pulse poling by XRD measurement. Figure 3(a) shows XRD patterns around (004)/(400) peaks of MPB-PZT thin films pulse poled at 40-100 V. The peak intensity at $2\theta = 99.1°$ gradually decreased at the unipolar pulse poling voltage of 40-80 V and abruptly decreased at the voltage of 100 V. In addition, the intensity of the shoulder around 97°-98° increased by unipolar pulse poling at 100 V. We have estimated the volume fraction of the c-domain ($V_c = V_{(004)}/V_{(004)} + V_{(400)}$) by the pseudo-Voigt function fitting.

Figure 3(b) shows, the volume fraction of c-domain V_c as a function of pulse poling voltage. As shown in fig. 3(b), V_c has increased from 37.4 to 41.6% with the unipolar pulse poling voltage from 40 to 100 V. Based on the results, we concluded that the increase in $|d_{31}|$ is attributed to alignment of polarization direction by disappearance of 180° domain at the pulse poling voltage of 40-80 V and to increase in volume fraction of the c-domain at the pulse poling voltage of 100 V.

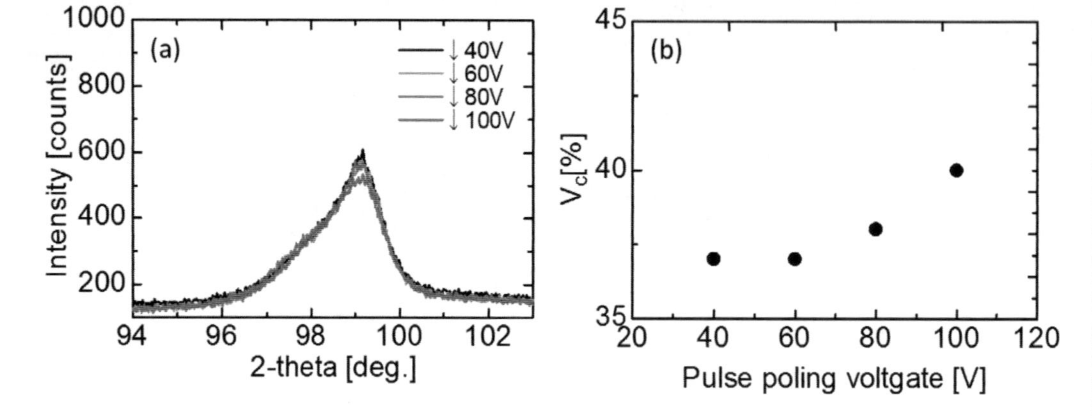

Figure 3. (a)XRD patterns around (004)/(400) peaks of MPB-PZT thin films pulse poled at 40-100 V, (b) volume fraction of *c*-domain V_c as a function of pulse poling voltage.

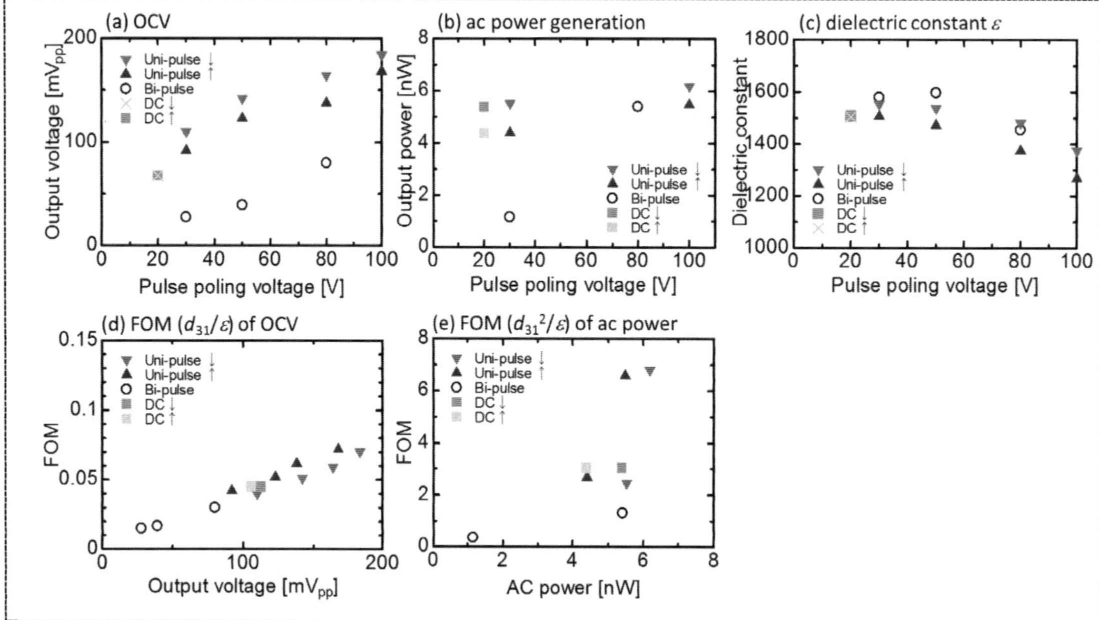

Figure 4. (a) Open circuit voltage (OCV), (b) ac power generation, (c) dielectric constant, (d) figure of merit of OCV, and (e) figure of merit of ac power generation as a function of pulse poling voltage.

3.3. Open circuit voltage and power generation

Figures 4(a,b) show the open circuit voltage (OCV) and ac power generation of the MPB-PZT thin films measured under resonant condition of the piezoelectric microcantilevers with proof mass, where ac power generation is characterized under impedance matching with a load resistance. The unipolar pulse poled MPB-PZT thin films have the larger OCV and ac power generation than the dc poled ones. The dielectric constant ε of the MPB-PZT thin films was measured to obtain the figure of merit (FOM[5]) of the OCV (d_{31}/ε) and the ac power generation (d_{31}^2/ε). The dielectric constant decrease with the increasing pulse poling voltage as shown in fig. 4(c).

Figures 2(d,e) show relation between the measured data and The measured data is well proportional to the FOM for the OCV, while ac power generation is not. Since the ac power generation is determined by *V*/2*R* at impedance matching condition, FOM for power generation should include the resistivity of PZT thin films as well as dielectric constant.

4. Summary

In conclusion, we have succeeded in activating the positive and inverse piezoelectric constant d_{31} of PZT thin films to 95 and 105 pm/V by applying unipolar triangle pulse voltage as pulse poling. It takes within 1 second, which is much shorter than dc poling leading to much better productivity of piezoelectric MEMS devices.

Acknowledgments

This research is granted by the Japan Society for the Promotion of Science (JSPS) through the ''Funding Program for World-Leading Innovative R&D on Science and Technology (FIRST Program)'', initiated by the Council for Science and Technology Policy (CSTP). This research is also granted by JSPS KAKENHI Grant No. 25820339.

References

[1] E. Fujii, R. Takayama, K. Nomura, A. Murata, T. Hirasawa, A. Tomozawa, S. Fujii, T. Kamada, and H. Torii, IEEE Trans. Ultrason. Ferroelectr. Freq. Control 54, 2431 (2007)

[2] T. Kobayashi, R. Maeda, T. Itoh, and R. Sawada, Appl. Phys. Lett. 90, 183514 (2007)

[3] A. Hajati, D. Latev, D. Gardner, A. Hajati, D. Imai, M. Torrey, and M. Schoeppler, Appl. Phys. Lett. 101, 253101 (2012)

[4] T. Kobayashi, S. Oyama, N. Makimoto, H. Okada, T. Itoh, and R. Maeda, Sensors and Actuators A 198, 87 (2013)

[5] H. Nogami, T. Kobayashi, H. Okada, N. Makimoto, R. Maeda, and T. Itoh, Smart Mater. Struct. 22, 095001 (2013)

[6] Y. Tomimatsu, H. Takahashi, T. Kobayashi, K. Matsumoto, I. Shimoyama, T. Itoh, and R. Maeda, J. Micromech. Microeng., 23, 125023 (2013)

[7] H. Liu, C. J. Tay, C. Quan, T. Kobayashi, and C. Lee, J. Microelectromechanical Syst. 20, 1131 (2011)

[8] H. Liu, S. Zhang, R. Kathiresan, T. Kobayashi, and C. Lee, Appl. Phys. Lett. 100, 223905 (2012)

[9] T. Kobayashi, H. Okada, T. Masuda, R. Maeda, and T. Itoh, Smart Mater. Struct., 19, 105030 (2010)

[10] H. Nogami, T. Koayashi, H. Okada, T. Masuda, R. Maeda, and T. Itoh, Jpn. J. Appl. Phys. 51, 09LD11 (2012)

[11] S. Hong and N. Setter, Appl. Phys. Lett. 81, 3437 (2002)

[12] M. Kohli, P. Muralt, and N. Setter, Appl. Phys. Lett. 72, 3217 (1998)

[13] T. Ogawa and K. Nakamura, Jpn. J. Appl. Phys. 37, 5241 (1998)

[14] T. Kobayashi, N. Makimoto, Y. Suzuki, H. Funakuobo, T. Oikawa, A. Wada, and R. Maeda, Jpn. J. Appl. Phys. 52, 09KA01 (2013)

[15] N. Ledermann, P. Muralt, J. Baborowski, S. Gentil, K. Mukati, M. Cantoni, A. Seifert, and N. Setter, Sensors and Actuators A 105, 162 (2003)

[16] T. Kobayashi, M. Ichiki, J. Tsaur, and R. Maeda, Thin Solid Films 489, 74 (2005)

The Power Generation Characteristics and Durability of a Vibration Power Generation Device using Piezoelectric Thick Film Formed Directly by Screen Printing on a Stainless Steel Substrate

A. Oishi[1], H. Okumura[1], H. Katsumura[1] and H. Kagata[1]
[1]Ecomaterials Development Center,
Corporate Engineering Division
Automotive & Industrial Systems Company
Panasonic Corporation
1006 Kadoma, Kadoma City, Osaka 571-8506, Japan

E-mail: oishi.akihiro@jp.panasonic.com

Abstract. This paper presents a small, high-performance and novel device that generates power from vibrations, made using screen-printing to form a piezoelectric thick film directly on a stainless steel substrate. This simple and cheap method realizes a 20 - 40 μm-thick piezoelectric film, otherwise difficult to achieve using thin-film techniques or ceramic sintering, on a stainless steel substrate. A maximum output power of 1.1 mW was recorded with acceleration of 0.1 G_{0p} (0.98 ms^{-2}) applied at a resonance frequency of 24 Hz. We also evaluated the durability of the device by repeatedly striking the tip of the element. Output power exceeding 100 mW under damped resonant vibration was generated at the instant of striking, with approximately 0.9 mJ of power measured per single damped vibration. No deterioration was seen in the integrity of the stainless steel substrate or the piezoelectric thick film after over 10 million strikes.

1. Introduction

We have been working on developing vibration power generation devices, which are one form of energy harvesting technology. Vibration energy has the second largest energy density after light. Efficient harvesting and power conversion of vibration energy would enable power to be generated in dark or cramped places where solar cells are impractical. Vibration power generation devices can be classified into four types based on their power generation mechanism: piezoelectric, electromagnetic, electrostatic, and magnetostrictive. The piezoelectric type can be made smaller and can generate higher voltages than the electromagnetic type, while the magnetostrictive type has certain disadvantages, such as very high internal resistance and low electric current. The piezoelectric type is thus generally regarded as the most suitable match for ICs and similar devices used in electronic circuits.

Piezoceramic materials, including PZT, are widely used in piezoelectric-type vibration power generation devices. However, piezoceramic materials need to be sintered at high temperatures to be able to exhibit their excellent piezoelectric properties. The resultant bulk materials are usually fragile and have durability concerns during harvesting of vibrational energy. The inner stress caused by the difference in thermal expansion coefficient with the electrode materials can also damage the device. Roundy *et al.* [1] have obtained an output power of 375 μW with 0.26 G (2.5 ms^{-2}) acceleration applied at a frequency of 120 Hz using a bimorph device with piezoelectric sheets integrally formed from piezoceramic and electrode materials; and Yu *et al.* [2] have achieved an output power of 66.75 μW with 0.5 G (5 ms^{-2}) acceleration at a frequency of 234.5 Hz using a MEMS device with the film thickness reduced to the level of several micrometers. However, thin-film fabricating processes such as sputtering are painstaking and require special equipment.

This report focuses on the new vibration power generation device we have developed and evaluates the materials' physical properties and their power generation characteristics. Our new vibration power generation device is able to obtain output power that is sufficient to drive a sensor circuit. It is also able to overcome the problems characteristic of conventional vibration power generation devices that use piezoceramic materials, such as mechanical durability and high manufacturing costs.

2. Preparation of the vibration power generation device

A screen-printing method is employed to fabricate our vibration power generation device: an inexpensive process that requires no special equipment. Stainless steel was selected as the base material of the vibration power generation device, as it is a readily available material, has a high resistance to fracture, and shows little loss of durability during high-temperature processing. As the piezoceramic material of the device, we used a hard PZT material with the composition shown below.

$$\mathbf{Pb_{1.02}Zr_{0.44}Ti_{0.46}(Zn_{1/3}Nb_{2/3})_{0.10}O_{3.02} + 0.25wt\%MnO_2}$$

This material was pre–sintered and then pulverized to give it an average particle diameter of less than 5 μm. A solvent, dispersant, and binder resin were added to the resultant powder and kneaded together with a triple roll mill into a piezoelectric paste.

As the bottom electrode, an AgPd (90:10) paste was applied, using screen-printing, to a 0.1 mm-thick stainless steel substrate with high heat resistance. It was then dried. The piezoelectric paste was applied on top of this layer using screen printing and then dried. Lastly, as the top electrode, the AgPd (90:10) paste was printed and dried. Sintering was then applied at 900 °C. The thickness of the piezoelectric layer after the sintering was approximately 30 μm. The layer was polarized by applying 120 V at 120 °C for ten minutes, and a fixing member was attached to one end of the element and a weight to the other to form the cantilever-type vibration power generation device. Figure 1 shows the appearance and structure of the device.

Beeby *et al.* prepared a vibration power generation device by forming the bottom electrode, insulation material, and PZT on stainless steel using screen-printing. A smartcard including the device realized an output power of 240 μW at 67 Hz under an acceleration of 0.4 G(3.9 ms^{-2}) [3]. Beeby *et al.* also reported, at Power MEMS 2013, on obtaining output powers of 0.1 to 0.86 μW using wind power with wind speeds of 1.5 - 8 ms^{-1} [4]. The device we have prepared does not require application of insulation materials: this has been achieved by effective matching of the materials with the sintering processes.

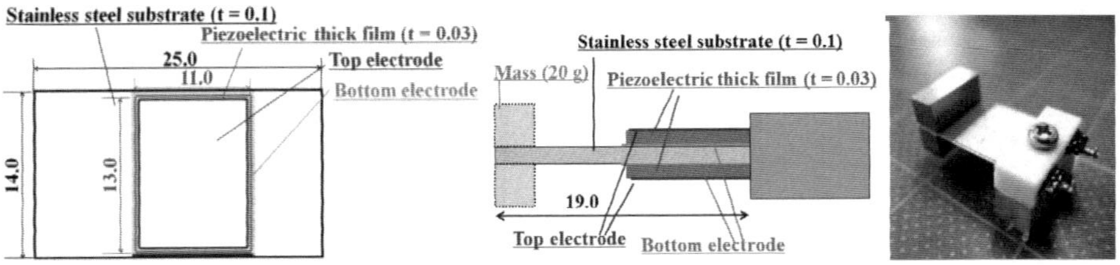

Figure 1. Appearance and structure of vibration power generation device.

3. Characteristics of PZT thick film

Table 1 shows the electric properties of the PZT thick films prepared using screen-printing and of the bulk PZT materials. The dielectric permittivity was determined using the capacity value (measured using an LCR meter), area, and thickness. The piezoelectric constant d_{31} of the thick films was measured on a cantilever device with the PZT thick film on only one surface. How much the tip of the beam was displaced when an alternating voltage was applied to the cantilever device was measured using a laser Doppler vibrometer (LDV), and the d_{31} and k_{31} value was calculated using the equation below [5]. Here, t_1 and t_2 denote the thickness of the PZT thick film and stainless steel, and s_1 and s_2 denote the elastic compliance of the PZT and stainless steel.

$$d_{31} = \frac{\delta(s_1^2 t_2^4 + 4s_1 s_2 t_1 t_1^3 + 6s_1 s_2 t_1^2 t_2^2 + 4s_1 s_2 t_1^3 t_2 + s_2^2 t_1^4)}{3(s_1 s_2 t_1 (t_1 + t_2) l 2V)}$$

$$k_{31} = d_{31} / (\varepsilon_{33}^T s_{11}^E)^{1/2}$$

Table 1. Piezoelectric properties of PZT thick films and bulk ceramics

	Thick film	bulk
$\varepsilon_{33}^T/\varepsilon_0$	400	1400
d_{31}	-45 pm/V	-115 pm/V
k_{31}	0.20	0.30

Figures 2 and 3 (SEM images) show a surface view and a cross-sectional view of the piezoelectric thick film. The images show that the piezoelectric thick film has been sintered to sufficient density, indicating that the decline in relative permittivity and piezoelectric constant was not caused by inadequate sintering.

Figure 2. SEM image of the surface view of PZT thick film.

Figure 3. SEM image of the cross-section of PZT thick film.

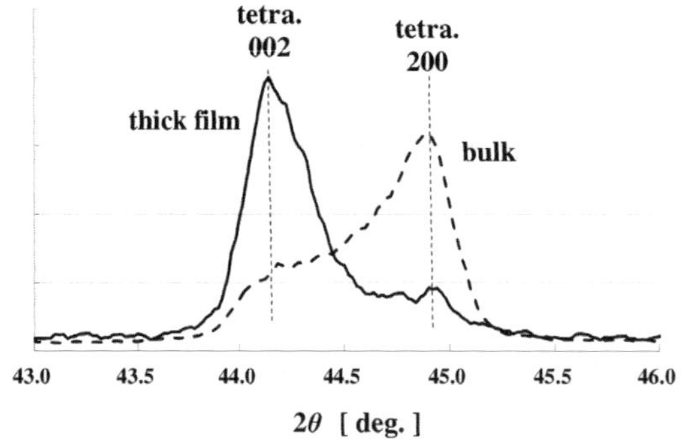

Figure 4. Comparison of PZT thick film and bulk PZT.

Figure 4 shows a comparison result between the XRD pattern of the bulk PZT and the piezoelectric thick-film device. Comparison between the (002) and (200) peak intensities indicates that the (002) peak intensity is higher in the piezoelectric thick film. This is likely due to the difference in thermal expansion coefficient between the stainless steel and piezoelectric materials, which produced high compression stress on the piezoelectric thick film during the temperature-lowering step of the sintering process. The large compression stress caused more PZT crystals to be oriented in the c-axis direction with respect to the film thickness direction, causing a lowering of dielectric permittivity and piezoelectric constant [6].

Measuring the residual stress of the piezoelectric film using x-ray diffractometry indicated a stress of approximately 500 MPa. Although causing a deterioration of electric characteristics, the large residual compression stress of the piezoelectric thick films is able to compensate for the fragility of the ceramic materials under tensile force.

4. Characteristics of vibration power generating device

Figure 5 shows the relationship between the output power and the frequency when sinusoidal wave vibration with acceleration of 0.1 G_{0p}(0.98 ms^{-2}) was applied to the cantilever-type vibration power generation device. The output power was determined by connecting a load resistance of 100 kΩ between the top and bottom electrodes and measuring the voltage at both ends. A maximum output power of approximately 1.1 mW was recorded at the device's resonance frequency of 24 Hz.

Figure 6 shows the relationship between the vibration acceleration and the output power at the resonance frequency. The output power increased with a rise in vibration acceleration, reaching approximately 5 mW or higher at 0.3 G_{0p}(2.9 ms^{-2}).

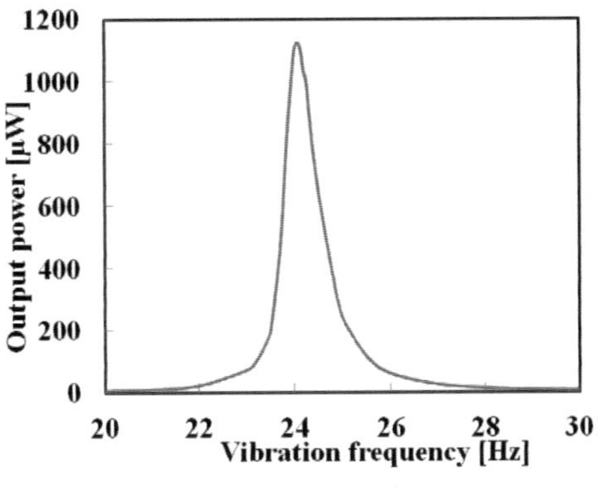

Figure 5. Frequency characteristics of vibration power generation device.

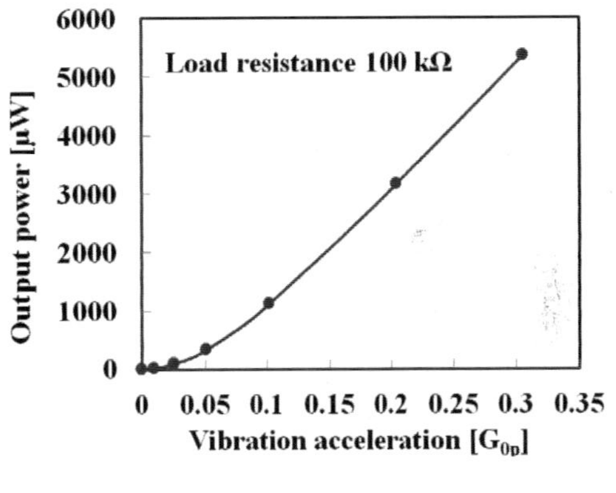

Figure 6. Relationship between vibration acceleration and output power at the resonance frequency.

Next, the durability of the vibration power generation device was evaluated. A 10 mm-wide and 17 mm-long vibration power generation device was prepared for the durability test as shown Figure 7. The device was fabricated from an element consisting of a 10 mm-wide, 0.2 mm-thick stainless steel substrate with a piezoelectric thick film measuring 9.4 mm wide, 10 mm long, and 0.03 mm thick, formed on both surfaces. The tip of the element was struck with a rotating bar to evaluate its durability based on how the power generation characteristics changed when damped resonant vibration was repeatedly applied to the device. Figure 8 shows the power generation characteristics under damped resonant vibration. At the striking instant, a maximum power of 100 mW or higher was generated, with approximately 0.9 mJ of power measured per single damped resonant vibration. Figure 9 shows how the power generation characteristics changed over 10 million strikes. The stainless steel substrate and piezoelectric thick films remained intact, with no change in the power generation characteristics.

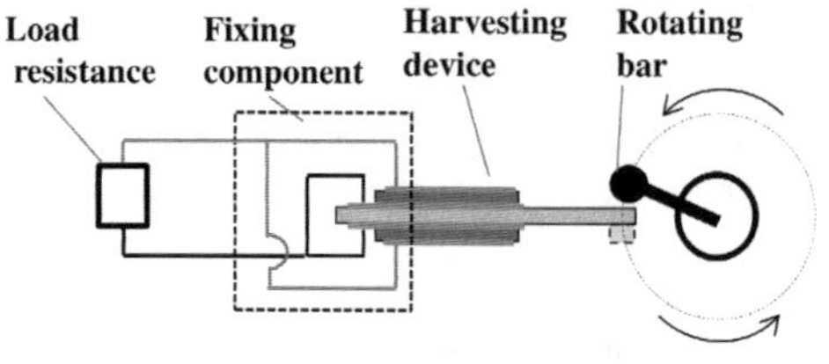

Figure 7. Appearance of the durability test by striking with rotating bar.

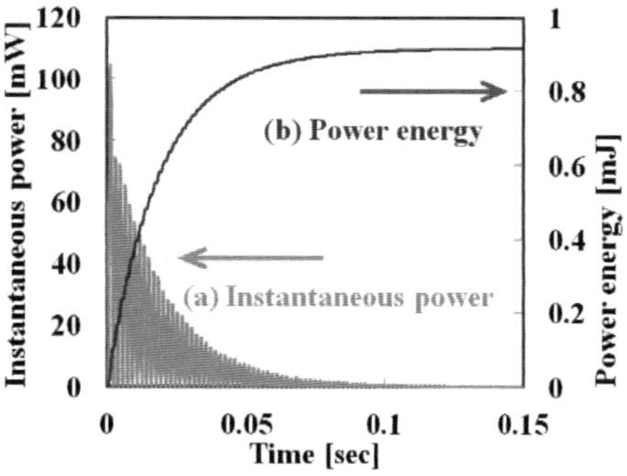

Figure 8. The behavior of instantaneous power and power energy with time.

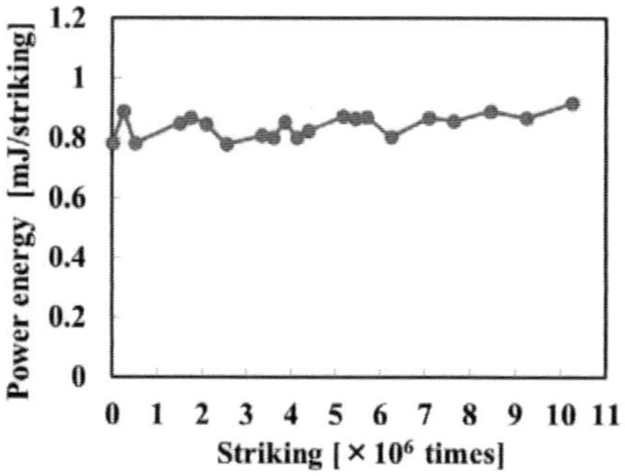

Figure 9. Power output with cumulative number of strikes.

The power obtained with the vibration power generation device was rectified, converted into 2.5 V direct voltage and used to successfully drive a temperature and acceleration sensor module. This indicates that continuous sensing and wireless transmission is possible over the 920 MHz radio band.

5. Conclusion

We prepared a simple and cheap cantilever-type vibration power generation device using PZT materials and screen-printing. The piezoelectric thick films obtained with screen-printing showed lower relative permittivity and piezoelectric constant than the bulk materials. These films are not prone to the fragility that is characteristic of conventional ceramic materials, and are inexpensive to prepare, and are thus likely to find applications in a wide range of fields as self-supporting power sources for sensor modules, for which there is expected to be a growing demand.

6. References

[1] S. Roundy, P.K. Wright, *Smart Mater. Struct.*, **13** (2004) 1131-1142
[2] Hua Yu, Jielin Zhou, Licheng Deng, Zhiyu Wen, *Sensors*, **14** (2014) 3323-3341
[3] D. Zhu, S.P. Beeby, M.J. Tudor, N.R. Harris, *Sensors and Actuators A*, (2011)
[4] H. Sun, D. Zhu, N.M. White, S.P. Beeby, *Sensors Journal of Physics: Conference Series,* **476** (2013)
[5] R. Takayama, E. Fuji *et al.*, *IEEJ Trans. SM*, **127**. 12 (2007) 553-558
[6] T. Miyoshi, M. Nakajima, H. Funakubo, *Jpn. J. Appl. Phys.*, **48** (2009) 09KD01-09

Low Temperature Sintering of PZT

A. Medesi[1,a], T. Greiner[1], M. Benkler[1], C. Megnin[1], and T. Hanemann[1,2]

[1]University of Freiburg, Department of Microsystems Engineering, IMTEK, Georges-Koehler-Allee 102, 79110 Freiburg, Germany
[2]Karlsruhe Institute of Technology, KIT, Institute for Applied Materials, Hermann von Helmholtz-Platz 1, 76344 Eggenstein-Leopoldshafen, Germany

Abstract. This paper describes the fabrication and characterization of lead zirconate titanate (PZT) films fired in a liquid-phase sintering process at 900 °C in air. In detail the manufacturing of piezoelectric multilayers with internal pure silver (T_m = 961 °C) electrodes are reported. The feasibility of ten sintering aids in two different volume fractions was investigated for a commercial hard PZT powder (PIC 181, PI Ceramics) with respect to density, microstructure, mechanical behaviour, and piezoelectric properties. Li_2O, Li_2CO_3, PbO, MnO_2, V_2O_5, CuO, Bi_2O_3, the eutectic mixtures $Cu_2O{\cdot}PbO$ and $PbO{\cdot}WO_3$ and the ternary system $Li_2CO_3{\cdot}Bi_2O_3{\cdot}CuO$ (LBCu) have been tested as liquid phase sintering aids. The combination of PZT with LBCu showed the best results. With 5 vol.% LBCu an average relative density of 97% and a characteristic breaking strength of 77 MPa was achieved. Composition of PZT with 2 vol.% LBCu exhibits the highest averaged piezoelectrical charge constant (d_{33}) of 181 pC/N.

1. Introduction

Piezoelectric devices based on PZT ceramic components have a current worldwide market of several billion dollars [1]. Despite lead-containing, PZT is still the most commonly used piezoelectric material, due to the unreached high electromechanical coupling constant. Especially the applications of piezoelectric materials as transducers are continuously growing. Transducers are devices that convert electrical energy into mechanical energy and vice versa. A special kind of transducers are energy harvesters, which use the energy of ambient vibrations to generate electrical energy. A high electrical output is expected for multi-layer bending transducers from hard PZT. For multilayered stacks the inner electrodes have to be co-fired with the ceramic. The electrical connection of PZT in multilayer devices is mostly realized by expensive high-temperature resistant metals, as Pt [2] or Ag/Pd [3]. The usage of cost-effective Ag-electrodes is hampered by the high sintering temperature of PZT, which is about 1200 °C. For sintering temperatures ≤ 900 °C pure Ag-electrodes can be used. Besides cost reduction for electrode material and energy consumption of the firing process, the volatility of PbO is reduced significantly. Due to that the stoichiometric composition of the PZT is more constant, which is necessary to realize stable electromechanical properties and reliable electrical components. Therefore a lot of effort was done and a variety of approaches, like sol-gel method [4], hot-pressing in oxygen [5], the vacuumed-air-venting process [6], or using fine ball-milled powders [7], has already been attempted. Nevertheless the liquid-phase sintering technique remains one of the most successful in lowering the sintering temperature of PZT by several hundred degrees.

[a] Corresponding author: anna.medesi@imtek.uni-freiburg.de

This technique is based on the acceleration of the densification progress by forming a liquid phase out of a lower melting sintering aid, which facilitates rearrangement of the PZT particles and hence lowers the required sintering temperature.

2. Fabrication Process

Piezoceramic green tapes based on a commercial hard PZT powder (PIC 181, PI Ceramics) were manufactured by tape casting. Slurries with a specially developed formulation were prepared with azeotropic mixture of ethanol and toluene as solvent and other organic additives. 3,6,9-trioxadecanoic acid (Clariant) was added as dispersant, polyvinyl butyral (Mowital-B45H, Kuraray) as temporary binder, and polyethylene glycol (Roth) as plasticizer. All slurries were ball-mixed in a planetary mill for 24 h and degased for 15 min directly before tape casting. The tape casted films were dried at ambient conditions. For each sintering aid two slurries with the same composition of PZT powder and organic additives were prepared containing 2 or 5 vol.% of the corresponding sintering aid in relation to the volume of PZT powder. The complex sintering aids were added in their eutectic mixtures. $PbO \cdot WO_3$ containing 16.5 mol% of WO_3 has a melting point of 730 °C [8], while $Cu_2O \cdot PbO$ (1:4) has a melting point of about 680 °C [9]. The ratio of Li_2CO_3, Bi_2O_3, and CuO (1:1:3.7) in the ternary system LBCu was chosen according to Wang *et al.*[10].

2.1. Single-Layer (SL)

Square-shaped green bodies (18 x 18 mm^2 and 12 x 12 mm^2) were punched out of dried green tapes and sintered in ambient air at 900 °C for 3 h. The samples were heated up to 400 °C with a slow heating rate of 1 K/min to ensure a complete removal of the organic additives (debinding). The sintered bodies (≈ 100 µm thick) have been metallized with a silver-conductive varnish (L100, Kemo-Electronic) to realize electrodes on both sides. Poling of each sample was performed in a silicon oil bath at 120 °C for 15 min in an electrical field of 3 kV/mm. Fig. 1 shows the intermediate steps of this fabrication process. Additional to the squared samples, beam-shaped specimens with respect to DIN EN 843-1 were sintered at the same temperature program for bending tests.

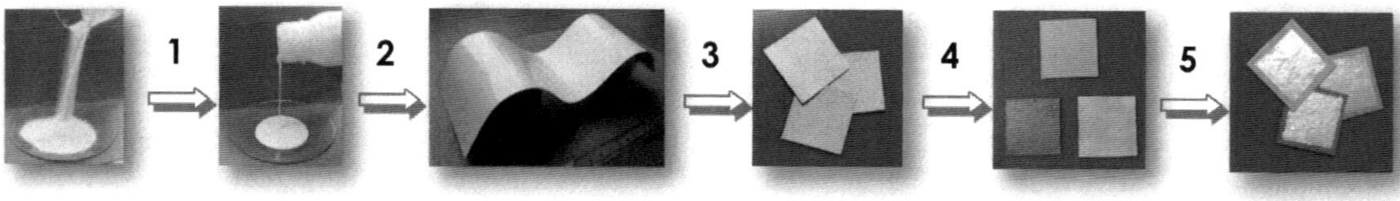

Figure 1. Fabrication process of investigated samples: 1. Preparation of a pourable slurry from PZT, organic and sintering additives, 2. Tape casting of films, 3. Punching out green bodies, 4. Debinding and sintering, 5. Metallizing with Ag varnish and poling with 3 kV/mm.

2.2. Multi-Layer (ML)

Beam-shaped pieces were punched out from green tapes. Metallization was performed by coating them with a conductive paste containing Ag-microparticles (Sigma-Aldrich). The metallized layers were stacked in a stepped manner to provide easy access to inner electrodes. The whole stack of alternating ceramic and metallic layers has been sintered in a co-firing process at 900 °C for 3 h. External electrodes (top and bottom) were applied after the co-firing process with a silver-conductive varnish (L100, Kemo-Electronic GmbH). Fig. 2 shows the sintered multi-layered stack including a magnified side view. In a new co-casting process the highly viscous Ag-paste (Sigma-Aldrich) was directly tape casted on a dried ceramic green tape. Subsequently, a second ceramic layer was tape casted on top of the dried Ag-film. This step is depicted in Fig. 3. Beam-shaped bimorphs were cut by laser (DPL *Smart* Marker, ACI Laser) from this co-casted three-layered stack and co-fired at 900 °C for 3 h. Access to the inner electrode was realized after the sintering process by removal of a squared area of the upper ceramic layer by laser treatment. This laser structured entry to the internal Ag-electrode is shown in a SEM micrograph in Fig. 4 (upper right corner).

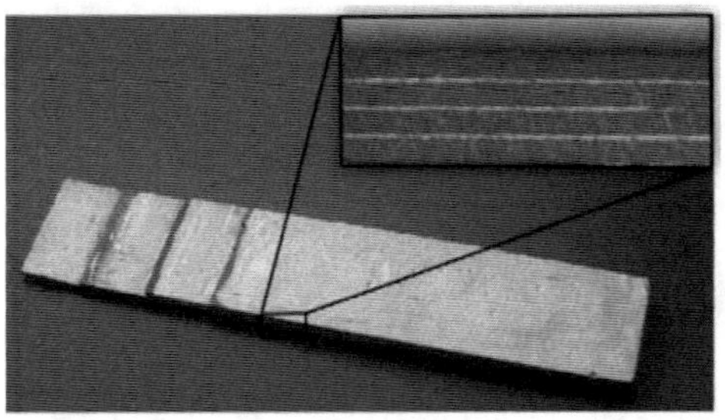

Figure 2. Co-fired multilayer with Ag-electrodes fabricated by stacking. No degradation of Ag or de-lamination of the layers was observed by SEM.

Figure 3. Co-casting plant (on lab scale) for manufacturing ML without stacking of SLs.

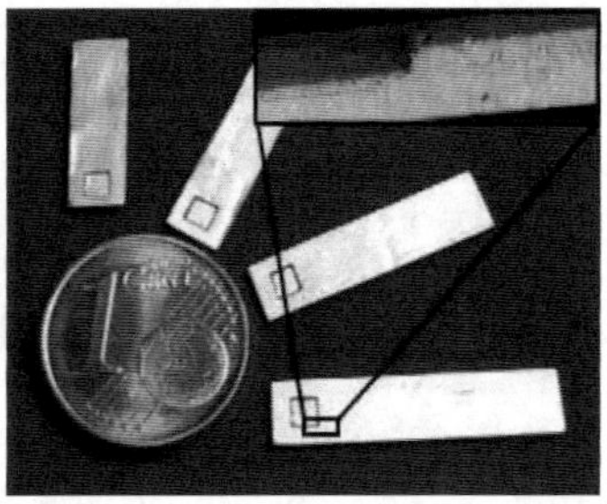

Figure 4. Co-casted bimorphs. Tailoring and access to the inner Ag-electrode was made by laser treatment.

3. Characterization

Densities of sintered bodies (single-layers), as depicted in Fig. 5, have been determined from their geometrical dimensions for assaying the potency of the sintering aids on densification progress of the PZT particles. Layer thicknesses have been measured with a digital measuring sensor system (MT 25, Heidenhain). Theoretical densities for each composition were calculated as mean values with respect to the volume fractions of the powder components. Relative densities were related to the theoretical densities of each composition. Microstructure was analyzed by SEM (Phenom G2 pro). For meeting a reliable statement concerning the mechanical stability of the sintered bodies from all PZT-based compositions 3-point-bending tests have been performed on specially prepared specimens from the same green tapes with dimensions referring to DIN EN 843-1. The characteristic breaking strengths were evaluated with Weibull statistic using the Maximum-Likelihood-Method. Longitudinal piezoelectric charge constants d_{33} have been determined by means of the piezoelectric d_{33} testing system PM300 (Piezotest).

4. Results and Discussion

Due to different efficiencies of added liquid sintering aids the sintered bodies show different thicknesses and sizes. A sample of each composition is pictured in Fig. 5. It must be taken into consideration that the investigated amounts of sintering aids are differently well suited for each PZT-aid combination. While PZT with PbO, Bi_2O_3 or $PbO{\cdot}WO_3$ shows improved mechanical and piezoelectric properties with increased amount of sintering aid, the Li-compounds suffer from significantly deteriorated piezoelectric properties (see Tab. 1). For usage of Li_2O or Li_2CO_3 as a sintering aid for PZT significantly lower amounts than 2 vol.% are recommended [11]. In this study we used these amounts to give a direct comparison for equal volume fractions of all sintering aids.

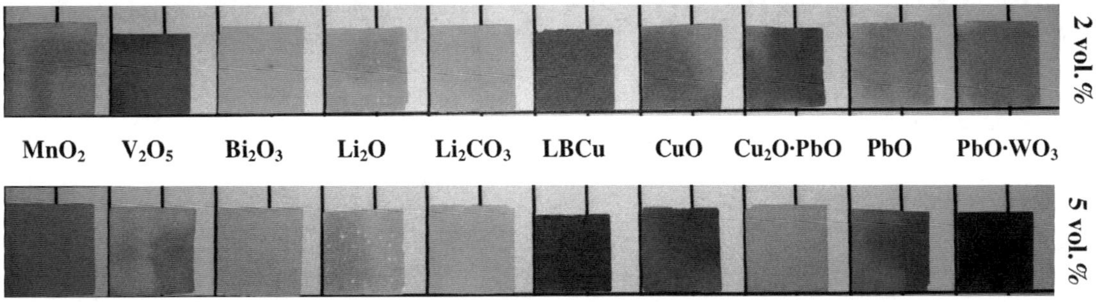

Figure 5. Sintered bodies from compositions of PZT with 2 and 5 vol.% of sintering aids.

4.1. Dimensional Reproducibility

A balanced and reliable shrinking behavior is crucial to dimensional reproducibility of future piezoelectric devices. Quite obvious is the imbalanced shrinkage of samples with $Cu_2O{\cdot}PbO$ or PbO as sintering aid (see Fig. 5). Samples with V_2O_5, Li_2O and CuO show also an imbalanced shrinkage, which causes distortion of the originally squared sample shape. The unequal discoloration of these samples is another hint for the inhomogeneous microstructure of these material composites. Shrinkage of the sintered bodies of PZT with MnO_2 after sintering at 900 °C is so low that it seems to be regular. A balanced shrinkage behavior was obtained for PZT with the sintering aids Li_2CO_3, Bi_2O_3 or the complex sintering aid LBCu, which is a combination of these aids and CuO.

4.2. Mechanical stability

Extremely brittle are samples with MnO_2 as sintering aid. Samples of PZT with Li_2O, Li_2CO_3 or Bi_2O_3 exhibit hardly any higher breaking strengths (see Tab. 1). Due to insufficient densification of the PZT particles in these composites the microstructure remains porous during sintering at 900 °C (see Fig. 6). PbO is also not able to effect an adequate densification of the PZT particles. Nevertheless, the samples with PbO revealed average breaking strengths above 40 MPa. Combination of PZT with V_2O_5 shows also low densification. However, V_2O_5 stabilizes the porous PZT matrix through a formation of V_2O_5-ligaments on the surface (see Fig. 6). For that reason breaking strengths above 50 MPa were measured for PZT with V_2O_5. The sintering aids CuO and Li_2O accumulate also on the surface of the sintered bodies. Unlike in the case of the Li-compounds, the compositions with CuO delivers mechanical very stable films with a high average breaking strength of 66 MPa for the addition of 2 vol.% (see Tab. 1). Significant densification of the PZT particles and formation of sintering necks during the sintering process at 900 °C were observed for samples with 5 vol.% of $Cu_2O{\cdot}PbO$ or $PbO{\cdot}WO_3$ (see Fig. 6). However, cracks within these materials obviously reduce their mechanical stability. Actually dense sintered and stable films with significant grain growth of the PZT particles are provided by the addition of 5 vol.% LBCu. Samples with this composition showed the highest average relative density of 97 ± 3 % and a characteristic breaking strength of 77 MPa.

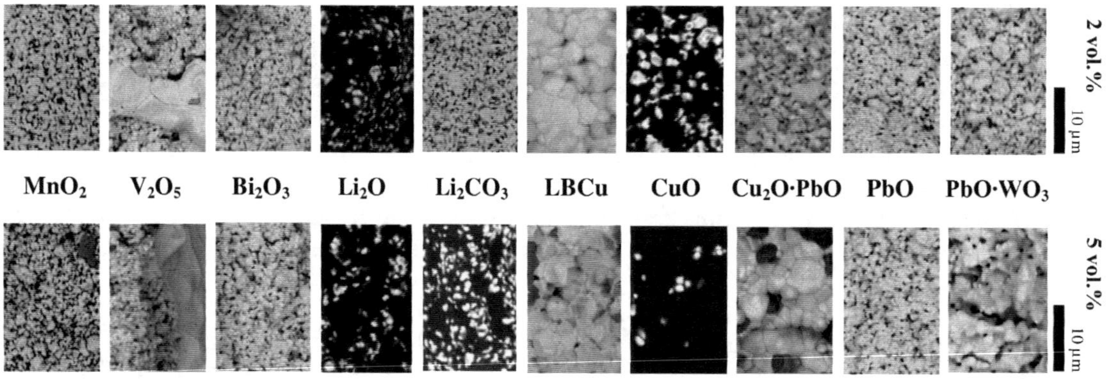

Figure 6. SEM micrographs of the surfaces of sintered films from compositions of PZT with 2 and 5 vol.% of sintering aids.

4.3. Piezoelectric charge constant

In principle, all sintering additives have to be considered as impurities which remain in the PZT matrix and cause a deterioration of the piezoelectric properties. Concurrently, sintering aids support the densification progress of the PZT particles. For most of the samples with 5 vol.% of sintering additive we measured lower d_{33} values compared to the corresponding samples with 2 vol.%. Strongly enhanced d_{33} values were obtained for samples with an increased amount of $PbO{\cdot}WO_3$ or V_2O_5 (see Tab. 1). The highest average d_{33} value of 181 pC/N was measured for films out of PZT and 2 vol.% LBCu.

We assume that the content of CuO in LBCu is responsible for this high piezoelectric constant, since all samples containing CuO (or Cu_2O in low amounts) show extraordinary high d_{33} values (see Tab. 1). Due to the increased d_{33} values of samples with 5 vol.% $PbO \cdot WO_3$ of 143 pC/N in average, we suggest that WO_3 could have a similar effect as CuO. This piezoelectric properties increasing effect could be related to the semiconductive nature of CuO [12] and WO_3 [13].

Table 1. Density, characteristic breaking strength and piezoelectric charge constant of PZT-based, at 900 °C sintered films in relation to sintering aid amount.

Sintering additive	2 vol.%					5 vol.%				
	ρ_{abs} [g/cm³]	ρ_{rel} [%]	σ_0 [MPa]	d_{33} (ø) [pC/N]	d_{33} (max) [pC/N]	ρ_{abs} [g/cm³]	ρ_{rel} [%]	σ_0 [MPa]	d_{33} [pC/N]	d_{33} (max) [pC/N]
MnO_2	4.2	55	12	13	14	4.5	59	10	4	5
V_2O_5	5.7	74	61	51	70	5.8	77	51	67	**140**
Li_2O	5.3	69	29	34	74	5.7	76	24	17	37
Li_2CO_3	4.9	63	31	79	**146**	5.7	75	34	53	82
Bi_2O_3	4.8	62	22	29	42	5.6	71	53	34	71
LBCu	6.3	**81**	49	**181**	**246**	7.4	**97**	**77**	98	113
CuO	5.8	70	**66**	136	**196**	6.0	78	58	134	**262**
$Cu_2O \cdot PbO$	5.7	74	45	126	**151**	5.6	72	36	46	52
PbO	5.4	69	40	30	46	6.4	81	48	34	46
$PbO \cdot WO_3$	5.4	69	26	59	67	7.0	89	55	**143**	**185**

5. Conclusion

CuO cannot be considered a liquid phase sintering aid in original sense, since its melting point is about 1326 °C. However, its addition increases significantly both, the piezoelectric properties and the mechanical stability of hard PZT films. LBCu, a combination of Li_2CO_3, Bi_2O_3 and CuO, acts as a great sintering aid for PZT. The highest average piezoelectric constant was measured for PZT with 2 vol.% LBCu. Additionally, LBCu provides the highest densification progress of all investigated PZT-sintering aid composites. With an addition of 5 vol.% LBCu a relative density of PZT-based films (≈100 μm thick) of 100% could be achieved after sintering at 900 °C for 3 h.

References

[1] Testino A 2013 *Int. J. Appl. Ceram. Technol.* **10** 723-730
[2] Xu R et al. 2012 *J. Micromech. Microeng.* **22** 094007-16
[3] Kühnlein T, Stiegelschmitt A, Roosen A, and Rauscher M 2013 *J. Eur. Cer. Soc.* **33** 991-1000
[4] Zhu W, Wang Z, Zhao C, and Tan O K 2003 *Mat. Sci. Eng. B* **99** 56
[5] Haertling G H and Land C E 1971 *J. Am. Cer. Soc.* **54**
[6] Chen C Y, Hu Y, and Lin H L 2006 *Mat. Chem. Phys.* **99** 26-29
[7] Lee J S et al. 2007 *Ceram. Int.* **33** 1283-86
[8] Nielsen E R, Ringgaard E, and Kosec M 2002 *J. Eur. Cer. Soc.* **22** 1847-55
[9] Corker D L, Whatmore R W, Ringgaard E, and Wolny W W 2000 *J. Eur. Cer. Soc.* **20** 2039-45
[10] Wang X X, Murakami K, Sugiyama O, and Kaneko S 2001 *J. Eur. Cer. Soc.* **21** 1367-70
[11] Han H-S, Park E-C, Lee J-S, Yoon J I, Ahn K K 2011 *Trans. Electr. Electron. Mater.* **12** 249-52
[12] Rahnama A, and Gharagozlou M 2012 *Opt. Quant.Electron.* **44** 313-22
[13] Vemuri R S, Kamala Bharathi K, Gullapalli S K, and Ramana C V 2010 *ACS Appl. Mater. Interfaces* **2** 2623-28

Stability and spring constant investigation for micromachined inductive suspensions: theoretical analysis vs. experimental results

K Poletkin[1], Z Lu[1], B den Hartogh[2], U Wallrabe[1], and V Badilita[1,3]
[1] Laboratory for Microactuators, Dept. of Microsystems Engineering – IMTEK, Univ. of Freiburg, Georges-Köhler-Allee 102, 79110 Freiburg, Germany
[2] FemtoTools AG, Furtbachstrasse 4, CH 8107 Buchs/ZH, Switzerland

E-mail: vlad.badilita@imtek.de

Abstract. We present a linear analytical model coupled with experimental analysis to discuss stability of a levitated proof mass (PM) in a micromachined inductive suspension (MIS), which has been previously introduced and characterized. The model is a function of the MIS geometry, describes the dynamics of a levitated disk-shaped PM near the equilibrium point, and predicts conditions for stable levitation. The experimental setup directly measures the lateral component of the Lorentz force, which has a stabilization role in the MIS structure, as well as the vertical levitation force. The experimental setup is further used to derive mechanical parameters such as stiffness values relative to lateral, vertical and angular displacements, proven to be in excellent agreement with the values predicted by the analytical model.

1. Introduction

Downscaling of micromachined inductive suspensions (MIS) has been enabled by the advances in the field of micro-electro-mechanical systems (MEMS) technologies. Due to their non-contact operation based on electromagnetic induction, MIS constitute the building blocks of a wide range of systems operating in a frictionless, dust-free environment, such as gyroscopes, micromotors, micro-positioning systems.

The first achievements in the field of MIS have been reported by Shearwood et al. in 1995, the technology being essentially based on surface micromachining, while the main application proposed was a rotating micro-gyroscope [1]. Shearwood et al. [2] reported 1000 rpm rotation speed and showed that this limit is attributed to a slight wobble during rotation that excited a lateral resonant mode. The performance was improved by Zhang et al. [3] by separating the so-called rotation and levitation coils, reporting a maximum speed of 3000 rpm. However, there was still plenty of room for improvement since the maximum theoretical rotation speed in air is of the order of 100000 rpm [2].

Most of the MIS structures reported so far, including the examples mentioned above, extensively make use of planar microfabrication techniques in order to define the coil structures required to achieve various functions such as levitation, rotation and stabilization of the proof mass (PM) to be levitated.

[3] To whom any correspondence should be addressed.

While defining the metal track for these coils is extremely straightforward for one single metal layer, any addition for multi-layered/multi-windings coils tremendously increases the fabrication complexity, at the same time decreasing the yield. Each additional metal layer requires additional masks corresponding to metallization, etching, insulation, vias. Moreover, the current carrying capabilities of metal layers obtained via planar microfabrication techniques (evaporation/sputtering, electroplating) are rather limited. In order to circumvent these problems we have developed a coil winding technology based on an automatic wirebonder [4] for the fabrication of 3D microcoils with a high degree of freedom in terms of shape, height, winding pitch, and number of windings.

Using this technology, we reported a 3D micromachined inductive suspension (3D MIS), which was preliminarily characterized [5, 6] demonstrating a dramatically reduced threshold current necessary to achieve levitation, along with increased levitation height compared to Shearwood et al.[2], due to the superior number of ampere-turns in the case of 3D solenoidal microcoils as opposed to the coils obtained through planar technology.

The purpose of this paper is to give a brief theoretical background as a basis to describe the stability and to calculate characteristic parameters of the system such as stiffness values. At the same time, this is the first direct verification of the theoretical model through direct experimental measurements of the forces acting on the proof mass.

2. Operating principle and fabrication

The basics of the 3D MIS proposed by our group have been previously introduced and the fabrication process has been presented in detail [6]. The device consists of two coaxial wirebonded microcoils that are excited with AC signals having 180° phase-shift with respect to each other. The inner coil is responsible for the levitation of the conductive disk (aluminium, 25 μm thick), while the outer coil is responsible for the stabilization of the proof mass. By applying an AC signal to the coils, a time-variable magnetic flux is created, thus inducing eddy currents in the PM. These eddy currents also generate a magnetic field that interacts with the excitation field producing the levitation and the stabilization effects respectively. By biasing the outer coil with an AC signal having 180° phase-shift with respect to the inner coil, a restoring force is introduced that pulls the PM back towards the equilibrium position, thus shaping a potential well which defines the stability of the system.

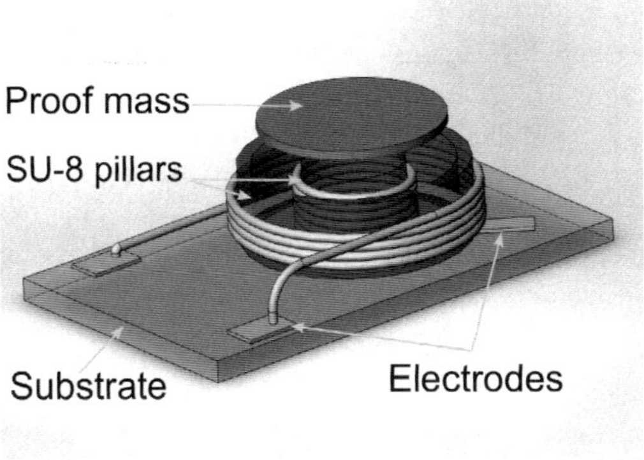

Figure 1. Schematic of the 3D MIS.

Figure 2. Photo of the 3D MIS without PM.

The fabrication process of the 3D MIS starts with the metallization of a Pyrex wafer in order to define the contact pads for microcoil connection. A CrAu layer is evaporated, followed by UV patterning and electroplating up to a thickness of 10 μm. After dissolving the photoresist mold used for electroplating and etching away the CrAu seed layer, tall SU-8 2150 structures are defined by UV photolithography to serve as support for the subsequent coil winding process. The height of these support structures is approximately 700 μm. The final step of the fabrication is the coil winding process, which is performed using an automatic coil-winding machine.

The PM to be levitated is laser-cut to a diameter of 3200 μm from a 25 μm thick Al foil.

Table 1. Parameters of the 3D MIS prototype.

Radius of the levitation coil, r_l	1000 μm
Radius of the stabilization coil, r_s	1900 μm
The coils pitch of winding, p	25 μm
Number of windings for stabilization coil, N	12
Number of windings for levitation coil, M	20
Radius of PM, r_{PM}	1600 μm
Thickness of PM	25 μm

3. Linear analytical model for MIS

3.1. Theoretical background

In the present paper, we adapt and apply the linear analytical model previously developed in [7, 8] to the MIS structure based on 3D wirebonded microcoils. The linear theoretical model can be seen as the analogy for the 3D MIS of the harmonic oscillator model, allowing us to predict the important parameters of the structure such as different stiffness components and, as a consequence, the corresponding resonant frequencies. The model starts from the dynamical equations written for the three degrees of freedom required to completely characterize the system: vertical displacement (l), lateral displacement (s), angular displacement (ϕ).

$$\begin{cases} m\ddot{l} + \mu_l\dot{l} + k_l \cdot l = F_l; \\ m\ddot{s} + \mu_s\dot{s} + k_s \cdot s + k_{s\varphi} \cdot \varphi = F_s; \\ J\ddot{\varphi} + \mu_\varphi\dot{\varphi} + k_{s\varphi} \cdot s + k_\varphi \cdot \varphi = M_\varphi. \end{cases} \quad (1)$$

where m is the mass of PM, J is the moment of inertia of the PM about the axis lying on its equatorial plane, μ_l, μ_s, and μ_ϕ, and are the damping coefficients of the PM relative to the appropriate velocities of the generalized coordinates l, s and ϕ, respectively; F_l, F_s, and M_ϕ are the generalized forces and torque acting on the PM relative to the generalized coordinates; k_l, k_s, k_ϕ and $k_{s\phi}$ are the stiffness coefficient corresponding to the generalized coordinates.

As we have previously demonstrated [7, 8], the stable levitation condition requires that the following inequalities are fulfilled simultaneously:

$$\begin{cases} k_l > 0; & (2a) \\ k_s > 0; & (2b) \\ k_\varphi > 0; & (2c) \\ k_s \cdot k_\varphi - k_{s\varphi}^2 > 0. & (2d) \end{cases}$$

3.2. Direct experimental measurement of stiffness coefficients

The strategy to measure the stiffness coefficients corresponding to vertical, lateral and angular displacement is to apply a force in a controlled manner and to measure the displacement determined by this force. In order to perform these measurements, we employed a microforce sensing probe (FT-S100, FEMTO-TOOLS AG Switzerland) and a mechanical probe (FT-FS1000, FEMTO-TOOLS AG Switzerland), simultaneously recording the applied force and the vertical and lateral displacement, respectively. The force resolution of the sensing probe was 0.005 μN and the displacement resolution 5 nm. Figure 3 shows the schematic of the experimental setup. Each coil was fed with a square wave AC current provided by a current amplifier (LCF A093R). The amplitude and the frequency of the current

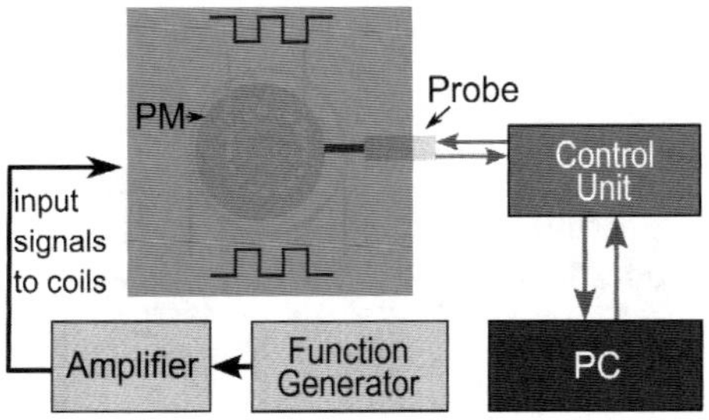

Figure 3. Schematic of the experimental setup.

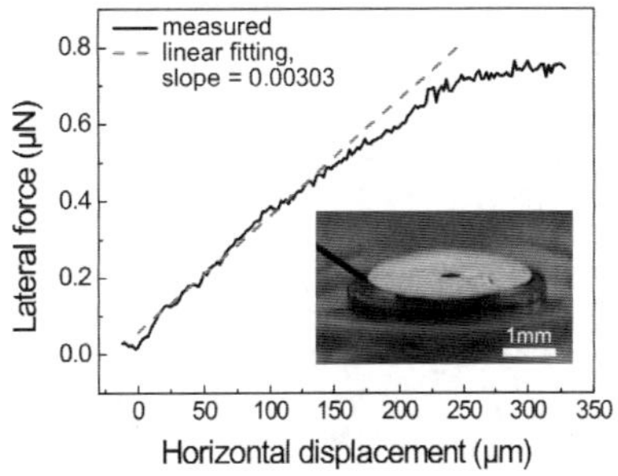

Figure 4. Lateral force versus displacement measurement.

in each coil were controlled by a function generator (Arbstudio 1104D) via a computer. In this experiment the levitation height, h, of the PM was 104 μm, measured with a laser distance sensor (LK-G32). Root mean square currents in the stabilization and levitation coils were to be I_S =0.106 A and I_L=0.11 A, respectively, at a frequency of 12 MHz. For force and displacement measurements the microforce sensing probe was moved towards the PM until mechanical contact, as shown in the inset of figure 4. Once in contact, the force applied to the PM and the linear displacement along the direction of action were recorded.

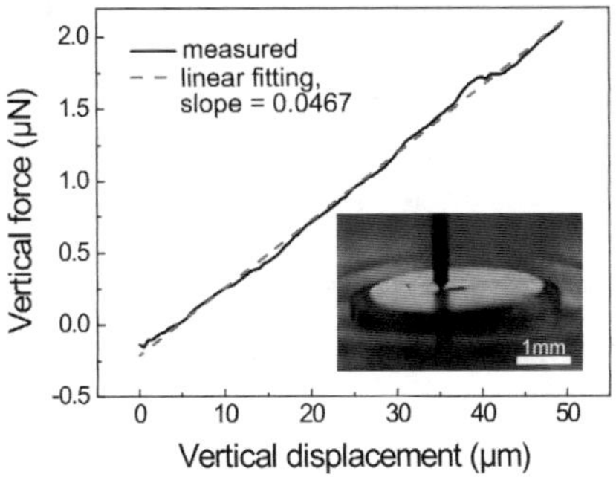

Figure 5. Vertical force versus displacement measurement for the tip placed at r_{p1}=300 μm.

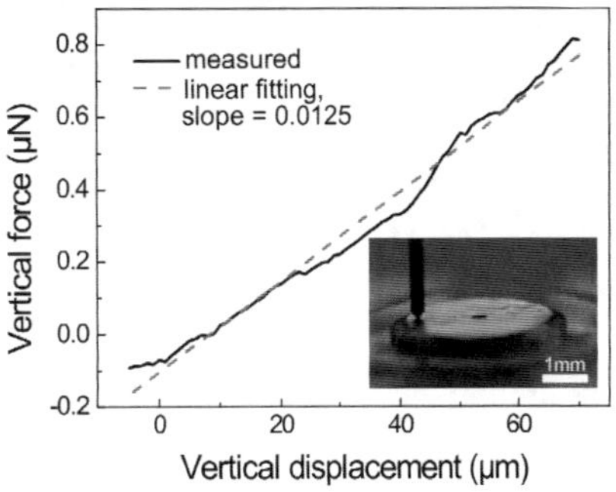

Figure 6. Vertical force versus displacement measurement for the tip placed at r_{p1}=1450 μm.

To evaluate the lateral stiffness, the tip of the sensing probe was tilted to an angle of 15° with respect to the PM plane due to topological restrictions. The lateral displacement measurement results are shown in figure 4, with a lateral stiffness coefficient, k_s, calculated to be 3.0×10^{-3} N·m^{-1}. Vertical and angular stiffness coefficients, k_l and k_ϕ respectively, have been evaluated indirectly by applying the sensing probe vertically at two different points located at r_{p1}=300 μm and r_{p2}=1450 μm from the PM center (insets of figure 5 and 6). Force versus displacement measurements are shown in figure 5 and 6. The corresponding stiffness values are: k_{p1}=4.5×10^{-2} N·m^{-1} and k_{p2}=1.2×10^{-2} N·m^{-1}. Substituting k_{p1}, k_{p2}, r_{p1} and r_{p2} into the set below:

$$\begin{cases} k_\varphi = \dfrac{\left(r_{p2}^2 - r_{p1}^2\right)k_{p1}}{k_{p1}/k_{p2} - 1}; \\ k_l = \dfrac{k_{p2}}{1 - r_{p2}^2\left(k_{p2}/k_\varphi\right)}, \end{cases} \tag{3}$$

the desired stiffness coefficients can be calculated. Hence we have the vertical stiffness k_l, relative to the generalized coordinated, l: k_l=5.3×10^{-2} N·m^{-1}, and the angular stiffness k_φ, relative to the generalized coordinate, ϕ: k_φ=3.3×10^{-8} N·m·rad^{-1}. On the other hand, we have also calculated the same stiffness coefficients using the linear analytical model. A comparison between the values measured experimentally and the calculated ones are shown in Table 2, showing a very good agreement between the experiment and the model. We also plot the set of inequalities (2), therefore providing a map of stability for the PM as shown in figure 7, where the variable *d* is the difference between the PM radius and the levitation coil radius, $d=r_{PM}-r_l$.

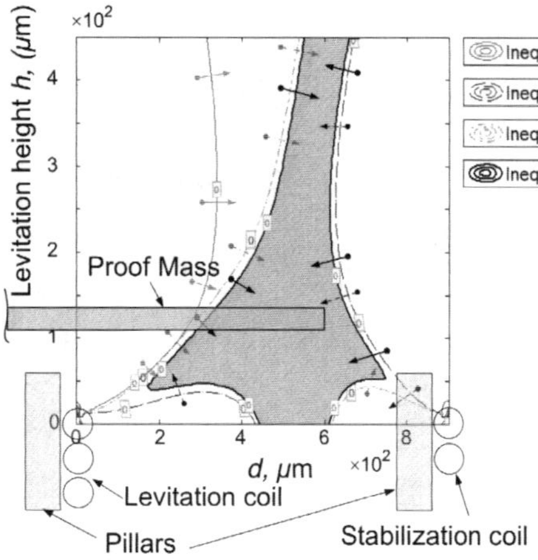

Table 2. Stiffness coefficients comparison: measured vs. calculated

Stiffness coefficients	Measured values	Calculated values
k_s, [N·m^{-1}]	3.0×10^{-3}	3.0×10^{-3}
k_l, [N·m^{-1}]	5.3×10^{-2}	4.0×10^{-2}
k_ϕ, [N·m·rad^{-1}]	3.3×10^{-8}	0.8×10^{-8}

Figure 7. Stability map of the levitating proof mass as derived from the set of inequalities (2). The arrows show the sign change when the boundary is crossed in the indicated direction.

4. Conclusions and outlook

The present work provides a theoretical insight in the 3D MIS reported previously [6], as well a comparison between the important parameters calculated by means of the theoretical model and the values obtained from direct experiment. This is the first time when the levitation and stabilization forces have been directly measured in such a structure. Moreover, this paper provides the means to describe the dynamics of the 3D MIS structure and clearly formulates the condition for stable levitation. The ability to theoretically predict the stability and the spring constants of the MIS enables future designs with improved dynamics for applications such as gyroscopes or high speed rotating micro-motors.

5. Acknowledgements

Zhiqiu Lu gratefully acknowledges support from the Siemens-DAAD Scholarship. Dr. Kirill Poletkin acknowledges with thanks support of the Alexander von Humboldt Foundation. Dr. Vlad Badilita kindly acknowledges support from the German Research Foundation (DFG) – project number BA 4275/2-1.

6. References

[1] Shearwood C, Williams C, Mellor P, Yates R, Gibbs M and Mattingley A 1995 Levitation of a micromachined rotor for application in a rotating gyroscope *Electronics Letters* **31** 1845-6

[2] Shearwood C, Ho K, Williams C and Gong H 2000 Development of a levitated micromotor for application as a gyroscope *Sensors and Actuators A* **83** 85-92

[3] Zhang W, Chen W, Zhao X, Wu X, Liu W, Huang X and Shao S 2006 The study of an elmg levitating μmotor for application in a rotating gyroscope *Sensors and Actuators A* **132** 651-7

[4] Kratt K, Badilita V, Burger T, Korvink J and Wallrabe U 2010 A fully MEMS-compatible process for 3D high aspect ratio μcoils with an automatic wire bonder *J. Micromech. Microeng.* **20** 015021

[5] Lu Z, Poletkin K, Wallrabe U and Badilita V 2014 Performance characterization of MIS based on 3D wirebonded microcoils *Micromachines* submitted - under review

[6] Badilita V, Rzesnik S, Kratt K and Wallrabe U 2011 Characterization of the 2nd generation magnetic microbearing with integrated stabilization for frictionless devices. In: *TRANSDUCERS, 2011 16th International*: IEEE) pp 1456-9

[7] Poletkin K, Chernomorsky A I, Shearwood C and Wallrabe U 2014 A qualitative analysis of designs of micromachined electromagnetic inductive contactless suspension *Intl J Mechanical Sciences* **82** 110-21

[8] Lu Z, Poletkin K, den Hartogh B, Wallrabe U and Badilita V 2014 3D micro-machined inductive contactless suspension: Testing and Modeling *Sensors and Actuators A* http://dx.doi.org/10.1016/j.sna.2014.09.017, in press

Experimental Analysis of a Coupled Energy Harvesting System with Monostable and Bistable Configuration

D Hoffmann[1], B Folkmer[1] and Y Manoli[1,2]

[1] HSG-IMIT – Institute of Micromachining and Information Technology, Wilhelm-Schickard-Str.10, 78052 Villingen-Schwenningen, Germany
[2] Fritz Huettinger Chair of Microelectronics, Department of Microsystems Engineering – IMTEK, University of Freiburg, Germany

daniel.hoffmann@hsg-imit.de

Abstract. In this paper we present experimental results from an energy harvesting system with two coupled energy harvesters. The energy conversion mechanism of the two coupled energy harvesters is based on the electromagnetic principle. The coupling is generated by two magnets in a repulsive arrangement. In this manner a bistable configuration can be obtained if the gap between the magnets is sufficiently small. We demonstrate that the total power output can be increased in comparison to a linear reference system, if specific conditions are fulfilled. In this respect, the highest power output occurs in the nonlinear region of a monostable system configuration, mostly near the transition to a bistable configuration. On the other hand, the results also indicate, that a bistable operating mode does not necessarily enhance the power output of the coupled system.

1. Introduction

The fundamental drawback of conventional vibration energy transducers is apparent when considering vibration profiles with a broad frequency spectrum or with a time-varying frequency content. For many years great effort was put into finding practical solutions to enhance the bandwidth of resonant vibration transducers [1]. In this respect the introduction of nonlinear systems with a bistable characteristic had a significant influence on the type of solutions so far [2, 3]. A further step towards a system with a coupled bistable structure was taken by Zhu et. al. [4]. In this system an assisting cantilever is coupled to a main cantilever in which only the main cantilever is able to generate energy. It was found that the power output is larger in comparison to a bistable structure without coupling. However, the assisting cantilever requires an additional volume, which could be used to accommodate another active energy harvesting structure.

In this paper, a coupled structure, which incorporates two cantilever-based electromagnetic energy harvesters coupled via a magnetic field, is investigated. The total power output of both energy harvesters together is considered and is compared to a linear reference system. The linear reference system incorporates two individual energy harvesters of the same size as in the coupled system. However, there is no coupling between the two energy harvesters in the linear case. In this respect, the volume occupied by the linear system is equal to the volume of the coupled system, which allows reasonable comparison between both systems.

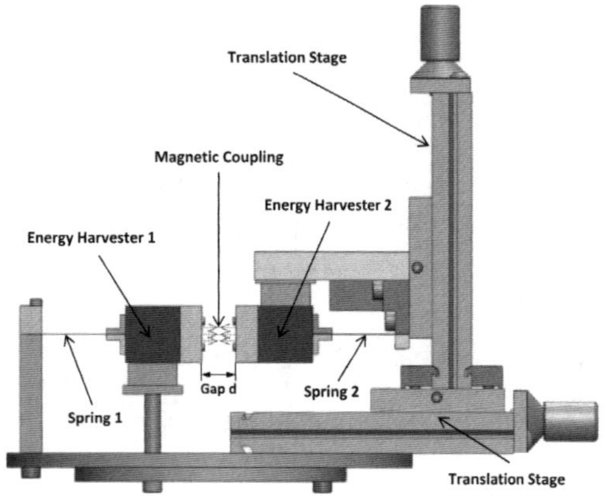

Figure 1. Schematic showing the experimental setup of the coupled energy harvester including two energy harvesters of the same size, which are coupled through a magnetic field

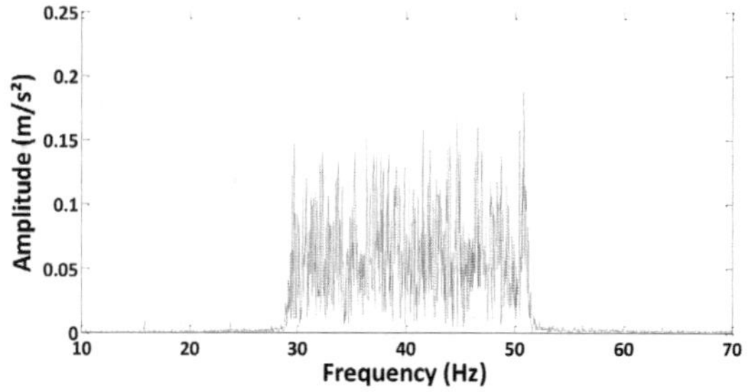

Figure 2. Frequency spectrum of the vibration profile 30-50-1 with a frequency content between 30 Hz and 50 Hz and with an RMS amplitude of 1 m/s^2

2. Experimental Setup

The coupled structure, which was investigated in this paper, is depicted in Figure 1. The structure comprises two cantilever-based energy converters with an identical design for magnetic circuit and coil. The influence of the Eigen-frequency of each energy harvester on the coupling dynamics and thus on the total power output was captured by varying the stiffness of the springs. In this respect, springs of two different thicknesses were used. The length and the width of the springs were kept equal. In this work two different spring configurations were investigated: An equal thickness of 0.7 mm was used for both springs in configuration one. As a result, both energy harvesters possessed the same Eigen-frequency. In a second configuration the thicknesses of the springs 1 and 2 were 0.8 mm and 0.7 mm. Consequently, the Eigen-frequency of the energy harvester 1 was larger in comparison to energy harvester 2.

The two different configurations of the coupled structure were excited using random vibrations with two different bandwidths and two different amplitudes resulting in four different vibration profiles. Vibration profiles 1 and 2 contain frequencies in the range of 30 Hz – 50 Hz. The RMS-amplitude was 1 m/s^2 for vibration profile 1 and 5 m/s^2 for vibration profile 2. The frequency spectrum of vibration profile 1 (30-50-1) is shown in Figure 2. Vibration profiles 3 and 4 contain frequencies in the range of 20 Hz – 60 Hz. The RMS-amplitude was varied in the same manner as for vibration profile 1 and 2.

For each combination (2 spring configurations times four vibration profiles) the RMS power of both energy harvesters was measured. For each measurement, the respective vibration profile was played for 1 min and the instantaneous power output at the load resistor (800 Ohm, matched) was recorded over time. The RMS power was then calculated and plotted as a function of distance between the two energy harvesters.

For better interpretation of the results the frequency response of each energy harvester was also investigated for varying distances d. In this regard, a harmonic excitation signal with an amplitude of 2 m/s^2 was applied. The voltage across a load resistance of 500 Ohm was measured.

3. Results and Discussion

Figure 3(a) and (b) show the frequency response of oscillator 1 and 2, respectively, for the spring configuration 1. For a gap of 14 mm both oscillators show an equal Eigen- frequency of approximately 40 Hz. The Eigen-frequency of both oscillators does not change much with varying distance d. For a gap of 4 mm the Eigen-frequency shifts down to a minimum value of 39.2 Hz. This system behavior

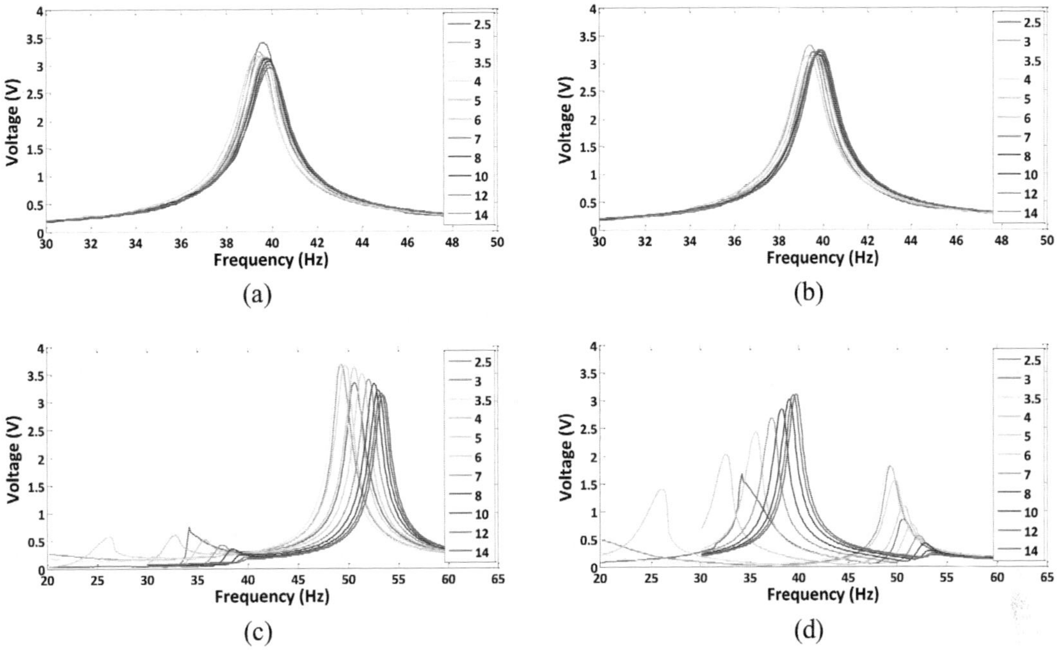

Figure 3. Frequency response of both energy harvesters for two different spring configurations: (a) and (b): Energy harvester 1 and 2 have an equal spring thickness of 0.7 mm. (c) and (d):The spring thickness of energy harvester 1 (c) and 2 (d) is 0.8 mm and 0.7 mm.

seems very interesting and was also observed for a spring configuration in which both springs had a thickness of 0.8 mm. From this it follows, that the power output will not change significantly with varying distances d. A bistable configuration is obtained for distances smaller than 3.7 mm.
A different behavior was observed for spring configuration two (Figure 3(c) and (d)). The initial Eigen-frequency of oscillator 1 (spring thickness of 0.8 mm) at the largest distance d is approximately 53.5 Hz (Figure 3(c)). When decreasing d from 14 mm to 3 mm the Eigen-frequency declines to 49 Hz. Moreover, small peaks are visible in the region between 30 Hz and 40 Hz. When looking at the frequency response of oscillator 2 (Figure 3(d)) it becomes clear, that the small peaks originate from this oscillator, which resonates in the respective frequency region. The magnetic coupling between the two oscillators becomes apparent. The frequency shift of oscillator 2 as a function of distance is much more significant. Here, the Eigen-frequency drops below 20 Hz for a distance of 3 mm. For distances smaller than 3.7 mm a bistable system configuration occurs. Therefore, the Eigen-frequency increases again if the distance is further reduced. At a distance of 2.5 mm the Eigen-frequency becomes 34 Hz. The oscillations of oscillator 1 are also coupled back to the energy harvester 2, which is apparent from Figure 3(d). In summary there are several important features to be considered for the coupled structure investigated in this paper: 1. there is a certain threshold distance at which a transition from a monostable towards a bistable configuration occurs. 2. If the stiffness of the two oscillators is unequal, the Eigen-frequency decreases when the distance is reduced from large values down to the threshold value. When the distance is further reduced the Eigen-frequency increases again. 3. The oscillations of each oscillator are coupled by means of the magnetic field. Therefore, the large amplitudes of a resonating oscillator lead to oscillations of the other.

Figure 4 shows the power output as a function of the distance d for spring configuration 1. In Each diagram a dotted vertical line indicates the threshold distance at which the transition between a monostable and a bistable configuration occurs. The power output is pictured for both energy

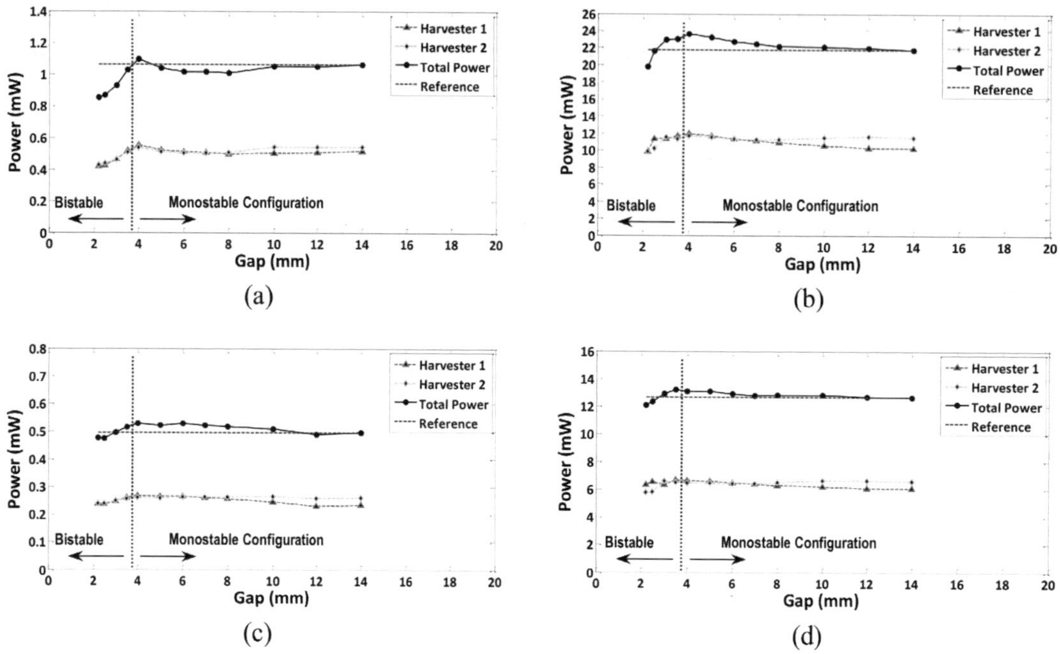

Figure 4. RMS Power output vs. gap for spring configuration 1 (equal spring stiffness): (a) Profile: 30-50-1. (b) Profile: 30-50-5. (c) Profile: 20-60-1. (d) Profile: 20-60-5.

harvesters and its summation. Moreover, the power obtainable with a linear reference system is also shown.

In general, the power output of both energy harvesters is very similar and does not change significantly with varying distance d. This is due to the fact that the frequency response, which is almost equal for both oscillators, does not depend on the distance d (Figure 3(a) and (b)). A very distinctive observation is that the largest power output occurs at a rather small distance d (ca. 4 mm). In comparison to the linear reference system, a larger power output is therefore achievable if a nonlinear monostable system configuration is chosen. Once the coupled system is configured bistable, the power output continuously declines if the distance is reduced further. Another characteristic of spring configuration 1 is that both energy harvesters always oscillate in a synchronous fashion even in the bistable configuration. As a consequence, no interwell oscillations occur. During experimental characterization oscillations always occurred within the potential well.

The power output of both energy harvesters differs significantly from each other when considering spring configuration 2 (Figure 5). For vibration profiles 1 and 2 the power output of energy harvester 1 is very low at a large distance of 14 mm. This is due to the fact that the Eigen-frequency is located outside the excitation frequency band (Figure 3(c)). In contrast, energy harvester 2 shows a much higher power output at the same distance since its Eigen-frequency is located in the center of the excitation frequency band (Figure 3(d)). When the distance is reduced, the power output of energy harvester 1 starts to increase while the power output of energy harvester 2 declines at the same time (Figure 5(a) and (b)). The total power output has its maximum near or at the transition towards a bistable system configuration. However, the power output of the linear reference system is not exceeded.

For vibration profiles 3 and 4 the initial power output of both energy harvesters is nearly equal at a distance of 14 mm since the Eigen-frequency of both oscillators is within the excitation frequency band (Figure 5(c) and (d)). Considering the total power output an increase is observed when reducing the distance. This again demonstrates the positive influence of the nonlinearity on the power output

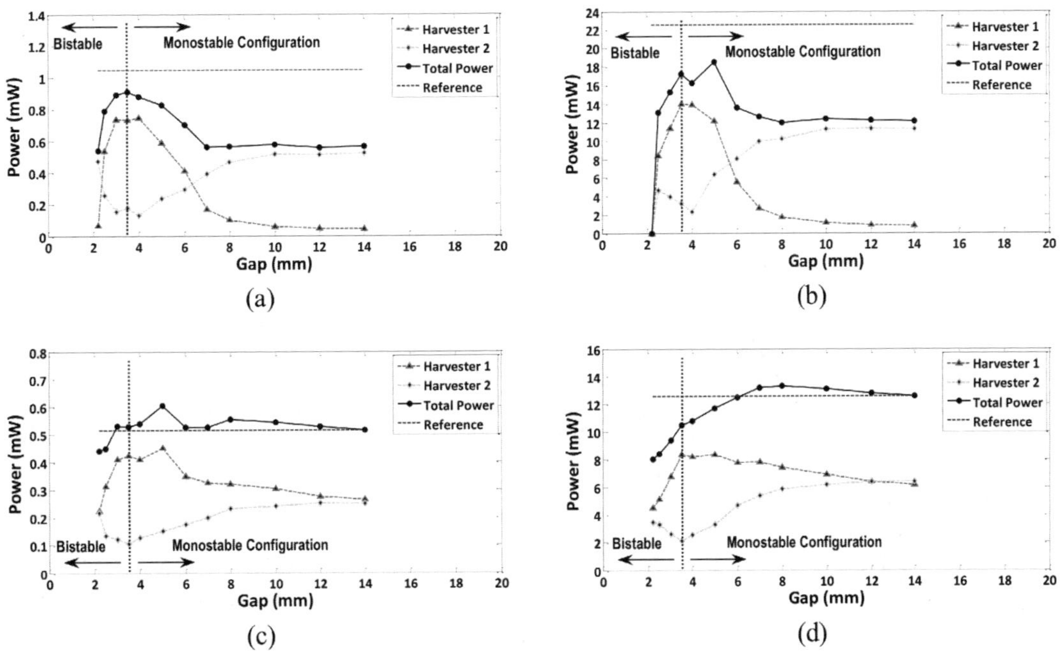

Figure 5. RMS Power output vs. gap for spring configuration 2. Spring 1: 0.8 mm, Spring 2: 0.7 mm. (a) Profile: 30-50-1. (b) Profile: 30-50-5. (c) Profile: 20-60-1. (d) Profile: 20-60-5.

in comparison to the linear reference system.

In case of vibration profile 2 and 4 interwell oscillations occur once the system is configured bistable. However, the total power output is still below the linear reference system.

4. Conclusion

In this work a coupled energy harvesting system including two cantilever-based electromagnetic energy harvesters was investigated experimentally. The results indicate that there is no benefit in terms of power output if a bistable system configuration is used. Even in case of bistable operation (interwell oscillations) the power output was lower than that of the linear reference system. However, it must be noted that this conclusion is only valid with respect to the four particular vibration profiles applied in this work.

On the other hand, a power gain over a linear system can be achieved if a monostable system configuration with a certain nonlinearity is chosen.

References

[1] D. Zhu, M. J. Tudor, and S. P. Beeby, "Strategies for increasing the operating frequency range of vibration energy harvesters: a review," *Meas. Sci. Technol,* vol. 21, no. 2, p. 22001, 2010.

[2] R. L. Harne and K. W. Wang, "A review of the recent research on vibration energy harvesting via bistable systems," *Smart Mater. Struct,* vol. 22, no. 2, p. 23001, 2013.

[3] S. P. Pellegrini, N. Tolou, M. Schenk, and J. L. Herder, "Bistable vibration energy harvesters: A review," *Journal of Intelligent Material Systems and Structures*, vol. 24, no. 11, pp. 1303–1312, 2013.

[4] D. Zhu and S. P. Beeby, "A broadband electromagnetic energy harvester with a coupled bistable structure," *J. Phys.: Conf. Ser,* vol. 476, p. 12070, 2013.

Mechanical Amplifier for Translational Kinetic Energy Harvesters

I Shahosseini and K Najafi

Center for Wireless Integrated MicroSensing and Systems (WIMS[2]), University of Michigan, Ann Arbor, MI, USA

E-mail: imanshah@umich.edu, najafi@umich.edu

Abstract. This paper reports the design, optimization, and test results of a mechanical amplifier coupled to an electromagnetic energy harvester to generate power from low-amplitude (±1 mm) and low-frequency (<5 Hz) vibrations in the presence of large static displacements. When coupled to a translational kinetic energy harvester, the amplifier boosts small vibration amplitudes by as much as 4x while accommodating translational displacements of more than 10x of vibration amplitudes. A complete electromagnetic energy harvester using this mechanical amplifier produces 16x improvement in output power (30 mW vs 1.9 mW without amplifier at 5 Hz), and a high power density of 170 μW/cm^3.

1. Introduction

There is a growing interest in using ambient vibrations as an alternative energy source for power consumer electronics. However, kinetic/vibration energy harvesting becomes challenging in low frequency environments (human motions, bridges, buildings, and some automotive/avionic systems) [1]. To address this challenge, different approaches have been proposed: using a resonant oscillating mechanism [2], deploying mechanical frequency up-conversion [3], or using rotational movement of an eccentric mass [4]. The performance of many of these systems degrades in the presence of non-axial loads, large static displacements, aperiodic vibrations, and low-amplitude stimuli [5].

While most of the aforementioned works propose inertial harvesters which transfer kinetic energy into a resonant proof mass tuned to resonate at source vibrations, an alternative is to deploy direct-force harvesters where source vibrations are directly applied on the proof mass. Although such systems tend to generate less power relative to resonant harvesters, they inherently remain operational over a wider excitation bandwidth. However, a real challenge to generate useful amounts of power arises when working at low-amplitude and low-frequency vibrations. To address this issue, this paper proposes a new approach of coupling a mechanical amplifier with direct-force kinetic energy harvesters to increase the amplitude of source vibrations. To investigate the impact of the mechanical amplifier on the generated power, a macro-scale electromagnetic energy harvester was used for.

This paper is organized as follows: in section 2 the design concept of the entire harvesting system is presented. Next in section 3 different design architectures are analyzed and refined to prototype the mechanical amplifier. After discussing experimental results in section 4, section 5 presents conclusions and future work.

2. Design conception

The operation principle of electromagnetic power generators is based on Faraday's law of induction which relates the electromotive force ε to the changing rate of magnetic flux $d\Phi_B/dt$ passing through a closed conductor as follows:

$$|\varepsilon| = d\Phi_B / dt \tag{1}$$

When considering Ohm's law and Kirchhoff's law for an electric circuit with source resistance of R_{coil}, the generator's output power P_{out} becomes proportional to:

$$P_{out} \propto \{(d\Phi_B / dz)^2/R_{coil}\}\ \{(dz / dt)^2\} \tag{2}$$

where dz/dt is the magnet-coil relative velocity.

To increase P_{out} both expressions on the right-hand side of (2) should be at their maximum. Regarding the first expression, previous work [6] presented an optimization procedure to adjust coil dimensions and position relative to a given magnetic structure in a way to meet the optimum ratio of flux linkage to coil resistance.

It is also essential to maximize the second expression, the vibration velocity which itself varies with vibration amplitude and frequency. When using frequency-up conversion or resonant mechanism in inertial harvesters, more power is generated via the increase in frequency and/or vibration amplitude. In direct-force energy harvesters other mechanisms should be implemented to enhance the output power.

As figure 1 illustrates, this energy harvester consists of two main parts: A) an electromagnetic transducer composed of an assembled stack of custom-made NdFeB magnets which remain stationary and a copper coil which moves axially relative to magnets, B) the mechanical amplifier consisting of three elements: 1) the base which is connected to the same frame as the transducer's stationary part, here the magnets, 2) the input member to receive ambient (source) vibrations, 3) the output member to drive the mobile part of the energy transducer, here the coil, at increased vibration amplitudes.

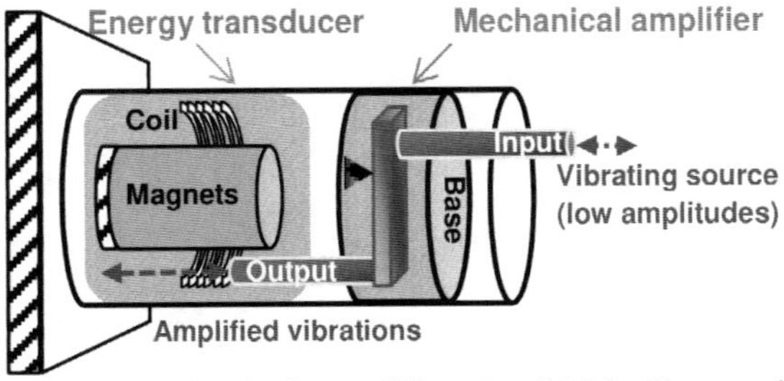

Figure 1. Schematic view of translational kinetic energy harvester: mechanical amplifier coupled to electromagnetic transducer to increase vibration amplitude of the coil moving axially relative to stationary magnets

The mechanical amplifier should ideally provide both a high mechanical gain for small dynamic vibrations, and a high capacity to accommodate large static/dynamic displacements to prevent structural damage of the harvester. In other words, the challenge is twofold: at the same time that small input vibrations are amplified as much as possible, very large static inputs, i.e. more than 10x of small vibrations, should be attenuated by the amplifier. Besides, depending on the application field, other criteria like weight, size, or damping force may also be taken into account in the design of the amplifier.

3. Amplifier architecture

In this section few designs of the amplifier based on mechanical linkage are examined and compared to each other. It should be mentioned that the objective of this work is to study the coupling effect of the mechanical amplifier with energy harvesters. The comprehensive study of all different configurable types of amplifiers lies beyond the scope of this paper.

In rotary energy harvesters higher output power can be generated by modifying the pitch diameter of the engaged gears [7]. As this concept remains incompatible with translational harvesters, other design architectures should be investigated. A very basic type of amplification is the lever mechanism composed of one link and one pivot point (figure 2-a). Despite its simple structure, it is not an ideal translational amplifier. Indeed, for input amplitudes much smaller than the link length the amplification can be estimated as translational; however, due to the rotational movement of the link about the fulcrum axis, it inherently provides rotational motions. Amplifiers based on the scissors linkage (figure 2-b) appear as a solution to obtain complete translational amplification. In spite of this advantage, the mechanical gain and the input capacity remain unchanged relative to the lever system.

By coupling two rack-pinion mechanisms together, linear motions are converted to rotational motions and vice versa. As Figure 2-c shows, the amplification takes place when both pinions are locked on the same shaft and the pitch diameter of the driven pinion is set larger than the driving pinion. This mechanism accommodates unlimited input amplitudes; however, the output amplitudes remain also unlimited. Except for applications where the harvester volume is not a design criterion, the unlimited output strokes become a major drawback to conceive a compact harvester unit.

Figure 2-d depicts a mechanical amplifier using one rack-pinion mechanism paired with one disk-crank mechanism providing piston-like motion. In such a system, after converting the linear motion into the rotational motion, while the angular velocity remains the same, the linear velocity increases for the disk of larger diameter than the pinion. Then, via a slider-crank linkage the rotational motion is converted back into linear motion. This assembly, furthermore, provides the advantage of accepting infinite input amplitudes while confining the output strokes within the disk diameter.

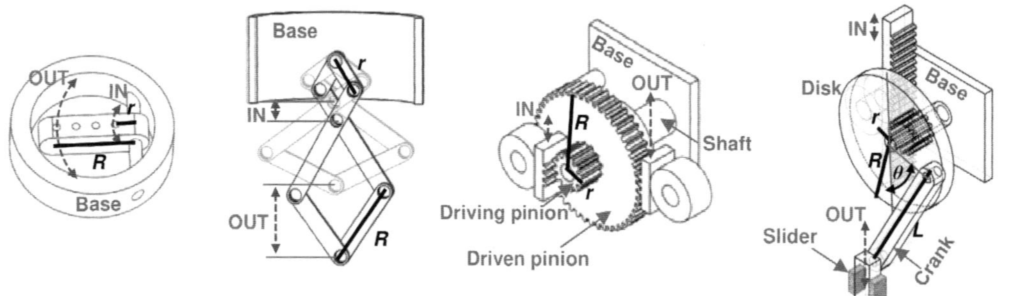

(a)	(b)	(c)	(d)
$G = R/r$	$G = R/r$	$G = R/r$	$G = \frac{R}{r}\left(Sin(\theta) + \frac{Sin(2\theta)}{2\sqrt{\left(\frac{L}{R}\right)^2 - Sin^2(\theta)}}\right)$
$Input_{Max} = 2\,r$	$Input_{Max} = 2\,r$	$Input_{Max} = \infty$	$Input_{Max} = \infty$
$Output_{Max} = 2\,R$	$Output_{Max} = 2\,R$	$Output_{Max} = \infty$	$Output_{Max} = 2\,R$

Figure 2. (a) Schematic view of different amplifier design conceptions with indicated mechanical gain G and input and output capacities: (a) simple lever, (b) scissors linkage, (c) two rack-pinion joints, (d) rack-pinion paired with piston-like motion

Figure 3-a shows detailed schemas of the developed mechanical amplifier based on figure 2-d design concept. Two solid plates are used as amplifier's backbone on which all components are mounted. Linear bearings/sliders are deployed to limit both the rack (the amplifier's input member) and the coil rod (the amplifier's output member) to one translational degree of freedom along the x-axis. To size the amplifier, trade-offs were set between device volume, mechanical gain, and friction. In this way the pinion pitch diameter and the disc diameter were set at 0.6 cm and 2.5 cm respectively ($R/r = 4$). As for the crank, a smaller length produces a higher mechanical gain. On the other hand, to meet the design requirement of high input load capacity, a minimum crank length equal to the disk radius R is necessary to succeed disk complete revolution (0–360°). Besides, as tested, very short-length cranks impede smooth functioning of the system unless using 4 cm or longer cranks.

Figure 3-b shows the finished amplifier prototype with a copper coil, 0.6 kΩ, attached to the output member and a spherical rod end screwed to the input member to decouple angular motions from linear motions of the vibrating source. The amplifier prototype was fabricated by mounting brass machined parts and off-the-shelf stainless steel rack & pinion with 120 diametral pitch and 20° pressure angle. Set screws and E-rings were also used to adjust and hold in position the movable components. This prototype weighs 130 gr and occupies 40 cm^3.

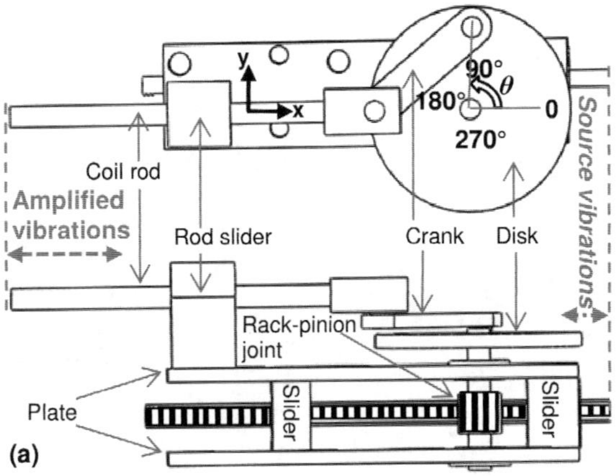

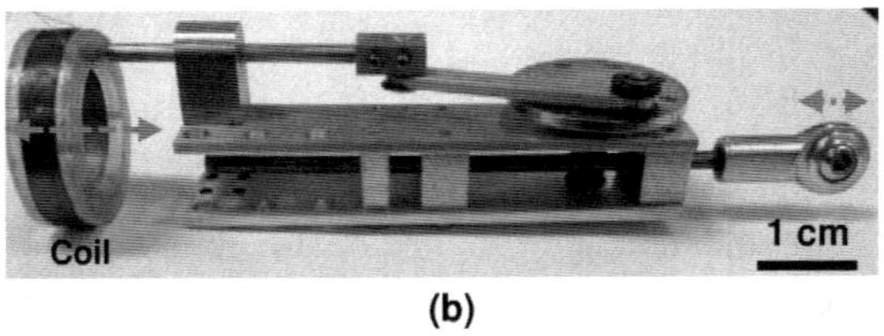

Figure 3. (a) Detailed structure of developed mechanical amplifier based on figure 2-d mechanism–top view and side view, (b) built amplifier prototype by assembling brass machined parts and off-the-shelf stainless steel rack & pinion

4. Results and discussion

To characterize the developed mechanical amplifier two series of tests were carried out: static measurements on the amplifier alone and dynamic tests on the complete energy harvesting system.

4.1. Static measurements

Referring to figure 2-d, it is noticed that the mechanical gain varies not only with disk-pinion radius ratio R/r (here 4) and the crank length L (here 4 cm), but also with the crank-disk angle θ. The mechanical gain between input and output displacements was measured and traced in figure 4 as a function of crank-disk angle. In good correlation with simulation results, the presence of two identical peaks defining the optimum operation zone is noticeable. The width of dotted zones (~30° in disk revolution) corresponds to ±1 mm input amplitude and their height to ±4 mm output (a gain of 4). Two dotted zones are identical because the rod slider and the disk are centered about the same x-axis.

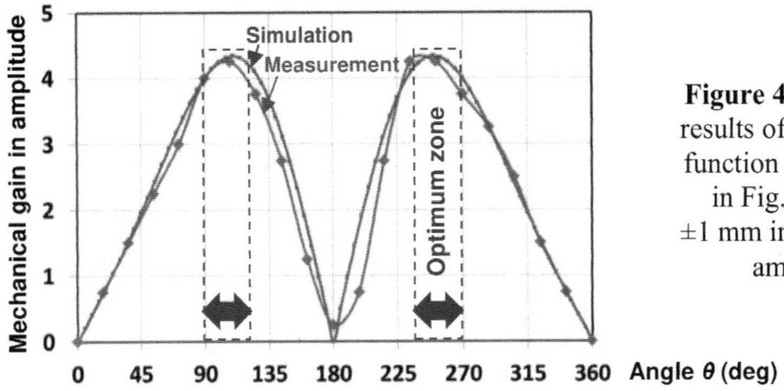

Figure 4. Simulation and experimental results of amplifier mechanical gain as a function of crank-disk angle θ indicated in Fig.3-a; within dotted rectangles ±1 mm inputs (~30° in disk rotation) are amplified to ±4 mm outputs

4.2. Dynamic measurements

Dynamic tests were performed at low-frequency low-vibration amplitudes (±1 mm) using APS 113 Electro-Seis shaker. While the energy transducer alone generates 1.9 mW power at 5 Hz, once incorporating the mechanical amplifier, the output power is boosted to 30 mW. As figure 5-a shows, after a transient regime, the generated power rises by as much as 16x when using a 4x-amplifier. The total damping force increases up to 8 N, 2.5 N of which is estimated to be caused by friction. Despite the increased total volume of the harvester when integrating the amplifier, the power density is quadrupled and reaches 170 μW/cm^3. The improvement can also be observed when comparing the resulting 43 mW/cm^3/g^2 normalized power density with other reported electromagnetic energy harvesters in figure 5-b. An inset of the complete energy harvesting system integrated into 4 cm wide tube is also illustrated in figure 5-b.

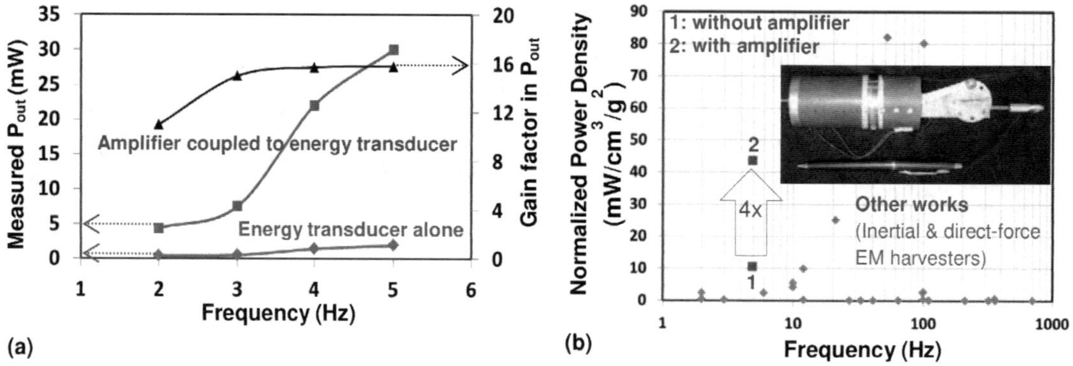

Figure 5. (a) Measured output power and gain factor at 1 mm vibration amplitude using source-load matched resistance, (b) normalized power density as a function of frequency for this work and other electromagnetic energy harvesters, an inset of fabricated energy harvester incorporating mechanical amplifier

5. Conclusion

This work studies the design of a translational mechanical amplifier and its impact on the generated power once coupled to an electromagnetic energy harvester. Several design concepts based on mechanical linkage were analyzed to obtain large mechanical gain and high load capacity to accommodate static/dynamic displacements much larger than vibration amplitudes. After refining the design architecture for the most attractive mechanism (a rack-pinion paired with a piston-motion mechanism), amplifier prototypes with 4x mechanical gain were built and tested.

Dynamic measurements showed when coupling the developed mechanical amplifier to a direct-force electromagnetic energy transducer, 30 mW power is generated from vibration amplitudes as low as 1 mm at low-frequencies. With mechanical amplifier integrated, the output power rises by as much as 16x and the power density quadruples up to 170 μW/cm^3. The design concept of the detailed mechanical amplifier remains compatible with other types of kinetic energy harvesters, e.g. piezoelectric or electrostatic. With the goal of size reduction and efficiency improvement, future work will investigate other mechanisms using pneumatic or hydraulic means.

Acknowledgements

The authors thank Pascal Sautier and Patrice Levallard at Hutchinson Research & Development Center, Montargis, for technical discussions, and acknowledge the support of Christophe Dominiak for the project. They also thank Robert Gordenker at WIMS^2 Research Center, University of Michigan, for his technical assistance with experimental setup. This work is funded and supported by Hutchinson SA, Centre de Recherche, France.

References

[1] Kwon S D, Park J and Law K 2013 *Smart Mater. Struct.* **22** pp 1–12
[2] Sardini E and Serpelloni M 2011 *Sensors and Actuators A* **172** pp 475–482
[3] Kulah H and Najafi K 2008 *IEEE Sens. J.* **8** pp 261–268
[4] Hara T and Kitahara J 2000 Electronic watch US6012838 A- PCT/JP1996/003419
[5] Tang Q C, Xia X Y and Li X X 2012 *Proc. Int. MEMS Conf.* pp 1273–1275
[6] Shahosseini I, Peterson R L, Aktakka E E and Najafi K 2014 *Proc. Sensors Conf.*
[7] Donelan J M, Li Q, Naing V et al. 2008 *Science* **319** pp 807–810

Electrodynamic Wireless Power Transmission to Rotating Magnet Receivers

A Garraud[1], J D Jimenez, N Garraud and D P Arnold
Interdisciplinary Microsystems Group,
Department of Electrical and Computer Engineering,
University of Florida, Gainesville, FL, 32611, USA

E-mail: agarraud@ufl.edu

Abstract. This paper presents an approach for electrodynamic wireless power transmission (EWPT) using a synchronously rotating magnet located in a 3.2 cm^3 receiver. We demonstrate wireless power transmission up to 99 mW (power density equal to 31 mW/cm^3) over a 5-cm distance and 5 mW over a 20-cm distance. The maximum operational frequency, and hence maximal output power, is constrained by the magnetic field amplitude. A quadratic relationship is found between the maximal output power and the magnetic field. We also demonstrate simultaneous, power transmission to multiple receivers positioned at different locations.

1. Introduction

A century ago, before the electrical wire grid, Tesla developed and demonstrated the concept of electromagnetic wireless power transmission (WPT) [1]. Despite this early vision, it was only in the past ten years that WPT has received a surge of interest [2–4] due to the increasingly numerous and diverse power demands of mobile electronics devices (laptops, smart phones, tablets, wearables) in everyday life. Other application areas for WPT include automotive vehicle recharging [5] and the *in vivo* charging of biomedical implants [6]. The majority of efforts have relied on the electromagnetic coupling between two coils to wirelessly transfer the power, called inductive WPT. In this approach, a transmitter coil produces an alternating magnetic field that directly induces a voltage and current in a receiver coil placed at some distance away.

We have recently presented an alternative method, named electrodynamic wireless power transmission (EWPT) [7–9]. In this approach, a transmitter produces an alternating magnetic field that moves a permanent magnet in a receiver, the motion of which generates a voltage in the receiver windings. If a load is connected to the receiver windings, EWPT is achieved. The electrodynamic coupling solution affords operation at much lower frequencies (e.g. 100's of Hz) than what is commonly used in inductively coupled WPT (e.g. 0.1 – 10's of MHz). This low-frequency range of operation facilitates higher safety margins for magnetic field amplitudes [10,11] and better penetration (less attenuation) through electrically conductive media. Therefore, the EWPT method provides the opportunity to safely transmit power in crowded environments, such as a home, office space, automobile, or to an implanted biomedical device within the human body.

We have previously demonstrated the EWPT concept using mechanical resonance in the receiver magnet. A system employing a cantilever-based oscillator yielded a power transmission of 0.15 mW at

[1] To whom any correspondence should be addressed.

2.2 cm [7]. A second prototype utilized a torsional resonator with an average transmitted power of 3.09 mW at 1 cm (equivalent to a power density of 143 μW/cm^3), and 0.126 mW at 7 cm [9]. System modeling has shown that a high mechanical quality factor in the receiver resonator is desirable for maximizing the power transmission and efficiency [8]. The inductive coupling between the transmitter coil and the receiver windings was previously shown to have a minor influence in the proposed EWPT scheme [9].

In this paper, an alternative motion in the receiver is explored. Rather than exciting a torsional resonance, the receiver magnet is continuously rotated by the transmitted B-field. Under steady-state operation, the rotating magnet acts as a synchronous machine, supplying electrical power to an external load. Compared to the resonant-type receivers, which only generated appreciable power when excited at their specific resonant frequencies, the rotational approach enables the receiver(s) to generate power across a wider frequency range of operation. As will be shown, this approach also enables much higher power transmission levels as well as simpler and more compact receiver construction. One limitation of this continuously rotating approach is that the receivers cannot instantaneously synchronize with the sinusoidal transmitter B-field. The machines take a certain amount of transient time to "start-up" from stand still, requiring a ramping of the transmitter frequency.

Section 2 first describes the concept of electrodynamic wireless power transmission through a continuously rotating magnet and it introduces the experimental setup used to perform EWPT. Section 3 discusses the experimental results. Conclusions are provided in section 4.

2. EWPT system design

2.1. Working principle

The EWPT system reported here includes two distinct parts: a coil-based transmitter and a receiver system featuring a permanent magnet rotor that is allowed to rotate within receiver windings, as illustrated in Figure 1.a. The transmitter coil is a thin solenoid with ~112 turns of 15-AWG copper wire, having an outer diameter of 30 cm, an inner diameter of 25 cm, and a length of 12 mm (L=8.5 mH, R=1.3 Ω). The receiver system comprises a diametrically magnetized, cylindrical N42 grade Nd–Fe–B permanent magnet rotor (K&J Magnetics, D36DIA, diameter = 4.8 mm, length = 9.5 mm) that rotates on a ferrofluid bearing (inspired by [12]) within a 3D-printed structure (diameter=25 mm, length=30 mm) with slots for coil windings, as presented by the schematic in Figure 1.b. In this prototype, the magnet and ferrofluid are held within a capped glass vial and centered via to plastic inserts. The receiver is wound with a single coil of 250 turns of 34-AWG copper wire. Figure 1.c presents a picture of the receiver system along with an example of capped glass vial containing a magnet covered by ferrofluid.

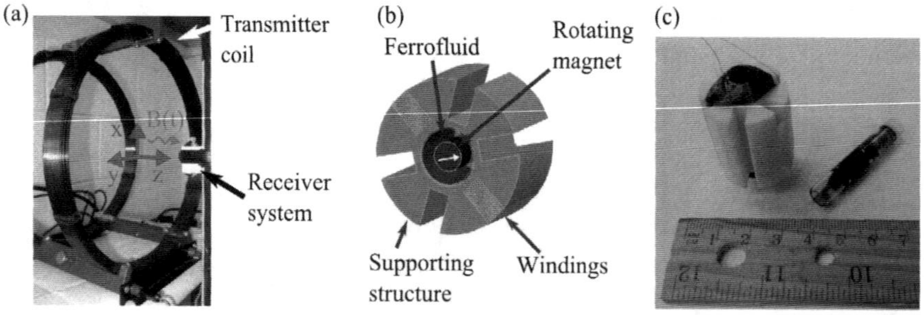

Figure 1: (a) Picture of the electrodynamic wireless power transmission system, (b) Cross-section schematic of the receiver system, (c) Picture of the receiver system and an example of the rotating magnet and capped glass vial containing the magnet and ferrofluid.

To transmit power, an alternating current is injected into the transmitter coil, which results in the generation of a time-varying magnetic flux density **B**(t). When the receiver system is placed nearby, the transmitted B-field applies an alternating torque that induces the synchronous rotation of the permanent magnet. The magnet motion produces a time-varying magnetic flux in the receiver windings thereby inducing a voltage. If an electrical load is connected to the two terminals of the receiver windings, power is delivered. Compared to a resonant-based system, the continuously-rotating EWPT has the potential to induce a larger variation of magnetic flux through the coil, since the magnet steadily rotates 360° around its axis at the frequency set by the transmitter AC current. Additionally, as described earlier, the receiver can operate across a large range of frequencies, the limits of which are discussed in Section 3.1.

2.2. Experimental setup

The EWPT system was characterized using the following setup. An HP33120A arbitrary waveform generator was used to supply an AC voltage to a Crown XLS2500 linear power amplifier. This produced an AC current input into the transmitter coil, monitored by a Tektronix TCP312A current probe connected to a Tektronix TCPA300 current probe amplifier. An Agilent DSO-X-2004A 4-channel oscilloscope was used to measure the input current to the transmitter, as well as the voltage measured across the receiver load. The resistance of the receiver windings was independently measured with a Keithley 2000 multimeter, using the 4-wire configuration. For a given AC excitation of the transmitter, the resultant B-field spatial distribution was measured with a Lakeshore XHMMA-1482 axial Hall probe connected to a Lakeshore 475DSP gaussmeter.

3. Results and discussion

First, the basic operation of the continuously rotating EWPT system was studied. Then, the output power as a function of the transmitted B-field was characterized along the central axis of the transmitter coil. Next, the off-axis wireless transmission was measured. Last, multi-receiver power transmission was demonstrated.

3.1. Basic Operation

After placing the receiver system at a certain distance from the transmitter, a fixed-amplitude 4 A_{RMS} sinusoidal current was injected in the transmitter coil, corresponding to 18.4 W. The resistive load was adjusted to 22.5 Ω to match the receiver windings resistance for maximum power transmission. The frequency was slowly increased from a low frequency, 40 Hz—where the receiver locked into synchronous, steady-state rotation—up to a point where the magnet lost synchronization with the transmitted B-field, resulting in no more magnet rotation. This upper frequency was called the maximal frequency and the output power at that frequency, the maximal output power. The frequency where synchronization is lost represents the "pull out torque" for a synchronous machine—the point where the magnetic torque acting on the magnet equals the sum of the opposing mechanical and magnetomotive torques [13].

At a distance of z=4 cm between the center of the transmitter and of the receiver, the output power was measured at representative frequencies, as illustrated in Figure 2. We observed that the output power increased quadratically with the frequency (load voltage proportional to frequency), as expected with a synchronous machine, up to the maximum frequency of 295 Hz. The test was reproduced at several distances between transmitter and receiver. The results obtained at z=10 cm and z=20 cm are also reported in Figure 2. For a given frequency, the output power was found to be independent of the distance. However the maximal frequency, hence the maximal output power, were strongly dependent on the distance. The farther the receiver was from the transmitter, the lower the magnetic field, and the lower the maximum output power that was transmitted. The maximum output power was 66.4 mW at z=4 cm, but only 5.1 mW at z=20 cm. Consequently, in this configuration, a 5-fold increase in distance resulted in only a 13-fold decrease in the maximal output power.

3.2. Power vs. B-field

Keeping the transmitter current at 4 A_{RMS}, the distribution of the B-field along the z-axis was measured. Based on these measurements, the maximal frequency and the maximal output power were plotted as a function of the peak B-field, at discrete receiver locations along the z-axis. Figure 3 summarizes the results. A linear dependence was found between the maximal frequency and the transmitted B-field. Consequently, the maximal output power was found to be quadratic with B-field. Based on these observations, for maximal power transfer, the B-field from the transmitter should be as high as possible to increase the maximal frequency, hence the maximal output power, as long as it remains within safety limits.

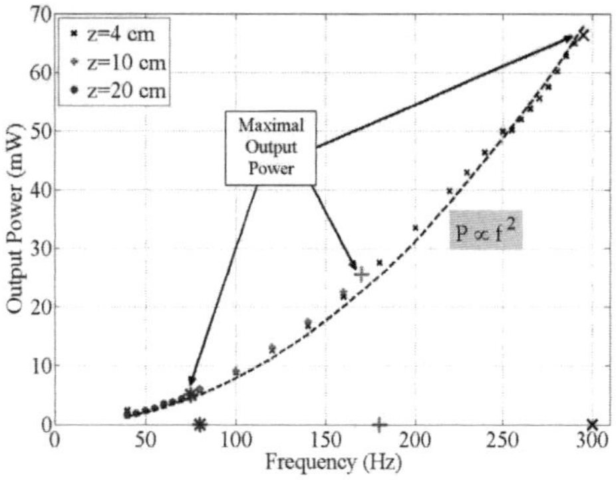

Figure 2: Output power vs. frequency, at different on-axis distances.

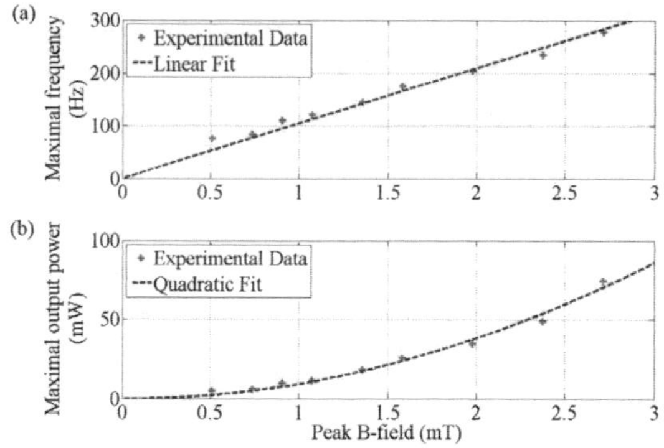

Figure 3: (a) Maximal frequency and (b) Maximal output power, vs. the transmitted peak B-field. Linear and quadratic relationships were found, respectively.

3.3. Off-axis wireless power transmission

Another capability for EPWT is off-axis power transmission. Due to symmetrical consideration, off-axis power was only measured for offsets in the positive y-axis. The receiver system was first placed at a distance z=6 cm away from the transmitter, and the off-axis distance y was modified. The resulting maximal output power as a function of y is reported in Figure 4. A maximal output power of 99 mW was measured at y=7 cm (z=6 cm). Numerical predictions of the expected output power levels at z=10 cm, z=15 cm, and z=20 cm are also plotted, based on the quadratic relationship with the peak B-field, found previously in Figure 3. The numerical estimations indicate that a reasonable output power can still be received off-axis, up to 20 cm away from the transmitter.

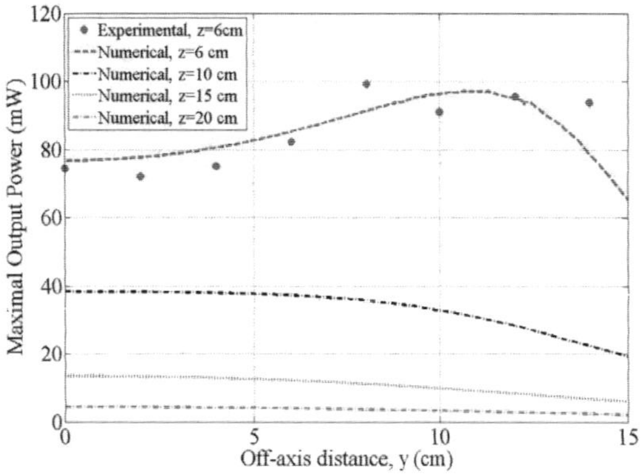

Figure 4: Experimental maximal output power vs. the distance y to the axis at z=6 cm and simulation results of the maximal output power vs. y at varying z distances.

3.4. Multi-receiver wireless power transmission

To test the ability to power multiple receivers, three similar receiver systems were placed in proximity to the transmitter. The EWPT system was shown to simultaneously power the three receivers distributed in space with no apparent limitation on their relative positions. As a worst-case scenario, the three receiver systems were placed one next to each other wherein maximum magnetic interaction could occur between the three receiver magnets. Even in this configuration, at a given frequency, each device received roughly the same amount of power as if powered alone. However, the maximal frequency was found to be lower (~ -30 %), causing a decrease in the maximal output power compared to the alone-receiver case.

4. Conclusion

In this paper, an EWPT scheme was presented using the continuous rotation of a permanent magnet, rather than resonant mechanical motion as presented in previous works. This approach resulted in a significant increase in the amount of power delivered to a resistive load, achieving up to 99 mW, with a power density up to 31 mW/cm^3. The system was shown capable of power a receiver off-axis from the transmitter, as well as powering multiple receivers at the same time. The maximal output power was demonstrated to be quadratic with the transmitted B-field. Future works will focus on improving the design of the receiver system to achieve a size reduction, as well as designing a transmitter coil with an improved B-field distribution.

Acknowledgments

This work was supported by the US Army Communications-Electronics Research, Development and Engineering Center (CERDEC) through the General Technical Services sub-contract GTS-S-13-226.

References

[1] Tesla N 1914 U.S. Patent 1119732 Apparatus for transmitting electrical energy
[2] Kurs A, Karalis A, Moffatt R, Joannopoulos J D, Fisher P and Soljacic M 2007 Wireless power transfer via strongly coupled magnetic resonances. *Science* **317** 83–6
[3] Wei X, Wang Z and Dai H 2014 A Critical Review of Wireless Power Transfer via Strongly Coupled Magnetic Resonances *Energies* **7** 4316–41
[4] Shinohara N 2014 *Wireless Power Transfer via Radiowaves* vol 2014 (Wiley)
[5] Gil A and Taiber J 2013 A Literature Review in Dynamic Wireless Power Transfer for Electric Vehicles: Technology and Infrastructure Integration Challenges *Sustainable Automotive Technologies* Lecture Notes in Mobility ed J Wellnitz, A Subic and R Trufin (Springer) pp 289–98
[6] Kim J-D, Sun C and Suh I-S 2014 A proposal on wireless power transfer for medical implantable applications based on reviews *2014 IEEE Wireless Power Transfer Conference* (IEEE) pp 166–9
[7] Challa V R, Mur-Miranda J O and Arnold D P 2012 Wireless power transmission to an electromechanical receiver using low-frequency magnetic fields *Smart Mater. Struct.* **21** 115017
[8] Mur-Miranda J O, Cheng S and Arnold D P 2013 Improving the efficiency of electrodynamic wireless power transmission *7th European Conference on Antennas and Propagation (EuCAP)* (Gothenburg, Sweden) pp 2848–52
[9] McEachern K M and Arnold D P 2013 Electrodynamic wireless power transmission to a torsional receiver *J. Phys. Conf. Ser.* **476** 012004
[10] C95.6 2002 *IEEE Standard for Safety Levels With Respect to Human Exposure to Electromagnetic Fields, 0-3 kHz*
[11] C95.1 2010 *IEEE Standard for Safety Levels With Respect to Human Exposure to Radio Frequency Electromagnetic Fields, 3 kHz to 300 GHz*
[12] Assadsangabi B, Tee M H, Wu S and Takahata K 2014 Ferrofluid-assisted micro rotary motor for minimally invasive endoscopy applications *2014 IEEE 27th International Conference on Micro Electro Mechanical Systems (MEMS)* (IEEE) pp 200–3
[13] Chapman S J 2002 *Electric Machinery and Power System Fundamentals* (McGraw-Hill Higher Education)

AUTHOR INDEX

AUTHOR INDEX

AUTHOR INDEX

AUTHOR INDEX